POWER ELECTRONICS

Devices, Circuits, and Industrial Applications

V.R. MOORTHI

Formerly, Professor and Head,
Department of Instrumentation and Control
Delhi College of Engineering, New Delhi

OXFORD
UNIVERSITY PRESS

OXFORD
UNIVERSITY PRESS

YMCA Library Building, Jai Singh Road, New Delhi 110001

Oxford University Press is a department of the University of Oxford.
It furthers the University's objective of excellence in research, scholarship,
and education by publishing worldwide in

Oxford New York
Auckland Cape Town Dar es Salaam Hong Kong Karachi
Kuala Lumpur Madrid Melbourne Mexico City Nairobi
New Delhi Shanghai Taipei Toronto

With offices in
Argentina Austria Brazil Chile Czech Republic France Greece
Guatemala Hungary Italy Japan Poland Portugal Singapore
South Korea Switzerland Thailand Turkey Ukraine Vietnam

Oxford is a registered trade mark of Oxford University Press
in the UK and in certain other countries.

Published in India
by Oxford University Press

First published 2005
Second impression 2005

ISBN-13: 978-0-19-567092-9
ISBN-10: 0-19-567092-2

Typeset in Times Roman
by Archetype, New Delhi 110063
Printed in India by Radha Press, Delhi 110031
and published by Manzar Khan, Oxford University Press
YMCA Library Building, Jai Singh Road, New Delhi 110001

CHAPTER 1

Thyristor Physics

1.1 Introduction

Power electronics is a branch of engineering that combines the fields of electrical power, electronics, and control. It started with the introduction of the mercury arc rectifier in 1900. The grid controlled vacuum tube rectifier, ignitron, and thyratron followed later. These found extensive application in industrial power control till 1950. In the meanwhile the invention of the transistor—a semiconductor device—in 1948 marked a revolution in the field of electronics. It also paved the way for the introduction of the silicon controlled rectifier (SCR), which was announced in 1957 by the General Electric Company. In due course it has come to be named as the 'thyristor'.

There is one important difference between the earlier electronic tubes and their semiconductor successors. On the low-current side the transistor is a base current controlled device whereas its predecessor, the vacuum tube, is a grid voltage controlled device. Similarly the thyristor, which is designed to carry high currents, is a gate current-controlled device; it has succeeded the thyratron, which is a grid voltage controlled gas tube.

After the inception of the thyristor, which is a *p-n-p-n* device, many other devices of its family came into existence. These include the DIAC, TRIAC, gate turn-off thyristor (GTO), and the MOS-controlled thyristor (MCT) among others. Subsequently, devices such as the power bipolar junction transistor (BJT), metal oxide semiconductor field effect transistor (MOSFET), and the IGBT have evolved as alternative power electronic devices, each of them being superior to the thyristor in one or more ways.

With its rapid development, power electronic equipment now forms an important part of modern technology. Power electronic applications can be broadly divided into the following five categories:

(a) Motor controls: ac (alternating current) and dc (direct current) drives used in steel, cement, and various other industries

(b) Consumer applications: heat controls, light dimmers, security systems, hand power tools, food mixers, and other home appliances

(c) Vehicle propulsions: electric locomotives used in railways, forklift trucks, and dc chopper based electric vehicles
(d) Power system applications: applications such as static VAR control, active harmonic filtering, flexible ac transmission system (FACTS) devices, and high-voltage dc (HVDC) systems
(e) Other industrial applications: uninterruptible power supplies (UPSs), switch mode power supplies (SMPSs), battery chargers, industrial heating and melting, arc welding, electrolysis, HV supplies for electrostatic precipitators, aerospace applications, electromagnets, etc.

It is important here to note that all the devices employed for power electronic applications are used in the 'switch' mode. The moments of switching on or off are controlled to fulfil the requirements of the circuit under consideration. Likewise, the BJT is operated as a switch.

The advantages of power electronic applications are (a) high efficiency because of low 'ON state' conduction losses when the power semiconductor is conducting and low 'OFF state' leakage losses when it is blocking the source voltage, (b) reduced maintenance, (c) long life, (d) compactness because of the facility of assembling the thyristors, diodes, and *RLC* elements in a common package, (e) faster dynamic response as compared to electromechanical equipment, and (f) lower acoustic noise as compared to electromagnetic controllers.

Thyristorized power controllers have some disadvantages; important among them are (i) they generate harmonics which adversely affect the loads connected to them and also get injected into the supply lines, (ii) controlled rectifiers operate at low power factors and cause derating of the associated rectifier transformers, and (iii) they do not have a short-time overload capacity. However, as their advantages outnumber their demerits, they are widely used in the various applications detailed above. They have also replaced conventional controllers.

After going through this chapter the reader should

- gain a good understanding of the physical processes that go behind the switching on of a thyristor,
- know the various methods of triggering a thyristor and different gate turn-on methods,
- develop a good idea about the turning-off mechanism of a thyristor and get acquainted with some methods of turning a thyristor off,
- realize the importance of the various ratings of a thyristor,
- understand how the current rating of a thyristor is arrived at, based on the thermal model of the thyristor,
- get acquainted with methods of protecting thyristors against overvoltages and overcurrents,
- become familiar with other members of the thyristor family as well as other power electronic devices which have been recently introduced, and
- get to know the characteristics of different power electronic devices.

This chapter lays a firm foundation for the chapters that follow, namely, rectifiers, choppers, inverters, and industrial applications of the thyristor.

To my late son V. Suryaprakash

Preface

Power electronics is a branch of engineering that combines the fields of electrical power, electronics, and control. The introduction of the mercury arc rectifier in 1900 and the invention of the transistor-a semiconductor device-in 1948 marked a revolution in the field of electronics. The inception of the thyristor during the 1960s as a power device led to the development of compact, reliable, and maintenance-free drive circuits. This application got a further boost with the advent of the gate-turn-off thyristor, power transistor, and MOSFET, among others, as power electronic devices. The utility of these devices spread to industrial applications such as uninterruptible power supplies, induction heating, high-voltage dc transmission, electrical welding, static reactive volt-ampere control of power systems, etc.

Microprocessors and microcomputers have made their impact on power electronics based industrial equipment since the 1970s. Their application for control of electric devices started in the shape of data acquisition systems. In due course (1980s), their involvement became full-fledged for the closed-loop control of drives.

With its rapid development, power electronic equipment now forms an integral part of modern technology. Some of the important applications of power electronics are electric traction, cranes, hoists, lathe machines, rolling mills, subway cars, and battery-operated vehicles.

About the Book

This book has mainly evolved out of the class notes of a course on power electronics taught by me at the undergraduate level at Delhi College of Engineering. Fora like winter schools, curriculum development workshops, and technical seminars conducted by me and others, both at Delhi College of Engineering and at IIT Delhi, have also provided useful inputs.

The book emphasizes the fundamental aspects of power electronics. It has a student-friendly approach, building up the discussion smoothly from chapter to chapter. Detailed mathematical derivations and illustrations augment a thorough understanding of the concepts presented. To complement the theory, a large number of solved numerical problems have been included. In addition, answers to all the unsolved numericals are provided to build up the confidence of students. The lucid treatment in the book will encourage self-study and motivate students towards independent problem solving.

Content and Structure

The prerequisite knowledge for this course is basic circuit theory, calculus, and simple differential equations. This book will serve as a valuable text for a one-semester course especially at the undergraduate level. Chapters 7-9, 13, and 14, which deal with control of modern electrical drives, and Chapter 11, which deals with recent industrial applications, should form a vital part of a typical post-graduate course on power electronic drives and other applications.

Chapter 1 is devoted to the theory of various power electronic devices: the thyristor and other members of the thyristor family, namely, the GTO, power transistor, MOSFET, and IGBT among others.

Chapter 2 provides a firm base for all the rectifier-based circuits. It discusses the various performance factors used by practising engineers to evaluate their equipment, such as the input power factor and rectifier efficiency.

Chapter 3 deals with the theory of dc choppers, which form the heart of chopper-based drives. These choppers also constitute the basis for buck-boost and flyback choppers, which are employed in switch mode power supplies.

Chapter 4 deals with single- and three-phase ac voltage controllers. Related operational features like reduction of distortion, improvement of the power factor, and production of harmonic torques are also presented.

Chapter 5 presents various types of inverters, including pulse width modulated, load commutated, voltage source, and current source inverters. Detailed derivations are provided wherever necessary to augment circuit functioning descriptions.

Chapter 6 develops the concept behind cycloconverters with the help of the dual converter.

Chapter 7 deals with dc drives, which are based on controlled converters as well as dc choppers. It also covers speed-torque curves and closed-loop control of converters.

Chapter 8 is devoted to ac drives, induction motor based drives being the most important among them. Speed control features of the induction motor using different strategies are dealt with at length, the Scherbius scheme being a recent one among them.

Chapter 9 elucidates the principle of the brushless dc motor, duly supported by appropriate diagrams and waveforms. A typical closed-loop control strategy is described in detail.

Chapter 10 is devoted to the description of control circuits for all the important devices introduced in Chapter 1. In addition, the necessity of zero voltage and zero current detectors is brought out with the help of appropriate circuits.

Chapter 11 deals with various industrial applications, the most important among them being high-voltage dc transmission, static-VAR control, and switch mode power supplies. The relevant theoretical aspects are also discussed.

Chapter 12 provides an introduction to microprocessor/microcomputer technology, starting from the basic digital electronic circuits. Thereafter, the Intel

8085A (8-bit) and 8086 (16-bit) microprocessors are described along with their assembly level programming aspects.

Chapter 13 is devoted to the microprocessor control of two of the most commonly used electrical drives, namely, a reversible dc drive and an induction motor drive. A few other related topics are also dealt with.

Finally, Chapter 14 deals with the relatively recent technique of field oriented control. The principle is explained with the help of mechanical analogies, followed by a thorough mathematical derivation. In addition, the microprocessor implementation of this technique in ac motors is elaborated.

This book is written so as to cater to the teachers and students of technical educational institutions, practising engineers, as well as students who wish to augment their engineering qualifications through professional bodies like the Institution of Engineers (India) and the Institution of Electronics and Telecom Engineers (India) by self-study.

Key Features

- Provides detailed and easy-to-understand derivations
- Includes detailed coverage of various kinds of electrical drives along with microcomputer implementations for a few of them
- Contains a separate chapter on recent industrial applications
- Includes a large number of illustrations, solved and unsolved problems, and review questions

Acknowledgements

I wish to thank the authorities of Netaji Subhash Institute of Technology for the facilities provided to me during my visiting assignment with them. My thanks are also due to Professor Bhim Singh of IIT Delhi for discussions on brushless dc motors and control circuits for modern power electronic devices. Among my family members, my eldest son V.N. Moorthy was instrumental in bringing out computer printouts of all the chapters, including the illustrations. The patience of my wife Laxmi Moorthi and the encouragement provided by my other children helped me throughout the preparation. My warm appreciation to all of them.

V.R. MOORTHI

Contents

List of Symbols

α firing angle
β thermal resistance, angle (used with extinction angle control)
γ extinction angle, ripple factor
δ derating factor (used for rectifier), ratio (τ_{ON}/τ)
μ commutation angle, distortion factor
λ input power factor
ω angular frequency (variable)
χ angle between stator and rotor current vectors of induction motor
θ temperature, angle
η intrinsic stand-off ratio (UJT), efficiency (used for rectifier)
ϕ flux (variable), phase angle
Ω angular frequency (constant)
Φ flux (constant)
ρ resistivity
σ ripple (voltage), leakage factor
τ time period, time constant

a harmonic coefficient, armature
b harmonic coefficient, battery, back emf (subscript)
d subscript used for direct current (also voltage and power)
f frequency (source, chopper), field, friction
g voltage reduction (due to commutation), air gap
i current (variable)
j junction (of a semiconductor)
k cathode
m maximum
n negative (layer), speed in rps
p positive (layer), pulse (time), primary (transformer), pole pairs, power (variable)
q charge (variable), quadrature
r reverse (current), reduction factor, receiving end
s settling (time), source (voltage, current), saturation (current), slip, sending end
t time (variable)
v voltage (variable)

A anode
B base (transistor)
B magnetic flux density
C collector (transistor)
C capacitance
D drain (CMOS device)
D diode, delay
E emitter (semiconductor device)
E voltage
G gate (CMOS device)
H holding (current), magnetic field strength
I current (dc), current (RMS), current (vector)
J junction (semiconductor device), inertia constant
K cathode
L inductance, latching (current for thyristor)
N number of turns, speed in rpm, negative current conducting rectifier
P electrical power, positive current conducting rectifier, pulse transformer
Q charge (constant), transistor (BJT)
R resistance, reverse (current), rise (time)
S rating of transformer
T period, tachometer, transformer, time constant (L/R)
$\mathcal{T}$ torque (constant)
$\bar{\mathcal{T}}$ torque (variable)
V voltage (dc), voltage (RMS), voltage (vector)
X inductive or capacitive reactance
Z impedance

1.2 Behaviour of Semiconductor Devices Under Biased Conditions

The silicon controlled rectifier (SCR) or thyristor comes under the family of semiconductor devices. Therefore, to understand its behaviour under biased conditions, it is necessary to recall the behaviour of a semiconductor *p-n* diode and a transistor under similar conditions.

p-n diode A *p-n* junction diode in which the two layers are equally doped is shown in Fig. 1.1(a). Due to the diffusion of the majority carriers (holes) of the *p*-layer towards the right and the majority carriers (electrons) of the *n*-layer towards the left, a depletion (or transition) region is formed at the junction. This depletion region consists of uncompensated negatively charged acceptor ions on the left side and uncompensated positive donor ions on the right side, as shown in Fig. 1.1(b). An acceptor ion is indicated by a minus sign because after an atom releases a hole (or equivalently, accepts an electron) it becomes a negative ion. Similarly the donor ion is represented by a plus sign because after the impurity atom donates an electron it becomes a positive ion. These uncompensated ions on either side constitute a potential barrier as shown in Fig. 1.1(c), which prevents the further diffusion of holes to the *n*-side and electrons to the *p*-side. The device is then said to be under equilibrium.

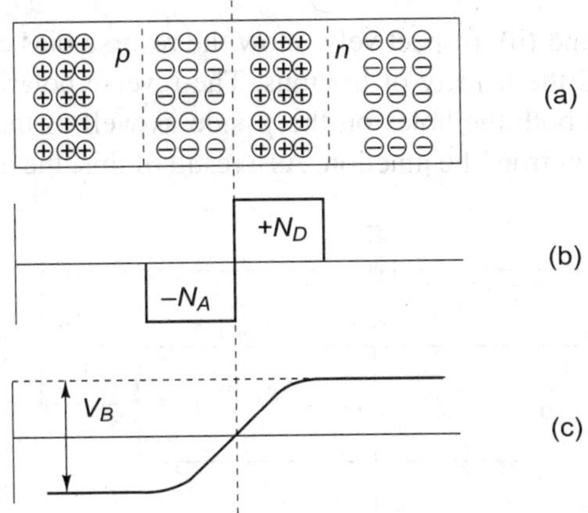

Fig. 1.1 *p-n* junction under the no bias condition: (a) schematic diagram, (b) density of uncompensated acceptor and donor ions, (c) built-in potential distribution

If a battery is now connected across the *p-n* junction as shown in Fig. 1.2(a), the device is said to be subjected to a forward bias. The battery potential disturbs the equilibrium and causes the holes to move to the right and the electrons to the left. This movement of the carriers under external bias conditions of the diode is known as the *drift*. It is to be noted that an electrical field of magnitude, say, *E* (volts/metre) must exist along the length of the device so as to initiate the drift current; as a result, electrons tend to move towards the anode and holes towards the cathode.

The movement of the charged carriers leads to neutralization of the originally uncompensated ions on either side; consequently the width of the depletion layer is reduced as shown in Fig. 1.2(b). Further, the potential barrier across the junction is reduced to the extent of the applied positive bias voltage, as shown in Fig. 1.2(c).

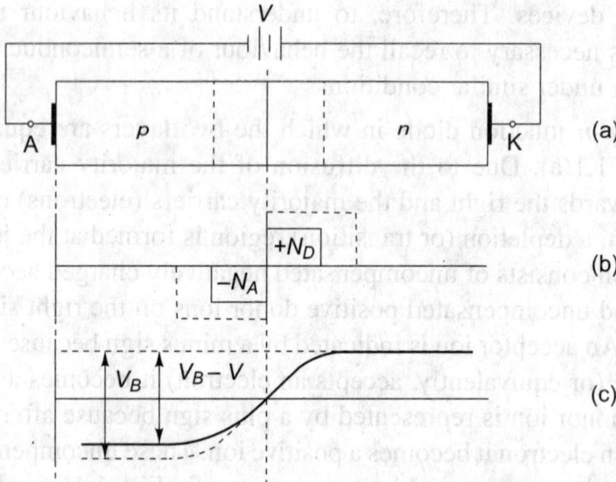

Fig. 1.2 *p-n* junction under the forward bias condition: (a) circuit diagram, (b) density of ions, (c) potential distribution

Figures 1.3(a) and (b), respectively, show the reverse bias condition of a *p-n* junction diode and the density of the ions. The reverse polarity of the external connection causes both the holes on the *p*-side as well as the electrons on the *n*-side to move away from the junction. As a result of this, the dense region of the

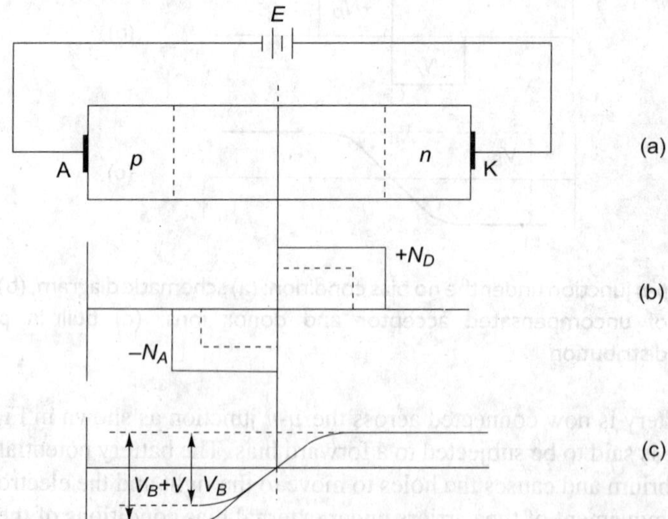

Fig. 1.3 *p-n* junction under the reverse bias condition: (a) circuit diagram, (b) density of ions, (c) potential distribution

negatively charged acceptor ions shifts to the left and that of the positively charged donor ions to the right. In other words, the depletion layer gets widened. In this condition the holes that are supplied from the n-side move across the junction to the left and get attracted to the negative terminal of the battery. Likewise the electrons from the p-side move to the right and go to the positive terminal of the battery. These two charge carrier currents crossing the junctions constitute the total current, which is called the *reverse saturation current* of the diode. Reverse bias causes widening of the potential barrier to the extent of the applied reverse voltage, as shown in Fig. 1.3(c), and prevents further drift of the charge carriers in either direction. Thus a new equilibrium condition is reached.

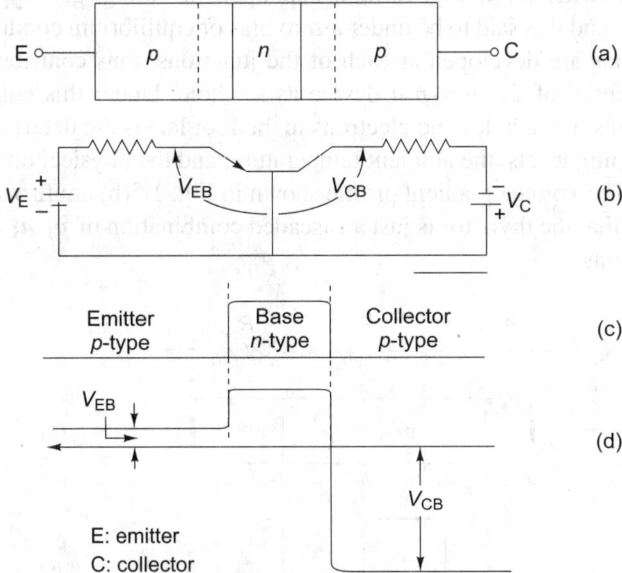

Fig. 1.4 (a) Schematic diagram of a p-n-p transistor, (b) forward bias of the emitter–base junction and reverse bias of the collector–base junction. Potential distributions under the (c) unbiased and (d) biased conditions.

Junction transistor The physical structure as well as the schematic diagram of a p-n-p transistor are shown in Fig. 1.4(a) and the bias condition of the junctions in Fig. 1.4(b). The three layers are, respectively, the emitter, base, and collector. Such a device consists of a very thin n-layer sandwiched between two wide p-layers. Under the equilibrium or no bias condition the variation of potential through the device will be as shown in Fig. 1.4(c). For transistor action to take place, the emitter–base junction, which is just a p-n junction, is forward-biased and the collector–base junction is reverse-biased. Under this condition the potential barrier at the emitter–base junction is reduced and that at the collector–base junction is augmented, as shown in Fig. 1.4(d). This causes more holes to be injected from the emitter layer to the base region. As the width of the centrally situated base (n) layer is small, a large fraction of the holes that are injected into the base region are sucked away by the collector due to the reverse bias action of the collector–base p-n junction. A very small fraction of these holes recombine in the base region to form a fraction of the base current which flows in the emitter–base circuit. The collector current constitutes 90–99 per cent of the emitter current.

1.2.1 Behaviour of a Thyristor Under Biased Conditions

The operation of the thyristor is now dealt with. The thyristor is a four-layer (p_1, n_1, p_2, n_2) semiconductor device as shown in Fig. 1.5(a) and has three *p-n* junctions, namely, p_1-n_1, n_1-p_2, and p_2-n_2. Hence there is considerable difference in its action as compared to that of the transistor. However, the basic principles that apply to a biased semiconductor diode hold good for its three *p-n* junctions.

When the switch Sw in Fig. 1.5(a) is kept open, no voltage gets applied across the thyristor, and it is said to be under a zero bias or equilibrium condition. Space charge regions are developed at each of the junctions, thus contributing to the built-in potential of the *p-n-p-n* device as a whole. Under this condition, the concentrations of the holes and electrons in the four layers are determined by the impurity doping levels, the ambient temperature, and the physical dimensions of the device. The voltage gradient profile shown in Fig. 1.5(b) confirms the earlier observation that the thyristor is just a cascaded combination of p_1-n_1, n_1-p_2, and p_2-n_2 junctions.

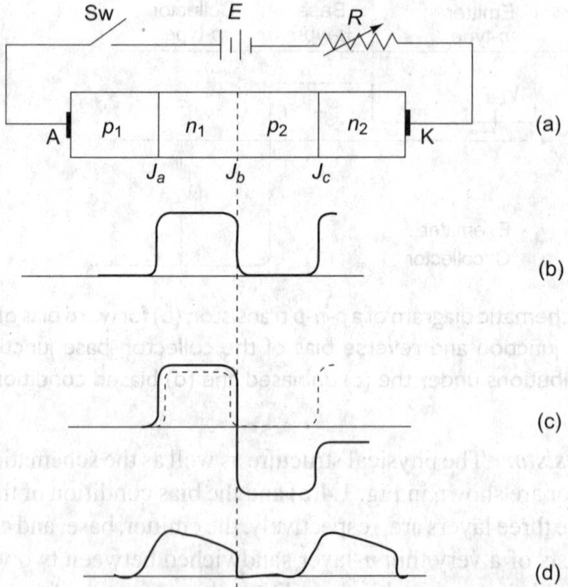

Fig. 1.5 Thyristor under forward bias condition: (a) circuit; potential diagrams under the (b) equilibrium, (c) forward blocking, and (d) forward conducting conditions

Now, if the switch Sw in Fig. 1.5(a) is closed, the p_1-layer is connected to the positive terminal of a battery and the n_2-layer is connected to its negative terminal through a resistor. Under this condition, called the forward bias condition of the thyristor, junctions J_a and J_c are forward-biased and junction J_b is reverse-biased. In order to understand the flow of the charge carriers prior to sudden conduction or 'firing' of the thyristor, it is first essential that the forward bias voltage be slightly less than that required to cause such conduction. In this condition, called the *forward blocking* condition, junction J_b is reverse-biased. The voltage gradient along the length of the device will be as shown in Fig. 1.5(c). However, the forward bias causes movement of the charge carriers in the device in three different ways as follows:

(a) The reverse-biased condition of junction J_b causes a widening of its space charge layer. A high electric field is created in it and a potential barrier is set up against the external bias. This barrier prevents the flow of electrons from the n_1-layer to the p_2-side and holes from the p_2-layer to the n_1-side. On the other hand holes, which are minority carriers in the n_1-region, tend to travel towards junction J_b and accumulate on its left side. Similarly electrons, which are minority carriers in the p_2-layer, move towards J_b and accumulate on its right side.

(b) Reverse-biasing of J_b leads to the movement of electrons in the n_1-layer towards junction J_a and causes them to cross this junction to the p_1-layer and traverse through it to the anode of the battery. Consequently, positively charged donor ions are left behind on the left side of the space charge region of J_b. These electrons increase the forward bias of J_a, narrow down its space charge region, and cause more holes in p_1 to cross J_a and go towards J_b. In a similar manner holes in the p_2-layer cross J_c into n_2 and through it to the cathode of the battery, thus increasing the forward bias of J_c. In so doing they release more electrons from n_2, which move to the left, cross J_c, and travel towards J_b. The net effects of all these actions are
   (i) the space charge region of J_b is widened and
   (ii) the accumulation of holes to the left of junction J_b and electrons to its right as discussed in (a) is augmented.

(c) Due to thermal agitation, hole–electron pairs are generated in and around the space charge layer of J_b. Thus more holes are thermally generated in the n_1-layer and move towards J_b and get accumulated to the left of J_b. Similarly thermally generated electrons in the n_2-layer move towards J_b and get accumulated on its right side. This action further strengthens the action discussed in (a) and (b).

Now, if the forward bias is increased to a value known as the *forward breakover* voltage rating of the thyristor, the following movements, which are a continuation of those discussed in (a), (b), and (c) above, respectively, take place.

(a′) The holes which are minority carriers in the n_1-layer and which are accumulated on the left side of J_b gain enough energy to cross junction J_b, go to its right side, and move towards junction J_c. This increases the forward bias of J_c and causes its space charge layer to narrow down. Similarly the electrons which are accumulated on the right side of J_b gain enough energy to cross it, go to its left side, and move towards the anode. This action increases the forward bias of J_a and causes its space charge layer to narrow down.

(b′) The action mentioned in (b) is further strengthened. The forward bias of J_a is further increased, releasing more holes from p_1 to the right to cross J_a and move toward the cathode. Similarly more holes in the p_2-layer go to the right, cross J_c, and release more electrons from the n_2-layer, which move to the left to cross J_c and J_b and go to the anode.

(c′) The thermally generated holes on the left of J_b and electrons on its right side cross it in either direction.

The actions in (a′), (b′), and (c′) aid each other causing copious movement of holes to the cathode and electrons to the anode. As a consequence, a large quantity of donor (positive) impurities which are on the left side of J_b and another large quality of acceptor (negative) impurities which are on the right side of J_b get neutralized by electrons and holes, respectively. Junction J_b no longer supports the blocking voltage and the device suddenly switches into the conduction state. Under this condition J_b becomes forward-biased and the device as a whole attains a low-impedance condition. Consequently, the voltage barriers at the three junctions get reduced considerably. The total resistance of the device as well as the voltage drop across it drop down to low values. Figure 1.5(d) shows the voltage gradient diagram under this condition.

A quantitative analysis of the flow of the electron and hole currents in the thyristor can be made at this stage. The currents flowing from anode to cathode consist of holes proceeding from the left side of junction J_b to the cathode and that of electrons moving from its right side towards the anode. A large portion of the current flowing from the anode terminal through the device to the cathode terminal is due to the combined movement of holes and electrons as detailed in (b′) above. These currents are shown as $\alpha_1 I_A$ and $\alpha_2 I_K$ in Fig. 1.6. Only a small portion of this current is contributed by the reverse saturation current discussed in (a′) on the one hand and the thermal current mentioned in (c′) on the other. The sum of the reverse saturation and thermal currents due to holes is denoted as I_{ps} and that due to electrons as I'_{ns}. It is also shown in Fig. 1.6 that a small fraction of the holes proceeding from p_1 into n_1 recombine in n_1; this is given by $(1-\alpha_1)I_A$. Likewise, a small fraction of the electrons moving from n_2 into p_2 recombine in p_2, this current being given by $(1-\alpha_2)I_K$.

The expression for the total current flowing from the anode to the cathode can be written as

$$I = \alpha_1 I_A + \alpha_2(-I_K) + I_{ps} + (-I'_{ns}) \tag{1.1}$$

Recognizing that

$$I_K = -I_A$$

and

$$I'_{ns} = -I_{ns}$$

where I_{ns} has the same direction as I_{ps}, the total current I becomes

$$I = (\alpha_1 + \alpha_2)I_A + I_{ps} + I_{ns} \tag{1.2}$$

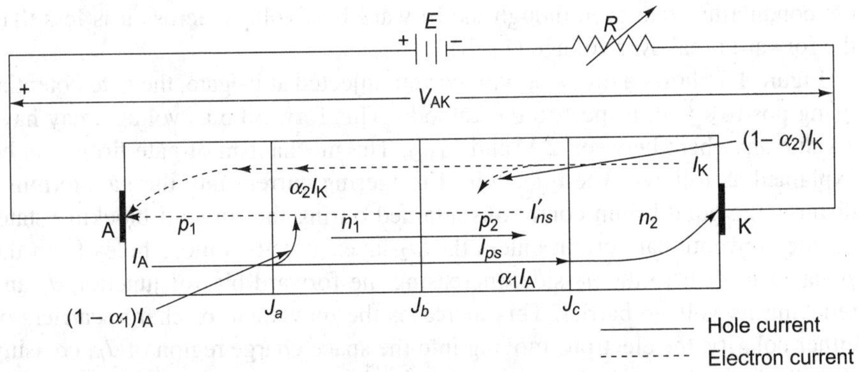

Fig. 1.6 Hole and electron currents in a forward-biased thyristor

Writing the total reverse saturation current as

$$I_s = I_{ps} + I_{ns}$$

and noting that $I = I_A$, Eqn (1.2) can be written as

$$I_A = (\alpha_1 + \alpha_2)I_A + I_s \tag{1.3}$$

Solving Eqn (1.3) for I_A gives

$$I_A = \frac{I_s}{1-(\alpha_1 + \alpha_2)} \tag{1.4}$$

Equation (1.4) can be interpreted as follows. As long as the thyristor is in the forward blocking state, the sum $(\alpha_1 + \alpha_2)$ will be much less than unity. Accordingly I_A will be very small in magnitude. When a voltage greater than or equal to the forward breakover voltage of the thyristor is applied across the thyristor, sudden switching takes place due to the regenerative action occurring in the thyristor. This can be interpreted as $(\alpha_1 + \alpha_2)$ approaching unity, causing I_A to shoot up, and the thyristor suddenly switches on. The device is then said to 'fire'. The large current that flows through the external circuit should be limited to a value less than the current rating of the thyristor by connecting an external series resistance in the circuit; the device will otherwise get damaged due to the heating effect consequent to the flow of this large current. In the absence of an external resistor this current will be large even when the external supply voltage is small, say, a few volts. It can be inferred from the above discussion that the thyristor acts like a closed switch when it gets fired and offers very little resistance to the flow of current. The drop across the thyristor in the conducting state will be of the order of 1–2 V.

1.2.2 Gate Firing of the Thyristor

In the above discussion, the gate terminal G of the thyristor or any current fed through it from an external source was not mentioned. Such a current has a catalytic effect on the firing of the thyristor as follows. If a small current, called the *triggering current*, is fed into this terminal, then the thyristor will switch onto the conducting state even though the forward bias voltage across it is less than the forward breakover voltage (V_{FBO}).

Figure 1.7 shows a thyristor with current injected at its gate, the gate potential being positive with respect to the cathode. This forward bias voltage may have a value anywhere between 2 V and V_{FBO}. The mechanism of gate firing can be explained as follows. The injection of triggering current into the gate terminal disturbs the equilibrium conditions attained during the forward blocking state. As the conventional current enters the p_2-layer, it causes more holes from the p_2-layer to drift to the n_2-side, increasing the forward bias of junction J_c and reducing its voltage barrier. This increases the movement of charge carriers of either polarity, the electrons moving into the space charge region of J_b, crossing it, and moving into p_1, thus increasing the forward bias of J_a. The rest of the action takes place exactly as discussed above with a forward bias voltage equal to or greater than V_{FBO} applied across the thyristor, in the absence of the gate signal. Avalanche multiplication takes place and the thyristor switches to the ON state.

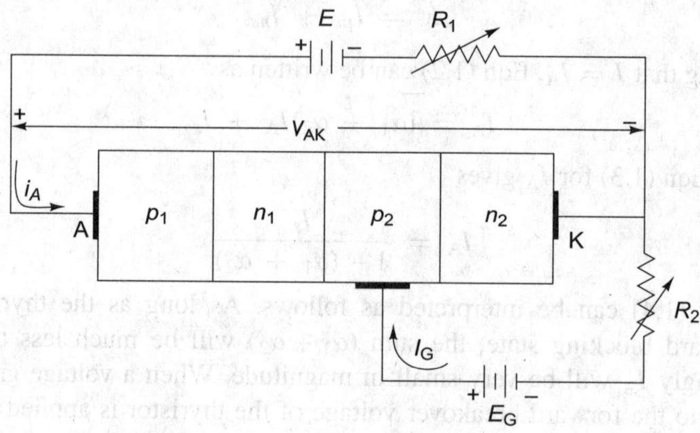

Fig. 1.7 Thyristor with a forward bias of V_{AK} volts and a gate current of I_{G} amperes

The switching action with a finite gate current may also be explained mathematically as follows. Equation (1.1) is modified as

$$I = \alpha_1 I_{\text{A}} + \alpha_2(-I_{\text{K}}) + I_{ps} + (-I'_{ns}) \tag{1.5}$$

With $I_{\text{K}} = -(I_{\text{A}} + I_{\text{G}})$ and $I'_{ns} = -I_{ns}$, Eqn (1.5) becomes

$$I = (\alpha_1 + \alpha_2)I_{\text{A}} + \alpha_2 I_{\text{G}} + I_{ps} + I_{ns} \tag{1.6}$$

Defining $I_s = I_{ps} + I_{ns}$ and noting that the current through the thyristor is $I = I_{\text{A}}$, Eqn (1.6) becomes

$$I_{\text{A}} = (\alpha_1 + \alpha_2)I_{\text{A}} + \alpha_2 I_{\text{G}} + I_s \tag{1.7}$$

Equation (1.7) can be solved for I_{A} as

$$I_{\text{A}} = \frac{\alpha_2 I_{\text{G}} + I_s}{1 - (\alpha_1 + \alpha_2)} \tag{1.8}$$

The regenerative action of the thyristor can be interpreted from Eqn (1.8) as follows: As I_A increases, α_1 increases; as I_K and I_G increase, α_2 increases. At a certain stage of this cumulative process, the sum $(\alpha_1 + \alpha_2)$ approaches unity and the thyristor switches to the ON state.

1.2.3 Two-transistor Analogy of the Thyristor

The switching action that takes place with gate current injection can also be explained with the help of the two-transistor analogy. Figures 1.8(a) and (b), respectively, show the back-to-back connection of *n-p-n* and *p-n-p* transistors as well as their equivalent circuits. The base current and collector current, respectively, of the *p-n-p* and *n-p-n* transistors can be expressed as

$$I_{B_1} = (1 - \alpha_1)I_A - I_{CB_1} \tag{1.9}$$

and

$$I_{C_2} = \alpha_2(-I_K) + I_{CB_2} \tag{1.10}$$

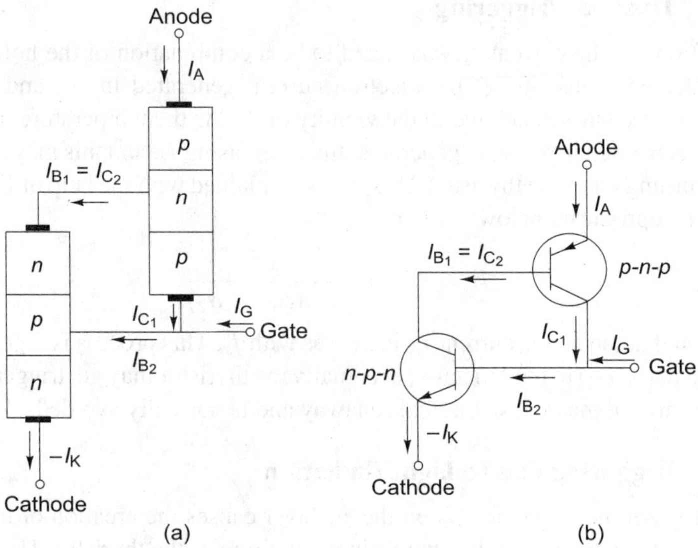

Fig. 1.8 Two-transistor analogy of a thyristor: (a) schematic diagram, (b) equivalent circuit

Here I_{CB_1} and I_{CB_2} are, respectively, the reverse saturation currents of *p-n-p* and *n-p-n* transistors. Recognizing that $I_{B_1} = I_{C_2}$ and $I_K = -(I_A + I_G)$, Eqns (1.9) and (1.10) can be combined as

$$(1 - \alpha_1)I_A - I_{CB_1} = \alpha_2(I_A + I_G) + I_{CB_2} \tag{1.11}$$

Solving Eqn (1.11) for I_A gives

$$I_A = \frac{\alpha_2 I_G + I_{CB_1} + I_{CB_2}}{1 - (\alpha_1 + \alpha_2)} \tag{1.12}$$

Defining $I_s = I_{CB_1} + I_{CB_2}$, Eqn (1.12) becomes

$$I_A = \frac{\alpha_2 I_G + I_s}{1 - (\alpha_1 + \alpha_2)} \tag{1.13}$$

If the sum $\alpha_1 + \alpha_2$ approaches unity, then I_A will increase to a very large value and the thyristor will switch to the conduction state. If the derivation is started with expressions for I_{B_2} and I_{C_1} instead of I_{B_1} and I_{C_1}, it can be shown that it leads to the same final result as in Eqn (1.13).

1.3 Methods for Triggering

It is stated above that with the slow increase of the positive bias voltage, the thyristor suddenly gets triggered at a certain value of the voltage called the forward breakover voltage (V_{FBO}). Though this is one way of triggering the thyristor, it is seldom resorted to. Other methods of triggering are described below, gate triggering being the most commonly used one.

1.3.1 Thermal Triggering

In Section 1.1 the current I_s was stated to be a combination of the hole current thermally generated in n_1, the electron current generated in p_2, and also the electron–hole pairs generated in the vicinity of J_b. As the temperature increases, more electron–hole pairs are generated, thus increasing I_s, and this may even lead to the turning on of the thyristor. This can be explained with the help of Eqn (1.4), which is reproduced below:

$$I = I_A = \frac{I_s}{1 - (\alpha_1 + \alpha_2)} \tag{1.14}$$

It is seen that the anode current I_A increases with I_s. This process is regenerative, causing $(\alpha_1 + \alpha_2)$ to tend to unity; eventually the thyristor may get triggered. This type of turn-on may cause thermal runaway and is normally avoided.

1.3.2 Triggering due to Light Radiation

Light focused near junction J_c on the p_2-layer causes the creation of holes and electrons, thus disturbing the equilibrium condition of the thyristor. The forward bias on J_c increases, and if the production of the charge carriers is sufficiently high, the device switches into conduction.

1.3.3 Gate Triggering

As stated in Section 1.2.2, a current injected into the p_2-layer with the gate terminal having positive polarity with respect to the cathode triggers the thyristor with a forward bias less than V_{FBO}. Different methods for such gate triggering exist and are named according to the auxiliary device that is employed; the more important among these are saturable core reactor triggering and unijunction transistor (UJT) triggering. These methods are described later.

Figure 1.9 gives a set of forward characteristics of a thyristor for various values of the gate current. Considering the characteristic with zero gate current, the portion OA is known as the *forward blocking region* and is a stable one. In the region AB the process of increased flow of charge carriers through J_b is initiated. The abscissa of point B gives the value of the forward breakover voltage.

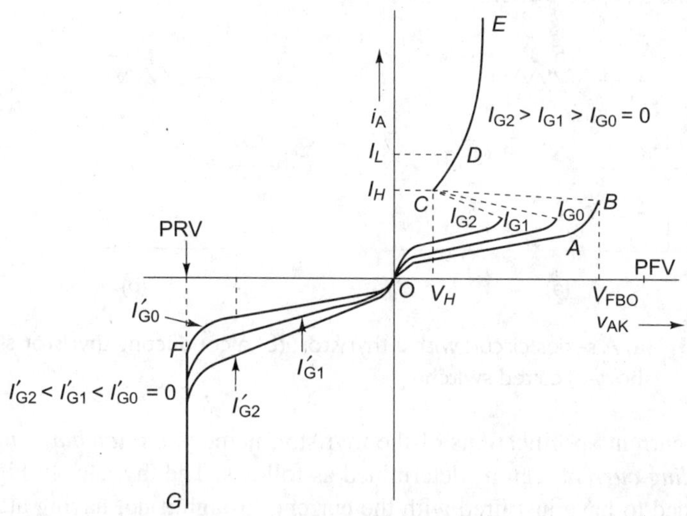

Fig. 1.9 Forward and reverse characteristics of a thyristor

The portion BC, which is shown dotted, is unstable and is a quick transition region from the OFF to the ON state. The characteristic CDE is the conducting region, which is also a stable one. It is seen from Fig. 1.9 that the breakover voltage decreases as the gate current is increased; this implies that with a fairly large gate current the thyristor gets fired even with negligible breakover voltage. The figure also shows that the forward characteristic of the thyristor then approaches that of a semiconductor diode. The relationship between the breakover voltage and the gate current shown in Fig. 1.10 confirms the fact that the forward voltage required to trigger the thyristor is negligibly small for a value of gate current equal to or exceeding I_{Gm}, this being the maximum allowable gate current rating.

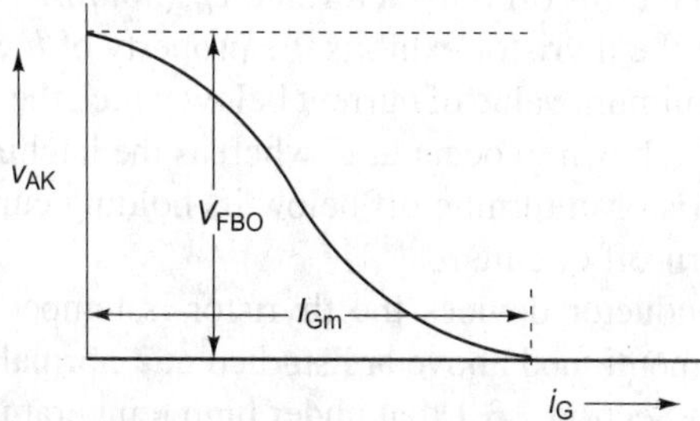

Fig. 1.10 Variation of breakover voltage with gate current

In the conducting state the current has to be limited by an external resistance connected in series with the thyristor, failing which, a very large current, which exceeds the continuous carrying capacity of the thyristor, flows. The device then gets damaged due to the excessive heating effect of this current. In Fig. 1.11(b) current I_A flowing through the device becomes V/R, with the conducting thyristor acting like a closed switch.

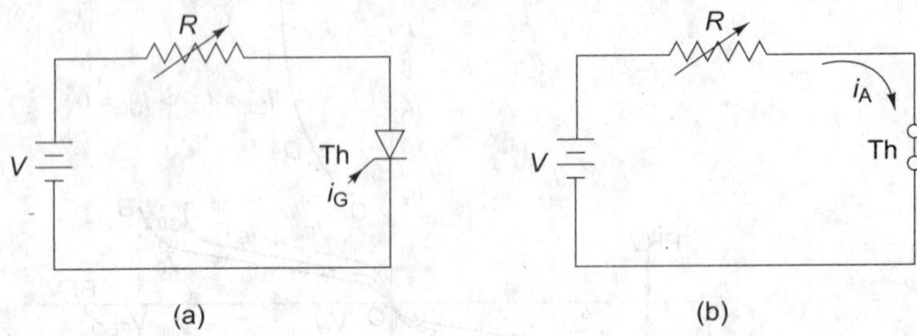

Fig. 1.11 (a) A series circuit with a thyristor; (b) a conducting thyristor shown as a short-circuited switch

Two current specifications of the thyristor, namely, the *latching current* and the *holding current*, can be determined as follows. The thyristor in Fig. 1.11(a) is assumed to have just fired with the current through it not having attained the *latching value*. However, when the resistance R is slowly decreased, at some stage the current through the thyristor reaches the latching value; that is, the device attains a stable conduction state at this value of current. The significance of the latching current can be further explained as follows. If the gate triggering current is removed before the thyristor current attains a value equal to I_L, the device will go back to the OFF state. This will not happen if the current exceeds the latching value I_L; on the other hand, the gate triggering signal loses its control over the device after the current attains the latching current value. It follows that the gate triggering signal is needed only till the thyristor current attains the latching value. The time taken to attain I_L is ordinarily of the order of a few microseconds, provided the load resistance is of appropriate value. If the thyristor current is initially above the latching value, the gate signal is disconnected and the thyristor current is slowly reduced, the thyristor will not turn off when the current is brought down to I_L; instead it turns off only at a value I_H (*holding value*) which is below I_L. This shows that the thyristor exhibits the property of *hysteresis*. The holding current I_H is the minimum value of current below which the thyristor turns off. In Fig. 1.9 this value is shown to occur at C whereas the latching value is at D. The property of the thyristor in turning off below its holding current value is utilized for designing its turn-off circuits.

Like all semiconductor devices the thyristor is temperature-dependent. All the characteristics mentioned above are studied at a normal temperature of, say, 25 °C. It is stated in Section 1.3.1 that under high temperatures the thyristor may get inadvertently triggered; this eventuality is normally prevented. Section 1.6 deals with a related topic, namely, the provision of an appropriate heat sink to the thyristor in order to limit the maximum junction temperature to 125 °C.

The height and duration of the gate current pulse required for firing a thyristor are decided by keeping in mind that the thyristor is a charge operated device. Unless sufficient charge is injected into the vicinity of junction J_c, no triggering takes place. If the current pulse is of height I and its duration h, then the charge injected into the p_2-layer will be Ih. This charge should be sufficient to fire the thyristors; sharp fronted gate pulses, called *trigger pulses*, are also found to successfully fire the thyristor. For injecting the same charge with such a narrow pulse, its height should be increased so that the area under it is the same as that for a rectangular pulse that successfully fires the thyristor.

Figure 1.12 shows the variation with time of the current i_A through the thyristor and the voltage v_A across it, as functions of time, during the turn-on process. The turn-on time t_{ON} is constituted of three components, namely, the delay time t_D, the rise time t_R, and the settling time t_S. Thus

$$t_{ON} = t_D + t_R + t_S \tag{1.15}$$

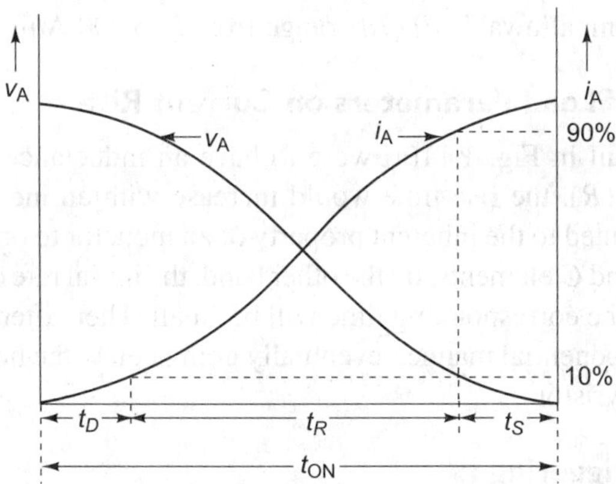

Fig. 1.12 Thyristor voltage and current during the turn-on process

The delay time t_D is the time required for the minority carriers (electrons) of p_2 to go to junction J_a, plus the time required for the holes released from p_1 to diffuse across n_1 and impinge into the space charge region of J_b. It is also equal to the time required for the minority carriers (holes) of n_1 to go to junction J_c plus that required for the electrons released from n_2 to diffuse through p_2 and finally to impinge on the space charge region of J_b. These two actions will simultaneously initiate the decrease of the voltage across the device, and the rise of the current through the device. The rise time t_R is governed by the time taken for the junction J_b to become forward-biased, this being due to the freely moving holes towards the cathode and the freely moving electrons towards the anode. The rise time t_R depends upon three quantities, namely, the gate current, the device temperature, and the elements (R, L, C) contained in the external circuit. A higher gate current leads to a lower rise time; again, a high temperature leads to copious production of charge carriers, which in turn leads to the reduction of rise time. The settling time t_S is the time taken for the conduction to spread over the entire area of cross section of the crystal. It depends upon this cross section as well as the dimensions of the electrodes. As the conduction reaches the equilibrium state, the voltage also settles down to the steady-state value, which is usually of the order of 1–2 V.

The turn-on time of the thyristor may range from 1 μs to a few hundred microseconds depending on the application for which the thyristor used. As it takes some time for the equilibrium condition of conduction to be attained, the device should not be subjected to a high rate of rise of the anode current (di_A/dt). This is because prior to reaching the equilibrium state, the voltage across the device is high and if it is also subjected to high di_A/dt it will be forced to carry large current in a short time. As a result of this, a hot spot develops due to current crowding, a large amount of heat dissipation occurs in the device, and temperatures approaching the melting point of the semiconductor device may be attained, thus leading to its damage. It can be inferred from the above that with a non-inductive load circuit, it is necessary to deliberately include an inductor of appropriate size to protect the thyristor from high di_A/dt values. The usual values of maximum allowable di_A/dt range from 5 to 200 A/μs.

1.3.4 Effect of Load Parameters on Current Rise

If the series circuit in Fig. 1.11(a) were to have an inductance (L) in addition to the resistance (R), the rise time would increase with an increase in di_A/dt. This can be attributed to the inherent property of an inductor to oppose any rise of current. With R and C elements, on the other hand, the initial rate of rise of current will be high and the corresponding time will be small. Thereafter the current will decrease in an exponential manner, eventually going below the holding value and turning off the thyristor.

1.3.5 dv/dt Triggering

If an alternating voltage v of high frequency and equal to $E\sin(2\pi ft)$ is applied to a circuit containing a capacitance and high resistance as shown in Fig. 1.13, the current flowing through the capacitor can be written as

$$i(t) = C\left(\frac{dv}{dt}\right) \tag{1.16}$$

$$= CE2\pi f\cos(2\pi ft) \tag{1.17}$$

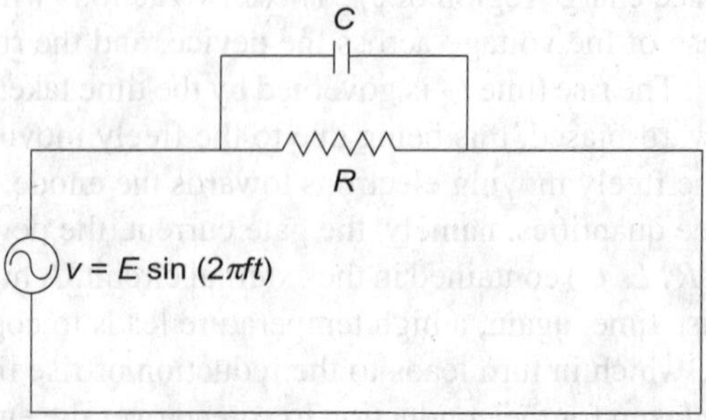

Fig. 1.13 *dv/dt* triggering

The RMS value of this current, given by $(CE2\pi f)/\sqrt{2}$, increases with frequency, that is, increases with a high dv/dt. In a thyristor the space charge region of each of the forward-biased junctions, namely, J_a and J_c, may be represented by a low resistance shunted by a capacitance, whereas that of the reverse-biased junction J_b may be represented by a high resistance shunted by a capacitance. Accordingly, in the high-frequency model of the thyristor shown in Fig. 1.14, the resistances R_{J_a} and R_{J_c} will be low and R_{J_b} will be high.

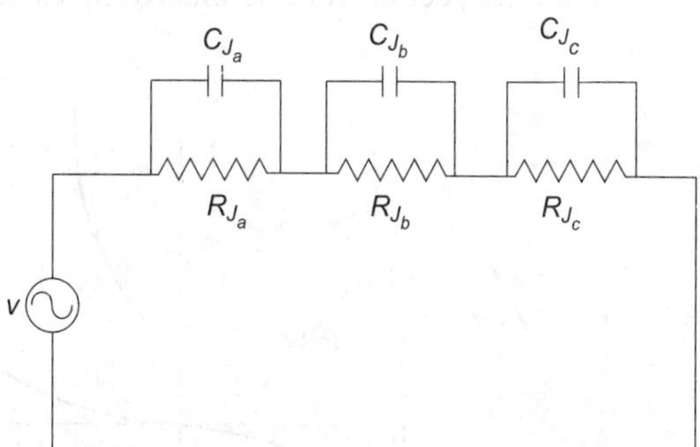

Fig. 1.14 High-frequency model of a thyristor

When a rapidly varying voltage is applied across the terminals of the thyristor for which junction J_b is initially biased, the current through its capacitance, given by $C_{J_b}(dv/dt)$, will be high and thus junction J_b almost becomes a short circuit. It is this current which initiates the turning on of the thyristor and plays the same part as (i) the application of a forward bias equal to or greater than V_{FBO} or (ii) the application of a triggering current at the gate, together with an appropriate forward bias voltage less than V_{FBO}. Regenerative action takes place as follows. I_a increases with α_1 and the sum $(\alpha_1 + \alpha_2)$ approaches unity, further increasing I_a as can be seen from Eqn (1.4). This cumulative process finally leads to the turning on of the thyristor. This type of switching on of the thyristor, being known as *dv/dt triggering*, evidently is an undesirable type of triggering. Section 1.7 deals with the design of circuits that prevent such high dv/dt.

An alternative interpretation of dv/dt triggering is as follows. If a rapidly varying voltage is applied across the terminals of a thyristor, the reverse-biased junction J_b experiences a rapid change in the charge contained in it. The resulting current through the junction is given by

$$i_{\text{b}} = \frac{d(q_{J_b})}{dt} = \frac{d(C_{J_b}v)}{dt} = C_{J_b}\frac{dv}{dt} + v\frac{dC_{J_b}}{dt} \tag{1.18}$$

The variation of the junction capacitance is usually very small; hence the second term can be neglected. Equation (1.18) now simplifies to

$$i_{\text{b}} = C_{J_b}\frac{dv}{dt} \tag{1.19}$$

The current i_b now initiates regenerative action, finally leading to the firing of the thyristor.

1.3.6 Reverse Characteristic

Figure 1.15(a) shows a reverse-biased thyristor; here the cathode is given a positive polarity with respect to that of the anode. As in the case of a reverse-biased diode, reverse saturation (or leakage) current flows through a reverse-biased thyristor. The junctions J_a and J_c therefore become reverse-biased and J_b becomes forward-biased.

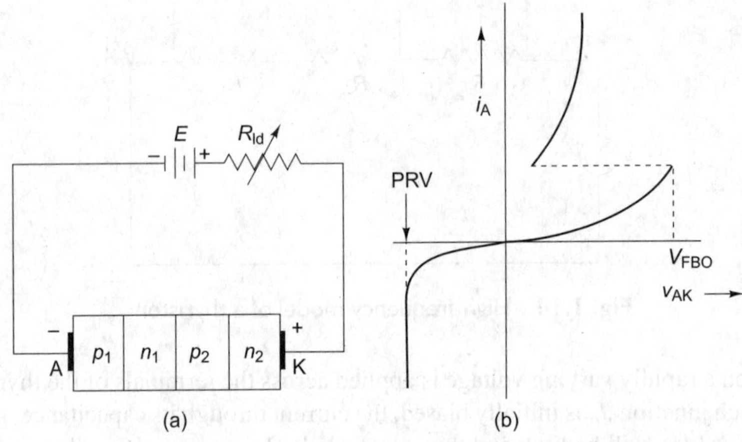

Fig. 1.15 A reverse-biased thyristor: (a) circuit, (b) characteristic curve showing the peak reverse voltage (PRV)

The holes in p_1 and the electrons in n_2 get attracted to the anode and the cathode, respectively. This leaves widened regions of negative acceptor ions to the left of J_a and positive donor ions to the right of J_c. Similarly the built-in potentials at these two reverse-biased junctions increase. Simultaneously the space charge region of the forward-biased junction J_b gets narrowed down. When the reverse voltage is increased, the junction J_c attains an avalanche breakdown condition because its built-in voltage is much small compared to that of J_a. The latter, however, does not attain the avalanche condition until the reverse voltage is increased to a high value called the *peak reverse voltage* (PRV) or *peak inverse voltage* (PIV) rating. This is one of the voltage specifications of a thyristor which are dealt with in Section 1.5. Figure 1.15(b) shows this feature on the SCR characteristic.

The PRV is a function of the maximum electric field intensity in junction J_a and also the *punch-through* across the n_1-layer. This in turn depends upon the width and doping level of this layer. The greater the width of this layer and the higher the doping level, the higher will be the value of the PRV. Punch-through occurs because the free carriers in the space charge region attain highly accelerated speeds, colliding with the lattice structure, thus releasing high-speed electron–hole pairs. This process multiplies and ultimately leads to the avalanche breakdown of J_a and the consequent sudden conduction (punch-through) of the thyristor. Another interpretation for the phenomenon of punch-through is that the space charge region of J_a extends and finally reaches that of J_b, making the thyristor a three-layered device. From Eqn (1.4) the equation for the reverse current I_R can be inferred to be

$$I_R = \frac{I_s}{1 - \alpha_1} \tag{1.20}$$

where I_s is the reverse saturation (or leakage) current of junction J_a and α_1 is the current gain parameter of the open base transistor p_1-n_1-p_2. With a small increase of the reverse voltage beyond the PRV, the charge carriers multiply, causing α_1 to approach unity. Thus the reverse characteristic beyond the PRV is very steep. In Fig. 1.9, *OF* is the reverse blocking region and *FG* is the avalanche conduction region.

1.4 Gate turn-on Methods

It is evident from the above discussion that injecting current through the gate terminal, which is called *gate triggering*, is the most appropriate method for turning a thyristor on. It also provides the means for controlling the forward breakover voltage, for the reason that this voltage decreases as the gate current increases. Figures 1.16(a) and (b), respectively, show the gate–cathode circuit and the v_G-i_G characteristics. The latter shows the region of successful turn-on in the i_G-v_G plane provided the anode current and the anode-to-cathode voltage, respectively, exceed the holding current and holding voltage values. The shaded portion is a region of points which will not turn on the thyristor under normal temperature conditions. If, however, the ambient temperature is increased, the thyristor gets triggered even for (i_G, v_G) points that lie in the shaded portion. The rectangular hyperbola on the right of this figure is the limitation imposed by the heat dissipation at the gate ($P_{G(max)} = i_G v_G$).

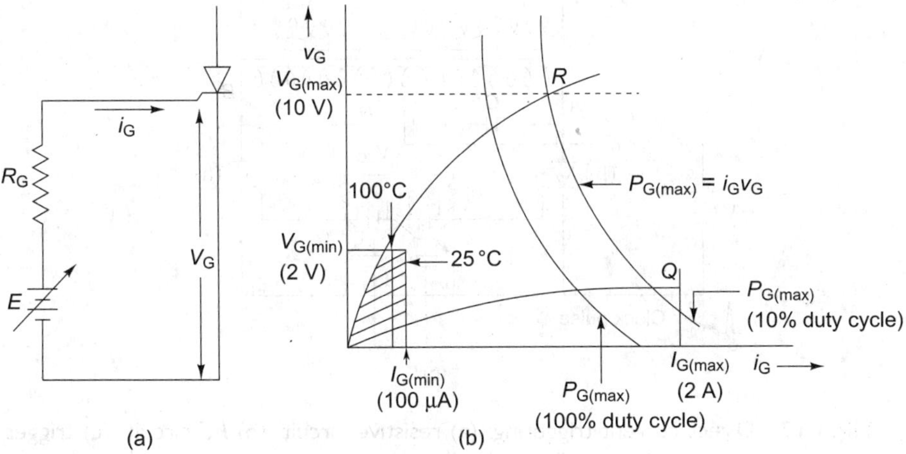

Fig. 1.16 (a) Gate–cathode circuit of a thyristor; (b) v_G-i_G characteristics

The three commonly used methods for gate triggering a thyristor are (a) ac triggering, (b) dc triggering, and (c) pulse triggering. Dc triggering and pulse triggering are adopted in chopper and inverter circuits whereas ac triggering is employed only in some single-phase rectifier circuits. Typical circuits for the above triggering methods are given below. The nature of the circuit depends upon the purpose for which the thyristor is used and the availability of direct or alternating voltage.

1.4.1 Dc Triggering

Dc triggering is adopted when the input supply to the thyristor is of the direct current type, as in the case of inverters and choppers; it is also used when the load is of an inductive nature. Three types of dc triggering are illustrated in Fig. 1.17 and these are described below.

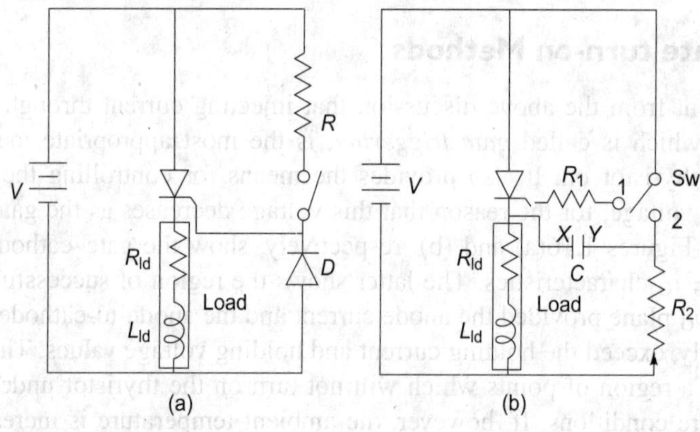

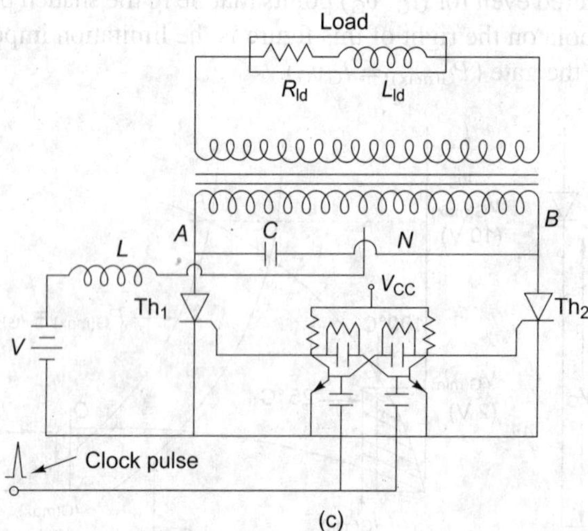

Fig. 1.17 Direct current triggering: (a) resistive circuit, (b) *RC* circuit, (c) trigger circuit with a bistable multivibrator

Figure 1.17(a) shows a thyristor feeding an inductive load, the gate signal being obtained through R. If the switch Sw is closed, the thyristor gets turned on by a dc signal. The diode D protects the gate from negative signals. This circuit may be made to work as a dc chopper if the switch Sw is alternately switched on and off; however, it does not incorporate a mechanism for turning off.

Figure 1.17(b) gives a circuit that can also be used as a dc chopper and is an improvement over that given in Fig. 1.17(a). When the thyristor is triggered and the switch is at position 1, the capacitor C is charged with the X-plate positive; with the switch at position 2, the capacitor discharges through the thyristor, thereby applying a reverse voltage to it and turning it off.

Figure 1.17(c) is a parallel capacitor inverter circuit in which the thyristors Th_1 and Th_2 are alternately triggered by the output pulses obtained from the multivibrator at each clock pulse. The duration of the dc pulses of the bistable multivibrator depends upon the frequency of the clock pulses. Other members of the multivibrator family can also be used for triggering purposes with suitable alterations in the circuitry.

In Fig. 1.18(a) anode current versus time characteristics are drawn for both resistive and inductive type of loads; the required lengths of time for the gate pulse in the two cases are shown in Fig. 1.18(b) as t_1 and t_2. It is evident from these figures that the gate current switch should be kept closed for a sufficiently long time so as to cause successful turn-on, this duration being longer for inductive loads.

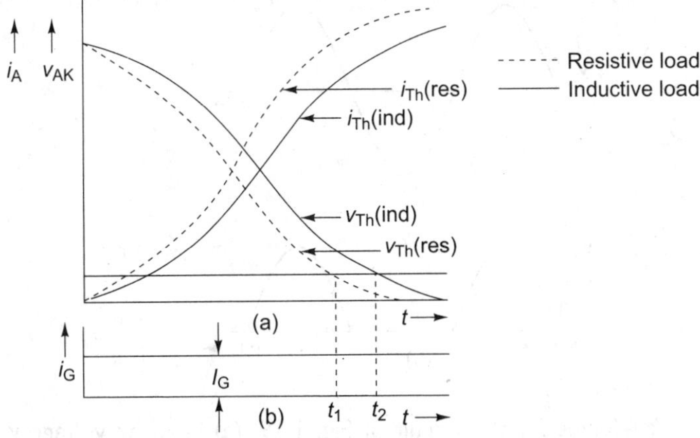

Fig. 1.18 (a) Anode current and v_{AK} vs time for a thyristor at turn-on, (b) gate triggering current vs time for resistive and inductive loads

1.4.2 Ac Triggering

Ac triggering is used where the source of supply is of the single- or three-phase ac type. Four methods of ac triggering are described below: (a) resistive, (b) RC, (c) saturated reactor, and (d) pulse triggering using a transistor as well as a UJT.

1.4.2.1 Resistance triggering Figure 1.19 shows a resistance triggering circuit for a single-phase, half-wave circuit, the firing angle being varied only during the positive half-cycle. The gate circuit consists of a fixed resistance R_f and a variable resistance R_v. Increasing R_v decreases the gate current and increases the angle of firing, the angle of firing being varied by this method in the range 0°–90°. The resistor R_b serves the purpose of giving a negative bias to the gate in the negative half-cycle of the ac and also of keeping the thyristor in the stable OFF state, when high temperatures are attained or when undesirable anode voltage transients occur. The input and load voltage waveforms for this circuit are given, respectively, in Figs 1.20(a) and (b).

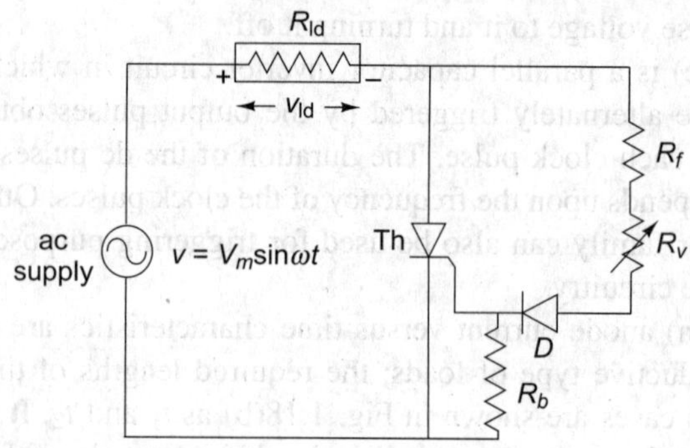

Fig. 1.19 Resistance triggering circuit

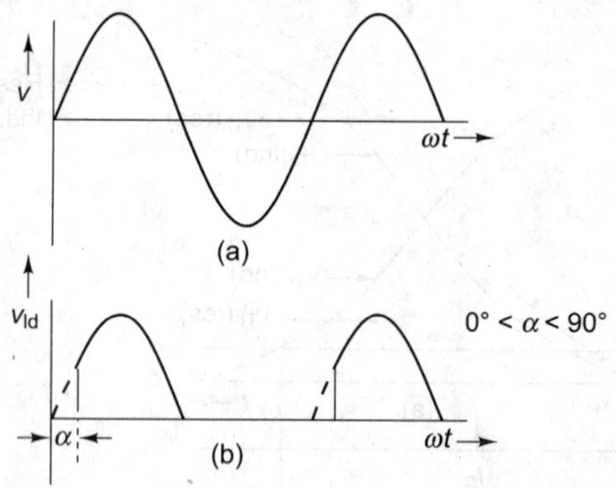

Fig. 1.20 Waveforms for the circuit of Fig. 1.19: (a) input ac voltage v, (b) load voltage v_{ld}

1.4.2.2 *RC* triggering Figure 1.21(a) shows a circuit in which the combination of a resistor and a capacitor is used to trigger the thyristor. Figure 1.21(b) shows that the capacitor voltage v_c lags the applied voltage V by an angle ϕ, which in turn varies with the value of R. The equivalent of the gate–cathode portion of the circuit given in Fig. 1.21(c) shows that the gate–cathode resistance R_{GK} is non-linear; its parallel connection with C helps in varying the triggering angle in the range 0°–180°.

The capacitor charges through R_{ld} and D_2 only after the ac voltage attains its negative maximum. This is because, in the duration t_0t_1 the supply voltage is

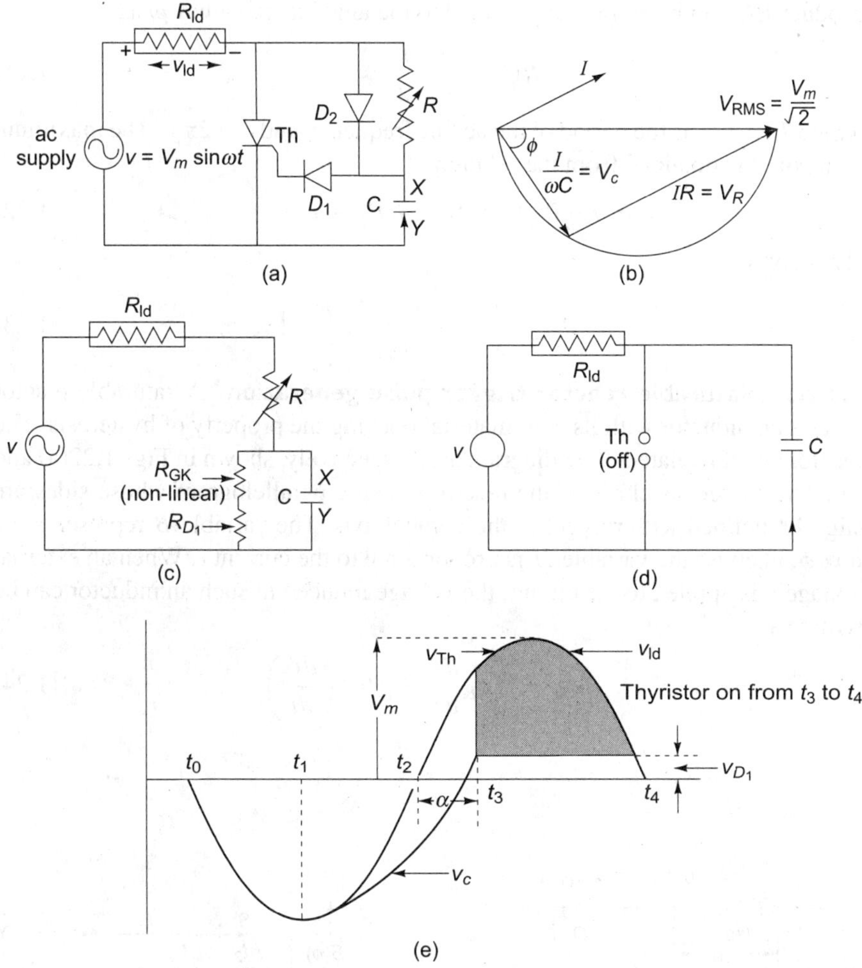

Fig. 1.21 *RC* triggering scheme: (a) circuit diagram, (b) phasor diagram, (c) equivalent gate–cathode circuit from t_1 to t_3, (d) Equivalent anode–cathode circuit from t_1 to t_3, (e) waveform showing the firing angle

negative, the current through the capacitor, $C(dv/dt)$, is also negative, and the Y-plate of the capacitor is positive. From time t_1 onwards the slope dv/dt and also the capacitor current are positive and the X-plate of the capacitor charges from $-V_{\max}$ towards $+V_{\max}$. From time t_2 onwards the equivalent circuit in Fig. 1.21(c) holds good. At a certain point in the duration of the positive half-cycle, the capacitor voltage v_c crosses zero towards the positive direction, but the thyristor fires only when its value exceeds that of the drop of the diode D_1, the firing angle being controlled by varying R. The SCR will conduct when the capacitor voltage v_c equals $v_{GT} + v_{D_1}$. The firing angle α is a non-linear function of R and C, thus being expressed as $\alpha = f(\omega RC)$. In the following derivation, it is assumed that the load resistance R_{ld} is negligible compared to the resistance R, which consists of R_f (the fixed part) and R_v (the variable part, not shown in the figure). The product RC can be empirically related to the angular frequency ω as

$$RC \geq 1.3\frac{T}{2} \approx \frac{4}{\omega} \tag{1.21}$$

where $T = 1/f$ is the period of the ac line frequency and $\omega = 2\pi f$. The maximum value of R is obtained from the relation

$$v \geq i_{GT}R + v_c = i_{GT}R + v_{GT} + v_{D_1} \tag{1.22}$$

This gives

$$R \leq \left(\frac{v - v_{D_1} - v_{GT}}{i_{GT}}\right) \tag{1.23}$$

1.4.2.3 Saturable reactor trigger pulse generation A saturable reactor (SR) is an inductor with its core material obeying the property of hysteresis. The reactor and the related B-H diagram are, respectively, shown in Figs 1.22(a) and (b). The hysteresis curve of the reactor is like a parallelogram whose sides are slightly inclined with respect to the vertical axis. The variable B represents the flux ϕ; likewise the variable H is proportional to the current i. When an external voltage v is applied to the circuit, the voltage induced in such an inductor can be written as

$$v_{SR} = L_{SR}\frac{di}{dt} = N_G\left(\frac{d\phi}{dt}\right) \tag{1.24}$$

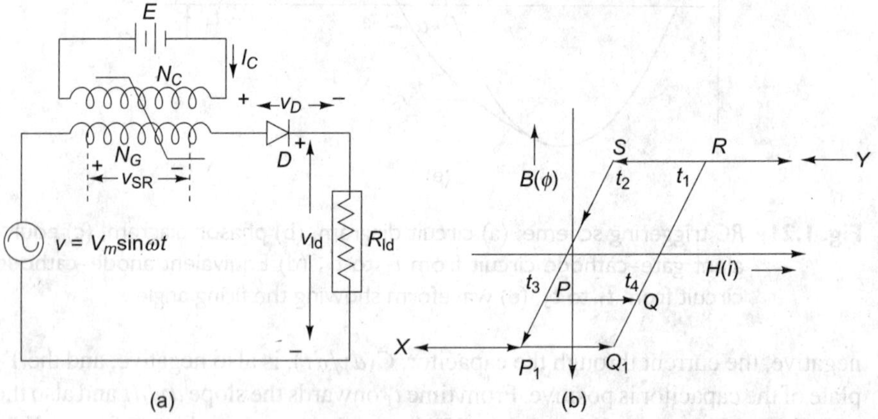

Fig. 1.22 (a) A saturable reactor connected in series with load; (b) B-H diagram (same as the ϕ-diagram)

Here, L_{SR} and N_G are, respectively, the inductance of the saturable reactor and the number of turns of the winding in the load circuit. Also, N_C is the number of turns of the control winding. Equation (1.24) can be solved for L_{SR} as

$$L_{SR} = N_G\left(\frac{d\phi}{di}\right) \tag{1.25}$$

Thus the inductance of the saturable reactor is obtained at any point on the ϕ-I curve as its slope at that point. It can be seen from Fig. 1.22(a) that in the unsaturated portions QR and PS, $d\phi/di$ is very high whereas in the flat portions SR and PQ, $d\phi/di = 0$. If the current were to be slowly increased, the reactor enters the saturated region after a certain value, the inductance collapses from a high value to zero. In the circuit of Fig. 1.22(b) this property is utilized to control the load voltage as follows. When the magnetizing current i_c of the reactor is zero, its voltage drop is high, leaving very little voltage at the load. When the reactor is saturated, the drop across it becomes zero and the load voltage equals the supply voltage. The waveforms of v, i_{ld}, and ϕ given in Fig. 1.23 illustrate this phenomenon.

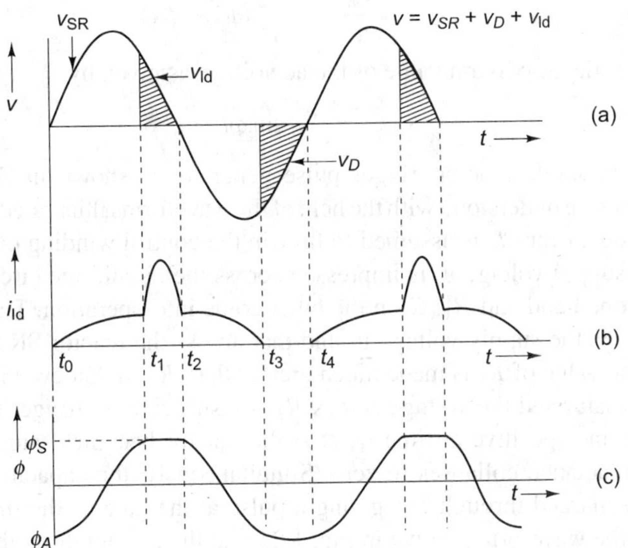

Fig. 1.23 Waveforms for the circuit of Fig. 1.22(a): (a) load voltage, (b) load current, (c) flux

At time $t_0 = 0$, a current I_c is assumed to pass through the control winding of the reactor causing a steady flux ϕ_A in the core. In the positive half-cycle, the voltage across, as well as the current in, the series circuit increase. At t_1 the core attains saturation and the potential across the reactor becomes zero, this condition continuing till t_2, at which point the supply voltage reverses. Though the load current i_{ld} is still positive, the exciting current is zero. This current, which is equal to the sum of $N_C I_{\text{ld}}/N_G$ and an instantaneous current (depending upon the instantaneous flux value), passes through the gate winding. At time t_2 the ϕ-i curve enters the high sloped region and remains so till t_3. The diode D is forward-biased as long as current flows into the load. At t_3 the load current, which lags the voltage, becomes zero. During the interval t_3 to t_4 the diode gets reverse-biased and its drop equals the full supply voltage. At t_4 the initial condition that occurs at t_0 is attained and a current equal to (EI_c/v_{SR}) flows in the gate winding, the flux now having a negative value. From t_4 onwards the cycle repeats. The angle of lag is seen to be equal to $\omega(t_2 - t_3)$. The average load voltage v_{dc} at the load can be written as

$$v_{dc} = \frac{V_m}{\pi} - \frac{1}{\pi}\int_{t_0}^{t} v_{SR}\,dt \tag{1.26}$$

Using Eqn (1.24), v_{dc} can be expressed as

$$v_{dc} = \frac{V_m}{\pi} - \frac{1}{T}\int_{\phi_A}^{\phi_S} N_G\,d\phi \tag{1.27}$$

Evaluating the integral in Eqn (1.27) gives

$$v_{dc} = \frac{V_m}{\pi} - \frac{N_G}{T}(\phi_S - \phi_A) \tag{1.28}$$

where V_m is the maximum value of the ac voltage as given by

$$v = V_m \sin \omega t$$

A typical saturable reactor trigger pulse generator is shown in Fig. 1.24; its operation can be understood with the help of the waveforms illustrated in Fig. 1.25. Initially a dc current I_c is assumed to flow in the control winding of the reactor. When the supply voltage v_s is impressed across the circuit, the circuits R_1, SR, R_2 on the one hand and R_1, C on the other come into operation. The voltage v_c across C lags the supply voltage v_s and that across the reactor SR (not shown) leads it. The value of R_1 is made much greater than R_2 so that even if the reactor were to be saturated the voltage across R_2 is insufficient to trigger the thyristor Th. v_c becomes positive at time t_1; at t_2 the reactor becomes saturated and its inductive reactance collapses to zero. Simultaneously, the capacitor voltage is quickly discharged through R_2, giving a pulse at the gate of the thyristor. It is seen from the waveforms shown in Fig. 1.25 that the current through R_2 (that is, i_{R_2}) is initially small, but once the reactor gets saturated at t_2 it shoots up to a high value. At t_3 the supply voltage as well as the reactor voltage become negative, and the reactor once again attains the unsaturated condition. The current being a lagging one, continues to flow through the diode D till t_4. During the interval t_4 to t_5 the diode gets reverse-biased. At t_5 the supply voltage waveform crosses zero to become positive and the cycle repeats. The net effect is that trigger pulses, which are synchronized with the supply voltage waveform, are generated periodically.

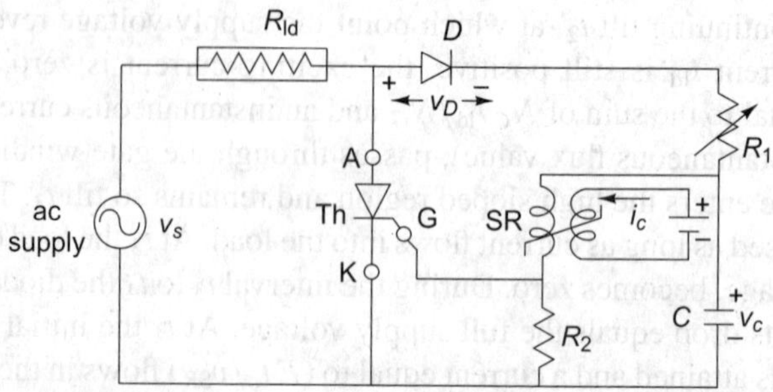

Fig. 1.24 Saturable reactor trigger pulse generator

The time instant at which firing takes place can be controlled by varying the resistance R_1.

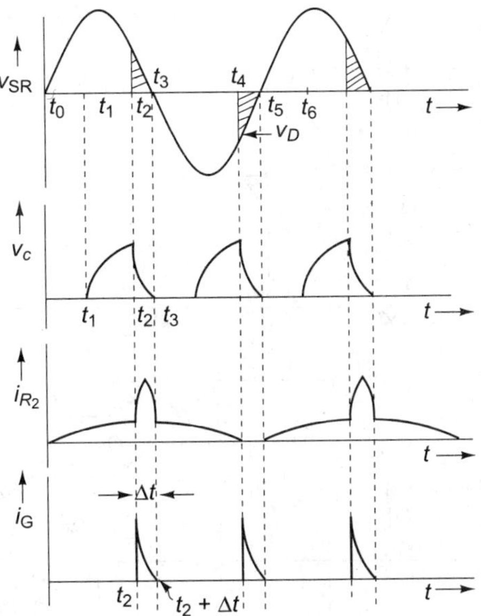

Fig. 1.25 Waveforms for the pulse generator circuit of Fig. 1.24

Figure 1.26 shows a transistorized circuit that generates rectangular pulses. The power and control circuits are fed from the same ac supply through the centre-tapped transformers S and T. The circuit operation is as follows. In the positive half-cycle, the secondary winding of the transformer T has its upper terminal positive with respect to the centre point n_T. Thus v_{t_1}, the voltage of the upper half of the secondary winding, is positive as shown in Fig. 1.27. Consequently, D_1 gets forward-biased and Q_1 turns off. The voltage v_c of the capacitor C gets charged to a negative values, with its upper plate having negative polarity. When v_c goes below the control voltage V_{con} (at the emitter terminal of Q_2), Q_2 conducts and v_c remains constant at a value slightly below V_{con} till the end of the positive half-cycle. Q_3 also switches on, giving a pulse at each of the two secondaries of the pulse transformers. Just when v_{t_1} passes through zero to the negative side, the emitter–base junction of Q_1 becomes forward-biased and Q_1 conducts, thus discharging the capacitor. Simultaneously, Q_2 and Q_3 switch off. In the negative half-cycle v_{t_2} becomes positive with respect to n_T; D_2 conducts, reducing the bias of the emitter–base junction of Q_1 and turning it off. The cycle repeats from this moment onwards. The magnitude of V_{con} controls the width of the pulses obtained at the secondary of the pulse transformer. It is seen from Fig. 1.27 that the width of the dc pulse obtained at the pulse transformer is $\pi - \alpha$. Such a rectangular pulse is suitable for firing a thyristor that supplies an inductive load. Using the same primary for both the power and the control circuit as in Fig. 1.26 helps in the production of firing pulses that have a definite delay with respect to the positive-going zero crossing of the ac supply waveform. This is termed as *synchronization* of pulses with respect to the input supply.

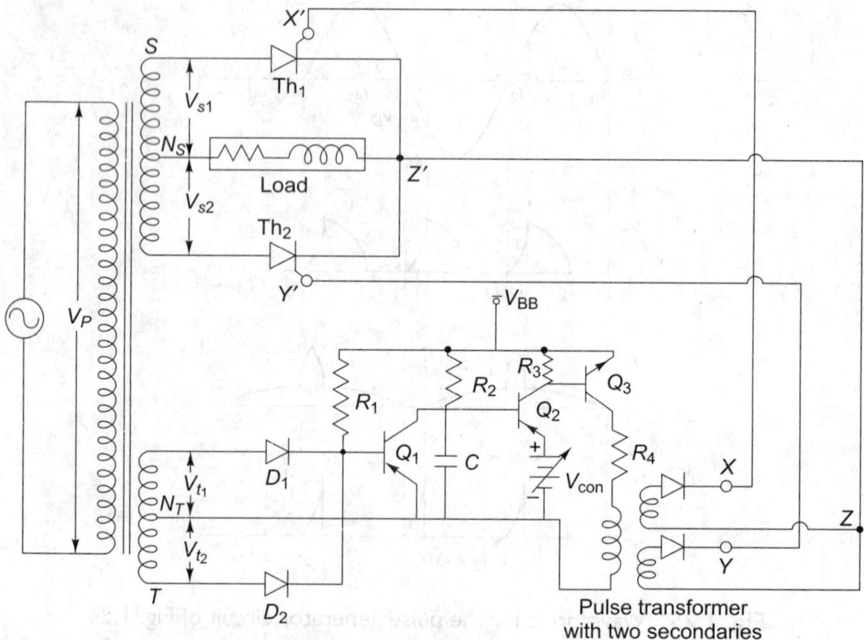

Fig. 1.26 Transistorized pulse generating circuit for a single-phase, centre-tapped type of full-wave rectifier

1.4.2.4 Pulse triggering Turning a thyristor off with the help of trigger pulses at the gate is the most economical method because (a) it is the charge that triggers the thyristor through the gate and (b) the gate loses its control once the thyristor anode current attains the latching value. A typical pulse generating circuit rigged up with transistor-based multivibrators and a unijunction transistor are described below.

A pulse generator circuit with multivibrators Figure 1.28 illustrates a pulse generating circuit employing multivibrators constituted of *p-n-p* and *n-p-n* transistors. Rectangular pulses are generated by an astable multivibrator or a Schmitt trigger and passed through a differentiating circuit; a diode rectifier then filters off the negative pulses. Synchronization is achieved by passing the ac waveform through a zero crossing detector whose output triggers a monostable multivibrator. The rectangular output pulse of the latter acts as a gate for the output pulses of the differentiator; that is, the output pulses of the differentiator are allowed only as long as the output of the monostable multivibrator exists. The firing angle is controlled by varying the period T of the astable multivibrator. A typical zero voltage-crossing detector circuit is described in Section 10.5.

1.4.3 Unijunction Transistor Trigger Pulse Generator

All devices having negative resistance characteristics are useful in oscillator circuits; the unijunction transistor (UJT) is one such device. It is eminently useful

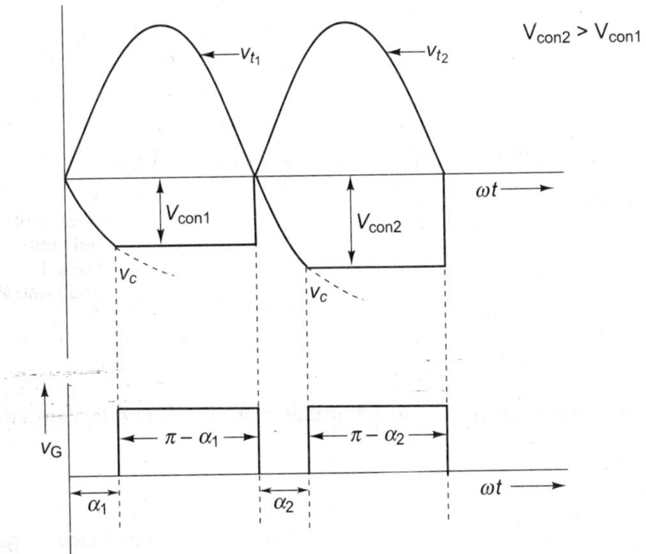

Fig. 1.27 Pulse voltage v_G with different control voltages

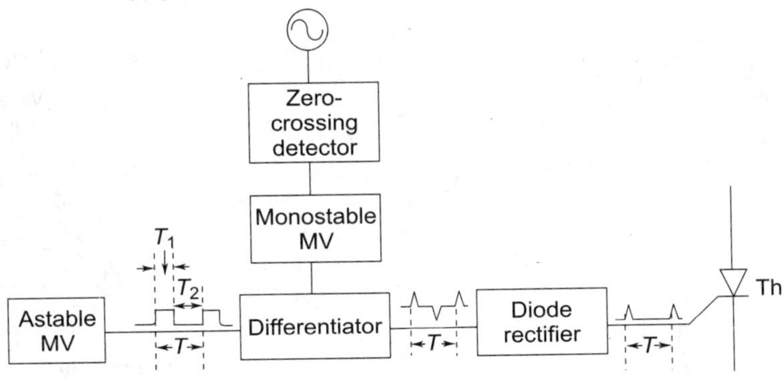

Fig. 1.28 Typical pulse generator circuit using multivibrators

for the production of trigger pulses for the gate firing of thyristors because of its simplicity and flexibility. The latter quality can be attributed to its compatibility with both ac and dc sources.

Figure 1.29(a) shows a UJT made of an n-type semiconductor bar with two base leads B_1 and B_2 at the two ends. At a point between the ends of this bar, a p-type emitter is developed by alloying. The circuit symbol for the device is shown in Fig. 1.29(b). Figure 1.30(a) shows a circuit for determining the emitter voltage versus emitter current characteristics of a UJT; Fig. 1.30(b) shows these characteristics.

Initially, as the resistance R_2 is decreased, both i_E as well as v_E will increase; this state is indicated by the region $O\,P_k$ in Fig. 1.30(b). In this state the inequality $v_E < \eta V_{BB} + V_D$ holds good, where η is the intrinsic stand-off ratio of the UJT.

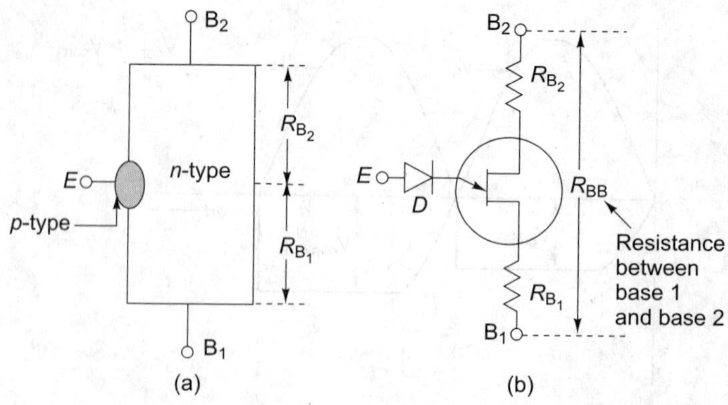

Fig. 1.29 Unijunction transistor: (a) schematic diagram, (b) equivalent circuit

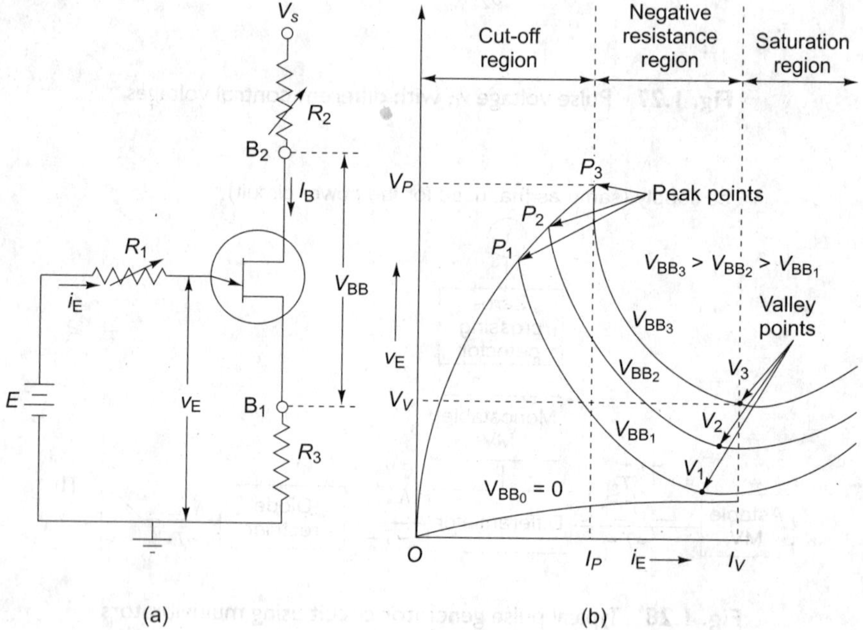

Fig. 1.30 (a) Circuit for determining the v_E-i_E characteristics of a UJT; (b) characteristic curves

It is related to the base resistances as

$$\eta = \frac{R_{B_1}}{R_{BB}} = \frac{R_{B_1}}{R_{B_1} + R_{B_2}} \tag{1.29}$$

v_E is the diode drop of the E-B$_1$ junction and v_{BB} is the voltage drop across R_{BB}, the *interbase resistance*. But once a typical peak point P_k, given by (i_P, v_P), is reached and v_E equals $(\eta V_{BB} + V_D)$, the E-B$_1$ junction of the UJT breaks down and the operating pointing (i, v) follows the negative resistance region as indicated by the region $P_k V_k$ ($k = 1, 2,$ or 3) of the characteristic. After the valley point V_k, the device regains its positive-resistance property. The value of V_{BB} can be varied by varying R_3, and a different characteristic is obtained for each value of V_{BB}.

Figures 1.31(a) and (b), respectively, show a UJT oscillator circuit without and with a pulse transformer at the base point B_1. When a battery of voltage V_s is connected to the circuit consisting of R_1, R_2, and C, the capacitor C gets charged with its X-plate positive. When the capacitor voltage v_c attains the value V_P, the E-B_1 junction breaks down causing a gush of current through base B_1 and resistance R_4. In this short duration of the discharge, a voltage spike is obtained at the upper end of R_4 and this pulse may be utilized for the gate firing of the thyristor. The resistance R_{B_1}, which is of the order of a few kilo-ohms prior to the breakdown of the E-B_1 junction, collapses to a few tens of ohms after the breakdown. This latter value of R_{B_1} is termed as $R_{B_1(con)}$. V_P is known as the peak point voltage and is expressed as

$$V_P = \eta V_{BB} + V_D \tag{1.30}$$

where

$$V_{BB} = I_B R_{BB} \tag{1.31}$$

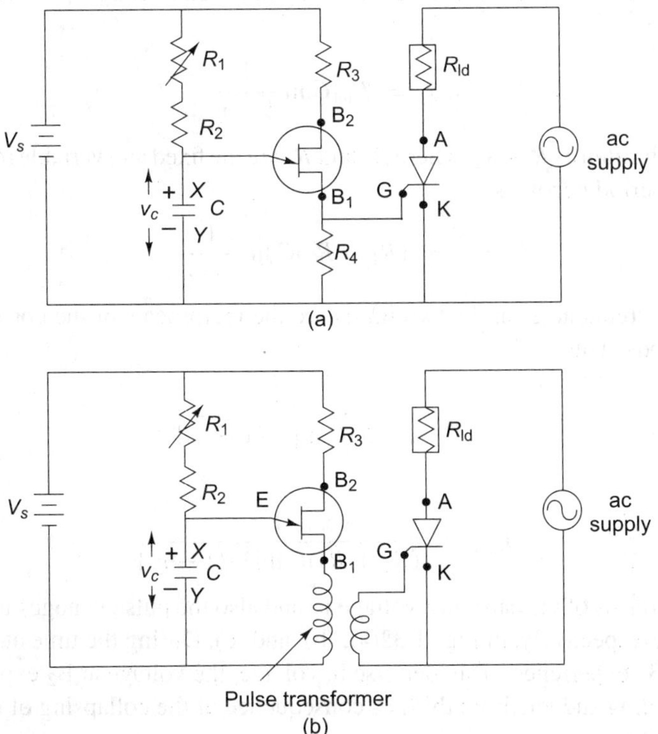

Fig. 1.31 UJT oscillator circuits: (a) pulse taken from the top terminal of R_4, (b) pulse taken from the secondary of the pulse transformer

Also,

$$I_B = \frac{V_s}{R_3 + R_{BB} + R_4} \quad (1.32)$$

As will be explained later, R_3 and R_4 will be much smaller than R_{BB}, and hence V_s can be approximated as

$$V_s \approx I_B R_{BB} = V_{BB} \quad (1.33)$$

The diode drop V_D in Eqn (1.30) is usually small and can be neglected. Using this approximation and combining Eqns (1.30) and (1.33) gives V_P as

$$V_P = \eta V_s \quad (1.34)$$

Considering the RC circuit, the capacitor voltage v_c is given by

$$v_c = V_s\left(1 - e^{-t/R_{tot}C}\right) \quad (1.35)$$

where $R_{tot} = R_1 + R_2$. When the E-B$_1$ junction breaks down at $t = \tau$, v_c in Eqn (1.35) equals V_P. With this condition for v_c, the combination of Eqns (1.34) and (1.35) yields

$$V_s\left(1 - e^{-\tau/R_{tot}C}\right) = \eta V_s \quad (1.36)$$

where τ is the time period of the generated pulses. Cancelling V_s and rearranging gives

$$\tau = R_{tot} C \ln \frac{1}{1-\eta} \quad (1.37)$$

With R_{tot} written as $R_1 + R_2$, where R_2 and R_1 are the fixed and variable resistances, the time period becomes

$$\tau = (R_1 + R_2) C \ln \frac{1}{1-\eta} \quad (1.38)$$

The pulse frequencies at the two limits are the reciprocals of the corresponding time periods. Thus,

$$f_{max} = \frac{1}{R_2 C \ln[1/(1-\eta)]} \quad (1.39)$$

and

$$f_{min} = \frac{1}{(R_1 + R_2) C \ln[1/(1-\eta)]} \quad (1.40)$$

The waveforms of the capacitor voltage v_c and also the pulse voltages v_{B_2} and v_{B_1} are given, respectively, in Figs 1.32(a), (b), and (c). During the time at which the terminal B$_1$ experiences a sudden rise in voltage, the voltage at B$_2$ experiences a corresponding sudden drop; this is a consequence of the collapsing of the base-1 resistance to $R_{B_1(con)}$.

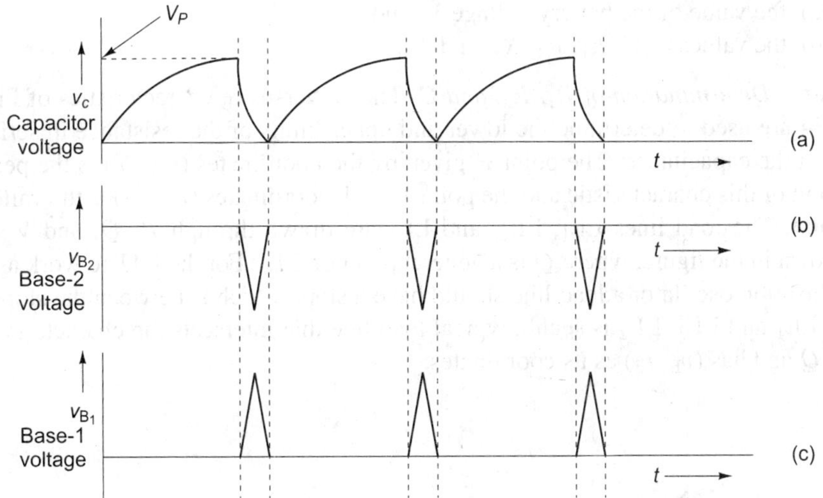

Fig. 1.32 Voltage waveforms of a UJT: (a) v_c, (b) v_{B_2}, (c) v_{B_1}

The pulse height can be obtained from Fig. 1.33, which gives the E-B_1 circuit in the conducting state as

$$v_{\text{pulse}} = v_{R_4} = V_P \frac{R_4}{R_4 + R_{B_1(\text{con})}} \tag{1.41}$$

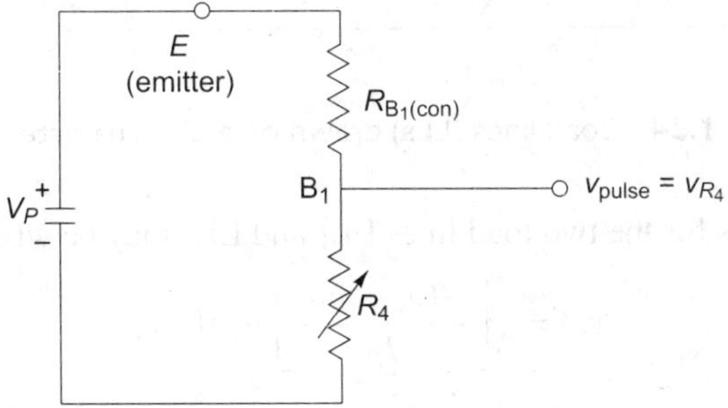

Fig. 1.33 Emitter-based circuit during conduction

Design of a UJT oscillator The values for the various circuit elements of a UJT relaxation oscillator can be chosen to obtain (i) a specified range of pulse frequencies and (ii) a given pulse height. The following data for the UJT can be obtained from its data sheets:

(a) V_P and I_P
(b) V_V and I_V
(c) η and R_{BB}

The output quantities required are (i) the frequency range and (ii) the pulse height. Using this data, it is required to determine

(a) the value of the battery voltage V_s and
(b) the values of C, R_1, R_2, R_3, and R_4.

Step 1 Determination of R_1, R_2, and C The v_E versus i_E characteristics of Fig. 1.34 are used to determine the lower and upper limits of the resistance in series with the capacitance. The point P given by the coordinates (v_P, i_P) is the peak point of this characteristic and the point V with coordinates (v_V, i_V) is the valley point. The load lines LL_1, LL_2, and LL_3 are drawn through P, Q, and V as shown in the figure, where Q is a general point on LL_2. For the UJT to work as a relaxation oscillator a load line should have a slope which is between the slopes of LL_1 and LL_3. LL_2 is such a typical load line that intersects the characteristic at Q and has (v_E, i_E) as its coordinates.

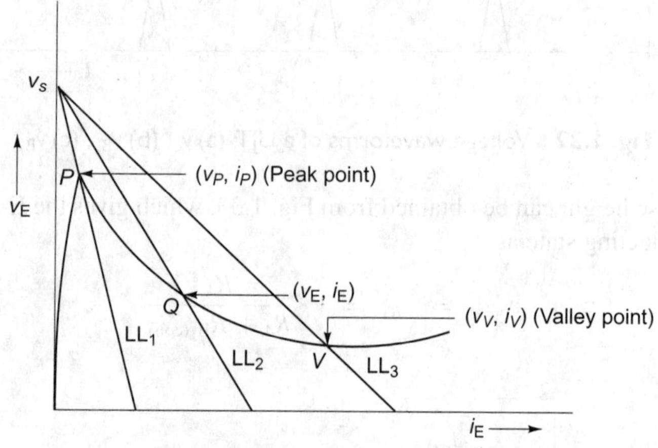

Fig. 1.34 Load lines (LLs) drawn on a UJT characteristic

The equations for the two load lines LL_1 and LL_3 may be written as

$$v_E = \left[\frac{-(v_s - v_P)}{I_P}\right] i_E + V_s \tag{1.42}$$

and

$$v_E = \left[\frac{-(v_s - v_V)}{i_V}\right] i_E + V_s \tag{1.43}$$

As the slope represents the resistance in the RC circuit, the resistance R_{tot} obeys the inequality

$$\frac{v_s - v_V}{i_V} \leq R_{tot} \leq \frac{v_s - v_P}{i_P} \tag{1.44}$$

With R_2 as the fixed resistance and R_1 as the variable one, the inequality in Eqn (1.44) can be split up as

$$R_2 \geq \frac{v_s - v_V}{i_V} \tag{1.45}$$

and

$$(R_1 + R_2) \leq \frac{v_s - v_P}{i_P} \qquad (1.46)$$

The value of C may be chosen to be between 0.01 and 0.5 μF. The procedure for design is to first fix up C and then determine R_1 and R_2 from the Eqns (1.45) and (1.46).

Step 2 Determination of R_3 and R_4 R_3 should be low enough such that V_{BB} is sufficiently high. This is to ensure a pulse of the required height. At the same time, R_3 should be high enough such that the permitted value of the interbase voltage V_{BB} is not exceeded. This means that I_B in Eqn (1.31) should not exceed a certain limit. A value in the range 0.4–1.0 kΩ is usually employed. As seen from Eqn (1.4) the value of R_4 depends upon $R_{B_1(con)}$ and the required pulse height. Its value usually ranges from 40 to 100 Ω.

Step 3 Choice of the dc supply voltage V_s The pulse height depends on v_P, which is approximately equal to ηV_{BB}, which in turn is nearly equal to ηV_s. V_s should be large enough to give a pulse of appropriate height as per the gate requirements [Fig. 1.16(b)]. This fixes its lower limit. On the high side, V_s should be so chosen that the maximum allowable interbase voltage V_{BB} is not exceeded. A value of 10–30 V is usually adopted for commonly available thyristors. A practical method for obtaining the required dc power for the UJT oscillator is to drop a part of the output voltage of a single-phase, half- or full-wave bridge rectifier in a resistor and connect a zener diode to truncate it, as shown in the circuit and the waveforms of the oscillator given in Figs 1.35(a) and (b). This method has the advantage that the trigger pulses are synchronized with the ac waveform. The thyristor rectifies the positive half-cycle during the period of conduction, which occurs from α to π. It is seen from Fig. 1.35(a) that the ac voltage is dropped in R_{ld} and R_5 and then truncated by the zener diode D_z so that a reduced voltage is fed to the UJT circuit. This is because the RMS magnitude of the ac supply voltage is usually

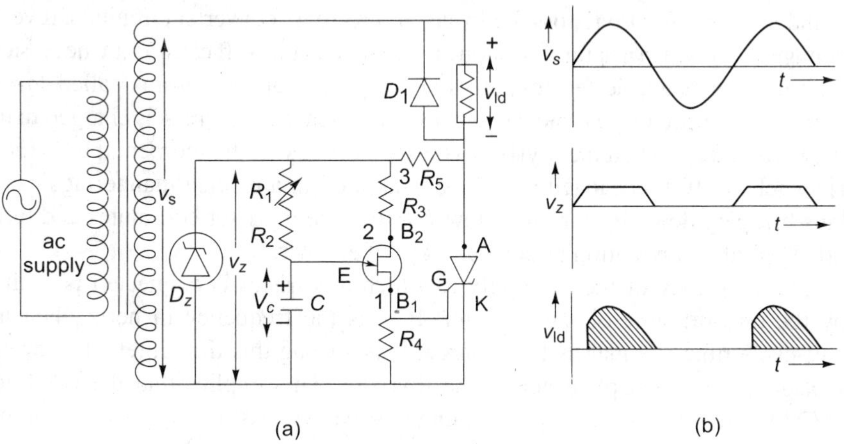

Fig. 1.35 Single-phase, half-wave controlled converter using a UJT: (a) circuit diagram, (b) waveforms

220 V whereas a voltage less than 30 V (dc) is required for the UJT oscillator. The frequency of the pulse obtained at the upper terminal of R_4 can be controlled by varying R_1, and hence a potentiometer can be used as R_1.

The UJT triggering circuit is inexpensive and is convenient as a gate triggering pulse generator. But its use is limited only to single-phase, half-wave and full-wave circuits. For more complicated circuits, namely, a single-phase bridge rectifier or a three-phase, half- or full-wave rectifier, digital circuits giving as many trigger outputs as the thyristors used in the rectifier are required. For instance, in a three-phase, full-wave fully controlled converter, six trigger pulses are needed; the digital electronic circuits are designed and optimized using Karnaugh maps for such circuits. Microprocessors are finding wide application in this regard because of their compactness and flexibility.

1.5 Thyristor Turn-off

The word 'turn-off' implies the sequence of events taking place from a forward conducting state to a forward blocking state. In the blocking state the thyristor will not fire without a gate triggering signal when an anode voltage less than V_{FBO} is applied.

The turn-off time is defined as the time interval between the moment the thyristor current attains a zero value and the moment the thyristor attains a forward blocking condition. The typical turn-off time of a thyristor varies from 5 μs for low-rated thyristors to about 200 μs for high-rated ones. For a fully conducting thyristor to be brought back to the forward blocking state it is necessary to sweep out the excess charge carriers in the device either by decay or recombination. The appropriate technique for turning off a thyristor in an electrical circuit is to reduce its current below the holding current value I_H. This may be accomplished by diverting the current through a low-resistance shunt or by any other means. Applying a reverse voltage across the thyristor for a sufficiently long time, namely, the turn-off time, also forces the current to become zero and leads to the turn-off of the thyristor. For thyristors fed by ac circuits (e.g., converter circuits) a reverse voltage is applied in a natural fashion and no special turn-off circuits are necessary. On the other hand, dc-fed thyristors need special turn-off means called *forced* commutation circuits, a typical one consists of a capacitor which is charged to the required voltage when the thyristor is turned on. Just at the moment the thyristor is switched off, this capacitor voltage is applied in a reverse manner against it, thus bringing down its current below I_H; thus the thyristor commutes and turns off. Typical commutation circuits are described later.

The frequency of the ac supply to which a thyristor is connected is limited by the turn-off time of the thyristor. If f is the frequency in hertz, then the half-cycle time is equal to $1/2f$ and at least during this time interval a reverse voltage has to be applied across the thyristor. This implies that the condition $(1/2f) > t_{OFF}$ has to be ensured, failing which the thyristor may conduct without a gate signal when a positive voltage is applied across it. For example, if the turn-off time of a particular thyristor is 80 μs this time is the minimum half-cycle time of the ac supply. Therefore the frequency of the ac supply should not exceed $1/(2 \times 80 \times 10^{-6}) = 6250$ Hz.

The turn-off time consists of two components:

(a) the time required for attaining the reverse blocking property (t_r) and
(b) the time required for attaining the forward blocking property (t_{OFF}).

To demonstrate the mechanism of turn-off, the thyristor is first subjected to a reverse voltage for sometime and then to a forward voltage. The waveforms in Figs 1.36(a) and (b) are drawn, respectively, for the voltage across the thyristor and the current through it. In Fig. 1.36(a) the dotted line represents the external voltage impressed across the thyristor; the continuous line shows the actual drop. The sequence of events that finally ends up in the turning off of the thyristor is given below.

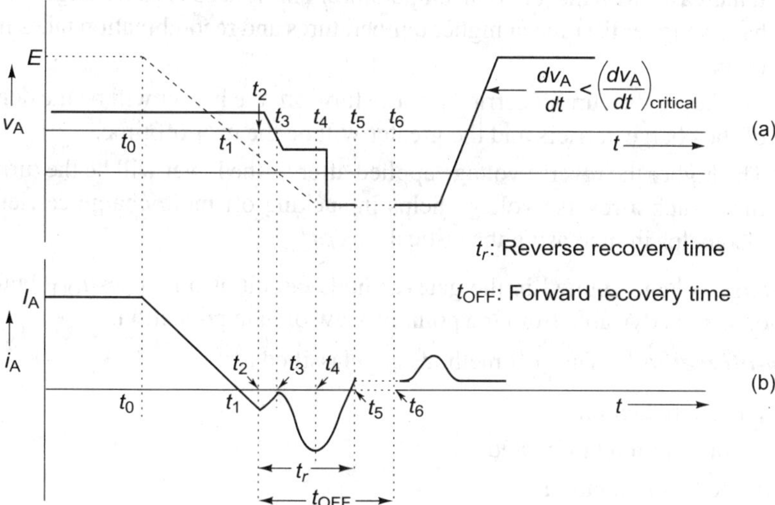

Fig. 1.36 Turn-off mechanism of a thyristor that supplies a resistive load: (a) voltage across the thyristor (v_A), (b) current through the thyristor (i_A)

When a thyristor is conducting, a large number of mobile charge carriers, namely, holes and electrons, exist in the device. The thyristor turn-off starts at t_0, when the forward voltage applied across it is reduced and then reversed. Simultaneously, the current in the device starts to decrease. This entails a decrease of the charge carriers across the junctions J_a and J_c, which are initially in a forward-biased condition. At t_1 the voltage drops to zero; the current also goes to zero at the same instant provided the load is resistive. On the other hand, with an inductive $(R + L)$ load, there will be a time lag between the voltage zero and the current zero. Though some of the charge carriers are absorbed by recombination, current continues to flow in the reverse direction. At t_2 the charge carriers in the vicinity of J_c start decreasing and go to zero at t_3; at this moment J_c regains its reverse blocking capability. The reverse current decreases in the interval $(t_3 - t_2)$ but again increases till the instant t_4. From t_4 onwards till t_5 the junction J_a also regains its reverse blocking capability and the current through it attains a zero value. The duration t_2 to t_5 is called the *reverse recovery time*. If a forward voltage is applied at t_5 across the thyristor, it will conduct again due to the fact that a large number of charge carriers in the vicinity of J_b still need some time to recombine and only then (say, at t_6) can the forward blocking capability of the thyristor be attained. The duration t_2 to t_6 is called the *forward recovery time* or simply *recovery time* and is denoted by t_{OFF}. If a positive voltage is applied across the thyristor after t_6, it will not fire unless a gate signal is applied. However, if the rate of increase of the anode voltage (dv_A/dt) exceeds a critical value, namely, $(dv_A/dt)_{critical}$, the thyristor sets triggered again as explained before, even in the absence of a gate signal.

The recovery time t_{OFF} is an important specification of the thyristor and depends on the following factors.

(a) It increases with the junction temperature. This is because the charge carriers have a higher lifetime at higher temperatures and recombination takes more time.

(b) The larger the current carried prior to turn-off, the higher will be the density of the charge carriers and the greater will be the turn-off time.

(c) The higher the reverse voltage applied after t_r, the lower will be the turn-off time; such a reverse voltage helps in sucking off more charge carriers. It also helps in increasing the critical dv_A/dt.

A reverse voltage applied in the gate–cathode circuit also reduces t_{OFF} but this method is not advisable from the point of view of gate protection.

Turn-off methods Turn-off methods are classified as

(a) current reduction,

(b) dc line commutation, and

(c) forced commutation.

1.5.1 Current Reduction

The current reduction method consists in directly applying the principle of bringing down the current below the holding current I_H. An on–off switch may be connected in series with the thyristor and load, as shown in Fig. 1.37(a). The switch has to be closed again only after a lapse of time equal to the recovery time of the thyristor, provided $(dv_A/dt)_{critical}$ is not exceeded. This method leads to heavy sparking and is used only for low-current thyristors. An alternative is to bypass the thyristor current through a transistor [Fig. 1.37(b)]. A pulse whose duration is greater than the recovery time of the thyristor is applied at the base of the transistor, thus turning off the thyristor, which remains so till another gate firing pulse is applied.

These two methods may be applied to thyristors used in dc choppers with low current ratings. However, for choppers with higher ratings as well as inverters, easily realizable circuits coming under the third category, namely, forced commutation, are available.

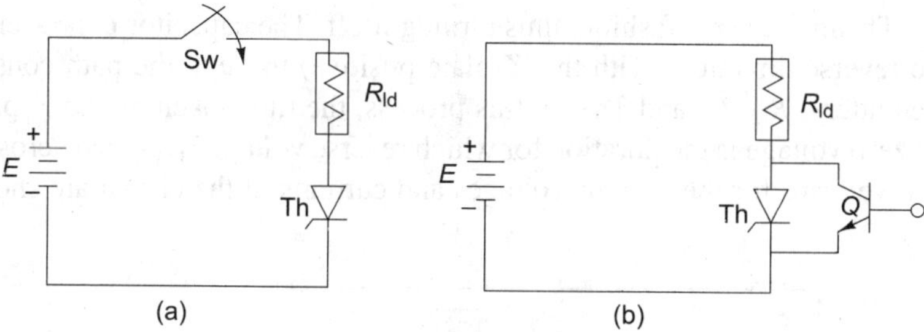

Fig. 1.37 Methods for current reduction: (a) series switch, (b) transistor connected in shunt

1.5.2 Ac Line Commutation

Ac line commutation is used in single- as well as three-phase converter circuits. The thyristor conducts during a portion of the positive half-cycle of the ac voltage waveform. As the voltage crosses zero towards the negative side, the thyristor current also goes to zero either at the same instant or after a certain lag, depending on whether the load is resistive or inductive. This feature together with the application of negative voltage across the thyristor in the negative half-cycle turns off the thyristor in a natural fashion. With the prevailing power frequencies (say, 50 Hz or 60 Hz) the duration for which reverse voltage is applied across the thyristor is many times greater than the turn-off time of any thyristor, thus making the thyristor a natural choice for use in single- or three-phase rectifier circuits and also in ac choppers.

1.5.3 Forced Commutation Circuits

As stated earlier, the method of forced commutation consists in charging a capacitor to the dc supply voltage when the thyristor is conducting, and applying this capacitor voltage in a reverse manner across the thyristor to facilitate its turn-off. The size of the capacitor is so chosen that its discharge time is much in excess of the t_{OFF} of the thyristor. Another method is that of passing a reverse current of magnitude greater than that of the thyristor current. Resonant circuits are commonly used for this purpose. The following four types of forced commutation are described here:

(i) parallel capacitor turn-off,

(ii) auxiliary resonant turn-off, and

(iii) self-commutation by resonance; two circuits are available under this category, namely, (a) the series circuit and (b) the parallel circuit.

1.5.3.1 Parallel capacitor turn-off The circuit for parallel capacitor turn-off is given in Fig. 1.38(a). Th_1 is the load thyristor and Th_2 is the auxiliary thyristor. R_1 is an auxiliary resistance whose value has to be properly determined as explained below. The operation of the circuit is initiated by firing Th_1, thus passing current through the load. Simultaneously, the capacitor C charges towards E (with its Y-plate positive) through the circuit, i.e., the battery, R_1, C, and Th_1. When Th_2 is triggered after a lapse of time, the capacitor voltage is impressed across Th_1 in a reverse fashion, thus turning it off. The capacitor C now charges in the reverse direction (with the X-plate positive) through the path consisting of the battery, R_{ld}, C, and Th_2. In this process, the time taken by the Y-plate to attain zero voltage is the duration for which reverse voltage is applied across Th_1. The waveforms for the various voltages and currents of the circuit are shown in Fig. 1.38(b).

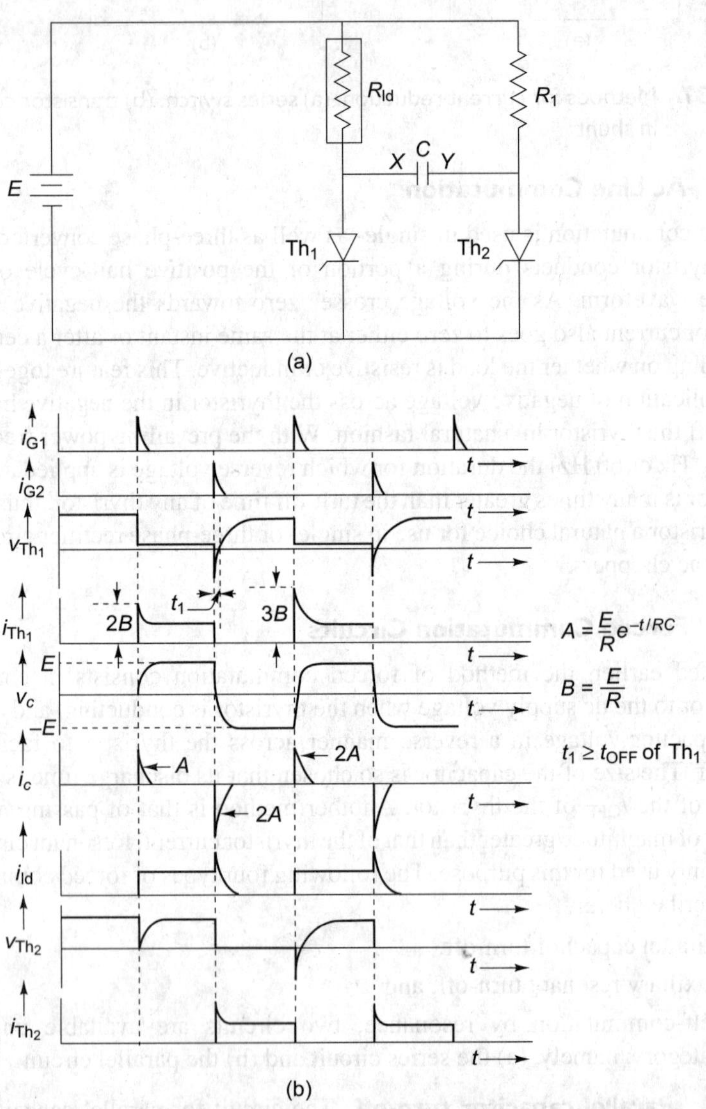

Fig. 1.38 Parallel capacitor turn-off schemes: (a) circuit and (b) waveforms

The auxiliary resistance R_1 should be large enough to ensure that the power loss in it is minimized during the conduction time of Th_2. At the same time it should be small enough to ensure that the current through it exceeds the leakage current of Th_2, which, being very small, flows through Th_2 in its OFF state. If this precaution is not taken, some current may leak through this thyristor and the current through the capacitor will not be sufficient to charge it to the value E in the other direction.

The value of capacitance that will cause successful turn-off of Th_1 can be determined as follows. Let the moment at which Th_2 is turned on be taken as $t = 0$. The equation of the circuit constituting the battery, R_{ld}, C, and Th_2 can now be written as

$$-E + R_{ld}i + \frac{1}{C}\int_0^t i\,dt = 0 \tag{1.47}$$

with the initial condition $v_c(0) = -E$ or $q(0) = -EC$. Equation (1.47) can be rewritten in terms of the charge q as

$$R_{ld}\frac{dq}{dt} + \frac{q}{C} = E \tag{1.48}$$

From Appendix B, the solution of Eqn (1.48) with the initial condition $q(0) = -EC$ is given by Eqn (B.11a), in which V_0 should be taken equal to E. Thus, $q(t)$ becomes

$$q(t) = CE\left(1 - 2e^{-t/R_{ld}C}\right) \tag{1.49}$$

Also,

$$i = \frac{dq}{dt} = \frac{2E}{R_{ld}}\exp\left(\frac{-t}{R_{ld}C}\right) \tag{1.50}$$

It is required to determine the circuit time taken to forward bias Th_1, after Th_2 has been turned on. This can be obtained by writing the expression for v_{Th_1} and determining the time taken by it to just become forward-biased:

$$v_{Th_1} = E - iR_{ld} \tag{1.51}$$

Substitution for i from Eqn (1.50) gives

$$v_{Th_1} = E - R_{ld}\left[\frac{2E}{R_{ld}}\exp\left(\frac{-t}{R_{ld}C}\right)\right] \tag{1.52}$$

$$= E\left[1 - 2\exp\left(\frac{-t}{R_{ld}C}\right)\right] \tag{1.53}$$

Let t_1 be the time taken by v_{Th_1} to just become zero:

$$0 = E\left[1 - 2e^{(-t_1/R_{ld}C)}\right] \tag{1.54}$$

Equation (1.54) can be solved for t_1 to get

$$t_1 = 0.7CR_{ld} \tag{1.55}$$

This time should at least be equal to t_{OFF}. That is,

$$t_1 \geq t_{OFF}$$

or

$$0.7CR_{ld} \geq t_{OFF}$$

The capacitance value is now obtained as

$$C \geq \frac{t_{\text{OFF}}}{0.7R_{\text{ld}}} \tag{1.56}$$

A practical method is to use a value of capacitance which is slightly more than that given by Eqn (1.56). Its value can also be experimentally found, by slowly increasing C from a very small value to a value slightly greater than that for which commutation just fails.

1.5.3.2 Auxiliary resonant turn-off The circuit of a voltage commutated chopper is given in Fig. 1.39(a). In addition to the main thyristor Th_1 it consists of an auxiliary thyristor Th_2, inductor L, capacitor C, and a diode D. Its operation with a purely resistive load is dealt with here and that with an inductive load is described in Chapter 3. The waveforms pertaining to the various voltages and currents of this circuit are given in Fig. 1.39(b).

It is assumed that initially the capacitor C is charged to the battery voltage E with the Y-plate having positive polarity. When Th_1 is turned on at t_0 the load current flows in the circuit consisting of the battery, Th_1, and R_{ld}. Simultaneously, an LC oscillation occurs in the closed circuit consisting of Th_1, C, L, and D. Thus, Th_1 has to carry the sum total of the load current i_{ld} and the oscillatory current i_c, whose peak value is i_{cm}. At t_1, one half-cycle of the oscillatory current is completed. At this moment, the current tries to reverse, but is prevented by the diode D. The capacitor is now charged to voltage E with the X-plate positive and remains so till t_2. When Th_2 is fired at t_2 the capacitor voltage is applied in a reverse manner across Th_1 and it turns off. Current now flows through the circuit consisting of the battery, Th_2, C, and R_{ld}, which is only an RC circuit; when this current goes below the holding current of Th_2, it turns off. The capacitor now charges back to the voltage E with the Y-plate positive. t_3 marks the time when the capacitor voltage just becomes zero, and hence the circuit time available for commutation of Th_1 is $(t_3 - t_2)$. During the interval t_3 to t_4, V_c rises exponentially to the battery voltage E and remains at that value. The cycle repeats when the firing signal is given to Th_1 at t_4.

The analysis of the circuit aims at determining the expressions for the values of L and C required to successfully commutate Th_1. The known quantities are the battery voltage E, load resistance R_{ld}, and the turn-off time t_{OFF}. For the purpose of analysis, t_2 is taken as the starting moment in time and accordingly the time

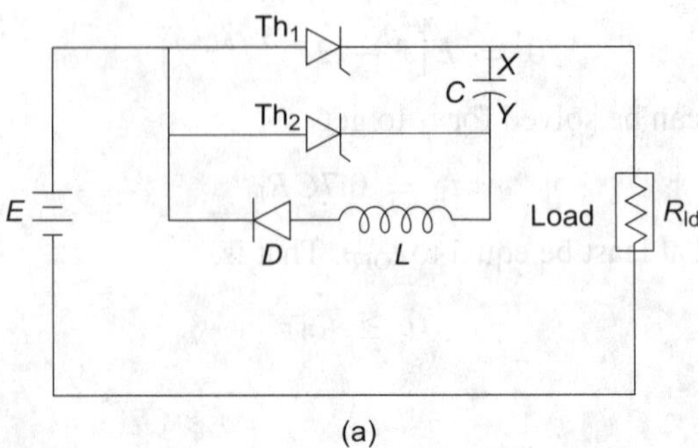

Fig. 1.39(a)

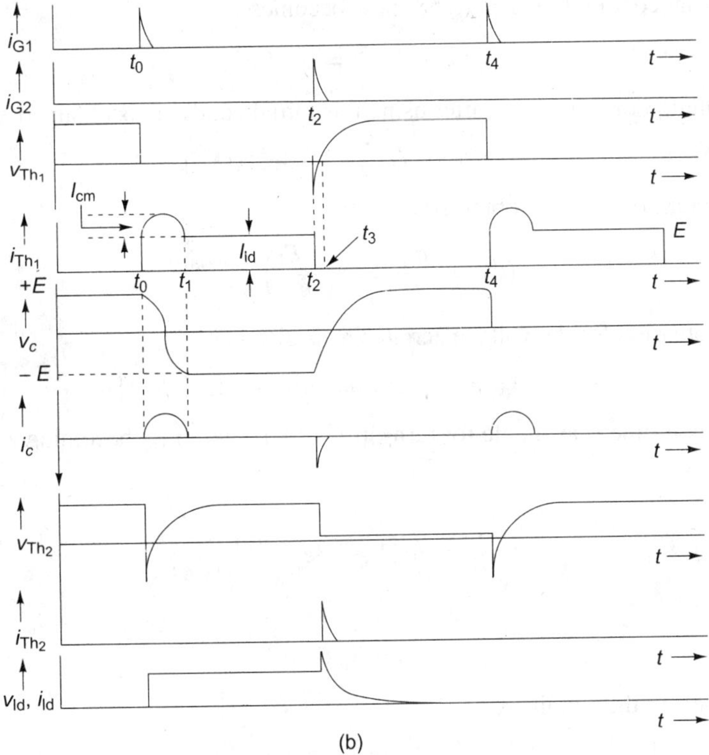

Fig. 1.39 Auxiliary resonant turn-off scheme: (a) circuit diagram, (b) waveforms

variable is taken as $t' = t - t_2$, where t is the actual time. The equation governing the circuit is that of an RC circuit charged through the thyristor Th_2 and is written as

$$E = \frac{1}{C}\int_{-\infty}^{t'} i\,dt' + i R_{ld} \tag{1.57}$$

with the Y-plate of the capacitor assumed to be of positive polarity (the capacitor is initially charged to the voltage $-E$). Hence

$$\frac{1}{C}\int_{-\infty}^{0} i\,dt' = -E \tag{1.58}$$

Equation (1.57) can be rewritten as

$$E = -E + \frac{1}{C}\int_{0}^{t'} i\,dt' + i R_{ld} \tag{1.59}$$

With $i = dq/dt'$, Eqn (1.59) becomes

$$\frac{dq}{dt'} + \frac{q}{R_{ld}C} = \frac{2E}{R_{ld}} \tag{1.60}$$

The initial condition in Eqn (1.58) now becomes

$$q(t')|_{t'=0} = -CE$$

The solution of Eqn (1.60) after using this initial condition is obtained as

$$q(t') = CE\left[2 - 3e^{-t'/R_{\text{ld}}C}\right] \tag{1.61}$$

The current $i(t')$ is now obtained as

$$i(t') = \frac{dq}{dt'} = \left(\frac{3E}{R_{\text{ld}}}\right) e^{-t'/R_{\text{ld}}C} \tag{1.62}$$

The expression for the voltage across Th_1 is given as

$$v_{\text{Th}_1} = E - iR_{\text{ld}} = E\left[1 - 3e^{-t'/R_{\text{ld}}C}\right] \tag{1.63}$$

The circuit time t_c available for turn-off is $t_3 - t_2$, with v_{Th_1} becoming zero at t_3. Thus

$$0 = E\left[1 - 3e^{-t_c/R_{\text{ld}}C}\right] \tag{1.64}$$

This gives

$$t_c = R_{\text{ld}}C \ln 3 \tag{1.65}$$

By imposing the condition

$$t_c \geq t_{\text{OFF}} \tag{1.66}$$

the capacitor value required for turn-off is given by the inequality

$$C \geq \frac{t_{\text{OFF}}}{R_{\text{ld}} \ln 3} \tag{1.67}$$

The value of the inductor L can be obtained from the fact that the current rating of the thyristor must be at least equal to $I_{\text{cm}} + I_{\text{ld}}$, where I_{cm} is the peak value of the LC oscillatory current. I_{cm} can be written as

$$I_{\text{cm}} = \omega CE = E\sqrt{\frac{C}{L}} \tag{1.68}$$

The maximum current capacity $I_{\text{Th}_1\text{M}}$ of the thyristor can now be expressed as

$$I_{\text{Th}_1\text{M}} \geq E\sqrt{\frac{C}{L}} + \frac{E}{R_{\text{ld}}} \tag{1.69}$$

This gives the value of L as

$$L \geq \frac{C}{\left(\dfrac{I_{\text{Th}_1\text{M}}}{E} - \dfrac{1}{R_{\text{ld}}}\right)^2} \tag{1.70}$$

For a resistive load, the load voltage v_{ld} and the load current i_{ld} are seen to have the same shape. It is shown in Chapter 3 that for a large inductive load, the load current will be nearly constant whereas the load voltage waveform remains the same as that given in Fig. 1.39(b).

1.5.3.3 Self-commutation by resonance Self-commutation by resonance for series and parallel circuits is explained below.

Series circuit Figure 1.40(a) gives a series resonance circuit consisting of the elements Th, C, and L in series with the supply (Ramshaw 1975). It is assumed that the load is purely resistive and that the value of the load resistance is negligible compared to the inductive and capacitive reactances in the series circuit. When the thyristor is switched on, the circuit equation for this condition, assuming it to be lossless, can be written as

$$L\frac{di}{dt} + \frac{1}{C}\int_{-\infty}^{t} i\,dt = E \tag{1.71}$$

with initial conditions

$$q(0) = \int_{-\infty}^{0} i\,dt = 0 \quad \text{and} \quad i(0) = 0$$

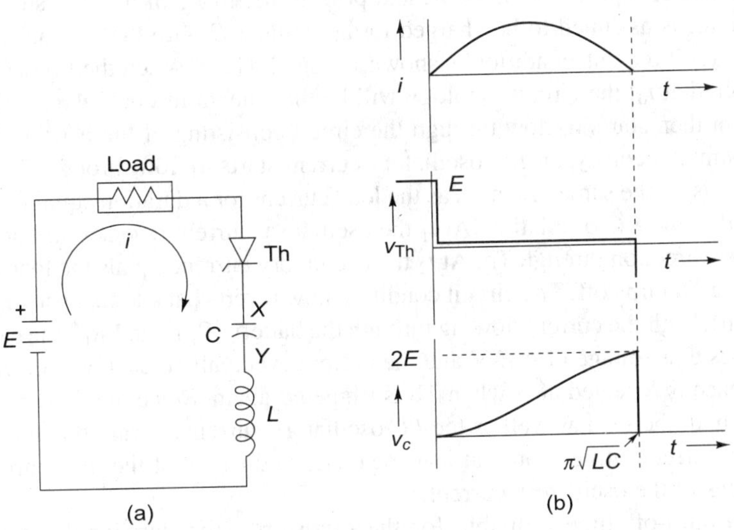

Fig. 1.40 Series resonant turn-off scheme: (a) circuit diagram, (b) waveforms

Splitting the integral as $\int_{-\infty}^{t} i\,dt = \int_{-\infty}^{0} i\,dt + \int_{0}^{t} i\,dt$, making use of the given initial condition, and using the equalities $i = dq/dt$ and $\int_0^t i\,dt = q$ gives

$$L\frac{d^2q}{dt^2} + \frac{q}{C} = E \tag{1.72}$$

or

$$\frac{d^2q}{dt^2} + \frac{q}{LC} = \frac{E}{L} \tag{1.73}$$

This is the equation of a forced LC oscillator. After using the initial condition, its solution becomes

$$q = CE(1 - \cos \omega t) \tag{1.74}$$

Also, the voltage across the capacitor becomes

$$v_c = \frac{q}{C} = E(1 - \cos \omega t) \tag{1.75}$$

As a check, it can be seen that Eqn (1.74) satisfies both the above initial conditions. The current i, voltage across the thyristor v_{Th}, and the voltage across the capacitor v_c are shown in Fig. 1.40(b). After the half-period $\pi\sqrt{LC}$, the X-plate of the capacitor attains a voltage $2E$ at the thyristor cathode and X-plate junction, whereas the thyristor anode is at the battery voltage E. This implies that a reverse voltage E is applied across the thyristor at $t = \pi\sqrt{LC}$. At this moment the current crosses zero to the negative side and the thyristor gets turned off. Assuming C and L to be lossless, the reverse voltage remains for sufficient time to commutate the thyristor.

Parallel circuit The parallel resonance circuit, circuit variations, and waveforms are all shown in Figs 1.41(a), (b) and (c), and (d), respectively. The load is assumed to consist of a pure resistance R, and prior to the firing of the thyristor Th, the capacitor is assumed to be charged to the voltage E with the X-plate positive. The circuit for this condition is shown in Fig. 1.41(b). When the thyristor Th is triggered at t_0, the circuit topology will be that shown in Fig. 1.41(c). The load current then starts to flow through the circuit consisting of the battery, Th, and R_{ld}; simultaneously, an LC oscillatory current starts to flow through C, Th, and L, that is, in the same direction as the load current for a duration equal to half the period of the LC oscillation. At t_1 the oscillatory current reverses and flows in a reverse direction through Th. At t_2 the oscillatory current equals the load current I_{ld}, and Th turns off. The circuit condition now reverts back to that shown in Fig. 1.41(b). With the current flowing through the battery, C, L, and R_{ld}, the capacitor charges to a voltage of $+E$ V and the current eventually dies down to zero. The sequence is repeated at t_4 when Th is triggered again. Since the thyristor has to carry both the load as well as the LC oscillatory currents during the interval (t_0, t_1), its current rating should at least be equal to the sum of the load current and the peak of the oscillatory current.

The turn-off time available for the thyristor is the duration from t_2 to t_3 because in this interval the load voltage exceeds E and the voltage across the thyristor becomes negative as shown by the waveform of v_{Th} given in Fig. 1.41(d). The magnitude of the negative peak depends upon the values of the R, L, and C elements. As the reverse oscillatory current causes reduction of the thyristor current to zero, this circuit comes under the category of forced current commutation. For successful turn-off, the amplitude I_{cm} of the oscillatory current should exceed the load current I_{ld}, that is, the inequality

$$\frac{E}{1/\omega C} \geq \frac{E}{R_{\text{ld}}} \tag{1.76}$$

should hold good. Usually I_{cm} is kept at a value of 1.5–2 times I_{ld}. *Assuming* $I_{\text{cm}} = 2I_{\text{ld}}$ the circuit elements L and C can be designed in the following manner. The

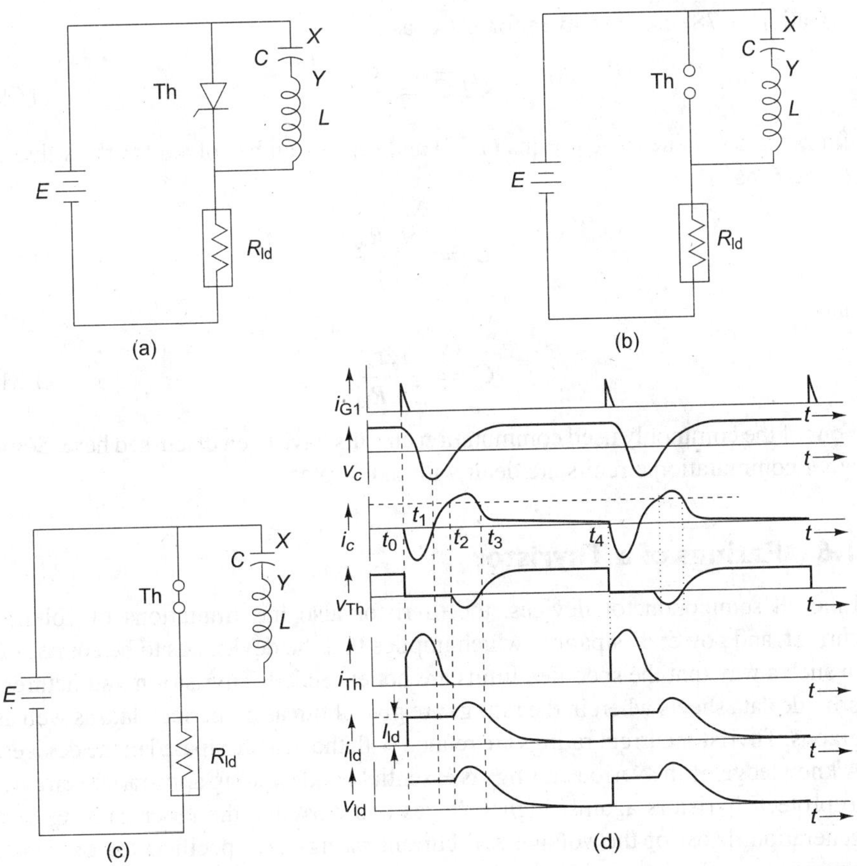

Fig. 1.41 Parallel resonant turn-off scheme: (a) circuit diagram, (b) circuit for charging of the capacitor, (c) circuit for the *LC* oscillation and load current condition, (d) waveforms

assumed relationship can be written as

$$E\omega C = 2\left(\frac{E}{R_{ld}}\right)$$

or

$$\sqrt{\frac{C}{L}} = \frac{2}{R_{ld}} \tag{1.77}$$

A second condition can be derived by specifying the duration of the current pulse in the load. The value of t_p, the duration of the current pulse, can be seen to be the sum total of the entire negative half-cycle plus the time between t_2 and t_1, which in this case becomes $\pi/6$. t_p can therefore be expressed as

$$t_p = \pi\sqrt{LC} + \frac{\pi\sqrt{LC}}{6} = \frac{7\pi}{6}\sqrt{LC} \tag{1.78}$$

Equation (1.78) can be solved for $\sqrt{LC}$ as

$$\sqrt{LC} = \frac{6t_p}{7\pi} \tag{1.79}$$

Knowing the value of R_{ld}, Eqns (1.77) and (1.79) can be solved for the value of L and C as

$$L = \frac{3t_p R_{\text{ld}}}{7\pi} \tag{1.80}$$

and

$$C = \frac{12t_p}{7\pi R_{\text{ld}}}. \tag{1.81}$$

Four of the commonly used commutation circuits have been discussed here. Some other commutation circuits are dealt with in Chapter 3.

1.6 Ratings of a Thyristor

Like all semiconductor devices, the thyristor also has limitations of voltage, current, and power dissipation, which implies that the device has to be connected in such a way that the specified limits are not exceeded. Thyristor manufacturers provide data sheets wherein the ratings are given both as numerical data as well as graphs. Thyristor current ratings are related with the heat dissipated in the devices. A knowledge of the various ratings is essential for designing appropriate circuits to protect thyristors against high voltages and currents, the latter causing heat generation. Most of the voltage and current ratings are specified for the worst operation condition, namely, for a maximum permissible junction temperature of 125 °C.

1.6.1 Voltage Ratings

While considering protection of thyristors against overvoltages, a distinction has to be made between slowly varying overvoltages and short-time surges. With the former type of overvoltages the thyristor will fire when the voltage exceeds the forward voltage rating V_{FBO}. Hence these voltages get reflected as overcurrents, which can be limited by external resistances. On the other hand, short-duration surges may cause permanent damage to the thyristor, and hence the device has to be safeguarded against them using proper protective circuits called *snubber circuits*. These are discussed in Section 1.9, which deals with protection of thyristors.

A thyristor may be subject to constant or repetitive overvoltage both in the forward as well as reverse directions. Direct overvoltages occur when the thyristor is used in the dc chopper and inverter circuits whose input is a direct voltage. Repetitive overvoltages may occur when it is connected in converter circuits which are supplied from an ac source. The following voltage ratings are usually given in the data sheets and are shown in Fig. 1.42; the nature of the overvoltages is evident from their nomenclature.

Forward voltage (or breakover voltage) (V_{FBO}) This is the forward voltage at which the thyristor will turn on with no gate triggering signal. As explained in Section 1.1, applying a voltage equal to this rating is one of the methods for triggering a thyristor.

Maximum repetitive peak forward blocking voltage (V_{FB} or V_{DRM}) This is the peak repetitive voltage, occurring under ac circuit conditions, which the thyristor can block in the forward direction.

Maximum non-repetitive transient forward voltage (V_{FBT} or PFV) This is the limiting anode-to-cathode voltage above which the thyristor may get damaged. As shown in Fig. 1.42, PFV is greater than V_{FBO}.

Maximum repetitive reverse voltage (V_{RB} or PRV or V_{RRM}) This is the peak repetitive voltage, occurring under ac circuit conditions, which the thyristor can block in the reverse direction. If this rating is exceeded, a very large reverse current may flow through the thyristor and damage it.

Maximum non-repetitive reverse voltage (V_{RBT} or V_{RSM}) This is the peak value of the transient reverse voltage which can be safely blocked by the thyristor. V_{RSM} is larger than V_{RRM} as shown in Fig. 1.42.

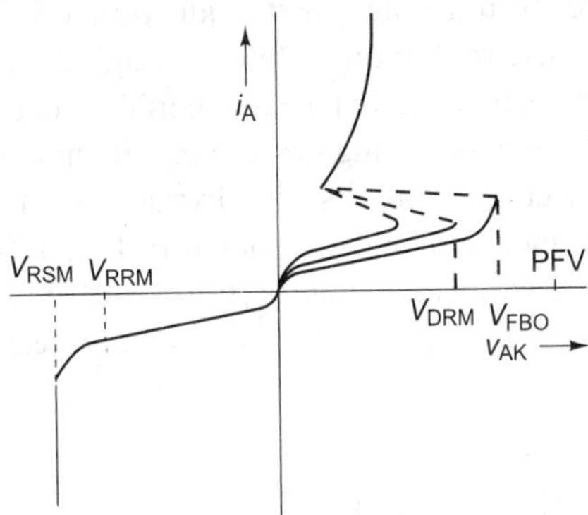

Fig. 1.42 Thyristor characteristics with various voltage ratings indicated thereon

1.6.2 Current Ratings

All semiconductor devices need to be guarded against high temperature by limiting their currents, the maximum temperature for which a device is designed being normally higher than the $I_A^2 R_d t$ heating, where I_A is the rated forward current, R_d is the resistance of the device, and t is the conduction time. Though this provides some factor of safety to the device, the thyristor is a unique semiconductor device for which the temperature attained may exceed the designed maximum under certain special conditions. In order to understand these contingencies and the phenomena underlying them, it is necessary to have a knowledge of the following aspects: (a) the heat dissipation under steady as well as transient states, (b) heating and cooling parameters of the junctions, (c) the nature of the fault current waveforms, and (d) the ambient temperature.

Heat dissipation in a thyristor is of five kinds, namely, forward dissipation, reverse dissipation, turn-on dissipation, turn-off dissipation, and dissipation at the gate. The first two kinds of dissipation are due to flow of forward current and reverse current, respectively. Turn-on dissipation is due to incomplete spread of the conduction in the cross section of the device and occurs when a device is frequently turned on and off. Turn-off dissipation is also associated with frequent switching. Out of these five losses, forward dissipation constitutes a major fraction and is discussed below in detail.

1.6.3 Average Forward Current ($I_{F(\mathrm{av})}$ or I_T)

The thyristor, being a controlled rectifier, conducts current either in a continuous or in an intermittent fashion. In the former case there may be ripples in the current waveform and in the latter it is necessary to know the conduction angle for computing the heat dissipation. It is the average current which is of interest in both the cases. The current rating is related to the heat (or power) dissipation in the device and is usually limited by a maximum junction temperature of 125 °C.

The detailed procedure adopted for arriving at the proper value of $I_{F(\mathrm{av})}$ for a thyristor under a particular set of operating conditions is now described.

Figures 1.43(a) and (b) illustrate how the junction temperature of a thyristor varies when a single pulse of current I flows through it. At the initiation of the pulse, the temperature starts rising in an exponential manner, this rise continuing till the pulse lasts; it then starts going down from the moment of termination of the pulse. In power electronic circuits, the thyristor has to conduct a series of current pulses. The sequence of pulses, shown in Fig. 1.44(a), will result in a cyclic variation of the junction temperature [Fig. 1.44(b)] in the steady state. The lower and upper limits of the junction temperature are, respectively, denoted by $\theta_{j1(\mathrm{min})}$ and $\theta_{j1(\mathrm{max})}$.

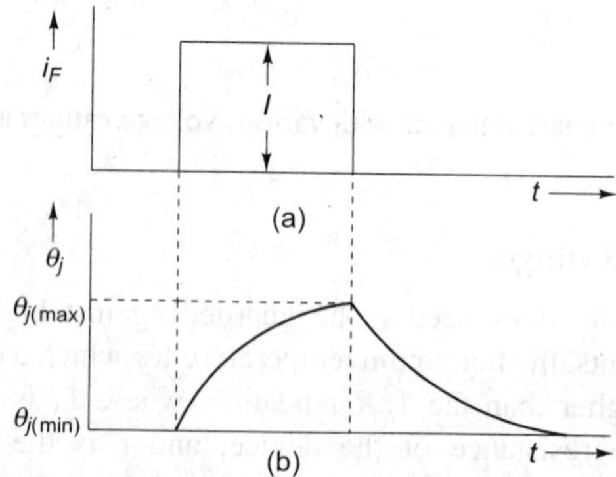

Fig. 1.43 (a) A typical current pulse; (b) curve showing the variation of junction temperature with time

If the duration of the pulse is reduced from t_1 to t_2 keeping the average current $I_{F(\mathrm{av})}$ constant, the pulse height has to be increased to I_2 as shown in Fig. 1.45(a). But in the latter case the steady-state variation of the junction temperature will be

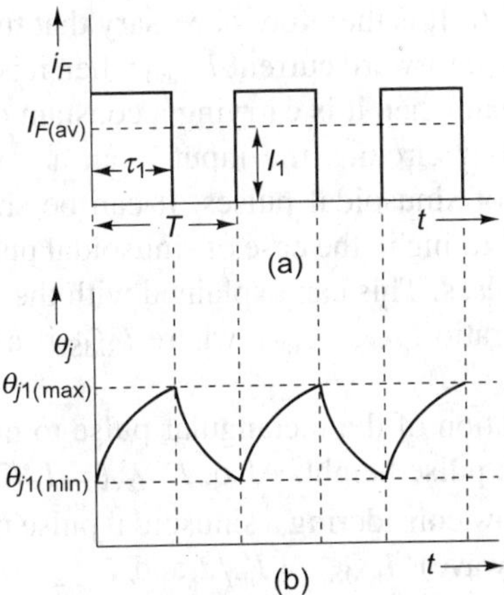

Fig. 1.44 (a) A typical current pulse with a sequence of duration τ_1 and period T; (b) curve showing the variation of junction temperature with time

different, as shown in Fig. 1.45(b); the new minimum and maximum temperature limits are denoted as $\theta_{j2(\text{min})}$ and $\theta_{j2(\text{max})}$, respectively. This difference in limits can be explained due to the fact that a higher pulse magnitude will give rise to higher heat dissipation, this in turn leading to a higher $\theta_{j(\text{min})}$. It can now be concluded that if the pulse duration is reduced, with constant pulse frequency, the $I_{F(\text{av})}$ rating has to be reduced so as to limit the maximum temperature attained by the thyristor. This is called the *derating of the thyristor for a reduced conduction angle*.

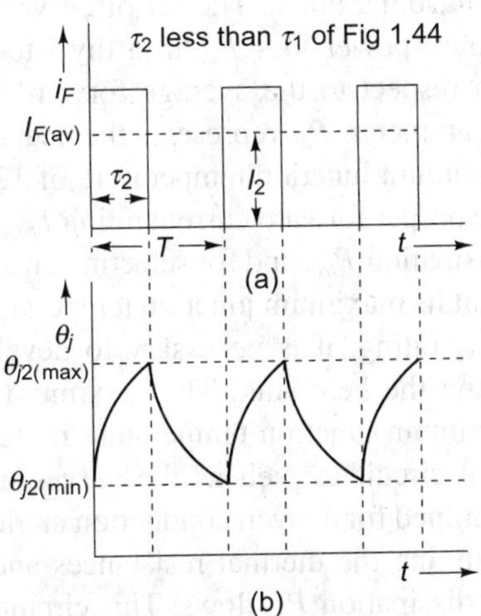

Fig. 1.45 (a) A typical current pulse with a sequence of duration τ_2 and period T; (b) curve showing variation of junction temperature

Due to a rapid increase of the junction temperature θ_j, when i_F begins to flow, the average value of i_F for a sequence of rectangular pulses that will just cause θ_j to reach 125 °C is less than the value of the constant direct current that would lead to the same rise in θ_j. It is therefore necessary that the thyristor be rated at a lower value of the average forward current $I_{F(\text{av})}$ when it is conducting a sequence of rectangular pulses than when it is carrying a constant direct current.

In controlled converter circuits, the input is an ac sinusoid, and hence the output is in the shape of sinusoidal pulses. It can be shown that for the same heating effect, the $I_{F(\text{av})}$ rating in the case of sinusoidal pulses is low compared to that with rectangular pulses. This can explained with the help of the form factor, which is defined as the ratio $I_{\text{RMS}}/I_{F(\text{av})}$, where I_{RMS} is a measure of the heating effect.

Considering the duration of the rectangular pulse to be $T/2$ or 180°, $I_{\text{RMS1}} = I_m/\sqrt{2}$, where I_m is the pulse height. Also, $I_{F1(\text{av})} = I_m/2$ and the form factor $= I_{\text{RMS1}}/I_{F1(\text{av})} = \sqrt{2}$. Now considering a sinusoidal pulse with a conduction angle of 180° (that is, a half-wave), $I_{\text{RMS2}} = I'_m/2$ and $I_{F2(\text{av})} = I'_m/\pi$, where I'_m is the peak of the sinusoid. If the heating effect were to be the same in both the cases, then $I_{\text{RMS1}} = I_{\text{RMS2}}$.

The ratio of the average values can be expressed as

$$\frac{I_{F2(\text{av})}}{I_{F1(\text{av})}} = \frac{I_{\text{RMS1}}}{I_{F1(\text{av})}} \div \frac{I_{\text{RMS2}}}{I_{F2(\text{av})}}$$

$$= \sqrt{2} \times \frac{2}{\pi} = \frac{2.828}{\pi}$$

This can be explained as follows. Due to the fact that, for the same I_{RMS}, the peak value is greater in the case of a sinusoidal pulse, the maximum junction temperature attained will also be higher. Therefore the average forward current will be smaller for a sinusoidal pulse. The set of curves in Fig. 1.46 shows the variation of the maximum power loss P_{av} in a thyristor which is subjected to sinusoidal pulses, with respect to the average forward current $I_{F(\text{av})}$, with the conduction angle as a parameter. P_{av} represents the heat dissipation occurring in the thyristor with a maximum junction temperature of 125 °C. These curves are useful for arriving at the proper forward current rating $I_{F(\text{av})}$, given the conduction angle and the power dissipation P_{av}, and for selecting an appropriate heat sink for the thyristor so as to limit its maximum junction temperature to 125 °C. However, for arriving at the $I_{F(\text{av})}$ rating, it is necessary to develop the thermal circuit of the thyristor including the heat sink. The maximum power dissipation P_{av} corresponding to a maximum junction temperature of 125 °C is arrived at with the help of this thermal circuit as well as the curves in Fig. 1.46. The $I_{F(\text{av})}$ rating can then be determined for a given conduction angle. Figure 1.47(a) shows a thermal circuit constituting the thermal resistances and thermal capacitances through which the heat dissipation P_{av} flows. This circuit bears analogy with an electrical network consisting of resistors and capacitors.

The quantity P_{av} can be shown to play the same part as a current source in an electrical circuit. The notations for the various terms are as follows: θ_j is the junction temperature, θ_c is the temperature of the casing, θ_s is the temperature of the heat sink, and θ_a is the ambient temperature. The thermal resistance between the junction and the casing is defined as

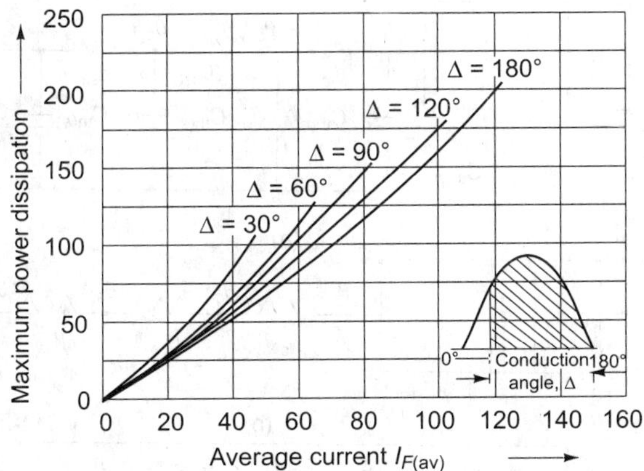

Fig. 1.46 P_{av} vs $I_{F(av)}$ curves for various conduction angles

$$\beta_{jc} = \frac{\theta_j - \theta_c}{P_{av}} \tag{1.82}$$

where P_{av} is the heat loss occurring at the junction and getting dissipated at the various resistances. The units of β_{jc} are °C/W. β_{cs} and β_{sa} are similarly defined, and $\beta_{th(a)}$ is the thermal resistance offered by a thin film of air surrounding the sink. $C_{th(j)}$ is the thermal capacitance at the junction, its units being W s/°C. $C_{th(c)}$, $C_{th(s)}$, and $C_{th(a)}$ are the thermal capacitances, respectively, of the casing, sink, and a thin film of air surrounding the sink. The circuit in Fig. 1.47(a) can be simplified to that in Fig. 1.47(b) with the understanding that the capacitances act as open circuits for the thermal current under steady-state conditions. The equation for this simplified circuit is written as

$$\theta_j - \theta_a = P_{av}(\beta_{jc} + \beta_{cs} + \beta_{sa}) \tag{1.83}$$

where θ_j is usually taken as 125 °C. The thermal resistances β_{jc} and β_{cs} are provided in the data sheet of the thyristor. The sink-to-ambient thermal resistance β_{sa} is supplied by heat sink manufacturers in the shape of curves drawn between $\theta_s - \theta_a$ and P_{av}. A typical set of such curves is shown in Fig. 1.47(c). These curves are drawn for heat sinks made with different metals with or without a black coating. A particular heat sink, say, curve c, may be selected and, guessing a value of P_{av}, the corresponding thermal resistance of the metal corresponding to curve c is

$$\beta_{sa} = \frac{(\theta_s - \theta_a)}{P_{av}} \tag{1.84}$$

This value of β_{sa} and the given β_{jc} and β_{cs} are substituted in Eqn (1.83), which is then solved for P_{av}. If the P_{av} so obtained tallies with the P_{av} estimated in the beginning, $I_{F(av)}$ is read off from Fig. 1.46, corresponding to the heat loss P_{av} and

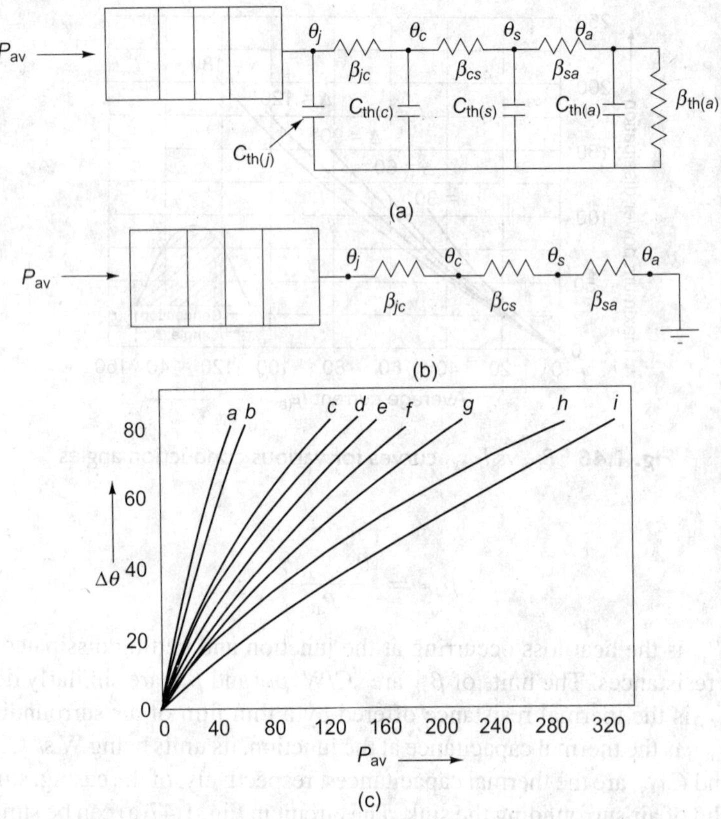

Fig. 1.47 (a) Thermal equivalent circuit of a thyristor; (b) simplified thermal circuit; (c) heat sink characteristics $\Delta\theta$ vs P_{av}

to the appropriate conduction angle. Otherwise the above procedure is repeated till both the values of P_{av} are very nearly equal.

The $I_{F(av)}$ arrived at as above is the average current rating of the thyristor which is required to limit the maximum junction temperature to 125°C for the given conduction angle and with the heat sink *c* fitted on the thyristor.

1.6.4 RMS Current Rating (I_{TR} or I_{RMS})

This is the specified maximum RMS forward current required to limit excessive heating in the resistive elements in the thyristor circuit, which includes leads, metallic junctions, and other such interfaces.

1.6.5 Peak Repetitive Forward Current Rating (I_{TRM})

This is the maximum repetitive instantaneous forward current permitted under stated conditions. This rating becomes relevant when the thyristor is used to supply a condenser load for which the ratio of the peak to average current is high.

1.6.6 Surge Current Rating (I_{TSM} or I_{FM})

This is the maximum current that can be permitted under fault conditions so that the junction temperature does not exceed 125 °C. Such abnormal conditions occur only a few times during the life of the thyristor. These surge currents are assumed to be sinusoidal with a frequency of 50 or 60 Hz and are categorized as half-cycle, three-cycle, or 10-cycle surge currents. The longer the duration, the lesser will be the value of this current rating. This phenomenon is now explained with the help of the thermal equivalent circuit of the thyristor shown in Fig. 1.47(a). The capacitances $C_{th(c)}$, $C_{th(s)}$, and $C_{th(a)}$ act as short circuits for a very short duration from the moment of application of a current surge, thereby conducting the heat to the surrounding air. The current through the resistances and the heat dissipation under such conditions will be minimum. Hence, very high half-cycle currents can be permitted. If the duration of the current is large, more and more of the capacitance comes into play, the capacitive impedance becomes appreciable, and more current passes through the resistances, thus increasing the resistive dissipation. This implies that the three-cycle current rating has to be less than the half-cycle rating, and in turn the 10-cycle current rating has to be smaller than the three-cycle rating.

1.6.7 I^2t Rating

The I^2t rating is specified for overloads of duration less than a half-cycle. It is useful in selecting a fast acting fuse or circuit-breaker that is employed to protect the thyristor.

1.6.8 Other Ratings

Some other ratings which are usually provided in the data sheet are as follows:

(a) dv/dt rating
(b) di/dt rating
(c) Turn-on time
(d) Turn-off time
(e) Gate current (average and maximum)
(f) Latching current
(g) Holding current

A detailed discussion of these ratings is given in the earlier sections.

1.7 Protection of Thyristors

The transient overvoltages occurring in a thyristorized circuit may be 5 to 10 times the normal forward and reverse voltages applied to the thyristor. These voltages cause permanent damage to the thyristor if means are not provided to absorb the energy associated with them. These surges can be broadly categorized into two types: (a) switching voltage surges and (b) atmospheric surges. As already stated, the thyristor has got an inherent protection against slowly increasing voltages because it turns on if the voltage exceeds V_{FBO}. Therefore the overvoltage problem gets converted into an overcurrent problem. This implies that the thyristor has to be mainly protected from switching and atmospheric surges.

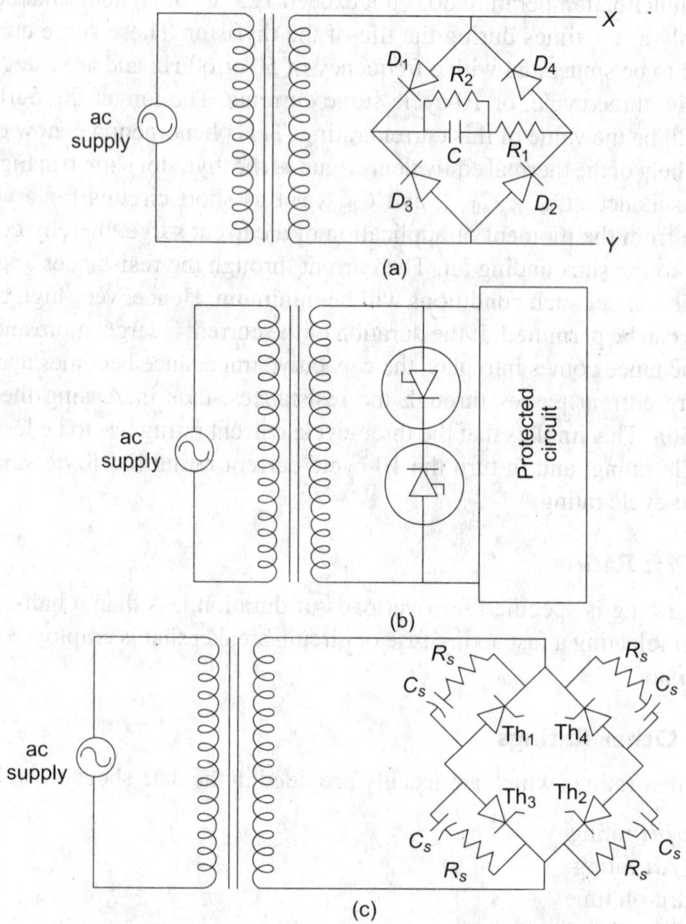

Fig. 1.48 Thyristor protection circuits against high voltages: (a) diode rectifier bridge, (b) selenium rectifier cells, (c) snubber circuits across each of the thyristors

1.7.1 Protection Against Voltage Surges

Voltage transients may be caused by any of the following conditions (Gentry et al. 1964).

(a) Sudden disconnection of circuits which have inductances and capacitances. This disconnection is caused by (i) blowing out of fuses, (ii) switching done on the primary side of a transformer, or (iii) the action of fast turn-off thyristors.

(b) Voltage surges due to causes external to the circuits, namely, atmospheric surges or those due to switching off of transformers under no load, where there is no low-impedance path for the surges; such switchings result in the occurrence of surges larger than those occurring in loaded transformers. The duration of occurrence of voltage surges may range from a fraction of a microsecond to a few milliseconds. The overvoltages that occur within the circuitry can be countered by using thyristors with voltage ratings of 1.5 to 2 times the voltages handled by them; alternatively surge absorbing circuits can be connected across the thyristors. A typical surge absorbing circuit consists of a diode rectifier bridge, selenium rectifier cells, or a resistance–capacitance combination known as a snubber circuit; these are shown respectively in Figs 1.48(a), (b), and (c). Figure 1.49 shows a full-wave fully controlled bridge converter protected by surge absorbing circuits. The protective circuit consists of a capacitor C_t in series with a resistor R_t. If a switching action occurs in the transformer, the magnetic energy in the transformer core and the $(1/2)Li^2$ energy in the transformer winding are initially absorbed by C_t and eventually dissipated in the R_t; the leakage inductance of the transformer, the capacitor C_t, and the resistor R_t constitute a damped oscillatory circuit.

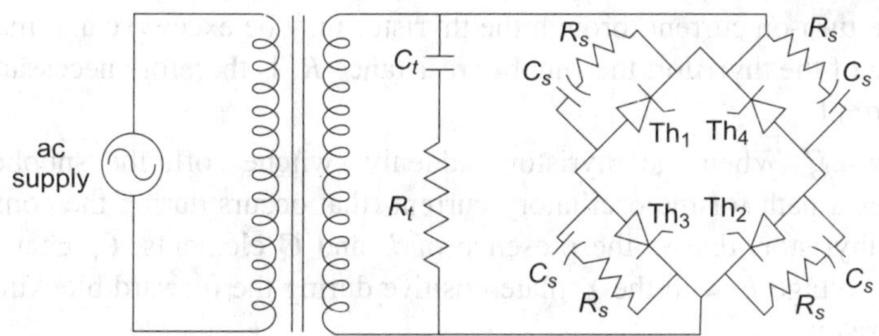

Fig. 1.49 Circuit showing surge absorbing elements and snubbers

The required value of the capacitor to absorb the surges may be calculated by the equation

$$\frac{1}{2}C_t(V_{\text{RSM}}^2 - V_m^2) = \frac{1}{2}L_\mu I_\mu^2 \tag{1.85}$$

where V_{RSM} is the peak non-repetitive reverse voltage of the thyristor and V_m is the peak value of the voltage handled by it; L_μ is the magnetizing inductance of the transformer and I_μ is the peak value of the magnetizing current, both these quantities being referred to the secondary of the transformer.

The series combination of the capacitor C_s and the resistor R_s across each of the thyristors in Fig. 1.49 constitutes the snubber circuit and its action is explained below with the help of Fig. 1.50.

C_s is required to prevent the unwanted dv/dt triggering of the thyristor. When the switch is closed, a sudden voltage appears across the circuit. C_s then acts like a short circuit, and in the absence of R_s the voltage across the thyristor will be zero. The voltage of the capacitor builds up at a slow rate and this helps in keeping down the dv/dt across the capacitor C_s as well as across the thyristor, to a value which is less than $(dv/dt)_{\text{max}}$ of the device. Eventually the voltage across C_s will

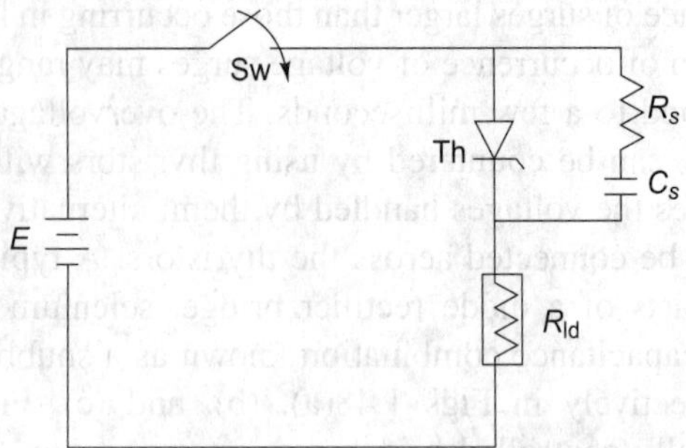

Fig. 1.50 Circuit explaining snubber action

become equal to the supply voltage E. The necessity of R_s can be explained as follows. When the thyristor is turned on with only C_s connected across it, C_s, which was charged earlier, discharges through it. The initial current will be equal to $V/(R_{\text{Th}} + R_{cs})$, where R_{Th} and R_{cs} are the resistances of the conducting thyristor and the capacitor C_s, respectively. As this combined resistance is quite low, the turn-on current through the thyristor may be excessive and may lead to damage of the thyristor; the snubber resistance R_s is therefore necessary to limit this current.

At turn-off When the thyristor suddenly switches off, the snubber circuit provides a path for the oscillatory current (that occurs during the commutation of the thyristor) due to the presence of L and C elements. C_s charges to the supply voltage E with the Y-plate positive during the forward blocking state of the thyristor.

The snubber circuit acts like a shock absorber for the high-voltage transients, which otherwise would cause damage to the thyristor. Figure 1.50 is redrawn as Fig. 1.51 with the extra element, namely, the inductance L connected in series with the thyristor. This circuit is useful in determining both the $(di/dt)_{\max}$ and $(dv/dt)_{\max}$ ratings of the thyristor. The approximate values of L, R_s, and C_s that will limit the actual di/dt and dv/dt can be determined with the help of these ratings, which are provided in the thyristor data sheet. In the following derivation it is assumed that the initial voltage on the capacitor (V_c) remains constant throughout and that the capacitor acts as a short circuit for a very short duration between the moment of closing of the switch Sw and the moment of firing of the thyristor. The thyristor remains in the forward blocking state during this interval.

The equation describing the circuit can be written as

$$L\frac{di}{dt} + (R_s + R_{\text{ld}})i = E - V_c$$

With the initial condition $i(0) = 0$ and defining R to be equal to R_s+R_{ld}, this equation becomes

$$L\frac{di}{dt} + Ri = E - V_c \tag{1.86}$$

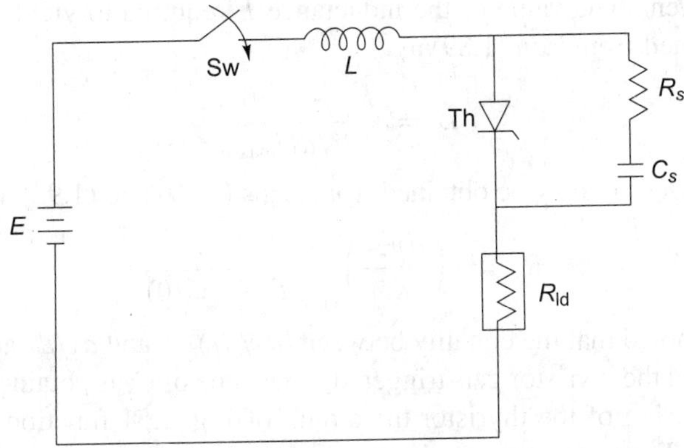

Fig. 1.51 Circuit for determining $(dv/dt)_{max}$ and $(di/dt)_{max}$ ratings

The solution of this first-order differential equation can be written as

$$i(t) = \left(\frac{E - V_c(0)}{R}\right)(1 - e^{-Rt/L}) \tag{1.87}$$

The differentiation of Eqn (1.87) gives

$$\frac{di}{dt} = \left(\frac{E - V_c(0)}{L}\right)e^{-Rt/L} \tag{1.88}$$

di/dt has its maximum at $t = 0$. Thus

$$\left(\frac{di}{dt}\right)_{max} = \left(\frac{di}{dt}\right)_{t=0} = \frac{E - V_c(0)}{L} \tag{1.89}$$

The voltage across the thyristor can be written as

$$v_{Th} = (E - V_c) - L\frac{di}{dt} - R_{ld}i = R_s i \tag{1.90}$$

Hence

$$\frac{dv_{Th}}{dt} = R_s\frac{di}{dt} \tag{1.91}$$

Hence the maximum values on both sides are related as

$$\left(\frac{dv_{Th}}{dt}\right)_{max} = R_s\left(\frac{di}{dt}\right)_{max} \tag{1.92}$$

It is evident from Eqn (1.92) that $(di/dt)_{max}$ and, hence, $(dv_{Th}/dt)_{max}$ occur at $t = 0$. Thus,

$$\left(\frac{dv_{Th}}{dt}\right)_{max} = \left(\frac{dv_{Th}}{dt}\right)_{t=0} = R_s\frac{E - V_c(0)}{L} \tag{1.93}$$

The $(dv/dt)_{max}$ rating of the thyristor can be experimentally determined as follows. The value of L is kept fixed and R_s varied till the thyristor fires without any gate current. The value of the inductance L required to yield $(dv/dt)_{max}$ is then determined from Eqn (1.89) as

$$L = \frac{E - V_c(0)}{(di/dt)_{max}} \tag{1.94}$$

The value of R_s can now be obtained from Eqns (1.89) and (1.92) as

$$R_s = \left(\frac{dv_{Th}}{dt}\right)_{max} \frac{L}{E - V_c(0)} \tag{1.95}$$

It should be noted that the equality between $(dv/dt)_{max}$ and dv/dt at $t = 0$ holds good provided the thyristor gets triggered. The value of C_s is obtained as follows. Prior to the firing of the thyristor the circuit of Fig. 1.51 functions as a forced RLC circuit. For such a circuit, the value of R_s for optimum damping (Csaki et al. 1975) is given as

$$R_s = \sqrt{\frac{L}{C_s}} \tag{1.96}$$

Equation (1.96) can be solved for C_s as

$$C_s = \frac{L}{R_s^2} \tag{1.97}$$

The following worked example explains the procedure for designing the values of L, R_s, and C_s, given the values of E, V_c, R_{ld}, $(di/dt)_{max}$, and $(dv/dt)_{max}$.

Worked Example

For the protective circuit given in Fig. 1.51, determine suitable values for L, C_s and R_s given the following data: $E = 130$ V, $V_c(0) = 30$ V, $R_{ld} = 20\ \Omega$, $(dv/dt)_{max} = 300$ V/μs, and $(di/dt)_{max} = 10$ A/μs.

Solution

From Eqn (1.94) the inductance L can be obtained as

$$L \geq \frac{130 - 30}{10} \times 10^{-6}\ \text{H}$$

This gives

$$L \geq 10\ \mu\text{H}$$

L is initially taken as 10 μH and R_s is obtained from Eqn (1.95) as

$$R_s \leq \frac{300}{10^{-6}} \times \frac{10 \times 10^{-6}}{130 - 30} = \frac{300 \times 10 \times 10^{-6}}{100 \times 10^{-6}}$$

$$\leq 30\ \Omega$$

Let R_s be taken as 30 Ω. C_s is obtained from Eqn (1.97) as

$$C_s = \frac{10 \times 10^{-6}}{30^2} = 0.011\ \mu\text{F}$$

The initial values of L, R_s, and C_s determined as above need to be modified in view of the following considerations (Bimbhra 1991).

When the thyristor is turned on, a total current $E/R_{ld} + E/R_s$ flows through the device. The waveform of E/R_s has the shape of a spike superimposed over the load current. It should be ensured that this total current is less than the peak repetitive forward current rating I_{TRM} of the thyristor.

Thus R_s has to be taken to be greater than that required to limit dv/dt. On the other hand the value of C_s has to be reduced so as to ensure that the discharge of the snubber does not harm the thyristor when it is turned on. Let the I_{TRM} rating in this case be 9 A. Thus R_s should be such that $E/R_s = 9 - (E/R_{ld}) = 9 - (130/20)$, which works out to 2.5 A. With R_s taken as 60 Ω, L has also to be recomputed to keep down the di/dt of the circuit. L is recomputed using Eqn (1.95) as

$$L \geq \frac{R_s(E - V_c)}{(dv/dt)_{max}} = \frac{60 \times 100 \times 10^{-6}}{300} = 20\,\mu\text{H}$$

Let L be taken as 25 μH. The actual $(di/dt)_{max}$ that is possible with this value of L is obtained from Eqn (1.89) as

$$\left(\frac{di}{dt}\right)_{max} = \frac{130 - 30}{25} = 4\,\text{A}/\mu\text{s}$$

This value of di/dt is well within the rated $(di/dt)_{max}$. Finally, C_s is obtained from Eqn (1.97) as

$$C_s = \frac{25}{60 \times 60}\,\mu\text{F} = 0.007\,\mu\text{F}$$

It is seen that this final value of C_s is less compared to the initially designed value. The final design values are $L = 25$ μH, $R_s = 60$ Ω, $C_s = 0.007$ μF.

1.7.2 Protection Against Direct Overcurrents

Overcurrents occur in thyristorized circuits due to (a) negligence, (b) failure of components on the load side, (c) short-circuiting of motor windings, and (d) failures due to the non-blocking (that is, not turning off) of the thyristors. An example of (d) is that of the parallel capacitor inverter shown in Fig. 1.52. The purpose of this circuit is to convert the dc of the battery into ac in the load, by alternately switching on the thyristors Th_1 and Th_2. It is to be ensured in this process that with one thyristor on, the other should be off, failing which there is a risk of damage to the battery. If, for example, Th_1 fails to turn off when Th_2 is on, the battery will get shorted through one half of the transformer primary winding, L, and Th_1 or Th_2.

The transient current ratings of the thyristor may be characterized as a one to three-cycle surge current rating and a half-cycle surge current rating, the former being much less than the latter. Overcurrents may also be categorized accordingly. The one to three-cycle overcurrents occur in circuits where the fault current is limited due to the presence of some amount of impedance, whereas half-cycle overcurrents are practically unlimited because of the presence of very little impedance. The lower fault current may be protected by fuses and circuit-breakers as in an ordinary power circuit, taking care to ensure that the

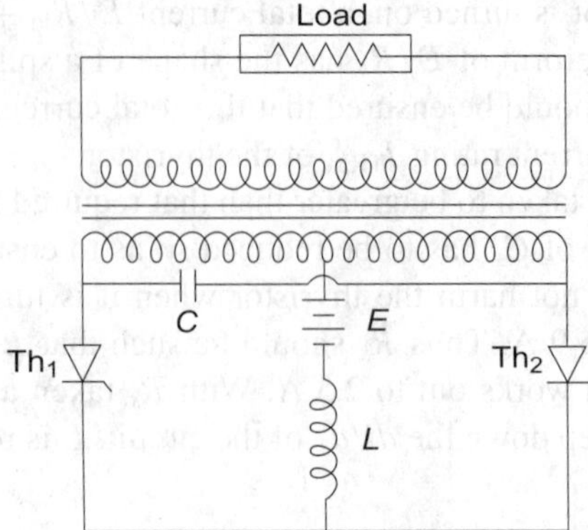

Fig. 1.52 A parallel capacitor type of inverter in which failure of blocking may occur

I^2t rating of the fuse is less than that of the series thyristor which is protected; also the short-circuit kilovolt-ampere rating of the circuit-breaker should be less than that of the thyristor protected by it. A scheme which is of particular interest in protecting the thyristor from such overloads is that of feeding back a signal proportional to the load current into the gate circuit, so as to increase the firing angle α and reduce the average load current.

For overcurrents exceeding the higher rating, namely, the half-cycle surge rating, all the schemes discussed above are too slow for satisfactory protection.

Figure 1.53 gives the curve for the overcurrent occurring in the case of a fault; the point P of the dashed curve is the peak value of the current that occurs in case no fuse is provided. On the other hand, the point B is the maximum value attained when the fuse is connected. The time taken from A to B is called the *melting time*. Once point B is reached, the fuse starts melting and, as a result, current flows through the fuse in the shape of an arc. This also introduces a high arcing resistance, which brings down the current. Finally, when the fuse completely melts, the arcing resistance becomes infinite and the current becomes

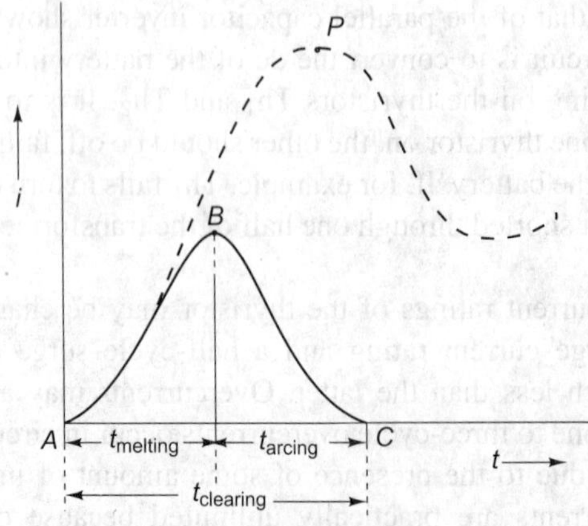

Fig. 1.53 Plot of i vs t illustrating the action of a current limiter

zero. This is represented by point C. The time taken from B to C is called the *arcing time*, and the total time, which is the sum of the melting and arcing times, is called the *clearing time*.

A problem with fast acting fuses is that of di/dt, which occurs due to the abrupt interruption of current during arcing. This leads to the production of overvoltage transients due to the presence of inductance in the circuit. Hence this protection scheme is to be accompanied by suitable overvoltage protection features, for limiting these surges to values less than the thyristor ratings.

As protective features normally cause interruptions to the supply, sometimes the overloading problem is solved by providing adequate redundancy in the circuit. As a result, when a fault occurs, the faulty section is isolated, and the load current is diverted through the healthy section. Though this ensures continuity of supply to the load, the cost of the additional thyristors and associated circuitry should be justified by other economic factors.

A third way of protecting thyristors from overcurrents is to provide circuits employing auxiliary thyristors as shown in Fig. 1.54. This circuit is a slight modification of the parallel capacitor commutation shown in Fig. 1.38(a). The action of the circuit is explained below.

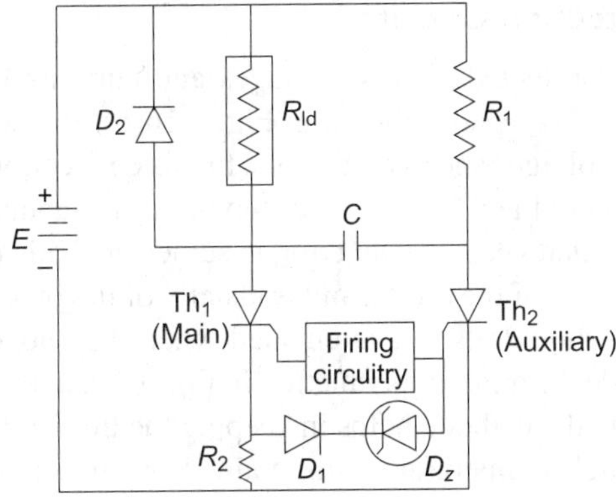

Fig. 1.54 Fast acting current limiter circuit using an auxiliary thyristor

Normally the circuit may be made to operate as a chopper by alternately firing the two thyristors. When the main thyristor Th_1 is conducting, the load is supplied through it and the load current flows through a low resistance (R_2). At normal load currents, the voltage across R_2 is negligible. But when abnormal current flows through the load, this voltage causes an avalanche breakdown of the zener diode D_z, fires the thyristor Th_2, and turns off Th_1. The fault current is interrupted and the diode D_1 helps in providing reverse current protection to the gate of Th_2. The diode D_2 serves to provide a free path for the inductive $[(1/2)Li^2]$ energy of the load when Th_1 is blocked either during the chopper action or due to the emergency action caused by the switching of Th_2.

1.7.3 Protection Against Alternating Overcurrents

Protection may be provided by means of a resonant circuit to limit the ac load current when it exceeds its normal value. A typical circuit is shown in Fig. 1.55.

The series LC elements provide a path here, with the saturable reactor in the shunt path offering very high impedance to the normally occurring alternating currents. Under heavy fault currents, the saturable reactor saturates and offers zero impedance, thus causing resonance between the parallel L and C elements. By forming an infinite impedance path, the LC combination provides protection to the thyristors from these fault currents.

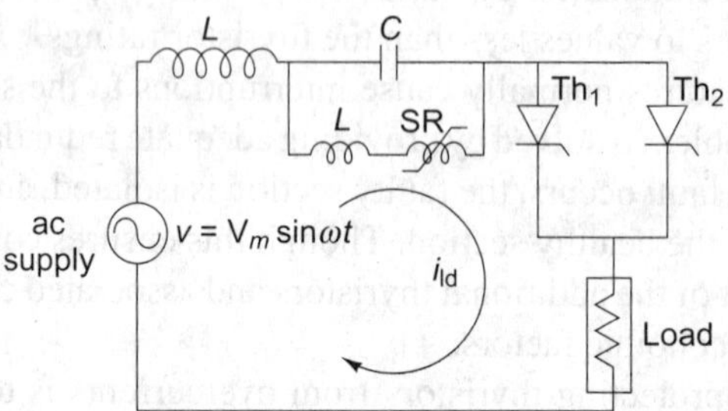

Fig. 1.55 Circuit for alternating overcurrent protection

1.7.4 Gate Protection Circuits

The gate has to be protected against voltages and currents falling outside the safe region of the (i_G, v_G) plane shown in Fig. 1.16(b). It also needs protection against reverse overvoltages and overcurrents. In this connection, the peak reverse gate voltage V_{GRM} is defined as the rated maximum instantaneous reverse bias voltage to which the gate–cathode junction is subjected. This bears similarity to the peak reverse voltage of the diode, and is usually of the order of 5 to 10 V. The positive gate signal should also be attenuated when the anode is negative with respect to the cathode. Similar to its function in Fig. 1.19, a resistor R_b connected between the gate and the cathode helps in keeping the thyristor in the stable OFF state when either high temperatures are attained or undesirable anode voltage transients occur.

The zener diode pair shown in Fig. 1.56(a) helps in clipping both the positive and negative overvoltage signals at the gate. In Fig. 1.56(b) the circuit consisting of D_1, R, and D_2 helps in attenuating positive gate signals when the anode is negative with respect to the cathode and also when the gate voltage is negative with respect to the cathode. Figure 1.56(c) shows a diode D commonly used to protect the gate from reverse currents.

1.8 Other Members of the Thyristor Family

The development of the silicon controlled rectifier (thyristor) in 1957 caused a revolution in power-control circuitry. It has replaced gas tube devices such as the thyratron and ignitron because of its operational superiority and compactness. However, continued research in device technology led to the development of a number of other *p-n-p-n* devices since 1970. Some of the important devices are the (i) DIAC, (ii) TRIAC, (iii) inverter grade thyristor, (iv) gate turn-off thyristor,

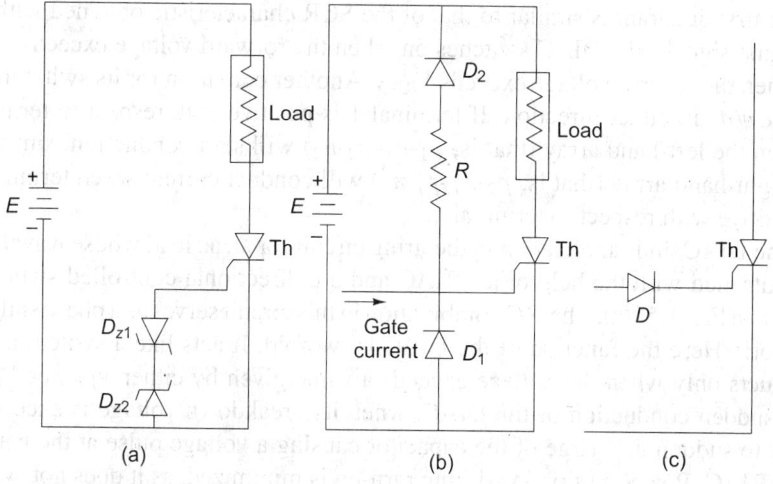

Fig. 1.56 Gate protection circuits: (a) against positive and negative gate-to-cathode voltages, (b) attenuation of gate overvoltage signals, (c) against reverse currents

(v) programmable unijunction transistor, (vi) reverse-conducting thyristor or asymmetrical thyristor, (vii) static induction thyristor, (viii) light-activated SCR, and the (ix) MOS-controlled thyristor. Each of these devices has one or more features that are either not present in the conventional thyristor or, if present, are superior to those of the thyristor. These are described below.

1.8.1 DIAC

The DIAC is a bidirectional diode switch. Its circuit symbol and schematic construction are given, respectively, in Figs 1.57(a) and (b). It consists of a *p-n* layer pair connected to each of the terminals 1 and 2, the pairs being located adjacent to each other. Its characteristic, shown in Fig. 1.57(c), consists of two parts, one in the first quadrant and the other in the third quadrant. The curve drawn

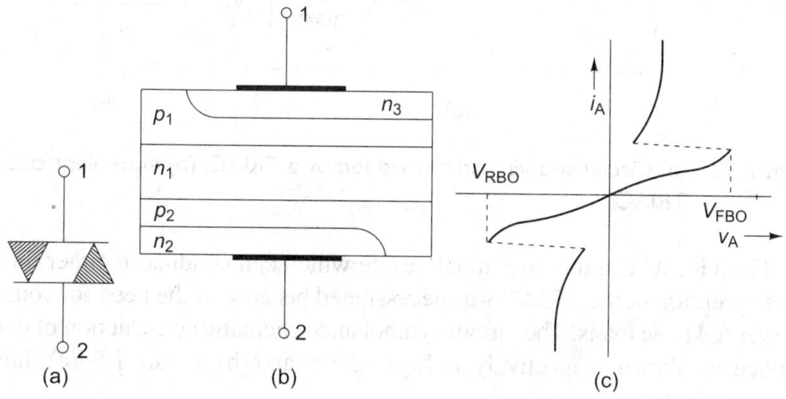

Fig. 1.57 DIAC: (a) circuit symbol, (b) schematic diagram, (c) characteristics

in the first quadrant is similar to that of the SCR characteristic obtained without any gate signal. The DIAC switches on when the forward voltage exceeds V_{FBO} or when the reverse voltage exceeds V_{RBO}. Another condition for its switching is high dv/dt in either direction. If terminal 1 is positive with respect to terminal 2, then the left-hand array (that is, p_1-n_1-p_2-n_2) will start conduction; similarly the right-hand array (that is, p_2-n_1-p_1-n_3) will conduct current when terminal 2 is positive with respect to terminal 1.

The DIAC finds application in the firing circuit for an ac load whose waveform is controlled with the help of a TRIAC and a bidirectional controlled switch as shown in Fig. 1.58(a). The RC combination in this circuit serves as a phase shifting network. Here the function of the DIAC is twofold. It acts like a switch, i.e., it conducts only when the voltage exceeds a value given by either V_{FBO} or V_{RBO}. The sudden conduction of the DIAC, when its breakdown voltage is exceeded, leads to sudden discharge of the capacitor causing a voltage pulse at the gate of the TRIAC. Power dissipation during turn-on is minimized, as it does not switch on for a lower voltage. It provides a positive- or negative-capacitance voltage at the gate depending on whether terminal 1 is positive with respect to terminal 2 or vice versa. The DIAC is also used in the speed control scheme of a universal motor.

1.8.2 TRIAC

The operation of the TRIAC is equivalent to that of a combination of two thyristors connected antiparallelly, as given in Fig. 1.58(b). Whereas it is possible to control the TRIAC with the help of a single gate [Fig. 1.58(a)], both the thyristor gates are required to be fired in the circuit of Fig. 1.58(b).

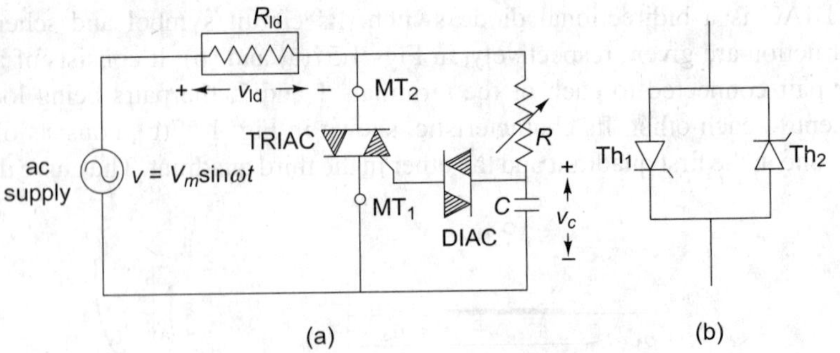

Fig. 1.58 (a) Circuit showing an ac load fed by a TRIAC; (b) equivalent circuit of a TRIAC

The TRIAC is a three-terminal device which can conduct in either direction. The invention of the TRIAC was necessitated because of the need for controlling power fed to ac loads. The circuit symbol and schematic construction of one such device are shown, respectively, in Figs 1.59(a) and (b). Figure 1.59(c) shows the characteristics.

Flexibility of operation in such a device consists of turning it on in any of the four modes given below.

Mode 1 The main terminal MT_2 is positive with respect to the main terminal MT_1 and the gate triggering signal is positive. The gate plays the part of the normal gate of a thyristor formed by the p_1-n_1-p_2-n_2 combination.

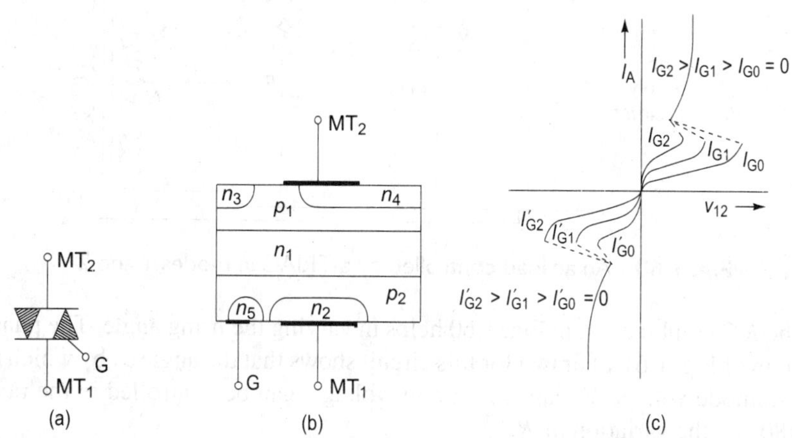

Fig. 1.59 (a) TRIAC: (a) circuit symbol, (b) schematic diagram, (c) characteristics

Mode 2 MT_2 is negative with respect to MT_1 and the gate triggering signal is positive. In this case the p_2-n_1-p_1-n_4 combination forms a thyristor and the p_2-n_2 combination forms an indirect gate, initiating electrons from n_2 to p_2, from p_2 to n_1, and finally from n_1 to p_1. Thus the combination n_2-p_2-n_1-p_1 fires the thyristor p_2-n_1-p_1-n_4.

Mode 3 MT_2 is positive with respect to MT_1 and the gate triggering signal is negative. The gate G forms the igniting cathode through the n_5-layer and the p_1-n_1-p_2-n_2 combination acts as the thyristor.

Mode 4 MT_2 is negative with respect to MT_1 and the gate triggering signal is negative. In this condition the gate G forms the indirect gate to the thyristor p_2-n_1-p_1-n_4 through the n_5 layer.

It can be seen that the gate connections to p_2 and n_5 facilitate firing, respectively, in the first and second modes.

The above discussion shows that the turn-on of the TRIAC is easier than that of the thyristor. Modes 1 and 4 can be classified as primary modes because in the one case both MT_2 and G are positive and in the other case both of them are negative. From this point of view, modes 2 and 3 are considered secondary modes. The sensitivity of triggering differs in the four modes, implying thereby that the charge required for firing varies from mode to mode. The TRIAC has the property that mode 4 is most sensitive, thus requiring least charge. Figure 1.58(a) is an example of a circuit in which the TRIAC works in the two primary modes. In case the *RC* circuit is connected as a phase shifting network as shown in Fig. 1.60, the TRIAC works in mode 1 (primary mode) in the positive half-cycle of the ac and in mode 2 (secondary mode) in the negative half-cycle.

It follows that the primary modes 1 and 4 are obtained naturally in an ac circuit but, for operating in the secondary modes, a separate supply is needed for the gate circuit.

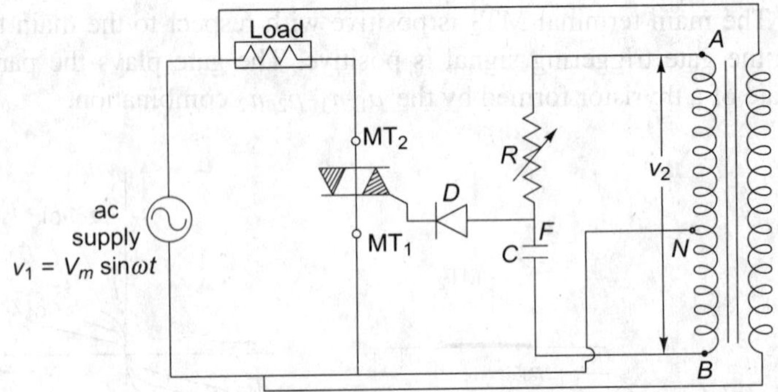

Fig. 1.60 An ac load controlled by a TRIAC in modes 1 and 2

The RC combination in Fig. 1.60 helps in varying the firing angle. The phasor diagram of Fig. 1.61(a) drawn for this circuit shows that the angle α, by which the gate–cathode voltage V_T lags the supply voltage, can be controlled in the range 0°–180° by the variation of R.

As the voltage V_2 in Fig. 1.60 is derived from the same ac source, there is an inherent synchronization of the gate triggering pulse with the ac supply waveform. Thus the firing angle of the TRIAC can be varied in both half-cycles as indicated in Fig. 1.61(b).

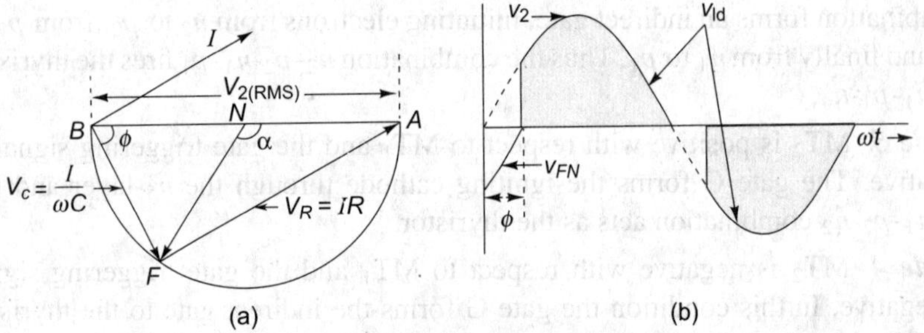

Fig. 1.61 (a) Phasor diagram for the RC circuit of Fig. 1.60; (b) load voltage waveform for the controlled ac circuit of Fig. 1.60

The turn-off of a TRIAC is more difficult relative to that of the thyristor because, if a negative voltage is applied across a TRIAC which is conducting with MT_2 positive with respect to MT_1, it will start conducting in the opposite direction. Other methods available for turn-off of a TRIAC are (i) current starvation or switching off the load current and (ii) ac line commutation. In both these cases, the current becomes zero and the TRIAC recovers its blocking state. The gate signal can then be used to turn it on in the other direction. However, in the case of an $R + L$ load, turn-off is not easy because the current attains a zero value only ϕ (power factor angle) electrical degrees after the voltage becomes zero, and the TRIAC tends to turn off. But in this short time the voltage attains a negative value and this amounts to the application of high dv/dt in the reverse direction and the TRIAC turns on. To prevent this undesirable turn on, an RC (snubber) circuit has to be connected across the TRIAC to absorb the energy accompanied with the high dv/dt.

The TRIAC finds the following applications: as an electronic switch, in spot and seam welding equipment, in light controlling circuits, single-phase ac series motors or universal motors, and in consumer appliances such as food mixers, portable drills, etc.

The limitation of the TRIAC is its low $(dv/dt)_{max}$ as compared to thyristors. This feature limits its use to only low-frequency circuits.

1.8.3 Inverter Grade Thyristor

The inverter grade thyristor is a conventional reverse blocking thyristor with enhanced dynamic characteristics to give improved performance in forced commutated inverter circuits. Rectifier circuits use conventional thyristors that are naturally commutated by the normal reversal of the ac supply voltage. The frequency of the ac supply should not exceed $1/2t_{OFF}$, where t_{OFF} is the turn-off time of the thyristor. The turn-off time is usually less than 100 μs and hence the frequency of the ac supply should not exceed 5000 Hz. The power supply frequency being usually 50–60 Hz, an adequate factor of safety is available and the turn-off time is not critical in these applications. These conventional thyristors called *converter grade* thyristors work satisfactorily in rectifier applications. However, choppers and inverters have a dc supply as their input, and hence the thyristors used with them must be turned off by forced commutation circuits which make use of the energy stored in a charged capacitor. Forced commutation involves the discharging and recharging of the capacitor with a consequent energy loss, this being prohibitively high at high switching frequencies. In these applications t_{OFF} is a critical parameter that must be minimized in order to limit the size and weight of the commutating capacitor, thereby reducing the commutation losses. The inverter grade thyristor has been developed to meet this requirement. In this device the silicon crystal is doped with gold atoms to reduce the lifetime of charge carriers in the n-base region. By accelerating the recombination of the electron–hole pairs during the turn-off, t_{OFF} is reduced to the range 5–50 μs, depending on the voltage rating.

Another method for achieving this accelerated recombination and reduced turn-off time is the *electron irradiation* of the crystal.

In high-frequency applications the turn-on time of the thyristor has also to be reduced. This is achieved by incorporating a high di/dt capability, which helps in eliminating a di/dt limiting inductor and reducing turn-on dissipation. However, this is accompanied by (a) low-voltage blocking capability, (b) low dv/dt rating, and (c) increased on-state voltage drop; the last feature results in higher conduction losses.

1.8.4 Gate turn-off Thyristor

Like the conventional thyristor, the gate turn-off thyristor (GTO) can be triggered into the conducting state by a pulse of positive gate current. However, unlike the ordinary thyristor a pulse of negative current at the gate terminal can cause its turn-off. This feature simplifies the commutation circuitry of a GTO and facilitates the construction of more compact inverter and chopper circuits using the GTO.

The operation of the GTO can be explained with the help of the two-transistor analogy shown in Fig. 1.62 and by comparison of its features with those of thyristors. As the current gains of the *p-n-p* and *n-p-n* transistors in a conventional thyristor are high, the device latches into conduction with a small gate current signal. But due to the same reason, it cannot be turned off by negative gate current pulses. The GTO differs from the thyristor in that the gain of the *p-n-p* transistor is at a low value, facilitating turn-off of the device with a negative gate current signal. A negative current at the terminal G reduces the base current of the *n-p-n* transistor as well as the collector current of the *p-n-p* transistor to zero. This explains the quick turn-off of the GTO. The reduced gain of the *p-n-p* transistor is obtained for large-current GTOs by constructing some anode-to-*n*-base short-circuiting spots as shown in Fig. 1.63(a).

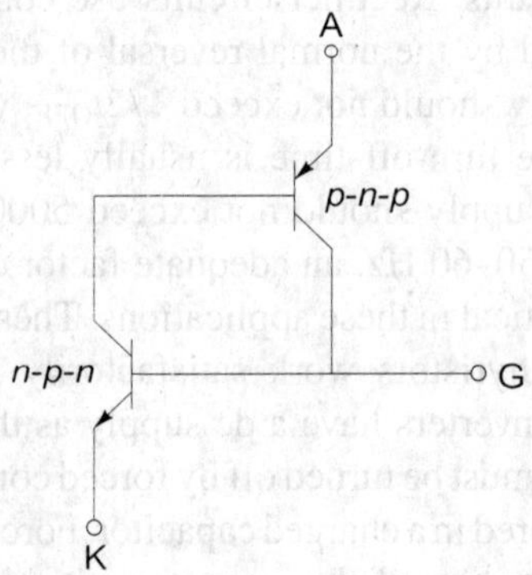

Fig. 1.62 Two-transistor analogy of a GTO

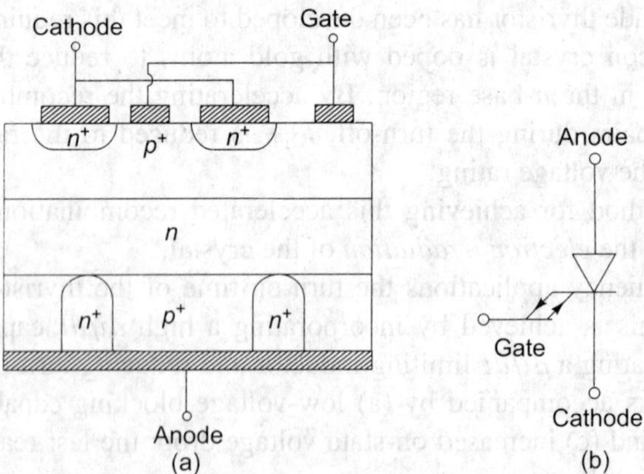

Fig. 1.63 (a) Schematic of a GTO showing *n*-base short-circuiting spots; (b) circuit symbol of a GTO

On the cathode side, a few cathode terminals constructed and interconnected to form a parallel combination of cathodes connected to *n*-type fingers in *p*-type regions. This type of construction for the *p-n-p-n* device helps in simultaneous turn-on or turn-off of the whole active area of the chip. Alternatively, gold or any other heavy metal can be diffused to reduce the carrier lifetime and consequently reduces the gain of *p-n-p* part of the GTO. The circuit symbol of a GTO is shown in Fig. 1.63(b).

Characteristics of a GTO The turn-off mechanism of a GTO can be explained with the help of the chopper circuit and its waveforms given, respectively, in Figs 1.64(a) and (b). In the former figure, a battery of voltage E_s supplies an inductive load through a GTO. The diode D_1 across the load serves as a freewheeling diode for circulating the load current at the turn-off of the GTO. The snubber circuit consisting of the elements R_s, D_2, and C_s helps in reducing the forward dv/dt during turn-off, thus improving the current interrupting capability of the GTO.

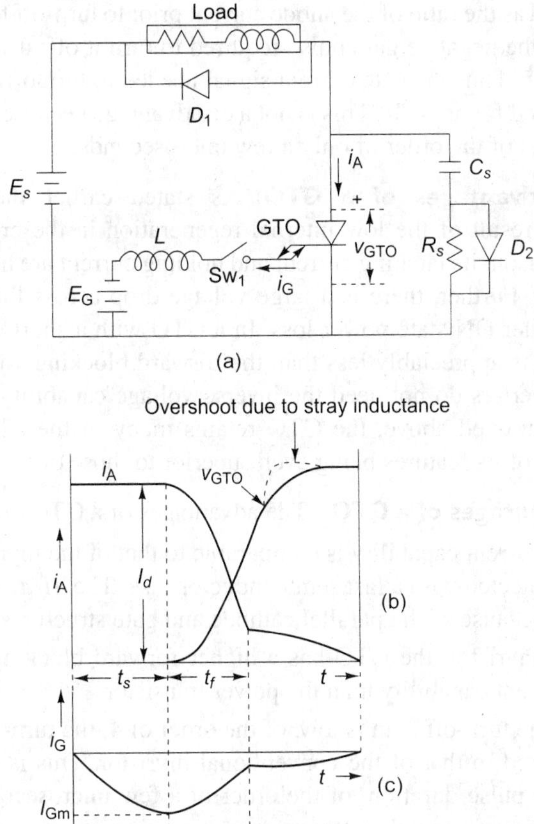

Fig. 1.64 (a) Circuit for turn-off of GTO; (b) plots of i_A and v_{GTO} during turn-off; (c) plot of i_G during turn-off

The turn-off mechanism can be understood by assuming that the circuit is initially in an ON state and that the GTO conducts a steady-state current I_d through the load. In this condition, the crystal is filled with plasma consisting of electron–hole pairs. The turn-off is initiated by passing a negative current pulse of amplitude I_{Gm} [Fig. 1.64(c)] through the gate–cathode circuit by turning on the switch Sw_1. During the interval called the *storage time* (t_s), excess holes in the plasma can be removed by this reverse gate current pulse, but the anode current remains constant at I_d. In this state the anode current is pinched through the narrow filaments which are formed under the cathode terminals. During the time period called the *fall time* (t_f), these current filaments quickly collapse and the anode current rapidly decreases to a value called the *tail current*. During this interval the load current is diverted through the snubber capacitor C_s and its voltage increases at a rate given by

$$\frac{dv}{dt} = \frac{I_d}{C_s} \tag{1.98}$$

The tail current of the anode then decays to zero. A current rating known as the *turn-off gain* of a GTO gives an idea of the gate current pulse required to turn it off. It is defined as the ratio of the anode current prior to turn-off to the magnitude of the peak of the negative gate pulse required to turn it off; its value is usually in the range 3–5. Thus the gate current signal needed to turn off a GTO is larger than that required for an SCR. This is not a disadvantage because the gate current pulse duration is of the order of only a few microseconds.

1.8.4.1 Disadvantages of a GTO As stated earlier the gate turn-off capability is a result of the low internal regeneration in the crystal. However, for the same reason, its latching current and holding current are higher than those of the thyristor. Further, there is a large voltage drop across the GTO and this results in a higher ON state power loss. In a GTO with a shorted anode emitter the PRV rating is appreciably less than the forward blocking voltage, but many GTO-based inverters do not need the inverse voltage capability. In spite of the drawbacks mentioned above, the GTO retains many of the advantages of the thyristor, some of its features being even superior to those of the thyristor.

1.8.4.2 Advantages of a GTO The advantages of a GTO are as follows.

(a) Its surge current capability is comparable to that of the thyristor and thus it can be protected with a fast semiconductor fuse. The di/dt limitation is not so acute because of the parallel cathode and gate structures.

(b) Like the thyristor, the GTO has a higher forward blocking voltage and a larger current capability than the power transistor.

(c) Though the turn-off gain is low, of the order of 4, the turn-off loss is small as compared to that of the conventional thyristor. This is due to the very short gate pulse duration, of the order of a few microseconds. As against this, the conventional thyristor requires a reverse current pulse of about 1.6 times the anode current with a duration of two to three times the thyristor turn-off time.

(d) The GTO has a faster switching speed than a conventional thyristor and in devices like converters this facilitates a smaller filter size. This is one of the reasons for its improved performance.

(e) The firing circuit of the GTO can be rigged up so as to generate both the turn-on and turn-off current pulses at the gate, and hence no separate commutation circuit is necessary. This is illustrated in Fig. 1.65, in which separate batteries are shown for turn-on and turn-off. Here, the transistor Q_1 is an auxiliary device used for turn-on purposes whereas an auxiliary thyristor Th_1 is included for facilitating turn-off. The presence of the inductor L helps in the turn-off process.

(f) A GTO circuit is smaller in size and lower in weight than a thyristor circuit. Moreover, it has a higher efficiency, because the increase in the gate drive power and ON state power loss is more than compensated by the elimination of the forced commutation losses.

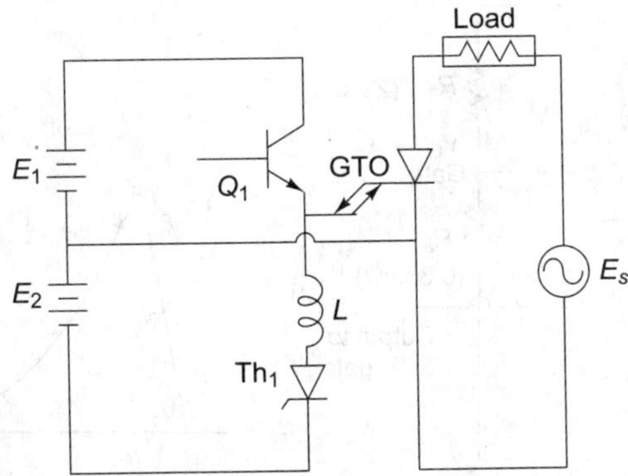

Fig. 1.65 Gate-drive circuit of a GTO for turn-on and turn-off

1.8.5 Programmable Unijunction Transistor

The programmable unijunction transistor (PUT) is a three-terminal planar *p-n-p-n* device like the SCR. Its terminals are also designated as anode, cathode, and gate. However, in the PUT the gate is connected to the *n*-type material near the anode as shown in Fig. 1.66(a). The circuit symbol is given in Fig. 1.66(b), the *V-I* characteristic in Fig. 1.66(d), and a typical relaxation oscillator circuit in Fig. 1.66(c). Its operation is somewhat similar to that of the conventional UJT.

The gate voltage V_G, which is variable, is obtained by the potential division of the supply voltage by the resistors R_4 and R_5. V_G determines the peak voltage V_P, whereas in a conventional UJT, V_P is fixed for a device by the dc supply voltage. If the anode voltage V_A is less than the gate voltage V_G the PUT remains in its OFF state. On the other hand, if V_A obeys the inequality

$$V_A \geq V_G + 0.7 \tag{1.99}$$

(where 0.7 V is the voltage drop across a conducting diode), the peak point is reached and the upper *p-n* junction starts conduction and the device turns on. The peak and valley currents I_P and I_V depend on the equivalent impedance $R_{eq} = R_4R_5/(R_4 + R_5)$ and the supply voltage V_{CC}, R_{eq} is usually kept at a value below 500 kΩ.

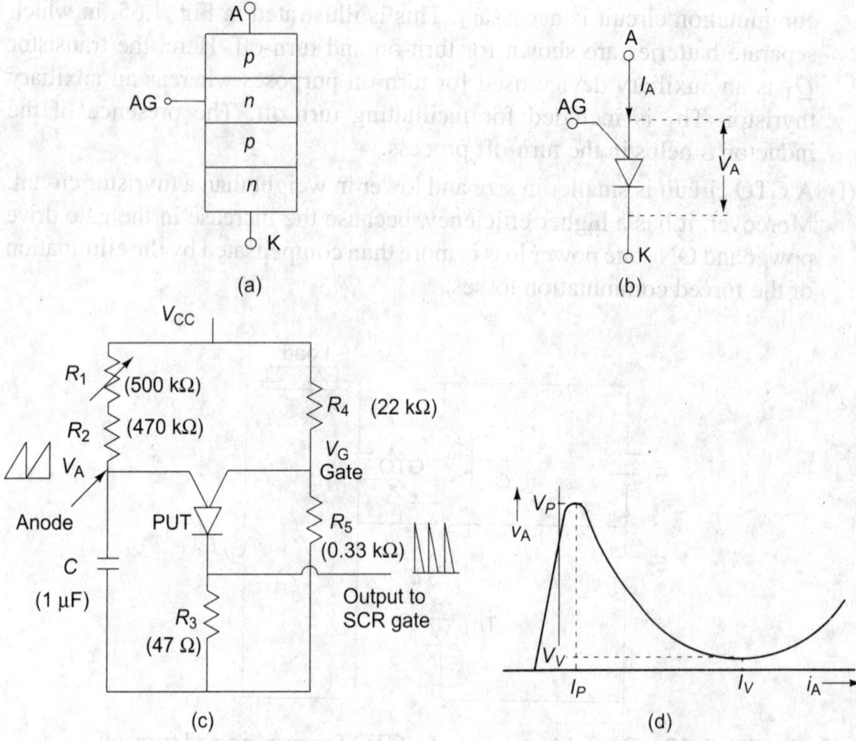

Fig. 1.66 Programmable unijunction transistor: (a) schematic diagram, (b) circuit symbol, (c) typical relaxation oscillator circuit, (d) v_A-i_A characteristics

$$V_P = \frac{R_5}{R_4 + R_5} V_{CC} \tag{1.100}$$

The intrinsic stand-off ratio is now obtained as

$$\eta = \frac{V_P}{V_{CC}} = \frac{R_5}{R_4 + R_5} \tag{1.101}$$

Equation (1.101) shows that the factor η can be varied at will by varying R_4 and R_5. The frequency of the output pulse is controlled by R_1, R_2, C, R_4, and R_5. The period of oscillation is given approximately by

$$T = \frac{1}{f} = (R_1 + R_2)C \ln \frac{V_{CC}}{(V_{CC} - V_P)}$$

$$= (R_1 + R_2)C \ln \left(1 + \frac{R_5}{R_4}\right) \tag{1.102}$$

The gate current I_G at the valley point is given by

$$I_G = (1 - \eta)\frac{V_{CC}}{R_{eq}} \tag{1.103}$$

where R_{eq} is as given above. R_4 and R_5 can be obtained as

$$R_4 = \frac{R_{eq}}{\eta} \quad \text{and} \quad R_5 = \frac{R_{eq}}{1-\eta}$$

The advantage of the PUT can be seen with the help of the relaxation oscillator circuit of Fig. 1.66(c), where R_1 and R_2 can be chosen to get any desired UJT characteristic such as R_{BB}, I_P, and I_V. The following inequalities have also to be satisfied for bringing the PUT into the ON and OFF states:

$$I_A \geq I_P \quad \text{for turning on} \tag{1.104}$$

$$I_A \leq I_V \quad \text{for turning off} \tag{1.105}$$

$$V_A = V_G + 0.7 \tag{1.106}$$

For the circuit given in Fig. 1.66(d), typical values of V_G, V_A, and I_A are calculated as below for a V_{CC} of 20 V:

$$V_G = \frac{20R_5}{R_4 + R_5} = \frac{20 \times 0.33}{22 + 0.33} = 0.3\text{ V}$$

$$V_A = 0.3 + 0.7 = 1.0\text{ V}$$

$$I_A = \frac{20 - 0.3 - 0.7}{R_1 + R_2} = \frac{19}{R_1 + R_2}$$

The potentiometer R_1 can be varied to give the required anode current. Similar to the UJT, if the conducting resistance R_{con} of the device is given for the PUT, the pulse height can be determined by the equation

$$v_{pulse} = \frac{V_P R_3}{R_{con} + R_3} \tag{1.107}$$

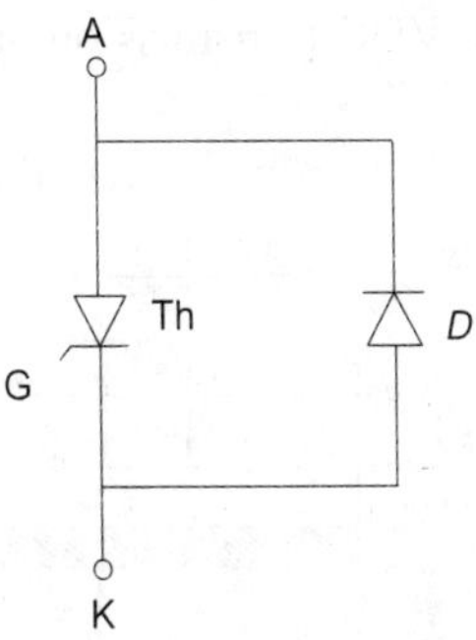

Fig. 1.67 Construction of an RCT

1.8.6 Reverse Conducting Thyristor

The reverse blocking capability of the reverse conducting thyristor (RCT) is much less than that of the SCR, of the order of 20–40 V. This makes it cheaper than the ordinary SCR. It can be used in inverter and dc chopper circuits in which a high reverse blocking capability is not needed. As shown in Fig. 1.67, its construction is equivalent to that of a thyristor with a built-in antiparallel diode connected across it. This feature clamps the reverse voltage to 1–2 V. Under steady-state conditions the diode helps in eliminating unwanted voltage transients. The forward blocking capability of an RCT is, however, comparable to that of a thyristor. The low reverse blocking voltage permits a reduction in the turn-off time of the RCT. This in turn reduces the cost of the commutation circuitry. The RCT is also called an *asymmetrical thyristor*.

1.8.7 Static Induction Thyristor

The static induction thyristor (SITH) is a minority carrier device that has the same characteristics as the metal oxide semiconductor field effect transistor (MOSFET), which is described in Section 1.9.2. It is turned on by applying a positive gate voltage like the enhancement-type MOSFET and is turned off by a negative gate voltage. It has a low resistance during conduction. Its voltage and current ratings are, respectively, limited to 2500 V and 500 A. It has fast switching speeds which are of the order of hundreds of kilohertz, typical switching times ranging from 1 to 6 μs. Its dv/dt and di/dt ratings are high and its characteristics are sensitive to the load conditions.

1.8.8 Light-activated SCR

The circuit symbol of the light-activated silicon controlled rectifier (LASCR) is given in Fig. 1.68(a) and its construction in Fig. 1.68(b). Electron–hole pairs are regenerated due to light radiation on the silicon wafer, and this produces a triggering current in the presence of an electric field. In order to obtain high dv/dt, high di/dt, and high gate sensitivity, the gate structure is designed to have good sensitivity to light. It finds application in high-voltage, high-current power systems such as the HVDC transmission and static reactive volt-ampere

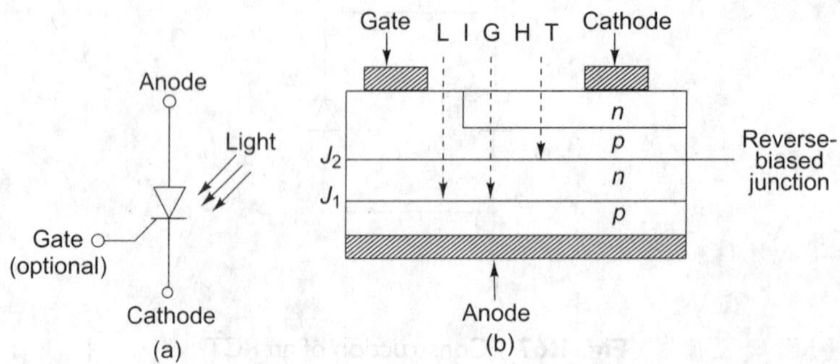

Fig. 1.68 LASCR: (a) circuit symbol, (b) schematic diagram

(VAR) compensation. It provides isolation between the light source, which is a low-voltage device (such as an LED), and the power circuitry operating at a few kilovolts (of the order of 4 kV).

1.8.9 MOS-controlled Thyristor

The construction of a MOS-controlled thyristor (MCT) is a blend of MOS and thyristor technologies. Like the conventional thyristor it is a four-layered device but has a MOS-type gate structure. A typical MCT cell and also the schematic diagram of its two-MOSFET–two-transistor analogy are shown, respectively, in Figs 1.69(a) and (b). Its equivalent circuit and circuit symbol are given in Figs 1.69(c) and (d), respectively.

It has an *n-p-n-p* structure as against the *p-n-p-n* structure of the conventional thyristor, and thus the gate voltage has to be measured with its anode as the reference. Assuming that the MCT is in its forward blocking state, a negative gate-to-anode voltage will turn it on as explained below. This is equivalent to a negative V_{G1S1}, and a *p*-channel is formed in the *p*-channel MOSFET M_p. The flow of conventional current (hole current) is from the emitter E_2 of transistor Q_2 to D_1 and then through the *p*-channel to the source S_1, which is connected to base B_1 of the transistor Q_1. This provides the hole current for the base–emitter junction of Q_1, which then emits electrons from E_1 to C_1 through B_1. As C_1 and B_2 are at the same potential, this electron current from B_2 to E_2 forward-biases Q_2. E_2 in turn injects holes into C_2 through D_2, thus turning it on and latching the MCT into conduction.

If the MCT is in the conducting state, it can be turned off with the help of a positive gate-to-anode voltage as explained below. This voltage is equivalent to a positive V_{G2S2}, and an *n*-channel is formed in the *n*-channel MOSFET M_n. Electrons will flow from D_2 through the *n*-channel, through S_2 and B_1, back to

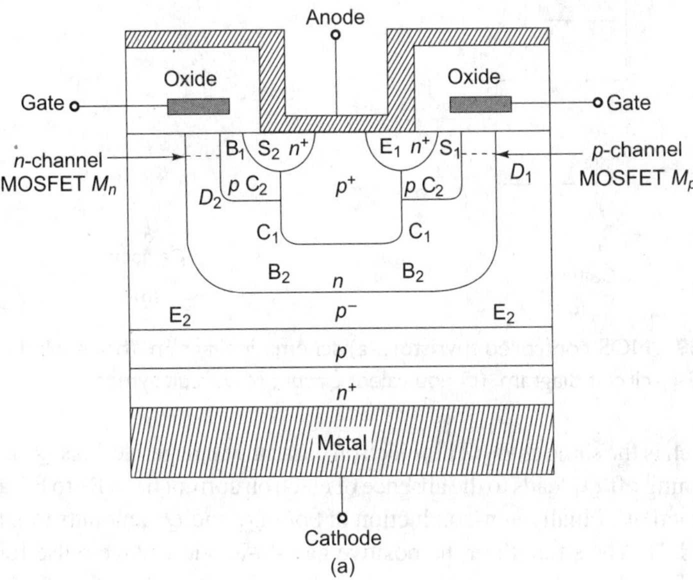

Fig. 1.69(a)

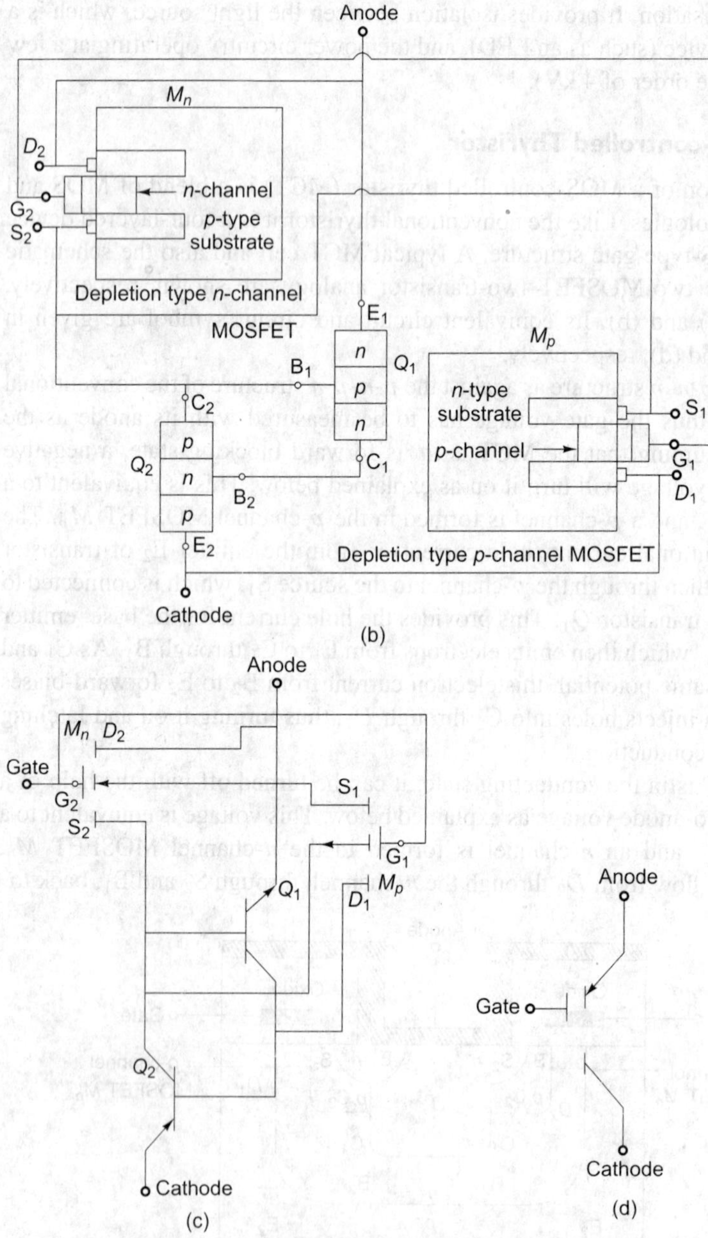

Fig. 1.69 MOS-controlled thyristor: (a) schematic diagram, (b) detailed analogous circuit diagram, (c) equivalent circuit, (d) circuit symbol

E_1, which is the same as D_2. This electron current will negative-bias Q_1 and turn it off. Turning off Q_1 leads to the absence of electron current from B_2 to E_2, and Q_2 is also turned off. Finally non-conduction of both Q_1 and Q_2 amounts to turning off of the MCT. The strength of the positive gate-to-anode voltage pulse for turning the MCT off depends upon the current conducted by it, provided this is within the rated value. Thus, larger currents need a wider pulse width. However, for turning off currents greater than the rated current of the MCT, forced commutation circuits are needed as in the case of the ordinary thyristor.

The MCT has many advantages over the conventional thyristor, namely, a low forward voltage drop, very small turn-on and turn-off times, less switching losses, and a high gate input impedance. Two or more devices can be paralleled when larger currents have to be conducted. Its switching frequency can be as high as 20 kHz.

1.9 Other Power Electronic Devices

Apart from the thyristor and other members of its family, high-power versions of some other electronic devices have been developed during the past 20 to 25 years. The power transistor and power MOSFET are two such semiconductor devices. The insulated gate bipolar transistor (IGBT) is a recent addition to this non-thyristor family of devices. This last device has the combined merits of the transistor and the MOSFET. A detailed discussion of these three power-control devices follows.

1.9.1 The Power Transistor

The power transistor, which is a magnified version of the bipolar junction transistor (BJT), can be advantageously used as a semiconductor switch in all power conversion circuits in place of the conventional thyristor. However, unlike the thyristor, it has to be kept in a conducting state by continuous current in the base–emitter circuit. The BJT is so called because holes and electrons flow in either direction, its two versions being the *p-n-p* and *n-p-n* transistors. Likewise there are two versions of the power transistor, the *n-p-n* type finding wider use in high-voltage, high-current applications because of the ease in its manufacture and lower cost. The circuit for this *n-p-n* type of power transistor is given in Fig. 1.70.

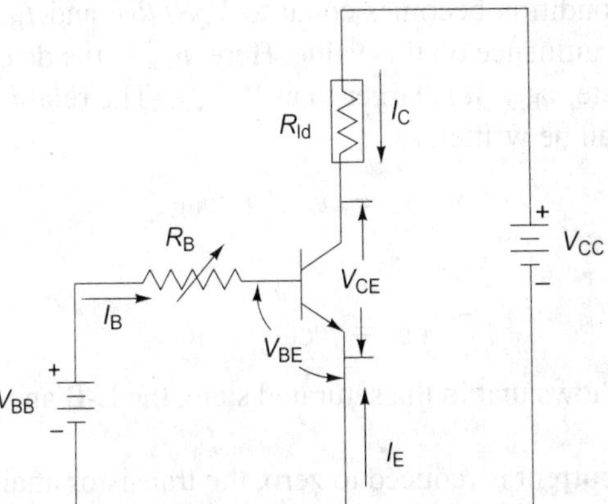

Fig. 1.70 Common-emitter connection of an *n-p-n* transistor

As in the case of other power electronic devices, the switching performance and associated losses are of importance for a power transistor.

1.9.1.1 Characteristics The characteristics of a power transistor are similar to the conventional BJT. In Fig. 1.71 the characteristics of an *n-p-n* type of the device are drawn on the i_C-v_{CE} plane.

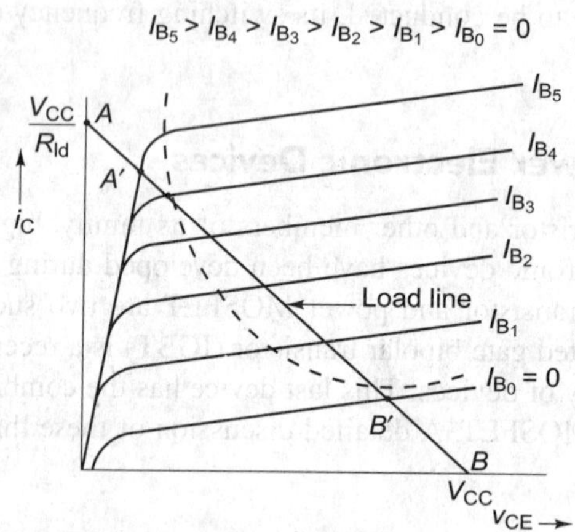

Fig. 1.71 i_C-v_{CE} characteristics of an *n-p-n* transistor in the common-emitter connection

For linear operation, the operating point lies anywhere between *A* and *B*; for switching operations, it lies at *A* or *B*. In practice, when a large base current is passed through the emitter–base circuit, the transistor gets saturated and the operating point lies at *A*′. The V_{CE} at this point is less than 1 V and the transistor is in a fully conducting state. It is then said to be under hard saturation, the losses in this condition being minimum. However, the ideal point is *A*, where $I_C = V_{CC}/R_{ld}$. For attaining this condition an excess of base current is driven through the base–emitter circuit. The transistor is then said to be overdriven. The $I_{C(sat)}$ under this condition becomes equal to V_{CC}/R_{ld}, and I_B, which is equal to $I_{C(sat)}/h_{FE}$, has no influence on this value. Here, h_{FE} is the dc current gain I_C/I_B. In the saturated state, $v_{BE(sat)}$ is larger than $V_{CE(sat)}$. The relationship between the various voltages can be written as

$$v_{CE} = v_{CB} + v_{BE} \tag{1.108}$$

or

$$v_{CB} = v_{CE} - v_{BE} \tag{1.109}$$

Equation (1.108) shows that in the saturated state, the E-B and B-C junctions are forward-biased.

When the base current is reduced to zero, the transistor attains its cut-off state and the operating point will be located at *B*′, which is the intersection point of the load line and the $I_B = 0$ characteristic. In this case a small reverse current I_{co} flows in the base–collector circuit. It can be reduced to a negligible value by applying a reverse bias in the base–emitter junction. In this condition both the junctions are reverse-biased and the operating point moves to a location very close to the point B.

When the transistor is used for switching operations, the operating point lies at A or B. When the switching condition changes from ON to OFF and vice versa, the operating point moves from A to B and back. Thus it passes through the active region between A and B, implying a high instantaneous power dissipation. This dissipation can be minimized (a) if the switching frequencies used for the transistor are fairly high and (b) if the points A' and B' lie below the dashed curve of Fig. 1.71, which signifies the limit for maximum allowable dissipation.

1.9.1.2 Switching performance The switching behaviour of the power transistor can be understood with the help of the waveforms of i_B, i_C, and v_{CE} shown in Fig. 1.72 for a resistive load. It is assumed that prior to $t = 0$ the device is in an OFF condition. That is, $i_B = 0$, i_C is very small, and $v_{CE} = V_{CC}$. At $t = 0$, a current that saturates the transistor is passed through the base. But the collector current does not immediately attain its full value; instead it takes time t_r for attaining it. This is called the *rise time*. During t_r, v_{CE} falls down to its conducting state value (about 0.7 V). When i_B is removed at time t_1, i_C does not fall down to zero but remains at its maximum value till the end of time t_s called the *storage time*. At the end of t_s, i_C starts falling and reaches its cut-off value after a time t_f called the *fall time*, with a small reverse current I_{co} flowing in the base–collector circuit in this condition. The net effect of all these time delays is to limit the switching frequency to which the transistor is subjected.

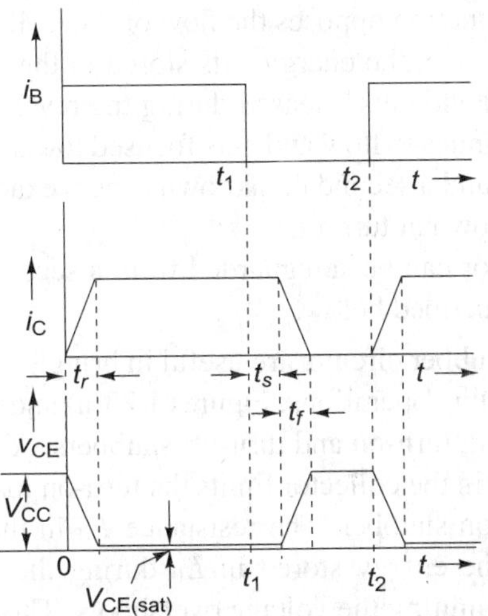

Fig. 1.72 Switching waveforms of a transistor with resistive load

Under the saturated state of the transistor, excess minority carriers exist in the base and collector regions. The storage time t_s is the time taken for these excess carriers to recombine, thus enabling the transistor to cross the boundary region between the active and saturated states. This suggests that t_s can be reduced if i_B is adjusted to the minimum value that just makes the transistor cross this boundary. In this state the transistor is said to be in a *quasi-saturation* condition. Another method of reducing t_s and also t_f is to apply a negative bias to the transistor, but this may lead to a breakdown of the transistor especially when it is supplying an inductive load.

The rise time can be reduced by overdriving the transistor, that is, by passing more base current than that required to saturate the transistor.

Phenomenon of secondary breakdown In a power transistor, two types of breakdowns occur: (a) the avalanche breakdown, also known as the first breakdown, and (b) the breakdown when the collector current is non-uniformly distributed over the emitter cross section and has high density in small areas, causing the development of hot spots—this is known as the second breakdown and can occur during turn-on as well as turn-off conditions.

During turn-on the base–emitter junction is forward-biased, and due to the uneven distribution of current, the emitter periphery is more positive than the centre. Any current entering the base region from the collector is focused towards a small area at the emitter periphery. Localized heating takes place in this region, leading to a rise in temperature. Due to the negative temperature coefficient of the ON state resistance of the transistor, there is a cumulative increase in the current concentration in the small area and this leads to a second breakdown.

During turn-off, the base potential is negative with respect to that of the emitter (that is, the base–emitter junction is reverse-biased) and the emitter centre more positive than its periphery. If any current crosses the collector–base junction, it gets directed to the small area at the emitter centre. Normally, the reverse-biased base–emitter junction opposes the flow of this collector current. But if the collector load is inductive, the energy gets stored in the load inductance during the conducting period and gets released during the reverse bias. As a result, the collector current continues to flow and gets focused towards the emitter centre. A hot spot is developed and a second breakdown occurs exactly as explained above for the second breakdown at turn-on.

The power transistor can be safeguarded from a second breakdown by using snubber circuits as described below.

Snubber circuits Snubber circuits are useful in bringing down the currents and voltages during switching operations. Figures 1.73(a) and (b), respectively, show an *n-p-n* transistor with turn-on and turn-off snubbers. As shown in Fig. 1.73(a) a small inductance L_0 in the collector limits the turn-on losses by limiting di/dt, thus acting as a turn-on snubber. The resistance R_0 in the freewheeling circuit helps in dissipating the energy stored in L_0 during the OFF condition of the transistor and also in limiting the voltage overshoots. The diode D_1 ensures that no current flows through the turn-on snubber resistance R_0 at turn-on.

At turn-off, the load current is diverted through D_2 and C_s, the latter helping in limiting the dv/dt applied to the device [Fig. 1.73(b)]. A large C_s also helps in keeping down V_{CE} and the turn-off losses. Just before turn-on, however, C_s charges to the dc supply voltage V_{CC}, and at turn-on it discharges through the snubber resistor R_s and the transistor. The resistance R_s serves to limit the discharge current of the capacitor at turn-on. Since the presence of R_s, however, is harmful at turn-off, it is shunted by a diode D_2. In a typical power electronic application, the transistor is protected by both turn-on and turn-off snubbers. They are also called *polarized* snubber circuits because of the inclusion of diodes.

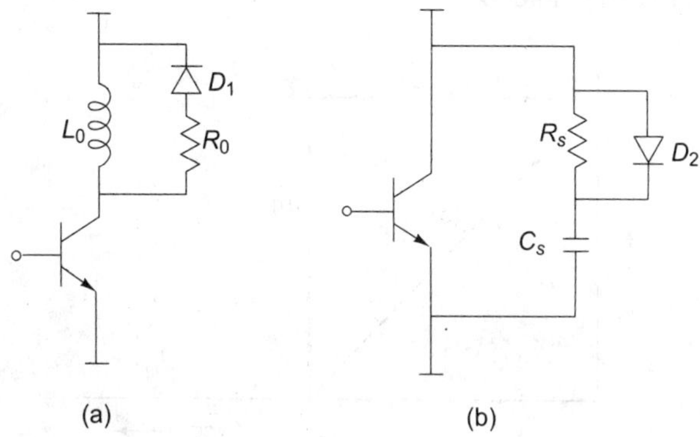

Fig. 1.73 Transistor with (a) a turn-on snubber and (b) a turn-off snubber

The above discussion shows that the snubbers keep down the switching losses. The switching performance is depicted in the form of a load line that indicates the locus of the operating point. For instance, the effect of the turn-off snubber on the load lines (Fig. 1.74) is that the overvoltage at turn-off is suppressed by this polarized snubber, the switching loss being substantially decreased as a consequence.

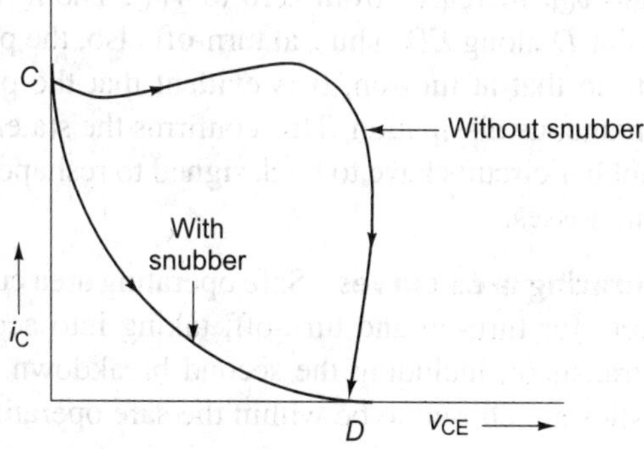

Fig. 1.74 Typical transistor load lines with an inductive load

In a typical switching cycle, energy is stored in the snubber inductance L_0 and the snubber capacitance C_s and then discharged. The switching power losses are therefore proportional to the switching frequency and are necessarily transferred to the snubber resistors. Hence the loss in a circuit protected by a snubber can be lower than that in an unprotected circuit.

Figure 1.74 also shows that excess currents and voltages that occur during switching operations can be assessed by means of load lines, thus helping their correction by appropriate snubbers. This technique of reduction of currents and voltages is known as *load shaping*. It is clear from Fig. 1.75 that the load changes are along the line *CD* for a resistive load, with the points *C* and *D*, respectively, depicting fully on and fully off conditions. The instantaneous power dissipation at these points is zero because either the voltage or the current is zero, but it has a finite value between *C* and *D*.

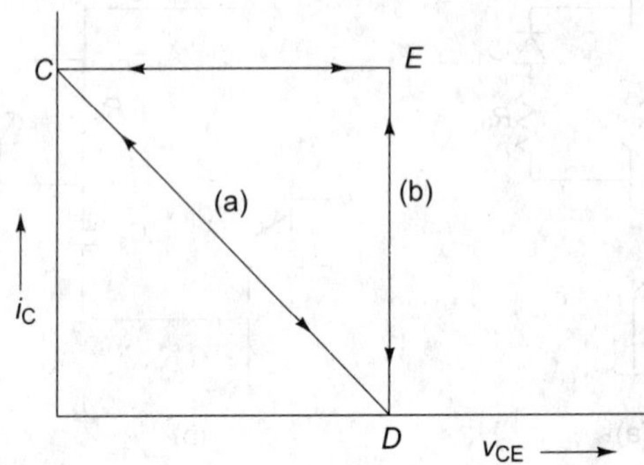

Fig. 1.75 Ideal transistor load lines with (a) a purely resistive load and (b) an inductive load

The dynamic load line will be different from that in Fig. 1.75 and may enter the region of high power dissipation when the power transistor caters to an inductive load. At turn-on, the current increases from *D* to *E* with a constant voltage characteristic (V_{CC}). The freewheeling diode across the load inductance recovers and v_{CE} decreases to $v_{CE(sat)}$ at *C*. Again, at turn-off, the current remains constant till *E* is reached and v_{CE} increases from zero to V_{CC}. The i_C then decreases to the cut-off value point *D* along *ED*. Thus, at turn-off also, the path is rectangular and is just opposite to that at turn-on. It is evident that the point *E* is a point of maximum instantaneous dissipation. This confirms the statement made above that appropriate snubber circuits have to be designed to reshape the load line and reduce the switching losses.

1.9.1.3 Safe operating area curves Safe operating area curves are supplied by the manufacturers for turn-on and turn-off, taking into account the various limitations of the transistor, including the second breakdown. The turn-on and turn-off characteristics are chosen to be within the safe operating area (SOA) so as to minimize the associated power losses. A typical safe operating curve called the *reverse bias safe operating area* (RBSOA) curve is shown in Fig. 1.76 from which it is seen that the overvoltage surge at turn-off is suppressed by a polarized turn-off snubber and the switching loss is drastically reduced.

1.9.1.4 Advantages of power transistors The advantages of power transistors are as follows.

(i) (a) Power transistors have a considerably higher switching frequency than thyristors; thus transistorized dc choppers and inverters have improved and more efficient operation. (b) Current limit protection can be provided by them through the base drive circuit. (c) There is a saving on the losses as

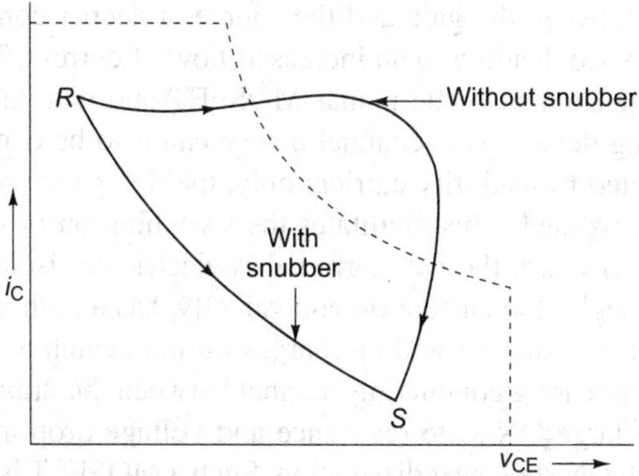

Fig. 1.76 Transistor turn-off load lines with and without snubber protection

well as the cost because of the absence of forced commutated circuitry. (d) They also have a low conduction drop.

(ii) Advantages (a) and (b) given above hold good even in comparison to GTOs. Like power transistors, GTOs do not require forced commutation circuits. Finally, these devices (power transistors and GTOs) are efficient, respectively, for short and long duty cycles.

1.9.1.5 Disadvantages of power transistors The disadvantages of power transistors are as follows.

(i) As compared to thyristors: (a) They require large and continuous base drives. (b) They cannot withstand reverse voltage. (c) The ratio of maximum to continuous current is low. (d) They have a negative temperature coefficient of resistance thus making the paralleling of the transistor-based drives difficult.

(ii) As compared to GTOs: In the case of a short circuit or a fault leading to overcurrent, transistors may fail due to the second breakdown. The chances of a GTO failing due to such a fault are less because the magnitude of the controllable current is much higher, a typical range being two to six times the rated rms current.

1.9.2 Power MOSFET

The power MOSFET is a FET in which MOS technology has been incorporated. Figures 1.77(a) and (b) give the schematic diagrams, respectively, of the conventional and power MOSFETs. The conventional MOSFET shown in Fig. 1.77(a) has a planar structure with three external terminals, namely, the source, drain, and gate. It consists of a p-type silicon substrate with two n^+-regions being diffused into it. An SiO_2 layer is spread on the top and serves as an insulator. The source, drain, and gate terminals are metallic in nature and are located on this layer. With zero gate-to-source bias, no current flows from the drain to the source. On the other hand, with a positive bias, the induced electronic field draws free electrons to the top of the p-layer, thus leading to the formation of an n-channel and causing flow of electrons from the source terminal to the drain terminal. This causes a lateral current to flow from the drain to the source. With increase of the positive voltage between the gate and the source, a deeper conducting channel of electrons is formed, leading to an increased flow of current. This is known as an *n*-channel enhancement mode planar MOSFET and is a voltage controlled, current-conducting device. A *p*-channel device can also be constructed. As the current is conducted by majority carriers only, the delay caused for removal or recombination is avoided. This facilitates the switching on of the device in the megahertz range. Second, the gate terminal is electrically isolated. Due to this feature, it draws negligible current or, equivalently, has a high input impedance. The input capacitance charges and discharges during switching operations. The planar structure has a long conducting channel between the drain and the source, thus leading to a large ON state resistance and voltage drop in the conducting state and also high ON state heat dissipation. Such a MOSFET is used for powers up to 1 W.

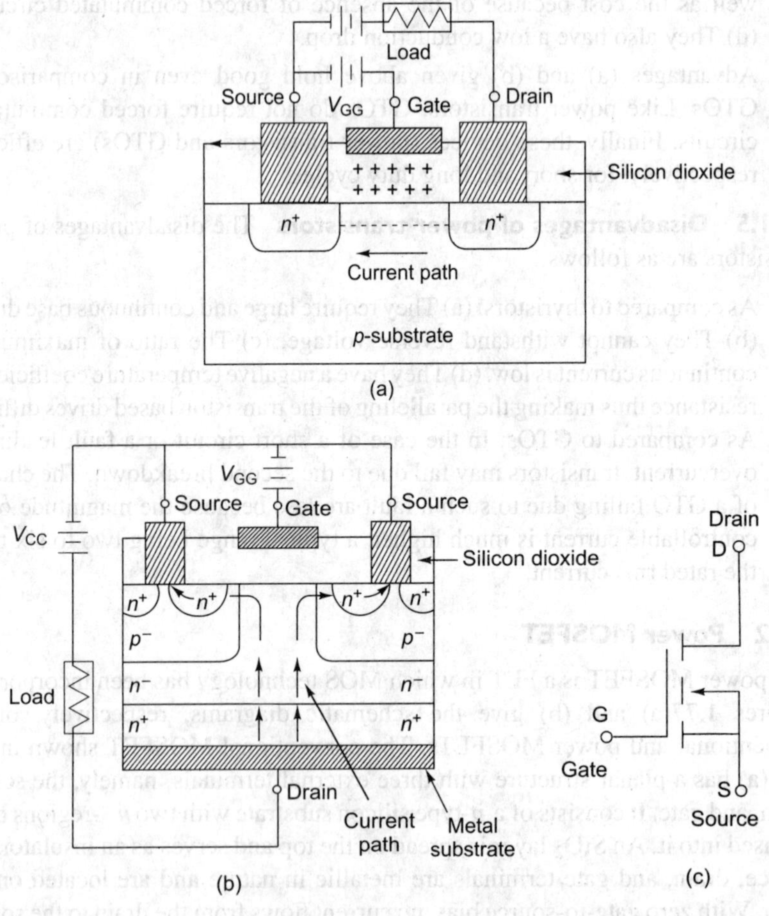

Fig. 1.77 (a) Schematic diagram of an *n*-channel planar MOSFET, (b) schematic diagram of a vertical channel diffused MOS power MOSFET, (c) circuit symbol of a MOSFET

The construction of an n-channel power MOSFET, which is a double diffused MOS device, is shown in Fig. 1.77(b). An n^--layer is epitaxially spread on a substrate of n^+. The resistivity and thickness of this n^--layer determines the blocking voltage capability. Next, a p^--layer is diffused on this n^--layer. Finally, the n^+-fingers are diffused on top of this p^--layer. The power MOSFET has a cellular construction, which means that several layers of such diffused MOS cells are placed one on top of the other. The conducting layer of electrons is formed in the p^--layer situated under the gate oxide layer. The current first flows in a vertical direction and then in a horizontal direction towards the source terminals. The ON state resistance depends on the channel length. This length is suitably adjusted by means of diffusion so as to reduce both this resistance and the ON state dissipation. The normal operation is with a positive drain-to-source voltage, in which the device behaves like a normal p-n junction with a negative drain-to-source voltage. The circuit symbol given in Fig. 1.77(c) shows the three terminals; the insulation of the gate terminal is highlighted by not having any electrical contact between the gate and the other two terminals.

1.9.2.1 MOSFET characteristics

Similar to the characteristics of the n-p-n transistor given in Fig. 1.71, the static characteristics of the power MOSFET are drawn by plotting v_{DS} versus i_D with a fixed V_{GS} by means of the circuit shown in Fig. 1.78. Figure 1.79(a) shows several such characteristics drawn with V_{GS} as a parameter. A load line can be drawn on these characteristics similar to that drawn on the i_C-v_{CE} characteristic of the n-p-n transistor. The MOSFET permits a rapid changeover from the ON state to the OFF state, thus permitting the employment of high switching frequencies.

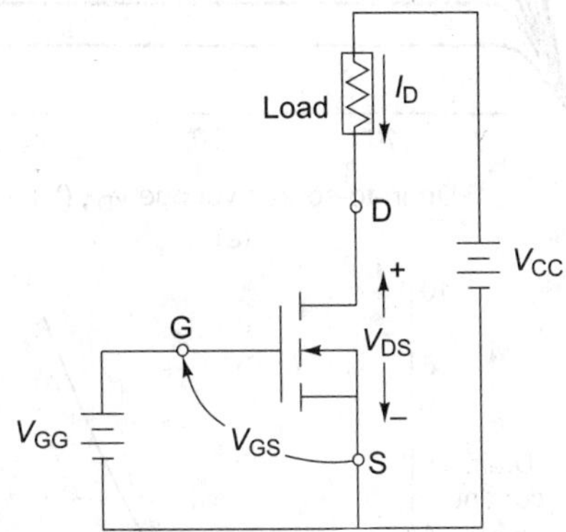

Fig. 1.78 Circuit for obtaining power MOSFET characteristics

The i_D-v_{GS} characteristic, called the transfer characteristic, of the MOSFET is given in Fig. 1.79(b). It is seen that I_D remains zero until a threshold voltage of 2–3 V is reached, the characteristic being nearly linear above this value. V_{GS} has to be nearly 10 V for attaining the ON condition. The switching frequency is determined by the following factors: (a) the RC time constant, where R is the source resistance and C is the input capacitance, (b) the time required for the threshold value of V_{GS} to be reached, (c) the time needed for the full conduction state to be reached, and (d) the discharge rate of the capacitance through the drive circuit.

The above discussion implies that the time delay can be reduced and the switching frequency can be increased by minimizing the source resistance and keeping the charging current of the capacitance at a few amperes. Typical switching times are of the order of 100 ns for high-current MOSFETs. The power MOSFET can be conveniently driven by complementary MOS or transistor–transistor logic circuits because of negligible gate power requirement.

1.9.2.2 Advantages of MOSFETs MOSFETs are preferred to power transistors because the fast switching times obtainable with them vary little with temperature. Second, because of the high input impedance of the device the drive circuitry can be of low power, and also compact and simple. Two or more MOSFETs can be connected in parallel to supply high-power loads because of the positive temperature coefficient of the device. This shows that the second breakdown phenomenon that occurs in transistors is absent in a MOSFET, thus permitting a wider safe operating area. Under high-frequency switching conditions the MOSFET proves to be more efficient because of lower losses compared to the transistor.

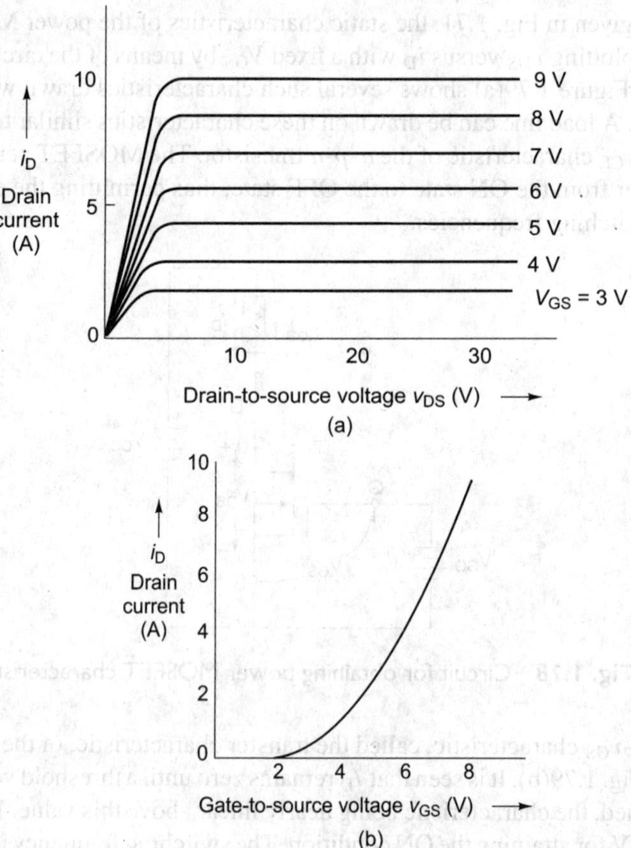

Fig. 1.79 Power MOSFET: (a) output characteristics, (b) transfer characteristic

1.9.2.3 Disadvantages of MOSFETs Under low-frequency switching conditions the losses in a conducting MOSFET are high, the performance of the transistor being superior under these conditions. The ON state drain-to-source resistance $R_{DS(ON)}$ at 150 °C is twice that at 25 °C. Third, the ON state voltage drop increases with its voltage rating irrespective of the current rating. For instance, for a 100-V, 30-A MOSFET, the ON state voltage drop is about 3 V, but this drop goes up to 8 V for a device with ratings of 500 V and 10 A. In spite of these disadvantages, the MOSFET proves to be the best high-frequency switching device for voltages below 500 V.

1.9.3 Insulated Gate Bipolar Transistor

The schematic diagram of an insulated gate bipolar transistor (IGBT), its equivalent circuit, simplified equivalent circuit, and circuit symbol are shown, respectively, in Figs 1.80(a), (b), (c), and (d). Its construction is identical to that of a MOSFET except for the p^+-substrate. Its equivalent circuit is similar to that of a combination of a thyristor and a MOSFET. Its performance features are nearer to that of a transistor but with no secondary breakdown phenomenon. It has a high input impedance like the MOSFET and a low saturation voltage like the transistor. For the same current rating the size of an IGBT is smaller than that of a MOSFET.

The IGBT can be turned on by applying a positive voltage between the gate and the emitter (source) and, similar to a MOSFET, it can be turned off by making the gate voltage zero or negative. It has much lower ON state resistance than the MOSFET.

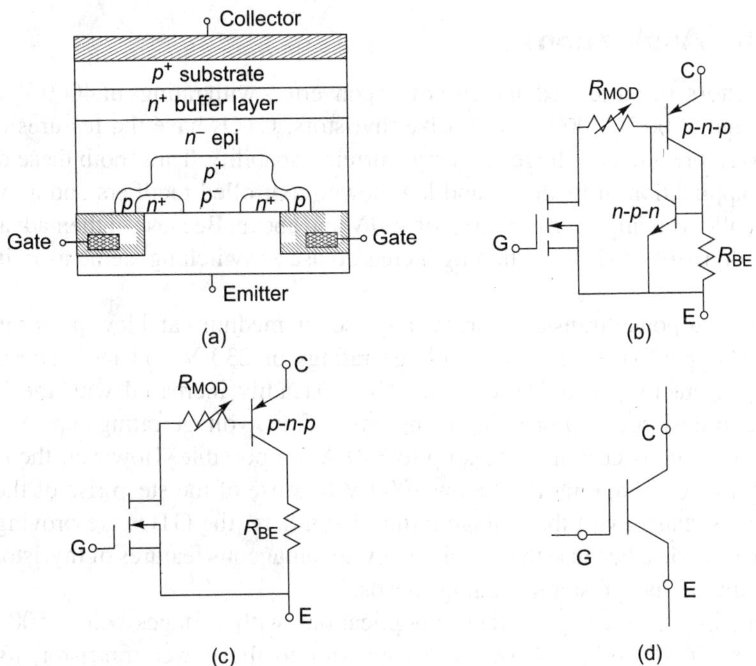

Fig. 1.80 IGBT: (a) cross section, (b) equivalent circuit, (c) simplified equivalent circuit, (d) circuit symbol

Once the IGBT has been latched up, the gate will have no control over the collector (drain) current. Like the thyristor, the only way to turn off a conducting IGBT is by forced commutation circuits. The ON state voltage of the IGBT depends on the gate voltage; also, a sufficiently high gate voltage is necessary to reduce the ON state voltage. The voltage across the conducting IGBT increases with temperature.

Though the secondary breakdown feature is absent in the IGBT, it is necessary to keep below the maximum power dissipation and also the specified maximum junction temperature of the device, so as to ensure its reliable operation under all conditions. The bipolar action in the IGBT slows down the speed of the device, its switching frequency being smaller than that of the MOSFET. It can be subjected to switching frequencies up to 20 kHz, and its maximum current and voltage ratings are, respectively, 400 A and 1200 V. IGBTs cannot be easily paralleled like the MOSFETs. The factors that prevent current sharing during parallel operation are: (a) the ON state current unbalance caused by the $V_{CE(sat)}$ distribution and the main circuit wiring resistance distribution and (b) the current imbalance at turn-on and turn-off caused by the switching time difference between the parallel connected devices, and the circuit wiring inductance distribution.

Two IGBT ratings that are provided by manufacturers are (i) the peak allowable drain current that can flow before the latch-up of the device and (ii) the corresponding gate-to-source voltage that permits this current to flow and which should not be exceeded. IGBTs find use in medium-power applications such as dc and ac motor drives, power supplies, solid-state relays, and solid-state contactors, UPS systems, FACTS devices, custom power products, etc.

1.10 Applications

Thyristors are employed in high-power converters with ratings of 4000 V, 1500 A (av) and 1200 V, 3000 A (av). Like thyristors, GTOs have the features of high forward breakover voltage and large current capability. Thus, both these devices find application in medium- and low-power controlled rectifiers and ac voltage controllers having voltage ratings of 460 V and above. Because of their advantages over thyristors, GTOs are finding increased use as switching elements in inverter drives.

Bipolar power transistors are widely used in medium- and low-power inverter and chopper drives that have voltage ratings of 230 V and less. The voltage and current ratings of these devices have steadily increased with the help of modern design and processing technology. Thus, voltage ratings up to 1200 V and continuous current ratings up to 300 A are possible. However, their use is confined to voltage ratings below 1000 V because of the steep rise of their ON state resistance with the voltage rating. Even here, the GTOs are proving to be the best choice because they retain many advantageous features of thyristors and, in addition, have faster switching speeds.

For high-frequency switching applications with voltages below 500 V, the power MOSFET has proved to be superior to the power transistor, its main drawback being the increase of the ON state voltage drop with the voltage rating.

Summary

The characteristics of a thyristor indicate that the voltage at which it fires can be controlled by varying the gate current. The turn-off of a thyristor needs a circuitry for either reducing the thyristor current below its holding current value or applying a reverse voltage across it. Like other semiconductor devices, the thyristor is sensitive to temperature and hence needs protection against high temperatures. The thyristor has to be safeguarded (a) against overvoltages and high dv/dt by connecting a snubber circuit across it and (b) against overcurrents and high di/dt by connecting an inductor in series with it.

A knowledge of the gate characteristics is necessary to know the strength of the gate pulse that can successfully fire a thyristor. The UJT relaxation oscillator can be designed to generate trigger pulses of appropriate strength; transistorized circuits can also used for this purpose. Digital ICs and microprocessors are proving to be convenient for producing firing sequences in the case of multiphase rectifiers and inverters.

Forced commutation circuits are necessary for the operation of thyristors in inverters, dc and ac choppers, and cycloconverters. A careful analysis is necessary to arrive at appropriate values of the commutation elements (L and C) which constitute the heart of such circuits.

The TRIAC is a bidirectional device which serves the same purpose as a thyristor pair connected antiparallel. Whereas the turn-on of the TRIAC is simple, its turn-off circuit needs careful design. The GTO has a faster switching capability than the thyristor and its commutation circuitry is quite simple. Though its latching current and holding current are high, it is replacing the thyristor for control of medium-power equipment because of the simplicity of its commutating circuit. The PUT serves the same purpose as the UJT but is flexible because of characteristics that allow alteration to suit particular needs. The MCT has a lower forward voltage drop, smaller turn-on and turn-off times, and smaller switching losses as compared to the thyristor.

The power transistor has of late become a strong contender for circuits that need high switching speeds. The pulse width modulation inverter is an example of such a circuit. It does not need separate forced commutation circuits and has a low ON state drop. However, it suffers from the phenomenon of secondary breakdown. Its other disadvantages are as follows: (i) it needs continuous base drive, (ii) it cannot withstand reverse voltages, and (iii) it does not permit paralleling with other transistors.

High switching frequencies are possible with a power MOSFET. Its drive circuitry is also simple. Paralleling of MOSFETs is permitted because of the positive temperature coefficient of the device. Its demerits are as follows: (a) its low-frequency switching losses are more than those of the power transistor and (b) its ON state voltage drop increases with the voltage rating.

The IGBT has performance features which approximate those of a power transistor, but it does not have the second breakdown phenomenon. Its switching frequency is smaller than that of the MOSFET, but unlike MOSFETs, IGBTs cannot be easily paralleled.

Worked Examples

1. The parallel resonance turn-off circuit of Fig. 1.81(a) has the following data: $E = 180$ V, $R_{ld} = 580\ \Omega$, pulse width $t_p = 1.3$ ms. The load current is 0.7 times $I_{c(max)}$. Determine the values of L and C.

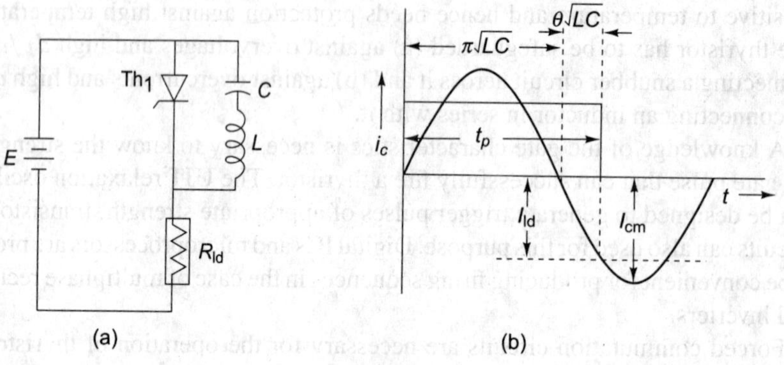

Fig. 1.81

Solution

The waveform is given in Fig. 1.81(b). The thyristor turns off in the negative half-cycle at an angle

$$\theta = \sin^{-1}\frac{I_{ld}}{I_{cm}} = \sin^{-1} 0.7 = 44.42° = 0.775 \text{ rad}$$

The pulse duration is given as

$$t_p = (\pi + \theta)\sqrt{LC} = 3.917\sqrt{LC}$$

Therefore

$$3.917\sqrt{LC} = 1.3 \times 10^{-3}$$

or

$$\sqrt{LC} = \frac{1.3 \times 10^{-3}}{3.917} = 0.3319 \times 10^{-3}$$

The inequality in Eqn (1.76) can be written as an equality. Thus,

$$\frac{E}{R_{ld}} = 0.7E\sqrt{\frac{C}{L}}$$

or

$$\sqrt{\frac{C}{L}} = \frac{1}{0.7R_{\text{ld}}} = \frac{1}{0.7 \times 580}$$

$$C = \sqrt{LC}\sqrt{\frac{C}{L}} = \frac{0.3319 \times 10^{-3}}{0.7 \times 580}\ \text{F}$$

$$= 0.817\ \mu\text{F}$$

$$L = \sqrt{LC}\sqrt{\frac{L}{C}} = 0.3319 \times 10^{-3} \times 0.7 \times 580$$

$$= 0.135\ \text{H}$$

2. The series resonance turn-off circuit of Fig. 1.82 has the following data: E = 160 V, L = 8 mH, resistance of inductor coil r_L = 0.2 Ω, R_{ld} = 0.6 Ω, and C = 65 μF. (a) Derive an expression for the current $i(t)$. Determine (b) the pulse width and (c) the time required for the capacitor voltage to attain a voltage equal to $1.7E$.

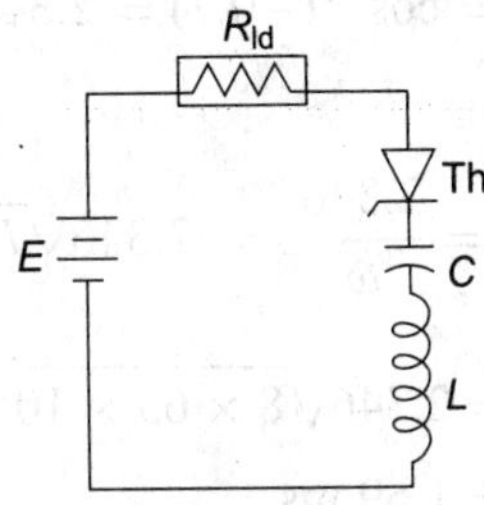

Fig. 1.82

Solution

(a) $R_{\text{ld}} + r_L = 0.6 + 0.2 = 0.8\ \Omega$. From Eqn (1.74),

$$q = CE(1 - \cos \omega t)$$

$$i = \frac{dq}{dt} = CE\omega \sin \omega t$$

$$= \frac{CE}{\sqrt{LC}\sin \omega t} = E\sqrt{\frac{C}{L}}\sin \omega t$$

Substitution of values gives

$$i(t) = 160\left(\sqrt{\frac{65 \times 10^{-6}}{8 \times 10^{-3}}}\right)\sin \omega t$$

$$= 14.42 \sin \omega t$$

where

$$\omega = \frac{1}{\sqrt{LC}} = \frac{1}{\sqrt{8 \times 65 \times 10^{-9}}} = 1386.8\ \text{rad/s}$$

The assumption made in Section 1.5.3.3 is realized in this circuit. That is, the total resistance in the circuit (0.8 Ω) is negligible when compared to the inductive or capacitive reactance, both of which have a value of 11.09 Ω.
(b) Pulse width = $\pi\sqrt{LC} = \pi\sqrt{8 \times 65 \times 10^{-9}} = 2.27$ ms. The capacitor voltage can be expressed as

$$v_c = \frac{q}{C} = E(1 - \cos\omega t)$$

Let t_1 be the time at which v_c attains the value $1.7E$. This gives

$$1.7E = E(1 - \cos\omega t_1)$$

$$1.7 = 1 - \cos\omega t_1$$

This gives

$$\cos\omega t_1 = 1 - 1.7 = -0.7$$

or

$$\omega t_1 = \cos^{-1}(-0.7) = 2.346 \text{ rad}$$

Finally,

$$t_1 = \frac{2.346}{\omega} = 2.346\sqrt{LC}$$

$$= 2.346\sqrt{8 \times 65 \times 10^{-9}}$$

$$= 1.69 \text{ ms}$$

3. Design the various elements of a UJT relaxation oscillator, given the following data for the UJT: V_P = 24 V, I_P = 48 μA, V_V = 0.75 V, I_V = 5.1 mA, η = 0.72, R_{BB} = 6.5 kΩ, and $R_{B_1(con)}$ = 60 Ω. Also compute (a) the maximum frequency range and (b) the pulse height obtained as per the design.
Solution
From Eqns (1.45) and (1.46) the values of the lower and higher values of resistance are

$$R_2 \geq \frac{V_s - V_V}{I_V}$$

$$R_1 + R_2 \leq \frac{V_s - V_P}{I_P}$$

Also from Eqn (1.29)

$$V_P \approx \eta V_s + V_D$$

V_P can be expressed approximately as

$$V_P \approx \eta V_s$$

or

$$V_s = V_P/\eta = 24/0.72 = 33.3\ \text{V}$$

V_s can be taken as 30 V. Substitution of values in the above inequalities give the marginal values as

$$R_2 = \frac{30 - 0.750}{5.1 \times 10^{-3}} = 5.74\ \text{k}\Omega \simeq 6\ \text{k}\Omega$$

$$R_1 + R_2 = \frac{30 - 24}{48 \times 10^{-6}} = 125\ \text{k}\Omega$$

Hence $R_1 = 125 - 6 = 119$ kΩ. A potentiometer of 100 kΩ can be chosen for R_1. R_3 is selected as 0.8 kΩ and R_4 as 70 Ω. From Eqns (1.34) and (1.41),

$$\text{Pulse height} = \frac{\eta V_s R_4}{R_{B_1(\text{con})} + R_4}$$

$$= \frac{0.72 \times 30 \times 70}{60 + 70} = 11.6\ \text{V}$$

Let the capacitor C be chosen to have a capacitance of 0.1 μF. Now f_{max} and f_{min} are computed using Eqns (1.39) and (1.40), respectively. Thus

$$f_{\text{max}} = \frac{1}{R_2 C \ln[1/(1-\eta)]}$$

$$= \frac{1}{6 \times 10^3 \times 0.1 \times 10^{-6} \ln[1/(1-0.72)]}$$

$$= \frac{10^4}{6 \ln[1/(1-0.72)]}$$

$$= 1309 \text{ pulses per second}$$

$$f_{\text{min}} = \frac{1}{(R_1 + R_2) C \ln[1/(1-\eta)]}$$

$$= \frac{1}{125 \times 10^3 \times 0.1 \times 10^{-6} \ln[1/(1-0.72)]}$$

$$= 63 \text{ pulses per second}$$

4. For the protective circuit shown in Fig. 1.83, determine the values of V, L, and C. Data: maximum load current is 1.2 A, frequency of ac supply is 50 Hz, $R_{\text{ld}} =$ 200 Ω. The Q-factor of the combination of the inductance and load resistance is 0.38. The resistance of the inductor is negligible.

Solution

$$Q = 0.38 = \frac{\omega L}{R_{\text{ld}}} = \frac{2\pi \times 50 \times L}{R_{\text{ld}}} = \frac{314L}{200}$$

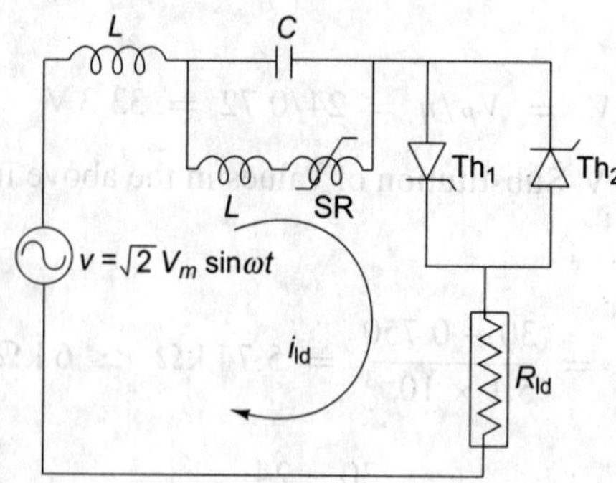

Fig. 1.83

This gives $L \simeq 0.24$ H.

$$\text{Maximum load current} = \frac{V_{max}}{R_{ld}}$$

$$\omega = \frac{1}{\sqrt{LC}} = 314$$

or

$$\sqrt{LC} = \frac{1}{314}$$

$$LC = \frac{1}{(314)^2}$$

$$C = \frac{1}{0.25 \times (314)^2}\ \text{F} = \frac{10^6}{0.25 \times (314)^2}\ \mu\text{F}$$

$$= 40.6\ \mu\text{F}$$

Maximum load current

$$I_{ld(max)} = \frac{V_{max}}{R}$$

Substitution of values gives

$$1.2 = \frac{\sqrt{2}V}{200}$$

or

$$V = \frac{200 \times 1.2}{\sqrt{2}} = \frac{240}{\sqrt{2}} = 169.7\ \text{V}$$

5. The circuit of Fig. 1.84 is used to determine the $(di/dt)_{max}$ and $(dv/dt)_{max}$ ratings of the thyristor. R_v is increased in steps and the switch Sw is closed to trigger the thyristor. In this process, at a value of $R_v = 12\ \Omega$, the thyristor gets triggered. Determine the values of $(di/dt)_{max}$ and $(dv/dt)_{max}$ given the following data: $E = 120$ V, $L = 10\ \mu$H, $V_c(0) = 0$ V, and $R_{ld} = 18\ \Omega$.

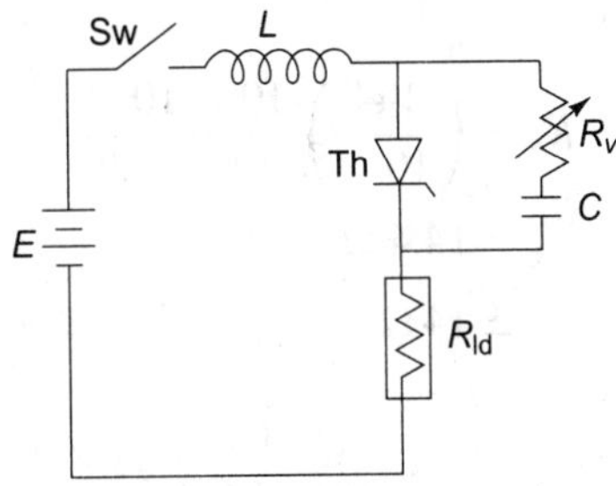

Fig. 1.84

Solution
From Eqn (1.88)

$$\left(\frac{di}{dt}\right)_{max} = \frac{E - V_c(0)}{L}$$

Substitution of the values gives

$$\left(\frac{di}{dt}\right)_{max} = \frac{120 - 0}{10 \times 10^{-6}} = 12\ \text{A}/\mu\text{s}$$

Also, from Eqn (1.90)

$$\left(\frac{dv}{dt}\right)_{max} = R_v\left(\frac{di}{dt}\right)_{max} = \frac{12 \times 12}{10^{-6}} = 144\ \text{V}/\mu\text{s}$$

6. Using the values of $(di/dt)_{max}$ and $(dv/dt)_{max}$ of Example 5, determine suitable values of L for overcurrent protection, and also R_v and C_s for protection against switching transients for the same thyristor having the same load current. Data: $E = 120$ V, $V_c(0) = 20$ V, $I_{TRM} = 10$ A.

Solution
The circuit is given in Fig. 1.84. The preliminary design is done as follows. From Eqn (1.94),

$$L = \frac{E - V_c(0)}{(di/dt)_{max}}$$

This gives

$$L \geq \frac{(120 - 20) \times 10^{-6}}{12}\ \mu\text{H}$$

$$\geq 8.33\ \mu\text{H}$$

L is initially taken as 10 μH, and with this value, R_v is found from Eqn (1.95) as

$$R_v = \left(\frac{dv}{dt}\right)_{max}\left[\frac{L}{E - V_c(0)}\right]$$

Substitution of values gives

$$R_v = \left(\frac{144}{10^{-6}}\right)\frac{10\times10^{-6}}{120-20}$$
$$= 14.4\,\Omega$$
$$\simeq 14\,\Omega$$

From Eqn (1.96),

$$C_s = \frac{L_s}{(R_v)^2} = \frac{10\times10^{-6}}{(14^2)} = 0.051\,\mu\text{F}$$

From the discussion of the worked example in Section 1.7.1,

$$\frac{E}{R_{\text{ld}}} + \frac{E}{R_v} = I_{\text{TRM}}$$

or

$$\frac{E}{R_v} = I_{\text{TRM}} - \frac{E}{R_{\text{ld}}}$$

Substitution of values gives

$$\frac{120}{R_v} = 10 - \frac{120}{18} = 3.33$$

This gives

$$R_v = \frac{120}{3.33} = 36\,\Omega$$

Taking R_v as 40 Ω, L is found using Eqn (1.93) as

$$L = \frac{40\times(120-20)\times10^{-6}}{144} = 28\,\mu\text{H}$$

The actual di/dt is now computed from Eqn (1.88) as

$$\frac{di}{dt} = \frac{120-20}{28\times10^{-6}} = 3.6\,\text{A}/\mu\text{s}$$

As the value of di/dt is less than 10 A/μs, the value of L is considered to be satisfactory. Substituting these values of R_v and L in Eqn (1.96), the value of C_s is finally determined as

$$C_s = \frac{28\times10^{-6}}{(40)^2}\,\text{F} = 0.018\,\mu\text{F}$$

This value of C_s is less than the initially computed value and hence is satisfactory. The final design values are $L = 28\,\mu$H, $R_v = 40\,\Omega$, and $C_s = 0.018\,\mu$F.

7. For the parallel capacitor turn-off circuit of Fig. 1.85(a) determine the size of the capacitor to successfully turn off the load thyristor. Sketch the waveforms of v_{Th_1}, i_{Th_1}, and i_{ld}. The data are E = 60 V, $R_1 = R_{\text{ld}} = 5\,\Omega$, t_{OFF} of Th_1 = 40 μs.

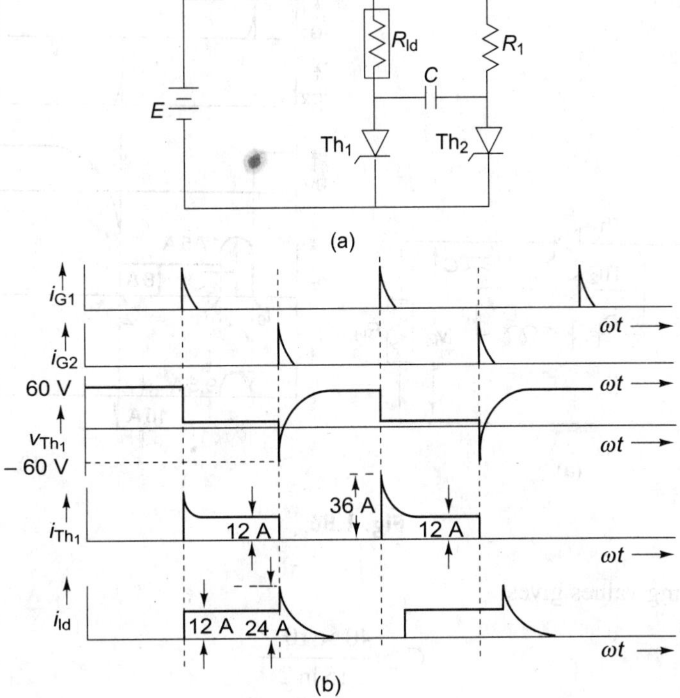

Fig. 1.85

Solution

From Eqn (1.56),

$$C \geq \frac{t_{OFF}}{0.7R_{ld}}$$

Substituting the given data yields

$$C \geq \frac{40 \times 10^{-6}}{0.7 \times 5}$$

$$\geq 11.43\,\mu F$$

Therefore, $C \simeq 12\,\mu F$. The waveforms are sketched in Fig. 1.85(b).

8. The auxiliary resonant turn-off circuit of Fig. 1.86(a) has a battery voltage of 120 V and a load resistance of 15 Ω. The maximum current that the thyristor Th_1 can carry is 18 A and the turn-off time of the load thyristor is 40 μs. Determine the values of L and C and sketch the waveforms of i_{Th_1}, i_c, and v_{Th_1}.

Solution

From Eqn (1.67),

$$C \geq \frac{t_{OFF}}{R_{ld} \ln 2}$$

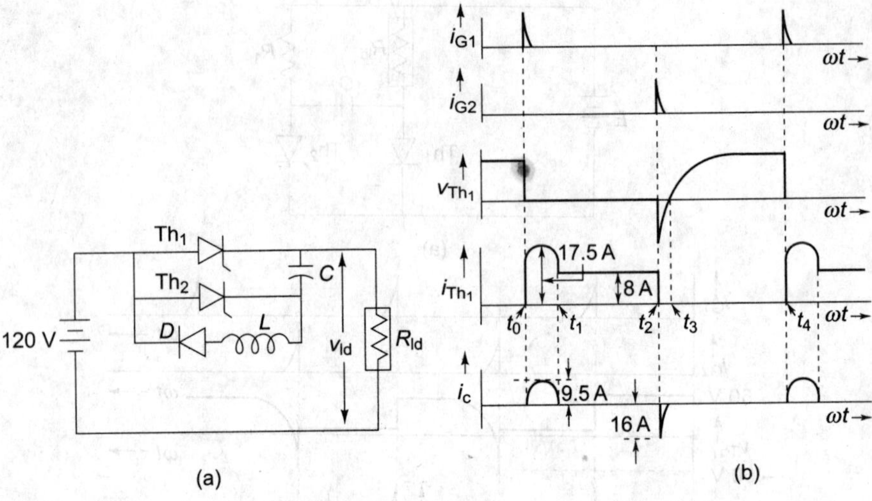

Fig. 1.86

Substituting values gives

$$C \geq \frac{40 \times 10^{-6}}{15 \ln 2}$$

$$\geq 3.85\ \mu\text{F}$$

Taking C as 5 μF, L is given by the inequality

$$L \geq \frac{C}{(I_{\text{ThM}}/E - 1/R_{\text{ld}})^2}$$

Substitution of values gives

$$L \geq \frac{5 \times 10^{-6}}{(18/120 - 1/15)^2}$$

$$\geq 0.72\ \text{mH}$$

$$\simeq 0.8\ \text{mH}$$

The waveforms are shown in Fig. 1.86(b). Maximum current through the thyristor $= E\sqrt{C/L} + E/R_{\text{ld}} = 17.5$ A. This is less than 18 A.

9. Figure 1.87 shows a resistance firing circuit. Given $V_{\text{RMS}} = 230$ V, $R_{\text{ld}} = 12\ \Omega$, and the maximum and minimum gate current ratings as 10 mA and 2 mA, respectively, estimate the values of R_f and R_v.

Solution

R_{ld} is usually very small when compared to R_f and R_v and is hence neglected. The maximum gate current will flow with $\alpha = 0$.

$$V_{\text{av}} = \frac{V_m(1 + \cos\alpha)}{2\pi}$$

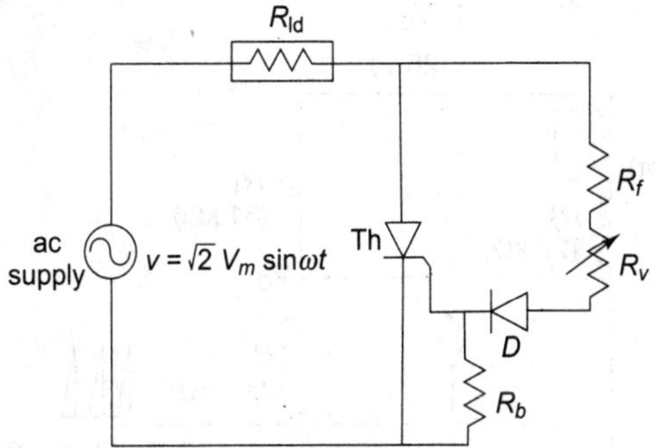

Fig. 1.87

With $\alpha = 0$, V_{av} $(\alpha = 0) = V_m/\pi = 230\sqrt{2}/\pi$. For maximum gate current to flow, there should be minimum resistance in the gate circuit. Thus

$$R_f = \frac{V_{av}(\alpha = 0°)}{10\,\text{mA}}$$
$$= \frac{230\sqrt{2}}{\pi \times 10 \times 10^{-3}}$$
$$= 10.35\,\text{k}\Omega$$
$$\simeq 10\,\text{k}\Omega$$

With the resistance firing circuit, the maximum value of α occurs at 90° with gate firing. Here,

$$R_f + R_v = \frac{V_{av}(\alpha = 90°)}{2\,\text{mA}} = \frac{V_{RMS}\sqrt{2}}{2\pi \times 2\,\text{mA}} = 25.9\,\text{k}\Omega \simeq 26\,\text{k}\Omega$$

This gives

$$R_v = 26 - 10 = 16\ \text{k}\Omega$$

R_f should be chosen as a fixed resistance of 10 kΩ and R_v as a potentiometer of 20 kΩ.

10. A PUT-based relaxation oscillator is to be designed to give a pulse frequency of 100 Hz and a pulse height of 8 V. The data are $V_{CC} = 28$ V, $I_G = 1.2$ mA, and the peak voltage V_P is 12 V. Determine the values of R_1, R_2, R_3, C, R_4, and R_5. The conducting resistance of the device, R_{con}, can be taken as 30 Ω.

Solution

The oscillator is shown in Fig. 1.88. From Eqns (1.101) and (1.102),

$$\eta = \frac{V_P}{V_{CC}} = \frac{12}{28} = 0.429$$

and

$$I_G = (1 - \eta)\frac{V_{CC}}{R_{eq}}$$

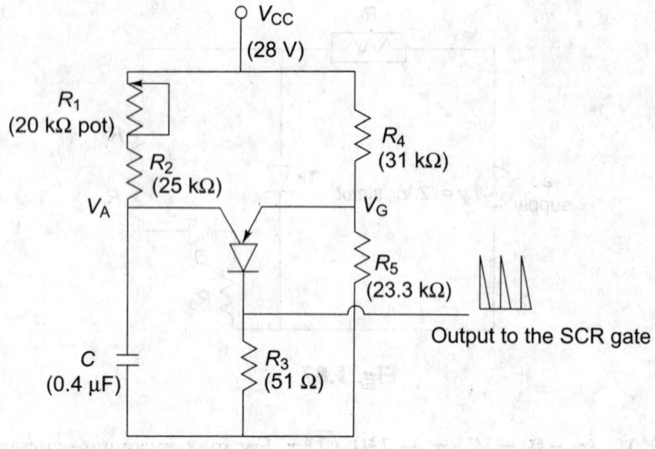

Fig. I.88

Substitution of values yields

$$1.2 \times 10^{-3} = (1 - 0.429)\frac{28}{R_{eq}}$$

This gives

$$R_{eq} = (1 - 0.429)\frac{28}{1.2 \times 10^{-3}} = 13.32 \text{ k}\Omega$$

Substituting values gives

$$R_4 = \frac{R_{eq}}{\eta} = \frac{13.32}{0.429} = 31 \text{ k}\Omega$$

$$R_5 = \frac{R_{eq}}{1 - \eta} = \frac{13.32}{1 - 0.429} = 23.3 \text{ k}\Omega$$

Let C be 0.4 μF. From Eqn (1.100) and the expression for time period T,

$$T = \frac{1}{f} = (R_1 + R_2)C \ln\left(1 + \frac{R_5}{R_4}\right)$$

Substitution of values gives

$$\frac{1}{100} = (R_1 + R_2) \times 0.4 \times 10^{-6} \ln\left(1 + \frac{23.3}{31.0}\right)$$

That is,

$$R_1 + R_2 = \frac{1}{100 \times 0.4 \times 10^{-6} \ln[1 + (23.3/31.0)]}$$

$$= \frac{10^4}{0.4 \ln(1 + 23.3/31.0)} = 44.6 \text{ k}\Omega \simeq 45 \text{ k}\Omega$$

R_2 can be taken as a 25 kΩ fixed resistance and R_1 can be taken as a 20-kΩ potentiometer. The anode voltage is obtained from Eqns (1.99) and (1.100) as

$$V_A = V_G + 0.7 = V_P + 0.7$$
$$= V_{CC}\frac{R_5}{R_4 + R_5} + 0.7$$
$$= \frac{28 \times 23.3}{31 + 23.3} + 0.7$$
$$= 12.71 \text{ V}$$

With the approximation that $V_A \approx V_P$, the pulse height is obtained as

$$\text{Pulse height} = 8 \text{ V} = \frac{V_A R_3}{30 + R_3} = \frac{12.71 R_3}{30 + R_3}$$

From this equation R_3 is obtained as 51 Ω.

11. For the RC firing of Fig. 1.89, determine the values of R_f, R_v, and C given the following data: V = 230 V, $I_{GT(max)}$ = 12 mA, $I_{GT(min)}$ = 2 mA, V_{GT} = 1.0 V, R_{ld} = 10 Ω. The firing angle range is from 0° to 120°. V_{D_1} can be taken as 0.7 V.

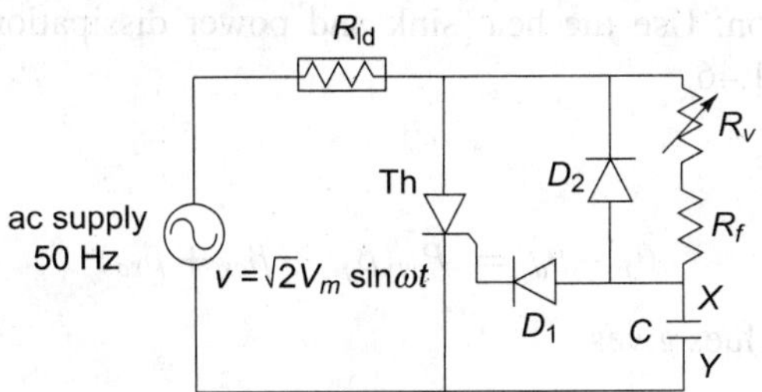

Fig. 1.89

Solution

It is assumed that the load resistance is negligible compared to R_f and R_v. Equation (1.23) gives

$$R \leq \frac{V_{av} - V_{D_1} - V_{GT}}{I_{GT}}$$

Substitution of $I_{GT(max)}$ and $I_{GT(min)}$ gives, respectively, R_f and $R_f + R_v$. A firing angle of 0° gives

$$V_{av} = \frac{V_m}{\pi} = \frac{230\sqrt{2}}{\pi} = 103.5 \text{ V}$$
$$V_{D_1} = 0.7 \quad \text{and} \quad I_{GT(max)} = 12 \text{ mA}$$

Hence

$$R_f \leq \frac{103.5 - 0.7 - 1.0}{12 \times 10^{-3}} = 8.5 \text{ k}\Omega$$

V_{av} for an α of 120° is found as

$$V_{av} = \frac{1}{2\pi}\int_{2\pi/3}^{\pi} V_m \sin\omega t \, d(\omega t) = \frac{V_m}{4\pi}$$

$$R_f + R_v = \frac{(230\sqrt{2}/4\pi) - 0.7 - 1.0}{2 \times 10^{-3}} = 12.1\,\text{k}\Omega$$

$$\simeq 12\,\text{k}\Omega$$

R_f can be taken as 8.5 kΩ and R_v as a potentiometer of 5 kΩ. From Eqn (1.21),

$$RC = \frac{4}{\omega}$$

Average value of $R = (12 + 8.5)/2$ kΩ = 10.25 kΩ.

$$C = \frac{4 \times 10^6}{314 \times 10.25 \times 10^3} = 1.24\,\mu\text{F}$$

12. Figure 1.90 shows a thermal circuit for a thyristor. The following data are given: $\theta_j = 100\,°\text{C}$, $\theta_a = 25\,°\text{C}$, $\beta_{jc} = 0.45\,°\text{C/W}$, $\beta_{cs} = 0.085\,°\text{C/W}$. Specify a heat sink and the appropriate power dissipation P_{av}; also find the $I_{F(av)}$ of the thyristor for 90° conduction. Use the heat sink and power dissipation curves given in Figs 1.47(c) and 1.46.

Solution

From Eqn (1.83)

$$\theta_j - \theta_a = P_{av}(\beta_{jc} + \beta_{cs} + \beta_{sa})$$

Substitution of values gives

$$100 - 25 = P_{av}(0.45 + 0.085 + \beta_{sa})$$

From Fig. 1.47(c), the heat sink h is selected, and with an estimated P_{av} of 80 W,

$$\beta_{sa} = \frac{32}{80} = 0.4$$

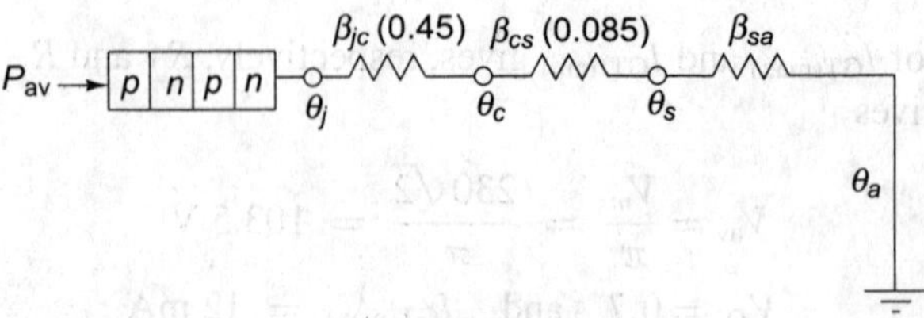

Fig. 1.90

The calculated power dissipation is

$$P_{av} = \frac{75}{0.45 + 0.085 + 0.4} = 80\,\text{W}$$

Hence the appropriate heat sink is h. The average power dissipation is 80 W. From Fig. 1.46, $I_{F(\text{av})}$ for 90° conduction is 57 A.

13. A thyristor has an $I_{F(\text{av})}$ rating of 25 A. The other data concerning its thermal model are $\theta_j = 90\,°\text{C}$, $\theta_a = 25\,°\text{C}$, $\beta_{jc} = 0.56\,°\text{C/W}$, $\beta_{cs} = 0.10\,°\text{C/W}$. If the thyristor is required to conduct for a duration of 180°, suggest an appropriate heat sink and determine the thermal resistance β_{sa}. [Make use of the curves in Figs 1.46 and 1.47(c).]

Solution

The thermal equivalent circuit is as given in Fig. 1.90. From the curves in Fig. 1.46, the power dissipation for an $I_{F(\text{av})}$ of 25 A and 180° conduction is approximately 30 W. From Eqn (1.82),

$$\theta_j - \theta_a = P_{\text{av}}(\beta_{jc} + \beta_{cs} + \beta_{sa})$$

On substitution of the given values, the equation becomes

$$90 - 25 = 30(0.56 + 0.10 + \beta_{sa})$$

This gives

$$\beta_{sa} = 1.5\,°\text{C/W}$$

With a P_{av} of 30 W, curve a in Fig. 1.47(c) gives

$$\Delta\theta = \theta_s - \theta_a = 45°$$

That is,

$$\beta_{sa} = \frac{45}{30} = 1.5$$

Hence the required heat sink is that corresponding to curve a and the thermal resistance is 1.5 °C/W.

14. The phase shifting circuit shown in Fig. 1.91(a) is utilized to fire the thyristor which rectifies the same ac supply. (a) Determine the range of α that can be obtained if R_1 is varied from 2 kΩ to 6 kΩ. The transformer ratio is 230 V/115 V for both the secondaries. The capacitance has a value of 0.3 μF. (b) If a firing angle of 45° is to be obtained, determine the value of R.

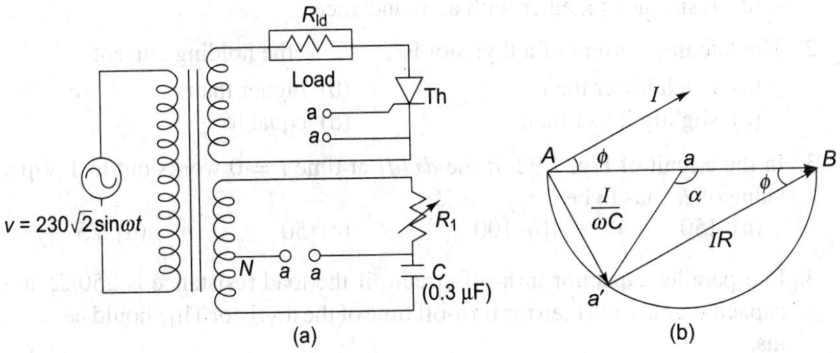

Fig. 1.91

Solution

(a) From the geometry of the phasor diagram shown in Fig. 1.91(b), $\alpha = 180 - 2\phi$.

$$\tan\phi = \frac{I/\omega C}{IR} = \frac{1}{\omega CR} \tag{1.110}$$

Let α_1 and α_2 be the lower and higher values of the firing angle. Also let ϕ_1 and ϕ_2 be the corresponding phase angles.

$$\tan\phi_1 = \frac{1}{314 \times 0.3 \times 10^{-6} \times 2000} = 5.3$$

So $\phi_1 = 79.3°$ and $\alpha_1 = 180 - 2\phi_1 = 21.4° \simeq 21°$.

$$\tan\phi_2 = \frac{1}{314 \times 0.3 \times 10^{-6} \times 6000} = 1.77$$

This gives

$$\phi_2 = 60.5° \qquad \text{and} \qquad \alpha_2 = 180 - 2\phi_2 = 59°$$

Therefore, the range of α is 21°–59°.

(b) With $\alpha_1 = 45°$,

$$\phi_1 = \frac{180 - 45}{2} = 67.5°$$

$$\tan\phi_1 = \frac{1}{314 \times 0.3 \times 10^{-6} \times R} = \tan\ 67.5° = 2.414$$

This gives R as 4.4 kΩ.

Exercises

Multiple Choice Questions

1. In a parallel capacitor turn-off circuit, it is desirable to use a ________ on the second leg of the circuit.
   (a) very high resistance
   (b) very low resistance
   (c) resistance which is neither high nor low
   (d) resistance together with an inductance

2. The latching current of a thyristor is ________ the holding current
   (a) much lower than (b) higher than
   (c) slightly lower than (d) equal to

3. In the circuit of Fig. 1.92, if the dv/dt at time $t = 0$ works out to 1 V/μs, the value of R_s has to be ________ Ω.
   (a) 150 (b) 100 (c) 50 (d) 25

4. In a parallel capacitor turn-off circuit, if the load resistance is 250 Ω and the capacitance is 4 μF, then the turn-off time of the thyristor Th_1 should be ________ ms.
   (a) more than 1.4 (b) less than 1.4
   (c) more than 0.7 (d) less than 0.7

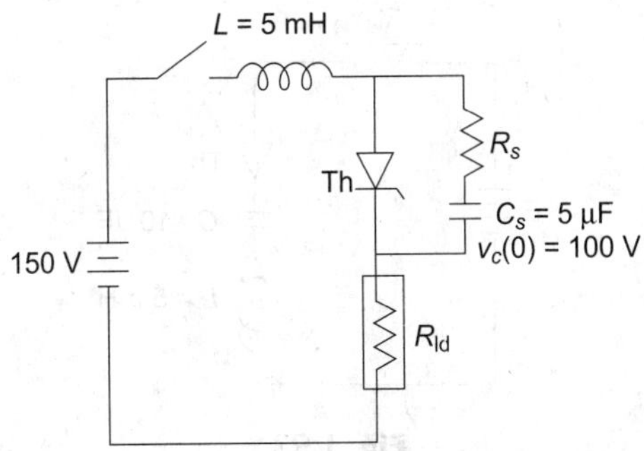

Fig. 1.92

5. The maximum permissible value of the frequency of the ac voltage source whose voltage is to be rectified by a thyristor with a t_{OFF} of 75 μ is ______ Hz.
   (a) 7500 (b) 3333 (c) 6667 (d) 9990

6. In a UJT firing circuit, the height of the output pulse will ______ by lowering the value of R_3.
   (a) be increased (b) be decreased
   (c) remain the same as before (d) become zero

7. The range of α in the case of an RC firing circuit will be in the range 0°–180° because of ______.
   (a) the presence of the capacitance
   (b) the non-linear nature of R_{GK}
   (c) the presence of the capacitance as well as R_{GK}
   (d) the presence of the diode D_1 in the gate–cathode circuit

8. In a UJT relaxation oscillator circuit, if a UJT of higher R_{BB} value than that of the original UJT is used, the pulse frequency ______.
   (a) increases (b) decreases
   (c) remains unaffected (d) becomes zero

9. For a certain UJT relaxation oscillator circuit, the following data are given: $\eta = 0.67$, $V_s = 21$ V, $R_4 = 50\ \Omega$, and the pulse height = 8 V. With these values, $R_{B_1(con)}$ can be computed to be nearly ______ Ω.
   (a) 37.5 (b) 75.0 (c) 85.0 (d) 42.5

10. In a thyristor, a lower value of the middle junction capacitance is ______ because it ______ the chance of dv/dt triggering.
    (a) undesirable, increases (b) desirable, increases
    (c) undesirable, decreases (d) desirable, decreases

11. In the circuit of Fig. 1.93, which provides self-commutation by series resonance turn-off, if the load resistance is negligible as compared to the inductive and capacitive reactances, the half-cycle time of the oscillatory current can be computed to be approximately ______ ms.
    (a) 0.70 (b) 0.065 (c) 0.003 (d) 0.054

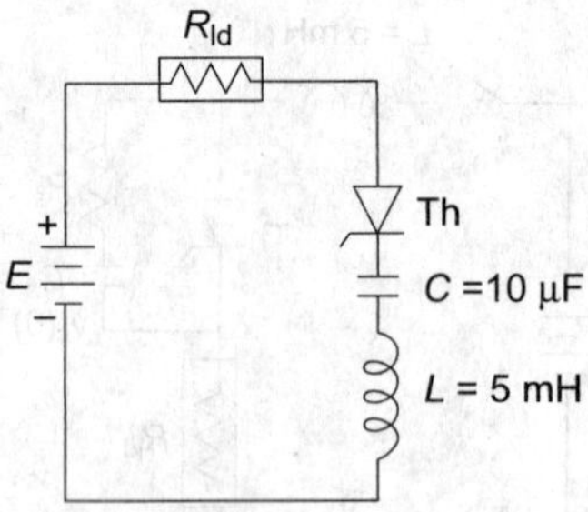

Fig. 1.93

12. In the protection circuit for a power transistor, the diode D_2 connected across the snubber inductor L_s ensures that ______.
    (a) current does not flow through R_s during the turn-off condition
    (b) current does not flow through R_s during the turn-on condition
    (c) current flows only in one direction during the turn-on condition
    (d) current flows only in one direction during the turn-off condition
13. In the turn-off snubber circuit of a power transistor, the diode D_2 is connected across the snubber resistor R_s to prevent current ______.
    (a) from flowing through R_s during the turn-on condition
    (b) from flowing through C_s in both directions
    (c) from flowing through C_s during the turn-off condition
    (d) from flowing through R_s during the turn-off condition
14. The storage time t_s of a power transistor can be minimized ______.
    (a) by keeping i_B at the maximum value
    (b) by keeping i_C at the minimum value
    (c) by keeping i_C at the maximum value required for saturation
    (d) by keeping i_B at the minimum value required for cut-off
15. The rise time t_R of a power transistor can be reduced ______.
    (a) by passing less forward base current than necessary
    (b) by short-circuiting the emitter and collector terminals
    (c) by passing more forward base current than necessary
    (d) by short-circuiting the emitter and base terminals
16. The MOSFET is a ______ -controlled ______ -conducting device.
    (a) current, current (b) voltage, voltage
    (c) current, voltage (d) voltage, current
17. The switching frequency of a MOSFET will be reduced with ______.
    (a) an increase in the output impedance of the device
    (b) an increase in the discharge rate of the input capacitance
    (c) an increase in the source resistance
    (d) a decrease in the discharge rate of the input capacitance
18. The ON state voltage of the MOSFET ______ as the voltage rating ______.
    (a) increases, decreases (b) decreases, decreases
    (c) decreases, increases (d) increases, increases

19. It is possible to connect two or more MOSFETs in parallel because ______.
    (a) the threshold value of the gate-to-source voltage is only 2–3 V
    (b) fast switching times are obtainable with it
    (c) the MOSFET has a very small power loss under high-frequency conditions
    (d) the MOSFET resistance has a positive temperature coefficient
20. During the fall time t_f at the turn-off of a GTO, the current falls down ______.
    (a) at a rapid rate to zero
    (b) at a rapid rate to the tail current value and then at a slow rate to zero
    (c) at a slow rate to the tail current value and then at a rapid rate to zero
    (d) at a slow rate to the tail current value and remains fixed at that value
21. The turn-off loss in a GTO is ______ that of a thyristor.
    (a) smaller than
    (b) greater than
    (c) of the same order as
    (d) double
22. The snubber resistance R_s across a GTO is necessary ______.
    (a) for keeping down the current both during turn-on and turn-off
    (b) for keeping down the current during turn-off
    (c) for keeping down the current during turn-on
    (d) for dissipating the trapped energy of the capacitor during turn-off
23. As a consequence of low internal generation in a GTO, it has ______.
    (a) an increased latching current value
    (b) an increased holding current value
    (c) both of the above
    (d) none of the above
24. The phenomenon of second breakdown in a transistor occurs at the time of turn-off because of ______.
    (a) uneven distribution of current in the collector–base junction
    (b) uneven distribution of current in the emitter cross section
    (c) uneven distribution of current in the base cross section
    (d) uneven distribution of current in the collector cross section

True or False

Indicate whether the following statements are *true* or *false*. Briefly justify your choice.

1. A high value of di/dt is more harmful for a thyristor than a high value of dv/dt.
2. As the ambient temperature increases, the turn-on time of a thyristor increases but the turn-off time decreases.
3. The advantages of a MOSFET as compared to a power transistor are the following: (i) under high-frequency conditions the losses of the MOSFET are very high and (ii) the ON state voltage drop of the MOSFET increases with a decrease in voltage rating.
4. In the turn-on snubber circuit provided with a power transistor, the diode D_2 connected across the inductor L_s ensures that the current does not flow through R_s during the turn-off of the device.

5. The magnitudes of the gate triggering voltage and gate triggering current for a thyristor are independent of the ambient temperature.
6. It is possible to determine the latching current value in a laboratory.
7. The maximum permissible magnitude of the ac supply frequency to which a thyristor with a turn-off time of 60 μs is connected is 8060 Hz.
8. In a UJT relaxation oscillator circuit, increasing the value of R_3 results in a decrease in the pulse frequency.
9. In the circuit of Fig. 1.94, if the initial dv/dt works out to 1 V/μs, then the value of R_s has to be 80 Ω.

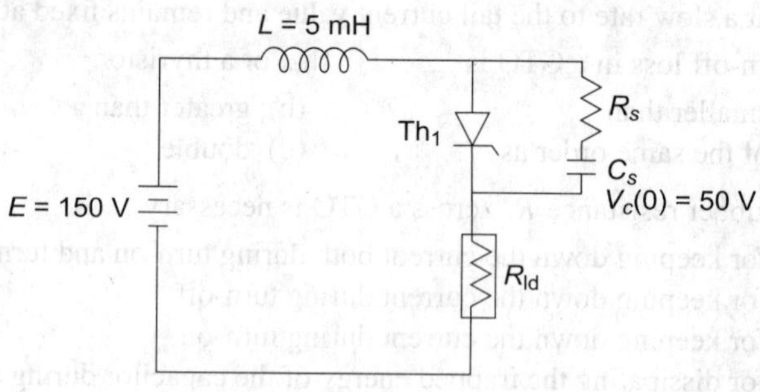

Fig. 1.94

10. When determining $I_{F(\mathrm{av})}$ from thermal considerations, the values of the thermal capacitances cannot be ignored, as this will lead to erroneous results.
11. In a UJT firing circuit, a low value of R_4 is always desirable from the gate protection point of view.
12. A reverse gate-to-cathode voltage is not desirable for a thyristor.
13. It is possible to design an automatic temperature control system for an oven using a single thyristor and the associated circuit.
14. The turn-off time of a thyristor decreases with increase in temperature.
15. In the parallel capacitor turn-off circuit, it is advantageous to use a very large resistance for the second leg.
16. The continuous current rating $I_{F(\mathrm{av})}$ of a thyristor increases when β_{jc}, the thermal resistance between the junction and the casing, is decreased.
17. It is not possible to determine the holding current of a thyristor in a laboratory.
18. In a UJT firing circuit, the higher the value of the capacitance, the larger will be the pulse height.
19. The turn-off gain of a GTO is defined as the ratio of the average current rating $I_{F(\mathrm{av})}$ of the thyristor to the minimum gate current required for its turn-off.
20. Power transistors can be employed for higher switching frequencies than thyristors but for lower switching frequencies than GTOs.
21. A power electronic circuit using a GTO is more reliable than a circuit using power transistors because GTOs have less chances of failure under a short-circuit fault.

22. Two power transistors cannot be connected in parallel whereas two MOSFETs can be.
23. The MOSFET is more efficient for use in low-frequency switching conditions, whereas the power transistor is more efficient under high-frequency conditions.
24. The IGBT has a higher ON state resistance than the MOSFET.
25. The IGBT has a lower switching frequency as compared to the MOSFET.
26. It is advantageous to use a DIAC and not a conventional diode in the gate circuit because a DIAC can conduct in both directions.
27. Using a PUT in a relaxation oscillator is advantageous over a UJT because it has a high peak current I_P and low values of R_{BB} and I.
28. In a UJT relaxation oscillator if the resistance R_4 is doubled, the pulse frequency is halved but the pulse height remains unaffected.
29. The MOSFET has a smaller safe operating area as compared to the power transistor.
30. In the parallel turn-off circuit of Fig. 1.95 with data $E = 60$ V, $R_{ld} = 4\ \Omega$, and t_{OFF} of the thyristor = 35 μs, the size of the capacitor should not be less than 8 μF.

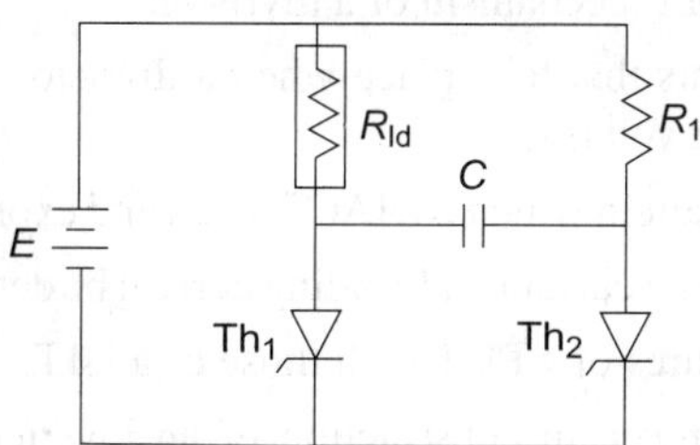

Fig. 1.95

31. In the TRIAC triggering circuit of Fig. 1.96, if the DIAC is replaced by two diodes connected antiparallel, the operation of the circuit will be inferior to that using the DIAC.

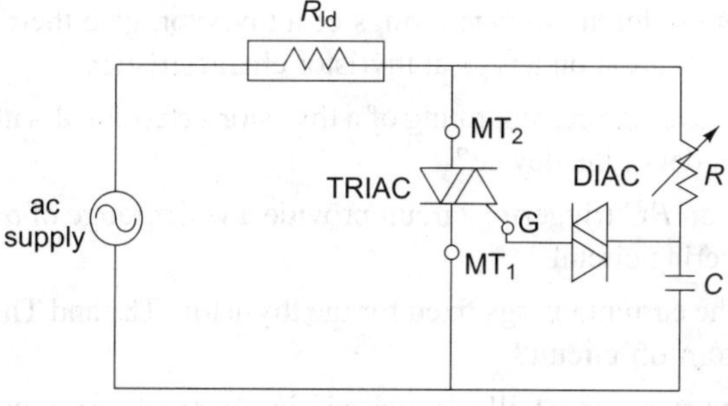

Fig. 1.96

32. The forward recovery time t_{OFF} is more than the reverse recovery time t_r because the junction J_a takes more time for forward recovery than the junctions J_b and J_c.

Short Answer Questions

1. Discuss briefly the voltage commutation and current commutation techniques used for the commutation of thyristors.
2. Explain the two-transistor analogy of a thyristor and derive an expression for the anode current I_A using this analogy.
3. Explain the role of the diode D_2 in an RC firing circuit. How does the synchronization of the firing pulse with the input ac supply take place in a thyristor?
4. Name some negative resistance devices other than the UJT and give their characteristics.
5. How is the negative resistance property of the UJT helpful for its operation as an oscillator?
6. Develop the thermal model of a thyristor and explain its use.
7. Discuss the turn-off mechanism of a thyristor.
8. Describe the events that take place when a thyristor is subjected to a slowly increasing reverse voltage.
9. Give a typical circuit in which a DIAC is used and explain its role in the circuit.
10. How can the latching current and holding current be determined in a laboratory?
11. Compare the features of a PUT with those of a UJT.
12. Explain the difference in construction of an inverter grade thyristor over a conventional one.
13. Compare the features of a GTO with those of a conventional thyristor.
14. Compare the features of a GTO with those of a power transistor.
15. Compare the features of a thyristor with those of a power transistor.
16. Compare and contrast an IGBT with a power transistor.
17. Explain how the second breakdown phenomenon occurs in a power transistor.
18. List out the different voltage ratings of a thyristor, give their definitions, and show their location on a typical thyristor characteristic.
19. How is the average current rating of a thyristor determined with the help of the thermal model of the device?
20. How does an RC triggering circuit provide a wider range of α as compared to an R triggering circuit.
21. How are the current ratings fixed for the thyristors Th_1 and Th_2 in an auxiliary resonant turn-off circuit?
22. Explain the events that will take place if, in a parallel capacitor turn-off circuit, Th_1 fails to turn off when Th_2 is triggered.
23. Explain how a power transistor can be safeguarded against current and voltage transients. What is their effect on the safe operating area of the device?

24. How is protection against direct overcurrents provided for a thyristor?
25. How is the gate of a thyristor protected against overcurrents and overvoltages?
26. Explain with the help of a diagram how the limitations of gate voltage and gate current arise for a thyristor.
27. Explain how the turn-on and turn-off snubber circuits reduce the power losses during the switching operations of a power transistor.
28. Compare the turn-off mechanisms of a TRIAC and a thyristor.
29. Write short notes on the (i) SITH, (ii) LASCR, and (iii) MCT.
30. Discuss the switching performance of a power transistor.
31. Explain the difference in construction of a power MOSFET as compared to a conventional MOSFET.
32. Discuss the protection afforded to the thyristor against (a) steady overcurrents and (b) current transients.
33. Give a firing circuit for a thyristor that uses a saturable reactor.
34. Explain how the thyristor gets triggered with a high dv/dt.
35. Describe a gate triggering circuit in which digital ICs are employed.
36. For the circuit given in Fig. 1.97, compute the value of R_s that gives a maximum dv/dt of 100 V/μs.

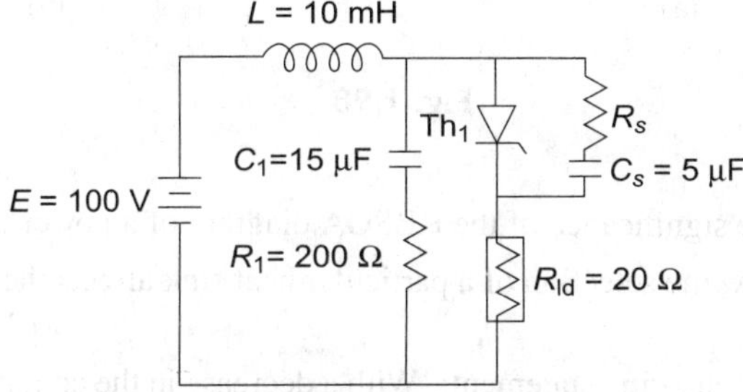

Fig. 1.97

37. Comment on the following statement: 'When subjected to slowly increasing overvoltages, the thyristor needs overcurrent protection but not overvoltage protection.'
38. Explain how the values of the R_f and R_v are determined for a resistance firing circuit. Explain the role played by the resistance R_B connected between the gate and the cathode.
39. Justify the following statement: 'The thyristor exhibits the property of hysteresis.'
40. Define the reverse recovery time of a thyristor and explain how it can be determined in the laboratory.
41. What are the factors governing the rise time during thyristor turn-on?
42. List out the factors on which the recovery time t_{OFF} of a thyristor depends.

43. Explain how the switching losses of a power transistor increase under inductive load conditions.
44. Explain the turn-on and turn-off processes of a GTO with the help of the two-transistor analogy.
45. Explain the mechanism of firing of the thyristor for an anode-to-cathode voltage less than V_{FBO} in the presence of a gate signal.
46. Why do the V_{AK} versus t as well as the I_A versus t characteristics of a thyristor with $R + L$ load differ from those of a thyristor with a purely resistive load?
47. Explain the role played by the junction J_b of a thyristor (a) during turn-on, (b) during turn-off, and (c) when a slowly increasing voltage is applied across it.
48. Compare the operation of the circuits given in Figs 1.98. Draw the load voltage waveforms in both the cases.

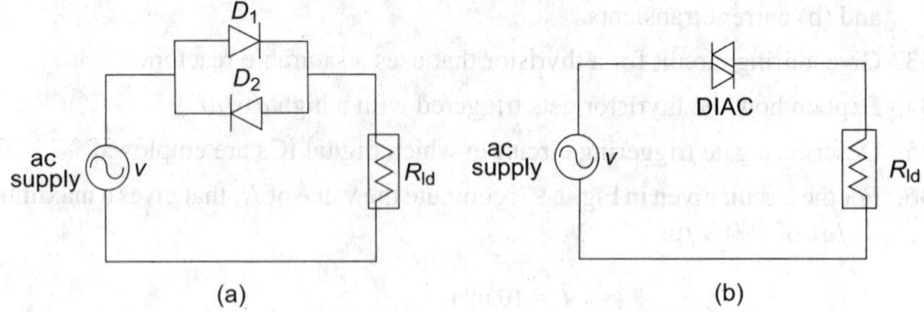

Fig. 1.98

49. Explain the significance of the RBSOA diagram of a power transistor.
50. Explain how the selection of a particular heat sink affects the $I_{F(av)}$ rating of a thyristor.
51. Justify the following statement: 'With a decrease in the conduction angle from 180° to 120°, its average thyristor current rating should also be reduced'.
52. What are the factors that govern the $I_{F(av)}$ rating of a thyristor?

Problems

1. For the circuit of Fig. 1.99, determine the values of L and C given the following data: $E = 200$ V, $R_1 = 664$ Ω, t_p (pulse duration) = 1.5 ms, and $I_{ld} = (\sqrt{3}/2)I_{c(max)}$.

   Ans. $L = 0.206$ H, $C = 0.62$ μF

2. The circuit of Fig. 1.100 has the following data: $E = 150$ V, $L = 10$ mH, r_L (resistance of the inductor) = 0.1 Ω, $R_{ld} = 0.7$ Ω, and $C = 70$ μF. Determine (i) the expression for the current $i(t)$, (ii) the pulse width, and (iii) the time for v_c to attain a value of $1.4E$.

   Ans. $i(t) = 12.56 \sin \omega t$ with $\omega = 1196$ rad/s, $t_p = 2.63$ ms, and the time for v_c to attain $1.4E$ is 1.66 ms

3. The following data are given for a UJT: $V_P = 22$ V, $I_P = 50$ μA, $V_V = 0.8$ V, $I_V = 4.8$ mA, $\eta = 0.73$, $R_{BB} = 6.95$ kΩ, and $R_{B_1(con)} = 65$ Ω. Determine the

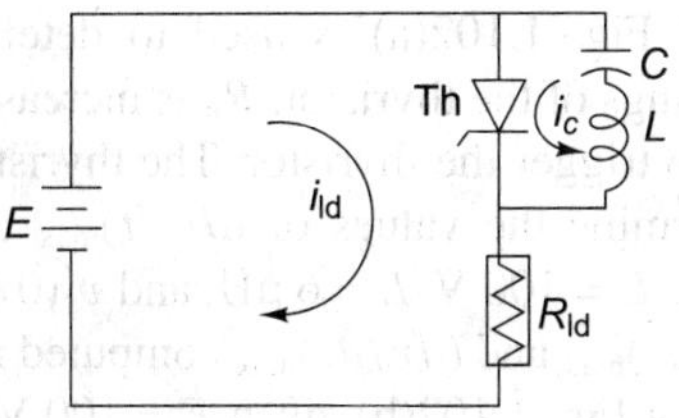

Fig. 1.99

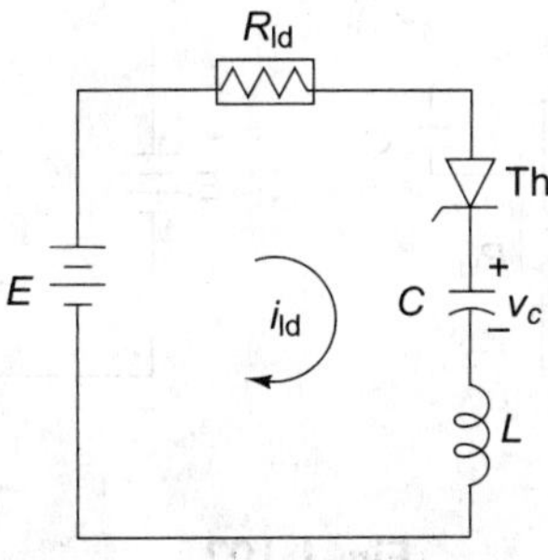

Fig. 1.100

various elements of the UJT relaxation oscillator for this device. Compute the maximum frequency range and pulse height obtainable with the designed oscillator.

Ans. $V_s = 30$ V, $C = 0.08$ μF, $R_1 = 154$ kΩ, $R_2 = 6$ kΩ, $R_3 = 700$ Ω, $R_4 = 70$ Ω, $f_{min} = 60$ Hz, $f_{max} = 1591$ Hz, and pulse height = 11.4 V

4. Determine the values of V, L, and C for the circuit of Fig. 1.101 with the following data: Load current is not to exceed 0.8 A, output frequency of the inverter = 50 Hz, Q factor of LR combination = 0.35, load resistance $R_{ld} = 220$ Ω, and the resistance of the inductor is to be neglected.

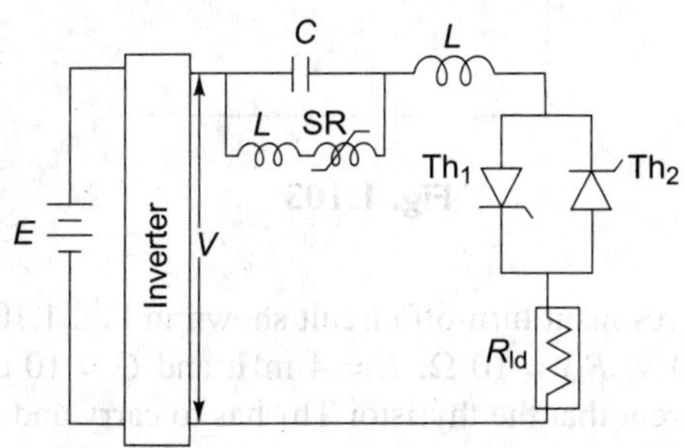

Fig. 1.101

Ans. $V = 176$ V, $L = 0.245$ H, $C = 41.4$ μF

5. (a) The circuit of Fig. 1.102(a) is used to determine the $(di/dt)_{max}$ and $(dv/dt)_{max}$ ratings of the thyristor. R_v is increased in steps and the switch Sw is closed to trigger the thyristor. The thyristor gets triggered with R_v = 10 Ω. Determine the values of $(di/dt)_{max}$ and $(dv/dt)_{max}$ given the following data: E = 100 V, L = 6 μH, and $v_c(0)$ = 50 V.
(b) Using the $(di/dt)_{max}$ and $(dv/dt)_{max}$ computed in (a), design the values of L, R_s, and C_s of Fig. 1.102(b) given E = 100 V and $V_c(0)$ = 50 V.

Ans. (a) 8.33 A/μs, 83.33 V/μs; (b) 20 μH, 25 Ω, 0.032 μF

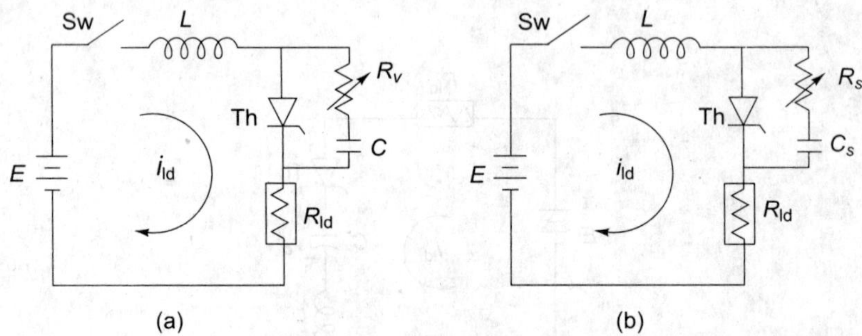

Fig. 1.102

6. For the circuit of Fig. 1.103 determine the size of the capacitor required and sketch the waveforms of v_{Th_1}, i_{Th_1}, and i_{ld}. The data for the circuit are E = 63 V, $R_1 = R_{ld}$ = 4 Ω, t_{OFF} of Th$_1$ = 33 μs.

Ans. C = 12 μF, I_{ld} is 15.75 A and shoots up to 31.5 A, i_{Th_1} has peaks of 31.5 A and 47.25 A

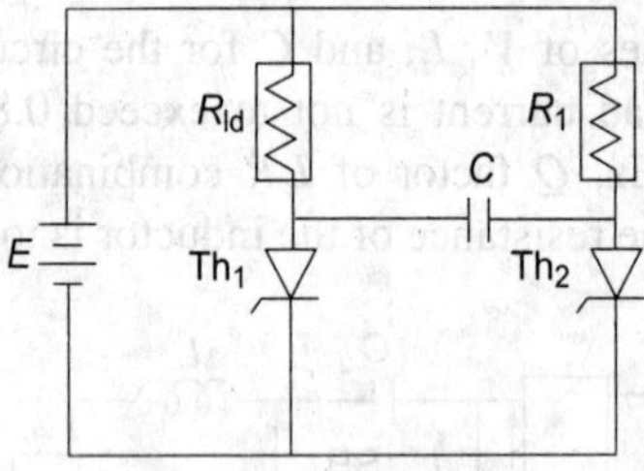

Fig. 1.103

7. The auxiliary resonant turn-off circuit shown in Fig. 1.104 has the following data: E = 100 V, R_{ld} = 10 Ω, L = 4 mH, and C = 10 μF. Compute (a) the maximum current that the thyristor Th$_1$ has to carry and (b) the turn-off time of Th$_1$. Sketch the waveforms of the voltage across Th$_1$, current through Th$_1$, and load voltage v_{ld}.

Ans. I_{Th_1M} = 15 A, t_{OFF} = 110 μs

8. For the UJT relaxation oscillator of Fig. 1.105, the following data are given: I_P = 65 μA, V_P = 19.5 V, I_V = 5.2 mA, V_V = 0.58 V, η = 0.65, R_{BB} = 7.5 kΩ, $R_{B_1(con)}$ = 60 Ω.

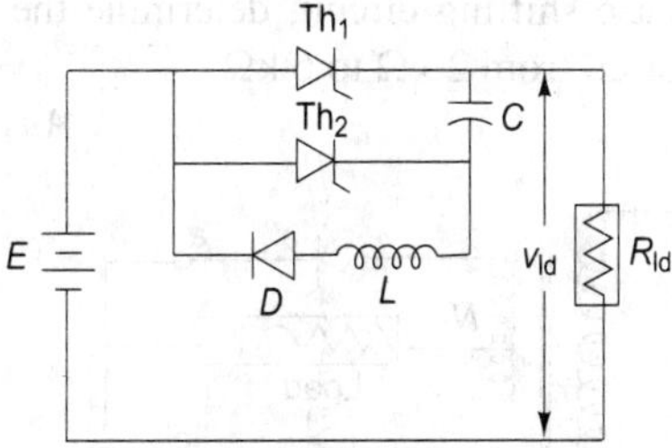

Fig. 1.104

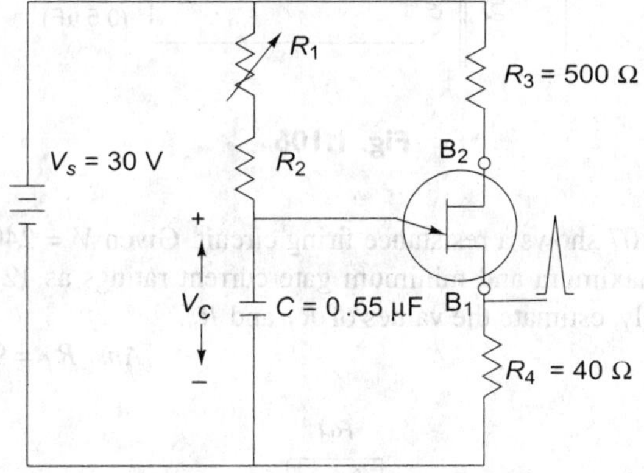

Fig. 1.105

(i) Determine the minimum value of R_2 and the corresponding pulse frequency.

(ii) By what factor does the pulse repetition rate increase or decrease if C is reduced from 0.55 μF to 0.40 μF?

(iii) By what factor does the pulse height increase or decrease if R_4 is decreased from 40 Ω to 25 Ω?

Ans. (i) $R_{2(\min)}$ = 5.66 kΩ and the corresponding $f_{\max}$ = 306 Hz,
(ii) increases to 1.37 times the original frequency,
(iii) decreases to 0.735 times the original height

9. For the thermal equivalent circuit of Fig. 1.47(b) the following data are given: θ_j = 85 °C, θ_a = 25 °C, β_{jc} = 0.41 °C/W, and β_{cs} = 0.09 °C/W. By choosing an appropriate heat sink from the curves of Fig. 1.47(c) and using the power dissipation versus average current curves of Fig. 1.46, determine the average current rating $I_{F(\text{av})}$ of the associated thyristor for 120° conduction.

Ans. Curve g is selected for the heat sink, P_{av} = 60 W, $I_{F(\text{av})}$ = 40 A

10. The centre-tapped phase shifting circuit shown at the bottom of Fig. 1.106 is utilized to fire the thyristor rectifier shown at the top. By drawing the phasor

diagram of the phase shifting circuit, determine the range of α that can be obtained if R is varied from 2 kΩ to 6 kΩ.

Ans. Range of α is 35°–87°

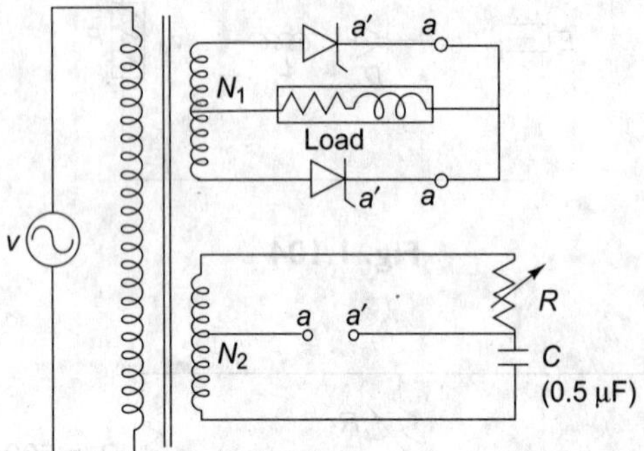

Fig. 1.106

11. Figure 1.107 shows a resistance firing circuit. Given $V = 240$ V, $R_{ld} = 15\ \Omega$, and the maximum and minimum gate current ratings as 12 mA and 3 mA, respectively, estimate the values of R_f and R_v.

Ans. $R_f = 9$ kΩ, $R_v = 9$ kΩ

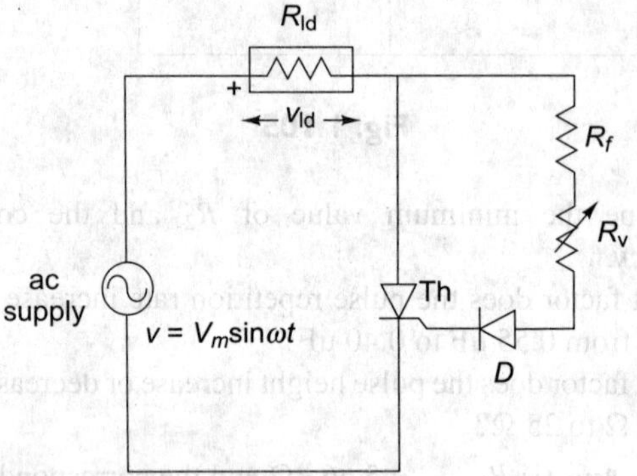

Fig. 1.107

12. In the circuit of Fig. 1.106, if the resistance is kept constant at 4 kΩ and the capacitance varied from 0.02 μF to 0.45 μF, determine the range of α that can be obtained.

Ans. 3°–59°

13. The phase shifting circuit shown in Fig. 1.108 is utilized to fire the thyristor which rectifies the same ac supply. (a) Determine the range of α that can be obtained if R_1 is varied from 1 kΩ to 5 kΩ. The transformer ratio is 220 V/110 V for both the secondaries; the supply frequency f is 50 Hz. (b) Find the value of R_1 to obtain a firing angle α of 30°.

Ans. (a) $\alpha_1 = 14°$, $\alpha_2 = 64°$; (b) $R_1 = 2.1$ kΩ

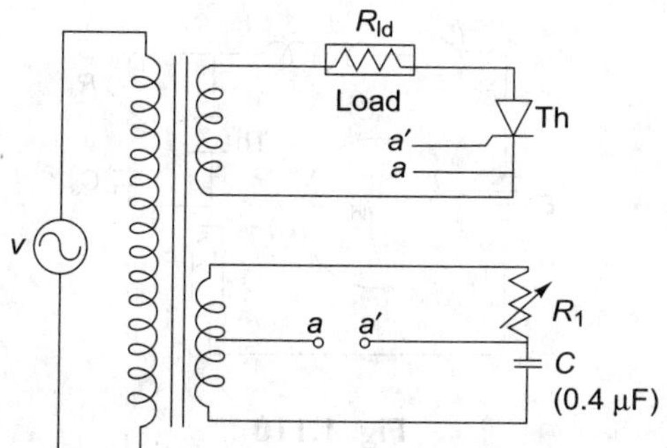

Fig. 1.108

14. (a) The parallel resonant turn-off circuit of Fig. 1.109 has the following data: E = 150 V, R_{ld} = 500 Ω, t_p (pulse width) = 1.8 ms, and $I_{ld} = I_{c(max)}$. Determine the values of L and C.

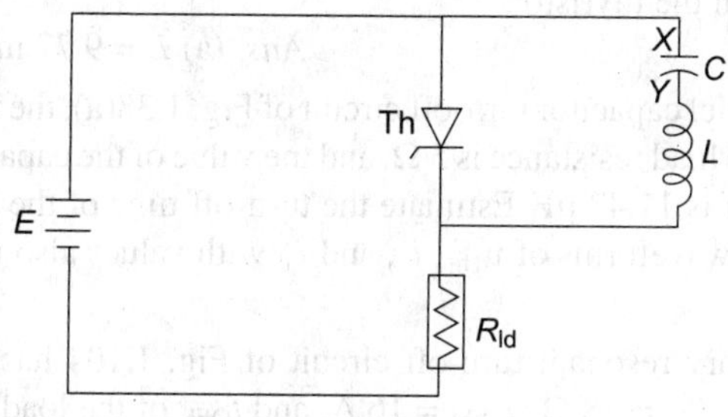

Fig. 1.109

(b) If now it is desired to provide reliable commutation by making $I_{ld} = 0.9I_{c(max)}$, by how much should the value of C be increased or decreased? Also compute the new pulse width.

Ans. (a) L = 0.19 H, C = 0.764 μF; (b) increase in C by 0.174 μF; new t_p = 1.8 ms

15. The following data pertain to the circuit given in Fig. 1.110: E = 100 V, $V_c(0)$ = 60 V, R_{ld} = 20 Ω, $(dv/dt)_{max}$ = 320 V/μs, $(di/dt)_{max}$ = 12 A/μs, and I_{ThM} = 12 A. Determine the values of L, R_s, and C_s. [Hint: See worked out problem in Section 1.7.1.]

Ans. The final designed values are R_s = 27 Ω, L = 7.5 μH, and C_s = 0.01 μF

16. Suggest a heat sink and estimate the thermal resistance of the heat sink, given the following data for a thyristor whose conduction angle is 120°, θ_j= 75 °C, θ_a= 25 °C, β_{jc}= 0.20 °C/W, β_{cs}= 0.10 °C/W, and $I_{F(av)}$ = 43 A.

Ans. The heat sink corresponding to curve d, β_{sa}= 0.50 °C/W

17. For the auxiliary resonant turn-off circuit of Fig. 1.104, the following data are given: E = 110 V, R_{ld} = 15 Ω, I_{ThM} = 10.5 A, and t_{OFF} of the thyristor is 40 μs. Determine the values of L and C for successful turn-off.

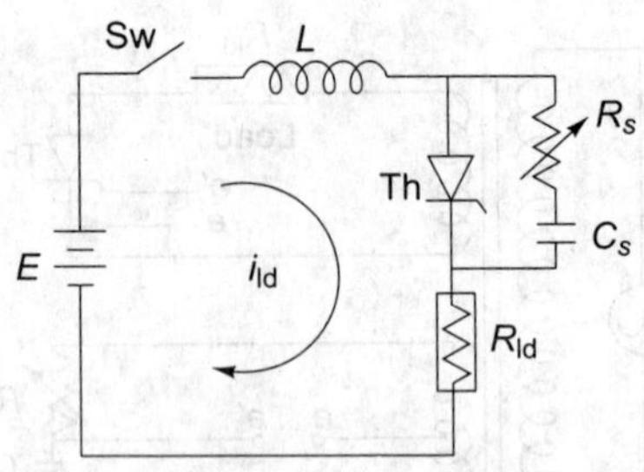

Fig. 1.110

Ans. $C = 2.5\ \mu F$ and $L = 3.0$ mH

18. The data for the auxiliary resonant turn-off circuit of Fig. 1.104 are as follows: $E = 120$ V, $R_{ld} = 14\ \Omega$, $I_{ThM} = 11$ A, and $C = 4\ \mu F$.
    (a) Compute the value of L for successful turn-off.
    (b) If now the value of C is doubled, compute the percentage of excess current through the thyristor.

    Ans. (a) $L = 9.77$ mH, (b) $\Delta I_{Th} = 9.1\%$

19. In the parallel capacitor turn-off circuit of Fig. 1.38(a), the battery has a voltage of 68 V, the load resistance is 5 Ω, and the value of the capacitance just required for turn-off is 11.43 μF. Estimate the turn-off time of the load thyristor. Also, sketch the waveforms of v_{Th_1}, i_c, and v_c with values also noted therein.

    Ans. 40 μs

20. The auxiliary resonant turn-off circuit of Fig. 1.104 has the following data: $E = 126$ V, $R_{ld} = 18\ \Omega$, $I_{ThM} = 16$ A, and t_{OFF} of the load thyristor = 42 μs.
    (a) Estimate the values of L and C. Sketch the waveforms of i_{Th_1}, v_c, i_{Th_2}, and i_c.
    (b) If the value of L is increased by 50% compute the new value of I_{ThM}.

    Ans. (a) $C = 2.12\ \mu F$, $L = 0.42$ mH; (b) 14.3 A

21. Design a PUT-based relaxation oscillator, given $V_{CC} = 26$ V, $I_G = 1.1$ mA, $V_P = 11$ V, the pulse frequency is 90 Hz, the pulse height is 7 V, and the conducting resistance of the device is 32 Ω.

    Ans. $R_1 = 20$-kΩ potentiometer, $R_2 = 21$ kΩ, $R_3 = 56\ \Omega$, $R_4 = 32.2$ kΩ, $R_5 = 23.6$ kΩ, and $C = 0.5\ \mu F$

22. For the RC firing circuit of Fig. 1.111, determine the values of R_f, R_v, and C given the following data: $V = 240$ V, $I_{GT(max)} = 13$ mA, $I_{GT(min)} = 1.5$ mA, $V_{GT} = 1.2$ V, $R_{ld} = 12\ \Omega$. The firing angle ranges from 0° to 110°. V_{D_1} can be taken as 0.7 V.

    Ans. $R_f = 8$ kΩ, $R_v = 20$-kΩ potentiometer, and $C = 0.91\ \mu F$

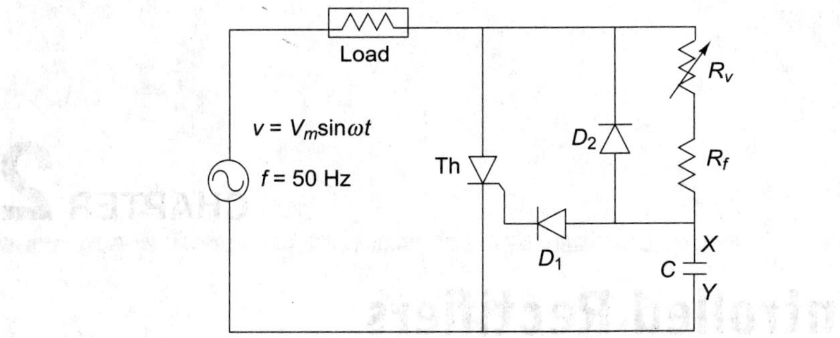

Fig. 1.111

CHAPTER 2

Controlled Rectifiers

2.1 Introduction

During the first half of this century, the ignitron, which is a voltage controlled gas tube, was widely used for controlled rectification. With the advent of the thyristor, the ignitron has been replaced not only in rectification circuits but also in various other power-control equipment. In all subsequent treatment, the words 'converter' and 'rectifier' will be treated as synonymous terms. The advantages of using the thyristor as an element for controlled rectification are as follows.

(a) A firing circuit (or a gate control circuit) can be constituted with the help of digital electronic devices, a recent trend being the use of microprocessors for this purpose. These IC chips are mounted on printed circuit cards which can be conveniently housed in card cages.

(b) Thyristorized circuits are compact and need very little maintenance.

(c) By properly changing the firing instant, thyristorized rectifier circuits can also be used for inversions. This mode of operation is adopted in applications, such as hoist control, that involve regenerative braking.

(d) A controlled rectifier is an important component of a dc drive, which consists of a dc motor, a rectifier, and thyristorized power-control circuits together with relevant firing circuitry. Dc drives are driven by rectifiers whereas ac drives are controlled by inverters, ac controllers, or cycloconverters. In dc drives the speed of the motor is controlled by varying the armature voltage. On the other hand, speed control of ac drives involves the variation of both the magnitude and frequency of the applied voltage. Though the commutator of a dc motor makes it a costlier proposition, the precision obtained with it for speed control is higher than that obtained with an ac drive. When used for rectification purposes, a thyristor does not need a separate circuit for turning off. This is because, in the negative half-cycle, the polarity of the anode-to-cathode voltage applied to the device is reversed and this turns the thyristor off in a natural manner.

(e) Thyristorized drives may be conveniently converted into closed-loop speed control systems, thus deriving all the advantages of such control. In these systems, a signal proportional to the actual shaft speed of the drive is compared with a signal that represents the desired speed. The error so obtained adjusts the firing angle to restore the speed to the nominal value.

The following are some of the drawbacks associated with thyristorized rectifiers.

(a) The waveform of the voltage obtained after rectification is of a pulsating nature and therefore contains harmonics. The harmonic content increases with the firing angle. Such a rectified voltage is not suitable for certain load circuits. To obtain a fairly constant voltage at the load, smoothing (or filter) circuits have to be employed, but these add to the cost of the rectifier.

(b) The harmonics cause excessive heating due to increased iron losses in the armature of dc motors. Therefore, special precautions like providing high-quality insulation and extra cooling arrangements become necessary.

(c) When the load current of a rectifier is made constant by means of smoothing inductors, the thyristors in each phase carry rectangular pulses of current. Consequently, the current in the primary of a rectifying transformer is usually not sinusoidal. This causes some derating of the rectifier transformer. It thus becomes imperative that a transformer of appropriate capacity, over and above the load power that is supplied, be installed after an assessment of the extent of derating.

(d) Thyristorized circuits are subject to overvoltages and overcurrents. This may be, to a large extent, due to switching transients caused by the presence of smoothing inductors, capacitors, and transformers in the circuit. For safeguarding the circuits against such transients, protective features and also some redundancy of devices are essential. However, these add to the cost of the circuitry.

(e) The presence of harmonics causes an increase in the RMS value of the input current and this causes additional heat dissipation in the conductors. The effective resistance also increases due to the skin effect caused by the harmonic frequencies. Thus there is an increase in the overall ohmic loss, resulting in a decrease of the power factor presented to the input supply system.

In spite of the above-mentioned drawbacks, thyristorized converters have become popular because of their compactness as well as other merits enumerated earlier. They have found extensive use in all types of industry involving speed and power control.

After going through this chapter the reader should

- know the different configurations of single- and three-phase, half-wave controlled rectifiers as well as the topologies of single- and three-phase, full-wave controlled rectifiers,
- understand how the concept of single- and three-phase rectifiers can be generalized to n-phase rectifiers and how analyses are made with different types of loads,

- become familiar with the inverting mode of a converter,
- know how a quantitative expression for the ripple factor can be obtained,
- realize that overlapping of conduction occurs for thyristors in an n-phase rectifier when the source inductance has a significant magnitude, and know how the average voltage output of the rectifier gets reduced because of this,
- know how to evaluate the *rectifier efficiency* and *derating factor* for different rectifier configurations,
- get acquainted with dual converters and their operation with simultaneous and non-simultaneous control,
- become familiar with different methods for braking of motors, and
- understand how the power factor of a rectifier can be improved by using common forced commutation methods.

2.2 Single-phase Rectifiers

A thyristor conducts only in one direction under the control of the gate signal and this feature makes it an ideally suited device for controlled rectification. Depending on the type of ac supply, such circuits are classified as single-phase or polyphase rectifiers. They are also called phase-controlled rectifiers because the firing angle, which is measured with reference to the zero crossing of the voltage waveform, can be controlled over the range 0°–180°.

The circuit and waveforms of a single-phase, half-wave rectifier feeding a resistive load are illustrated in Fig. 2.1. The thyristor starts conduction only in the positive half-cycle at an angle α, which is determined by the gate current signal applied by means of the firing circuitry. The load current waveform also starts at the angle α; its instantaneous magnitude is given by $(v_{ld} - v_{Th(con)})/R_{ld}$. Here $v_{Th(con)}$ is the voltage drop across the conducting thyristor and is of the order of 1–2 V. For analysis purposes, an ideal thyristor with negligible drop is assumed and $v_{Th(con)}$ is taken to be negligible, thus approximating the instantaneous value of current as v_{ld}/R_{ld}. For a resistive load, the thyristor stops conduction at the angle π, at which the voltage waveform attains its zero.

The circuit and waveforms of a single-phase, half-wave rectifier feeding an inductive load are shown in Fig. 2.2. The thyristor stops conduction at an angle $\alpha_e > \pi$, but not necessarily equal to $\pi + \alpha$; its actual value depends on the L_{ld}/R_{ld} ratio. It can be stated that conduction beyond π is caused by the inductive energy stored during the period α to π. The load voltage becomes negative from π to α_e whereas the current remains positive. Thus, the power becomes negative during this interval and can be interpreted as power being fed back from the load to the ac source.

The average load voltage V_α can be obtained as follows:

$$V_\alpha = \frac{1}{2\pi} \int_\alpha^{\alpha_e} V_m \sin \omega t \, d(\omega t) \tag{2.1}$$

which works out to give

$$V_\alpha = \frac{V_m}{2\pi}(\cos\alpha - \cos\alpha_e) \tag{2.2}$$

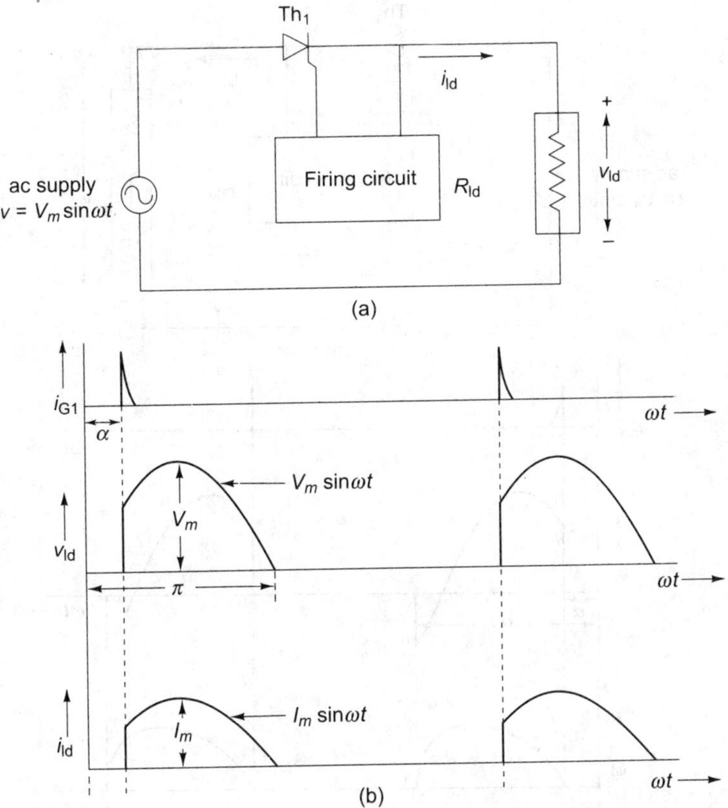

Fig. 2.1 Single-phase, half-wave controlled rectifier with resistive load: (a) circuit, (b) waveforms

For a resistive load, $\alpha_e = \pi$ and V_α becomes $(V_m/2\pi)(1 + \cos\alpha)$. On the other hand, for $\alpha_e > \pi$, say $\alpha_e = \pi + \beta$, the average voltage becomes $(V_m/2\pi)(\cos\alpha + \cos\beta)$, its value decreasing with increasing β. It is evident from this that the negative part of the voltage waveform (after π) causes a reduction in the average voltage.

The single-phase, half-wave circuit is usually not employed because the ripple content in the load voltage is high and the load current becomes discontinuous even with normal values of load inductance.

2.2.1 Single-phase, Full-wave Circuit with Centre-tapped Secondary

The circuit shown in Fig. 2.3(a) consists of a single-phase controlled rectifier fed by a centre-tapped transformer and supplying an inductive load. By virtue of the centre tap, v_1 and v_2 are each equal to half the secondary voltage, but have opposite polarities. Conduction of Th_1 is initiated by a firing pulse i_{G1} at an angle α from the zero crossing of v_1. Th_1 continues to conduct up to $\pi + \alpha$ because of the load inductances. Th_2 starts conduction at $\pi + \alpha$ because (i) v_2 is in its

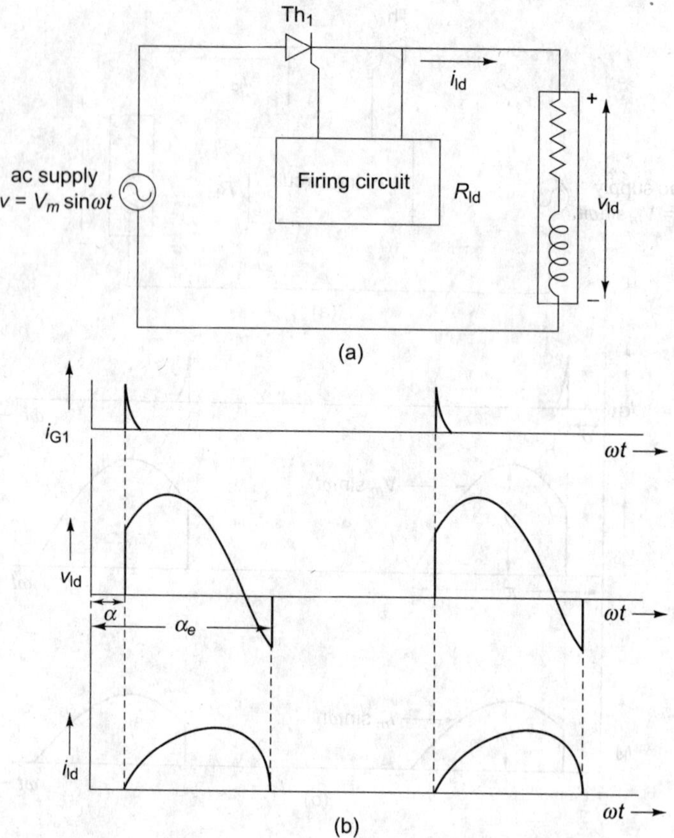

Fig. 2.2 Single-phase, half-wave controlled rectifier with inductive load: (a) circuit, (b) waveforms

positive half-cycle and (ii) the trigger pulse i_{G2} starts its (Th_2) conduction. With sufficient inductance in the load, the current i_{ld} becomes continuous and attains a nearly flat shape as shown in Fig. 2.3(b). It is evident from this figure that the load current is conducted alternately by the thyristors Th_1 and Th_2. Neglecting the drop of the conducting thyristor, the voltage v_{Th_1} across Th_1 becomes zero in the interval α to $\pi + \alpha$. Also, when Th_2 is on, the full secondary voltage $2V_m$ is impressed against Th_1.

The average voltage V_α across the load is given by

$$V_\alpha = \frac{2}{2\pi}\int_{\alpha}^{\pi+\alpha} V_m \sin\omega t \, d(\omega t) = \frac{2V_m}{\pi}\cos\alpha \qquad (2.3)$$

If the drop across the conducting thyristor is significant, the actual average load voltage is obtained by subtracting this drop from V_α. The conduction will be discontinuous when the load is either purely resistive or purely inductive with a small L/R ratio. In the purely resistive case, simple integration shows that V_α equals $(V_m/2\pi)(1 + \cos\alpha)$. This average is greater than that obtained with an inductive load.

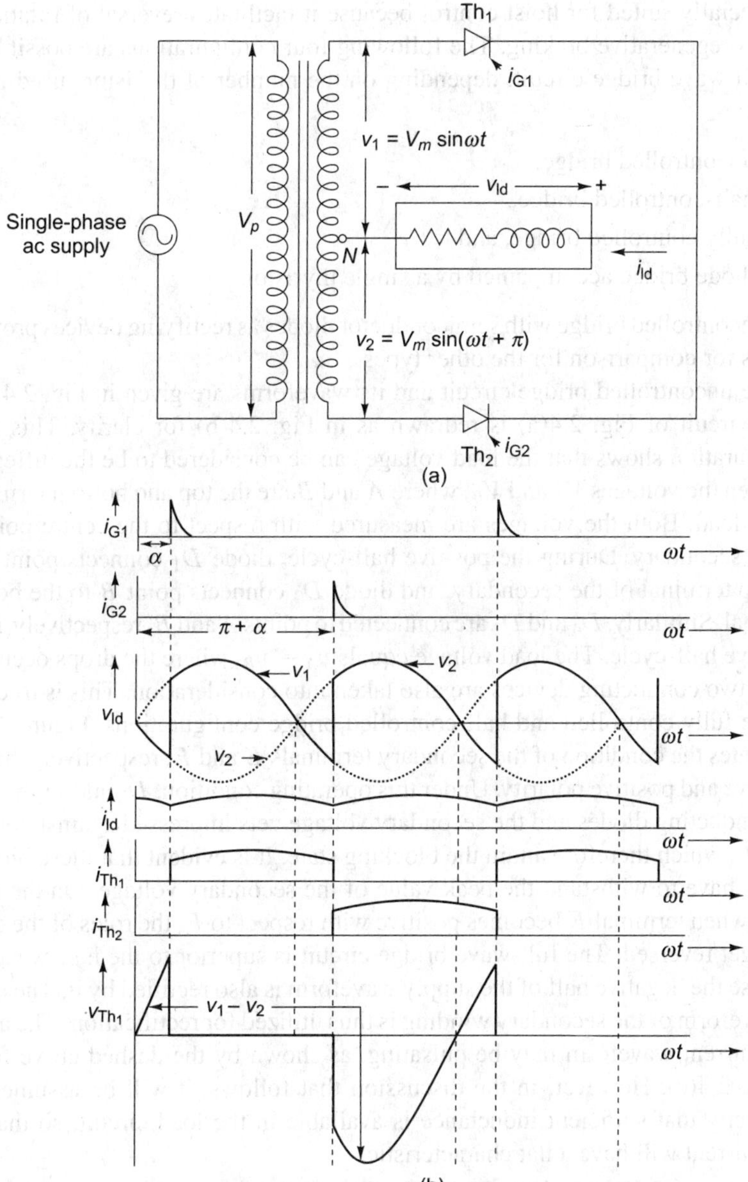

Fig. 2.3 Single-phase centre-tapped secondary type of full-wave rectifier: (a) circuit, (b) waveforms

2.2.2 Single-phase, Full-wave Bridge Rectifiers

The bridge type of controlled rectifier is widely used because of its versatility. Whereas the application of the half-wave converter is restricted to motors of 2 kW and below, the full-wave bridge converter is used to drive dc motors having capacities up to 20 kW. It is shown in a later section that a dual converter consisting of two bridge converters connected in a back-to-back configuration is especially suited for hoist control because it facilitates reversal of rotation as well as regenerative braking. The following four configurations are possible for the full-wave bridge circuit, depending on the number of thyristors used in the circuit:

(a) uncontrolled bridge,
(b) half-controlled bridge,
(c) fully controlled bridge, and
(d) diode bridge accompanied by a single thyristor.

The uncontrolled bridge with semiconductor diodes as rectifying devices provides a basis for comparison for the other types.

The uncontrolled bridge circuit and its waveforms are given in Fig. 2.4. The basic circuit of Fig. 2.4(a) is redrawn as in Fig. 2.4(b) for clarity. This latter configuration shows that the load voltage can be considered to be the difference between the voltages V_A and V_B, where A and B are the top and bottom terminals of the load. Both the voltages are measured with respect to the centre point N of the secondary. During the positive half-cycle, diode D_1 connects point A to the top terminal of the secondary, and diode D_3 connects point B to the bottom terminal. Similarly, D_4 and D_2 are connected to points A and B, respectively, in the negative half-cycle. The load voltage equals $v_A - v_B$, where the drops occurring in the two conducting devices are also taken into consideration. This is true also for the fully controlled and half-controlled bridge configurations. Figure 2.4(d) illustrates the condition of the secondary terminals E and F, respectively, having negative and positive polarity. Under this operating condition, D_2 and D_4 become the conducting diodes and the secondary voltage gets impressed against both D_1 and D_3, which therefore attain the blocking state. It is evident that these blocked diodes have to withstand the peak value of the secondary voltage. On the other hand, when terminal E becomes positive with respect to F, the roles of the diode pairs get reversed. The full-wave bridge circuit is superior to the half-wave one because the negative half of the supply waveform is also rectified by it. The entire ac waveform of the secondary winding is thus utilized for rectification. The actual load current waveform may be pulsating, as shown by the dashed curve for i_{ld} in Fig. 2.4(c). However, in the discussion that follows, it will be assumed for simplicity that sufficient inductance is available in the load circuit, so that the load current will have a flat characteristic.

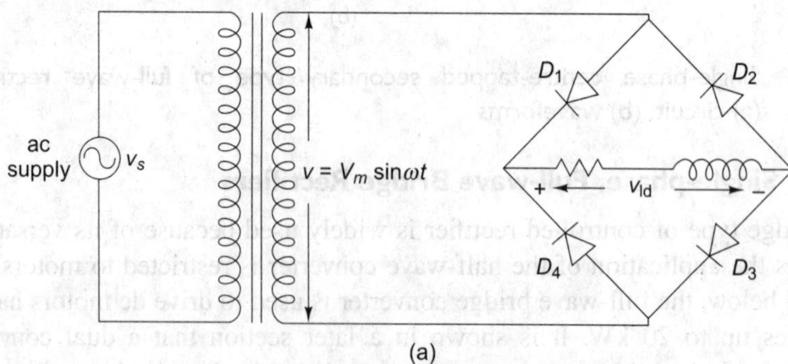

Fig. 2.4(a)

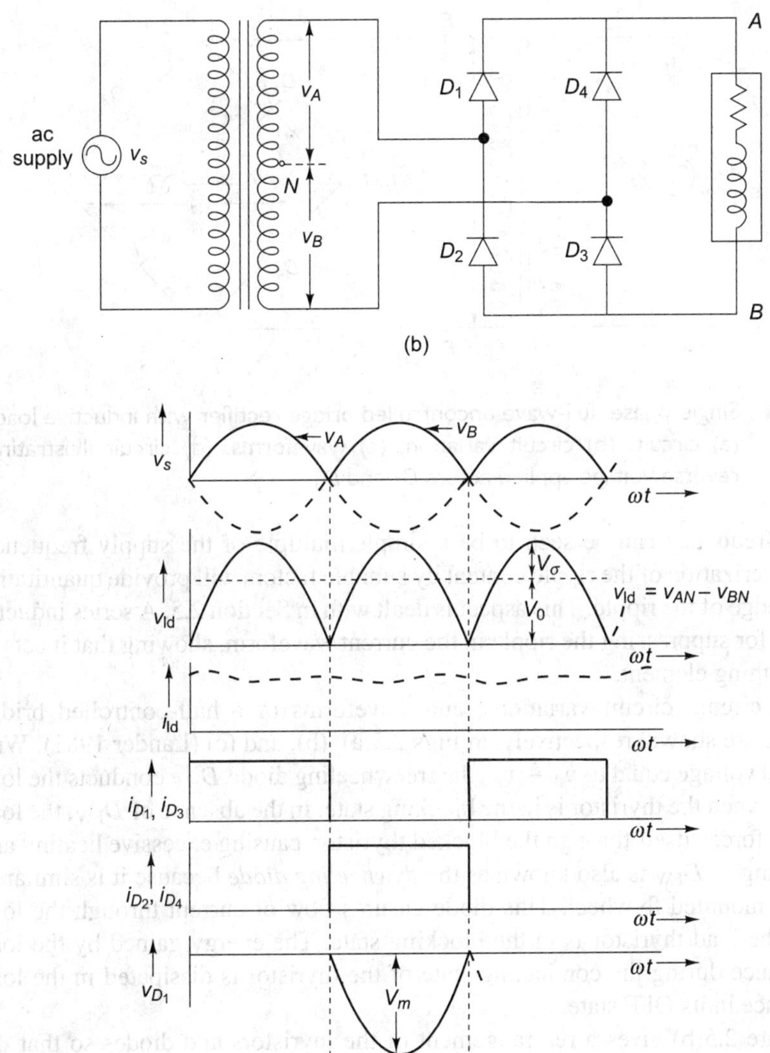

Figs 2.4(b), (c)

The effect of the presence of a large inductor in the load circuit can be explained with the help of the time constant concept as follows. In a series RL circuit, a large value for the inductor makes the time constant (L/R) large, and correspondingly the droop of the current curve is large. In the limit when L is infinitely large, the current attains a flat characterstic.

The instantaneous load voltage is seen from Fig. 2.4(c) to be the sum of the average voltage V_o and the ripple voltage V_σ, the latter fluctuating about V_o with α taken as zero. V_o is obtained from Eqn (2.3), with $\alpha = 0$, as $2V_m/\pi$. The load voltage waveform has a two-pulse characteristic showing thereby that the ripple frequency is twice that of the supply frequency. It will be shown in a later section that the load voltage waveform gets smoothed with an increase in the ripple frequency. In both single-phase as well as multiphase rectifiers, the

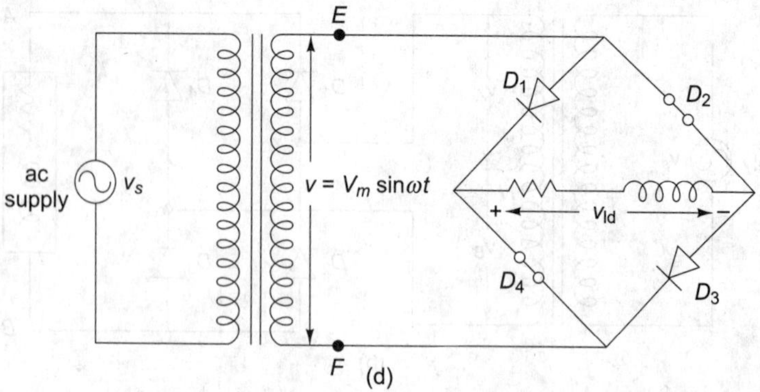

Fig. 2.4 Single-phase, full-wave uncontrolled bridge rectifier with inductive load: (a) circuit, (b) circuit variation, (c) waveforms, (d) circuit illustrating reverse voltage applied across D_1 and D_3

ripple frequency can be seen to be a simple multiple of the supply frequency. Characterization of the ripple content by suitable factors will provide quantitative knowledge of the ripple. This aspect is dealt with in Section 2.5. A series inductor is used for suppressing the ripple in the current waveform, showing that it acts as a smoothing element.

The circuit, circuit variations, and waveforms of a half-controlled bridge rectifier are shown, respectively, in Figs 2.5(a), (b), and (c) (Lander 1981). With the load voltage equal to $v_A - v_B$, the freewheeling diode D_{FW} conducts the load current when the thyristor is in the blocking state. In the absence of D_{FW}, the load current forces itself through the blocked thyristor, causing excessive heating and damaging it. D_{FW} is also known as the *flywheeling diode* because it is similar to a shaft mounted flywheel. This diode ensures flow of current through the load when the load thyristor is in the blocking state. The energy gained by the load inductance during the conducting state of the thyristor is dissipated in the load resistance in its OFF state.

Figure 2.5(b) gives a rearrangement of the thyristors and diodes so that the diodes D_1 and D_2 can provide a freewheeling path for the load current. If D_{FW} is now connected across the load, it provides a preferred path because the forward resistance of a single diode is less than that of two diodes. The rectifier circuit of Fig. 2.5(a) also has the following freewheeling paths: during the interval *OP* [Fig. 2.5(c)], the thyristor Th_2 and D_2 form a freewheeling path; also in the interval *QR*, the devices Th_1 and D_1 constitute the freewheeling path. However, a preferential path exists through the freewheeling diode D_{FW} as explained above. Though the $i_{D_{FW}}$ pulses are shown to be flat, they are actually slightly sloping curves because of the dissipation of the inductive energy of the load. It can be shown that the average voltage V_α in this case works out to $(V_m/\pi)(1 + \cos\alpha)$.

A fully controlled bridge circuit and its waveforms are given in Fig. 2.6, in which either the thyristor pair Th_1, Th_3 or the pair Th_2, Th_4 conducts at a time. The resistors forming the pair have to be fired at the same instant.

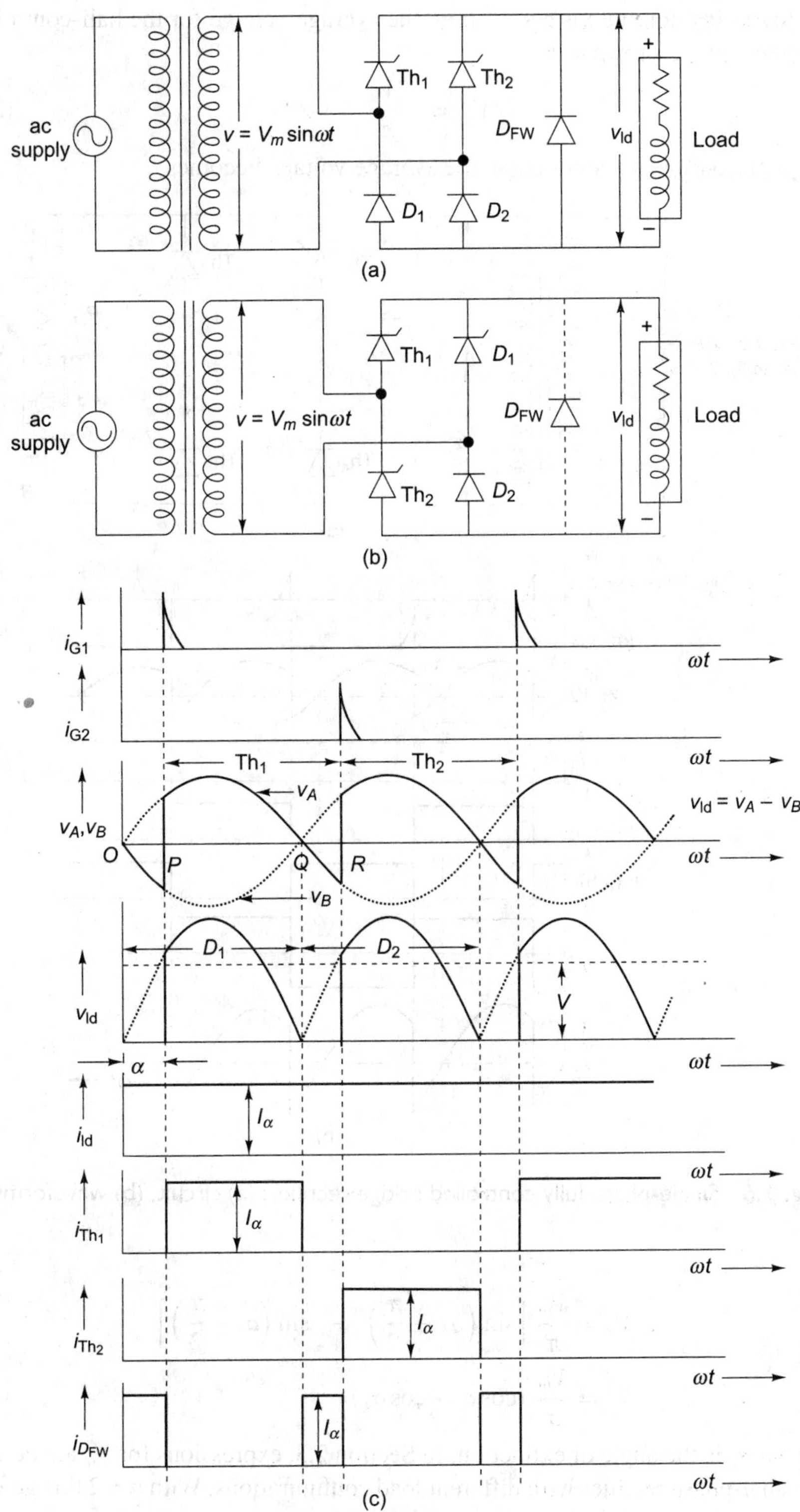

Fig. 2.5 Single-phase half-controlled bridge rectifier with inductive load and a freewheeling diode: (a) circuit, (b) circuit variation, (c) waveforms

Assuming continuous conduction, the average voltage for the half-controlled rectifier can be shown to be

$$V_\alpha = \frac{2V_m}{\pi}\cos\alpha \tag{2.4}$$

With discontinuous conduction, the average voltage becomes

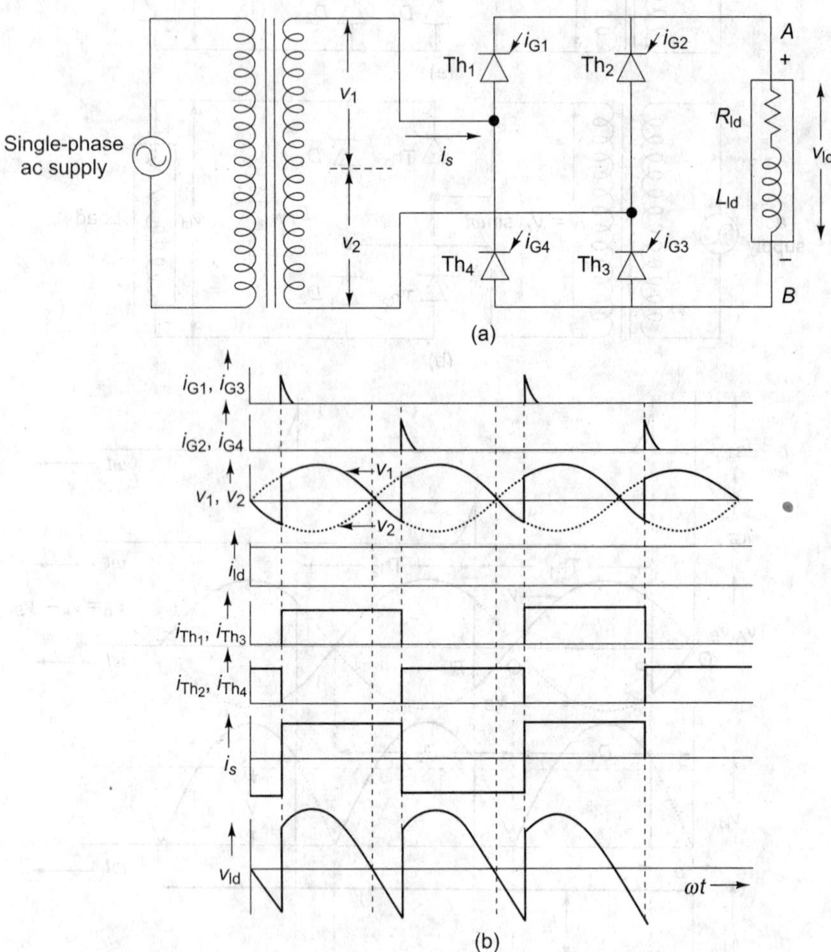

Fig. 2.6 Single-phase fully controlled bridge rectifier: (a) circuit, (b) waveforms

$$\begin{aligned} V_\alpha &= \frac{V_m}{\pi}\left[\sin\left(\alpha_e - \frac{\pi}{2}\right) - \sin\left(\alpha - \frac{\pi}{2}\right)\right] \\ &= \frac{V_m}{\pi}(\cos\alpha - \cos\alpha_e) \end{aligned} \tag{2.5}$$

where α_e is the angle of extinction. In Section 2.3, expressions for V_α are derived for an n-phase rectifier with different load configurations. With $n = 2$ this general expression can then be interpreted to be that of the average voltage for a single-phase rectifier. Likewise, with $n = 3$, the expression gives the average load voltage for a three-phase rectifier.

A unique feature of the fully controlled bridge converter is that, with α greater than 90° (electrical), or $\pi/2$ rad, the circuit functions as an inverter. Hence this mode of operation of a fully controlled converter is called the *inverting mode*; Section 2.5 is devoted to a detailed discussion on it.

A circuit that is commonly employed for the speed control of small dc motors consists of a diode bridge rectifier followed by a single thyristor as shown in Fig. 2.7(a). The thyristor controls the conduction angle in each of the half-cycles; the symmetry of conduction is ensured by spacing the triggering pulses π radians apart. With a supply frequency of 50 Hz, the pulse frequency for a unijunction

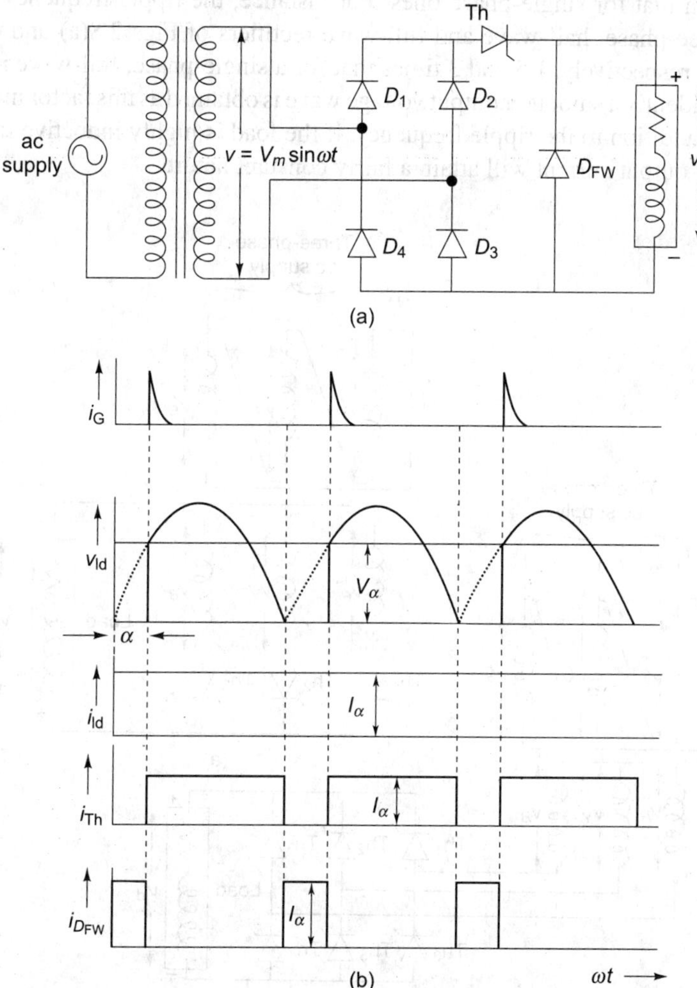

Fig. 2.7 Single-phase diode bridge with a single thyristor in the load circuit: (a) circuit, (b) waveforms

transistor relaxation oscillator has to be fixed at 100 Hz for this purpose. The continuity or otherwise of the load current waveform depends on the L/R ratio of the load; the freewheeling diode D_{FW} serves the same purpose. The waveforms for this circuit are given in Fig. 2.7(b).

2.3 Three-phase Rectifiers

Three-phase rectifiers are more commonly used because of the following reasons.

(i) Three-phase ac power is readily available.

(ii) It is economical to provide dc supply to dc motors of capacity 20 kW and more from a three-phase rectifier rather than single-phase one.

(iii) The ripple frequency of the output current of three-phase rectifiers is higher than that for single-phase ones. For instance, the ripple frequency for the three-phase, half-wave and full-wave rectifiers of Figs 2.8(a) and (b) will be, respectively, 1.5 and 3 times that for a single-phase, full-wave rectifier. Evidently a smoother output voltage wave is obtained as this factor increases. In addition to the ripple frequency, if the load is highly inductive in nature the output current will attain a fairly constant nature.

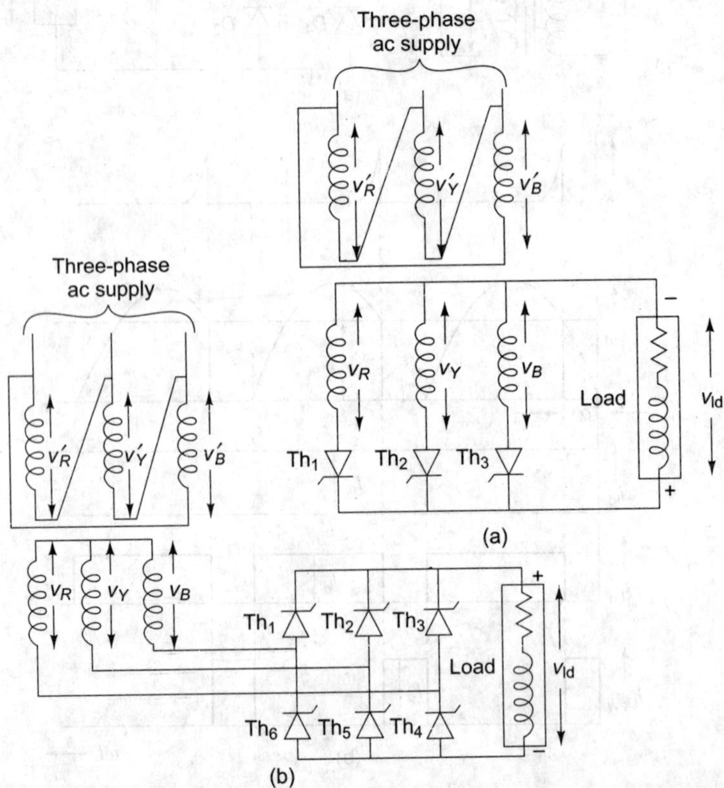

Fig. 2.8 (a) Three-phase, half-wave controlled bridge rectifier; (b) three-phase, full-wave fully controlled bridge rectifier

The three-phase rectifier circuits are examined in respect of the following operational features: (a) average load voltage, (b) peak inverse voltage across a single rectifying device, and (c) symmetry of the current waveform on the ac supply side.

2.3.1 Three-phase, Half-wave Controlled Rectifier

A three-phase, half-wave rectifier feeding an inductive load and its waveforms are shown in Fig. 2.9. The source for the rectifier is usually a delta–star transformer as shown in Fig. 2.9(a) or a star–star one.

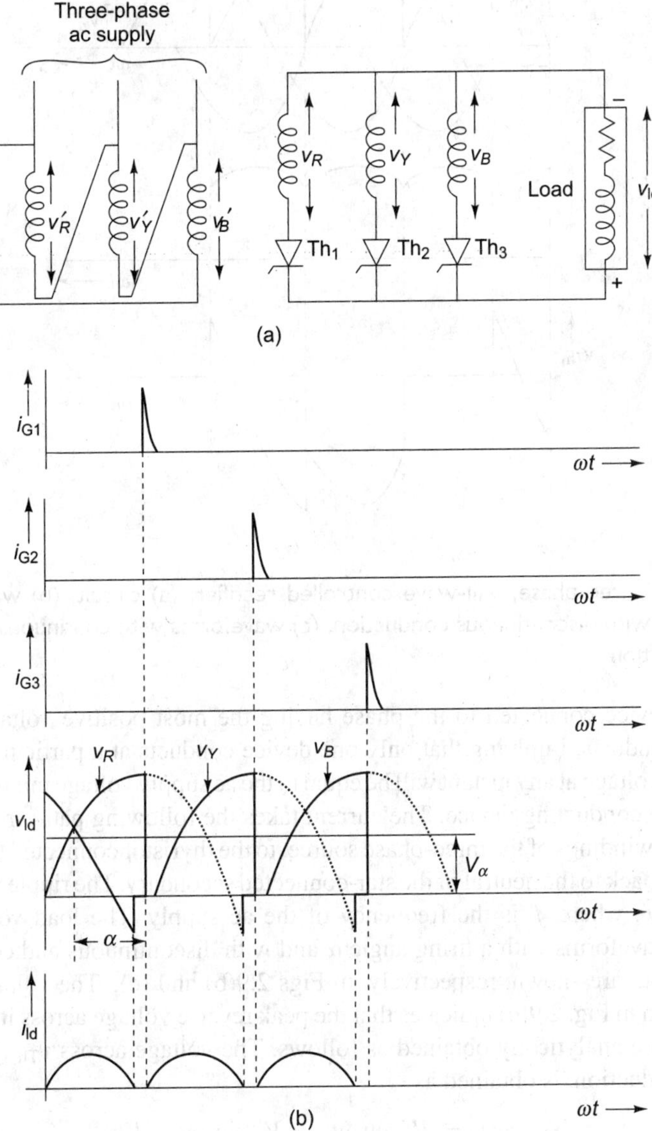

Figs 2.9(a), (b)

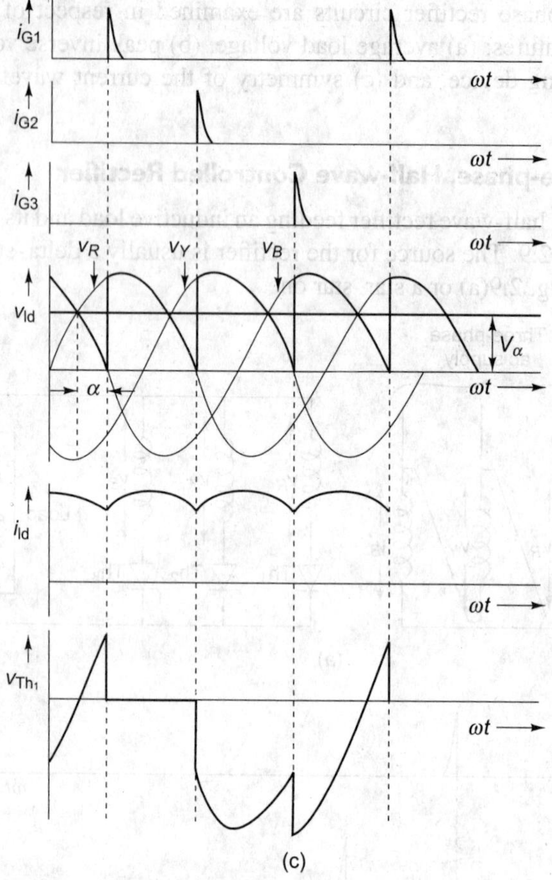

Fig. 2.9 Three-phase, half-wave controlled rectifier: (a) circuit, (b) waveforms with discontinuous conduction, (c) waveforms with continuous conduction

The device connected to the phase having the most positive voltage at that instant conducts, implying that only one device conducts at a particular instant. The load voltage at any instant will be equal to the ac supply voltage minus the drop across the conducting device. The current takes the following path: from one of the phase windings of the three-phase source, to the thyristor connected to it, to the load, and back to the neutral of the star-connected secondary. The ripple frequency will be $3f$, where f is the frequency of the ac supply. The load voltage and current waveforms with a firing angle α and with discontinuous and continuous conduction, are shown, respectively, in Figs 2.9(b) and (c). The voltage across Th_1 shown in Fig. 2.9(c) indicates that the peak reverse voltage across it is $\sqrt{3}V_m$. This can be analytically obtained as follows. The voltage across v_{Th_1}, after Th_2 starts conduction, is obtained as

$$v_{\mathrm{Th}_1}(\omega t) = V_m \sin \omega t - V_m \sin(\omega t - 120°) \tag{2.6}$$

Setting $dv_{\text{Th}}/dt = 0$ gives $\omega t = 60°$ and $v_{\text{Th(max)}}$ works out to be $\sqrt{3}V_m$. The average voltage can be shown to be

$$V_\alpha = \frac{3\sqrt{3}}{2\pi} V_m \cos\alpha \tag{2.7}$$

where V_m is the peak value of the phase-to-neutral voltage of the input ac waveform.

At any instant, the secondary current is unidirectional, and hence the flux in the core is also unidirectional, causing saturation of the core. The zigzag configuration shown in Fig. 2.10(a) overcomes this drawback and provides an alternating waveform on the primary side (i'_R), as can be seen from the current waveforms shown in Fig. 2.10(b) (Lander 1981).

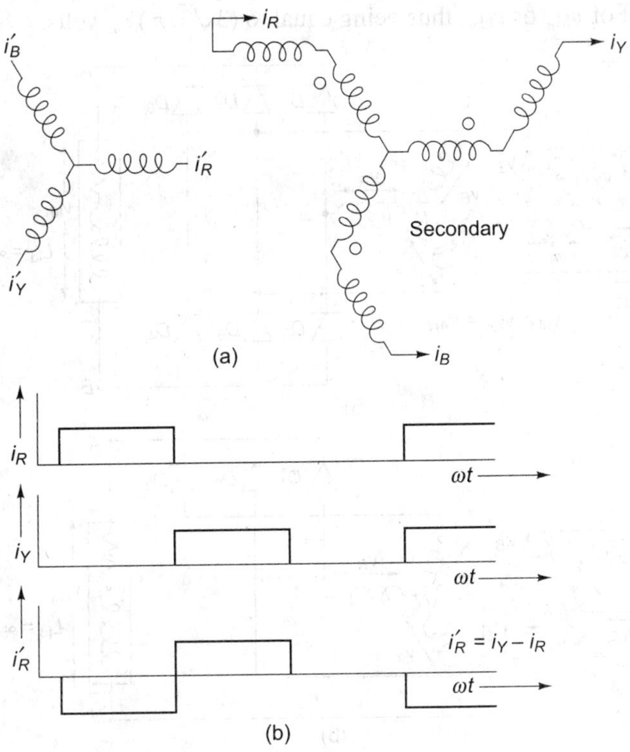

Fig. 2.10 (a) Primary and secondary windings of a star/zigzag transformer; (b) current waveform

2.3.2 Three-phase, Full-wave Rectifiers

The full-wave bridge configuration shown in Fig. 2.11(a) with diodes as the rectifying devices is commonly used, as it gives an alternating current waveform in the transformer primary and also has the merit of utilizing the negative half-waves of the secondary voltage. The load is considered to be highly inductive so as to make the diode as well as secondary current pulses rectangular. This full-wave circuit can be considered to be composed of two half-wave circuits as shown in Figs 2.11(b) and (c). In Fig. 2.11(b), the load terminal *A* is connected between the most positive phase and the neutral, each diode conducting for a period of 120° or $2\pi/3$ rad. Similarly, in Fig. 2.11(c), the diode connections are such that the load terminal *B* is connected to the most negative phase and the neutral. The load voltage waveform v_{ld} shown in Fig. 2.11(d) is therefore the algebraic difference between the top caps of v_{ld_A} and the bottom caps of v_{ld_B}, which are measured with respect to the neutral of the secondary winding. The device pairs that conduct in each interval are also indicated in the figure.

The net load voltage $v_{\mathrm{ld}} = v_{AB} = v_{AN} - v_{BN}$ is a pulsating one and is seen to have a ripple frequency of $6f$. The waveforms of i_1, i_2, and i_3 confirm that the primary current is alternating, although they are stepped in nature and hence have harmonic content. Neglecting the diode drops, the average load voltage v_{ld} is twice that of v_{ld_A} or v_{ld_B}, thus being equal to $(3\sqrt{3}/\pi)V_m$ volts.

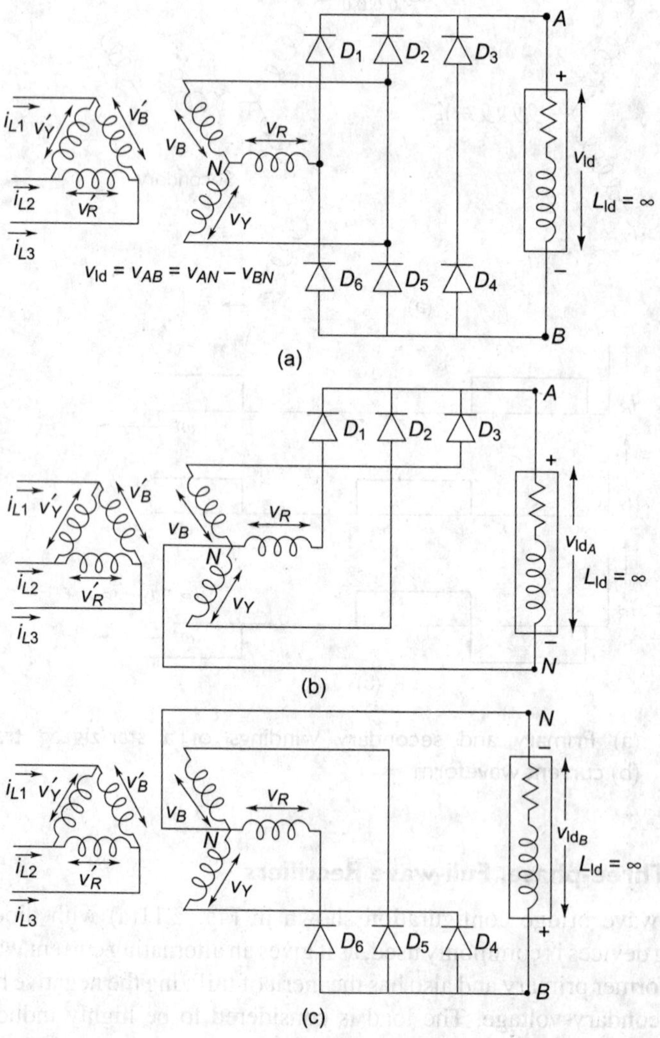

Figs 2.11 (a)–(c)

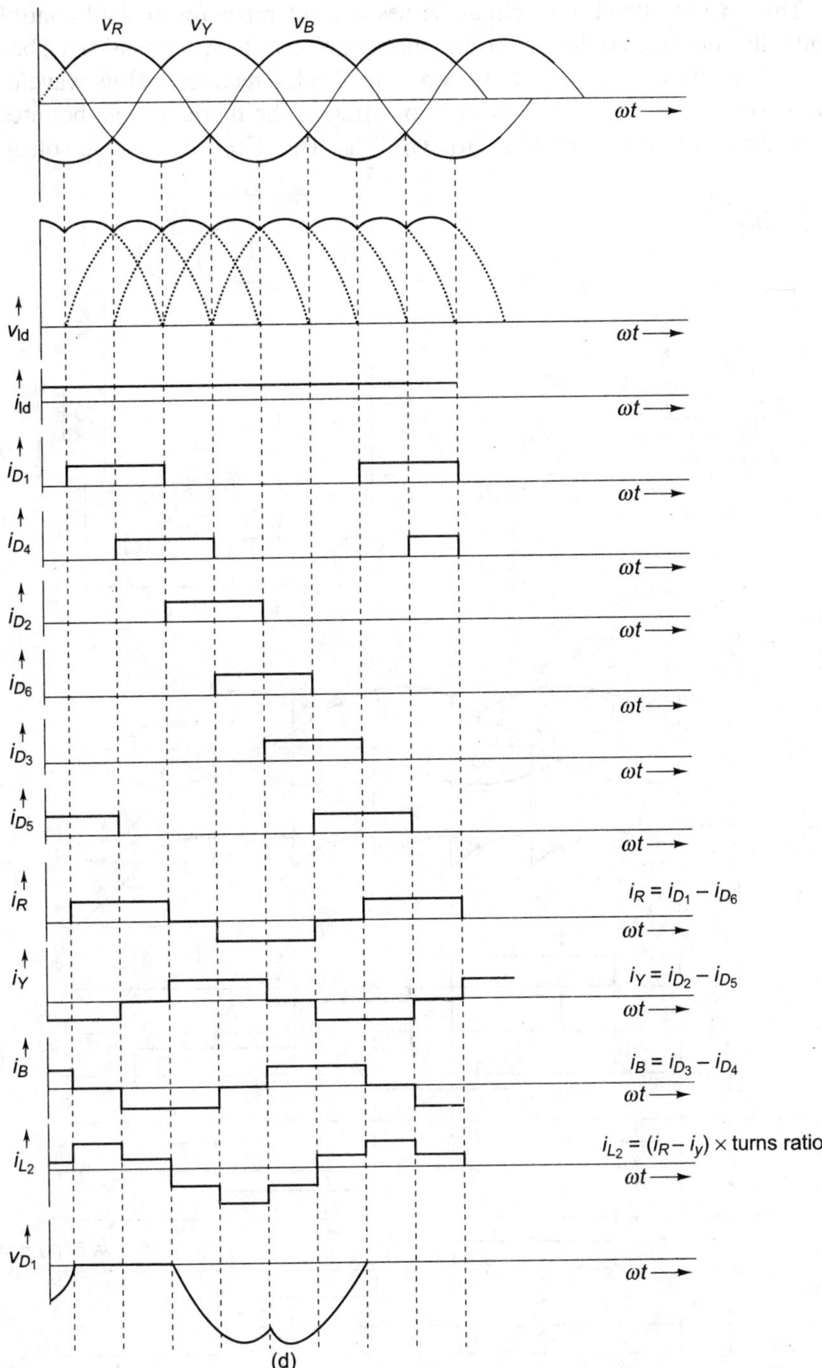

Fig. 2.11 Three-phase, full-wave uncontrolled bridge rectifier: (a) circuit, (b) circuit with positive terminal of load, (c) current with negative terminal of load, (d) waveforms

This uncontrolled bridge circuit serves as a reference for the half-controlled and fully controlled full-wave bridge configurations which are considered below.

A three-phase, full-wave fully controlled bridge rectifier and its waveforms are given in Fig. 2.12. The sequence of firing of the thyristors can be inferred from the waveforms to be Th_1, Th_4, Th_2, Th_6, Th_3, Th_5. The average voltage at

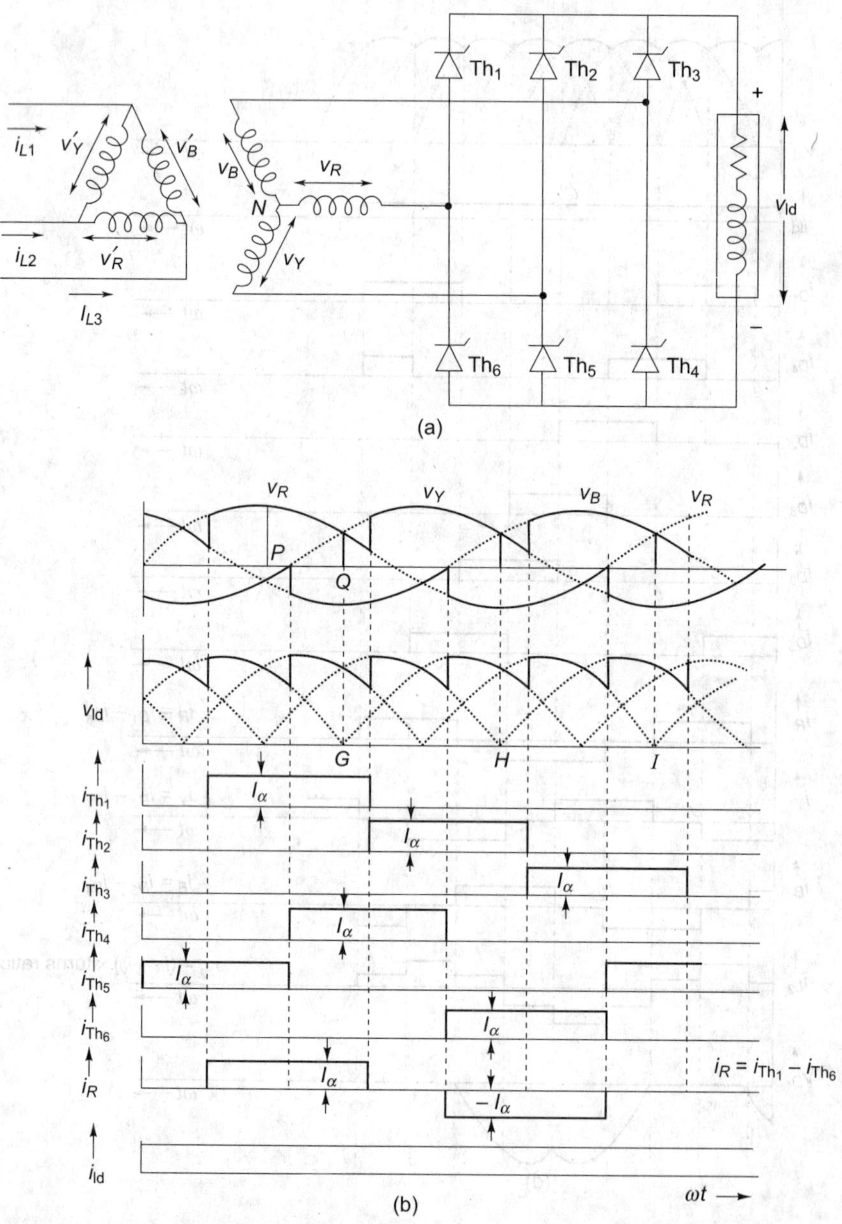

Fig. 2.12 Three-phase, full-wave fully controlled bridge rectifier: (a) circuit, (b) waveforms

the load can be shown to be $(3\sqrt{3}/\pi)V_m \cos\alpha$, with the firing angle α measured from the point G, which is the point of natural commutation for the thyristor Th_1. Likewise, the points H and I are, respectively, the points of natural commutation for Th_2 and Th_3. The peak inverse voltage occurring against any thyristor is $\sqrt{3}V_m$, which is the same as the peak magnitude of the ac line voltage. The ripple frequency remains at $6f$ as in the case of the uncontrolled bridge rectifier, but the ripple content increases with α. The continuity or otherwise of the load current depends on two factors, namely, the firing angle α and the load power factor, or equivalently, the L/R ratio of the load. If the load were a dc motor, the back emf of the motor would also influence this feature. Like the single-phase fully controlled bridge, the three-phase, full-wave fully controlled bridge can be operated in the inversion mode by increasing α to a value above 90°.

A unique requirement of this circuit is that two firing pulses spaced 60° apart have to be provided for each of the thyristors to ensure proper functioning of this circuit. This can be explained as follows. At any moment, two thyristors conduct the load current, say, one conducting the positive half-wave of the R-phase and the other conducting the negative half-wave of the B-phase. If the circuit is switched on at the peak of the R-phase given by the point P, the next firing pulse to be given is that for Th_4 at Q, for the negative half-wave of the B-phase. However, thyristor Th_4 will not conduct unless Th_1 (which conducts the positive R-phase waveform) is on, because at any instant, one device from the positive group and another from the negative group has to conduct the load current. Hence, for starting the operation of the circuit, the firing circuit must generate a second firing pulse at an instant which is 60°after the first pulse for thyristor Th_1; the pulses will then be at P and Q, where ωt, respectively, attains values of 60° and 120° from the point of natural commutation of the R-phase. Likewise, two firing pulses spaced by 60° have to be provided for each of the remaining five thyristors.

The circuit and waveforms of a three-phase, full-wave half-controlled bridge rectifier feeding a highly inductive load are given in Fig. 2.13. This circuit employs three thyristors and three diodes and hence is less costly compared to the fully controlled rectifier considered above. However, the presence of the diodes does

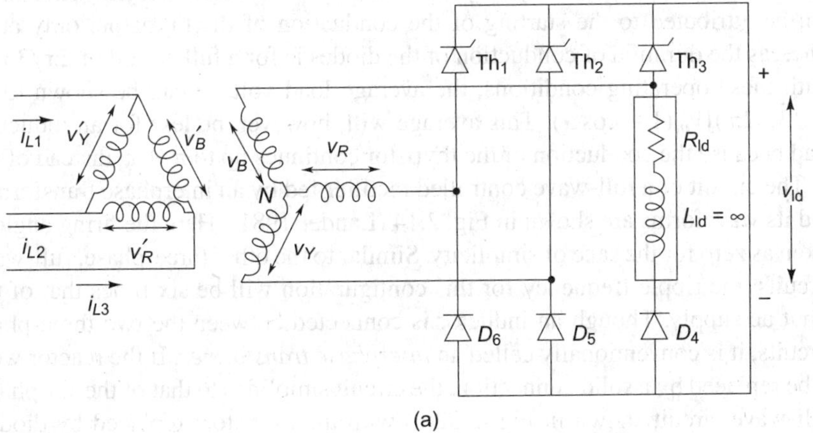

Fig. 2.13(a)

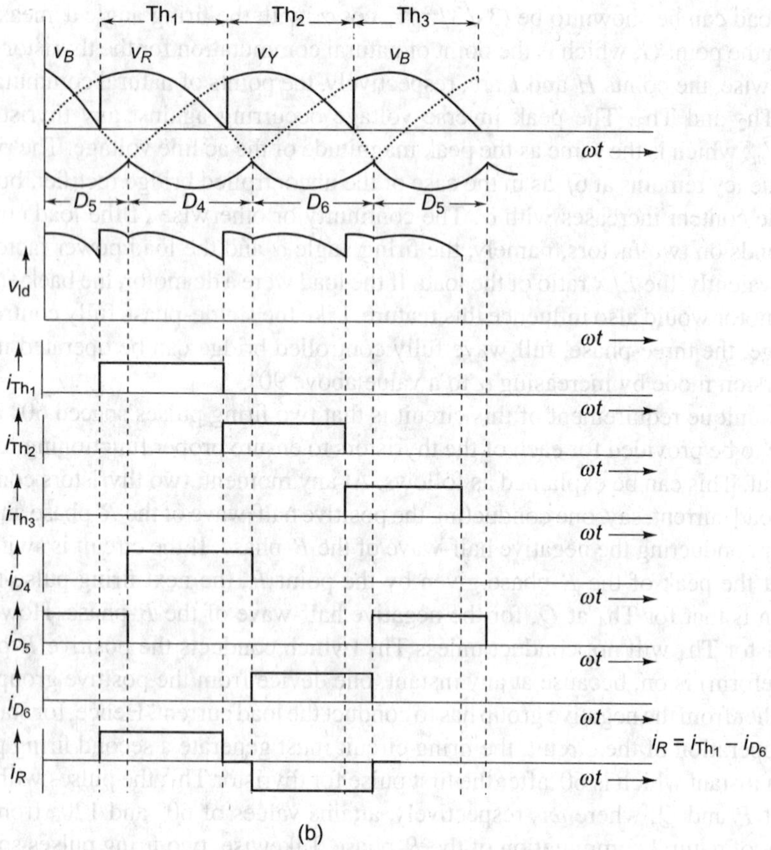

Fig. 2.13 Three-phase, full-wave half-controlled bridge rectifier: (a) circuit, (b) waveforms

not permit the operation of this rectifier in the inversion mode. The ripple content in the output voltage (v_{ld}) of this circuit is higher than that in the uncontrolled circuit of Fig. 2.11. The unsymmetrical nature of the output current waveforms can be attributed to the starting of the conduction of the thyristors only at α, whereas the duration of conduction of the diodes is for a full period of $2\pi/3$ rad. With these operating conditions, the average load voltage can be shown to be $[3\sqrt{3}/(2\pi)]V_m(1 + \cos\alpha)$. This average will, however, be less for an inductive load because the conduction of the thyristor continues up to $\pi + \phi$ instead of π.

The circuit of a full-wave controlled rectifier fed by an interphase transformer and its waveforms are shown in Fig. 2.14 (Lander 1981). Here the firing angle is taken as zero for the sake of simplicity. Similar to the other three-phase, full-wave circuits, the ripple frequency for this configuration will be six times that of the input ac supply. Though an inductor is connected between the two three-phase circuits, it is conventionally called an *interphase transformer*. If the reactor were to be replaced by a solid connection, the circuit simplifies to that of the six-phase, half-wave circuit shown in Fig. 2.32(a) with the thyristors replaced by diodes. The physical dimensions of the reactor can be small because no unidirectional flux is allowed to pass through the core.

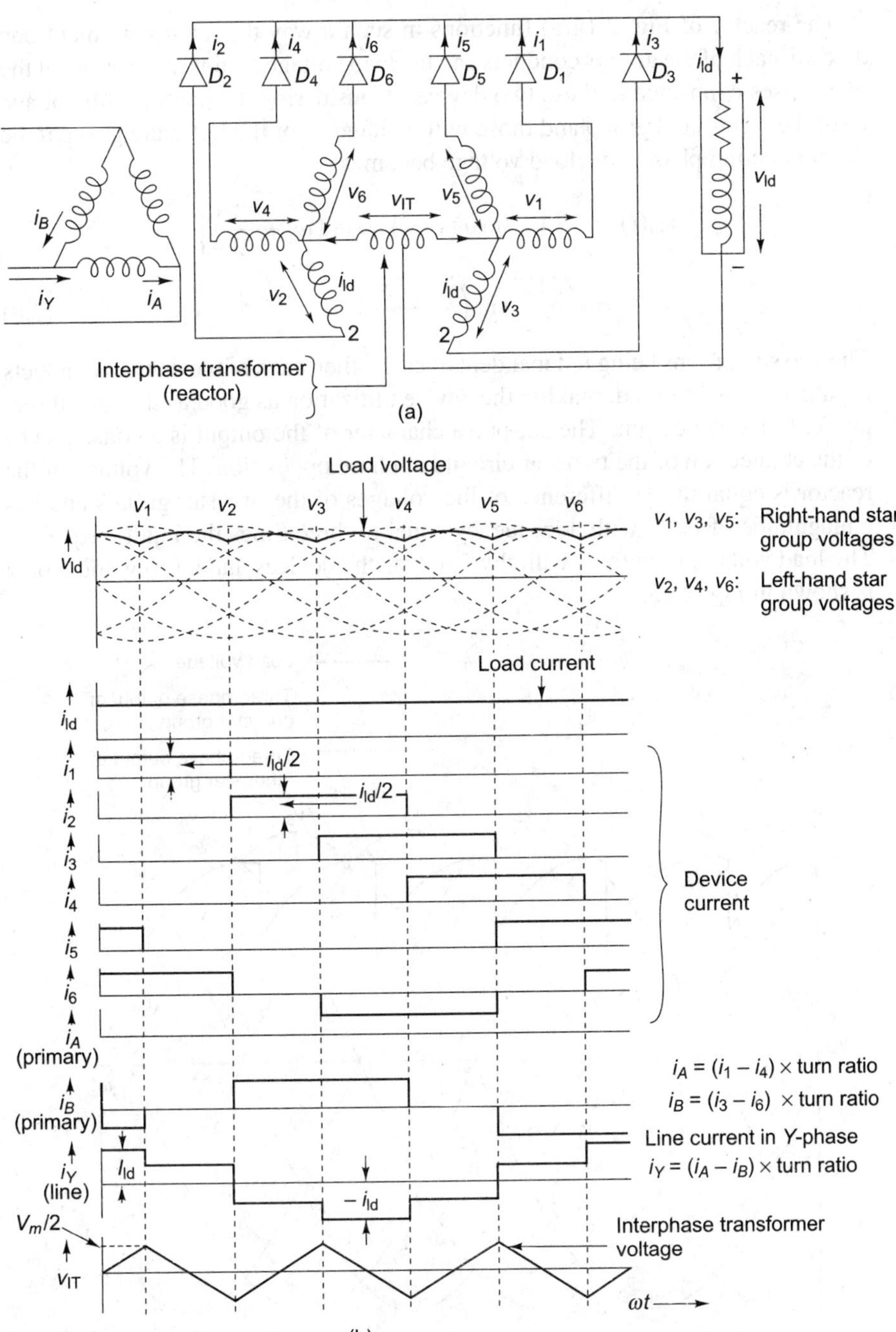

Fig. 2.14 Three-phase six-pulse rectifier with an interphase transformer (IT): (a) circuit, (b) waveforms

The reactor of Fig. 2.14(a) functions in such a way that at any moment one diode in each of the groups conducts and the load voltage attains the average of the two phases connected to these two devices. Considering the phases with voltage v_1 of the right-hand group and those with voltage v_2 of the left-hand group to be the conducting phases, the load voltage becomes

$$v_{\mathrm{ld}}(t) = \frac{1}{2}\left[V_m \sin \omega t + V_m \sin\left(\omega t - \frac{\pi}{3}\right)\right]$$

$$= \frac{\sqrt{3}V_m}{2} \sin\left(\omega t - \frac{\pi}{6}\right) \tag{2.8}$$

The two star groups being independent of each other, each of the devices conducts for a period of $2\pi/3$ rad, making the device utilization as good as that in a three-phase, full-wave circuit. The six-pulse character of the output is a consequence of the connection of the two star circuits in phase opposition. The voltage of the reactor is equal to the difference of the voltages of the two star groups and has a magnitude of $v_m/2$ with a frequency equal to three times the input frequency. The load voltage waveform with thyristors as the devices and a firing angle of α is shown in Fig. 2.15.

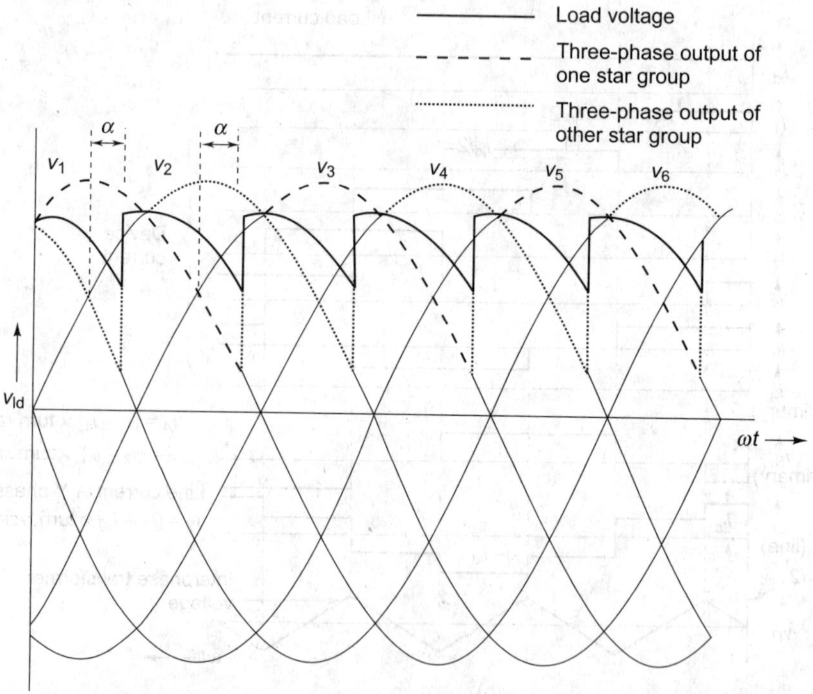

Fig. 2.15 Load waveforms for the circuit of Fig. 2.14(a) with a firing angle α

2.4 Voltage and Current Expressions for an n-phase Controlled Rectifier

The desire to obtain the smoothest possible output voltage waveform led to the construction of multiphase rectifiers, the number of phases usually being 3 or its multiples. Figure 2.32(a) gives the circuit diagram of a six-phase, half-wave rectifier in which the phase difference between any two phases is $2\pi/6$ or $60°$. Twelve-pulse rectifiers are used for high-voltage dc transmission, which is dealt with in Chapter 11. A separate analysis of single-phase, three-phase, and other multiphase rectifiers is not usually done because it proves to be monotonous. A better approach is to derive the expressions for the average load voltage and current for an n-phase rectifier. The results can then be interpreted for a single-phase bridge rectifier by taking $n = 2$, for a three-phase, half-wave rectifier by taking $n = 3$, and so on.

The load voltage and current waveforms depend on the type of load that is connected. The following discussion is centred on the operation of an n-phase rectifier with the following four kinds of loads (Csaki 1975):

(i) a purely resistive load (R),
(ii) a resistive load in series with a battery of voltage E_b, which is taken to be an $R + E_b$ load,
(iii) an inductive load ($R + L$), and
(iv) an inductive load in series with a battery of voltage E_b, which is taken to be an $R + L + E_b$ load.

The last case covers the dc motor which is usually modelled as an $R + L + E_b$ type of load.

2.4.1 n-phase Controlled Rectifier Feeding a Purely Resistive Load

The circuit and waveform for this rectifier are given in Fig. 2.16. The thyristors being considered are fired at an angle α with respect to the point of natural commutation (point A on the ωt-axis).

Had the firing angle been zero, Th_1, Th_2, or Th_3 would have conducted continuously, that is, from A to B, B to C, and C to D, respectively, with a conduction period of $2\pi/n$ rad for each device. If the firing angle of Th_1 had been between A and E, the angle of extinction (or turn-off) would have extended to a value beyond B; the conduction period would remain at $2\pi/n$, with the thyristor Th_2 starting immediately. However, if the firing angle is beyond the point E, that is, with $\alpha > \pi/2 - \pi/n$, the angle of extinction remains at π because with a resistive load, Th_1 ceases conduction at the zero crossing of the voltage. Defining the angle of natural commutation, $\alpha_r = \pi/2 - \pi/n$, it can be concluded that the conduction is continuous when $\alpha < \alpha_r$ and discontinuous when $\alpha > \alpha_r$. The points A, B, C, and D are therefore called the points of natural commutation.

The average voltage at the load can be arrived at as follows, with the origin shifted to F (which is at the peak of the waveform of v_R) for this purpose. The instantaneous value of voltage for the R-phase can be expressed as

$$v(t) = V_m \cos \omega t$$

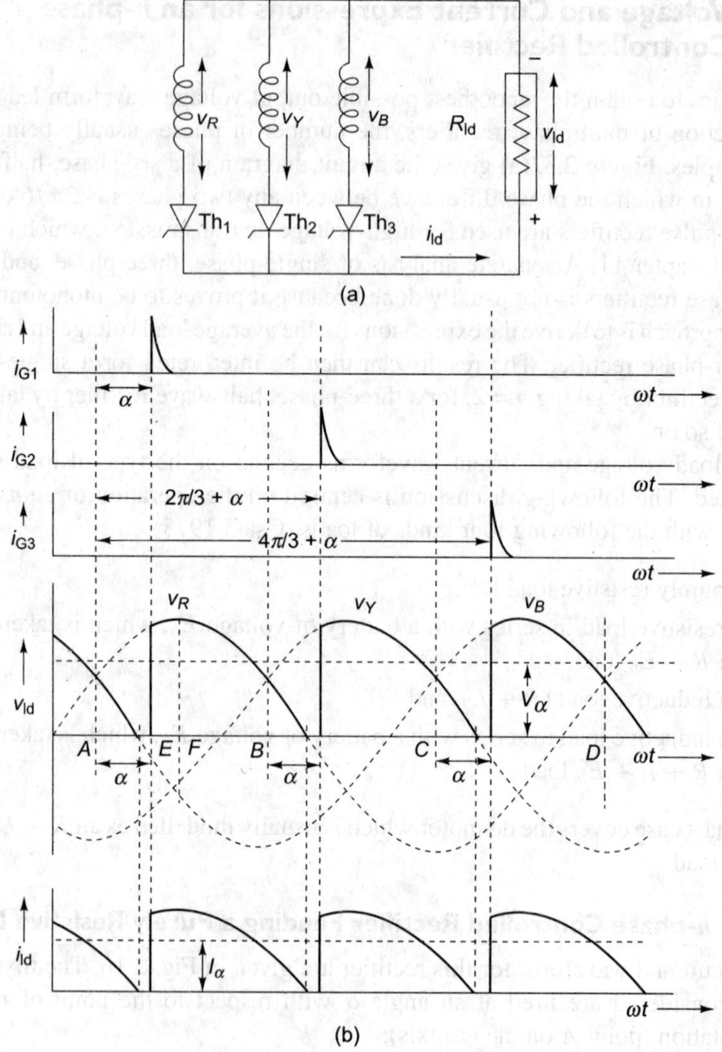

Fig. 2.16 *n*-phase, half-wave controlled rectifier feeding a resistive load: (a) circuit, (b) waveforms

The average voltage V_α is obtained as

$$V_\alpha = \frac{n}{2\pi}\int_{-(\pi/n)+\alpha}^{\pi/2} V_m \cos \omega t \, d(\omega t)$$

$$= \frac{n}{2\pi} V_m \left[1 - \sin\left(\alpha - \frac{\pi}{n}\right)\right] \tag{2.9}$$

and

$$I_\alpha = \frac{V_\alpha}{R_{ld}} = \frac{n}{2\pi}\frac{V_m}{R_{ld}}\left[1 - \sin\left(\alpha - \frac{\pi}{n}\right)\right] \tag{2.10}$$

For continuous conduction, the angle of extinction α_e becomes equal to $2\pi/n + \alpha$ and the average voltage becomes

$$V = \frac{n}{2\pi}\int_{(-\pi/n)+\alpha}^{(-\pi/n)+\alpha_e} V_m \cos\omega t\, d(\omega t)$$

where $\alpha_e = (2\pi/n) + \alpha$. Evaluation of the integral gives

$$V_\alpha = \frac{n}{2\pi}V_m\left[\sin\left(\alpha + \frac{\pi}{n}\right) - \sin\left(\alpha - \frac{\pi}{n}\right)\right]$$

Using the well-known trigonometrical identity, V_α simplifies as

$$V_\alpha = \frac{n}{\pi}V_m \sin\left(\frac{\pi}{n}\right)\cos\alpha \tag{2.11}$$

Defining V_0 as

$$V_0 = \frac{n}{\pi}V_m \sin\left(\frac{\pi}{n}\right)$$

Eqn (2.11) can be written as

$$V_\alpha = V_0 \cos\alpha \tag{2.12}$$

The load current is then given as

$$I_\alpha = \frac{V_0\cos\alpha}{R_{ld}} \tag{2.13}$$

2.4.2 n-phase Rectifier Feeding a Resistive Load in Series with a Battery of Voltage E_b

The circuit as well as waveforms for this rectifier are illustrated in Fig. 2.17. Here, the continuity or otherwise of the current depends upon the magnitude of E_b and the value of the firing angle α. With α_r defined as before and $\alpha_s = \pi/2 - \pi/n - \sin^{-1}(E_b/V_m)$, the following four alternatives have to be considered.

(a) $E_b \le V_m \sin\alpha_r = V_m \cos\pi/n$ ***and*** $\alpha \le \alpha_s$ In this case there is continuous conduction as shown in Fig. 2.17(b), because the value of E_b is below the point on the sinusoidal waveform at which the previous thyristor gets extinguished. The angle α_s is arrived at with the help of Fig. 2.17(f), which is an enlargement of the crossing point of the v_B and v_R waveforms. E_b cuts the v_R and v_B waveforms at P and Q, respectively, their projections on the ωt-axis being R and T, respectively. If α is less than the duration ST, then the current will never become zero because the load voltage waveform v_{ld} is always above E_b. Thus, the limit for continuous conduction is with $\alpha = ST$, where ST can be expressed as

$$ST = SU - TU$$

By symmetry,

$$TU = OR$$

Hence,

$$ST = SU - OR$$

$$= \alpha_r - \sin^{-1}\left(\frac{E_b}{V_m}\right) = \frac{\pi}{2} - \frac{\pi}{n} - \sin^{-1}\left(\frac{E_b}{V_m}\right)$$

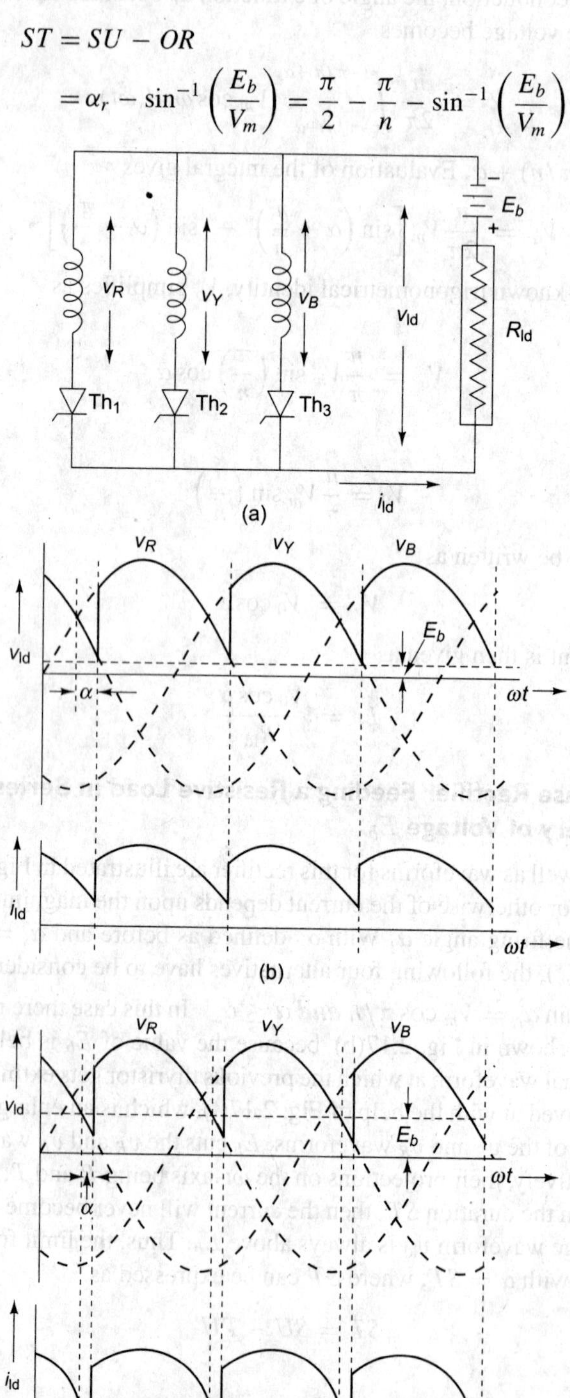

Figs 2.17(a)–(c)

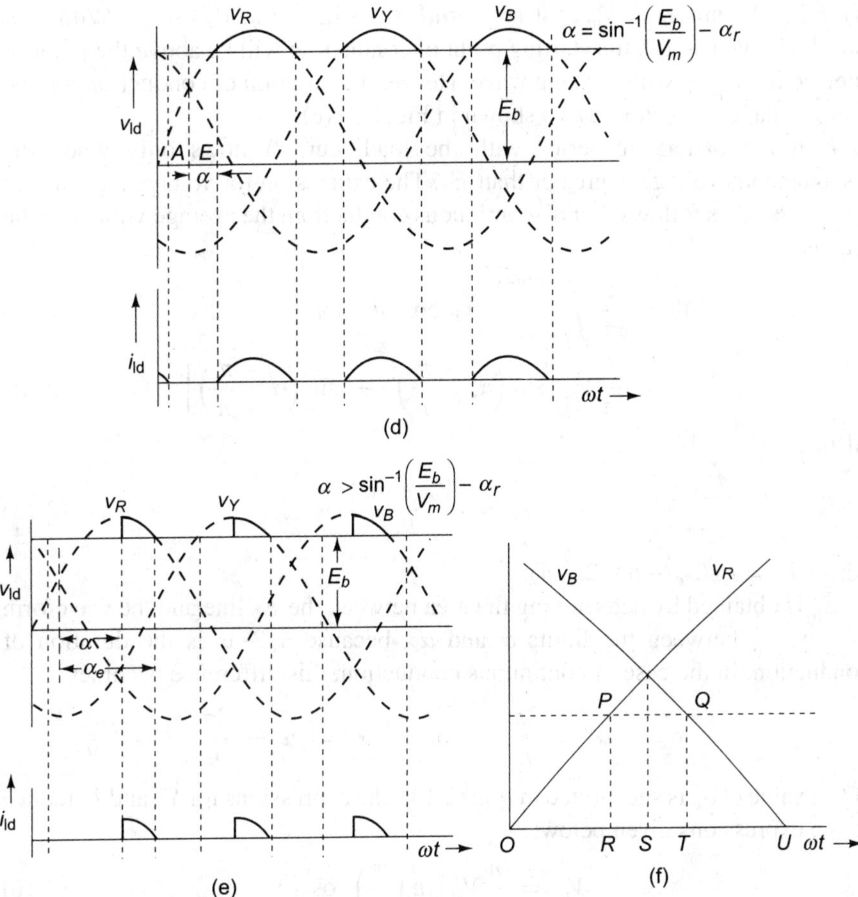

Fig. 2.17 (a) Circuit of an n-phase controlled rectifier feeding a resistive load in series with a battery of voltage E_b. Voltage and current waveforms for (b) $E_b < V_m \sin(\pi/2 - \pi/n)$ and $\alpha \leq \alpha_s$, (c) $E_b < V_m \sin(\pi/2 - \pi/n)$ and $\alpha > \alpha_s$, (d) $E_b > V_m \sin(\pi/2 - \pi/n)$ and $\alpha = \sin^{-1}(E_b/V_m) - \alpha_r$, and (e) $E_b > V_m \sin(\pi/2 - \pi/n)$ and $\alpha > \sin^{-1}(E_b/V_m) - \alpha_r$. (f) Enlarged waveform around the crossing of v_B and v_R.

(b) $E_b \leq V_m \sin \alpha_r = V_m \cos(\pi/n)$ ***and*** $\alpha > \alpha_s$ In this case, conduction continues up to a point on the v_R waveform which is below E_b. The current is therefore discontinuous as shown in Fig. 2.17(c).

(c) $E_b > V_m \sin \alpha_r = V_m \cos(\pi/n)$ ***and*** α ***satisfying the inequality*** $0 < \alpha < \sin^{-1}(E_b/V_m) - \alpha_r$ From Fig. 2.17(d) it is seen that A is the point of natural commutation, implying that had α been equal to zero and had there been no battery in series with the load, Th_1 would have started conduction at A. With a battery in series, its voltage E_b cuts the input voltage waveform above the point A, whose ordinate has the value $V_m \sin \alpha_r$. Hence, conduction will be discontinuous even with $\alpha = \sin^{-1}(E_b/V_m) - \alpha_r$.

(d) $E_b > V_m \sin\alpha_r = V_m \cos(\pi/n)$ ***and*** $\alpha > \sin^{-1}(E_b/V_m) - \alpha_r$ With $\alpha > [\sin^{-1}(E_b/V_m) - \alpha_r]$, the starting point of conduction will be above the point of intersection of E_b with the sine wave. Hence, the duration of conduction will be shorter than that in case (d), as shown in Fig. 2.17(e).

With the battery in series with the load, current flows only when the instantaneous voltage is greater than E_b. The expression for voltage and current are arrived at as follows. For discontinuous conduction, the average voltage at the load is

$$V_\alpha = \frac{n}{2\pi}\int_{-\pi/n+\alpha}^{-\pi/n+\alpha_e} V_m \cos\omega t \, d(\omega t)$$

$$= \frac{n}{2\pi}V_m\left[\sin\left(\alpha_e - \frac{\pi}{n}\right) - \sin\left(\alpha - \frac{\pi}{n}\right)\right] \quad (2.14)$$

Also,

$$I_\alpha = \frac{V_\alpha - E_b^*}{R_{\mathrm{ld}}} \quad (2.15)$$

where $E_b^* = E_b(\alpha_e - \alpha)/(2\pi/n)$.

E_b^* is obtained by determining the area between the E_b line and the waveform of, say, v_R between the limits α and α_e, because $\alpha_e - \alpha$ is the duration of conduction. In the case of continuous conduction, this difference becomes

$$\alpha_e - \alpha = \frac{2\pi}{n} \qquad \text{or} \qquad \alpha_e = \alpha + \frac{2\pi}{n}$$

If this value of α_e is substituted in Eqn (2.14), the expressions for V_α and I_α reduce to the expressions given below:

$$V_\alpha = \frac{n}{\pi}V_m \sin\left(\frac{\pi}{n}\right)\cos\alpha \quad (2.16)$$

and

$$I_\alpha = \frac{V_\alpha - E_b}{R_{\mathrm{ld}}} \quad (2.17)$$

2.4.3 *n*-phase Rectifier Feeding an Inductive Load

Figure 2.18(a) shows an n-phase rectifier feeding a load consisting of an inductor L_{ld} in series with a resistance. As stated before, conduction may be discontinuous or otherwise, respectively, depending upon the L/R ratio being low or high.

Figures 2.18(b) and (c) show the waveforms, respectively, for discontinuous and continuous modes of operation. In the former case, the current becomes zero between the instant at which the device connected to voltage v_B stops conduction and the moment at which the device connected to voltage v_R starts conduction. In the continuous mode, the current waveform does not touch zero but is in the form of a ripple which momentarily dips at the instant of transition. The expressions for average voltage and current with discontinuous conduction are given in Eqns (2.14) and (2.15), respectively.

An alternative approach which facilitates the determination of α_e is to write the differential equation for the circuit and then solve for the current. This is

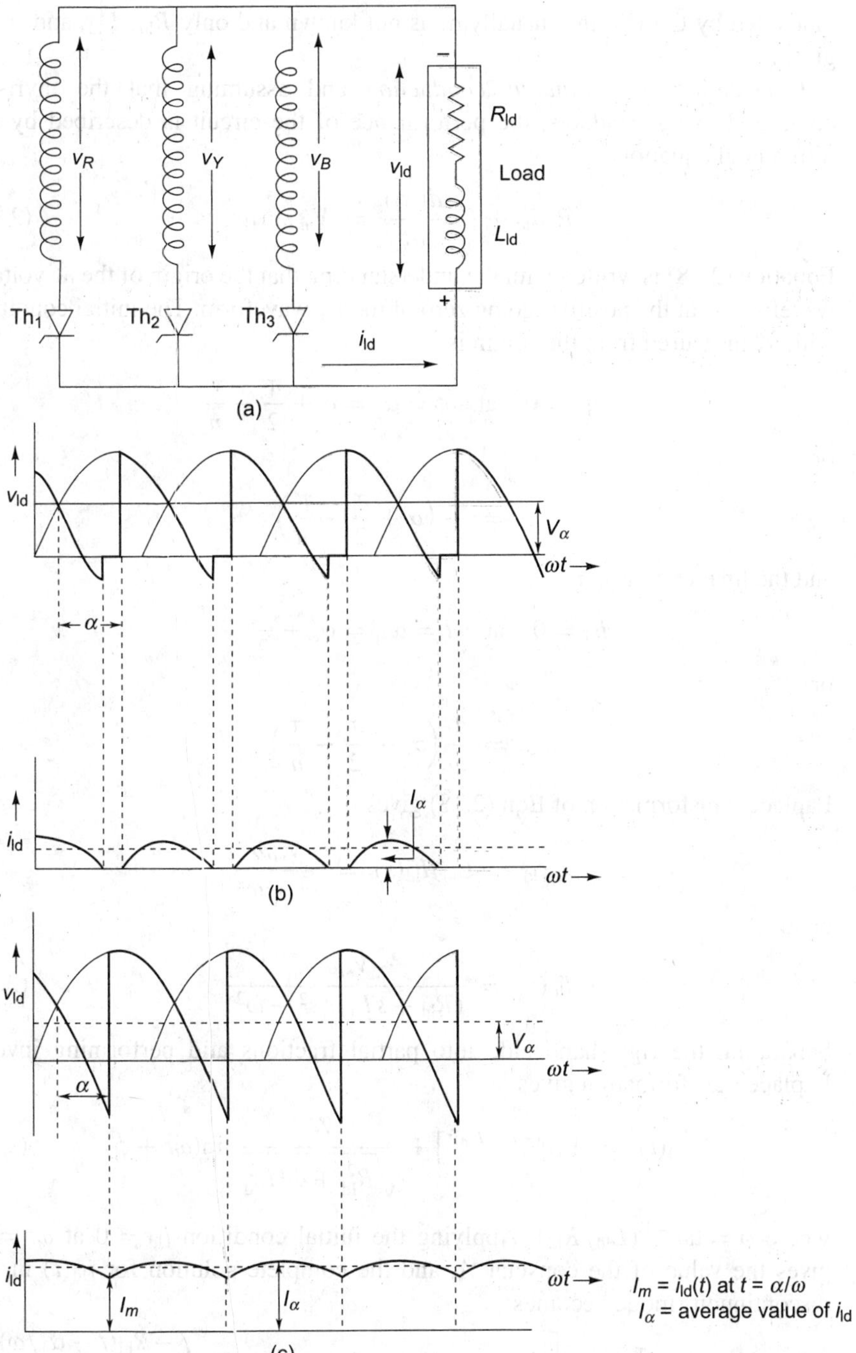

Fig. 2.18 n-phase, half-wave controlled rectifier with an inductive load: (a) circuit, (b) waveforms for discontinuous conduction, (c) waveforms for continuous conduction

motivated by the fact that usually α_e is not known and only R_{ld}, L_{ld}, and α are given.

Considering *discontinuous conduction* and assuming that the thyristor connected to v_R conducts, the performance of the circuit is described by the differential equation

$$R_{\text{ld}} i_{\text{ld}} + L_{\text{ld}} \frac{d(i_{\text{ld}})}{dt} = V_m \sin \omega t \tag{2.18}$$

Equation (2.18) is written with the understanding that the origin of the ac voltage waveform is at the positive-going zero of the v_R waveform. The initial condition with α_1' measured from the origin is

$$i_{\text{ld}} = 0 \quad \text{at } \omega t = \alpha_1' = \alpha + \frac{\pi}{2} - \frac{\pi}{n}$$

or

$$t = \frac{1}{\omega}\left(\alpha + \frac{\pi}{2} - \frac{\pi}{n}\right)$$

and the final condition is

$$i_{\text{ld}} = 0 \quad \text{at } \omega t = \alpha_{e1}' = \alpha_e + \frac{\pi}{2} - \frac{\pi}{n}$$

or

$$t = \frac{1}{\omega}\left(\alpha_e + \frac{\pi}{2} - \frac{\pi}{n}\right)$$

Laplace transformation of Eqn (2.18) gives

$$(R_{\text{ld}} + sL_{\text{ld}}) I_{\text{ld}}(s) = \frac{V_m \omega}{s^2 + \omega^2}$$

or

$$I_{\text{ld}}(s) = \frac{V_m \omega}{(R_{\text{ld}} + sL_{\text{ld}})(s^2 + \omega^2)} \tag{2.19}$$

Separating the right-hand side into partial fractions and performing inverse Laplace transformation gives

$$i_{\text{ld}}(t) = A\left[e^{(-R_{\text{ld}}/L_{\text{ld}})t}\right] + \frac{V_m}{\sqrt{R_{\text{ld}}^2 + \omega^2 L_{\text{ld}}^2}} \sin(\omega t + \phi) \tag{2.20}$$

where $\phi = \tan^{-1}(L_{\text{ld}}/R_{\text{ld}})$. Applying the initial condition $i_{\text{ld}} = 0$ at $\omega t = \alpha_1$ gives the value of the constant A, and the complete solution for $i_{\text{ld}}(t)$ in this discontinuous mode becomes

$$i_{\text{ld}}(t) = \frac{V_m}{\sqrt{R_{\text{ld}}^2 + \omega^2 L_{\text{ld}}^2}} \left[\sin(\omega t - \phi) - \sin(\alpha_1 - \phi) \exp\left(\frac{-R_{\text{ld}}(t - \alpha_1/\omega)}{L_{\text{ld}}}\right)\right] \tag{2.21}$$

which can be split up as

$$i_{\text{ld}}(t) = i_1(t) + i_2(t)$$

where $i_1(t)$ and $i_2(t)$ are, respectively, the steady-state and transient solutions. Knowing the load power factor $\cos\phi$ and the firing angle α_1, the instantaneous value of $i_{ld}(t)$ may be determined from Eqn (2.21). The average value of the load current is now obtained from the steady-state part as

$$I_\alpha = \frac{n\omega}{2\pi}\int_{\alpha/\omega}^{\alpha_{e1}/\omega} i_1(t)dt \tag{2.22}$$

and the load voltage V_α can be written as

$$V_\alpha = I_\alpha R_{ld} \tag{2.23}$$

The above method of arriving at I_α and V_α requires the solution of the differential equation and is hence not convenient. If a method can be evolved to determine α_{e1}, then V_α may be directly determined from Eqn (2.14) or (2.16) depending on the conduction being discontinuous or continuous. The final condition can be used for this purpose. Substitution of the condition

$$i_{ld} = 0 \qquad \text{at } \omega t = \alpha_{e1}$$

or

$$t = \alpha_{e1}/\omega$$

in Eqn (2.21) results in the equation

$$e^{\alpha_{e1}/\tan\phi}\sin(\alpha_{e1} - \phi) - e^{\alpha_1/\tan\phi}\sin(\alpha - \phi) = 0 \tag{2.24}$$

Equation (2.24), which is of a transcendental nature, can be solved explicitly for α_{e1}. A much simpler method is to use the chart given in Fig. 2.19 to determine α_{e1}, with the knowledge of the firing angle α_1 and the load power factor $\cos\phi$. The 45° lines in Fig. 2.19 are the demarcating or border case lines between continuous and discontinuous conduction for various values of n. Given α_1 and $\cos\phi = R_{ld}/(R_{ld}^2 + L_{ld}^2)^{1/2}$, the value of α_{e1} may be read from the chart. Depending upon whether the point lies above or below the border case lines, the mode of conduction is also ascertained and the average values of the load voltage and load current computed using the appropriate expression.

A special case of Eqn (2.24) is that obtained by taking $\alpha_{e1} = \alpha + 2\pi/n$, which is the borderline case of $\alpha_1 = \alpha_{CD}$. It can be obtained as

$$\alpha_{CD} = \tan^{-1}\left(\frac{\sin(2\pi/n - \phi) + e^{-2\pi/n\,\tan\phi}\sin\phi}{e^{-2\pi/n\,\tan\phi}\cos\phi - \cos(2\pi/n - \phi)}\right) \tag{2.25}$$

In the *continuous conduction mode*, the value of the current I_m of Fig. 2.16(c) can be determined as follows. Substituting the value of I_m for $i_{ld}(t)$ at $\omega t = \alpha_1$ or $t = \alpha_1/\omega$ in Eqn (2.20) gives the value of the constant A. On substituting for A in Eqn (2.20), i_{ld} is obtained as

$$i_{ld}(t) = \left[I_m - \frac{V_m}{\sqrt{R_{ld}^2 + \omega^2 L_{ld}^2}}\sin(\alpha_1 - \phi)\right]\exp\left[\frac{R_{ld}}{L_{ld}}\left(\frac{\alpha_1}{\omega} - t\right)\right]$$
$$+ \frac{V_m}{\sqrt{R_{ld}^2 + \omega^2 L_{ld}^2}}\sin(\omega t + \phi) \tag{2.26}$$

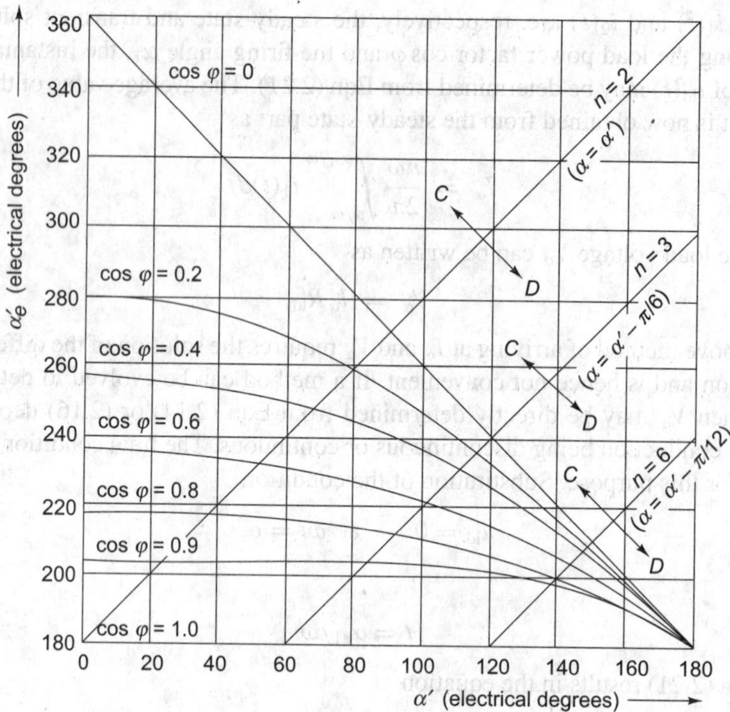

Fig. 2.19 Chart for determining α_e' for the rectifier of Fig. 2.18(a)

Again by substitution of the final condition $i_{\rm ld}(t) = I_m$ at $\omega t = \alpha_{e1} = \alpha_1 + 2\pi/n$, I_m is obtained as

$$I_m = \frac{V_m}{R_{\rm ld}}\cos\phi\left[\frac{\sin(\alpha_1 + 2\pi/n - \phi) - \sin(\alpha_1 - \phi)e^{-2\pi/n\tan\phi}}{1 - e^{-2\pi/n\tan\phi}}\right] \tag{2.27}$$

Now the average load current I_α may be determined from Eqn (2.22) by taking the expression given for $i_{\rm ld}$ in either Eqn (2.21) or Eqn (2.26), respectively, for discontinuous or continuous modes, with the expression for I_m as in Eqn (2.27). Thus,

$$I_\alpha = \frac{n}{2\pi}\int_{\alpha_1}^{\alpha_{e1}} i_{\rm ld}\,d(\omega t) \tag{2.28}$$

where $\alpha_{e1} = \alpha_1 + 2\pi/n$ for continuous conduction. Also,

$$V_\alpha = E_b^* + I_\alpha R_{\rm ld} \tag{2.29}$$

where E_b^* is equal to $[(\alpha_{e1} - \alpha_1)/(2\pi/n)]E_b$ for discontinuous conduction. For continuous conduction, $(\alpha_{e1} - \alpha_1)$ becomes equal to $2\pi/n$ and E_b^* equals E_b.

2.4.4 *n*-phase Rectifier Feeding an Inductive Load in Series with a Battery of Voltage E_b

The circuit as well as waveforms for this case are given in Fig. 2.20. Similar to the case of the inductive load, the expression for the average current can be obtained in two ways. The sequence of computation is to first obtain the average voltage and then calculate the average current. For discontinuous conduction, the expressions for V_α and I_α can be obtained as

$$V_\alpha = \left(\frac{n}{2\pi}\right) V_m \left[\sin\left(\alpha_e - \frac{\pi}{n}\right) - \sin\left(\alpha - \frac{\pi}{n}\right)\right] + E_b\left(1 - \frac{\alpha_e - \alpha}{2\pi/n}\right) \quad (2.30)$$

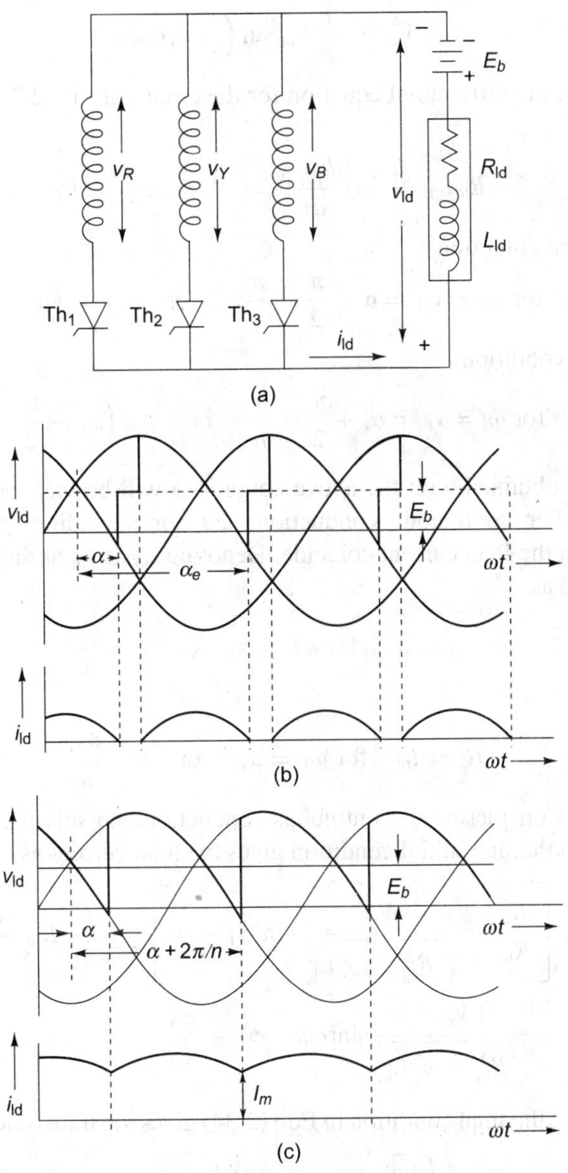

Fig. 2.20 n-phase, half-wave controlled rectifier feeding an $R + L + E_b$ load: (a) circuit, (b) waveforms for discontinuous conduction, (c) waveforms for continuous conduction

and

$$I_\alpha = \frac{V_\alpha - E_b}{R_{\mathrm{ld}}} \tag{2.31}$$

where α and α_e are assumed to be given as data. Substitution of the relation $\alpha_e = \alpha + 2\pi/n$ in the above expression gives the average voltage for continuous conduction as

$$V_\alpha = \frac{n}{\pi} V_m \sin\left(\frac{\pi}{n}\right) \cos\alpha \tag{2.32}$$

Alternatively, the differential equation for the circuit of Fig. 2.20 can be written as

$$R_{\mathrm{ld}} i_{\mathrm{ld}} + L_{\mathrm{ld}} \frac{d(i_{\mathrm{ld}})}{dt} = V_m \sin\omega t - E_b \tag{2.33}$$

with the initial condition

$$i_{\mathrm{ld}}(t) = 0 \quad \text{for } \omega t = \alpha_1 = \alpha + \frac{\pi}{2} - \frac{\pi}{n} \qquad \text{or} \qquad t = \left(\alpha + \frac{\pi}{2} - \frac{\pi}{n}\right)/\omega$$

and the final condition

$$i_{\mathrm{ld}}(t) = 0 \quad \text{for } \omega t = \alpha_{e1} = \alpha_e + \frac{\pi}{2} - \frac{\pi}{n} \qquad \text{or} \qquad t = \left(\alpha_e + \frac{\pi}{2} - \frac{\pi}{n}\right)/\omega$$

The solution obtained with the above conditions will be that for discontinuous conduction. For continuous conduction, the corresponding initial and final conditions on the load current coincide. Denoting i_{ld} as I_m at these points, these are expressed as

$$i_{\mathrm{ld}} = I_m \quad \text{for } \omega t = \alpha_1 \quad \text{or} \quad t = \frac{\alpha_1}{\omega}$$

and

$$i_{\mathrm{ld}} = I_m \quad \text{for } \omega t = \alpha_{e1} \quad \text{or} \quad t = \frac{\alpha_{e1}}{\omega}$$

respectively. Considering discontinuous conduction and solving the differential equation with the first initial condition gives the load current as

$$i_{\mathrm{ld}}(t) = \left[\frac{E_b}{R_{\mathrm{ld}}} - \frac{V_m}{\sqrt{R_{\mathrm{ld}}^2 + \omega^2 L_{\mathrm{ld}}^2}} \sin(\alpha_1 - \phi)\right] \exp\left(-R_{\mathrm{ld}} \frac{\omega t - \alpha_1}{\omega L_{\mathrm{ld}}}\right) + \frac{V_m}{\sqrt{R_{\mathrm{ld}}^2 + \omega^2 L_{\mathrm{ld}}^2}} \sin(\omega t - \phi) - \frac{E_b}{R_{\mathrm{ld}}} \tag{2.34}$$

Substitution of the final condition in Eqn (2.34) gives the transcendental equation

$$\left[\cos\phi \sin(\alpha_{e1} - \phi) - \frac{E_b}{V_m}\right] e^{\alpha_{e1}/\tan\phi} = \left[\frac{E_b}{V_m} - \cos\phi \sin(\alpha_1 - \phi)\right] e^{\alpha_1/\tan\phi} \tag{2.35}$$

The chart in Fig. 2.21 represents this equation. It gives α_{e1} versus α_1 with the parameters $a = E_b/V_m$ and $\cos\phi = R_{\mathrm{ld}}/(R_{\mathrm{ld}}^2 + \omega^2 L_{\mathrm{ld}}^2)^{1/2}$. With $a = 0$, it gives the

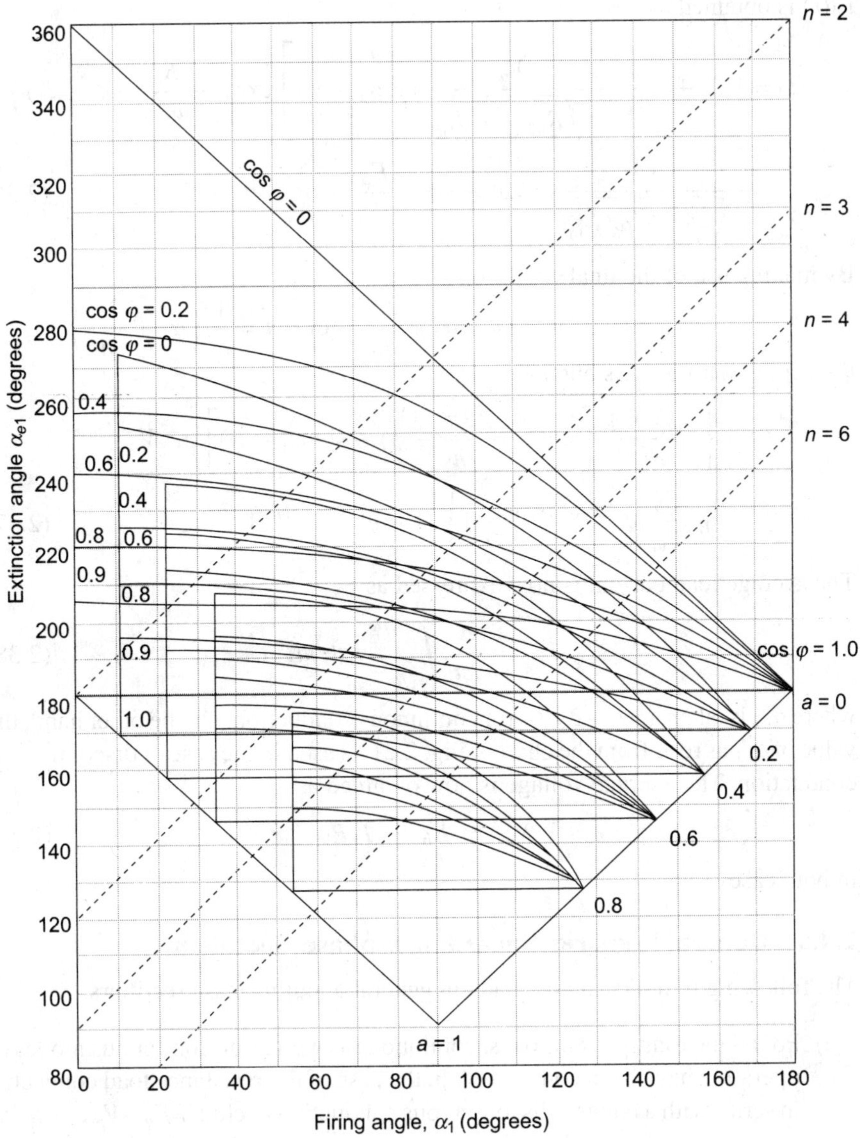

Fig. 2.21 Chart for determining α_{e1} for the rectifier of Fig. 2.20(a)

same information as that contained in the chart of Fig. 2.19. As before, I_m for the continuous current i_{ld} of Fig. 2.20(b) can be obtained with the help of Eqn (2.34) as follows. Using the initial conditions

$$i_{ld} = 0 \quad \text{for } \omega t = \alpha_1 \quad \text{or} \quad t = \frac{\alpha_1}{\omega}$$

$i_{\mathrm{ld}}(t)$ is obtained as

$$i_{\mathrm{ld}}(t) = \left[I_m + \frac{E_b}{R_{\mathrm{ld}}} - \frac{V_m}{\sqrt{R_{\mathrm{ld}}^2 + \omega^2 L_{\mathrm{ld}}^2}} \sin(\alpha_1 - \phi) \right] \exp\left(-\frac{R_{\mathrm{ld}}}{\omega L_{\mathrm{ld}}} (\omega t - \alpha_1) \right) + \frac{V_m}{\sqrt{R_{\mathrm{ld}}^2 + \omega^2 L_{\mathrm{ld}}^2}} \sin(\omega t - \phi) - \frac{E_b}{R_{\mathrm{ld}}} \tag{2.36}$$

By making use of the final condition

$$i_{\mathrm{ld}} = 0 \quad \text{for} \quad \omega t = \alpha_1 + \frac{2\pi}{n} \quad \text{or} \quad t = \frac{1}{\omega}\left(\alpha_1 + \frac{2\pi}{n}\right)$$

the expression for I_m is obtained as

$$I_m = \left\{ \frac{V_m \cos\phi}{R_{\mathrm{ld}}} \left[\sin(\alpha_1 + \frac{2\pi}{n} - \phi) - \sin(\alpha_1 - \phi) \right] e^{-2\pi/(n \tan\phi)} + \frac{E_b}{R_{\mathrm{ld}}} [e^{-2\pi/(n \tan\phi)} - 1 \right\} / 1 - e^{-2\pi/(n \tan\phi)} \tag{2.37}$$

The average load current is now expressed as

$$I_\alpha = \frac{n}{2\pi} \int_{\alpha_1/\omega}^{\alpha_{e1}/\omega} i_{\mathrm{ld}}(t) dt \tag{2.38}$$

where α_{e1} is taken as $\alpha_1 + 2\pi/n$ for continuous conduction. On the other hand, the value of α_{e1} as read from the chart of Fig. 2.21, is used in the case of discontinuous conduction. The average voltage is now obtained as

$$V_\alpha = E_b + I_\alpha R_{\mathrm{ld}} \tag{2.39}$$

in both cases.

2.4.5 General Remarks Regarding n-phase Rectifiers

The following remarks can be made in general about n-phase rectifiers.

(a) For all the four types of loads, continuous as well as discontinuous modes of operation have been considered. In the case of the resistance load connected in series with a battery, discontinuous conduction occurs if $E_b > V_m \cos(\pi/n)$ irrespective of the value of α. For the case of $\alpha = 0$, the angle β at which conduction starts and the angle δ at which the device stops conduction, can be obtained from Fig. 2.22 as

$$\beta = \sin^{-1}\left(\frac{E_b}{V_m}\right) - \left(\frac{\pi}{2} - \frac{\pi}{n}\right) \tag{2.40}$$

and

$$\delta = \left[\pi - \sin^{-1}\left(\frac{E_b}{V_m}\right)\right] - \left(\frac{\pi}{2} - \frac{\pi}{n}\right)$$

Simplification gives δ as

$$\delta = \left(\frac{\pi}{2} + \frac{\pi}{n}\right) - \sin^{-1}\left(\frac{E_b}{V_m}\right) \tag{2.41}$$

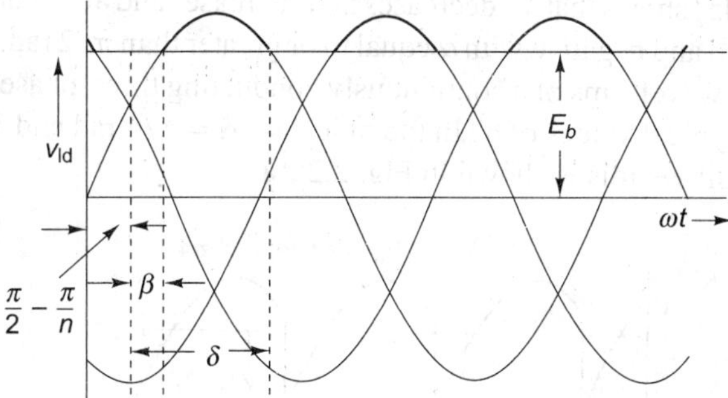

Fig. 2.22 Determination of β and δ for an n-phase, half-wave controlled rectifier when $E_b > V_m \cos(\pi/n)$

(b) For the case of the inductive load and that of the inductive load in series with a back emf, the angles α_1 and α_{e1} are measured with respect to the origin of the sinusoid of the ac supply. These are related to α and α_{e1} as

$$\alpha_1 = \alpha + \frac{\pi}{2} - \frac{\pi}{n} \tag{2.42}$$

and

$$\alpha_{e1} = \alpha_e + \frac{\pi}{2} - \frac{\pi}{n} \tag{2.43}$$

For a single-phase bridge rectifier where $n = 2$, α_1 and α_{e1} in Eqns (2.42) and (2.43) become equal to α and α_e, respectively. If R_{ld} and L_{ld} are known, $\cos\phi$ may be obtained as

$$\cos\phi = \frac{R_{\text{ld}}}{\sqrt{R_{\text{ld}}^2 + \omega^2 L_{\text{ld}}^2}} \tag{2.44}$$

Given α_1, a, and $\cos\phi$, the value of α_{e1} may be read from the chart.

(c) The appropriate value of n has to be substituted in the expressions for the voltage and current. Thus, in the case of a three-phase, half-wave rectifier, n is taken as 3.

(d) The expressions derived above cannot be used for a full-wave, n-phase rectifier. However, in this case the average voltage is twice that of the corresponding half-wave rectifier.

2.5 Inverting Mode of a Converter

It is stated before that fully controlled single- and three-phase rectifiers operate in the inverting mode when the firing angle α is kept at a value greater than 90° (electrical) or $\pi/2$ rad. This aspect will be examined here in greater detail. From Eqn (2.16), the average output voltage of a continuously conducting n-phase converter is seen to be a cosine function of α and is given as

$$V_\alpha = V_0 \cos\alpha \tag{2.45}$$

where $V_0 = (n/\pi)V_m \sin(\pi/n)$.

Equation (2.45) shows that V_α decreases as α increases and also that it becomes, respectively, zero and negative with α equal to or greater than $\pi/2$ rad. In Fig. 2.23 the load voltage waveforms of a continuously conducting three-phase rectifier are given for three typical values of α. In the first case, $\alpha = \pi/4$ rad and $V_\alpha = V_0/\sqrt{2}$ and is hence positive; this is shown in Fig. 2.23(a).

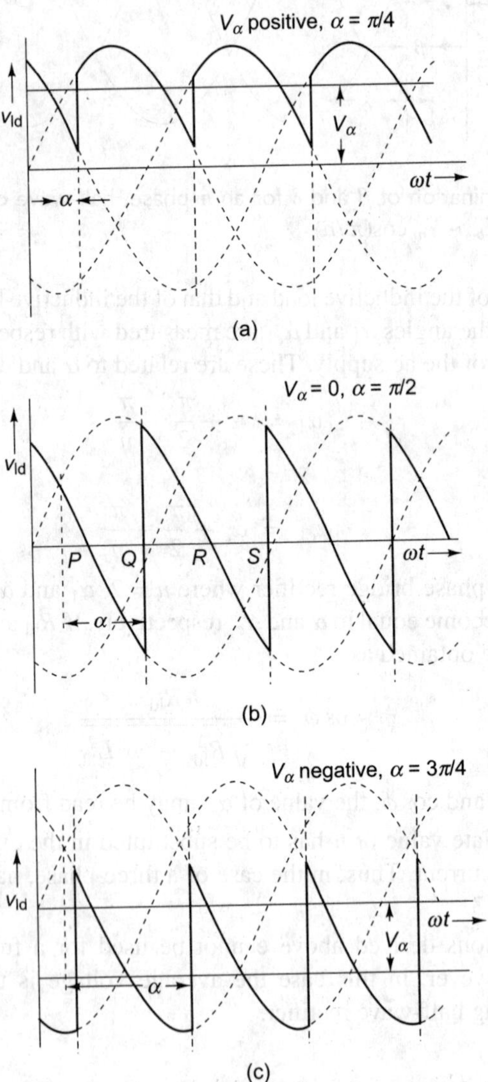

Fig. 2.23 Load voltage waveforms for a three-phase, half-wave controlled rectifier with (a) $\alpha = \pi/4$, (b) $\alpha = \pi/2$, and (c) $\alpha = 3\pi/4$

It can be seen from Fig. 2.23(b) that with $\alpha = \pi/2$, the areas under the positive and negative parts of the waveform become equal. Equation (2.45) shows that this is true for any n-phase rectifier. Considering the R-phase waveform, the firing angle α is measured from the point P, the angular displacement from P to Q

being $\pi/2$. Hence the conduction period on the positive side is given as

$$QR = \pi - \left(\frac{\pi}{2} - \frac{\pi}{n} + \frac{\pi}{2}\right) = \frac{\pi}{n}$$

As the continuously conducting rectifier has a total conduction period of $2\pi/n$, *RS* is also equal to π/n, *R* being the zero crossing point. This confirms that the average voltage is zero.

In the last case [Fig. 2.23(c)], $\alpha = 3\pi/4$ and with this value of α the average load voltage V_α becomes negative. The thyristor conducts in only one direction; hence the current through the load continues to be positive, showing that the power delivered to the load is negative. This can be interpreted as power flowing from the load back to the supply and this mode of operation is termed as the *inverting mode*.

The conditions favourable for the inverting mode of operation can now be summarized with the help of Figs 2.24(a) and (b), in which a battery in series with a resistance constitutes the load. If the firing angle of the thyristor is less than 90° as shown in Fig. 2.24(a), a positive voltage will be applied at the terminals AA', enabling the converter to operate in the rectifying mode. In Fig. 2.24(b) the converter is assumed to be operating in the inverting mode, *the terminals of the battery now being reversed.*

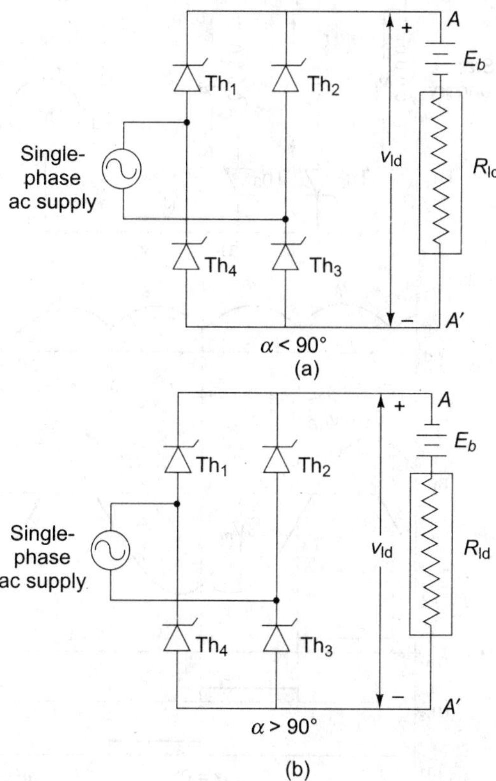

Fig. 2.24 Single-phase fully controlled bridge rectifier: (a) rectifying mode, (b) inverting mode

If $|E_b| > |V_\alpha|$, then current can flow from the battery to the source as shown. This suggests that if (a) the load is a separately excited dc motor, (b) the armature terminals are reversed, and (c) the inequality $|E_b| > |V_\alpha|$ is ensured, then the motor converts its kinetic energy to electrical energy and returns it back to the source through the rectifier. As a consequence, the motor comes to a stop after a short time. This mode of operation is called the *regenerative braking* mode and is dealt with in detail in Section 2.12.

2.5.1 Extinction Angle and its Significance

The significance of an angle called the *extinction angle*, which is defined in connection with the inverting mode of a converter, is explained below. Figure 2.25(a) shows a three-phase, half-wave controlled rectifier which is assumed to be operating in the inverting mode. The load is assumed to be highly inductive so as to make the load current continuous and ripple-free.

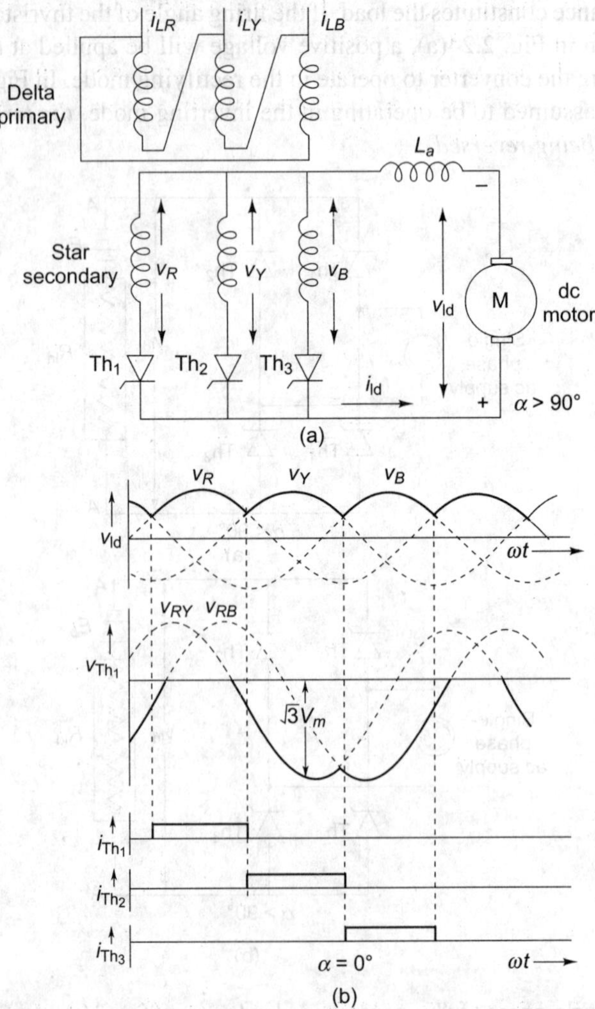

Figs 2.25(a), (b)

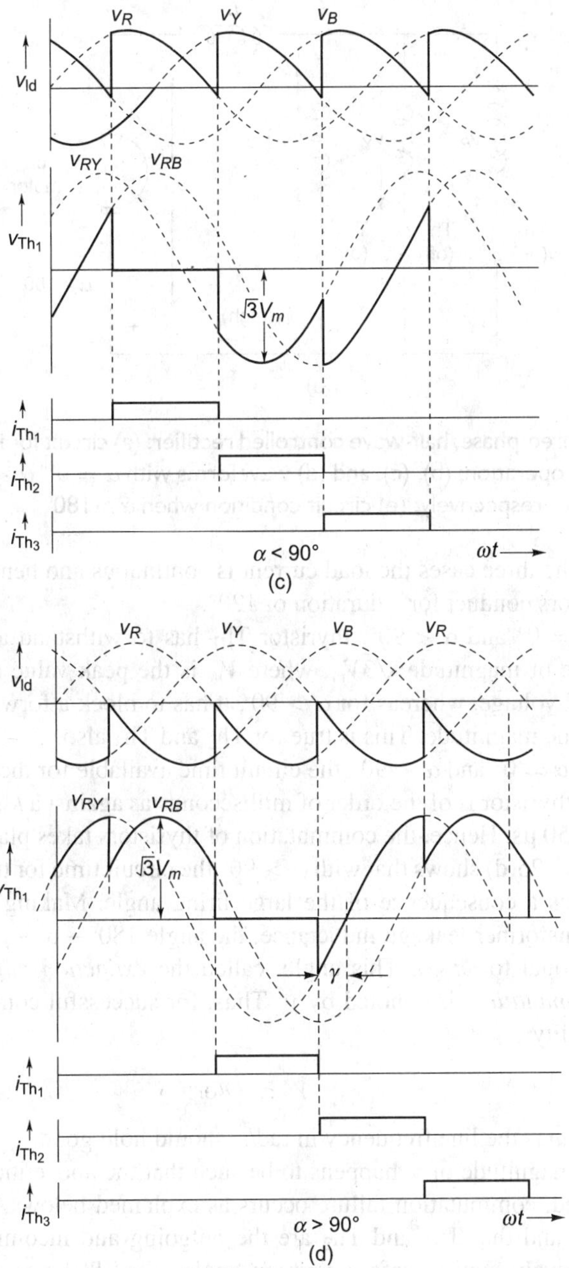

Figs 2.25(c), (d)

Figure 2.25(d) gives the waveforms of load voltage, supply voltage, load current, and the thyristor currents for this mode of operation ($\alpha > 90°$). For the sake of comparison, the same set of waveforms are drawn with $\alpha = 0°$ and $\alpha < 90°$ in Figs 2.25(b) and (c), respectively. The following remarks can be made after a comparison of Figs 2.25(b), (c), and (d).

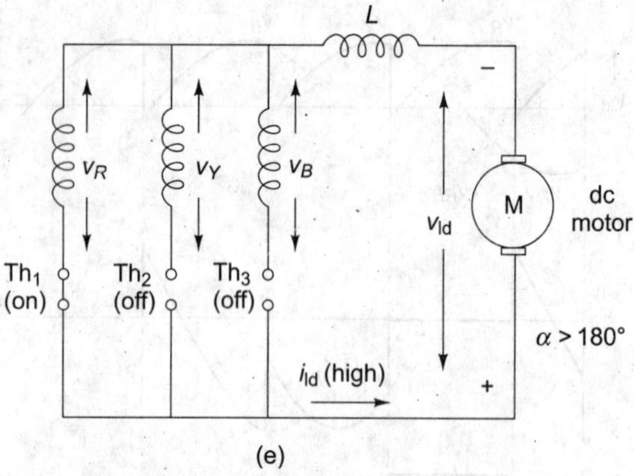

Fig. 2.25 Three-phase, half-wave controlled rectifier: (a) circuit for inverting mode of operation; (b), (c), and (d) waveforms with $\alpha = 0°$, $\alpha < 90°$, and $\alpha > 90°$, respectively; (e) circuit condition when $\alpha > 180°$

1. In all the three cases the load current is continuous and hence each of the thyristors conduct for a duration of 120°.
2. For $\alpha = 0°$ and $\alpha < 90°$, thyristor Th_1 has to withstand a peak reverse voltage of magnitude $\sqrt{3}V_m$, where V_m is the peak value of the line-to-neutral voltage, whereas for $\alpha > 90°$, it has to block a forward voltage of the same magnitude. This is true for Th_2 and Th_3 also.
3. When $\alpha = 0°$ and $\alpha < 90°$, the circuit time available for the commutation of the thyristor is of the order of milliseconds as against a t_{OFF} of the order of 20–50 μs. Hence, the commutation of thyristors takes place naturally.
4. Figure 2.25(d) shows that with $\alpha > 90°$ the circuit time for turn-off is very small as a consequence of the large firing angle. Making an allowance for transformer leakage inductance, the angle $180° - \alpha - \mu$ should be at least equal to ωt_{OFF}. This angle, called the *extinction angle* or *margin of commutation*, is denoted by γ. Thus, for successful commutation, the inequality

$$\gamma \geq \omega t_{OFF}$$

where ω is the line frequency in rad/s, should hold good.
5. If the magnitude of γ happens to be such that the above inequality is not satisfied, commutation failure occurs as explained below. Assuming that $\mu = 0$ and that Th_1 and Th_2 are the outgoing and incoming thyristors, respectively, with a safe extinction angle $\gamma = 180° - \alpha$, the following operating conditions occur. It is seen from Fig. 2.25(d) that at the time of changeover, the voltage v_{RY} is negative or equivalently v_{YR} is positive. This implies that the anode of the incoming thyristor Th_2 is more positive than that of the outgoing thyristor Th_1, and natural commutation takes places for Th_1 by, the application of a reverse voltage. The instantaneous voltage difference between the phases Y and R causes the circulation of a commutating current as usual. However, with $\alpha > 180°$, this commutating voltage v_{RY} is positive (or v_{YR} is negative) and natural commutation does

not occur because conduction of Th_1 continues into the positive half-cycle of v_R. The rectifier output voltage becomes positive, its average V_α aiding the load voltage v_{ld} and causing a high direct current to flow through the armature resistance, thus leading to a short circuit as shown in Fig. 2.25(e).

6. This principle is important, and the above-mentioned condition on γ [that is, $\gamma \geq \omega t_{\text{OFF}}$] has to be ensured during the inverting mode of operation in converters used for HVDC transmission as well as those used for cycloconverters.

2.6 Ripple Factor

The output voltage waveforms of single- and three-phase controlled rectifiers are non-sinusoidal and hence contain considerable harmonic content. However, the ripple frequency increases with the number of supply phases and a smoother load voltage waveform can be obtained. Again, for a particular kind of ac source, say, a three-phase supply, the ripple content increases with the firing angle. Here, we attempt a quantitative characterization of the ripple content by means of a factor known as the *ripple factor*. With the number of phases taken as n, general expressions are derived for the ripple factor under different firing angle conditions. It is assumed throughout that conduction is continuous for the controlled rectifier.

At any instant, the load voltage $v_{\text{ld}}(t)$ for an n-phase rectifier can be written as

$$v_{\text{ld}}(t) = V_\alpha + v_\sigma(t) \tag{2.46}$$

where V_α is the average load voltage and $v_\sigma(t)$ is the instantaneous value, also called the *ripple voltage*. A typical output waveform shows that $v_\sigma(t)$ fluctuates about V_α, thus contributing to the ripple. Figure 2.26(a) shows one such waveform. The ripple factor γ is defined as the ratio of the net harmonic content of the output voltage to the average voltage. Thus, for an n-phase controlled rectifier, γ can be written as

$$\gamma = \frac{1}{V_\alpha}\sqrt{\frac{n}{2\pi}\int_\beta^\rho [v_{\text{ld}}(t) - V_\alpha]^2 d(\omega t)} \tag{2.47}$$

The angles β and ρ, indicated in Fig. 2.26(a), depend upon the shape of the load voltage waveform as elaborated below.

The exact expression for the ripple factor differs for the two different operating conditions on V_α, namely,
(a) $\sin^{-1}(V_\alpha/V_m) \geq \alpha + (\pi/2 - \pi/n)$ or $\alpha \leq \sin^{-1}(V_\alpha/V_m) - (\pi/2 - \pi/n)$
(b) $\sin^{-1}(V_\alpha/V_m) < \alpha + (\pi/2 - \pi/n)$ or $\alpha > \sin^{-1}(V_\alpha/V_m) - (\pi/2 - \pi/n)$
Accordingly, two separate expressions are derived here for this factor.

2.6.1 $\alpha \leq \sin^{-1}(V_\alpha/V_m) - (\pi/2 - \pi/n)$

Figure 2.26(b) shows the load voltage waveform for this condition. Defining

$$\theta = \sin^{-1}\left(\frac{V_\alpha}{V_m}\right) \tag{2.48}$$

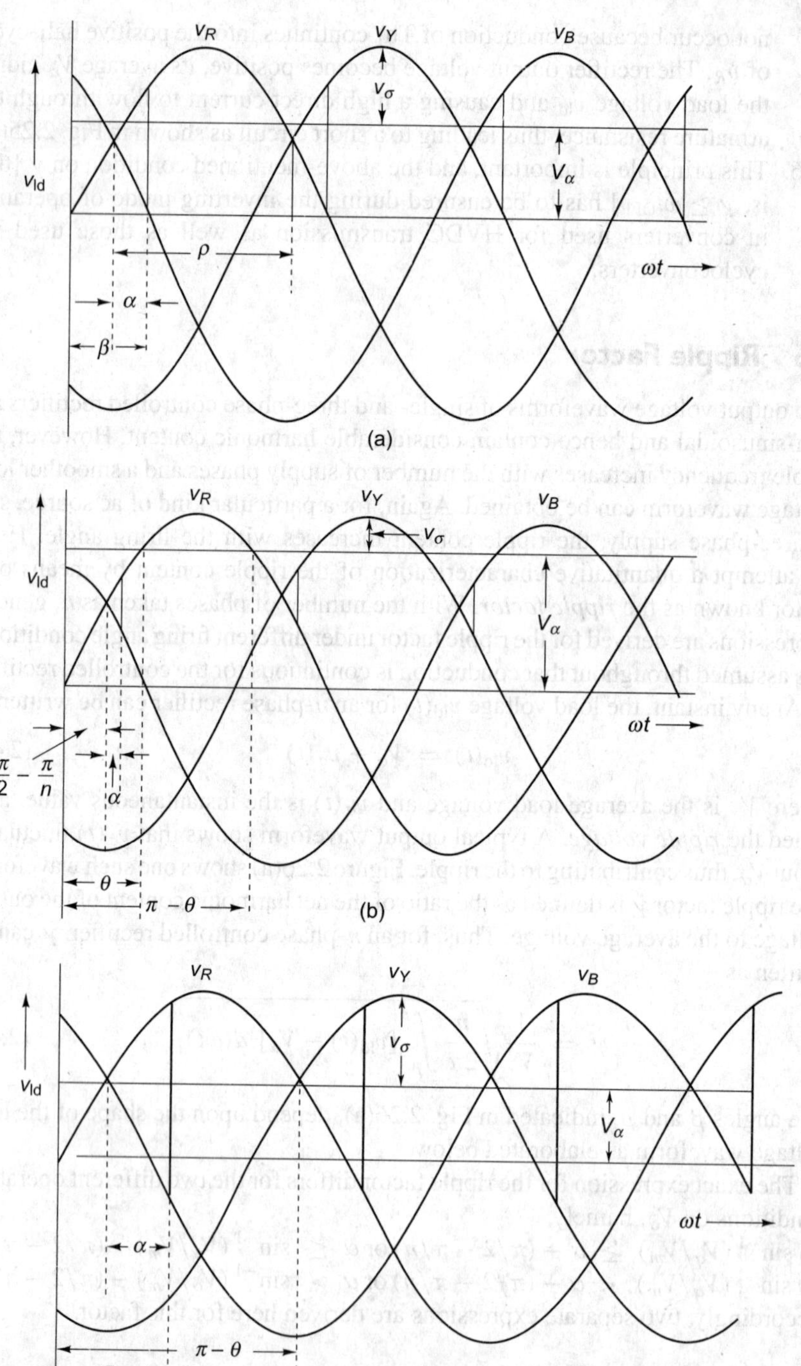

Fig. 2.26 Output of a three-phase, half-wave controlled rectifier with (a) $\alpha > 0°$, (b) $\alpha \leq \sin^{-1}(V_\alpha/V_m) - (\pi/2 - \pi/n)$, and (c) $\alpha > \sin^{-1}(V_\alpha/V_m) - (\pi/2 - \pi/n)$

this condition can be expressed as

$$0 \le \alpha \le \theta - \left(\frac{\pi}{2} - \frac{\pi}{n}\right) \tag{2.49}$$

In this range the shape of the load voltage waveform above V_α is seen to be a sinusoidal one and symmetrical about the peak. Figure 2.26(b) shows that $v_\sigma(t)$ exceeds V_α in the range $\theta \le \omega t \le \pi - \theta$; thus $\beta = \theta$ and $\rho = \pi - \theta$ for this condition. Equation (2.47) can now be rewritten by taking $v_{\text{ld}}(t) = V_m \sin \omega t$ as

$$\gamma = \frac{1}{V_\alpha}\sqrt{\frac{n}{2\pi}\int_\theta^{\pi-\theta} [V_m \sin \omega t - V_\alpha]^2 d(\omega t)} \tag{2.50}$$

Now V_α can be expressed as

$$V_\alpha = \frac{n}{\pi} V_m \sin\frac{\pi}{n} \cos\alpha = \frac{V_m}{k_1}$$

where

$$1/k_1 = \frac{n}{\pi} \sin\frac{\pi}{n} \cos\alpha$$

This gives

$$V_m = k_1 V_\alpha$$

Also

$$\gamma = \frac{1}{V_\alpha}\sqrt{\frac{n}{2\pi}\int_\theta^{\pi-\theta} [k_1 V_\alpha \sin \omega t - V_\alpha]^2\, d(\omega t)}$$

$$= \frac{V_\alpha}{V_\alpha}\sqrt{\frac{n}{2\pi}\int_\theta^{\pi-\theta} [k_1 \sin \omega t - 1]^2\, d(\omega t)} \tag{2.51}$$

Cancelling V_α in Eqn (2.51) and expanding the expression under the integral gives

$$\gamma = \sqrt{\frac{n}{2\pi}\int_\theta^{\pi-\theta} [k_1^2 \sin^2 \omega t - 2k_1 \sin \omega t + 1] d(\omega t)}$$

$$= \sqrt{\frac{n}{2\pi}\left[\int_\theta^{\pi-\theta} k_1^2 \sin^2 \omega t\, d(\omega t) - 2k_1 \int_\theta^{\pi-\theta} \sin \omega t\, d(\omega t) + \int_\theta^{\pi-\theta} d(\omega t)\right]} \tag{2.52}$$

with $\theta \le \alpha \le \theta - (\pi/2 - \pi/n)$. Defining

$$I_1 = k_1^2 \int_\theta^{\pi-\theta} \sin^2 \omega t\, d(\omega t)$$

$$I_2 = -2k_1 \int_\theta^{\pi-\theta} \sin \omega t\, d(\omega t)$$

and

$$I_3 = \int_{\theta}^{\pi-\theta} d(\omega t)$$

the ripple factor can be expressed as

$$\gamma = \sqrt{\frac{n}{2\pi}\, k_2} \tag{2.53}$$

where

$$k_2 = I_1 + I_2 + I_3$$

With $\alpha = 0$, the ripple factors for single-, three-, and six-phase supply waveforms work out to 0.31, 0.116, and 0.046, respectively. These values confirm the fact that with the increase in the number of secondary phases of the rectifier transformer, the dc load voltage approaches the direct voltage.

2.6.2 $\alpha > \sin^{-1}(V_\alpha/V_m) - (\pi/2 - \pi/n)$

The waveform for this condition is illustrated in Fig. 2.26(c). It is seen from this figure that the part of the waveform above V_α is no longer symmetrical about the peak. The ripple factor for the waveform is now expressed as

$$\gamma = \frac{1}{V_\alpha}\sqrt{\frac{n}{2\pi}\int_{\alpha+\pi/2-\pi/n}^{\pi-\theta} [V_m \sin\omega t - V_\alpha]^2 d(\omega t)} \tag{2.54}$$

Thus, $\beta = \alpha + \pi/2 - \pi/n$ and $\rho = \pi - \theta$ in this case. A comparison of Eqns (2.50) and (2.54) shows that the expression for γ in the latter is different in only one aspect, namely, the lower limit is now $\alpha + \pi/2 - \pi/n$ instead of θ. Proceeding along the same lines as in Section 2.6.1, γ can be expressed, after the cancellation of V_α, as

$$\gamma = \left\{\frac{n}{2\pi}\left[\int_{\alpha+\pi/2-\pi/n}^{\pi-\theta} k_1^2 \sin^2\omega t\, d(\omega t) - \int_{\alpha+\pi/2-\pi/n}^{\pi-\theta} 2k_1 \sin\omega t\, d(\omega t) + \int_{\alpha+\pi/2-\pi/n}^{\pi-\theta} d(\omega t)\right]\right\}^{1/2} \tag{2.55}$$

Now, taking

$$I_A = \int_{\alpha+\pi/2-\pi/n}^{\pi-\theta} k_1^2 \sin^2\omega t\, d(\omega t)$$

$$I_B = -\int_{\alpha+\pi/2-\pi/n}^{\pi-\theta} 2k_1 \sin\omega t\, d(\omega t)$$

and

$$I_C = \int_{\alpha+\pi/2-\pi/n}^{\pi-\theta} d(\omega t)$$

and also defining $k_3 = I_A + I_B + I_C$, γ can be expressed as

$$\gamma = \sqrt{\frac{n}{2\pi}\,k_3} \tag{2.56}$$

Remarks:

(a) For computing the ripple factor, it is necessary to first check whether the given firing angle α is greater or less than the border case firing angle given by

$$\alpha_b = \sin^{-1}\left(\frac{V_\alpha}{V_m}\right) - \left(\frac{\pi}{2} - \frac{\pi}{n}\right)$$

For single-, three-, and six-phase, half-wave bridge rectifiers, α_b works out to 32.65°, 20.7°, and 10.1°, respectively. The appropriate expression from amongst those in Eqns (2.52) and (2.55) is then chosen.

(b) Equations (2.52) and (2.55) are derived for continuous conduction. However, the derivation can be extended to discontinuous conduction by taking the following aspects into account.

(i) The expression for the average load voltage should be taken as

$$V_\alpha = \frac{n}{2\pi}V_m\left[\sin\left(\alpha_e - \frac{\pi}{n}\right) - \sin\left(\alpha - \frac{\pi}{n}\right)\right]$$

with an appropriate value for n. $1/k_1$ is now defined as

$$\frac{1}{k_1} = \frac{n}{2\pi}\left[\sin\left(\alpha_e - \frac{\pi}{n}\right) - \sin\left(\alpha - \frac{\pi}{n}\right)\right]$$

(ii) After checking whether α is greater or lesser than α_b, the appropriate expression from amongst Eqns (2.52) and (2.55) is chosen.

(iii) The upper limit of integration should be taken as $\alpha_e + \pi/2 - \pi/n$ if $\pi - \theta$ is greater than $\alpha_e + \pi/2 - \pi/n$, and as $\pi - \theta$ if $\pi - \theta$ obeys the inequality $(\pi - \theta) \le (\alpha_e + \pi/2 - \pi/n)$.

2.7 Transformer Leakage Reactance and its Effects on Converter Performance

In Section 2.4 the expressions for average voltages and currents have been derived for an n-phase controlled rectifier with the assumption that the converter transformer is ideal and that its leakage reactance as well resistance are negligible. While it is reasonable to consider the resistance to be negligible, ignoring the inductance is not possible and leads to serious error because the transformer windings are predominantly inductive in nature. The following treatment gives the expression for the load voltage of an n-phase rectifier with a leakage inductance L_s in each phase of the transformer, referred to the secondary. Sufficient inductance is assumed to be present in the load, so that the load current I_{ld} can be considered to be constant. The analysis here is first carried out for an n-phase uncontrolled rectifier or equivalently an n-phase controlled rectifier with firing angle $\alpha = 0$.

2.7.1 Effect of Leakage Reactance with $\alpha = 0$

Figures 2.27(a) and (b), respectively, give the circuit as well as load voltage waveforms for such a rectifier with $\alpha = 0$ and $\alpha > 0$.

An inspection of the waveforms shows that thyristor Th_3 conducting the B-phase current stops conduction at the point P on the ωt-axis, and Th_1 connected to the phase R starts conduction at the same moment. However, as the inductor L_s prevents sudden change of current, Th_3 continues to conduct the load current till the point Q. At the same time, Th_1 starts conduction at P because the R-phase voltage becomes most positive at that point. Thus, both the devices conduct during the interval PQ, called the *period of commutation*, and the load voltage v_{ld} for this duration becomes the mean of the voltages v_R and v_B. Consequently, there is a drop in the load voltage during the interval of commutation, which is indicated by the striped area in Fig. 2.27. The average load voltage $V_{\mu 0}$ will be less than that obtained in the absence of transformer leakage inductance. An inspection of the current waveforms shows that during the commutation interval, current i_B slowly decreases and i_R slowly increases, their sum remaining at I_{ld} throughout this period. Also, the cathodes of Th_1 and Th_3 are shorted and the voltage $v_R - v_B$ is absorbed by the inductances L_s in the B- and R-phase. A short-circuit current i_{sc} flows from the R-phase winding to the B-phase winding. To facilitate the analysis of this condition, it is assumed that the two cathodes are at a common voltage v_T. Considering $v_B - v_T$ and $v_R - v_T$ to be, respectively, the drops in the inductances of the B- and R-phase windings, they can be expressed as

$$v_R - v_T = L_s \frac{di_R}{dt} \tag{2.57}$$

$$v_B - v_T = L_s \frac{di_B}{dt} \tag{2.58}$$

where $v_R = v_m \sin \omega t$ and $v_B = V_m \sin(\omega t + 2\pi/n)$.

The difference between the R- and B-phase voltages is obtained from Eqns (2.57) and (2.58) as

$$v_R - v_B = L_s \left(\frac{di_R}{dt} - \frac{di_B}{dt} \right) \tag{2.59}$$

If the reference for the two sinusoids is shifted to the point P where commutation starts, v_R and v_B can be expressed as

$$v_R = V_m \sin \left(\omega t + \frac{\pi}{2} - \frac{\pi}{n} \right) \tag{2.60}$$

and

$$v_B = V_m \sin \left(\omega t + \frac{\pi}{2} + \frac{\pi}{n} \right) \tag{2.61}$$

Thus,

$$v_R - v_B = 2V_m \sin \omega t \sin(\pi/n) \tag{2.62}$$

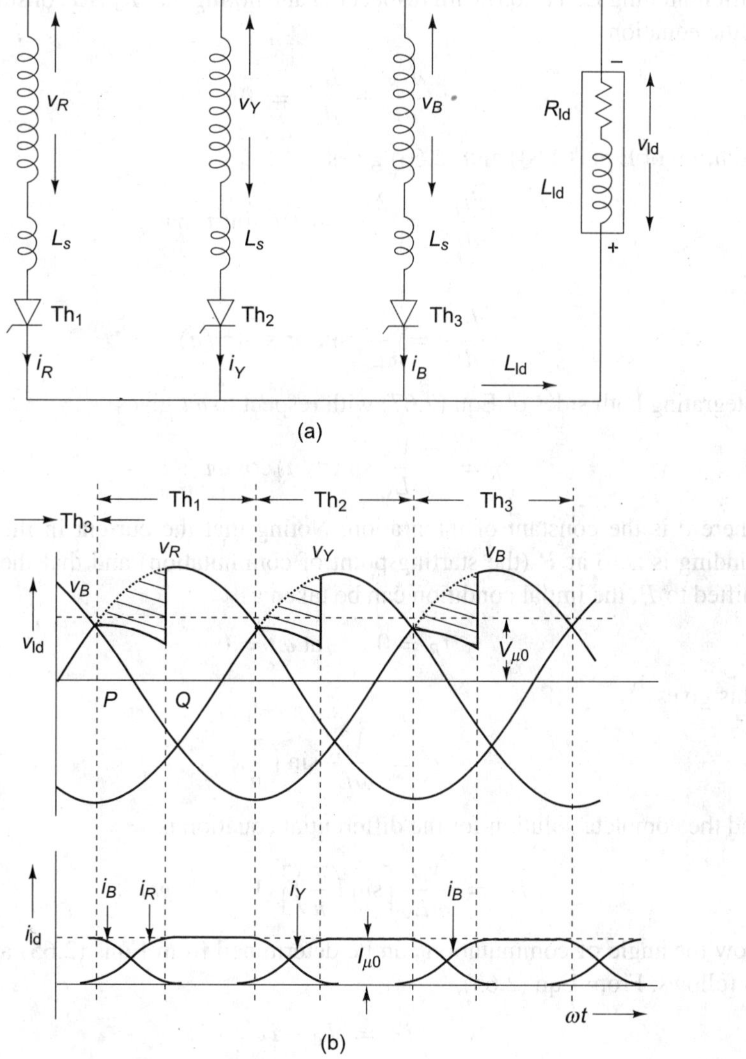

Fig. 2.27 n-phase, half-wave controlled converter with transformer inductance L_s: (a) circuit, (b) waveforms with $\alpha = 0°$

The right-hand sides of Eqns (2.59) and (2.62) can now be equated to get

$$L_s \left(\frac{di_R}{dt} - \frac{di_B}{dt} \right) = 2V_m \sin \omega t \, \sin (\pi/n) \tag{2.63}$$

or

$$\left(\frac{di_R}{dt} - \frac{di_B}{dt} \right) = \frac{2V_m}{L_s} \sin \omega t \sin(\pi/n) \tag{2.64}$$

The R- and B-phase currents sum up to I_{ld}. Thus,

$$i_R + i_B = I_{ld} \tag{2.65}$$

Differentiating Eqn (2.65) with respect to t and noting that I_{ld} is a constant results in the equation

$$\frac{di_R}{dt} + \frac{di_B}{dt} = 0 \tag{2.66}$$

Addition of Eqns (2.64) and (2.66) gives

$$2\frac{di_R}{dt} = \frac{2V_m}{L_s} \sin \omega t \, \sin(\pi/n)$$

or

$$\frac{di_R}{dt} = \frac{V_m}{L_s} \sin \omega t \, \sin(\pi/n) \tag{2.67}$$

Integrating both sides of Eqn (2.67) with respect to ωt gives

$$i_R = \frac{-V_m}{\omega L_s} \sin(\pi/n) \cos \omega t + c \tag{2.68}$$

where c is the constant of integration. Noting that the current in the R-phase winding is zero at P (the starting point of commutation) and that the origin is shifted to P, the initial condition can be taken as

$$i_R = 0 \qquad \text{at } \omega t = 0$$

This gives

$$c = \frac{V_m}{\omega L_s} \sin\left(\frac{\pi}{n}\right)$$

and the complete solution for the differential equation is

$$i_R = \frac{V_m}{\omega L_s} \left[\sin\left(\frac{\pi}{n}\right)\right] (1 - \cos \omega t) \tag{2.69}$$

Now the angle of commutation can be determined from Eqns (2.65) and (2.69) as follows. From Eqn (2.65),

$$i_B = I_{\mathrm{ld}} - i_R \tag{2.70}$$

Substitution for i_R from Eqn (2.69) for gives

$$i_B = I_{\mathrm{ld}} - \frac{V_m}{\omega L_s} \left[\sin\left(\frac{\pi}{n}\right)\right] (1 - \cos \omega t) \tag{2.71}$$

The final condition, namely, that the B-phase current goes to zero at the end of the period of commutation, is now invoked. Thus,

$$i_B = 0 \qquad \text{at } \omega t = \mu$$

Substituting this condition in Eqn (2.71) gives

$$0 = I_{\mathrm{ld}} - \frac{V_m}{\omega L_s} \left[\sin\left(\frac{\pi}{n}\right)\right] (1 - \cos \mu) \tag{2.72}$$

or

$$\cos\mu = 1 - \frac{I_{\text{ld}}\omega L_s}{V_m \sin(\pi/n)} \tag{2.73}$$

The average voltage (V_α) at the load with no transformer leakage inductance is given in Eqn (2.16) as

$$V_\alpha = \frac{n}{\pi} V_m \sin\left(\frac{\pi}{n}\right)\cos\alpha \tag{2.74}$$

With $\alpha = 0$, this becomes

$$V_0 = \frac{n}{\pi} V_m \sin\left(\frac{\pi}{n}\right) \tag{2.75}$$

The reduction in average voltage due to the phenomenon of commutation is obtained by computing the striped area and averaging it over the period $2\pi/n$. For this purpose, the instantaneous value of voltage reduction can be written as

$$v_\Delta = v_R - \frac{v_R + v_B}{2} \tag{2.76}$$

$$= \frac{v_R - v_B}{2} \tag{2.77}$$

$$= V_m \sin(\pi/n)\sin\omega t \tag{2.78}$$

using Eqn (2.62). The reduction in the average voltage denoted as $g_{\mu 0}$ is now obtained by averaging this instantaneous reduction over the period $2\pi/n$:

$$g_{\mu 0} = \frac{n}{2\pi}\int_0^\mu v_\Delta d(\omega t) \tag{2.79}$$

$$= \frac{n}{2\pi}\int_0^\mu V_m \sin(\pi/n)\sin\omega t \; d(\omega t) \tag{2.80}$$

$$= \frac{n}{2\pi} V_m \left[\sin\left(\frac{\pi}{n}\right)\right](1 - \cos\mu) \tag{2.81}$$

Substituting for $\cos\mu$ from Eqn (2.73) gives $g_{\mu 0}$ as

$$g_{\mu 0} = \frac{n}{2\pi} V_m \left[\sin\left(\frac{\pi}{n}\right)\right]\frac{I_{\text{ld}}\omega L_s}{V_m \sin(\pi/n)} = \frac{n}{2\pi} I_{\text{ld}}\omega L_s = \frac{n I_{\text{ld}} X_s}{2\pi} \tag{2.82}$$

where $X_s = \omega L_s$. The net voltage at the load can now be written as

$$V_{\mu 0} = V_\alpha - g_{\mu 0} \tag{2.83}$$

2.7.2 Effect of Leakage Reactance with $\alpha > 0$

The derivation given above can be extended to an n-phase controlled rectifier with $\alpha > 0$, the circuit and waveforms for which are, respectively, shown in Figs 2.28(a) and (b). The analysis made earlier up to Eqn (2.68) remains the same for $\alpha > 0$. However, the constant c is obtained by using the initial condition $i_R = 0$ at $\omega t = \alpha$ because the R-phase thyristor starts conduction only at α.

Thus,

$$c = \frac{V_m}{\omega L_s} \sin\left(\frac{\pi}{n}\right) \cos\alpha$$

and the complete solution for i_R becomes

$$i_R = \frac{V_m}{\omega L_s}\left[\sin\left(\frac{\pi}{n}\right)\right](\cos\alpha - \cos\omega t) \tag{2.84}$$

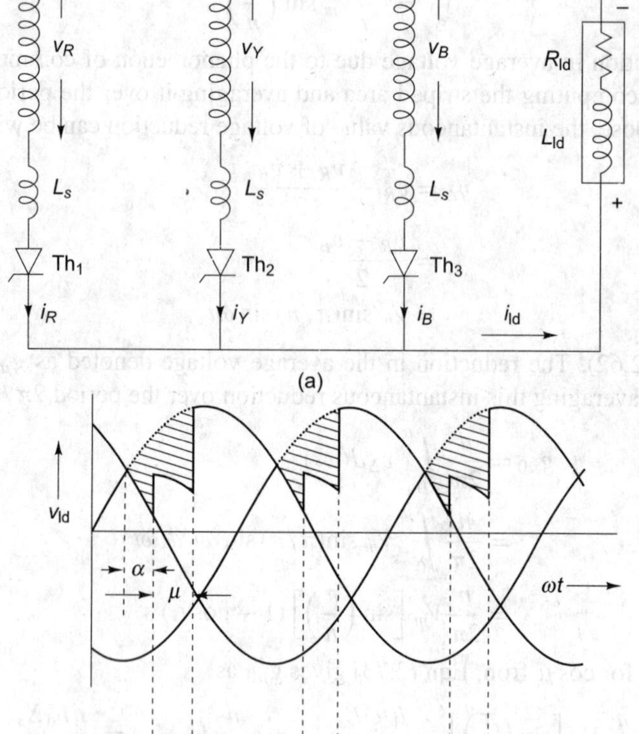

Fig. 2.28 n-phase, half-wave controlled converter with transformer inductance L_s: (a) circuit, (b) waveforms with $\alpha > 0°$

As before, the reference point is shifted to the point of natural commutation, making the duration of commutation α to $\alpha + \mu$. The commutation angle μ is arrived at as follows. From Eqn (2.65),

$$i_B = I_{ld} - i_R \tag{2.85}$$

Substituting the expression for i_R from Eqn (2.84) gives

$$i_B = I_{\mathrm{ld}} - \frac{V_m}{\omega L_s}\left[\sin\left(\frac{\pi}{n}\right)\right](\cos\alpha - \cos\omega t) \tag{2.86}$$

Applying the final condition $i_B = 0$ at $\omega t = \alpha + \mu$ yields

$$0 = I_{\mathrm{ld}} - \left\{\frac{V_m}{\omega L_s}\left[\sin\left(\frac{\pi}{n}\right)\right][\cos\alpha - \cos(\alpha+\mu)]\right\} \tag{2.87}$$

or

$$\cos(\alpha+\mu) = \cos\alpha - \frac{I_{\mathrm{ld}}\omega L_s}{V_m \sin(\pi/n)} \tag{2.88}$$

Equation (2.88) is an implicit one and can now be solved for μ by trial and error, using the known values of I_{ld}, L_s, and α. The reduction in average voltage, which is denoted as $g_{\mu\alpha}$, is obtained as before by considering the reduction in area over the interval α to $\alpha+\mu$ and averaging the area over $2\pi/n$. Thus,

$$g_{\mu\alpha} = \frac{n}{2\pi}\int_{\alpha}^{\alpha+\mu} \frac{v_R - v_B}{2}\, d(\omega t) \tag{2.89}$$

$$= \frac{n}{2\pi}\int_{\alpha}^{\alpha+\mu} V_m \sin(\pi/n)\sin\omega t\, d(\omega t) \tag{2.90}$$

Evaluation of the integral gives

$$g_{\mu\alpha} = \frac{n}{2\pi}V_m\left[\sin\left(\frac{\pi}{n}\right)\right][\cos\alpha - \cos(\alpha+\mu)] \tag{2.91}$$

Using Eqn (2.88), $g_{\mu\alpha}$ can be expressed as

$$g_{\mu\alpha} = \frac{n}{2\pi}V_m\left[\sin\left(\frac{\pi}{n}\right)\right]\frac{I_{\mathrm{ld}}\omega L_s}{V_m \sin(\pi/n)} = \frac{n}{2\pi}I_{\mathrm{ld}}\omega L_s = \frac{n}{2\pi}I_{\mathrm{ld}}X_s \tag{2.92}$$

A comparison of Eqns (2.82) and (2.92) shows that the expression for the reduction in the average voltage remains unaltered irrespective of the value of α. The net average voltage at the load is now given by

$$V_{\mu\alpha} = V_\alpha - g_{\mu\alpha} \tag{2.93}$$

Upon substituting for $g_{\mu\alpha}$ from Eqn (2.92), the net average voltage becomes

$$V_{\mu\alpha} = \frac{n}{\pi}V_m\left[\sin\left(\frac{\pi}{n}\right)\right]\frac{[\cos\alpha + \cos(\alpha+\mu)]}{2} \tag{2.94}$$

2.8 Rectifier Efficiency and Derating Factor of Rectifier Transformers

The output voltage waveform of a controlled rectifier contains harmonics, the content of which increases with the firing angle α. These harmonics cause additional hysteresis and eddy current losses in the rectifier transformer. A second drawback of using a controlled rectifier is that the load current is unidirectional, and as a consequence the primary current waveform may not be sinusoidal. The combined effect of both these features is that the capacity of the rectifier transformer has to be higher than that of the dc load connected at the output of the rectifier. In this section two quantities, namely, *rectifier efficiency* for the controlled rectifier and *derating factor* for the rectifier transformer, are defined. A third quantity, the *line power factor*, is also derived in the case of transformers having a delta primary winding. It is assumed that the devices used in the controlled rectifiers are either diodes or, equivalently, thyristors with zero firing angle.

2.8.1 General Principles

Though there is a direct relationship between the rectifier current and the transformer primary current, it is not possible to arrive at a general expression for the RMS value of the primary current. The waveforms of the output currents of the rectifier depend upon the type of the load, the number of secondary phases, and, in the case of a thyristor controlled rectifier, the firing angle. In the following, expressions for rectifier efficiency are derived for rectifiers fed by single-phase and three-phase transformers. A relationship between the magnetomotive forces (mmfs) of the primary and secondary windings is established rather than those between the primary and secondary currents. It is assumed that the magnetizing mmfs are negligible and that the rectifier elements are ideal.

2.8.1.1 Rectifier fed by single-phase transformers In a single-phase transformer, all the windings are traversed by the same flux irrespective of whether the magnetic circuit is of the core or the shell type. The polarity of the windings is indicated by the dots shown in Fig. 2.29.

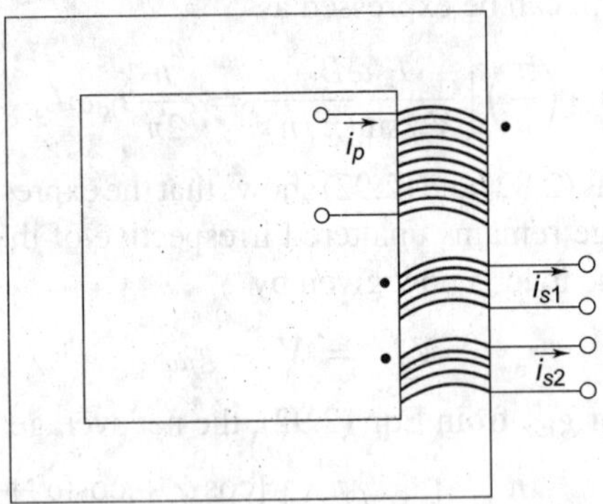

Fig. 2.29 Magnetic circuit of a single-phase transformer

The phenomenon occurring in these transformers can be explained as follows. If the mean of the secondary ampere turns is non-zero, this dc component cannot be compensated by the primary ampere turns, because the primary current is necessarily an ac current and has a zero mean. The uncompensated secondary ampere turns do not effect the computation of the primary current but saturate the magnetic circuit, thereby leading to an increase in the primary magnetizing current as well as the core losses.

If the mean value is non-zero and the magnetizing current is ignored, the ampere-turn balance is written as

$$N_1 i_p = N_2 i_{s1} + N_2 i_{s2} - N_2 I_d \tag{2.95}$$

where $N_2 I_d$ is the mean value of the sum of the ampere turns. With a zero mean value, Eqn (2.95) becomes

$$N_1 i_p = N_2 i_{s1} + N_2 i_{s2} \tag{2.96}$$

The primary current i_p can now be computed from Eqn (2.96), given the secondary currents i_{s1} and i_{s2}.

2.8.1.2 Rectifiers fed by three-phase and six-phase transformers Three-phase transformers usually have a three limbed core. To ensure that the primary ampere turns form a balanced system, the secondary windings are distributed so that the mmfs are identical at an electrical angle of $2\pi/3$ rad. The following discussion centres around various kinds of three-limbed, three-phase transformers, each of the limbs having one primary winding and one or more secondary windings. The core in Fig. 2.30 has a secondary winding on each limb; however, the treatment can be extended to transformers having more than one secondary winding and hence there is no loss of generality in such an approach. The windings on both sides can be connected in a delta or a star, with the star point either isolated or grounded. Hence, a number of winding combinations are possible. Some commonly occurring combinations are considered here. If the mean value of the secondary ampere turns is non-zero, it is not possible to compensate for this dc component. This is true for all three-phase, half-wave rectifiers (both controlled and uncontrolled configurations). Thus, in cases 1 to 5 considered below, there is a dc component in the transformer secondary current. This component may lead to core saturation. It is assumed in the expressions derived later that the sum of the secondary ampere turns excludes this dc component. Fortunately, the effects of such a dc component are minimized in a three-phase, three-limbed magnetic circuit because

(i) the uncompensated ampere turns on the three limbs are equal and in the same direction and
(ii) the extraneous fluxes generated by them on the three limbs are of very small magnitudes because the flux paths have to pass through very high reluctance regions outside the magnetic circuit. On the other hand, in a single-phase transformer, the extraneous fluxes have their low-reluctance paths in the magnetic circuit itself and hence the dc components cannot be ignored.

Case 1:* Δ/Y *transformer As the primary windings are delta-connected here, the condition that the primary ampere turns should be alternating is imposed. Also, when the secondary ampere turns form a balanced system, the sum of their ampere turns is zero. Thus,

$$N_2(i_{sR} + i_{sY} + i_{sB}) = 0 \tag{2.97}$$

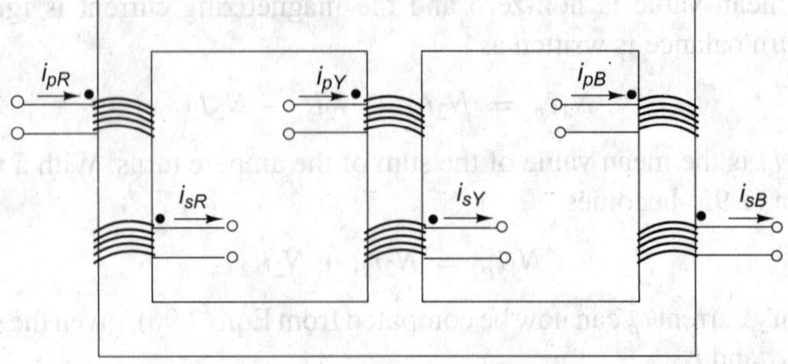

Fig. 2.30 Three-limbed transformer core with primary and secondary windings

The above condition on the primary ampere turns together with Eqn (2.97) implies that ampere-turn balance exists on each of the limbs. This can be expressed as

$$\begin{aligned} i_{pR}N_1 &= i_{sR}N_2 \\ i_{pY}N_1 &= i_{sY}N_2 \\ i_{pB}N_1 &= i_{sB}N_2 \end{aligned} \tag{2.98}$$

The line currents can now be written as the difference of the primary currents:

$$\begin{aligned} i_{L1} &= i_{pR} - i_{pY} \\ i_{L2} &= i_{pY} - i_{pB} \\ i_{L3} &= i_{pB} - i_{pR} \end{aligned} \tag{2.99}$$

Case 2: Δ/Y-with-neutral connection As in case 1, it is assumed that the primary ampere turns are alternating. However, two alternatives exist here. If the secondary ampere turns form a balanced system, this becomes identical to case 1. On the other hand, when the secondary ampere turns have a zero sequence component, this component I_{s0} is given as

$$i_{s0} = \frac{1}{3}N_2(i_{sR} + i_{sY} + i_{sB}) \tag{2.100}$$

The corresponding zero sequence component in the primary winding circulates around the delta circuit and will not be present in the line currents. The primary currents can now be determined from the set of equations (2.103) given under case 4. The set of equations (2.99) hold good in this case also and this confirms the fact that the zero sequence components are absent in the line currents.

Case 3: Y/Y connection As the primary windings are connected in a star configuration without a neutral conductor, the instantaneous currents are constrained to sum up to zero. Thus, their zero sequence component I_{p0} can be written as

$$i_{p0} = \frac{1}{3}(i_{pR} + i_{pY} + i_{pB}) = 0 \tag{2.101}$$

When the secondary ampere turns also form a balanced system, it can be concluded that alternating ampere-turn balance exists on each of the limbs. Hence the set of equations (2.98) holds good here.

Case 4: Y/Y-with-neutral connection Similar to case 3, the primary star here is without a neutral conductor. Accordingly, the zero sequence primary current is zero as per Eqn (2.101). When the secondary ampere turns do not have a zero sequence component, this case is identical to case 3. On the other hand, when the secondary ampere turns have a zero sequence component, this component is expressed by Eqn (2.100). The zero sequence component of the secondary, i_{s0}, cannot be compensated because the star-connected primary ampere turns cannot have a zero sequence component. Hence, Eqn (2.98) has to be written for all primary currents as

$$
\begin{aligned}
i_{pR}N_1 &= i_{sR}N_2 - \frac{1}{3}N_2(i_{sR} + i_{sY} + i_{sB}) \\
i_{pY}N_1 &= i_{sY}N_2 - \frac{1}{3}N_2(i_{sR} + i_{sY} + i_{sB}) \qquad (2.102) \\
i_{pB}N_1 &= i_{sB}N_2 - \frac{1}{3}N_2(i_{sR} + i_{sY} + i_{sB})
\end{aligned}
$$

These equations can be solved for primary currents as

$$
\begin{aligned}
i_{pR} &= \frac{N_2}{N_1}\left[\frac{2}{3}i_{sR} - \frac{1}{3}i_{sY} - \frac{1}{3}i_{sB}\right] \\
i_{pY} &= \frac{N_2}{N_1}\left[\frac{2}{3}i_{sY} - \frac{1}{3}i_{sB} - \frac{1}{3}i_{sR}\right] \qquad (2.103) \\
i_{pB} &= \frac{N_2}{N_1}\left[\frac{2}{3}i_{sB} - \frac{1}{3}i_{sR} - \frac{1}{3}i_{sY}\right]
\end{aligned}
$$

The set of equations (2.103) is applicable for the core as well as shell type of magnetic circuits. The uncompensated zero sequence component in the core induces undesirable voltages in the three primary windings. As in the case of dc components, these zero sequence components are negligible for three-limbed transformers but not for four- or five-limbed ones. When the condition of Eqn (2.97) is not satisfied, it is advantageous to connect the primary windings in a delta.

Case 5: Y-with-neutral/Y-with-neutral connection In this case, all the secondary alternating ampere turns are compensated by primary ampere turns. The set of equations (2.98) can be used to determine the primary currents. However, undesirable currents of frequency $3f$ and their multiples would flow in each of the phase windings and also in the neutral conductor.

Case 6: Y/zigzag connection For the Y/zigzag transformer shown in Fig. 2.31(b), the primary currents are constrained to sum up to zero and thus Eqn (2.101) holds good. The zigzag connection ensures that the secondary ampere turns sum up to zero and the dc component, if any, is nullified. Hence, Eqn (2.97) is satisfied and there is a perfect balance of the ampere turns on each of the three limbs. The winding disposition on the core is given in Fig. 2.31(a). The ampere-turn balance can be obtained from this figure as

$$N_1 i_{pR} = \frac{N_2}{2}(i_{sY} - i_{sR})$$

$$N_1 i_{pY} = \frac{N_2}{2}(i_{sB} - i_{sY}) \qquad (2.104)$$

$$N_1 i_{pB} = \frac{N_2}{2}(i_{sR} - i_{sB})$$

The waveforms in Fig. 2.31(c) show that each of the primary current waveforms is also an alternating one. Thus, the zigzag winding ensures the satisfaction of both Eqns (2.97) and (2.101).

Case 7: Y/interstar connection For the star–interstar connection [Fig. 2.32(b)], the secondary can be considered to be a six-phase connection. As in case 3, the instantaneous primary ampere turns are constrained to sum up to zero. Corresponding to the set of equations (2.98), the ampere-turn balance on each limb can be written as

$$N_1 i_{pR} = N_2(i_{s1} - i_{s4})$$

$$N_1 i_{pY} = N_2(i_{s3} - i_{s6}) \qquad (2.105)$$

$$N_1 i_{pB} = N_2(i_{s5} - i_{s2})$$

It can be inferred from the set of equations (2.105) and Fig. 2.32(a) that the secondary ampere turns have a zero mean value. For example, in the first of Eqns (2.105), the right-hand side equals + $N_2 I_d$ for an interval of one-sixth of the time period when Th_1 conducts and $-N_2 I_d$ for another interval of one-sixth of the time period during the ON time of Th_4. However, an alternating zero sequence component exists for the secondary ampere turns because, at any instant, only one device conducts and the sum of the secondary ampere turns of the three limbs is

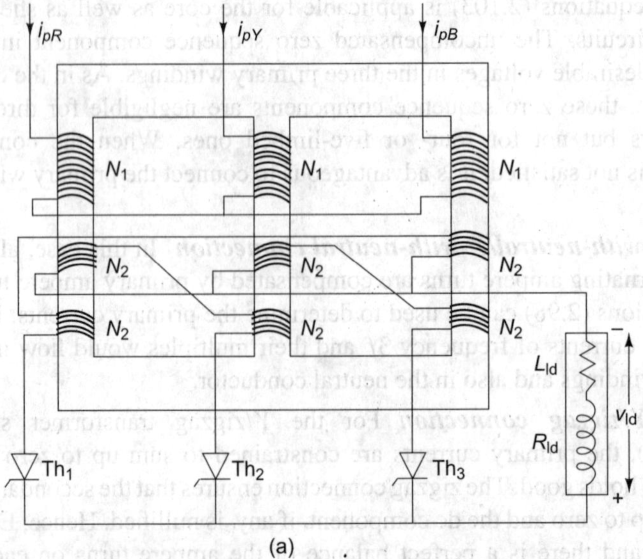

Fig. 2.31(a)

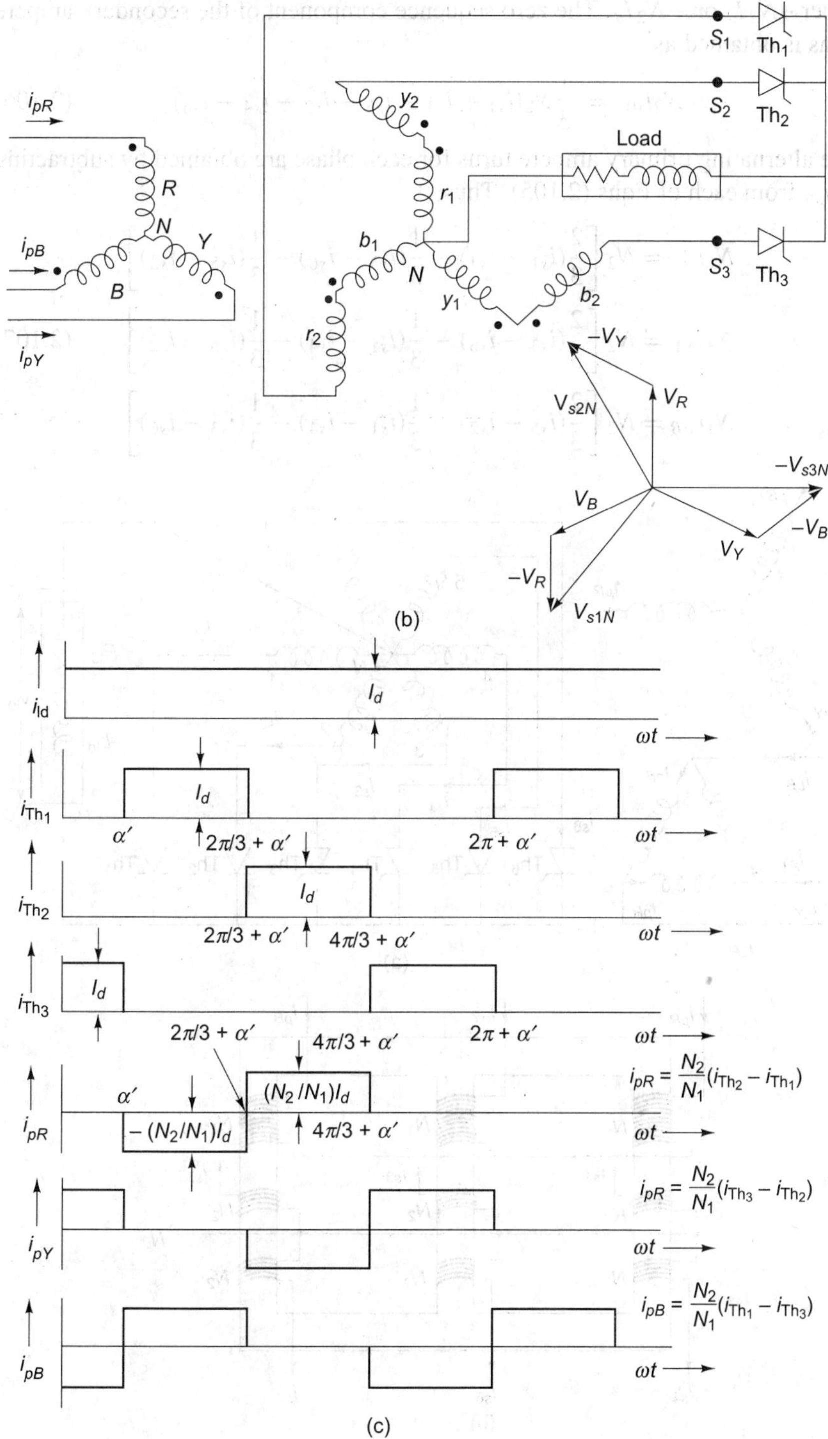

Fig. 2.31 Three-phase, half-wave rectifier supplied by a star/zigzag transformer and feeding an inductive load: (a) schematic diagram, (b) primary and secondary windings and phasor diagram, (c) Waveforms

either $+N_2I_d$ or $-N_2I_d$. The zero sequence component of the secondary ampere turns is obtained as

$$N_2 i_{s0} = \frac{1}{3}N_2(i_{s1} + i_{s3} + i_{s5} - i_{s2} - i_{s4} - i_{s6}) \tag{2.106}$$

The alternating primary ampere turns for each phase are obtained by subtracting $N_2 i_{s0}$ from each of Eqns (2.105). Thus,

$$\begin{aligned} N_1 i_{pR} &= N_2\left[\frac{2}{3}(i_{s1} - i_{s4}) - \frac{1}{3}(i_{s3} - i_{s6}) - \frac{1}{3}(i_{s5} - i_{s2})\right] \\ N_1 i_{pY} &= N_2\left[\frac{2}{3}(i_{s3} - i_{s6}) - \frac{1}{3}(i_{s1} - i_{s4}) - \frac{1}{3}(i_{s5} - i_{s2})\right] \\ N_1 i_{pB} &= N_2\left[\frac{2}{3}(i_{s5} - i_{s2}) - \frac{1}{3}(i_{s1} - i_{s4}) - \frac{1}{3}(i_{s3} - i_{s6})\right] \end{aligned} \tag{2.107}$$

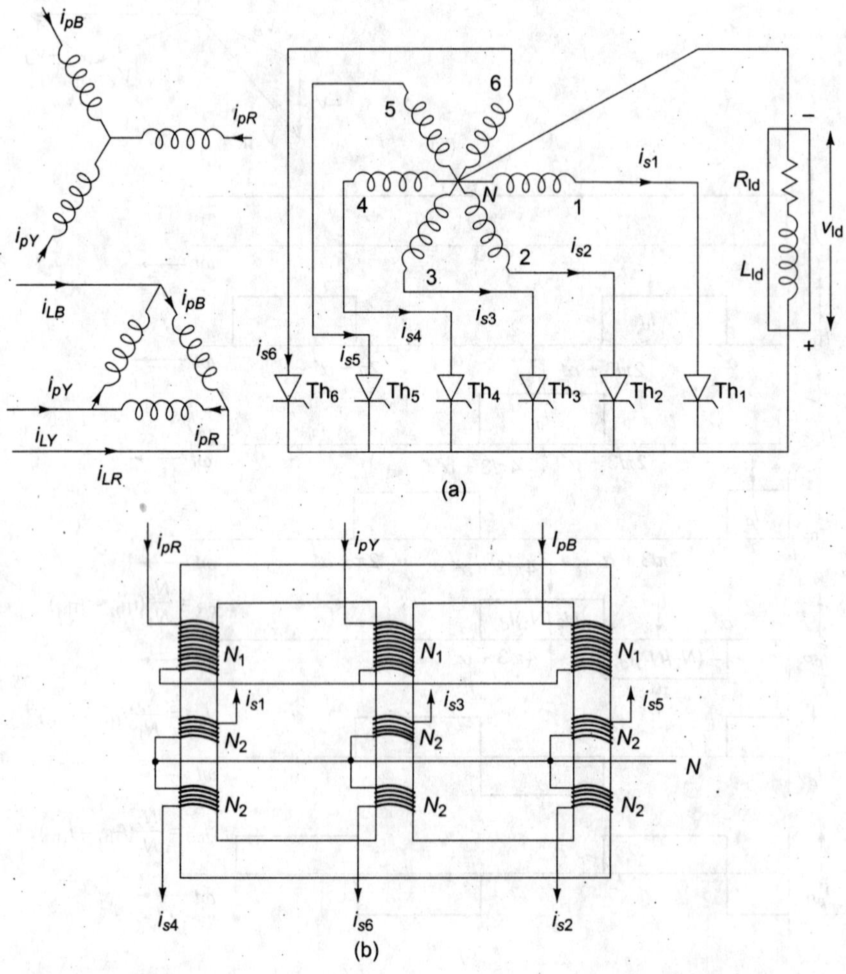

Figs 2.32(a), (b)

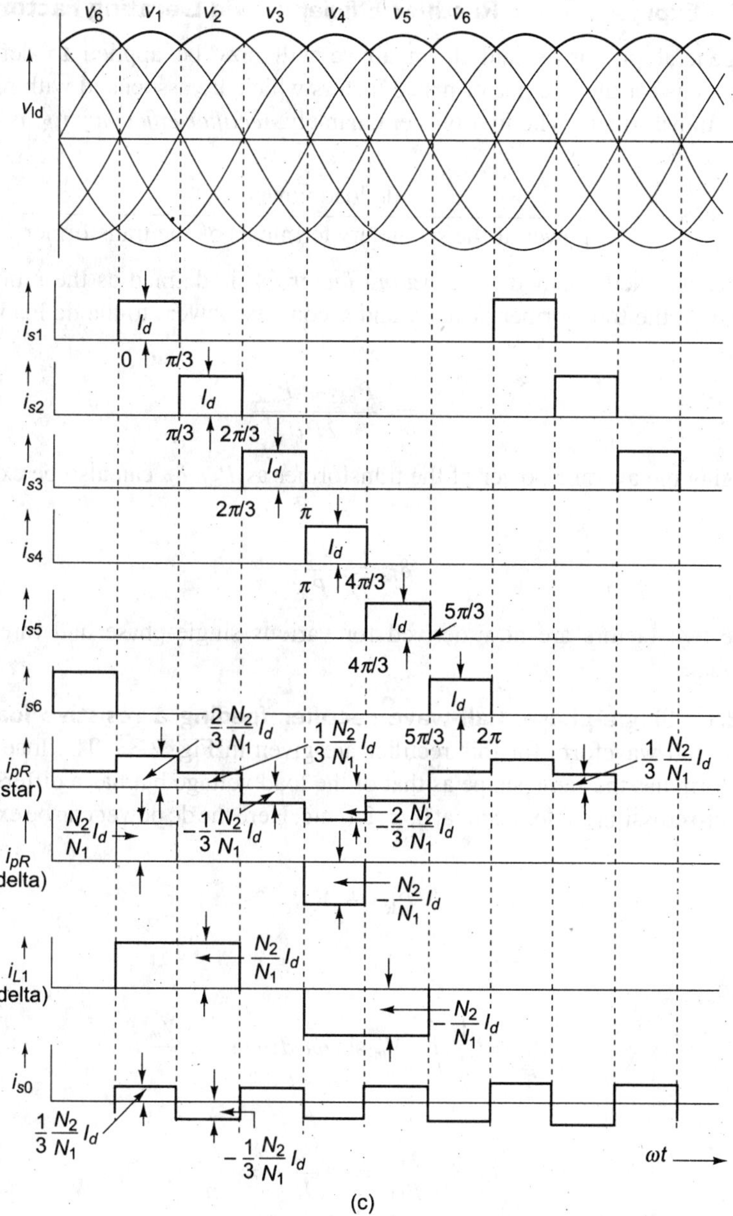

Fig. 2.32 Six-phase, half-wave rectifier feeding an inductive load (a) circuit, (b) winding disposition, (c) waveforms

It is seen from the waveforms of Fig. 2.32(c) that for a star primary, the primary phase current waveform is of a stepped nature and is nearly sinusoidal in shape. Therefore, this vector group connection is superior to others. Further, the waveform is of an alternating nature for the delta winding also.

2.8.2 Expressions for Rectifier Efficiency and Derating Factor

The general principles laid down above will now be applied to derive the expressions for the two performance factors which are associated with rectifiers fed by transformers. The first of them, namely, *rectifier efficiency* η_R, is defined as

$$\eta_R = \frac{\text{dc load power}}{\text{ac power at the secondary terminals of the transformer}}$$

The second factor, called the *derating factor* δ_R, is defined as the ratio of the average of the transformer primary and secondary powers to the dc load power. Thus,

$$\delta_R = \frac{P_{pac} + P_{sac}}{2P_{dc}} \tag{2.108}$$

Denoting the average power of the transformer as P_{eq}, δ_R can also be expressed as

$$\delta_R = \frac{P_{eq}}{P_{dc}} \tag{2.109}$$

These two factors are now derived for various single-phase and three-phase rectifiers.

2.8.2.1 Single-phase, half-wave rectifier feeding a resistive load The circuit and waveform for this rectifier are given in Fig. 2.33. The load current waveform has the same shape as that of the load voltage but has a different size. The zero crossing angles occur at $0, \pi, 2\pi$, etc. Here the dc power can be expressed as

$$P_{dc} = V_d I_d \tag{2.110}$$

where

$$V_d = \frac{1}{2}\int_0^{\pi} V_m \sin \omega t \; d(\omega t) = \frac{V_m}{\pi}$$

and

$$I_d = \frac{V_d}{R_{ld}} = \frac{V_m}{\pi R_{ld}} = \frac{I_m}{\pi}$$

where

$$I_m = \frac{V_m}{R_{ld}}$$

Thus,

$$P_{dc} = \frac{V_m I_m}{\pi^2} = \frac{V_m^2}{\pi^2 R_{ld}}$$

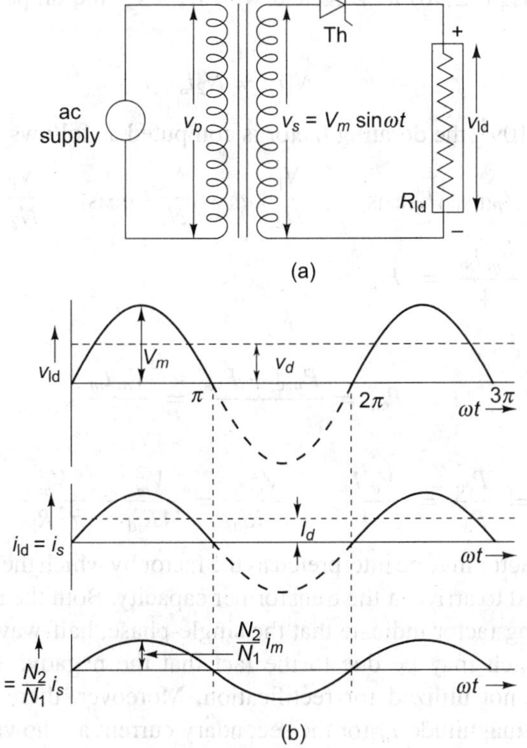

Fig. 2.33 Single-phase, half-wave rectifier feeding a resistive load: (a) circuit, (b) waveforms

Also,

$$P_{sac} = V_{s(\text{RMS})} I_{s(\text{RMS})} \tag{2.111}$$

From Fig. 2.33, the right-hand side quantities of Eqn (2.111) can be seen to be

$$V_{s(\text{RMS})} = \sqrt{\frac{1}{2\pi}\int_0^{\pi} (V_m \sin\omega t)^2 d(\omega t)} = \frac{V_m}{2}$$

and

$$I_{s(\text{RMS})} = \sqrt{\frac{1}{2\pi}\int_0^{\pi} (I_m \sin\omega t)^2 d(\omega t)} = \frac{I_m}{2}$$

Thus,

$$P_{sac} = \frac{V_m I_m}{4} = \frac{V_m^2}{4R_{\text{ld}}} \tag{2.112}$$

Now,

$$\eta_R = \frac{P_{\text{dc}}}{P_{sac}} = 100 \times \frac{V_m^2}{\pi^2 R_{\text{ld}}} \div \frac{V_m^2}{4R_{\text{ld}}} = 40.5\%$$

Taking i_{s2} in Eqn (2.96) as zero and with $i_{s1} = i_s$, the ampere-turn balance is expressed as

$$N_1 i_p = N_2 i_s \tag{2.113}$$

Using Eqn (2.109), the derating factor is computed as follows

$$P_{pac} = V_{p(\text{RMS})} I_{p(\text{RMS})} = \frac{N_1}{N_2} V_{s(\text{RMS})} \frac{N_2}{N_1} I_{s(\text{RMS})} = \frac{N_1}{N_2} \frac{V_m}{\sqrt{2}} \frac{N_2}{N_1} \frac{I_m}{2}$$

$$= \frac{V_m I_m}{4} = P_{sac}$$

Thus,

$$P_{\text{eq}} = \frac{P_{pac} + P_{sac}}{2} = \frac{V_m I_m}{4} \tag{2.114}$$

$$\delta_R = \frac{P_{\text{eq}}}{P_{\text{dc}}} = \frac{V_m I_m}{4} \div \frac{V_m^2}{\pi^2 R_{\text{ld}}} = \frac{V_m^2}{4 R_{\text{ld}}} \div \frac{V_m^2}{\pi^2 R_{\text{ld}}} = 2.47$$

The derating factor may be interpreted as the factor by which the dc load power has to be multiplied to arrive at the transformer capacity. Both the rectifier efficiency and the derating factor indicate that the single-phase, half-wave configuration is inefficient, which may be due to the fact that the negative half-waves of the ac supply are not utilized for rectification. Moreover, there is a non-zero dc component of magnitude I_d for the secondary current as shown in Fig. 2.33(b).

2.8.2.2 Single-phase, full-wave bridge rectifier feeding a resistive load

The circuit and waveforms for this configuration are given in Fig. 2.34. The dc quantities V_d and I_d are arrived at as follows:

$$V_d = \frac{2}{2\pi} \int_0^{\pi} V_m \sin \omega t \, d(\omega t) = \frac{2V_m}{\pi}$$

Similarly,

$$I_d = \frac{2I_m}{\pi} = \frac{2V_m}{\pi R_{\text{ld}}}$$

$$P_{\text{dc}} = \frac{4V_m^2}{\pi^2 R_{\text{ld}}} = \frac{4V_m I_m}{\pi^2} \tag{2.115}$$

$$V_{s(\text{RMS})} = \sqrt{\frac{2}{2\pi} \int_0^{\pi} (V_m \sin \omega t)^2 d(\omega t)} = \frac{V_m}{\sqrt{2}}$$

Similarly,

$$I_{s(\text{RMS})} = \frac{I_m}{\sqrt{2}}$$

This gives

$$P_{\text{sac}} = \frac{V_m I_m}{2} = \frac{V_m^2}{2R_{\text{ld}}} \tag{2.116}$$

$$\eta_R = 100 \times \frac{P_{\text{dc}}}{P_{\text{sac}}} = 100 \times \frac{4}{\pi^2}\frac{V_m^2}{R_{\text{ld}}} \div \frac{V_m^2}{2R_{\text{ld}}} = 81\%$$

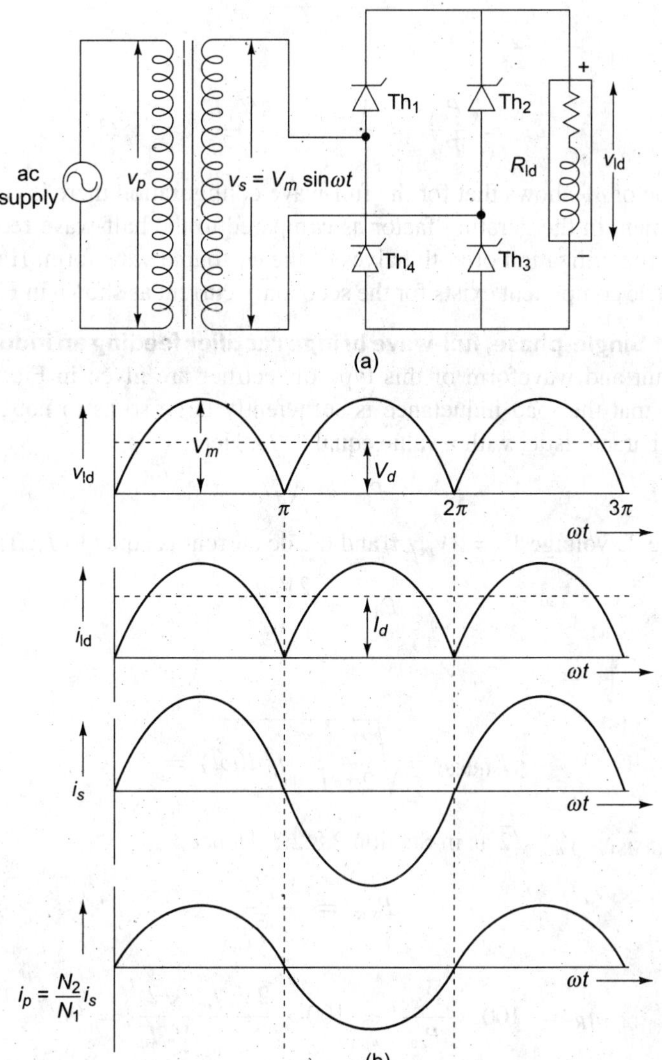

Fig. 2.34 Single-phase, full-wave bridge rectifier feeding a resistive load: (a) circuit, (b) waveforms

The ampere-turn balance in this case is the same as in Section 2.8.2.1:

$$N_1 i_p = N_2 i_s \tag{2.117}$$

or

$$i_p = \frac{N_2}{N_1} i_s$$

$$P_{pac} = V_{p(\text{RMS})} I_{p(\text{RMS})} = \frac{N_1}{N_2}\frac{V_m}{\sqrt{2}}\frac{N_2}{N_1}\frac{I_m}{\sqrt{2}} = \frac{V_m I_m}{2}$$

$$P_{\text{eq}} = \frac{P_{pac} + P_{sac}}{2} = \frac{V_m I_m}{2} \tag{2.118}$$

Hence,

$$\delta_R = \frac{P_{\text{eq}}}{P_{\text{dc}}} = \frac{V_m I_m}{2} \div \frac{4 V_m I_m}{\pi^2} = 1.23$$

This value of δ_R shows that for the full-wave configuration there is considerable improvement in the derating factor as compared to the half-wave rectifier. This is due to the utilization of both halves of the ac source waveform. However, the non-zero dc component exists for the secondary current as shown in Fig. 2.34(b).

2.8.2.3 Single-phase, full-wave bridge rectifier feeding an inductive load

The circuit and waveform of this type of rectifier are given in Fig. 2.35. It is assumed that the load inductance is sufficiently large so as to flatten the load current characteristic, with a value equal to I_d. Here,

$$P_{\text{dc}} = V_d I_d$$

where the dc voltage $V_d = 2V_m/\pi$ and the dc current is equal to I_d. Thus,

$$P_{\text{dc}} = \frac{2 V_m I_d}{\pi} \tag{2.119}$$

$$I_{s(\text{RMS})} = \sqrt{\frac{2}{2\pi}\int_0^{\pi} I_d^2 \, d(\omega t)} = I_d$$

Also, $V_{s(\text{RMS})} = V_m/\sqrt{2}$ as in Section 2.8.2.2. Hence,

$$P_{sac} = \frac{V_m I_d}{\sqrt{2}} \tag{2.120}$$

$$\eta_R = 100 \times \frac{P_{sdc}}{P_{sac}} = 100 \times \frac{2 V_m I_d}{\pi}\frac{\sqrt{2}}{V_m I_d} = 90\%$$

The primary quantities can be determined with the help of Fig. 2.35 as

$$I_{p(\text{RMS})} = \frac{N_2}{N_1} I_d \tag{2.121}$$

and

$$V_{p(\text{RMS})} = \frac{N_1}{N_2}\frac{V_m}{\sqrt{2}}$$

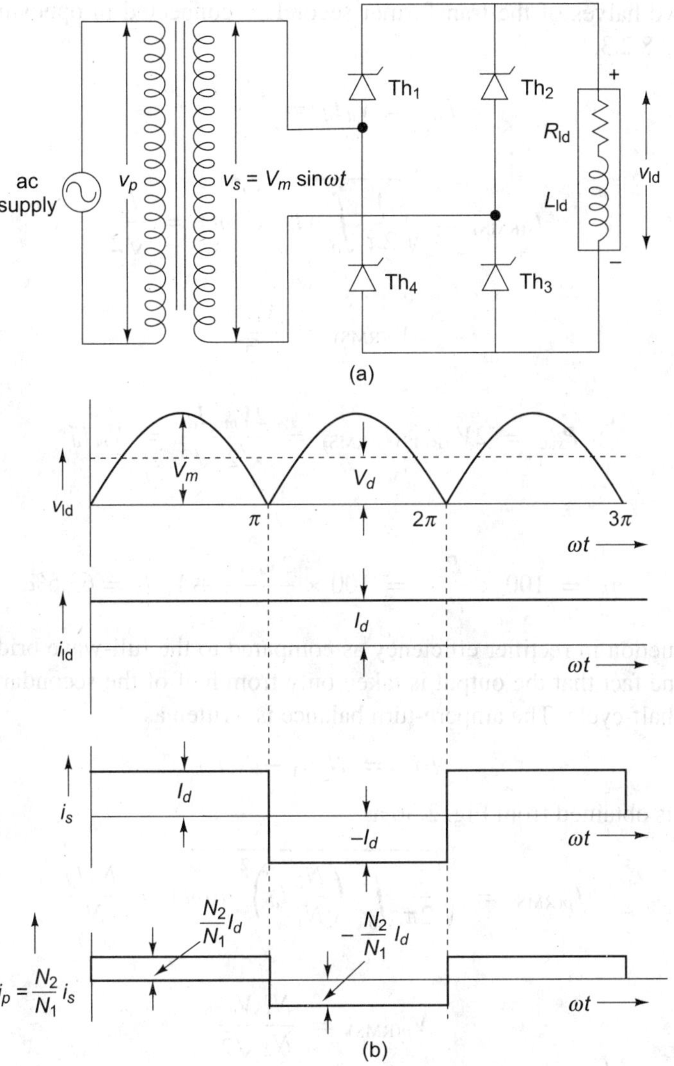

Fig. 2.35 Single-phase, full-wave bridge rectifier feeding an inductive load: (a) circuit, (b) waveforms

Thus,

$$P_{eq} = P_{pac} = P_{sac} = \frac{V_m I_d}{\sqrt{2}} \tag{2.122}$$

and

$$\delta_R = \frac{P_{eq}}{P_{dc}} = \frac{V_m I_d}{\sqrt{2}} \div \frac{2V_m I_d}{\pi} = 1.11$$

2.8.2.4 Single-phase, full-wave rectifier supplied by a centre-tapped secondary and feeding an inductive load The circuit and waveforms of the rectifier are given in Fig. 2.36. The transformer core has two circuits corresponding to the two halves of the transformer secondary connected in opposition. As in Section 2.8.2.3,

$$P_{dc} = V_d I_d = \frac{2V_m I_d}{\pi} \tag{2.123}$$

$$I_{s(RMS)} = \sqrt{\frac{1}{2\pi}\int_0^{\pi}(I_d)^2 d(\omega t)} = \frac{I_d}{\sqrt{2}}$$

$$V_{s(RMS)} = \frac{V_m}{\sqrt{2}}$$

$$P_{sac} = 2V_{s(RMS)} I_{s(RMS)} = \frac{2V_m}{\sqrt{2}}\frac{I_d}{\sqrt{2}} = V_m I_d \tag{2.124}$$

Hence,

$$\eta_R = 100 \times \frac{P_{dc}}{P_{sac}} = 100 \times \frac{2V_m I_d}{\pi} \div V_m I_d = 63.6\%$$

The reduction in rectifier efficiency as compared to the full-wave bridge case is due to the fact that the output is taken only from half of the secondary winding in each half-cycle. The ampere-turn balance is written as

$$N_1 i_p = N_2 i_{s1} - N_2 i_{s2} \tag{2.125}$$

$I_{p(RMS)}$ is obtained from Fig. 2.36 as

$$I_{p(RMS)} = \sqrt{\frac{2}{2\pi}\int_0^{\pi}\left(\frac{N_2}{N_1}I_d\right)^2 d(\omega t)} = \frac{N_2 I_d}{N_1}$$

and

$$V_{p(RMS)} = \frac{N_1}{N_2}\frac{V_m}{\sqrt{2}}$$

Now,

$$P_{pac} = V_{p(RMS)} I_{p(RMS)}$$

Using the expressions for $V_{p(RMS)}$ and $I_{p(RMS)}$, the primary power can be expressed as

$$P_{pac} = \frac{N_1}{N_2}\frac{V_m}{\sqrt{2}}\frac{N_2}{N_1}I_d = \frac{V_m I_d}{\sqrt{2}}$$

and

$$P_{eq} = \frac{1}{2}[P_{pac} + P_{sac}] = \frac{1}{2}\left[\frac{V_m I_d}{\sqrt{2}} + V_m I_d\right] \tag{2.126}$$

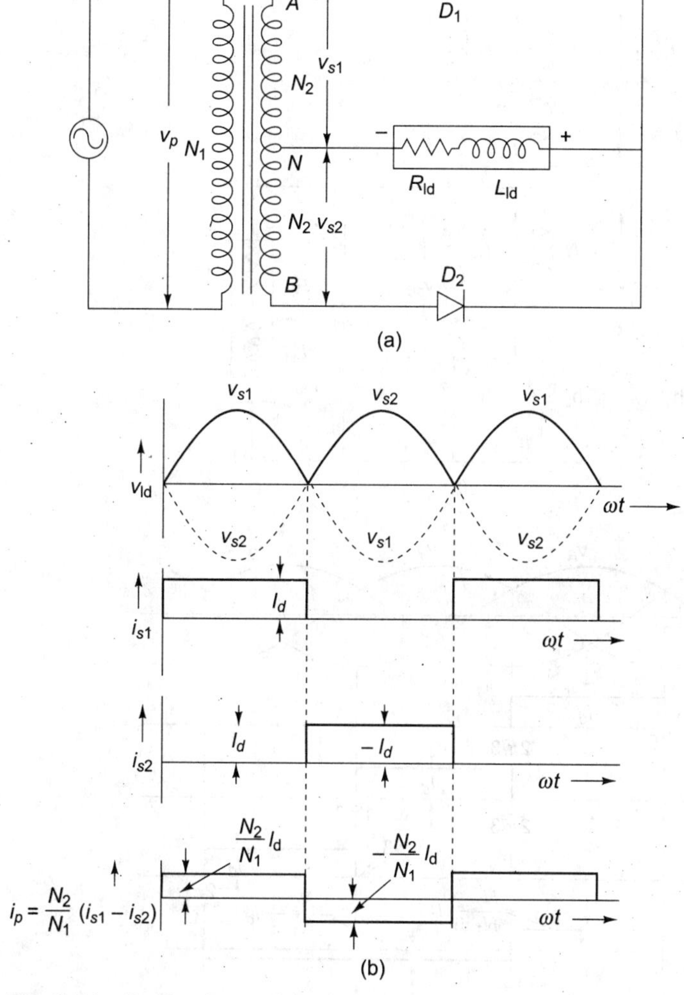

Fig. 2.36 Single-phase, full-wave rectifier fed by a centre-tapped secondary: (a) circuit, (b) waveforms

Finally,

$$\delta_R = \frac{P_{\text{eq}}}{P_{\text{dc}}} = \frac{1}{2}\frac{\left[V_m I_d/\sqrt{2} + V_m I_d\right]}{2V_m I_d/\pi} = \frac{1.707\pi}{2 \times 2} = 1.34$$

2.8.2.5 Three-phase, half-wave rectifier supplied by a Y/Y-with-neutral transformer and feeding an inductive load The circuit and waveforms are given in Fig. 2.37. As stated in case 4 of Section 2.8.1.2, the secondary ampere turns have a zero sequence component which cannot be compensated by the primary ampere turns.

$$P_{\text{dc}} = V_d I_d = \frac{3}{\pi} V_m \sin\left(\frac{\pi}{3}\right) I_d = \frac{3\sqrt{3}}{2\pi} V_m I_d \qquad (2.127)$$

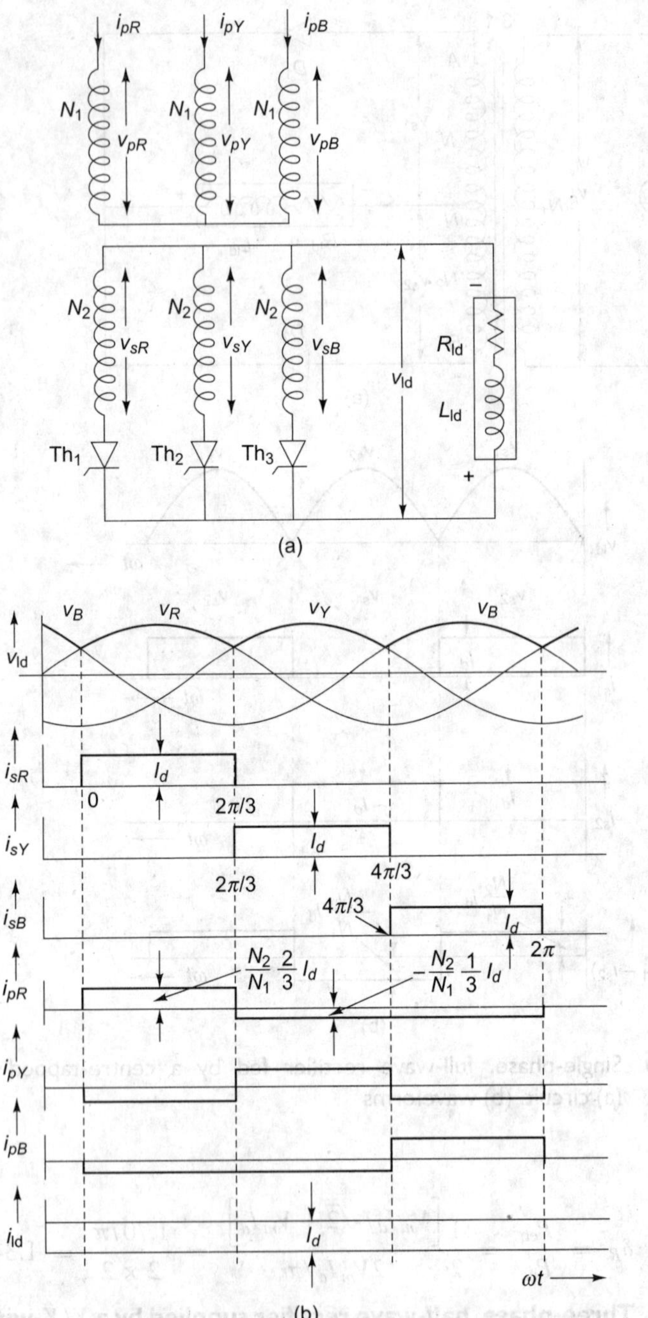

Fig. 2.37 Three-phase, half-wave controlled rectifier fed by a star/star transformer and feeding an inductive load: (a) circuit, (b) waveforms

From Fig. 2.37(b),

$$I_{s(\mathrm{RMS})} = \sqrt{\frac{1}{2\pi}\int_0^{2\pi/3} (I_d)^2 d(\omega t)} = \frac{I_d}{\sqrt{3}}$$

$$V_{s(\mathrm{RMS})} = \frac{V_m}{\sqrt{2}}$$

Hence,

$$P_{s\mathrm{ac}} = \frac{3V_m}{\sqrt{2}}\frac{I_d}{\sqrt{3}} = \frac{3}{\sqrt{6}}V_m I_d \tag{2.128}$$

Thus,

$$\eta_R = 100 \times \frac{P_{\mathrm{dc}}}{P_{s\mathrm{ac}}} = 100 \times \left(\frac{3\sqrt{3}}{2\pi}V_m I_d\right) \div \left(\frac{3}{\sqrt{6}}V_m I_d\right) = 67.5\%$$

Referring to Fig. 2.37 and using Eqns (2.103) gives $I_{p(\mathrm{RMS})}$ as

$$I_{p(\mathrm{RMS})} = \sqrt{\frac{1}{2\pi}\int_0^{2\pi/3}\left(\frac{N_2}{N_1}\frac{2I_d}{3}\right)^2 d(\omega t) + \int_{2\pi/3}^{2\pi}\left(-\frac{N_2}{N_1}\frac{I_d}{3}\right)^2 d(\omega t)} = \frac{N_2}{N_1} I_d \frac{\sqrt{2}}{3}$$

Also,

$$V_{p(\mathrm{RMS})} = \frac{N_1}{N_2}\frac{V_m}{\sqrt{2}}$$

Hence,

$$P_{p\mathrm{ac}} = \frac{3N_1}{N_2}\frac{V_m}{\sqrt{2}}\frac{N_2}{N_1}\frac{I_d\sqrt{2}}{3} = V_m I_d$$

$$P_{\mathrm{eq}} = \frac{1}{2}\left(\frac{3}{\sqrt{6}} + 1\right)V_m I_d = 1.11 V_m I_d \tag{2.129}$$

Finally,

$$\delta_R = \frac{P_{\mathrm{eq}}}{P_{\mathrm{dc}}} = 1.11 V_m I_d \div \frac{3\sqrt{3}}{2\pi}V_m I_d = 1.34$$

2.8.2.6 Three-phase, half-wave rectifier supplied by a delta/star transformer and feeding an inductive load The circuit and waveforms for this configuration are given in Fig. 2.38. It is assumed that the secondary ampere turns have a zero sequence component. The delta-connected winding gives rise to three new current variables i_{LR}, i_{LY}, and i_{LB}. However, it is seen from Fig. 2.38(b) that the phase current waveforms both in the primary and the secondary windings are identical to those of Fig. 2.37(b). Hence, the values for η_R and δ_R remain at 67.5% and 1.34, respectively.

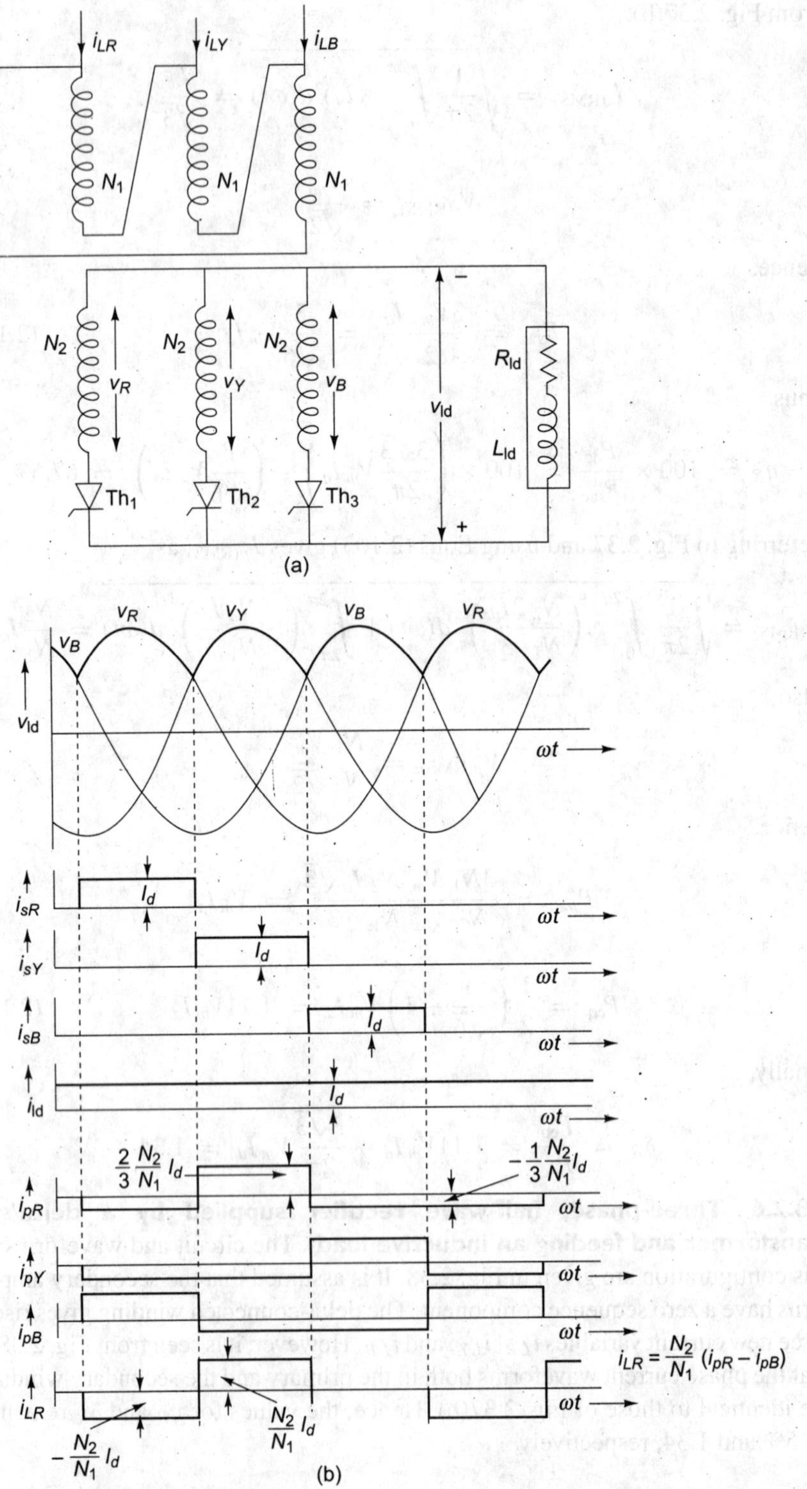

Fig. 2.38 Three-phase, half-wave rectifier supplied by a delta/star transformer and feeding an inductive load: (a) circuit, (b) waveforms

A factor called the *line power factor* is denoted for this case as PF_{line} and is defined as

$$PF_{line} = \frac{P_{dc}}{\text{input power}} = \frac{P_{dc}}{3V_L I_L}$$

where V_L and I_L are the RMS values of the line voltage and line current, respectively. They are determined as

$$V_L = V_{pR} = \frac{N_1}{N_2}\frac{V_m}{\sqrt{2}}$$

From Fig. 2.38, the instantaneous value of line current is

$$I_{LR} = \frac{N_2}{N_1}(i_{pR} - i_{pB}) = \frac{N_2}{N_1}I_d$$

The RMS value of line current is obtained as

$$I_L = \sqrt{\frac{2}{2\pi}\int_0^{2\pi/3}\left(\frac{N_2}{N_1}I_d\right)^2 d(\omega t)} = \frac{N_2}{N_1}I_d\sqrt{\frac{2}{3}}$$

Now,

$$P_{ac(line)} = V_L I_{LR} = \frac{\sqrt{3}N_1}{N_2}\frac{V_m}{\sqrt{2}}\frac{N_2}{N_1}I_d\sqrt{\frac{2}{3}} = V_m I_d \quad (2.130)$$

Finally, with P_{dc} as in Section 2.8.2.5, the input power factor is obtained as

$$PF_{line} = \frac{P_{dc}}{P_{ac(line)}} = \frac{(3\sqrt{3}/2\pi)V_m I_d}{V_m I_d} = \frac{3\sqrt{3}}{2\pi} = 0.827$$

2.8.2.7 Three-phase, half-wave rectifier supplied by a star/zigzag transformer and feeding an inductive load The transformer connections considered in Sections 2.8.2.5 and 2.8.2.6 above suffer from the drawback that the secondary ampere turns give rise to dc components; also, they do not sum up to zero and hence have zero sequence components. These two drawbacks are eliminated in the zigzag connection of Fig. 2.31(a). The zigzag secondary requires 15% more turns than a star secondary. The primary current waveforms in Fig. 2.31(c) have an alternating nature and confirm that the dc component is nullified. As stated in case 6 of Section 2.8.1.2, there is a perfect balance of the primary ampere turns in this case and thus the set of equations (2.101) holds good. The RMS values of the primary currents sum up to zero as given by Eqn (2.104). The rectifier efficiency can be computed with the help of the waveforms of Fig. 2.31(c). The dc output voltage with $\alpha = 0$ is

$$V_{d0} = \frac{3}{\pi}V_m \sin\left(\frac{\pi}{3}\right) = \frac{3\sqrt{3}}{2\pi}V_m$$

where V_m is the peak value of v_{s1} as well as that of v_{s2} and v_{s3}. The dc current is equal to I_d. Hence,

$$P_{dc} = V_d I_d = \frac{3\sqrt{3}}{2\pi}V_m I_d \quad (2.131)$$

From the phasor diagram of Fig. 2.31(b),

$$V_{s1N(\text{RMS})} = V_{s2N(\text{RMS})} = V_{s3N(\text{RMS})} = \frac{V_m}{\sqrt{2}}$$

Hence the RMS voltage of each of the six windings becomes

$$V_{R(\text{RMS})} = V_{Y(\text{RMS})} = V_{B(\text{RMS})} = \frac{1}{\sqrt{3}}\frac{V_m}{\sqrt{2}}$$

The current in each of the six windings is

$$I_{s(\text{RMS})} = \sqrt{\frac{1}{2\pi}\int_0^{2\pi/3} I_d^2 d(\omega t)} = \frac{I_d}{\sqrt{3}}$$

Thus,

$$P_{\text{sac}} = V_{s(\text{RMS})}I_{s(\text{RMS})} = \frac{6}{\sqrt{3}}\frac{V_m}{\sqrt{2}}\frac{I_d}{\sqrt{3}} = \sqrt{2}V_m I_d \tag{2.132}$$

Also,

$$\eta_R = \frac{P_{\text{dc}}}{P_{\text{sac}}} = 100 \times \left(\frac{3\sqrt{3}}{2\pi}V_m I_d\right) \div \left(\sqrt{2}V_m I_d\right) = 58.5\%$$

The RMS value of the primary current is computed with the help of the waveforms in Fig. 2.31(c) as follows:

$$I_{p(\text{RMS})} = \sqrt{\frac{1}{2\pi}\int_0^{2\pi} i_{pB}^2 d(\omega t)}$$

That is,

$$I_{p(\text{RMS})} = \sqrt{\frac{1}{2\pi}\left[\int_{4\pi/3}^{2\pi}\left(-\frac{N_2}{N_1}I_d\right)^2 d(\omega t) + \int_{2\pi}^{8\pi/3}\left(\frac{N_2}{N_1}I_d\right)^2 d(\omega t)\right]}$$

$$= \frac{N_2}{N_1}I_d\sqrt{\frac{2}{3}}$$

Also,

$$V_{p(\text{RMS})} = \frac{N_1}{N_2}\frac{1}{\sqrt{3}}\frac{V_m}{\sqrt{2}}$$

Thus,

$$P_{\text{pac}} = 3V_{p(\text{RMS})}I_{p(\text{RMS})}$$

$$= 3\frac{N_1}{N_2}\frac{1}{\sqrt{3}}\frac{V_m}{\sqrt{2}}\frac{N_2}{N_1}I_d\sqrt{\frac{2}{3}}$$

$$= V_m I_d$$

$$P_{\text{eq}} = \frac{1}{2}(\sqrt{2}+1)V_m I_d = 1.21V_m I_d \tag{2.133}$$

Finally,

$$\delta_R = \frac{P_{\text{eq}}}{P_{\text{dc}}} = 1.21 V_m I_d \div \frac{3\sqrt{3}}{2\pi} V_m I_d$$
$$= 1.46$$

2.8.2.8 Six-phase, half-wave rectifier feeding an inductive load The circuit and winding disposition of the rectifier are given in Fig. 2.32(a) and the waveforms in Fig. 2.32(b). The first configuration with the star-connected primary winding helps in obtaining a nearly sinusoidal primary current waveform. Equations (2.105) and (2.107) in case 7 of Section 2.8.1.2 are applicable here, respectively, for the ampere-turn balance and alternating primary ampere turns. The rectifier efficiency and derating factor for the two types of primary windings are arrived at as follows.

Star-connected primary From the waveforms of secondary currents given in Fig. 2.32(c), $I_{s(\text{RMS})}$ is obtained as

$$I_{s(\text{RMS})} = \sqrt{\frac{1}{2\pi}\int_0^{2\pi/6} I_d^2 d(\omega t)} = \frac{I_d}{\sqrt{6}}$$

Also,

$$V_{s(\text{RMS})} = \frac{V_m}{\sqrt{2}}$$

The apparent power at the secondary terminals is given as

$$P_{s\text{ac}} = 6 V_{s(\text{RMS})} I_{s(\text{RMS})} = 6 \frac{V_m}{\sqrt{2}} \frac{I_d}{\sqrt{6}} = \sqrt{3} V_m I_d \tag{2.134}$$

From the waveform of i_{pR} (star),

$$I_{p(\text{RMS})} = \left[\frac{2}{2\pi}\int_0^{\pi/3}\left(\frac{N_2}{N_1}\frac{1}{3}I_d\right)^2 d(\omega t) + \int_{\pi/3}^{2\pi/3}\left(\frac{N_2}{N_1}\frac{2}{3}I_d\right)^2 d(\omega t)\right.$$
$$\left. + \int_{2\pi/3}^{\pi}\left(\frac{N_2}{N_1}\frac{1}{3}I_d\right)^2 d(\omega t)\right]^{1/2} = \frac{N_2}{N_1}\frac{\sqrt{2}}{3} I_d \tag{2.135}$$

The apparent power at the primary terminals is given as

$$P_{p\text{ac}} = 3 V_{p(\text{RMS})} I_{p(\text{RMS})} = 3 \frac{N_1}{N_2}\frac{V_m}{\sqrt{2}}\frac{N_2}{N_1}\frac{\sqrt{2}}{3} I_d = V_m I_d$$

V_{dc} for this six-phase rectifier can be obtained by taking $n = 6$ and $\alpha = 0$ in Eqn (2.16). Thus,

$$V_{\text{dc}} = \frac{6}{\pi} V_m \sin\left(\frac{\pi}{6}\right) = 3\frac{V_m}{\pi}$$

Also,

$$I_{\text{dc}} = I_d$$

Hence,

$$P_{dc} = V_{dc} I_{dc} = \frac{3V_m I_d}{\pi} \tag{2.136}$$

$$\eta_R = \frac{P_{dc}}{P_{sac}} = 100 \times \frac{3V_m I_d/\pi}{\sqrt{3} V_m I_d} = 55\%$$

$$P_{eq} = \frac{1}{2}(P_{sac} + P_{pac}) = \frac{1}{2}(\sqrt{3}+1)V_m I_d = 1.37 V_m I_d \tag{2.137}$$

Finally,

$$\delta_R = \frac{P_{eq}}{P_{dc}} = \frac{1.37 V_m I_d}{(3/\pi) V_m I_d} = 1.43$$

Delta-connected primary For this case, P_{sac} and P_{dc} remain as before and hence η_R is equal to 55%. $i_{p(RMS)}$ is arrived at from the waveform of i_{pR} as follows:

$$I_{p(RMS)} = \frac{2}{2\pi} \int_{\pi/3}^{2\pi/3} \left(\frac{N_2}{N_1} I_d\right)^2 d(\omega t) = \frac{N_2}{N_1} \frac{I_d}{\sqrt{3}}$$

Also,

$$V_{p(RMS)} = \frac{N_1}{N_2} \frac{V_m}{\sqrt{2}}$$

$$P_{pac} = 3V_{p(RMS)} I_{p(RMS)} = 3 \frac{N_1}{N_2} \frac{V_m}{\sqrt{2}} \frac{N_2}{N_1} \frac{I_d}{\sqrt{3}} = \frac{3V_m I_d}{\sqrt{6}}$$

$$P_{eq} = \frac{1}{2}(P_{sac} + P_{pac}) = \frac{1}{2}\left(\sqrt{3} + \frac{3}{\sqrt{6}}\right) V_m I_d = 1.48 V_m I_d \tag{2.138}$$

$$\delta_R = \frac{P_{eq}}{P_{dc}} = \frac{1.48 V_m I_d}{(3/\pi) V_m I_d} = 1.55$$

The RMS value of line current can be computed as follows. i_{LR} can be written as

$$i_{LR} = i_{pR} - i_{pB} = \frac{N_2}{N_1}(i_{s1} - i_{s4} - i_{s5} + i_{s2})$$

Hence,

$$I_{L(RMS)} = \sqrt{\frac{2}{2\pi} \int_0^{2\pi/3} \left(\frac{N_2}{N_1} I_d\right)^2 d(\omega t)} = \frac{N_2}{N_1} \sqrt{\frac{2}{3}} I_d$$

$$V_{L(RMS)} = V_{p(RMS)} = \frac{N_1}{N_2} \frac{V_m}{\sqrt{2}}$$

Now, the input power

$$P_{\text{ac(line)}} = 3V_{L(\text{RMS})}I_{L(\text{RMS})} = \sqrt{3}\frac{N_1}{N_2}\frac{V_m}{\sqrt{2}}\frac{N_2}{N_1}\sqrt{\frac{2}{3}}I_d = V_m I_d \qquad (2.139)$$

Finally, the line power factor is

$$\text{PF}_{\text{line}} = \frac{P_{\text{dc}}}{P_{\text{ac(line)}}} = \frac{(3/\pi)V_m I_d}{V_m I_d} = 0.955$$

2.8.3 Summary of the Expressions for Rectifier Efficiency and Derating Factor

The rectifier efficiencies and derating factors for single-phase rectifiers are given in Table 2.1. The low η_R for S_1 is due to the low ac component of current [$I_{s(\text{RMS})} = I_m/2$]. As stated earlier, the unidirectional ampere turns on the secondary cannot be compensated by the primary ampere turns and thus they contribute to the core losses. For the case S_2, the secondary ac component of current has a higher magnitude [$I_{s(\text{RMS})} = I_m/\sqrt{2}$] and hence there is better utilization of transformer capacity. The high values of η_R for case S_3 are due to the absence of uncompensated dc ampere turns on the secondary side. In case S_4, only half the transformer voltage is utilized in each half-cycle; this feature together with the dc ampere turns results in poor performance factors.

Table 2.1

Factor	Single-phase, half-wave (resistive load) (S_1)	Single-phase, full-wave (resistive load) (S_2)	Single-phase, full-wave (inductive load) (S_3)	Single-phase, full-wave, centre-tapped secondary (inductive load) (S_4)
η_R	40.5%	81%	90%	63.6%
δ_R	2.47	1.23	1.11	1.34

Table 2.2 sums up the results for the rectifiers fed by three-phase and six-phase transformers. Due to the dc ampere turns present in cases T_1 and T_2, their rectifier efficiencies are low at 67.5%. In the case T_2, the line power factor is equal to the primary power factor [$P_{pac} = P_{\text{ac(line)}}$] because both the systems of currents are balanced. Even though there is ampere-turn balance for case T_3, the primary current is zero for one-third of the period. Hence η_R is low at 58.5%. The alternating zero sequence components in cases T_4 and T_5 are the cause of the low rectifier efficiency at 55%. The nearly sinusoidal primary current waveform in case T_4 is advantageous and hence the derating factor for case T_4 is less than that for case T_5. For the latter case, the primary power factor is 0.78 as against the line power factor of 0.955. The reason for this will be evident from the waveforms of i_{pR} and i_{L1}. Whereas in the former case the no current intervals have a duration of $4\pi/3$ rad, in the latter case the duration is only $2\pi/3$ rad. In spite of the poor performance factors for the three cases T_3, T_4, and T_5 as compared to those of

cases T_1 and T_2, the assumption of flatness of the current waveform is more justified because of the six-pulse nature of the current.

Table 2.2

Factor	Three-phase Y/star-with-neutral (T_1)	Three-phase Δ/star-with-neutral (T_2)	Three-phase Y/zigzag (T_3)	Six-phase Y/interstar (T_4)	Six-phase Δ/interstar (T_5)
η_R	67.5%	67.5%	58.5%	55%	55%
δ_R	1.34	1.34	1.46	1.43	1.55
PF_{line}		0.827			0.955

2.9 Input Power Factor

In all phase-controlled rectifiers, the firing angle delay causes the thyristor current to lag the phase voltage. With a highly inductive load, the load current that is carried by the thyristor attains the shape of a rectangular pulse which is delayed by an angle of α.

The fundamental power factor, also called the *displacement factor*, is defined as $\cos\phi_1$, where ϕ_1 (termed the *displacement angle*) is the phase angle difference between the phase voltage and the fundamental component of the transformer secondary current. It is shown below that for all full-wave fully controlled connections, ϕ_1 will be equal to the firing angle α.

The *input power factor* λ for a rectifier fed by the secondary of a transformer is defined as

$$\lambda = \frac{\text{mean power at the secondary}}{\text{apparent power}} \tag{2.140}$$

The transformer secondary current for a full-wave controlled rectifier consists of alternating rectangular waveforms with a time period of 2π. Denoting this current as i_s and using Fourier analysis, it can be decomposed as

$$i_s = \sum_{n=1}^{\infty} a_n \cos n\omega t + b_n \sin n\omega t \tag{2.141}$$

where a_n and b_n are the nth-order Fourier coefficients. Equation (2.140) now becomes

$$\lambda = \frac{V_{1(\mathrm{RMS})} I_{1(\mathrm{RMS})} \cos\phi_1}{V_{\mathrm{RMS}} I_{\mathrm{RMS}}} \tag{2.142}$$

where $V_{1(\mathrm{RMS})}$ and $I_{1(\mathrm{RMS})}$ are, respectively, the fundamental components of thyristor voltage and current. Here,

$$V_{1(\mathrm{RMS})} = V_{\mathrm{RMS}} \tag{2.143}$$

for the sinusoidal secondary voltage. Hence, Eqn (2.142) becomes

$$\lambda = \frac{I_{1(\mathrm{RMS})} \cos\phi_1}{I_{\mathrm{RMS}}} \tag{2.144}$$

The ratio $I_{1(\text{RMS})}/I_{\text{RMS}}$ is called the *distortion factor* denoted by μ. Equation (2.144) can now be expressed as

$$\lambda = \mu \cos \phi_1 \tag{2.145}$$

Thus the input power factor is just the product of the distortion and displacement factors. It will always be less than unity for a secondary current containing harmonic components. This is true even when the fundamental component of current is in phase with the voltage, with the diode as the rectifying device.

Expressions for input power factor are derived below with respect to commonly occurring transformer fed rectifiers.

2.9.1 n-phase, Full-wave Fully Controlled Rectifier

For a continuously conducting n-phase, full-wave rectifier, the secondary current of the transformer is an alternating rectangular waveform of amplitude I_d as shown in Fig. 2.12. If the secondary current of the transformer is denoted by i_s, it is expressed in terms of a Fourier series as

$$i_s = \sum_{m=1}^{\infty} (a_m \cos m\omega t + b_m \sin m\omega t) \tag{2.146}$$

As explained in Section 2.4.1, the firing angle α is measured from the point of natural commutation, which is at an angle of $\pi/2 - \pi/n$ rad from the origin of the n-phase sinusoid. The Fourier coefficients are now expressed as

$$a_m = \frac{2}{\pi} \int_{\pi/2-\pi/n+\alpha}^{\pi/2-\pi/n+\alpha+2\pi/n} I_d \cos m\omega t \; d(\omega t) \tag{2.147}$$

and

$$b_m = \frac{2}{\pi} \int_{\pi/2-\pi/n+\alpha}^{\pi/2-\pi/n+\alpha+2\pi/n} I_d \sin m\omega t \; d(\omega t) \tag{2.148}$$

The Fourier coefficients a_1 and b_1 associated with the fundamental component of current are therefore given as

$$a_1 = \frac{2}{\pi} \int_{\pi/2-\pi/n+\alpha}^{\pi/2-\pi/n+\alpha+2\pi/n} I_d \cos \omega t \; d(\omega t) \tag{2.149}$$

and

$$b_1 = \frac{2}{\pi} \int_{\pi/2-\pi/n+\alpha}^{\pi/2-\pi/n+\alpha+2\pi/n} I_d \sin \omega t \; d(\omega t) \tag{2.150}$$

a_1 and b_1 are obtained by evaluating the integrals, as

$$a_1 = -\frac{4I_d}{\pi} \sin\left(\frac{\pi}{n}\right) \sin \alpha \tag{2.151}$$

and

$$b_1 = \frac{4I_d}{\pi} \sin\left(\frac{\pi}{n}\right) \cos \alpha \tag{2.152}$$

The RMS magnitude of the fundamental component is obtained as

$$I_{1(\mathrm{RMS})} = \sqrt{\frac{a_1^2 + b_1^2}{2}} = \frac{2\sqrt{2}I_d}{\pi} \sin\left(\frac{\pi}{n}\right) \tag{2.153}$$

Also, the displacement angle ϕ_1 is obtained as

$$\phi_1 = \tan^{-1}\left(\frac{a_1}{b_1}\right) = \tan^{-1}(-\tan\alpha) = -\alpha \tag{2.154}$$

For single- and three-phase rectifiers $I_{1(\mathrm{RMS})}$ works out, respectively, to $2\sqrt{2}I_d/\pi$ and $\sqrt{6}I_d/\pi$. The RMS magnitude of the current pulse is obtained as

$$I_{\mathrm{RMS}} = \sqrt{\frac{2}{2\pi}\int_{\pi/2-\pi/n+\alpha}^{\pi/2-\pi/n+\alpha+2\pi/n} I_d^2 d(\omega t)} \tag{2.155}$$

$$= I_d\sqrt{\frac{2}{n}} \tag{2.156}$$

Finally from Eqn (2.145), the input power factor is obtained as

$$\lambda = \frac{(2\sqrt{2}/\pi)I_d \sin(\pi/n)}{I_d\sqrt{2/n}} \cos(-\alpha) \tag{2.157}$$

$$= \frac{2\sqrt{n}}{\pi} \sin\left(\frac{\pi}{n}\right) \cos\alpha \tag{2.158}$$

The input power factors for single- and three-phase rectifiers work out, respectively, to $(2\sqrt{2}/\pi)\cos\alpha$ and $(3/\pi)\cos\alpha$.

2.9.2 Three-phase, Half-wave Controlled Rectifier

As stated in Section 2.8.1, a dc component exists for the transformer secondary current for all configurations of three-phase, half-wave controlled rectifiers. Two of the commonly occurring connections are (a) Y/Y-with-neutral [Fig. 2.8(a)] and (b) Δ/Y-with-neutral [Fig. 2.9(a)]. In both of these cases, an uncompensated zero sequence component flows through the secondaries, causing the saturation of the core. Hence, the use of three-phase, half-wave controlled rectifiers is limited to low-power applications where such undesirable currents can be ignored.

2.9.3 Three-phase, Half-wave Controlled Rectifier Fed by a Star/Zigzag Transformer

The drawbacks mentioned above for the three-phase, half-wave controlled rectifiers are eliminated in the star/zigzag transformer, because here the secondary ampere turns sum up to zero. However, it employs three more secondary windings, which implies that it requires 15% more turns than a star secondary; this is with the assumption that the turn ratios between the primary winding R′ and the secondary windings r_1 and r_2 are in the ratio $\sqrt{3}$:1:1. Figures 2.31(a), (b), and (c), respectively, show the circuit, phasor diagram, and waveforms. The ratio N_2/N_1 given in the waveforms is $1/\sqrt{3}$.

In Fig. 2.31(b), the windings b_1 and r_2 connected to Th_1 are mutually displaced by a space angle of 120°. Hence, the voltage $V_{s1N} = (V_{b_1} + V_{r_2})\cos 30 = (2 \times V_{R'}/\sqrt{3})(\sqrt{3}/2) = V_{R'}$, where $V_{R'}$ is assumed to be the phase voltage of the primary. Hence the voltage ratio of the transformer is 1.

Taking the primary star as the source for the associated rectifier and assuming the firing angle to be measured from the origin of the supply voltage waveform, the input power factor is arrived at as follows. The RMS value of the primary current i_{pR} is evaluated as

$$I_{R(\mathrm{RMS})} = \left\{ \frac{1}{2\pi} \left[\int_{\alpha'}^{2\pi/3+\alpha'} \left(\frac{-I_d}{\sqrt{3}} \right)^2 d(\omega t) + \int_{2\pi/3+\alpha'}^{4\pi/3+\alpha'} \left(\frac{I_d}{\sqrt{3}} \right)^2 d(\omega t) + \int_{4\pi/3+\alpha'}^{2\pi+\alpha'} [0]^2 d(\omega t) \right] \right\}^{1/2} \tag{2.159}$$

$$= \frac{I_d\sqrt{2}}{3} \tag{2.160}$$

i_{pR} can be expressed in terms of a Fourier series as

$$i_{pR} = \sum_{n=1}^{\infty} (a_n \cos n\omega t + b_n \sin n\omega t) \tag{2.161}$$

The fundamental component coefficients a_1 and b_1 are obtained as

$$a_1 = \frac{1}{\pi} \left[\int_{\alpha'}^{2\pi/3+\alpha'} \left(\frac{-I_d}{\sqrt{3}} \right) \cos \omega t \, d(\omega t) + \int_{2\pi/3+\alpha'}^{4\pi/3+\alpha'} \left(\frac{I_d}{\sqrt{3}} \right) \cos \omega t \, d(\omega t) + \int_{4\pi/3+\alpha'}^{2\pi+\alpha'} 0 \, d(\omega t) \right] \tag{2.162}$$

$$= -\frac{\sqrt{3} I_d}{\pi} \sin \left(\frac{\pi}{3} - \alpha' \right) \tag{2.163}$$

and

$$b_1 = \frac{1}{\pi} \left[\int_{\alpha'}^{2\pi/3+\alpha'} \left(\frac{-I_d}{\sqrt{3}} \right) \sin \omega t \, d(\omega t) + \int_{2\pi/3+\alpha'}^{4\pi/3+\alpha'} \left(\frac{I_d}{\sqrt{3}} \right) \sin \omega t \, d(\omega t) \right] \tag{2.164}$$

$$= -\frac{\sqrt{3} I_d}{\pi} \cos \left(\frac{\pi}{3} - \alpha' \right) \tag{2.165}$$

Now,

$$I_1 = \frac{c_1}{\sqrt{2}} = \sqrt{\frac{a_1^2 + b_1^2}{2}} = \sqrt{\frac{3}{2}} \frac{I_d}{\pi} \tag{2.166}$$

and

$$\phi_1 = \tan^{-1}\left(\frac{a_1}{b_1}\right) = \tan^{-1}\left[\tan\left(\frac{\pi}{3} - \alpha'\right)\right] = \frac{\pi}{3} - \alpha' \tag{2.167}$$

Finally, the input power factor can be derived as

$$\lambda = \mu \cos\phi_1 = \left(\frac{I_{1(\text{RMS})}}{I_{R(\text{RMS})}}\right)\cos\phi_1$$

Finally, the expression for λ can be written as

$$\lambda = \frac{\sqrt{3/2}(I_d/\pi)}{(\sqrt{2}/3)I_d}\cos\left(\frac{\pi}{3} - \alpha'\right)$$

$$= \frac{3\sqrt{3}}{2\pi}\cos\left(\frac{\pi}{3} - \alpha'\right) \tag{2.168}$$

2.9.4 Six-phase, Half-wave Rectifier Fed by a Star/Interstar Transformer

The winding connections and waveforms of this rectifier are shown, respectively, in Figs 2.32(a) and (c). The primary star is considered to be the source (input) because the corresponding phase current is an alternating waveform of stepped nature. The input power factor is now derived with the simplifying assumption that the turn ratio N_2/N_1 is 1:1.

$$I_{R(\text{RMS})} = \frac{2}{2\pi}\left[\int_{\alpha'}^{\pi/3+\alpha'}\left(\frac{I_d}{3}\right)^2 d(\omega t) + \int_{\pi/3+\alpha'}^{2\pi/3+\alpha'}\left(\frac{2I_d}{3}\right)^2 d(\omega t) + \int_{2\pi/3+\alpha'}^{\pi+\alpha'}\left(\frac{I_d}{3}\right)^2 d(\omega t)\right] \tag{2.169}$$

Simplification gives

$$I_{R(\text{RMS})} = \frac{I_d\sqrt{2}}{3} \tag{2.170}$$

The Fourier coefficients corresponding to the fundamental component are arrived at as follows:

$$a_1 = \frac{2}{\pi}\left[\int_{\alpha'}^{\pi/3+\alpha'}\frac{I_d}{3}\cos\omega t\, d(\omega t) + \int_{\pi/3+\alpha'}^{2\pi/3+\alpha'}\frac{2I_d}{3}\cos\omega t\, d(\omega t) + \int_{2\pi/3+\alpha'}^{\pi+\alpha'}\frac{I_d}{3}\cos\omega t\, d(\omega t)\right] \tag{2.171}$$

$$= -\frac{2I_d}{\pi}\sin\alpha' \tag{2.172}$$

Also,

$$b_1 = \frac{2}{\pi}\left[\int_{\alpha'}^{\pi/3+\alpha'} \frac{I_d}{3}\sin\omega t\, d(\omega t) + \int_{\pi/3+\alpha'}^{2\pi/3+\alpha'} \frac{2I_d}{3}\sin\omega t\, d(\omega t) + \int_{2\pi/3+\alpha'}^{\pi+\alpha'} \frac{I_d}{3}\sin\omega t\, d(\omega t)\right] \tag{2.173}$$

$$= \frac{2I_d}{\pi}\cos\alpha' \tag{2.174}$$

The corresponding RMS value of current and its phase angle are given as

$$I_{R_1(\mathrm{RMS})} = \frac{\sqrt{a_1^2 + b_1^2}}{\sqrt{2}} = \frac{\sqrt{2}I_d}{\pi} \tag{2.175}$$

and

$$\phi_1 = \tan^{-1}\left(\frac{a_1}{b_1}\right) = \tan^{-1}(-\tan\alpha') = -\alpha' \tag{2.176}$$

The input power factor is obtained as

$$\lambda = \frac{I_{R_1(\mathrm{RMS})}}{I_{R(\mathrm{RMS})}}\cos\phi_1 = \left(\frac{I_d\sqrt{2}/\pi}{I_d\sqrt{2}/3}\right)\cos(-\alpha') = \left(\frac{3}{\pi}\right)\cos\alpha' \tag{2.177}$$

2.9.5 Six-phase, Half-wave Rectifier Fed by a Delta/Interstar Transformer

The winding disposition and waveforms for this rectifier are also given in Figs 2.32(a) and (c). The primary delta is considered to be the source by virtue of its current waveform being an alternating rectangular one. As before, the turn ratio N_2/N_1 is assumed to be 1:1 for the sake of simplicity.

$$I_{p(\mathrm{RMS})} = \sqrt{\frac{2}{2\pi}\int_{\pi/3+\alpha'}^{2\pi/3+\alpha'} I_d^2 d(\omega t)} = \frac{I_d}{\sqrt{3}} \tag{2.178}$$

As before, the Fourier coefficients a_1 and b_1 are obtained as follows:

$$a_1 = \frac{2}{\pi}\int_{\pi/3+\alpha'}^{2\pi/3+\alpha'} I_d\cos\omega t\, d(\omega t) = -\frac{2I_d}{\pi}\sin\alpha' \tag{2.179}$$

and

$$b_1 = \frac{2}{\pi}\int_{\pi/3+\alpha'}^{2\pi/3+\alpha'} I_d\sin\omega t\, d(\omega t) = \frac{2I_d}{\pi}\cos\alpha' \tag{2.180}$$

Hence,

$$I_{R_1(\mathrm{RMS})} = \sqrt{\frac{a_1^2 + b_1^2}{2}} = \frac{\sqrt{2}I_d}{\pi} \tag{2.181}$$

and

$$\phi_1 = \tan^{-1}\left(\frac{a_1}{b_1}\right) = \tan^{-1}(-\tan\alpha') = -\alpha' \tag{2.182}$$

Finally, the input power factor is given as

$$\lambda = \frac{I_d\sqrt{2}/\pi}{I_d/\sqrt{3}}\cos(-\alpha') = \frac{\sqrt{6}}{\pi}\cos\alpha' \tag{2.183}$$

2.10 Dual Converters

The operation of a fully controlled converter in both the rectifying and inverting modes is termed as *two-quadrant* operation because it can be represented by points in the first and fourth quadrants, respectively, on the v-i plane. Typical (v, i) points in the two quadrants are shown in Fig. 2.39(a).

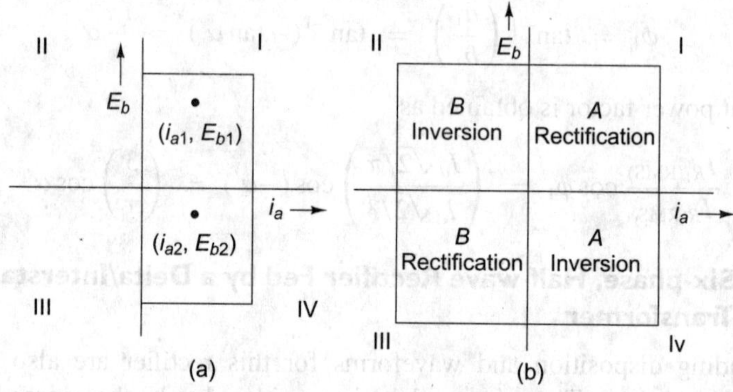

Fig. 2.39 Multiquadrant operation of a separately excited dc motor: (a) two-quadrant operation, (b) four-quadrant operation

If a second fully controlled converter which conducts current in the reverse direction is connected as shown in Fig. 2.40(a), the combination gives four-quadrant operation. Figure 2.39(b) illustrates the rectifying and inversion modes of the P and N converters. This combined circuit is called a *dual converter* and can be employed as a source for a reversible separately excited dc motor. With the positive direction of motor current taken to be from C to D, converters P and N act, respectively, as the positive and negative conducting converters. With the thyristors of converter N being oriented in the reverse direction, its rectification mode will be in the third quadrant and inversion mode in the fourth quadrant.

Two schemes of dual converter control are considered here: (a) non-simultaneous control and (b) simultaneous control.

2.10.1 Non-simultaneous Control

In non-simultaneous control [Fig. 2.40(a)], when one of the converters, say, converter P, is in operation, the other converter (N) is kept in a blocked condition by disconnecting its triggering signals. Assuming that the motor initially operates in the first quadrant, braking of the motor involves changing the operating point

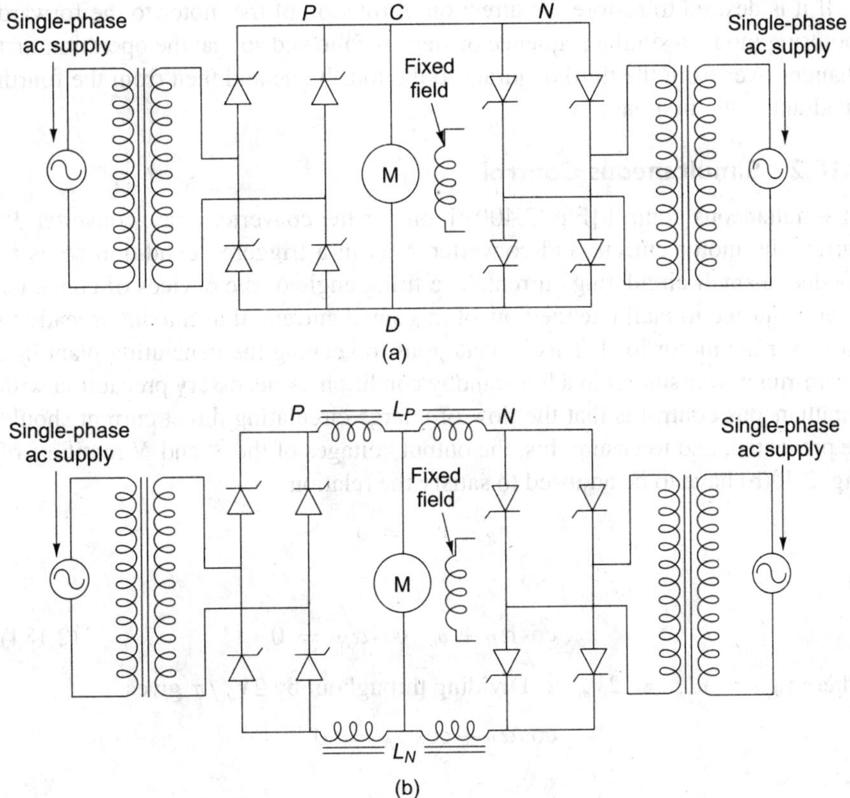

Fig. 2.40 Dual converter configurations: (a) circuit for non-simultaneous control, (b) circuit for simultaneous control

from the first quadrant to the second one. The changeover is carried out in a smooth manner as follows.

(i) The armature current is forced to zero by increasing the firing angle α_P of the devices of converter P to 180°.

(ii) After zero current is sensed, a dead time of 2–10 ms is allowed, to ensure the turn-off of all the thyristors of converter P, and at this stage its triggering signals are disconnected. No drastic change occurs in the motor speed during this period because of inertia.

(iii) The firing angle α_N of the devices of converter N are kept at a value which is slightly less than 180° and then its firing signals released.

(iv) α_N is then reduced to fall in the range $90° < \alpha_N < 180°$, thus ensuring the operation of converter N in the inversion mode (second quadrant). After a short time the motor speed become zero and this completes the braking of the motor.

If now it is required to reverse the rotation, α_N is further reduced to fall in the range $(0° < \alpha_N < 90°)$. This facilitates shifting of the operation to the third quadrant.

If it is desired to restore the direction of rotation of the motor to the forward motoring mode, a similar sequence of steps is followed so that the operation first changes over from the third quadrant to the fourth one and then from the fourth quadrant to the first one.

2.10.2 Simultaneous Control

In simultaneous control [Fig. 2.40(b)], one of the converters, say, converter P, carries the motor current and converter N is in a triggered condition so as to conduct a small circulating current. The firing angle of the devices of converter N are adjusted to facilitate the flow of this small current, thus making it ready to take over the motor load. This is analogous to keeping the generating plant in a steam-run power station in a hot standby condition. A necessary precaution with simultaneous control is that the flow of a large circulating direct current should be prevented, and to ensure this, the output voltages of the P and N rectifiers of Fig. 2.40(b) have to be adjusted to satisfy the relation

$$v_P + v_N = 0$$

or

$$v_{\alpha_P} \cos \alpha_P + v_{\alpha_N} \cos \alpha_N = 0 \tag{2.184}$$

where $v_{\alpha_P} = v_{\alpha_N} = 2V_m/\pi$. Dividing throughout by $2V_m/\pi$ gives

$$\cos \alpha_P + \cos \alpha_N = 0$$

or

$$\alpha_P + \alpha_N = 180^\circ \tag{2.185}$$

Equation (2.185) implies that, with the converter P working as a rectifier ($0 < \alpha_P < 90^\circ$), the converter N has to be operated in the inverting mode ($90 < \alpha_N < 180^\circ$) and vice versa. However, as their instantaneous voltages are not equal, an ac circulating current flows through the two rectifiers. The inductors L_P and L_N are necessary to limit this ac current. Assuming that converter P carries the motor current, motor control is possible either in the first or fourth quadrant. Converter N carries only the circulating current. Now, if converter N takes over the motor current, operation takes place only in the second or third quadrant, with converter P carrying the circulating current. Section 6.4 deals with the sequence of steps for a smooth changeover of conduction from one converter to the other for a simultaneously controlled dual converter.

2.10.3 Circulating Current

The following analysis is aimed at deriving an expression for the circulating current flowing in a simultaneously controlled single-phase dual converter. It is assumed that rectifier P carries the load current and rectifier N carries only the circulating current. In Fig. 2.41(a), v_P and v_N are the output voltages of rectifiers P and N, respectively, with the firing angle α_P kept at 60°. As per Eqn (2.185), α_N has to be adjusted at 120°, implying that rectifier N has to work in the inversion mode. The difference between the voltages v_{AB} and $-v_{dc}$ equals the drop across the inductors L_P and L_N. This is expressed as

$$L\frac{di_c}{dt} = v_{AB} - v_{CD} = v_P - (-v_N) = v_P + v_N \tag{2.186}$$

where i_c is the circulating current and $L = L_P + L_N$. It is seen from the waveforms of Fig. 2.41(b) that the sum $v_P + v_N$ is zero during the intervals α_P to $\pi - \alpha_P$ and $\pi + \alpha_P$ to $2\pi - \alpha$. Considering the interval $\pi - \alpha_P$ to $\pi + \alpha_P$, v_P and v_N are related as

$$v_P + v_N = 2V_m \sin \omega t$$

with $i_c(\omega t) = 0$ at $\pi - \alpha_P$. Equation (2.186) can be solved for i_c as

$$i_c = \frac{1}{L}\int_{(\pi-\alpha_P)/\omega}^{t} (v_P + v_N)dt = \frac{2V_m}{L}\int_{(\pi-\alpha_P)/\omega}^{t} \sin \omega t \, dt + c_1$$

The integration constant c_1 becomes zero because of the initial condition on i_c. Performing the integration gives the circulating current as

$$i_c = -\frac{2V_m}{\omega L}(\cos\alpha_P + \cos\omega t) \qquad \text{for } \pi - \alpha_P \le \omega t \le \pi + \alpha_P \tag{2.187}$$

Figure 2.41(c) shows the plot of the circulating current for the interval $\pi - \alpha_P \le \omega t \le \pi + \alpha_P$, which becomes 120° (elec) $\le \omega t \le$ 240° (elec) for α_P = 60°; the plot is shown as a continuous curve. For the intervals $0 \le \omega t \le \alpha_P$ and $2\pi - \alpha_P \le \omega t \le 2\pi$, $v_P + v_N$ becomes $-2V_m \sin \omega t$. However, for integration, it is convenient to take the former interval as $-\alpha_P \le \omega t \le \alpha_P$, the latter interval remaining as it is. The per-unit magnitude for I_c for the interval $-\alpha_P \le \omega t \le \alpha_P$ is drawn only for the half intervals 0° to 60° and 300° to 360°. Proceeding as above, the circulating current for these intervals is obtained as

$$i_c = \frac{2V_m}{\omega L}(\cos\omega t - \cos\alpha_P)$$

The plots for these two intervals also are given in Fig. 2.41(c). It is easy to see that for α_P = 0°, the circulating current is constant at zero as shown by the dash-dot-dash line, and equal to $(2V_m/\omega L)\cos\omega t$ when α_P = 90°, which is shown as the dashed cosine curve. It can be concluded that the plot of the circulating current depends on α_P on the one hand and the inductance L on the other.

The voltage that causes the circulating current is the algebraic sum of the instantaneous voltages and is itself an alternating quantity. The path of the circulating current is through the conducting SCRs, the transformer windings, and the two inductors L_P and L_N. It is assumed in the above analysis that the leakage inductances of the transformer windings are negligible as compared to the combined inductance L. The following remarks are relevant in connection with the operation of the dual converter.

(a) Though the above discussion has centred around a dual converter consisting of single-phase bridge rectifiers, it can be extended to a converter consisting of three-phase, full-wave fully controlled bridge rectifiers.

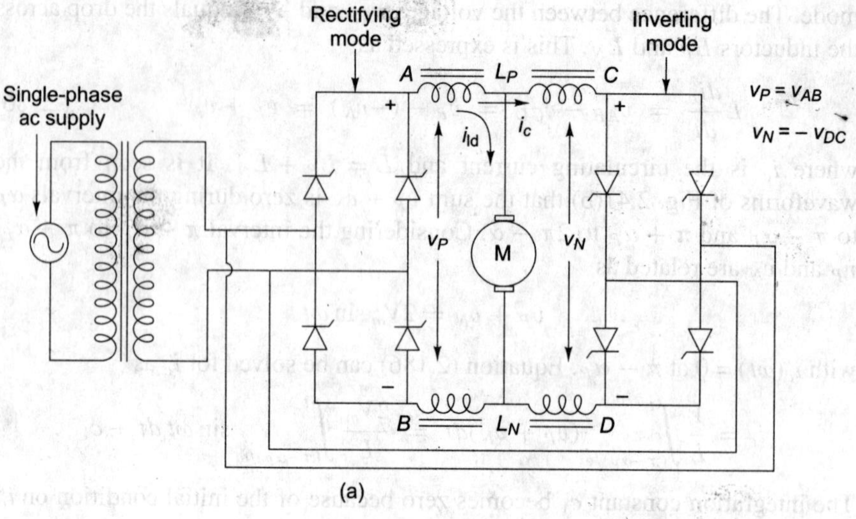

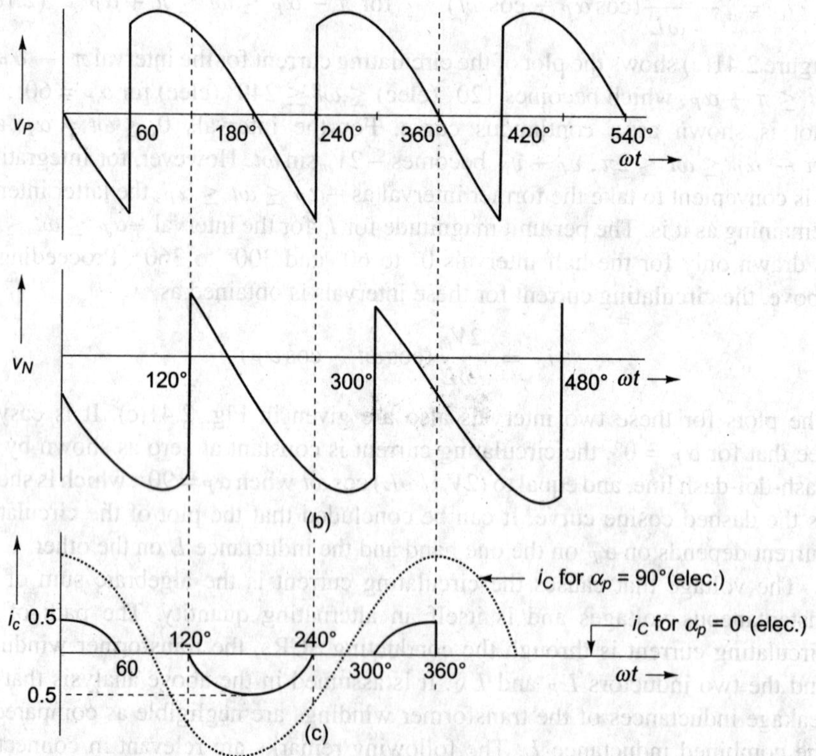

Fig. 2.41 Single-phase simultaneously controlled dual converter: (a) circuit, (b) waveforms of v_P and v_N, (c) waveforms of circulating currents with $\alpha = 0°, 60°$, and $90°$

(b) Though the details of the firing circuitry have not been dealt with, it is to be noted that a carefully designed firing circuitry is essential to provide the proper sequence of triggering signals for both the converters.

(c) A comparative assessment of the two types of control shows that the presence of reactors in simultaneous control leads to higher cost, lower efficiency, and lower power factor as compared to the non-simultaneous type. However, it has the following merits: (i) continuous conduction is guaranteed because of the freedom for the motor to rotate in either direction, the gain being constant; (ii) simpler control and better speed regulation are achieved; and (iii) continuous conduction of both the converters is facilitated over the whole control range, irrespective of the magnitude of load current. This in turn facilitates faster changeover of operation from one quadrant to another. The main drawback of non-simultaneous control is that speed regulation is poor because of discontinuous conduction at light loads. This feature, combined with the presence of a dead zone during reversal makes its gain characteristic non-linear. A third disadvantage is that additional logic circuitry is needed to detect zero current as well as implement dead time. The availability of ICs and hardware for detecting zero current at lower price are making the implementation of non-simultaneous control easy and also facilitating faster response. This type of dual converter is finding wider acceptance because of these trends.

(d) The motor employed with the dual converter has been assumed to be a separately excited one. Though using the dc series motor is advantageous for some kinds of hoists, one needs to ensure that the field current remains constant during the reversal of armature current. For accomplishing this, a diode bridge has to be interposed in the armature circuit, the field is connected at the output terminal of the diode bridge.

Chapter 6 deals with cycloconverters in which dual converters find extensive application.

2.11 Hoist Operation

A ready application of the four-quadrant operation of a dual converter is found in hoist operation. Hoists are used in hydroelectric power stations and mines, which are situated 30–300 m below the ground level, to move men and material up to the ground level. It is assumed that a load has to be hoisted up by means of a hook. After this load is removed from the hook, the light hook has to be lowered down to the bottom so that it gets ready to hoist a second load.

The operation of a hoist using a separately excited dc motor controlled by a dual converter can be explained with the help of Figs 2.42(a)–(c). The torque and speed of a dc motor are assumed to be positive in the hoisting mode, the equation for power being given as

$$P = KT_d\Omega \qquad (2.188)$$

Equation (2.188) shows that the shaft power is positive when torque and speed are positive. It is shown below that there is a one-to-one correspondence between the

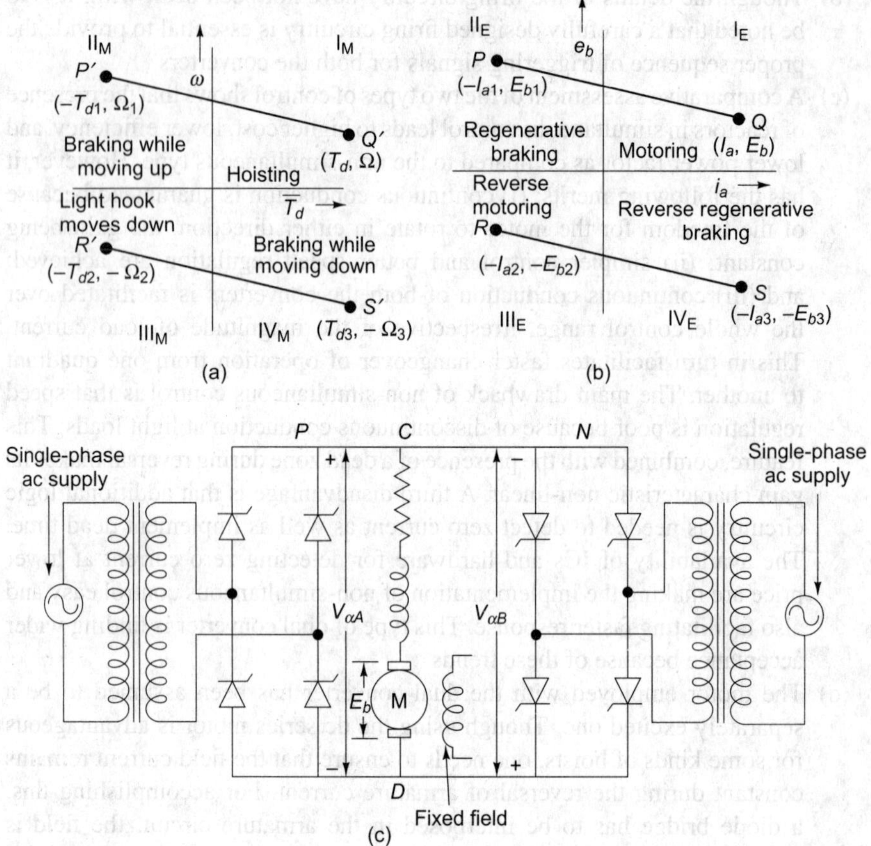

Fig. 2.42 Four-quadrant operation of a dc separately excited motor: (a) ω-$\overline{T}$ diagram, (b) e_b-i_a diagram, (c) circuit of a dual converter

corresponding quadrants of the $\overline{T}_d$-ω plane and the e_b-i_a plane drawn, respectively, in Figs 2.42(a) and (b). The relations for a separately excited dc motor are

$$\phi = K_\phi i_f \tag{2.189}$$

$$v_a = e_b + i_a R_a \tag{2.190}$$

$$e_b = K_b \phi \omega \tag{2.191}$$

$$\overline{T}_d = K \phi i_a \tag{2.192}$$

where ϕ is the field flux, I_f is the field current, v_a is the terminal voltage of either of the rectifiers, e_b is the back emf of the motor, i_a is the armature current, R_a is the armature resistance, ω is the variable angular speed of the motor, and $\overline{T}_d$ is the variable motor torque developed at the shaft.

With constant field current I_f and hence constant field flux Φ, e_b is proportional to ω and $\overline{T}_d$ is proportional to i_a. The positive motoring shown in the first quadrant of the e_b-i_a diagram corresponds to the hoisting mode. The correspondence of the other quadrants of the e_b-i_a and $\overline{T}_d$-ω diagrams are also shown in the figures. The line PQ in Fig. 2.42(b) represents Eqn (2.190), with V_a as the intercept on the e_b-axis and R_a as its slope. As the point Q falls in the I_E quadrant, torque is positive; the corresponding (T_d, Ω) point is shown as Q' in Fig. 2.42(a). Assuming non-simultaneous control of the dual converter, converter P of Fig. 2.42(c) will be in the rectifying mode and converter N will be off. In this mode, shown in Fig. 2.43, the hoist moves from the bottom and proceeds up towards the ground level.

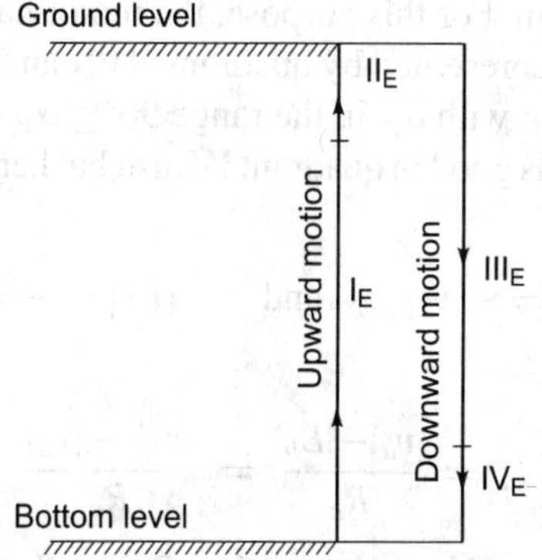

Fig. 2.43 Range of hoist motion during four-quadrant operation

When the hoist has reached a point slightly below the ground level, it is necessary to brake the motor at this level to facilitate its stopping at the ground level. This involves the sequence of steps given in Section 2.10 for the non-simultaneously controlled dual converter. The motor operation will now shift to a new (i_a, E_b) point located in quadrant II_E, which represents regenerative braking and implies eventual stopping of the motor. The voltage will be positive as viewed from the terminals of the converter N. The relation between E_b and i_a in quadrant II_E can be written as

$$v_a = e_b + i_a R_a \tag{2.193}$$

which gives i_a as

$$i_a = \frac{v_a - e_b}{R_a} \tag{2.194}$$

As v_a is less than E_b in the second quadrant, i_a is negative and the point P has the coordinates $(-i_{a1}, E_{b1})$. The torque is also negative, with ω having a positive sign, and therefore the point on the $\overline{T}_d$-ω diagram falls in quadrant II_M, with P' as a representative point.

For lowering the light hook, the motor has to rotate in a reverse direction. This is shown as the reverse motoring mode and the operation is represented by a point in quadrant III_E. Converter N operates as a rectifier with α_N in the range $90° \leq \alpha_N < 180°$. Equation (2.190) holds good but with both v_a and e_b

negative. Writing $v_a = -|v_1|$ and $e_b = -|v_2|$, i_a can be written as

$$i_a = \frac{-|v_1| + |v_2|}{R_a} \tag{2.195}$$

As $|v_1| > |v_2|$ in the III_E quadrant, the armature current i_a will be negative, R being a typical operating point. The transition from quadrant II_E to quadrant III_E is carried out smoothly as described earlier. The corresponding point in quadrant III_M of the τ_d-ω diagram is shown as R'.

As the light hook reaches a point slightly above the bottom of the mine (or the level of a hydropower house), it is necessary to initiate action to bring it to a stop exactly at the bottom. For this purpose, the motor has to be brought into the reverse braking mode, represented by quadrants IV_M and IV_E. Rectifier P has to be brought into operation with α_P in the range $90° \leq \alpha_P \leq 180°$.

Equation (2.190) holds good in quadrant IV also but here E_b and v_a are negative and written as

$$v_a = -|v_2| \quad \text{and} \quad |E_b| = -|v_3|$$

and

$$i_a = \frac{v_a - E_b}{R_a} = \frac{|v_3| - |v_2|}{R_a} \tag{2.196}$$

As $|v_3| > |v_2|$ in quadrant IV_E, i_a is positive. Points S and S' are typical points, respectively, in quadrants IV_E and IV_M.

2.12 Braking of Dc Motors

Single-phase and three-phase motors have to be braked at some stage of their operation and it is desirable that this be done with minimum power loss. Three electrical methods of braking of dc motors, along with their relative merits and demerits, are discussed here.

When the terminals of a motor are disconnected from the dc source, the machine continues to run as a generator. Because of the opposing load torque, it eventually slows down, finally coming to rest when the kinetic energy is completely dissipated. If either the load torque is small or the inertia is large, the motor takes a long time to come to a stop. It follows that if it is desired to brake the motor within a short interval, additional opposing torque has to be applied to the rotor. By applying a mechanical brake an opposing frictional torque can be generated but this causes wear and tear of mechanical parts. Hence, electrical braking, which is smoother, is preferred, three of the usually implemented methods being (i) dynamic or rheostatic braking, (ii) plugging or reverse voltage braking, and (iii) regenerative braking.

2.12.1 Dynamic Braking

Dynamic braking consists in first disconnecting the dc supply and then connecting a resistance, called the *braking resistance*, across the terminals of the machine. The kinetic energy of the machine, which now runs as a generator, is converted

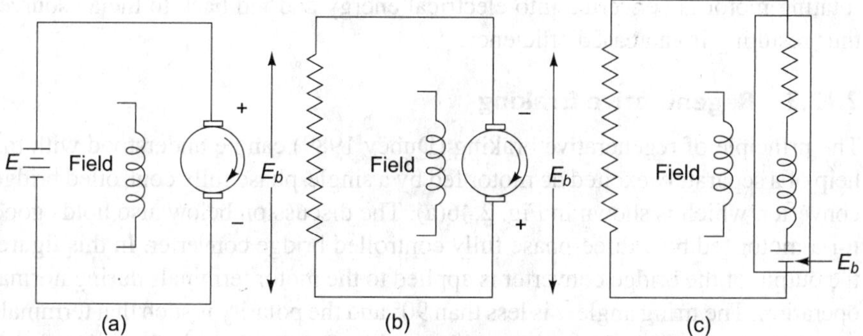

Fig. 2.44 Dynamic braking of a separately excited dc motor: (a) machine in normal motoring mode, (b) machine running as a generator feeding the braking resistor, (c) equivalent circuit

into electrical energy and dissipated in the braking resistor. Figures 2.44(a)–(c) illustrate the three steps of this method of braking.

2.12.2 Plugging

In the plugging mode connection shown in Fig. 2.45(b), the motor armature terminals of Fig. 2.45(a) are reversed and a braking resistor is connected in the circuit. As a result of the reversal, the supply voltage and the back emf combine to pass a large reverse current through the armature. Thus a braking torque is generated and the motor quickly comes to a stop.

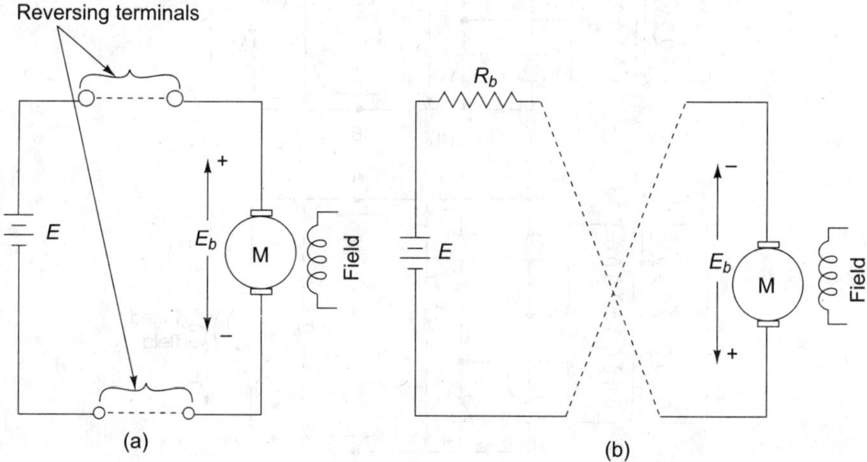

Fig. 2.45 Plugging of a separately excited dc motor: (a) normal motoring mode, (b) plugging mode

As power is wasted in both of the above methods, techniques that do not involve such wastage are preferred. In regenerative braking, which is described below, no power is spent for braking; on the other hand, the kinetic energy of the rotating motor is converted into electrical energy and fed back to the ac source, thus resulting in increased efficiency.

2.12.3 Regenerative Braking

The principle of regenerative braking (Dubey 1989) can be understood with the help of a separately excited dc motor fed by a single-phase fully controlled bridge converter, which is shown in Fig. 2.46(a). The discussion below also holds good for a motor fed by a three-phase fully controlled bridge converter. In this figure, the output of the bridge converter is applied to the motor terminals during normal operation. The firing angle α is less than 90° and the polarity is such that terminals A and B are, respectively, positive and negative. Current flows in the series circuit consisting of the positive terminal of rectifier, A, M, M_1, B, and back to the negative terminal of rectifier, and this is assumed to be the positive direction. If it is desired to reverse the flow of current, the following conditions have to be fulfilled:

(a) $|E_b| > |V_\alpha|$, and
(b) the devices used in the bridge should be able to conduct current in the reverse direction.

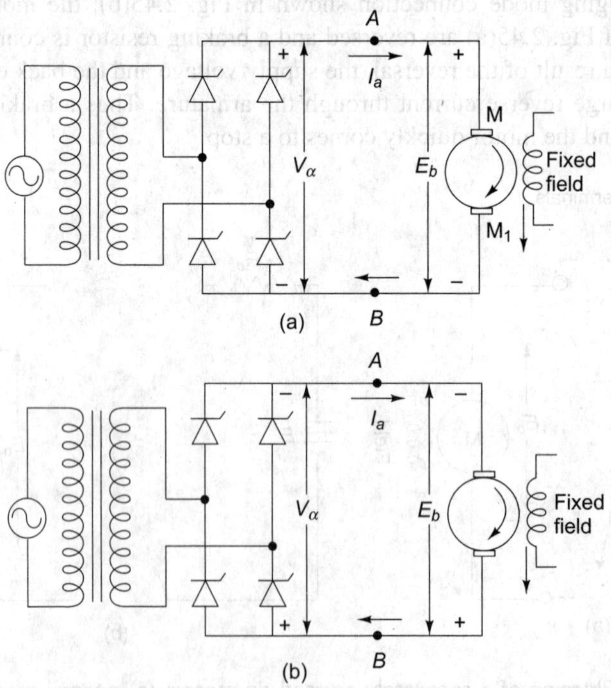

Fig. 2.46 Regenerating braking of a separately excited dc motor: (a) normal motoring mode, (b) regenerative braking mode

The first condition can be satisfied by adjusting α to be greater than 90°. The second condition cannot be fulfilled because thyristors can conduct current only in one direction. Hence, reversal of current is not possible with this circuit. The only alternative left is to reverse the polarities of V_α and E_b, as shown in Fig. 2.46(b), and also to ensure that $|E_b| > |V_\alpha|$. The polarity of V_α can be reversed as stated above; however, there are four methods for reversal of the polarity of E_b. These are

(i) reversal of the direction of rotation of the prime mover,
(ii) reversal of the armature terminals,
(iii) reversal of the field terminals, and
(iv) using a dual converter in place of a single converter for providing dc supply to the motor.

The reversal of the direction of rotation may not be practically feasible. The armature terminals can be reversed by connecting a reversing switch as shown in Fig. 2.47(a) or by means of a semiconductor bridge circuit as shown in Fig. 2.47(b). The disadvantages of using a reversing switch are that it is slow and there is wear and tear of the moving contacts. In the semiconductor bridge arrangement shown in Fig. 2.47(b), the thyristor pairs Th_A, Th_C on the one hand and Th_B, Th_D on the other conduct at a time. A necessary precaution consists of firing the new pair only after the conducting pair of thyristors is successfully commutated. This, however, requires zero current detection which adds to the overall cost.

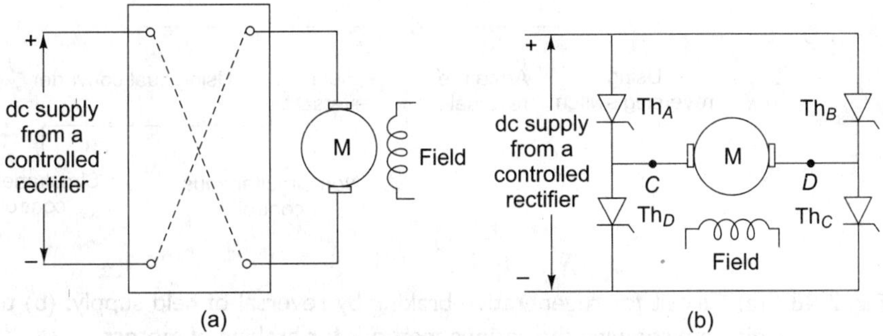

Fig. 2.47 (a) dc motor connected to the supply through a reversing switch; (b) dc motor connected across a semiconductor bridge

In the third method, the supply to the field coil is reversed, whereas that of the armature circuit remains unchanged. As shown in Fig. 2.48(a), a thyristor bridge can be used as a source for the field so as to facilitate reversal. The drawback of this method is that the reversal of field current affects the commutation of the armature current. Second, as the field time constant is of the order of 1–2 s, natural decay and rise of the field current may take 5–10 s. Consequently, this method suffers from the disadvantage of reduction of speed of operation.

The fourth method, namely, reversal of E_b using a dual converter, is by far the most superior among all the methods because of the fast response and flexibility of control obtained with it. It can be accomplished either by simultaneous control or by non-simultaneous control, as elaborated in Section 2.10.

The various methods given above for braking are summarized in the tree diagram of Fig. 2.48(b). Chapter 7 gives yet another method, namely, employing a dc chopper for this purpose.

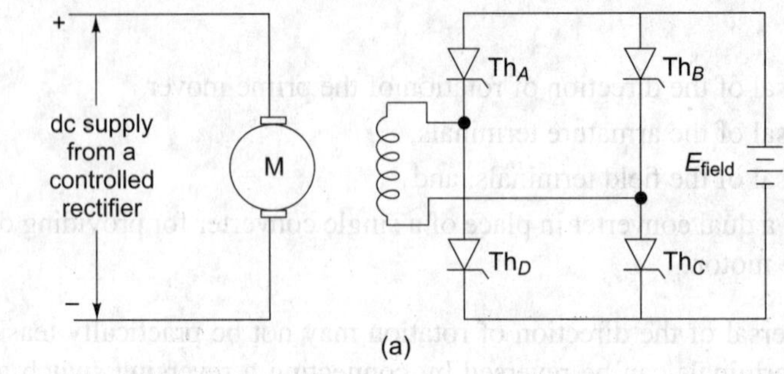

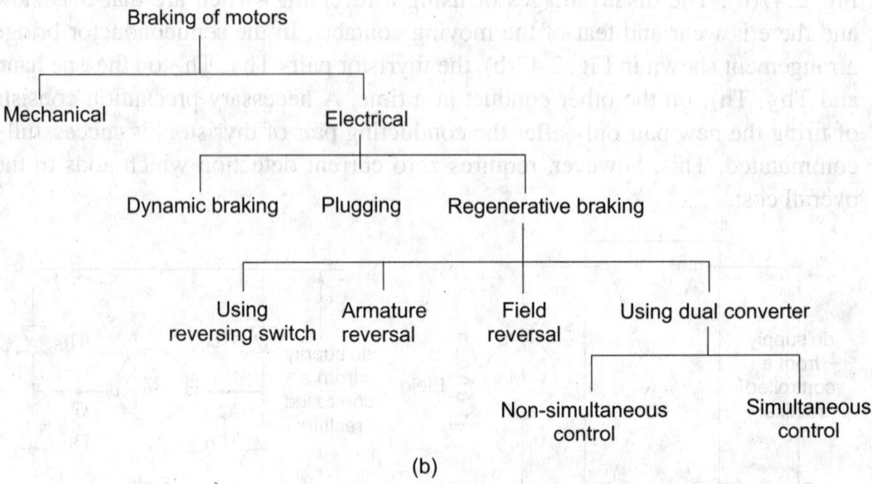

Fig. 2.48 (a) Circuit for regenerative braking by reversal of field supply; (b) tree diagram showing the various methods for braking of motors

2.13 Power Factor Improvement

Conventional half-controlled and fully controlled rectifiers present a low power factor to the input supply system and hence need PF improvement. Some forced commutation methods that help in improving the power factor of rectifiers are

(a) extinction angle control,
(b) symmetrical pulse width modulation,
(c) selective harmonic elimination using pulse width modulation (PWM),
(d) sinusoidal PWM, and
(e) sequence control of rectifiers.

2.13.1 Extinction Angle Control

Figure 2.49(a) shows a single-phase half-controlled rectifier for which the technique of extinction angle control is to be implemented. Though thyristors are shown as the controlled devices, GTOs can also be used instead. The operation of the circuit can be understood with the help of the waveforms given in Fig. 2.49(b).

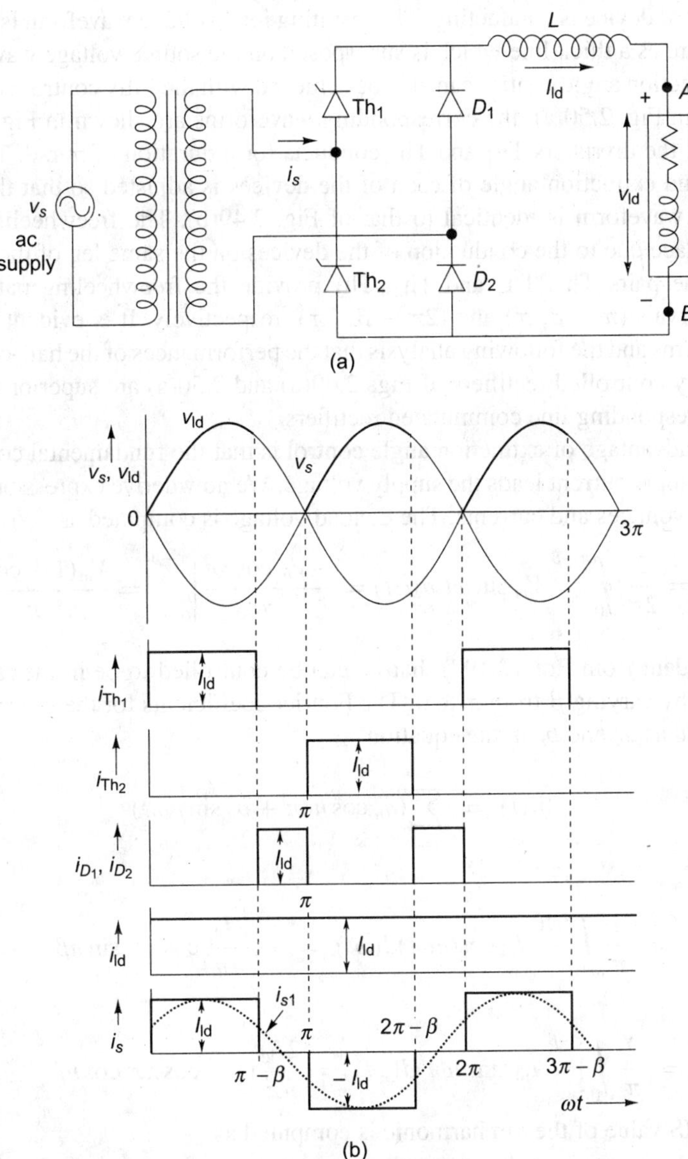

Fig. 2.49 Single-phase half-controlled bridge converter with forced commutation: (a) circuit, (b) waveforms for extinction angle control

An external inductor L of appropriate size is assumed to be provided to make the load a highly inductive one, thus ensuring a constant load current characteristic. This allows the load proper to be resistive or slightly inductive.

As shown in Fig. 2.49(b), Th_1 is turned on at $\omega t = 0$ and off at $\omega t = \pi - \beta$; again, Th_2 is turned on and off, respectively, at angles π and $2\pi - \beta$. The devices that conduct during the various parts of the cycle are also given in the figure. The diode pair D_1, D_2 provides a freewheeling path for the load current when no controlled device is conducting. The resulting load voltage waveform is drawn in the figure as a thick line which is superposed on the source voltage waveform.

Extinction angle control can also be achieved with the fully controlled rectifier shown in Fig. 2.50(a); the corresponding waveforms are shown in Fig. 2.50(b). Each of the thyristors Th_1 and Th_2 conducts for a duration of π rad. The firing angle and extinction angle of each of the devices is adjusted so that the source current waveform is identical to that of Fig. 2.49(b). The freewheeling action takes place due to the conduction of the devices on the same leg of the rectifier. Thus the pairs Th_1, Th_4 and Th_2, Th_3 provide the freewheeling path during the intervals $(\pi - \beta, \pi)$ and $(2\pi - \beta, 2\pi)$, respectively. It is evident from the waveforms and the following analysis that the performances of the half-controlled and fully controlled rectifiers of Figs 2.49(a) and 2.50(a) are superior to that of the corresponding line commutated rectifiers.

The advantage of extinction angle control is that the fundamental component of the supply current leads the supply voltage. We now derive expressions for the various voltages and currents. The dc load voltage is computed as

$$v_\beta = \frac{2}{2\pi}\int_0^{\pi-\beta} V_m \sin\omega t \, d(\omega t) = \frac{-V_m\cos\omega t}{\pi}\Bigg|_0^{\pi-\beta} = \frac{V_m(1+\cos\beta)}{\pi} \tag{2.197}$$

It is evident from Eqn (2.197) that v_β can be controlled to be in the range 0 to $2V_m/\pi$ by varying β from π to 0. The Fourier coefficients for the source current are given as a_n and b_n in the equation

$$i_s(t) = \sum_{n=1}^{\infty}(a_n\cos n\omega t + b_n \sin n\omega t)$$

where

$$a_n = \frac{2}{\pi}\int_0^{\pi-\beta} I_{\text{ld}}\cos(n\omega t)d(\omega t) = -\frac{2I_{\text{ld}}}{n\pi}\cos n\pi \sin n\beta \tag{2.198}$$

and

$$b_n = \frac{2}{\pi}\int_0^{\pi-\beta} I_{\text{ld}}\sin(n\omega t)d(\omega t) = \frac{2I_{\text{ld}}}{n\pi}(1-\cos n\pi\cos n\beta) \tag{2.199}$$

The RMS value of the nth harmonic is computed as

$$I_{sn} = \left[\frac{a_n^2+b_n^2}{2}\right]^{1/2} = \begin{cases}(2\sqrt{2}I_{\text{ld}}/n\pi)\sin(n\beta/2) & \text{for } n \text{ even}\\ (2\sqrt{2}I_{\text{ld}}/n\pi)\cos(n\beta/2) & \text{for } n \text{ odd}\end{cases} \tag{2.200}$$

$$\phi_n = \tan^{-1}\left(\frac{a_n}{b_n}\right) = \frac{-\cos n\pi \sin n\beta}{1-\cos n\pi\cos n\beta}$$

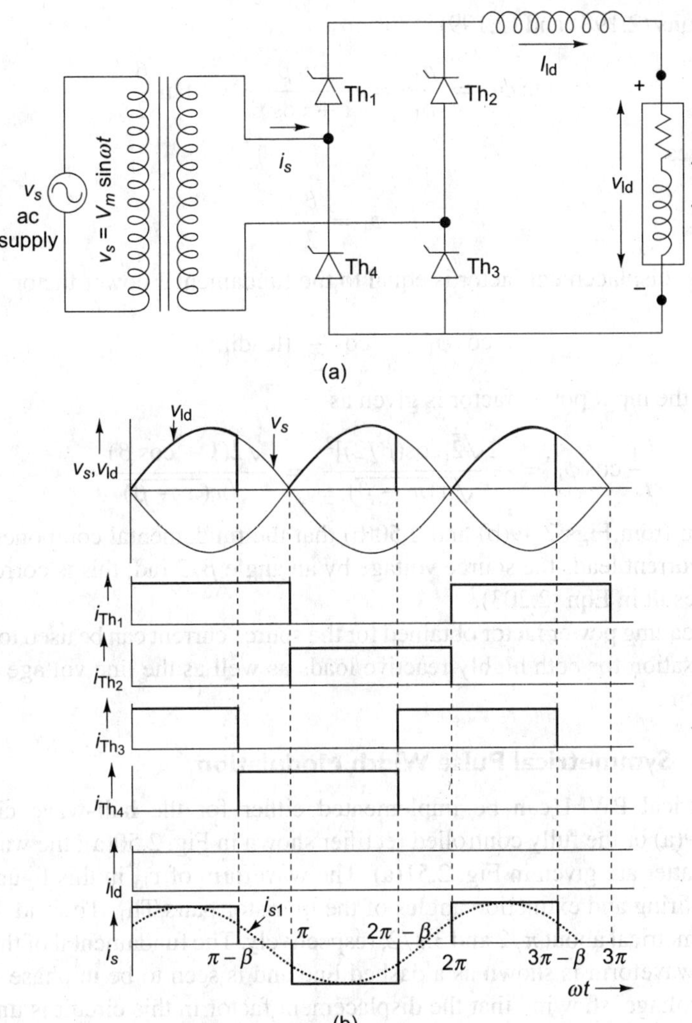

Fig. 2.50 Single-phase fully controlled bridge converter with forced commutation: (a) circuit, (b) waveforms for extinction angle control

The RMS value of the source current is given by

$$I_s = \sqrt{\frac{2}{2\pi}\int_0^{\pi-\beta} I_{\rm ld}^2 d(\omega t)} = I_{\rm ld}\sqrt{\frac{\pi - \beta}{\pi}} \tag{2.201}$$

An expression for the input power factor can also be derived as follows. From Eqns (2.200) and (2.201), the distortion factor is obtained (with $n = 1$) as

$$\frac{I_{s1}}{I_s} = \frac{2\sqrt{2}\cos(\beta/2)}{\sqrt{\pi(\pi - \beta)}} \tag{2.202}$$

From Eqns (2.198) and (2.199),

$$\tan\phi_1 = \frac{a_1}{b_1} = \frac{\sin\beta}{1+\cos\beta} = \tan\frac{\beta}{2}$$

This gives

$$\phi_1 = \frac{\beta}{2} \tag{2.203}$$

Here, the displacement factor is equal to the fundamental power factor. Thus,

$$\cos\phi_1 = \cos\frac{\beta}{2} \text{ (leading)} \tag{2.204}$$

Finally, the input power factor is given as

$$\frac{I_{s1}}{I_s}\cos\phi_1 = \frac{2\sqrt{2}[\cos(\beta/2)]^2}{\sqrt{\pi(\pi-\beta)}} = \frac{\sqrt{2}(1+\cos\beta)}{\sqrt{\pi(\pi-\beta)}} \tag{2.205}$$

It is seen from Figs 2.49(b) and 2.50(b) that the fundamental component of the source current leads the source voltage by an angle $\beta/2$ rad; this is corroborated by the result in Eqn (2.203).

The leading power factor obtained for the source current can be used to provide compensation for both highly reactive loads as well as the line voltage drops in the system.

2.13.2 Symmetrical Pulse Width Modulation

Symmetrical PWM can be implemented either for the half-wave circuit of Fig. 2.49(a) or the fully controlled rectifier shown in Fig. 2.50(a); the waveforms for the latter are given in Fig. 2.51(a). The waveform of v_{ld} in this figure shows that the firing and extinction angles of the thyristor pairs Th_1, Th_3 and Th_2, Th_4 are symmetrical about $\pi/2$ and $3\pi/2$, respectively. The fundamental of the source current waveform is shown as a dashed line and is seen to be in phase with the source voltage, showing that the displacement factor in this circuit is unity. The turn-on and extinction angles can be varied by comparing a triangular carrier voltage v_c with a variable dc reference voltage v_r. Figure 2.51(b) shows these two waveforms and also the resulting gate pulses.

The various voltages and currents are computed as follows:

$$V_{\text{av}} = \frac{2}{2\pi}\int_{\pi/2-\beta/2}^{\pi/2+\beta/2} V_m \sin\omega t\, d(\omega t) = \frac{2V_m}{\pi}\sin\frac{\beta}{2} \tag{2.206}$$

The Fourier series expression of $i_s(t)$ can be written as (Appendix A)

$$i_s(t) = \sum_{n=1}^{\infty}(a_n\cos n\omega t + b_n \sin n\omega t)$$

where

$$a_n = \frac{2}{\pi}\int_{\pi/2-\beta/2}^{\pi/2+\beta/2} I_{\text{ld}}\cos(n\omega t)d(\omega t) = 0 \tag{2.207}$$

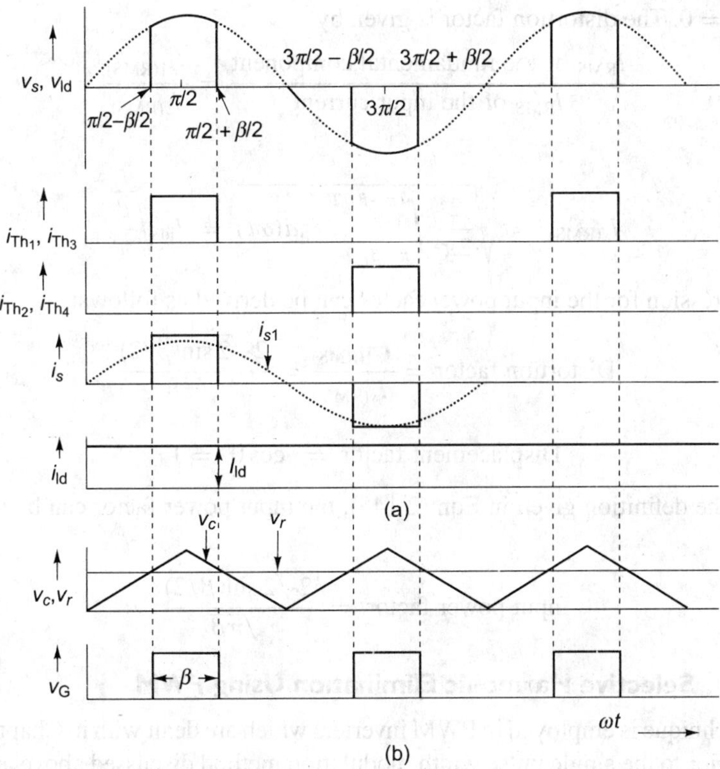

Fig. 2.51 (a) Waveforms for symmetrical single pulse width modulation; (b) carrier and reference waveforms and resulting gate pulses

and

$$b_n = \frac{2}{\pi}\int_{\pi/2-\beta/2}^{\pi/2+\beta/2} I_{ld}\sin(n\omega t)d(\omega t) = \frac{4I_{ld}}{n\pi}\sin\frac{n\pi}{2}\sin\frac{n\beta}{2} \tag{2.208}$$

Thus the RMS value of the nth harmonic of the input current can be computed as

$$I_{sn(\mathrm{RMS})} = \begin{cases} \left(\dfrac{a_n^2+b_n^2}{2}\right)^{1/2} = \dfrac{2\sqrt{2}I_{ld}}{n\pi}\sin\dfrac{n\beta}{2} & \text{for } n = 1, 3, \ldots \\ 0 & \text{for } n = 2, 4, \ldots \end{cases} \tag{2.209}$$

and

$$\phi_n = \tan^{-1}\left(\frac{a_n}{b_n}\right) = 0 \tag{2.210}$$

From Eqn (2.209), the RMS value of the fundamental component of the input current

$$I_{s1(\mathrm{RMS})} = \frac{2\sqrt{2}I_{ld}}{\pi}\sin\frac{\beta}{2}$$

and $\phi_1 = 0$. The distortion factor is given by

$$\frac{I_{\text{RMS}} \text{ of the fundamental component}}{I_{\text{RMS}} \text{ of the input current}} = \frac{I_{s1(\text{RMS})}}{I_{s(\text{RMS})}}$$

Here,

$$I_{s(\text{RMS})} = \sqrt{\frac{2}{2\pi}\int_{(\pi-\beta)/2}^{(\pi+\beta)/2} I_{\text{ld}}^2 d(\omega t)} = I_{\text{ld}}\sqrt{\frac{\beta}{\pi}}$$

An expression for the input power factor can be derived as follows:

$$\text{Distortion factor} = \frac{I_{s1(\text{RMS})}}{I_{s(\text{RMS})}} = \frac{2\sqrt{2}\sin(\beta/2)}{\sqrt{\pi\beta}} \tag{2.211}$$

$$\text{Displacement factor} = \cos 0° = 1 \tag{2.212}$$

Using the definition given in Eqn (2.145), the input power factor can be written as

$$\text{Input power factor} = \frac{2\sqrt{2}\sin(\beta/2)}{\sqrt{\pi\beta}} \tag{2.213}$$

2.13.3 Selective Harmonic Elimination Using PWM

This technique is employed in PWM inverters which are dealt with in Chapter 5. It is superior to the single pulse width modulation method discussed above and can be applied either to a half-controlled converter or to a fully controlled converter. As shown in Fig. 2.52(a), the method consists in (i) placing of, say, N_1 number of current pulses per quarter-cycle to eliminate N_1 harmonics and (ii) variation of the magnitude of the output voltage by controlling the pulse width (2ρ) symmetrically about the pulse position. Here, N_1 is taken as 2 and all the pulse widths are arranged to be equal. β_1 and β_2 denote the pulse positions in the first quarter-cycle, (0, $\pi/2$). Thus there are two positive pulses on either side of $\pi/2$ and similarly two negative pulses on each side of $3\pi/2$. Figure 2.52(b) gives the waveforms of v_c and v_r whose intersection points determine the pulse widths. The block diagram of Fig. 2.52(c) shows the method of generating the trigger pulses to be fed to the gates of the thyristors. The following derivation is made with the assumption that the load current has a varying nature but a very small ripple content and an average value I_{ld}. It is assumed here that the constant term in the expression for the Fourier series is zero. The supply current i_s can be expressed as

$$i_s = \sum_{n=1,3}^{N_1} (a_n \cos n\omega t + b_n \sin n\omega t) \tag{2.214}$$

The coefficient a_n can be obtained as

$$a_n = \frac{1}{\pi}\int_0^{2\pi} i_s(t)\cos n\omega t \, d(\omega t) \tag{2.215}$$

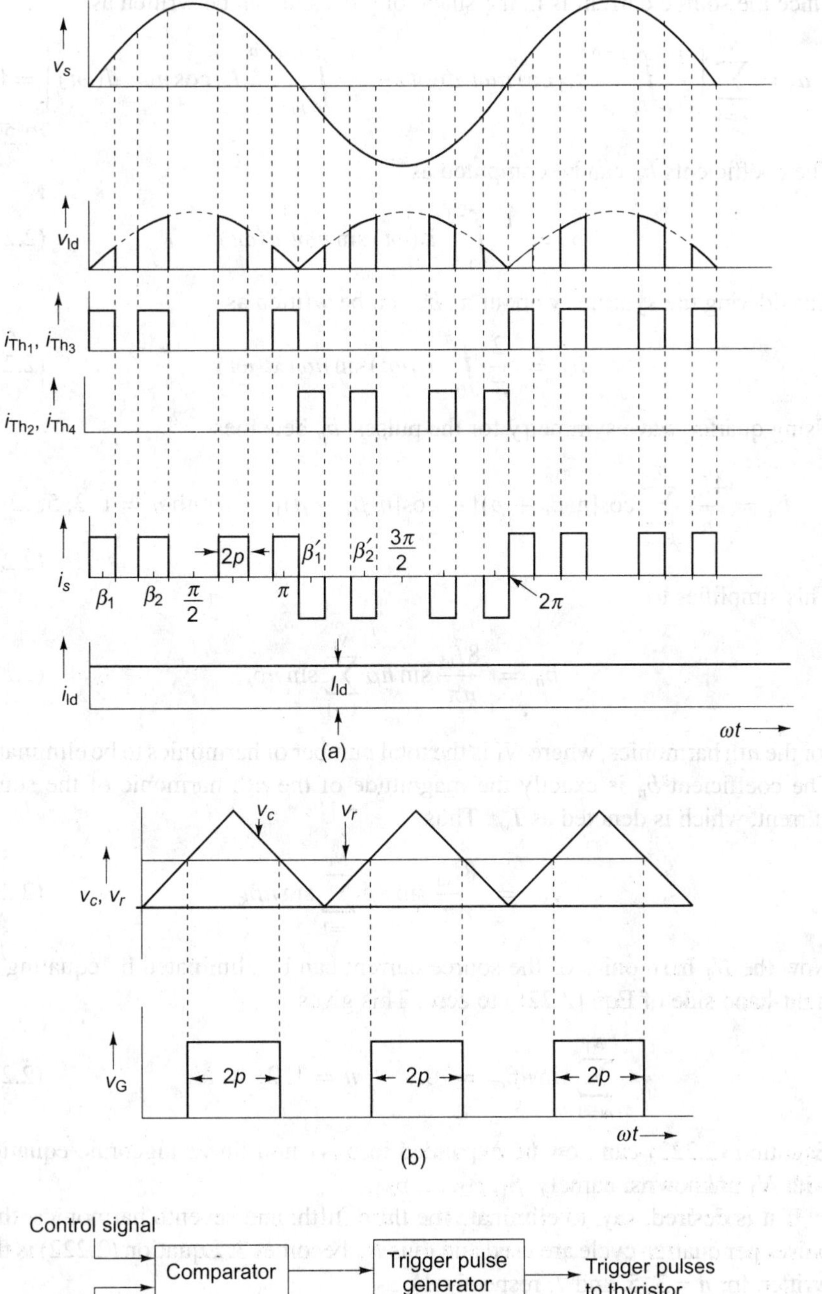

Fig. 2.52 (a) Waveforms for symmetrical multiple PWM; (b) carrier and reference waveforms and resulting gate pulses; (c) block diagram for generating gate trigger pulses

Since the source current is in the shape of pulses, it can be written as

$$a_n = \sum_{k=1}^{N_1}\left[\frac{1}{\pi}\int_{\beta_k-\rho}^{\beta_k+\rho} I_{\rm ld}\cos n\omega t\, d(\omega t) - \frac{1}{\pi}\int_{+\beta_k-\rho}^{+\beta_k+\rho} I_{\rm ld}\cos n\omega t\, d(\omega t)\right] = 0 \tag{2.216}$$

The coefficients b_n can be computed as

$$b_n = \frac{1}{\pi}\int_0^{2\pi} i_s(\omega t)\sin n\omega t\, d(\omega t) \tag{2.217}$$

Considering the symmetry about π, b_n can be written as

$$b_n = \frac{2}{\pi}\int_0^{\pi} i_s(\omega t)\sin n\omega t\, d(\omega t) \tag{2.218}$$

Using quarter-wave symmetry for the pulses, b_n becomes

$$b_n = \frac{4I_{\rm ld}}{n}\sum_{k=1}^{N_1}\cos[n(\beta_k+\rho)] - \cos[n(\beta_k-\rho)] \qquad \text{with } n = 1, 3, 5, \ldots \tag{2.219}$$

This simplifies to

$$b_n = \frac{8I_{\rm ld}}{n\pi}\sin n\rho \sum_{k=1}^{N_1}\sin n\beta_k \tag{2.220}$$

for the nth harmonics, where N_1 is the total number of harmonics to be eliminated. The coefficient b_n is exactly the magnitude of the nth harmonic of the source current, which is denoted as I_{sn}. Thus,

$$I_{sn} = \frac{8I_{\rm ld}}{n\pi}\sin n\rho \sum_{k=1}^{N_1}\sin n\beta_k \tag{2.221}$$

Now the N_1 harmonics of the source current can be eliminated by equating the right-hand side of Eqn (2.221) to zero. This gives

$$\sum_{m=1}^{N_1}\sin n\beta_m = 0, \qquad n = 1, 2, \ldots, N_1 \tag{2.222}$$

Equation (2.222) can now be expanded into N_1 non-linear algebraic equations with N_1 unknowns, namely, $\beta_1, \beta_2, \ldots, \beta_{N_1}$.

If it is desired, say, to eliminate the third, fifth, and seventh harmonics, three pulses per quarter-cycle are used and thus N_1 becomes 3. Equation (2.222) is then written for n = 3, 5, and 7, respectively, as

$$\begin{aligned}
\sin 3\beta_1 + \sin 3\beta_2 + \sin 3\beta_3 &= 0\\
\sin 5\beta_1 + \sin 5\beta_2 + \sin 5\beta_3 &= 0\\
\sin 7\beta_1 + \sin 7\beta_2 + \sin 7\beta_3 &= 0
\end{aligned} \tag{2.223}$$

These non-linear equations are solved on the computer with the help of numerical techniques. The solution gives the values of β_1, β_2, and β_3, which are the centres of the pulses. These are obtained as

$$\beta_1 = 33^\circ, \qquad \beta_2 = 62^\circ, \qquad \text{and} \qquad \beta_3 = 78^\circ$$

Evidently, placing the pulses in the first quarter-cycle at these angles will give the supply current waveform in which the above three harmonics, namely, 3, 5, and 7, are eliminated and only harmonics of the order of 9 and above will be present. The magnitudes of such higher harmonics usually being quite small, the harmonic content in the source current will be substantially reduced.

It is evident from the above discussion that elimination of unwanted harmonics is the main advantage of the technique of selective harmonic elimination. An instance is that of electric traction in which the unwanted harmonics occur in a range close to the track circuit frequency. This method has the merit that the operation of the signals is not disturbed due to the harmonics generated by the converters. It, however, has the drawback that elimination of harmonics causes a reduction in the maximum available voltage.

2.13.4 Sinusoidal PWM

The technique of sinusoidal PWM is based on the fact that a nearly sinusoidal input current is very helpful in minimizing the lower order harmonics. A half-sinusoidal reference waveform is compared with a triangular carrier waveform to generate the current pulses. The widths of the pulses so obtained will be proportional to the sines of the angles at their locations. Figure 2.53(a) shows these two waveforms (v_r and v_c) as also thyristor current pulses, input current pulses, and the load current for the single-phase fully controlled bridge converter of Fig. 2.50(a). The block diagram of a scheme for generating such current pulses is given in Fig. 2.53(b).

The following two factors are associated with sinusoidal PWM:

$$\text{Carrier ratio } (p) = \frac{\text{frequency of carrier signal waveform}}{\text{frequency of reference sinusoidal waveform}}$$

$$\text{Modulation index } (M) = \frac{\text{amplitude of reference waveform}}{\text{amplitude of carrier waveform}}$$

The amplitudes of the input current harmonics also vary with the modulating index M. The carrier ratio p determines the order of the predominant harmonics in the output voltage waveform. Fourier analysis shows that the harmonic order is given by $k = np \pm m$, where m is odd if n is even or vice versa. Thus, for $n = 2$, the harmonics present are $2p\pm1$, $2p\pm3$, $2p\pm3$, $2p\pm5$. Similarly, for $n = 1$, the harmonics that exist are $p\pm2$, $p\pm4$, etc. In the latter case, the harmonic content may increase with m. For instance, with respect to the harmonics of the order $p\pm2$, the harmonic content increases from 0 to 0.2. However, with p usually being a multiple of 3 (such harmonics are called *triplen harmonics*), harmonics which are sidebands of such a multiple are not applied to a three-phase load.

The number of pulses for a carrier ratio of p will be $p + 1$ if the zero crossing of both the waveforms is at the origin. On the other hand, there will be p pulses if the peak of the carrier waveform and the positive-going zero crossing of the reference waveform occur at the origin.

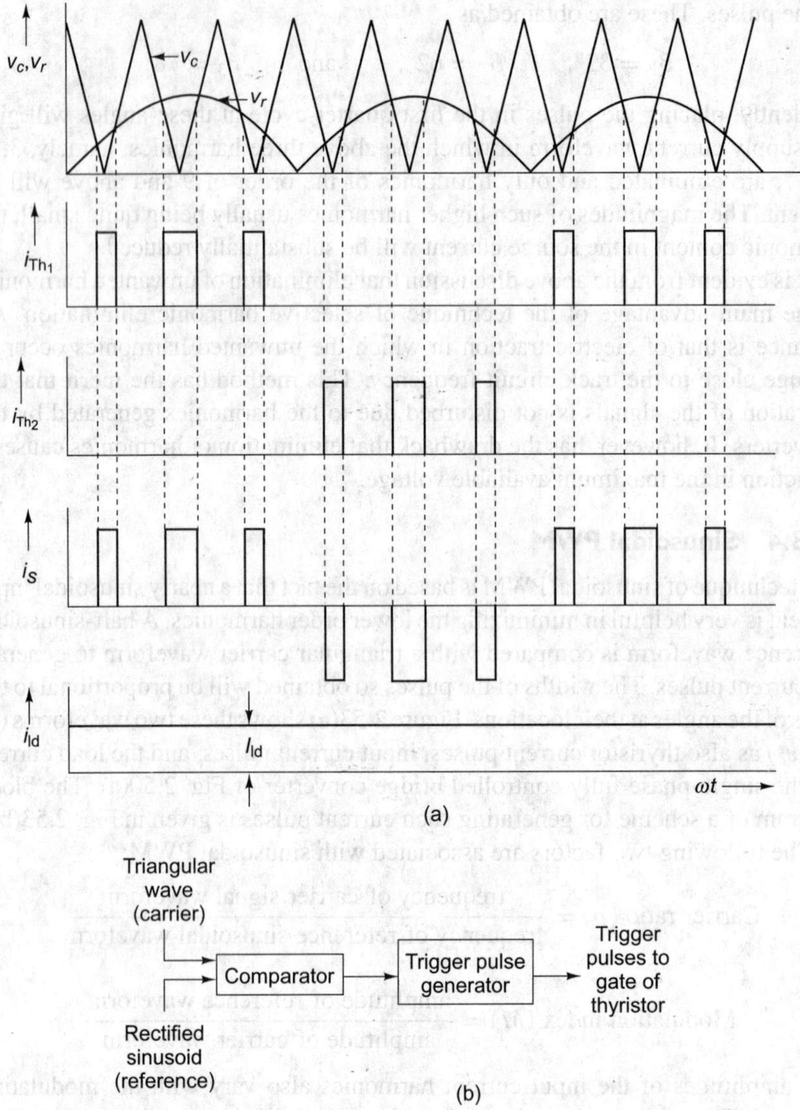

Fig. 2.53 (a) Waveforms for sinusoidal PWM; (b) block diagram for generating gate trigger pulses

The advantage of the sinusoidal PWM is that for the modulation index in the range $0 \leq M \leq 1$, harmonics other than those given above are absent. However, if M is increased to a value greater than 1, low-frequency harmonics will get introduced in the source current waveform. The principle of selective harmonic elimination can also be applied to sinusoidal PWM so as to derive the advantages of both the techniques. This helps in improving the power factor and also in keeping the displacement factor at unity. Thus, sinusoidal PWM is the best among all the PWM methods and can be applied for both half-controlled as well as fully controlled rectifiers.

Selective harmonic elimination was preferred earlier due to difficulties in the implementation of sinusoidal PWM. However, with the advent of microprocessors, generation of sinusoidal waveforms has become possible with the help of software techniques. Such a software-oriented sinusoidal PWM is called *regular sampled PWM* and is available as a package. This topic is elaborated in Chapter 13.

2.13.5 Sequence Control of Rectifiers

Power factor improvement can also be achieved by connecting two half-controlled or fully controlled rectifiers in series and connecting the load across the combination [Fig. 2.54(a)]. For facilitating the analysis, the load is assumed to be highly inductive so as to give a nearly constant load current. Figure 2.54(b) shows that, by employing two single-phase half-controlled rectifiers, the load voltage is obtained as the sum of the output voltages of these two converters.

The common primary winding of the transformer has N_1 turns, whereas each rectifier is supplied by a separate secondary with N_2 turns. The turn ratio N_1/N_2 is taken as 2. The flexibility available with this circuit is as follows: (i) either or both of the rectifiers can be put into operation and (ii) with both of them under operation, the firing angle of one rectifier can be kept at zero, while that of the second is kept at an appropriate angle which is greater than zero. An alternative to the second method of operation is to vary the firing angles of both the rectifiers.

When one of the rectifiers is off, the freewheeling diode across it will conduct the load current. For instance, if the thyristors of rectifier 2 are off and those of rectifier 1 are on, the diode D_{FW2} will conduct the load current because the voltage at the top bus is positive and will reverse bias D_{FW1}.

Figure 2.54(b) shows that rectifier 2 does not conduct current in the ranges $0 \le \omega t \le \alpha_2$ and $\pi \le \omega t \le \pi + \alpha_2$. This confirms that the presence of D_{FW2} is a necessity to bypass the load current during these intervals. If the roles of the rectifiers are reversed, then rectifier 1 will also have no current periods. Hence the inclusion of D_{FW1} is necessary to facilitate the bypassing of this rectifier. The range of voltage control obtainable will be $0 \le V_{ld} \le 4V_m/\pi$, where V_m is the peak of the voltage across the secondary. The higher limit occurs when both α_1 and α_2 are zero. The load voltage can be expressed as

$$V_{ld} = V_{\alpha_1} + V_{\alpha_2}$$

With $\alpha_1 = 0$ and $0 \le \alpha_2 \le \pi/2$, the outputs of the rectifiers become

$$V_{\alpha_1} = \frac{2V_m}{\pi} \quad \text{and} \quad V_{\alpha_2} = \frac{V_m}{\pi}(1 + \cos\alpha_2)$$

The load voltage then becomes

$$V_{ld} = \frac{V_m}{\pi}(3 + \cos\alpha_2) \tag{2.224}$$

The various performance factors for this circuit can be arrived at from Fig. 2.54(b) as follows. The input current waveform (i_p), being periodic but non-sinusoidal,

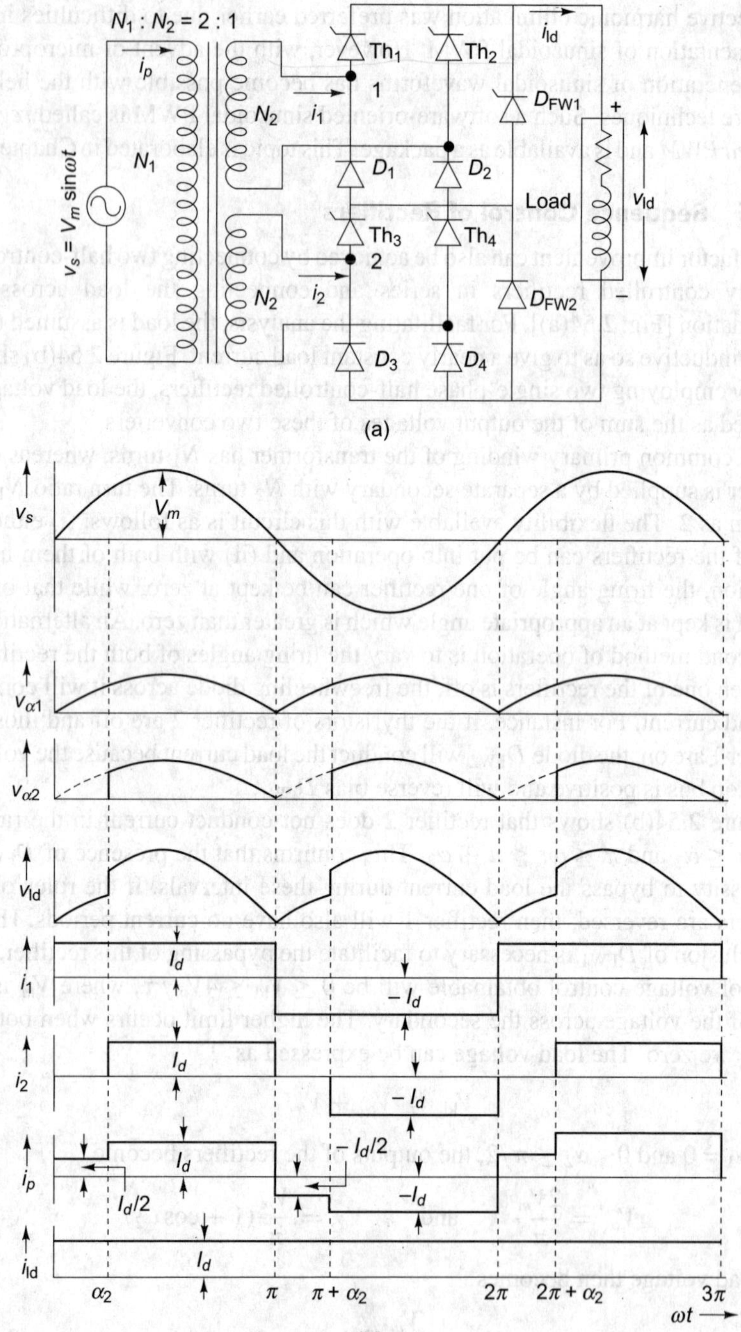

Fig. 2.54 Sequence control of two single-phase half-controlled rectifiers: (a) circuit, (b) waveforms

can be expressed as

$$i_p(\omega t) = I_{p0} + \sum_{n=1}^{\infty}(a_n \cos n\omega t + b_n \sin n\omega t)$$

where the dc component I_{p0} can be expressed as

$$\begin{aligned} I_{p0} &= \frac{1}{2\pi}\int_0^{2\pi} i_p(\omega t) d(\omega t) \\ &= \frac{1}{2\pi}\left[\int_0^{\alpha_2} \frac{I_d}{2} d(\omega t) + \int_{\alpha_2}^{\pi} I_d\, d(\omega t)\right. \\ &\quad \left. + \int_{\pi}^{\pi+\alpha_2} \frac{I_d}{2} d(\omega t) + \int_{\pi+\alpha_2}^{2\pi} (-I_d)\, d(\omega t)\right] \\ &= 0 \end{aligned}$$

Because of symmetry of the waveforms about π, a_n can be obtained as

$$\begin{aligned} a_n &= \frac{1}{\pi}\int_0^{2\pi} i_p(\omega t) \cos n\omega t\, d(\omega t) \\ &= \frac{1}{2\pi}\left[\int_0^{\alpha_2} \frac{I_d}{2} \cos n\omega t\, d(\omega t) + \int_{\alpha_2}^{\pi} I_d \cos n\omega t\, d(\omega t)\right] \\ &= \frac{I_d}{\pi n} \sin n\alpha_2 \end{aligned} \tag{2.225}$$

Similarly, symmetry gives b_n as

$$\begin{aligned} b_n &= \frac{1}{\pi}\int_0^{2\pi} i_p(\omega t) \sin n\omega t\, d(\omega t) \\ &= \frac{1}{2\pi}\left[\int_0^{\alpha_2} \frac{I_d}{2} \sin n\omega t\, d(\omega t) + \int_{\alpha_2}^{\pi} I_d \sin n\omega t\, d(\omega t)\right] \\ &= \frac{I_d}{\pi n}(3 + \cos n\alpha_2) \end{aligned} \tag{2.226}$$

The RMS value of the nth component of the input current is

$$I_{pn(\text{RMS})} = \frac{1}{\sqrt{2}}(a_n^2 + b_n^2)^{1/2} = \frac{I_d}{\pi n}(5 + 3\cos n\alpha_2)^{1/2} \tag{2.227}$$

Thus,

$$\frac{I_{pn(\text{RMS})}}{I_d} = \frac{1}{\pi n}(5 + 3\cos n\alpha_2)^{1/2} \tag{2.228}$$

The RMS value of the input current is obtained as

$$I_{p(\text{RMS})} = \left\{\frac{2}{2\pi}\left[\int_0^{\alpha_2}\left(\frac{I_d}{2}\right)^2 d(\omega t) + \int_{\alpha_2}^{\pi} I_d^2 d(\omega t)\right]\right\}^{1/2} = I_d\left(1 - \frac{3\alpha_2}{4\pi}\right)^{1/2}$$

Thus,

$$\frac{I_{p(\text{RMS})}}{I_d} = \left(1 - \frac{3\alpha_2}{4\pi}\right)^{1/2} \tag{2.229}$$

The fundamental component I_{p1} is obtained as

$$\begin{aligned} I_{p1(\text{RMS})} &= \frac{1}{\sqrt{2}}(a_1^2 + b_1^2)^{1/2} \\ &= \frac{1}{\sqrt{2}}\left[\left(\frac{I_d}{\pi}\sin\alpha_2\right)^2 + \left\{\frac{I_d}{\pi}(3 + \cos\alpha_2)\right\}^2\right]^{1/2} \\ &= \frac{I_d}{\pi}(5 + 3\cos\alpha_2)^{1/2} \end{aligned}$$

$$\begin{aligned} \text{Displacement factor} = \cos\phi_1 &= \frac{I_{p1}\cos\phi_1}{I_{p1}} = \frac{b_1/\sqrt{2}}{(a_1^2 + b_1^2)^{1/2}/\sqrt{2}} \\ &= \frac{(I_d/\pi)(3 + \cos\alpha_2)}{(\sqrt{2}I_d/\pi)(5 + 3\cos\alpha_2)^{1/2}} \\ &= \frac{3 + \cos\alpha_2}{\sqrt{2}(5 + 3\cos\alpha_2)^{1/2}} \end{aligned} \tag{2.230}$$

$$\begin{aligned} \text{Distortion factor} &= \frac{I_{p1(\text{RMS})}}{I_{p(\text{RMS})}} = \frac{I_d(5 + 3\cos\alpha_2)^{1/2}}{\pi I_d\,(1 - 3\alpha_2/4\pi)^{1/2}} \\ &= \frac{(5 + 3\cos\alpha_2)^{1/2}}{\pi\,(1 - 3\alpha_2/4\pi)^{1/2}} \end{aligned} \tag{2.231}$$

The input power factor, being the product of the displacement and distortion factors, is obtained as

$$\text{Input power factor} = \frac{3 + \cos\alpha_2}{\sqrt{2}\pi\,(1 - 3\alpha_2/4\pi)^{1/2}} = \frac{3 + \cos\alpha_2}{\pi\,(2 - 3\alpha_2/2\pi)^{1/2}} \tag{2.232}$$

The circuit for sequence control of two single-phase fully controlled converters and the resulting waveforms are, respectively, shown in Figs 2.55(a) and (b). There is no necessity of bypass diodes in this circuit, because at any instant one pair of devices in each bridge conducts the load current. The paths for the load current during the various intervals are as follows. (a) Intervals $0 \le \omega t \le \alpha_2$ and $\pi \le \omega t \le \pi + \alpha_2$: A, Th_1, load, Th_{IV}, C, D, Th_{II}, Th_3, B, and back to A. (b) Intervals $\alpha_2 \le \omega t \le \pi$ and $\pi + \alpha_2 \le \omega t \le 2\pi$: A, Th_1, load, Th_{III}, D, C, Th_{I}, Th_3, B, and back to A.

Assuming that the conduction is continuous in both the rectifiers, the range of load voltage control in this case is $0 \le v_{\text{ld}} \le 4V_m/\pi$, which is the same as that for a half-controlled converter. Again, α_1 is set at zero and α_2 is varied from zero to 180°. This gives the two average voltages as

$$V_{\alpha_1} = \frac{2V_m}{\pi}$$

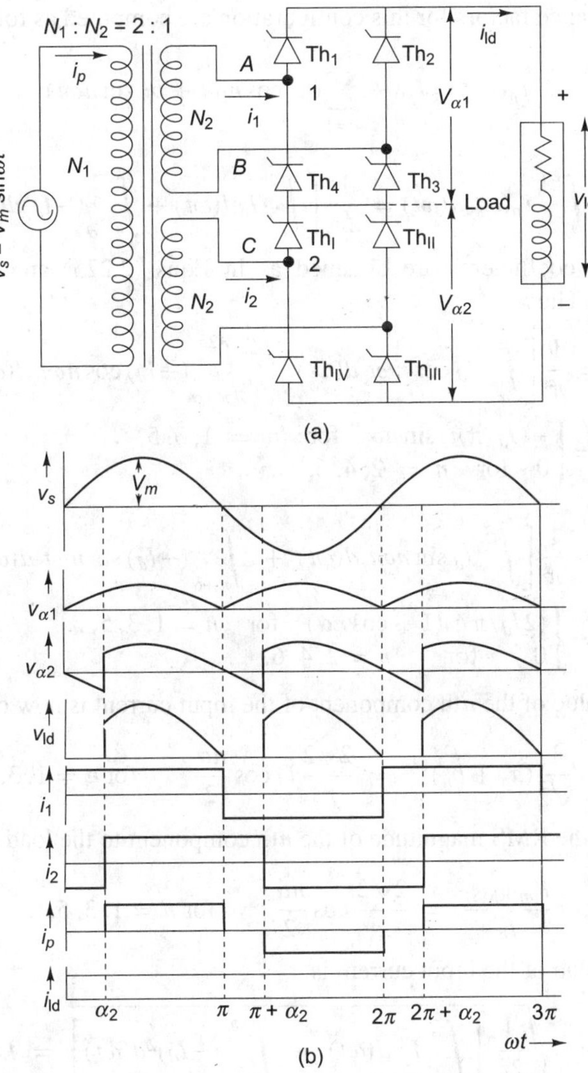

Fig. 2.55 Sequence control of two single-phase fully controlled rectifiers: (a) circuit, (b) waveforms

and

$$V_{\alpha_2} = \frac{2V_m}{\pi} \cos \alpha_2$$

The load voltage is given as

$$v_{ld} = V_{\alpha_1} + V_{\alpha_2} = \frac{2V_m}{\pi}(1 + \cos \alpha_2) \quad (2.233)$$

The performance factors for this configuration are computed as follows:

$$i_p(\omega t) = I_{p0} + \sum_{n=1}^{\infty}(a_n \cos n\omega t + b_n \sin n\omega t)$$

$$I_{p0} = \frac{1}{2\pi}\int_0^{2\pi} i_p(\omega t)d(\omega t) = \frac{1}{2\pi}\left[\int_\alpha^\pi I_d d(\omega t) + \int_{\pi+\alpha}^{2\pi}(-I_d)d(\omega t)\right] = 0$$

The Fourier coefficients are obtained as in Eqns (2.225) and (2.226) from Fig. 2.55(b). Thus

$$a_n = \frac{1}{\pi}\left[\int_\alpha^\pi I_d \cos n\omega t\, d(\omega t) + \int_{\pi+\alpha}^{2\pi}(-I_d)\cos n\omega t\, d(\omega t)\right]$$

$$= \begin{cases} -(I_d/\pi n)\sin n\alpha & \text{for } n = 1, 3, 5, \ldots \\ 0 \quad \text{for } n = 2, 4, 6, \ldots \end{cases} \tag{2.234}$$

$$b_n = \frac{1}{\pi}\left[\int_\alpha^\pi I_d \sin n\omega t\, d(\omega t) + \int_{\pi+\alpha}^{2\pi}(-I_d)\sin n\omega t\, d(\omega t)\right]$$

$$= \begin{cases} (2I_d/\pi n)(1+\cos n\alpha) & \text{for } n = 1, 3, 5, \ldots \\ 0 & \text{for } n = 2, 4, 6, \ldots \end{cases} \tag{2.235}$$

The RMS value of the *n*th component of the input current is now obtained as

$$I_{pn(\text{RMS})} = \frac{1}{\sqrt{2}}(a_n + b_n)^{1/2} = \frac{2\sqrt{2}}{\pi} I_d \cos\frac{n\alpha}{2} \qquad \text{for } n = 1, 3, 5, \ldots \tag{2.236}$$

The ratio of the RMS magnitude of the *n*th component to the load current is

$$\frac{I_{pn(\text{RMS})}}{I_d} = \frac{2\sqrt{2}}{\pi}\cos\frac{n\alpha}{2} \qquad \text{for } n = 1, 3, 5, \ldots \tag{2.237}$$

The RMS value of the input current is

$$I_{p(\text{RMS})} = \sqrt{\frac{1}{2\pi}\left[\int_\alpha^\pi I_d^2\, d(\omega t) + \int_{\pi+\alpha}^{2\pi}(-I_d)^2 d(\omega t)\right]} = I_d\sqrt{\frac{\pi-\alpha}{\pi}} \tag{2.238}$$

The distortion factor is obtained as

$$\frac{I_{p1(\text{RMS})}}{I_{p(\text{RMS})}} = \frac{(2\sqrt{2}I_d/\pi)\cos(\alpha/2)}{I_d\sqrt{(\pi-\alpha/\pi)}} = \frac{2\sqrt{2}\cos(\alpha/2)}{\sqrt{\pi(\pi-\alpha)}} \tag{2.239}$$

The displacement factor is obtained from Eqns (2.235) and (2.236) with $n = 1$ as

$$\cos\phi_1 = \frac{b_1/\sqrt{2}}{(a_1^2 + b_1^2)/\sqrt{2}}$$

$$= \frac{\sqrt{2}I_d(1+\cos\alpha)/\pi}{(2\sqrt{2}I_d/\pi)\cos(\alpha/2)} = \cos\frac{\alpha}{2} \tag{2.240}$$

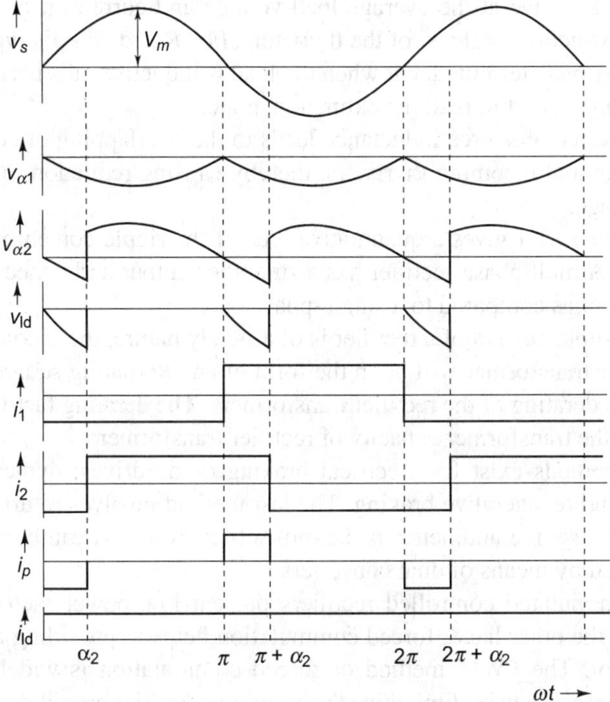

Fig. 2.56 Waveforms for the inverting mode of sequentially controlled single-phase rectifiers

Finally, using the identity $\cos^2(\alpha/2) = (1 + \cos\alpha)/2$,

$$\text{Input power factor} = \frac{2\sqrt{2}\cos(\alpha/2)}{[\pi(\pi-\alpha)]^{1/2}} \cos\frac{\alpha}{2}$$

$$= \frac{\sqrt{2}(1+\cos\alpha)}{[\pi(\pi-\alpha)]^{1/2}} \tag{2.241}$$

Similar to a single fully controlled rectifier bridge, sequentially operated fully controlled converters can work in the inverting mode. For this purpose, rectifier 1 is operated with $\alpha_1 = \pi$ and rectifier 2 with an arbitrary firing angle, as before. The load voltage will then be on the negative side of the ωt-axis as shown in Fig. 2.56. However, the currents i_1 and i_2 remain positive, as per the orientation of the thyristors.

Summary

For transformers which serve as ac sources for all types of half-wave rectifiers, their secondary currents contain a dc component. If no corrective steps are taken, this component causes saturation of the transformer core. However, a zigzag type of secondary for the transformer helps in eliminating this undesirable feature.

In order to arrive at the average load voltage and current, it is necessary to know the extinction angle α_e of the thyristors. For R and $R + E_b$ types of loads, α_e can easily be determined, but when the load is inductive, it is necessary to use the firing angle chart to read the extinction angle.

The presence of source inductance leads to the overlapping of conduction of the outgoing and incoming thyristors, thereby causing reduction in the average output voltage.

The ripple factor gives a quantitative idea of the ripple content in the output waveform. A multiphase rectifier has a smoother output and hence it has lower ripple content as compared to a single-phase one.

As the output current of a rectifier is of a steady nature, the secondary current in a rectifier transformer will be in the form of an alternating square waveform. This causes derating of the rectifier transformers. The derating factor is useful in estimating the transformer capacity of rectifier transformers.

Three methods exist for electrical braking of dc drives: dynamic braking, plugging, and regenerative braking. The last method involves returning the load energy to the source and hence is the most efficient one. It can be conveniently implemented by means of dual converters.

Line commutated controlled rectifiers present low power factors to the ac source. On the other hand, forced commutation helps in providing an improved power factor. The PWM method of forced commutation is widely employed because of its efficacy in eliminating the harmonics besides providing an improved power factor.

Worked Examples

1. A single-phase, half-wave rectifier with an ac voltage of 150 V has a pure resistive load of 9 Ω. The firing angle α of the thyristor is $\pi/2$. Determine the (a) rectification efficiency, (b) form factor, (c) transformer derating factor, (d) peak inverse voltage of the SCR, and (e) ripple factor of the output voltage. Assume that the transformer ratio is 2:1.

Solution

The circuit is shown in Fig. 2.57.

(a) $$\text{Rectification efficiency} = \frac{\text{dc load power}}{\text{secondary ac power}}$$

$$\text{Average voltage at the load} = \frac{1}{2\pi}\int_{\pi/2}^{\pi} \sqrt{2} \times 150 \sin \omega t \, d(\omega t)$$

$$= \frac{\sqrt{2} \times 150}{2\pi} = 33.8 \text{ V}$$

Dc load current = $33.8/R_{\text{ld}} = 33.8/9 = 3.75$ A. Dc load power = $33.8 \times 3.75 =$ 127 W. Ac power at the secondary = $V_{\text{RMS}} I_{\text{RMS}}$. Here $V_{\text{RMS}} = 150$ V.

$$I_{\text{RMS}} = \sqrt{\frac{1}{2\pi}\int_{\pi/2}^{\pi} (I_m \sin \omega t)^2 d(\omega t)}$$

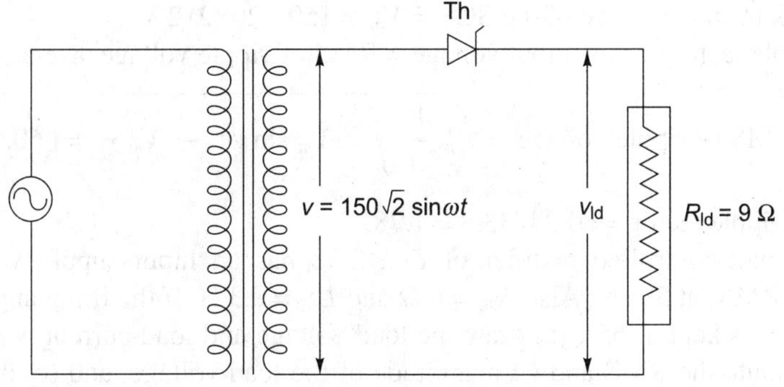

Fig. 2.57

where $I_{\mathrm{RMS}} = V_m / R_{\mathrm{ld}} \;=\; \sqrt{2} \times 150/9 = 23.6$ A. After evaluating the integral,

$$I_{\mathrm{RMS}} = \frac{I_m}{2\sqrt{2}} = \frac{23.6}{2\sqrt{2}} = 8.34 \text{ A}$$

Thus,

$$\text{Rectification efficiency} = \frac{33.8 \times 3.75 \times 100}{150 \times 8.34} = 10.13\%$$

(b) $$\text{Form factor} = \frac{\text{RMS output voltage}}{\text{average output voltage}}$$

where

$$\text{RMS output voltage} = \sqrt{\frac{1}{2\pi}\int_{\pi/2}^{\pi}(V_m \sin \omega t)^2 d(\omega t)}$$

$$= \frac{V_m}{2\sqrt{2}} = \frac{150\sqrt{2}}{2\sqrt{2}} = 75 \text{ V}$$

Hence, the form factor = 75/33.8 = 2.22.

(c) The transformer derating factor

$$\delta_r = \frac{P_{\mathrm{pac}} + P_{\mathrm{sac}}}{2P_{\mathrm{dc}}}$$

$$P_{\mathrm{sac}} = 150 \times 8.34 = 1251 \text{ W}$$

$$P_{\mathrm{pac}} = \frac{N_1}{N_2}\frac{V_m}{\sqrt{2}}\frac{N_2}{N_1}\frac{I_m}{2}$$

$$= \frac{V_m I_m}{2\sqrt{2}}$$

$$= P_{\mathrm{sac}}$$

$$\delta_r = \frac{2P_{\mathrm{sac}}}{2P_{\mathrm{dc}}} = \frac{P_{\mathrm{sac}}}{P_{\mathrm{dc}}} = \frac{1251}{33.8 \times 3.75} = 9.87$$

(d) Peak inverse voltage of the SCR = $V_m = 150\sqrt{2} = 212$ V
(e) Ripple factor of the output voltage = RMS of ripple voltage/average voltage

$$\text{RMS of ripple voltage} = \sqrt{\frac{1}{2\pi}\int_{\pi/2}^{\pi}(V_m \sin\omega t - V_{av})^2} = 60.24$$

Hence, ripple factor = 60.24/33.8 = 1.78.

2. The half-controlled rectifier of Fig. 2.58 has an input supply voltage of 115 V (RMS) at 50 Hz. Also $R_{ld} = 6\ \Omega$ and $L_{ld} = 0.3$ H. If the firing angle of the thyristors is kept at 65°, (a) draw the load voltage and load current waveforms, (b) compute the RMS and dc magnitude of the load voltage, and (c) determine the magnitude of the dc load current.

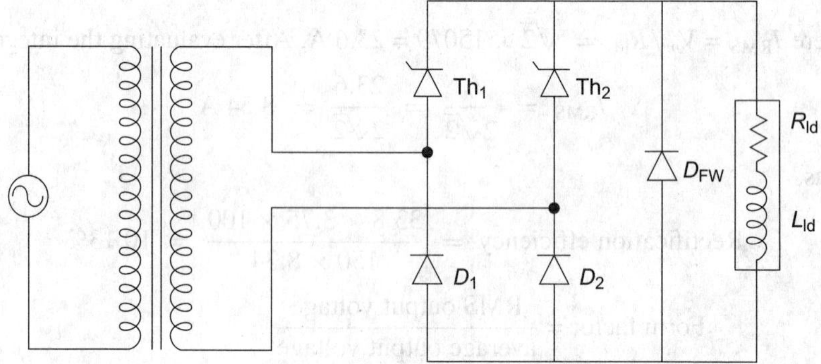

Fig. 2.58

Solution

(a) The load voltage and load current waveforms are given in Fig. 2.5(c). As the inductive reactance is very high at 314 × 0.3 = 94.2 Ω, the current has a flat characteristic.

$$\text{RMS value of load voltage} = \sqrt{\frac{2}{2\pi}\int_{65\pi/180}^{\pi}(115\sqrt{2}\ \sin\omega t)^2 d(\omega t)}$$
$$= 0.617V_m$$
$$= 0.617 \times 150\sqrt{2}$$
$$= 130.8 \text{ V}$$

$$\text{Dc magnitude of load voltage} = \frac{2}{2\pi}\int_{65\pi/180}^{\pi}(115\sqrt{2}\ \sin\omega t)d(\omega t)$$
$$= \frac{115\sqrt{2}(1+\cos 65°)}{\pi}$$
$$= 73.65 \text{ V}$$

$$\text{Dc load current} = \frac{\text{dc load voltage}}{R_{ld}} = \frac{73.65}{6} = 12.28 \text{ A}$$

3. A three-phase, half-wave rectifier is operated by a three-phase, star-connected, 220-V, 50-Hz supply. The load resistance is 12 Ω. Load inductance is negligible. If it is required to obtain 75% of the maximum possible output voltage, calculate the (a) firing angle, (b) average and RMS load currents, (c) average and RMS thyristor currents, (d) rectification efficiency, and (e) transformer derating factor.

Solution

(a) The maximum value of the average voltage is given as

$$V_0 = \frac{3}{\pi} V_m \sin\left(\frac{\pi}{3}\right) \cos 0^\circ$$

$$= \frac{3\sqrt{3}}{2\pi} \times \sqrt{2} \times 220 \;=\; 257.3 \text{ V}$$

$$V_\alpha = \frac{3\sqrt{3}}{2\pi}\sqrt{2} \times 220 \;\cos\alpha \;=\; 0.75 \times 257.3$$

$$\cos\alpha = \frac{0.75 \times 257.3 \times 2\pi}{3\sqrt{6} \times 220} \;=\; 0.75$$

Hence

$$\alpha = 41.4^\circ$$

(b) Average load current $I_\alpha = V_\alpha/12 = 0.75 \times 257.3/12 \;=\; 16.08$ A.

$$\text{RMS load current} = \frac{1}{R_{\text{ld}}}\sqrt{\frac{3}{2\pi}\int_{41.4^\circ}^{120^\circ} (V_m \sin\omega t)^2 d(\omega t)}$$

$$= \frac{1}{12}\sqrt{\frac{3}{2\pi}\int_{41.4^\circ}^{120^\circ} (220\sqrt{2}\;\sin\omega t)^2 d(\omega t)}$$

$$= \frac{220\sqrt{2} \times 0.741}{12} \;=\; 19.2 \text{ A}$$

(c) Average thyristor current = $I_\alpha/3 \;=\; 16.08/3 \;=\; 5.36$ A.

$$\text{RMS value of thyristor current} = \frac{1}{R_{\text{ld}}}\sqrt{\frac{1}{2\pi}\int_{41.4^\circ}^{120^\circ} (V_m \sin\omega t)^2 d(\omega t)}$$

$$= \frac{19.2}{\sqrt{3}} = 11.09 \text{ A}$$

(d) Rectification efficiency = $\dfrac{P_{\text{dc}}}{P_{sac}} \;=\; \dfrac{0.75 \times 257.3 \times 16.08}{220 \times 19.2} \;=\; 73.5\%$

(e) Derating factor = $\dfrac{P_{sac} + P_{pac}}{P_{\text{dc}}} \;=\; \dfrac{2P_{sac}}{2P_{\text{dc}}} \;=\; 1.36$

4. A single-phase fully controlled bridge rectifier supplies an $R + L + E_b$ load. The data are V_{s1} = 230 V (RMS), $E_b = 130$ V, L_{ld} = 12 mH, α = 30°, and frequency of ac supply = 50 Hz. (a) Of what value will the load resistance be if the conduction is required to be just continuous? (b) Determine the average values of the load voltage and load current. (c) Sketch the waveforms of the load voltage and load current.

Solution

(a) $\alpha' = \alpha + 0° = 30° + 0° = 30°$. The parameter $a = E_b/V_m = 130/\sqrt{2} \times 230 = 0.4$. With the value of $a = 0.4$ and $\alpha' = 30°$, $\cos\phi$ for just starting conduction is obtained from the chart of Fig. 2.21 as 0.4.

$$\tan\phi = \tan(\cos^{-1} 0.4) = 2.29$$

As $\omega L_{ld}/R_{ld} = 2.29$,

$$R_{ld} = \frac{314 \times 12 \times 10^{-3}}{2.29} = 1.65$$

(b) The conduction is marginally continuous. Hence the expression for continuous conduction can be used:

$$V_\alpha = \frac{2}{\pi} V_m \cos\alpha = \frac{2}{\pi} \times 230\sqrt{2} \cos 30° = 179.4$$

$$I_\alpha = \frac{V_\alpha}{R_{ld}} = \frac{179.4}{1.65} = 108.74 \text{ A}$$

(c) The waveforms are shown in Fig. 2.59.

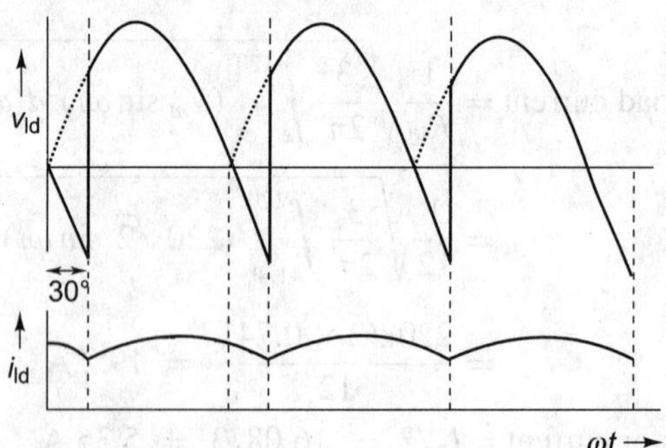

Fig. 2.59

5. Find the rectifier efficiency for a single-phase bridge rectifier for which the data are V_s = 120 V (RMS) at 50 Hz, $\alpha = 40°$, $R_{ld} = 12\ \Omega$, L_{ld} = 50 mH. Sketch the voltage across a thyristor and the current through it. Also calculate the ripple factor of the output voltage.

Solution

The waveforms are given in Fig. 2.60. For the single-phase rectifier,

$$\alpha' = \alpha = 40°$$

$$\tan\phi = \frac{\omega L}{R} = \frac{314 \times 50}{1000 \times 12} = 1.308$$

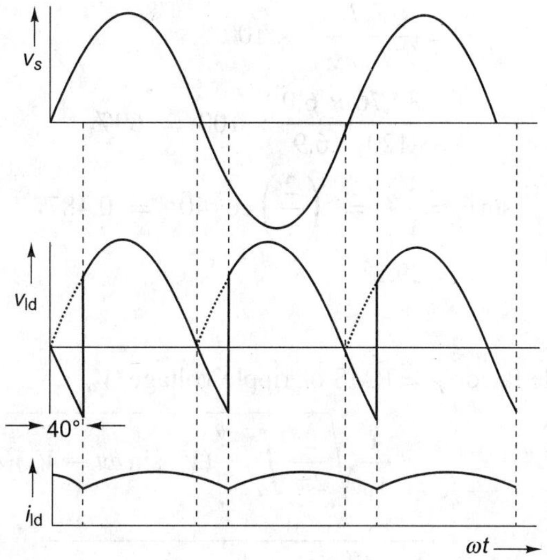

Fig. 2.60

$$\phi = \tan^{-1}(1.308) = 52.6°$$

$$\cos\phi = 0.607$$

From the chart of Fig. 2.19, α'_e is determined as follows: For $\cos\phi$ = 0.607 and α = 40°,

$$\alpha_e = \alpha'_e = 233°$$

$\alpha_e - \alpha$ = 233°– 40°= 193°, which is greater than 180°. Hence the conduction is continuous.

$$V_\alpha = \frac{2}{\pi} V_m \sin\left(\frac{\pi}{2}\right) \cos 40° = \frac{2}{\pi}\sqrt{2} \times 120 \cos 40°$$

This gives

$$V_\alpha = 82.76 \text{ V}$$

Also,

$$I_\alpha = \frac{V_\alpha}{R} = \frac{82.76}{12} = 6.9 \text{ A}$$

Assuming a flat characteristic for I_{ld}

$$I_{ld(RMS)} = \sqrt{\frac{2}{2\pi}\int_0^\pi (6.9)^2 d(\omega t)} = 6.9 \text{ A}$$

$$\text{Rectifier efficiency} = \frac{P_{dc} \times 100}{P_{sac}}$$

$$= \frac{V_\alpha I_\alpha}{V_s I_{\text{ld(RMS)}}} \times 100$$

$$= \frac{82.76 \times 6.9}{120 \times 6.9} \times 100 = 69\%$$

$$\sin\theta = \frac{V_\alpha}{V_m} = \left(\frac{2}{\pi}\right)\cos 40^\circ = 0.4877$$

$$\theta = 29.2^\circ$$

Hence $\alpha > \theta$.

$$\text{Ripple factor } \gamma = \text{RMS of ripple voltage}/V_\alpha$$

$$= \frac{1}{V_\alpha}\sqrt{\frac{2}{2\pi}\int_\alpha^{\pi-\theta}(V_m \sin\omega t - V_\alpha)^2 d(\omega t)}$$

$$= \frac{1}{82.76}\sqrt{\frac{1}{\pi}\int_{40^\circ}^{150.8^\circ}(V_m \sin\omega t - V_\alpha)^2 d(\omega t)}$$

$$= \frac{1}{82.76}\left(\frac{1}{\pi}\left\{(120\sqrt{2})^2\frac{1}{2}\left[\omega t - \frac{\sin 2\omega t}{2}\right]_{40^\circ}^{150.8^\circ} - 2V_m V_\alpha[\cos 40^\circ - \cos 150.8^\circ] + V_\alpha^2 \frac{\pi}{180}[150.8^\circ - 40^\circ]\right\}\right)^{1/2}$$

$$= \frac{1}{82.76}\sqrt{\frac{8280.5}{\pi}}$$

$$= 0.62$$

That is, the ripple factor is 0.62.

6. Find the mean values of load voltage and current for a centre-tapped type of rectifier with the following data: $V_{s1} = V_{s2} = 135$ V (RMS) at 50 Hz, $E_b = 98$ V, $R_{\text{ld}} = 3.5\ \Omega$, $L_{\text{ld}} = 14.5$ mH, $\alpha = 50^\circ$. Also calculate the ripple factor of the output voltage.

Solution

The circuit and waveforms are given in Fig. 2.3. The rectifier will work as a single-phase bridge rectifier with an ac input voltage of 135 V. It is first required to find out whether the conduction is continuous or discontinuous:

$$\alpha' = \alpha = 50^\circ$$

$$\tan\phi = \frac{314 \times 14.5}{1000 \times 3.5} = 1.3$$

$$\phi = 52.44^\circ$$

$$\cos\phi = 0.609$$

With $\cos\phi$ of 0.609 and $\alpha' = 50°$, α'_e is found from the chart of Fig. 2.19 as

$$\alpha_e = \alpha'_e = 232°$$

This gives

$$\alpha_e - \alpha = 232° - 50° = 182°$$

which is greater than 180°. Hence the conduction is continuous.

$$V_\alpha = \frac{2}{\pi} V_m \sin\left(\frac{\pi}{2}\right) \cos\alpha$$

$$= \frac{2}{\pi} \times 135\sqrt{2} \times 1 \times \cos 50° = 78.13 \text{ V}$$

$$I_\alpha = \frac{V_\alpha}{R} = \frac{78.13}{35} = 22.32 \text{ A}$$

$$\text{Ripple factor of output voltage} = \frac{\text{RMS of ripple voltage}}{V_\alpha}$$

$$= \frac{\sqrt{\frac{2}{2\pi}\int_\alpha^{\alpha+\pi}(V_m \sin\omega t - V_\alpha)^2 d(\omega t)}}{V_\alpha}$$

Substitution of values gives

$$\text{Ripple factor} = \frac{110.10}{78.13} = 1.41$$

7. A three-phase, half-wave rectifier supplies a load with $R_{\text{ld}} = 5.0\ \Omega$ and $L_{\text{ld}} = 20$ mH. If the firing angle $\alpha = 70°$ and the ac input voltage is 160 V (RMS), determine V_α and I_α.

Solution

It is first required to find α_e to decide whether the conduction is discontinuous or continuous.

$$\tan\phi = \frac{314 \times 20}{1000 \times 5} = 1.256$$

$$\phi = 51.5°$$

$$\cos\phi = 0.623$$

Here $\alpha' = \alpha = 70°$. With α' as 70° and $\cos\phi$ as 0.623, α'_e is found to be 230°.

$$\alpha_e = \alpha'_e = 230°$$

$$\alpha_e - \alpha = 230° - 70° = 160°$$

which is less than 180°. Hence conduction is discontinuous. From Eqn (2.14),

$$V_\alpha = \frac{2}{2\pi} V_m \left\{\sin\left(\alpha_e - \frac{\pi}{2}\right) - \sin\left(\alpha - \frac{\pi}{2}\right)\right\}$$

$$= \frac{160\sqrt{2}}{\pi}\{\sin(230^\circ - 90^\circ) - \sin(70^\circ - 90^\circ)\}$$

$$= 70.95 \text{ V}$$

$$I_\alpha = \frac{V_\alpha}{R_{ld}} = \frac{70.95}{5} = 14.19 \text{ A}$$

8. A three-phase, half-wave rectifier is supplied by a transformer with a secondary voltage of 180 V (RMS) at 50 Hz. Other data are $R_{ld} = 10\ \Omega$, $L_{ld} = 10$ mH, and back emf $E_b = 153$ V. Determine V_α and I_α for a firing angle of 60°. Also sketch the waveforms.

Solution

The waveforms are given in Fig. 2.9(b).

$$V_m = 180\sqrt{2} = 254.53 \text{ V}$$

$$a = \frac{E_b}{V_m} = \frac{153}{254.53} = 0.6$$

$$\cos\phi = \cos\left\{\tan^{-1}\left(\frac{10 \times 314}{1000 \times 10}\right)\right\} = 0.954$$

$$\alpha' = \alpha + 30^\circ = 60^\circ + 30^\circ = 90^\circ$$

From the chart of Fig. 2.21, α'_e for $a = 0.6$, $\alpha' = 90°$, and $\cos\phi = 0.954$ is 153°.

$$\alpha_e = \alpha'_e - 30^\circ = 153^\circ - 30^\circ = 123^\circ$$

$$\alpha_e - \alpha = 123^\circ - 60^\circ = 63^\circ$$

which is less than 120°. Hence the conduction is discontinuous.

From Eqn (2.30), the average voltage for discontinuous conduction is

$$V_\alpha = \left(\frac{n}{2\pi}\right) V_m \left[\sin\left(\alpha_e - \frac{\pi}{n}\right) - \sin\left(\alpha - \frac{\pi}{n}\right) + E_b\left(1 - \frac{\alpha_e - \alpha}{2\pi/n}\right)\right]$$

Substituting values gives the average voltage as

$$V_\alpha = \left(\frac{3}{2\pi}\right) \times 180\sqrt{2}[\sin(123^\circ - 60^\circ) - \sin(60^\circ - 60^\circ)]$$

$$+ 153\left(1 - \frac{\pi(123 - 60)/180}{2\pi/n}\right)$$

$$= 108.30 + 72.68$$

$$= 181 \text{ V}$$

$$I_\alpha = \frac{181}{R_{ld}} = \frac{181}{10} = 18.1 \text{ A}$$

9. If, in Example 8, the load inductance is increased to 156 mH and E_b is decreased to 51 V, other quantities remaining the same, determine the new V_α and I_α.

Solution

The waveforms are given in Fig. 2.9(c).

$$\cos\phi = \cos\left[\tan^{-1}\left(\frac{314 \times 156}{1000 \times 10}\right)\right] = 0.2$$

With $\alpha' = 90°$, $\cos\phi = 0.2$, and $a = 0.6$, α'_e is found to be 212°.

$$\alpha_e = \alpha'_e - 30° = 212° - 30° = 182°$$

$$\alpha_e - \alpha = 182° - 60° = 122°$$

which is greater than 120°. Hence the conduction is continuous.

The average voltage is

$$V_\alpha = \left(\frac{n}{\pi}\right) V_m \sin\left(\frac{\pi}{n}\right) \cos\alpha$$

$$= \frac{3}{\pi} \times 180\sqrt{2} \times \frac{\sqrt{3}}{2} \cos 60° = 105.25 \text{ V}$$

Also,

$$I_\alpha = \frac{V_\alpha}{R_{ld}} = \frac{105.25}{10} = 10.53 \text{ A}$$

10. A three-phase, half-wave rectifier is supplied by a 220-V 3ϕ supply and feeds an $R + E_b$ load with $R_{ld} = 15\ \Omega$ and $E_b = 120$ V. Determine the ranges of α for which conduction is (a) continuous and (b) discontinuous.

Solution

The waveforms are as given in Fig. 2.17(b) for continuous conduction and Fig. 2.17(c) for discontinuous conduction. The range of α for continuous conduction or otherwise is demarcated by the E_b line.

From Fig. 2.17(b), the range of α for continuous conduction is

$$\alpha \le \frac{\pi}{2} - \frac{\pi}{n} - \alpha_r$$

where $\alpha_r = \sin^{-1}(E_b/V_m) = \sin^{-1}(120/220\sqrt{2}) = 22.6°$. Thus,

$$\frac{\pi}{2} - \frac{\pi}{n} - \alpha_r = 30° - 22.6° = 7.4°$$

Hence the range for continuous conduction is $0° \le \alpha \le 7.4°$ and discontinuous conduction is $7.4° \le \alpha \le 120°$.

11. For the three-phase rectifier in Example 10, find V_α and I_α for (a) $\alpha = 5°$ and (b) $\alpha = 35°$.

Solution

(a) From the solution of Example 10, the conduction is continuous as $\alpha = 5°$, which is less than 7.4°. The expression for continuous conduction can be used:

$$V_\alpha = \left(\frac{3}{\pi}\right) V_m \sin\left(\frac{\pi}{3}\right) \cos\alpha = \frac{3}{\pi} \times 220\sqrt{2} \times \frac{\sqrt{3}}{2} \cos 5° = 256.3 \text{ V}$$

$$I_\alpha = \frac{V_\alpha}{R} = \frac{256.3}{15} = 17.1 \text{ A}$$

(b) For $\alpha = 35°$ the conduction is discontinuous. Hence V_α is determined from Eqn (2.14):

$$V_\alpha = \frac{nV_m}{2\pi}\left[\sin\left(\alpha_e - \frac{\pi}{n}\right) - \sin\left(\alpha - \frac{\pi}{n}\right)\right]$$

Here $\alpha_e = 180° - \sin^{-1}(E_b/V_m) = 180° - \sin^{-1}(120/220\sqrt{2}) = 180° - 22.6° = 157.4°$. Thus,

$$V_\alpha = \frac{3}{2\pi} \times 220\sqrt{2}[\sin(157.4° - 60°) - \sin(35° - 60°)] = 210 \text{ V}$$

From Eqn (2.15),

$$I_\alpha = \frac{V_\alpha - E_b^*}{R_{\text{ld}}}$$

where $E_b^* = E_b(\alpha_e - \alpha)/(2\pi/3)$. Here

$$E_b^* = \frac{120(157.4 - 35) \times \pi \times 3}{180 \times 2\pi} = 122.4 \text{ V}$$

Thus,

$$I_\alpha = \frac{210 - 122.4}{15} = 5.84 \text{ A}$$

12. For a three-phase, half-wave rectifier the transformer inductance L_s per phase is given as 4 mH. Other data are $V_{sR} = V_{sY} = V_{sB} = 180$ V (RMS), $f = 50$ Hz, $R_{\text{ld}} = 2.5\ \Omega$, $L_{\text{ld}} \approx \infty$, $\alpha = 55°$. Compute the overlap angle μ, reduction in voltage due to source inductance $g_{\mu\alpha}$, and load current I_{ld}.

Solution

The circuit and waveforms are given in Fig. 2.28. From Eqns (2.92) and (2.93),

$$V_{\mu\alpha} = V_\alpha - g_{\mu\alpha} = V_\alpha - \frac{n}{2\pi} I_{\text{ld}} X_s$$

Also,

$$I_{\text{ld}} = \frac{V_{\mu\alpha}}{R_{\text{ld}}} = \frac{V_\alpha - (n/2\pi) I_{\text{ld}} X_s}{R_{\text{ld}}}$$

$$= \frac{V_\alpha}{R_{\text{ld}} + (n/2\pi) X_s}$$

$$= \frac{(n/2\pi) V_m \sin(\pi/n) \cos\alpha}{R_{\text{ld}} + (n/2\pi) X_s}$$

Substitution of the values gives

$$V_\alpha = \frac{3}{\pi} V_m \sin\left(\frac{\pi}{n}\right) \cos\alpha = \frac{3}{\pi} \times 180\sqrt{2} \sin\left(\frac{\pi}{3}\right) \cos 55° = 120.7 \text{ V}$$

$$I_{\rm ld} = \frac{\frac{3}{\pi}\frac{\sqrt{3}}{2}180\sqrt{2}\cos 55^\circ}{2.5 + \frac{3}{2\pi} \times \frac{314}{1000} \times 4}$$

$$= \frac{120.7}{3.1} = 38.95 \text{ A}$$

From Eqn (2.88),

$$\cos(\mu + \alpha) = \cos\alpha - \frac{I_{\rm ld} X_s}{V_m \sin(\pi/n)}$$

Substitution of values yields

$$\cos(\mu + 55^\circ) = \cos 55^\circ - \frac{38.95 \times 314 \times 4/1000}{180\sqrt{2} \times \sqrt{3}/2}$$

$$= 0.3517$$

Therefore,

$$\mu + 55^\circ = 69.4^\circ$$

This gives the overlap angle as

$$\mu = 14.4^\circ$$

The reduction in voltage is

$$g_{\mu\alpha} = \frac{n}{2\pi} I_{\rm ld} X_s = \frac{3}{2\pi} \times 38.95 \times \frac{314 \times 4}{1000} = 23.36 \text{ V}$$

Hence

$$V_{\mu\alpha} = 120.7 - 23.36 = 97.34 \text{ V}$$

$$I_{\rm ld} = \frac{V_{\mu\alpha}}{R_{\rm ld}} = \frac{97.34}{2.5} = 38.94 \text{ A}$$

13. A three-phase, half-wave rectifier has an input voltage of 185 V (RMS) at 50 Hz at the secondary and its transformer leakage inductance is 2 mH. Other data are $\mu = 20^\circ$, $R_{\rm ld} = 3.5\ \Omega$, $L_{\rm ld} = \infty$, and $\alpha = 63^\circ$. Compute the load current $I_{\rm ld}$, reduction in voltage $g_{\mu\alpha}$, and net load voltage $V_{\mu\alpha}$.

Solution

The circuit and waveforms are given in Fig. 2.28. The load current $I_{\rm ld}$ can be found from Eqn (2.88):

$$\cos(\alpha + \mu) = \cos\alpha - \frac{I_{\rm ld} X_s}{V_m \sin(\pi/n)}$$

Substitution of values gives

$$\cos(63^\circ + 20^\circ) = \cos 63^\circ - \frac{I_{\rm ld} \times 314 \times 2/1000}{185\sqrt{2} \times (\sqrt{3}/2)}$$

which gives I_{ld} = 119.82 A. Also,

$$g_{\mu\alpha} = \frac{n}{2\pi} I_{\text{ld}} X_s = \frac{3}{2\pi} \times 119.82 \times \frac{314 \times 2}{1000} = 35.93 \text{ V}$$

$$\begin{aligned} V_{\mu\alpha} &= V_\alpha - 35.93 \\ &= \frac{3\sqrt{3}}{2\pi} \times 185\sqrt{2} \cos 63^\circ - 35.93 \\ &= 98.23 - 35.93 = 62.3 \text{ V} \end{aligned}$$

14. A three-phase, half-wave rectifier is supplied by a three-phase ac input of 175 V (RMS). It supplies a dc motor load with a back emf of 90 V. Other data are $R_a = R_{\text{ld}} = 1.2\ \Omega$, $L_{\text{ld}} = \infty$, $\alpha = 58^\circ$, and $L_s = 1.8$ mH. Determine I_{ld}, $g_{\mu\alpha}$, and $V_{\mu\alpha}$.

Solution

$V_{\mu\alpha}$ can expressed in two ways. From Eqn (2.93),

$$\begin{aligned} V_{\mu\alpha} &= V_\alpha - g_{\mu\alpha} \\ &= \left(\frac{n}{2\pi}\right) V_m \sin\left(\frac{\pi}{n}\right) \cos\alpha - \frac{n}{2\pi} I_{\text{ld}} X_s \end{aligned} \tag{i}$$

Also,

$$V_{\mu\alpha} = E_b + I_{\text{ld}} R_a \tag{ii}$$

Equating Eqn (i) with Eqn (ii) and solving for I_{ld} gives

$$I_{\text{ld}} = \frac{(n/\pi) V_m \sin(\pi/n) \cos\alpha - E_b}{R_a + (n/2\pi) X_s}$$

On substitution of the numerical values, the current is obtained as

$$\begin{aligned} I_{\text{ld}} &= \frac{\dfrac{3\sqrt{3}}{2\pi} 175\sqrt{2} \cos 58^\circ - 90}{1.2 + \dfrac{3}{2\pi} \times \dfrac{314}{1000} \times 1.8} \\ &= \frac{18.459}{1.470} = 12.56 \text{ A} \end{aligned} \tag{2.242}$$

The load voltage is given as

$$\begin{aligned} V_{\mu\alpha} &= E_b + I_{\text{ld}} R_a \\ &= 90 + 12.56 \times 1.2 \\ &= 105.1 \text{ V} \end{aligned}$$

Finally, the reduction in voltage is

$$\begin{aligned} g_{\mu\alpha} &= \frac{3}{2\pi} I_{\text{ld}} X_s \\ &= \frac{3}{2\pi} \times 12.56 \times \frac{314}{1000} \times 1.8 \\ &= 3.39 \text{ V} \end{aligned}$$

15. A three-phase, half-wave rectifier supplies a dc motor load consisting of $R_a + L_a + E_b$, where $R_a = 3\ \Omega$, $E_b = 130$ V, and $\alpha = 50°$. If the rectifier operates on the border of continuous and discontinuous conduction, determine the load power factor and the value of L_a of the motor armature.

Solution

The parameter a is calculated as

$$a = \frac{130}{230\sqrt{2}} = 0.4$$

$$\alpha' = \alpha + 30° = 50° + 30° = 80°$$

From the firing angle chart in Fig. 2.21, $\cos\phi$ is read as 0.45.

$$\tan\phi = \frac{\omega L_a}{R_a} = \tan(\cos^{-1} 0.45) = 1.9845$$

$$L_a = \frac{R_a \times 1.9845}{\omega} = \frac{3 \times 1.9845}{314}$$

$$= 0.019 \text{ H or } 19 \text{ mH}$$

16. A single-phase dual converter is used for controlling the speed of a separately excited dc motor as shown in Fig. 2.40(a). It is supplied by a single-phase ac source of 240 V at 50 Hz. Normally, rectifier P is used for motor operation. It is required to brake the motor in a regenerative manner by operating rectifier N in the inverting mode (quadrant II) with $\alpha_N = 120°$. If the motor is initially running at 1500 rpm, the combined inertia of the rotor and load (J) is 1.2×10^{-3} kg m^2, and the average current through the motor is 0.2 A, determine the time required for the braking of the motor.

Solution

The output voltage of rectifier N just at the start of regenerative braking is

$$V_{\alpha_N} = \left(\frac{n}{\pi}\right) V_m \sin\left(\frac{\pi}{n}\right) \cos\alpha$$

For the single-phase rectifier N,

$$V_{\alpha_N} = \left(\frac{2}{\pi}\right) V_m \sin\left(\frac{\pi}{2}\right) \cos 120° = -\frac{n}{\pi} \times 240\sqrt{2} \times 0.5 = -108 \text{ V}$$

The average value of the terminal voltage is $V_{av} = -108/2 = -54$ V. Hence $E_{b(av)} = 54$ V. The kinetic energy dissipated for braking purposes $= (1/2)J\omega^2$ $= 1/2 \times 1.2 \times 10^{-3} \times [2\pi \times 1500/60]^2 = 14.8$ J. The electrical energy returned back to the source is equal to the kinetic energy dissipated. That is, electrical power of the motor $\times$ time required for braking = kinetic energy dissipated. Thus,

$$E_{b(av)} I_{av} t = 14.8$$

or

$$54 \times 0.2 \times t = 14.8$$

This gives

$$t = \frac{14.8}{54 \times 0.2} = 1.37 \text{ s}$$

17. A three-phase, half-wave controlled rectifier is fed by a 254-V (RMS) (line-to-neutral), 50-Hz supply and provides an adjustable dc voltage at the terminals of a separately excited dc motor. The back emf is related to ω as $E_b = 1.05\,\omega$ and the full load armature current $I_a = 400$ A. The motor operates at 50% of the full load at a speed of 180 rad/s. If $\alpha = 50°$ and the conduction is marginal, determine the values of R_a and L_a.

Solution

$$50\% \text{ of full load} = 400/2 = 200 \text{ A}$$

$$E_b = 1.05\omega = 1.05 \times 180 = 189 \text{ V}$$

$$V_\alpha = 3\sqrt{3}/2\pi \times 254\sqrt{2} \cos 50° = 191 \text{ V}$$

$$= E_b + I_a R_a$$

This gives

$$R_a = \frac{V_\alpha - E_b}{I_a} = \frac{191 - 189}{200} = 0.01\ \Omega$$

$$\alpha' = \alpha + 30° = 50° + 30° = 80°$$

$$a = 189/254\sqrt{2} = 0.526$$

From the chart of Fig. 2.21, for $a = 0.526$ and $\alpha' = 80°$, $\cos\phi = 0.2$.

$$\frac{\omega L}{R} = \tan(\cos^{-1} 0.2) = 4.899$$

$$L = \frac{4.899 \times 0.01 \times 1000}{314} \text{ mH}$$

$$= 0.156 \text{ mH}$$

18. A single-phase bridge rectifier is operated with extinction angle control with $\beta = 45°$. The ac supply voltage is 160 V (RMS) and a load resistance of 23 Ω is connected at the output of the rectifier. Determine the (a) distortion factor, (b) displacement factor, (c) input power factor, (d) average load voltage, and (e) average load current.

Solution

The waveforms are given in Fig. 2.50(b). From Eqns (2.202)–(2.204), the various factors are determined as below.

(a) Distortion factor $\mu = \dfrac{2\sqrt{2}\cos(\beta/2)}{\sqrt{\pi(\pi - \beta)}}$

$$= \frac{2\sqrt{2}(\cos 22.5°)}{\sqrt{\pi(\pi - \pi/4)}} = 0.96$$

(b) Displacement factor $\cos\phi_1 = \cos(\beta/2) = \cos 22.5° = 0.924$

(c) Input power factor $= \mu \cos\phi_1 = 0.96 \times 0.924 = 0.887$

(d) Average load voltage $= \dfrac{V_m(1+\cos\beta)}{\pi} = \dfrac{160\sqrt{2}(1+\cos 45^\circ)}{\pi} = 123$ V

(e) Average load current = 123/23 = 5.35 A

19. A single-phase, half-controlled rectifier is operated with symmetrical pulse width modulation with the angle $\beta = 56^\circ$. The ac supply voltage is 220 V (RMS) and the load resistance is 21 Ω. Compute the (a) distortion factor, (b) input power factor, (c) average load voltage, and (d) average load current.

Solution

The waveforms are given in Fig. 2.51.

(a) From Eqn (2.162),

$$\text{distortion factor } \mu = \frac{2\sqrt{2}\sin(\beta/2)}{\sqrt{\pi\beta}}$$

$$= \frac{2\sqrt{2}\,\sin 28^\circ}{\sqrt{\pi \times 56 \times \pi/180}} = \frac{1.328}{1.752} = 0.76$$

From Eqn (2.212), the displacement factor $= \cos\phi_1 = 1$.

(b) Input power factor $= \mu\ \cos\phi_1 = 1 \times 0.76 = 0.76$

(c) Average load voltage $= (2V_m/\pi)\sin(\beta/2) = (2 \times 220\sqrt{2}/\pi)\sin 28^\circ = 93$ V

(d) Average load current $= 93/R_{\text{ld}} = 93/21 = 4.43$ A

20. The specifications of a rectifier supplied by asymmetrical voltage [Fig. 2.61(a)] are $V_{s1} = 230$ V and $V_{s2} = 115$ V at 50 Hz. Other data are $R_{\text{ld}} = 15$ Ω and $\alpha = 70^\circ$. Find the average load current I_α with (a) $L_{\text{ld}} = 0$ and (b) $L_{\text{ld}} = 63.7$ mH.

Solution

The waveforms for an inductive load are shown in Fig. 2.61(b).

(a) With no load inductance, the load current waveform is in phase with the load voltage waveform.

$$\alpha_e = \pi \text{ rad or } 180^\circ$$

$$V_\alpha = \frac{1}{2\pi}\left[\int_{0\text{ rad}}^{\pi\text{ rad}} V_{s1}(\omega t)d(\omega t) + \int_{0\text{ rad}}^{\pi\text{ rad}} V_{s2}(\omega t)d(\omega t)\right]$$

$$= \frac{1}{2\pi}\left[\int_{70^\circ}^{180^\circ} 230\sqrt{2}\ \sin\omega t\ d(\omega t) + \int_{70^\circ}^{180^\circ} 115\sqrt{2}\ \sin\omega t\ d(\omega t)\right]$$

$$= \frac{1}{2\pi}[230\sqrt{2}(\cos 70^\circ - \cos 180^\circ) + 115\sqrt{2}(\cos 70^\circ - \cos 180^\circ)]$$

$$= \frac{1}{2\pi}[436.52 + 218.26]$$

$$= 104.2 \text{ V}$$

$$I_\alpha = \frac{V_\alpha}{R_{\text{ld}}} = \frac{104.2}{15}$$

$$= 6.95 \text{ A}$$

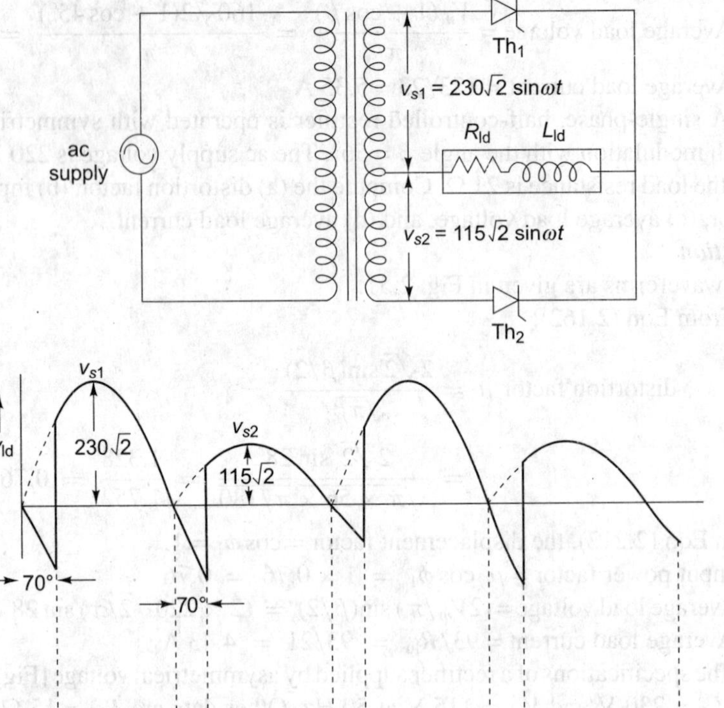

Fig. 2.61

(b) With load inductance

$$\tan\phi = \frac{\omega L}{R} = \frac{314 \times 63.7}{1000 \times 15} = 1.33$$

Hence,

$$\phi = 53.12°, \qquad \cos\phi = 0.6$$

Here $\alpha' = \alpha = 70°$. From the chart in Fig. 2.19, with α' of 70° and $\cos\phi = 0.6$,

$$\alpha_e = 232° = \alpha'_e$$

Also,

$$\alpha_e - \alpha = 232° - 70° = 162° < 180°$$

Hence the conduction is discontinuous.

$$V_\alpha = \frac{1}{2\pi}\left[\int_\alpha^{\alpha_e} V_{s1}(\omega t)d(\omega t) + \int_\alpha^{\alpha_e} V_{s2}(\omega t)d(\omega t)\right]$$

where $V_{s1} = 230\sqrt{2}\ \sin\omega t$ and $V_{s2} = 115\sqrt{2}\ \sin\omega t$. Thus,

$$V_\alpha = \frac{1}{2\pi}[230\sqrt{2}(\cos\alpha - \cos\alpha_e) + 115\sqrt{2}(\cos\alpha - \cos\alpha_e)$$
$$= \frac{1}{2\pi}[230\sqrt{2}(\cos 70^\circ - \cos 232^\circ) + 115\sqrt{2}(\cos 70^\circ - \cos 232^\circ)]$$
$$= 74.4\ \text{V}$$

$$I_\alpha = \frac{V_\alpha}{R_{\text{ld}}} = \frac{74.4}{15} \approx 4.96\ \text{A}$$

21. The following data pertain to the single-phase, half-wave controlled rectifier circuit of Fig. 2.62. Neglecting the ON state drop across the thyristor, determine the mean load current and the mean load voltage. Draw the current and voltage waveforms. The data are ac supply voltage = 300 V (RMS), supply frequency = 50 Hz, load resistance $R_{\text{ld}} = 15\ \Omega$, load inductance $L_{\text{ld}} = 0.2$ H, and firing angle $\alpha = 90^\circ$.

Solution

The waveforms are given in Fig. 2.62. The peak of ac supply voltage = $300\sqrt{2}$ = 424.3 V; X_{ld} of reactor = $\omega L_{\text{ld}} = 100\ \pi \times 0.2 = 62.8\ \Omega$. The expression for the current is obtained from Eqn (2.21) as

$$i_{\text{ld}}(t) = \frac{V_m}{\sqrt{R_{\text{ld}}^2 + X_{\text{ld}}^2}}[\sin(\omega t - \phi) - \sin(\alpha - \phi)e^{-(R_{\text{ld}}/L_{\text{ld}})(t-\alpha/\omega)}]$$

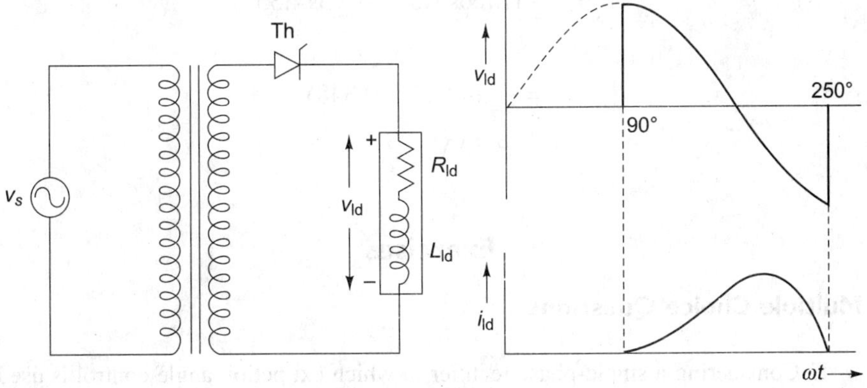

Fig. 2.62

Here

$$\phi = \tan^{-1}(62.8/15) = 1.336$$

$$\alpha = \pi/2\ \text{rad}$$

$$\frac{\alpha}{\omega} = \frac{1.57}{2\pi \times 50} = 0.005\ \text{s}$$

Thus,

$$i_{ld}(t) = \frac{424.3}{\sqrt{15^2 + 62.8^2}}[\sin(\omega t - 1.336) - \sin(1.57 - 1.336)e^{-(15/0.2)(t-0.005)}]$$

$$= 6.57[\sin(\omega t - 1.336) - 0.337e^{-75t}]$$

$i_{ld}(t)$ becomes zero at $t = 0.01387$ or at an angle of 250°. Thus $\alpha_e = 250°$ or 4.357 rad.

The mean load current can be obtained as

$$I_\alpha = \frac{\omega}{2\pi}\int_{\alpha/\omega}^{\alpha_e/\omega} i_{ld}(t)dt$$

$$= \frac{100\pi}{2\pi}\int_{0.005}^{0.01387}[6.57\sin(\omega t - 1.336) - 2.214e^{-75t}]dt$$

This gives

$$I_\alpha = 1.57 \text{ A}$$

Also,

$$V_\alpha = I_\alpha R_{ld} = 1.57 \times 15 = 23.55 \text{ V}$$

The value of V_α may be checked by direct computation as

$$V_\alpha = \frac{1}{2\pi}\int_{1.57}^{4.357} V_m \sin \omega t \, d(\omega t)$$

$$= \frac{1}{2\pi}V_m[\cos 1.57 - \cos 4.357]$$

$$= \frac{1}{2\pi}(424.3 \times 0.348)$$

$$= 23.50 \text{ V}$$

Exercises

Multiple Choice Questions

1. Considering a single-phase rectifier in which extinction angle control is used for improving the power factor, if the angle β is 50°, the input power factor will be ______.

   (a) 0.75 (b) 0.87
   (c) 0.96 (d) 1.13

2. In the case of a single-phase rectifier in which symmetrical PWM is used for improving the power factor, if the angle β is 60°, the input power factor will be nearly ______.

   (a) 0.85 (b) 0.67
   (c) 1.12 (d) 0.78

3. In a single-phase rectifier in which sinusoidal PWM is used for improving the power factor, the modulation index M should not be allowed to go above unity because ______.
   (a) some new high-frequency harmonics will get introduced
   (b) the harmonic content will remain the same as for $M \leq 1$
   (c) some low-frequency harmonics will get introduced
   (d) both low-frequency and high-frequency harmonics will get introduced
4. The circuit for selective harmonic elimination method not only helps in improving the power factor but also eliminates ______ with it.
   (a) all the high-frequency harmonics which are multiples of the carrier frequency
   (b) all the low-frequency harmonics which are submultiples of the carrier frequency
   (c) some high-frequency harmonics which are multiples of 4 with respect to the input supply frequency
   (d) some selected harmonics
5. In the sinusoidal PWM method of improving the power factor, if the carrier ratio is 6 and $n = 1$, then the four nearest sidebands will be ______.
   (a) (2, 4, 8, 10) (b) (3, 5, 9, 11)
   (c) (1, 3, 7, 9) (d) (2, 4, 12, 14)
6. In the sinusoidal PWM method of improving the power factor, if the carrier ratio is 8, then the pulses per half-cycle will also be 8, provided ______.
   (a) the zero crossings of both the waveforms occur at the origin
   (b) the peak of the reference waveform and the zero crossing of the carrier waveform occur at the origin
   (c) the peaks of both the waveforms occur at the origin
   (d) the peak of the carrier waveform and the zero crossing of the reference waveform occur at the origin
7. In the sinusoidal PWM method of improving the power factor, if the carrier ratio is 9, then the pulses per half-cycle will be 9, provided ______.
   (a) the peak of the carrier waveform and the zero crossing of the reference waveform occur at the origin
   (b) the zero crossings of both the waveforms occur at the origin
   (c) the peak of the reference waveform and the zero crossing of the carrier waveform occur at the origin
   (d) the peaks of both the waveforms occur at the origin
8. The disadvantage of the method of selective harmonic elimination is that ______.
   (a) some required harmonics are also eliminated along with unwanted ones
   (b) some spikes may occur in the output voltage waveform
   (c) the peak of the available voltage exceeds the rated value
   (d) the peak of the available voltage becomes limited
9. The ripple frequency of a three-phase fully controlled full-wave rectifier is ______ times that of the supply frequency.
   (a) 6 (b) 3
   (c) 9 (d) 12

10. The peak inverse voltage of a thyristor in a centre-tapped secondary type of transformer is ________ times that of the maximum voltage read at one of the secondaries.
    (a) 4 (b) 2
    (c) 8 (d) 6

11. In a three-phase transformer supplying a three-phase rectifier, it is advantageous to use a zigzag winding for the secondary because ________.
    (a) the losses are minimum with it
    (b) it helps in preventing the saturation of the iron core
    (c) it helps in getting a nearly sinusoidal output voltage
    (d) it helps in preventing the eventuality of the transformer being overloaded

12. The maximum output voltage obtainable with a fully controlled full-wave 3ϕ rectifier is ________, where V is the RMS magnitude of the phase-to-neutral voltage of the secondary.
    (a) $4\sqrt{3}V/3\pi$ (b) $6\sqrt{2}V/\pi$
    (c) $2\sqrt{6}V/\pi$ (d) $3\sqrt{6}V/\pi$

13. The condition suitable for the inverting mode of operation of a single-phase bridge rectifier is ________.
    (a) α being greater than 90°
    (b) an extra inductance in series with a dc motor load, with $\alpha > 120°$
    (c) a battery in series with a dc motor, with α in the range $90° < \alpha < 120°$
    (d) a battery and an extra inductance in series with the load, with $\alpha > 120°$

14. A three-phase, half-wave controlled rectifier is supplied by a transformer with a secondary voltage of 200 V (RMS) and feeds an $R + E_b$ load with $R_{ld} = 20\ \Omega$ and $E_b = 50$ V. If the firing angle measured from the point of natural commutation is 12°, then the extinction angle α_e of the thyristor will be ________.
    (a) 155° (b) 162°
    (c) 167° (d) 171°

15. A three-phase controlled rectifier feeds an $R + E_b$ load. The data for the rectifier is $V_s = 220$ V (RMS), $E_b = 140$ V, and $R_{ld} = 15\ \Omega$. If the firing angle α measured from the point of natural commutation is 8°, then the extinction angle α_e of the thyristor will be ________.
    (a) 140.5° (b) 135.5°
    (c) 123.3° (d) 160°

16. A three-phase controlled rectifier feeds an $R + E_b$ load. The data for the rectifier is $V_s = 240$ V (RMS), $E_b = 130$ V, and $R_{ld} = 24\ \Omega$. If the firing angle α measured from the point of natural commutation is 15°, then the duration of conduction of a thyristor will be ________.
    (a) 80.5° (b) 110.4°
    (c) 104.8° (d) 102.2°

17. A three-phase, half-wave controlled rectifier feeds a purely resistive load. The data are $V_s = 200$ V (RMS) and $R_{ld} = 20\ \Omega$. If the firing angle α is 25°, then the extinction angle α_e of the rectifier is ________.
    (a) 135° (b) 145°
    (c) 155° (d) 165°

18. A three-phase controlled rectifier feeds a purely resistive load. The data are $V_s = 220$ V (RMS) and $R_{ld} = 15\ \Omega$. If the firing angle α is 45°, then the duration of a conduction of a thyristor will be ________.

    (a) 110° (b) 125°
    (c) 105° (d) 100°

19. A three-phase controlled rectifier feeds a purely resistive load. The data are $V_s = 240$ V (RMS) and $R = 24\ \Omega$. If the firing angle α is 90°, then the average current delivered to the load is ________.

    (a) 8.50 A (b) 9.65 A
    (c) 3.38 A (d) 6.75 A

20. If the n-phase rectifier for which transformer leakage inductance is to be considered has a firing angle $\alpha = 0°$, then the commutation angle μ can be determined from the expression ________.

    (a) $\cos\mu = 1 + \dfrac{V_m \sin(\pi/n)}{I_d X_s}$ (b) $\cos\mu = 1 - \dfrac{I_d V_m}{X_s \sin(\pi/n)}$
    (c) $\sin\mu = 1 - \dfrac{V_m X_s}{I_d \sin(\pi/n)}$ (d) $\cos\mu = 1 - \dfrac{I_d X_s}{V_m \sin(\pi/n)}$

21. For an n-phase rectifier in which the transformer leakage inductance L_s is significant, the expression for the reduction in average voltage at the load with a firing angle α will be ________.

    (a) $(n/2\pi)I_{ld}\omega L_s$ (b) $(2\pi/n)I_{ld}\omega L_s$
    (c) $(3n/2\pi)V_m I_{ld}\omega L_s$ (d) $(2\pi/n)(V_m\omega L_s/I_{ld})$

22. For an n-phase rectifier in which the transformer leakage inductance has to be taken into consideration, the commutation angle μ can be found from the expression ________.

    (a) $\cos\alpha = \cos(\alpha + \mu) - I_d X_s/V_m \sin(\pi/n)$
    (b) $\cos(\alpha + \mu) = \cos\alpha + V_m \sin(\pi/n)/I_d X_s$
    (c) $\cos(\alpha + \mu) = \cos\alpha - I_d X_s/V_m \sin(\pi/n)$
    (d) $\cos(\alpha + \mu) = \cos\alpha + V_m \sin(\pi/n)/I_d X_s$

23. The rectifier efficiency of a single-phase bridge rectifier feeding an $R + L$ load will be ________ than that with a pure resistive load; also its derating factor will be ________ than that with a pure resistive load.

    (a) more, more (b) less, more
    (c) less, less (d) more, less

24. The rectifier efficiency of a single-phase centre-tapped secondary type of rectifier feeding an $R + L$ load will be ________ than that of a bridge rectifier feeding an $R + L$ load; also its derating factor will be ________ than that of the bridge rectifier.

    (a) less, more (b) less, less
    (c) more, more (d) more, less

25. A dc motor fed by a dual converter will operate in the ________ quadrant for forward braking and the ________ quadrant for reverse braking.

    (a) first, fourth (b) second, fourth
    (c) third, fourth (d) fourth, second

26. A dc motor fed by a dual converter will operate in the ________ quadrant for reverse motoring and the ________ quadrant for forward motoring.
    (a) second, first
    (b) fourth, first
    (c) second, third
    (d) third, first

27. The reversal of rotation of a separately excited dc motor using the field reversal method is disadvantageous because ________.
    (a) it affects the commutation of the armature current
    (b) it causes voltage spikes due to the presence of armature inductance
    (c) it leads to excessive current in the armature
    (d) its implementation is not as easy as the method of armature reversal

28. The average load voltage that is obtained for the circuit of Fig. 2.63(a) will be ________ that obtained with the circuit of Fig. 2.63(b).
    (a) less than
    (b) more than
    (c) equal to
    (d) much more than

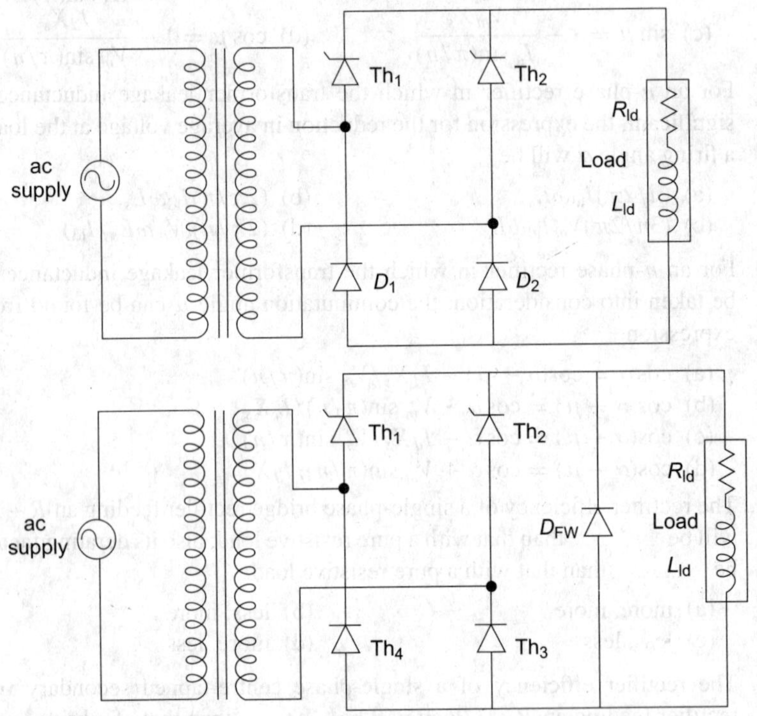

Fig. 2.63

29. In a single-pulse circuit using an interphase transformer, the peak magnitude of the voltage of the reactor will be equal to ________, where V_m is the peak of the phase-to-neutral voltage, and the frequency of this reactor voltage is ________, where f is the ac supply frequency.
    (a) $3V_m/2, 6f$
    (b) $V_m/2, 3f$
    (c) $3V_m/2, 3f$
    (d) $V_m/2, 6f$

30. In a single-phase bridge type of controlled rectifier supplying an $R + L$ load, discontinuous conduction occurs when the magnitude of L is ______ and the firing angle α is ______.
    (a) small, large
    (b) small, small
    (c) large, large
    (d) large, small

31. The presence of a freewheeling diode is essential in the circuit of Fig. 2.64 because it ______.
    (a) improves the power factor of the circuit
    (b) helps in preventing the flow of large current through the thyristors when they are just triggered
    (c) prevents the flow of reverse current through the thyristors which are in the OFF state
    (d) prevents the flow of large forward current through the thyristors which are in the OFF state

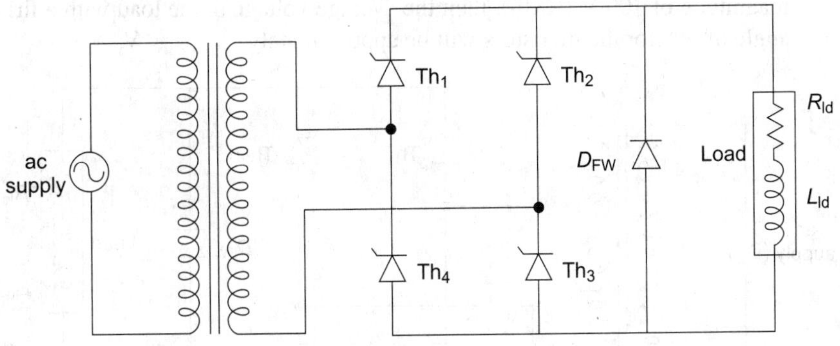

Fig. 2.64

32. A three-phase rectifier supplies a current of 20 A to an $R + L$ load, and its transformer leakage inductance is 5 mH. If the ac supply frequency is 50 Hz and the firing angle α is 30°, then the reduction in the average voltage at the load will be ______.
    (a) 25 V
    (b) 15 V
    (c) 12.5 V
    (d) 35 V

33. As compared to a simultaneously controlled dual converter, a non-simultaneously controlled converter presents ______ power factor to the input supply.
    (a) a lower
    (b) a higher
    (c) the same
    (d) a much higher

34. For an n-phase rectifier with an $R + E_b$ load, if the back emf obeys the inequality $E_b > V_m \cos(\pi/n)$, where V_m is the peak value of the phase-to-neutral voltage, there will be discontinuous conduction ______.
    (a) for low values of α
    (b) for large values of α
    (c) for all values of α
    (d) only when $\alpha > \sin^{-1}(E_b/V_m)$

35. For a three-phase, half-wave controlled rectifier in which the transformer inductance L_s is significant and for which the following data are given, the commutation angle μ will be equal to ______. The data are $\alpha = 0°$, $I_{ld} = 2$ A, $X_s = 2\ \Omega$, and $V_m = 100\sqrt{2}$ V.

   (a) 10.3° (b) 8.8°
   (c) 15.4° (d) 14.7°

36. In the case of a three-phase, half-wave controlled rectifier feeding a purely resistive load, if two values of the firing angle α measured from the point of natural commutation are 20° and 40°, respectively, then the extinction angles measured from the origin will be ______ and ______, respectively.

   (a) 170°, 190° (b) 170°, 180°
   (c) 140°, 200° (d) 140°, 160°

37. In the case of the single-phase, half-controlled rectifier bridge of Fig. 2.65, if the load is highly inductive and the ac supply voltage has a line-to-neutral magnitude of 100 V (RMS), then the average voltage at the load with a firing angle of 60° for the thyristors will be approximately ______ V.

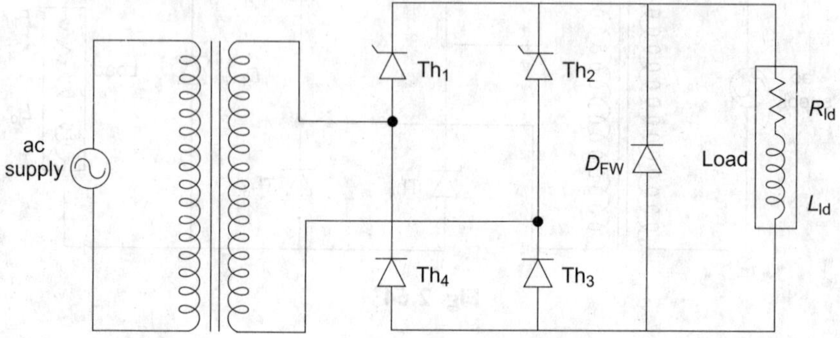

Fig. 2.65

   (a) 67.5 (b) 75.0
   (c) 87.5 (d) 80.0

38. For the three-phase, full-wave half-controlled rectifier of Fig. 2.66, if the load is highly inductive and the magnitude of the ac input voltage is 100 V (RMS), then the average voltage at the load with a firing angle of 60° for the thyristors will be approximately ______ V.

   (a) 212.5 (b) 195.0
   (c) 175.5 (d) 162.0

True or False

Indicate whether the following statements are *true* or *false*. Briefly justify your choice. Refer to the charts in Figs 2.19 and 2.21 wherever necessary.

1. For the three-phase rectifier of Fig. 2.67, the maximum reverse voltage that is applied across any thyristor is $3.5V_m$, where V_m is the peak of the phase-to-neutral voltage of the transformer secondary winding.

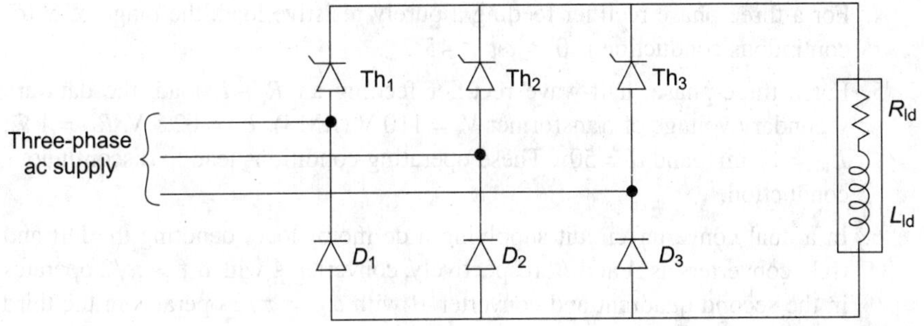

Fig. 2.66

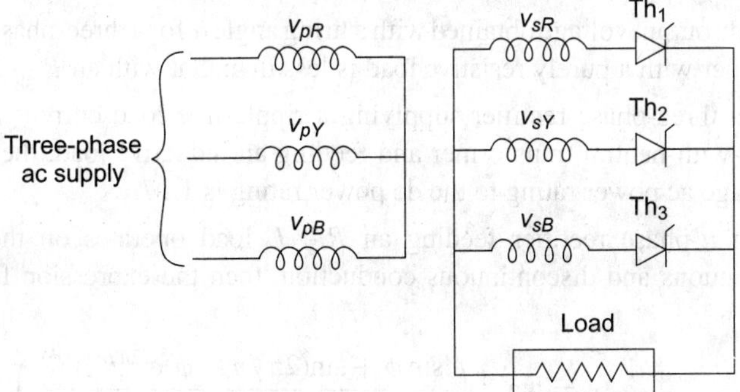

Fig. 2.67

2. The operation of the half-controlled rectifier circuit of Fig. 2.68 gets affected if the diode D_{FW} is removed from the circuit.

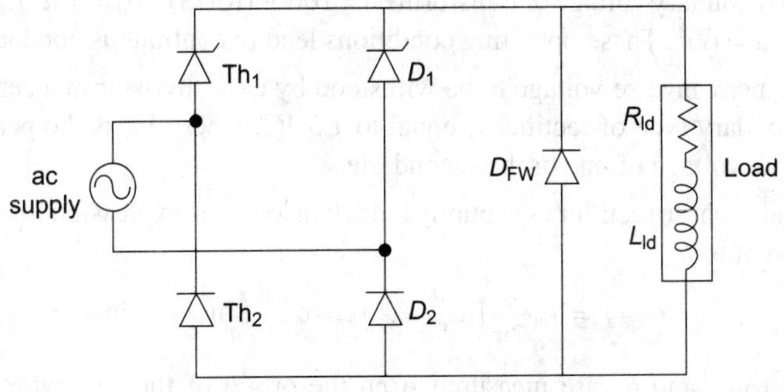

Fig. 2.68

3. Considering hoist operation using a dual converter, while the hoist is going down, the operation is in quadrant IV for a short time before the hoist reaches the bottom.

4. For a three-phase rectifier feeding a purely resistive load, the range of α for continuous conduction is $0° \leq \omega t \leq 45°$.
5. For a three-phase, half-wave rectifier feeding an $R + L$ load, the data are secondary voltage of transformer V_s = 110 V (RMS), E_b = 62.5 V, R_{ld} = 1 Ω, L_{ld} = 17 mH, and α = 50°. These operating conditions lead to discontinuous conduction.
6. In a dual converter circuit supplying a dc motor load, denoting the left and right converters as A and B, respectively, converter A with $\alpha_A > \pi/2$ operates in the second quadrant and converter B with $\alpha_B < \pi/2$ operates in the third quadrant.
7. In an n-phase rectifier in which transformer inductance L_s is considered in all phases, the reduction in the average voltage at the load end is independent of the firing angle α.
8. The dc output voltage obtained with a firing angle α for a three-phase, half-wave rectifier with a purely resistive load is less than that with an $R + L$ load.
9. For a three-phase rectifier supplying a ripple-free load current I_d, fed by a Y/Y-with-neutral transformer and feeding an inductive load, the ratio of the average ac power rating to the dc power rating is 1.17.
10. If an n-phase rectifier feeding an $R + L$ load operates on the border of continuous and discontinuous conduction, then the expression for α_{CD} will be

$$\alpha_{CD} = \tan^{-1}\left(\frac{\sin\phi + \sin(2\pi/n - \phi)e^{(2\pi/n\tan\phi)}}{\cos\phi - \cos(2\pi/n - \phi)e^{(2\pi/n\tan\phi)}}\right)$$

11. Among (a) the single-phase, half-wave rectifier, (b) the centre-tapped secondary type of rectifier, and (c) the single-phase, full-wave bridge rectifier, the centre-tapped secondary type of rectifier gives the best performance.
12. For a centre-tapped secondary type of single-phase rectifier, the data are V_s (secondary voltage of transformer) = 100 V (RMS), R_{ld} = 5 Ω, L_{ld} = 20 mH, and α = 60°. These operating conditions lead to continuous conduction.
13. The peak inverse voltage to be withstood by each thyristor in a centre-tapped secondary type of rectifier is equal to 1.5 V_m, where V_m is the peak value of the ac voltage of one of the secondaries.
14. In an n-phase rectifier with purely inductive load, the expression for the average current is

$$I_d = I_m\frac{n}{2\pi}[(\alpha'_e - \alpha')\cos\alpha' + \sin\alpha' - \sin\alpha'_e]$$

where α' and α'_e are measured from the origin of the sine wave and $I_m = V_m/\omega L_{ld}$.
15. Considering a three-phase, half-wave controlled rectifier circuit feeding an $R + E_b$ load, if the load current waveform is as shown in Fig. 2.69, it can be inferred that $E_b < V_m\cos(\pi/n)$ and the firing angle obeys the inequality $\alpha > \pi/2 - \pi/n$.
16. The method of extinction angle control gives a better input power factor than the symmetrical PWM method.

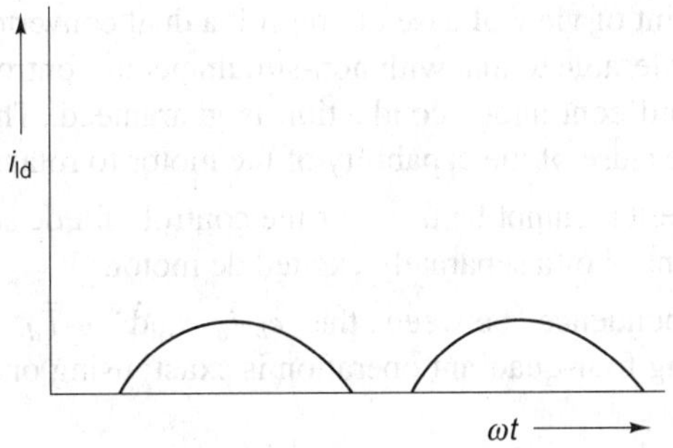

Fig. 2.69

17. In the sinusoidal PWM method of improving the power factor, the value of the modulation index can be either less than or more than unity but cannot be less than 0.5.
18. Though the extinction angle control method gives a leading power factor, it cannot be used to correct the power factor of highly lagging loads.
19. Symmetrical pulse width modulation cannot be applied to the half-wave controlled single-phase rectifier because the duration of conduction of a diode and that of a thyristor are not equal.
20. The sinusoidal PWM method cannot be practically implemented because of limitations of hardware modules.
21. The rectifier efficiency of a centre-tapped secondary type of single-phase rectifier with $R + L$ load is less than that of a single-phase, full-wave bridge rectifier with $R + L$ load, because in the former case each secondary carries current for only half the time, whereas in the latter case the secondary winding carries current all the time.
22. The rectifier efficiency of a single-phase, full-wave bridge rectifier with a purely resistive load will be greater than that with an $R + L$ load, because in the former case, the load current will be in phase with the load voltage.
23. In the case of transformers feeding rectifiers, it can be stated in general that as the derating factor increases, the rectifier efficiency increases.
24. For a three-phase rectifier fed by a transformer having inductance L_s, the data are $V_m = 110$ V (RMS), $I_{ld} = 5$ A, $L_s = 3$ mH, and $\alpha = 0°$. With this data the commutation angle μ will be equal to 10°.
25. In a three-phase rectifier fed by a transformer with inductance L_s, the data are $V_s = 180$ V (RMS), $I_{ld} = 6$ A, $L_s = 2$ mH, and $\alpha = 45°$. With this data the rectifier commutation angle μ will have a value of 12°.
26. Braking a separately excited dc motor using the method of plugging is faster than the method of dynamic braking because more power is dissipated by the former method.
27. Regenerative braking using the field reversal method is preferable to the armature reversal method because the field current is very small and hence reversal can take place much faster.

28. From the point of view of ease of control, a dual converter with simultaneous control is preferable to that with non-simultaneous control because the former is simpler and continuous conduction is guaranteed. The latter condition is facilitated because of the capability of the motor to rotate in either direction.
29. A dual converter cannot be used for the control of a dc series motor as easily as for the control of a separately excited dc motor.
30. The correspondence between the e_b-i_a and ω-$\overline{T}_d$ diagrams used for understanding four-quadrant operation is exact; using one is as good as using the other.
31. For a rectifier fed by a star/zigzag type of transformer, the primary current waveform approximates more closely to a sinusoidal waveform than in the case of a star/star-with-neutral type of transformer, but the rectifier efficiency of the former will be lower than that of the latter.
32. The average voltage of a three-phase, full-wave rectifier will be three times that of a three-phase, half-wave rectifier, provided the firing angle α has the same value in both the cases.

Short Answer Questions

1. For the half-controlled bridge rectifier of Fig. 2.70. determine the (a) output RMS voltage and (b) average output voltage. Assume a firing angle of α.
2. Repeat the derivations (a) and (b) of Question 1 with D_1 and D_2 of Fig. 2.70 replaced by thyristors Th_3 and Th_4.
3. Derive an expression for the firing angle α when an n-phase rectifier feeding an $R + L$ load operates on the border of continuous and discontinuous conduction.
4. Explain why the load voltage for an n-phase rectifier in which the leakage inductance of the transformer has a significant value is less than that for a rectifier with a negligible leakage inductance.
5. What are the possible alternatives to be considered in the case of a three-phase controlled rectifier that has to feed an $R + E_b$ load where E_b is the back emf? Explain how V_α and I_α can be obtained in each case.

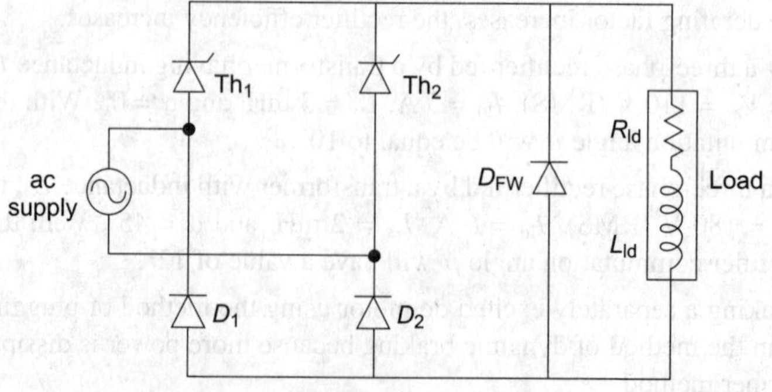

Fig. 2.70

6. Sketch the waveforms of v_{ld} and i_{ld} for a three-phase half-controlled rectifier feeding an $R + E_b$ load; also derive the expressions for their mean values.

7. Describe the four-quadrant operation of a hoist using a thyristorized dual converter.
8. Give the circuit and waveforms of a three-phase, full-wave half-controlled bridge rectifier.
9. Derive the expression for the load current $i_{ld}(t)$ for a continuously conducting n-phase rectifier with firing angle α. Assume that the ac voltage is sinusoidal in nature and the load is inductive.
10. What are the merits and demerits of the selective harmonic elimination method of improving the power factor?
11. Justify the following statement: 'The control circuitry of the symmetrical PWM method of power factor improvement is more complicated than that of the method of extinction angle control.'
12. What are the merits of the sinusoidal PWM method of power factor improvement over the multiple PWM method?
13. Given a choice between the multiple PWM method and that of sequence control of rectifiers, which is preferable and why?
14. Given a choice between any one of the conventional methods of power factor improvement and the method of extinction angle control, which is preferable and why?
15. List out the various methods of improving the input power factor using rectifier circuits and describe the best among them.
16. Explain how unwanted harmonics can be eliminated using the selective harmonic elimination method.
17. Discuss the various features of the inverting mode of a single-phase fully controlled bridge converter, with special reference to load conditions conducive to this mode.
18. For a three-phase, half-wave controlled rectifier with $R + E_b$ load, discuss the conditions for which conduction is discontinuous irrespective of the value of α.
19. With the help of a circuit diagram, discuss the operation of a six-phase, half-wave circuit with an interphase transformer.
20. Derive an expression for determining the overlap angle, given the firing angle α, input supply voltage, load resistance, and load inductance, for an n-phase rectifier in which the transformer inductance L_s has a significant value.
21. Define the ripple factor and derive an expression for it. Explain how the ripple factor decreases with an increase in the number of phases of a rectifier transformer.
22. How is four-quadrant operation made possible with the help of a dual converter? Describe an application of it.
23. Give the circuit and waveforms of a three-phase, full-wave half-controlled bridge rectifier assuming the load to be highly inductive and derive an expression for the average load voltage.
24. Discuss the simultaneous and non-simultaneous types of control of a dual converter. Give their merits and demerits.
25. Define the (i) rectifier efficiency and (ii) derating factor of a rectifier. Derive

the expressions for these two factors for a single-phase bridge rectifier feeding an $R + L$ load.

26. Show that a transformer with star/star-with-neutral and one with delta/star-with-neutral windings and with the same turn ratios have the same rectifier efficiency and derating factor.
27. Explain why a zigzag type of secondary is preferred for a three-phase transformer with a star primary. Support your argument with suitable waveforms.
28. Establish the approximate correspondence between the e_b-i_a plane and ω-$\overline{T}_d$ plane for dc motor operation when it is supplied by a dual converter.
29. Explain why the dual converter method is superior to the rest of the methods for the purpose of regenerative braking.
30. Discuss the conditions under which dynamic braking will become a necessity for a dc motor.
31. Explain why a narrow trigger pulse may fail to trigger the thyristors used in a single-phase bridge rectifier with $R + L$ load.
32. Assuming continuous conduction, derive an expression for the load current I_o at the point of transition from one thyristor to another for a single-phase controlled bridge rectifier.

Problems

1. The single-phase controlled bridge rectifier of Fig. 2.71 has an input supply voltage of 120 V (RMS) at 50 Hz. $R_{ld} = 5\ \Omega$ and $L_{ld} = 0.02$ H. If the firing angle α of the thyristors is kept at 60°, (a) draw the load voltage and load current waveforms, (b) compute the RMS and dc magnitudes of the load voltage, (c) determine the magnitude of the dc load current.

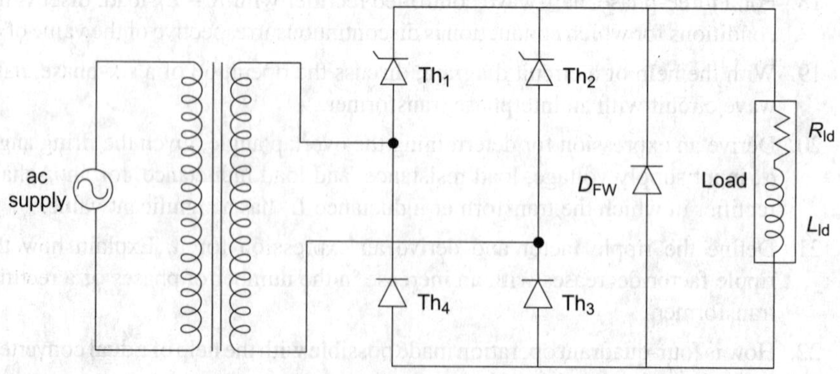

Fig. 2.71

Ans. (b) 107.6 V, 81.0 V; (c) 16.2 A

2. If the single-phase, half-wave converter of Fig. 2.72 has a purely resistive load with R_{ld} as 8 Ω and the firing angle α of the thyristor as $\pi/2$ rad, determine the

(a) rectification efficiency, (b) form factor, (c) transformer utilization factor, (d) peak inverse voltage of the SCR, and (e) ripple factor of the output voltage.

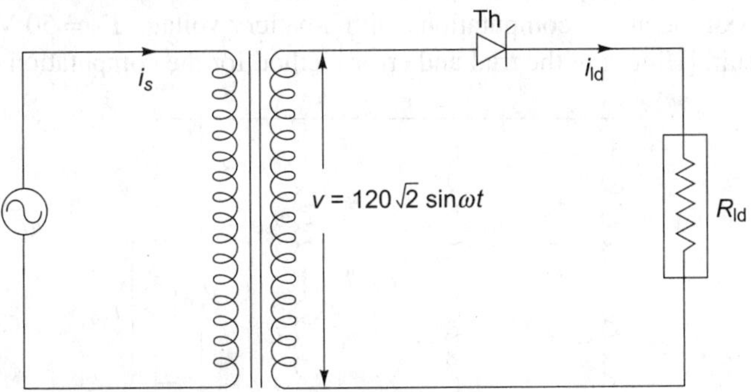

Fig. 2.72

Ans. (a) 20.3%, (b) 2.22, (c) 4.93, (d) 170 V, (e) 1.784

3. The 3ϕ half-wave converter shown in Fig. 2.73 is operated by a 3ϕ, Y-connected, 210-V (line-to-neutral), 50-Hz supply. The load resistance $R_{ld} = 10\ \Omega$, the load inductance is zero. If it is required to obtain an average output voltage of 50% of the maximum possible output voltage, calculate the (a) firing angle α, (b) RMS and average load currents, (c) RMS and average thyristor currents, (d) rectifier efficiency, and (e) transformer derating factor. [Hint: See Worked Example 3.]

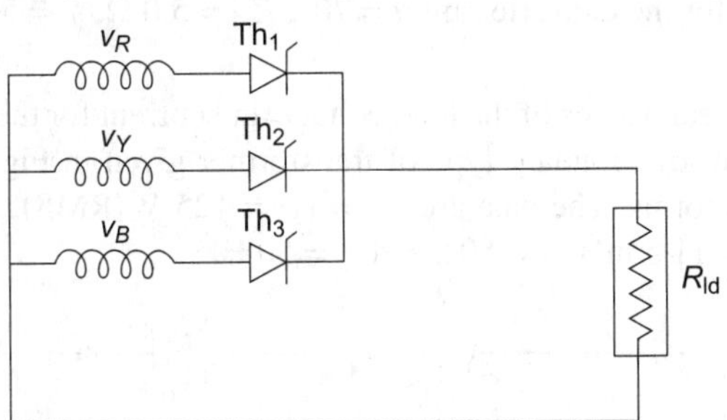

Fig. 2.73

Ans. (a) 67.7°; (b) 16.54 A, 12.28 A; (c) 9.55 A, 4.1 A; (d) 14.47%; (e) 6.91

4. A single-phase fully controlled bridge rectifier supplies a load composed of R_{ld}, L_{ld}, and back emf E_b. The given data are V_s= 220 V (RMS), E_b= 62.5 V, L_{ld} = 15.6 mH, $\alpha = 40°$, and f = 50 Hz.
   (a) What value will the resistance R_{ld} have if the conduction is required to be just continuous.
   (b) Determine the average values of the load voltage and current.
   (c) Sketch the waveforms of the load voltage and current.

Ans. (a) 3.67 Ω; (b) 151.7 V, 24.3 A

5. For the 3ϕ rectifier of Fig. 2.74, assuming that the battery is absent and the load consists of R_{ld} and L_{ld} (a) compute the overlap angle μ, reduction in load

voltage due to the commutation inductance L_s, and the load current. Sketch the load voltage and current waveforms. The given data are $V_{sR} = V_{sY} = V_{sB} =$ 160 V (RMS), $L_s = 2$ mH, $f = 50$ Hz, $R_{ld} = 1.8\ \Omega$, $L_{ld} = \infty$, and $\alpha = 40°$. (b) Repeat the above computations with a battery voltage $E_b = 50$ V in the load circuit. [Hint: Use the trial and error method for the computation of μ.]

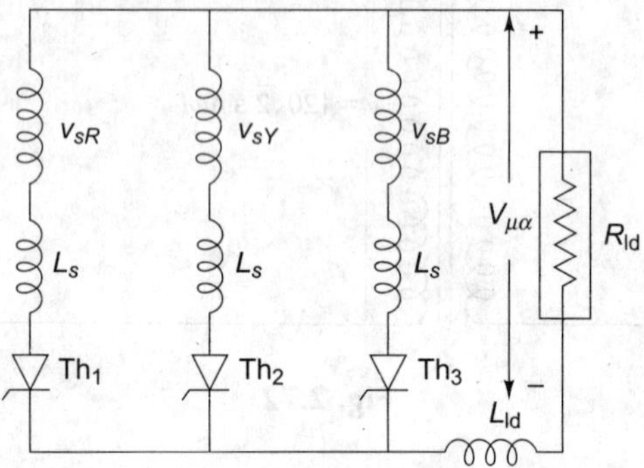

Fig. 2.74

Ans. (a) $\mu = 16.8°$, $g_{\mu\alpha} = 20.5$ V, $I_{ld} = 68.2$ A; (b) $\mu = 11.4°$, $g_{\mu\alpha} = 13.3$ V, $I_{ld} = 44.5$ A

6. A 3ϕ half-wave converter has a 50-Hz ac source and supplies an inductive load. It operates on the border of continuous and discontinuous conduction. If all the elements of the rectifier are ideal, determine the value of the load inductance L_{ld}. Data for the converter are $\alpha = 70°$, $R_{ld} = 5.0\ \Omega$, $f = 50$ Hz.

Ans. 17.2 mH approx.

7. Find the mean values of the load voltage and current for the rectifier fed by the centre-tapped secondary type of transformer given in Fig. 2.75. Also sketch their waveforms. The data are $v_{s1} = v_{s2} = 125$ V (RMS), $E_b = 88.0$ V, $R_{ld} =$ 3 Ω, $L_{ld} = 11.5$ mH, $\alpha = 50°$, and $f = 50$ Hz.

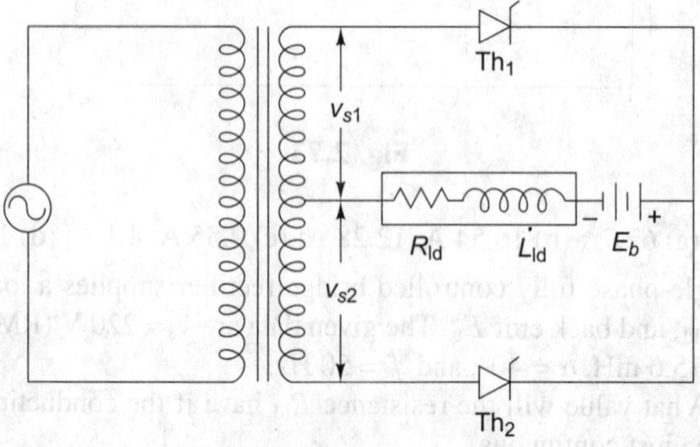

Fig. 2.75

Ans. $V_\alpha = 92$ V, $I_\alpha = 1.33$ A

8. Find the rectifier efficiency for a phase-controlled single-phase bridge rectifier with the following data: $V_s = 100$ V (RMS), $\alpha = 30°$, $R_{ld} = 15\ \Omega$, and $L_{ld} = \infty$. Draw the waveforms of the rectified voltage and current, and also that of the current through a thyristor. Determine the ripple factor of the output voltage.

   Ans. $\eta_R = 78\%$, ripple factor = 0.58

9.(a) A three-phase, half-wave rectifier feeding a resistive load is supplied by a star/star type of transformer. The data are $V_s = 230$ V (RMS) (line-to-neutral), $V_{ld} = 160$ V (dc), $R_{ld} = 1.2\ \Omega$. Determine the value of α.

   (b) If the firing angle is kept at the value computed in (a) and a load inductance of 0.6 mH is included in the load, find the new values of average voltage and current at the load. Also determine the RMS value of the current through a thyristor.

   Ans. (a) 58.3°; (b) 160 V, 133.3 A, 97.6 A

10. The specifications of a rectifier supplied by an asymmetric voltage as shown in Fig. 2.76 are $v_{s1} = 220$ V (RMS), $v_{s2} = 110$ V (RMS) (line-to-neutral), $f = 50$ Hz, $R_{ld} = 12\ \Omega$, $\alpha = 80°$. Find the average load voltage and current I_α with (a) $L_{ld} = 0$ and (b) $L_{ld} = 10$ mH.

    Ans. (a) $V_\alpha = 87.2$ V, $I_\alpha = 7.27$ A; (b) $V_\alpha = 86.4$ V, $I_\alpha = 7.2$ A

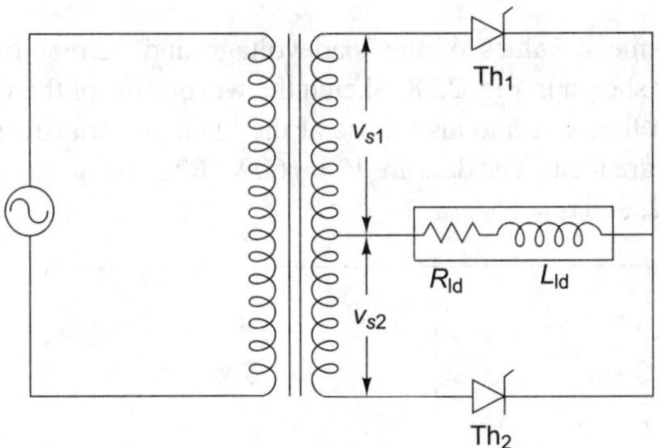

Fig. 2.76

11. In the 3ϕ rectifier of Fig. 2.77, the values for the various quantities are V_s= 108 V (RMS) (line-to-neutral), $f = 50$ Hz, $R_{ld} = 1.5\ \Omega$, $L_{ld} = \infty$, $\alpha = 38°$, and commutation inductance L_s= 1.2 mH. Determine the (a) angle of overlap μ, (b) mean value of load voltage $V_{\mu\alpha}$, and (c) mean value of load current I_{ld}.

    Ans. (a) 13.7°, (b) 88.87 V, (c) 59.24 A

12. A three-phase, half-wave controlled rectifier is fed by a 277-V (RMS) (line-to-neutral), 50-Hz supply and provides an adjustable dc voltage at the terminals of a separately excited dc motor. The motor specifications are R_a = unknown, L_a = unknown, $E_b = 1.2\ \omega$, where ω is the angular speed of the motor, and full load armature current $I_a = 500$ A. The motor operates with full load current at a speed of 200 rad/s. If $\alpha = 40°$, and conduction is just continuous, determine the values of R_a and L_a.

    Ans. $R_a = 0.0164\ \Omega$, $L_a = 0.344$ mH

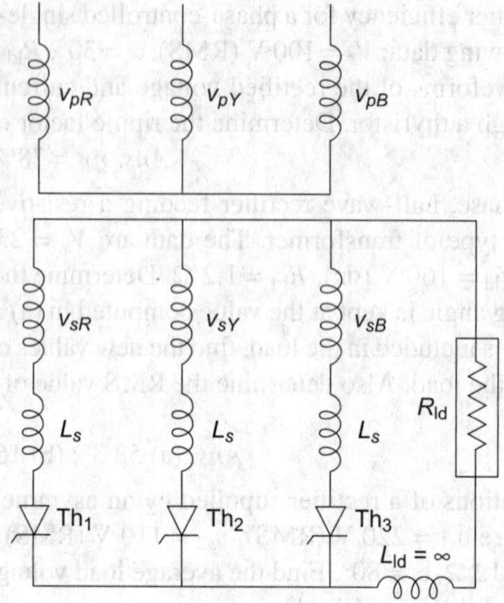

Fig. 2.77

13. Find the mean values of the load voltage and current for the three-phase converter shown in Fig. 2.78. Sketch the waveforms of the voltage and current through a thyristor and also those at the load. The transformer and converter elements are ideal. The data are $V_s = 160$ V (RMS) (line-to-neutral), $E_b = 80$ V, $R_{ld} = 2\ \Omega$, and $\alpha = 65°$.

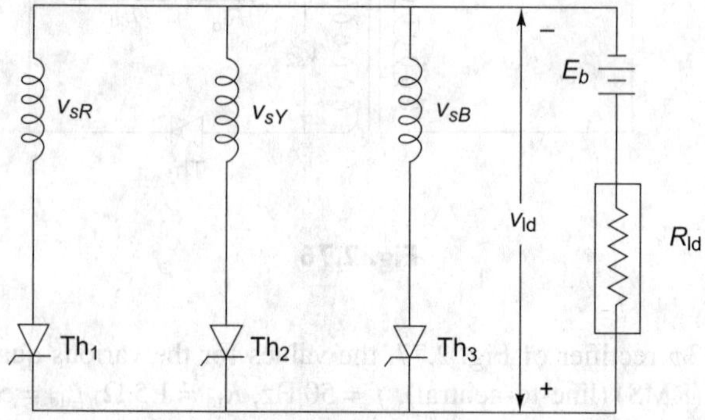

Fig. 2.78

Ans. 61.8 V, 9.47 A

14.(a) A circuit is connected as shown in Fig. 2.79(a) to a 230-V (line-to-neutral), 50-Hz supply. Determine the average load voltage and current if the load comprises $R_{ld} = 12.5\ \Omega$, $L_{ld} = 0.15$ H, and the firing angle $\alpha = 70°$.

(b) Repeat the computations of (a) with a diode connected as in Fig. 2.79(b). Neglect the drops across conducting devices.

Ans. (a) 2.57 V, 0.205 A; (b) 52.76 V, 4.22 A

15.(a) For the rectifier circuit of Fig. 2.80, determine the (i) RMS values of current in the secondary winding of the transformer, (ii) average current in the load,

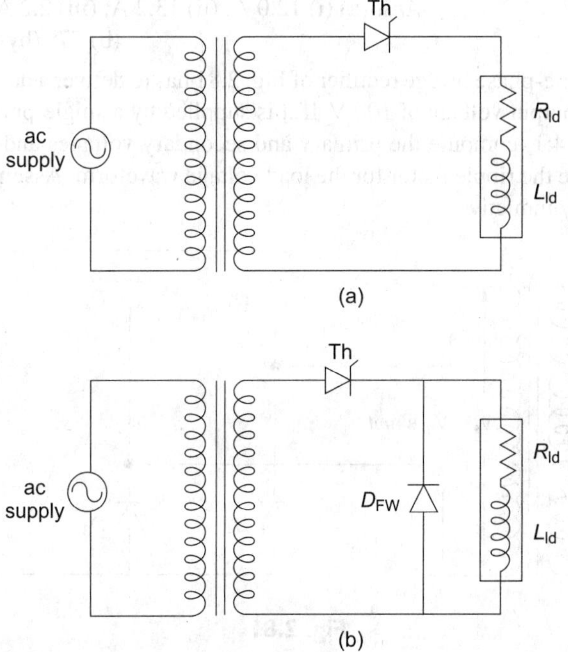

Fig. 2.79

(iii) average value of current in the bypass diode, and (iv) mean value of rectified voltage. The data are $\alpha = 50°$, $V_s = 200$ V (RMS) (line-to-neutral), $f = 50$ Hz, $R_{ld} = 12\ \Omega$, and $L_{ld} = \infty$.

(b) At what firing angle does the conduction become just discontinuous if the bypass diode D is removed and the infinite inductance replaced by a finite conductance $L_{ld} = 65$ mH?

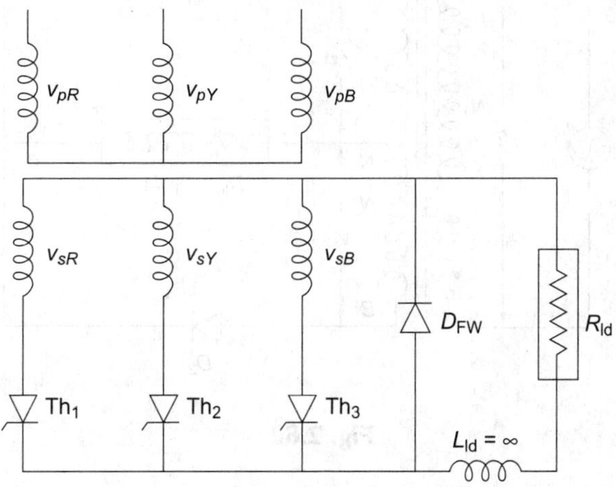

Fig. 2.80

Ans. (a) (i) 12.0 A, (ii) 13.2 A, (iii) 2.2 A, (iv) 158.5 V; (b) 77° (by trial and error)

16. The single-phase bridge rectifier of Fig. 2.81 has to deliver a dc current of 30 A at a dc output voltage of 100 V. If it is supplied by a single-phase transformer of ratio 4:1, compute the primary and secondary voltages and currents. Also calculate the ripple factor for the load voltage waveform. Assume that the load is highly inductive.

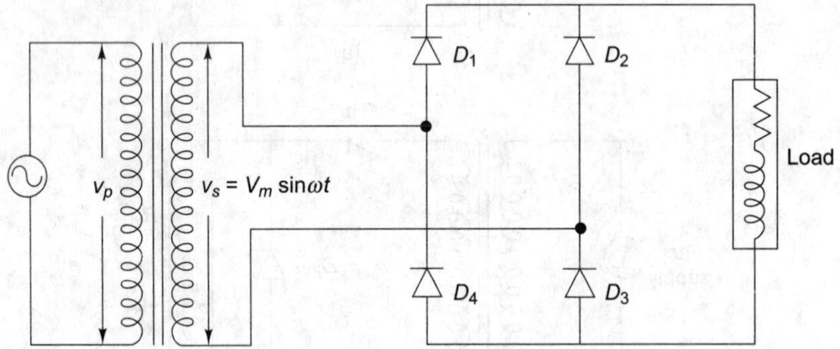

Fig. 2.81

Ans. Primary—444 V, 7.5 A; secondary—111 V, 30 A; ripple factor = 0.48

17. The centre-tapped secondary type of rectifier of Fig. 2.82 delivers a dc current of 15 A at a dc output voltage of 110 V. If it is supplied by a single-phase transformer of ratio 3:1, compute the primary and secondary voltages and currents. Also determine the ripple factor for the load voltage. Assume α to be zero and that the load is highly inductive.

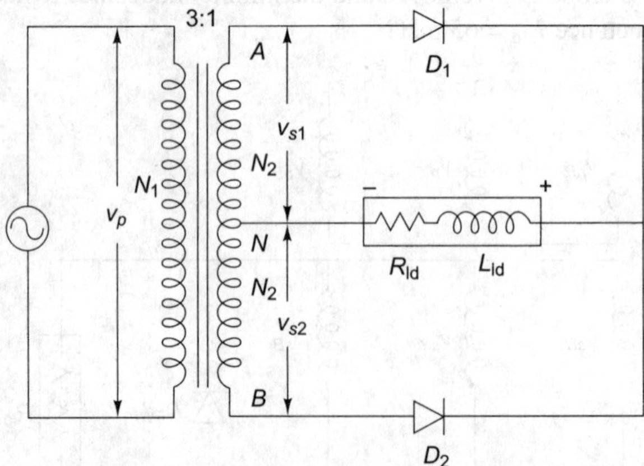

Fig. 2.82

Ans. Primary—366.6 V, 5.55 A; secondary—122.0 V, 16.67 A; ripple factor = 0.48

18. The three-phase, half-wave rectifier of Fig. 2.83 delivers a maximum dc current of 20 A at a maximum dc output voltage of 150 V to a load which is highly

inductive. If the transformer ratio is 5:1, compute (a) the primary and secondary voltages and currents and the (b) ripple factor. Assume $\alpha = 0°$.

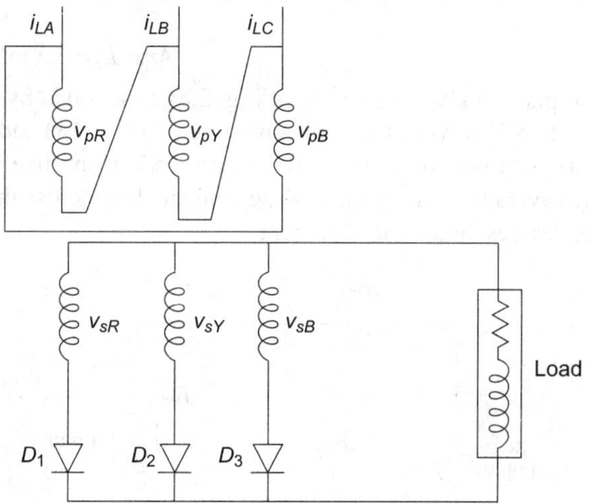

Fig. 2.83

Ans. (a) V_s(L to L) = 128.3 V, I_s = 11.55, V_p(L to L) = 641.3 V, I_p = 1.88 A; (b) ripple factor = 0.54

19. Compute V_α and I_α for a three-phase, half-wave controlled rectifier feeding an $R + E_b$ load and having the following data: input voltage V_s = 200 V (L to N) (RMS), E_b= 141.4 V, R_{ld} = 3 Ω, and $\alpha = 30°$.

Ans. 184.5 V, 26.2 A

20. A single-phase bridge rectifier is operated with extinction angle control with the angle $\beta = 50°$. The ac supply voltage is 150 V (RMS) and a load resistance of 20 Ω is connected at the output side of the rectifier. Compute the (a) distortion factor, (b) displacement factor, (c) input power factor, (d) average load voltage, and (e) average load current. Draw the load voltage and load current waveforms.

Ans. (a) 0.95, (b) 0.9, (c) 0.86, (d) 110.9 V, (e) 5.55 A

21. A single-phase, half-controlled rectifier is operated with symmetrical pulse width modulation with the angle $\beta = 60°$. The ac supply voltage is 200 V (RMS) and the load resistance is 25 Ω. Compute the (a) distortion factor, (b) input power factor, (c) average load voltage, and (d) average load current. Draw the load voltage and load current waveforms.

Ans. (a) 0.78, (b) 0.78, (c) 90 V, (d) 3.6 A

22. A three-phase, half-wave controlled rectifier is fed by a 282-V (RMS) (line-to-neutral), 50-Hz supply and has a separately excited dc motor as its load; R_a= 0.015 Ω. The relationship between E_b and ω is $E_b = k\omega$, where k = 1.2 V/(rad/s), $\alpha = 50°$, and E_b/V_m= 0.4. If the conduction is just continuous, determine the value of L_a and the speed at which the motor rotates.

Ans. 0.1 mH, 1270 rpm

23. A three-phase, half-wave rectifier is fed by a star/star transformer whose leakage inductance is to be considered. If the commutation angle μ and the

firing angle α are 10° and 35°, respectively, determine the value of L_s for the transformer and the mean value of the load current. The other data are V_s (RMS) (L to N) = 110 V, $R_{ld} = 2\ \Omega$, $L_{ld} = \infty$, and $f = 50$ Hz. Assume E_b= 60 V.

Ans. L_s= 2.5 mH, I_{ld} = 19.1 A

24. The single-phase, half-wave circuit of Fig. 2.84 is supplied by a single-phase ac source of 15 V (RMS), the load consisting of $R_{ld} = 2\ \Omega$ and L_{ld} = 15 mH. If the diode and thyristor drops are 0.7 V and 1.6 V, respectively, and $\alpha = 50°$, find the (a) average load voltage, (b) percentage drop across the load voltage due to the devices, and (c) load current.

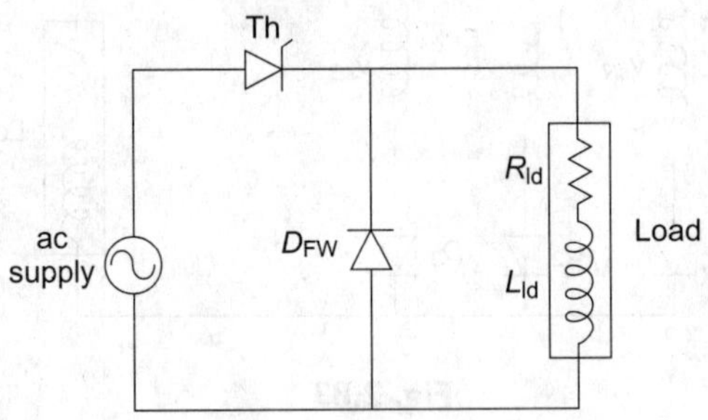

Fig. 2.84

Ans. (a) 4.97 V, (b) thyristor drop 10.4% (the diode contribution is for the negative part of the voltage waveform and of duration 67%), (c) 2.49 A

25. The single-phase, half controlled bridge rectifier shown in Fig. 2.85 supplies a load consisting of R_{ld} = 1.5 Ω and L_{ld} = 12 mH. The other data are V_s(RMS) = 20 V, diode drop = 0.7 V, thyristor drop = 1.6 V, and $\alpha = 45°$. Determine the (a) average load voltage, (b) percentage drop across the load voltage due to the devices, and (c) load current.

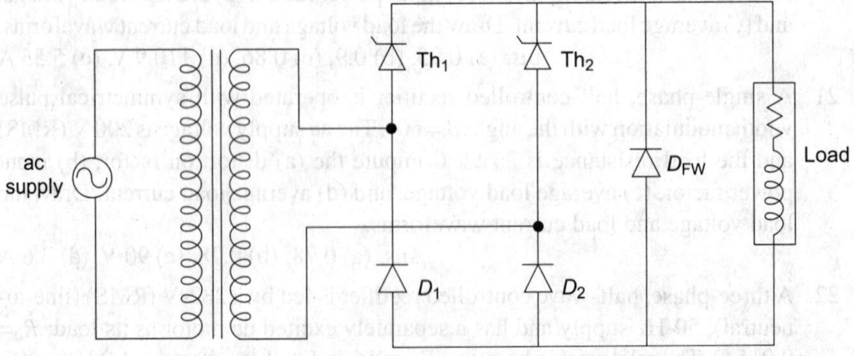

Fig. 2.85

Ans. (a) 13.07 V, (b) 14.96%, (c) 8.71 A

26. A three-phase, half-wave controlled rectifier is supplied by an ac source with a voltage (L to N) of 100 V (RMS) and feeds an $R + E_b$ load. If the load resistance is 5 Ω and the back emf is 70.7 V, what is the range of values of α for which the conduction will be discontinuous? Find the average load voltage and the load current for the load voltage waveform with this border case. Draw the waveforms of load voltage and current.

Ans. $\alpha > 0°$, $V_\alpha = 116.95$ V, $I_\alpha = 9.25$ A

27. The three-phase, half-wave controlled rectifier of Fig. 2.86 feeds an $R + E_b$ load with a load resistance of 8 Ω and E_b of 100 V. If the supply voltage (L to N) is 200 V (RMS), state the value of α in the range $0 \le \alpha \le 120°$ for which there will be border case conduction. Determine the average voltage and current at the load for this value of α. Also calculate the ripple factor for the load voltage waveform.

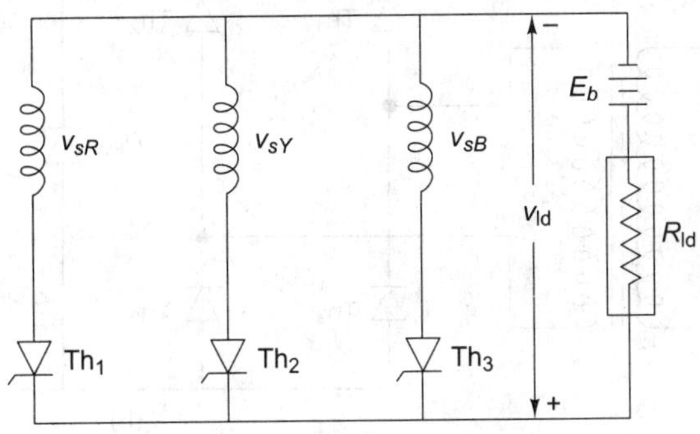

Fig. 2.86

Ans. $\alpha = 9.3°$, $V_\alpha = 230.8$ V, $I_\alpha = 16.35$ A, ripple factor = 0.124

28.(a) A single-phase controlled bridge rectifier feeds an inductive load as shown in Fig. 2.87(a). The data are $V_s = 30$ V (RMS), drop across a conducting thyristor = 1.5 V, $\alpha = 45°$, $R_{ld} = 2$ Ω, and $X_{ld} = 2.5$ Ω. Determine the average voltage and current at the load and the percentage mean voltage drop due to the thyristors.

(b) Now, if a freewheeling diode is connected at the load as shown in Fig. 2.87(b), repeat the computations of (a) assuming that the drop across a conducting diode is 0.7 V. Also find the percentage mean voltage drop due to the freewheeling diode.

Ans. (a) 16.1 V, 8.05 A, 15.7%; (b) 20.63 V, 10.31 A, 10.52%,
% mean drop due to the freewheeling diode = 0.76%

29. For the six-phase rectifier of Fig. 2.88, determine the net average voltage and the current at the load. Also compute the ripple factor, percentage voltage drop due to thyristors, and the maximum RMS current through the thyristors. The data are V_s(L to N) (RMS) = 100 V, $R_{ld} = 2$ Ω, $L_{ld} = 15$ mH, $f = 50$ Hz, and $\alpha = 30°$. Drop across a conducting thyristor = 1.5 V. Assume that the load is highly inductive.

Ans. $V_\alpha = 115.5$ V, $I_\alpha = 57.75$ A, $r_v = 0.115$,
% drop due to a thyristor = 1.3%, $I_{ThM(RMS)} = 27.58$ A

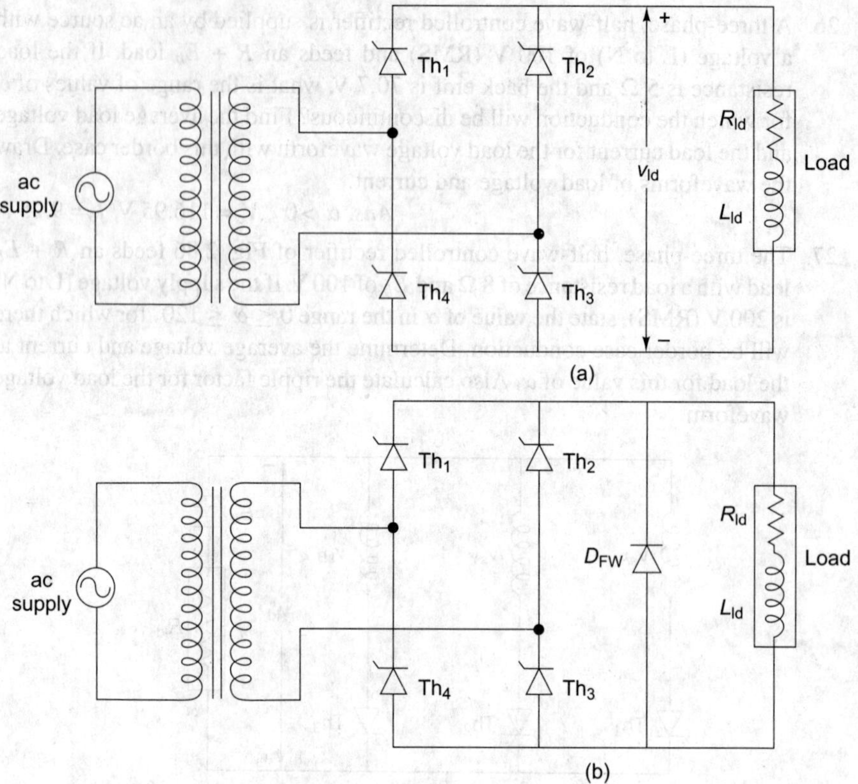

Fig. 2.87

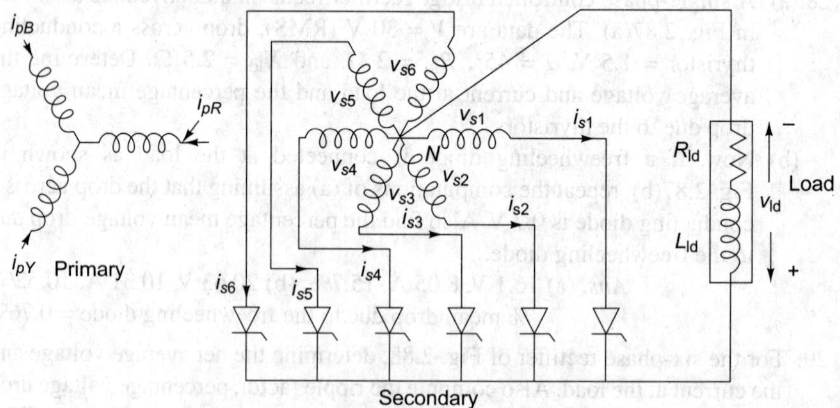

Fig. 2.88

CHAPTER 3

Dc Choppers

3.1 Introduction

A dc chopper is a power converter whose input is a fixed dc voltage and output a variable dc voltage. Dc choppers can be realized using thyristors, gate turn-off power transistors (GTOs), metal oxide semiconductor field effect transistors (MOSFETs), or integrated gate bipolar transistors. Dc choppers are used in variable speed dc drives because of high efficiency, flexibility, quick response, and capability of regeneration. Their applications include subway cars, machines hoists, forklift trucks, and mine haulers among others. Earlier, for traction purposes, series dc motors controlled by dc choppers were preferred because of the high starting torque provided by them. However, a series motor has the following limitations:

(i) its field voltage cannot be easily controlled by power electronic converters,

(ii) when field control is not employed the motor base speed has to be set equal to the highest desired speed of the drive; this implies using fewer field ampere turns and therefore lower torque per ampere at low speeds, and

(iii) the implementation of regenerative braking is not easy.

Besides being free from these drawbacks, a separately excited dc motor can also be operated to give the characteristics of a series motor. Hence the present trend is to use separately excited motors fed from dc choppers for traction as well as other applications.

After going through this chapter the reader should

- know the principle of a dc chopper,
- understand the working of a step-down chopper and its application to a separately excited dc motor,
- know the derivation of the expressions for its current and developed torque and also the determination of its speed–torque characteristics,

- get acquainted with the principle of a step-up chopper and its practical utility in the regenerative braking of dc motors,
- become familiar with two kinds of two quadrant choppers,
- get acquainted with the various methods of operation of a four-quadrant chopper,
- understand how the speed–torque characteristic of a dc series motor can be drawn with the duty ratio as a parameter, and
- become familiar with the various turn-off circuits employed in choppers.

3.1.1 Principle of a Dc Chopper

The principle of a chopper can be understood with the help of the circuit shown in Fig. 3.1(a). The switch Sw is repeatedly kept on for τ_{ON} seconds and off for τ_{OFF} seconds. Thus the time period of the chopper is $\tau = \tau_{ON} + \tau_{OFF}$ and its frequency is $1/\tau$ cycles per second. The average load voltage V_{ld} can be obtained as

$$V_{ld} = \frac{1}{\tau}\int_0^{\tau} v_{ld}\,dt = \frac{1}{\tau}\left[\int_0^{\tau_{ON}} E\,dt + \int_{\tau_{ON}}^{\tau} 0\,dt\right]$$

$$= \frac{E\tau_{ON}}{\tau} \tag{3.1}$$

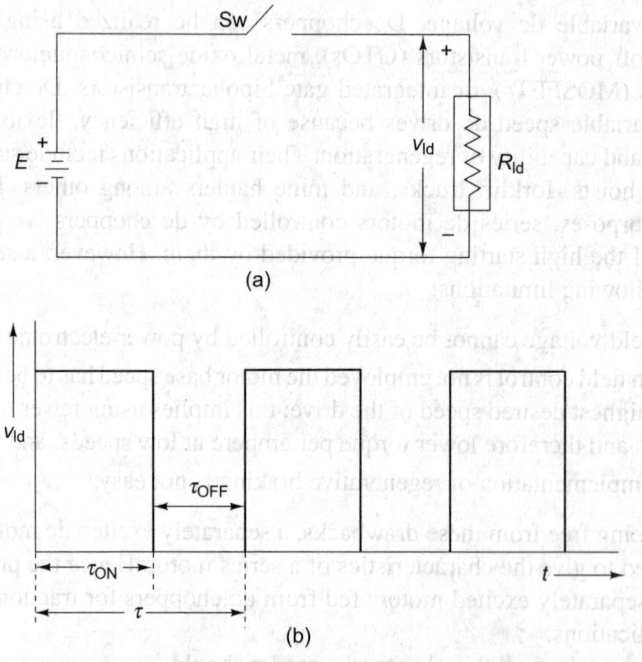

Fig. 3.1 (a) Simple circuit explaining the principle of a chopper; (b) waveform of load voltage

Thus, if τ is kept constant and τ_{ON} is varied from 0 to τ, V_{ld} varies from 0 to E. Though the principle of a basic chopper is explained with the help of a mechanical switch, in a practical chopper the switch consists of a semiconductor device, namely, a thyristor, transistor, GTO, etc. As these devices need a separate commutation circuit, the device along with this commutation circuitry is termed as a dc chopper. It is symbolically represented by a thyristor enclosed in a dashed square as shown in Fig. 3.2.

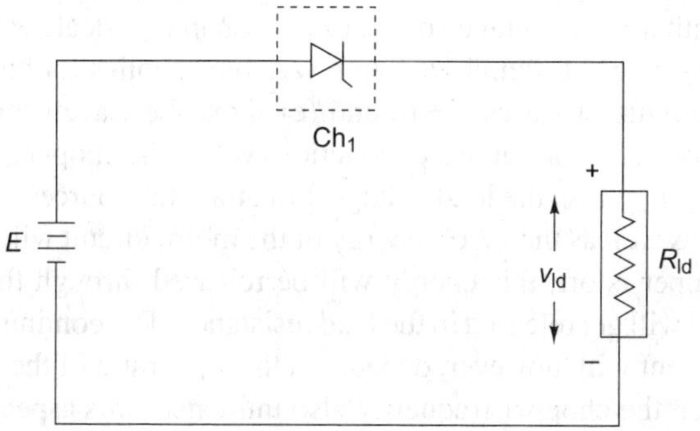

Fig. 3.2 A chopper with resistance load

Figure 3.1(b) gives the output of the switch type of chopper shown in Fig. 3.1(a). As stated above, the turn-on time of the device can be varied but the frequency of switching ($f_{Ch} = \tau$) is kept constant. For the chopper shown in Fig. 3.2, the output voltage profile will be similar to that given in Fig. 3.1(b); the output current will be a scaled version of the voltage because the load is resistive. By Fourier analysis it can be shown that this load current is composed of an ac voltage superimposed over a dc component. With a dc motor, which can be modelled as an $R + L + E_b$ type of load, the load current waveform will no longer be rectangular in nature but consist of exponentially rising and falling curves as shown in Fig. 3.3. In addition, the load voltage may not be zero during τ_{OFF} and the conduction may be continuous or discontinuous. The waveforms of the practical chopper given in Figs 3.3(b) and (c) clearly demonstrate these features.

3.2 Step-down and Step-up Choppers

Two types of classifications exist for dc choppers. The first one depends upon whether the load voltage is lesser or greater than the source voltage. In a step-down chopper the load voltage is always a fraction of the source voltage. The chopper considered in the previous section is a step-down one. A step-up chopper, like a step-up transformer, provides an output voltage which is greater than the source voltage. The second classification is based on the quadrant of the v-i plane in which the operating point of the chopper falls; such choppers will be considered in the next section.

3.2.1 Step-down Chopper

A practical application of the step-down chopper is the speed control of a separately excited dc motor. Here, an external inductor is usually connected in series with the motor load and serves to smoothen the load current waveform. In all future discussion, it will be assumed that L is the sum total of L_{ext} and L_{arm}, where L_{ext} is the external inductance and L_{arm} is the small inductance of the armature.

Figure 3.3(a) shows the circuit of a such a step-down chopper with a dc motor load. Though for clarity the motor field is not shown, it is assumed that a separate field circuit with a fixed voltage source exists. Chapter 7 deals with dc drives for which dc choppers can be employed with variable supplies for both the armature and the field circuits. Figures 3.3(b) and (c) show the waveforms, respectively, for discontinuous and continuous conduction. When the chopper is on, that is, in the interval $0 \leq t \leq \tau_{ON}$, the load voltage will attain the source voltage E and the motor current as well as the $\frac{1}{2}LC^2$ energy of the motor circuit will increase. Next, when the chopper is off, this energy will be released through the freewheeling diode D_{FW} and will get released in the load resistance. The continuity or otherwise of the load current will, however, depend on the L/R ratio of the load circuit. As elaborated later, the chopper frequency also influences this aspect.

The load voltage of the chopper depends on the continuity or otherwise of conduction of the load current. In the case of continuous conduction [Fig. 3.3(c)], the load current freewheels through D_{FW} as stated above, in which case the load voltage becomes nearly zero. Hence the average load voltage is governed by Eqn (3.1). On the other hand, for discontinuous conduction the load voltage as well as current will be those shown in Fig. 3.3 (b). During the interval (τ_{ON}, τ_1) the load current freewheels through D_{FW} and falls down in an exponential manner, finally decaying to zero. Though the load current stops at τ_1, the motor continuous to run due to inertia. As a consequence of this, during the interval (τ_1, τ), the load voltage attains a value equal to E_b, which is the back emf of the motor.

The source current will be discontinuous under both discontinuous and contionuous conduction states. This discontinuous but periodical source current gives rise to harmonic ac components through the other loads connected to the source. When the source is not an ideal one, the harmonic current leads to electromagnetic interference (EMI) by means of conduction and radiation. LC filters may be properly designed and employed to minimize these effects. Harmonics and filters are, however, beyond the scope of this book.

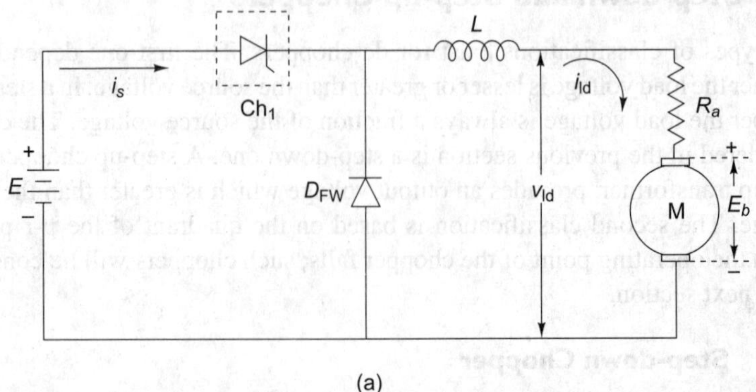

Fig. 3.3(a)

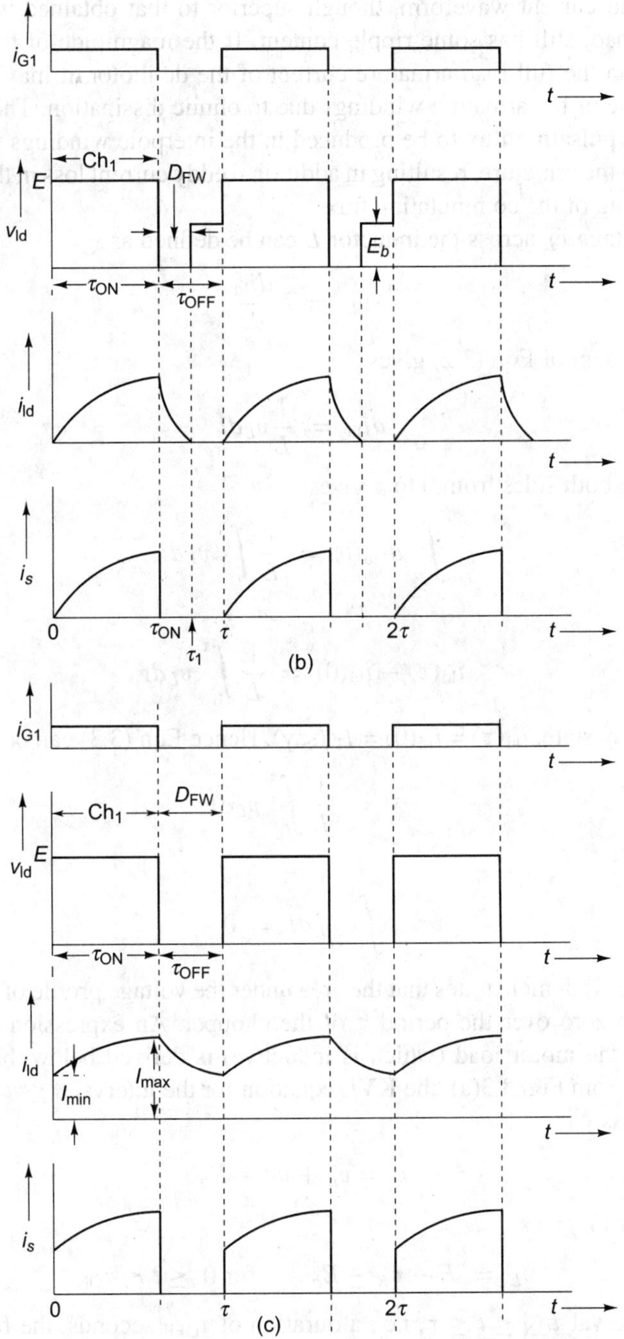

Fig. 3.3 Step-down chopper: (a) circuit, (b) waveforms for discontinuous conduction, (c) waveforms for continuous conduction

The load current waveform, though superior to that obtained with a purely resistive load, still has some ripple content. If the magnitude of this current is greater than the full load armature current of the dc motor, it may increase the temperature of the armature windings due to ohmic dissipation. The ripple may also cause pulsating flux to be produced in the interpole windings which are in series with the armature, resulting in additional eddy current loss in the interpoles and damping of the commutating flux.

The voltage v_L across the inductor L can be defined as

$$v_L = L\frac{di_{\text{ld}}}{dt} \tag{3.2}$$

Rearrangement of Eqn (3.2) gives

$$di_{\text{ld}} = \frac{1}{L}v_L dt$$

Integrating both sides from 0 to τ gives

$$\int_0^{\tau} di_{\text{ld}}(t) = \frac{1}{L}\int_0^{\tau} v_L dt$$

or

$$i_{\text{ld}}(\tau) - i_{\text{ld}}(0) = \frac{1}{L}\int_0^{\tau} v_L dt \tag{3.3}$$

In the steady state, $i_{\text{ld}}(\tau) = i_{\text{ld}}(0) = I_1$ (say). Hence Eqn (3.3) can be written as

$$0 = \frac{1}{L}\int_0^{\tau} v_L dt$$

or

$$\int_0^{\tau} v_L dt = 0 \tag{3.4}$$

Equation (3.4) demonstrates that the area under the voltage profile of the inductor sums up to zero over the period τ of the chopper. An expression for the load voltage of the motor load (which is inductive) is derived below, based on this principle. From Fig. 3.3(a), the KVL equation for the interval $0 \le t \le \tau_{\text{ON}}$ can be written as

$$E = v_L + v_R + E_b \tag{3.5}$$

Solving for v_L gives

$$v_L = E - v_R - E_b \qquad \text{for } 0 \le t \le \tau_{\text{ON}} \tag{3.6}$$

For the interval $\tau_{\text{ON}} \le t \le \tau$, i.e., a duration of τ_{OFF} seconds, the freewheeling diode conducts and v_L becomes

$$v_L = -(v_R + E_b) \tag{3.7}$$

The application of Eqn (3.4) to the values of v_L in Eqns (3.6) and (3.7) now gives

$$(E - v_R - E_b)\tau_{\text{ON}} = (v_R + E_b)\tau_{\text{OFF}} \tag{3.8a}$$

The relationship in Eqn (3.8a) is illustrated in Fig. 3.4. This equation can be written as

$$(E - v_R - E_b)\tau_{ON} = (v_R + E_b)(\tau - \tau_{ON})$$

or

$$E\tau_{ON} - (v_R + E_b)\tau_{ON} = (v_R + E_b)\tau - (v_R + E_b)\tau_{ON}$$

That is,

$$v_R + E_b = \frac{E\tau_{ON}}{\tau} \tag{3.8b}$$

If τ_{ON} and τ are fixed and if the circuit attains a steady state, v_R in Eqn (3.8b) can be considered as the average voltage across the motor resistance and written as v_R. Also, the average load voltage can be written as V_{ld} and taken to be equal to $v_R + E_b$. Thus, Eqn (3.8b) becomes

$$V_{ld} = \frac{E\tau_{ON}}{\tau} \tag{3.9}$$

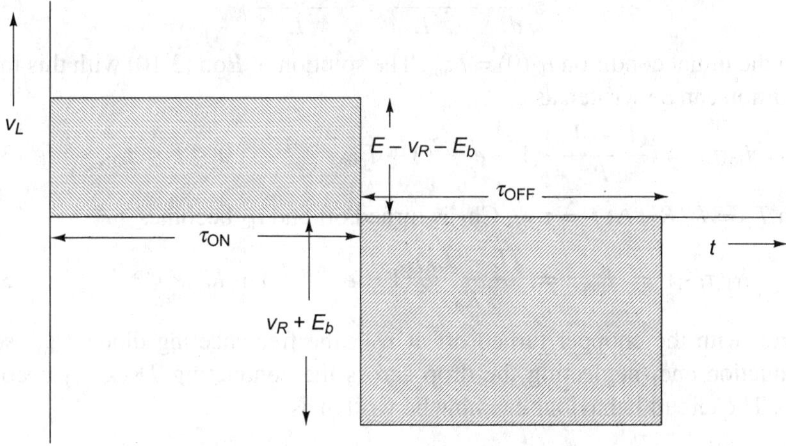

Fig. 3.4 Voltage across the inductor during one period

Equation (3.9) shows that with continuous conduction and with the circuit attaining a steady state, the basic chopper equation remains as in Eqn (3.1) and that the average load voltage can be controlled from 0 to E volts by varying the ratio τ_{ON}/τ from 0 to 1. The control of the average load voltage V_{ld} as explained above is called *time ratio control*. It can be further subclassified depending on the type of control—*fixed frequency control* or *variable frequency control*; the chopper period τ is kept constant and τ_{ON} is varied in the former and vice versa in the latter. Variable frequency control is seldom resorted to because, with a large τ (or with a low chopper frequency), the load current waveform becomes discontinuous as shown in Fig. 3.3(b) and this may deteriorate the motor performance; secondly, it complicates the design of the input filter.

A drawback of the step-down chopper is that the load current waveform varies between low and high limits and hence has a ripple content. Though this may be somewhat reduced by the inductor L, increasing the chopper frequency has a marked effect in smoothing the current waveform. The switching frequency obtainable with a thyristorized chopper is of the order of hundreds of hertz, whereas with a power transistor, frequencies of the order of tens of kilohertz are realizable. With MOSFETs, frequencies up to hundreds of kilohertz have been used.

3.2.2 Analysis with Dc Motor Load

The step-down chopper of Fig. 3.3(a) is analysed with the assumptions that the load current is continuous and the circuit operation has attained a steady state. With the chopper Ch_1 turned on at $t = 0$, the circuit behaviour can be expressed as

$$L\frac{di_{ld}}{dt} + R_a i_{ld} = E - E_b \quad \text{for } 0 \le t \le \tau_{ON}$$

This can be rewritten as

$$\frac{di_{ld}}{dt} + \frac{R_a i_{ld}}{L} = \frac{E - E_b}{L} \tag{3.10}$$

with the initial condition $i_{ld}(0) = I_{min}$. The solution of Eqn (3.10) with this initial condition can be written as

$$i_{ld}(t) = \frac{(E - E_b)}{R_a}(1 - e^{-t/T_a}) + I_{min}e^{-t/T_a}, \quad 0 \le t \le \tau_{ON} \tag{3.11}$$

with $T_a = L/R_a$. At $t = \tau_{ON}$, Ch_1 is turned off and i_{ld} becomes

$$i_{ld}(\tau_{ON}) = I_{max} = \frac{(E - E_b)}{R_a}(1 - e^{-\tau_{ON}/T_a}) + I_{min}e^{-\tau_{ON}/T_a} \tag{3.12}$$

Again, with the chopper turned off at τ_{ON}, the freewheeling diode D_{FW} starts conduction and, neglecting the drop across the conducting D_{FW}, v_{ld} becomes zero. The circuit behaviour can now be written as

$$\frac{di_{ld}}{dt'} + \frac{R_a}{L}i_{ld} = -\frac{E_b}{L}, \quad 0 \le t' \le \tau_{OFF} \ (\text{or } \tau_{ON} \le t \le \tau) \tag{3.13}$$

where $t' = t - \tau_{ON}$. The initial condition for the differential equation is $i_{ld}(t') = I_{max}$ at $t' = 0$. The solution of Eqn (3.13) is given as

$$i_{ld}(t') = -\frac{E_b}{R_a}(1 - e^{-t'/T_a}) + I_{max}e^{-t'/T_a}, \quad 0 \le t' \le \tau_{OFF} \ (\text{or } \tau_{ON} \le t \le \tau) \tag{3.14}$$

At $t' = \tau_{OFF}$, i_{ld} attains the value of I_{min}. Thus,

$$I_{min} = -\frac{E_b}{R_a}(1 - e^{-\tau_{OFF}/T_a}) + I_{max}e^{-\tau_{OFF}/T_a} \tag{3.15}$$

Using the relation $\tau_{OFF} = \tau - \tau_{ON}$ in Eqn (3.15) gives

$$I_{min} = -\frac{E_b}{R_a}(1 - e^{-(\tau - \tau_{ON})/T_a}) + I_{max}e^{-(\tau - \tau_{ON})/T_a} \tag{3.16}$$

Equations (3.12) and (3.16) can now be solved for I_{max} and I_{min} as

$$I_{max} = \frac{E}{R_a}\frac{(1 - e^{-\tau_{ON}/T_a})}{(1 - e^{-\tau/T_a})} - \frac{E_b}{R_a} \tag{3.17}$$

and

$$I_{min} = \frac{E}{R_a}\frac{(e^{\tau_{ON}/T_a} - 1)}{(e^{\tau/T_a} - 1)} - \frac{E_b}{R_a} \tag{3.18}$$

The current ripple can now be obtained as

$$\Delta i_{ld} = \frac{I_{max} - I_{min}}{2} = \frac{E}{2R_a}\left[\frac{1 + e^{\tau/T_a} - e^{\tau_{ON}/T_a} - e^{\tau_{OFF}/T_a}}{e^{\tau/T_a} - 1}\right] \tag{3.19}$$

The average load current denoted as i_{ld}(av) can now be expressed as

$$i_{ld(av)} = \frac{1}{\tau}\left[\int_0^{\tau_{ON}} i_{ld}(t)dt + \int_0^{\tau_{OFF}} i_{ld}(t')dt'\right] \tag{3.20}$$

Substituting the expressions for $i_{ld}(t)$ and $i_{ld}(t')$ in Eqn (3.20) and performing the integrations gives

$$i_{ld(av)} = \frac{E}{R_a}\frac{\tau_{ON}}{\tau} - \frac{e_b}{R_a} \tag{3.21}$$

where e_b is taken as a variable. This back emf is proportional to the speed. Hence it can be written as $e_b = K_1'\omega$, $i_{ld(av)}$ can be expressed as a function of ω. Thus,

$$i_{ld}(\omega) = \frac{E}{R_a}\frac{\tau_{ON}}{\tau} - \frac{K_1'\omega}{R_a} \tag{3.22}$$

The average torque developed by the motor, which is proportional to the flux as well as to the load current, is written as

$$\overline{T}_d = K_t\Phi_f i_{ld}(\omega) \tag{3.23}$$

where K_t is the torque constant, numerically equal to K_b, the voltage constant associated with the back emf [refer to Eqn (7.2)]. Φ_f being constant, $K_t\Phi_f$ can be replaced by K_1', with units N m/A. However, when associated with e_b, the units of K_1' are V/(rad/s). Now $\overline{T}_d$ becomes

$$\overline{T}_d = K_1' i_{ld}(\omega)$$

Using Eqn (3.22), $\overline{T}_d$ can be expressed as

$$\overline{T}_d = K_1'\left[\frac{E}{R_a}\frac{\tau_{ON}}{\tau} - \frac{K_1'\omega}{R_a}\right] \tag{3.24}$$

Equation (3.24) can now be solved for ω as

$$\omega = \frac{E}{K_1'}\frac{\tau_{ON}}{\tau} - \frac{R_a}{(K_1')^2}\overline{T}_d \tag{3.25}$$

Figure 3.5 shows the speed–torque curves with low chopper frequency taking the per-unit (p.u.) value of the speed, $\omega/\omega_{\text{rated}}$, as the ordinate. The regions of continuous and discontinuous conduction are demarcated by a dashed line.

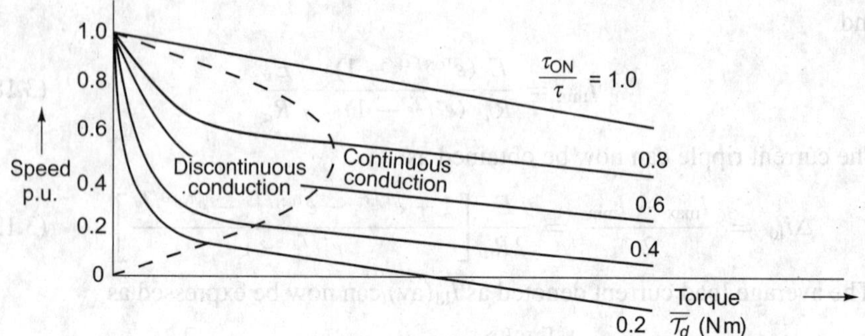

Fig. 3.5 Speed–torque curves with low chopper frequency

The speed–torque characteristics of Fig. 3.6, drawn with a high chopper frequency, show that the region of discontinuous conduction is reduced compared to that in Fig. 3.5. It can therefore be concluded that increasing the chopper frequency has the effect of smoothing the speed–torque curves in the region of continuous conduction.

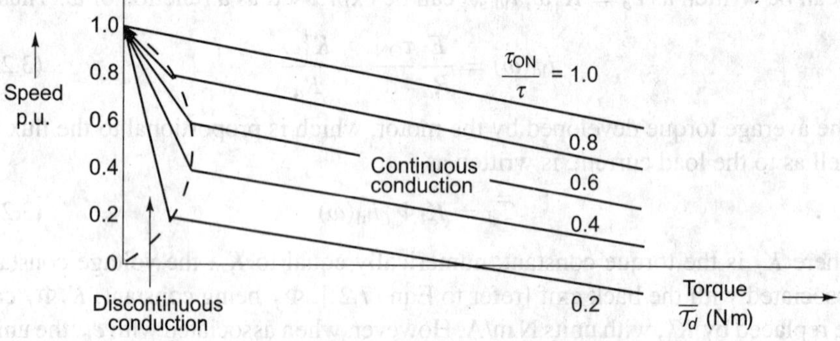

Fig. 3.6 Speed–torque curves with high chopper frequency

3.2.3 Step-up Chopper

The circuit and waveforms of a step-up chopper are, respectively, shown in Figs 3.7(a) as well as (b) and (c). The source voltage V_s in Fig. 3.7(a) can be a motor working in the inverting mode. Ch_2 is alternately switched on and off for the durations τ_{ON} and τ_{OFF}, respectively. The waveforms shown in Fig. 3.7(b) are obtained for the chopper of Fig. 3.7(a) after the circuit attains the steady state. In mode 1 the chopper is switched on and the inductance L_1 gains energy. The equation governing this mode is

$$V_s = L_1 \frac{di_1}{dt} \tag{3.26}$$

where V_s is the average source voltage. The solution of Eqn (3.26) can be written as

$$i_1(t) = \frac{V_s t}{L_1} + I_{\min} \tag{3.27}$$

where $I_{\min}$ is taken to be the inital current (at $t = 0$) for this mode. The quantity V_s/L_1 being positive, i_1 has a positive slope. At $t = \tau_{\text{ON}}$,

$$i_1(\tau_{\text{ON}}) = I_{\max} = \frac{V_s \tau_{\text{ON}}}{L_1} + I_{\min} \tag{3.28}$$

In mode 2, Ch_2 is turned off and the equation governing this mode is

$$V_s - E = L_1 \frac{di_2}{dt'} \tag{3.29}$$

where $t' = t - \tau_{\text{ON}}$. With the initial condition $i_2(0) = I_{\max}$, $i_2(t')$ can be obtained as

$$i_2(t') = \left(\frac{V_s - E}{L_1}\right) t' + I_{\max} \tag{3.30}$$

$i_2(t')$ must have a negative slope; otherwise the current continues to rise, leading to an unstable situation. Therefore the condition for power transfer from the inductor to the load is

$$\frac{V_s - E}{L_1} < 0$$

or

$$V_s < E$$

At $t' = \tau_{\text{OFF}}$, $i_2(t')$ attains the value $I_{\min}$. Equation (3.30) can now be written as

$$I_{\min} = \left(\frac{V_s - E}{L_1}\right) \tau_{\text{OFF}} + I_{\max} \tag{3.31}$$

A combination of Eqns (3.28) and (3.31) yields

$$V_s \tau_{\text{ON}} = (E - V_s)\tau_{\text{OFF}} \tag{3.32}$$

Equation (3.32) can also be obtained from the condition given in Eqn (3.4), which states that, under steady-state conditions, the net area under the curve of inductor voltage is zero over the interval τ and can be graphically represented by the striped figure in Fig. 3.7(b). Equation (3.32) can be rewritten as

$$E = \frac{V_s \tau}{\tau_{\text{OFF}}} = \frac{V_s \tau}{\tau - \tau_{\text{ON}}} \tag{3.33}$$

where

$$\tau_{\text{OFF}} = \tau - \tau_{\text{ON}}$$

Dividing both the numerator and denominator of Eqn (3.33) by τ and defining

$$\frac{\tau_{\text{ON}}}{\tau} = \delta$$

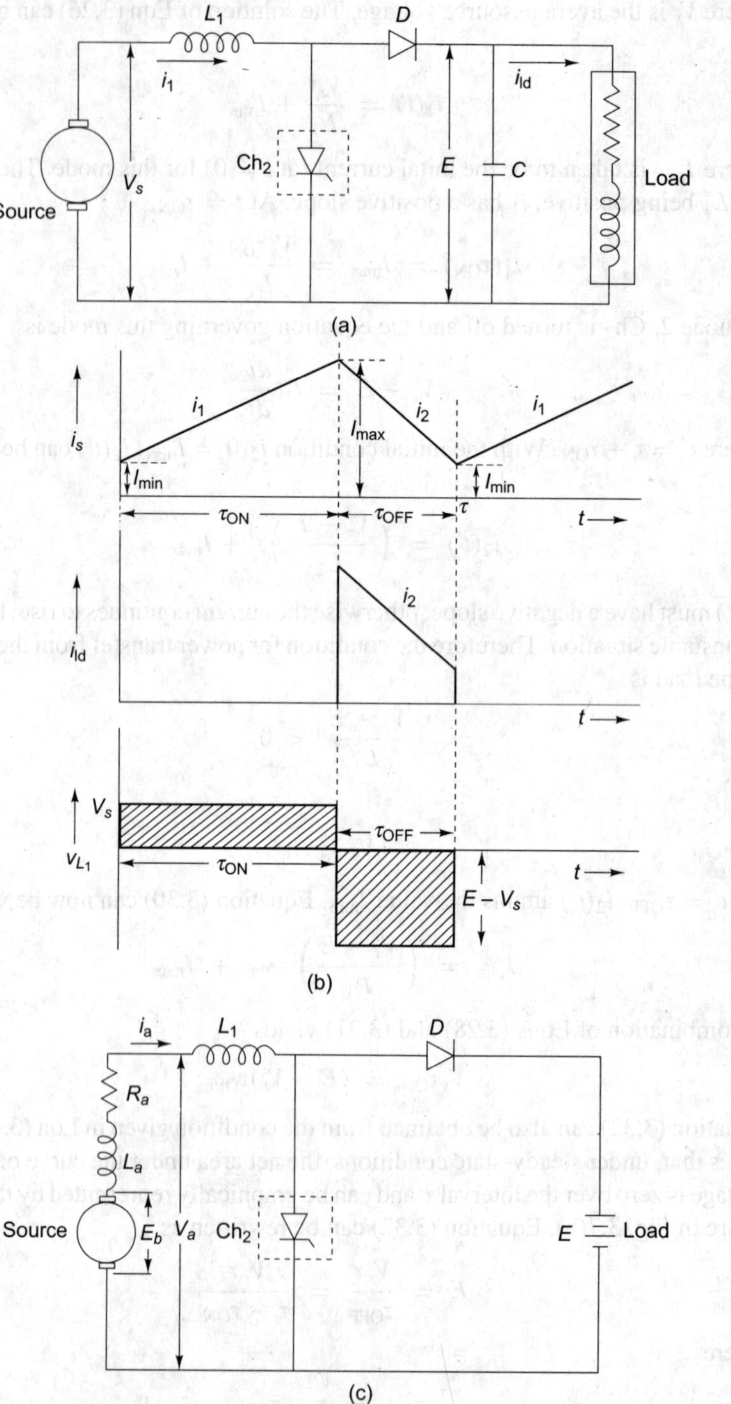

Fig. 3.7 Step-up chopper: (a) circuit, (b) waveforms, (c) circuit for regenerative braking

results in the equation

$$E = \frac{V_s}{1 - \delta} \tag{3.34}$$

where δ is in the range $0 \leq \delta \leq 1$.

Equation (3.34) shows that power transfer takes place with the load voltage being greater than that of the source. Though theoretically the upper value of the load voltage can be very large, it will be limited by any filter that may be used in the chopper circuit. If the load is considered to be a battery as shown in Fig. 3.7(c) then the operation described above can be interpreted to be that of a motor operating as a generator (source) and supplying power to a load (battery), with the back emf E_b (source) being less than the battery voltage E (load). This aspect is elaborated below.

The capacitor C in Fig. 3.7(a) serves two purposes. Just at the moment of switching on the chopper Ch_2, the source current i_s and the load current i_{ld} may not be the same. In the absence of the capacitor C, the turn-off of Ch_2 will force the two currents to have the same value, causing high induced voltages in the inductance L and as well as in the load inductance. Secondly, C will reduce the load voltage ripple. The diode D prevents any flow of current from the load side to the source or into the chopper Ch_2.

The main advantage of the step-up chopper is that the source current waveform will have a low ripple content. Whereas most applications require a step-down chopper, the step-up chopper finds application in low-power battery-driven vehicles such as golf carts, trolleys, etc. The principle of the step-up chopper is also used for regenerative braking of motors. Figure 3.7(c) shows the chopper circuit along with a separately excited dc motor which, with its supply cut off, works as a generator. For clarity, the field circuit, which is assumed to provide constant field current, is not shown in the figure. Here, the back emf E_b serves the same purpose as that of the voltage V_s of Fig. 3.7(a) and the battery with voltage E can be interpreted as the load. When the dc motor runs without the input supply, its kinetic energy gets converted into electrical energy, which is transferred to the battery (E) by the step-up chopper action. Consequently, the motor speed falls down and E_b also gets reduced. It can be inferred from this that the mechanical power of the shaft can be converted and transferred to the battery as long as the motor has a non-zero speed. This justifies the term regenerative braking mode for this mode of operation. The next section gives a detailed analysis of the dc motor working under such a regenerative mode.

3.3 Choppers Based on the Quadrants of Operation

The step-down chopper discussed in Section 3.2.1 operates in the first quadrant of the V-I plane and, hence, can be called a *first-quadrant* chopper. The applied voltage and current of a normal motor are considered to be positive and hence can be controlled by a first-quadrant chopper. Some other choppers—second-quadrant, two-quadrant, and four-quadrant choppers—are now dealt with in detail.

3.3.1 Second-quadrant Chopper

The second-quadrant chopper gets its name from the fact that the flow of current is from the load to the source, the voltage remaining positive throughout the range of operation. Such a reversal of power can take place only if the load is active, implying thereby that it should be capable of providing continuous power output. A dc motor which, after disconnection of its input supply, works in the regenerative braking mode, is an example of such a load. Figure 3.8(a) shows the circuit of a second-quadrant chopper which consists of a source that is a battery of voltage E, a chopper connected across the source and a diode, a separately excited dc motor (load), and an additional inductance L_1. This extra inductance is necessary for successful operation of the second-quadrant chopper. Figure 3.8(c) gives the waveforms for this chopper, under the assumption that its operation has attained steady state.

To bring out the feature of the regenerative mode, Fig. 3.8(a) is redrawn as Fig. 3.8(b) in which the motor is replaced by its equivalent circuit (back emf E_b, armature resistance R_a, and inductance L_a). For convenience, L_a is merged with the external inductance L_1 and the following are defined:

$$L = L_1 + L_a$$

and

$$R = R_a$$

A comparison of the step-up chopper of Fig. 3.7(c) with that of Fig. 3.8(b) reveals that the latter is just a swapped version of the former (the source in one becomes the load in the other and vice versa). The following analysis aims at showing that the operation of the second-quadrant chopper of Fig. 3.8(a) is identical to that of the step-up chopper of Fig. 3.7(a) with some approximations. Thus, it will be shown that the circuit of Fig. 3.8(b) is appropriate for being operated in the regenerative braking mode.

3.3.1.1 Analysis of the second-quadrant chopper circuit of Fig. 3.8(b)

The chopper Ch_2 is turned on at $t = 0$ and kept in this condition during the interval $0 \le t \le \tau_{\mathrm{ON}}$; the inductance L gains energy. The circuit behaviour in this interval is described by the equation

$$Ri_{\mathrm{ld}} + L\frac{di_{\mathrm{ld}}}{dt} = E_b, \quad 0 \le t \le \tau_{\mathrm{ON}} \tag{3.35}$$

with the initial condition $i_{\mathrm{ld}}(0) = I_{\min}$. The solution of Eqn (3.35) is given as

$$i_{\mathrm{ld}}(t) = I_{\min}e^{-Rt/L} + \frac{E_b}{R}(1 - e^{Rt/L}), \quad 0 \le t \le \tau_{\mathrm{ON}} \tag{3.36}$$

At $t = \tau_{\mathrm{ON}}$, i_{ld} attains the value $I_{\max}$. Thus, Eqn (3.36) becomes

$$I_{\max} = I_{\min}e^{-(R/L)\tau_{\mathrm{ON}}} + \frac{E_b}{R}(1 - e^{-(R/L)\tau_{\mathrm{ON}}}) \tag{3.37}$$

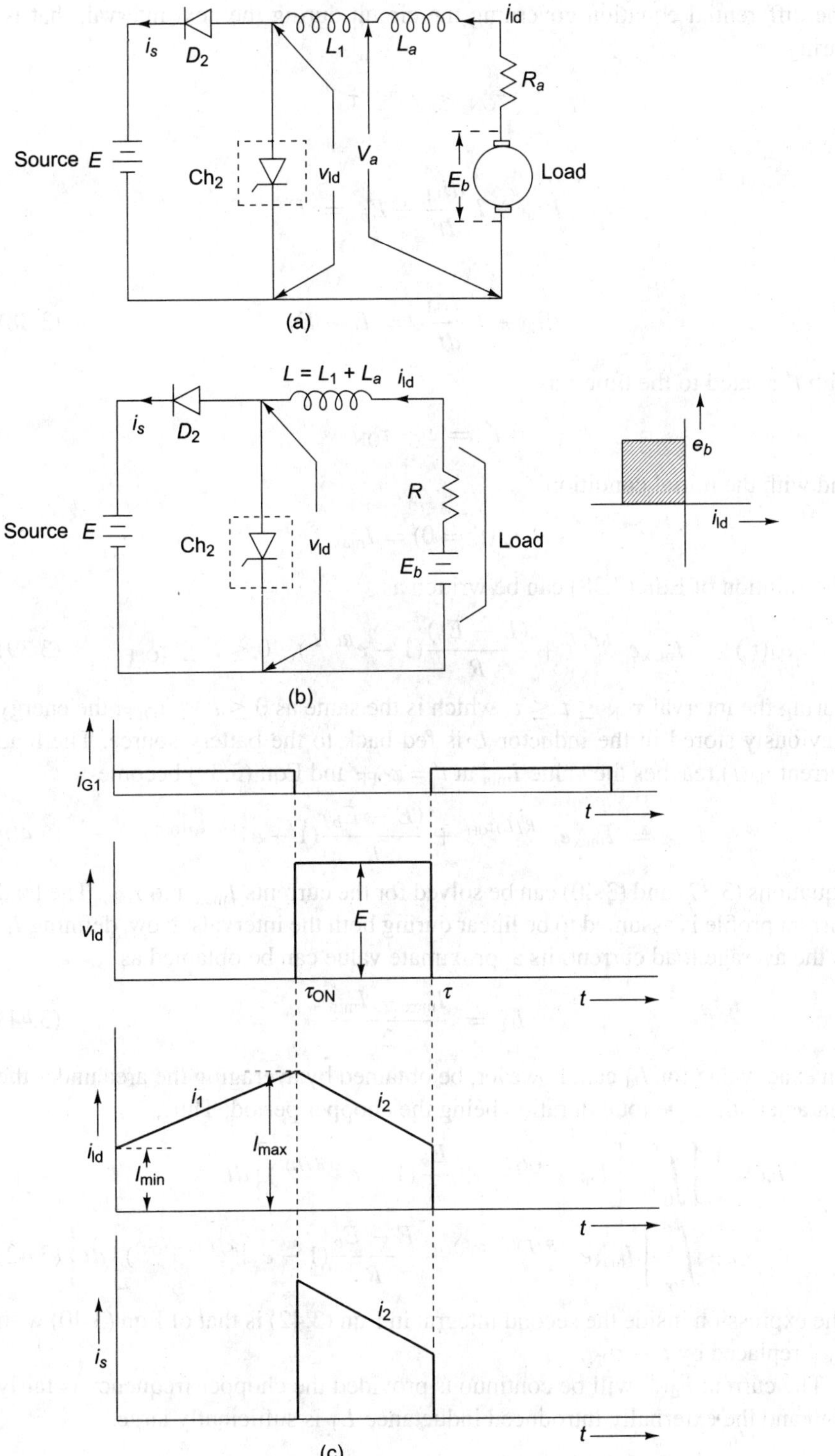

Fig. 3.8 Second-quadrant chopper: (a) circuit, (b) equivalent circuit, (c) waveforms

The differential equation governing the circuit during the next interval, that is, during

$$\tau_{ON} \le t \le \tau$$

is

$$Ri_{ld} + L\frac{di_{ld}}{dt'} + E_b = E$$

or

$$Ri_{ld} + L\frac{di_{ld}}{dt'} = E - E_b \tag{3.38}$$

with t' related to the time t as

$$t' = t - \tau_{ON}$$

and with the initial condition

$$i_{ld}(t' = 0) = I_{max}$$

The solution of Eqn (3.38) can be written as

$$i_{ld}(t') = I_{max}e^{-Rt'/L} + \frac{(E - E_b)}{R}(1 - e^{Rt'/L}), \;\; 0 \le t' \le \tau_{OFF} \tag{3.39}$$

During the interval $\tau_{ON} \le t \le \tau$, which is the same as $0 \le t' \le \tau_{OFF}$, the energy previously stored in the inductor L is fed back to the battery source. The load current $i_{ld}(t)$ reaches the value I_{min} at $t' = \tau_{OFF}$ and Eqn (3.39) becomes

$$I_{min} = I_{max}e^{-(R/L)\tau_{OFF}} + \frac{(E - E_b)}{R}(1 - e^{-(R/L)\tau_{OFF}}) \tag{3.40}$$

Equations (3.37) and (3.40) can be solved for the currents I_{max} and I_{min}. The load current profile is assumed to be linear during both the intervals. Now, defining I_{ld} as the average load current, its approximate value can be obtained as

$$I_{ld} = \frac{I_{max} + I_{min}}{2} \tag{3.41}$$

An exact value for I_{ld} can, however, be obtained by averaging the area under the characteristics, the total duration being the chopper period. Thus,

$$I_{ld} = \frac{1}{\tau}\left\{\int_0^{\tau_{ON}} \left[I_{min}e^{-(R/L)t} + \frac{E_b}{R}(1 - e^{-(R/L)t})\right] dt \right.$$
$$\left. + \int_{\tau_{ON}}^{\tau} \left[I_{max}e^{-(R/L)(t-\tau_{ON})} + \frac{E - E_b}{R}(1 - e^{-(R/L)(t-\tau_{ON})})\right] dt\right\} \tag{3.42}$$

The expression inside the second integral in Eqn (3.42) is that of Eqn (3.40) with τ_{OFF} replaced by $t - \tau_{ON}$.

The current $i_{ld}(t)$ will be continuous provided the chopper frequency is fairly high and the externally introduced inductance L_1 is sufficiently large.

3.3.1.2 Equivalence of Figs 3.8(b) and 3.7(a) In the analysis made above, the resistance R is just the armature resistance of the dc motor, which is usually very small. Similarly, the armature inductance L_a is small, and L can be considered to be nearly equal to L_1. This approximation is now incorporated in the expressions for I_{max} and I_{min}. Thus, Eqn (3.37) can be approximated as

$$I_{max} = I_{min}e^0 + \frac{E}{R}\left[1 - \left(1 - \frac{R\tau_{ON}}{L_1}\right)\right] \tag{3.43}$$

$$= I_{min} + \frac{E\tau_{ON}}{L_1} \tag{3.44}$$

Similarly Eqn (3.40) can be approximated as

$$I_{min} = I_{max} + \frac{E - E_b}{L_1}\tau_{OFF} \tag{3.45}$$

If the voltages V_s and E of Fig. 3.7(a) are, respectively, replaced by E and E_b, then Eqns (3.44) and (3.45), respectively, become the same as Eqns (3.28) and (3.31), which are derived for the step-up chopper. It is now evident that the second-quadrant chopper of Fig. 3.8(b) will function as a step-up chopper. Corresponding to Eqn (3.34) of the step-up chopper, the following equation can be shown to hold good for the second-quadrant chopper by manipulating Eqns (3.44) and (3.45):

$$E_b = \frac{E}{1 - \delta} \tag{3.46}$$

where δ is the duty ratio τ_{ON}/τ of this chopper. This confirms the remarks made earlier that under the regenerative braking mode, power transfer takes place from the load (motor) to the source if the load back voltage E_b is greater than the source voltage E.

3.3.2 Two-quadrant Chopper

The two-quadrant chopper (Dubey 1989) is just a combination of the first- and second-quadrant choppers. Two such choppers are dealt with here, one operating in the first and second quadrants and the other in the first and fourth quadrants. Evidently there is flexibility for the motor to operate in any one of the quadrants, instantaneous changeover being possible in both the cases.

3.3.2.1 Two-quadrant type-A chopper Figure 3.9 shows the circuit diagram of a two-quadrant chopper providing motor (load) operation in the first and second quadrants. With Ch_2 off, D_2 not conducting, but with Ch_1 on, the circuit works as a first-quadrant chopper. Again, with Ch_1 off, Ch_2 on, and D_1 not forward-biased, it works as a second-quadrant chopper. An important precaution to be taken is that the choppers Ch_1 and Ch_2 should not be switched on at the same time, as this will cause a direct short circuit across the source. To ensure this, the control signal i_{G1} for Ch_1 is kept on for a duration τ_{ON} and then i_{G2}, the firing signal of Ch_2, is applied during the time interval τ_{ON} to τ, this duration being equal to τ_{OFF}. Depending on the load time constant $T_a = L_a/R_a$, the chopping period τ, and the conducting period τ_{ON} of Ch_1, the following four possibilities exist. As before, the positive direction of the current is taken to be the direction from the source to the motor.

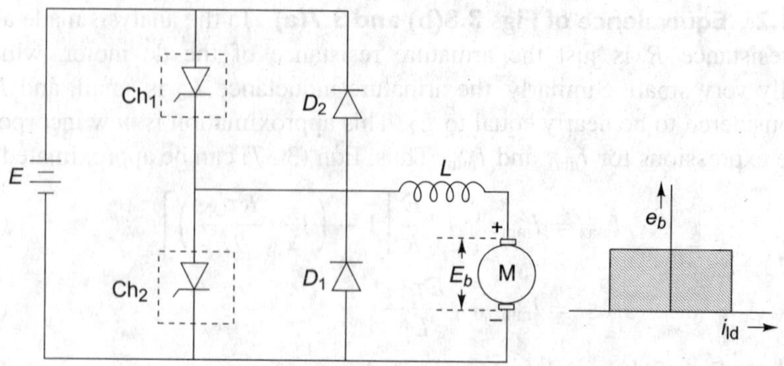

Fig. 3.9 Two-quadrant type-A chopper

(i) $\boldsymbol{I_{\text{min}} > 0}$ ***and both*** $\boldsymbol{I_{\text{max}}}$ ***and*** $\boldsymbol{I_{\text{min}}}$ ***are positive*** The dc motor works only in the motoring mode with only the chopper Ch_1 and the diode D_1 in operation. When Ch_1 is switched on at time $t = 0$, current flows from the source to the motor and the inductor L gains energy. At time τ_{ON}, Ch_1 is turned off but the current continues to flow in the same direction and finds a closed path through the motor, the freewheeling diode D_1, and the inductor L. Evidently the instantaneous load current i_{ld} is positive throughout, and hence the average current I_{ld} is also positive. Therefore the chopper operates in the first quadrant. Figure 3.10(a) gives the waveforms for this condition.

(ii) $\boldsymbol{I_{\text{max}} > 0, I_{\text{min}} < 0}$***, and*** $\boldsymbol{I_{\text{ld}}}$ ***is positive*** In this case, the instantaneous motor current can be positive or negative as shown in Fig. 3.10(b) but its profile is such that the average load current is positive. It is assumed that the circuit operation has attained a steady state. When i_{G1} is applied to Ch_1 at $t = 0$ the load current is negative and D_2 conducts it. The drop across D_2 reverse-biases the main thyristor of Ch_1, thus preventing conduction. The supply voltage E will be greater than E_b, and hence di_{ld}/dt will be positive. When i_{ld} reaches zero, D_2 stops conduction and Ch_1 gets forward-biased. As i_{G1} is still present, Ch_1 starts conduction and continues to do so till τ_{ON}. The firing signal i_{G2} of the thyristor Ch_2 is started at τ_{ON} but the chopper cannot conduct because the current is in the positive direction. As the source is isolated, D_1 freewheels the inductive current. The slope di_{ld}/dt being negative, i_{ld} becomes zero after some time and D_1 stops conduction. When i_{ld} crosses to the negative side, Ch_2 starts conduction because its firing signal is still present. This condition remains till time τ at which instant i_{G1} is applied again. The sequence is now repeated. The quantities τ_{ON}, τ, and T_a are such that the average load current i_{ld} is positive; the motor operation is therefore represented in the first quadrant. The presence of the chopper Ch_2 and the diode D_2 facilitates continuous flow of current irrespective of its direction.

(iii) $\boldsymbol{I_{\text{max}} > 0, I_{\text{min}} < 0}$***, and*** $\boldsymbol{I_{\text{ld}}}$ ***is negative*** The sequence of events for this case is exactly as given above except that τ_{ON}, τ, and T_a are such that the average load current I_{ld} is negative. Hence the motor operates in the second quadrant. The waveforms for this condition are given in Fig. 3.10(c).

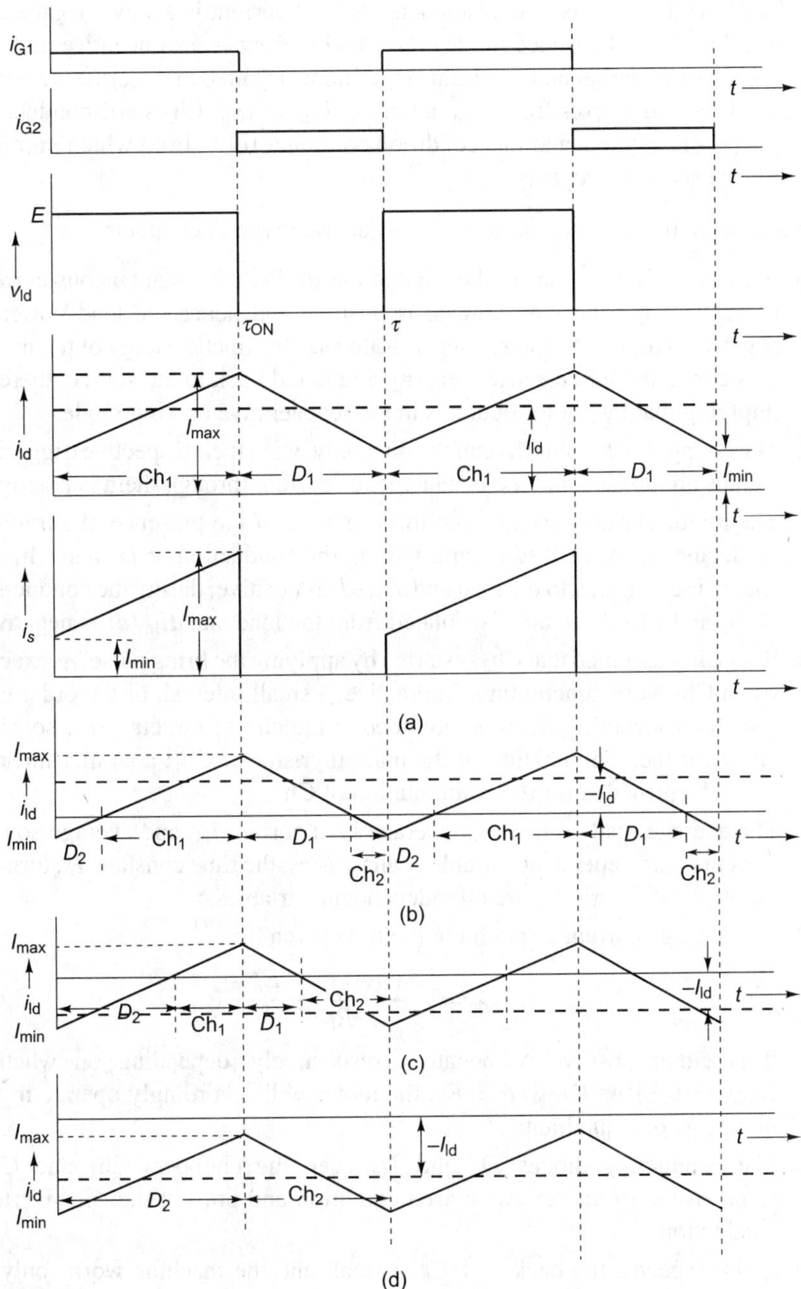

Fig. 3.10 Waveforms with different values of duty cycle (τ_{ON}/τ) for the two-quadrant type-A chopper. (a) $I_{min} > 0$, I_{ld} positive; (b) $I_{min} < 0$, $I_{max} > 0$, I_{ld} positive; (c) $I_{min} < 0$, $I_{max} > 0$, I_{ld} negative; (d) $I_{max} < 0$, I_{ld} negative

(iv) $I_{max} < 0$ In this case the instantaneous load current is always negative as shown in Fig. 3.10(d). Hence the average load current is also negative and the motor operates in the second quadrant. D_2 conducts from 0 to τ_{ON}, di_{ld}/dt being positive. The current rises from I_{min}, attaining I_{max} at τ_{ON}. Ch_2 starts conduction at τ_{ON}, as i_{G2} is applied, and this condition continues till τ, from which moment onwards the sequence repeats.

The following remarks pertain to the two-quadrant type-A chopper:

(a) For cases (ii)–(iv), during the conduction of D_2, the instantaneous current is negative but the load voltage is positive and hence the load power is negative. This can be interpreted as follows. The kinetic energy of the motor gets converted into electrical energy and is fed back to the source, thereby implying that the motor operates in the regenerative braking mode.

(b) The choppers Ch_1 and Ch_2 can conduct only when their respective triggering signals are present and the instantaneous current through them is positive.

(c) The motor current flows all the time because of the presence of sufficient inductance L in the load circuit. During the conduction of D_2 and Ch_1 the source is connected to the load and di_{ld}/dt is positive; during the conduction of D_1 and Ch_2 the source is isolated from the load and di_{ld}/dt is negative.

(d) It is tacitly assumed that Ch_2 is started by applying the firing pulse i_{G2} exactly when Ch_1 stops functioning. In practice, a small interval, of the order of a few microseconds, has to be provided at the end of the pulse i_{G1} so as to allow for the commutation of the main thyristor of Ch_1; a similar interval has to be provided for the commutation of Ch_2.

(e) The average load voltage v_{ld}, average load current i_{ld}, and average source current i_d are dependent variables, whereas e_b, the time constant T_a, turn-on time τ_{ON}, and period τ are all independent variables.

(f) The average current through the motor is given by

$$i_{ld} = \frac{E(\tau_{ON}/\tau) - E_b}{R_{ld}} \tag{3.47}$$

It is either positive or negative, respectively, depending on whether $E\tau_{ON}/\tau > E_b$ or $E\tau_{ON}/\tau < E_b$; the motor will accordingly operate in the first or second quadrant.

(g) The conducting diodes D_1 and D_2 keep the choppers Ch_2 and Ch_1, respectively, in the reverse-biased condition and prevent their inadvertent conduction.

(h) At low speeds, the back emf E_b is small and the machine works only in the motoring mode. However, both motoring and regenerative braking are possible when E_b is large, that is, when the motor speed is high.

3.3.2.2 Two-quadrant type-B chopper Figure 3.11(a) gives the circuit of a two-quadrant type-B chopper that can operate in the first and fourth quadrants. This is achieved by the two methods of control described below.

Method 1 It is seen from the waveforms of Fig. 3.11(b) that both the choppers are turned on and off, respectively, during τ_{ON} and τ_{OFF} $(= \tau - \tau_{ON})$. The load voltage has the values $+E$ and $-E$, respectively, for these intervals. When the two choppers are turned off, the current through the inductor L continues to flow in the same direction, making the diode pair D_1, D_2 conduct, thus feeding the load energy back to the dc source. The average load voltage is obtained as

$$V_{ld} = \frac{1}{\tau}\left[\int_0^{\tau_{ON}} E dt + \int_{\tau_{ON}}^{T} (-E)\, dt\right] = \frac{E}{\tau}(\tau_{ON} - \tau_{OFF}) \tag{3.48}$$

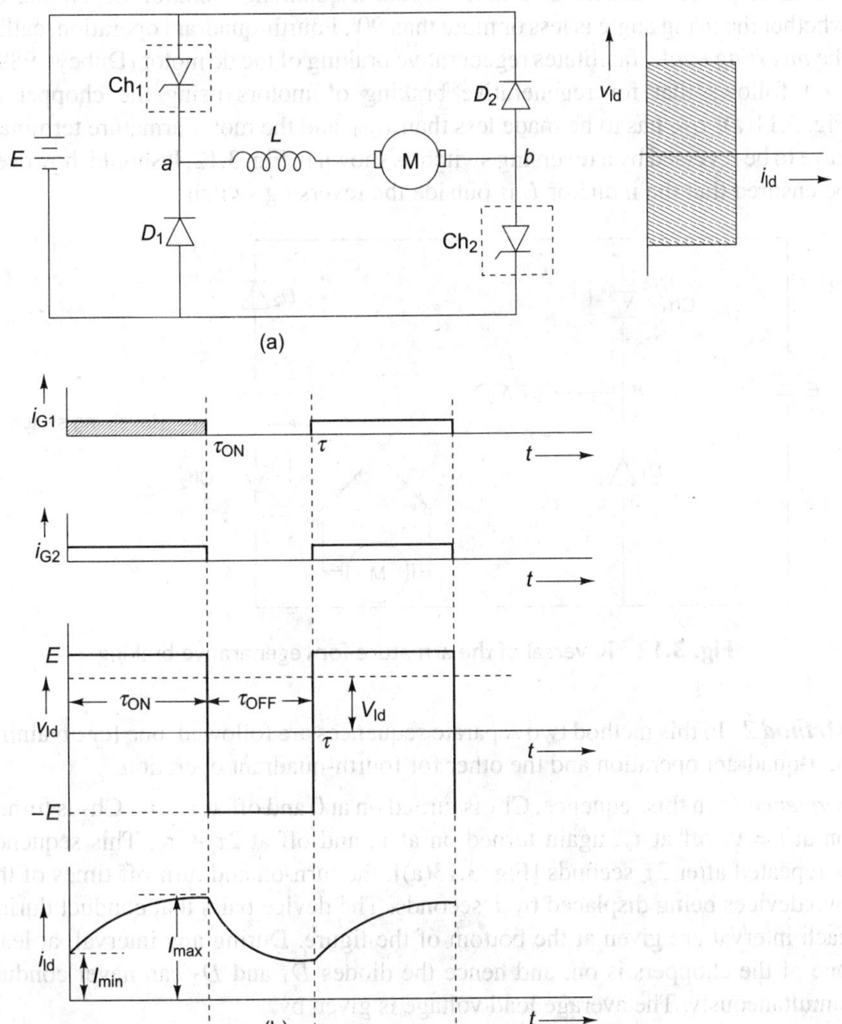

Fig. 3.11 Two-quadrant type-B chopper (method 1): (a) circuit, (b) waveforms

Equation (3.48) shows that for the condition $\tau_{ON} > \tau_{OFF}$, v_{ld} is positive and the current flows in the circuit constituted by the positive terminal of the battery, Ch_1,

L, the motor, Ch_2, and back to the negative terminal of the battery. Both the average load voltage v_{ld} and the average load current i_{ld} being positive, the operation of the motor can be represented as a point in the first quadrant; this is just the motoring mode. When $\tau_{ON} < \tau_{OFF}$, v_{ld} is negative but i_{ld} is positive because it flows from terminal a to terminal b, the direction of the current remaining unchanged even when the diode pair conducts. Hence, this operation can be depicted as a point in the fourth quadrant, implying thereby that the motor operates in the inverting mode.

Regenerative braking of a dc motor It is shown in Chapter 2 that single-phase and three-phase rectifiers give first- or fourth-quadrant operation depending on whether the firing angle is less or more than 90°. Fourth-quadrant operation, called the *inverting mode*, facilitates regenerative braking of the dc motor (Dubey 1989).

It follows that for regenerative braking of motors using the chopper of Fig. 3.11(a), τ_{ON} has to be made less than τ_{OFF} and the motor armature terminals have to be reversed by a reversing switch as shown in Fig. 3.12. It should, however, be ensured that the inductor L is outside the reversing switch.

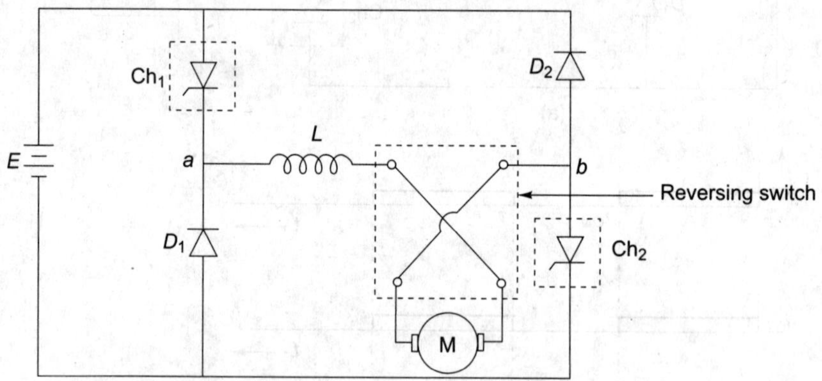

Fig. 3.12 Reversal of the armature for regenerative braking

Method 2 In this method two separate sequences are followed, one for obtaining first-quadrant operation and the other for fourth-quadrant operation.

Sequence 1 In this sequence, Ch_1 is turned on at 0 and off at $\tau + \tau_x$. Ch_2 is turned on at $t = 0$, off at τ_x, again turned on at τ, and off at $2\tau + \tau_x$. This sequence is repeated after 2τ seconds [Fig. 3.13(a)], the turn-on and turn-off times of the two devices being displaced by τ seconds. The device pairs that conduct during each interval are given at the bottom of the figure. During any interval, at least one of the choppers is on, and hence the diodes D_1 and D_2 can never conduct simultaneously. The average load voltage is given by

$$V_{ld} = \frac{1}{\tau}\int_0^{\tau_x} E\,dt$$

$$= \frac{E\tau_x}{\tau} \tag{3.49}$$

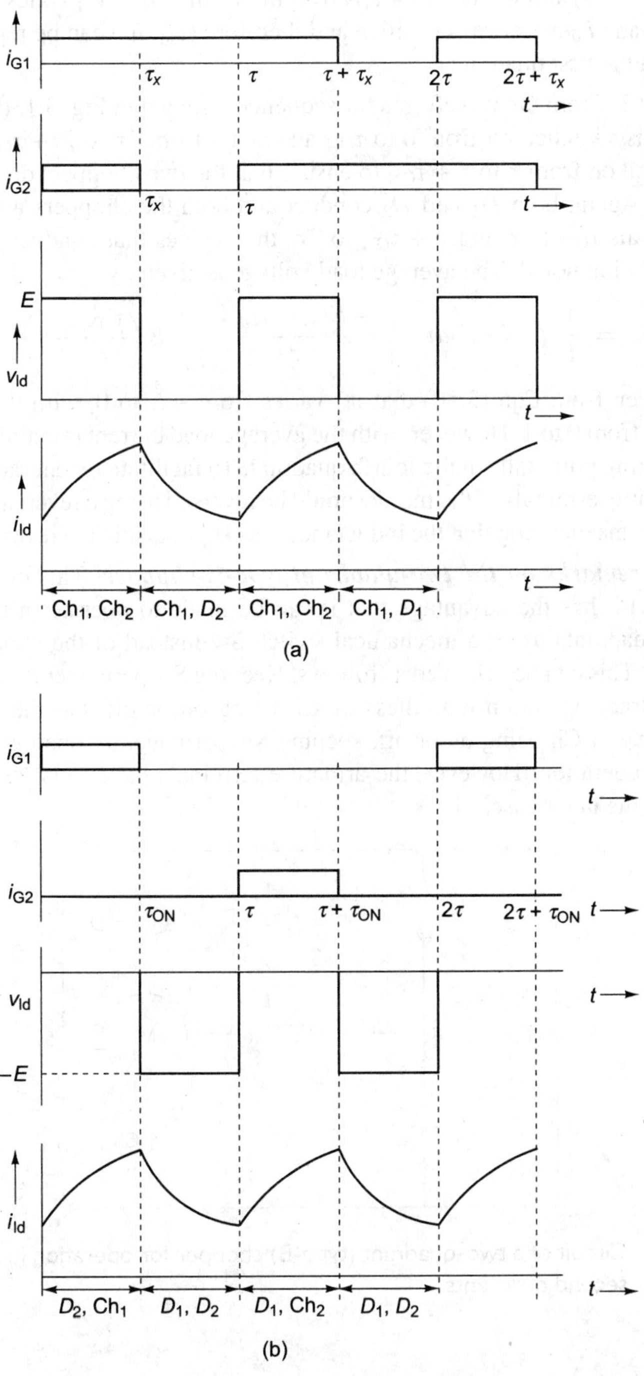

Fig. 3.13 Two-quadrant (type-B) chopper (method 2): (a) waveforms for sequence 1, (b) waveforms for sequence 2

Equation (3.49) shows that, with τ_x/τ varying from 0 to 1, V_{ld} varies from 0 to E volts. v_{ld} and i_{ld} are always positive and therefore (v_{ld}, i_{ld}) can be represented as a point in the first quadrant.

Sequence 2 From the waveforms for sequence 2 shown in Fig. 3.13(b), it is seen that Ch_1 is switched on from 0 to τ_{ON} and again from 2τ to $2\tau + \tau_{ON}$; also Ch_2 is switched on from τ to $\tau + \tau_{ON}$ to ensure that the two choppers do not conduct together. Again, both D_1 and D_2 conduct and both the choppers are off during the intervals τ_{ON} to τ and $\tau + \tau_{ON}$ to 2τ, the devices that conduct during each duration being noted. The average load voltage is given by

$$V_{ld} = \frac{1}{\tau}\int_{\tau_{ON}}^{\tau}(-E)dt = \frac{-E(\tau - \tau_{ON})}{\tau} = E\left(\frac{\tau_{ON}}{\tau} - 1\right) \tag{3.50}$$

It is seen from Eqn (3.50) that v_{ld} varies from $-E$ to 0, with the duty ratio changing from 0 to 1. However, with the average load current remaining positive, the operating point falls in the fourth quadrant. To facilitate regenerative braking, the armature terminals of the motor should be reversed using a reversing switch, at the same time ensuring that the inductance L is kept outside the reversing switch.

General remarks on the two-quadrant type-B chopper This circuit, shown in Fig. 3.14, has the advantage that it can be made to operate in the first and second quadrants using a mechanical switch Sw instead of the chopper Ch_1 in Fig. 3.12. This can be achieved as follows. Keeping Sw permanently closed gives first-quadrant operation regardless of Ch being on or off. On the other hand, irrespective of Ch being on or off, keeping Sw permanently open gives second-quadrant operation. However, the armature terminals have to be reversed by a switch in the latter case.

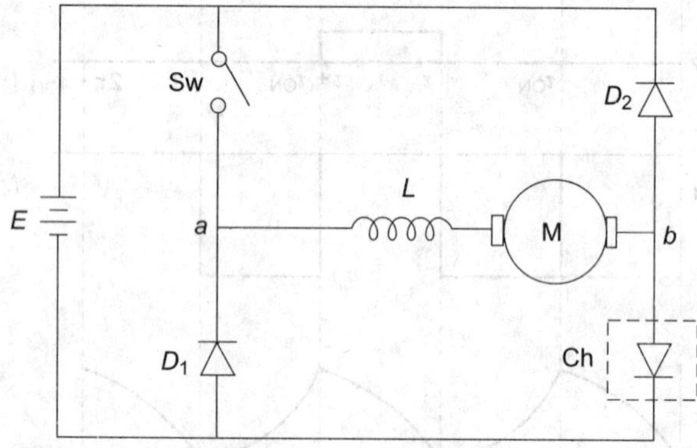

Fig. 3.14 Circuit of a two-quadrant (type-B) chopper for operation in the first and second quadrants

3.3.3 Four-quadrant Chopper

The circuit of a four-quadrant chopper is shown in Fig. 3.15 in which the inductor L is assumed to be composed of the armature inductance and an external inductor.

Three methods of control are possible for the operation of this chopper as elaborated below.

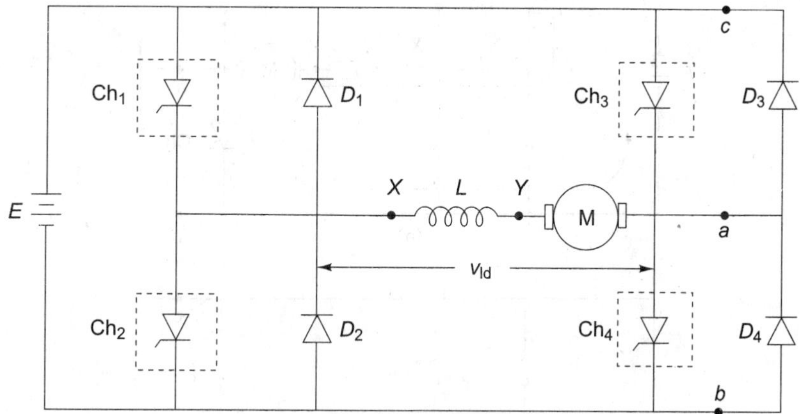

Fig. 3.15 Circuit of a four-quadrant chopper

Method 1 The circuit is operated as a two-quadrant chopper to obtain (a) first- and second-quadrant operation as well as (b) third- and fourth-quadrant operation.

Sequence 1 To obtain mode (a), Ch_4 is permanently kept on; terminals a and b are always kept shorted by ensuring conduction by either Ch_4 or D_4 and terminals a and c are always kept open. The choppers Ch_1 and Ch_2 are controlled as per the following four steps.

(a) If Ch_1 and Ch_4 are turned on at $t = 0$, the battery voltage E will be applied to the load circuit and current will flow from X to Y as shown in Fig. 3.16(a); this direction is the positive one. Thus the load voltage during this interval is kept at $+E$.

(b) When Ch_1 is turned off at τ_{ON}, the current due to the stored $(1/2)Li^2$ energy of the inductor L drives the current through D_2 and Ch_4 as shown in Fig. 3.16(b). Ch_2, which is turned on at τ_{ON}, does not conduct because it is shorted by D_2.

(c) Ch_2, which is on, conducts the current when it reverses, as shown in Fig. 3.16(c).

(d) Finally, when Ch_2 is turned off at τ, current flows through the path consisting of the negative of the battery, D_4, the motor, L, D_1, and the positive of the battery as shown in Fig. 3.16(d). If the machine were to be operated as a generator, this circuit facilitates regenerative braking. The zero crossing instants of the current waveform depend upon the values of E, E_b, L, and the armature resistance R_a of the motor. It is seen that Ch_1 does not conduct till i_{ld} becomes positive and Ch_2 does not conduct till i_{ld} flows in the negative direction. Also, D_4 conducts the reverse current and applies a reverse bias against Ch_4. The devices that conduct during each of the intervals are shown in Fig. 3.17(a), which gives the waveforms of this mode.

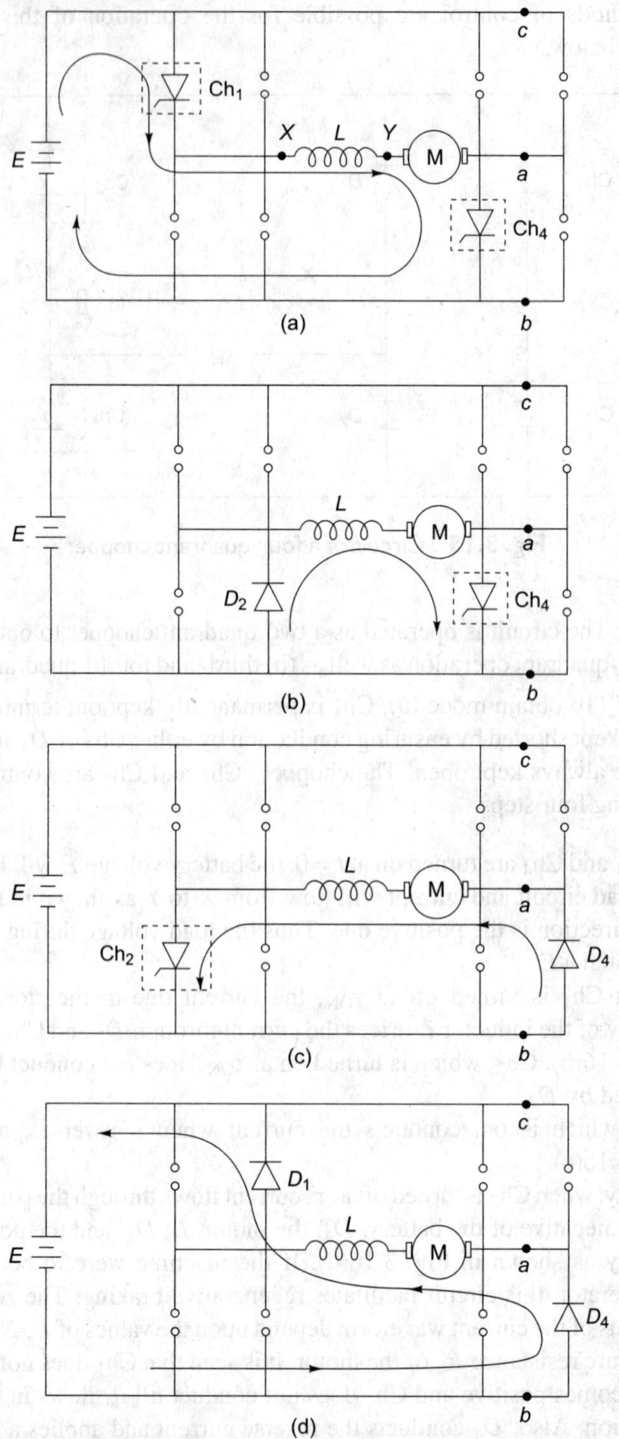

Fig. 3.16 Circuit conditions of a four-quadrant chopper with method 1: (a) Ch_1 and Ch_4 turned on; (b) Ch_1 turned off, Ch_4 remaining on; (c) Ch_2 turned on, Ch_4 shorted by D_4; (d) Ch_2 turned off, Ch_4, shorted by D_4

Sequence 2 For the circuit to provide third- and fourth-quadrant operation, Ch_3 is permanently kept on. Terminals *a* and *c* remain shorted due to conduction by either Ch_3 or D_3; terminals *a* and *b* remain open throughout. The relevant waveforms are shown in Fig. 3.17(b) (in the figure, I_B denotes the current through the battery).

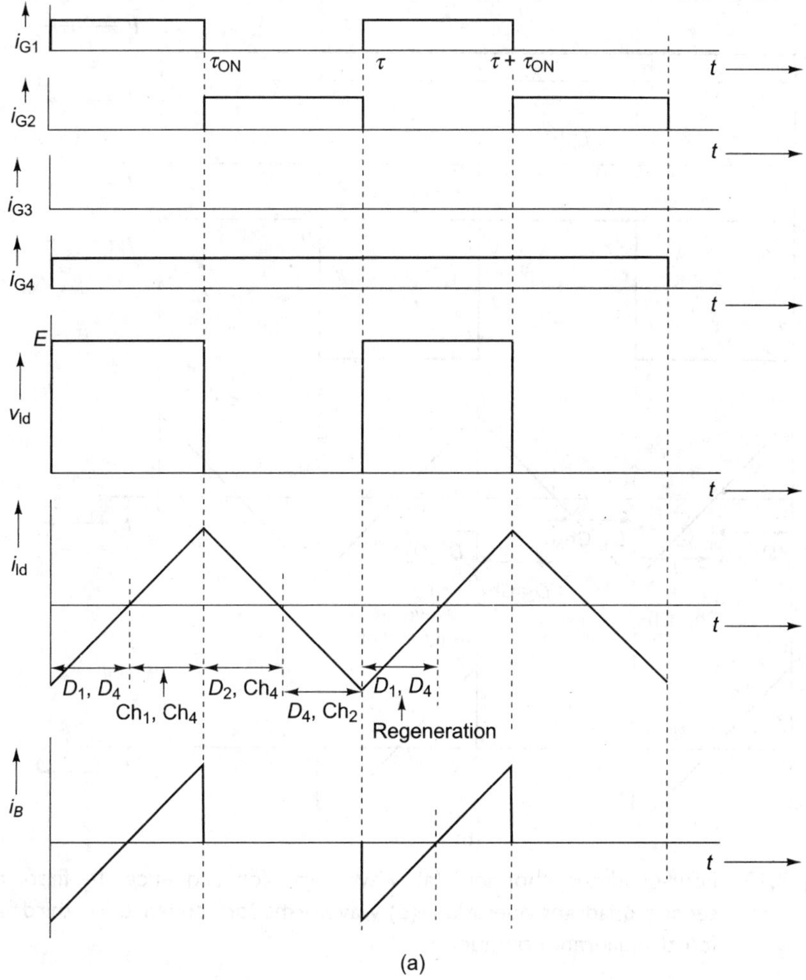

Fig. 3.17(a)

(a) Ch_2 is triggered on at $t = 0$ but starts conduction only when a reverse current flows through the path consisting of the positive of the battery, Ch_3, the motor, L, Ch_2, and the negative of the battery. A voltage equal to $-E$ is applied at the load terminals.

(b) When Ch_2 is turned off at τ_{ON}, the inductor continues to drive the current in the reverse direction through the path consisting of Ch_3, the motor, L, and D_1. The load voltage then becomes zero.

(c) Ch_1 is triggered at τ_{ON} but starts conduction only when the current flows in the positive direction, flowing through the closed circuit consisting of Ch_1, L, the motor, and D_3.

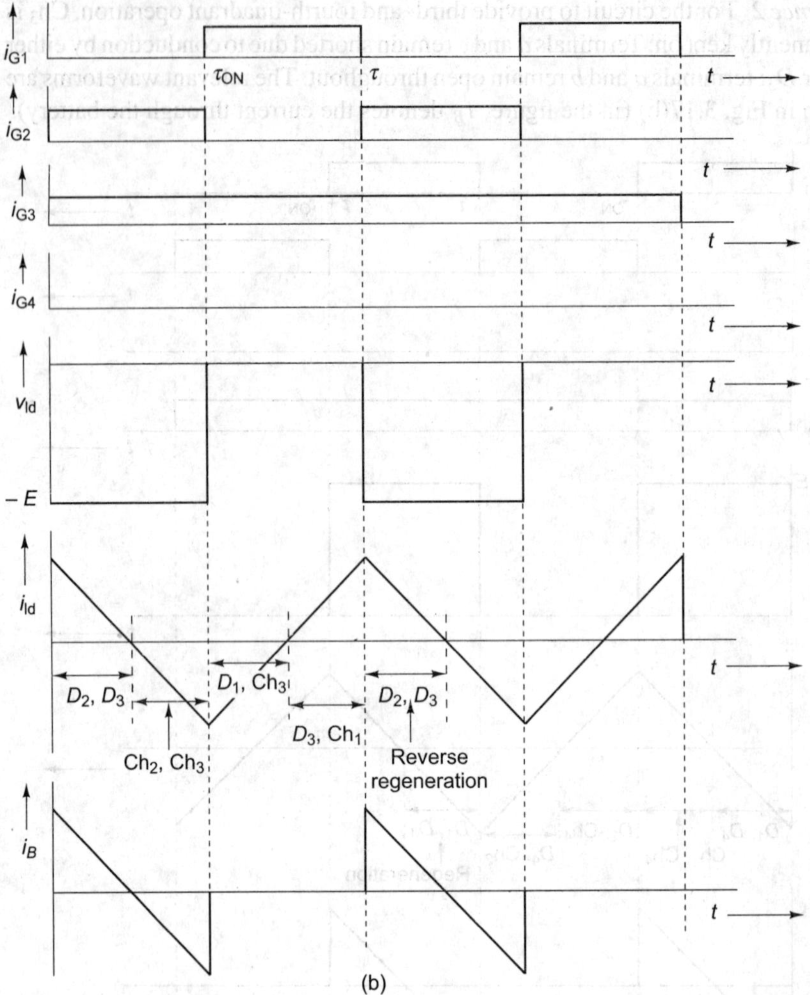

Fig. 3.17 Four-quadrant chopper: (a) waveforms for sequence 1—first- and second-quadrant operation, (b) waveforms for sequence 2—third- and fourth-quadrant operation

(d) When Ch_1 is turned off at τ, a negative battery voltage is applied to the load but positive current flows through the negative terminal of the battery, D_2, L, the motor, D_3, and back to the positive terminal of the battery.

It is seen that either Ch_1 or Ch_2 conducts current when i_{ld} becomes positive or negative, respectively. This is because, even though their control signals are present prior to the zero crossing of the load current, the conducting diodes D_1 and D_2 apply a reverse bias, respectively, across Ch_1 and Ch_2. The devices that conduct during each interval are given in Fig. 3.17(b).

This circuit suffers from the disadvantage that either Ch_3 or Ch_4 are kept on for a long time, which may lead to commutation problems. An important precaution to be taken is that the choppers Ch_1 and Ch_2 should not conduct simultaneously, as otherwise the source gets shorted through them. To ensure this, a small interval of time has to be provided between the turn-off of Ch_2 and the turn-on of Ch_1 and vice versa; this feature, however, limits the maximum chopper frequency.

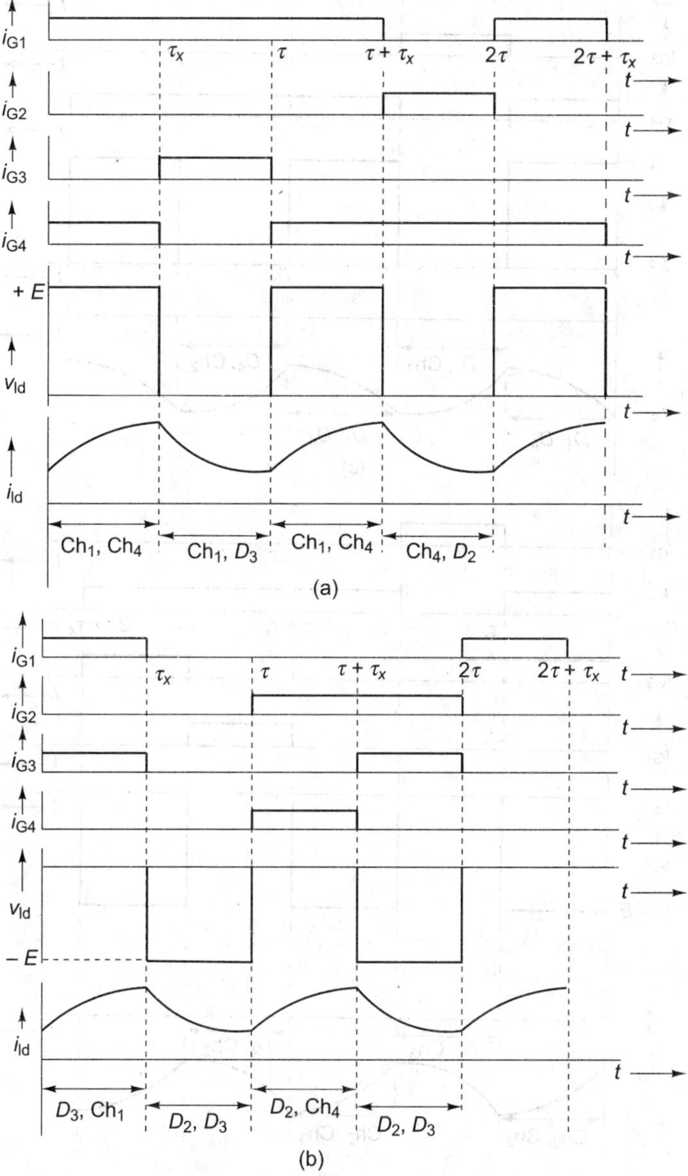

Figs 3.18(a) and (b)

Method 2 In this method, the four-quadrant chopper provides first- and fourth-quadrant operation similar to method 2 of the two-quadrant type-B chopper. Thus the chopper pair Ch_1, Ch_4 and the diode pair D_2, D_3 conduct alternately; the other chopper pair is permanently kept off. Accordingly the waveforms will be identical to those of Figs 3.13(a) and (b), respectively. Likewise, for obtaining second- and

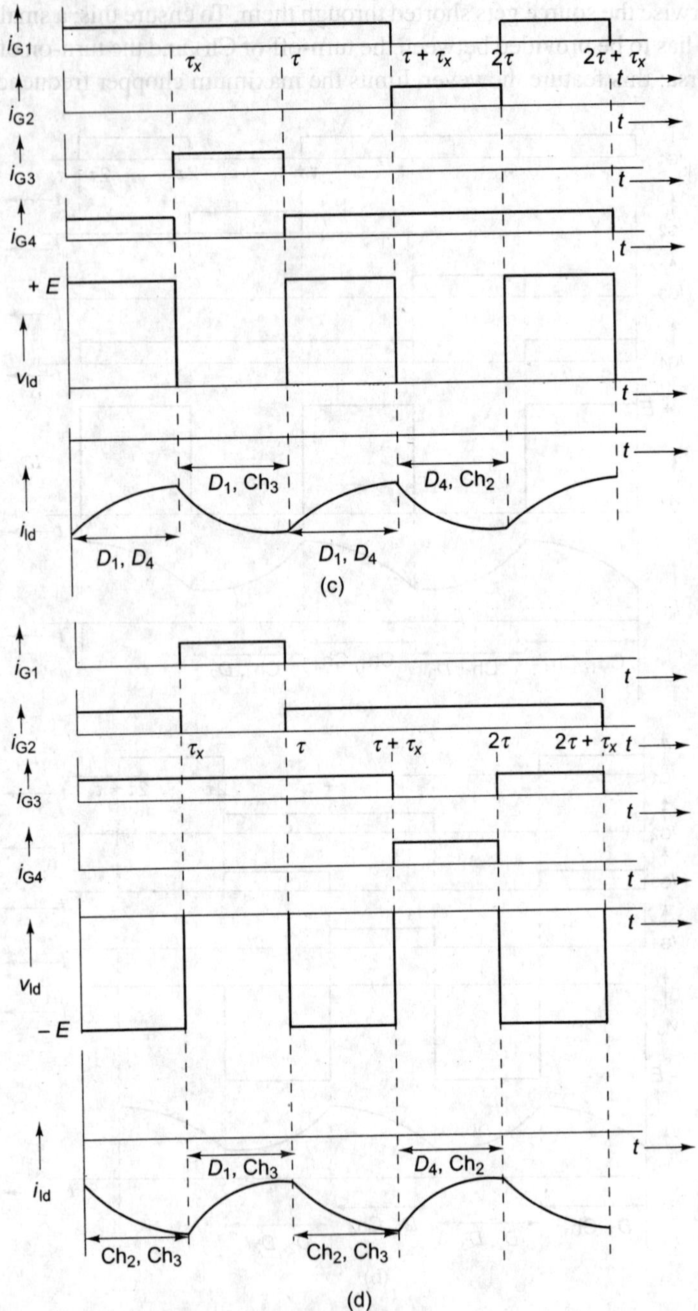

Fig. 3.18 Waveforms for four-quadrant chopper operation (method 3): (a) first-quadrant operation, (b) fourth-quadrant operation, (c) second-quadrant operation, (d) third-quadrant operation

third-quadrant operation, the chopper pair Ch_2, Ch_3 and the diode pair D_1, D_4 of Fig. 3.15 conduct in alternate intervals with the chopper pair Ch_1, Ch_4 always kept off. The waveforms in this case will be similar to those of Figs 3.13(a) and (b) except for the fact that the instantaneous current in both cases is always negative. Thus the operating point will be in either the second or the third quadrant.

Method 3 This method consists of operating the same combinations of chopper pairs, as in method 2, to provide four-quadrant operation. However, the chopper pairs are controlled in such a way that if one of them conducts during some interval, the other pair is off. The waveforms for the first, fourth, second, and third quadrants are given, respectively, in Figs 3.18(a), (b), (c), and (d).

3.4 Speed Control of a Chopper-controlled Dc Series Motor

The inductive load for the choppers discussed so far is taken as a separately excited dc motor. However, in some applications such as traction and hoist control, the dc series motor (Dubey 1989) is preferred because of the high starting torque provided by it. As its name implies, the series motor has its field in series with the armature, implying that the field current remains the same as the armature current. Figures 3.19(a) and (b), respectively, give the circuit and speed–torque characteristics of a dc series motor. In Fig. 3.19(a) R_a and L_a are, respectively, the armature resistance and inductance; R_f and L_f are, respectively, the field resistance and inductance. Since the total inductance (L_a+ L_f) is large, it is assumed that the current ripple is small and the inductive drop is negligible. The aim of the following analysis is to (a) derive a steady-state relationship between speed and torque and (b) examine how a chopper can be employed to obtain a set of speed–torque characteristics.

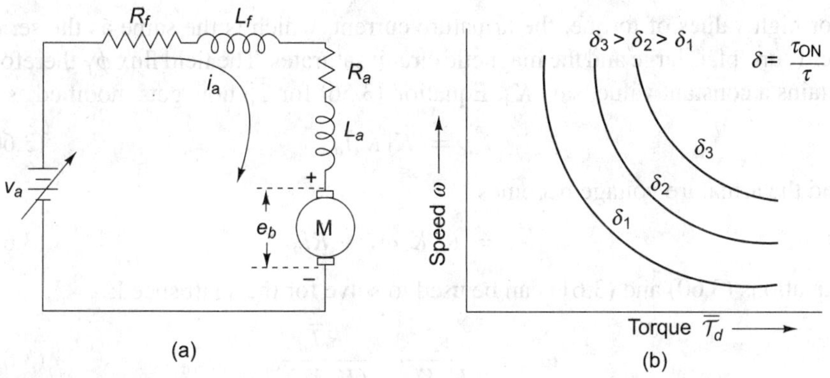

Fig. 3.19 dc series motor: (a) circuit, (b) speed–torque characteristics

The basic equation is that of the applied voltage, which is given as

$$v_a = e_b + Ri_a \tag{3.51}$$

where $R = R_a + R_f$. In turn, e_b can be expressed as

$$e_b = K_1 \phi_f \omega_m \tag{3.52}$$

where K_1 is a constant. The ϕ_f-I_a characteristic for a series motor is just the magnetization characteristic. In the linear region the field flux is proportional to the field current i_f, which, here, is the same as the armature current i_a. Thus,

$$\phi_f = K_2 i_a \tag{3.53}$$

where K_2 is a constant. Using Eqn (3.53), the back emf becomes

$$e_b = K_1 K_2 i_a \omega_m \tag{3.54}$$

This expression for e_b can now be substituted in Eqn (3.51), giving

$$v_a = K_1 K_2 i_a \omega_m + R i_a \tag{3.55}$$

The torque developed by the machine is given as

$$\overline{T}_d = K_1 \phi_f i_a = K_1 K_2 (i_a)^2 \tag{3.56}$$

where Eqn (3.53) is made use of for replacing ϕ_f. Combining Eqns (3.55) and (3.56) and solving for ω_m gives

$$\omega_m = \frac{v_a}{\sqrt{\overline{T}_d K_1 K_2}} - \frac{R}{K_1 K_2} \tag{3.57}$$

The resistance R is usually very small and hence can be neglected, the speed now becomes

$$\omega_m = \frac{v_a}{\sqrt{\overline{T}_d K_1 K_2}} \tag{3.58}$$

If a first-quadrant chopper is interposed between the source (with a fixed voltage E) and the chopper, Eqn (3.57) can be written as

$$\omega_m = \frac{E(\tau_{\mathrm{ON}}/\tau)}{\sqrt{\overline{T}_d K_1 K_2}} \tag{3.59}$$

For high values of torque, the armature current, which is the same as the series field current, is large and the magnetic circuit saturates. The field flux ϕ_f therefore attains a constant value, say, K_4. Equation (3.56) for $\overline{T}_d$ now gets modified as

$$\overline{T}_d = K_1 K_4 i_a \tag{3.60}$$

and the armature voltage becomes

$$v_a = K_1 K_4 \omega_m + R i_a \tag{3.61}$$

Equations (3.60) and (3.61) can be used to solve for the shaft speed:

$$\omega_m = \frac{v_a}{K_1 K_4} - \frac{R\overline{T}_d}{(K_1 K_4)^2} \tag{3.62}$$

Neglecting R as before and assuming that a chopper (supplied by a battery of E volts) controls the armature voltage of the motor, Eqn (3.62) gets modified as

$$\omega_m = \frac{E(\tau_{\mathrm{ON}}/\tau)}{K_1 K_4} \tag{3.63}$$

Thus, Eqns (3.59) and (3.63) define the speed–torque characteristics, respectively, for small and large values of torque. Accordingly, it is seen from Fig. 3.19(b) that for a fixed duty ratio τ_{ON}/τ and for low and medium values of $\overline{T}_d$, the speed–torque characteristics attain a very inverse shape as given by

$$\overline{T}_d \omega_m^2 = \text{constant} \tag{3.64}$$

It is also seen that Eqn (3.64) is just another version of Eqn (3.59). Figure 3.19(b) also shows that for high values of $\overline{T}_d$, the characteristics become parallel to the torque axis, implying that, for a fixed duty ratio, the speed becomes constant for high values of torque. Thus, each of the speed–torque characteristics in Fig. 3.19(b) corroborates the relations in Eqns (3.59) and (3.63).

Braking of a dc series motor The two-quadrant type-B chopper can also be used for regenerative braking of a dc series motor, taking the necessary precautions, to ensure that the series field is kept outside the reversing operation.

3.5 Commutation Circuits

It is stated in Section 3.2 that the basic chopper consists of one or more semiconductor devices along with commutation circuitry. Choppers differ from one another in the type of commutation circuitry employed. Commonly occurring commutation circuits can be broadly categorized into the following groups.

Voltage commutation This consists of applying a negative voltage across the conducting thyristor for a period long enough to successfully turn off the thyristor. The Jones, Morgan, and voltage commutated choppers come under this category.

Current commutation In current commutation a reverse current equal to or greater than the current in the conducting thyristor is made to pass through the thyristor to turn it off. Turn-off is then ensured by applying a negative voltage against it. The modified parallel resonant chopper and the special current-commutated chopper are examples of this type of commutation.

Load commutation This consists of transferring the load current flowing through one pair of thyristors to a second one. A typical load commutated chopper employs voltage commutation with the help of a capacitor connected in the centre of a bridge circuit that has a thyristor in each of its arms. Four simple commutation circuits have been discussed in Section 1.5. Some more commonly occurring circuits are now described here.

3.5.1 Modified Parallel Resonant Turn-off Circuit

The parallel resonant turn-off circuit given in Fig. 1.41(a) can be modified to that of a first-quadrant chopper circuit, shown in Fig. 3.20, by means of additional elements, namely, a diode D_1, an inductance L, and a freewheeling diode D_{FW}. The various circuit conditions occurring during the operation of this chopper are given in Figs 3.21(a)–(c) and the waveforms are given in Fig. 3.22. For D_{FW} to conduct the load current, the conditions to be fulfilled are that the thyristor should be off and the capacitor current should be below the load current I_{ld}. The sequence of operations that take place is now described.

At time t_0, a gate-triggering signal is applied to the thyristor Th_1 to turn it on. This sets up an LC-oscillatory current in the loop consisting of C, L, and Th_1;

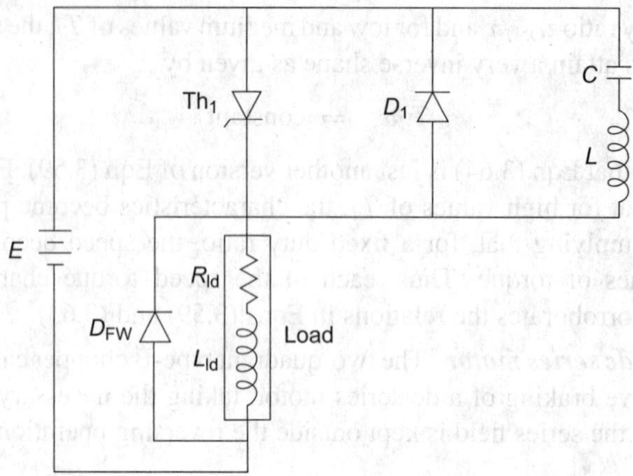

Fig. 3.20 Circuit for modified parallel resonant turn-off

simultaneously, current I_{ld} flows in the load circuit consisting of E, Th$_1$, and the load. At t_1, i_c reverses and v_c attains a negative maximum of $-E$ volts. Again, at t_2, i_c attains a value equal to I_{ld} and Th$_1$ turns off. As long as Th$_1$ conducts (that is, prior to its turn-off), a reverse voltage equal to the forward conducting drop of the thyristor is applied across the diode D_1, keeping it off. After Th$_1$ is turned off, the load current flows in the circuit consisting of E, C, L, and the load. However, from t_2 to t_3 the magnitude of i_c is in excess of I_{ld} and this excess current flows through D_1. The duration from t_2 to t_3 is the commutation time available for the thyristor, because the conducting diode applies a negative voltage equal to its forward drop across Th$_1$. At t_3, i_c drops back to I_{ld} and D_1 stops conduction; v_c attains a positive value E_1, which is less than E. The new circuit condition, as shown in Fig. 3.21(c), comes into operation and the capacitor subsequently charges to E with the X-plate having positive polarity. This condition remains till t_4, at which moment Th$_1$ is fired so as to start a second cycle of operation. As the thyristor conducts both the load current as well as oscillatory current during t_0 to t_1, its rating should exceed the sum I_{ld} + I_{cm}. During t_1 to t_2, i_c gradually rises and i_{Th_1} gradually falls to zero, their sum being equal to I_{ld}. The diode D_{FW} starts conduction at t_3 and the current through it rises exponentially to reach I_{ld} and remains at this value till Th$_1$ is fired again at t_4. Because the reverse LC-oscillatory current stops the conduction of the thyristor, this chopper comes under the category of current-commutated choppers.

The analysis consists of three stages corresponding to three modes of operation. For the capacitor charging mode shown in Fig. 3.21(a) the results of the forced RLC circuit with the non-zero initial condition [case (vii) of Appendix B] hold good. The unforced LC circuit with an initial condition [case (iii) of Appendix B] applies for the closed LC circuit of Fig. 3.21(b). Figure 3.21(c) shows the circuit in which the L and C elements carry a part of the load current I_{ld}, the diode D_{FW} conducting the other part. The duration of conduction of D_1 determines the commutation time. If t_c is the circuit time available for commutation, it can be

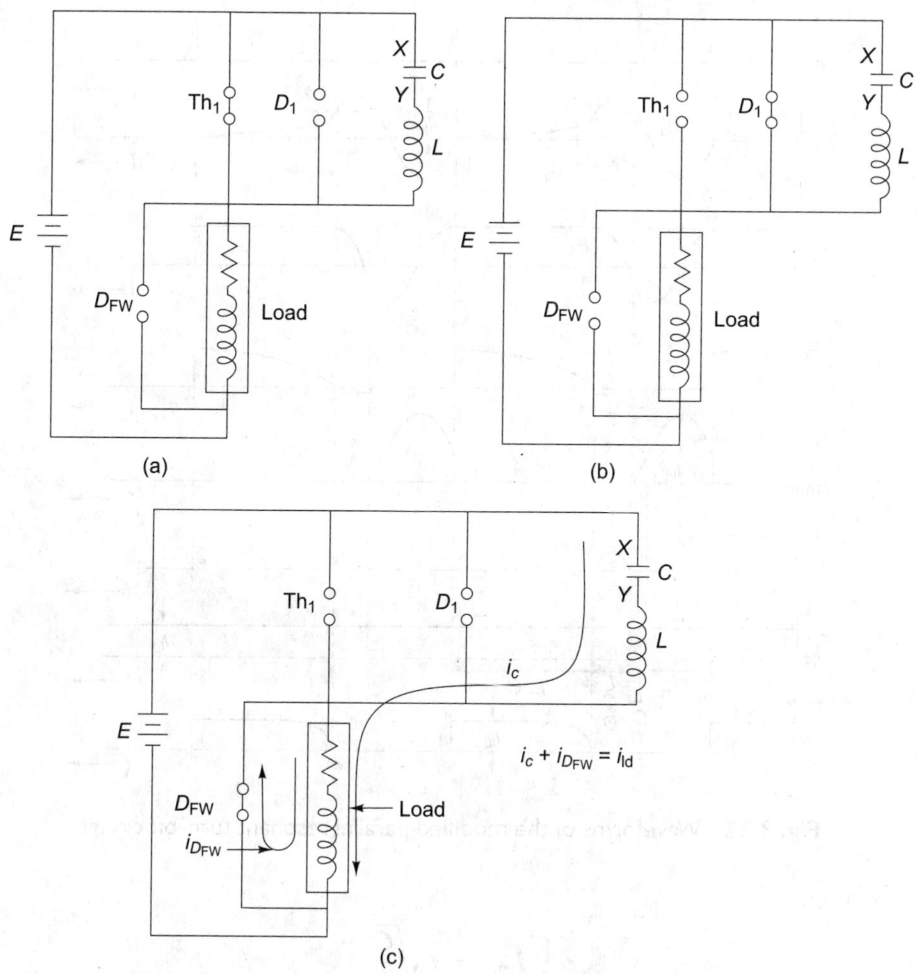

Fig. 3.21 Circuit conditions for the modified parallel resonant turn-off chopper: (a) just after t_0, (b) just after t_2, (c) just after t_3

expressed as

$$t_c = \frac{1}{\omega}\left[\pi - 2\sin^{-1}\left(\frac{I_{\text{ld}}}{I_{\text{cm}}}\right)\right] \tag{3.65}$$

For successful commutation of the thyristor the inequality

$$t_c \geq t_{\text{OFF}} \tag{3.66}$$

has to be satisfied. The maximum continuous current of the thyristor is equal to the sum of the load current and *LC* oscillatory current. Thus,

$$I_{\text{Th}_1\text{M}} = \frac{E}{R_{\text{ld}}} + I_{\text{cm}} \tag{3.67}$$

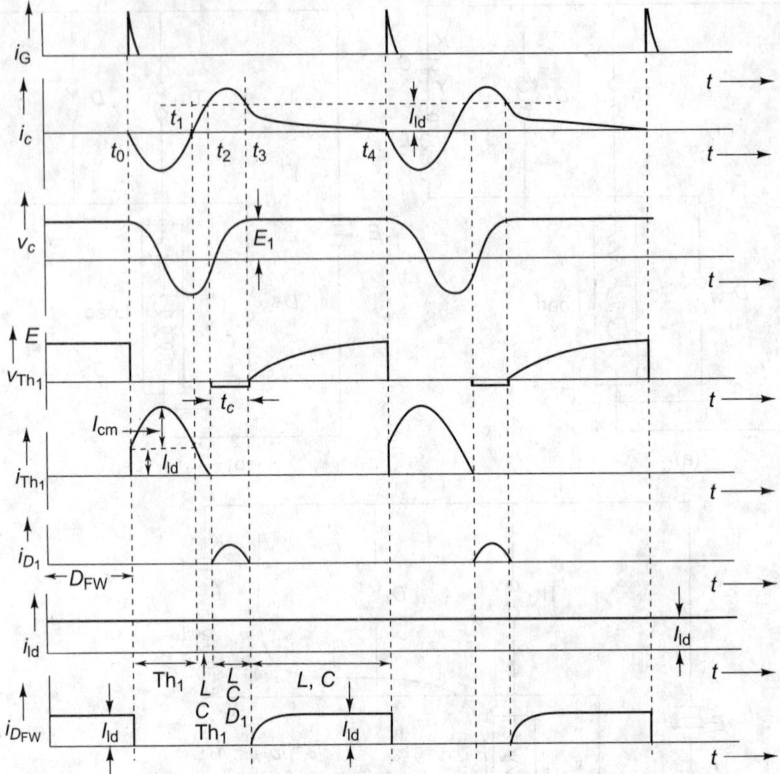

Fig. 3.22 Waveforms of the modified parallel resonant turn-off circuit

where

$$I_{cm} = E\sqrt{\frac{C}{L}}$$

Given the battery voltage E, load resistance R_{ld}, and the turn-off time t_{OFF}, the elements L and C can be designed to provide sufficient circuit time t_c for reliable commutation. From Eqn (3.67), I_{cm} can be written as

$$I_{cm} = I_{Th_1M} - \frac{E}{R_{ld}} \tag{3.68}$$

Writing I_{cm} in terms of E, C, and L and using Eqn (3.67) gives

$$E\sqrt{\frac{C}{L}} = I_{Th_1M} - \frac{E}{R_{ld}}$$

Finally,

$$\sqrt{\frac{C}{L}} = \frac{I_{Th_1M}}{E} - \frac{1}{R_{ld}} \tag{3.69}$$

From Eqns (3.65), (3.66), and (3.68), the following inequality is obtained:

$$\sqrt{LC}\left[\pi - 2\sin^{-1}\left\{\frac{I_{\mathrm{ld}}}{(I_{\mathrm{Th_1M}} - E/R_{\mathrm{ld}})}\right\}\right] \geq t_{\mathrm{OFF}} \tag{3.70}$$

Considering only the equality in Eqn (3.70) and solving for $\sqrt{LC}$ gives

$$\sqrt{LC} = \frac{t_{\mathrm{OFF}}}{\left[\pi - 2\sin^{-1}\left\{\dfrac{I_{\mathrm{ld}}}{(I_{\mathrm{Th_1M}} - E/R_{\mathrm{ld}})}\right\}\right]} \tag{3.71}$$

Multiplication of Eqns (3.69) and (3.71) yields

$$C = \frac{t_{\mathrm{OFF}}\left[I_{\mathrm{Th_1M}}/E - 1/R_{\mathrm{ld}}\right]}{\left[\pi - 2\sin^{-1}\left\{\dfrac{I_{\mathrm{ld}}}{(I_{\mathrm{Th_1M}} - E/R_{\mathrm{ld}})}\right\}\right]} \tag{3.72}$$

Also, the division of Eqn (3.71) by Eqn (3.69) gives

$$L = \frac{t_{\mathrm{OFF}}}{\left[\dfrac{I_{\mathrm{Th_1M}}}{E} - \dfrac{1}{R_{\mathrm{ld}}}\right]\left[\pi - 2\sin^{-1}\left\{\dfrac{I_{\mathrm{ld}}}{(I_{\mathrm{Th_1M}} - E/R_{\mathrm{ld}})}\right\}\right]} \tag{3.73}$$

The actual value of C should be higher by 10–20 % than that given by Eqn (3.72) to ensure successful commutation.

3.5.2 Morgan Chopper

The circuit, three circuit conditions, and waveforms of the Morgan chopper are, respectively, given in Figs 3.23, 3.24, and 3.25. The uniqueness of this circuit is that a saturable core reactor (SR) replaces the inductor for the purpose of LC oscillation. The load is assumed to consist of a dc motor running at constant speed. The latter depends upon the load voltage, which in turn depends upon the frequency of the gate triggering signal i_G, the values of the capacitor C, and the size of the SR. As shown in Fig. 3.23, the cathode of the thyristor is connected to

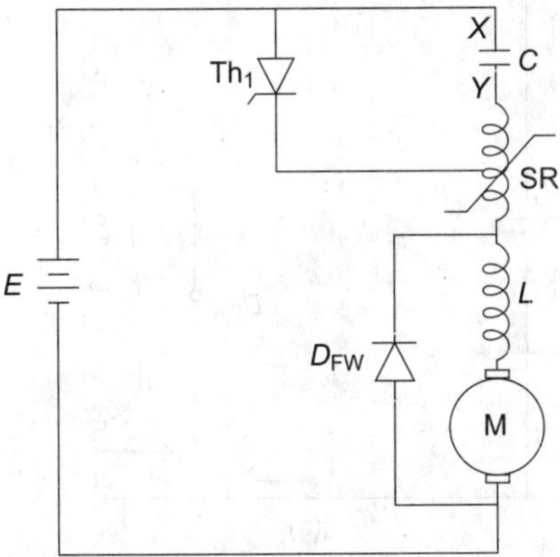

Fig. 3.23 Circuit of a Morgan chopper

the centre of the SR. The unsaturated reactance of the SR is assumed to be very large; hence the corresponding current will be negligibly small.

Prior to the starting of the sequence of operations, the capacitor is assumed to be charged to the battery voltage E with positive polarity at the X-plate. The sequence of operations starts with the thyristor Th$_1$ being triggered at t_0. The load current then flows through the battery E, Th$_1$, the lower half of the SR, the external inductance L, and the motor; simultaneously a current i_c flows in the closed circuit comprising of the capacitor C, Th$_1$, and the upper half of the SR as shown in Fig. 3.24(b). The high unsaturated reactance of SR causes a very small current to flow through this circuit from t_0 to t_1 and thus slightly decreases the capacitor voltage v_c. At t_1, the SR saturates and an LC oscillatory current sets in, in the anticlockwise direction with a frequency of $(1/2)\pi\sqrt{L_sC}$, where L_s is the low post-saturation inductance of the SR. From t_1 to t_2, which is the half-cycle time of the oscillatory current, v_c changes from $+E$ to $-E$. At t_2, the LC oscillatory current starts flowing in the clockwise direction. However, because

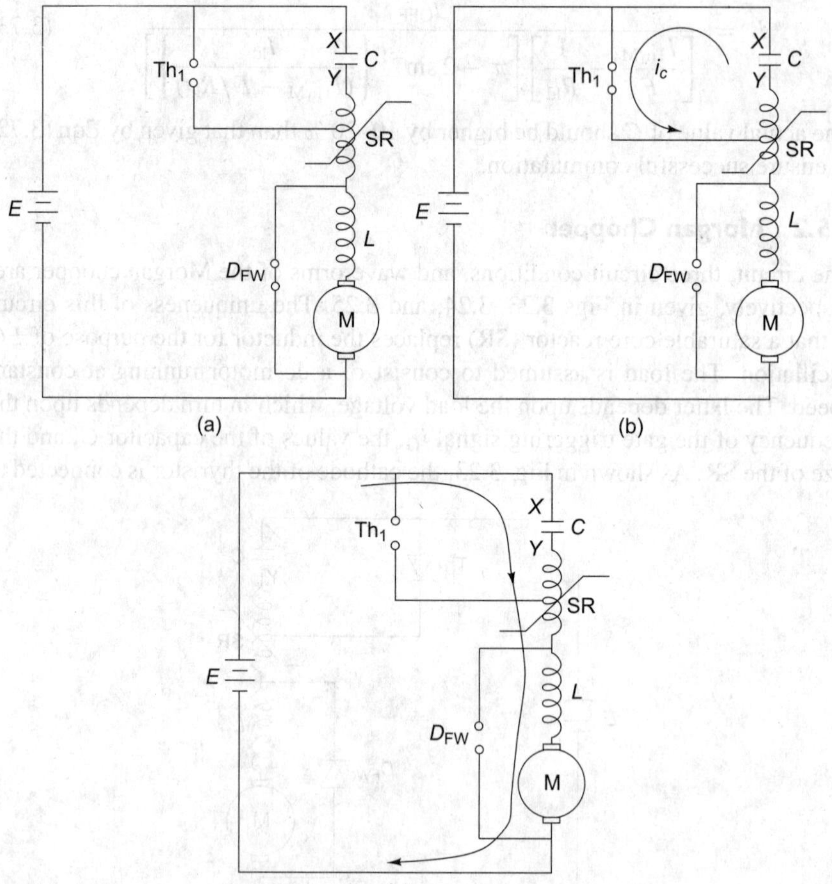

Fig. 3.24 Circuit conditions for a Morgan chopper

of the reversal of the current in the SR, it comes out of saturation and offers a very high impedance to the current. This situation exists till t_3, at which time the reactor flux builds up in the opposite direction. At t_4, the SR gets saturated and then along with C an oscillation is set up in the opposite sense through Th_1. When at t_5, this oscillatory current exceeds I_{ld}, Th_1 turns off; the circuit condition at this moment is shown in Fig. 3.24(c). At t_6, i_c falls below I_{ld} and, due to the freewheeling effect of the load (that is, the motor) inductance together with L, $i_{D_{FW}}$ slowly starts increasing as shown in Fig. 3.25. i_c gradually falls to zero and correspondingly $i_{D_{FW}}$ gradually increases to I_{ld} and remains so till the gate firing signal for Th_1 occurs at t_7, thus turning it on. The cycle repeats from t_7 onwards.

When D_{FW} starts conduction at t_6 the load voltage drops down to zero, whereas the voltage across the SR and C combination increases. As a consequence of this, the capacitor charges quickly to $+E$.

The duration from t_2 to t_5 is the extra time available for the conduction of the thyristor due to the presence of the saturable reactor. This interval can be split up into two subintervals as follows: (a) the duration after saturation when current flows in the reverse direction (t_2 to t_3) and (b) the duration before saturation when current flows in the forward direction (t_3 to t_5).

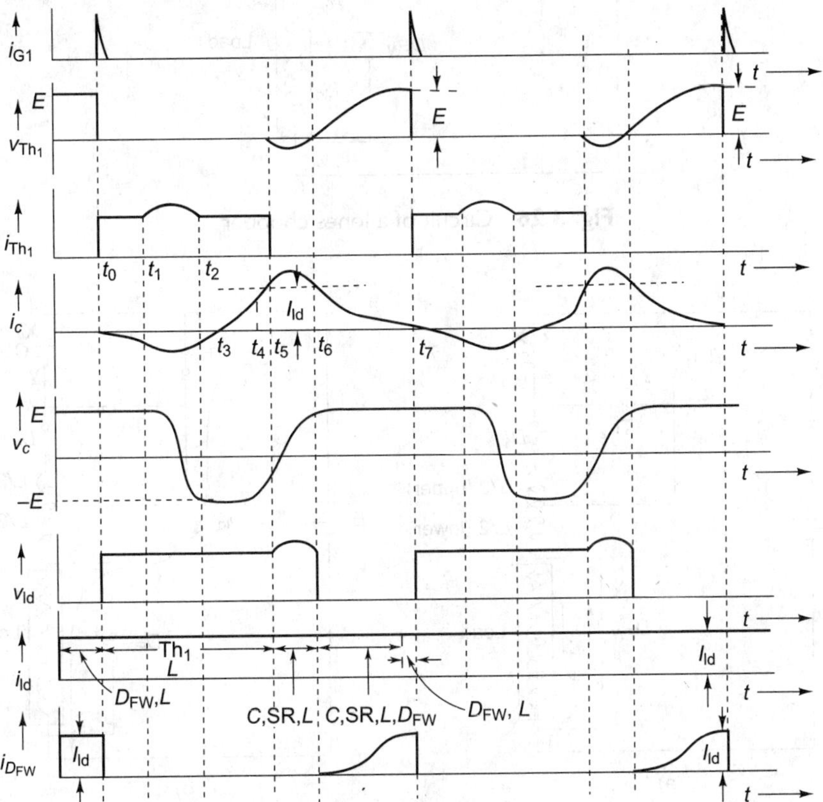

Fig. 3.25 Waveforms of a Morgan chopper

The saturable reactor together with the inductor L and the freewheeling diode D_{FW} enables a fairly constant current to be maintained in the motor load. Since the load current is terminated by the LC oscillatory current, the Morgan chopper can be categorized as a current-commutated chopper.

3.5.3 Jones Chopper

The circuit diagram, two circuit conditions, and waveforms of a Jones chopper (Ramshaw 1975) are given, respectively, in Figs 3.26, 3.27, and 3.28. Here, an autotransformer replaces the saturable core reactor of the Morgan circuit and helps the process of commutation of the main thyristor Th_1. An auxiliary thyristor Th_2

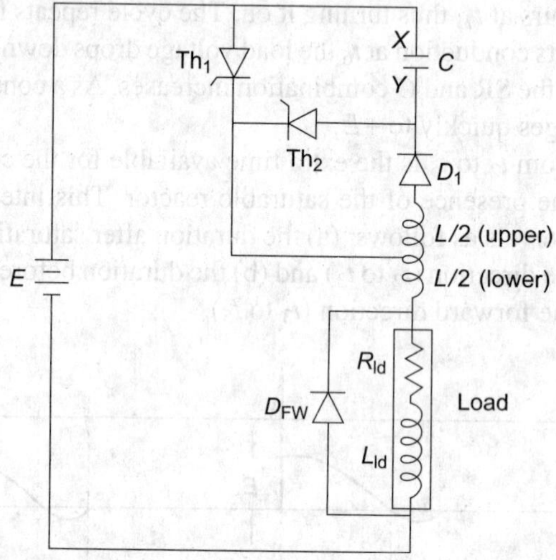

Fig. 3.26 Circuit of a Jones chopper

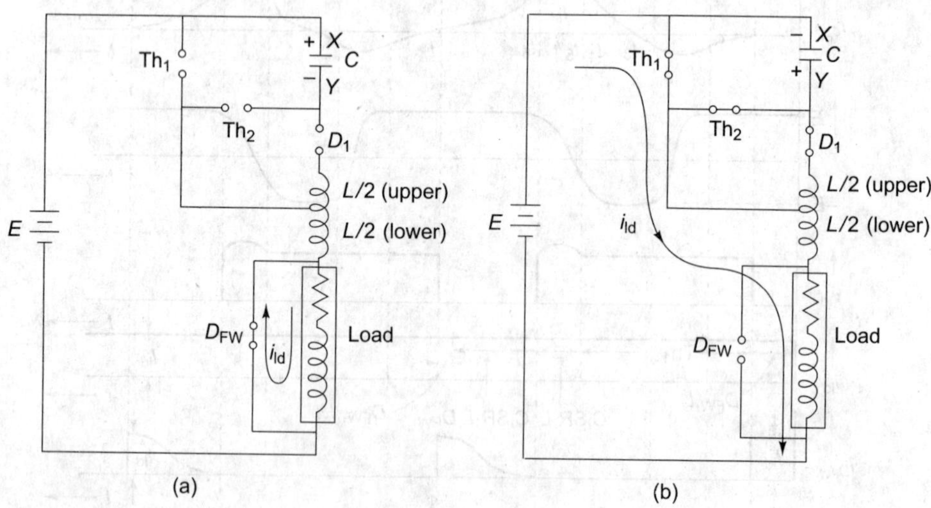

Fig. 3.27 Circuit conditions of a Jones chopper: (a) just before firing of Th_1, (b) just before turn-off of Th_1

and a diode D_1 also form part of this chopper circuit. The circuit operation starts with the assumption that prior to the turning on of the main thyristor Th_1, the capacitor C is charged to a voltage E_0, which is slightly less than the battery voltage E. At t_0, Th_1 is fired and the following two circuits come into operation simultaneously: (a) $E, Th_1, L/2$ (lower), and the load and (b) the oscillatory circuit consisting of $C, Th_1, L/2$ (upper), and D_1. At t_1 the oscillatory current i_c becomes zero and tends to reverse, but its reversal is prevented by D_1. During the period t_0 to t_1, v_c reverses and attains a value $-E'$, where $|E'| > |E|$. This is because the flow of load current in $L/2$ (lower) induces a voltage equal to double its voltage, above $L/2$ (upper), due to autotransformer action. The Y-plate of C also attains the value $-E$ and remains so till t_2 because D_1 becomes open-circuited due to reverse-biasing. Th_2 is fired at t_2 and the capacitor C discharges through it, thereby applying a reverse voltage across Th_1 and turning it off at t_2. The autotransformer action of the inductor ensures that the magnitude of E' is greater than that of the supply voltage E.

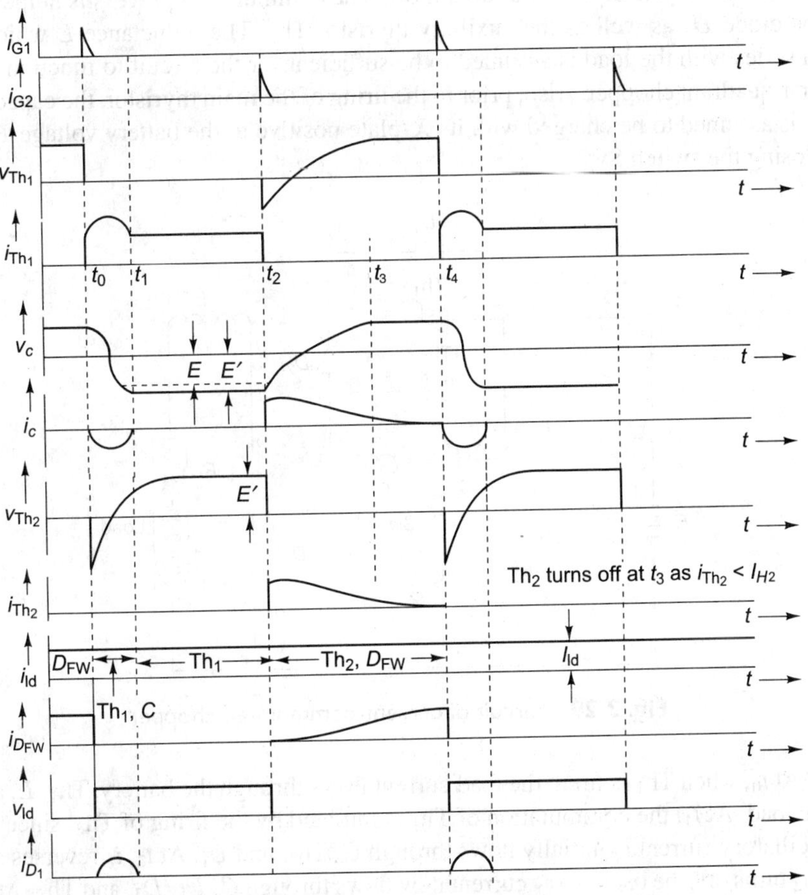

Fig. 3.28 Waveforms of a Jones chopper

From t_2 onwards, the series circuit consisting of the battery, C, Th_2, and $L/2$ (lower) comes into operation and the capacitor gets charged, with its X-plate positive. At t_3 the capacitor current goes below the holding current value (I_H) of Th_2, thus turning it off. v_c then attains a value equal to $+E_0$ and remains so till t_4. The cycle repeats from t_4 onwards. As the capacitor voltage causes Th_1 to turn-off, the Jones circuit comes under the category of voltage commutated choppers.

The Jones circuit is also called a *coupled-pulse commutation circuit* because of the coupling of the upper and lower halves of L. The combination works as an autotransformer whose effect increases with the increase of load current. The circuit turn-off time provided by this chopper increases with the increase of v_c.

3.5.4 A Special Current-commutated Chopper

Figure 3.29 shows a chopper circuit which works with the principle of forced current commutation. The waveforms for the circuit operation are given in Fig. 3.30. It is the LC oscillatory current that opposes the load current flowing in the main thyristor Th_1 and turns it off. The commutation process is helped by the diode D_1 as well as the auxiliary thyristor Th_2. The inductance L which is in series with the load is assumed to be sufficient for the circuit to function as a first-quadrant chopper. Also, prior to the firing of the main thyristor, the capacitor C is assumed to be charged with its X-plate positive to the battery voltage E by closing the switch Sw.

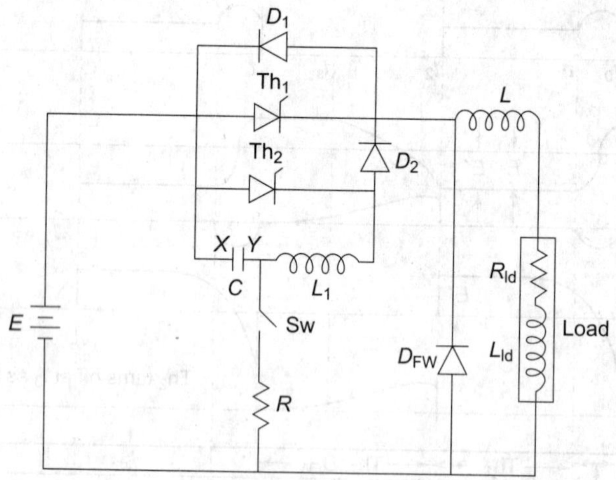

Fig. 3.29 Circuit of current-commutated chopper

At t_0, when Th_1 is fired, the load current flows through the battery, Th_1, L, and the load. At t_1, the commutation of Th_1 is initiated by the firing of Th_2, since the oscillatory current i_c initially flows through C, Th_2, and L_1. At t_2, i_c reverses and Th_2 turns off; the oscillatory current now flows through C, L_1, D_2, and Th_1. At t_3, i_c becomes equal to the load current I_{ld} and Th_1 turns off. The elements L_1 and C conduct the load current from this moment onwards. During the interval t_3 to t_4 the oscillatory current, which is in excess of I_{ld}, flows through D_1, thus applying a reverse voltage across Th_1 and commutating it. At t_4, D_1 stops conduction, and if the freewheeling diode D_{FW} is not forward-biased, the oscillatory current flows through the battery, C, L_1, D_2, L, and the load. From t_5 onwards i_c decreases to zero and the current through D_{FW} increases to I_{ld}, implying thereby that the load is short-circuited through D_{FW}. By this time the capacitor C charges to $+E$ through the charging circuit comprising the battery, C, and R and remains at this level till Th_1 is fired at t_7. The next cycle of operations is initiated from t_7 onwards. The commutation time provided for Th_1 is the interval from t_3 to t_4 during which a negative voltage equal to the forward drop of D_1 is applied across it. This duration depends upon L and C as well as the load current. For reliable commutation of Th_1, I_{cm}, which is the peak oscillatory current, must be at least twice the load current I_{ld}. As usual, the analysis of the circuit aims at designing the commutating elements L_1 and C. The behaviour of the circuit during the period of LC oscillation can be written as

$$L_1\frac{di_c}{dt} + \frac{1}{C}\int i_c dt = 0 \tag{3.74}$$

with the initial conditions

$$v_c(0) = E$$

and

$$i_c(0) = C\frac{dv_c}{dt}\bigg|_{t=0} = 0$$

Noting that $i_c = C(dv_c/dt)$, Eqn (3.74) becomes

$$L_1C\frac{d^2v_c}{dt^2} + v_c = 0 \tag{3.75}$$

or

$$\frac{d^2v_c}{dt^2} + \frac{1}{L_1C}v_c = 0 \tag{3.76}$$

Equation (3.76) is seen to be the equation of a harmonic oscillator and hence v_c can be written as

$$v_c(t) = E\cos\omega t \tag{3.77}$$

where $\omega = 1/\sqrt{L_1C}$. Thus,

$$i_c = C\frac{dv_c}{dt} = -\omega CE\sin\omega t$$

For the equation for i_c, maximum oscillatory current is

$$I_{cm} = \omega CE = \frac{CE}{\sqrt{L_1C}} = E\sqrt{\frac{C}{L_1}} \tag{3.78}$$

To determine L_1 and C, I_{cm} and I_{ld} are assumed to be related as

$$I_{cm} = kI_{ld}$$

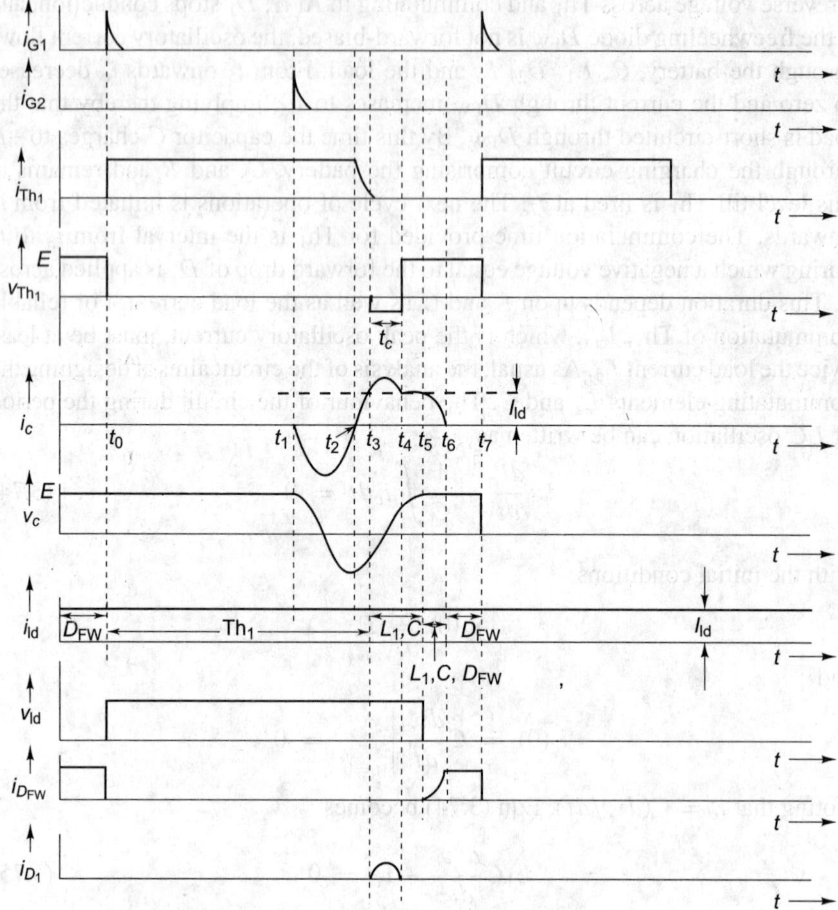

Fig. 3.30 Waveforms of the current-commutated chopper

where k is any real number. I_{ld} can be expressed as

$$I_{ld} = \frac{E}{R_{ld}}$$

where R_{ld} is the load resistance. Thus, assuming that

$$I_{cm} = E\sqrt{\frac{C}{L_1}} = kI_{ld} = \frac{kE}{R_{ld}} \tag{3.79}$$

Eqn (3.79) provides one relation for the unknowns C and L_1, namely,

$$\sqrt{\frac{C}{L_1}} = \frac{k}{R_{ld}} \tag{3.80}$$

The second relation is obtained from the fact that the duration from t_3 to t_4 should be greater than the turn-off time of Th_1. Thus with

$$\frac{I_{ld}}{I_{cm}} = \frac{1}{k}, \quad \omega(t_3 - t_2) = \sin^{-1}\left(\frac{1}{k}\right), \quad \text{and} \quad \omega(t_4 - t_2) = \pi - \sin^{-1}\left(\frac{1}{k}\right)$$

Thus,

$$\omega(t_4 - t_3) = \pi - 2\sin^{-1}\left(\frac{1}{k}\right)$$

or

$$t_4 - t_3 = \frac{1}{\omega}\left[\pi - 2\sin^{-1}\left(\frac{1}{k}\right)\right] \tag{3.81}$$

Taking $1/\omega = \sqrt{L_1 C}$, the requirement that $t_4 - t_3 \geq t_{\mathrm{OFF}}$ can be used to write the inequality

$$\sqrt{L_1 C}\left[\pi - 2\sin^{-1}\left(\frac{1}{k}\right)\right] \geq t_{\mathrm{OFF}}$$

or

$$\sqrt{L_1 C} \geq \frac{t_{\mathrm{OFF}}}{\left[\pi - 2\sin^{-1}\left(\frac{1}{k}\right)\right]} \tag{3.82}$$

Taking the inequality in Eqn (3.82) to be an equality and solving Eqns (3.80) and (3.82) gives the values of C and L_1. Thus,

$$C = \frac{k}{R_{\mathrm{ld}}} \times \frac{t_{\mathrm{OFF}}}{\left[\pi - 2\sin^{-1}\left(\frac{1}{k}\right)\right]} \tag{3.83}$$

and

$$L_1 = \frac{R_{\mathrm{ld}}}{k} \times \frac{t_{\mathrm{OFF}}}{\left[\pi - \sin^{-1}(1/k)\right]} \tag{3.84}$$

The actual value of C is now taken to be slightly greater than that computed above. L_1 can then be suitably corrected using Eqn (3.80) or Eqn (3.82).

3.5.5 Load commutated Chopper

A special type of chopper in which four thyristors are connected in a bridge configuration and a capacitor is connected in the centre of the bridge is the *load commutated chopper*, the circuit of which is shown in Fig. 3.31. Three circuit variations for this chopper are given in Fig. 3.32 and the waveforms in Fig. 3.33. As in the other cases discussed above, the inductance L and freewheeling diode D_{FW} are essential for maintaining the load current at a constant value. During the operation of the circuit, the load current is switched from one thyristor pair (Th_1, Th_3) to the other (Th_2, Th_4), thus justifying its name. It is indeed a voltage commutated chopper, since the capacitor voltage is applied in a reverse fashion across each of the thyristors in a pair to turn them off. The following sequence of operations gives a detailed description of the circuit.

It is assumed that v_c, the capacitor voltage, is the voltage of the X-plate with respect to the Y-plate and is equal to $-E$ volts prior to the firing of the first device pair, namely, Th_1 and Th_3. The firing of these thyristors initiates the reverse charging of the X-plate of the capacitor by the load current I_{ld}; v_c thus charges

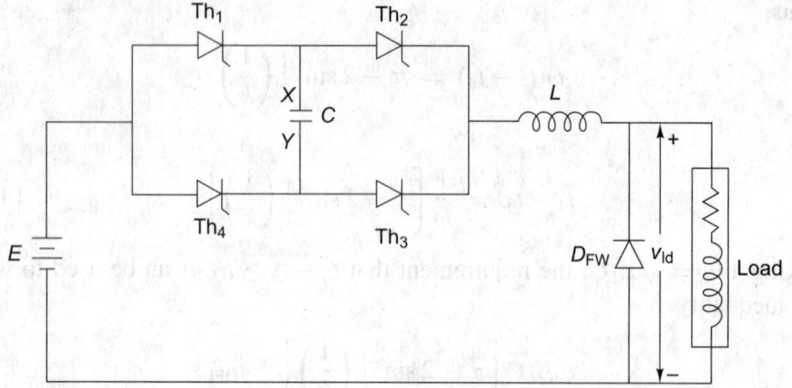

Fig. 3.31 Circuit of a load commutated chopper

from $-E$ to $+E$. The corresponding circuit condition is given in Fig. 3.32(a). When V_c attains the voltage $+E$ at t_1, the capacitor C stops conducting and D_{FW} starts conducting the entire load current. This condition is shown in Fig. 3.32(b).

At t_2 the thyristors Th_2 and Th_4 are fired and, through these devices, a reverse voltage equal to E is applied across each of the thyristors Th_3 and Th_1, turning them off. v_c now starts charging from $+E$ to $-E$ and reaches zero at t_3. Th_1 and Th_3 will successfully commutate if the circuit time $t_3 - t_2$ is greater than the t_{OFF} of the thyristor. At t_4, v_c attains a value of $-E$ and remains in this state till t_5, at which moment Th_1 and Th_3 are fired again to repeat the next cycle.

The time for which the main thyristor conducts, τ_{ON}, is determined from the equation

$$\frac{1}{C_1}\int_0^{\tau_{ON}} I_{ld}dt = 2E \tag{3.85}$$

where C_1 is the value of the capacitor that is used in the bridge. Performing the integration in Eqn (3.85) gives

$$\frac{I_{ld}\tau_{ON}}{C_1} = 2E$$

or

$$\tau_{ON} = \frac{2C_1E}{I_{ld}} \tag{3.86}$$

From Eqn (3.9) of Section 3.2.1,

$$V_{ld} = \frac{E\tau_{ON}}{\tau} = E\tau_{ON}f_{ch} \tag{3.87}$$

where $f_{ch} = 1/\tau$ is the chopper frequency. Substitution of the expression for τ_{ON} from Eqn (3.86) in Eqn (3.87) gives

$$V_{ld} = \frac{2E^2C_1}{I_{ld}}f_{ch} \tag{3.88}$$

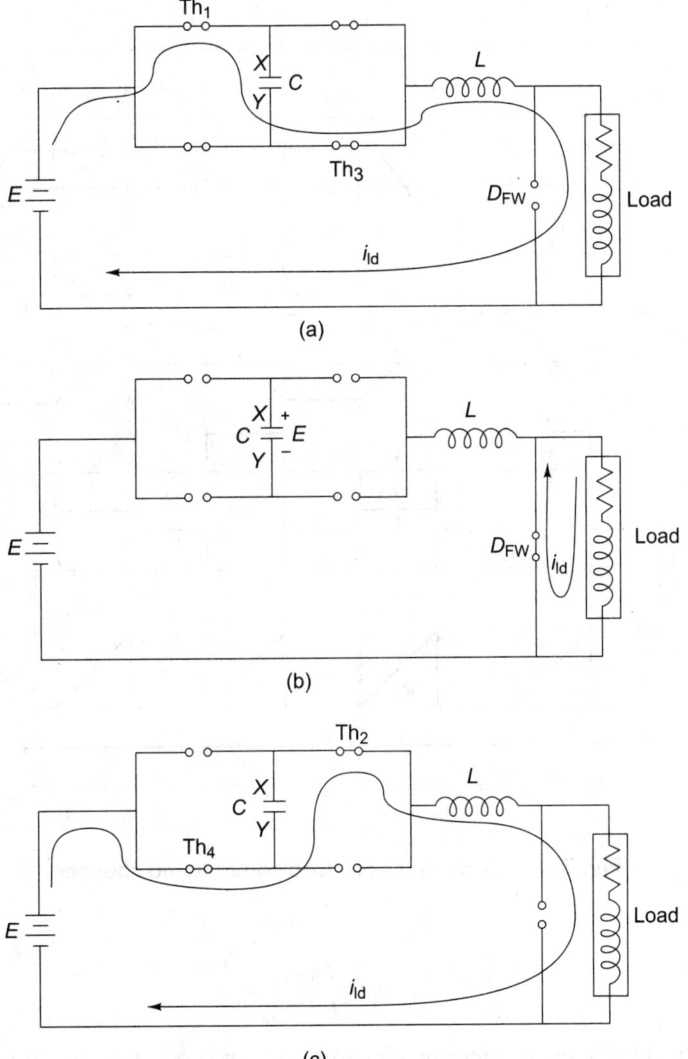

Fig. 3.32 Circuit conditions of a load commutated chopper: (a) Th_1 and Th_3 conducting load current, (b) all thyristors turned off, (c) Th_2 and Th_4 conducting load current

For determining the value of C, the maximum value of f_{ch}, namely, $f_{ch(max)}$, is taken and correspondingly I_{ld} is taken to be equal to $I_{ld(max)}$. But

$$V_{ld}|_{f_{ch(max)}} = E$$

Thus,

$$E = \frac{2E^2C_1}{I_{ld(max)}} f_{ch(max)}$$

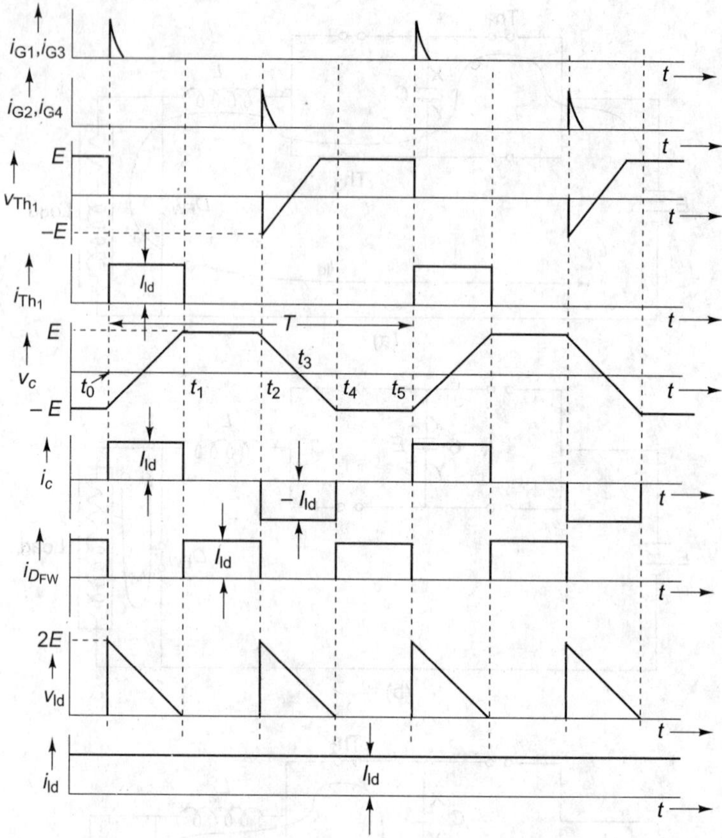

Fig. 3.33 Waveforms of a load commutated chopper

or

$$C_1 = \frac{I_{ld(max)}}{2Ef_{ch(max)}} \tag{3.89}$$

Alternatively, C can be determined as follows: Let $t_{OFF} = t_3 - t_2$. An inspection of the waveform of v_{Th1} suggests the equation

$$\frac{1}{C_2}\int_{t_2}^{t_3} I_{ld}dt = E$$

or

$$\frac{I_{ld}(t_3 - t_2)}{C_2} = E$$

This gives

$$C_2 = \frac{I_{ld}(t_3 - t_2)}{E} \tag{3.90}$$

Now C is chosen from among the capacitances given by Eqns (3.89) and (3.90) as

$$C = \max(C_1, C_2) \tag{3.91}$$

3.5.6 Voltage commutated Chopper

The simple voltage commutation circuit of Fig. 1.39(a) can be modified to that of the voltage commutated first-quadrant chopper shown in Fig. 3.34 to facilitate the voltage control of an inductive load. As before, D_{FW} conducts the current of the inductive load circuit when neither of the thyristors Th_1 and Th_2 conduct. The auxiliary thyristor Th_2 is turned off when the capacitor charges to $+E$ with the Y-plate having positive polarity.

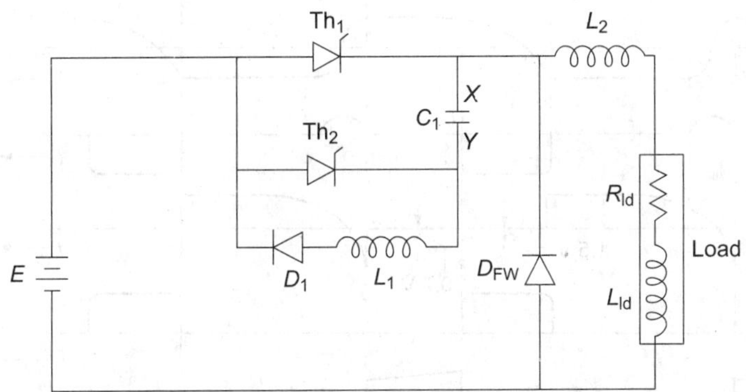

Fig. 3.34 Circuit of a voltage commutated chopper

The sequence of operations of the circuit can be understood with the help of the waveforms given in Fig. 3.35, which is drawn for the circuit after it reaches the steady state. Prior to the firing of Th_1, the capacitor is initially charged with the X-plate positive with respect to the Y-plate. When Th_1 is fired at t_0, it conducts both the oscillatory current of the L_1 and C_1 elements as well as the load current. After completing one half-cycle at t_1, the oscillatory current is reversed and stopped by D_1. At this moment the Y-plate of the capacitor attains a potential of $-E$ volts. When Th_2 is triggered at t_2, v_c is applied in a reverse manner across Th_1 and it turns off; also, the Y-plate of C starts charging from $-E$ to $+E$, with the current through C remaining constant at $I_{ld(max)} = E/R_{ld}$ because of the large value of L_2.

At t_4, v_c attains a voltage of $+E$ with the Y-plate positive with respect to the X-plate, and C gets open-circuited. The cycle repeats when Th_1 is fired again at t_5. The diode D_{FW} conducts the load current from t_4 to t_5. This current is highly inductive due to the load inductance L_{ld} as well as the external inductance L_2. The capacitor conducts the load current during the interval t_2 to t_4, which is assumed to be maintained at its maximum value of $I_{ld(max)} = E/R_{ld}$. Also, during the interval t_2 to t_3, v_c charges from $-E$ to 0. Assuming the duration $t_3 - t_2$ to be greater than the turn-off time of Th_1, the charge equation is written as

$$\frac{1}{C}\int_{t_2}^{t_3} I_{ld(max)}dt = 0 - (-E) = E \tag{3.92}$$

Performing the integration on the left-hand side yields

$$CE = I_{ld(max)}(t_3 - t_2)$$

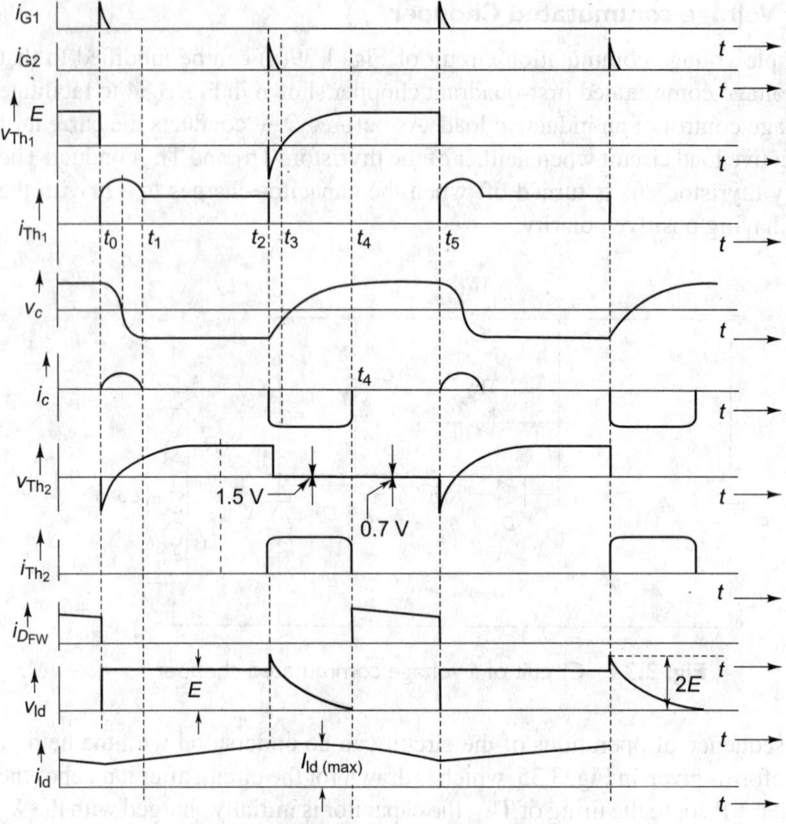

Fig. 3.35 Waveforms of a voltage commutated chopper

or

$$C = \frac{I_{ld(max)}(t_3 - t_2)}{E} \tag{3.93}$$

Imposing the condition $t_3 - t_2 \geq t_{OFF}$ and using Eqn (3.93) gives the inequality

$$C \geq \frac{I_{ld(max)}t_{OFF}}{E} \tag{3.94}$$

The value of the commutating inductor L_1 has to determined based on the following two conflicting requirements: (a) the peak of the capacitive current should be less than or equal to $I_{Th_1M} - I_{ld(max)}$, where I_{Th_1M} is the maximum current rating of Th_1 and (b) the half-cycle time of the LC oscillation should be as small a fraction of the chopper time period τ as possible. These two conditions can be quantitatively expressed as follows.

The first requirement on I_{cm} gives

$$E\sqrt{\frac{C}{L}} \leq I_{Th_1M} - I_{ld(max)}$$

or

$$L \geq C\left[\frac{E}{I_{\text{Th}_1\text{M}} - I_{\text{ld(max)}}}\right]^2 \tag{3.95}$$

To convert Eqn (3.95) into an equality, $I_{\text{Th}_1\text{M}} - I_{\text{ld(max)}}$ can be multiplied by a factor b, where $0 < b < 1$. Thus,

$$L = C\left[\frac{E}{b(I_{\text{Th}_1\text{M}} - I_{\text{ld(max)}})}\right]^2 \tag{3.96}$$

This gives the lower bound for L. The second requirement can be interpreted as follows. As per Eqn (3.9) the average output voltage of the chopper is

$$V_{\text{ld}} = E\frac{\tau_{\text{ON}}}{\tau} \tag{3.97}$$

To get as large a variation for v_{ld} as possible it should be possible to vary τ_{ON}/τ in the range

$$d < \frac{\tau_{\text{ON}}}{\tau} < 1$$

where d should be as small a fraction as possible. A limitation with the voltage commutation circuit is that a minimum interval equal to $t_1 - t_0$ is necessary for the capacitor voltage to change from $+E$ to $-E$. Thus, τ_{ON}/τ cannot become less than $t_1 - t_0/\tau$, or $d = (t_1 - t_0)/\tau$. This in turn implies that $t_1 - t_0$, which is the half-cycle time of the LC oscillation, should be as small a fraction of the chopper period as possible. Taking d to be equal to, say, 0.1, the inequality

$$0.1 \leq \frac{\pi\sqrt{LC}}{\tau}$$

is obtained. This gives the second inequality for L as

$$L \geq \frac{1}{C}\left(\frac{0.1\tau}{\pi}\right)^2$$

Taking the equality sign gives

$$L = \frac{1}{C}\left(\frac{0.1\tau}{\pi}\right)^2 \tag{3.98}$$

Equation (3.98) gives the upper bound for L. A compromise must be made between the values obtained in Eqns (3.96) and (3.98) to arrive at the proper value of L.

3.6 Applications

Dc choppers are useful in surface as well as underground traction, where the source voltage is required to be converted into different levels with high efficiency. In servos, quick response is a desired feature, and choppers fit this requirement, provided they are used for the armature control of these servos; the field can be either of the separately excited type or of the permanent magnet type.

Two-quadrant choppers find application in low-power battery-operated vehicles because they can also be operated in the regenerative braking mode, which enables increased mileage per charge of the battery. A four-quadrant chopper can be used in reversible drives, a ready application of which is in hoists used in mines and hydropower stations.

3.7 Advantages and Drawbacks of Dc Choppers

With a high chopper frequency, a fairly smooth output can be obtained with dc choppers. As against this, the ripple content in controlled rectifiers can be decreased only by increasing the number of output pulses. For example, in high-voltage dc transmission a 12-pulse output is obtained by connecting star–star and star–delta transformers in series. This shows that in multiphase rectifiers, the complexity of the equipment increases with the number of output pulses. However, there is a practical limitation on the number of output pulses. On the other hand, the frequency of dc choppers can be increased to hundreds of kilohertz and even megahertz, respectively, with power transistors and MOSFETs. The size of the inductor also decreases considerably with such high chopper frequencies. As a result, the region of discontinuity for the ω-$\overline{T}$ curves will be smaller in the case of dc choppers than for controlled rectifiers. The drop in speed with increase in torque is lower for a dc chopper than for a phase-controlled rectifier.

The main advantage of a step-up chopper is the low ripple content in the source current waveform. The two-quadrant chopper provides flexibility for the dc motor to operate both in the motoring mode as well the regenerative mode. A four-quadrant chopper has still more flexibility by way of application to reversible motors also.

The drawback of a step-down chopper is that the source current is discontinuous and hence a high harmonic content is presented to the source. The cost of the commutation circuit forms a significant part of the cost of this chopper, the additional inductor in the load circuit also contributing to the cost.

A particular problem encountered during the regenerative braking of a separately excited motor is violation of the safe value of the motor current. This can be seen from Eqn (7.55) in which δ cannot be decreased below a certain value because of commutation problems. On the other hand, the motor speed may exceed the maximum limit, thus increasing the value of E_b. Such a contingency is possible for a traction vehicle going down a steep slope.

Although regeneration can be provided by multiquadrant choppers, the source cannot normally store this energy. The only alternative left is to resort to dynamic braking, which involves power loss and decrease in efficiency.

Summary

The step-down chopper is a viable alternative to the controlled rectifier as a source for the separately excited dc motor. The step-up chopper, on the other hand, facilitates regenerative braking of a dc motor. If the resistance on the load side of a second-quadrant chopper is neglected, this chopper becomes identical to a step-up chopper. This implies that the second-quadrant chopper is appropriate for regenerative braking of a dc motor. When a changeover of conduction from the first quadrant to the second quadrant or vice versa is desired, a small interval has to be provided for the commutation of the thyristor in the outgoing chopper.

The two-quadrant chopper contains the features of both first- and second-quadrant choppers and hence provides flexibility of operation. By properly changing the sequence, first- and fourth-quadrant operation is also possible with a two-quadrant chopper. To ensure the continuity of current in chopper-based drives, the inductance L on the load side should be fairly large.

In four-quadrant operation, choppers can be made to conduct continuously or partially, depending on whether the motoring or braking operation is desired. The diodes across the thyristors help in returning inductive energy from the load to the source and also facilitate successful commutation of the thyristors.

For facilitating the two-quadrant operation of a dc series motor, the series field has to be kept outside the reversing operations.

A chopper, in all chopper circuits, consists of one of the turn-off circuits described in Sections 1.5 and 3.5. The inductor and capacitor of the chopper have to be properly designed to enable reliable turn-off. The provision of a second thyristor, called the auxiliary thyristor, helps in efficient turn-off of the main device. The voltage and current waveforms of the various circuit elements help in understanding the circuit operation as well as the determination of the inductance and capacitance values.

Worked Examples

1. Determine the commutating elements for a voltage commutated chopper with the following data: $E = 80$ V, $I_{\text{Th}_1\text{M}} = 11.0$ A, and t_{OFF} of the main thyristor is 25 μs. The chopper frequency is 600 Hz and $R_{\text{ld}} = 15\ \Omega$.

Solution

$$I_{\text{ld(max)}} = \frac{E}{R_{\text{ld}}} = \frac{80}{15} = 5.33 \text{ A}$$

From Eqn (3.94),

$$C \geq \frac{I_{\text{ld(max)}} t_{\text{OFF}}}{E}$$

Substitution of values gives

$$C \geq \frac{5.33 \times 25 \times 10^{-6}}{80} = 1.67\ \mu\text{F}$$

Therefore, C can be chosen as 1.7 μF. Here,

$$I_{\text{Th}_1\text{M}} - I_{\text{ld(max)}} = 11.00 - 5.33 = 5.67 \text{ A}$$

From Eqn (3.96), the lower bound for L with, say, $b = 0.7$ is

$$L = C\left(\frac{E}{0.7 \times 5.67}\right)^2 = 1.7 \times 10^{-6}\left(\frac{80}{0.7 \times 5.67}\right)^2 = 0.69 \text{ mH}$$

From Eqn (3.98), the upper bound for L is

$$L = \frac{1}{C}\left(\frac{0.1\tau}{\pi}\right)^2$$

$$= \frac{1}{1.7 \times 10^{-6}}\left(\frac{0.1}{600\pi}\right)^2$$

$$= 1.66 \text{ mH}$$

Therefore, L can be chosen as 1.2 mH. Hence, L and C should be taken as 1.2 mH and 1.7 μF, respectively.

2. Design the commutating elements for a current-commutated chopper with the following data: $E = 180$ V, $R_{ld} = 30\ \Omega$, t_{OFF} of $Th_1 = 40$ μs. Also, I_{cm} is 1.5 times I_{ld}.

Solution

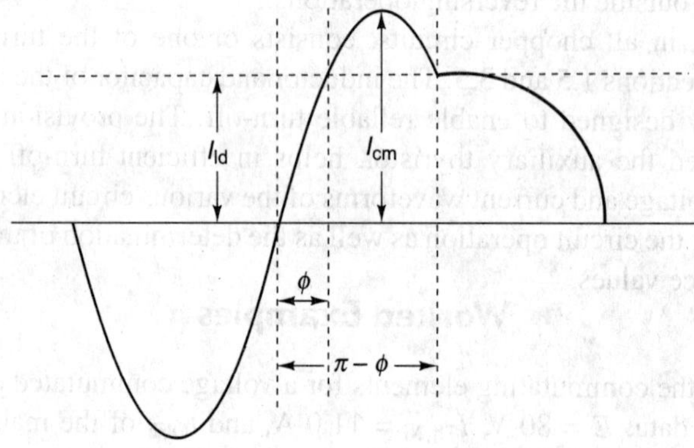

Fig. 3.36

$$\phi = \sin^{-1}\left(\frac{I_{ld}}{I_{cm}}\right) = \frac{1}{1.5} = 41.8° \text{ or } 0.73 \text{ rad}$$

$$\pi - 2\phi = \pi - 2 \times 0.73 = 1.68 \text{ rad}$$

From the relation between I_{cm} and I_{ld} given as data and the relations between L and C given in Eqns (3.80) and (3.82),

$$\sqrt{\frac{C}{L}} = \frac{1.5}{R_{ld}} = \frac{1.5}{30}$$

and

$$\sqrt{LC} = \frac{t_{OFF}}{\pi - 2\phi} = \frac{40 \times 10^{-6}}{1.68}$$

Multiplying $\sqrt{C/L}$ by $\sqrt{LC}$ gives C. Thus,

$$C = \frac{1.5}{30} \times \frac{40}{1.68} \times 10^{-6}\ \text{F} = 1.19\ \mu\text{F}$$

Also

$$L = \frac{40 \times 10^{-6}}{1.68} \times \frac{30}{1.5} = 0.48\ \text{mH}$$

The commutating elements L and C are, respectively, 0.48 H and 1.19 μF.

3. For a modified parallel resonant turn-off circuit, I_{cm} is 1.2 times I_{ld}. Other data are E = 120 V, R_{ld} = 60 Ω, t_{OFF} of Th = 45 μs. If I_{ThM} is 4.2 A, determine the value of L and C.

Solution

$$\frac{I_{ld}}{I_{cm}} = \frac{1}{1.2}$$

Hence $\sin^{-1}(I_{ld}/I_{cm})$ = 56.44° or 0.985 rad. From Eqn (3.71),

$$\sqrt{LC} = \frac{t_{OFF}}{[\pi - 2\sin^{-1}(I_{ld}/I_{cm})]} = \frac{45 \times 10^{-6}}{\pi - 2 \times 0.985}$$

This gives

$$\sqrt{LC} = 38.42 \times 10^{-6}$$

Also, from Eqn (3.69),

$$\sqrt{\frac{C}{L}} = \frac{I_{ThM}}{E} - \frac{1}{R_{ld}}$$

That is,

$$\sqrt{\frac{C}{L}} = \frac{4.2}{120} - \frac{1}{60} = 0.0183$$

Therefore,

$$C = \sqrt{\frac{C}{L}} \times \sqrt{LC} = 0.0183 \times 38.42 \times 10^{-6}\ \text{F} = 0.7\ \mu\text{F}$$

and

$$L = \sqrt{\frac{L}{C}} \times \sqrt{LC} = 2.1\ \text{mH}$$

4. The following data are provided for a modified parallel resonant turn-off circuit: E = 150 V, C = 0.35 μF, L = 2.5 mH, R_{ld} = 100 Ω. Determine the turn-off time and the maximum current capacity of the thyristor.

Solution

From Eqn (3.69),

$$\sqrt{\frac{C}{L}} = \frac{I_{ThM}}{E} - \frac{1}{R_{ld}}$$

This gives

$$I_{\text{ThM}} = E\sqrt{\frac{C}{L}} + \frac{E}{R_{\text{ld}}}$$

$$= 150\sqrt{\frac{0.35 \times 10^{-6}}{2.5 \times 10^{-3}}} + \frac{150}{100}$$

$$= 3.28 \text{ A}$$

Also, from Eqn (3.70),

$$t_{\text{OFF}} = \sqrt{LC}\left[\pi - 2\sin^{-1}\left(\frac{I_{\text{ld}}}{I_{\text{cm}}}\right)\right]$$

Here,

$$I_{\text{cm}} = E\sqrt{\frac{C}{L}} = 150\sqrt{\frac{0.35 \times 10^{-6}}{2.5 \times 10^{-3}}} = 1.775 \text{ A}$$

$$I_{\text{ld}} = \frac{E}{R_{\text{ld}}} = \frac{150}{100} = 1.5 \text{ A}$$

Therefore,

$$\pi - 2\sin^{-1}\left(\frac{I_{\text{ld}}}{I_{\text{cm}}}\right) = \pi - 2\sin^{-1}\left(\frac{1.5}{1.775}\right) = 1.128$$

$$\sqrt{LC} = \sqrt{0.35 \times 10^{-6} \times 2.5 \times 10^{-3}} = 2.96 \times 10^{-5}$$

Hence,

$$t_{\text{OFF}} = 2.96 \times 10^{-5} \times 1.128 = 33 \text{ µs}$$

5. The following data pertain to a load commutated chopper: E = 150 V, R_{ld} = 45 Ω, $I_{\text{ld(max)}}$ = 6 A. The maximum chopper frequency is 1200 Hz and the turn-off time is 36 µs. Determine the value of C.

Solution

From Eqn (3.89),

$$C_1 = \frac{I_{\text{ld(max)}}}{2Ef_{\text{ch(max)}}} = \frac{6}{2 \times 150 \times 1200} = 16.67 \text{ µF}$$

Also, from Eqn (3.90),

$$C_2 = \frac{I_{\text{ld(max)}}(t_3 - t_2)}{E} = \frac{I_{\text{ld(max)}}t_{\text{OFF}}}{E} = \frac{6 \times 36 \times 10^{-6}}{150} = 1.44 \text{ µF}$$

Finally,

$$C = \max(C_1, C_2) = 16.67 \text{ µF}$$

6. A step-down chopper supplies a separately excited dc motor with a supply voltage E = 240 V and back emf E_b = 100 V. Other data are total inductance L

= 30 mH, armature resistance R_a = 2.5 Ω, chopper frequency = 200, and duty cycle = 50%. Assuming continuous current determine I_{max}, I_{min}, and the current ripple.

Solution

The circuit and waveforms are given in Fig. 3.3. From Eqn (3.17),

$$I_{max} = \frac{E}{R_a}\left(\frac{1-e^{-\tau_{ON}/T_a}}{1-e^{-\tau/T_a}}\right) - \frac{E_b}{R_a}$$

$$\frac{\tau_{ON}}{\tau} = 0.5 \quad \text{and} \quad \tau = \frac{1}{200} = 0.005$$

Hence,

$$\tau_{ON} = \frac{0.5}{200} = 0.0025$$

$$T_a = \frac{L_a}{R_a} = \frac{30 \times 10^{-3}}{2.5} = 0.012$$

$$\frac{\tau_{ON}}{T_a} = \frac{0.0025}{0.012}$$

$$\frac{\tau}{T_a} = \frac{0.005}{0.012}$$

$$(1 - e^{-\tau_{ON}/T_a}) = 0.188, \quad (e^{-\tau_{ON}/T_a} - 1) = 0.2316$$

$$(1 - e^{-\tau/T_a}) = 0.34, \quad (e^{-\tau/T_a} - 1) = 0.5169$$

Hence,

$$I_{max} = \frac{240}{2.5} \times \frac{0.188}{0.34} - \frac{100}{2.5} = 13.0 \text{ A}$$

From Eqn (3.18),

$$I_{min} = \frac{E}{R_a}\left[\frac{e^{\tau_{ON}/T_a} - 1}{e^{\tau/T_a} - 1}\right] - \frac{E_b}{R_a}$$

$$= \frac{240}{2.5} \times \frac{0.2316}{0.5169} - \frac{100}{2.5} = 3.0 \text{ A}$$

The current ripple is

$$\Delta i_{ld} = \frac{I_{max} - I_{min}}{2} = \frac{13-3}{2} = 5 \text{ A}$$

7. A step-down chopper feeds a dc motor load. The data pertaining to this chopper-based drive is E = 210 V, R_a = 7 Ω, L (including armature inductance) = 12 mH. Chopper frequency = 1.5 kHz, duty cycle = 0.55, and E_b = 55 V. Assuming continuous conduction, determine the (a) average load current, (b) current ripple,

(c) RMS value of current through chopper, (d) RMS value of current through D_{FW}, and (e) effective input resistance seen by the source, and (f) RMS value of load current.

Solution

(a) The circuit and waveforms are given in Fig. 3.3. The average load current is given by Eqn (3.21) as

$$I_{ld} = \frac{E}{R_a}\frac{\tau_{ON}}{\tau} - \frac{E_b}{R_a}$$

Substitution of values gives

$$I_{ld} = \frac{210}{7} \times 0.55 - \frac{55}{7}$$
$$= 8.64 \text{ A}$$

(b) The current ripple is given by Eqn (3.19) as

$$\Delta i_{ld} = \frac{E}{2R_a}\left\{\frac{1 + e^{\tau/T_a} - e^{\tau_{ON}/T_a} - e^{\tau_{OFF}/T_a}}{e^{\tau/T_a} - 1}\right\}$$

Here,

$$T_a = \frac{L_a}{R} = \frac{12 \times 10^{-3}}{7}$$

$$\tau = \frac{1}{1.5 \times 10^3} = \frac{1 \times 10^{-3}}{1.5}$$

$$\frac{\tau}{T_a} = \frac{7}{1.5 \times 12} = 0.389$$

$$\frac{\tau_{ON}}{T_a} = \frac{0.55 \times 10^{-3}}{1.5} \times \frac{7}{12} \times 10^{-3} = 0.214$$

$$\frac{\tau_{OFF}}{T_a} = \frac{0.45 \times 10^{-3}}{1.5} \times \frac{7}{12} \times 10^{-3} = 0.175$$

Substitution of values gives

$$\Delta i_{ld} = \frac{210}{2 \times 7} \times \frac{1 + 1.475 - 1.239 - 1.191}{1.475 - 1} = 1.42 \text{ A}$$

(c) It is assumed that the load current increases linearly from I_{min} to I_{max} during $(0, \tau_{ON})$. Thus the instantaneous current i_{ld} can be expressed as

$$i_{ld} = I_{min} + \frac{I_{max} - I_{min}}{\tau_{ON}}t, \quad 0 \le t \le \tau_{ON}$$

The RMS value of the current through the chopper can now be found as

$$I_{ch(RMS)} = \sqrt{\frac{1}{\tau}\int_0^{\tau_{ON}} (i_{ld})^2 dt}$$

Here,

$$I_{ch(RMS)} = \left\{\sqrt{\frac{\tau_{ON}}{\tau}\left[I_{min}^2 + I_{min}(I_{max} - I_{min}) + \frac{(I_{max} - I_{min})^2}{3}\right]}\right\}$$

where

$$I_{min} = \frac{E}{R_a}\left[\frac{e^{\tau_{ON}/T_a} - 1}{e^{\tau/T_a} - 1}\right] - \frac{E_b}{R_a}$$

$$= \frac{210}{7}\left[\frac{1.239 - 1}{1.475 - 1}\right] - \frac{55}{7} = 15.095 - 7.857 = 7.24 \text{ A}$$

$$I_{max} = \frac{E}{R_a}\left[\frac{1 - e^{-\tau_{ON}/T_a}}{1 - e^{-\tau/T_a}}\right] - E_b/R_a$$

$$= \frac{210}{7}\frac{(1 - 0.807)}{(1 - 0.678)} - \frac{55}{7} = 17.981 - 7.857 = 10.12 \text{ A}$$

Thus,

$$I_{ch(RMS)} = \sqrt{\frac{\tau_{ON}}{\tau}\left[7.24^2 + 7.24(10.12 - 7.24) + \frac{(10.12 - 7.24)^2}{3}\right]}$$

$$= \sqrt{0.55 \times 76.03}$$

$$= 6.46 \text{ A}$$

(d) The RMS value of current through D_{FW} can be found as

$$I_{D_{FW}(RMS)} = \sqrt{\frac{1}{\tau}\int_0^{\tau_{OFF}} (i_{ld})^2 dt'} \quad (t' = t - \tau_{ON})$$

$$= \sqrt{\frac{\tau_{OFF}}{\tau}\left[(10.12)^2 + \frac{(2.88)^2}{3} - 10.12 \times 2.88\right]}$$

(where $I_{max} - I_{min} = 10.12 - 7.24 = 2.88$ A)

$$= \sqrt{0.45 \times 76.02} = 5.85 \text{ A}$$

(e) Average source current $= (\tau_{ON}/\tau)\times$ average load current

$$= 0.55 \times \frac{(7.24 + 10.12)}{2} = 0.55 \times 8.68 = 4.77 \text{ A}$$

Hence, the effective input resistance seen by the source = 210/4.77 = 44 Ω.

(f) RMS value of load current $= \sqrt{\frac{1}{\tau}\left[\int_0^{\tau_{ON}} \{i_{ld}(t)\}^2 dt + \int_0^{\tau_{OFF}} \{i_{ld}(t')\}^2 dt'\right]} = 8.72$ A

with $t' = t - \tau_{ON}$. This is seen to be nearly equal to the average load current, namely, 8.68 A.

8. A step-down chopper has the following data: $R_{\rm ld} = 0.40\,\Omega$, $E = 420$ V, $E_b = 25$ V. The average load current is 175 A and the chopper frequency is 280 Hz. Assuming the load current to be continuous, and linearly rising to the maximum and then linearly falling, calculate the inductance L which would limit the maximum ripple in the load current to 12% of the average load current.

Solution

The circuit is given in Fig. 3.3(a). The expression for the current ripple in Eqn (3.19) can be written with substitutions $\tau_{\rm ON} = \delta\tau$ and $\tau_{\rm OFF} = (1-\delta)\tau$, where δ is in the range $0 < \delta < 1$. Thus,

$$\Delta i_{\rm ld} = \frac{E}{2R_a}\left[\frac{1 + e^{\tau/T_a} - e^{\delta\tau/T_a} - e^{(1-\delta)\tau/T_a}}{e^{\tau/T_a} - 1}\right]$$

with $\delta = \tau_{\rm ON}/\tau$. Differentiating the ripple current with respect to δ and equating this to zero gives the value of δ for maximum ripple:

$$\frac{d(\Delta i_{\rm ld})}{d\delta} = \left(-\frac{\tau}{T_a}e^{\delta\tau/T_a}\right) + \frac{\tau}{T_a}e^{(1-\delta)\tau/T_a} = 0$$

This yields

$$e^{\delta\tau/T_a} = e^{(1-\delta)\tau/T_a}$$

or

$$\delta = 1 - \delta$$

This gives $\delta = 0.5$. Substituting this value of δ in the expression for $\Delta i_{\rm ld}$ gives

$$\Delta i_{\rm ld} = \frac{E}{2R_a}\left[\frac{1 + e^{\tau/T_a} - 2e^{0.5\tau/T_a}}{e^{\tau/T_a} - 1}\right]$$

$$= E/2R_a\left[\frac{(e^{0.5\tau/T_a} - 1)^2}{(e^{0.5\tau/T_a} - 1)(e^{0.5\tau/T_a} + 1)}\right]$$

$$= \frac{E}{2R_a}\left[\frac{e^{0.5\tau/T_a} - 1}{e^{0.5\tau/T_a} + 1}\right]$$

$$= \frac{E}{2R_a}\tanh\frac{R_a}{4fL}$$

where $f = 1/\tau$ is the chopper frequency. If $4fL >> R_a$, then $\tanh(R_a/4fL) \approx R_a/4fL$. Thus,

$$\Delta I_{\rm ld(max)} = \frac{E}{2R_a}\frac{R_a}{4fL} = \frac{E}{8fL}$$

The condition $\Delta I_{\rm ld(max)} = 12\% I_{\rm ld}$ gives

$$\frac{E}{8fL} = 0.12 \times 175$$

Hence,

$$L = \frac{E}{8f \times 0.12 \times 175} = \frac{420}{8 \times 280 \times 0.12 \times 175} = 0.0089 \text{ H}$$
$$= 8.9 \text{ mH}$$

9. A 240-V, separately excited dc motor has an armature resistance of 2.2 Ω and an inductance of 4 mH. It is operated at constant load torque. The initial speed is 600 rpm and the armature current is 28 A. Its speed is now controlled by a step-down chopper with a frequency of 1 kHz, the input voltage remaining at 240 V. (a) If the speed is reduced to 300 rpm, determine the duty cycle of the chopper. (b) Compute the current ripple with this duty cycle.

Solution

(a) The equation for the motor is

$$V = E_b + I_a R_a$$

Substitution of values gives

$$240 = E_b + 28 \times 2.2$$

Thus,

$$E_b = 240 - 28 \times 2.2 = 178.4 \text{ V}$$

Other quantities on the right-hand side of the expression for E_b remaining constant, it can be expressed as

$$E_b = kN$$

or

$$178.4 = k600$$

$$k = \frac{178.4}{600} = 0.297$$

The new speed is 300 rpm. Hence,

$$E_{b(\text{new})} = 0.297 \times 300 = 89.1 \text{ V}$$

The new applied voltage with the same load torque, that is, the same armature current, is

$$V_{\text{new}} = E_b + I_a R_a = 89.1 + 28 \times 2.2 = 150.7 \text{ V}$$

The duty cycle of the chopper (τ_{ON}/τ) can be determined from the relation

$$V_{\text{new}} = \frac{\tau_{\text{ON}}}{\tau} \times \text{input voltage}$$

This gives

$$\frac{\tau_{\text{ON}}}{\tau} = \frac{V_{\text{new}}}{\text{input voltage}} = \frac{150.7}{240} = 0.628$$

(b) $T_a = L/R_a = 4 \times 10^{-3}/2.2$; $\tau = 1/1000 = 10^{-3}$. Therefore,

$$\frac{\tau}{T_a} = \frac{2.2}{4} = 0.55$$

$$\frac{\tau_{\rm ON}}{T_a} = \frac{\tau_{\rm ON}}{\tau}\frac{\tau}{T_a} = 0.628 \times 0.55 = 0.345$$

$$\frac{\tau_{\rm OFF}}{T_a} = \frac{\tau}{T_a} - \frac{\tau_{\rm ON}}{T_a} = 0.55 - 0.345 = 0.205$$

The current ripple is given as

$$\begin{aligned}\Delta i_{\rm ld} &= \frac{E}{2R_a}\frac{(1 + e^{\tau/T_a} - e^{\tau_{\rm ON}/T_a} - e^{\tau_{\rm OFF}/T_a})}{e^{\tau/T_a} - 1}\\ &= \frac{150.7}{2 \times 2.2} \times \frac{1 + 1.733 - 1.412 - 1.227}{1.733 - 1}\\ &= 4.39 \text{ A}\end{aligned}$$

10. A 220-V, 80-A, separately excited dc motor operating at 800 rpm has an armature resistance of 0.18 Ω. The motor speed is controlled by a chopper operating at 1000 Hz. If the motor is regenerating, (a) determine the motor speed at full load current with a duty ratio of 0.7, this being the minimum permissible ratio (b) Repeat the calculation with a duty ratio of 0.1.

Solution

(a) When the machine is working as a motor, E_b is obtained from the equation

$$E_b = E - I_a R_a = 220 - 80 \times 0.18 = 220 - 14.4 = 205.6 \text{ V}$$

From the equation

$$E_b = kN$$

$$k = \frac{205.6}{800} = 0.257$$

When it is regenerating, the step-up configuration of Fig. 3.7(c) holds good. Thus,

$$E_b = E(1 - \delta) + I_a R_a = 220(1 - 0.7) + 80 \times 0.18 = 66 + 14.4 = 80.4 \text{ V}$$

$$N = \frac{E_b}{k} = \frac{80.4}{0.257} = 313 \text{ rpm}$$

(b) The speed for $\delta = 0.1$ is obtained as follows:

$$E_b = 220(1 - 0.10) + 80 \times 0.18 = 198 + 14.4 = 212.4 \text{ V}$$

Therefore the speed is

$$N = \frac{212.4}{0.257} = 827 \text{ rpm}$$

11. A 250-V, 105-A, separately excited dc motor operating at 600 rpm has an armature resistance of 0.18 Ω. Its speed is controlled by a two-quadrant chopper with a chopping frequency of 550 Hz. Compute (a) the speed for motor operation with a duty ratio of 0.5 at 7/8 times the rated torque and (b) the motor speed if it regenerates at $\delta = 0.7$ with rated current.

Solution

(a) The initial back emf is to be determined from the equation

$$E_b = E - I_a R_a = 250 - 105 \times 0.18 = 231.1 \text{ V}$$

Hence the back emf constant $k = 231.1/600 = 0.385$. A fraction 7/8 of the rated current $I'_a = 7/8 \times 105 = 91.875$ A. The new E_b is obtained as

$$E'_b = E\delta - I'_a R_a$$

where $\delta = 0.5$ and $I'_a = 91.875$. Its numerical value is

$$E'_b = 250 \times 0.5 - 91.875 \times 0.18 = 108.46\text{V}$$

The new speed is

$$N = \frac{E'_b}{k} = \frac{108.46}{0.385} = 282 \text{ rpm}$$

(b) When it is regenerating, Eqn (7.54) is to be used. Thus,

$$I_a = \frac{E_b - E(1-\delta)}{R_a}$$

or

$$E_b = E(1-\delta) + I_a R_a$$

Substituting values gives

$$E_b = 250(1 - 0.7) + 105 \times 0.18 = 93.9 \text{ V}$$

From this, the speed $N = 93.9/0.385 = 244$ rpm

12. A 300-V, 100-A, separately excited dc motor operating at 600 rpm has an armature resistance and inductance of 0.25 Ω and 16 mH, respectively. It is controlled by a four-quadrant chopper with a chopper frequency of 1 kHz. (a) If the motor is to operate in the second quadrant at 4/5 times the rated current, at 450 rpm, calculate the duty ratio. (b) Compute the duty ratio if the motor is working in the third quadrant at 500 rpm and at 60% of the rated torque.

Solution

(a) $E_b - I_a R_a = 300 - 100 \times 0.25 = 275$ V. Back emf constant $k = E_b/N = 275/600 = 0.458$. Operation in the second quadrant implies that the motor works as a generator. Hence the motor terminal voltage V_a is written as

$$V_a = E(1-\delta)$$

New current

$$I_a = \frac{4}{5} \times 100 = 80 \text{ A}$$

Hence,

$$E_b = V_a + I_a R_a$$

$$kN = E(1-\delta) + I_a R_a$$

Substitution of values gives

$$0.458 \times 450 = 300(1 - \delta) + 80 \times 0.25$$

This yields $\delta = 0.38$.
(b) In the third quadrant, the machine works in the motoring mode but with reverse voltage and reverse current. The voltage equation relevant in this case is

$$E_b = V_a - I_a R_a$$

where

$$E_b = kN = 0.458 \times 500$$

$$V_a = E\delta = 300\delta$$

and

$$I_a = 0.6 \times 100 = 60 \text{ A}$$

By substituting numerical values, the equation becomes

$$0.458 \times 500 = 300 \times \delta - 60 \times 0.25$$

This gives $\delta = 0.813$.

Exercises

Multiple Choice Questions

1. In a step-down chopper the load current ripple can be reduced by _______.
   (a) increasing L (b) increasing f_{ch}
   (c) both the above (d) none of the above
2. The regions of discontinuous conduction in the ω-T curves of a motor controlled by a first-quadrant chopper can be decreased by _______.
   (a) increasing f_{ch} (b) decreasing f_{ch}
   (c) increasing R_{ld} (d) decreasing R_{ld}
3. For the two-quadrant type-A chopper under operating conditions (ii) and (iii) (Section 3.3.2.1), the devices _______ are connected to the source for one part of the period τ and the devices _______ are connected for the other part.
   (a) D_1 and Ch_1, D_2 and Ch_2 (b) D_2 and Ch_1, D_1 and Ch_2
   (c) Ch_2, D_2 and Ch_1 (d) D_1 and D_2, Ch_1 and Ch_2
4. Considering both the methods of the two-quadrant type-B chopper, _______.
   (a) v_{ld} can only be positive and i_{ld} can be positive or negative
   (b) v_{ld} can only be negative and i_{ld} can be positive or negative
   (c) v_{ld} can be positive or negative and i_{ld} can only be negative
   (d) v_{ld} can be positive or negative and i_{ld} can only be positive
5. For a two-quadrant chopper type-A, regenerative braking is _______.
   (a) possible at low speeds
   (b) possible at high speeds

(c) possible at both low and high speeds
(d) not possible at all

6. In a two-quadrant chopper type-A, the average voltage v_{ld} depends on ______.
   (a) τ_{ON} and τ
   (b) τ_{OFF} and τ
   (c) only τ_{ON}
   (d) only τ_{OFF}

7. The two-quadrant ______ chopper cannot work in the third and fourth quadrants; the two-quadrant ______ chopper can work in the first and fourth quadrants.
   (a) type-A, type-A
   (b) type-B, type-A
   (c) type-A, type-B
   (d) type-B, type-B

8. For ______ values of torque, the ω-$\overline{T}$ characteristics of a dc series motor become parallel to the torque axis.
   (a) low
   (b) medium
   (c) high
   (d) very low

9. Apart from the two halves of the centre-tapped inductor, the various other elements that support the load current in the Jones chopper, over a time period are ______.
   (a) Th_1, C, and Th_2
   (b) C, Th_2, and D_1
   (c) Th_1, C, Th_2, and D_1
   (d) Th_1, Th_2, C, and D_{FW}

10. In the Jones chopper the voltage E' is greater than E because ______, where v_L is the voltage of the whole inductor.
    (a) $v_L/2$ (upper) adds to v_c
    (b) the diode drop across D_1 adds to $v_L/2$ (upper)
    (c) $v_L/2$ (lower) induces a voltage at $v_L/2$ (upper)
    (d) the load voltage adds to v_c

11. In the Jones chopper the current $i_{D_{\text{FW}}}$ rises slowly because the current through ______ falls slowly.
    (a) Th_2 and C
    (b) Th_2 and D_1
    (c) Th_1 and Th_2
    (d) Th_1 and D_1

12. In the Jones chopper the diode D_1 conducts only when ______ exceeds the load current.
    (a) the current through $L/2$ (upper) and Th_1
    (b) the current through Th_1 and $L/2$ (lower)
    (c) the current through $L/2$ (upper) and Th_2
    (d) the current through C and Th_2

13. In a load commutated chopper, during the conduction of Th_1 and Th_3, the load voltage v_{ld} falls from ______.
    (a) E to 0
    (b) $2E$ to $-E$
    (c) E to $-E$
    (d) $2E$ to 0

14. In a load commutated chopper, the freewheeling diode D_{FW} conducts only when ______ do not conduct.
    (a) Th_1, Th_2, Th_3, Th_4, and C
    (b) Th_1, Th_3, and C
    (c) Th_2 and Th_4
    (d) Th_2, Th_4, and C

15. A load commutated chopper is so called because ______.
    (a) the load current freewheels through D_{FW}
    (b) the load current alternately flows through the two thyristor pairs
    (c) the load current through one thyristor pair commutates the other thyristor pair
    (d) the voltage of the capacitor which is charged by the load current commutates the thyristor pairs.
16. As per the waveforms, the various components that support the load current in a current-commutated chopper over a period of time are ______.
    (a) Th_1, Th_2, L, and C (b) Th_1, L, C, and D_{FW}
    (c) Th_1, Th_2, C, and D_{FW} (d) Th_1, Th_2, L, and D_{FW}
17. In a current-commutated chopper, the current $i_{D_{FW}}$ increases slowly because the current through ______ falls slowly.
    (a) Th_1 and C (b) Th_2 and L_1
    (c) Th_2 and D_1 (d) L_1 and C
18. In a current-commutated chopper, the circuit turn-off time is facilitated by the reverse voltage applied by the conducting element, namely, ______.
    (a) Th_2 (b) D_2
    (c) C (d) D_1
19. In a current-commutated chopper, the diode D_1 conducts the current only when ______ are not conducting.
    (a) Th_1 and D_{FW} (b) Th_2, D_2, and D_{FW}
    (c) Th_1, D_2, and D_{FW} (d) Th_1, Th_2, and D_{FW}
20. In the Morgan chopper, the current $i_{D_{FW}}$ slowly increases because the current through ______ slowly decreases.
    (a) Th_1 and C (b) SR and Th_1
    (c) C and SR (d) Th_1, SR, and C
21. In the Morgan chopper, the circuit turn-off time occurs when the current flows through the components ______.
    (a) D_{FW}, L, and the load (b) SR, L, and the load
    (c) D_{FW}, L, C, and the load (d) C, SR, L, and the load
22. In a voltage commutated chopper, the current through C is ______ for some time after the thyristor turns on and is ______ for some time after it turns off.
    (a) nearly flat, sinusoidal (b) sinusoidal, sinusoidal
    (c) nearly flat, nearly flat (d) sinusoidal, nearly flat
23. In a voltage commutated chopper, current flows through C when the thyristor Th_1 turns off at t_2 and it stops flowing at t_4 because ______.
    (a) i_c falls down naturally
    (b) the load current i_{ld} cancels i_c
    (c) the voltage v_c of the capacitor attains the value E
    (d) the load voltage exactly opposes v_c
24. In a voltage commutated chopper, the voltage across Th_2 just after it stops conducting i_c will be nearly ______ V.

(a) 1.5
(b) 0.7
(c) 1.1
(d) 0.3

25. In a modified parallel resonant turn-off circuit, the current through D_{FW} rises slowly because the current through ________ falls slowly.

(a) C and L
(b) C and the load
(c) L and the load
(d) L and D_1

26. In a modified parallel resonant turn-off circuit, the components that conduct the load current over a period τ are ________.

(a) Th_1, L, C, and D_{FW}
(b) Th_1, L, C, and D_1
(c) Th_1, C, and D_{FW}
(d) Th_1, L, C, D, and and D_{FW}

27. In a modified parallel resonant turn-off circuit, the circuit turn-off time is due to the current flow through ________.

(a) L and C
(b) L, C, and D_1
(c) D_{FW}
(d) the load

28. In a modified parallel resonant turn-off circuit, the diode D_1 conducts only when ________.

(a) the load current exceeds the peak oscillatory current
(b) the peak oscillatory current exceeds half the load current
(c) the load current exceeds half the peak oscillatory current
(d) the oscillatory current exceeds the load current

29. In a modified parallel resonant turn-off circuit with data $E = 120$ V, $R_{ld} = 40\ \Omega$, $C = 2\ \mu$F, and $L = 10$ mH, the maximum current through the thyristor (I_{ThM}) will be nearly ________ A.

(a) 10.3
(b) 4.7
(c) 8.5
(d) 7.8

30. In a modified parallel resonant turn-off circuit, if $t_{OFF} = 20\ \mu$s and the relationship $I_{ld} = 0.4 I_{cm}$ holds good, the quantity $\sqrt{LC}$ works out to nearly ________.

(a) 5.43×10^{-3}
(b) 10.84×10^{-6}
(c) 8.62×10^{-6}
(d) 16.31×10^{-6}

True or False

Indicate whether the following statements are *true* or *false*. Briefly justify your choice.

1. The role played by the saturable reactor in a Morgan chopper is that commutation of the thyristor is made more reliable.
2. The Jones chopper is more reliable than the Morgan chopper because of the inductor present in it.
3. In a modified resonant turn-off circuit, the diode D_1 which is connected across the thyristor carries three-fourths of the load current for a short time and zero current for the rest of the time.
4. The waveform of $i_{D_{FW}}$ in a Jones chopper will be as shown in Fig. 3.37; the diode D_{FW} will conduct when both Th_1 and Th_2 are not conducting.

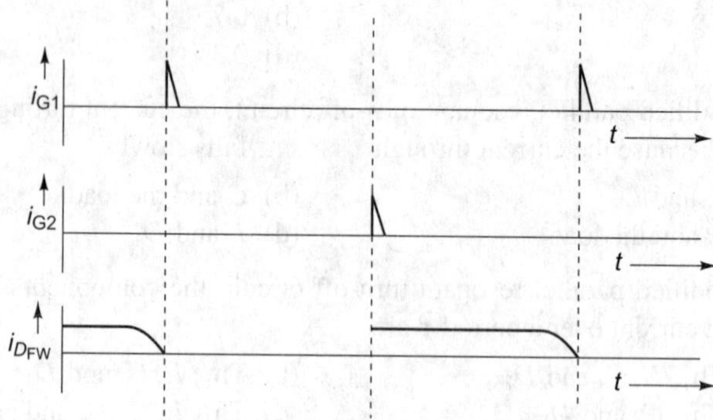

Fig. 3.37

5. In a current-commutated chopper, the magnitudes of L and C determine the circuit turn-off time of the thyristor.
6. In the current-commutated chopper of Fig. 3.29, the purpose of the diode D_1 is to carry the load current when no other device is conducting.
7. In the voltage commutated chopper of Fig. 3.34, the function of the diode D_{FW} is to conduct load current when the auxiliary thyristor Th_2 is conducting and the main thyristor Th_1 is in the blocked condition.
8. For better operation of the Morgan chopper, it is desirable that the freewheeling diode dissipate the stored energy before the thyristor is turned on.
9. The advantage of the Morgan chopper as compared to the Jones chopper is that the voltage to which the capacitor will get charged is higher than that in the Jones chopper.
10. In the current-commutated chopper circuit, the purpose of the diode D_1 across the load thyristor is twofold: (a) to carry the excess oscillatory current and (b) to keep the main thyristor reverse-biased.
11. In the load commutated chopper circuit, the purpose of the freewheeling diode is to carry the load current when the current carried by the capacitor goes to zero.
12. In an auxiliary resonant turn-off circuit, care is necessary to see that the capacitor does not discharge through the main thyristor Th_1 and the diode D_1.
13. Considering two-quadrant choppers, the type-A chopper operates in the first and fourth quadrants whereas the type-B chopper operates in the first and second quadrants.
14. A chopper-controlled, separately excited dc motor gives better performance than a rectifier-controlled motor.
15. Power transistors are superior to MOSFETs and thyristors as switching devices in choppers.
16. The source current in a step-down chopper has a higher ripple content than that in a step-up chopper.

17. In the circuit of a four-quadrant chopper, if one semiconductor switch is kept closed continuously and the other three switches are controlled, it is possible to obtain two-quadrant operation.
18. Four-quadrant operation is also possible with field control of a motor but the response in this case is not as fast as that obtained with armature control.
19. The performance of the modified parallel resonant turn-off circuit is inferior to the conventional parallel resonant turn-off circuit.
20. In the current-commutated chopper, the switch Sw_1 has to be kept open throughout the operation of the circuit, except when the load current becomes excessive.

Short Answer Questions

1. Derive an expression for the commutating capacitance used in a load commutated chopper.
2. Explain the role played by the saturable reactor in a Morgan chopper.
3. With the help of the load current waveform, list out the various devices that conduct this current in a Morgan chopper.
4. Discuss the relative merits and demerits of Morgan and Jones choppers.
5. Explain why the waveform of i_c in a load commutated chopper is a straight line parallel to the ωt-axis and not exponentially decreasing.
6. Explain the role played by the diode D_1 across the thyristor in a modified parallel resonant turn-off circuit. Give the waveforms for i_c, v_c, and i_{D_1} for this circuit.
7. Give the classification of choppers based on quadrants; also give their circuit diagrams along with a brief description of operation.
8. Give arguments in support of the following statement: 'With a certain range of τ_{ON}, the operating point of the two-quadrant chopper (type-B) falls in the fourth quadrant.'
9. Explain how the modified parallel resonant turn-off circuit gives better performance than the normal parallel resonant turn-off circuit.
10. Explain the factors on which the circuit turn-off time depends in the case of a modified parallel resonant turn-off circuit.
11. With the help of a waveform indicate the various devices that conduct the load current during a time period τ for a current-commutated chopper.
12. Show that the relation $v_{\mathrm{ld}} = E\tau_{\mathrm{ON}}/\tau$ holds good, even for an $R + L$ type of load, in a case of a step-down chopper with continuous conduction.
13. How can the regions of discontinuous conduction be reduced for the speed–torque curves of a separately excited dc motor fed from a first-quadrant chopper?
14. Explain how the two-quadrant type-B chopper can be made to operate in the first and second quadrants.
15. Derive an expression for ω_m, as a function of the torque $\overline{T}_d$, for a dc series motor.

16. Explain the necessity of a large inductance L in series with the motor armature for a separately excited dc motor controlled by a four-quadrant chopper.
17. Explain how regenerative braking takes place for a separately excited dc motor controlled by a four-quadrant chopper.
18. Derive expressions for the inductor L and the capacitor C used in a modified parallel resonant turn-off circuit.
19. Give suitable diagrams for the various devices which conduct in a current-commutated chopper when the oscillatory current slowly reduces from a value equal to the load current.
20. For a current-commutated chopper, state what components help in the turn-off of the main thyristor?
21. Explain the part played by the saturable reactor in the Morgan chopper circuit. Support your answer with waveforms.
22. What are the devices that conduct the load current in a Morgan chopper circuit during a time period τ? When does the diode D_{FW} start conduction?
23. Compare the v_c and i_c waveforms obtained in the modified parallel resonant turn-off and Morgan chopper circuits.
24. Compare the v_c and i_c waveforms occurring in the Morgan and Jones chopper circuits.
25. Explain how a four-quadrant chopper will be useful in motor control.
26. Explain how the step-up chopper will be useful for regenerative braking of motors.
27. Derive the expressions for L_1 and C for a current-commutated chopper with the load current related to the maximum oscillatory current as $I_{ld} = 0.4I_{cm}$.

Problems

1. Determine the values of L_1 and C_1 for the voltage commutated chopper of Fig. 3.34. The data are $E = 100$ V, $I_{ThM} = 7.5$ A, t_{OFF} of Th = 30 μs, chopper frequency = 500 Hz. [Hint: Take $b = 0.5$; assume $I_{Th_1M} = 2I_{ld(max)}$.]

   Ans. $C = 1.7$ μF, $L = 2.3$ mH
2. The following data pertain to a current-commutated chopper: $E = 200$ V, $R_{ld} = 25\ \Omega$, t_{OFF} of $\text{Th}_1 = 50$ μs, $I_{cm} = 1.6I_{ld}$. Determine the values of L and C.

   Ans. $C = 1.8$ μF, $L = 0.44$ mH
3. The following data pertain to a modified parallel resonant turn-off circuit: $E = 150$ V, $t_{OFF} = 50$ μs, $I_{ThM} = 5.0$ A, $I_{ld}/I_{cm} = \sqrt{3}/2$. Determine the values of L and C.

   Ans. $C = 0.78$ μF, $L = 2.35$ mH
4. The following data pertain to a modified parallel resonant turn-off circuit: $E = 120$ V, $t_{OFF} = 45$ μs. If $I_{ThM} = 4.2$ A, and L and C are 1.98 mH and 0.66 μF, respectively, find (a) the ratio I_{ld}/I_{cm} and (b) R_{ld}.

   Ans. (a) 0.812, (b) 60 Ω
5. The following data are provided for a modified parallel resonant turn-off circuit: $E = 140$ V, $C = 1.2$ μF, $L = 0.45$ mH, $R_{ld} = 50\ \Omega$. Determine the t_{OFF} and I_{ThM} of the thyristor.

   Ans. $I_{ThM} = 10.03$ A, $t_{OFF} = 54.5$ μs

6. A step-down chopper supplies a separately excited dc motor. The data for the circuit are E = 220 V, back emf E_b of motor = 120 V, total inductance (including armature inductance) L = 20 mH, armature resistance R_a = 2 Ω, chopper frequency = 100 Hz, and duty cycle = 60%. Assuming continuous current, determine I_{max}, I_{min}, and the current ripple.

   Ans. I_{max} = 18.52 A, I_{min} = −7.37 A, Δi_{ld} = 12.93 A

7. A second-quadrant chopper supplies a dc motor. The relevant data are E = 240 V, E_b = 130 V, L = 25 mH, R_a = 2.5 Ω, chopper frequency = 120 Hz, and τ_{ON}/τ = 0.4. Derive the two expressions for the load current i_{ld} during intervals τ_{ON} and τ_{OFF} and find the average load current I_{ld}.

   Ans. (i) $i_{ld} = 46.48e^{-100t} + 52(1 - e^{-100t}),\ 0 \le t \le \tau_{ON}$;
   (ii) $i_{ld} = 48.06^{-100(t-\tau_{ON})} + 44[1 - e^{-100(t-\tau_{ON})}],\ \tau_{ON} \le t \le \tau$; 47.46 A

8. For a current-commutated chopper, the values for the commutating elements L and C are given as 15 mH and 0.5 μs, respectively. If the ratio I_{ld}/I_{cm} is $\sqrt{3}/2$, determine (a) the circuit turn-off time and (b) the load resistance that can be supplied by the circuit.

   Ans. τ_{OFF} = 90.7 μs, R_{ld} = 200 Ω

9. The following data pertain to a load commutated chopper: E = 160 V, R_{ld} = 40 Ω, $I_{ld(max)}$ = 5 A, maximum chopper frequency = 1000 Hz, and turn-off time = 30 μs. Determine a suitable value for the capacitor C.

   Ans. C = 16 μF

10. A voltage commutated chopper has the following data: E = 150 V, $I_{ld(max)}$ = 15 A, L = 10 mH, and C = 0.5 μF. Compute the turn-off time and the maximum chopper frequency.

    Ans. τ_{OFF} = 4 μs, $f_{ch(max)}$ = 450 Hz

11. A step-down chopper feeds a dc motor load. The data are E = 200 V, R_a = 6 Ω, L (including armature inductance) = 10 mH, chopper frequency = 1.2 kHz, duty cycle δ = 0.6, and E_b = 60 V. Assuming continuous conduction, determine the (a) average load current, (b) current ripple, (c) RMS value of current through chopper, (d) RMS value of current through D_{FW}, (e) effective input resistance seen by the source, and (f) RMS value of load current.

    Ans. (a) 10.0 A, (b) 2.03 A, (c) 7.78 A, (d) 6.37 A, (e) 33.3 A, (f) nearly 10.0 A

12. (a) The step-down chopper of Fig. 3.3(a) has the following data: R_a = 0.35 Ω, E = 480 V, and E_b = 20 V. I_{ld} = 190 A and chopper frequency = 300 Hz. Assuming the load current waveform to be linear, calculate the inductance L which would limit the maximum load current ripple to 14% of the average load current.
    (b) Derive the expression for the duty cycle ratio that gives the maximum load current ripple used in part (a). [Hint: See Worked Example 8.]

    Ans. L = 15 mH

13. A 220-V, separately excited dc motor has an armature resistance of 1.8 Ω and an inductance of 5 mH. When driven at 500 rpm, the armature takes 25 A. The motor is now controlled by a step-down chopper with a frequency of 500 Hz, the input voltage remaining at 220 V. (a) Determine the duty cycle if the speed is to be reduced from the original value of 500 rpm to 300 rpm, the load torque remaining unchanged. (b) Calculate the current ripple with this duty cycle.

    Ans. (a) 68%, (b) 9.45 A

14. A 250-V, 100-A, separately excited dc motor operating at 600 rpm has an armature resistance and inductance of 0.15 Ω and 15 mH, respectively. The motor speed is controlled by a chopper operating at 500 Hz. (a) If the motor is regenerating, find the motor speed with a duty ratio of 0.6 with rated current. (b) Compute the speed for a duty ratio of 0.1 with rated current, this being the minimum permissible ratio.

Ans. (a) 294 rpm, (b) 613 rpm

15. A 240-V, 95-A, separately excited dc motor operating at 600 rpm has an armature resistance of 0.2 Ω and an armature inductance of 12 mH. Its speed is controlled by a two-quadrant chopper (type A) with a chopper frequency of 450 Hz. (a) Compute the speed of the motor for a duty ratio of 0.6 at two-thirds the rated torque. (b) Determine the motor speed if it regenerates at $\delta = 0.65$ and rated current.

Ans. (a) 357 rpm, (b) 280 rpm

16. A 250-V, 90-A, separately excited dc motor operating at 500 rpm has an armature resistance of 0.18 Ω and an inductance of 14 mH. It is controlled by a four-quadrant chopper with a chopper frequency of 450 Hz as in method 1 of Section 3.3.3. (a) If the motor is to be operated in the second quadrant at three-fourths the rated torque at a speed of 300 rpm, calculate the duty ratio. (b) Compute the duty ratio if the motor is working in the third quadrant at a speed of 400 rpm with 40% of the rated torque.

Ans. (a) 0.49, (b) 0.77

17. A load commutated chopper has the following data: $E = 120$ V, $C = 5$ μF, $I_{\text{ld(max)}} = 40$ A. Compute the chopper frequency and turn-off time of the thyristor.

Ans. $f_{\text{ch}} = 33$ kHz, $t_{\text{OFF}} = 15$ μs

18. A modified parallel resonant turn-off circuit has the following data: $E = 200$ V, $R_{\text{ld}} = 600$ Ω, $C = 0.8$ μF, and $L = 0.2$ mH. Compute the values of I_{ThM} and t_{OFF}.

Ans. $I_{\text{ThM}} = 12.98$ A, $t_{\text{OFF}} = 39$ μs

CHAPTER 4

Ac Line Voltage Control

4.1 Introduction

Ac line voltage control involves controlling the angle of conduction in both the half-waves of the ac voltage waveform. It is economical compared to other power conversion equipment for the variation of the average power through a load. Some of the applications of such ac voltage control are as follows.

(a) Control of heating in industrial heating loads.

(b) Control of illumination in a lighting control system.

(c) Control of the RMS magnitude of the voltage in a transformer tap changing system.

(d) Speed control of industrial ac drives.

The impetus for the use of ac voltage control has arisen from the fact that ac is employed for subtransmission and distribution all over the world. The following methods have evolved for the control of the RMS magnitude of ac voltage in the past.

(i) Obtaining a continuously variable ratio of voltages using a toroid transformer or an induction regulator [Fig. 4.1(a)].

(ii) Fully switching on or off the ac voltage in an $[Nx\%, N(100-x)\%]$ manner, where N is the number of integral cycles of the ac voltage; this method is called on–off control [Fig. 4.1(b)].

(iii) Employing time-ratio control, which involves the variation of the ratio of the on and off periods [Fig. 4.1(c)].

(iv) Periodically switching the supply once per half-cycle [Figs 4.1(d)–(f)].

(v) Periodically switching the supply several times per half-cycle [Figs 4.1(d), (e), and (g)].

This chapter introduces single- and three-phase ac controllers to which fixed ac supply is given and from which a variable ac output is obtained by the phase control of the semiconductor switches.

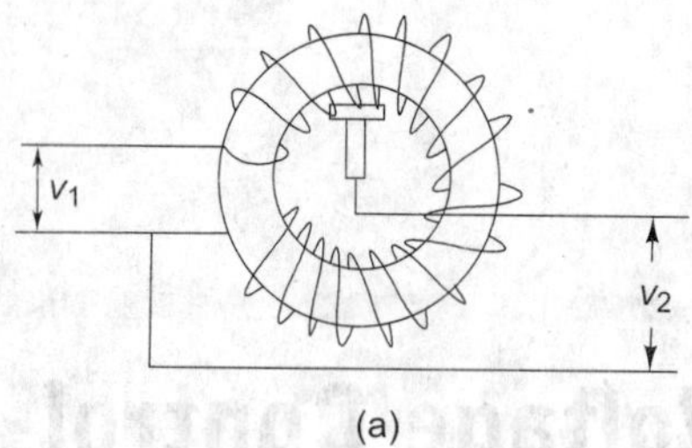

(a)

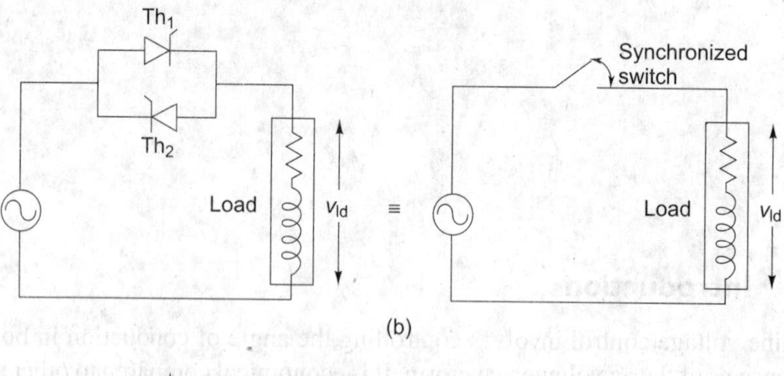

(b)

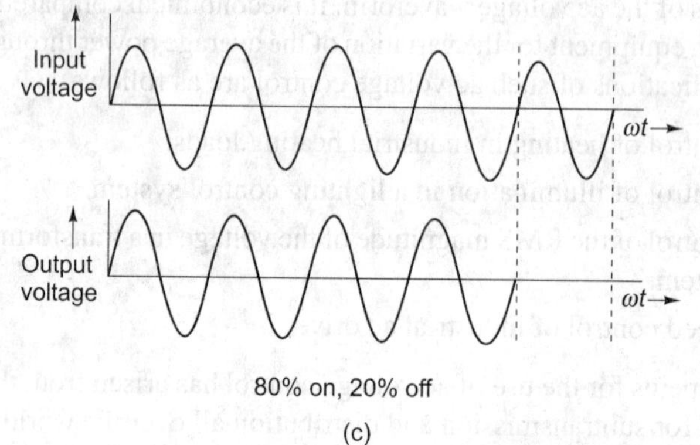

(c)

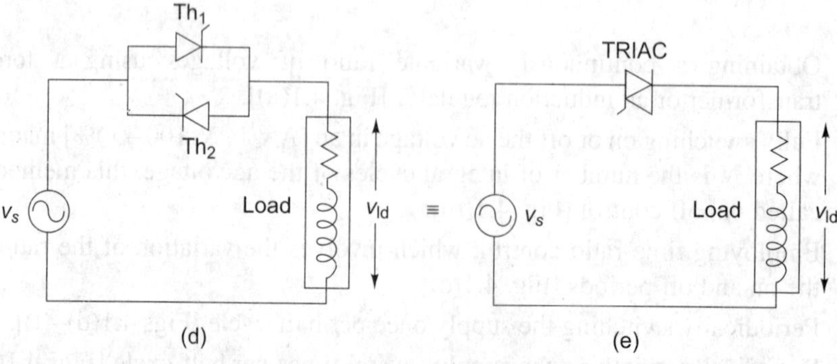

(d) (e)

Figs 4.1(a)–(e)

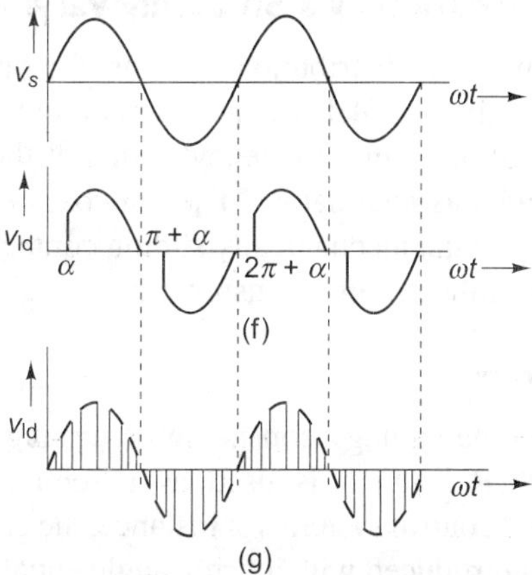

Fig. 4.1 Sources of variable voltage: (a) toroid transformer; (b) switching in an on–off manner over integral cycles; (c) time-ratio control; (d)–(f) switching once in a half-cycle; (d), (e), (g) switching several times in a half-cycle

After going through this chapter the reader should

- understand the different techniques employed in the industry for obtaining a variable ac output from a fixed ac input,
- compute the RMS and half-cycle average values of load voltage and current for single-phase ac choppers supplying resistive and inductive loads,
- know the different classifications of three-phase choppers and the methods of deriving the expressions for the output voltage for different ranges of α,
- gain acquaintance with the concept of transformer tap changing using ac controllers, and
- understand the problems that arise consequent to the use of ac controllers as a source of supply for ac motors, and know the derivation of expressions for harmonic torques.

After a discussion of the methods for voltage variation in Section 4.2, single-phase and three-phase ac voltage controllers are dealt with in Sections 4.3 and 4.4, respectively. For the three-phase controllers, different connections will be examined with regard to their performance. The sequence control of ac regulators used in transformer tap changing will be discussed in Section 4.5. Finally, the application of ac controllers, respectively, to sequence controllers and drives will be investigated in Sections 4.5 and 4.6.

4.2 Methods for Ac Voltage Variation

Some methods which are commonly used for ac voltage control are examined below in respect of their efficacy, which depends upon the number of switchings needed and the harmonic content introduced into the output.

4.2.1 Ac Voltage Variation by a Smoothly Varying Transformer

In this method, used in power distribution systems, the magnitude of the output can be varied continuously [Fig. 4.1(a)]. This output, obtained at the secondary, will be sinusoidal and of the same frequency as that of the source, provided the primary voltage is of a sinusoidal nature. The drawbacks of this method are (a) wear and tear occurring in the mechanically sliding contacts used for varying the turn ratio and (b) losses inherent in a transformer.

4.2.2 On–off Control

On–off control is a simple strategy consisting of turning on or off of a silicon controlled rectifier (SCR), which is used as a semiconductor switch. In a closed-loop temperature control system, for instance, the gate signal to the SCR is, removed altogether or introduced with a firing angle equal to zero, depending on whether the temperature of the heater is above or below the set value [Fig. 4.1(b)]. The disadvantage of such a control is the constant fluctuation of the actual temperature about the desired value. However, because of its simplicity, this circuit can be rigged up with minimum cost.

4.2.3 Time-ratio Control

In this technique [Fig. 4.1(c)], which is a modification of on–off control, the ratio of the on and off periods is varied depending on the actual magnitude of the output, say the temperature in a closed-loop temperature control system. Because of the precision obtained by this technique, it is more accurate than on–off control. Thyristors may be employed for switching the supply at zero crossing points.

4.2.4 Switching the Supply Once in a Half-cycle

This method is known as ac chopping, and the associated circuit is called an ac chopper. It is widely used in single-phase circuits because only two thyristors are needed for a single-phase ac chopper. The devices are connected in an antiparallel manner and the load is connected on the output side of the chopper circuit. A single TRIAC may as well be used [Figs 4.1(d)–(f)]. It gives the same performance as two thyristors, with a simpler gate triggering circuit. The circuit and waveforms of a typical single-phase ac chopper are, respectively, given in Figs 4.2(a) and (b). Employing a zero voltage detector circuit facilitates the synchronization of the firing signals with the ac supply waveform. A major disadvantage of the single-phase ac chopper is that its output is non-sinusoidal.

The principle described above can be extended to construct three-phase ac choppers, which are widely used in the industry. A certain amount of flexibility exists in the configuration of three-phase choppers, as will be elaborated later. Like single-phase choppers, three-phase circuits also give a non-sinusoidal, though periodic, output. The magnitude of the various harmonics can be determined using Fourier analysis. A feature of the waveforms is that the fundamental frequency remains the same as that of the input supply. Though the harmonics are beneficial for a few applications such as industrial heating, their presence is harmful for a majority of industrial loads. The advantages of ac choppers are: the absence

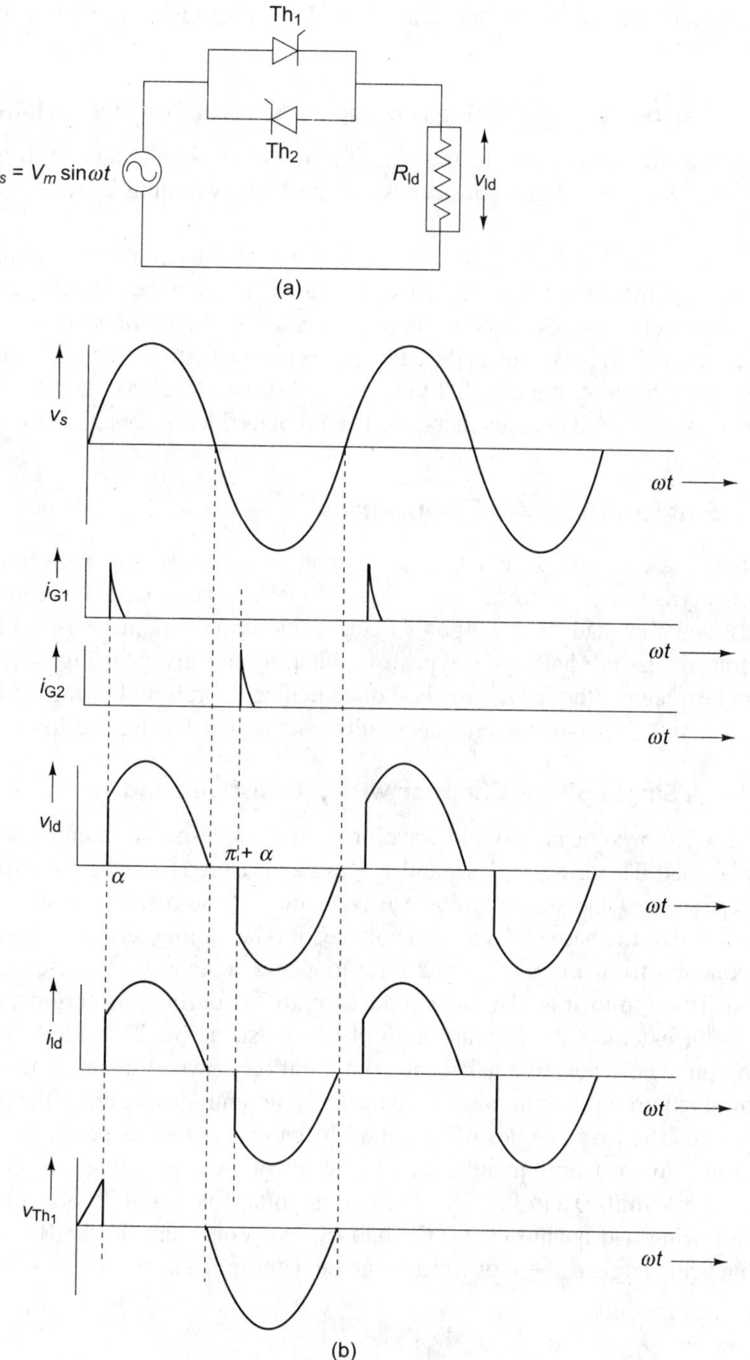

Fig. 4.2 Single-phase ac chopper: (a) circuit, (b) waveforms

of moving parts, reliability, and efficiency. They also facilitate high switching speeds.

4.2.5 Periodically Switching the Supply Several Times in a Half-cycle

With this method, which is illustrated in Fig. 4.1(g), the RMS value of the output can be varied; it is rarely used because of the high switching losses associated with it.

A modification of the single-phase chopper is the asymmetrical chopper, in which one of the thyristors is replaced by a diode. This can be extended to three-phase choppers also. Such controllers are called *half-controlled ac choppers*. These are found to have almost the same performance features as fully controlled choppers. Their advantage is that the circuit cost is reduced to the extent of the number of controlled devices that have been replaced by diodes.

4.3 Single-phase Ac Choppers

A single-phase ac chopper consists of either an antiparallel connection consisting of two thyristors or, equivalently, a single TRIAC interposed between the ac supply and the load. The output voltage waveforms for once-in-a-half-cycle and four-times-in-a-half-cycle types of switching are given in Fig. 4.1(g). As already explained, the former method of switching is preferred because of lesser complexity of the associated firing circuitry and reduced switching losses.

4.3.1 A Single-phase Chopper with a Resistive Load

Figure 4.2 shows the circuit and waveforms of a single-phase ac chopper with a resistive load. The firing angles α and $\pi + \alpha$ are measured from the positive-going zero crossing of the supply voltage waveform. For the resistive load, the load current will be in phase with the load voltage, thus becoming zero at an angle of π. Th_1 conducts from α to π, the duration π to $\pi + \alpha$ being a no-device-conducting period. Th_2 is fired at $\pi + \alpha$ and conducts up to 2π; during this period a reverse bias to the extent of the forward drop of Th_2 exists across Th_1. Th_1 is kept off during the whole negative half-cycle, the duration of which is much more than the time required to commutate it; likewise Th_2 commutates during the positive half-cycle. The firing angles of Th_1 and Th_2 can be varied, respectively, from 0 to π and π to 2π. Consequently, the RMS value of the load voltage can be varied between the limits 0 and $V_m/\sqrt{2}$. The output voltage is usually characterized by the following two quantities: (a) the half-cycle average and (b) the RMS value. With a firing angle α these quantities can be determined as follows.

Half-cycle average

$$V_{\text{ld}} = \frac{1}{\pi}\int_{\alpha}^{\pi} V_m \sin\omega t \; d(\omega t)$$

$$= \frac{V_m}{\pi}(1 + \cos\alpha) \tag{4.1}$$

The RMS value

$$V_{\text{ld(RMS)}} = \left[\frac{1}{\pi}\int_{\alpha}^{\pi} V_m^2 \sin^2 \omega t \, d(\omega t)\right]^{1/2}$$

$$= \frac{V_m}{\sqrt{2}}\left[1 - \frac{\alpha}{\pi} + \frac{\sin 2\alpha}{2\pi}\right]^{1/2} \tag{4.2}$$

As the output voltage is periodical but non-sinusoidal, it consists of the fundamental as well as higher harmonic components. However, the dc component will be zero as a consequence of the symmetry of the positive and negative waveforms. Employing Fourier analysis, the load voltage can be written as

$$v_{\text{ld}} = \sum_{n=1}^{\infty}[a_n \cos(n\omega t) + b_n \sin(n\omega t)] \tag{4.3}$$

where

$$a_n = \frac{1}{\pi}\int_0^{2\pi} v_{\text{ld}} \cos(n\omega t) d(\omega t) \tag{4.4}$$

and

$$b_n = \frac{1}{\pi}\int_0^{2\pi} v_{\text{ld}} \sin(n\omega t) d(\omega t) \tag{4.5}$$

where $v_{\text{ld}} = V_m \sin \omega t$ for the intervals (α, π) and $(\alpha + \pi, 2\pi)$. In the case of an induction motor load, the fundamental component $(v_{\text{ld},1})$ is of interest. The coefficients for this case are obtained as

$$a_1 = \frac{V_m}{\pi}\frac{(\cos 2\alpha - 1)}{2} \tag{4.6}$$

and

$$b_1 = \frac{V_m}{\pi}\left[(\pi - \alpha) + \frac{\sin 2\alpha}{2}\right] \tag{4.7}$$

$v_{\text{ld},1}$ can now be written as

$$v_{\text{ld},1}(t) = a_1 \cos \omega t + b_1 \sin \omega t$$

$$= V_{\text{ld},1m} \sin(\omega t + \phi_{\text{ld},1}) \tag{4.8}$$

where

$$V_{\text{ld},1m} = (a_1^2 + b_1^2)^{1/2} \tag{4.9}$$

and

$$\phi_{\text{ld},1} = \tan^{-1}\left(\frac{a_1}{b_1}\right) \tag{4.10}$$

Also,

$$V_{\text{ld},1(\text{RMS})} = \frac{1}{\sqrt{2}}(a_1^2 + b_1^2)^{1/2} \tag{4.11}$$

where a_1 and b_1 are given in Eqns (4.6) and (4.7), respectively.

The second and higher order harmonics are computed with the help of Eqns (4.3)–(4.5) by substituting $n = 2,3,\ldots$. These higher order harmonics are harmful for an induction motor load and cause torque pulsations, cogging, and other undesirable phenomena. However, in the case of heater type of loads, all the components including the fundamental contribute to the heating effect. The RMS value of the load voltage, $V_{\mathrm{ld(RMS)}}$, given by Eqn (4.2) is of interest in such heating applications. The determination of the average and RMS values of current for purely resistive loads is a straightforward computation. Thus,

$$I_{\mathrm{ld}} = \frac{V_m}{\pi R}(1 + \cos\alpha) \tag{4.12}$$

$$I_{\mathrm{ld(RMS)}} = \frac{V_m}{\sqrt{2}R}\left[1 - \frac{\alpha}{\pi} + \frac{\sin 2\alpha}{2\pi}\right]^{1/2} \tag{4.13}$$

The amplitude of the fundamental component of the output current is

$$I_{\mathrm{ld},1m} = \frac{1}{R}(a_1^2 + b_1^2)^{1/2} \tag{4.14}$$

The output power is characterized by the power factor, which can be defined as

$$\mathrm{PF} = \frac{\text{real power at the load}}{\text{apparent input power}}$$

$$= \frac{V_{\mathrm{ld(RMS)}}^2/R}{V_m/\sqrt{2}(V_{\mathrm{ld(RMS)}}/R)}$$

On simplification, this becomes

$$\mathrm{PF} = \frac{\sqrt{2}V_{\mathrm{ld(RMS)}}}{V_m} \tag{4.15}$$

4.3.1.1 A half-controlled $s\phi$ ac chopper A half-controlled single-phase ac chopper feeding a resistive load and its waveforms are, respectively, given in Figs 4.3(a) and (b). Because of asymmetry in the voltage and current waveforms, the dc component of the Fourier series for these waveforms will be non-zero. Moreover, the even harmonics will also have non-zero magnitudes. To this extent, it differs from the two-thyristor circuit of Fig. 4.2(a). If it is supplied by a finite impedance source such as a single-phase transformer, the dc component may saturate the transformer core. This effect is, however, negligible when the core is large. The use of the single-phase ac chopper is therefore limited to ideal sources whose impedance is zero. The even harmonics will also limit the application of the circuit to a few kilowatts.

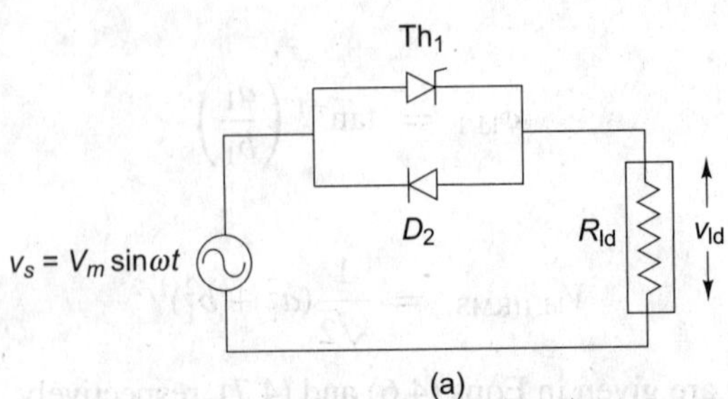

Fig. 4.3(a)

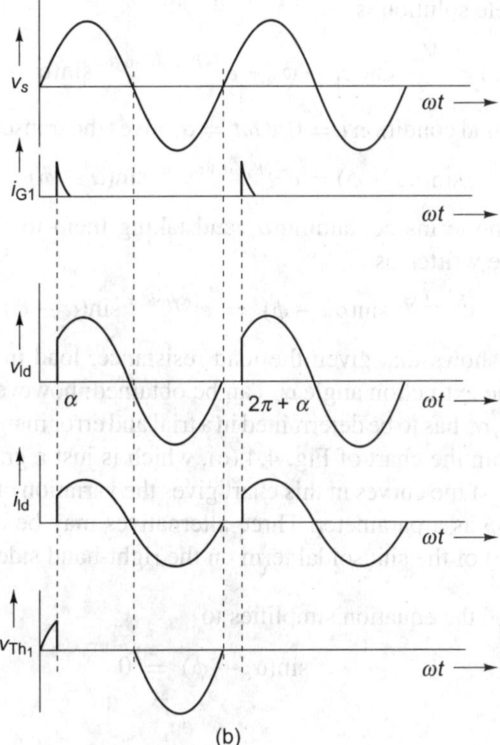

Fig. 4.3 Single-phase half-controlled ac chopper with resistive load: (a) circuit diagram, (b) waveforms

4.3.2 A Single-phase Chopper with an Inductive Load

The circuit of a single-phase ac chopper feeding an inductive or $R + L$ load is given in Fig. 4.4(a). The gating signal as well as load voltage and current waveforms are shown in Fig. 4.4(b). The circuit equation for the chopper of Fig. 4.4(a) is

$$L\frac{di}{dt} + Ri = V_m \sin \omega t \quad \text{for } \alpha < \omega t < \alpha_e \text{ and } \pi + \alpha < \omega t < \pi + \alpha_e \tag{4.16}$$

The solution of Eqn (4.16) is of the form

$$i(t) = \frac{V_m}{Z}\sin(\omega t - \phi) + A\exp\left(-\frac{R}{L}t\right) \tag{4.17}$$

where ϕ is the power factor angle. Imposing the condition $i = 0$ at $\omega t = \alpha$ (or $t = \alpha/\omega$) gives

$$0 = \frac{V_m}{Z}\sin(\alpha - \phi) + A\exp\left(\frac{-R}{L}\right)\frac{\alpha}{\omega} \tag{4.18}$$

Equation (4.18) can be solved for A as

$$A = -\frac{V_m}{Z}\sin(\alpha - \phi)e^{(R\alpha/\omega L)} \tag{4.19}$$

Thus, the complete solution is

$$i(t) = \frac{V_m}{Z}\sin(\omega t - \phi) - e^{-(R/L)(t-\alpha/\omega)}\sin(\alpha - \phi) \tag{4.20}$$

Substituting the final condition $i = 0$ at $\omega t = \alpha_e$ gives the transcendental equation

$$\sin(\alpha_e - \phi) = e^{-(R/\omega L)(\alpha_e - \alpha)}\sin(\alpha - \phi) \tag{4.21}$$

By sorting out the terms containing α_e and taking them to the left-hand side, Eqn (4.21) can be written as

$$e^{(R/\omega L)\alpha_e}\sin(\alpha_e - \phi) = e^{(R/\omega L)\alpha}\sin(\alpha - \phi) \tag{4.22}$$

Equation (4.22) shows that given the load resistance, load inductance, and the firing angle α, the extinction angle α_e can be obtained; however, as the equation is an implicit one, α_e has to be determined in a trial and error manner. Alternatively, it can be read from the chart of Fig. 4.4.(c), which is just a graphical version of Eqn (4.22). Each of the curves in this chart gives the variation of α_e, given α, with the phase angle ϕ as a parameter. Three alternatives may be considered for the argument $(\alpha - \phi)$ of the sinusoidal term on the right-hand side of Eqn (4.22).

(i) With $\alpha = \phi$, the equation simplifies to

$$\sin(\alpha_e - \phi) = 0$$

or

$$\alpha_e - \phi = \pi \tag{4.23}$$

Equation (4.23) shows that α_e equals $\pi + \phi$, or the duration of conduction $\alpha_e - \alpha$ becomes π radians. Th_1 conducts from α to $\pi + \alpha$ and Th_2 from $\pi + \alpha$ to $2\pi + \alpha$. This is the borderline case between continuous and discontinuous conduction.

(ii) With $\alpha > \phi$ the sinusoidal term on the left-hand side obeys the inequality

$$\sin(\alpha_e - \phi) > 0 \tag{4.24}$$

This gives two values of α_e, namely,

$$\alpha_e - \phi < \pi \quad \text{or} \quad \alpha_e > \phi \tag{4.25}$$

The former inequality together with the inequality $\alpha > \phi$ leads to the inequality

$$\alpha_e - \alpha < \pi \tag{4.26}$$

Equation (4.26) implies that the conduction is discontinuous [Fig. 4.4(b)]; it also shows that load voltage and current control is possible only for the condition $\alpha > \phi$. The second inequality in Eqn (4.25) does not help in deciding whether the conduction is continuous or discontinuous.

(iii) For $\alpha < \phi$ the waveforms for the load voltage and load current will be exactly those obtained with the load connected directly across the supply voltage. This may be explained as follows. When Th_1 is given a gate

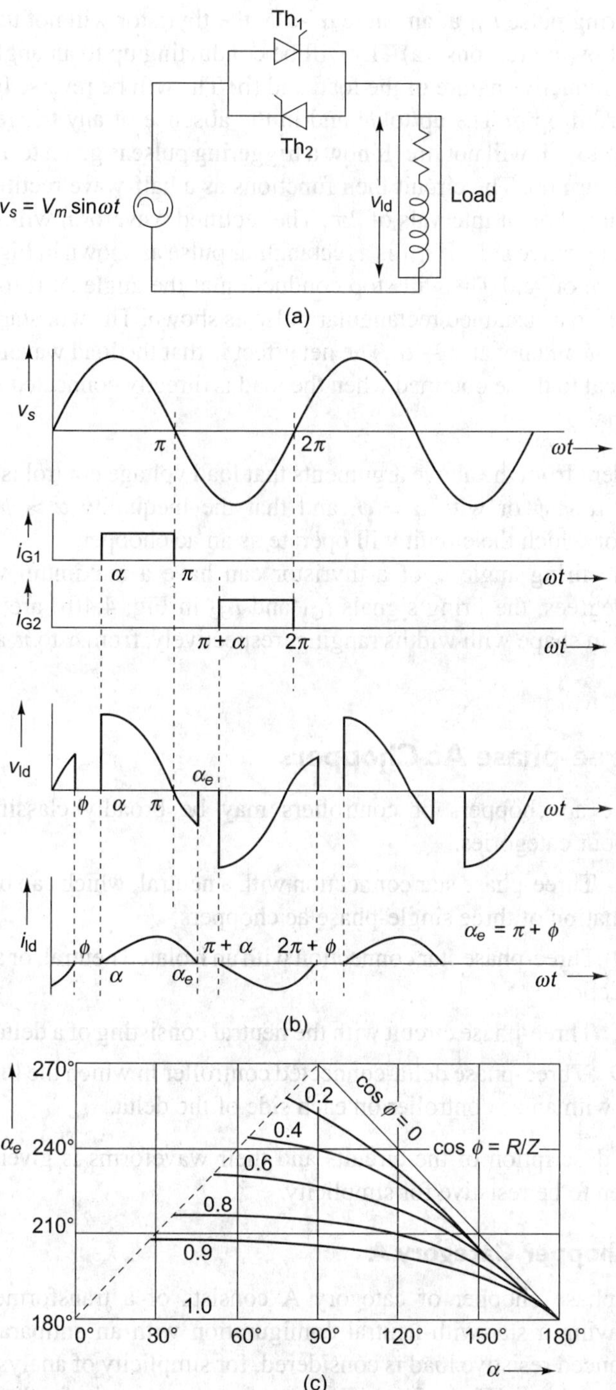

Fig. 4.4 Single-phase fully controlled ac chopper with inductive load: (a) circuit diagram, (b) waveforms, (c) chart showing α_e vs α with $\cos\phi$ as a parameter

triggering pulse i_{G1} at an angle $\alpha < \phi$, the thyristor will not turn on due to the following reasons. (a) Th_2 will be conducting up to an angle ϕ because of the inductive nature of the load and (b) Th_1 will be reverse biased by the forward drop of Th_2 up to ϕ; and in the absence of any triggering signals from ϕ to π it will not fire. If now a triggering pulse is given to Th_2 at $\pi + \alpha$, it will turn on. The circuit then functions as a half-wave rectifier with only Th_2 turned on at intervals of 2π. The rectified waveform will, however, be on the negative side. If i_{G1} is a rectangular pulse as shown in Fig. 4.4(b), Th_1 will turn on and Th_2 will stop conduction at the angle ϕ. It follows that if i_{G2} is also a sustained rectangular pulse as shown, Th_2 will start conduction at $\pi + \phi$ and not at $\pi + \alpha$. The net effect is that the load waveforms will be identical to those obtained when the load is directly connected to the supply terminals.

It is evident from the above arguments that load voltage control is not possible either with $\alpha < \phi$ or with $\alpha = \phi$, and that the inequality $\alpha > \phi$ is the only condition for which the circuit will operate as an ac chopper.

Since the firing angle α of a thyristor can have a maximum value of 180 electrical degrees, the firing signals i_{G1} and i_{G2} in Fig. 4.4(b) are shown to be rectangular in shape with widths ranging, respectively, from α to π and $\pi + \alpha$ to 2π radians.

4.4 Three-phase Ac Choppers

Three-phase ac choppers or controllers may be broadly classified into the following four categories.

Category A Three-phase star connection with a neutral, which can be considered as a combination of three single-phase ac choppers.

Category B Three-phase star connection with an isolated neutral, or a three-phase delta connection.

Category C Three-phase circuit with the neutral consisting of a delta connection.

Category D Three-phase delta-connected controller in which the load resistance is in series with an ac controller on each side of the delta.

A detailed description of the circuits and their waveforms is given below. The load is taken to be resistive for simplicity.

4.4.1 Chopper Category A

The three-phase chopper of category A consists of a transformer secondary winding having a star-with-neutral configuration with an antiparallel thyristor pair. A balanced resistive load is considered, for simplicity of analysis, in each of the secondary phases [Fig. 4.5(a)]. The gate triggering circuits for the thyristor pair in each of the phases will be identical to those of a single-phase ac chopper. The current ratings of the thyristors are arrived at under the condition of maximum duration of conduction (i.e., with $\alpha = 0$). The behaviour of this circuit can be understood by plotting the average value of the third-harmonic current i_0 flowing through the neutral as a function of the firing angle α. Accordingly, the waveforms of the phase voltages as well as those of i_0 for three possible ranges of α are plotted in Figs 4.5(b)–(d). The magnitude of the instantaneous value of the current i_0 can be arrived at as follows. When all the three phases are conducting, $i_0 = 0$. If just one phase does not conduct, i_0 becomes equal in magnitude, but opposite in sign, to the current that would flow in the non-conducting phase. Finally, if only one phase conducts, i_0 will be equal to the current flowing in that phase.

For the first range of α, namely, 0 to $\pi/3$, either two or three phases conduct. Application of the criteria laid down above and an inspection of the figure reveals that the duration 0 to $\pi/3$ can be split up into two subintervals, namely, 0 to α and α to $\pi/3$. During the first period 0 to α, only the R-phase does not conduct and hence i_0, the neutral current, will be exactly the negative of the R-phase current. In the second subinterval α to $\pi/3$, all the three phases conduct and hence $i_0 = 0$. The resultant waveforms for the different phase voltages as well as for i_0 are given in Fig. 4.5(b). Considering the symmetry of the phases, the RMS magnitude of the current in the three phases can be written as

$$\frac{|V_R'|_{(\text{RMS})}}{R} = \frac{|V_Y'|_{(\text{RMS})}}{R} = \frac{|V_B'|_{(\text{RMS})}}{R} = |I|_{(\text{RMS})} \tag{4.27}$$

The average value of i_0 can be determined as

$$\begin{aligned} I_{0(\text{av})} &= \frac{1}{\pi/3}\int_0^{\alpha} I_m \sin\omega t \; d(\omega t) \\ &= \frac{3I_m}{\pi}(1-\cos\alpha), \qquad 0 \le \alpha \le \frac{\pi}{3} \end{aligned} \tag{4.28}$$

In the second range, that is $\pi/3$ to $2\pi/3$, either one or two phases will conduct. The averaging interval remains $\pi/3$ in this case also, its duration being α to $\alpha + \pi/3$. This interval can be split up into two subintervals: (a) α to $2\pi/3$ and (b) $2\pi/3$ to $\alpha + \pi/3$. In the first subinterval, the phases R and Y conduct and i_0 in this case becomes equal to the negative of the instantaneous value of current in phase B. During the second subinterval, only the R-phase conducts and i_0

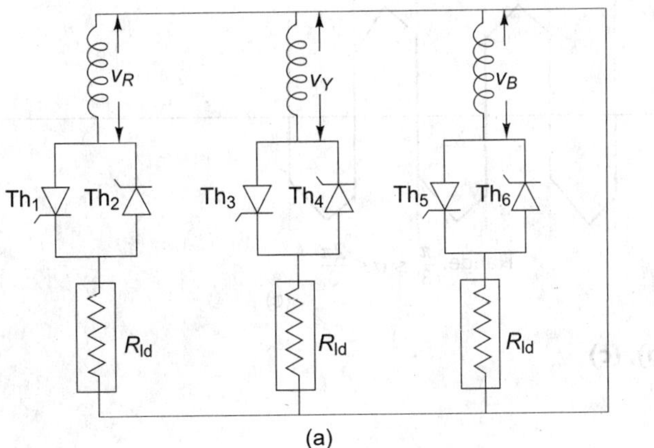

Fig. 4.5(a)

becomes equal to the current in this phase. The load voltage and neutral current waveforms are given in Fig. 4.5(c).

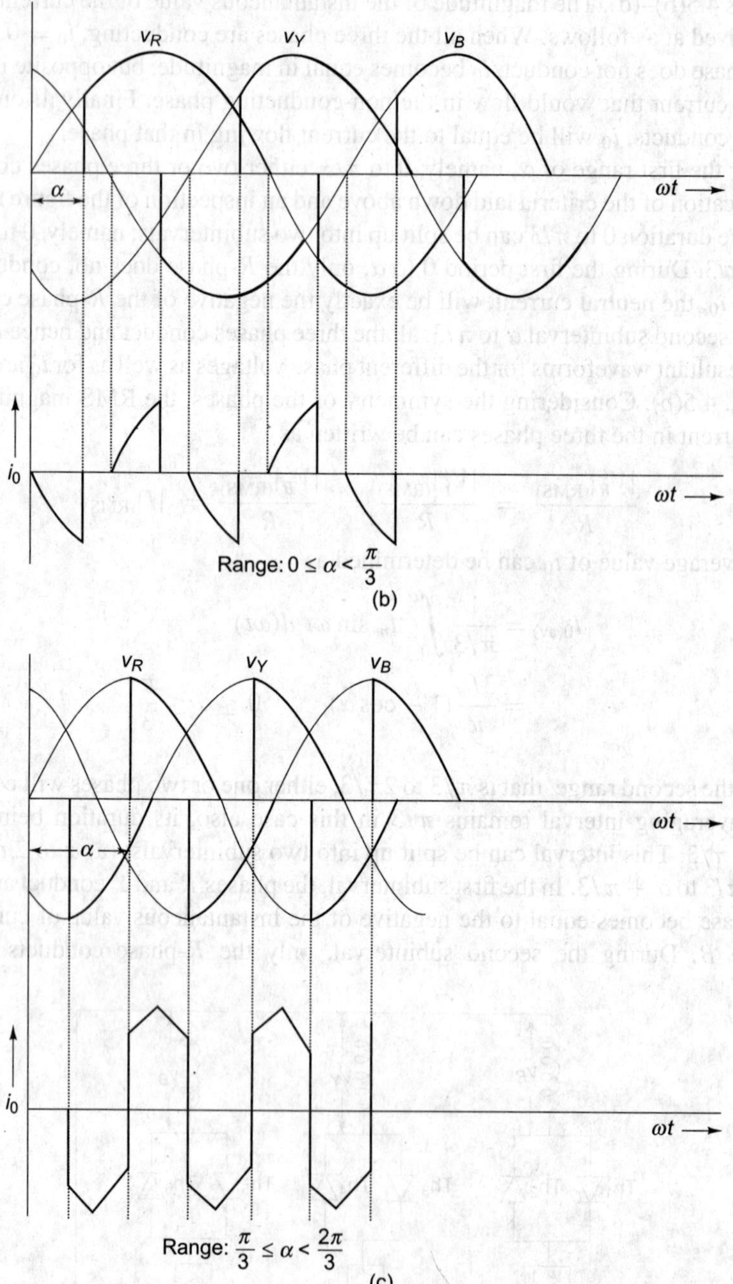

Figs 4.5(b), (c)

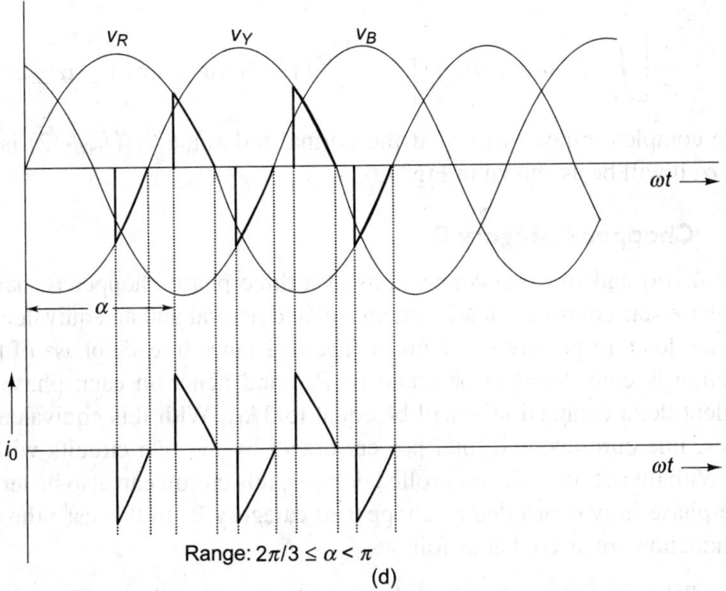

Fig. 4.5 **Three-phase ac controller with a star-connected load along with a neutral: (a) circuit diagram; waveforms for (b) $0 \le \alpha < \pi/3$, (c) $\pi/3 \le \alpha < 2\pi/3$, and (d) $2\pi/3 \le \alpha < \pi$**

$I_{0(\mathrm{av})}$ can be expressed as

$$I_{0(\mathrm{av})} = \frac{1}{\pi/3}\left[\int_{\alpha}^{2\pi/3} I_m \sin\left(\omega t - \frac{\pi}{3}\right) d(\omega t) + \int_{2\pi/3}^{\alpha+\pi/3} I_m \sin \omega t \, d(\omega t)\right]$$

$$= \frac{3I_m}{\pi}(\sqrt{3}\,\sin\alpha - 1), \quad \pi/3 \le \alpha \le 2\pi/3 \tag{4.29}$$

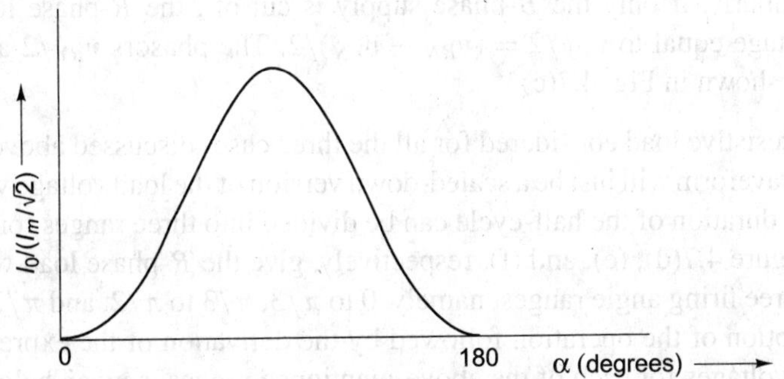

Fig. 4.6 **Variation of the normalized value of $I_{0(\mathrm{av})}$ with α**

In the third range, with duration $2\pi/3$ to π, only one phase conducts, as shown in Fig. 4.5(d). The average of the neutral current for this range may be determined as

$$I_{0(\text{av})} = \frac{3}{\pi}\left[\int_{\alpha}^{\pi} I_m \sin\omega t \; d(\omega t)\right] = \frac{3I_m}{\pi}(1+\cos\alpha), \; 2\pi/3 \le \alpha \le \pi \quad (4.30)$$

For the complete range, 0 to π, if the normalized value $I_0/(I_m/\sqrt{2})$ is drawn versus α, it will be as shown in Fig. 4.6.

4.4.2 Chopper Category B

In Figs 4.7(a) and (b), two connections of a three-phase chopper B, namely, a three-phase star-connected load with an isolated neutral and an equivalent delta-connected load, respectively, are given. The resistance in each phase of the star connection is considered to be equal to R_{ld}, and hence in each phase of the equivalent delta connection it will be equal to $3R_{\text{ld}}$. With this equivalence, the voltages, line currents, and total powers drawn by the two circuits will all be equal. With thyristors as the controlled devices, this circuit can also be termed as a three-phase fully controlled ac chopper of category B. In this case the criteria for conduction are arrived at as follows.

(a) If all the three phases conduct, the phase load receives its own line-to-neutral voltage. Thus, the R-phase resistive load gets the R line-to-neutral voltage, etc.

(b) If the phase voltage pertaining to a certain phase is switched off, the current through the load resistance of that phase as well as the voltage across it will become zero.

(c) Finally, considering a particular phase, say R, if the supply to only one of the other phases, say Y, is switched off, then the resistive load in this phase receives one-half of the difference of the line-to-neutral voltages of the R- and B-phase. For instance, if v_{RN} is the line-to-neutral voltage of the R-phase, v_{BN} that of the B-phase, and v_{RB} the difference of these two voltages, then the R-phase load gets a voltage $v_{RB}/2 = (v_{RN} - v_{BN})/2$. Similarly, if only the B-phase supply is cut off, the R-phase load gets a voltage equal to $v_{RY}/2 = (v_{RN} - v_{YN})/2$. The phasors $v_{RY}/2$ and $v_{RB}/2$ are shown in Fig. 4.7(c).

With resistive load considered for all the three cases discussed above, the load current waveform will just be a scaled-down version of the load voltage waveform. The total duration of the half-cycle can be divided into three ranges for the firing angle. Figure 4.7(d), (e), and (f), respectively, give the R-phase load waveforms for the three firing angle ranges, namely, 0 to $\pi/3$, $\pi/3$ to $\pi/2$, and $\pi/2$ to $5\pi/6$. A description of the operation followed by the derivation of the expressions for the load voltages for each of the above-mentioned ranges is given below.

(i) $0 \le \alpha < \pi/3$: It is assumed that the thyristors in all the phases are fired at the angle α measured from the corresponding zero crossing. This holds good for both positive- as well as negative-conducting thyristors. The duration α to π, which is the positive-half-wave duration of the R-phase is divided into five subintervals. In the subinterval α to $\pi/3$ all the three phases conduct and hence, applying the criteria given above, the resistance in the R-phase

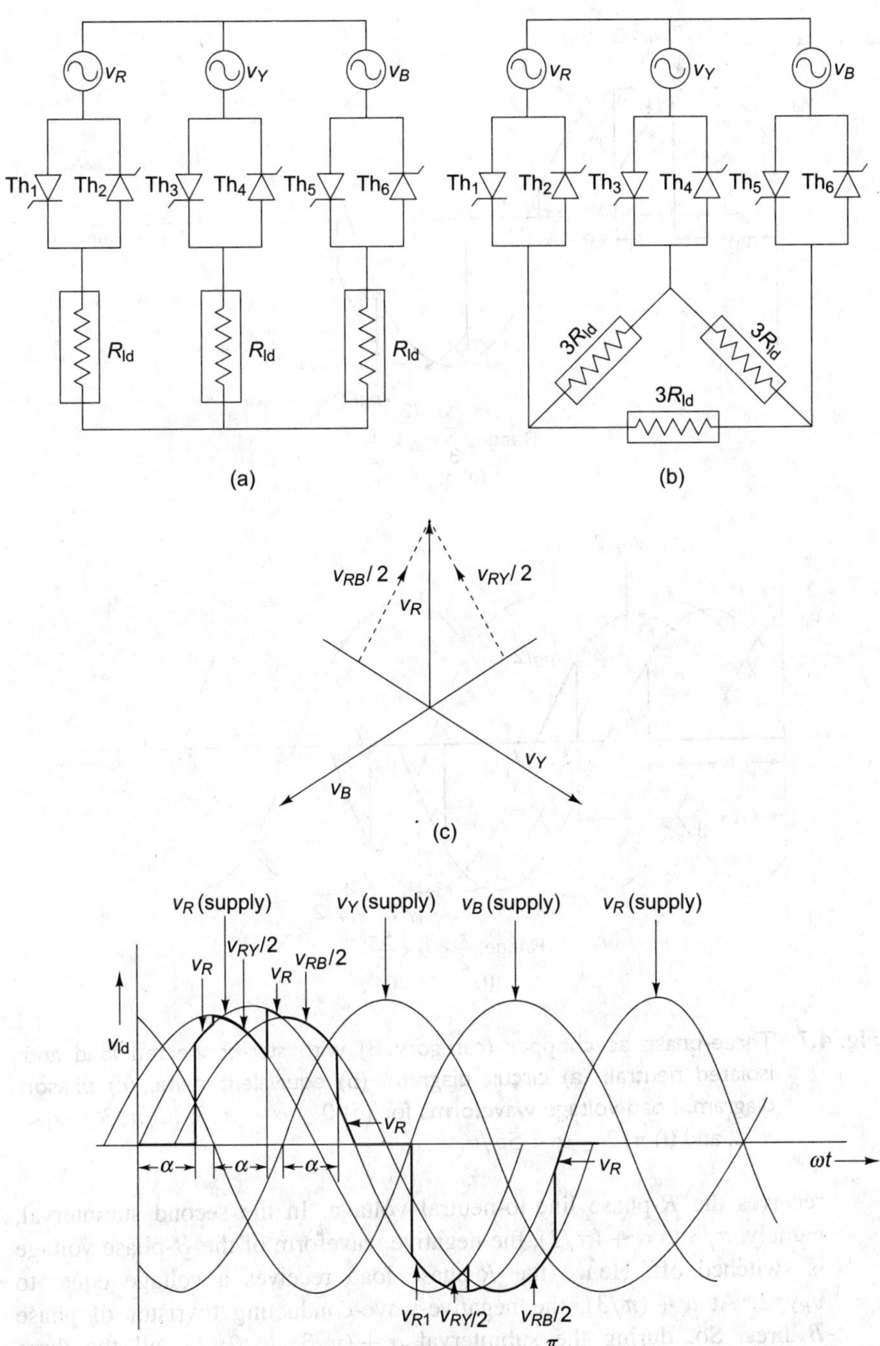

v_R
v_Y
v_B
Th1 Th2 Th3 Th4 Th5 Th6
R_{ld}
$3R_{ld}$
(a)
(b)
$v_{RB}/2$
$v_{RY}/2$
v_R
v_Y
v_B
(c)
v_R (supply)
v_Y (supply)
v_B (supply)
v_R (supply)
$v_{RY}/2$
$v_{RB}/2$
v_{ld}
α
ωt
v_{R1}
$v_{RY}/2$
$v_{RB}/2$
Range: $0 \leq \alpha < \frac{\pi}{3}$
(d)

Figs 4.7(a)–(d)

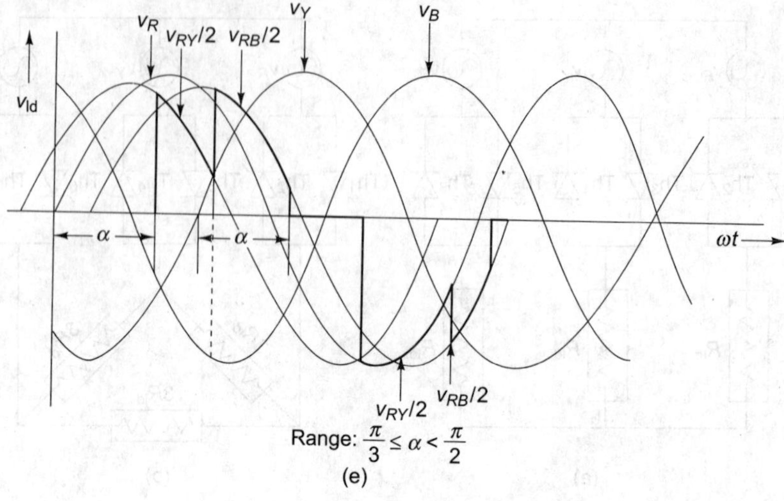

(e)

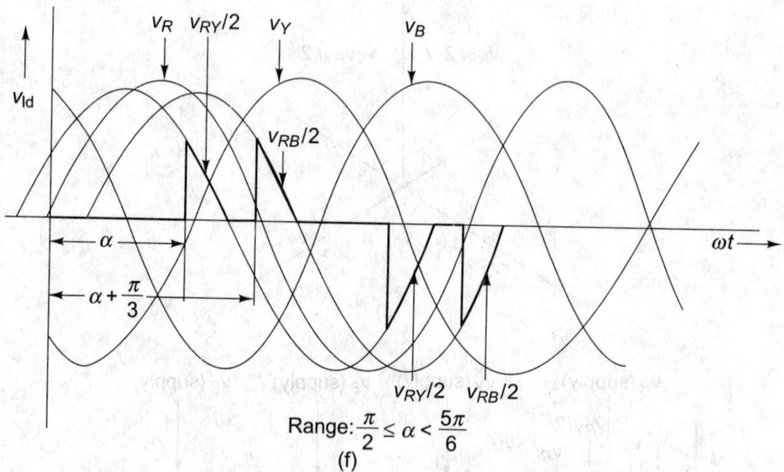

(f)

Fig. 4.7 Three-phase ac chopper (category B) with star-connected load and isolated neutral: (a) circuit diagram, (b) equivalent delta, (c) phasor diagram. Load voltage waveforms for (d) $0 \leq \alpha < \pi/3$, (e) $\pi/3 \leq \alpha < \pi/2$, and (f) $\pi/2 \leq \alpha < 5\pi/6$.

receives the R-phase line-to-neutral voltage. In the second subinterval, namely, $\pi/3$ to $\alpha + (\pi/3)$, the negative waveform of the B-phase voltage is switched off. Hence the R-phase load receives a voltage equal to $v_{RY}/2$. At $\alpha + (\pi/3)$, the negative-wave-conducting thyristor of phase B fires. So, during the subinterval $\alpha + (\pi/3)$ to $2\pi/3$, all the three phases conduct and the R-phase load receives its own phase voltage. At $2\pi/3$, the Y-phase positive wave is cut off and hence from $2\pi/3$ to $(2\pi/3) + \alpha$ the voltage across the R-phase load is $v_{RB}/2$. At $(2\pi/3) + \alpha$, the positive-wave-conducting thyristor of the Y-phase fires; all the phases now conduct and hence in the subinterval $(2\pi/3) + \alpha$ to π the R-phase load gets the R-phase voltage. Thus the total duration of conduction in the positive half-cycle of the R-phase is $\pi - \alpha$. Similarly, in the negative half-cycle of the R-phase, the conduction starts at $\pi + \alpha$ and ends at 2π. The duration still remains $(\pi - \alpha)$, and the voltage segments of the R-phase load in the negative half-cycle are obtained by the same criteria described above. It is seen from the waveforms that during the interval $\pi/3$ to $(\pi/3) + \alpha$ there is a deficit of area for the waveform of $v_{RY}/2$ with respect to the v_R waveform. Similarly, during the interval $2\pi/3$ to $(2\pi/3) + \alpha$ there is excess area for the waveform of $v_{RB}/2$ with respect to v_R. Though these positive and negative areas do not appear to be identical, their half-cycle averages remain the same. This is because the excess area obtained when the Y-phase is turned off is equal to the deficit of area obtained when the B-phase is turned off. Using this simplification, the half-cycle average can be obtained by integrating the R-phase voltage waveform over the duration α to π. Thus,

$$V_{R(\mathrm{av})} = \frac{V_m}{\pi}\int_{\alpha}^{\pi} \sin\omega t \, d(\omega t)$$

$$= \frac{V_m}{\pi}(1 + \cos\alpha), \quad 0 < \alpha < \pi/3 \tag{4.31}$$

The RMS voltage of the R-phase load is determined based on the above discussion as follows:

$$\begin{aligned} V_{R(\mathrm{RMS})} = {} & \left[\frac{2}{2\pi}\left\{\int_{\alpha}^{\pi/3} (V_m \sin\omega t)^2 d(\omega t) + \int_{\pi/3}^{\alpha+(\pi/3)} \left[\frac{3V_m}{2}\sin\left(\omega t + \frac{\pi}{6}\right)\right]^2 \right.\right. \\ & + \int_{\alpha+(\pi/3)}^{2\pi/3} (V_m \sin\omega t)^2 d(\omega t) \\ & + \int_{2\pi/3}^{(2\pi/3)+\alpha} \left[\frac{3V_m}{2}\sin\left(\omega t - \frac{\pi}{6}\right)\right]^2 d(\omega t) \\ & \left.\left. + \int_{(2\pi/3)+\alpha}^{\pi} (V_m \sin\omega t)^2 d(\omega t)\right\}\right]^{1/2} \\ = {} & \left[\frac{V_m^2}{\pi}\left\{\frac{1}{2}\left(\frac{\pi}{3} - \alpha - \frac{\sqrt{3}}{4} + \frac{\sin 2\alpha}{2}\right) + \frac{3}{8}\left(\alpha + \frac{\sin 2\alpha}{2}\right)\right.\right. \\ & + \frac{1}{2}\left(\frac{\pi}{3} - \alpha + \frac{\sqrt{3}}{4} - \frac{\sin 2\alpha}{4} + \frac{\sqrt{3}}{4}\cos 2\alpha\right) + \frac{3}{8}\left(\alpha + \frac{\sin 2\alpha}{2}\right) \\ & \left.\left. + \frac{1}{2}\left(\frac{\pi}{3} - \alpha - \frac{\sin 2\alpha}{4} - \frac{\sqrt{3}}{4}\cos 2\alpha\right)\right\}\right]^{1/2} \end{aligned}$$

By sorting out similar terms, $V_{R(\mathrm{RMS})}$ gets simplified to

$$V_{R(\mathrm{RMS})} = V_m\left[\frac{1}{2} - \frac{3\alpha}{4\pi} + \frac{3}{8\pi}\sin 2\alpha\right]^{1/2}, \quad 0 \le \alpha < \pi/3 \tag{4.32}$$

Figure 4.7(d) illustrates the load voltage waveform obtained for this range of α. The load current waveform will be a scaled-down replica of it.

(ii) $\pi/3 \le \alpha < \pi/2$: Figure 4.7(e) gives the R-phase load voltage waveform for this range of α. It is seen that there are always two simultaneously conducting phases in this case. The region of conduction for the positive half-wave of the R-phase can be divided into the two subintervals α to $(\pi/3)+\alpha$ and $(\pi/3)+\alpha$ to $(2\pi/3)+\alpha$. In the former interval, phase B is cut off and phases R and Y conduct. The voltage across the R-phase load is that of $v_{RY}/2$. At $(\pi/3)+\alpha$ the phase-B thyristor is fired and, as the instantaneous value of $v_{RB}/2$ is greater than that of $v_{RY}/2$, the Y-phase thyristor turns off. Thus the voltage of the R-phase load becomes $v_{RB}/2$. When the Y-phase thyristor starts conduction, at $(2\pi/3 + \alpha)$, the condition of the R-phase voltage is arrived at as follows. Since all the three phases conduct, the R-phase load gets its own phase voltage but the instantaneous value of the R-phase voltage is negative. The thyristor that conducts positive current is therefore reverse biased; moreover, the negative-current-conducting thyristor has not yet been fired. As a result, the voltage across the R-phase load collapses to zero. The total conduction period in the positive half-cycle for the R-phase load is therefore $2\pi/3$. Because of symmetry, the same condition holds good for the Y- and B-phase also. The half-cycle average of the R-phase load voltage can be computed as twice that of one subinterval, i.e., $\pi/3$:

$$V_R = \frac{2}{\pi}V_m\left(\frac{\sqrt{3}}{2}\right)\int_{\alpha}^{\alpha+\pi/3}\sin\left(\omega t+\frac{\pi}{6}\right)d(\omega t)$$

$$= \frac{\sqrt{3}V_m}{\pi}\cos\left(\alpha-\frac{\pi}{6}\right), \qquad \frac{\pi}{3}\le\alpha<\frac{\pi}{2} \tag{4.33}$$

The RMS magnitude of the R-phase load voltage is obtained as

$$V_{R(\mathrm{RMS})} = \left[\frac{2}{2\pi}\left\{\int_{\alpha}^{(\pi/3)+\alpha}\left(\frac{\sqrt{3}}{2}V_m\sin\left(\omega t+\frac{\pi}{6}\right)\right)^2 d(\omega t)\right.\right.$$
$$\left.\left.+\int_{(\pi/3)+\alpha}^{(2\pi/3)+\alpha}\left(\frac{\sqrt{3}V_m}{2}\sin\left(\omega t-\frac{\pi}{6}\right)\right)^2 d(\omega t)\right\}\right]^{1/2}$$

Simplification gives

$$V_{R(\mathrm{RMS})} = V_m\left[\frac{1}{4}+\frac{9}{16\pi}\sin 2\alpha+\frac{3\sqrt{3}}{16\pi}\cos 2\alpha\right]^{1/2}, \quad \pi/3\le\alpha<\pi/2 \tag{4.34}$$

(iii) $\pi/2 \le \alpha < 5\pi/6$: The two alternatives in this case are either the two phases conduct simultaneously or none of the phases conducts. This range differs from the above two cases in that the conduction is not continuous. As seen from Fig. 4.7(f), conduction stops at the zero crossing of the voltage

$v_{RY}/2$. The two thyristors Th_1 and Th_4 become reverse biased at this point. Conduction resumes again at $\alpha + (\pi/3)$ when the negative-conducting thyristor Th_6 of phase B is fired. The R-phase load now gets a voltage equal to $v_{RB}/2$. A unique feature of this range is that the thyristors Th_1 and Th_4 in the R- and Y-phase have to be fired simultaneously at the angle α, and the thyristors Th_1 and Th_6 in the R- and B-phase have to be fired at the angle $\alpha + (\pi/3)$. The subintervals in this range are decided as follows. The voltage $v_{RY}/2$ goes to zero at $5\pi/6$ and hence the duration of this first subinterval becomes $(5\pi/6) - \alpha$, the averaging for this duration being relevant for the waveform of $v_{RY}/2$. The waveform of $v_{RB}/2$ shows conduction at $\alpha + (\pi/3)$ and goes up to $7\pi/6$, this duration also being equal to $(5\pi/6) - \alpha$. Hence the half-cycle average of the total duration is obtained as twice that of one subinterval, i.e., $(5\pi/6) - \alpha$.

The half-cycle average and RMS magnitudes of load voltage for this interval are arrived at as below:

$$V_{R(\mathrm{av})} = \left(\frac{2}{\pi}\right)\left(\frac{\sqrt{3}}{2}\right) V_m \int_{\alpha}^{5\pi/6} \sin\left(\omega t + \frac{\pi}{6}\right) d(\omega t)$$

$$= \frac{\sqrt{3}V_m}{\pi}\left[1 - \sin\left(\alpha - \frac{\pi}{3}\right)\right], \qquad \pi/2 \le \alpha \le 5\pi/6 \qquad (4.35)$$

The RMS value of the R-phase load is determined as

$$V_{R(\mathrm{RMS})} = \left[\frac{2}{2\pi}\left\{\int_{\alpha}^{5\pi/6} \left(\frac{\sqrt{3}V_m}{2} \sin\left(\omega t + \frac{\pi}{6}\right)\right)^2 d(\omega t) + \int_{\alpha+(\pi/3)}^{7\pi/6} \left(\frac{\sqrt{3}V_m}{2} \sin\left(\omega t - \frac{\pi}{6}\right)\right)^2 d(\omega t)\right\}\right]^{1/2}$$

After simplification, $V_{R(\mathrm{RMS})}$ is obtained as

$$V_{R(\mathrm{RMS})} = V_m\left[\frac{5}{8} - \frac{3\alpha}{4\pi} + 3\frac{\sin 2\alpha}{16\pi} + 3\sqrt{3}\frac{\cos 2\alpha}{16\pi}\right]^{1/2}, \qquad \pi/2 \le \alpha \le 5\pi/6 \qquad (4.36)$$

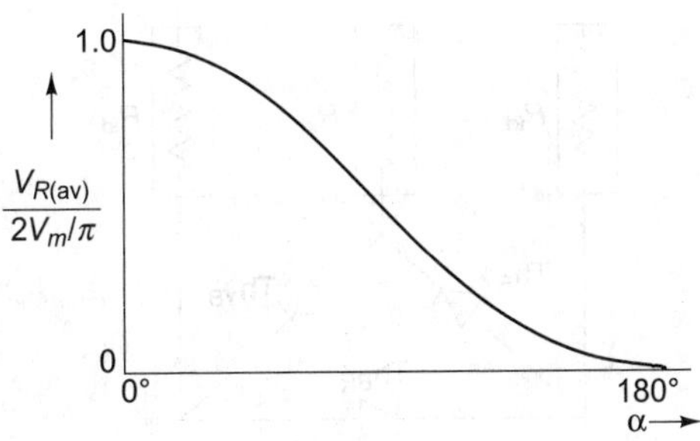

Fig. 4.8 Variation of the normalized value of $V_{R(\mathrm{av})}$ with α

The normalized values $[V_R(\text{av})/(2V_m/\pi)]$ for various values of α are plotted in Fig. 4.8. The range in this case is 0 to $5\pi/6$ and not 0 to π, the latter duration holding good only for the single-phase case.

4.4.3 Chopper Category C

The three-phase half-controlled chopper circuit and its equivalent delta connection shown in Figs 4.9(a) and (b) come under this category. The operation of the three-phase load circuit with the delta-connected neutral shown in Fig. 4.9(c) is identical to the above two circuits and so grouped with them. The circuit of Fig. 4.9(a) is considered here for discussion. The essential difference between this circuit and that of Fig. 4.7(a) is that the negative-current-conducting thyristors Th_2, Th_4, and Th_6 in the latter circuit have been replaced here by diodes D_2, D_4, and D_6, which conduct throughout the negative half-cycle of their respective phase voltages, thus justifying the name of this circuit. The operation of this circuit can be explained with the help of the same criteria as for the fully controlled circuit described in the previous section, any difference between the two cases being due to the diodes. Another special feature of this circuit is that the firing angle α exceeds π rad or 180° [Fig. 4.9(f)]. As before, the waveforms of the R-phase load voltage are plotted assuming a resistive load.

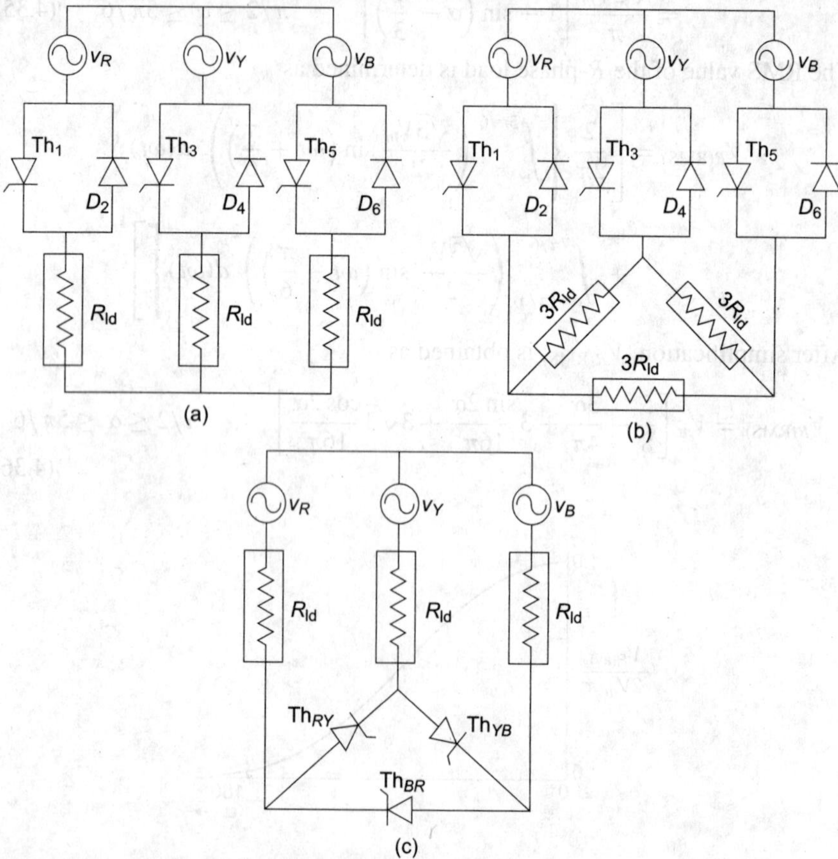

Figs 4.9(a)–(c)

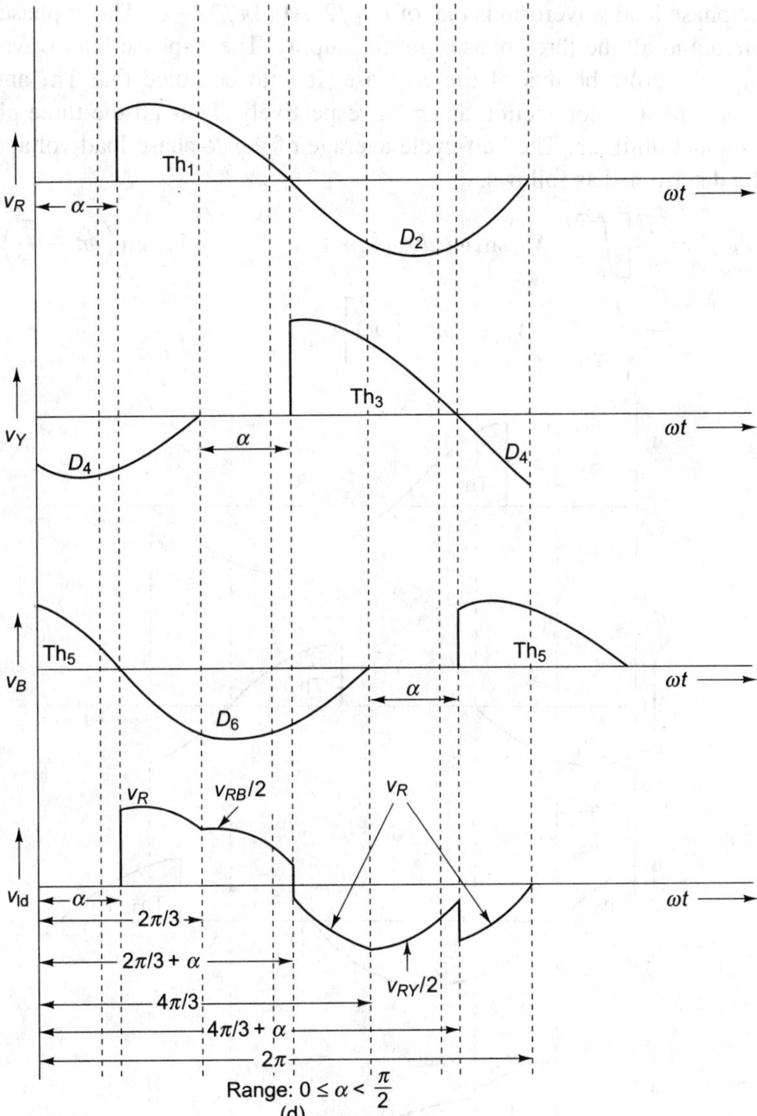

Fig. 4.9(d)

(i) $0 \leq \alpha < \pi/2$: The R-phase load voltage remains zero till the R-phase thyristor Th_1 is fired at α. After α, all the three phases get connected to the supply, and hence the R-phase load gets the R-phase voltage. At an angle of $2\pi/3$, the Y-phase is cut off and Th_3 does not conduct till $\alpha + (2\pi/3)$. During this interval the load waveform is that of $v_{RB}/2$. At $\alpha + (2\pi/3)$, the Y-phase thyristor Th_3 is switched on and all the three phases conduct; thereby the R-phase load gets its own phase voltage. This continues in the negative half-cycle also (that is, beyond π) because the diode D_2 starts conduction after this angle. At an angle of $4\pi/3$, the diode D_6 of phase B stops conduction and Th_5 of the same phase (B) switches off. Hence, the

R-phase load waveform is that of $v_{RY}/2$. At $(4\pi/3) + \alpha$, Th_5 of phase B is fired and all the three phases get the supply. The R-phase load waveform will therefore be that of the R-phase. It is to be noted that Th_3 and D_4 stop and start conduction at $5\pi/3$, respectively. Thus all the three phases conduct until 2π. The half-cycle average of the R-phase load voltage can be determined as follows:

$$V_{R(\text{av})} = \frac{1}{\pi}\left[\int_{\alpha}^{2\pi/3} V_m \sin\omega t \; d(\omega t) + \int_{2\pi/3}^{(2\pi/3)+\alpha} \frac{\sqrt{3}}{2} V_m \sin\left(\omega t - \frac{\pi}{6}\right) d(\omega t) + \int_{(2\pi/3)+\alpha}^{\pi} V_m \sin\omega t \; d(\omega t)\right]$$

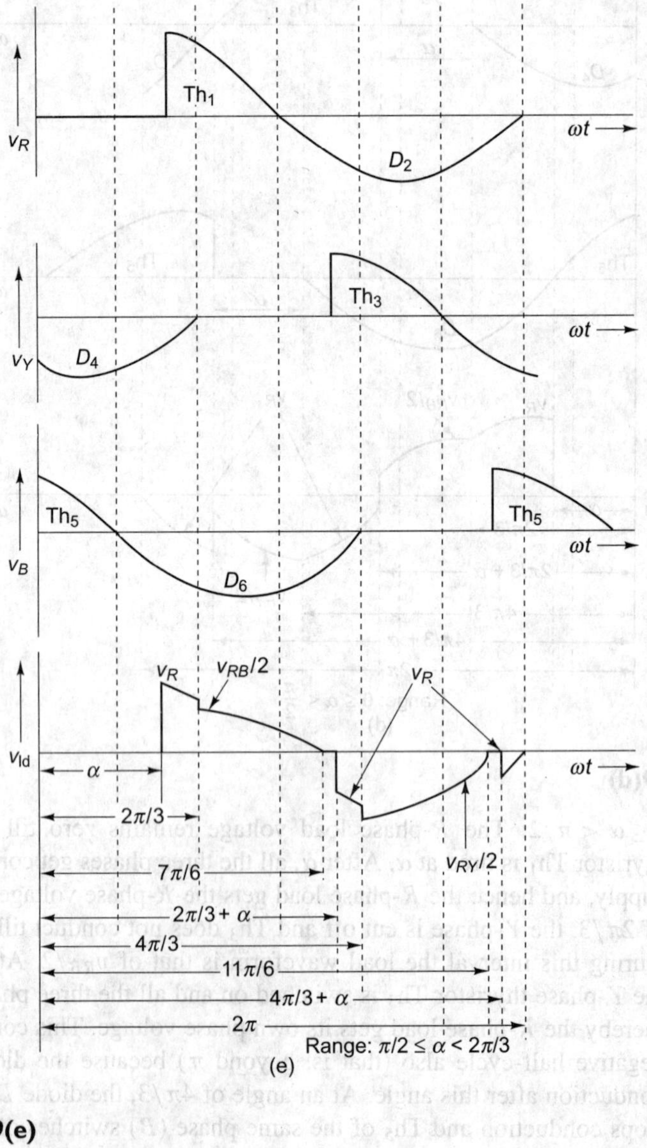

Fig. 4.9(e)

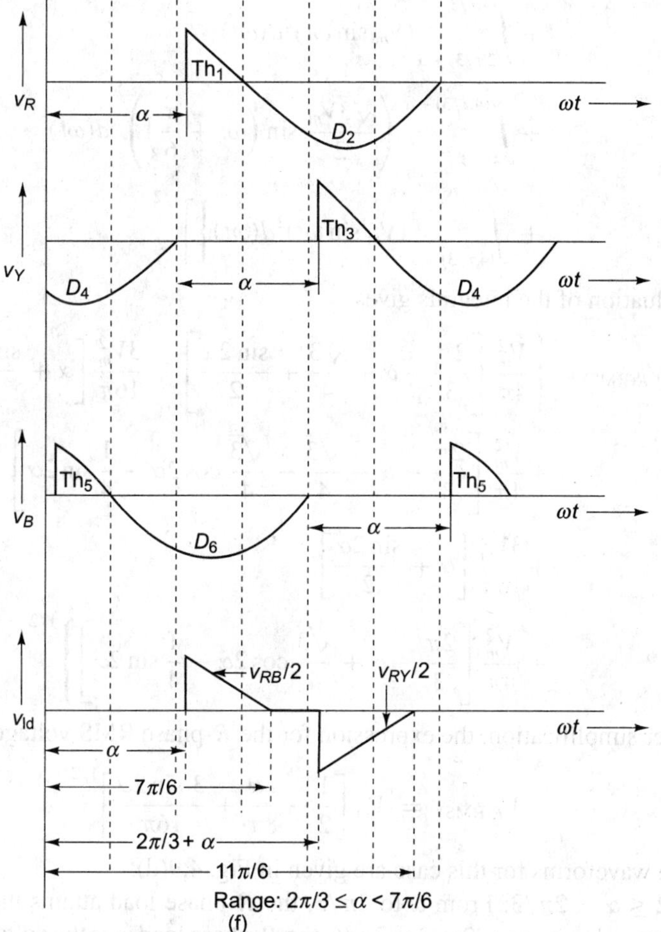

Fig. 4.9 Three-phase half-controlled chopper (category C): (a) star connection, (b) delta connection, (c) equivalent circuit with delta-connected thyristors. Waveforms with (d) $0 \le \alpha < \pi/2$, (e) $\pi/2 \le \alpha < 2\pi/3$, and (f) $2\pi/3 \le \alpha < 7\pi/6$.

$$= \frac{V_m}{2\pi}(\cos\alpha + 3) \tag{4.37}$$

The RMS voltage of the R-phase for this case is arrived at as follows:

$$V_{R(\text{RMS})} = \left[\frac{1}{2\pi}\left\{\int_{\alpha}^{(2\pi/3)} (V_m \sin\omega t)^2 d(\omega t)\right.\right.$$

$$\left. + \int_{2\pi/3}^{(2\pi/3)+\alpha} \left(\frac{\sqrt{3}V_m}{2}\sin\left(\omega t - \frac{\pi}{6}\right)\right)^2 d(\omega t)\right.$$

$$+\int_{(2\pi/3)+\alpha}^{4\pi/3} (V_m \sin \omega t)^2 d(\omega t)$$

$$+\int_{4\pi/3}^{(4\pi/3)+\alpha} \left(\frac{\sqrt{3}V_m}{2}\sin\left(\omega t+\frac{\pi}{6}\right)\right)^2 d(\omega t)$$

$$+\left.\left.\int_{(4\pi/3)+\alpha}^{2\pi} (V_m \sin \omega t)^2 d(\omega t)\right\}\right]^{1/2}$$

Evaluation of the integrals gives

$$V_{R(\text{RMS})} = \left\{\frac{V_m^2}{4\pi}\left[\frac{2\pi}{3}-\alpha+\frac{\sqrt{3}}{4}+\frac{\sin 2\alpha}{2}\right]+\frac{3V_m^2}{16\pi}\left[\alpha+\frac{\sin 2\alpha}{2}\right]\right.$$

$$+\frac{V_m^2}{4\pi}\left[\frac{2\pi}{3}-\alpha-\frac{\sqrt{3}}{4}-\frac{\sqrt{3}}{4}\cos 2\alpha-\frac{1}{4}\sin 2\alpha\right]$$

$$+\frac{3V_m^2}{16\pi}\left[\alpha+\frac{\sin 2\alpha}{2}\right]$$

$$\left.+\frac{V_m^2}{4\pi}\left[\frac{2\pi}{3}-\alpha+\frac{\sqrt{3}}{4}\cos 2\alpha-\frac{1}{4}\sin 2\alpha\right]\right\}^{1/2}$$

After simplification, the expression for the R-phase RMS voltage becomes

$$V_{R(\text{RMS})} = V_m\left[\frac{1}{2}-\frac{3\alpha}{8\pi}+\frac{3\sin 2\alpha}{16\pi}\right]^{1/2} \tag{4.38}$$

The waveforms for this case are given in Fig. 4.9(d).

(ii) $\pi/2 \le \alpha < 2\pi/3$: From α to $2\pi/3$, the R-phase load attains the R-phase voltage. Again, from $2\pi/3$ to $7\pi/6$, the R-phase load gets the voltage $v_{RB}/2$. From $7\pi/6$ to $(2\pi/3)+\alpha$, the voltage across the R-phase is zero because Th_1 is reverse biased and D_6 also gets reverse biased by virtue of v_{RB} becoming negative after its zero crossing. From $(2\pi/3)+\alpha$ to $4\pi/3$, the R-phase load again attains the R-phase voltage because all the phases conduct in this interval. For the duration $4\pi/3$ to $11\pi/6$, it gets the voltage $v_{RY}/2$. From $11\pi/6$ to $(4\pi/3)+\alpha$, the R-phase voltage is zero because of D_2 getting reverse biased. At $(4\pi/3)+\alpha$, Th_5 is fired and the B-phase starts conduction. Thus, all phases get the supply and the R-phase load gets its own phase voltage. The waveforms for this condition are given in Fig. 4.9(e).

The half-cycle average voltage of the R-phase is expressed as

$$V_{R(\text{av})} = \frac{1}{\pi}\left[\int_{\alpha}^{2\pi/3} V_m \sin \omega t \, d(\omega t) + \int_{2\pi/3}^{7\pi/6} \frac{\sqrt{3}}{2} V_m \sin\left(\omega t-\frac{\pi}{6}\right) d(\omega t)\right]$$

$$= \frac{V_m}{\pi}\left(\frac{1+\sqrt{3}}{2}+\cos\alpha\right) \tag{4.39}$$

The RMS voltage of the R-phase for this case can be written as

$$V_{R(\text{RMS})} = \left[\frac{1}{2\pi}\left\{\int_{\alpha}^{2\pi/3} (V_m \sin\omega t)^2 d(\omega t) + \int_{2\pi/3}^{7\pi/6}\left(\frac{\sqrt{3}V_m}{2}\sin\left(\omega t - \frac{\pi}{6}\right)\right)^2 d(\omega t) + \int_{2\pi/3+\alpha}^{4\pi/3} (V_m \sin\omega t)^2 d(\omega t) + \int_{4\pi/3}^{11\pi/6}\left(\frac{\sqrt{3}V_m}{2}\sin\left(\omega t + \frac{\pi}{6}\right)\right)^2 d(\omega t) + \int_{4\pi/3+\alpha}^{2\pi} (V_m \sin\omega t)^2 d(\omega t)\right\}\right]^{1/2}$$

After simplification of the right-hand side, the R-phase voltage is obtained as

$$V_{R(\text{RMS})} = V_m\left[\frac{11}{16} - \frac{3\alpha}{4\pi}\right]^{1/2} \tag{4.40}$$

(iii) $2\pi/3 \le \alpha \le 7\pi/6$: From α to $7\pi/6$, the R-phase load attains the waveform of $v_{RB}/2$. From $7\pi/6$ to $(2\pi/3) + \alpha$, it is zero because Th_1 and D_6 get reverse biased at $7\pi/6$. From $(2\pi/3) + \alpha$ to $11\pi/6$, the R-phase load gets the waveform of $v_{RY}/2$, because the B-phase voltage gets discontinued from an angle of $4\pi/3$ due to the crossing of the voltage carried by D_6 onto the positive side. From $11\pi/6$ to 2π, D_6 is reverse biased by a voltage equal to $v_{RY}/2$ and the R-phase voltage is zero. As shown in Fig. 4.9(f) the waveforms on both sides of the ωt-axis are identical.

The half-cycle average voltage of the R-phase for this case is obtained from the equation

$$V_{R(\text{av})} = \frac{1}{\pi}\int_{\alpha}^{7\pi/6} \frac{\sqrt{3}V_m}{2}\sin\left(\omega t - \frac{\pi}{6}\right) d(\omega t) \tag{4.41}$$

Upon performing the integration, the expression for the voltage gets simplified to

$$V_{R(\text{av})} = \frac{\sqrt{3}V_m}{\pi}\left[\frac{1}{2} + \frac{\sqrt{3}}{4}\cos\alpha + \frac{1}{4}\sin\alpha\right] \tag{4.42}$$

The RMS voltage of the R-phase for this case is expressed as

$$V_{R(\text{RMS})} = \left[\frac{1}{2\pi}\left\{\int_{\alpha}^{7\pi/6}\left(\frac{\sqrt{3}V_m}{2}\sin\left(\omega t - \frac{\pi}{6}\right)\right)^2 d(\omega t) + \int_{(2\pi/3)+\alpha}^{11\pi/6}\left(\frac{\sqrt{3}V_m}{2}\sin\left(\omega t + \frac{\pi}{6}\right)\right)^2 d(\omega t)\right\}\right]^{1/2} \tag{4.43}$$

After simplification of right-hand side, $V_{R(\text{RMS})}$ can be expressed as

$$V_{R(\text{RMS})} = V_m \left[\frac{7}{16} - \frac{3\alpha}{8\pi} + \frac{3}{32\pi} \sin 2\alpha - \frac{3\sqrt{3}\cos 2\alpha}{32\pi} \right]^{1/2} \tag{4.44}$$

4.4.4 Chopper Category D

As shown in Fig. 4.10(a), the circuit of a category-D chopper consists of a delta configuration of three single-phase ac choppers along with their resistive loads. The star-connected three-phase chopper circuit shown in Fig. 4.10(b) is its electrical equivalent.

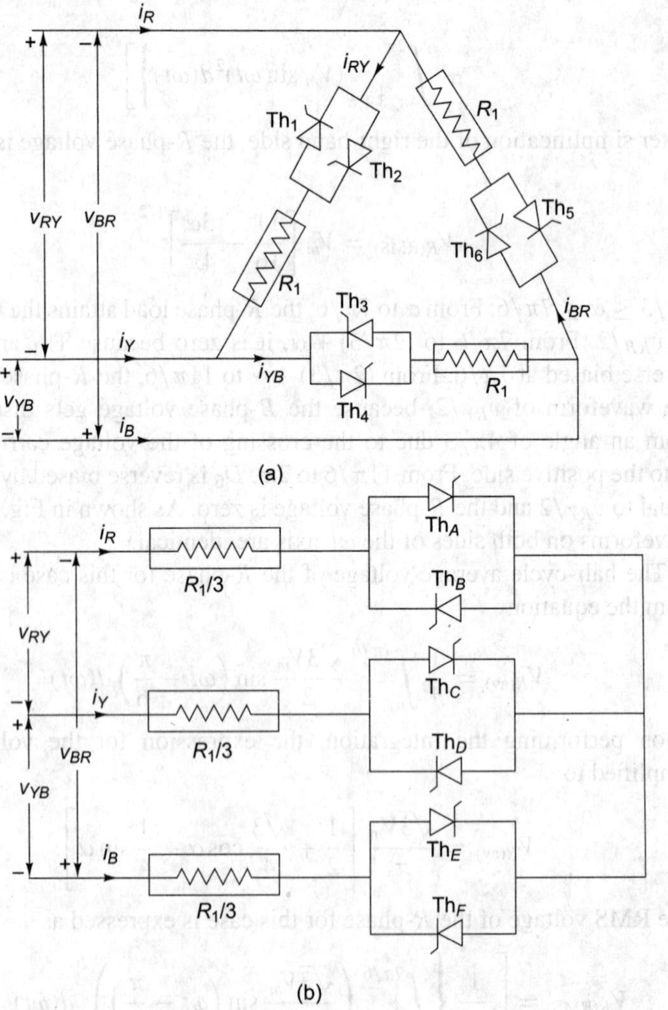

Fig. 4.10 Class-D ac chopper: (a) circuit diagram, (b) equivalent circuit

The instantaneous line-to-line voltages which are also the phase voltages for the delta connection are assumed to be

$$v_{RY} = V_m \sin \omega t$$

$$v_{YB} = V_m \sin\left(\omega t - \frac{2\pi}{3}\right) \tag{4.45}$$

and

$$v_{BR} = V_m \sin\left(\omega t - \frac{4\pi}{3}\right)$$

where V_m is the peak of the line-to-line voltage. The phase currents i_{RY}, i_{YB}, and i_{BR} in the delta-connected circuit are related to the line currents as

$$i_R = i_{RY} - i_{BR}$$

$$i_Y = i_{YB} - i_{RY} \tag{4.46}$$

$$i_B = i_{BR} - i_{YB}$$

with $i_{RY} = v_{RY}/R_1$, $i_{YB} = v_{YB}/R_1$, and $i_{BR} = v_{BR}/R_1$. The load being resistive, these currents are in phase with their respective voltages. Thus the expressions for i_R, i_Y, and i_B work out to be

$$i_R = \frac{\sqrt{3}V_m}{R_1} \cos\left(\omega t - \frac{2\pi}{3}\right) \tag{4.47a}$$

$$i_Y = -\frac{\sqrt{3}V_m}{R_1} \cos\left(\omega t - \frac{\pi}{3}\right) \tag{4.47b}$$

and

$$i_B = -\frac{\sqrt{3}V_m}{R_1} \cos(\omega t - \pi) \tag{4.47c}$$

Equations (4.47a)–(4.47c) show that the line current peaks have a magnitude of $\sqrt{3}V_m/R_1$, whereas the phase current peaks are all equal to V_m/R_1, with the load resistance in each of the legs of the delta being equal to R_1. The equivalent resistance in each leg of the star circuit of Fig. 4.10(b) works out to $R_1/3$ ohms. For this resistive case, the half-cycle average load voltage and the RMS line-to-line voltage, which is the same as the voltage of each phase of the delta circuit, can be worked out, from the waveforms given in Fig. 4.11, as follows (Rashid 1994.) The half-cycle average load voltage is

$$V_{\text{ld(av)}} = \frac{1}{\pi}\int_{\alpha}^{\pi} V_m \sin \omega t \, d(\omega t)$$

$$= \frac{V_m}{\pi}(1 + \cos\alpha) \tag{4.48}$$

Also,

$$V_{\text{ld(RMS)}} = \left[\frac{1}{2\pi}\int_0^{2\pi} v_{RY}^2 d(\omega t)\right]^{1/2}$$

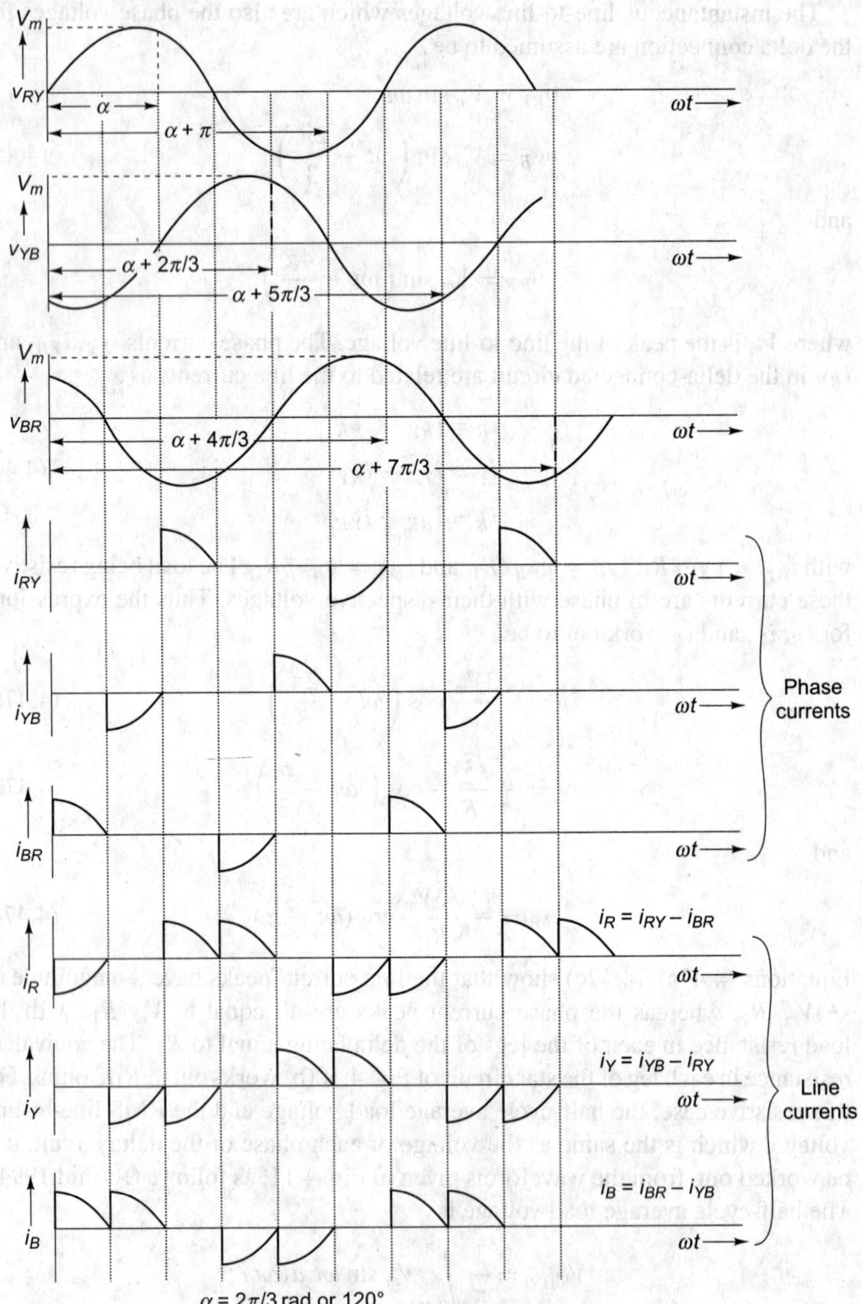

Fig. 4.11 Waveforms of a class-D ac chopper

$$= \left[\frac{2}{2\pi}\int_{\alpha}^{\pi} V_m^2 \sin^2 \omega t \, d(\omega t)\right]^{1/2}$$

$$= V_m \left[\frac{1}{2\pi}\left(\pi - \alpha + \frac{\sin 2\alpha}{2}\right)\right]^{1/2} \quad (4.49)$$

The maximum load voltage for the delta-connected load occurs at $\alpha = 0$ and its value is equal to $V_m/\sqrt{2}$, the range of α being $0 \le \alpha \le \pi$. The half-cycle average as well as RMS load currents are obtained as ratios of the corresponding voltages to the load resistance. Thus,

$$I_{\mathrm{ld(av)}} = \frac{V_{\mathrm{ld}}}{R_{\mathrm{ld}}} \quad (4.50)$$

and

$$I_{\mathrm{ld(RMS)}} = \frac{V_{\mathrm{ld(RMS)}}}{R_{\mathrm{ld}}} \quad (4.51)$$

The RMS current through any single thyristor can be determined from the fact that the thyristor conducts for half the total time of that phase. A simple calculation gives the RMS current as

$$I_{\mathrm{Th(RMS)}} = \frac{I_{\mathrm{ld(RMS)}}}{\sqrt{2}} \quad (4.52)$$

The line currents shown in Fig. 4.11 may be continuous or discontinuous depending on the value of α. Because of the irregular shapes of the current waveforms, the phase and line currents contain harmonics. The RMS value of the phase current can now be written as

$$I_{RY\mathrm{(RMS)}} = (I_1^2 + I_3^2 + I_5^2 + I_7^2 + \cdots + I_k^2)^{1/2} \quad (4.53)$$

where all odd harmonics up to the kth one are taken into consideration. The harmonics that are multiples of 3 circulate around the delta-connected load winding and do not appear in the line current (because the zero sequence currents are in phase in all the three branches). Hence a typical line current (here the R-phase current) can be expressed as

$$I_{R\mathrm{(RMS)}} = \sqrt{3}(I_1^2 + I_5^2 + I_7^2 + \cdots + I_n^2)^{1/2} \quad (4.54)$$

where n is not a multiple of 3. That is,

$$I_{R\mathrm{(RMS)}} < \sqrt{3} I_{RY\mathrm{(RMS)}} \quad (4.55)$$

Equation (4.55) shows that the current ratings of the thyristors in the delta will be more than $1/\sqrt{3}$ times the line currents. However, their peak inverse voltage is equal to the peak line voltage.

4.4.5 A Comparative Study of Three-phase Choppers

To provide a basis for the comparison of the performance of the four categories presented here, an input voltage of 250 V (L to L) is taken with an R_{ld} of 10 Ω (star) and an α of $\pi/3$ rad. Table 4.1 gives a summary of the results obtained.

Whereas the power factor is computed for category B, C, and D choppers, the input power factor is determined for the category A chopper. The results shows that category A and D choppers give the highest PF of 0.9. The category B chopper follows with a PF of 0.84 and the category C chopper gives the lowest PF of 0.6.

For a comparison of the voltages and currents, only categories A, B, and C are considered. The category C chopper gives the highest average load voltages and currents. Then categories A and B follow, giving the same average voltage and current. On the other hand, category A provides the highest RMS voltage followed by categories B and C in that order.

The above results cannot be generalized for all ranges of α, the ranges in turn being different for different categories. However, a general idea can be obtained by studying the various connections. The reader is advised to compute the performance features for other values of α.

Table 4.1

[Input voltage (L to L) = 250 V, $\alpha = \pi/3$ rad, $R_{\mathrm{ld}}(\mathrm{star}) = 10\ \Omega$]

Connection	Category A Star	Category B Star	Category C Star	Category D Delta-connected load
$V_{\mathrm{ld(av)}}$	97.45 V	97.45 V	113.6 V	168.8 V
$V_{\mathrm{ld(RMS)}}$	129.4 V	121.3 V	87.2 V	224.2 V
$I_{\mathrm{ld(av)}}$	9.75 A	9.75 A	11.36 A	5.63 A
$I_{\mathrm{ld(RMS)}}$	12.94 A	12.13 A	8.72 A	7.47 A
Power factor for categories B, C, D and input PF for category A	Input PF $= (I_1/I)\cos\phi_1 = \frac{12.10}{12.94} \times 0.959 \approx 0.9$	0.84	0.6	0.9

4.5 Single-phase Ac Choppers used as Sequence Controllers

The drawbacks of ac choppers highlighted in the previous sections are (a) high harmonic distortion and (b) low power factor for the output even with a resistive load. Distortion of waveforms causes torque pulsations, cogging, and other such undesirable phenomena in induction motor loads. Hence it should be minimized to improve the power factor. Interposing two ac controllers between the secondary of the transformer and the load is one such method that permits the control of the thyristors in a sequential manner and results in a smooth variation of the voltage across the secondary terminals. Though the resulting waveform may not be sinusoidal, it has the merit that all low-order harmonics are eliminated. Second, it can be extended to more than two ac controllers so as to further reduce the harmonic content and make the waveform nearly sinusoidal. A detailed treatment of the two- and multi-controller configurations as tap-changers for a single-phase transformer is now described.

4.5.1 Single-phase Transformer Tap-changer

The combination of two ac controllers connected to the secondary of a transformer (Fig. 4.12) may be operated as a continuous tap-changer. The antiparallel circuit consisting of thyristors Th_3 and Th_4 is connected to the centre point of the secondary, whereas that with Th_1 and Th_2 is connected at the top of the secondary. Assuming a resistive load, three methods of operation for these ac controllers are presented below and thus three corresponding ranges of load voltage are obtained.

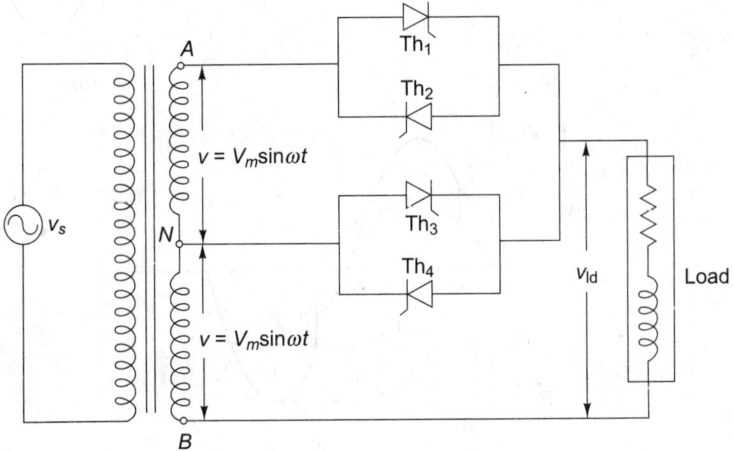

Fig. 4.12 Single-phase transformer tap-changer using two ac controllers

(a) Th_1 and Th_2 are turned off; Th_3 and Th_4 are turned on at α and $\pi+\alpha$, respectively. The variation of voltage obtained in this case is given by $0 \le V_{ld} \le V_m/\sqrt{2}$. The load voltage waveform for a typical value of α is given in Fig. 4.13(a).

(b) Th_3 and Th_4 are turned off; Th_1 and Th_2 are turned on at α and $\pi+\alpha$, respectively. The range of load voltage obtained with this mode of operation is given by the inequality $0 \le V_{ld} \le V_m\sqrt{2}$. A typical load voltage waveform is given in Fig. 4.13(b). With a scale change, this waveform also represents the load current waveform.

(c) Both the pairs of thyristors are employed.

(i) Resistive load: This is the most general case for the tap-changer of Fig. 4.12. To start with, the thyristors Th_1 and Th_2 are turned off. The thyristor Th_3 is fired at zero angle and Th_4 at an angle of π so that, together, they conduct the load current throughout the positive and negative half-cycles, respectively. Now if Th_1 is turned on at $\alpha < \pi$ in the positive half-cycle, the voltage induced across the upper half AN of the secondary winding is applied in a reverse fashion across Th_3 and it turns off. The load voltage waveform from α to π will be the same as that of the full secondary. At π, Th_1 is turned off and Th_4 is turned on. When Th_2 is turned on at $\pi+\alpha$, the voltage across AN, which is negative, is applied in a reverse direction across Th_4 and it turns off.

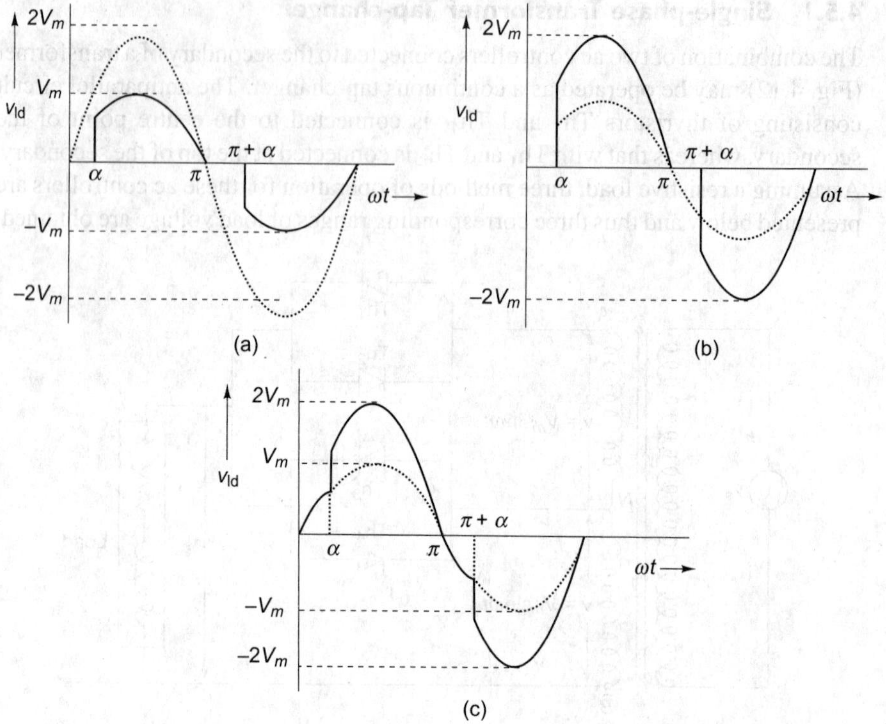

Fig. 4.13 Waveforms for the tap-changer of Fig. 4.12 with resistive load

The load voltage now is the same as that of the full secondary. Th_2 is turned off at 2π and Th_3 is turned on again to repeat the sequence. The resulting waveforms are shown in Fig. 4.13(c). For a variation of α in the range $0 \leq \alpha \leq \pi$, the load voltage will be in the range $V_m/\sqrt{2}$ to $2V_m/\sqrt{2}$.

(ii) Inductive load: The circuit configuration for this case is the same as that of Fig. 4.13(c). As in case (i), Th_1 and Th_2 are fired at α and $\pi + \alpha$, whereas Th_3 and Th_4 are fired at 0 and π, respectively. This implies that this latter pair conducts the full positive and negative half-cycles. Though the output voltage variation obtained in this case is also given by $V_m/\sqrt{2} \leq V_{ld} \leq 2V_m/\sqrt{2}$, the current lags the voltage as shown in Fig. 4.14. In the following treatment, the firing pulses of all the thyristors will be assumed to be rectangular in shape and their duration will extend up to the entire interval of conduction of the thyristors. The reason for this remains the same as that given in Section 4.3.2 for a single-phase chopper feeding an inductive load. Thus, the ranges of gate pulses for Th_1, Th_2, Th_3, and Th_4 will be α to π, $\pi + \alpha$ to 2π, 0 to π, and π to 2π, respectively.

The sequence of operation starts when Th_3 is fired at $\omega t = 0$, which is the start of the positive half-cycle. When Th_1 is fired at α, a reverse bias of v_1 (= $V_m \sin\alpha$) volts is applied across Th_3 and it turns off. The load current now flows through the whole of the transformer winding, the thyristor Th_1, and the load. The load

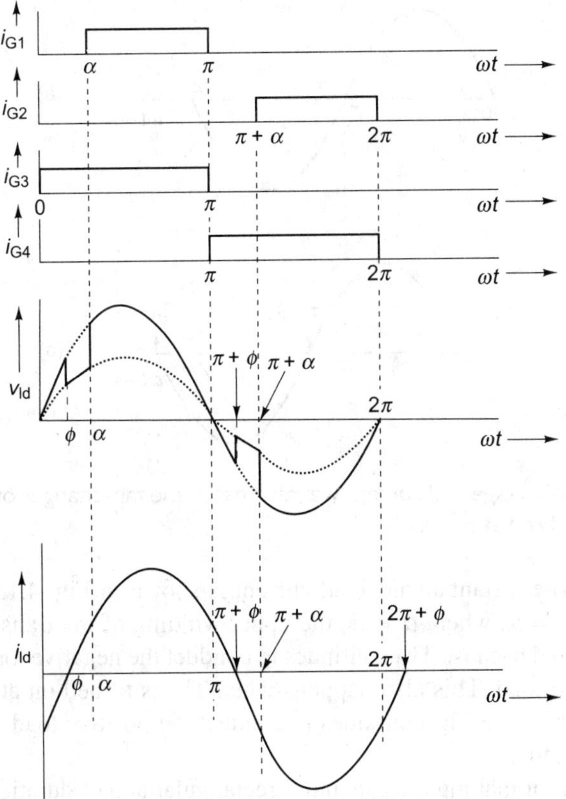

Fig. 4.14 Waveforms for the tap-changer of Fig. 4.12 with inductive load

voltage therefore jumps to $2v_1$ ($= 2V_m \sin\alpha$). At $\omega t = \pi$ the gate pulses of the thyristors Th_1 and Th_3 are terminated. However, Th_1 continues to conduct till $\pi + \phi$ because of the inductive nature of the load. Th_4 receives a firing signal at π but cannot turn on because the voltage of the upper half of the winding becomes negative even after π and gets applied in a reverse direction across Th_4. When the load current attains a zero value at an angle $\pi + \phi$, where ϕ depends on the $\omega L/R$ ratio, Th_1 gets reverse biased by a voltage $2v_2$ ($= -2V_m \sin\phi$) and turns off. Th_4 now gets forward biased and provides the load with a voltage v_2 ($= -V_m \sin\phi$). This negative voltage also helps in commutating Th_1. At $\omega t = \pi + \alpha$, Th_2 is turned on and the load voltage jumps to $2v$ ($= -2V_m \sin\alpha$). Half of this voltage, namely, $-V_m \sin\alpha$, is applied across Th_4 in a reverse manner and it turns off. Th_2 continues to conduct beyond $\omega t = 2\pi$ because of the inductive nature of the load.

One precaution to be observed with an inductive load is that Th_1 should not be given a firing signal at $\alpha < \phi$. The reason for this can be ascertained with the help of Fig. 4.15 as follows. It is assumed that the control range is $0 \le V_{ld(RMS)} \le 2V_m/\sqrt{2}$, where V_m is the peak of the secondary voltage of the transformer at the centre. Also, it is assumed that Th_1 and Th_2 are turned off while Th_3 and Th_4 are turned on during the alternative half-cycles at the zero crossings of the

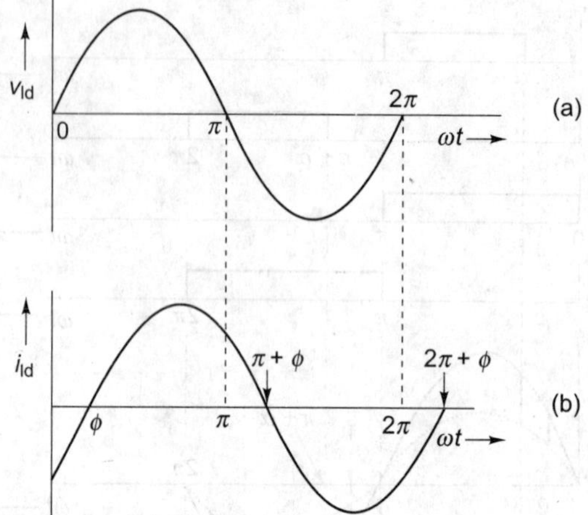

Fig. 4.15 Load voltage and current waveforms for the tap-changer of Fig. 4.12 with Th_1 fired at $\alpha < \phi$

load current. The instantaneous load current is shown in Fig. 4.15(b). If Th_1 is turned on at $\omega t = \alpha$, where $\alpha < \phi$, the upper winding of the transformer would be short-circuited because Th_4 continues to conduct the negative load current due to the inductive load. This also happens when Th_2 is turned on at $\omega t = \pi + \alpha$, where $\alpha < \theta$, because Th_3 continues to conduct the positive load current due to the inductive load.

The reasons for making the gate pulse rectangular and of duration equal to the conduction interval are as follows. With the load being inductive, Th_1 and Th_2 conduct beyond π and 2π, respectively. The thyristor Th_4 has to start conduction as soon as Th_1 stops conduction; similarly, Th_3 has to begin conducting when Th_2 stops conduction. With the power factor of the load varying between 0 and 1, the angles of discontinuation of conduction of thyristors Th_1 and Th_2 can, respectively, be the duration between π and 2π for Th_1 and 0 and π for Th_2.

The method of sequence control or tap changing can be extended to more than two controllers, say n controllers (with $n > 2$). These must be situated at equal distances along the secondary, as shown in Fig. 4.16, each contributing to an RMS voltage of $V_m/n\sqrt{2}$. Here, V_m is the peak value of the voltage induced across the whole of the secondary winding. A better method of constructing a tap-changer would be to split the secondary winding into n subsections which provide voltages in the geometric progression $v, 2v, \ldots, 2^{n-1}v$. The advantages of such a scheme are (a) possibility of continuous voltage control over a wide range, (b) low harmonic content, and (c) smaller number of thyristor pairs being used in the tap-changer compared to that used in the equal voltage scheme.

For the resistive load alternative under case (c) considered above, the RMS magnitude of the load voltage can be expressed as

$$V_{ld(RMS)} = \left[\frac{1}{2\pi}\int_0^{2\pi} (V_{ld}\omega t)^2 d(\omega t)\right]^{1/2}$$

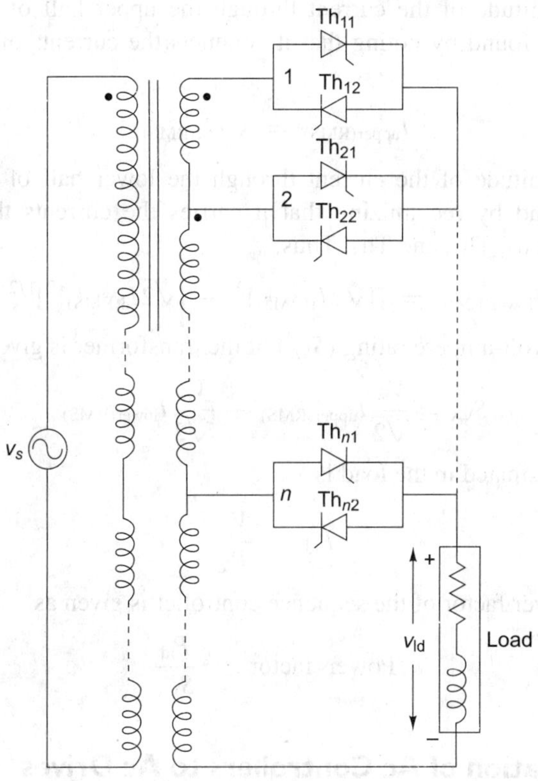

Fig. 4.16 Single-phase transformer with n ac controllers as tap-changers

$$= \left[\frac{2}{2\pi}\left\{\int_0^{\alpha}(V_m \sin \omega t)^2 d(\omega t) + \int_{\alpha}^{\pi}(2V_m \sin \omega t)^2 d(\omega t)\right\}\right]^{1/2}$$

$$= V_m\left[2 - \frac{3\alpha}{2\pi} + \frac{3\ \sin 2\alpha}{4\pi}\right]^{1/2} \tag{4.56}$$

The RMS magnitude of current through either Th_1 or Th_2 is given by

$$I_{1(\text{RMS})} = \left[\frac{1}{2\pi}\int_{\alpha}^{\pi}\left(\frac{2V_m \sin \omega t}{R_{\text{ld}}}\right)^2 d(\omega t)\right]^{1/2}$$

$$= \frac{V_m}{R_{\text{ld}}}\left[1 - \frac{\alpha}{\pi} + \frac{\sin 2\alpha}{2\pi}\right]^{1/2} \tag{4.57}$$

The RMS magnitude of current through either Th_3 or Th_4 can be expressed as

$$I_{3(\text{RMS})} = \left[\frac{1}{2\pi}\int_0^{\alpha}\left(\frac{V_m \sin \omega t}{R_{\text{ld}}}\right)^2 d(\omega t)\right]^{1/2}$$

$$= \frac{V_m}{2R_{\text{ld}}}\left[\frac{\alpha}{\pi} - \frac{\sin 2\alpha}{2\pi}\right]^{1/2} \tag{4.58}$$

The RMS magnitude of the current through the upper half of the secondary winding can be found by noting that it conducts the currents of both Th_1 and Th_2. Thus,

$$I_{\text{upper(RMS)}} = \sqrt{2}I_{1\text{(RMS)}} \tag{4.59}$$

The RMS magnitude of the current through the lower half of the secondary winding is found by recognizing that it carries the currents through all the thyristors Th_1, Th_2, Th_3, and Th_4. Thus,

$$I_{\text{lower(RMS)}} = [\{\sqrt{2}I_{1\text{(RMS)}}\}^2 + \{\sqrt{2}I_{3\text{(RMS)}}\}^2]^{1/2} \tag{4.60}$$

The secondary volt-ampere rating (S_{VA}) of the transformer is given as

$$S_{\text{VA}} = \frac{V_m}{\sqrt{2}}I_{\text{upper(RMS)}} + \frac{V_m}{\sqrt{2}}I_{\text{lower(RMS)}} \tag{4.61}$$

Also, power dissipated in the load is

$$P_{\text{ld}} = \frac{V_{\text{ld}}^2}{R_{\text{ld}}} \tag{4.62}$$

Finally, the power factor of the sequence controller is given as

$$\text{Power factor} = \frac{P_{\text{ld}}}{S_{\text{VA}}} \tag{4.63}$$

4.6 Application of Ac Controllers to Ac Drives

Ac controllers are usually employed for speed control of induction and synchronous motors. A three-phase synchronous or induction motor which has a two-pole three-phase winding has its three stator coils displaced in space by 120° from each other. These coils are excited by a three-phase system of voltages which are displaced from each other by 120° (elec.) in time. The input supply being a sinusoidal voltage in usual operation, each coil carries a sinusoidal current which establishes a pulsating magnetic field of magnetomotive force (mmf). Since the phase winding is usually distributed in a number of slots on an iron surface, the spatial distribution of the mmf is non-sinusoidal. However, by proper distribution it can be made to provide a nearly sinusoidal distribution of mmf. The mmf wave corresponding to each phase current pulsates in magnitude, the spatial distribution remaining sinusoidal in nature. The output voltage of an ac controller is non-sinusoidal but periodic in nature. Such a non-sinusoidal voltage applied to the stator winding of a motor gives rise to a corresponding non-sinusoidal current waveform which can be decomposed into various harmonics. The frequency of the fundamental waveform, however, remains the same as that of the input of the controller. If, for simplicity of analysis, any slight magnetic saturation that may exist in the motor is neglected, in respect of the harmonic currents it can be considered as a linear system which obeys the principle of superposition. The effect of each of the harmonic inputs can be evaluated and all such results summed up to obtain the net effect of the non-sinusoidal input given to the motor. Thus, the net motor current and torque are, respectively, equal to the sum of the currents and that of the average torques provided by each of the harmonic components.

The time harmonic currents set up in an ac motor are of the order of $6n\pm1$, where n is an integer. These currents give rise to harmonic waves of mmf of the same order. While the mmf harmonics of the order $6n+1$ rotate in the same direction as the main field, those of the order $6n-1$ rotate in an opposite direction. These are, respectively, called positive-sequence and negative-sequence harmonics. The reaction of a rotor harmonic mmf with a harmonic air gap flux of the same order gives rise to a steady-state harmonic torque. Expressions are now derived for these positive- and negative-sequence torques.

For the purpose of determining the effect of harmonics on the induction motor, the equivalent circuit of the motor has to be modified to reflect the effect of higher frequency input on its elements. Figures 4.17(a) and (b), respectively, give the equivalent circuits of an induction motor with fundamental and kth harmonic inputs.

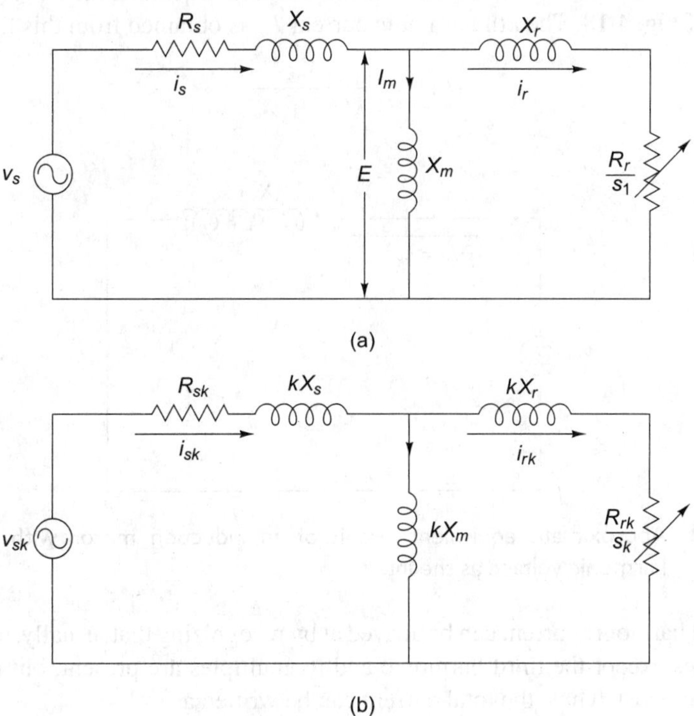

Fig. 4.17 Equivalent circuit of an induction motor with a sinusoidal harmonic input, with rotor parameters referred to the stator: (a) fundamental and (b) kth harmonic

The slip with respect to the fundamental rotating field is defined as

$$s_1 = \frac{N_s - N}{N_s} = \frac{\omega_s - \omega_m}{\omega_s} \tag{4.64}$$

where N_s and ω_s are the synchronous speeds of the rotor in rpm and the corresponding rotating field in rad/s, respectively. Likewise, N and ω_m are, respectively, the actual rotor speed in rpm and the actual speed of the corresponding rotating field in rad/s. The slip for the kth harmonic forward rotating field is given as

$$s_k = \frac{kN_s - N}{kN_s} = \frac{k\omega_s - \omega_m}{k\omega_s} \tag{4.65}$$

It is assumed that R_r and jX_r are, respectively, the equivalent rotor resistance per phase and the equivalent reactance per phase, at standstill. Likewise, the quantities R_{rk} and jkX_r are the equivalent resistance and reactance corresponding to the kth harmonic equivalent circuit.

The harmonic resistance R_{rk} is usually different from R_r due to the skin effect. However, the reactance jkX_r is much greater than R_{rk}. Hence the latter can be neglected. Moreover, the kth harmonic shunt magnetizing reactance jX_m is much greater than the rotor leakage reactance and hence can be treated as an infinite impedance and omitted. With these considerations the equivalent circuit simplifies to that of Fig. 4.18. The kth harmonic current I_{sk} is obtained from this figure as

$$|I_{sk}| = \frac{V_{sk}}{k(X_s + X_r)} \tag{4.66}$$

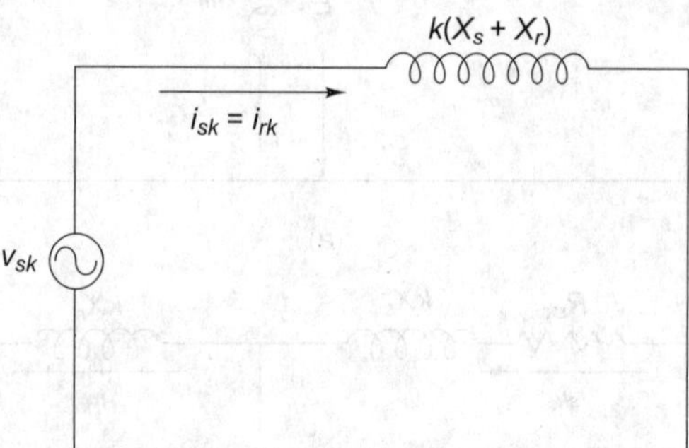

Fig. 4.18 Approximate equivalent circuit of an induction motor with the kth harmonic voltage as the input

The total harmonic current can be arrived at by recognizing that, usually, only odd harmonics except the third harmonic and its multiples are present, but no even harmonics exist. Thus, the total current can be written as

$$I_{s(\text{RMS})} = [I_{s1}^2 + I_{s5}^2 + I_{s7}^2]^{1/2} \tag{4.67}$$

From the theory of the induction motor, the torque developed due to the fundamental frequency voltage is given as

$$T_{d1} = \frac{3I_r^2 R_r}{\omega_m}\left(\frac{1 - s_1}{s_1}\right) \tag{4.68}$$

This torque is greater than the shaft torque by an amount equal to the windage and friction losses. The fundamental frequency f_1 in hertz is related to speed as follows:

$$f_1 = \frac{N_s p}{60} = \frac{\omega_s}{2\pi} p \tag{4.69}$$

where p is the number of pole pairs. From Eqn (4.64), ω_s can be written in terms of ω_m and s_1 as

$$\omega_s = \frac{\omega_m}{1 - s_1} \tag{4.70}$$

ω_s can also be obtained from Eqn (4.69) as

$$\omega_s = \frac{2\pi f_1}{p} \tag{4.71}$$

The right-hand sides of Eqns (4.70) and (4.71) can be equated to give

$$\frac{\omega_m}{1 - s_1} = \frac{2\pi f_1}{p} \tag{4.72}$$

Equation (4.72) can be solved for ω_m as

$$\omega_m = \frac{2\pi f_1 (1 - s_1)}{p} \tag{4.73}$$

Using Eqn (4.73), Eqn (4.68) for the steady fundamental torque can now be written as

$$T_{d1} = \frac{1.5p}{\pi f_1} \frac{I_r^2 R_r}{s_1} \tag{4.74}$$

By a similar argument, the expression for the kth harmonic torque can be derived as

$$T_{dk} = \pm \left(\frac{1.5p}{k\pi f_1} \right) \frac{I_{rk}^2 R_{rk}}{s_k} \tag{4.75}$$

Taking the positive value in Eqn (4.75), the ratio T_{dk}/T_{d1} can now be expressed as

$$\frac{T_{dk}}{T_{d1}} = \left(\frac{I_{rk}}{I_r} \right)^2 \frac{R_{rk}}{R_r} \frac{s_1}{k s_k} \tag{4.76}$$

Now, using Eqns (4.64) and (4.65), s_k can be written in terms of s_1 as

$$s_k = \frac{(k-1) + s_1}{k} \tag{4.77}$$

s_1 is usually very small for normal full load operation of the induction motor. Therefore, s_k can be approximated as

$$s_k \approx \frac{k-1}{k} \tag{4.78}$$

Using this approximation, T_{dk} of Eqn (4.76) can be expressed in terms of T_{d1} as

$$\frac{T_{dk}}{T_{d1}} \approx \left(\frac{I_{rk}}{I_r} \right)^2 \frac{R_{rk}}{R_r} \frac{s_1}{k-1} \tag{4.79}$$

A similar derivation can be done for the torque developed due to the kth harmonic backward field:

$$\frac{T_{dk}}{T_{d1}} = -\left(\frac{I_{rk}}{I_r}\right)^2 \frac{R_{rk}}{R_r} \frac{s_1}{k+1} \tag{4.80}$$

Fifth harmonic currents, being negative-sequence ones, generate negative torques. On the other hand, seventh harmonic currents are positive-sequence currents and hence give rise to positive torques. The values of the resistance, currents, and slip on the right-hand sides of Eqns (4.79) and (4.80) are such that the negative fifth harmonic torque will be nearly equal to the positive seventh harmonic torque. Therefore, their combined effect on the steady-state fundamental torque is negligible.

In addition to the steady components of torque discussed above, pulsating torques, produced by the reaction of harmonic rotor mmfs with harmonic rotating air gap fluxes of different orders, are generated in an ac machine. For instance, the fifth harmonic stator currents, which produce a space fundamental mmf rotating at five times the fundamental synchronous speed in the opposite direction, react with the fundamental rotating field to produce a pulsating torque at six times the fundamental frequency. This is because the relative speed of the fifth harmonic rotor mmf wave and the fundamental air gap field is six times the synchronous speed in the reverse direction. Likewise, the seventh harmonic currents, which rotate at a speed of seven times the fundamental frequency in the positive direction, also produce a pulsating torque at six times the fundamental frequency. This is because their relative speed with respect to the fundamental is also six times that of the synchronous speed in the forward direction. The above-mentioned two pulsating torques at six times the fundamental frequency combine to produce a fluctuation in the electromagnetic torque developed by the motor. Similarly, the 11th and 13th harmonics produce a 12th harmonic pulsating torque.

Though these pulsating torques have an average value equal to zero, they cause the angular velocity of the rotor to change during a revolution. As a consequence, the motor rotation may consist of jerks, or may stop at very low speeds. For avoiding such irregular motion called *cogging*, the motor has to be operated above a certain lower speed limit. This limit depends on the inertia of the rotor and associated parts of the rotating system. Such a performance is unacceptable in machine tool drives. In the case of geared systems these pulsating torques lead to excessive wear and tear of the gear teeth, especially when the torque pulsating frequency coincides with the resonant frequency of the mechanical shaft. A remedy for this behaviour consists in the suppression of the low-order harmonics from the output of the ac controller.

The fundamental equivalent circuit of Fig. 4.17(a) applies only to the three-phase induction motor. However, the harmonic equivalent circuit of Fig. 4.18 is also valid for the harmonic behaviour of the field excited and permanent magnet excited type of synchronous motors because these machines have cage windings or damper bars, and they operate asynchronously with respect to time-harmonic mmf waves.

Summary

Single-phase and three-phase ac choppers provide variable voltage inputs to ac motors and other loads with a much simpler circuit compared to inverters and cycloconverters. Ac controllers have the merit that the fundamental frequency of output voltage obtained with them remains the same as the input frequency. However, the output contains a high harmonic content and this will cause problems for various kinds of loads as follows. (a) In induction motor loads, phenomena such as cogging and crawling will deteriorate their performance. (b) In geared systems, the coincidence of the torque pulsations due to low-order harmonic torques with the resonant frequency of the shaft may lead to excessive wear and tear of gear teeth. These examples emphasize the necessity of suppressing low-order harmonics.

In the *B* and *C* categories of three-phase ac controllers, a careful assessment of the exact voltage that is applied to the individual phases is necessary. This is to arrive at the RMS and average values of the output voltage. For all the categories of three-phase controllers, the power factor serves as a yardstick for the assessment of the power loss due to harmonics.

A ready application for ac controllers can be found in lighting and heating loads, where harmonics do not cause any problems. On the other hand, harmonic currents augment the heat and light generated by the fundamental component. In other applications, namely, thyristor controlled reactors and thyristor switched capacitors, dealt with in Chapter 11, ac controllers provide a quick control over reactive power. Because of their simple circuitry these controllers are proving to be less costly and more compact than conventional power-control circuitry. Transformer tap changing is a third area in which single-phase ac controllers are finding wide usage. In this application, increasing the number of sections of the transformer winding is conducive to smoother control.

Worked Examples

1. The single-phase ac controller circuit of Fig. 4.19(a) is operated with multiple-cycle on–off control. The thyristors are switched on with $\alpha = 0$ for one-sixth of a second and switched off for one-fourth of a second. Given $R_{\rm ld} = 5\ \Omega$, $L_{\rm ld} = 15$ mH, and $E_m = 141.4$ V. Find (a) the ratio of the load voltage with on–off control to that with the direct connection of the load to the ac supply, α remaining the same in both the cases, and (b) the ratio of powers.

Solution

Here $p = 1/6 \times 50 = 8.33$ cycles, $q = 50/4 = 12.5$ cycles. The load voltage waveforms with on–off control, are given in Fig. 4.19(b). The RMS value of the load voltage with on–off control,

$$V_1 = \sqrt{\frac{1}{2\pi}\left(\frac{p}{p+q}\right)\int_0^{2\pi}(E_m \sin\omega t)^2 d(\omega t)}$$

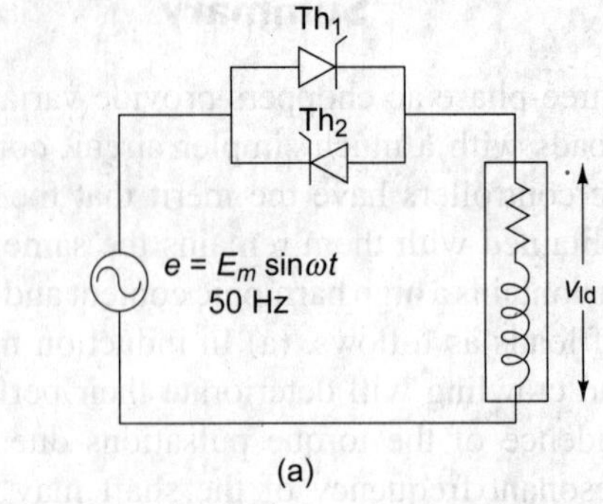

(a)

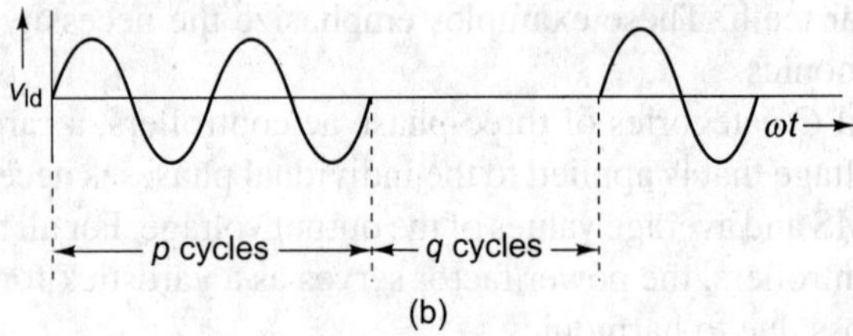

(b)

Fig. 4.19

$$= \frac{E_m}{\sqrt{2}}\sqrt{\frac{p}{p+q}}$$

$$= \frac{141.4}{\sqrt{2}}\sqrt{\frac{8.33}{20.83}}$$

$$= 63.24 \text{ V}$$

$$Z = \sqrt{5^2 + \left(\frac{314 \times 15}{1000}\right)^2} = 6.87\ \Omega$$

Current with on–off control,

$$I_1 = \frac{V_1}{Z} = \frac{63.24}{6.87} = 9.2 \text{ A}$$

Power delivered to the load,

$$P_1 = I_1^2 R = (9.2)^2 \times 5 = 424 \text{ W}$$

Voltage with direct connection,

$$V_0 = \frac{141.4}{\sqrt{2}} = 100 \text{ V}$$

Therefore, the ratio of voltages

$$\frac{V_1}{V_0} = \frac{63.2}{100} = 0.632$$

The load current with direct connection = 100/6.87 = 14.56. Therefore, the power delivered to the load with direct connection,

$$P_0 = (14.56)^2 \times 5 = 1060 \text{ W}$$

Ratio of powers

$$\frac{P_1}{P_0} = \frac{424}{1060} = 0.4$$

2. A single-phase ac controller is supplied by a source of RMS voltage 210 V at 50 Hz. A resistive load of 5.6 Ω is connected at its output. Determine (a) the value of α that gives an RMS voltage of 100 V, (b) the amplitude of the seventh harmonic current with $\alpha = \pi/6$, and (c) the input power factor.

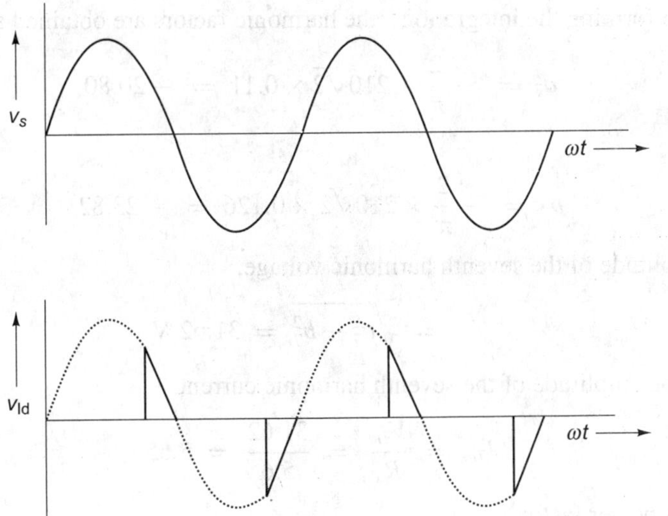

Fig. 4.20

Solution

(a) The circuit is shown in Fig. 4.4(a) and waveforms in Fig. 4.20. α is found by trial and error from the equation

$$100 = \frac{210\sqrt{2}}{\sqrt{2}}\sqrt{1 - \frac{\alpha}{\pi} + \frac{\sin 2\alpha}{2\pi}}$$

and is equal to 2.03 rad or 116.3°.

(b) The seventh harmonic voltage is first determined as follows:

$$v_{ld}(t) = \sum_{n=1}^{\infty}(a_n \cos n\omega t + b_n \sin n\omega t)$$

$$a_7 = \frac{1}{\pi}\int_0^{2\pi} v_{ld}(t)\cos(7\omega t)\, d(\omega t)$$

$$= \frac{2}{\pi}\int_{\alpha}^{\pi} V_m \sin \omega t \cos(7\omega t)\, d(\omega t)$$

$$= \frac{2}{\pi}\int_{116.3}^{180} 210\sqrt{2}\sin\omega t\cos(7\omega t)d(\omega t)$$

Also,

$$b_7 = \frac{1}{\pi}\int_{0}^{2\pi} v_{\text{ld}}(t)\sin(7\omega t)\,d(\omega t)$$

$$= \frac{2}{\pi}\int_{\alpha}^{\pi} V_m\sin\omega t\sin(7\omega t)d(\omega t)$$

$$= \frac{2}{\pi}\int_{116.3}^{180} 210\sqrt{2}\sin\omega t\sin(7\omega t)\,d(\omega t)$$

Upon performing the integrations, the harmonic factors are obtained as

$$a_7 = -\frac{2}{\pi}\times 210\sqrt{2}\times 0.11 = -20.80$$

and

$$b_7 = -\frac{2}{\pi}\times 210\sqrt{2}\times 0.126 = -23.82$$

The amplitude of the seventh harmonic voltage,

$$V_{7m} = \sqrt{a_7^2 + b_7^2} = 31.62 \text{ V}$$

Hence the amplitude of the seventh harmonic current,

$$I_{7m} = \frac{V_{7m}}{R_{\text{ld}}} = \frac{31.62}{5.6} = 5.65$$

(c) Input power factor:

$$\mu\cos\phi_1 = \frac{I_1}{I}\cos\phi_1$$

For the ac controller of Fig. 4.3,

RMS value of input current = RMS value of load current

and

$$\begin{pmatrix}\text{RMS value of fundamental}\\ \text{component of input current}\end{pmatrix} = \begin{pmatrix}\text{RMS value of fundamental}\\ \text{component of load current}\end{pmatrix}$$

The RMS value of the fundamental component of voltage is

$$V_1 = \frac{\sqrt{a_1^2 + b_1^2}}{\sqrt{2}}$$

where

$$a_1 = \frac{2}{\pi}\int_{\alpha}^{\pi} V_m\sin\omega t\cos\omega t\ d(\omega t) = -76.0$$

and

$$b_1 = \frac{2}{\pi}\int_{\alpha}^{\pi} V_m \sin^2\omega t \, d(\omega t) \; = \; 67.6$$

$$\phi_1 = \tan^{-1}\left(\frac{a_1}{b_1}\right) \; = \; \tan^{-1}\left(\frac{-76.0}{67.6}\right) \; = \; -48.3^\circ$$

$$\cos\phi_1 = 0.665$$

$$V_1 = \frac{\sqrt{(-76.0)^2 + (67.6)^2}}{\sqrt{2}} = 71.92$$

The RMS value of the fundamental component of the input current is

$$I_1 \; = \; \frac{V_1}{R} \; = \; \frac{71.92}{5.6} \; = \; 12.84 \text{ A}$$

The RMS value of input current $I = 100/5.6 = 17.86$ A. Hence,

$$\mu \; = \; \frac{I_1}{I} \; = \; \frac{12.84}{17.86} \; = \; 0.72$$

Now, the input power factor becomes

$$\mu\cos\phi_1 \; = \; 0.72 \times 0.665 \; = \; 0.478$$

3. A single-phase ac controller has the following data: $V = 280$ V, $f = 50$ Hz, $X_L = 40\ \Omega$, and $R_{\text{ld}} = 0.1\ \Omega$. Determine (a) the range of α for the load current in the chopper mode, (b) the maximum value of the average thyristor current over the permissible range of α, (c) the maximum RMS value of the thyristor current over the permissible range of α, (d) the range of α for which the greatest forward voltage is applied to the thyristors, and (e) the range of α for which the maximum reverse voltage is applied to the thyristors.

Solution

The circuit and waveforms are given in Figs 4.4(a) and (b).

(a) The load power factor angle

$$\phi \; = \; \tan^{-1}\left(\frac{X_{\text{ld}}}{R_{\text{ld}}}\right) \; = \; \tan^{-1}\left(\frac{40}{0.1}\right) \; = \; 89.86^\circ$$

This gives

$$\cos\phi \; \approx \; 0$$

Hence the range of α in the chopper mode is given as

$$89.86^\circ \; < \; \alpha \; < \; 180^\circ$$

(b) From Eqn (4.20),

$$i \; = \; \frac{V_m}{Z}\sin(\omega t - \phi) - e^{-(R/L)(t-\alpha/\omega)}\sin(\alpha - \phi)$$

i_{max} occurs with $\alpha = \phi$. Hence,

$$i_{max} = \frac{V_m}{Z}\sin(\omega t - \phi)$$

From the chart of Fig. 4.4(c), for $\alpha = 90°$and $\cos\phi = 0°$, $\alpha_e \approx 270° = \pi + \phi$. The maximum average value of i_{max} is given as

$$I_{max} = \frac{1}{\pi}\int_{\phi}^{\pi+\phi}\frac{V_m}{Z}\sin(\omega t - \phi)d(\omega t)$$

$$= \frac{1}{\pi}\frac{V_m}{Z}[-\cos(\omega t - \phi)]_{\phi}^{\pi-\phi}$$

$$= \frac{1}{\pi}\frac{V_m}{Z}[-\cos(\pi + \phi - \phi) + \cos(\phi - \phi)]$$

$$= \frac{1}{\pi}\frac{V_m}{Z}\left[-\cos\pi + \cos 0°\right]$$

$$= \frac{2V_m}{\pi Z} = \frac{2 \times 280\sqrt{2}}{\pi \times 40} = 6.3 \text{ A}$$

Here $Z = \sqrt{0.1^2 + 40^2} \approx 40\ \Omega$.

(c) The maximum RMS value of current

$$= \sqrt{\frac{1}{\pi}\int_{\phi}^{\pi+\phi}\left[\frac{V_m}{Z}\sin(\omega t - \phi)\right]^2 d(\omega t)}$$

$$= \sqrt{\frac{1}{\pi}\int_{\phi}^{\pi+\phi}\frac{V_m^2}{Z^2}\sin^2(\omega t - \phi)d(\omega t)}$$

$$= \frac{V_m}{\sqrt{2}Z}$$

$$= \frac{280\sqrt{2}}{40\sqrt{2}} = 7.0 \text{ A}$$

Hence, the maximum RMS value of the thyristor current is equal to

$$\text{Maximum RMS current}/\sqrt{2} = 7.0/\sqrt{2} = 4.95 \text{ A}$$

(d) The range of α for which the greatest forward voltage is applied to the thyristors is given by $\pi/2 \le \alpha \le \pi$.

(e) The range of α for which the greatest reverse voltage is applied across the thyristors is also given by $\pi/2 \le \alpha \le \pi$.

4. For the three-phase ac chopper of category B the line-to-line supply voltage is 300 V and the load consists of a 6 Ω resistance. Determine (a) the value of α required to give a line-to-neutral voltage of 110 V at the load, (b) the maximum forward and reverse thyristor voltages, (c) the maximum RMS thyristor current, and (d) the maximum average thyristor current.

Solution

(a) It is estimated that to give a line-to-neutral voltage of 110 V, α should be in the range $\pi/2 \leq \alpha \leq 5\pi/6$. Equation (4.36) gives the RMS value of voltage for this range. Thus,

$$110 = V_m\sqrt{\frac{5}{8} - \frac{3\alpha}{4\pi} + \frac{3}{16\pi}\sin 2\alpha + \frac{3\sqrt{3}}{16\pi}\cos 2\alpha}$$

By trial and error, α is found to be 82°.

(b) The maximum thyristor forward voltage $= 300\sqrt{2}/\sqrt{3} = 245$ V. The maximum thyristor reverse voltage $= 300\sqrt{2}/\sqrt{3} = 245$ V.

(c) The maximum RMS thyristor current is equal to

$$\frac{\text{Maximum RMS load voltage that occurs at } \alpha = 0}{\text{Resistance} \times \sqrt{2}} = \frac{300}{6\sqrt{2}} = \frac{50}{\sqrt{2}}\text{ A} = 35.35\text{ A}$$

(d) The maximum average thyristor current is

$$\frac{1}{\pi}\int_0^{\pi} I_m \sin \omega t \, d(\omega t) = \frac{2I_m}{R\pi} = \frac{2 \times 50\sqrt{2}}{\pi} = 45.0\text{ A}$$

5. A 12-kW, three-phase, 230-V, star-connected induction motor which has a full load power factor of 0.85 and efficiency of 88% is fed by a three-phase ac controller of category B. The ac controller is fed in turn by a three-phase supply with a line-to-line voltage of 240 V. Determine the (a) firing angle α of the thyristor, (b) RMS thyristor current, (c) average thyristor current, and (d) maximum thyristor forward and reverse voltage ratings.

Solution

(a) The line-to-neutral input voltage of an ac controller $= 240/\sqrt{3} = 138.56$ V (RMS). The output voltage $= 230/\sqrt{3} = 132.79$ V (RMS). Assuming α to be in the range $0 \leq \alpha \leq \pi/3$, it can be determined using Eqn (4.32):

$$132.79 = 138.56\sqrt{2}\sqrt{\frac{1}{2} - \frac{3\alpha}{4\pi} + \frac{3}{8\pi}\sin 2\alpha}$$

By trial and error, α is obtained as 38°.

(b) Efficiency = output/input. Therefore,

$$0.88 = \frac{12{,}000}{\sqrt{3} \times 230 \times I \times 0.85}$$

From this,

$$I = \frac{12{,}000}{\sqrt{3} \times 230 \times 0.88 \times 0.85} = 40.27\text{ A}$$

This is the RMS output current per phase of the ac controller.

$$\text{RMS current per thyristor} = \frac{40.27}{\sqrt{2}} = 28.48\text{ A}$$

Now,

$$I_{\text{RMS}} = I_m \sqrt{\frac{1}{2} - \frac{3\alpha}{4\pi} + \frac{3}{8\pi}\sin 2\alpha}$$

That is,

$$40.27 = I_m \sqrt{\frac{1}{2} - \frac{3 \times 0.663}{4\pi} + \frac{3}{8\pi} \times 0.97}$$

This gives I_m as 59.54 A.

(c) The half-cycle average of the motor current is computed as

$$I_{\text{av}} = \frac{1}{\pi}\int_{38^\circ}^{180^\circ} I_m \sin\omega t \; d(\omega t) = \frac{1}{\pi}\int_{38^\circ}^{180^\circ} 59.54 \sin\omega t \; d(\omega t) = 33.89 \text{ A}$$

This gives the average thyristor current as 33.89/2 = 16.94 A.

(d) The maximum thyristor forward voltage in the entire range of α is $240\sqrt{2}/\sqrt{3}$ = 196 V.

The maximum thyristor reverse voltage in the entire range of α is $240\sqrt{2}/\sqrt{3}$ = 196 V.

6. A three-phase ac controller of category B supplies a star-connected resistive load of 8 Ω resistance and the input line-to-line voltage V_{LL} is 200 V (RMS) at 50 Hz; α is given as $3\pi/8$ rad. Determine the (a) RMS magnitude of the load voltage, (b) half-cycle average magnitude of the load voltage, and (c) power factor of the controller.

Solution

(a) The given firing angle falls in the range $\pi/3 \leq \alpha < \pi/2$. From Eqn (4.34), the RMS phase voltage is

$$V_{R(\text{RMS})} = V_m \sqrt{\frac{1}{4} + \frac{9}{16\pi}\sin 2\alpha + \frac{3\sqrt{3}}{16\pi}\cos 2\alpha}$$

where V_m is the peak of the per phase voltage. Thus,

$$V_{R(\text{RMS})} = \frac{200\sqrt{2}}{\sqrt{3}} \sqrt{\frac{1}{4} + \frac{9}{16\pi}\sin\left(2 \times \frac{3\pi}{8}\right) + \frac{3\sqrt{3}}{16\pi}\cos\left(2 \times \frac{3\pi}{8}\right)} = 90 \text{ V}$$

(b) The half-cycle average is given by the expression in Eqn (4.33) as

$$V_R = \frac{\sqrt{3}V_m}{\pi}\cos\left(\alpha - \frac{\pi}{6}\right) = \frac{\sqrt{3} \times 200\sqrt{2}}{\pi\sqrt{3}}\cos\left(\frac{3\pi}{8} - \frac{\pi}{6}\right)$$

$$= 90.0 \times 0.793 = 71.4 \text{ V}$$

(c) The RMS value of the line-to-neutral current

$$I_{R(\text{RMS})} = \frac{V_{R(\text{RMS})}}{R_{\text{ld}}} = \frac{90}{8} = 11.25 \text{ A}$$

Output power

$$P_o = 3I_{R(\text{RMS})}^2 R_{\text{ld}}$$
$$= 3 \times 11.25^2 \times 8$$
$$= 3038 \text{ W}$$

The input power with $\cos\phi = 1$,

$$P_i = 3\frac{V_{\text{LL(RMS)}}}{\sqrt{3}} I_{R(\text{RMS})}$$
$$= 3 \times \frac{200}{\sqrt{3}} \times 11.25$$
$$= 3897 \text{ W}$$

The power factor of the controller is

$$\frac{P_o}{P_i} = \frac{3038}{3897} = 0.78 \text{ (lagging)}$$

7. The three-phase, delta-connected ac controller of Fig. 4.10 (class-D chopper) has a resistive load of 10 Ω. The line-to-line voltage V_{LL} is 208 V (RMS) at 50 Hz and the firing angle of the thyristors is 95°. Determine the (a) RMS and half-cycle average values of the load voltage, (b) expressions for the instantaneous currents i_Y, i_B, and i_{YB}, (c) RMS and half-cycle average values of the load current i_{RY}, (d) RMS half-cycle magnitude of the line current i_R, (e) power factor of the controller, and (f) RMS current through a thyristor.

Solution

(a) The firing angle $\alpha = 95° = 1.658$ rad. The RMS value of the load voltage is given as

$$V_{\text{ld(RMS)}} = V_m \left[\frac{1}{2\pi}(\pi - \alpha) + \frac{\sin 2\alpha}{2}\right]^{1/2}$$
$$= 208\sqrt{2}\left[\frac{1}{2\pi}(\pi - 1.658) + \frac{\sin 190°}{2}\right]^{1/2}$$
$$= 208\left[1 - \frac{1.658}{\pi} + \frac{\sin 190°}{2\pi}\right]^{1/2}$$
$$= 138.7 \text{ V}$$

The half-cycle average value of the load voltage

$$V_{\text{ld}} = \frac{1}{\pi}\int_0^{\pi} V_m \sin\omega t \, d(\omega t)$$
$$= \frac{V_m}{\pi}(1 + \cos\alpha)$$
$$= \frac{208\sqrt{2}}{\pi}(1 + \cos 95°)$$
$$= 85.46 \text{ V}$$

(b) The expressions for the currents are

$$i_Y = -\frac{\sqrt{3}}{R_{\text{ld}}} V_m \cos\left(\omega t - \frac{\pi}{3}\right)$$

$$= -\frac{\sqrt{3}}{10} \times 208\sqrt{2} \cos\left(\omega t - \frac{\pi}{3}\right)$$

$$= -51 \cos\left(\omega t - \frac{\pi}{3}\right)$$

$$i_B = -\frac{\sqrt{3} V_m}{R_{\text{ld}}} \cos(\omega t - \pi)$$

$$= -\frac{\sqrt{3} \times 208 \times \sqrt{2}}{10} \cos(\omega t - \pi)$$

$$= -51 \cos(\omega t - \pi)$$

$$i_{YB} = \frac{V_m}{R_{\text{ld}}} \sin\left(\omega t - \frac{2\pi}{3}\right)$$

$$= \frac{208\sqrt{2}}{10} \sin\left(\omega t - \frac{2\pi}{3}\right)$$

$$= 29.4 \sin\left(\omega t - \frac{2\pi}{3}\right)$$

(c) The RMS value of the load current i_{RY} is given as

$$I_{RY(\text{RMS})} = \frac{V_m}{R_{\text{ld}}}\left[\frac{1}{2\pi}\left(\pi - \alpha + \frac{\sin 2\alpha}{2}\right)\right]^{1/2}$$

$$= \frac{V_{RY}}{R_{\text{ld}}}$$

$$= \frac{138.7}{10} = 13.87 \text{ A}$$

The half-cycle average of the load current is

$$I_{\text{ld}} = \frac{V_{\text{ld}}}{R_{\text{ld}}}$$

$$= \frac{85.46}{10} = 8.55 \text{ A}$$

(d) $$i_R = i_{RY} - i_{BR} = \frac{v_{RY}}{R_{\text{ld}}} - \frac{v_{BR}}{R_{\text{ld}}}$$

The RMS value of the line current i_R is obtained as

$$I_{R(\text{RMS})} = \sqrt{\frac{2}{2\pi}\left\{\int_{\alpha}^{\pi}\left(\frac{V_m}{R_{\text{ld}}} \sin \omega t\right)^2 d(\omega t) + \int_{\alpha+\pi/3}^{4\pi/3}\left[\frac{V_m}{R_{ld}} \sin\left(\omega t - \frac{4\pi}{3}\right)\right]^2 d(\omega t)\right\}}$$

$$= \frac{V_m}{R_{\text{ld}}}\left[\frac{2}{2\pi}\left(\pi - \alpha + \frac{\sin 2\alpha}{2}\right)\right]^{1/2}$$

$$= \frac{V_{\text{ld(RMS)}}\sqrt{2}}{R_{\text{ld}}}$$

$$= \frac{138.7\sqrt{2}}{10} = 19.6 \text{ A}$$

The half-cycle average of the line current I_{RY} is

$$I_{RY(\text{av})} = \frac{2}{2\pi}\left[\int_{\alpha}^{\pi} \frac{V_m}{R_{\text{ld}}} \sin\omega t \, d(\omega t) + \int_{\alpha+\pi/3}^{4\pi/3} \frac{V_m}{R_{\text{ld}}} \sin\omega t \, d(\omega t)\right]$$

$$= \frac{2V_m}{\pi R_{\text{ld}}}(1 + \cos\alpha)$$

$$= \frac{85.46 \times 2}{10} = 17.1 \text{ A}$$

(e) The output power is given as

$$P_o = 3I_{\text{ld(RMS)}} V_{\text{ld(RMS)}}$$

$$= 3 \times 13.87 \times 138.7$$

$$= 5771 \text{ W}$$

The input power with $\cos\phi = 1$ is

$$P_i = \sqrt{3} V_{\text{LL(RMS)}} I_{\text{line(RMS)}}$$

$$= \sqrt{3} \times 208 \times 19.6$$

$$= 7061 \text{ W}$$

The power factor of the controller is

$$\frac{P_o}{P_i} = \frac{5771}{7061} = 0.82 \text{ (lagging)}$$

(f) The RMS current through a thyristor

$$I_{\text{Th(RMS)}} = \frac{I_{RY(\text{RMS})}}{\sqrt{2}} = \frac{13.87}{\sqrt{2}} = 9.8 \text{ A}$$

8. The circuit of Fig. 4.12 is operated as a single-phase transformer tap-changer with the transformer ratio 1:1. The primary voltage is 240 V (RMS) at 50 Hz. The secondary winding is tapped at the centre and each half is connected to an ac controller. If R_{ld} is 15 Ω and the RMS load voltage is 167.5 V, determine the (a) firing angle of the thyristors Th_1 and Th_2, (b) RMS and half-cycle average magnitudes of the currents through either Th_3 or Th_4, (c) RMS and half-cycle average magnitudes of the currents through either Th_1 or Th_2, and (d) the power factor of the controller.

Solution

(a) From Eqn (4.56),

$$V_{\text{ld(RMS)}} = V_m\sqrt{2 - \frac{3\alpha}{2\pi} + \frac{3\sin 2\alpha}{4\pi}}$$

$$167.5 = 120\sqrt{2}\sqrt{2 - \frac{3\alpha}{2\pi} + \frac{3\sin 2\alpha}{4\pi}}$$

This gives α_{Th_1} as 107°. Also, $\alpha_{\mathrm{Th}_2} = 287°$.
(b) The RMS value of current through Th_3 or Th_4 is determined as follows:

$$I_{3,4(\mathrm{RMS})} = \sqrt{\frac{1}{2\pi}\int_0^{\alpha}\left(\frac{V_m}{R_{\mathrm{ld}}}\sin\omega t\right)^2 d(\omega t)}$$

$$= \frac{V_m}{R_{\mathrm{ld}}}\sqrt{\frac{\alpha}{4\pi} - \frac{\sin 2\alpha}{8\pi}}$$

$$= \frac{120\sqrt{2}}{15}\sqrt{\frac{1.8675}{4\pi} + \frac{0.559}{8\pi}}$$

$$= 4.67 \text{ A}$$

The half-cycle average of the current through Th_3 or Th_4 is

$$I_{3,4(\mathrm{av})} = \frac{1}{\pi}\int_0^{\alpha}\frac{V_m}{R_{\mathrm{ld}}}\sin\omega t \, d(\omega t)$$

$$= \frac{1}{\pi}\frac{120\sqrt{2}}{15}\int_0^{107°}\sin\omega t \, d(\omega t)$$

$$= 4.66 \text{ A}$$

(c) The RMS value of the current through Th_1 or Th_2 is found as follows:

$$I_{1,2(\mathrm{RMS})} = \left[\frac{1}{2\pi}\int_{\alpha}^{\pi}\left(\frac{2V_m}{R_{\mathrm{ld}}}\sin\omega t\right)^2 d(\omega t)\right]^{1/2}$$

$$= \frac{V_m}{R_{\mathrm{ld}}}\left(1 - \frac{\alpha}{\pi} + \frac{\sin 2\alpha}{2\pi}\right)^{1/2}$$

$$= \frac{120\sqrt{2}}{15}\left(1 - \frac{1.8675}{\pi} + \frac{\sin 214°}{2\pi}\right)^{1/2}$$

$$= 6.37 \text{ A}$$

The half-cycle average of current through Th_1 or Th_2 is found as follows:

$$I_{1,2(\mathrm{av})} = \frac{1}{\pi}\int_{\alpha}^{\pi}\frac{2V_m}{R_{\mathrm{ld}}}\sin\omega t \, d(\omega t)$$

$$= \frac{1}{\pi} \times 2 \times \frac{120\sqrt{2}}{15}\int_{107°}^{180°}\sin\omega t \, d(\omega t)$$

$$= 5.1 \text{ A}$$

(d) The RMS current through the upper half of the secondary winding is $\sqrt{2}I_{1,2(\mathrm{RMS})} = 9.0$ A. The RMS current through the lower half of the secondary winding is the total RMS current of thyristors Th_1, Th_2, Th_3, and Th_4. It is given as

$$\begin{aligned} I_{\text{lower(RMS)}} &= [(\sqrt{2} I_{3,4(\text{RMS})})^2 + (\sqrt{2} I_{1,2(\text{RMS})})^2]^{1/2} \\ &= [(\sqrt{2} \times 4.67)^2 + (\sqrt{2} \times 6.38)^2]^{1/2} \\ &= 11.18 \text{ A} \end{aligned}$$

The volt-ampere capacity is given as

$$V_{\text{lower}} I_{\text{lower}} + V_{\text{upper}} I_{\text{upper}} = (120 \times 11.18 + (120 \times 9.02)$$

That is,

$$\text{VA capacity} = 120 \times 11.18 + 120 \times 9.0 = 2422 \text{ V A}$$

$$\text{Output power} = \frac{(167.5)^2}{15} = 1870 \text{ W}$$

$$\text{Power factor of the controller} = \frac{1870}{2422} = 0.77 \text{ (lagging)}$$

9. A three-phase, half-wave ac controller of category C has the following data: ac supply voltage (line-to-line) is 280 V at 50 Hz, load resistance for delta connection $R_{\text{ld(delta)}} = 21\ \Omega$, α is $5\pi/8$. Determine the (a) RMS values of load voltage and current, (b) half-cycle average values of load voltage and current, and (c) power factor of the controller.

Solution

(a) The line-to-neutral supply voltage is

$$V_{R(\text{RMS})} = \frac{v_{RY}}{\sqrt{3}} = \frac{280}{\sqrt{3}} = 161.66 \text{ V}$$

The load resistance for the star connection is

$$R_{\text{ld(star)}} = \frac{21}{3} = 7\ \Omega$$

The given value of α falls in the second range, namely, $\pi/2 \le \alpha \le 2\pi/3$. From Eqn (4.40), the RMS load voltage is

$$V_{R(\text{RMS})} = V_m \sqrt{\frac{11}{16} - \frac{3}{4\pi} \frac{5\pi}{8}}$$

Substituting the value of V_m gives this load voltage as

$$V_{R(\text{RMS})} = 161.66\sqrt{2} \sqrt{\frac{11}{16} - \frac{3}{4\pi} \frac{5\pi}{8}} = 106.93 \text{ V}$$

The RMS value of load current is

$$\frac{V_{R(\text{RMS})}}{R_{\text{ld(star)}}} = \frac{106.93}{7} = 15.28 \text{ A}$$

(b) From Eqn (4.39), the average load voltage is

$$V_R = \frac{V_m}{\pi}\left[\frac{1+\sqrt{3}}{2} + \cos\alpha\right] = 71.53 \text{ V}$$

The average load current is

$$\frac{V_R}{R_{\text{ld(star)}}} = \frac{71.53}{7} = 10.22 \text{ A}$$

(c) The output power is

$$\begin{aligned} P_{\text{ld}} &= 3V_{\text{ld}}I_{\text{ld}}\cos\phi \\ &= 3 \times 106.93 \times 15.28 \times 1 \\ &= 4902 \text{ W} \end{aligned}$$

$$\begin{aligned} \text{Input VA capacity} &= \sqrt{3}v_{RY}I_{\text{ld}} \\ &= \sqrt{3} \times 280 \times 15.28 \\ &= 7410 \text{ W} \end{aligned}$$

$$\text{Power factor of controller} = \frac{4902}{7410} = 0.66 \text{ (lagging)}$$

10. A three-phase ac controller of category A [Fig. 4.5(a)] supplies a star-connected resistive load of 8 Ω. The input voltage (line-to-line) is 200 V (RMS) at 50 Hz. Determine the (a) value of α that gives a line-to-neutral output voltage of 103.6 V, (b) maximum permissible RMS value of the current obtainable, (c) amplitude of the fifth harmonic current when $\alpha = 45°$, and (d) input power factor with $\alpha = 45°$.

Solution

(a) The ac controller is considered to be a combination of three single-phase choppers. The line-to-neutral voltage of the R-phase is

$$V_{R\text{(RMS)}} = \frac{200}{\sqrt{3}} = 115.47 \text{ V}$$

α is found from the equation

$$\begin{aligned} 103.6 &= \frac{115.47\sqrt{2}}{\sqrt{2}}\sqrt{1 - \frac{\alpha}{\pi} + \frac{\sin 2\alpha}{2\pi}} \\ &= 115.47\sqrt{1 - \frac{\alpha}{\pi} + \frac{\sin 2\alpha}{2\pi}} \end{aligned}$$

by trial and error as 60° or $\pi/3$ rad.

(b) The maximum RMS value of current obtainable is obtained as

$$\frac{\text{Maximum } R\text{-phase voltage which occurs at } \alpha = 0}{\text{Resistance}} = \frac{115.47}{8} = 14.43 \text{ A}$$

(c) From Appendix A, the load voltage can be written in terms of its harmonic components as

$$v_{\text{ld}}(t) = \sum_{n=1}^{\infty}(a_n \cos n\omega t + b_n \sin n\omega t)$$

The coefficient a_5 is determined as

$$a_5 = \frac{1}{\pi}\int_{\alpha}^{2\pi} V_m \sin \omega t \cos(5\omega t) d(\omega t)$$

$$= \frac{2}{\pi}\int_{\alpha}^{\pi} V_m \sin \omega t \cos(5\omega t) d(\omega t)$$

$$= \frac{2}{\pi}\int_{\alpha}^{\pi} 103.6\sqrt{2} \sin \omega t \cos(5\omega t) d(\omega t)$$

$$= 15.54 \text{ V}$$

Also,

$$b_5 = \frac{2}{\pi}\int_{\alpha}^{\pi} 103.6\sqrt{2} \sin \omega t \sin(5\omega t) d(\omega t) = 7.77 \text{ V}$$

Hence, the amplitude of the fifth harmonic current is

$$I_{5m} = \frac{(a_5^2 + b_5^2)^{1/2}}{R}$$

$$= \frac{(15.54^2 + 7.77^2)^{1/2}}{8}$$

$$= 2.17 \text{ A}$$

The input power factor is given as

$$\lambda = \mu \cos \phi_1 = \frac{I_1}{I} \cos \phi_1$$

The fundamental harmonic component of the load voltage is obtained as

$$a_1 = \frac{2}{\pi}\int_{\pi/4}^{\pi} V_m \sin \omega t \cos \omega t \, d(\omega t) = -23.33 \text{ V}$$

$$b_1 = \frac{2}{\pi}\int_{\pi/4}^{\pi} V_m \sin^2 \omega t \, d(\omega t) = 133.2 \text{ V}$$

$$\phi_1 = \tan^{-1}\left(\frac{a_1}{b_1}\right) - \tan^{-1}\left(\frac{-23.33}{133.2}\right) = -9.93°$$

$$I_1 = \frac{(a_1^2 + b_1^2)^{1/2}}{\sqrt{2}R} = \frac{135.23}{8\sqrt{2}} = 11.95 \text{ A}$$

$$\cos \phi_1 = \cos(-9.93°) = 0.985$$

$$I = \frac{103.6}{R} = \frac{103.6}{8} = 12.95 \text{ A}$$

Hence, the input power factor is

$$\lambda = \frac{11.95}{12.95} \times 0.985 = 0.909$$

Exercises

Multiple Choice Questions

1. In a single-phase ac chopper supplying a purely inductive load, the firing angle should be greater than ________ if switching is to take place in both half-waves.
   (a) $3\pi/4$ (b) $\pi/4$
   (c) $\pi/2$ (d) $2\pi/3$

2. In a single-phase ac chopper supplying an $R + L$ load, control is possible for ________, where ϕ is the power factor angle.
   (a) $\alpha = \phi$ (b) $\alpha > \phi$
   (c) $\alpha = \pi/2 + \phi$ (d) $\alpha < \phi$

3. In a three-phase ac chopper of category B, all the three phases conduct for some part of the range ________, whereas only two phases conduct for the entire range ________.
   (a) $\pi/2 < \alpha < 5\pi/6, 0 < \alpha < \pi/3$
   (b) $\pi/3 < \alpha < \pi/2, \pi/2 < \alpha < 5\pi/6$
   (c) $\pi/3 < \alpha < \pi/2, 0 < \alpha < \pi/3$
   (d) $0 < \alpha < \pi/3, \pi/3 < \alpha < \pi/2$

4. One criterion for arriving at the load waveforms in a three-phase ac chopper of category B is that the R-phase resistive load receives the line-to-neutral voltage of the R-phase ________.
   (a) when the Y- and B-phase conduct
   (b) when the B- and R-phase conduct
   (c) when the R- and Y-phase conduct
   (d) when the R-, Y-, and B-phase conduct

5. In the case of a three-phase ac chopper of category B, with α in the range ________, the average voltage remains at $(V_m/\pi)(1 + \cos\alpha)$, which is the same as when the load is connected directly to the ac supply.
   (a) $0 < \alpha < \pi/3$ (b) $\pi/3 < \alpha < \pi/2$
   (c) $\pi/2 < \alpha < 5\pi/6$ (d) all the above

6. In the sequence control circuit of Fig. 4.12, with a peak secondary voltage V_m of $200\sqrt{2}$ V and with only Th_1 and Th_2 as the conducting thyristors, the range of load voltage obtainable is ________.
   (a) $0 \le v_{ld} \le 300$ (b) $0 \le v_{ld} \le 100$
   (c) $0 \le v_{ld} \le 400$ (d) $0 \le v_{ld} \le 600$

7. If a load voltage variation of $0 \le V_{ld} \le 300$ V (RMS) is required in the sequence control circuit of Fig. 4.12 with a peak secondary voltage V_m of $150\sqrt{2}$ V, then the firing angles of thyristors Th_1 and Th_2 will be somewhere in the ranges ________ and ________, respectively.
   (a) $0 \le \alpha \le \pi/2, 3\pi/2 \le \alpha < 2\pi$

(b) $\pi \leq \alpha \leq 3\pi/2, 0 \leq \alpha \leq 2\pi$
(c) $0 \leq \alpha \leq \pi, \pi \leq \alpha \leq 2\pi$
(d) $\pi/2 \leq \alpha \leq \pi, 3\pi/2 \leq \alpha \leq 4\pi/3$

8. If a load voltage variation of $200 < V_{ld} < 400$ V (RMS) is required in the sequence control circuit of Fig. 4.12 with a peak secondary voltage V_m of $200\sqrt{2}$ V, then the firing angles of the thyristors Th_3 and Th_4 have to be kept fixed at ______ and ______, respectively, and the firing angles of Th_1 and Th_2 have to be varied in the range $0 \leq \alpha \leq \pi$ and $\pi \leq \alpha \leq 2\pi$, respectively.
   (a) $\alpha, \pi/2 + \alpha$ (b) $\pi, 2\pi$
   (c) $\pi/3 + \alpha, 2\pi/3 + \alpha$ (d) $0, \pi$

9. The three ranges of voltage variation obtained in the case of the sequence control circuit of Fig. 4.12 with $R + L$ load are ______ the ranges obtained with a purely resistive load.
   (a) exactly half of (b) same as
   (c) exactly twice of (d) three-fourths of

10. The precaution to be taken for the sequence circuit of Fig. 4.12 with an inductive load is that Th_2 should not be fired at an angle ______ when Th_3 is the only conducting thyristor.
   (a) $\alpha < \phi$ (b) $\pi + \alpha < \pi + \phi$
   (c) $2\pi - \alpha < 2\pi - \phi$ (d) $\pi/2 + \alpha < \pi/2 + \phi$

11. In a three-phase, delta-connected ac chopper (category D), the maximum RMS load voltage will be obtained with α equal to ______.
   (a) $\pi/2$ (b) π
   (c) $\pi/4$ (d) 0

12. In a three-phase, delta-connected ac chopper (category D), the current ratings of the thyristors will be ______ the line currents.
   (a) more than $\sqrt{3}$ times (b) less than $1/\sqrt{3}$ times
   (c) less than $\sqrt{3}$ times (d) more than $1/\sqrt{3}$ times

13. If a three-phase, delta-connected ac chopper (category D) has a balanced load resistance of 90 Ω, then in the equivalent star connection the load resistance will be equal to ______ Ω.
   (a) 75 (b) 30
   (c) 60 (d) 45

14. In a three-phase, delta-connected ac chopper (category D), if the phase currents have a magnitude of 20 A (RMS), then the line current peaks will have an approximate magnitude of ______ A.
   (a) 7 (b) 18
   (c) 49 (d) 63

15. In the case of a three-phase induction motor supplied by a three-phase ac controller, if the motor speed is 1500 rpm and the slip is 0.04, then the slip of the motor for the fifth harmonic will be approximately ______.
   (a) 0.35 (b) 0.81
   (c) 0.72 (d) 0.63

16. For a three-phase induction motor supplied by a three-phase ac controller, the _______ harmonic current is a negative-sequence one, whereas the _______ harmonic current is a positive-sequence one.
    (a) fundamental, seventh (b) fundamental, fifth
    (c) seventh, fifth (d) fifth, seventh

17. Figure 4.21(a) shows the equivalent circuit for the fundamental components of voltages in the case of a three-phase induction motor supplied with non-sinusoidal waveforms and Fig. 4.21(b) shows the approximate equivalent circuit with the seventh harmonic voltage applied to the stator and rotor windings. With the motor parameters given in the first figure, the value of the equivalent inductance X_{eq} in the second figure will be _______ Ω.
    (a) 35 (b) 21
    (c) 14 (d) 28

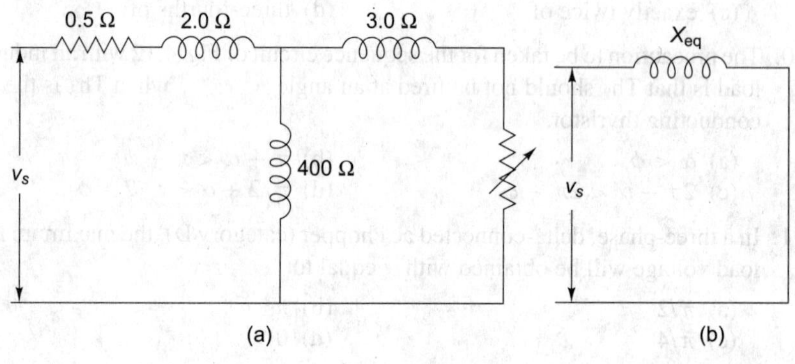

Fig. 4.21

18. For a three-phase induction motor supplied by a three-phase ac controller, if the fundamental stator winding current I_1 has an RMS value of 10 A and the harmonic part of the stator current, I_{har}, has an RMS value of 7 A, then the total RMS current through the stator winding will be approximately _______ A.
    (a) 11.7 (b) 12.9
    (c) 12.2 (d) 11.4

True or False

Indicate whether the following statements are *true* or *false*. Briefly justify your choice.

1. In a single-phase ac chopper supplying a purely inductive load, the firing angle α should be greater than $3\pi/4$ if switching is to take place in both half-waves.
2. In a single-phase ac chopper supplying an $R + L$ load, the control of the load voltage is not possible for $\alpha \leq \phi$, where ϕ is the power factor angle.
3. In a single-phase ac chopper supplying an $R + L$ load, if $\alpha = \phi$, where ϕ is the power factor angle, then the chopper works in the discontinuous conduction mode.
4. Time-ratio control is superior to on–off control.

5. Periodically switching the supply several times in a half-cycle gives better performance than switching it once in a half-cycle.
6. In a three-phase ac chopper of category B, all the three phases conduct for some part of the range $\pi/3 < \alpha < \pi/2$, whereas only one phase conducts for the range $\pi/2 < \alpha < 5\pi/6$.
7. The criteria for arriving at the load waveforms in a three-phase ac chopper of type C are the same as those in a three-phase ac chopper of type B.
8. In the case of two single-phase choppers being sequentially controlled, the following precautions, namely, the firing signal should not be given to Th_2 for $\alpha < \phi$ and Th_1 should not be fired for $\pi + \alpha < \pi + \phi$, are necessary.
9. In the case of a single-phase ac chopper supplying an $R + L$ load, the gate pulses can be just triggering pulses. These pulses will ensure reliable operation of the circuit.
10. Considering the sequence control of single-phase ac choppers for transformer tap changing, it is more advantageous to divide the secondary into a large number of sections and using a single-phase ac chopper for each section, than dividing the secondary into only two sections and accordingly providing two single-phase ac choppers.
11. In the case of a three-phase ac chopper of category B with α in the range $0 \le \alpha \le \pi/3$, the average voltage remains at $(V_m/\pi)(1 + \cos\alpha)$, which is the same as when the load is connected directly to the ac supply.
12. The criterion used for arriving at the load voltage in a category-B type chopper is that the R-phase load voltage gets the Y line-to-neutral voltage when all phases conduct.
13. In a single-phase ac chopper supplying a purely resistive load, the extinction angle α_e of the positive-half-wave-conducting thyristor remains at $2\pi/3$ irrespective of the value of α.
14. For analysis purposes the three-phase, star-connected load with an isolated neutral can be considered to be equivalent to a three-phase, star-connected load with the neutral earthed.
15. In the case of the delta connection of three single-phase ac controllers: (i) the line currents do not depend upon the firing angle α and (ii) these are always discontinuous.

Short Answer Questions

1. Explain how time-ratio control is superior to the on–off control.
2. How is the method of multiple switching in each half-cycle disadvantageous over the method of chopping used in an ac chopper?
3. Why are the higher order harmonics harmful for an induction motor load?
4. Why is the half-controlled, single-phase chopper inferior in performance to a fully controlled one?
5. What is the condition that makes the conduction discontinuous in the case of an ac chopper supplying an $R + L$ load?
6. Under what condition does the ac chopper work as a half-wave rectifier?

7. Give the different ranges of α which have to be considered in the case of a three-phase ac chopper of category B.
8. Give the different ranges of α which have to be considered in the case of a three-phase ac chopper of category C.
9. What is the principle behind the tap changing of a transformer with single-phase ac choppers?
10. What is the criterion for arriving at the load waveforms in the case of a three-phase ac chopper of category B?
11. In the case of the single-phase transformer tap-changer using two ac controllers, why is it necessary for both the thyristor pairs to be turned on to get maximum RMS load voltage?
12. Work out the RMS and average values of the load voltage for the three ranges of α in the case of a three-phase chopper of category D.
13. Work out the RMS and average values of the load voltage for the three ranges of α in the case of a three-phase chopper of category C.
14. Derive an expression for the input power factor in the case of a single-phase chopper supplying an $R + L$ load.
15. Derive an expression for the input power factor in the case of a three-phase category C chopper supplying an $R + L$ load.
16. Why must the firing pulses for a single-phase ac chopper supplying an $R + L$ load be rectangular with duration 0 to π and π to 2π in the positive and negative half-cycles, respectively?

Problems

1. Using the circuit given in Fig. 4.22, multicycle type of on–off control is implemented. The thyristors are switched on with $\alpha = 0$ for one-fifth of a second and switched off for one-third of a second. Find (a) the ratio between the power delivered to the load and the power delivered if the thyristors were to be kept on continuously and (b) the ratio of load currents for the two conditions.

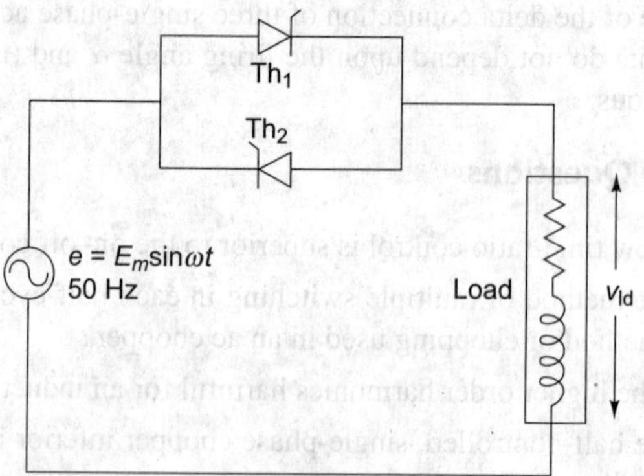

Fig. 4.22

Ans. (a) 0.375, (b) 0.612

2. The circuit shown in Fig. 4.22 has a purely resistive load. The given data are E_s = 250 V, f = 50 Hz, and R_{ld} = 4.5 Ω. Determine the (a) value of α that gives an RMS output voltage of 125 V, (b) maximum RMS value of the current obtainable over the permissible range of α, (c) maximum average value of the thyristor current obtainable over the permissible range of α, (d) amplitude of the seventh harmonic current when $\alpha = \pi/6$, (e) input power factor with $\alpha = \pi/6$, (f) angle for which the forward blocking voltage occurs for the thyristors, and (g) angle for which the reverse blocking voltage occurs for the thyristors.
Ans. (a) 114°, (b) 39.3 A, (c) 25 A, (d) 6.44 A, (e) 0.986, (f) 114°, (g) 180°

3. For the ac chopper circuit of Fig. 4.22, the data are supply voltage E_s = 300 V, f = 50 Hz, X_{ld} = 50 Ω, and R_{ld} = 0.05 Ω. Determine the (a) range of α for the load current in the chopper mode, (b) maximum value of the average thyristor current over the permissible range of α, (c) maximum value of the RMS thyristor current over the permissible range of α, (d) range of α for which the maximum forward voltage is applied to the thyristors, and (e) range of α for which the maximum reverse voltage is applied to the thyristors.
Ans. (a) $94° \le \alpha \le 180°$, (b) 5.4 A, (c) 4.24 A, (d) $90° \le \alpha \le 180°$, (e) $90° \le \alpha \le 180°$

4. The data for the chopper circuit of Fig. 4.22 are given as X_{ld} = 4 Ω, R_{ld} = 4 Ω, supply voltage E_s = 200 V, and f = 50 Hz. Determine the (a) range of α for the load current in the chopper mode, (b) maximum value of the average thyristor current over the permissible value of α, (c) maximum value of the RMS thyristor current over the permissible value of α, (d) range of α for which the greatest forward voltage is applied to the circuit, and (e) the range of α for which the greatest reverse voltage is applied to circuit.
Ans. (a) $45° \le \alpha \le 180°$, (b) 31.83 A, (c) 45.92 A, (d) $90° \le \alpha \le 180°$, (e) $90° \le \alpha \le 180°$

5. For the circuits of Problems 3 and 4, determine the maximum value of di/dt.
Ans. 2664 A/s, 1570 A/s

6. For the three-phase ac chopper of Fig. 4.7, the line-to-line supply voltage is 220 V and the load resistance is 5 Ω. Neglecting the inductance, compute the (a) value of α to give a line-to-neutral voltage of 115 V at the load, (b) half-cycle average of the thyristor current, and (c) RMS magnitude of the thyristor current.
Ans. (a) 50°, (b) 18.79 A, (c) 16.3 A

7. A 10-kW, three-phase, 250-V induction motor, which has a full load power factor of 0.866 and efficiency of 0.9, is fed by a three-phase ac controller of category B [Fig. 4.7(a)], which in turn is supplied by a source with a line-to-line voltage of 540 V. Compute the (a) firing angle α, (b) half-cycle average of the thyristor current, (c) RMS thyristor current, and (d) maximum thyristor forward and reverse voltage ratings.
Ans. (a) 65°, (b) 9.46 A, (c) 21.0 A, (d) 440.9 V, both forward and reverse

8. A three-phase ac controller of category B (Fig. 4.7) supplies a star-connected resistive load. The RMS magnitude of the input line-to-line voltage is 220 V

at 50 Hz. α is given as $\pi/6$. Determine the (a) RMS magnitude of the load voltage, (b) half-cycle average magnitude of the load voltage, and (c) power factor of the controller.

Ans. (a) 124 V, (b) 106.7 V, (c) 0.98

9. A three-phase, half-controlled ac controller of category C supplies a resistive load of $R_{\text{ld}} = 15\ \Omega$ connected in each phase with a star configuration as shown in Fig. 4.9(a). The input line voltage is 230 V (RMS) at 50 Hz. α is given as $\pi/6$. Determine the (a) RMS magnitudes of the load voltage and load current, (b) half-cycle average magnitude of the load voltage, and (c) power factor of the controller.

   Ans. (a) 91.93 V, 6.13 A; (b) 133.4 V, 8.9 A; (c) 0.69

10. A three-phase, delta-connected ac controller of category D [Fig. 4.10(a)] has a resistive load of value 12 Ω. The RMS magnitude of the supply voltage V_{LL} is 215 V at 50 Hz and the firing angle α of the thyristors is 120°. Determine the (a) RMS and half-cycle average values of the load voltage, (b) expression for instantaneous currents i_R, i_{RY}, and i_{BR}, (c) RMS and half-cycle average values of the load current through the RY branch, (d) RMS and half-cycle average magnitude of the line current i_R, (e) input power factor of the controller, (f) RMS current i_R through a thyristor.

    Ans. (a) 95.05 V, 48.39 V; (b) 107.5 $\cos(\omega t - 2\pi/3)$, 25.34 $\sin \omega t$, 25.34 $\sin(\omega t - 4\pi/3)$; (c) 7.92 A, 4.03 A; (d) 11.20 A, 8.07 A; (e) 0.54; (f) 5.6 A

11. The circuit of Fig. 4.12 is operated as a single-phase transformer tap-changer. The primary voltage is 220 V (RMS) at 50 Hz. The secondary winding is tapped at the centre and each half is connected to an ac controller. If the load resistance R_{ld} is 12 Ω and the RMS load voltage is 195 V, determine the (a) firing angle of the thyristors Th_1 and Th_2, (b) RMS and half-cycle average magnitudes of the currents through thyristors Th_1 and Th_2, (c) RMS and half-cycle average magnitudes of the currents through the thyristors Th_3 and Th_4, and (d) input power factor.

    Ans. (a) 70°; (b) 10.95 A, 11.08 A; (c) 5.35 A, 12.96 A; (d) 0.88

12. The circuit of Fig. 4.12 is operated as a single-phase transformer tap-changer. The primary voltage is 240 V at 50 Hz. The secondary winding is tapped unequally in the ratio 1:2 so that the lower part of the winding has half the number of turns in the upper part. If the load resistance is 10 Ω and the firing angle of the thyristor Th_1 is $\pi/4$ and that of Th_2 is $5\pi/4$, determine the (a) RMS and half-cycle average magnitudes of the currents through the thyristors Th_1 and Th_2, (b) RMS and half-cycle average magnitudes of the currents through the thyristors Th_3 and Th_4, and (c) power factor of the tap-changer.

    Ans. (a) 11.44 A, 18.44 A; (b) 5.14 A, 1.06 A; (c) 0.756

13. A three-phase, half-controlled ac controller of category C supplies a resistive load of 13 Ω connected in each phase of the delta connection. The input line-to-line voltage is 260 V (RMS). If the load voltage is 81.19 V, determine the (a) value of α and the range in which it occurs, (b) half-cycle average of the load voltage and current, and (c) power factor of the controller.

    Ans. (a) 130° approx., $2\pi/3 \le \alpha \le 7\pi/6$; (b) 48.4 V, 11.12 A; (c) 0.54

14. A three-phase, half-controlled ac controller of category C supplies a resistive load of 18 Ω connected in each phase of the delta-connected load. If the RMS

magnitude of the input line-to-line voltage is 270 V at 50 Hz and the load current is 19.12 A, compute the (a) firing angle, (b) RMS magnitude of the load voltage, (c) half-cycle average magnitudes of the load voltage and current, and (d) power factor of the controller.

Ans. (a) 100°, (b) 114.73 V, (c) 83.67 V, 13.95 A; (d) 0.736

15. A three-phase ac controller of category B supplies a delta-connected resistive load of 24 Ω. If the RMS magnitude of the input line-to-line voltage is 210 V at 50 Hz and the load voltage is 58.3 V, determine the (a) firing angle and the range in which it occurs, (b) half-cycle average values of the load voltage and current, and (c) power factor of the controller.

    Ans. (a) 62°, $\pi/3 \le \alpha < \pi/2$; (b) 80.2 V, 10.0 A; (c) 0.48

16. A three-phase ac controller of category A is supplied by a source of line voltage 300 V, the load being resistive with $R_{ld} = 9\ \Omega$. If the firing angle is 85°, compute the (a) RMS values of the load voltage and current, (b) amplitude of the seventh harmonic current, and (c) input power factor of the controller.

    Ans. (a) 129.1 V, 14.34 A; (b) 12.82 A; (c) 0.746

17. A three-phase ac controller of category A supplies a three-phase load consisting of a 11 Ω resistance in each phase. The input ac supply has an RMS magnitude of 250 V (line-to-line). If the RMS magnitude of the load voltage is 224.2 V (line-to-line), determine the (a) firing angle of the thyristors and the range in which it occurs, (b) half-cycle average values of the load voltage and current, and (c) input power factor.

    Ans. (a) $\alpha = 60°$; (b) 97.46 V, 8.86 A; (c) 0.9

CHAPTER 5

Inverters

5.1 Introduction

Inverters are used for conversion of dc power into ac power of variable voltage and frequency. The phase-controlled rectifiers dealt with in Chapter 2 can be operated in the inverting mode with a firing angle greater than 90° so as to return power from the output (dc) side to the ac supply network; the type of commutation provided for the thyristors therein is ac line commutation. This operation, known as the *inverting mode of a converter*, is useful for limited applications such as cranes and hoists (which operate in all the four quadrants), as well as in load commutated inverters. It is evident that the above-mentioned mode is one of the modes of operation for a converter. As against this, the inverters discussed in this chapter are based on a different principle: they have a dc source and therefore need forced commutation circuits for the thyristors. The loads catered by them include, among others, various types of ac motors such as induction motors, synchronous motors, etc. In order to control the speed and also provide constant torque operation for these motors, it is necessary that the inverters incorporate voltage- as well as frequency control features. The output frequency of such static inverters is determined by the switching rate of the semiconductor devices, and hence they can be independently controlled. Constant torque operation can be implemented in an ac machine by maintaining a constant V/f ratio, and thus voltage control is an essential part of such a mode of operation. Voltage control also facilitates the regulation of ac machines operating under widely ranging load conditions.

The ac output voltage of a power electronic inverter is usually non-sinusoidal and hence has a high harmonic content. These harmonics can be eliminated by means of appropriate filters, but the cost of the inverter increases with the sophistication demanded in the output. When the output frequency of the inverter varies over a wide range, the design of the filter becomes a formidable task. A decision regarding the incorporation of a filter should therefore be supported by economical justification. The technique of pulse width modulation (PWM) is beneficial for reducing harmonics and obtaining an output which is very nearly equal to the fundamental component of the desired output.

Dc choppers and inverters come under the broad title of *dc line commutation* because both have dc sources and need forced-commutation circuits. It is advantageous to use devices such as gate turn-off thyristors (GTOs), power transistors, and IGBTs for inversion purposes in view of the high switching speeds obtainable with them and also because no special commutation circuits are necessary as in the case of thyristors. For low-power applications that demand very high switching speeds, however, the metal oxide semiconductor field effect transistor (MOSFET) continues to be the preferred device.

Apart from ac machines, other important applications of inverters are induction heating, uninterrupted power supplies (UPSs), and high-voltage dc transmission. Though dc drives are still being used in the industry, ac drives have an edge over them because of the absence of the costly and cumbersome commutator that is essential for the former.

After going through this chapter the reader should

- understand the principle of single-phase and three-phase inverters and gain acquaintance with their configurations,
- know the operation as well as design of commutating elements for the simplest inverter, namely, the single-phase, parallel capacitor inverter,
- gain acquaintance with the construction of voltage waveforms for a three-phase, six-step voltage source inverter (VSI) as well as the design of commutating elements for its McMurray and McMurray–Bedford versions.
- become familiar with the principle and sequence of operation of the 120°-mode VSI and the input circuit commutated VSI,
- know the different methods for control of the VSI output,
- become familiar with PWM inverters and their features,
- understand the principles of operation of single-phase and three-phase current source inverters and their merits, and
- gain acquaintance with the principle underlying the load commutated inverter and the control of its output quantities.

5.2 Classification of Inverters

Inverters can be classified as (a) voltage source inverters (VSIs), (b) current source inverters (CSIs), and (c) load commutated inverters (LCIs). Pulse width modulated inverters form a subclassification under VSIs. Section 5.3 deals with the single-phase, parallel capacitor inverter, which is the simplest among single-phase VSIs. Section 5.4 together with Sections 5.5, 5.6, and 5.7 deal, respectively, with VSIs, CSIs, and LCIs.

As their name implies, VSIs have either a battery or more commonly a controlled or uncontrolled rectifier as their source. When a diode rectifier is used, a dc chopper is interposed between this rectifier and the VSI for obtaining the voltage control feature. On the other hand, a large inductor is interposed in the dc link between the rectifier and the CSI, the latter also including a control loop for regulating the current. The load current of a VSI depends upon the load impedance whereas the load voltage is independent of the load. In contrast with this, the load voltage in a CSI is governed by the load impedance but the load current is not influenced by the load. In an LCI the load current leads the voltage; thus, when the current crosses zero, the load voltage has a finite magnitude. This voltage is applied in a reverse manner across the thyristor and turns it off provided the duration from the current zero to the voltage zero is larger than the turn-off time of the thyristor.

Inverter configurations Figure 5.1 gives three commonly encountered schemes of inverters. The configurations (Dewan & Straughen 1975) depend upon the unit in which voltage control is incorporated. In Fig. 5.1(a) the uncontrolled rectifier gives a constant voltage output. The *LC* filter in the dc link smooths the dc voltage; the voltage control function is also incorporated in the inverter. In Fig. 5.1(b) the controlled rectifier provides a variable dc link voltage. In Fig. 5.1(c) the dc chopper gives an output with a wide range of variation for conversion by the inverter. The inverter is sometimes termed as an adjustable frequency ac voltage source because its output frequency depends on the switching rate of the semiconductor devices.

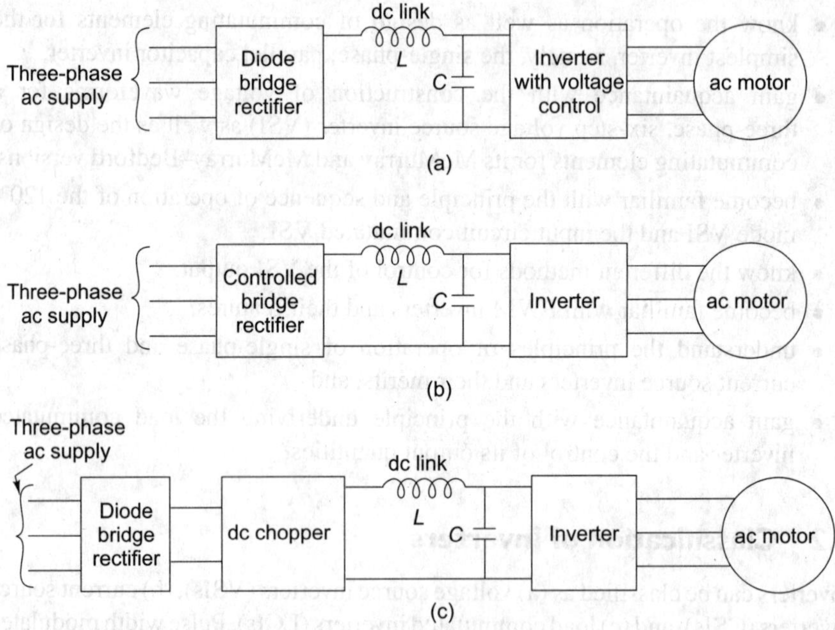

Fig. 5.1 Different configurations of three-phase inverters. Voltage control feature in the (a) inverter proper, (b) controlled bridge rectifier, and (c) dc chopper

5.3 Single-phase, Parallel Capacitor Inverter

Figure 5.2(a) shows a single-phase, parallel capacitor inverter, which is the simplest among single-phase VSIs. The thyristors Th_1 and Th_2 are alternately turned on; the turning on of Th_1 turns off Th_2, and similarly the firing of Th_2 helps in the commutation of Th_1. When Th_1 is turned on, current flows from *N* to *A* in the primary, and correspondingly, the secondary circuit current flows in a clockwise direction. Similarly when the primary current flows from N to B upon firing of Th_2, the direction of the secondary current becomes anticlockwise. The net effect is that the load connected in the secondary circuit gets an alternating current. The firing circuitry has been excluded from the figure for the sake of simplicity.

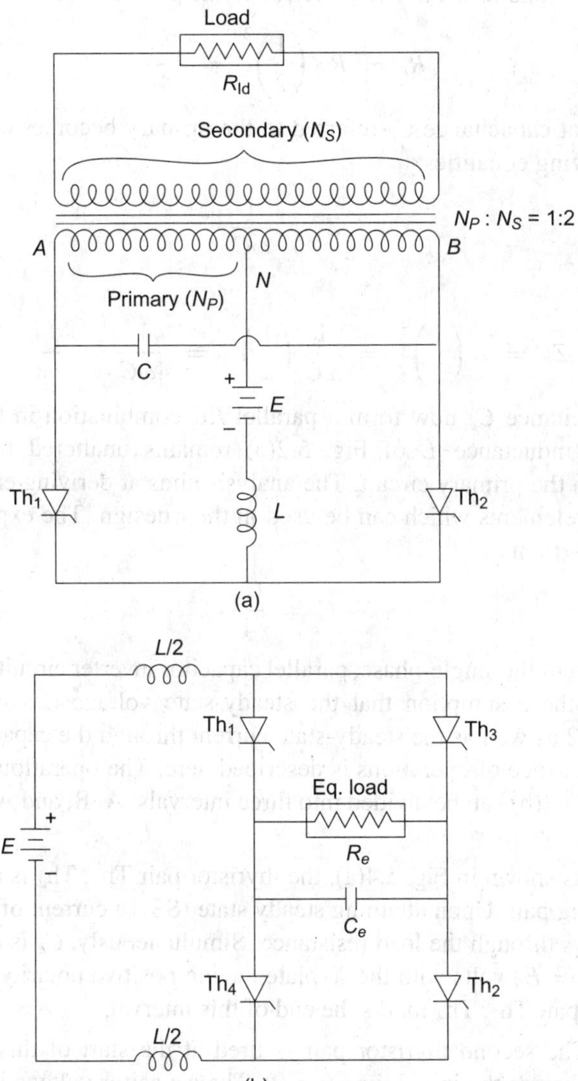

Fig. 5.2 Single-phase, parallel capacitor inverter: (a) circuit diagram, (b) equivalent circuit

Assuming Th_1 to be fired initially, current will flow in the primary and secondary windings as explained above. The capacitor C gets charged to a voltage of magnitude $2E$ due to autotransformer action. If Th_2 is fired at this stage, the capacitor voltage is applied in a reverse manner across Th_1 and it turns off.

Similar operations take place when Th_2 is initially conducting and Th_1 is fired subsequently. The elements L and C have to be properly designed to ensure successful turn-off of the thyristors. For simplicity of analysis and design, the load is taken to be purely resistive; Fig. 5.2(b), which is a simpler equivalent of Fig. 5.2(a) after referring all quantities to the primary, is taken up for discussion. The equivalent load resistance R_{ld} referred to the primary becomes

$$R_e = R_{\text{ld}}\left(\frac{1}{2}\right)^2 = \frac{R_{\text{ld}}}{4} \tag{5.1}$$

The equivalent capacitance C_e referred to the primary becomes equal to $4C$ as per the following equalities:

$$Z = \frac{1}{\omega C}$$

and

$$Z_e = Z\left(\frac{1}{2}\right)^2 = \frac{1}{\omega C}\left(\frac{1}{2}\right)^2 = \frac{1}{4\omega C} = \frac{1}{\omega C_e} \tag{5.2}$$

R_e and capacitance C_e now form a parallel RC combination in the equivalent circuit. The inductance L of Fig. 5.2(a) remains unaltered because of its connection in the primary circuit. The analysis aims at deriving expressions for the L and C elements which can be used in their design. The expression for C can then be written as

$$C = \frac{C_e}{4} \tag{5.3}$$

The operation of the single-phase, parallel capacitor inverter circuit of Fig. 5.2(b) is based on the assumption that the steady-state voltage across each of the inductors $L/2$ as well as the steady-state current through the capacitance C_e are zero. The sequence of operations is described here. The operation of the circuit given in Fig. 5.2(b) can be divided into three intervals: A, B, and A′, as shown in Fig. 5.3.

Interval A As shown in Fig. 5.4(a), the thyristor pair Th_1, Th_2 is assumed to be the conducting pair. Upon attaining steady state (SS), a current of I_{SS} $(= E/R_e)$ amperes flows through the load resistance. Simultaneously, C_e is charged to the voltage $V_{c\text{SS}}$ $(= E)$ volts with the X-plate having positive polarity. The firing of the thyristor pair Th_3, Th_4 marks the end of this interval.

Interval B The second thyristor pair is fired at the start of this interval; this moment is taken as the initial time ($t = 0$). The capacitor voltage is then applied in a reverse manner across Th_1 and Th_2 causing the devices to turn off. Also, the current through the inductors gets diverted through the thyristors Th_3 and Th_4 as illustrated in Fig. 5.4(b). As the capacitor current now flows from left to right, it starts getting charged in the reverse direction, with the Y-plate having positive polarity. Its voltage changes from the initial value $+E$ to the final steady-state value $-E$, the time for reversal depending on the circuit elements L, C_e, and R_e. Since Th_1 and Th_2 remain in parallel with C_e throughout this interval, they

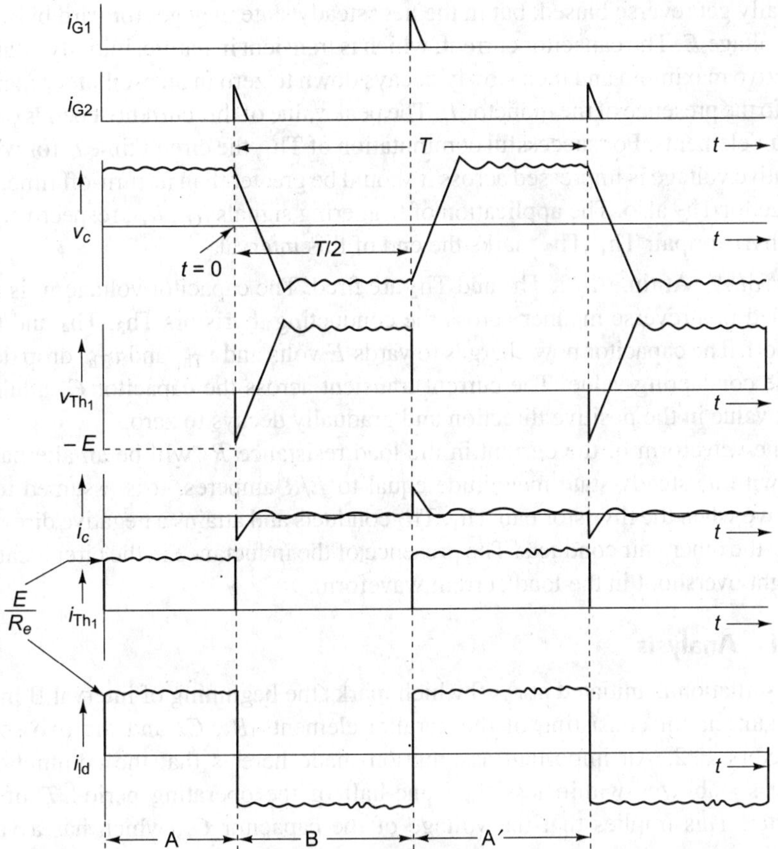

Fig. 5.3 Waveforms of a single-phase, parallel capacitor inverter

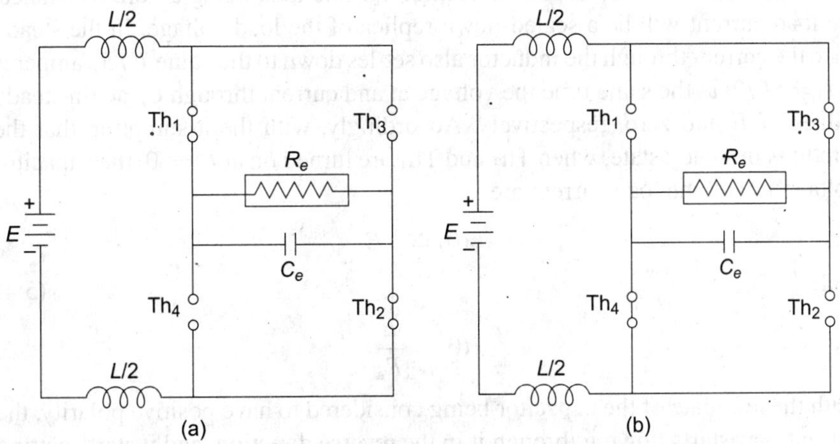

Fig. 5.4 Circuit conditions for the inverter of Fig. 5.2(b): (a) Th_1, Th_2 on, (b) Th_3, Th_4 on

initially get reverse biased, but in the new steady state they get forward biased to the voltage E. The capacitor current, which is transient in nature, initially attains a negative maximum and then slowly decays down to zero in an oscillatory manner due to the presence of the inductor L. The peak value of this current depends on the circuit elements. For successful commutation of Th_1, the circuit time t_c for which negative voltage is impressed across it should be greater than its turn-off time; this is true for Th_2 also. The application of triggering signals i_{G1}, i_{G2}, respectively, to the thyristor pair Th_1, Th_2 marks the end of this interval.

Interval A′ At time $T/2$, Th_1 and Th_2 are fired. The capacitor voltage v_c is now applied in a reverse manner across the conducting thyristors Th_3, Th_4 and they turn off. The capacitor now charges towards E volts and v_{Th_1} and v_{Th_2} drop down to the conducting value. The current transient across the capacitor C_e attains a peak value in the positive direction and gradually decays to zero.

The waveform of the current in the load resistance R_e will be an alternating one with its steady-state magnitude equal to E/R amperes. It is assumed to be positive when the thyristor pair Th_1, Th_2 conducts and attains a negative direction when the other pair conducts. The presence of the inductance in the circuit causes a slight overshoot in the load current waveform.

5.3.1 Analysis

An oscillation is initiated at $t = 0$ which marks the beginning of interval B in the resonant circuit consisting of the parallel elements R_e, C_e and the two series inductors $L/2$. An important assumption made here is that the commutation process stabilizes within less than one-half of the operating period T of the inverter. This implies that the voltage of the capacitor C_e, which has a value of $+E$ volts at $t = 0$, gets reversed and attains a steady-state value of $-E$ volts slightly before $T/2$ seconds.

The capacitance C_e and the load resistance R_e being connected in parallel, the load voltage follows the capacitor voltage v_c. The load being a pure resistance, the load current will be a scaled-down replica of the load voltage. In the steady state the current through the inductor also settles down to the value E/R_e amperes at $t = T/2$; at the same time the voltage at and current through C_e attain steady values of E and zero, respectively. Accordingly, with the assumption that the circuit is in steady state, when Th_3 and Th_4 are turned on at $t = 0$, the capacitor voltage v_c and the load current are

$$v_c(0) = +E$$

and

$$i(0) = \frac{E}{R_e} \tag{5.4}$$

With the left plate of the capacitor being considered to have positive polarity, the current now starts flowing through it in the reverse direction, and it starts getting charged to a negative voltage. The capacitor voltage can now be expressed as

$$-v_c = E - L\frac{di}{dt} \tag{5.5}$$

where the current i flows through the series circuit consisting of the positive terminal of the battery, inductor L, Th_3, the parallel combination of R_e and C_e, Th_4, and back to the negative terminal of the battery. This current is equal to the sum of the currents flowing through the elements R_e and C_e. Thus,

$$i = \frac{-v_c}{R_e} - C_e\frac{dv_c}{dt} \tag{5.6}$$

Substituting for $-v_c$ from Eqn (5.5) in Eqn (5.6) and rearranging the terms gives the differential equation

$$LC_e\left(\frac{d^2i}{dt^2}\right) + \frac{L}{R_e}\left(\frac{di}{dt}\right) + i = \frac{E}{R_e} \tag{5.7}$$

with the initial conditions given in Eqn (5.4). Now using Eqns (5.4) and (5.5), the initial condition on v_c is obtained as

$$\left.\left(E - L\frac{di}{dt}\right)\right|_{t=0} = -v_c(0) = -E \tag{5.8}$$

This gives the initial condition on di/dt as

$$\left.\frac{di}{dt}\right|_{t=0} = \frac{2E}{L} \tag{5.9}$$

With the initial conditions in Eqns (5.4) and (5.9), the solution of the above differential equation becomes

$$i(t) = \frac{E}{R_e}\left[1 + \frac{4(R_e^2C_e/L)}{\sqrt{4(R_e^2C_e/L)-1}}e^{-t/2R_eC_e}\sin\omega t\right] \tag{5.10}$$

Also, using Eqn (5.5), $v_c(t)$ is obtained as

$$v_c(t) = \frac{2E}{\cos\phi}e^{-t/2R_eC_e}\cos(\omega t + \phi) - E \tag{5.11}$$

where $\tan\phi = 1/2\omega R_eC_e$ and $\omega = \sqrt{1/LC_e - 1/4R_e^2C_e^2}$.

5.3.2 Design of Commutating Elements

It is required to determine the values of L and C in Fig. 5.2(a) given, the values of the load resistance R_{ld} and the turn-off time t_{OFF} of the thyristors. The guiding principle is that the commutating elements L and C_e in the equivalent circuit must facilitate the application of a reverse voltage across the thyristor Th_1 of Fig. 5.4(b) so as to successfully commutate it. This implies that the circuit turn-off time t_c should be greater than or equal to the turn-off time (t_{OFF}) of the thyristor. This design procedure applies to the other three thyristors of Fig. 5.2(b) as well as to the thyristor pair of the original inverter given in Fig. 5.2(a). For determining the value of C_e, $v_{Th_1}(t)$ [$= -v_c(t)$] is plotted as shown in Fig. 5.5 (Ramshaw 1975).

The value of capacitor C is now arrived at as follows. It is seen from this figure that the initial part of the plot of $v_{Th_1}(t)$ approximates to a straight line, and hence

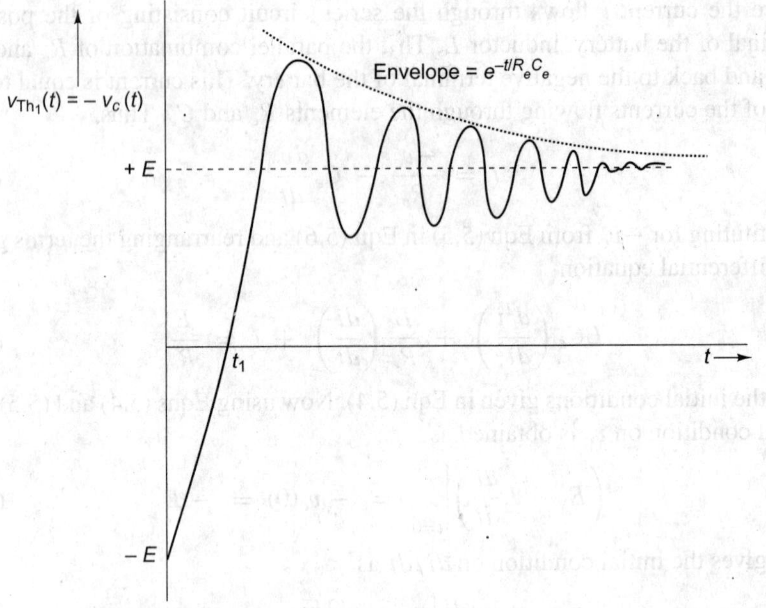

Fig. 5.5 Waveform of voltage across Th_1 in the single-phase, parallel capacitor inverter

its initial slope can be written as

$$\left.\frac{d[v_{Th_1}(t)]}{dt}\right|_{t=0} = \frac{-E}{t_1} \tag{5.12}$$

With $v_c(t)$ as given in Eqn (5.11), differentiating $-v_c(t)$ gives the initial rate of change of the thyristor voltage as

$$\left.\frac{d[v_{Th_1}(t)]}{dt}\right|_{t=0} = \frac{d[-v_c(t)]}{dt}\bigg|_{t=0}$$

$$= \frac{-2E}{\cos\phi}\left[\frac{1}{2R_eC_e}e^{-t/2R_eC_e}\cos(\omega t+\phi) + \omega e^{-t/2R_eC_e}\sin(\omega t+\phi)\right]\bigg|_{t=0}$$

Simplifying the right-hand side gives

$$\left.\frac{d[v_{Th_1}(t)]}{dt}\right|_{t=0} = -\frac{2E}{R_eC_e} \tag{5.13}$$

Equating Eqns (5.12) and (5.13) gives

$$\frac{E}{t_1} = \frac{2E}{R_eC_e}$$

or

$$t_1 = \frac{R_eC_e}{2} \tag{5.14}$$

Using the criterion $t_c \geq t_{OFF}$ and approximating t_c as t_1, the inequality to be satisfied is

$$\frac{R_e C_e}{2} \geq t_{\text{OFF}}$$

or

$$C_e \geq \frac{2t_{\text{OFF}}}{R_e} \tag{5.15}$$

Using Eqn (5.3), C is obtained as

$$C = \frac{C_e}{4} = \frac{2t_{\text{OFF}}}{4R_e} = \frac{t_{\text{OFF}}}{2R_e}.$$

L can be designed as follows. For damped oscillation to occur, the roots of Eqn (5.7) should be complex. This equation is rewritten as

$$\frac{d^2 i}{dt^2} + \frac{1}{R_e C_e}\frac{di}{dt} + \frac{i}{LC_e} = \frac{E}{R_e L C_e}$$

Its roots are

$$i_1, i_2 = -\frac{1}{2R_e C_e} \pm \sqrt{\frac{1}{4R_e^2 C_e^2} - \frac{1}{LC_e}}$$

For the roots to be complex, $1/LC_e$ has to be greater than $1/4R_e^2 C_e^2$. That is,

$$\frac{1}{LC_e} \geq \frac{1}{4R_e^2 C_e^2}$$

Solving for L gives

$$L \leq 4R_e^2 C_e$$

This gives the upper bound for the value of the inductor. The lower bound is determined as follows. The expression for current $i(t)$ in Eqn (5.10) is of the form

$$i(t) = a + b\exp\left(\frac{-t}{2R_e C_e}\right)\sin\omega t \tag{5.16}$$

Differentiation of $i(t)$ gives

$$\frac{di}{dt} = \frac{-b}{2R_e C_e}\exp\left(\frac{-t}{2R_e C_e}\right)\sin\omega t + b\omega\exp\left(\frac{-t}{2R_e C_e}\right)\cos\omega t$$

Defining $\cos\phi = b\omega$ and $\sin\phi = b/2R_e C_e$, di/dt can be written in a compact form as

$$\frac{di}{dt} = \exp\left(\frac{-t}{2R_e C_e}\right)\cos(\omega t + \phi) \tag{5.17}$$

Equation (5.17) shows that the maximum value of di/dt occurs at $t = 0$. This can be expressed as

$$\left(\frac{di}{dt}\right)_{\max} = \left.\left(\frac{di}{dt}\right)\right|_{t=0} \tag{5.18}$$

The combination of Eqns (5.9) and (5.18) yields

$$\left(\frac{di}{dt}\right)_{\text{max}} = \frac{2E}{L} \tag{5.19}$$

The right-hand side of Eqn (5.19) should be less than $(di/dt)_{\text{rated}}$ of the thyristor. Thus the inequality

$$\frac{2E}{L} \leq \left(\frac{di}{dt}\right)_{\text{rated}}$$

has to be fulfilled, and hence

$$L \geq \frac{2E}{(di/dt)_{\text{rated}}} \tag{5.20}$$

Thus, Eqn (5.20) gives the lower bound for L. The design procedure is to first select a value of L in the range

$$\frac{2E}{(di/dt)_{\text{rated}}} \leq L \leq 4R_eC_e \tag{5.21}$$

The values of ω and ϕ are then computed, respectively, from the expressions $\sqrt{1/LC_e - 1/4R_e^2C_e^2}$ and $\tan\phi = 1/2\omega R_eC_e$. Figure 5.5 shows the successful turn-off of the thyristor. The inequality

$$-v_c(t_{\text{OFF}}) \geq 0 \tag{5.22}$$

has to be satisfied because, in Fig. 5.5, negative voltage is applied across Th_1 till time t_1, which should be greater than t_{OFF} of Th_1. Using Eqn (5.11), this implies that the inequality

$$-\frac{2E}{\cos\phi}\exp\left(\frac{-t_{\text{OFF}}}{2R_eC_e}\right)\cos(\omega t_{\text{OFF}} + \phi) + E \geq 0 \tag{5.23}$$

has to be satisfied. The quantities E, R_e, and t_{OFF} in this last equation are known. Also, ω, $\cos\phi$, and $\cos(\omega t_{\text{OFF}} + \phi)$ can be calculated from the trial value of L. If the substitution of these quantities in Eqn (5.23) does not satisfy it, a slightly different value of L is taken, all the above quantities recalculated, and the inequality in Eqn (5.23) rechecked. This procedure may be repeated if necessary to find the appropriate value of L.

Worked Example

For the parallel capacitor inverter of Fig. 5.2(a), the data are $R_{\text{ld}} = 25\ \Omega$, turn ratio $= N_S/N_P = 2$, $E = 100$ V, t_{OFF} of thyristor $= 30\ \mu$s, and $(di/dt)_{\text{rated}} = 3$ A/μs. Determine the values of L and C.

Solution

$$R_e = \frac{R_{\text{ld}}}{(N_S/N_P)^2} = \frac{25}{4} = 6.25$$

$$C_e \geq \frac{2t_{\text{OFF}}}{R_e} = \frac{2 \times 30 \times 10^{-6}}{6.25}$$

$$L \le 4R_e^2C_e = 1500 \times 10^{-6}\ \text{H} = 1.5\ \text{mH}$$

Also,

$$L \ge \frac{2E}{(di/dt)_{\text{rated}}} = \frac{2 \times 100}{3 \times 10^6}\ \text{H} = 0.67 \times 10^{-4}\ \text{H} = 0.067\ \text{mH}$$

Thus,

$$0.067\ \text{mH} \le L \le 1.5\ \text{mH}$$

Choosing L as 0.32 mH, ω becomes

$$\omega = \sqrt{\frac{10^9}{0.32 \times 9.6} - \frac{10^{12}}{4 \times 6.25^2 \times 9.6^2}} \approx 16{,}000\ \text{rad/s}$$

and

$$\tan\phi = \frac{1}{2\omega R_e C_e} = 0.5208$$

This gives $\phi = 0.4802$, $\cos\phi = 0.8869$, and $\omega t_c = 0.48$.

$$\cos(\omega t_c + \phi) = \cos(0.48 + 0.4802) = \cos(0.9602) = 0.5733$$

The expression

$$\left[-\left(\frac{2E}{\cos\phi}\right)\exp\left(\frac{t_{\text{OFF}}}{2R_eC_e}\right)\cos(\omega t_c + \phi) + E\right]$$

is now evaluated as

$$-\frac{2 \times 100}{0.8869} \times 0.7788 \times 0.5733 + 100 = -0.11\ \text{V}$$

As the right-hand side value of – 0.11 V is near zero, it can be concluded that the chosen value of L is an appropriate one. Also,

$$C = \frac{C_e}{4} = 2.4\ \mu\text{F}$$

5.4 Voltage Source Inverters

The source for this type of inverter is usually a battery; it can be a controlled or an uncontrolled rectifier. The single-phase, parallel capacitor inverter discussed earlier can be categorized as a single-phase VSI because the output is single-phase ac voltage. Figure 5.6 shows the half-bridge and full-bridge types of single-phase VSIs in which the inductor and capacitor are dispensed with and each thyristor is shunted by a diode. The half-bridge inverter has only two thyristors which are controlled so as to connect point A of the load to the positive bus (through Th_1) for one half-cycle and to the negative bus (through Th_2) for the other.

Such an operation provides a square wave of amplitude $E/2$ for the load voltage as shown in Fig. 5.7(a). On the other hand, in the full-bridge inverter, Th_1, Th_2

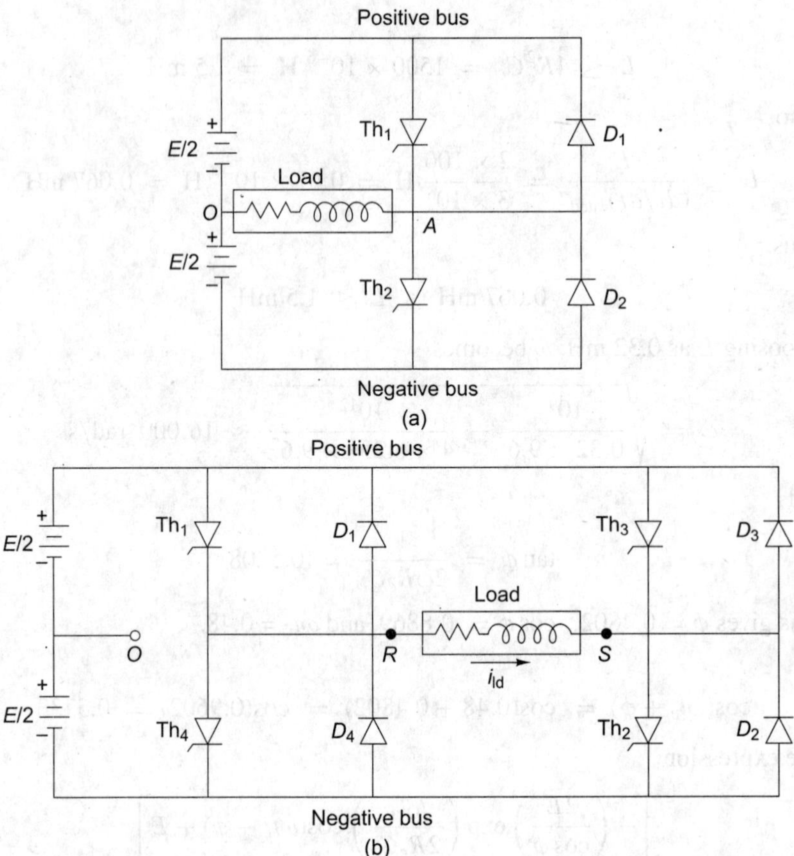

Fig. 5.6 Voltage source inverters: (a) half-bridge circuit, (b) full-bridge circuit

conduct the positive half-cycle of the output voltage waveform and Th_3, Th_4 conduct the other half-cycle. The circuit operation can be understood as follows. The load terminal R is considered to be connected to the positive terminal of the upper battery in the first positive half-cycle, making the voltage v_{RO} equal to $+E/2$. Also, the load terminal S is assumed to be connected to the negative terminal of the lower battery, making v_{SO} equal to $-E/2$. This situation reverses in the negative half-cycle. As shown in Fig. 5.7(b), the load voltage v_{RS} becomes $v_{RS} = v_{RO} - v_{SO} = +E$ during the positive half-cycle and $-E$ during the other half-cycle. Thus the load voltage is a square wave of amplitude E. It is evident that the output voltage contains harmonics; the fundamental harmonic has the same frequency as that of the square wave output.

The load current waveform for this single-phase, full-bridge VSI depends on the type of the load. Whereas it is a scaled-down replica of the voltage waveform for a resistive load, it lags the fundamental component of the load voltage for an inductive load. The lagging angle in the latter case depends on the L/R ratio of the load [Fig. 5.7(c)]. It rises exponentially like the current in a series RL circuit which is initially connected to a battery for a small duration, and then falls when this battery is shorted for another interval. For a small duration after the voltage

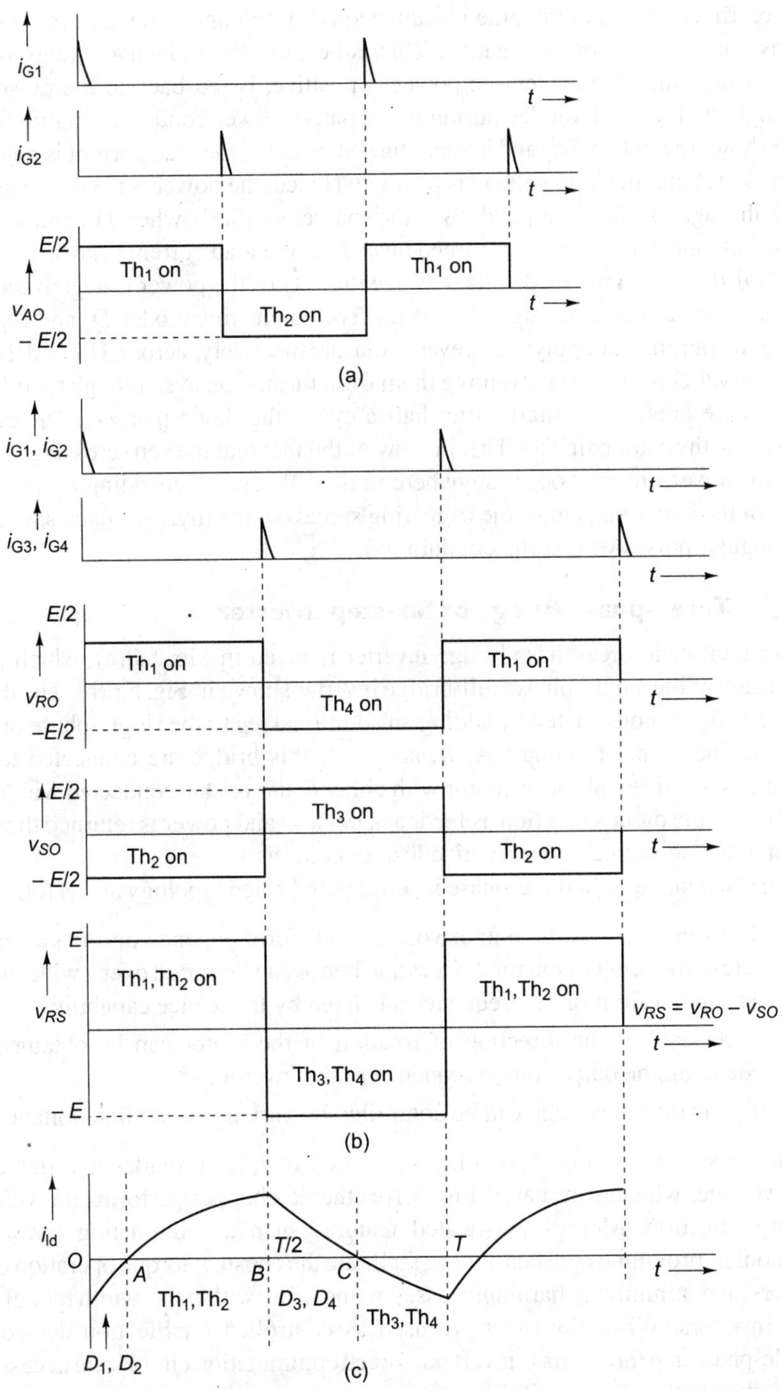

Fig. 5.7 Waveforms of the (a) half-bridge inverter load voltage, (b) full-bridge inverter load voltage, and (c) load current in one phase of the full-bridge inverter

waveform changes polarity, the instantaneous current and voltage have opposite signs, making the power negative. This is because the inductive energy of the load, which increases when the power is positive, is fed back to the dc source through the feedback diodes during the negative power condition. Figure 5.7(c) also shows that when Th_1 and Th_2 are turned on at 0°, the load current is negative from 0 to A and the load voltage is positive. Hence, the power is negative and fed back through the diodes D_1 and D_2 to the source. Similarly when Th_1 and Th_2 are turned off and Th_3 and Th_4 are turned on at $T/2$, the load current is positive in the interval B to C but the load voltage is negative. Thus the power is negative and is fed back to the source through D_3 and D_4. The conducting diodes D_1 and D_2 also serve the purpose of applying a reverse bias, respectively, across Th_1 and Th_2 in the interval O to A, thus preventing them from turning on even though their firing signals are present. Similarly after half a cycle, the diode pair D_3, D_4 reverse biases the thyristor pair Th_3, Th_4. In view of the fact that the zero crossings of the current waveform may occur anywhere in the half-cycle depending on the L/R ratio of the load impedance, the triggering signals of the thyristor pairs should be rectangular pulses with sufficient duration.

5.4.1 Three-phase Bridge or Six-step Inverter

The circuit of a three-phase bridge inverter is given in Fig. 5.8(a), which is an extension of the single-phase, full-bridge inverter shown in Fig. 5.6(b). The three-phase bridge is constructed by adding an additional leg to the single-phase bridge circuit. The output terminals A, B, and C of this bridge are connected to the terminals of a three-phase ac motor with either delta- or star-connected windings. As before, the diodes function as feedback devices and power is returned through them to the dc source under reactive load conditions.

The advantages of a three-phase inverter with bridge topology are as follows.

(a) The frequency of the output voltage waveform depends on the switching rate of the semiconductor devices and hence can be varied over a wide range. The upper limit of the frequencies is fixed by the device capability.

(b) A reversal of the direction of rotation of the motor can be obtained by changing the output phase sequence of the inverter.

(c) The ac output voltage can be controlled by varying the dc link voltage.

In the scheme of Fig. 5.1(b) the controlled rectifier provides a variable dc link voltage, whereas in that of Fig. 5.1(c) the dc chopper performs the voltage control function. More sophisticated features such as eliminating unwanted harmonics, providing constant V/f to facilitate the constant torque operation of ac motors, and minimizing harmonic losses are possible with pulse width modulated VSI inverters. When thyristors are used as controlled rectification devices in single-phase and three-phase inverters, forced commutation circuits are necessary. In addition to this, the circuit should also incorporate protective features. All these features increase the cost of the inverter circuit. Of late, in the high-, medium-, and low-power ranges, respectively, GTOs, IGBTs, and MOSFETs are used because of (a) the high switching rates obtainable with them and (b) the simplicity of their base drives. GTOs are evolving as a good compromise because their power capabilities are comparable with conventional thyristors. Moreover, their switching rates are higher than those of thyristors and even comparable with those of transistors. As explained earlier, the firing signal or the gate drive, as the case may be, has to be a sustained one as in the case a power transistor or should be a rectangular pulse as in the case of a thyristor.

The operation of the three-phase bridge inverter is similar to that of a single-phase inverter with the difference that the line-to-line voltages are applied across any two terminals. Hence, across any two lines, the load consists of two parallel branches of a delta-connected load winding or two series branches of a star-connected load winding. Each semiconductor device conducts for a duration of 180°. A displacement of 120° is to be maintained between successive phases by adjusting the switching sequence suitably. To achieve this, the thyristors have to be turned on or off as per the sequence noted in Fig. 5.8(j), namely, Th_1, Th_2, Th_3, Th_4, Th_5, and Th_6.

Voltage waveforms As before, a fictitious centre point O of the battery is considered to exist and the pole voltages v_{RO}, v_{YO}, and v_{BO} are, respectively, drawn in Figs 5.8(b), (c), and (d). It is evident from these waveforms that at any instant, three thyristors, one from each leg, have to be fired. The pole voltage is $E/2$ or $-E/2$ according to whether the upper or lower thyristor is turned on. Each of these thyristors is turned on for 180° or half the period of an alternating cycle; this is true of the other two thyristors also. The line voltages are obtained as the difference of the pole voltages. For instance, $v_{RY} = v_{RO} - v_{YO}$, and accordingly the waveform for v_{RY} [Fig. 5.8(e)] is obtained by subtracting the ordinates of Fig. 5.8(c) from those of Fig. 5.8(b). The other two line voltages v_{YB} and v_{BR} are also obtained in a similar manner. Because of the zero voltage intervals, these line voltage waveforms are known as *quasi-square waveforms*. Each of these waveforms is periodic, though non-sinusoidal, and hence can be expressed as a Fourier series. The Fourier series for the voltage v_{RY} can be obtained by shifting it to the right by 30° to obtain quarter-wave symmetry as shown in Fig. 5.8(i). v_{RY} can now be expressed as

$$v_{RY}(t) = \Sigma b_n \sin \omega t \qquad \text{where } n = 1, 5, 7, 11, \ldots \tag{5.24}$$

The coefficients b_n are obtained by referring to Fig. 5.8(i). The coefficient b_1 is given as

$$b_1 = \frac{2}{\pi}\int_{\pi/6}^{5\pi/6} E \sin \omega t \, d(\omega t) = \frac{2E}{\pi}[-\cos \omega t]_{\pi/6}^{5\pi/6} = \frac{2\sqrt{3}E}{\pi} \tag{5.25}$$

$$b_5 = \frac{2}{\pi}\int_{\pi/6}^{5\pi/6} E \sin 5\omega t \, d(\omega t) = \frac{2E}{\pi}\left[\frac{-\cos 5\omega t}{5}\right]_{\pi/6}^{5\pi/6} = -\frac{2\sqrt{3}E}{5\pi} \tag{5.26}$$

Similarly b_7, b_{11}, and b_{13} can be, respectively, obtained as $-2\sqrt{3}E/7\pi$, $2\sqrt{3}E/11\pi$, and $2\sqrt{3}E/13\pi$. The expression for v_{RY} now becomes

$$v_{RY}(t) = \frac{2\sqrt{3}E}{\pi}\left[\sin \omega t - \frac{1}{5}\sin 5\omega t - \frac{1}{7}\sin 7\omega t + \frac{1}{11}\sin 11\omega t + \frac{1}{13}\sin 13\omega t - \cdots\right] \tag{5.27}$$

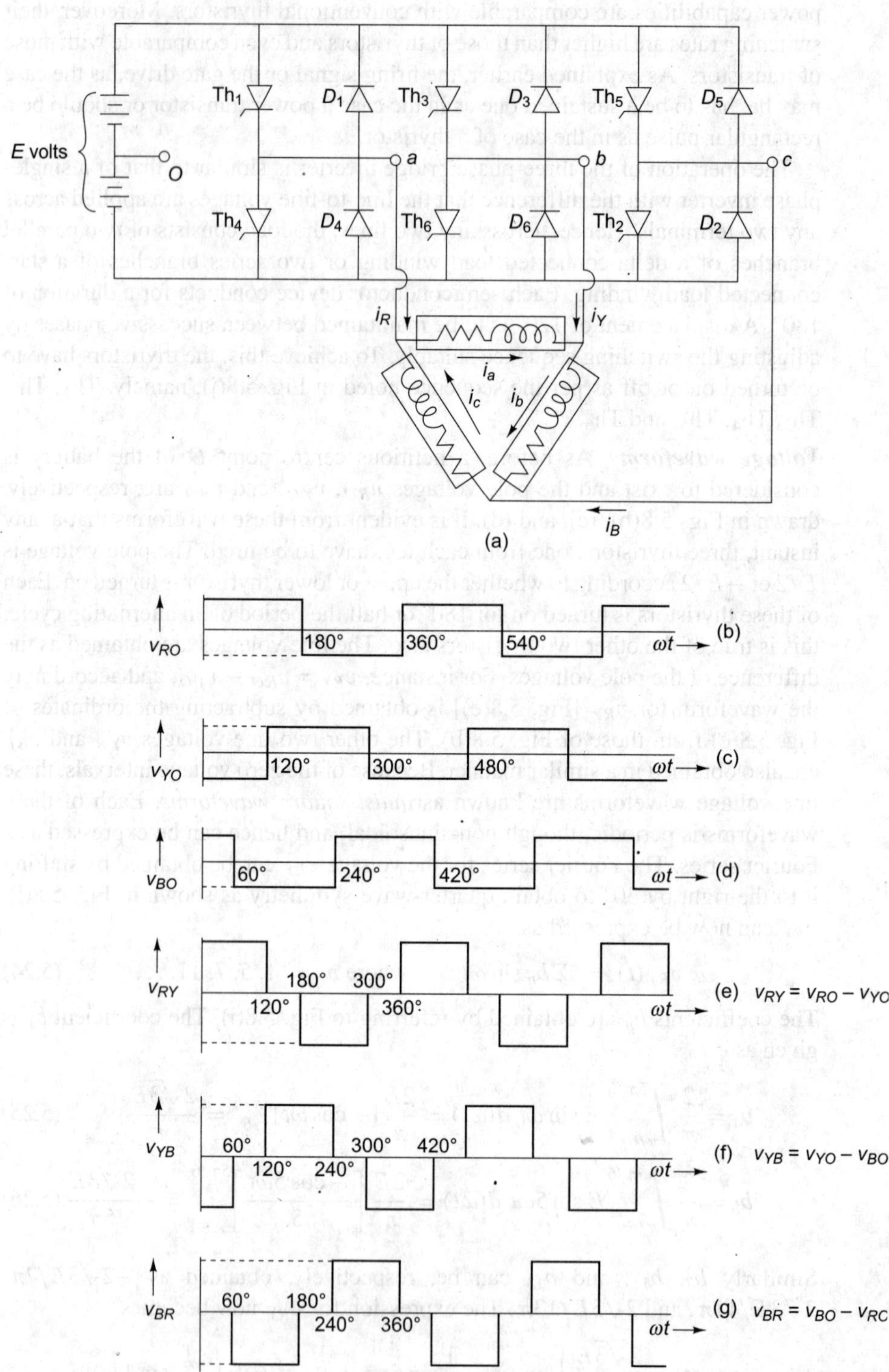
Th1
D1
Th3
D3
Th5
D5
E volts
O
a
b
c
Th4
D4
Th6
D6
Th2
D2
iR
iY
ia
ic
ib
iB
(a)
vRO
180°
360°
540°
ωt
(b)
vYO
120°
300°
480°
ωt
(c)
vBO
60°
240°
420°
ωt
(d)
vRY
180°
300°
120°
360°
ωt
(e) vRY = vRO − vYO
vYB
60°
300°
420°
120°
240°
ωt
(f) vYB = vYO − vBO
vBR
60°
180°
240°
360°
ωt
(g) vBR = vBO − vRO

Figs 5.8(a)–(g)

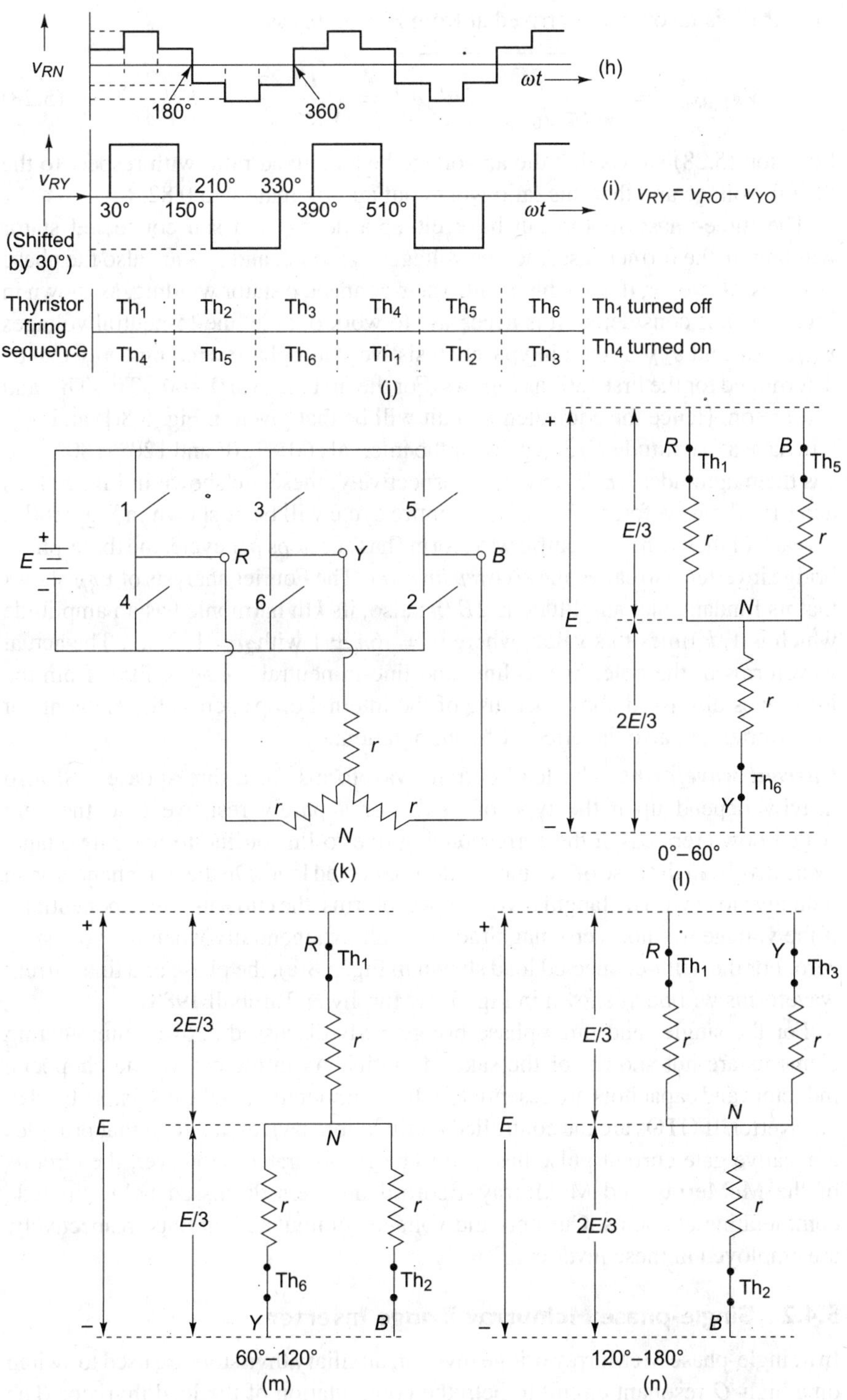

Fig. 5.8 Three-phase, six-step VSI: (a) circuit diagram, (b)–(d) pole voltages, (e)–(g) line voltages, (h) line-to-neutral voltage v_{RN}, (i) v_{RY} shifted to the right by 30°, (j) thyristor firing sequence, (k) equivalent circuit with six switches and resistive load, (l)–(n) equivalent circuits at different intervals

The RMS value of v_{RY} is arrived at from Fig. 5.8(e) as

$$V_{RY(\text{RMS})} = \sqrt{\frac{2}{2\pi}\int_0^{2\pi/3} E^2 d(\omega t)} = \sqrt{\frac{E^2}{\pi}\frac{2\pi}{3}} = E\sqrt{\frac{2}{3}} \qquad (5.28)$$

Equation (5.28) shows that the ac voltage bears a fixed ratio with respect to the dc link voltage and that this ratio works out approximately to 0.82.

The three-phase motor can have either a delta- or a star-connected stator winding. In the former case, the line voltages v_{RY}, v_{YB}, and v_{BR} are also the phase voltages. However, if a machine with a star-connected stator winding, as shown in Fig. 5.8(k), is considered, it is necessary to work out the line-to-neutral voltages v_{RN}, v_{YN}, and v_{BN}. Assuming purely resistive stator phase windings, v_{RN} can be determined for the first 180° as follows. For the first interval 0°–60°, Th_1, Th_5, and Th_6 are on. Hence the equivalent circuit will be that given in Fig. 5.8(l) and v_{RN} will have a magnitude $E/3$. Again for the intervals 60°–120° and 120°–180°, v_{RN} has the magnitudes $2E/3$ and $E/3$, respectively; these are shown in Figs 5.8(m) and (n). The waveform of v_{RN} for an entire cycle will be as shown in Fig. 5.8(h). As each of these line-to-neutral waveforms has six steps per cycle, the three-phase bridge inverter also called the *six-step inverter*. The Fourier analysis of v_{RN} shows that its fundamental amplitude is $2E/\pi$; also, its kth harmonic has an amplitude which is $1/k$ times this value, where $k = 6n \pm 1$ with $n = 1, 2, \ldots$. The actual waveforms of the pole, line-to-line, and line-to-neutral voltages differ from the ideal ones discussed above because of the internal drops across the elements of the inverter and also the effects of commutation.

Current waveforms The load current waveforms for a three-phase VSI also heavily depend upon the type of load. For a purely resistive load they are scaled-down replicas of the corresponding line-to-line or line-to-neutral voltage, respectively, in the case of a delta- or star-connected load. On the other hand, for an inductive load with rectangular voltage waveforms, the currents rise exponentially if the voltage has non-zero magnitude but fall exponentially when the voltage is zero. For the delta-connected load shown in Fig. 5.8(a), the phase and line current waveforms will be as shown in Fig. 5.9 (Murphy & Turnbull 1988).

For the single- and three-phase bridge VSIs discussed above, commutating elements are not shown for the sake of clarity. As in the case of dc choppers, inductors and capacitors are essential for the commutation of the thyristors used in an inverter. If GTOs are the controlled rectification devices, a circuit that provides a negative gate current pulse has also to be incorporated. However, the circuits of the McMurray and McMurray–Bedford inverters discussed below include commutating elements. Current- and voltage commutated circuits, respectively, are employed in these inverters.

5.4.2 Single-phase McMurray Bridge Inverter

In a single-phase McMurray bridge inverter, auxiliary thyristors are used to switch on a high-Q resonant circuit to help the commutation of the load thyristor. The circuit diagram and waveforms of this inverter are given in Fig. 5.10. The dc source may consist of a controlled rectifier, the output of which is smoothed by the filter elements L_0 and C_0. The inverter is assumed to feed an inductive load.

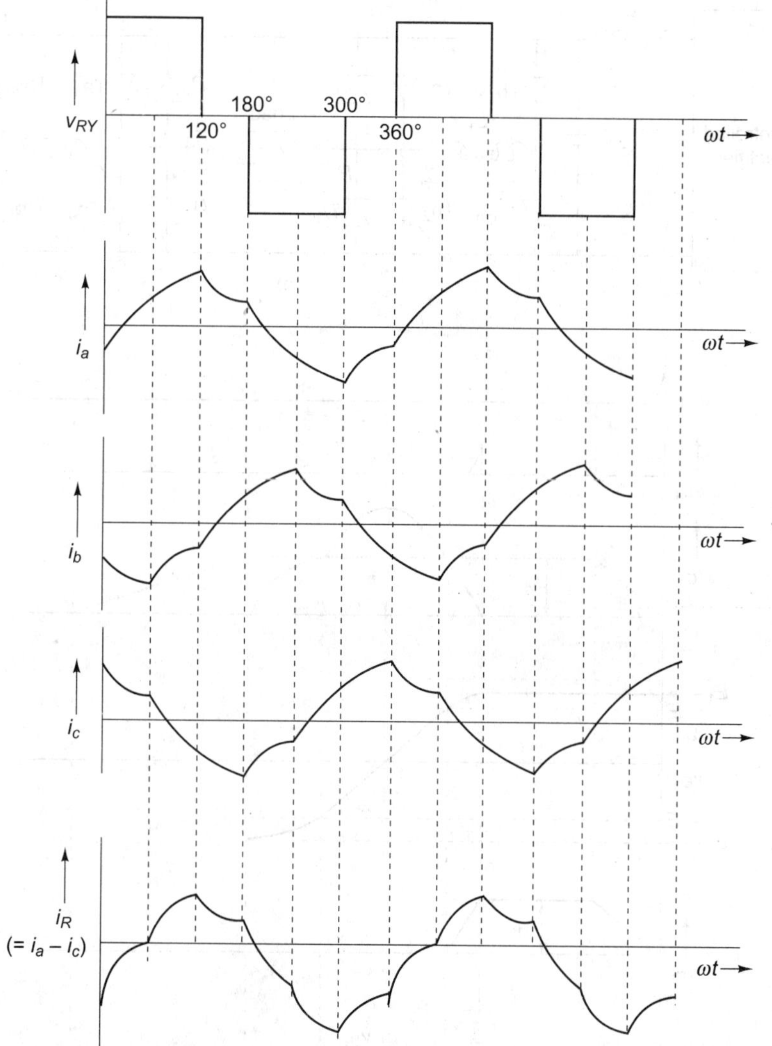

Fig. 5.9 Waveforms of the phase and line currents for the inverter of Fig. 5.8(a)

The thyristor pairs Th_1, Th_4 and Th_2, Th_3 conduct, respectively, in the positive and negative half-cycles. Considering the symmetry of the circuit about the load, only the left-hand side elements are taken up for discussion, with the understanding that the corresponding elements on the right-hand side undergo the same operating conditions.

The following assumptions, which are nearly true for practical circuits, facilitate the analysis of this inverter.

(a) The load inductance is assumed to be sufficiently large to give a constant load current till the commutation process is complete. This is justified by the fact that the commutation process is completed in 10–100 μs.

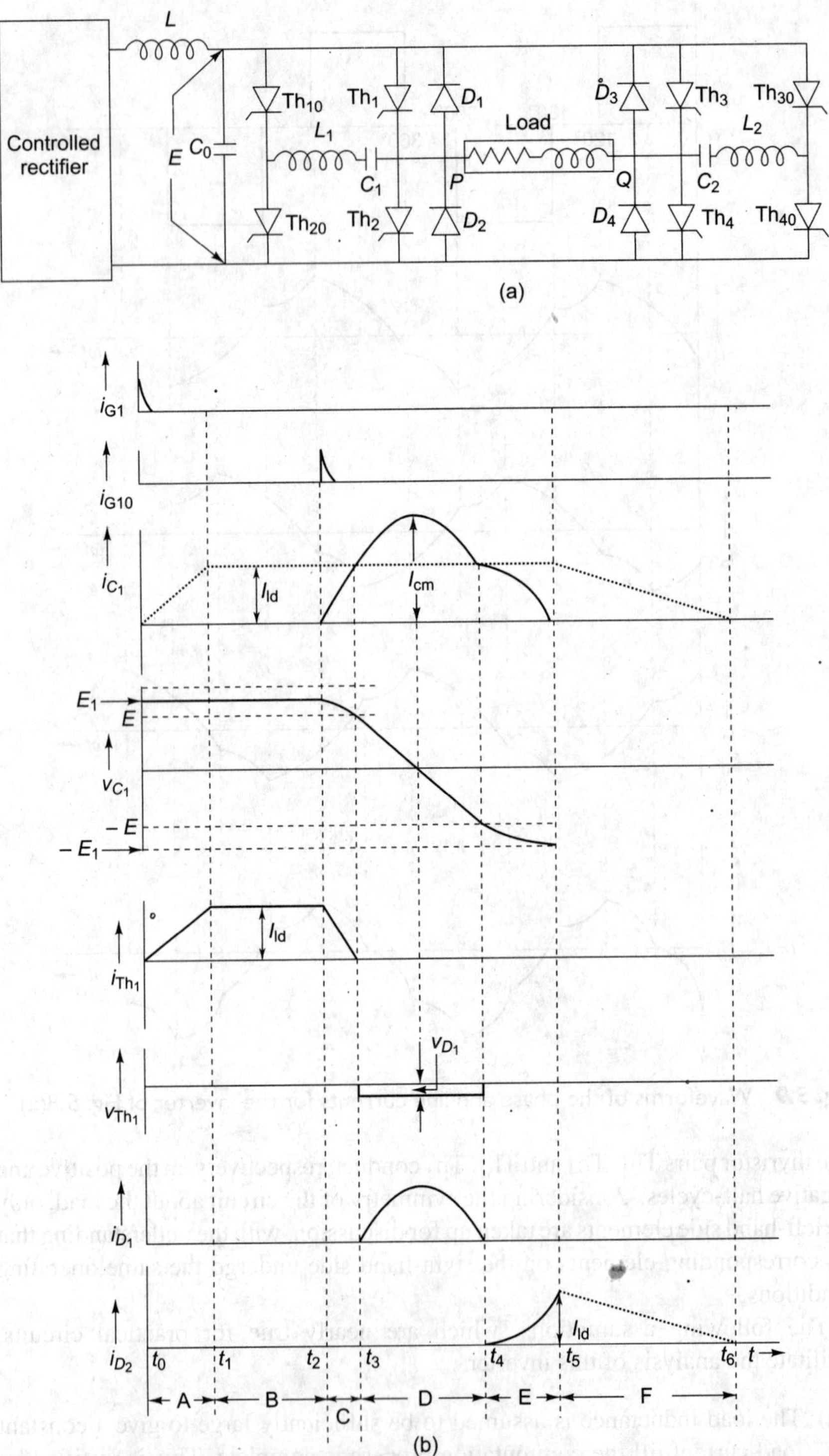

Fig. 5.10 Single-phase McMurray bridge inverter: (a) circuit diagram, (b) waveforms

(b) The elements L_1, L_2, C_1, and C_2 are taken to be lossless. In order to obtain a symmetrical output, the following equalities are assumed to hold good:

$$L_1 = L_2 = L$$

$$C_1 = C_2 = C$$

(c) The ON state voltage drops of the thyristors and diodes are considered to be negligible. This is justified by the fact that the drops are quite small compared to the supply and load voltages.

(d) It is assumed that the peak of the LC-oscillatory current (I_{cm}) is greater than that of the load current (I_{ld}). This is essential for inverter operation.

(e) As shown in Fig. 5.10(b), the sequence of operations is divided into the six intervals A–F.

Interval A At $t = t_0$, Th_1 is turned on so as to connect terminal P of the load to the positive dc bus. The capacitor C_1 is assumed to be already charged during the previous commutation to a voltage $E_1 > E$. The load current i_{ld} slowly increases and attains its full value of I_{ld} at t_1.

Interval B This interval starts at t_1. The full load current I_{ld} continues to flow through the Th_1, the load, and the corresponding right-hand thyristor. This interval terminates when the auxiliary thyristor Th_{10} is turned on at t_2. Figure 5.11(a) shows the circuit conditions at the start of this interval.

Interval C The turning on of Th_{10} at t_2 starts the commutation process. An LC-oscillatory current i_{C_1} flows through the circuit consisting of Th_{10}, L_1, C_1, and Th_1, and has an angular frequency of $\omega_0 = 1/\sqrt{L_1 C_1}$ rad/s. I_{C_1} attains its maximum (I_{cm}) after π/ω_0 seconds from t_2. At $t_3 < \pi/\omega_0$, I_{C_1} equals the load current I_{ld} and Th_1 turns off, marking the end of this interval.

Interval D The load being highly inductive, the load current i_{ld} continues to flow through Th_{10}, L_1, C_1, the load, and the corresponding elements on the right-hand side. However, the current in excess of I_{ld} flows through D_1. The forward drop V_{D_1} of diode D_1 is applied in a reverse manner across Th_1 and this ensures its commutation. When i_{C_1} attains its peak value of I_{cm}, v_{C_1} passes through zero because it has a phase difference of 90° (elec.) with respect to i_{C_1}. This interval terminates at t_4 when i_{C_1} goes just below the load current I_{ld} and D_1 stops conduction. The circuit time available for Th_1 to turn off is $t_4 - t_3$. Figure 5.11(b) shows the circuit conditions at the beginning of this interval.

Interval E As i_{C_1} falls short of the load current from t_4 onwards, this shortfall is offset by D_2. At t_4, D_2 gets forward biased because v_{C_1} attains the value $-E$. Consequently, the LC circuit experiences a step voltage E through Th_{10} and D_2. The inductive energy stored in L_1 gets transferred to the capacitor C_1 and overcharges it through this charging circuit consisting of Th_{10}, L_1, D_2, and back to Th_{10}. The capacitor gets charged to a negative maximum $-E_1$, thus becoming ready for the commutation of Th_2 in the next half-cycle. The load current is constant till t_5 ($< T/2$), where T is the time for one cycle of the output voltage waveform. The circuit conditions at the beginning of this interval is given in Fig. 5.11(c).

Interval F From t_5 to t_6, all the load current flows through the diode D_2, thus returning back the inductive energy of the load to the source. In this process the load current flows against the supply voltage E. Consequently, it gradually decreases, attains a zero value at t_6, and then reverses. At $t_6 = T/2$ a trigger pulse, already available at the gate of Th_2, turns it on; Th_2 now carries the reverse load current. The circuit conditions at the beginning of this interval is shown in Fig. 5.11(d).

The commutation of the McMurray inverter comes under the category of current commutation because the LC oscillatory current flows through the load thyristors in a reverse manner, thus turning them off.

Commutating time of Th_1 The commutation of Th_1 starts when Th_{10} is fired at t_1. Because of the fact that a reverse voltage is applied across Th_1 during the interval t_2 to t_3, the analysis of the circuit condition during the intervals t_1 to t_2 and t_2 to t_3 provides a method for determination of the circuit time t_c available for the commutation of Th_1. The KVL equation for the circuit consisting of Th_{10}, L, C, and Th_1 is given as

$$L\frac{di_c}{dt'} + \frac{1}{C}\int_0^{t'} i_c dt' - E = 0 \tag{5.29}$$

where $t' = t - t_1$.

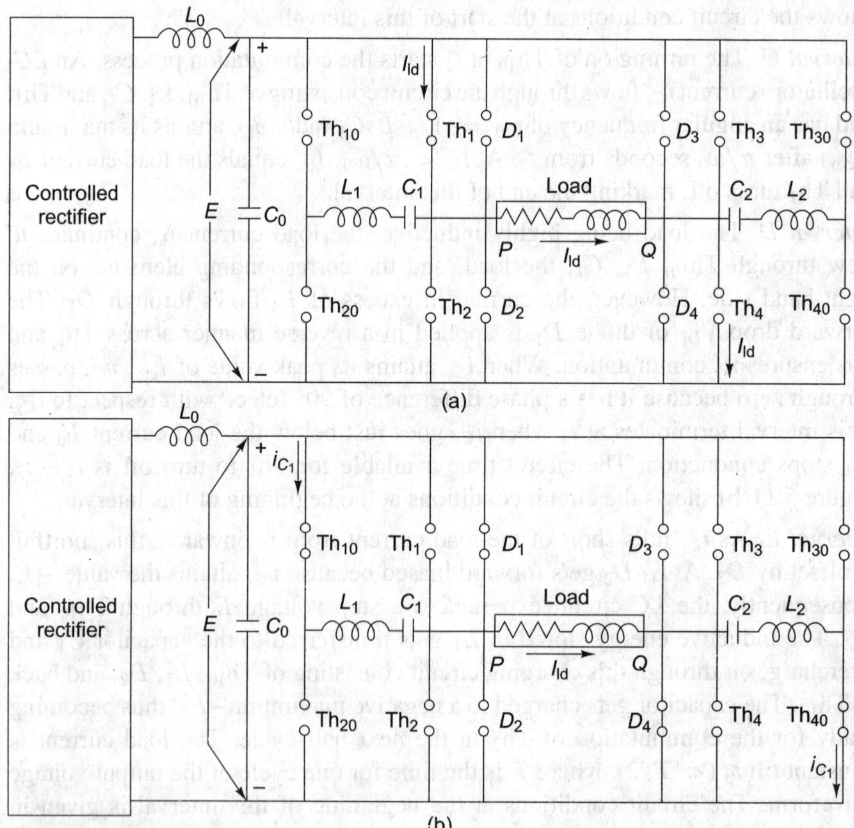

Figs 5.11(a), (b)

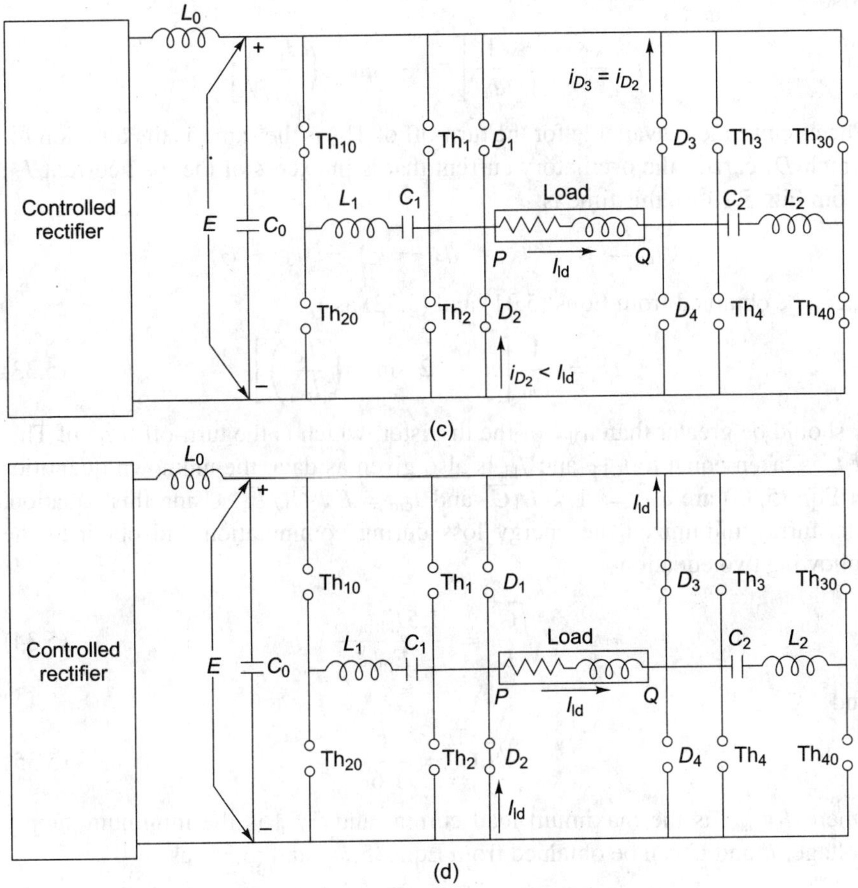

Fig. 5.11 Circuit conditions of the McMurray bridge inverter at the beginning of (a) interval B, (b) interval D, (c) interval E, and (d) interval F

One of the initial conditions is $i_c(0) = 0$; the second can be obtained from the fact that the initial voltage drop across the inductor L_1 must equal that across C_1, the latter being equal to E. Thus the second initial condition can expressed as

$$L_1 \frac{di_c}{dt'}|_{t'=0} = E$$

The solution of Eqn (5.29) with these two initial conditions is

$$i_c = I_{cm} \sin \omega_0 t' \tag{5.30}$$

where $I_{cm} = E\sqrt{C_1/L_1}$ and $\omega_0 = 1/\sqrt{L_1 C_1}$. From Fig. 5.10(b),

$$t_3 - t_2 = \frac{1}{\omega_0} \sin^{-1}\left(\frac{I_{ld}}{I_{cm}}\right) \tag{5.31}$$

Also,

$$t_4 - t_2 = \frac{1}{\omega_0}\left[\pi - \sin^{-1}\left(\frac{I_{ld}}{I_{cm}}\right)\right] \tag{5.32}$$

The circuit time t_c available for the turn-off of Th_1 is the same as the duration for which D_1 carries the oscillatory current that is in excess of the load current I_{ld}. From Fig. 5.10(b), this time is

$$t_c = t_4 - t_3 = (t_4 - t_2) - (t_3 - t_2)$$

t_c can be obtained from Eqns (5.31) and (5.32) as

$$t_c = \frac{1}{\omega_0}\left[\pi - 2\sin^{-1}\left(\frac{I_{ld}}{I_{cm}}\right)\right] \tag{5.33}$$

t_c should be greater than t_{OFF} of the thyristor, which is the turn-off time of Th_1. If t_c is taken equal to t_{OFF} and I_{ld} is also given as data, the unknown quantities in Eqn (5.33) are $\omega_0 = 1/\sqrt{L_1C_1}$ and $I_{cm} = E\sqrt{C_1/L_1}$. Using this equation, McMurray minimized the energy loss during commutation and obtained the following two equations:

$$\sqrt{\frac{C}{L}} = \frac{1.5 I_{ld(max)}}{E_{min}} \tag{5.34}$$

and

$$\sqrt{LC} = \frac{t_c}{1.68} \tag{5.35}$$

where $I_{ld(max)}$ is the maximum load current and E_{min} is the minimum supply voltage. L and C can be obtained from Eqns (5.34) and (5.35) as

$$L = 0.397\frac{E_{min} t_c}{I_{ld(max)}} \tag{5.36}$$

and

$$C = 0.893\frac{I_{ld(max)} t_c}{E_{min}} \tag{5.37}$$

5.4.3 Three-phase McMurray Bridge Inverter

The operation of the three-phase McMurray inverter can be understood from Fig. 5.12 as follows. The sequence of steps given for the single-phase inverter should be interpreted as the steps to be followed for the operation between any two phases of the three-phase load. A phase difference of 120° (elec) should be maintained between any two phases. The current waveforms depend both on the type of the load as well as the commutating elements used in the circuit.

5.4.4 Single-phase McMurray–Bedford Inverter

Figure 5.13(a) shows the circuit of a single-phase McMurray–Bedford inverter (Rajput 1993). It has half the number of thyristors in a single-phase McMurray inverter, this reduction is made possible because of the principle of complementary

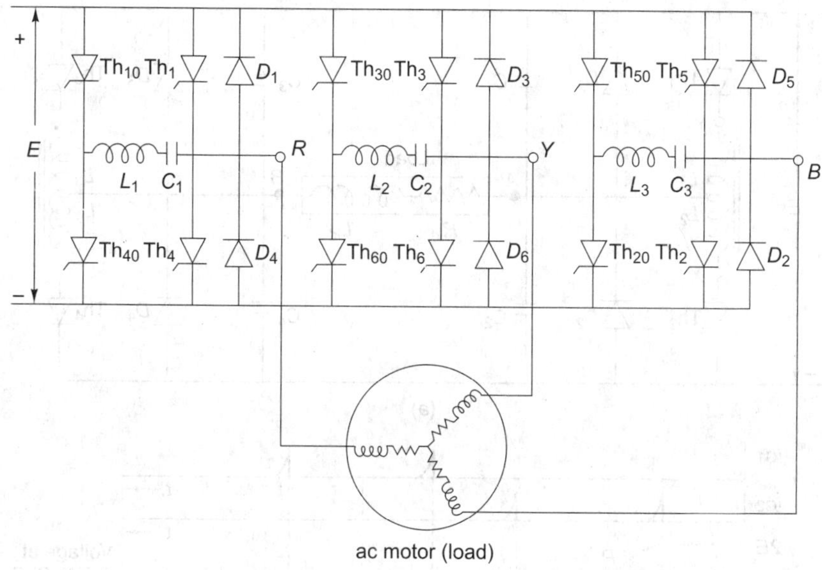

Fig. 5.12 Three-phase McMurray bridge inverter

commutation. In this technique, one device out, of the pair of thyristors on a leg causes the turn-off of the other. For example, Th_1 on the first leg initiates the turn-off of Th_2 and vice versa. Similarly thyristors Th_3 and Th_4 constitute a complementary pair. The centre-tapped inductors L_1, L_2 and L_3, L_4, as also the four capacitors C_1 to C_4 help in the commutation of the thyristors. According to the usual practice, diodes are employed to feed the load power back to the source. The operation of the circuit depends on whether the load is purely resistive or inductive. For each of these cases, in turn, the firing of the thyristors may be sequenced such that the load is either connected to the source or isolated from it during the post-commutation period. Thus, four alternative modes of operation are possible. For the purpose of this discussion, all the four inductors (L_1 to L_4) are assumed to be equal; that is, $L_1 = L_2 = L_3 = L_4 = L$. Similarly all capacitors are assumed to have the same capacitance, or $C_1 = C_2 = C_3 = C_4 = C$.
The four modes of operation are now taken up for discussion.

(i) Resistive load (first sequence): Load connected to the source in the post-commutation interval It is assumed that initially Th_1 and Th_4 are conducting the full positive load current I_{ld}. Because of the symmetry of the circuit, the operating conditions of Th_4, D_4, and C_4 will be, respectively, identical to those of Th_1, D_1, and C_1. Similarly, the operating conditions of the devices Th_3, D_3, and C_3 will be, respectively, identical to those of Th_2, D_2, and C_2. It is assumed that the load current remains at its full value till the commutation is completed.

As shown in Fig. 5.13(b) the sequence of operations is divided into six intervals, three each for the positive and negative half-cycles. The circuit conditions are given in Figs 5.13(c).

Interval A Thyristor Th_1 is turned on at t_0 and conducts the load current I_{ld}. Capacitor C_2 is charged to E prior to t_0 with its upper plate having positive

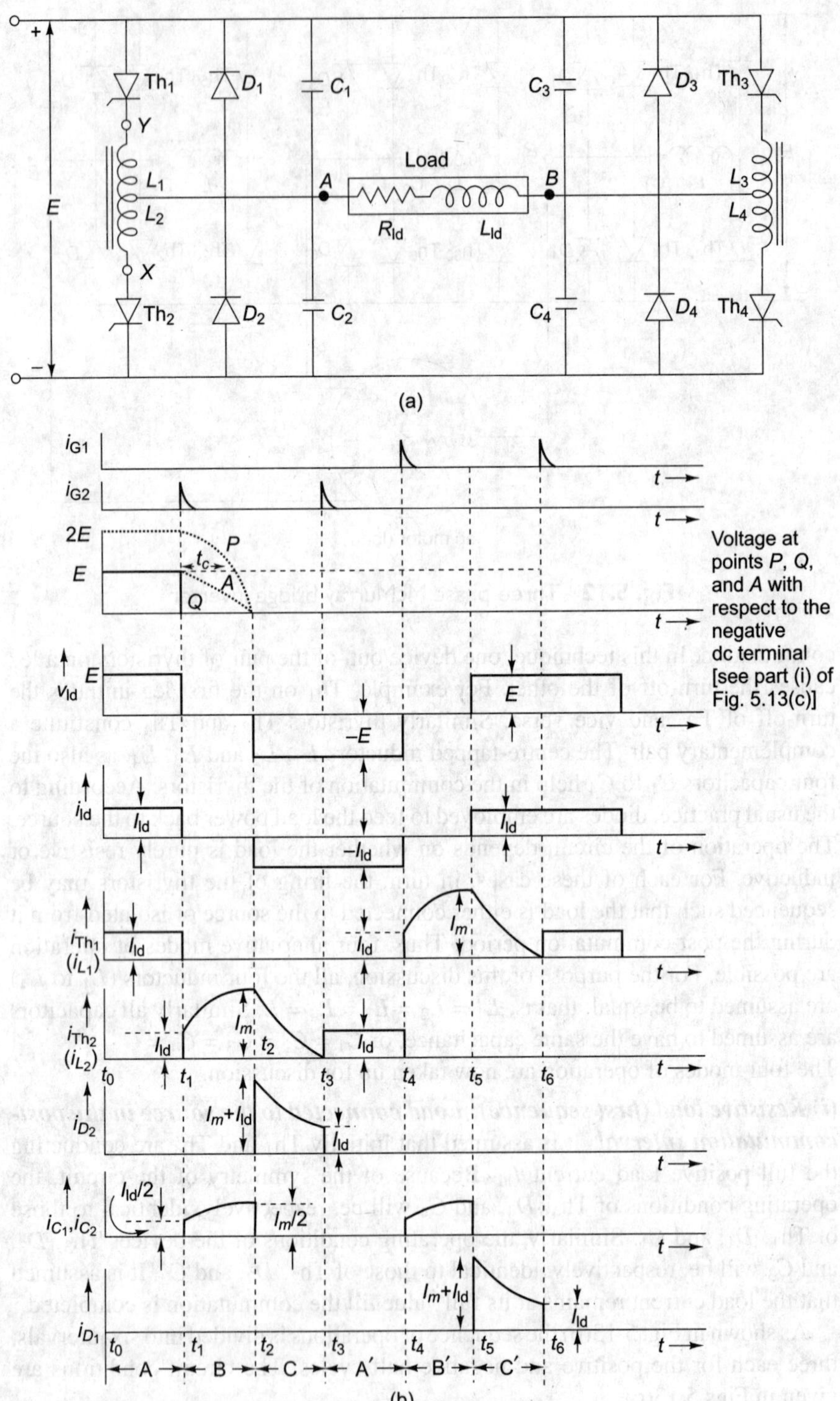
+
−
E
Th1
Y
L1
L2
X
Th2
D1
D2
C1
C2
A
Load
Rld
Lld
B
C3
C4
D3
D4
Th3
Th4
L3
L4
(a)
iG1
iG2
2E
E
P
tc
A
Q
Voltage at points P, Q, and A with respect to the negative dc terminal [see part (i) of Fig. 5.13(c)]
vld
E
−E
ild
Ild
−Ild
iTh1 (iL1)
Im
iTh2 (iL2)
iD2
Im+Ild
Ild/2
iC1, iC2
Im/2
iD1
t0
t1
t2
t3
t4
t5
t6
t
A
B
C
A′
B′
C′
(b)

Figs 5.13(a), (b)

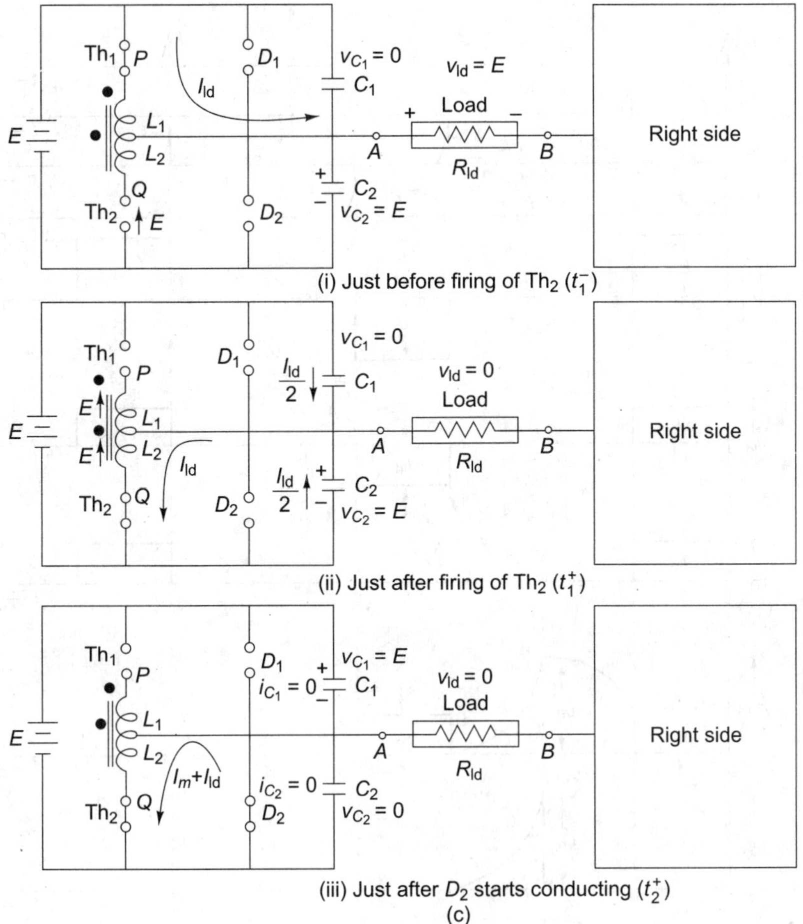

Fig. 5.13(c)

polarity. The voltage across the commutating inductors L_1 and L_2 will be very small compared to the battery voltage E because their inductances are assumed to be of the order of tens of microhenries. The voltage across the load terminals is therefore nearly equal to E, making the voltage across C_1 nearly zero. I_{ld} flows through Th_1, L_1, and the load as well as through the corresponding elements on the right-hand side. The circuit condition at the end of this interval, that is, prior to the firing of Th_2 is given in circuit (i) of Fig. 5.13(c).

Interval B Commutation of Th_1 is initiated when the other thyristor Th_2 is triggered at t_1. Since the load is resistive, the flow of current through it stops at t_1; likewise, the load voltage also becomes zero at t_1. As the voltage across the capacitors C_1 and C_2 cannot be changed instantaneously and the lower terminal (Q) of L_2 as well as the lower plate of C_2 are connected to the negative of battery, a voltage E appears across L_2. Due to autotransformer action, a voltage of $+2E$ volts appears at the top terminal of L_1, which is electrically at the same potential as the cathode of Th_1. But the anode of Th_1 is connected to the positive terminal of the battery; hence the device experiences a reverse bias of E volts and turns off.

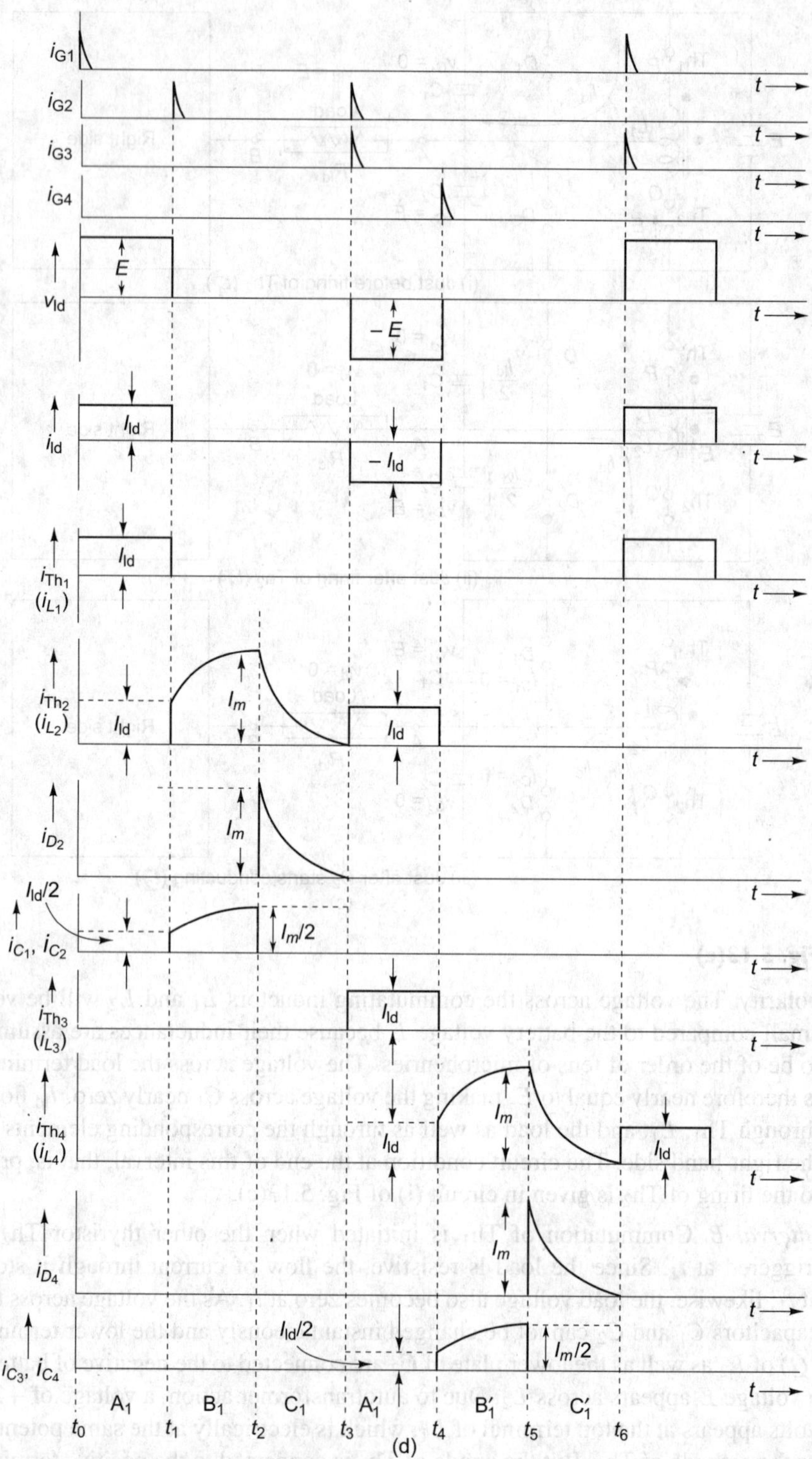
i_{G1}
i_{G2}
i_{G3}
i_{G4}
t
v_{ld}
E
$-E$
i_{ld}
I_{ld}
$-I_{ld}$
i_{Th_1} (i_{L_1})
i_{Th_2} (i_{L_2})
I_m
i_{D_2}
$I_{ld}/2$
i_{C_1}, i_{C_2}
$I_m/2$
i_{Th_3} (i_{L_3})
i_{Th_4} (i_{L_4})
i_{D_4}
i_{C_3}, i_{C_4}
A_1
B_1
C_1
A_1'
B_1'
C_1'
t_0
t_1
t_2
t_3
t_4
t_5
t_6
(d)

Fig. 5.13(d)

The current I_{ld} which was flowing through thyristor Th_1 and inductor L_1 gets transferred immediately to Th_2 and L_2 so as to preserve the ampere-turn balance in the autotransformer L_1-L_2. Accordingly, the inductive $(1/2)Li^2$ energy originally in L_1 is maintained. This current is equally contributed by capacitors C_1 and C_2 with $i_{C_1} = i_{C_2} = I_{ld}/2$ amperes, provided $C_1 = C_2 = C$.

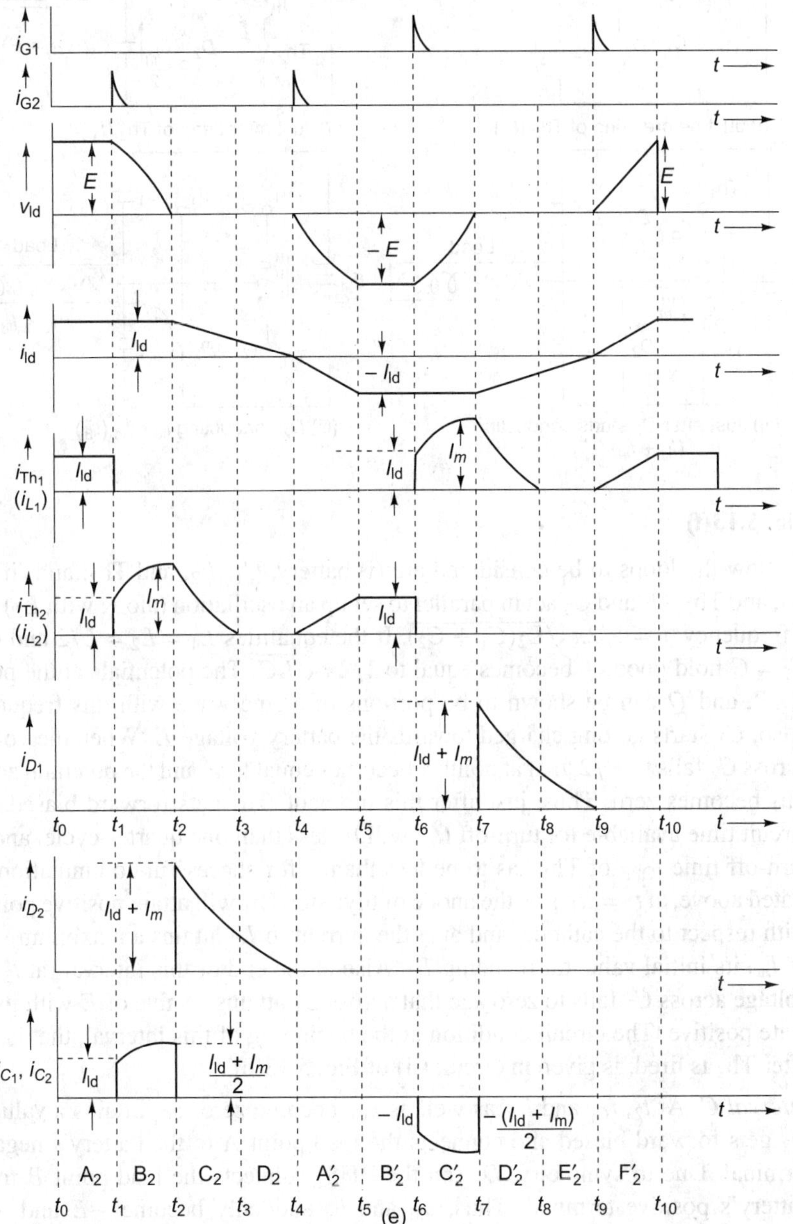

Fig. 5.13(e)

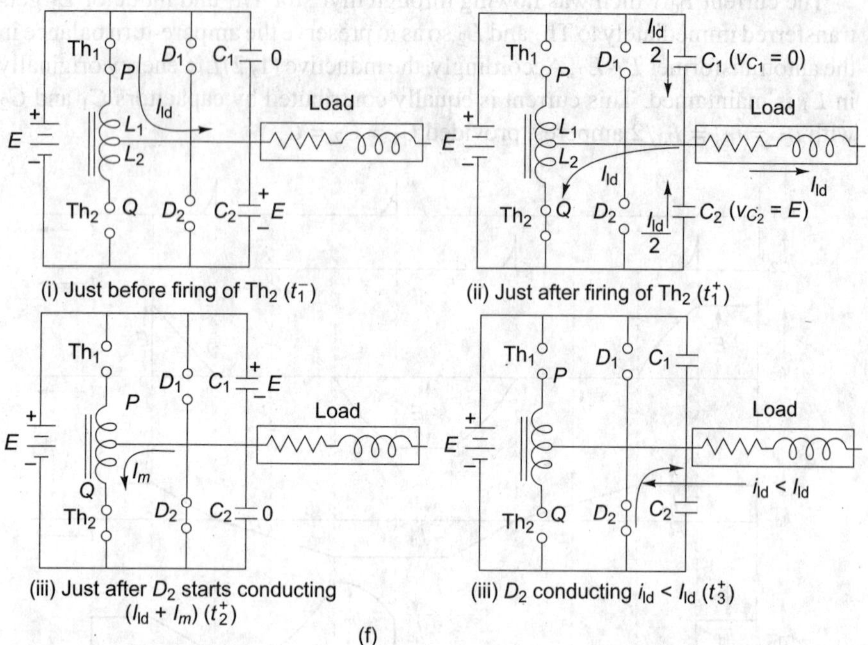

Fig. 5.13(f)

Now the loops to be considered are (i) battery, C_1, L_2, and Th$_2$, and (ii) C_2, L_2, and Th$_2$. C_1 and C_2 act in parallel to set up an oscillation (along with L_2) with a frequency $f = 1/2\pi\sqrt{L_2(C_1 + C_2)}$. If the equalities $L_1 = L_2 = L/2$ and $C_1 = C_2 = C$ hold good, f becomes equal to $1/2\pi\sqrt{LC}$. The potentials at the points A, P, and Q can be shown to be portions of a sine wave with this frequency. Also, C_1 starts getting charged towards the battery voltage E. When the voltage across C_2 falls to $E/2$, that at point A becomes equal to E and the potential across Th$_1$ becomes zero. Thus, just after this moment, Th$_1$ gets forward biased. The circuit time available for turn-off (t_c) will be less than one quarter-cycle, and the turn-off time t_{OFF} of Th$_1$ has to be less than t_c for successful commutation. As stated above, at $t = t_1 + t_c$, the anode of thyristor Th$_1$ will attain positive polarity with respect to the cathode, and at t_2 the current in L_2 attains a maximum value of I_m, its initial value (at t_1) being I_{ld}. Also at the end of this interval (at t_2) the voltage across C_2 falls to zero and that across C_1 attains a value of E with its top plate positive. The circuit condition at the beginning of this interval, that is, just after Th$_2$ is fired, is given in circuit (ii) of Fig. 5.13(c).

Interval C At t_2, i_{C_1} and i_{C_2} as well as v_{C_2} become zero. v_{C_1} attains a value E. D_2 gets forward biased and connects the load point A to the battery's negative terminal. Due to symmetry, D_3 (on the RHS) connects the load point B to the battery's positive terminal. Thus, v_{ld} and i_{ld} suddenly become $-E$ and $-I_{\text{ld}}$, respectively. D_2 now starts conducting a total current of $I_m + I_{\text{ld}}$. The trapped energy in L_2 circulating through D_2, L_2, and Th$_2$ gets dissipated through their resistances and becomes zero at t_3. Thus, i_{D_2} becomes equal to I_{ld} at t_3. v_{ld} and i_{ld} remain at $-E$ and $-I_{\text{ld}}$ till D_2 continues to conduct, that is, till t_3. The voltage drop

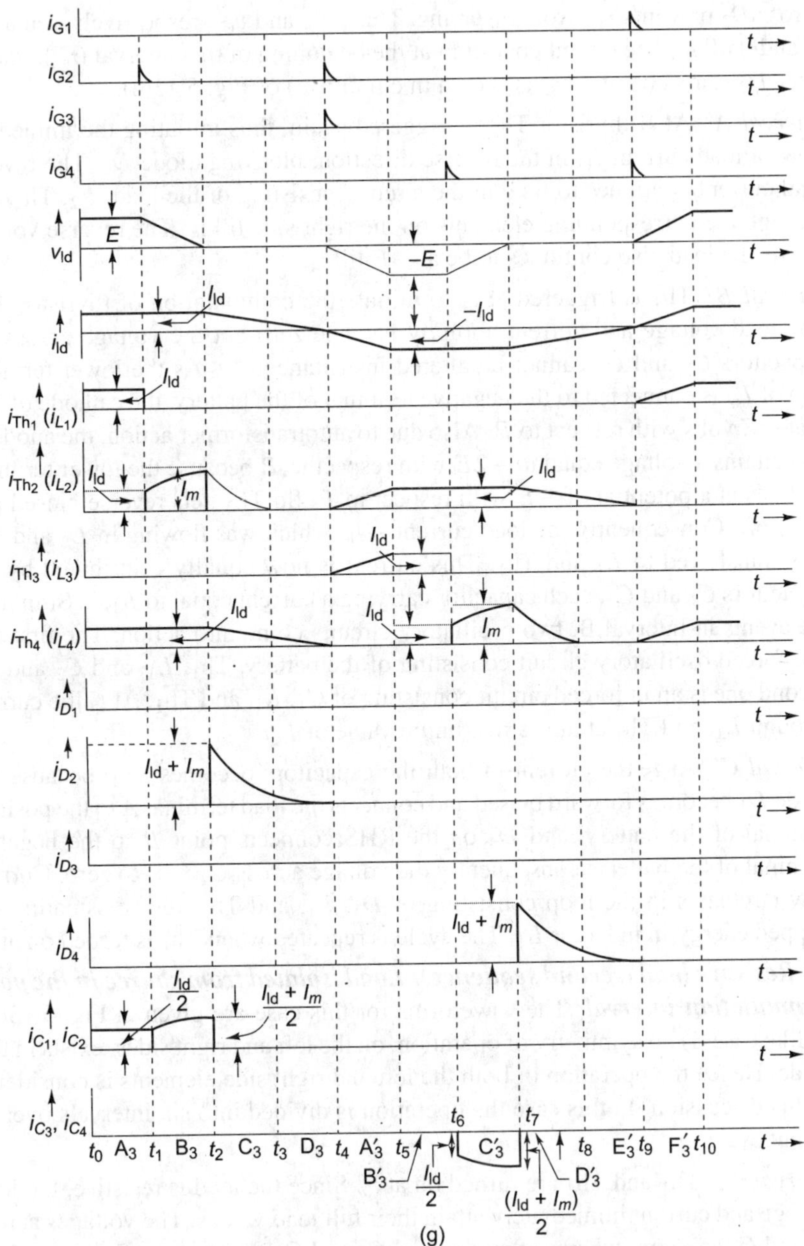

Fig. 5.13 Single-phase McMurray–Bedford impulse commutated bridge inverter: (a) circuit diagram, (b) waveforms with resistive load, load connected to the source after commutation. (c) Circuit conditions for resistive load as in (b): (i) at the end of interval A, (ii) at the beginning of interval B, (iii) at the beginning of interval C. (d) Waveforms for pure resistive load, load isolated from the source after commutation. (e) Waveforms with inductive load, load connected to the source after commutation. (f) Circuit condition for inductive load as in (e): (i) at the end of interval A_2, (ii) at the beginning of interval B_2 (iii) at the beginning of interval C_2, (iv) at the beginning of interval D_2. (g) Waveforms for inductive load, load isolated from the source after commutation

across D_2 prevents v_{C_2} from reversing. Thus, v_{C_1} and v_{C_2}, respectively, remain at E and 0 till t_2. The circuit condition at the beginning of this interval (t_2^+), that is, after D_2 starts conducting, is shown in circuit (iii) of Fig. 5.13(c).

Interval A′ At t_3 thyristor Th_2 is triggered again, thus initiating the immediate flow of load current I_{ld} in the reverse direction, blocking diode D_2. The reverse load current continues to flow in the circuit consisting of the load, L_2, Th_2, and through the corresponding elements on the right side till t_4. The reverse voltage across the load also continues to be at $-E$ till t_4.

Interval B′ Th_1 is triggered at t_4 to initiate the commutation of thyristor Th_2. The load voltage and current abruptly become zero but the voltage across the capacitors C_1 and C_2 cannot be altered instantaneously. As the lower terminal (Q) of L_2 is connected to the negative terminal of the battery, the cathode of Th_2 is at $-E$ volts with respect to P. Also due to autotransformer action, the anode of Th_2 attains a voltage equal to $-2E$ with respect to P because the lower terminal of L_1 is at a potential of $-E$ with respect to P. So Th_2 gets reverse biased and turns off. Consequently the load current $-I_{\text{ld}}$ which was flowing in L_2 and Th_2 gets transferred to L_1 and Th_1. This current is now equally contributed by the capacitors C_1 and C_2, each capacitor carrying a current equal to $I_{\text{ld}}/2$. Similar to the events in interval B, two oscillatory circuits come into action. The first one is a forced oscillatory circuit consisting of the battery, Th_1, L_1, and C_2 and the second one is an unforced circuit consisting of C_1, L_1, and Th_1. At t_5 the current through L_1, and Th_1 attains a maximum value of I_m.

Interval C′ At t_5 the current in both the capacitors becomes zero because the diode D_1 becomes forward biased and connects the load terminal A to the positive terminal of the battery, and D_4 on the RHS connects point B to the negative terminal of the battery; consequently the voltage across C_2 falls to zero. Current now circulates in the loop consisting of D_1, L_1, and Th_1, thus dissipating the trapped energy in inductor L_1. The cycle is repeated when Th_1 is turned on at t_6.

(ii) Resistive load (second sequence): Load isolated from source in the post-commutation interval The waveforms for this case are given in Fig. 5.13(d). Unlike case (i) no symmetry of operation, on the left and right sides, exists in this mode. Hence the operation of both the left and right side elements is considered in this discussion. In this case the operation is divided into six intervals over an output cycle.

Interval A_1 Th_1 and Th_4 are turned on at t_0. Since the load is resistive, the load voltage and current immediately attain their full load values. The voltages across C_1 and C_4 are zero, whereas those across C_2 and C_3 are equal to E.

Interval B_1 The commutation of Th_1 is initiated with the firing of Th_2 at t_1 and the load voltage and current abruptly become zero. As in interval B of case (i), Th_1 turns off because it experiences a reverse bias of E volts. The load current is transferred from L_1 to L_2 to maintain ampere-turn balance in the autotransformer. Two oscillating current loops, the battery, C_1, L_2, and Th_2 on the one hand and C_2, L_2, and Th_2 on the other, come into operation. The commutation of Th_1 will be successful if the reverse voltage is applied across it for a duration $t_c \geq t_{\text{OFF}}$.

At t_2, v_{C_2} becomes zero and v_{C_1} becomes equal to E. On the right-hand side, Th_4 stops conduction at t_1 itself.

Interval C_1 At t_2 the diode D_2 becomes forward biased and the inductive energy in L_2 is dissipated by the closed circuit consisting of D_2, L_2, and Th_2. At t_3 this circulating current becomes zero and this marks the end of this interval. The load gets isolated from the source during this interval unlike that in case (i) because of dissymmetry of operation of the circuit.

Interval A_1' Th_3 is fired at t_3. If Th_2 is turned off during the interval C_1 it is fired again. The load voltage and current suddenly attain their negative maximum values and remain so till t_4. The load current flows in the reverse direction in the circuit consisting of the positive of the battery, Th_3, L_3, the load, L_2, Th_2, and the negative of the battery. At t_4 both C_3 and C_2 become zero whereas C_4 and C_1 attain E volts each.

Interval B_1' When Th_4 is fired at t_4, Th_3 turns off and the load voltage and current become zero. The current in L_3 is transferred to L_4 to maintain the ampere-turn balance. The oscillatory current loops, battery, C_3, L_4, and Th_4 on the one hand and C_4, L_4, and Th_4 on the other, come into operation. At t_5, v_{C_4} becomes zero and v_{C_3} equals E and this marks the end of this interval. On the left-hand side Th_2 stops conduction at t_4 and remains in that state.

Interval C_1' Diode D_4 becomes forward biased at t_5. The remaining inductive energy of L_4 is dissipated in the circuit consisting of D_4, L_4, and Th_2. The cycle is repeated by turning on Th_1 and Th_4 at t_6. The load gets isolated from the source even during intervals B_1' and C_1'.

(iii) Inductive load (first sequence): Load connected to the source after commutation The sequence of operations in the case of inductive load differs from that of resistive load because the inductive energy in the load takes some time either to get dissipated or to be fed back to the source. As before, it is assumed that the load current remains at its full value till the commutation is completed. The waveforms and circuit conditions for this mode are given, respectively, in Figs 5.13(e) and (f). There is symmetry in this mode of operation as follows. When thyristor Th_1 conducts the load current on the left-hand side, the corresponding thyristor Th_4 on the right side conducts the same current. Similarly when the left-hand side elements L_1, C_1, and D_1 undergo certain operating conditions, the corresponding right-hand side elements L_4, C_4, and D_4 also experience the same conditions. The following sequence of operations is divided into ten intervals, and as before the discussion is confined to the left side elements.

Interval A_2 This interval starts at t_0. It is assumed that Th_1 is fired in the beginning of the previous interval and that its current has attained the full load value of I_{ld}. As in mode (i), L_1 is assumed to be small so that the load voltage equals the battery voltage E. v_{C_1} and v_{C_2} are assumed to have attained steady-state values, namely, zero and E volts, respectively. The circuit condition at the end of this interval is given in part (i) of Fig. 5.13(f).

Interval B_2 The triggering of Th_2 at t_1 initiates the commutation of Th_1. As the voltages across C_1 and C_2 cannot be altered simultaneously and as the lower terminal Q of inductor L_2 is connected to the negative of the battery, a voltage E appears across L_2. Due to autotransformer action, a voltage $2E$ is induced at the top terminal of L_1, which is electrically the same as the cathode of Th_1. As the anode of Th_1 is at a voltage E (it is connected to the positive of the battery), it experiences a reverse bias of E volts and turns off. The current I_{ld} in L_1 abruptly becomes zero and, to keep the magnetizing ampere-turns constant, an equal current is set up in L_2. A forced oscillation occurs in the circuit consisting of the battery, C_1, L_2, and Th_2 and an unforced one in the closed circuit consisting of C_2, L_2, and Th_2. Due to its inductive nature, the load current (I_{ld}) continues to flow through the load on the right side till t_2. At t_1 the capacitors C_1 and C_2 equally share a current totalling to $2I_{ld}$. The current in L_2, which has an initial magnitude of I_{ld} at t_1, rises to I_m at t_2. The current increases in the interval t_1 to t_2, and at t_2, C_1 and C_2 equally share a current equal to $I_{ld} + I_m$; also at t_2, v_{C_1} and v_{C_2}, respectively, attain voltages of magnitudes E and 0. Consequently the load voltage also becomes zero and D_2 becomes forward biased at t_2 and this marks the end of this interval. D_2 now conducts both the load current as well as the current I_m that is conducted by L_2 and Th_2. The circuit condition at the beginning of this interval, that is, just after Th_2 is fired, is given in part (ii) of Fig. 5.13(f).

Interval C_2 At t_2, the current I_m through L_2 and Th_2 suddenly falls to zero, consequent to the start of conduction of D_2. The load current, which was originally at the value I_{ld} at t_2, falls down to a lower value at t_3. The current in D_2 also decays through the closed circuit consisting of D_2, L_2, and Th_2; at t_3, i_{D_2} attains a value equal to a part of the load current I_{ld}. The circuit conditions at the beginning of this interval are given in part (iii) of Fig. 5.13(f).

Interval D_2 During this interval which starts at t_3, the load current i_{ld} ($< I_{ld}$) decays and becomes zero at t_4; this decaying current i_{ld} is conducted by D_2 on the left-hand side and D_3 on the right-hand side. The circuit condition at the beginning of this interval is given in part (iv) of Fig. 5.13(f).

Interval A'_2 At t_4, which marks the beginning of this interval, Th_2 is turned on if it is already in an off condition; consequently the load current reverses. This reverse current slowly builds up during this interval and attains the full value $-I_{ld}$ at t_5; the interval also ends at t_5.

Interval B'_2 The load voltage becomes nearly equal to $-E$ and v_{C_2} becomes zero at t_5. The load current flows through the closed circuit consisting of the battery, Th_2, L_2, the load, and the right-hand side elements. Finally Th_1 is triggered at the end of this interval (t_6).

Interval C'_2 The commutation of Th_2 starts upon the firing of Th_1 at t_6, which is also the start of this interval. Forced and unforced oscillations are set up in the following loops: battery, Th_1, L_1, and C_2 on the one hand and Th_1, L_1, and C_1 on the other. C_1 and C_2 each carry a current of $-I_{ld}$ amperes. The anode of Th_2 attains a voltage of $-2E$ with respect to P because the centre point of the L_1-L_2 combination is at a voltage of $-E$. This is a consequence of the autotransformer action of L_1 and L_2. Also, the cathode of Th_2 is at a voltage of $-E$ volts with respect to P, which is connected to the top of the battery. Hence Th_2 gets reverse biased and turns off. At t_7, which is the end of this interval, v_{C_1} becomes zero and v_{C_2} attains E volts; consequently D_1 becomes forward biased. The current in each of the capacitors C_1 and C_2 rises to $-(I_{ld} + I_m)/2$.

Interval D'_2 In this interval, which starts at t_7, current flows in the circuit consisting of D_1, L_1, and Th_1. Its initial value is $I_{ld} + I_m$. Its final value at t_8 will be i_{ld} ($< I_{ld}$).

Interval E'_2 During this interval, which starts at t_8, the load current which was flowing through D_1, the load, and the corresponding right-hand side diode reduces and becomes zero at the end of the interval. At t_9, Th_1 is turned on and the load current starts flowing in the positive direction, thereby starting the positive cycle.

Interval F'_2 During this interval the load current flowing through Th_1, L_1, and the load slowly increases and finally attains the full load value at t_{10}. v_{C_1} and v_{C_2} attain zero and E volts, respectively. The cycle is repeated from t_{10} onwards.

(iv) Inductive load (second sequence): Load isolated from the source after commutation Figure 5.13(g) gives the waveforms for this mode of operation. The symmetry of operation that exists in mode (iii) is not present here. Hence, both left- and right-hand side elements have to be considered while dealing with the operation of the inverter. The sequence of operations is divided into ten intervals as in mode (iii).

Interval A_3 This interval starts at t_0. The thyristors Th_1 and Th_4 are assumed to have been triggered earlier and thus conduct the full load current I_{ld}. As before, L_1 and L_4 are assumed to be small so that the load voltage is nearly equal to E. The voltages across C_1 and C_4 are zero and those across C_2 and C_3 are equal to E.

Interval B_3 At t_1, Th_2 is fired, initiating the commutation of Th_1. All the sequences of operations as given under interval B_2 of mode (iii) take place on the left-hand side. However, there is no change for the right-hand side elements and I_{ld} continues to flow through Th_4.

Interval C_3 The current through L_2 and Th_2, which attains a value of I_m prior to t_2, starts decaying at t_2 and becomes zero at t_3. The load current which has a value of I_{ld} up to t_2 also falls down to a lower value. D_2 conducts both the circulating current flowing through L_2 and Th_2 as well as this falling load current. On the right-hand side the same load current flows through Th_4.

Interval D_3 The load current i_{ld} ($< I_{ld}$) which was flowing through D_2, the load, and Th_4 decays to zero at t_4. A unique feature of this mode is that the load is isolated from the source during the intervals B_3, C_3, and D_3.

Interval A'_3 At t_4, thyristors Th_2 and Th_3 are fired to facilitate the flow of load current in the reverse direction, thus starting the negative half-cycle. The load current gradually rises and attains a value $-I_{ld}$ at t_5 and this marks the end of this interval.

Interval B'_3 At t_5, v_{C_3} becomes zero and v_{C_4} attains E volts. The full reverse load current, $-I_{ld}$, flows from t_5 onwards through Th_3, L_3, the load, L_2, and Th_2 till the end of this interval.

Interval C'_3 On the right-hand side, the commutation of Th_3 is initiated at t_6 by turning on Th_4. The current i_{Th_3} abruptly becomes zero, and for maintaining the ampere-turn balance of the autotransformer L_3-L_4, the current in L_3 gets transferred to L_4. Also, the forced oscillatory current in the circuit consisting of the battery, C_3, L_4, and Th_4 as well as the unforced oscillatory current in the circuit consisting of C_4, L_4, and Th_4 start circulation. At t_6 the load current $-I_{ld}$ is equally shared by C_3 and C_4. At the end of this interval, which occurs at t_7, I_{L_4} and I_{Th_4} attain the value I_m; also i_{C_3} and i_{C_4} each attain the value $-(I_{ld} + I_m)/2$. The reverse load current having the full value I_{ld} continues to flow through Th_2 on the left-hand side.

Interval D'_3 v_{C_4} and v_{C_3} attain values of zero and E volts, respectively, at t_7. Consequently diode D_4 gets forward biased. The load current, which is equal to the full value of I_{ld} at t_7, gradually falls down due to the dissipation occurring in the circuit consisting of D_4, the load, L_2, and Th_2.

Interval E'_3 The conditions occurring in interval D'_3 are maintained; also, reverse load current flows through D_4, the load, and Th_2, finally becoming zero at t_9. The load gets isolated from the source during the intervals C'_3, D'_3, and E'_3.

Interval F'_3 A new positive half-cycle is started at t_9 by firing Th_1 and Th_4. Consequently the load current slowly rises from zero to I_{ld}; it flows through the battery, Th_1, the load, L_1, L_4, Th_4, and back to the battery. At t_{10}, which marks the end of this interval, i_{ld} attains its full value I_{ld}. The cycle is repeated from t_{10} onwards.

5.4.5 Analysis of the Single-phase McMurray–Bedford Inverter

(i) Purely resistive load The inverter is analysed with reference to the circuit shown in Fig. 5.13(a) and the waveforms given in Fig. 5.13(b). The moment at which Th_2 is fired is taken as the initial time ($t = 0$). The circuit time available for turn-off is the time needed for v_{C_2} to drop from E to $E/2$.

The KVL equations for the two loops (a) battery, C_1, L_2, and Th_2 and (b) C_2, L_2, and Th_2 are written as

$$\frac{1}{C}\int i_1(t)dt + \frac{L}{2}\frac{d[i_1(t) + i_2(t)]}{dt} = E \tag{5.38}$$

and

$$\frac{1}{C}\int i_2(t)dt + \frac{L}{2}\frac{d[i_1(t) + i_2(t)]}{dt} = 0 \tag{5.39}$$

where i_1(t) and $i_2(t)$ are the currents flowing, respectively, in the first and the second loop. The initial conditions are

$$i_{L_2}(0) = (i_1 + i_2)|_{t=0} = I_{ld}$$

$$v_{C_1}(0) = \frac{1}{C}\int_{-\infty}^{0} i_1(t)dt = 0$$

and

$$v_{C_2}(0) = \frac{1}{C}\int_{-\infty}^{0} i_2(t)dt = -E \tag{5.40}$$

Laplace transformation of Eqns (5.38) and (5.39) results in the following equations:

$$\left(\frac{1}{Cs} + \frac{Ls}{2}\right) I_1(s) + \frac{Ls}{2} I_2(s) = \frac{E}{s} + \frac{L}{2} I_{\text{ld}} \tag{5.41}$$

$$\frac{Ls}{2} I_1(s) + \left(\frac{1}{Cs} + \frac{Ls}{2}\right) I_2(s) = \frac{E}{s} + \frac{L}{2} I_{\text{ld}} \tag{5.42}$$

Equations (5.41) and (5.42) can be solved to give the Laplace transforms of the currents as

$$I_1(s) = I_2(s) = \frac{E}{s(Ls + 1/Cs)} + \frac{(L/2)I_{\text{ld}}}{Ls + 1/Cs} \tag{5.43}$$

Inverse Laplace transformation of Eqn (5.43) gives the time functions as

$$i_1(t) = i_2(t) = E\sqrt{\frac{C}{L}} \sin \omega t + \frac{I_{\text{ld}}}{2} \cos \omega t \tag{5.44}$$

Defining

$$\frac{I_m}{2} \cos \phi = E\sqrt{\frac{C}{L}} \tag{5.45}$$

and

$$\frac{I_m}{2} \sin \phi = \frac{I_{\text{ld}}}{2} \tag{5.46}$$

Equation (5.44) can be written in a compact form as

$$i_1(t) = i_2(t) = \frac{I_m}{2} \sin(\omega t + \phi) \tag{5.47}$$

where

$$\tan \phi = \frac{I_{\text{ld}}/2}{E\sqrt{C/L}} \tag{5.48}$$

and

$$\omega = \frac{1}{\sqrt{LC}}$$

v_{C_2} is written as

$$v_{C_2}(t) = \frac{1}{C}\int i_2(t)dt \tag{5.49}$$

Using Eqns (5.47) and (5.49), $v_{C_2}(t)$ is obtained as

$$v_{C_2}(t) = -\sqrt{\frac{L}{C}}\frac{I_m}{2}\cos(\omega t + \phi) \tag{5.50}$$

Design of commutating elements The elements L and C which help in the commutation of the thyristor are now designed with the help of Eqns (5.44) and (5.50). The following quantities are assumed to be given:

- I_m, assumed to have a value which is k times I_{ld}, where k can be taken as a constant between 1.5 and 2,
- battery voltage E, and
- load resistance R_{ld}.

The expression for $\sqrt{C/L}$ can be obtained by manipulation of Eqns (5.44) and (5.45) as

$$\sqrt{\frac{C}{L}} = \frac{1}{2E}(I_m^2 - I_{\text{ld}}^2)^{1/2} \tag{5.51}$$

The second equation relating L and C is obtained from the fact that the circuit turn-off time t_c, which is the time required for v_{C_2} to attain a value of $-E/2$, should be at least equal to the turn-off time of the thyristor. Thus,

$$v_{C_2}(t_c) = -\frac{E}{2} = -\sqrt{\frac{L}{C}}\frac{I_m}{2}\cos(\omega t_c + \phi) \tag{5.52}$$

Equation (5.52) can be rewritten as

$$\frac{E}{2}\sqrt{\frac{C}{L}} = \frac{I_m}{2}\cos(\omega t_c + \phi)$$

Combining the above equation with Eqn (5.45) gives

$$\frac{\cos\phi}{2} = \cos(\omega t_c + \phi)$$

This can be solved for ω as

$$\omega = \frac{1}{t_c}\left[\cos^{-1}\left(\frac{\cos\phi}{2}\right) - \phi\right]$$

With ω expressed as $1/\sqrt{LC}$, an expression for $\sqrt{LC}$ can now be obtained as

$$\frac{1}{\sqrt{LC}} = \frac{1}{t_c}\left[\cos^{-1}\left(\frac{\cos\phi}{2}\right) - \phi\right]$$

or

$$\sqrt{LC} = \frac{t_c}{\cos^{-1}(\cos\phi/2) - \phi} \tag{5.53}$$

Taking $t_c = t_{\text{OFF}}$, L and C are obtained from Eqns (5.51) and (5.53) as

$$C = \frac{1}{2E}\frac{(I_m^2 - I_{\text{ld}}^2)^{1/2} t_{\text{OFF}}}{[\cos^{-1}(\cos\phi/2) - \phi]} \tag{5.54}$$

and

$$L = \frac{2Et_{OFF}}{(I_m^2 - I_{ld}^2)^{1/2}[\cos^{-1}(\cos\phi/2) - \phi]} \tag{5.55}$$

Worked Example

Design the commutating elements for a single-phase McMurray–Bedford inverter with the following data: $E = 200$ V dc, $I_{ld} = 40$ A (with resistive load), $t_{OFF} = 40$ μs, $I_m = 1.5I_{ld}$.

Solution

$$\sin\phi = \frac{I_{ld}/2}{I_m/2} = \frac{40/2}{1.5 \times 40/2} = \frac{2}{3} = 0.667$$

This gives $\phi = 0.73$ rad, $\cos\phi = 0.7454$, and $\cos^{-1}(\cos\phi/2) - \phi = 0.4589$. From Eqn (5.55),

$$L = \frac{t_{OFF}}{[\cos^{-1}(\cos\phi/2) - \phi]} \frac{E}{[(I_m/2)^2 - (I_{ld}/2)^2]^{1/2}}$$

where

$$\sqrt{\left(\frac{I_m}{2}\right)^2 - \left(\frac{I_{ld}}{2}\right)^2} = \sqrt{\left(\frac{60}{2}\right)^2 - \left(\frac{40}{2}\right)^2} = \sqrt{500} = 22.36$$

$$L = \frac{40 \times 10^{-6}}{0.4589} \times \frac{200}{22.36} \text{ H}$$

On simplification, L is obtained as

$$L = 0.78 \text{ mH}$$

or

$$\frac{L}{2} = 0.39 \text{ mH}$$

Also from Eqn (5.54),

$$C = \frac{t_{OFF}}{\cos^{-1}(\cos\phi/2) - \phi} \frac{[(I_m/2)^2 - (I_{ld}/2)^2]^2}{2E}$$

$$= \frac{40 \times 10^{-6}}{0.4589} \times \frac{22.36}{2 \times 200} \text{ F} = 9.75 \text{ μF}$$

(ii) Inductive load It is assumed that the load inductance is sufficiently high so that the load current flows for a while after the successful commutation of the conduction thyristor. Equations (5.38) and (5.39) hold good here also because the same KVL loops, namely, the battery, C_1, L_2, and Th_2 on the one hand and C_2, L_2, and Th_2 on the other, are considered for analysis. Thus,

$$\frac{1}{C}\int i_1 dt + \frac{L}{2}\frac{d(i_1 + i_2)}{dt} = E \tag{5.56}$$

and

$$\frac{1}{C}\int i_2 dt + \frac{L}{2}\frac{d(i_1+i_2)}{dt} = 0 \tag{5.57}$$

The moment of turn-on of Th_2 is taken as the initial time ($t = 0$). The initial current through L_2 and Th_2 is $2I_{ld}$ because the capacitors C_1 and C_2 share the sum of the load current I_{ld} and also the current transferred from L_1 to L_2, which is equal to I_{ld}. This can be expressed as

$$(i_1 + i_2)|_{t=0} = 2I_{ld}$$

As before, C_2 is assumed to be already charged to the battery voltage E. With i_2 taken to be in the reverse direction relative to this charging polarity of C_2, this capacitor voltage becomes

$$v_{C_2}|_{t=0} = \frac{1}{C}\int_{-\infty}^{0} i_2(t)dt = -E$$

The solution of Eqns (5.56) and (5.57) for the resistive load case gives

$$i_1(t) = i_2(t) = E\sqrt{\frac{C}{L}}\sin\omega t + I_{ld}\cos\omega t \tag{5.58}$$

Defining

$$E\sqrt{\frac{C}{L}} = \frac{I_{ld}+I_m}{2}\cos\phi \tag{5.59}$$

and

$$I_{ld} = \frac{I_{ld}+I_m}{2}\sin\phi \tag{5.60}$$

Equation (5.58) can be written in a compact manner as

$$i_1(t) = i_2(t) = \frac{I_{ld}+I_m}{2}\sin(\omega t+\phi) \tag{5.61}$$

where

$$\frac{I_{ld}+I_m}{2} = \sqrt{\frac{E^2C}{L}+I_{ld}^2} \tag{5.62}$$

$$\phi = \tan^{-1}\left(\frac{I_{ld}}{E\sqrt{C/L}}\right) \tag{5.63}$$

and

$$\omega = \frac{1}{\sqrt{LC}}$$

$v_{C_2}(t)$ is obtained as

$$v_{C_2}(t) = -\sqrt{\frac{L}{C}}\frac{(I_{ld}+I_m)}{2}\cos(\omega t+\phi) \tag{5.64}$$

With t_c as the time for v_{C_2} to attain a value of $-E/2$, this voltage at t_c can be expressed as

$$v_{C_2}(t_c) = -\frac{E}{2} = -\sqrt{\frac{L}{C}}\frac{(I_{\text{ld}} + I_m)}{2}\cos(\omega t_c + \phi) \qquad (5.65)$$

Design of commutating elements Squaring Eqn (5.62) yields

$$\frac{(I_{\text{ld}} + I_m)^2}{4} = \frac{E^2 C}{L} + I_{\text{ld}}^2 \qquad (5.66)$$

Equation (5.66) can be solved for C/L as

$$\frac{C}{L} = \frac{1}{E^2}\left\{\frac{(I_{\text{ld}} + I_m)^2}{4} - I_{\text{ld}}^2\right\} = \frac{1}{E^2}\left\{\frac{I_m^2}{4} + \frac{I_{\text{ld}} I_m}{2} - \frac{3I_{\text{ld}}^2}{4}\right\} \qquad (5.67)$$

Taking the square root of both sides of Eqn (5.67) gives

$$\sqrt{\frac{C}{L}} = \frac{1}{E}\sqrt{\frac{I_m^2}{4} + \frac{I_{\text{ld}} I_m}{2} - \frac{3I_{\text{ld}}^2}{4}} \qquad (5.68)$$

I_m is assumed to be l times I_{ld}, where $1.5 \le l \le 2$. Using Eqns (5.59) and (5.65) and proceeding along the same lines as in the resistive case, the second relation for L and C is obtained as

$$\sqrt{LC} = \frac{t_{\text{OFF}}}{\cos^{-1}(\cos\phi/2) - \phi} \qquad (5.69)$$

Equations (5.68) and (5.69) can now be solved for L and C as

$$L = \frac{t_{\text{OFF}}}{[\cos^{-1}(\cos\phi/2) - \phi]}\frac{E}{\sqrt{I_m^2/4 + I_{\text{ld}} I_m/2 - 3I_{\text{ld}}^2/4}} \qquad (5.70)$$

$$C = \frac{t_{\text{OFF}}}{\cos^{-1}(\cos\phi/2) - \phi}\frac{\sqrt{I_m^2/4 + I_{\text{ld}} I_m/2 - 3I_{\text{ld}}^2/4}}{E} \qquad (5.71)$$

Worked Example

Design the commutating elements for the McMurray–Bedford single-phase bridge inverter of Fig. 5.13(a) with the following data: load is inductive, $I_{\text{ld}} = 42$ A, $E = 280$ V, $t_{\text{OFF}} = 46$ μs.

Solution

Assuming I_m to be 1.8 times I_{ld},

$$I_m = 75.6 \text{ A}$$

The initial value of the inductor current is equal to I_{ld}. The initial capacitor currents are both equal to I_{ld}. Hence,

$$\frac{I_{\text{ld}} + I_m}{2}\sin\phi = I_{\text{ld}}$$

or

$$(I_{\text{ld}} + I_m)\sin\phi = 2I_{\text{ld}}$$

$$\sin\phi = \frac{2I_{ld}}{I_{ld} + I_m} = \frac{2 \times 42}{42 + 75.6} = \frac{84}{117.6}$$

$$\phi = 45.6° \text{ or } 0.7956 \text{ rad}$$

$$\cos\phi = 0.7$$

Substituting all values in Eqns (5.67) and (5.68) gives

$$L = \frac{46 \times 10^{-6}}{[\cos^{-1}(0.7/2) - 0.7956]} \frac{280}{\sqrt{\frac{75.6^2}{4} + \frac{75.6 \times 42}{2} - \frac{3 \times (42)^2}{4}}} = 0.75 \text{ mH}$$

$$C = \frac{46 \times 10^{-6}}{[\cos^{-1}(0.7/2) - 0.7956]} \frac{\sqrt{\frac{75.6^2}{4} + \frac{75.6 \times 42}{2} - \frac{3 \times (42)^2}{4}}}{280} = 16.2 \text{ μF}$$

5.4.6 Three-phase McMurray–Bedford Inverter

The circuit of a three-phase Murray–Bedford inverter is given in Fig. 5.14. The sequence of operations given for the single-phase McMurray–Bedford inverter holds good here if it is considered as the sequence of steps for the operation between any two phases of the three-phase load. Also, a phase difference of 120° has to be kept between any two phase currents. The line-to-line voltage should be taken into consideration for design purposes.

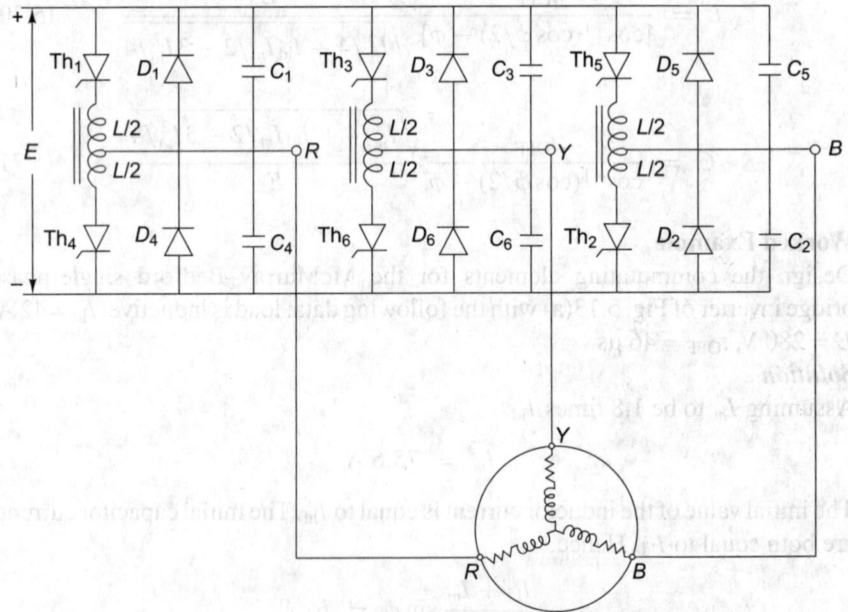

Fig. 5.14 Circuit diagram of a three-phase McMurray–Bedford bridge inverter

5.4.7 Three-phase 120°-mode VSI

For the three-phase voltage source bridge inverter shown Fig. 5.8(a), each of the six main thyristors requires a continuous 180° firing signal, so as to allow the devices to be turned on at any time in this duration. Such an arrangement has the disadvantage that if there is a delayed turn-off of a thyristor on one of the legs and if the complementary thyristor on the same leg is fired at exactly 180°, the dc supply will get short-circuited. Consequently, there is a possibility of damage to both the thyristors as well as the dc power supply. To avoid such a contingency, the 120°-mode VSI has been devised. In this approach each device on a leg conducts for a period for 120° so that there is a 60° interval between the turning off of the previous thyristor and the turning on of the complementary thyristor. However, the firing as well as turn-off sequences of the thyristors remain the same as in the six-step inverter. The drawback of the 120°-mode VSI is that during no-voltage intervals, the potentials of the terminals connected to the load depend mainly on the load; the fundamental voltage applied to the machine depends on the shaft

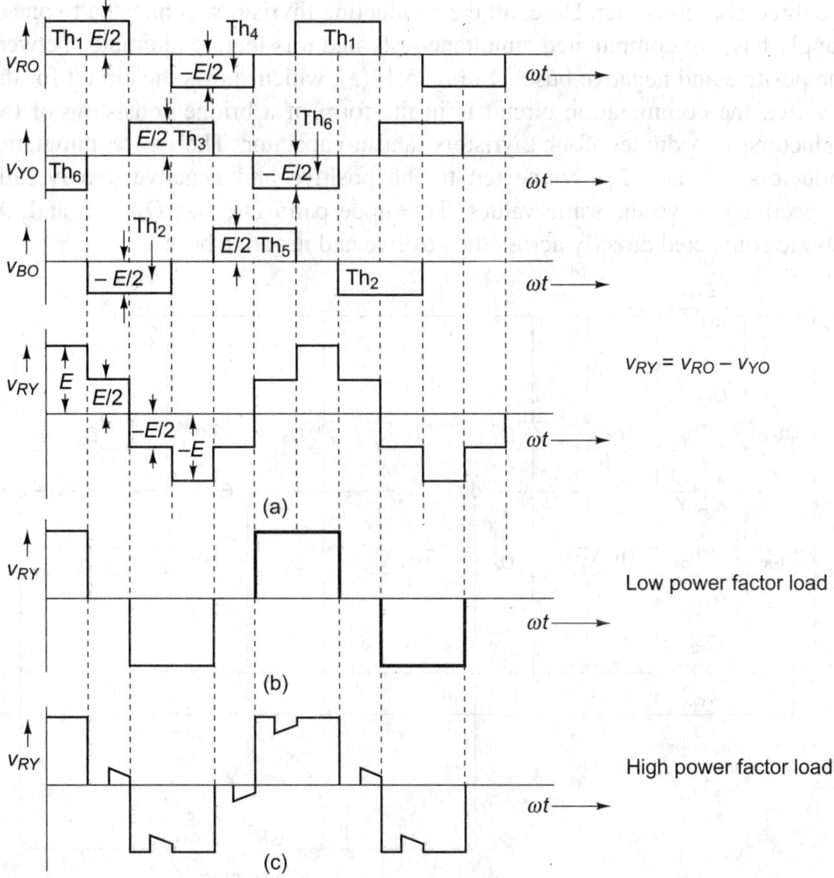

Fig. 5.15 Waveforms of the 120°-mode voltage source inverter: (a) pole voltages and line voltage V_{RY}, (b) line voltage with a low power factor load, (c) line voltage with a high power factor load

load, which determines the effective load power factor seen by the inverter. The circuit of Fig. 5.8(a) can function as a 120°-mode VSI, but with modified operation conditions as follows. Th_1 is turned off when Th_3 is turned on and the latter is turned off when Th_5 is turned on. In general, it can be stated that the commutation of the already conducting thyristor is ensured by the firing of another thyristor connected to the same dc bus. Figure 5.15(a) illustrates this feature.

The effect of load power factor on the line voltage can be seen from the waveforms shown in Figs 5.15(b) and (c). The former is the waveform for V_{RY} obtained for a low load power factor (of the order of 0.5 and below). Its well-defined nature can be attributed to the constancy of the load current during the 60° interval. The distortion in the latter can be attributed to the line current becoming zero at some time during the 60° interval.

5.4.8 Three-phase Input Circuit Commutated Inverter

This inverter (Finney 1988) is so called due to the fact that the commutating inductors, thyristors, and the capacitor are connected between the dc supply and the three-phase inverter. Here, all the conducting thyristors, connected to one dc supply bus, are commutated simultaneously and this feature alternates between the positive and negative buses. In Fig. 5.16(a), which shows the circuit for this inverter, the commutation circuit is in the form of a bridge consisting of two inductors, two diodes, four thyristors, and a capacitor. The two commutation inductors L_{c1} and L_{c2} connected to the positive and negative supply rails, respectively, have the same values. The diode pairs D_1, D_4; D_3, D_6; and D_5, D_2 are connected directly across the positive and negative buses.

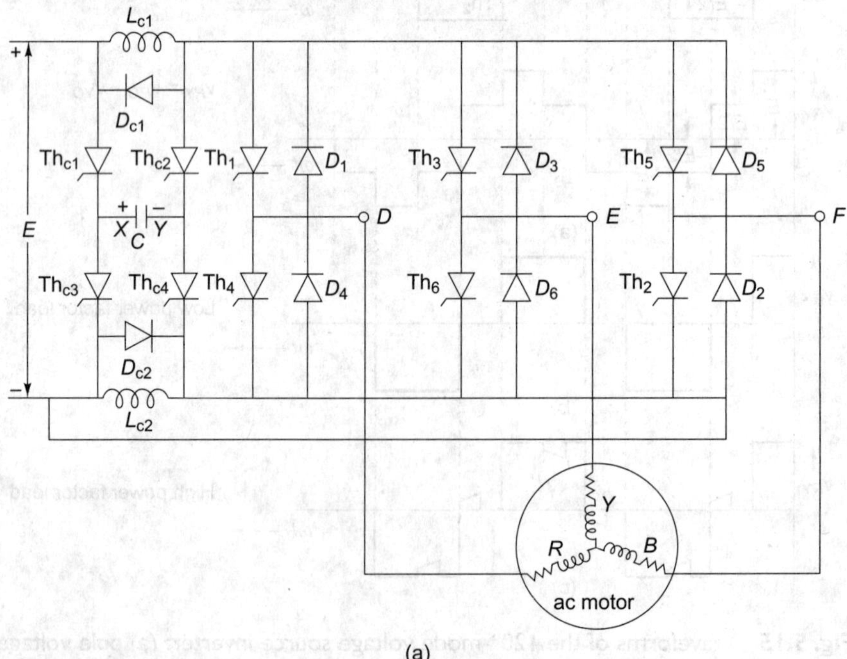

Fig. 5.16(a)

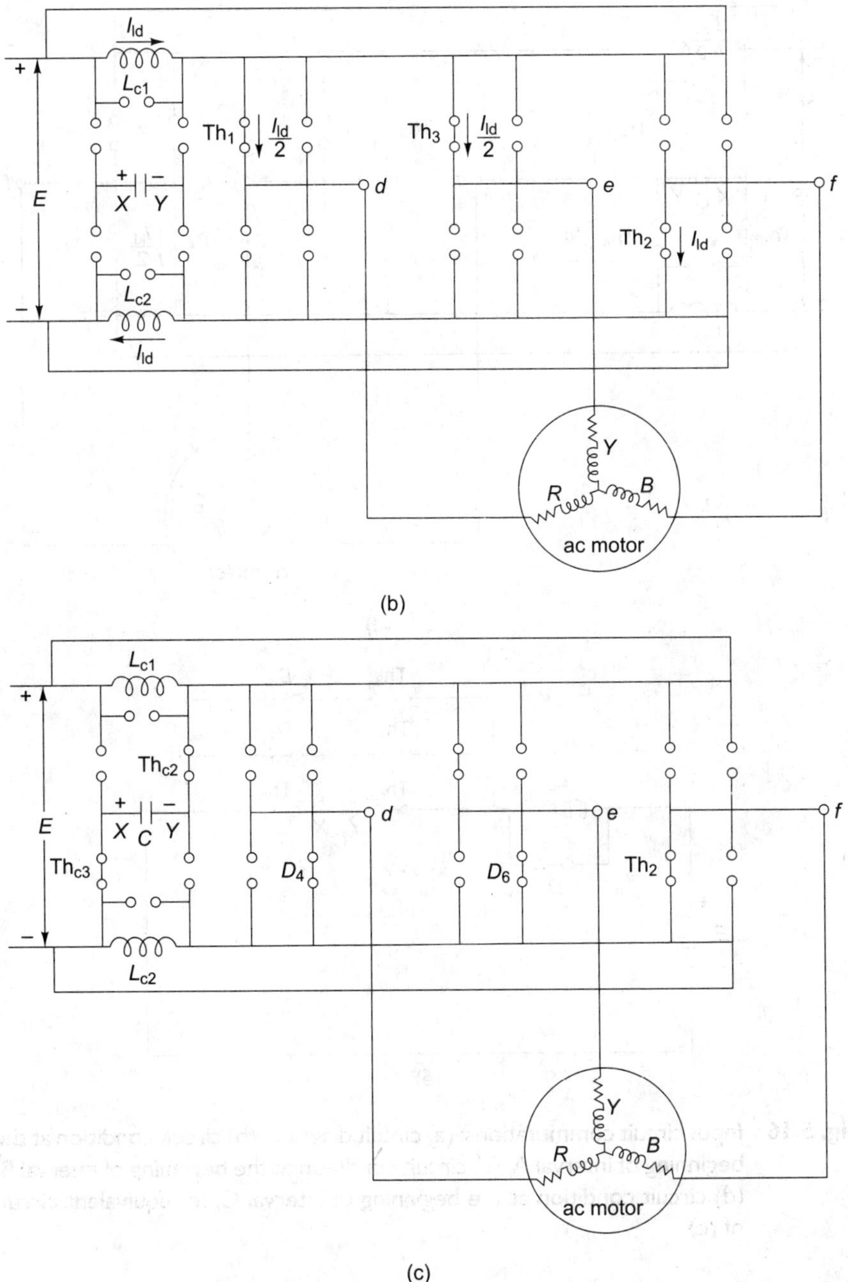

Figs 5.16(b), (c)

The sequence of operations can be divided into three intervals. The circuit conditions for the three intervals A, B, and C are given in Figs 5.16(b), (c), and (d) and the waveforms in Fig. 5.17.

Interval A It is assumed that the three thyristors Th_1, Th_2, and Th_3 are the conducting devices so that the load terminals d and e, respectively, of the R-

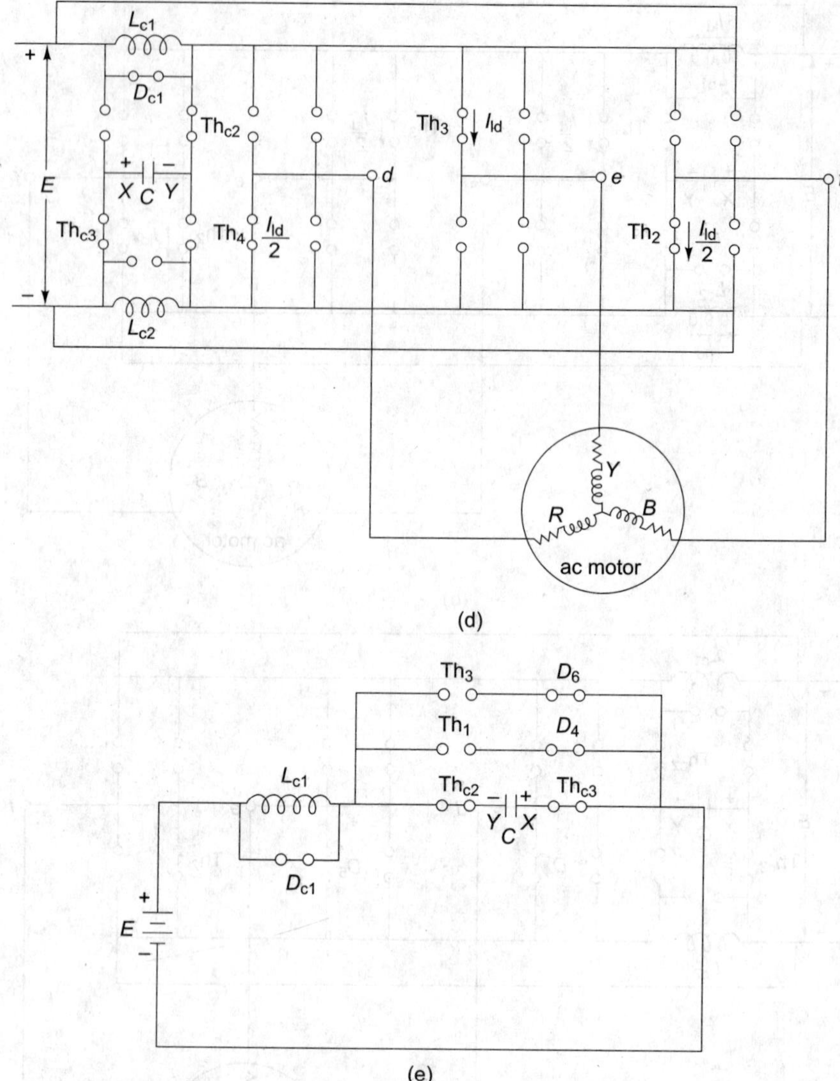

Fig. 5.16 Input circuit commutations: (a) circuit diagram, (b) circuit condition at the beginning of interval A, (c) circuit condition at the beginning of interval B, (d) circuit condition at the beginning of interval C, (e) equivalent circuit of (c)

and Y-phase windings are connected to the positive bus, and terminal f of the B-phase winding is connected to the negative bus. A steady dc current I_{ld} flows in the two inductors L_{c1} and L_{c2}, which in turn leads to the conduction of steady currents through the three thyristors. It is assumed that the capacitor C is charged to the dc supply voltage E with the X-plate positive; this is shown in Fig. 5.16(b). The sequence of turn-on of the thyristors is Th_1, Th_2, Th_3, Th_4, Th_5, and Th_6, and hence the next set of thyristors to conduct are Th_2, Th_3, and Th_4. As Th_4 is in the same leg as that of Th_1, it is necessary that Th_1 be commutated before Th_4 is

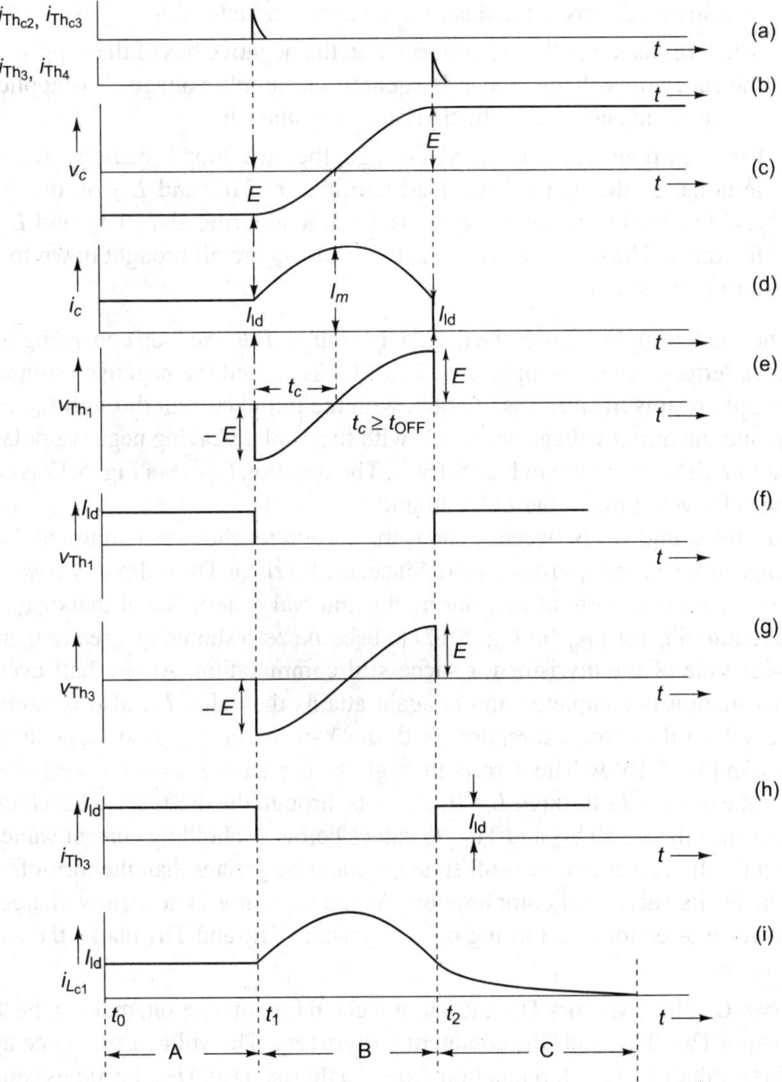

Fig. 5.17 Waveforms for input circuit commutation

turned on. However, as per the technique of input circuit commutation, both Th_1 and Th_3, which are connected to the positive bus, are turned off. This marks the end of interval A.

Interval B The commutation of Th_1 and Th_3 is initiated by disconnecting their gate currents and turning on the commutation thyristors Th_{c2} and Th_{c3}. The Y-plate of capacitor C, which is of negative polarity, is connected to the positive bus through Th_{c2} and the X-plate to the negative bus through Th_{c3}. As the load current in L_{c1} cannot change instantaneously, it gets diverted through Th_{c2}, C, Th_{c3}, and back to the negative terminal of the dc supply. The currents through the three legs of the star-connected motor (load) winding now find a path through the diodes D_4 and D_6. This leads to the following conditions.

(a) A voltage $2E$ gets applied across the upper inductor L_{c1}.
(b) The terminals d and e are connected to the negative bus of the supply. Also the capacitor voltage, which is equal to the supply voltage, E is applied in a reverse manner across the thyristors Th_1 and Th_3.
(c) Load currents are circulated through the two loops, namely, D_4, load terminal d, the motor load, load terminal f, Th_2, and L_{c2} on one hand, and D_6, load terminal e, the motor load, load terminal f, Th_2, and L_{c2} on the other. The line voltages, v_{de}, v_{ef}, and v_{fd} are all brought down to zero during interval B.

The sudden application of the dc supply voltage E to the loop consisting of the positive terminal of the supply, L_{c1}, Th_{c2}, C, Th_{c3}, and the negative terminal of the supply results in an LC oscillation, with the initial current through L_{c1} equal to I_{ld} and the initial voltage across C (with the Y-plate having negative polarity) equal to E. This is shown in Fig. 5.16(c). The duration (t_2-t_1) in Fig. 5.17 is equal to the half-cycle time of the LC oscillation.

At a point midway between t_1 and t_2 the current reaches a maximum of I_m and the capacitor voltage v_c crosses zero. Since, the thyristor Th_1 is directly connected across C, the waveform of v_{Th_1} during this interval is a replica of that of v_c. The time required t_c for v_{Th_1} (in Fig. 5.17) to become zero should be greater than the turn-off time of the thyristor for successful commutation. At t_2 a half-cycle of LC oscillation is completed and i_c again attains the value I_{ld}; also v_c attains a value $+E$ in the reverse direction, with the Y-plate having positive polarity, as shown in Fig. 5.16(d). The current through the capacitor is now reduced to zero; hence the current I_{ld} through L_{c1} freewheels through the diode D_{c1}. This reduces the currents through Th_{c2} and Th_{c3} to values below the holding current value and they turn off. The circuit turn-off time t_c should be greater than the turn-off time of Th_1 for its successful commutation. As the capacitor is now fully charged in the reverse direction, the turning on of thyristors Th_3 and Th_4 marks the end of interval B.

Interval C The thyristors Th_3 and Th_4 are gated for conduction, making the three thyristors Th_2, Th_3, and Th_4 conducting thyristors. The voltage v_{Th_3} once again attains a value of 1–2 V (conduction drop of a thyristor) and i_{Th_3} becomes equal to I_{ld}. This current I_{ld} circulates through L_{c1} and D_{c1} which is initially I_{ld} and slowly goes down to zero because of dissipation in the resistive part of L_{c1}. The interval C lasts till Th_{c1} and Th_{c4} are turned on to initiate the commutation of the thyristors Th_4 and Th_2, which are connected to the negative bus. The commutating elements L and C are designed to minimize the energy trapped during the commutation interval B.

The following analysis is made with the help of a circuit equivalent to that of Fig. 5.16(e). An expression is derived for the circuit time available for the successful commutation of Th_1 and Th_3. The differential equation for the forced LC circuit consisting of the battery, L_{c1}, and C can be written as

$$L_{c1}\frac{di_c}{dt'} + \frac{1}{C}\int_0^{t'} i_c dt' = E \tag{5.72}$$

The initial time is taken as $t' = 0$, which is the same as $t = t_1$, which is the firing instant of Th_{c2} and Th_{c3}. Equation (5.72) can be recast as

$$\frac{d^2q}{dt'^2} + \frac{q}{L_{c1}C} = \frac{E}{L_{c1}} \tag{5.73}$$

The initial conditions are

$$\left.\frac{dq}{dt'}\right|_{t'=0} = I_{ld}$$

and

$$q(0) = -CE$$

Laplace transformation of Eqn (5.73) gives

$$s^2Q(s) - sq_0 - \left.\frac{dq}{dt'}\right|_{t'=0} + \frac{Q(s)}{CL_{c1}} = \frac{E}{sL_{c1}}$$

or

$$s^2Q(s) + CEs - I_{ld} + \frac{Q(s)}{L_{c1}C} = \frac{E}{sL_{c1}}$$

This can be written as

$$Q(s)\left[s^2 + \frac{1}{L_{c1}C}\right] = \frac{E}{sL_{c1}} - CEs + I_{ld}$$

Solving for $Q(s)$ yields

$$Q(s) = \frac{E}{sL_{c1}(s^2 + 1/L_{c1}C)} - \frac{CEs}{s^2 + 1/L_{c1}C} + \frac{I_{ld}}{s^2 + (1/L_{c1}C)} \tag{5.74}$$

Inverse Laplace transformation of Eqn (5.74) now gives

$$q(t') = CE + \frac{I_{ld}}{\omega}\sin\omega t' - 2CE\cos\omega t' \tag{5.75}$$

with $\omega^2 = 1/L_{c1}C$. Differentiation of Eqn (5.75) with respect to time and taking $L_{c1} = L$ and $\omega = 1/\sqrt{LC}$ gives the current through the capacitor as

$$i_c(t') = I_{ld}\cos\omega t' + 2\omega CE\sin\omega t' \tag{5.76}$$

The voltage across the capacitor can now be expressed as

$$v_c(t') = E - L\frac{di_c}{dt'} \tag{5.77}$$

That is,

$$v_c(t') = E + \omega LI_{ld}\sin\omega t' - 2\omega^2LCE\cos\omega t'$$

Since $\omega^2 = 1/\sqrt{LC}$, the above equation can be written, after taking ω^2LC as 1, as

$$v_c(t') = E + \omega LI_{ld}\sin\omega t' - 2E\cos\omega t' \tag{5.78}$$

Initially, at t_1, $v_{\mathrm{Th}_1} = -E$. The circuit time t_c for the thyristor voltage to become zero can be obtained by equating it to zero. Thus,

$$v_{\mathrm{Th}_1}(t_1 + t_c) = v_c(t_1 + t_c) = 0 = E + \omega L I_{\mathrm{ld}} \sin \omega t_c - 2E \cos \omega t_c \tag{5.79}$$

or

$$\cos \omega t_c \; - \; \frac{\omega L I_{\mathrm{ld}}}{2E} \sin \omega t_c \; = \; \frac{1}{2}$$

Defining $\tan \phi \; = \; \omega L I_{\mathrm{ld}}/2E$, this can be written as

$$\sqrt{1 + \frac{\omega^2 L^2 I_{\mathrm{ld}}^2}{4E^2}} [\cos(\omega t_c + \phi)] \; = \; \frac{1}{2}$$

giving

$$\omega t_c + \phi \; = \; \cos^{-1}\left[\frac{1}{2\sqrt{1 \, + \, \omega^2 L^2 I_{\mathrm{ld}}^2/4E^2}}\right]$$

Finally, the expression for t_c can now be obtained as

$$t_c \; = \; \frac{1}{\omega}\left\{\cos^{-1}\left[\frac{1}{2\sqrt{1 \, + \, \omega^2 L^2 I_{\mathrm{ld}}^2/4E^2}}\right] \; - \; \phi\right\} \tag{5.80}$$

Given t_{OFF}, L, I_{ld}, and E, the value of C required to successfully commutate the thyristor can be obtained from the criterion that t_c should be greater than t_{OFF}.

5.5 Pulse Width Modulated VSIs

Pulse width modulated (PWM) VSIs are widely used because of the advantages highlighted in the previous section. This section gives a detailed treatment of two types of PWM waveforms that have been used in industry, namely, the square and sinusoidal ones. Generation as well as other operational features of both the single- as well as three-phase versions of both types are discussed at length.

5.5.1 Single Pulse Width Modulated Inverters

The circuit of a single-phase bridge inverter is shown in Fig. 5.18(a). The waveforms of the potentials v_{RO} and v_{SO}, respectively, of the load terminals R and S with respect to the centre tap O of the battery are shown in Fig. 5.18(b) along with the waveform of v_{RS} $(= v_{RO} - v_{SO})$. Such an output is obtained when the device pairs Th_1, Th_4 and Th_2, Th_3 conduct alternately for half-cycles, that is, for 180°. The waveform of v_{SO} lags that of v_{RO} by an angle $\beta = 180°$. With these conditions the output voltage waveform will have a square shape with amplitude E.

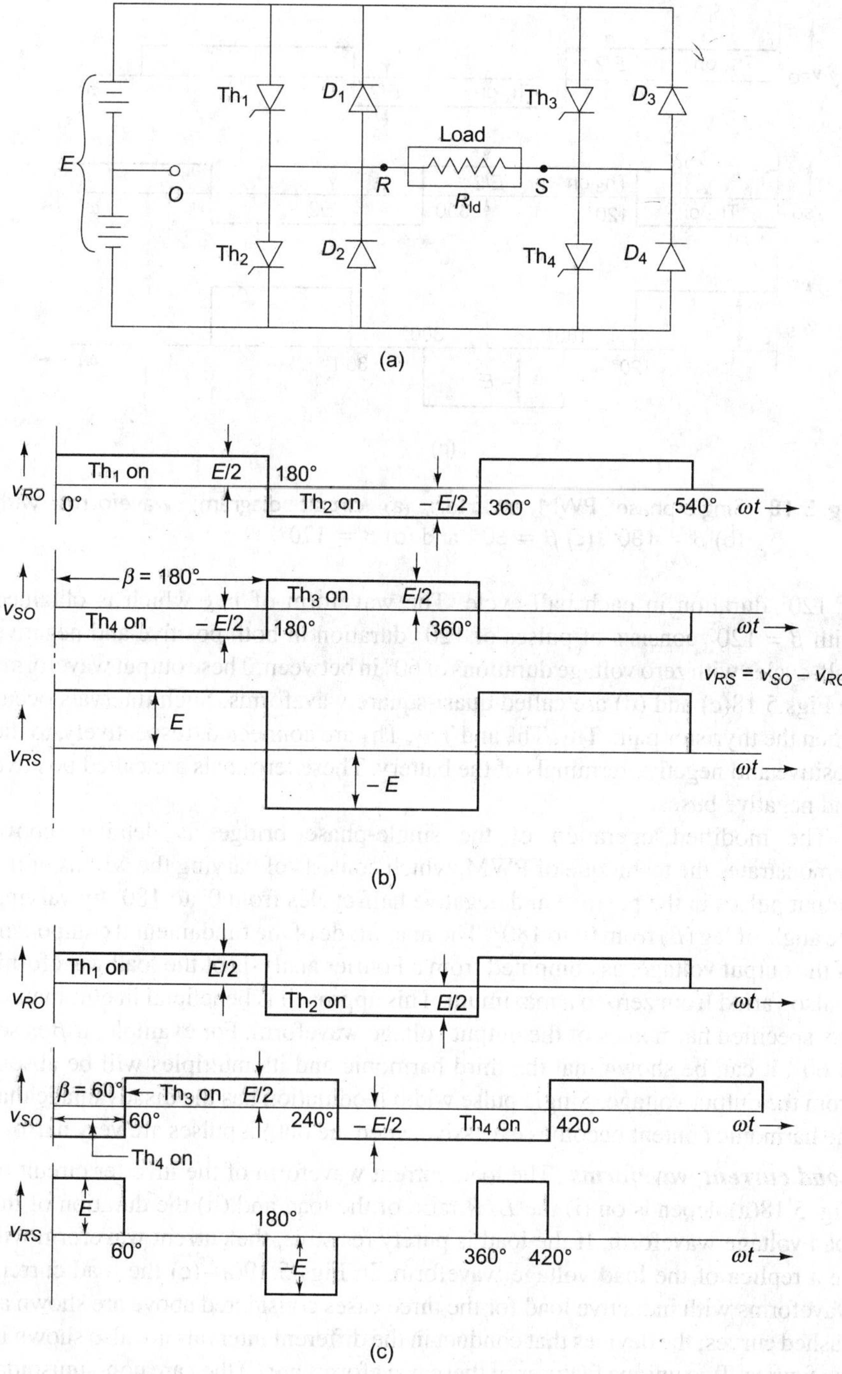

Figs 5.18(a), (b), and (c)

Figure 5.18(c) shows the waveforms v_{RO}, v_{SO}, and v_{RS} with the angle $\beta = 60°$. The resulting load voltage v_{RS} consists of rectangular pulses of 60° duration both in the positive and negative half-cycles. Further, there are zero voltage intervals

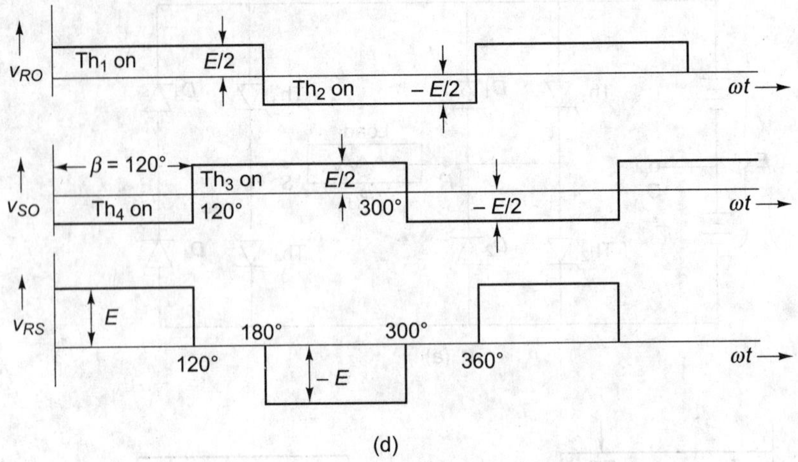

Fig. 5.18 Single-phase PWM inverter: (a) circuit diagram; waveforms with (b) $\beta = 180°$, (c) $\beta = 60°$, and (d) $\beta = 120°$

of 120° duration in each half-cycle. The waveform of v_{RS} which is obtained with β = 120° consists of pulses of 120° duration in both positive and negative half-cycles with zero voltage durations of 60° in between. These output waveforms in Figs 5.18(c) and (d) are called quasi-square waveforms. Such intervals occur when the thyristor pairs Th_1, Th_3 and Th_2, Th_4 are connected, respectively, to the positive and negative terminals of the battery. These terminals are called positive and negative buses.

The modified operation of the single-phase bridge as detailed above demonstrates the technique of PWM, which consists of varying the widths of the output pulses in the positive and negative half-cycles from 0° to 180° by varying the angle of lag (β) from 0° to 180°. The magnitude of the fundamental component of the output voltage, as computed from a Fourier analysis of the load waveform, is also varied from zero to a maximum. This approach is beneficial in eliminating any specified harmonics of the output voltage waveform. For example, if β is set at 60°, it can be shown that the third harmonic and its multiples will be absent from the output voltage. Single pulse width modulation has the disadvantage that the harmonic content becomes excessive when the output pulses are very narrow.

Load current waveforms The load current waveform of the inverter circuit of Fig. 5.18(a) depends on (i) the L/R ratio of the load and (ii) the duration of the load voltage waveform. If the load is purely resistive, the current waveform will be a replica of the load voltage waveform. In Figs 5.19(a)–(c) the load current waveforms with inductive load for the three cases considered above are shown as dashed curves; the devices that conduct in the different intervals are also shown in the figures. Two unique features of these waveforms are (i) they are non-sinusoidal and hence contain harmonics and (ii) they may have zero current intervals.

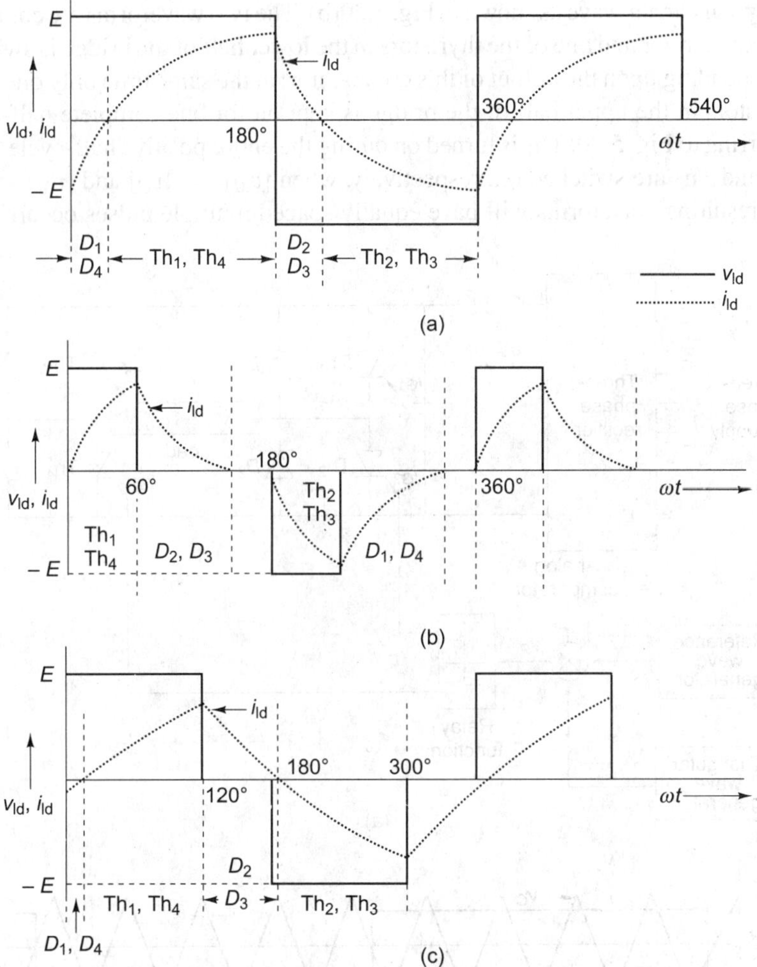

Fig. 5.19 Current waveforms for the inverter of Fig. 5.18(a) with (a) $\beta = 180°$, (b) $\beta = 60°$, (c) $\beta = 120°$

5.5.2 Multiple Pulse Width Modulated Inverters

The single-phase bridge circuit of Fig. 5.18(a) may be operated to give multiple pulses in each half-cycle of the inverter by comparing an alternating wave (called the *modulating* or *reference wave*) of amplitude V_M with a triangular wave, called the *carrier wave*, of amplitude V_C. The corresponding instantaneous values are denoted as v_M and v_C, respectively. The frequency of the modulating wave is the same as that of the fundamental of the output voltage waveform and is called the *reference frequency*. The frequency of the carrier wave, called the *carrier frequency*, is usually an odd multiple of the reference frequency, and can take the value 3 or its multiples. The reasons for this will be given later.

Single-phase square wave PWM Figure 5.20(a) shows a single-phase multiple pulse width modulated inverter that generates a square modulating wave and a triangular carrier wave as shown in Fig. 5.20(b). The two waveforms are compared in a comparator and one of the thyristors in the lower half of the bridge is switched on depending upon the output of this comparator; at the same time only one of the thyristors in the upper half of the bridge is kept on for one complete half-cycle. Referring to Fig. 5.20, Th_1 is turned on during the entire positive half-cycle. Also, Th_4 and Th_2 are switched on, respectively, when $|v_M| > |v_C|$ and $|v_M| < |v_C|$. The resulting waveforms will have equally spaced multiple pulses occurring on

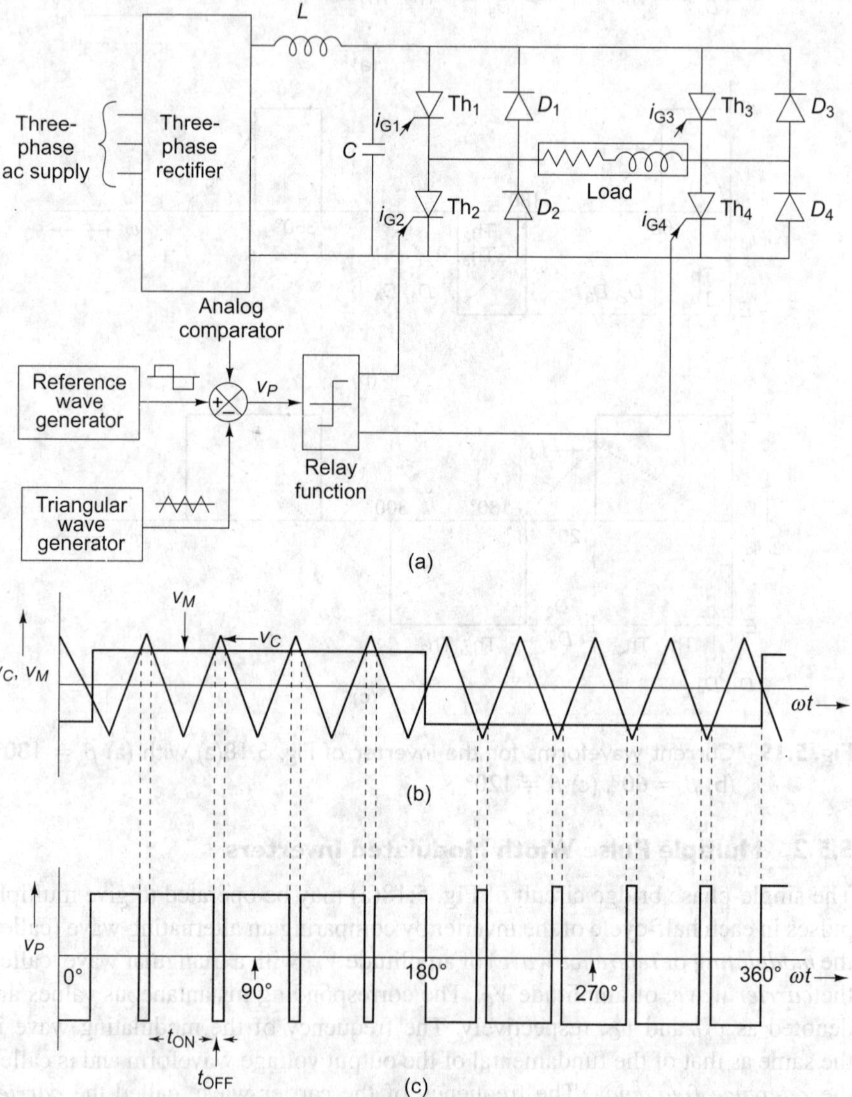

Fig. 5.20 Single-phase square wave PWM inverter: (a) circuit diagram, (b) waveforms of modulating (v_M) and carrier (v_C) voltages, (c) waveform of comparator output voltage (v_P)

the positive side of the ωt-axis as shown in Fig. 5.20(c). During the negative half-cycle, Th_3 is kept on continuously but Th_2 is switched on when $|v_M| > |v_C|$ and Th_4 is switched on when $|v_M| < |v_C|$. The resulting output will also have equal-width pulses located, however, on the negative side of the ωt-axis.

The widths of the output pulses are equal, except at both ends, where their widths are lesser than those of the other pulses. They can be varied by altering the ratio $t_{ON}/(t_{ON} + t_{OFF})$, where t_{ON} and t_{OFF} are the time periods of the pulse and notch, respectively. The variation is accomplished by raising or lowering the reference waveform.

The term pulse width modulation is derived from the nature of its reference waveform. Accordingly, the PWM in Fig. 5.20 is called a square wave PWM because the modulating wave has a square shape. The sinusoidal PWM, which has a sinusoidal reference waveform, overcomes some of the drawbacks of the square wave PWM.

5.5.3 Sinusoidal PWM (SPWM)

Single-phase SPWM In sinusoidal PWM, also called sine wave PWM, the resulting pulse widths are varied throughout the half-cycle in such a way that they are proportional to the instantaneous value of the reference sine wave at the centre of the pulses. However, the distance between the centres of the pulses is kept constant as in square wave PWM. Voltage control is achieved by varying the widths of all the pulses without disturbing the sinusoidal relationship. Figure 5.21 shows a typical circuit for single-phase sinusoidal PWM along with its waveforms.

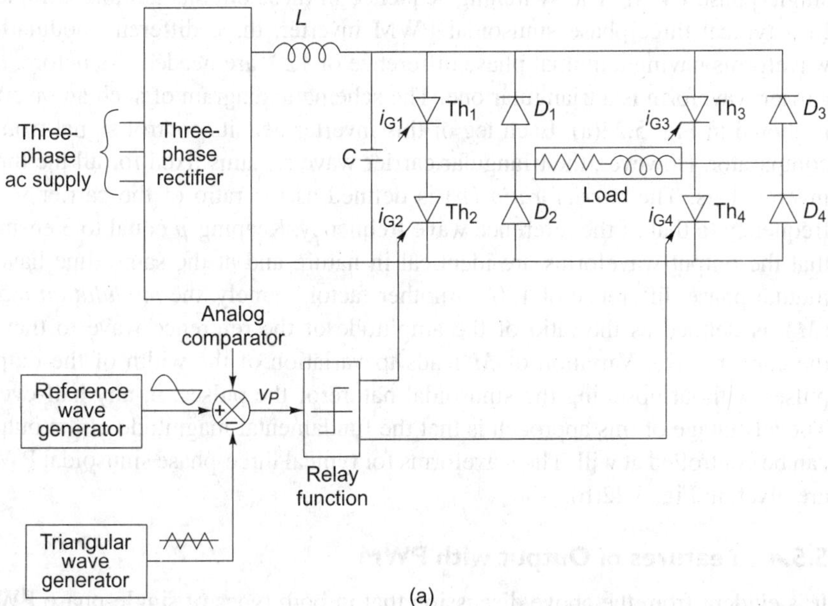

Fig. 5.21(a)

Three-phase SPWM High capacity drives are rigged up as three-phase configurations and hence three-phase PWM inverters are of considerable interest (Murphy & Turnbull 1988). These are based on an extension of the principle of

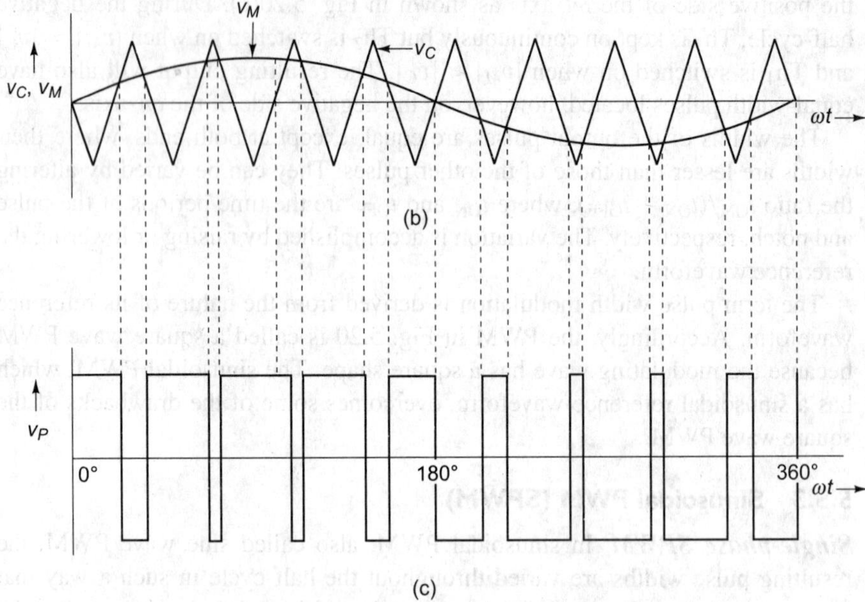

Fig. 5.21 Single-phase sinusoidal PWM inverter: (a) circuit diagram, (b) waveforms of modulating (v_M) and carrier (v_C) voltages, (c) waveform of comparator output voltage (v_P)

single-phase PWM. The switching sequence in these circuits is more complex. In a typical three-phase sinusoidal PWM inverter, three different modulating waveforms having a mutual phase difference of 120° are needed. As before, the carrier waveform is a triangular one. The schematic diagram of such an inverter is shown in Fig. 5.22(a). Each leg of this inverter gets its control signal from a comparator. However, the triangular carrier wave remains fixed for all the three inverter legs. The *carrier ratio* (p) is defined as the ratio of the carrier wave frequency to that of the reference wave frequency. Keeping p equal to 3 ensures that the output waveforms are identical in nature and at the same time have a mutual phase difference of 120°. Another factor, namely, the *modulation index* (M), is defined as the ratio of the amplitude of the reference wave to that of the carrier wave. Variation of M leads to variation of the width of the output pulses without upsetting the sinusoidal nature of the pulses, in any half-cycle. The advantage of this approach is that the fundamental magnitude of the output can be controlled at will. The waveforms for typical three-phase sinusoidal PWM are given in Fig. 5.22(b).

5.5.4 Features of Output with PWM

It is evident from the above discussion that in both types of single-phase PWM inverters, the switching frequencies of the thyristors on the lower half of each leg are higher than the switching frequencies of those situated on the upper half. As a consequence, switching losses in these devices are higher in comparison to those used in the bridge inverter. Another feature of the PWM output is that dominant harmonics occur at relatively higher frequencies. Though the inductance of the motor windings (load) filters these harmonics and limits the load current, the low-order harmonic content of the output is greater than that of the bridge type of VSI. For a motor load these low-order harmonics are harmful because they cause torque pulsations and irregular shaft rotation. It is, however, possible to eliminate the unwanted harmonics using digital PWM techniques and this feature makes PWM inverters superior to other voltage source inverters.

5.5.5 Harmonics in Three-phase PWM Inverters

Square wave PWM Because the output waveforms of the three-phase square wave PWM inverter are derived from a square wave reference, they contain all the harmonics of such reference waves. In addition, they contain switching harmonics of the form $lp \pm m$, where l is always even and m always odd. For example, with a carrier ratio of 6, l equal to 2, and m equal to 1, harmonics of orders 11 and 13 will be present. On the other hand, for the same p and l but with $m = 5$, the harmonic frequencies will be of orders 7 and 17. This implies that dominant harmonics

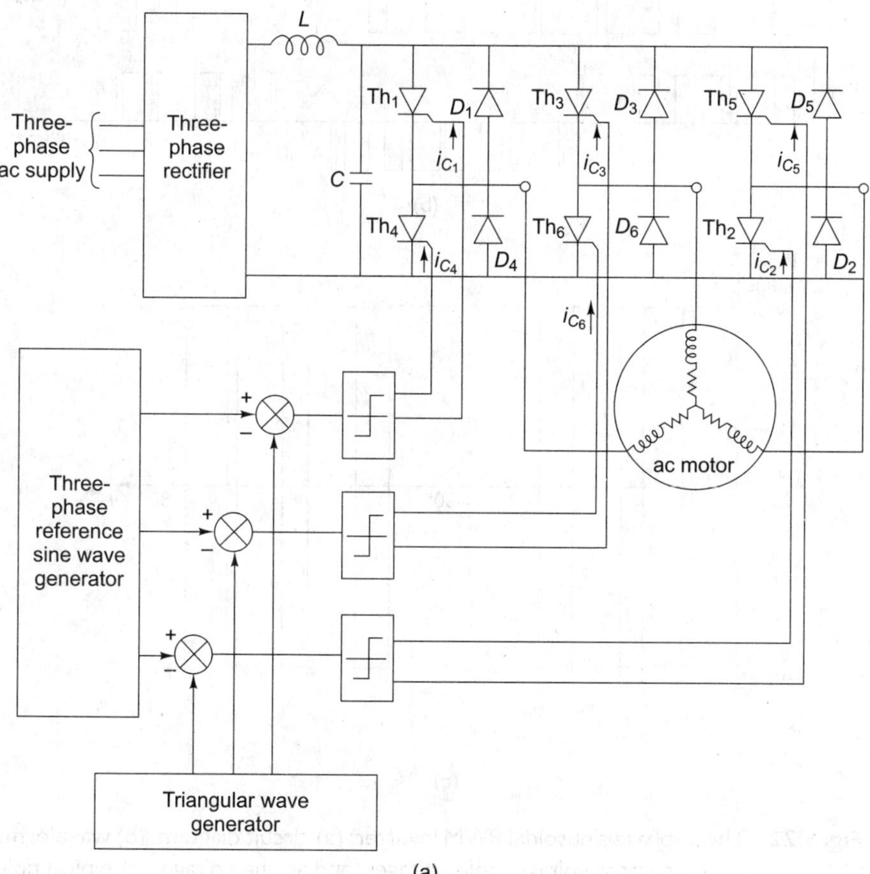

Fig. 5.22(a)

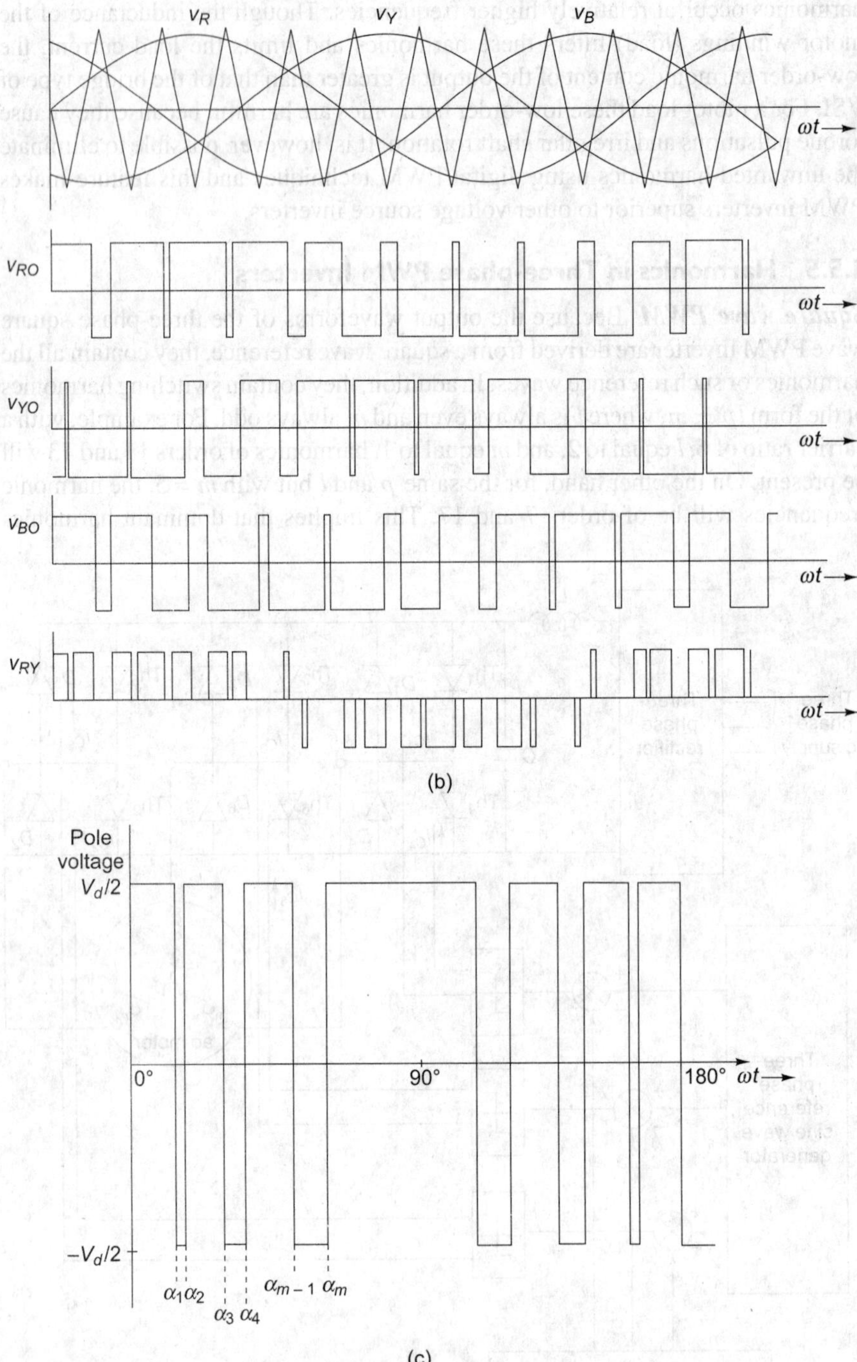

Fig. 5.22 Three-phase sinusoidal PWM inverter: (a) circuit diagram, (b) waveforms of comparator voltage, pole voltages, and ac line voltage, (c) typical pole voltage waveform

attain large magnitudes with a high value of p; the inductance of the motor windings filters these harmonics and limits the output current. In spite of this, the low-order harmonic content is greater than that of the bridge inverter and is undesirable as elaborated above. The switching frequency is high with PWM operation as compared to that of the bridge inverter; hence switching losses are higher for motors supplied by PWM inverters.

Sine wave PWM The harmonics in this case are sidebands of the carrier ratio and its multiples, and are of the form $kp \pm l$, where k is even and l is odd or vice versa. The dominant harmonics in this case will therefore be of the order $2p \pm 1$ or $p \pm 2$. If k and l are both even or both odd, the harmonics get absorbed.

For both square as well as sinusoidal PWM, the harmonics are independent of p for $p > 9$. Such values of p are especially beneficial for sinusoidal PWM because, with these values, the high-order harmonics are dominant, and are absorbed by the machine inductance. Consequently torque pulsations are eliminated and there is smooth motor rotation even at low speeds. Harmonic losses are therefore lower for sinusoidal PWM. This shows its superiority over both square wave PWM as well as bridge inverters.

5.5.6 Advantages and Disadvantages of PWM Inverters

The source for the three-phase bridge inverter consists of a three-phase bridge rectifier followed by an *LC* filter (Figs 5.12 and 5.13). There is a fixed ratio between the ac output voltage and the dc input voltage. It is therefore necessary to vary the rectifier output in order to control the output voltage of the inverter. This implies that the three-phase rectifier has to be a thyristorized one. On the other hand, in the case of PWM inverters, voltage control is achieved within the inverter and hence a diode bridge rectifier can be used as the input rectifier. Such a rectifier provides a constant dc link voltage, and this has many advantages as compared to an inverter based on a thyristorized rectifier.

First, the thyristor rectifier used in a bridge inverter is more costly. Further, the delayed firing of thyristors in the converter causes the inverter to present a low power factor to the incoming three-phase ac supply. As against this, the power factors presented by the PWM inverters are high (about 0.95). They are independent of the motor power factor because of the use of an uncontrolled rectifier on the input side. Third, the dc link inductance and capacitance for a PWM inverter based on a diode rectifier are less costly because they are designed to match the fixed output of the rectifier. On the other hand, large filter components used in a bridge inverter cause time lag and hence affect the stability of the speed control system of the motor at low speeds.

Another positive feature of the PWM inverter is that the fixed dc link voltage permits parallel operation of a number of PWM inverters connected to the same dc link. The overall transient response of PWM inverters is also better than that of the bridge inverter. A recent development in PWM inverters is that with both the square wave as well as the sine wave, the switching angles can be adjusted using digital techniques so as to eliminate unwanted harmonics. All these features go to show the supremacy of PWM inverters over bridge inverters.

PWM inverters, however, suffer from the following disadvantages. Switching takes place in the presence of load current, thus subjecting the devices to a high

di/dt, in turn leading to high dv/dt stresses and electromagnetic interference. In resonant pulse inverters, described in Appendix D, these drawbacks are overcome by ensuring switching during zero current intervals.

5.5.7 Constant Voltage-to-Frequency Operation of PWM Inverters

Constant torque and constant horsepower operation of motors fed by PWM inverters is carried out in a manner similar to those fed by a three-phase bridge inverter. However, greater flexibility of operation is available in PWM inverters in the shape of variation of the carrier ratio p. At low speeds, p is maintained at a high value to restrict the dominant harmonics to high values. On the other hand, at high speeds they are kept at low values so as to limit the switching losses. In the latter case, the low-order harmonics resulting from low values of p do not present problems. Normally, p is taken to be a multiple of 3 in order to produce identical phase voltage waveforms in the three phases. In the case of sinusoidal PWM, this restriction is removed for high reference wave frequencies, implying thereby that non-integral values are also allowed. Though such an operation gives rise to beat harmonics and associated losses, these do not cause problems unless p is less than 9. In the latter case, p is made a multiple of 3 and the reference and carrier waveforms are synchronized.

As the modulation index M is proportional to the fundamental component of the output voltage, constant $|V_1|/f$ ratio can be accomplished for PWM inverters by maintaining constant M/f ratio.

5.5.8 Reduction of Harmonics by the PWM Technique

The main advantages of the square wave and sinusoidal PWM schemes applied to three-phase systems are that (a) the magnitude of the fundamental harmonic of the output voltage can be controlled and (b) certain selected harmonics can be eliminated. This latter feature can be explained with the help of Fig. 5.22(b) as follows. The figure gives typical pole voltage waveforms v_{RO}, v_{YO}, and v_{BO} for a three-phase sinusoidal PWM inverter feeding a star-connected load with an isolated neutral [Fig. 5.22(a)]. For mathematical tractability it is assumed that there is half-wave as well as quarter-wave symmetry for a typical pole voltage waveform. Further, it is assumed that there are m switchings per quarter-cycle [Fig. 5.22(c)]. This means that in one quarter-cycle the waveform crosses the ωt-axis m times, either from the positive to the negative side or vice versa. A Fourier analysis of this waveform reveals that, because of quarter-wave symmetry, the coefficients a_k of the cosinusoidal terms become zero and the pole voltage can be expressed as

$$v_{AO}(t) = \sum_{k=1,3,5,\ldots}^{\infty} b_k \sin k\omega t \tag{5.81}$$

The third and other triple harmonics do not produce any motor current because of the star connection having an isolated neutral. Thus the expression for v_{AO} gets simplified to

$$v_{AO}(t) = b_1 \sin \omega t + b_5 \sin 5\omega t + b_7 \sin 7\omega t + \cdots \tag{5.82}$$

Each of the coefficients b_k are obtained [Appendix A and Fig. 5.22(c)] as

$$b_k = \frac{4}{\pi}\int_0^{\pi/2} v_{AO}\sin(k\omega t)d(\omega t)$$

or

$$b_k = \frac{4}{\pi}\left[\int_0^{\alpha_1}\left(+\frac{V_d}{2}\right)\sin k\omega t + \int_{\alpha_1}^{\alpha_2}\left(-\frac{V_d}{2}\right)\sin k\omega t + \int_{\alpha_2}^{\alpha_3}\left(+\frac{V_d}{2}\right)\sin k\omega t + \cdots + \int_{\alpha_m}^{\pi/2}(-1)^{m-1}\frac{V_d}{2}\sin k\omega t\right]$$

That is,

$$b_k = \frac{2V_d}{\pi}\left[\int_0^{\alpha_1}\sin k\omega t + \int_{\alpha_1}^{\alpha_2}(-1)\sin k\omega t + \int_{\alpha_2}^{\alpha_3}(-1)^2\sin k\omega t + \cdots + \int_{\alpha_m}^{\pi/2}(-1)^n\sin k\omega t\right]$$

$$= \frac{2V_d}{k\pi}[1 - 2\cos k\alpha_1 + 2\cos k\alpha_2 + \cdots + (-1)^m 2\cos k\alpha_m]$$

where b_k is the amplitude of the kth harmonic voltage [Eqn (5.82)]. Writing b_k in a compact manner gives

$$b_k = \frac{2V_d}{k\pi}\left[1 + 2\sum_{j=1}^{m}(-1)^j\cos k\alpha_j\right] \tag{5.83}$$

In order to suppress three of the harmonics, namely, the 5th, 7th, and 11th, keeping the fundamental at a prescribed fraction of $b_{k(\max)}$, it is necessary to solve a set of simultaneous equations. For instance, with $m = 4$, the following four equations have to be solved:

$$b_1 = \frac{2V_d}{\pi}(1 - 2\cos\alpha_1 + 2\cos\alpha_2 - 2\cos\alpha_3 + 2\cos\alpha_4) \tag{5.84a}$$

$$b_5 = \frac{2V_d}{5\pi}(1 - 2\cos 5\alpha_1 + 2\cos 5\alpha_2 - 2\cos 5\alpha_3 + 2\cos 5\alpha_4) \tag{5.84b}$$

$$b_7 = \frac{2V_d}{7\pi}(1 - 2\cos 7\alpha_1 + 2\cos 7\alpha_2 - 2\cos 7\alpha_3 + 2\cos 7\alpha_4) \tag{5.84c}$$

and

$$b_{11} = \frac{2V_d}{11}(1 - 2\cos 11\alpha_1 + 2\cos 11\alpha_2 - 2\cos 11\alpha_3 + 2\cos 11\alpha_4) \tag{5.84d}$$

Equations (5.84a)–(5.84d) are non-linear in nature and can be solved by numerical techniques for the four variables α_1, α_2, α_3, and α_4. These angles will then give the value of b_1 as a specified fraction of $b_{k(\max)}$, at the same time ensuring the elimination of the 5th, 7th, and 11th harmonics. The 13th harmonic is therefore the lowest harmonic that will appear in the output voltage waveform. For a fundamental frequency of 50 Hz this harmonic will be equal to 650 Hz and the leakage inductance of the ac motor can filter this and other higher harmonics. The output current waveform will therefore be a fairly smooth one.

This method of achieving voltage control together with elimination of unwanted harmonics is termed as the *optimum PWM* method. Because of the requirement of numerical techniques for solving the associated equations, optimum PWM is realizable by software programming on modern microcomputers.

5.6 Some Important Aspects of VSIs

The two aspects of VSIs, namely, voltage control and braking of VSI-based drives, are dealt with in detail here.

5.6.1 Voltage Control of VSIs

Output voltage control of VSIs is necessitated by the following.

(a) In applications like uninterrupted power supplies (or standby supplies) a constant ac voltage is needed, in spite of regulation within the inverter.

(b) In most adjustable speed drives, constant torque has to be maintained over the entire speed range. This is possible if the air gap flux is kept constant in this range. This in turn implies that the voltage-to-frequency ratio has to be kept constant; that is, the output voltage of the inverter has to be varied along with the motor speed. However, a voltage boost is necessary at low frequencies to overcome the effects of winding resistance.

The three methods that are commonly used for output voltage control of thyristors are (i) control of dc link voltage, (ii) control using multiple inverters, and (iii) pulse width modulation. These methods are described below.

Control of dc link voltage Figures 5.1(b) and (c) show two methods for controlling the dc link voltage of a three-phase bridge inverter. It is seen from Eqn (5.28) that the ac output voltage bears a fixed ratio with respect to the dc link voltage. The detailed circuits that correspond to the above two block diagrams are given in Figs 5.23(a) and (b), respectively. In the former, which uses a fully controlled rectifier as the dc source, the average dc voltage V_α of the rectifier is varied by changing its firing angle α. The pulsating voltage obtained at the output is smoothed by the *LC* filter shown in the figure.

The second method of changing the dc link voltage consists of a diode bridge followed by a dc chopper. The resulting voltage waveform, which is in the form of rectangular pulses, is filtered by the *LC* filter; the output of this filter in turn forms the input supply for the VSI. This circuit is less costly than the first one, as it uses only one thyristor in the chopper block as against six devices used in the controlled rectifier of the first method. Moreover, the filter elements can have much smaller sizes, because of the fact that the chopper can be switched at a high frequency. Third, a higher power factor is possible with a smaller output voltage of the dc chopper.

The drawbacks of the circuit using a chopper are as follows.

(a) It is non-regenerative because of the diodes being used in the uncontrolled rectifier; this means that the energy cannot be sent back to the input side. On

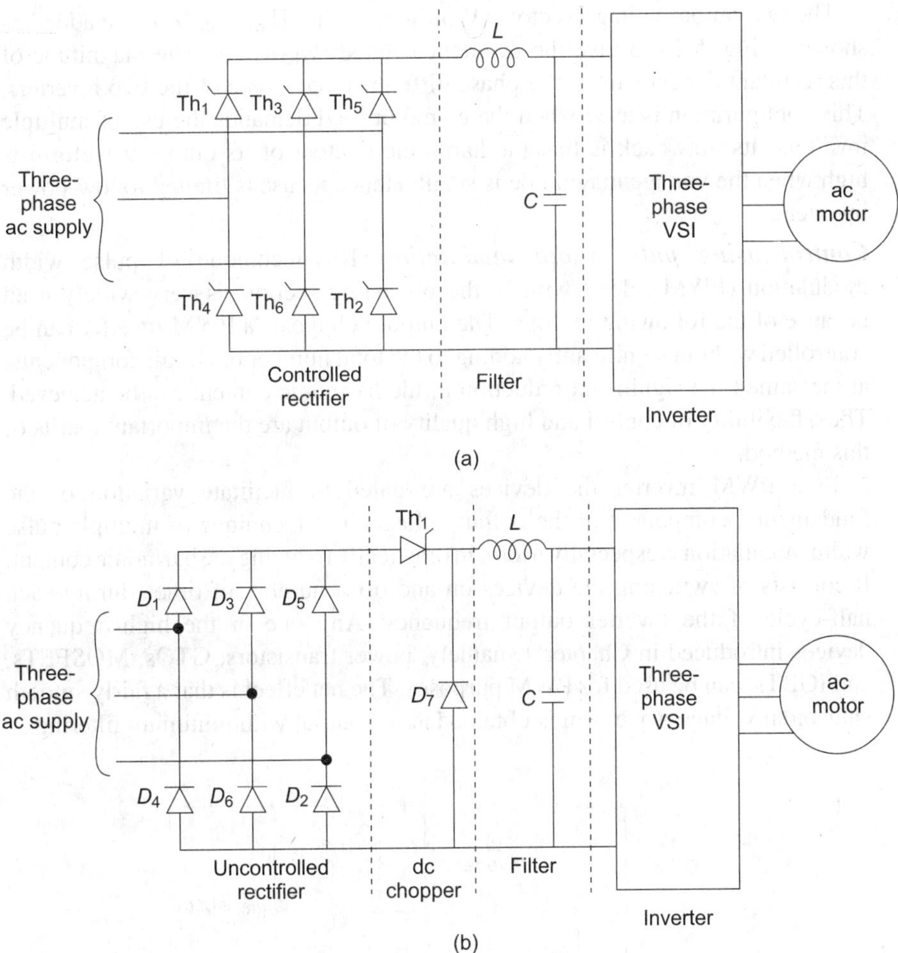

Fig. 5.23 Methods for control of dc link voltage using (a) a controlled rectifier and (b) an uncontrolled rectifier and a dc chopper

the other hand, the controlled rectifier used in the first method can operate in the inversion mode by making $\alpha > 90°$, thus facilitating reversal of power.

(b) It is necessary to connect an energy storage filter on the input side of the dc chopper when it is fed by a single-phase source. This is to prevent the input voltage from becoming zero whenever the output voltage falls to zero.

One great disadvantage of controlling the dc link voltage is that the *LC* filter causes a time lag which adversely affects the transient response of the system. The second drawback is that it requires an auxiliary dc supply for the commutation circuit if it is required to supply rated torque at low speeds.

Control using multiple inverters In this method, illustrated in Fig. 5.24, two or more inverters are connected in parallel and their output voltages are added up.

The two output voltage vectors $|V_{\alpha_1(\text{RMS})}|\angle\phi_1$ and $|V_{\alpha_2(\text{RMS})}|\angle\phi_2$ are added as shown in Fig. 5.24 to give the resultant voltage $|V_{(\text{RMS})}|\angle\phi$. The magnitude of this resultant depends upon the phase difference $(\phi_2 - \phi_1)$ of the two inverters. This configuration is used when the ac motor load demands the use of multiple inverters. Its drawback is that the harmonic content of its output waveform is high when the voltage magnitude is small. Hence its use is limited to low-power inverters.

Control using pulse width modulation The technique of pulse width modulation (PWM), dealt with in the preceding section, is very widely used because of the following reasons. The output voltage of a PWM inverter can be controlled without significantly adding to the total number of circuit components; at the same time significant reduction in the harmonic content can be achieved. Thus, flexibility of control and high quality of output are the important merits of this method.

In a PWM inverter the devices are gated to facilitate variation of the fundamental component of the output voltage. The technique of multiple pulse width modulation is especially useful in drastically reducing the harmonic content. It consists of switching the devices on and off a number of times during each half-cycle of the inverter output frequency. Any one of the high-frequency devices introduced in Chapter 1, namely, power transistors, GTOs, MOSFETs, and IGBTs, can be used for PWM purposes. The net effect is that a fairly smooth sinusoidal voltage waveform is obtained at the output with minimum filtering.

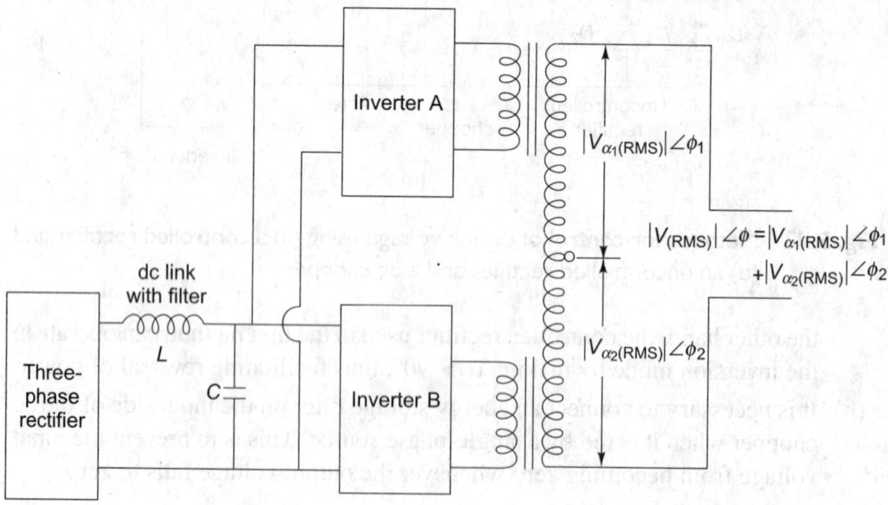

Fig. 5.24 Voltage control with multiple inverters

5.6.2 Braking of VSI-based Drives

A motor driven by a VSI is said to be *braked* when its speed is gradually reduced and it is finally brought to a standstill condition. A typical application for an inverter drive is a hoist, which involves four-quadrant operation. Figures 5.25(a) and (b), respectively, show a hoist and the four quadrants in which it operates. Here an ac machine is braked in the second quadrant whereas it works as a motor in the first quadrant. As the hoist moves up from D to B, the machine works in the (positive) motoring mode; the operating point is in the first quadrant of the i-v plane, where i is the armature current and v is the excitation voltage of the motor. At B the motor has to be decelerated or braked so that it comes to a standstill condition at A, which is at the ground level; this operation can be represented as a point in the second quadrant. Similarly, while the hoist goes down from A to C, reverse motoring takes place, this operation being depicted as a point in the third quadrant. As the hoist has to land at D, which is the bottom point, the last lap of movement is from C to D, which is made possible by the reverse braking of the motor, which comes under fourth-quadrant operation. The discussion below will confine to *braking* or second-quadrant operation. The braking of the motor can be achieved in the following ways:

(a) regenerative braking,
(b) dynamic braking, and
(c) braking that involves prevention of rapid deceleration.

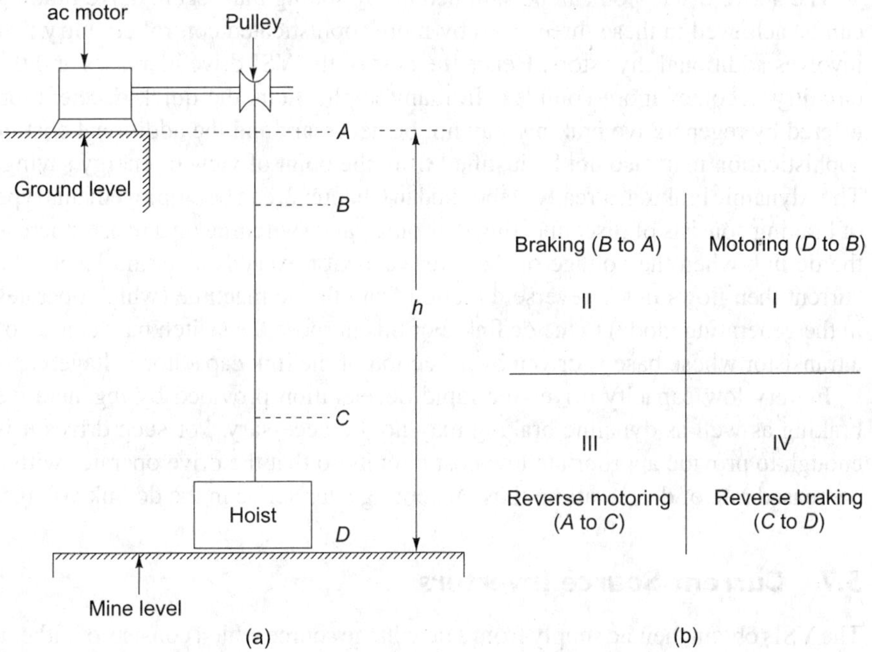

Fig. 5.25 A mine hoist: (a) schematic diagram (b) illustration of four-quadrant operation

An ac drive is said to be *regeneratively braked* when the kinetic energy is converted into electrical energy and returned to the source. This process causes deceleration of the motor and finally brings it to a stop. For an ac machine supplied by a VSI, when the output frequency of the inverter is reduced to a value less than the motor frequency, power flows in a reverse direction from the machine, which acts as a generator, to the dc link through the feedback diodes of the VSI. Regenerative braking involves utilizing this power usefully either by absorbing it in the dc link capacitor or by transferring it to the ac mains. The three configurations of Fig. 5.1 will now be taken up for discussion and necessary modifications explored to facilitate regenerative braking.

Considering the schematic diagram of Fig. 5.1(a), for the energy to be returned through the dc link to the ac source, it is necessary to replace the uncontrolled diode, that is, the rectifier bridge, by a dual converter which permits four-quadrant operation. The second quadrant is of interest because a typical point in it indicates power transfer from the dc link to the source. For the diagram shown in Fig. 5.1(b), the reverse dc link current cannot flow through the controlled rectifier bridge; only another fully controlled rectifier bridge connected in antiparallel with the existing one can permit reverse current to flow to the ac source, provided it is operated in the inverting mode.

Finally, for the configuration shown in Fig. 5.1(c), the first-quadrant dc chopper has to be replaced by a two-quadrant chopper, operating in the first and second quadrants to permit reversal of power. The diode rectifier bridge has also to be replaced by a dual converter.

The above discussion can be summed up by stating that regenerative braking can be achieved in these three cases by more sophisticated control circuitry that involves additional thyristors. Hence the cost of the VSI drive increases and the circuitry becomes more complex. In many applications the quick deceleration offered by regenerative braking may not be necessary and the additional cost of sophistication may also not be justified from the point of view of energy saving. Thus dynamic braking, already elaborated in Chapter 2, can be employed; this type of braking consists of disconnecting the source and switching on a resistor across the dc link when the voltage of the filter capacitor exceeds a certain level. The current then flows in the reverse direction from the ac machine (which operates in the generating mode) to the dc link. For this purpose the switch may consist of a transistor whose base is driven by a fraction of the link capacitor voltage.

In very low capacity drives the rapid deceleration provided by regenerative braking as well as dynamic braking may not be necessary. For such drives it is enough to provide appropriate low-cost circuits so that the drive operates with a decreased rate of deceleration, thus preventing a large rise in the dc link voltage.

5.7 Current Source Inverters

The VSIs obtain their ac supply from a dc voltage source which consists of either a diode bridge rectifier followed by a dc chopper and an *LC* filter or a fully controlled rectifier followed by a filter. The output voltage of a VSI is independent of the load impedance. In contrast to this, for a current source inverter (CSI) the shape of the output voltage depends on the load whereas the output current is independent of load. The source for a CSI consists of a phase-controlled rectifier which can be made to operate as a current source by means of a large series inductor at its output and a current regulating loop. A CSI has the following merits: (a) reversal of motor current, which is needed for regenerative braking, is possible without any additional components whereas these are essential in a VSI, (b) the large filter inductor prevents the short-circuiting of the output terminals when there is a commutation failure, (c) CSIs which are used in drives of medium or high power levels employ relatively less costly converter grade thyristors. CSIs are not without drawbacks and these will be dealt with later. A detailed description of single-phase and three-phase CSIs follows.

5.7.1 Single-phase Current Source Inverter

Figure 5.26(a) shows a single-phase autosequentially commutated inverter (ASCI) which consists of a constant current source rigged up with a dc supply. By alternate firing of the thyristor pairs Th_1, Th_4 and Th_2, Th_3 an alternating current wave of amplitude I_d is obtained at the load. The frequency of the output current is determined by the triggering frequency of the thyristor pairs. The sequence of operation for this circuit can be split up into four intervals as follows.

Interval 1 It is assumed that the thyristors Th_1 and Th_4 are already conducting the load current $I_{\text{ld}} = I_{\text{dc}}$ which flows in the circuit consisting of the positive terminal of the supply, smoothing inductor L, Th_1, D_1, the load, D_4, Th_4, and the negative terminal of the supply. The commutating capacitors C_1 and C_2 are assumed to be charged to $+E_1$ volts during the previous half-cycle, with the X_1- and X_2-plate attaining positive polarity. The voltage E_1 will be greater than the load voltage Ri_{ld}. The triggering of the other thyristor pair marks the end of this interval.

Interval 2 At time t_1 the thyristors Th_2 and Th_3 are triggered into conduction, enabling the capacitors C_1 and C_2 to apply a reverse voltage, respectively, across Th_1 and Th_4 and turn them off. Consequently the voltages across the capacitors decay down. At t_2 these capacitor voltages attain the load voltage RI_{ld}, and the diodes D_3 and D_2 start conduction. This marks the end of interval 2.

Interval 3 During this interval, which starts at t_2, the thyristors Th_2 and Th_3 as well as all the four diodes connect the load in parallel with the commutating capacitors. The smoothing inductor, capacitors C_1 and C_2, and the load resistance and inductance along with the current source form an *RLC* circuit which is driven by the current source. A current oscillation sets in, and the load current, which was originally equal to I_{ld}, becomes zero, then reverses and attains the value $-I_{\text{ld}}$. At t_3 the diodes D_1 and D_4 get reverse biased, terminating the commutation process.

Interval 4 The load current now flows through Th_3, D_3, the load (in a reverse direction), D_2, and Th_2. The capacitor voltage attains the value $-E_1$ (with X_1 and X_2 attaining negative polarity) and remains at this value till the commutation of Th_2 and Th_3.

If Th_1 and Th_4 are now triggered, Th_2 and Th_3 turn off and the load current and voltage reverse again, thus becoming positive. The cycle repeats.

When Th_2 and Th_3 are turned on at t_1 the input voltage (v_{in}) gets reversed whereas the input current remains constant at the positive value of I_{ld}. The power delivered to the load is $-v_{\text{in}}I_{\text{ld}}$. The negative sign indicates that the inductive energy of the load flows back to the dc source. After some time (interval 3) the load becomes ready to conduct current in the reverse direction.

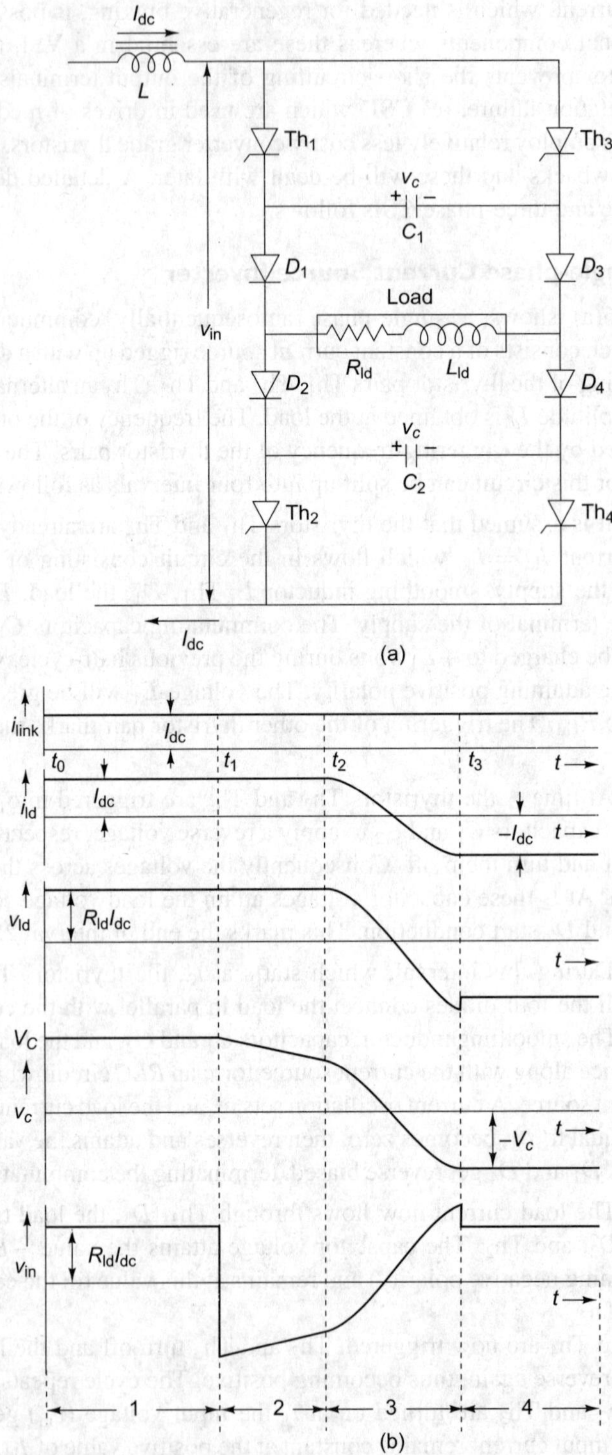

Fig. 5.26 Single-phase autosequentially commutated current source bridge inverter: (a) circuit diagram, (b) waveforms

5.7.2 Three-phase Bridge Type of Current Source Inverter

The single-phase bridge CSI is simple to construct and serves the purpose of demonstrating the principle of a current source inverter. However, because of the wider use of CSI-driven three-phase ac motors, the three-phase CSI (Murphy & Turnbull 1988) is of practical importance. Figures 5.27(a) and (b), respectively, show a three-phase (ASCI) feeding a star-connected balanced load (say, an induction motor) and its output waveforms. The thyristors Th_1 to Th_6 are fired in sequence, each device conducting for a period of 120°. The circuit can be seen to be an extension of the circuit of the single-phase ASCI and has three legs instead of two. It also has six capacitors C_1 to C_6 to store the energy necessary for commutation purposes and as many diodes, D_1 to D_6, to isolate the load from the capacitors.

The thyristors are gated in such a manner that one thyristor each from the top and bottom groups connects the load to the positive and the negative bus, respectively, thus conducting the current through it. It is assumed that initially the circuit is in the steady state and that the motor speed is constant. The sequence of operation of the inverter is divided into four intervals as before, detailed below.

Interval 1 The thyristors Th_1 and Th_2 are fired and kept on for some time; thus, only the R- and B-phase of the load conduct current. To start with, the capacitors C_1, C_3, and C_5 are charged with voltages E_1, 0, and $-E_1$, respectively, with the polarities shown, by a precharging circuit which is not needed for subsequent operation. This interval terminates at t_1 when Th_3 is fired. The circuit condition at the beginning of the interval is given in part (i) of Fig. 5.27(c).

Interval 2 The firing of Th_3 causes v_{C_1} to be applied in a reverse manner across Th_1, turning it off. The load current now flows through Th_3, the capacitor combination C_1 in parallel with the series combination of C_3, C_5, then

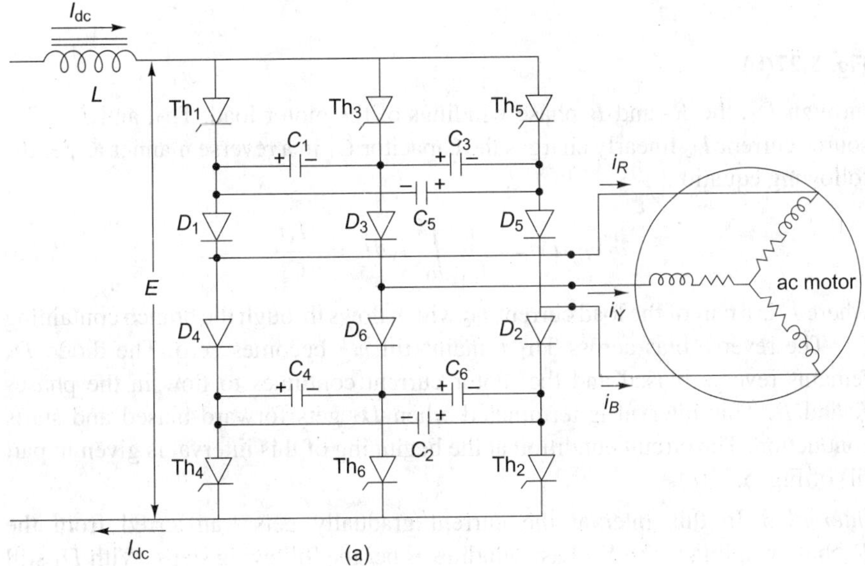

Fig. 5.27(a)

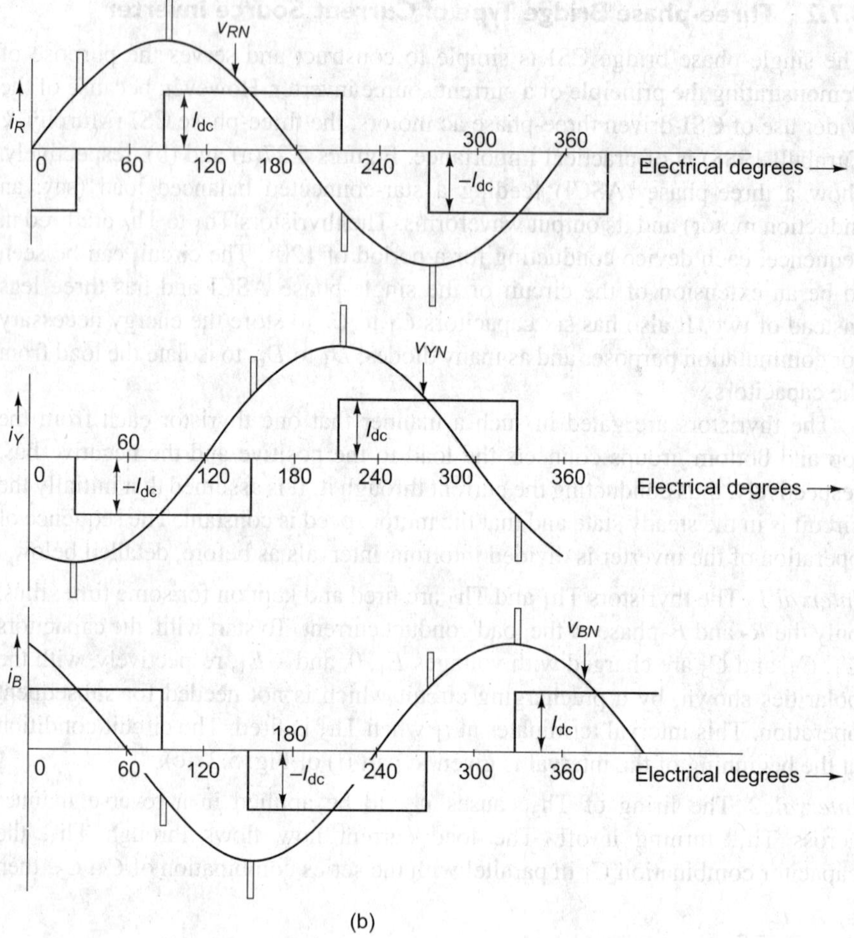

Fig. 5.27(b)

through D_1, the R- and B-phase windings of the motor load, Th_2, and D_2. The source current I_{dc} linearly charges the capacitor C_1 in a reverse manner as per the following equation:

$$v_{C_1}(t) = \frac{1}{C_1}\int_0^t I_x dt = \frac{I_x t}{C_1} \tag{5.85}$$

where I_x is a part of the load current I_{dc} which flows through the branch containing C_1. The reverse bias across Th_1 remains till v_{C_1} becomes zero. The diode D_3 remains reverse biased and the motor current continues to flow in the phases R and B. This interval is terminated when D_3 gets forward biased and starts conduction. The circuit condition at the beginning of this interval is given in part (ii) of Fig. 5.27(c).

Interval 3 In this interval the current gradually gets transferred from the R-phase winding to the Y-phase winding as per the following steps. With D_1 still

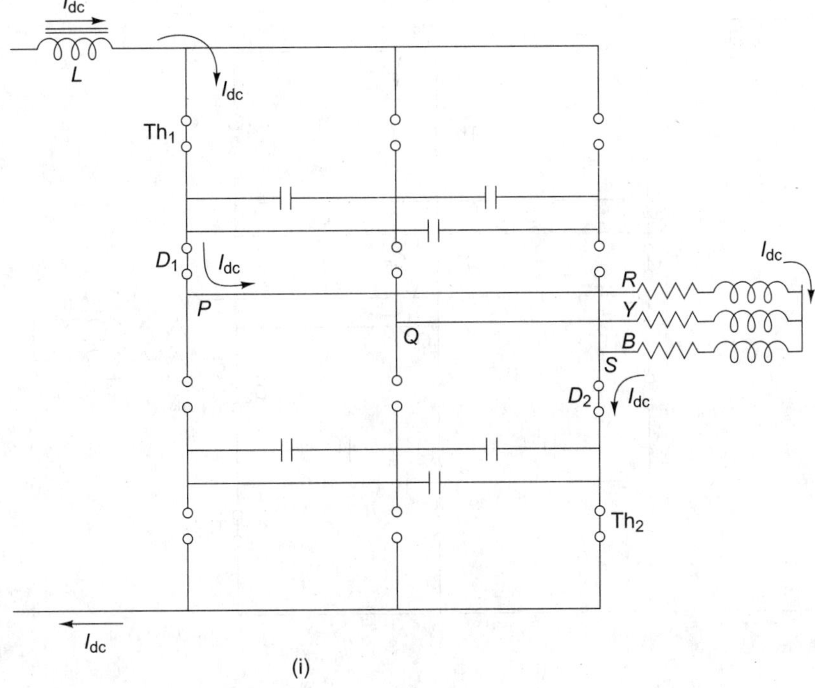

(i)

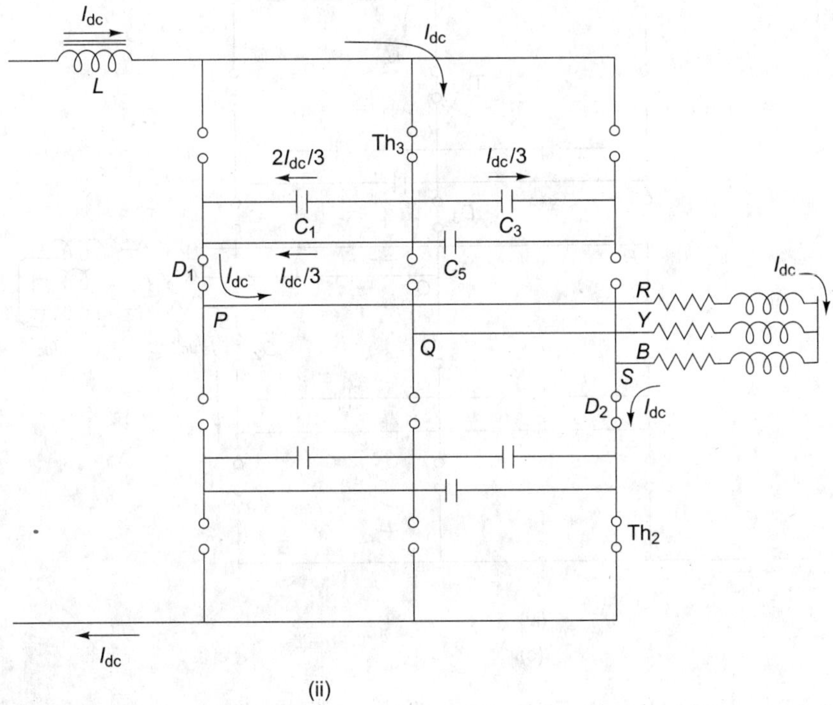

(ii)

Figs 5.27(c): (i) and (ii)

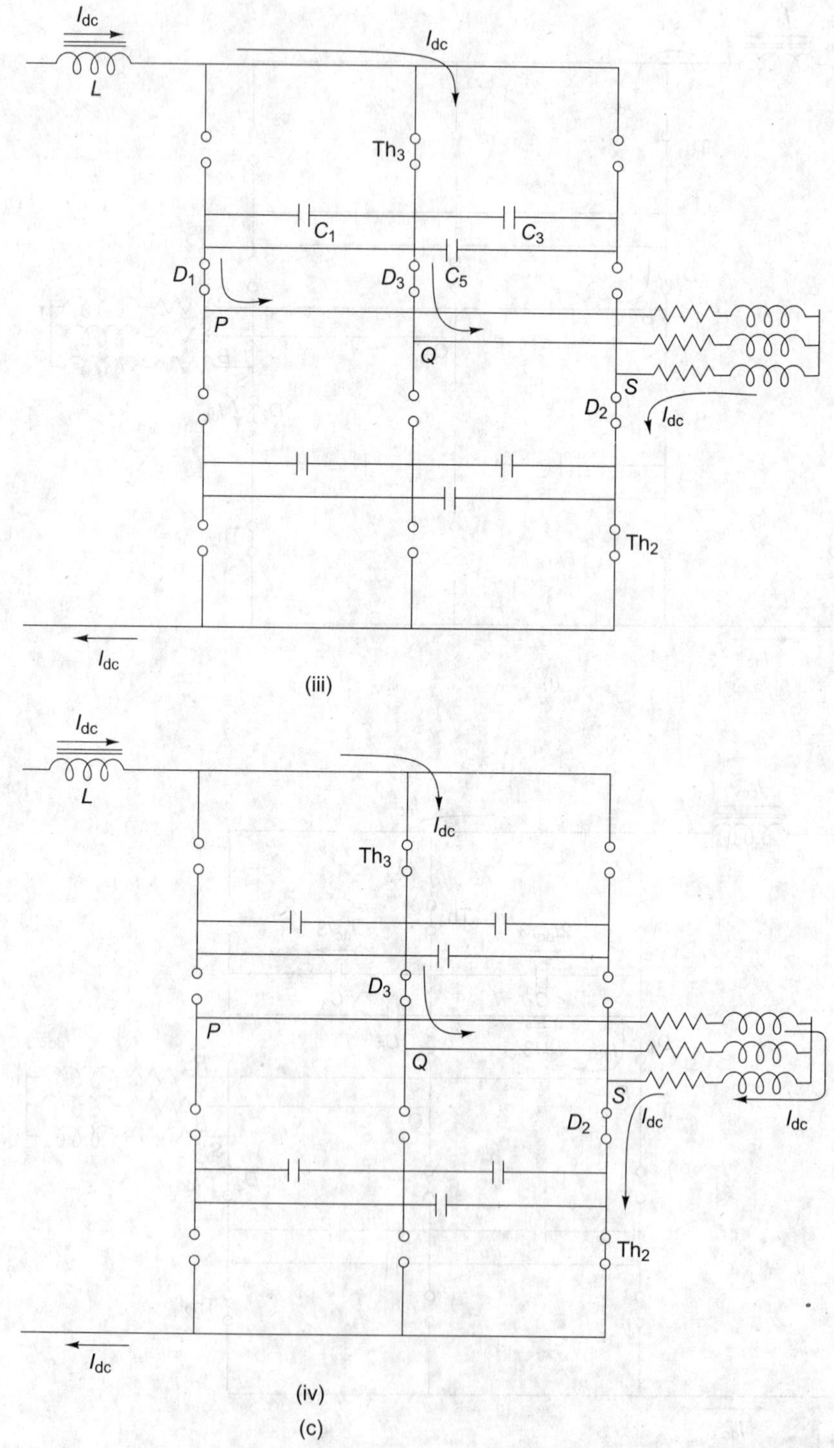

Fig. 5.27 Three-phase ASCI: (a) circuit diagram, (b) idealized voltage and current waveforms of ac motor load, (c) circuit conditions during the four intervals of commutation of Th_1

conducting, the capacitor combination C_1 in parallel with the series combination of C_3 and C_5 gets connected in parallel with the motor windings, through D_1 and D_3. The *RLC* circuit consisting of the smoothing inductor L, capacitor bank, and motor resistance and inductance forms a resonating circuit. In a quarter-cycle the current in the R-phase reduces from I_{dc} to 0 and that in the Y-phase increases from 0 to I_{dc}. D_1, which is connected to phase R, finally stops conduction, marking the end of this interval. The circuit condition at the beginning of this interval is given in part (iii) of Fig. 5.27(c).

Interval 4 The source current I_{dc} now flows through Th_3, the Y- and R-phase windings of the motor, and Th_2. This condition remains till Th_4 is gated, to initiate the commutation of Th_2. As the only conducting device D_3 is connected at the junction of C_1 and C_3, there is no path for their discharge. Hence these capacitors retain the charge built up during intervals 2 and 3. The circuit condition at the beginning of this interval is given in part (iv) of Fig. 5.27(c).

Voltage spikes and their reduction Figure 5.27(b) shows that the waveforms of the phase currents are quasi-square in nature and have a duration of 120°. The fundamental components of these currents and the leakage inductances of the motor winding determine the fundamental line-to-neutral voltages v_R, v_Y, and v_B at the motor terminals. Another feature is that due to the rectangular nature of the current waveforms, the $L(di/dt)$ voltages become zero during the flat portions and appear as spikes at the instants of the abrupt rise or fall of the current waveform. Hence the terminal voltages (line-to-neutral) are a composition of the fundamental voltages and the $L(di/dt)$ voltage transients which occur across the leakage inductances. It is assumed that the total inductance of the motor and stator windings, referred to the stator side, is L henries per phase. The three line-to-neutral voltages will be as shown in Fig. 5.27(b). The duration of the current transfer interval (interval 3) determines the magnitude of the $L(di/dt)$ spikes. The magnitude of the voltage wave, with spikes superimposed over it, should not exceed the blocking voltage capability of the inverter thyristors Th_1 to Th_6. This is ensured by increasing the duration of interval 3 and using induction motor windings with small leakage reactances. One way of increasing the duration of interval 3 is to use a large value for the capacitance C. This has the benefits of (i) reducing the magnitude of the transient voltage spikes and (ii) facilitating the use of converter grade thyristors instead of the costly inverter grade devices. A drawback of having a large value for the commutation capacitors is the reduction of the operating frequency at which the current bypass effect occurs. This last phenomenon is elaborated below.

Multiple commutation effect and current bypassing It has been stated before that the commutation cycle is so arranged that at any instant one thyristor each from the upper and lower groups conducts. It can be seen from Fig. 5.27(b) that an interval of 60° (or one-sixth of a cycle of the output frequency) is maintained between the turning off of one thyristor and the turning on of the other thyristor (called the complementary thyristor) on the same leg. This precaution is necessary because (a) it should be ensured that the dc supply is not shorted by the four devices of any particular leg and (b) the diversion of a portion of the load current I_{ld} from the motor windings should be prevented. However, if the output frequency is increased beyond a certain limit, partial overlap occurs when interval 2 commences in the upper group before interval 3 is completed in the lower group and vice versa. If it is further increased, the capacitor charging interval (interval 2) occurs simultaneously in both the groups. This is called *full commutation overlap*; in such a case, the load current will be bypassed from the load and circulates in the path consisting of the positive terminal of the dc supply, smoothing inductor L, two thyristors and two diodes on the same leg, and back to the negative of the dc supply.

5.7.3 Regenerative Braking of CSI Drives

Regenerative braking is possible in a CSI drive which uses a fully controlled rectifier bridge as a dc power source. To explain this feature, the schematic of Fig. 5.1(b) is modified to that in Fig. 5.28, which incorporates a current source inverter.

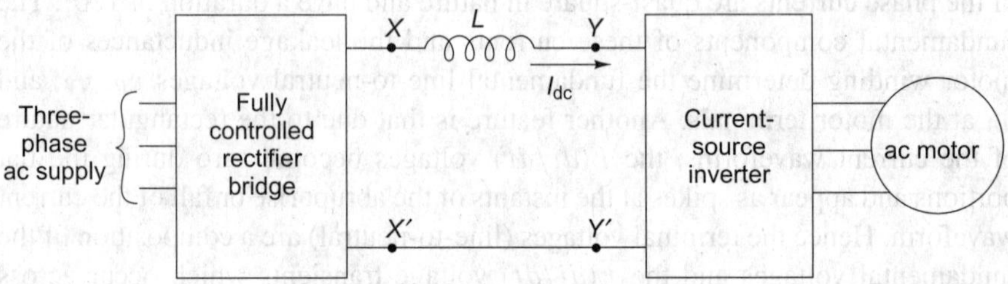

Fig. 5.28 Current source inverter fed from a controlled rectifier through a dc link

It was stated earlier that the output current in a CSI is independent of the load whereas the output voltage is governed by the load condition. Under the no-load condition the load voltage as well as the dc link voltage are nearly zero; on the other hand they attain their maximum values under load conditions. Under normal motor operation the terminals Y and Y' of Fig. 5.28, respectively, attain positive and negative polarities. The current regulating loop and gating circuits (Fig. 5.29) automatically adjust the thyristor firing angles of the fully controlled rectifier bridge to maintain positive and negative polarities at X and X', respectively.

When the ac motor is hauled by its load, it works as an ac generator and returns power to the three-phase ac source through the CSI, dc link, and the fully controlled bridge. This implies a reversal of direction of the ac motor rotation, which can be obtained by electrically reversing the output phase sequence of the CSI. This is known as the *regenerative mode*, in which the dc link current is unidirectional, but power reversal is made possible by means of a voltage reversal. The terminals Y and Y', respectively, attain negative and positive polarities in the regenerative mode. The current regulating loop and gating circuits accordingly adjust the firing angles of the controlled converter so that X and X', respectively, attain negative and positive polarities. This is in contrast to the braking of a VSI-controlled ac motor, where a second controlled converter is necessary under power reversal conditions. Thus a CSI has inherent regenerative capability and no

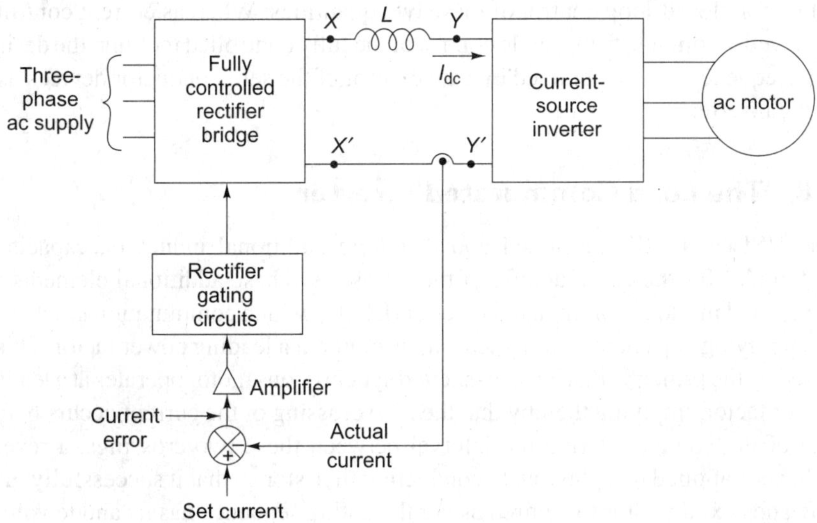

Fig. 5.29 Current regulating loop and rectifier gating circuits used in a CSI

additional components are needed for braking; hence it is less costly as compared to a VSI that incorporates the braking feature.

The CSI has some more beneficial features as follows. The magnitude of the filter inductor of the dc link is usually kept at a much greater value than the total leakage inductance of the induction motor, so that it offers protection under high di/dt conditions which may occur either due to commutation failure or because of the short-circuiting of the output terminals. Such a fault is cleared by the emergency stoppage of rectifier gate signals. In this way the drive can ride through transient conditions without any need to shut down the inverter. However, (a) this makes the inductor bulky and costly, (b) it leads to stability problems under high-speed operation with light loads, and (c) diodes and thyristors of high blocking voltages have to be employed because of the large voltage spikes occurring during commutation. A trade-off is therefore necessary while designing the size of the inductor.

In addition to these protective features, each semiconductor device is provided with a snubber circuit and a series inductor to protect it, respectively, from high dv/dt and high di/dt.

The quasi-square waveform of current contains low-order harmonic components which cause torque pulsations and irregular shaft rotation at low speeds called *cogging*. Consequently they may excite mechanical resonance for mechanical loads. For example, the fifth and seventh stator current harmonics, which result in the sixth harmonic component for the pulsating torque, lead to this phenomenon. Cogging may occur even during reversal of rotation. Fortunately, similar to PWM inverters it is possible to employ harmonic elimination techniques with a CSI to eliminate undesirable current harmonics.

The induction motor torque and air gap flux are uniquely determined by the stator current and rotor slip frequency. Hence separate loops may have to be set up for closed-loop control of these two quantities. Whereas current control is achieved by suitable firing angle control of the fully controlled rectifier, the desired slip frequency can be obtained by gate control of the semiconductor devices used in the inverter.

5.8 The Load Commutated Inverter

The VSIs and CSIs discussed above require additional inductors, capacitors, and diodes for the commutation of the thyristors. These additional elements are eliminated in a load commutated inverter (LCI) in which commutation is achieved by supplying current to an ac synchronous motor at a leading power factor. This is based on the principle that an over-excited synchronous motor operates at a leading power factor, implying thereby that the zero crossing of the current occurs before that of the voltage. During the interval between these two crossings, a reverse voltage is applied to a previously conducting thyristor so that it successfully turns off. The next thyristor to be fired as per the gating sequence has its anode voltage greater than that of the cathode; hence if a firing signal is given, it can take over the load current. The LCI, however, needs a forced commutation circuit whenever the machine speed is below 10% of its rated speed. This is because the low voltages generated by the synchronous machine at such low speeds are inadequate for the natural commutation of the thyristors of the LCI. Figure 5.30 gives the circuit of an LCI which consists of a line commutated converter on the left, a dc link with a series inductor in the centre, and a bridge inverter on the right. The ac machine is connected to the output terminals of the bridge inverter. A series commutating inductor L_c on each line connection from the inverter to the machine helps in the commutation process.

5.8.1 Line Commutated Converter

The source for the line commutated converter is a three-phase ac supply. Together with the dc link inductor, this converter supplies a regulated dc current at its output; hence this combination constitutes the current source for the LCI. As implied by its name, natural commutation is implemented for the thyristors. Figures 5.31(a), (b), (c), and (d), respectively, give the voltage waveforms of the positive and negative buses of the dc link, the dc link voltage to the left of inductor, the voltage across Th_1, and the current through Th_1. Since the converter has to be operated in the rectifying mode, the firing angle α should be less than 90°. With the orientation of the thyristors shown, the top and bottom buses of the dc link will, respectively, have positive and negative polarities. The conduction period of each thyristor is 120°. As shown in Fig. 5.31(c), the duration of application of reverse voltage across Th_1 can be as high as 240° and is adequate for commutation.

5.8.2 Bridge Inverter

The three-phase, full-wave thyristor bridge is made to work as an inverter by ensuring that α is always more than 90°. It therefore converts the dc link voltage into ac voltage, which becomes the input for the synchronous motor. The star neutral point of the machine windings becomes the reference for the machine-induced line-to-neutral voltages. Figure 5.32(a) gives these voltages as

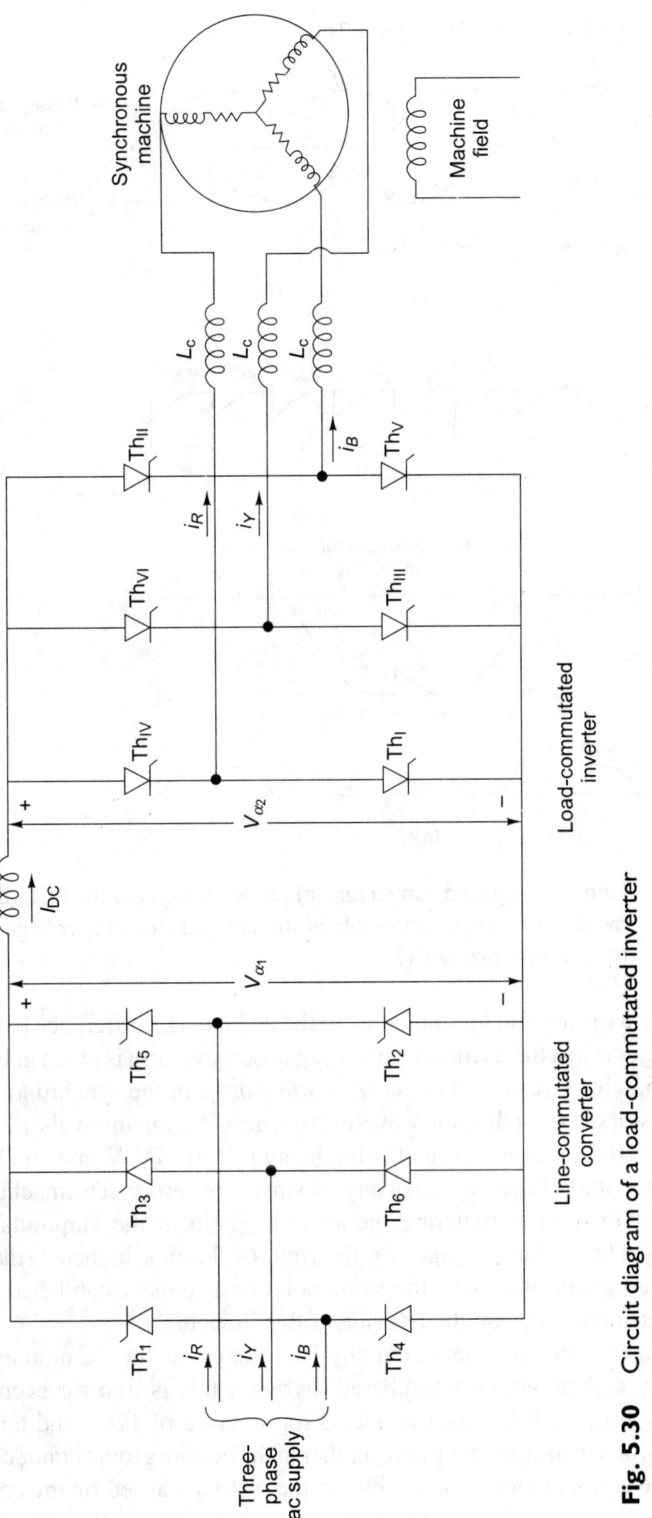

Fig. 5.30 Circuit diagram of a load-commutated inverter

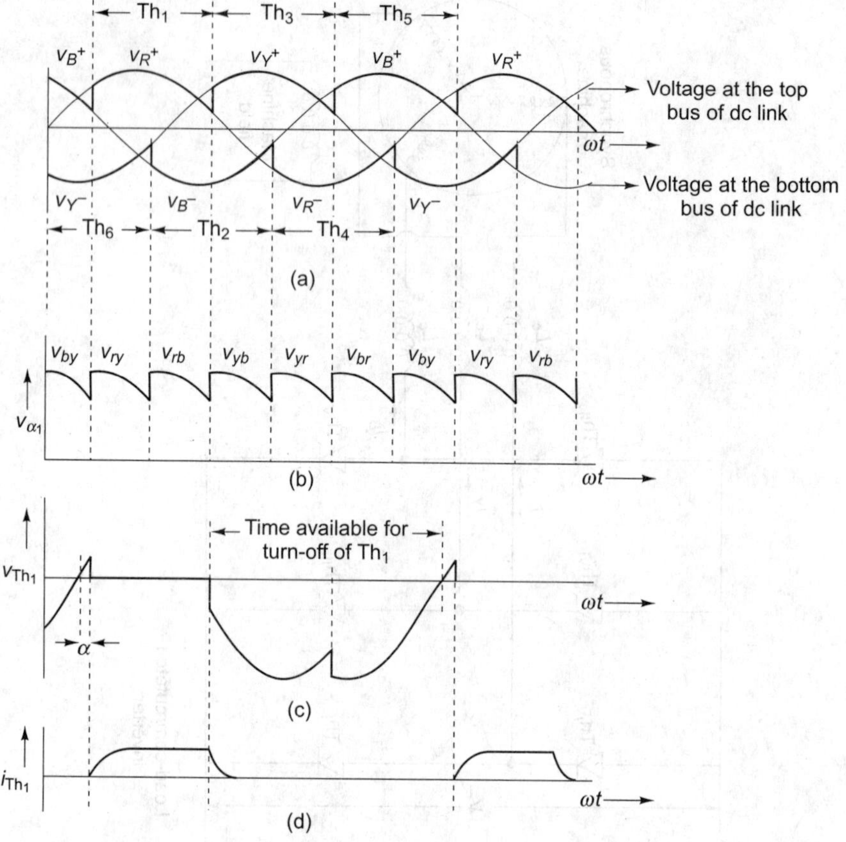

Fig. 5.31 Line commutated converter: (a) phase voltages of the LCI, (b) waveform of dc link voltage to the left of the link inductor, (c) voltage across Th_1, (d) current through Th_1

measured at the top and bottom buses of the dc link. The difference between these bus voltages is just the dc link voltage v_{α_2} whose waveform is given in Fig. 5.32(b). Finally, 5.32(c) gives the phase current waveforms of the synchronous machine. The thyristors that conduct the waveform during different intervals are also shown in Fig. 5.32(b), their sequence of firing being I, II, III, IV, V, and VI. The interval between the start of conduction of any two thyristors is 60°. The machine voltages together with the commutating inductors L_c help in the commutation of the thyristors. The dc link voltages on the right of the link inductor (that is, at the inverter terminals) will have the same polarity as those established by the line commutated converter on the left side of this inductor.

Turning on an odd numbered thyristor helps in the commutation of the previously conducting odd numbered thyristor; this is true for even-numbered thyristors also. Each thyristor conducts for a period of 120°, and normally two thyristors, one in the top group and another in the bottom group, conduct. However, during the commutation process, due to the overlap caused by the commutating inductors L_c, two thyristors of one group and one from another group conduct.

Though the corresponding dc link polarities of v_{α_1} and v_{α_2} are the same, the polarity of v_{α_2} will be negative with respect to the inverter. This is because, if α

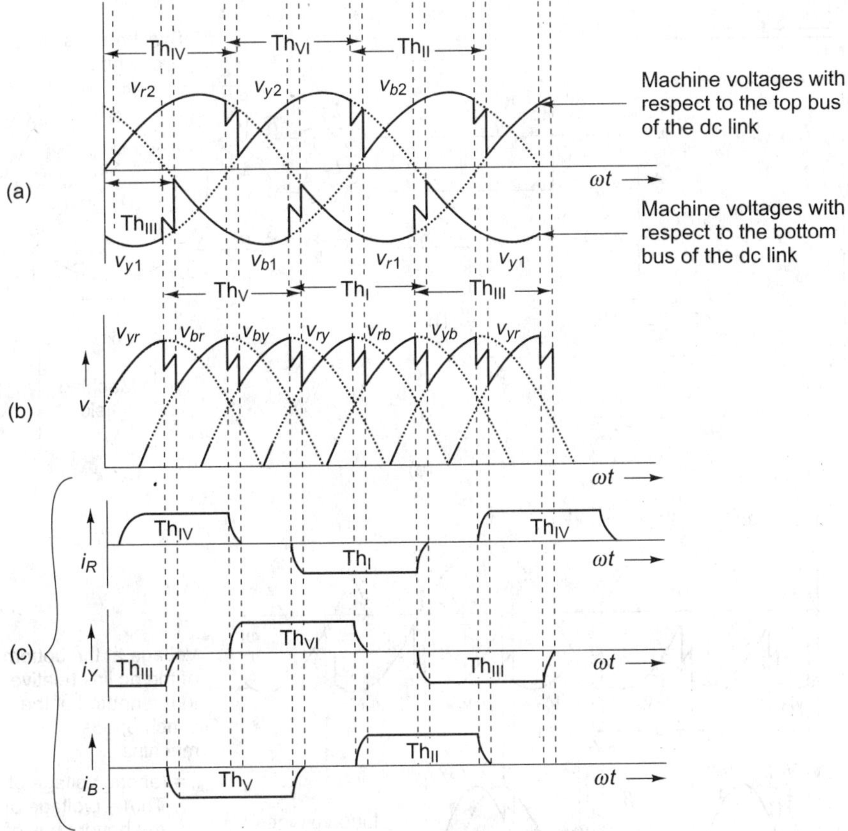

Fig. 5.32 Load commutated inverter: (a) phase voltages of a synchronous machine, (b) voltage of dc link to the right of the link inductor, (c) machine currents flowing through LCI

were to be less than 90° for its thyristors, the thyristor bridge of the bridge inverter gives a positive polarity to the bottom bus of the dc link and a negative polarity to the top bus. In the LCI circuit, however, the bridge operates in the inverter mode, that is, α is always kept greater than 90°. Accordingly, the bottom stop buses will, respectively, have negative and positive polarities. Thus v_{α_2} will have the same polarity as v_{α_1}.

The *commutation* of $\mathrm{Th_V}$, which is a bottom group thyristor, can be understood with the help Fig. 5.33. From the sequence of firing it is seen that with $\mathrm{Th_V}$ as the outgoing thyristor, $\mathrm{Th_I}$ will be the incoming one. Just before the commutation of $\mathrm{Th_V}$, the dc link voltage v_{α_2} will be equal to v_{BY} $(= v_{B1} - v_{Y2})$. When $\mathrm{Th_I}$ is fired with its $\alpha > 90°$, the R-phase machine terminal is connected to the bottom bus of the dc line. The current cannot be transferred from $\mathrm{Th_V}$ to $\mathrm{Th_I}$ because of the presence of the commutating inductors L_c. The time (in electrical degrees) taken for the transfer of conduction is the commutation interval μ.

The following discussion is confined to the commutation interval. The equivalent circuit of the bridge inverter, the waveforms of the bottom link voltage

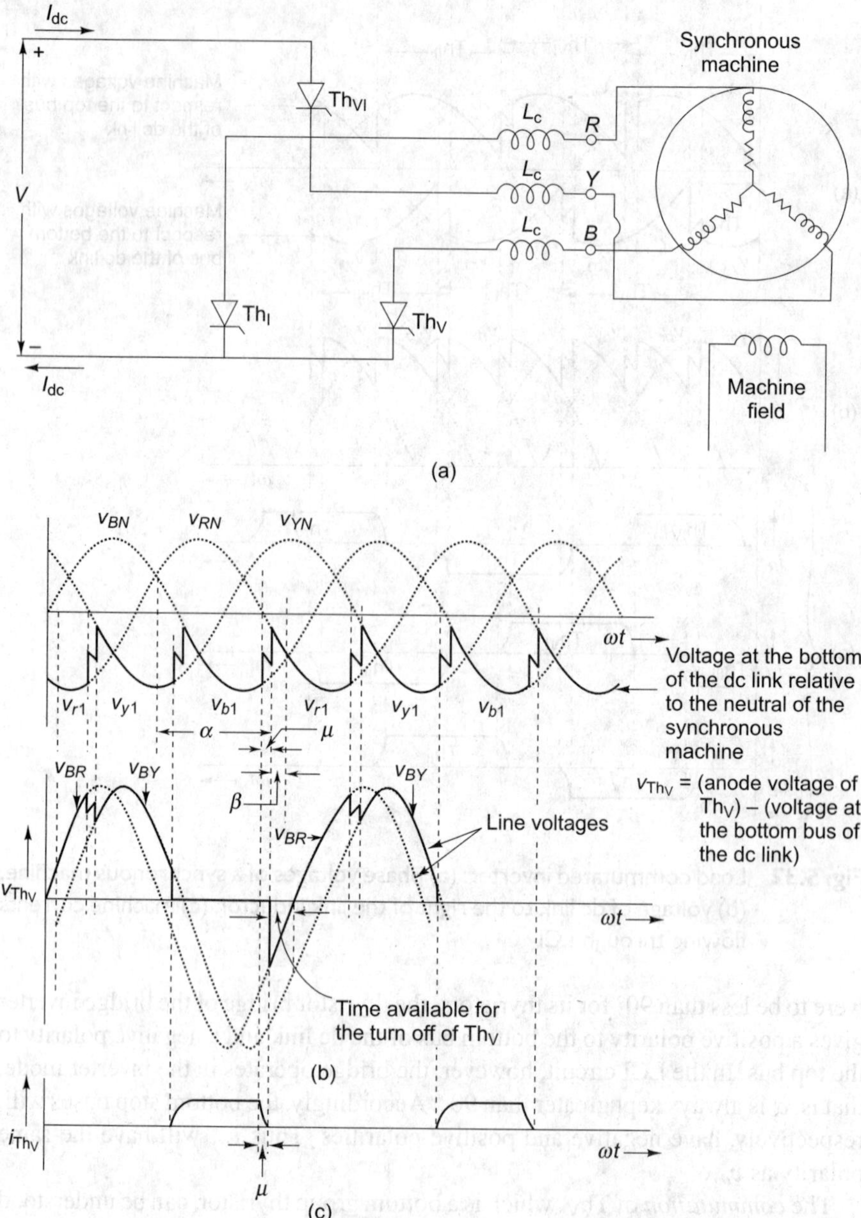

Fig. 5.33 Load commutated inverter: (a) circuit diagram during commutation of Th_V, (b) voltage across Th_V, (c) current through Th_V

as well as that of Th_V, and the current through Th_V are given, respectively, in Figs 5.33(a), (b), and (c). It is assumed that the dc link current i_d remains constant during this interval. Since the line voltage v_{RB} $(= v_{R1} - v_{B2})$ is impressed against Th_V in a reverse manner, the current through it gradually decreases and simultaneously the current through Th_I slowly increases. Because of the assumption made above, these two currents sum up to i_d at any instant of the commutation interval.

Th_V turns off at the end of μ and the new pair Th_I, Th_VI now conducts the link current i_d. The link voltage v_{α_2}, which now is the same as v_{RY} $(= v_{R1} - v_{Y2})$, will be negative as viewed from the inverter terminals. Thus the polarity of the top bus to the right of the link inductor will be positive with respect to the bottom bus. However, the link current i_d continues to be positive. Thus the product $v_{\alpha_2} i_d$ will be negative as viewed from the inverter terminals, implying thereby that power flows from the dc link to the synchronous machine through the bridge inverter. This makes the machine operate as a synchronous motor.

5.8.3 Other Features of the LCI

During the motoring mode of the synchronous machine, the current is adjusted by variation of the firing angle of the line commutated converter whereas the inverter frequency is controlled by gating its thyristors in synchronization with the machine frequency. This can be explained as follows. As the firing instants for the thyristors of the inverter are synchronized with the machine voltage waveforms, the frequency of operation of the inverter is the same as that of the voltages induced in the machine. Further, the control system is operated in such a way that the machine operates at a leading power factor. The average value of v_{α_1} is adjusted to be slightly greater than v_{α_2} to facilitate the flow of dc current through the resistance of the link inductor L. This inductor performs the following three functions: (a) it filters out the voltage harmonics of the line commutated converter in the rectifier mode, (b) it filters the harmonics of the inverter in the regenerative braking mode as elaborated below, and (c) it prevents the line commutated converter and the inverter from interfering with each other.

With $\alpha < 90°$, the bridge inverter operates as a rectifier, giving a reverse polarity to v_{α_2}. This means that the bottom and top buses of the dc link, respectively, have positive and negative polarities. As the direction of i_d, which depends on the orientation of the thyristors of the inverter, remains positive, power flows from the synchronous motor to the dc link through the inverter bridge and the machine operates as a generator. Now, if the firing angle of the line commutated converters is adjusted to be greater than 90°, it works in the inverting mode and transfers the dc link power to the ac source, thus facilitating regenerative braking of the synchronous machine. It is therefore to be concluded that the LCI can be operated in the regenerative braking mode by appropriately adjusting the firing angles of the left- and right-hand bridges.

5.8.4 Merits and Demerits of LCI

The major advantage of the LCI is that no forced commutation elements are needed for machine speeds nearing rated speeds. The cost of the LCI is reduced due to this and any losses in such elements are eliminated.

The drawbacks of an LCI are the following: (i) the three commutation inductors (L_c) which are needed for successful commutation add to the cost of the circuitry, (ii) the LCI needs a current regulating loop, which enables it to function as a precisely operated closed-loop control system, and (iii) additional forced commutation elements are needed for machine speeds below 10% of rated speed.

5.8.5 Applications of LCI

The LCI is used as a source of supply for high-power synchronous motors used for driving large fans, blowers, and compressors. Second, it is employed for starting and accelerating to full speed the large synchronous condensers that are used on ac transmission lines for voltage and power factor control. In this application the LCI is disconnected once the machine (synchronous condensers) attains full speed, thereby connecting it directly to the supply network. The LCI is also used for starting large pumped storage hydrogenerators for which it facilitates regenerative braking by causing a changeover of operation from the motoring mode to the generating mode.

Summary

Inverters are widely used for voltage and frequency control of ac motors. They involve forced commutation circuits that consist mainly of inductors, capacitors, and diodes. The design of these commutating elements needs a careful analysis of the circuit conditions. The McMurray and McMurray–Bedford types of commutation are well established because of their excellent performance. In the latter type, there exists flexibility for the operation to take place either with the load connected to the source after commutation or isolated from it.

While the six-step inverter is popular, other types of three-phase inverters such as the 120°-mode inverter and input-commutated inverter have some exclusive merits. While the possibility of short-through fault is eliminated in the 120°-mode inverter, multi-thyristor commutation is the positive feature of the input circuit commutated inverter. Reliable voltage control schemes have been developed with all the VSIs, the most attractive method being that of pulse width modulation (PWM). Harmonic minimization is the major benefit of PWM, other advantages being improved input power factor, faster transient response, and highly flexible voltage-to-frequency operation. The advent of microprocessors and associated software has made PWM inverters a ready alternative to the six-step inverter. They, however, suffer from the disadvantage of being subjected to high di/dt and dv/dt because of switching devices while conducting load current. The resonant pulse inverters discussed in Appendix D are helpful in overcoming these drawbacks.

Current source inverters (CSIs) have the following merits: (a) ability to reverse the motor current without using additional components and (b) protection, using the filter inductor, against the short-circuiting of output terminals in the eventuality of commutation failure. The first feature facilitates regenerative braking, thus improving the efficiency of this inverter. Harmonic elimination is also possible with CSIs.

The LCI facilitates natural commutation of thyristors, thus eliminating the need for the expensive commutating circuits used in VSIs. However, the commutating and current regulating loop needed for its operation add to its cost. In addition, forced commutation circuits are necessary under low machine speeds. LCIs find application in high-power equipment, which include both motors and generators.

Worked Examples

1. The single-phase, half-bridge inverter of Fig. 5.6(a) supplies a resistive load of 10 Ω. If the supply voltage $E = 200$ V, determine the (a) RMS output voltage at the fundamental frequency, (b) output power, (c) half-cycle average and RMS and peak currents of each thyristor, (d) input power factor, and (c) distortion factor.

Solution

The output waveform is shown in Fig. 5.7(a). Because of the odd symmetry of the output voltage waveform, the Fourier series for the instantaneous output voltage is obtained as

$$v(t) = \sum_{n=1}^{\infty} b_n \sin n\omega t$$

where

$$b_n = \begin{cases} \dfrac{2}{\pi}\displaystyle\int_0^{\pi} \dfrac{E}{2} \sin n\omega t \, d(\omega t) = \dfrac{2E}{n\pi} \quad \text{for } n = 1, 3, \ldots \\ 0 \text{ for } n = 2, 4, \ldots \end{cases}$$

Thus,

$$v(t) = \sum_{n=1,3,\ldots}^{\infty} \frac{2E}{n\pi} \sin n\omega t$$

The fundamental component becomes

$$v_1(t) = \frac{2E}{\pi} \sin \omega t$$

The RMS value of this fundamental component of voltage is

$$v_{1(\text{RMS})} = \frac{2E}{\sqrt{2}\pi} = \frac{\sqrt{2}E}{\pi} = \frac{\sqrt{2} \times 200}{\pi} = 90 \text{ V}$$

The fundamental component of current with purely resistive load is

$$i_1(t) = \frac{2E}{\pi R} \sin \omega t$$

The RMS value of the fundamental component of load current is

$$I_{\text{RMS}} = \frac{v_{1(\text{RMS})}}{R} = \frac{\sqrt{2}E}{\pi R} = \frac{\sqrt{2} \times 200}{\pi \times 10} = 9 \text{ A}$$

The fundamental component of output power

$$P_1 = V_1 I_1 \cos\phi_1 = 90 \times 9 \times 1 = 810 \text{ W}$$

The waveforms of currents through the thyristors are given in Fig. 5.34.
The half-cycle average current through Th_1 is

$$\frac{1}{\pi}\int_0^{\pi} \frac{E}{2R} d(\omega t) = \frac{E}{2R} = \frac{200}{2 \times 10} = 10 \text{ A}$$

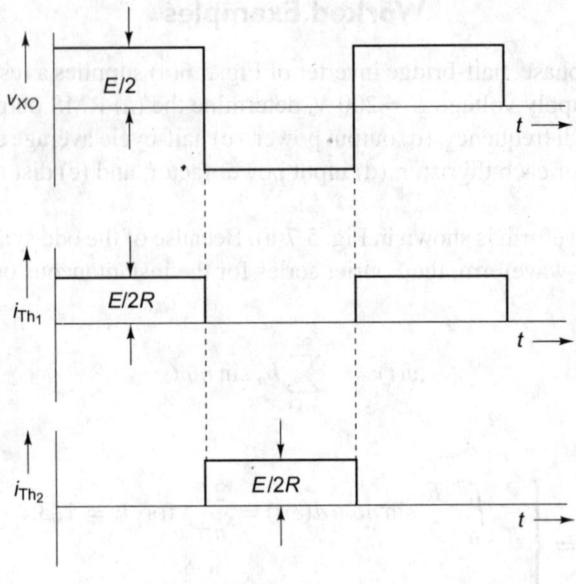

Fig. 5.34

The RMS value of the current through Th_1 is

$$\sqrt{\frac{1}{2\pi}\int_0^{\pi}\left(\frac{E}{2R}\right)^2 d(\omega t)} = \frac{E}{2R}\frac{1}{\sqrt{2}} = \frac{200}{2\times 10\times\sqrt{2}} = \frac{10}{\sqrt{2}} = 7.1\text{ A}$$

The peak value of thyristor current is

$$\frac{E}{2R} = \frac{200}{2\times 10} = 10\text{ A}$$

The RMS value of the output voltage is

$$\sqrt{\frac{2}{2\pi}\int_0^{\pi}\left(\frac{E}{2}\right)^2 d(\omega t)} = \frac{E}{2} = 100\text{ V}$$

The RMS value of the current through the load is

$$\sqrt{\frac{2}{2\pi}\int_0^{\pi}\left(\frac{E}{2R}\right)^2 d(\omega t)} = \frac{E}{2R} = \frac{200}{2\times 10} = 10\text{ A}$$

$$\begin{aligned}\text{Input power} &= \text{input voltage (RMS)} \times \text{input current (RMS)}\\ &= \text{input voltage (RMS)} \times \text{load current (RMS)}\\ &= 100\times 10 = 1000\text{ W}\end{aligned}$$

$$\begin{aligned}\text{Input power factor} &= \frac{\text{fundamental output power}}{\text{input power}}\\ &= \frac{V_1 I_1\cos\phi}{V_{\text{RMS}} I_{\text{RMS}}}\end{aligned}$$

$$= \frac{810}{100 \times 10} = 0.81$$

$$\text{Distortion factor} = \frac{\text{RMS of fundamental current } (I_1)}{\text{RMS of input current } (I)} = \frac{9}{10} = 0.9$$

2. For the inverter of Example 1, repeat the above calculations by considering an inductive reactance of 8 Ω at the load in addition to the resistance.

Solution

The waveform of the fundamental load current is given in Fig. 5.35. $V_1 = 90$ (same as in Example 1). For the lagging power factor load, the Fourier series for the instantaneous load current is given as

$$i_L = \sum_{n=1,3,5} \frac{2E}{n\pi\sqrt{R^2 + (nX)^2}} \sin(n\omega t - \theta_n)$$

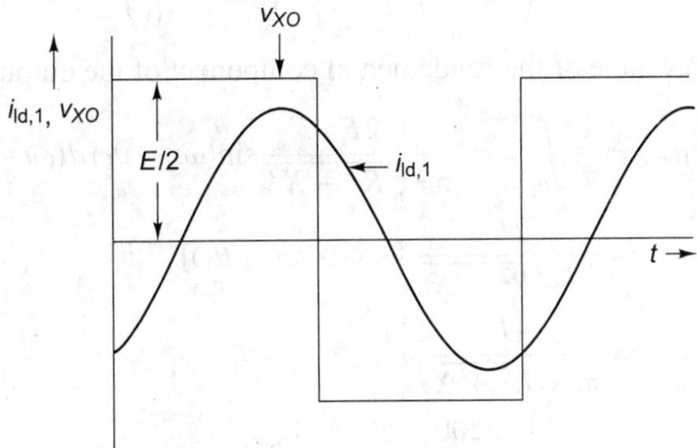

Fig. 5.35

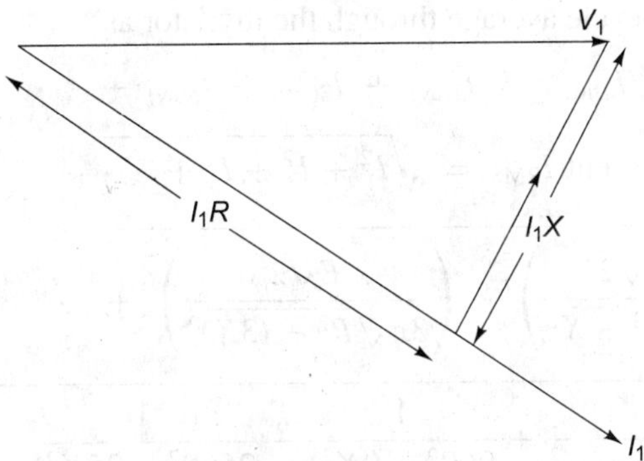

Fig. 5.36

where

$$\theta_n = \tan^{-1}\left(\frac{nX}{R}\right)$$

Hence the RMS value of the fundamental current

$$I_1 = \frac{\sqrt{2}E}{\pi\sqrt{R^2+X^2}} = \frac{\sqrt{2}\times 200}{\pi\sqrt{10^2+8^2}} = \frac{90}{12.8} = 7.03 \text{ A}$$

From Fig. 5.36, the fundamental output power is

$$P_1 = V_1 I_1 \cos\theta_1 = I_1^2 R$$

$$= \left(\frac{\sqrt{2}E}{\pi\sqrt{R^2+X^2}}\right)^2 \times 10$$

$$= \left(\frac{\sqrt{2}\times 200}{\pi\sqrt{10^2+8^2}}\right)^2 \times 10$$

$$= 494 \text{ W}$$

$$\cos\theta_1 = \cos\left(\tan^{-1}\frac{X}{R}\right) = \cos\left(\tan^{-1}\frac{8}{10}\right) = 0.78$$

The half-cycle average of the fundamental component of the output current is

$$I_{1(\text{av})} = \frac{1}{\pi}\int_{\theta_1}^{(\pi+\theta_1)} \frac{2E}{\pi^2\sqrt{R^2+X^2}}\sin(\omega t-\theta_1)\,d(\omega t)$$

$$= \frac{2E}{\pi^2\sqrt{R^2+X^2}}[-\cos(\omega t-\theta_1)]_{\theta_1}^{\pi+\theta_1}$$

$$= \frac{4E}{\pi^2\sqrt{R^2+X^2}}$$

$$= \frac{4\times 200}{\pi^2\times\sqrt{10^2+8^2}} = 6.3 \text{ A}$$

Similarly the half-cycle average of other components are calculated and added to give the total half-cycle average through the thyristor as

$$I_{\text{Th(av)}} = I_{1(\text{av})} + I_{3(\text{av})} + I_{5(\text{av})} + \cdots$$

$$\text{Total RMS load current } I_{\text{RMS}} = \sqrt{I_1^2 + I_3^2 + I_5^2 + \cdots}$$

$$= \sqrt{\left(\frac{E\sqrt{2}}{\pi\sqrt{R^2+X^2}}\right)^2 + \left(\frac{E\sqrt{2}}{3\pi\sqrt{R^2+(3X)^2}}\right)^2 + \cdots}$$

$$= \frac{E\sqrt{2}}{\pi}\sqrt{\frac{1}{R^2+X^2} + \frac{1}{9(R^2+9X^2)} + \frac{1}{25(R^2+25X^2)} + \cdots}$$

$$= \frac{200\sqrt{2}}{\pi}\sqrt{\frac{1}{10^2+8^2} + \frac{1}{9(10^2+9\times 8^2)} + \frac{1}{25(10^2+25\times 8^2)} + \cdots}$$

$$= \frac{200\sqrt{2}}{\pi}\sqrt{0.0061 + 0.00016 + \text{very small terms}}$$

That is,

$$I_{\text{RMS}} = \frac{200\sqrt{2}}{\pi}\sqrt{0.00626} = 7.1 \text{ A}$$

The RMS value of current through each thyristor is

$$\frac{\text{Total RMS value of load current}}{\sqrt{2}} = \frac{7.1}{\sqrt{2}} = 5.0 \text{ A}$$

The peak current through each thyristor is

$$\sqrt{2}(\text{RMS value of current through Th}_1) = \sqrt{2} \times 5 = 7.1 \text{ A}$$

The RMS value of the voltage at the load is

$$\frac{E}{2} = 100 \text{ V}$$

$$\text{Distortion factor } (\mu) = \frac{I_1}{I} = \frac{7.03}{7.1} = 0.99$$

$$\text{Displacement factor} = \cos\phi_1 = \cos\left[\tan^{-1}\left(\frac{8}{10}\right)\right] = 0.78$$

$$\text{Input power factor} = \mu \cos\phi_1 = 0.99 \times 0.78 = 0.77$$

3. The three-phase, six-step inverter of Fig. 5.3(a) has a star-connected resistive load with $R = 8\ \Omega$ and $L = 20$ mH. The inverter frequency is 50 Hz and the dc input voltage E is 200 V. Derive the expression for $i(t)$ as a Fourier series. Determine the (a) RMS line voltage V_{RY}, (b) RMS phase voltage V_{RN}, (c) RMS line voltage at fundamental frequency $V_{RY(1)}$, (d) RMS phase voltage at fundamental frequency, (e) load power factor, (f) RMS line current, and (g) average and RMS thyristor currents.

Solution

For a star-connected load, the phase voltage is $V_{RN} = V_{RY}/3$. From the Fourier series expression for V_{RY} of Eqn (5.23), the Fourier series for $i_R(t)$ for an R-L load is given by

$$i_R(t) = \frac{2E\sqrt{3}}{\sqrt{3}\pi}\left[\frac{\sin(\omega t - \theta_1)}{\sqrt{R^2 + (\omega L)^2}} - \frac{\sin(5\omega t - \theta_5)}{5\sqrt{R^2 + (5\omega L)^2}} - \frac{\sin(7\omega t - \theta_7)}{7\sqrt{R^2 + (7\omega L)^2}} + \frac{\sin(11\omega t - \theta_{11})}{11\sqrt{R^2 + (11\omega L)^2}} - \cdots\right]$$

Here,

$$|Z_{Ln}| = \sqrt{R^2 + (n\omega L)^2}$$

and

$$\angle Z_{Ln} = \tan^{-1}\left(\frac{n\omega L}{R}\right)$$

$$\omega = 2\pi \times 50 = 314 \text{ and } \omega L = \frac{314 \times 20}{1000} = 6.28\ \Omega$$

Thus,

$$|Z_{Ln}| = \sqrt{8^2 + (6.28n)^2} \text{ and } \theta = \tan^{-1}\left(\frac{6.28n}{8}\right)$$

Using the above equation for $i_R(t)$, the expression for the instantaneous line (or phase) current is given by

$$i_R(t) = 12.52 \sin(\omega t - 38°) - 0.785 \sin(5\omega t - 75.7°)$$
$$-0.41 \sin(7\omega t - 79.7°) + 0.17 \sin(11\omega t - 83.4°) + \cdots$$

(a) RMS line voltage

$$V_{RY(\text{RMS})} = 200\sqrt{\frac{2}{3}} = 163.3 \text{ V}$$

(b) RMS phase voltage

$$\frac{V_{RY(\text{RMS})}}{\sqrt{3}} = \frac{163.3}{\sqrt{3}} = 94.27 \text{ V}$$

(c) Fundamental component of RMS line voltage

$$V_{RY(1)(\text{RMS})} = \frac{2\sqrt{3}E}{\pi\sqrt{2}} = 156 \text{ V}$$

(d) Fundamental component of RMS phase voltage

$$V_{RN(1)(\text{RMS})} = \frac{V_{RY(1)(\text{RMS})}}{\sqrt{3}} = 90 \text{ V}$$

(e) RMS line current

$$I_{\text{line}} = \frac{\sqrt{12.52^2 + 0.785^2 + 0.41^2 + 0.17^2}}{\sqrt{2}} = 8.88 \text{ A}$$

The line current is shared by two thyristors. Hence the RMS thyristor current

$$I_{\text{Th(RMS)}} = \frac{8.88}{\sqrt{2}} = 6.28 \text{ A}$$

Load power

$$P_{\text{ld}} = 3I_L^2 R = 3 \times 8.88^2 \times 8 = 1893 \text{ W}$$

Average supply current

$$I_s = \frac{1893}{200} = 9.47 \text{ A}$$

In a three-phase circuit each phase conducts for a period of $2\pi/3$. Hence the average thyristor current is

$$\frac{I_s}{3} = \frac{9.47}{3} = 3.16 \text{ A}$$

4. The single-phase bridge inverter of Fig. 5.22 uses square wave PWM with two pulses per half-cycle. The modulation is adjusted to get a pulse width of $\pi/4$ and a distance of $\pi/4$ between the pulses. The load is inductive with $R = 8\ \Omega$ and $L = 20$ mH. The dc input voltage E is 180 V. (a) Determine the RMS output voltage and (b) express the instantaneous load current $i_{\text{ld}}(t)$ as a Fourier series with a frequency of 50 Hz for the inverter.

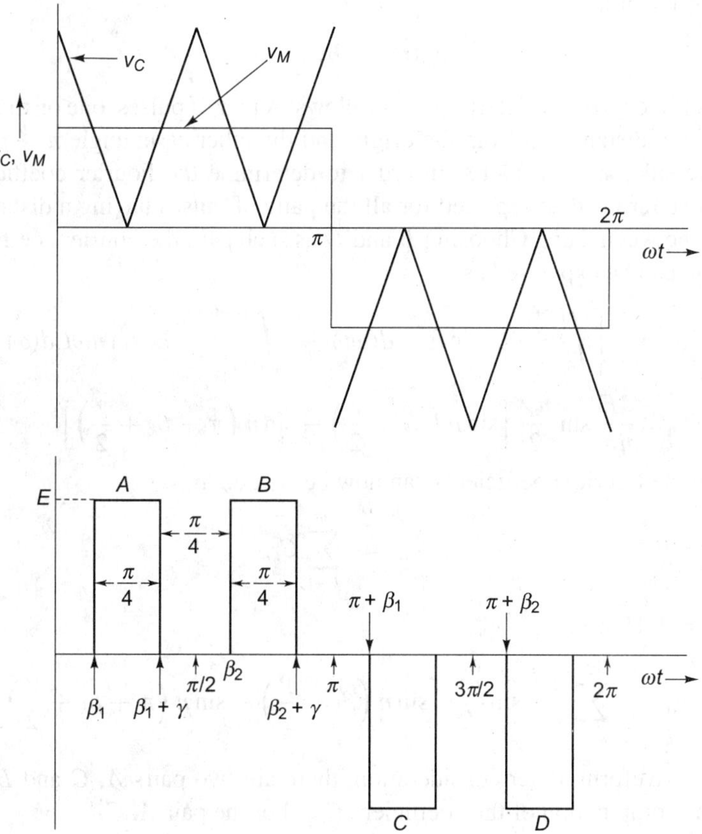

Fig. 5.37

Solution

The output waveform is shown in Fig. 5.37 with the pulse width γ, and q pulses in each half-cycle. The RMS load voltage is determined as

$$V_{\text{ld(RMS)}} = \left[\frac{2q}{2\pi}\int_{\beta_k}^{\beta_k+\gamma} E^2 d(\omega t)\right]^{1/2} = E\sqrt{\frac{q\gamma}{\pi}}$$

Here $q = 2$ and $\gamma = \pi/4$. Hence,

$$V_{\text{ld(RMS)}} = 180\sqrt{\frac{2\pi}{4\pi}} = 127.3 \text{ V}$$

The Fourier series expression for the PWM voltage waveform is

$$V_{\mathrm{ld(RMS)}} = \sum_{n=1}^{\infty}(a_n \cos n\omega t + b_n \sin n\omega t)$$

Because of the odd symmetry of the waveform, the coefficients a_n $(n = 1, 2, \dots)$ will be zero; also the odd (sine) harmonics with even coefficients become zero. Thus v_{ld} becomes

$$v_{\mathrm{ld}}(t) = b_n \sin n\omega t$$

The coefficients b_n are determined as follows. A pair of pulses, one on the positive side with a distance β_k from the origin and the other at an angle $\pi + \beta_k$ on the negative side, are considered in order to determine the Fourier coefficient C_k. The procedure is then repeated for all the pairs of pulses having a distance of π radians between them. Choosing B and D as such pair, the Fourier coefficient C_k for them can be expressed as

$$C_{kn} = \frac{1}{\pi}\left[\int_{\beta_k}^{\beta_k+\gamma} E \sin n\omega t \, d(\omega t) - \int_{\pi+\beta_k}^{\pi+\beta_k+\gamma} E \sin n\omega t \, d(\omega t)\right]$$

$$= \frac{2E}{n\pi}\sin\frac{n\gamma}{2}\left[\sin n\left(\beta_k + \frac{\gamma}{2}\right) - \sin n\left(\pi + \beta_k + \frac{\gamma}{2}\right)\right]$$

Each of the Fourier coefficients can now be written as

$$b_n = \sum_{k=1}^{q} C_{kn}$$

Here $q = 2$. Hence,

$$b_n = \sum_{k=1}^{2}\frac{2E}{n\pi}\sin\frac{n\gamma}{2}\left[\sin n\left(\beta_k + \frac{\gamma}{2}\right) - \sin n\left(\pi + \beta_k + \frac{\gamma}{2}\right)\right]$$

For the waveform under consideration, there are two pairs A, C and B, D that need the computation of the coefficient C_{kn}. For the pair A, C,

$$\beta_k = \frac{\pi}{8}, \quad \gamma = \frac{\pi}{4}, \quad \text{and } \pi + \beta_k = \frac{9\pi}{8}$$

With these values,

$$C_{1n} = \frac{2E}{n\pi}\sin\left(\frac{n\pi}{8}\right)\left\{\sin\left[n\left(\frac{\pi}{8} + \frac{\pi}{8}\right)\right] - \sin\left[n\left(\frac{9\pi}{8} + \frac{\pi}{8}\right)\right]\right\}$$

$$= \frac{2E}{n\pi}\sin\left(\frac{n\pi}{8}\right)\left[\sin\frac{n\pi}{4} - \sin\frac{5n\pi}{4}\right]$$

For the second pair B, D,

$$\beta_k = \frac{5\pi}{8}, \quad \gamma = \frac{\pi}{4}, \quad \text{and } \pi + \beta_k = \frac{13\pi}{8}$$

Hence,

$$C_{2n} = \frac{2E}{n\pi} \sin \frac{n\pi}{8} \left\{ \sin\left[n\left(\frac{5\pi}{8} + \frac{\pi}{8}\right)\right] - \sin\left[n\left(\frac{13\pi}{8} + \frac{\pi}{8}\right)\right]\right\}$$

$$= \frac{2E}{n\pi} \sin \frac{n\pi}{8} \left[\sin \frac{3n\pi}{4} - \sin \frac{7n\pi}{4} \right]$$

Now $b_n = C_{1n} + C_{2n}$. Thus,

$$b_n = \frac{2E}{n\pi} \sin \frac{n\pi}{8} \left[\sin \frac{n\pi}{4} - \sin \frac{5n\pi}{5} + \sin \frac{3n\pi}{4} - \sin \frac{7n\pi}{4} \right]$$

The load voltage $v_{ld}(t)$ now becomes

$$v_{ld}(t) = \sum_{n=1,3,5}^{\infty} \frac{2E}{n\pi} \sin \frac{n\pi}{8} \left[\sin \frac{n\pi}{4} - \sin \frac{5n\pi}{4} + \sin \frac{3n\pi}{4} - \sin \frac{7n\pi}{4} \right]$$

The first four coefficients b_1, b_3, b_5, and b_7 are worked out as follows:

$$b_1 = 2 \times \frac{180}{\pi} \sin \frac{\pi}{8} \left[\sin \frac{\pi}{4} - \sin \frac{5\pi}{4} + \sin \frac{3\pi}{4} - \sin \frac{7\pi}{4} \right] = 123.8 \text{ V}$$

$$b_3 = 2 \times \frac{180}{3\pi} \sin \frac{3\pi}{8} \left[\sin \frac{3\pi}{4} - \sin \frac{15\pi}{4} + \sin \frac{9\pi}{4} - \sin \frac{21\pi}{4} \right] = 99.8 \text{ V}$$

$$b_5 = 2 \times \frac{180}{5\pi} \sin \frac{5\pi}{8} \left[\sin \frac{5\pi}{4} - \sin \frac{25\pi}{4} + \sin \frac{15\pi}{4} - \sin \frac{53\pi}{4} \right] = -59.8 \text{ V}$$

$$b_7 = 2 \times \frac{180}{7\pi} \sin \frac{7\pi}{8} \left[\sin \frac{7\pi}{4} - \sin \frac{35\pi}{4} + \sin \frac{21\pi}{4} - \sin \frac{49\pi}{4} \right] = -17.7 \text{ V}$$

v_{ld} can now be obtained as

$$v_{ld}(t) = 123.8 \sin \omega t + 99.8 \sin 3\omega t - 59.8 \sin 5\omega t - 17.7 \sin 7\omega t$$

The Fourier coefficients for the load current can now be obtained as

$$i_{ld}(t) = \sum_{n=1}^{\infty} \frac{b_n}{Z_{Ln}} \sin(n\omega t - \theta_n)$$

where

$$Z_{Ln} = \sqrt{R^2 + (n\omega L)^2}$$

and

$$\theta_n = \tan^{-1}\left(\frac{n\omega L}{R}\right)$$

Here $R = 8\ \Omega$ and $L = 20$ mH. The harmonic impedances Z_{Ln} ($n = 1, 3, 5, 7$) and the corresponding angles are now obtained as

$$Z_{L1} = \sqrt{8^2 + \left(314 \times \frac{20}{1000}\right)^2} = \sqrt{8^2 + (6.28)^2} = 10.17$$

$$\theta_1 = 38°$$

$$Z_{L3} = \sqrt{8^2 + (3 \times 6.28)^2} = 20.4$$

$$\theta_3 = 67°$$

$$Z_{L5} = \sqrt{8^2 + (5 \times 6.28)^2} = 32.4$$

$$\theta_5 = 75.7°$$

$$Z_{L7} = \sqrt{8^2 + (7 \times 6.28)^2} = 44.68$$

$$\theta_7 = 79.68°$$

The Fourier series for the current can now be written as

$$i_{\text{ld}}(t) = \frac{123.8}{10.17}\sin(\omega t - 38°) + \frac{99.8}{20.4}\sin(3\omega t - 67°) - \frac{59.8}{32.4}\sin(5\omega t - 75.7°) - \frac{17.7}{44.68}\sin(7\omega t - 79.68°)$$

$$= 12.2\sin(\omega t - 38°) + 4.9\sin(3\omega t - 67°) - 1.85\sin(5\omega t - 75.7°) - 0.4\sin(7\omega t - 79.68°) + \cdots$$

5. For the input circuit commutated inverter of Fig. 5.16(a), the following data are given: source voltage $E = 200$ V, load current $I_{\text{ld}} = 20$ A, inductance $L_{c1} = L_{c2} = L = 10$ mH, and t_{OFF} of thyristor = 6.7 μs. Determine the value of C required to successfully commutate the thyristors of the inverter.

Solution

The circuit turn-off time t_c is given by Eqn (5.80) as

$$t_c = \frac{1}{\omega}\left[\cos^{-1}\left\{\frac{-1}{2\sqrt{1 + \omega^2 L^2 I_{\text{ld}}^2/4E^2}}\right\} - \phi\right]$$

where

$$\tan\phi = \frac{\omega L I_{\text{ld}}}{2E}$$

Using the above two equations, t_c can be obtained in a simplified form as

$$t_c = \frac{1}{\omega}\left[\cos^{-1}\left\{\frac{-\cos\phi}{2}\right\} - \phi\right]$$

Assuming $\omega = 12{,}200$,

$$\tan\phi = \frac{12{,}200 \times 10 \times 10^{-3} \times 20}{2 \times 200}$$

This gives $\tan\phi = 6.1$ and $\phi = 80.69°$ or 1.4083 rad. Also, $\cos\phi = 0.1618$. t_c can be computed as

$$t_c = \frac{1}{12{,}200}\left\{\cos^{-1}\left(-\frac{0.1618}{2}\right) - 1.4083\right\}$$

That is, $t_c = 6.68$ μs. As this value of t_c approximately tallies with the given t_{OFF} of 6.7 μs, C can be obtained from ω as follows:

$$\sqrt{LC} = \frac{1}{12{,}200}$$

Substitution of L gives

$$C = \frac{10^6}{(12.2)^2 \times 10^6} \times \frac{1}{10 \times 10^{-3}} = 0.67\ \mu\text{F}$$

Exercises

Multiple Choice Questions

1. A motor fed by an LCI has to be operated at a ________ power factor.
   (a) unity (b) lagging
   (c) leading (d) zero
2. In a single-phase voltage source bridge inverter, the shape of the load current depends on the ________.
   (a) source voltage
   (b) duration of conduction of thyristors
   (c) load impedance
   (d) duration of conduction of the feedback diodes
3. In a voltage source inverter, the purpose of the diodes across the thyristors is to ________.
   (a) help the commutation of the thyristors
   (b) see that excessive current does not pass through the thyristors
   (c) protect the thyristors from excessive voltages
   (d) feed the energy back from the load to the source under negative power conditions
4. For an input circuit commutated inverter, the voltage v_c of the capacitor C of the input bridge will charge to a maximum voltage, and this will be ________.
   (a) equal to E (b) less than E
   (c) greater than E (d) much greater than E
5. Considering an input circuit commutated inverter, a voltage equal to ________ will be impressed across the upper inductor L_{c1} in the process of commutation of the thyristors Th_1 and Th_3.
   (a) $1.5E$ (b) $2E$
   (c) E (d) $E/2$

6. For successful commutation of the thyristors of a single-phase McMurray type of inverter, it is necessary that the condition ______ be satisfied, where I_{cm} is the peak of the oscillatory current.
   (a) $I_{ld} > I_{cm}$
   (b) $I_{ld} < I_{cm}$
   (c) $I_{ld} \geq I_{cm}$
   (d) $I_{ld} = I_{cm}$
7. High switching losses occur in a ______ inverter as compared to other inverters.
   (a) three-phase, six-step
   (b) square wave PWM
   (c) current source
   (d) 120°-mode voltage source
8. In a sinusoidal PWM inverter the harmonics are of the form ______, where p is the carrier ratio.
   (a) $kp \pm l$ (where k and l are integers)
   (b) $kp \pm l$ (where k is always even and l is always odd)
   (c) $kp \pm l$ (where k is always odd and l is always even)
   (d) $kp \pm l$ (where k is odd and l is even or k is even and l is odd)
9. Reduction of harmonic content in the output of a system is possible in a ______ inverter.
   (a) current source
   (b) voltage source
   (c) square wave PWM
   (d) load commutated
10. In a McMurray bridge type of inverter, the reverse voltage applied across the thyristor during commutation is equal to ______.
   (a) the supply voltage E
   (b) half the supply voltage E
   (c) 1.5 times the supply voltage E
   (d) none of the above
11. As the load current decreases, the circuit turn-off time of the thyristor in a McMurray bridge inverter ______.
   (a) slightly increases
   (b) slightly decreases
   (c) remains unchanged
   (d) drastically decreases
12. The circuit turn-off time of the thyristor in a McMurray–Bedford inverter is ______ of an LC oscillation.
   (a) more than a half-cycle
   (b) less than a quarter-cycle
   (c) more than a quarter-cycle
   (d) less than a half-cycle
13. For a McMurray–Bedford inverter, in general, the output frequency with an $R + L$ load will be ______ that with a purely resistive load, the load resistance remaining the same in both cases.
   (a) greater than
   (b) lesser than
   (c) equal to that of
   (d) negligible when compared to
14. The thyristor turn-off in a McMurray bridge inverter can be categorized as ______ turn-off.
   (a) an auxiliary resonant
   (b) a series resonant
   (c) an impulse
   (d) a natural
15. The thyristor turn-off in a McMurray–Bedford inverter can be categorized as ______ turn-off.

(a) an auxiliary resonant (b) a series resonant
(c) an impulse (d) a natural

16. The performance features of a McMurray–Bedford type of inverter are ______ those of the McMurray bridge inverter.

    (a) slightly better than (b) inferior to
    (c) almost of the same order as (d) much better than

17. The ______ inverter is best suited for use as a PWM type of VSI.

    (a) input circuit commutated (b) McMurray
    (c) CSI (d) McMurray–Bedford

18. The major advantage of an input circuit commutated inverter is that ______.

    (a) it is suitable as a PWM inverter
    (b) it uses minimum number of components
    (c) it permits multiple thyristor turn-off
    (d) its switching losses are very low

19. If the size of the commutation capacitor in a three-phase CSI is increased, then it has the disadvantage of ______.

    (a) increased switching losses
    (b) decreased overall efficiency
    (c) increased harmonic content in the output
    (d) reducing the output frequency at which the bypass effect occurs

20. A CSI with a motor load is capable of operating ______.

    (a) only in the first quadrant
    (b) in all the four quadrants
    (c) only in the first and fourth quadrants
    (d) only in the first and second quadrants

21. Increasing the size of the filter inductor in the dc link of a CSI leads to ______.

    (a) stability problems (b) large voltage spikes
    (c) both of the above problems (d) none of the above problems

22. One of the drawbacks of an LCI is that forced commutation is needed for ______.

    (a) speeds greater than 50% of the normal speed
    (b) speeds nearing 90% of the normal speed
    (c) normal speeds
    (d) speeds lower than 10% of the normal speed

23. In a load commutated inverter the current of the connected motor is adjusted by ______ whereas the inverter frequency is controlled by ______.

    (a) gating the inverter thyristors in synchronization with machine speed, variation of α of the line commutated converter
    (b) gating the inverter thyristors in synchronization with the ac line frequency, variation of α of the line commutated converter
    (c) variation of α of the line commutated converter, gating of the inverter thyristors in synchronization with the ac line frequency
    (d) variation of α of the line commutated converter, gating of the inverter thyristors in synchronization with machine speed.

24. The phenomenon of commutation overlap occurs in the ________.
    (a) input circuit commutated inverter
    (b) CSI
    (c) McMurray inverter
    (d) LCI

True or False

Indicate whether the following statements are *true* or *false*. Briefly justify your choice.

1. In a voltage source inverter, the shape of the load current depends on the load parameters.
2. For successful commutation of the thyristors of the McMurray type of inverter, the load current must always be greater than the peak of the *LC*-oscillatory current I_{cm}.
3. The lower order harmonic content of a three-phase, six-step VSI is greater than that of a three-phase sine wave PWM inverter.
4. In a PWM inverter the modulating index is defined as the ratio of the carrier wave frequency to the reference wave frequency.
5. Torque pulsations occur at low speeds in an induction motor fed by a three-phase bridge inverter but not in a motor fed by a square wave PWM.
6. Torque pulsations at low speeds occur in an induction motor fed by a square wave PWM but not in a motor fed by a sine wave PWM.
7. The sequence of firing of thyristors in a three-phase 120° VSI is the same as that in a three-phase, six-step VSI.
8. The reverse voltage applied during commutation across the thyristor in a McMurray–Bedford inverter is more than that in a McMurray inverter.
9. The turn-off in a McMurray inverter can be categorized as an impulse type.
10. The turn-off in a McMurray–Bedford inverter can be categorized as a series resonant type of turn-off.
11. The McMurray–Bedford inverter has better performance features and higher efficiency than the McMurray inverter.
12. The shape of the load voltage of a McMurray type of inverter depends on the type of load it caters to.
13. The shape of the load current of a McMurray type of inverter depends on the type of load connected to it.
14. The shape of the load voltage in a PWM inverter depends on the type of load connected to it.
15. The shape of the load current in a CSI greatly depends on the type of load connected to it.
16. Increasing the size of the commutating capacitors in a three-phase CSI causes a decrease in the output frequency.
17. The CSI has the inherent capability of operating in the first and second quadrants but not in the third and fourth quadrants.

18. The CSI is a multiple-motor drive whereas the six-step VSI is a single-motor drive.
19. Regenerative braking can be done without additional components in a CSI whereas it involves additional circuitry in the case of a six-step VSI.
20. In a 120°-mode VSI the fundamental voltage applied to the load is independent of the shaft loading.
21. In a three-phase CSI increasing the size of the filter inductor in the dc link helps in riding through fault conditions but leads to high dc voltage surges.
22. The low-order harmonic content in a CSI is low and hence the problem of torque pulsations does not arise.
23. The transient response of a square wave PWM inverter is superior to that of a six-step inverter.
24. The performance under constant V/f operation of a PWM inverter is inferior to that of a six-step VSI.
25. One of the drawbacks of an LCI is that external commutation is needed for high machine speeds.
26. Regenerative braking of an LCI is possible but needs additional circuitry, thus involving extra cost.

Short Answer Questions

1. Draw the circuit of a McMurray–Bedford inverter and describe its operation. Compare and contrast its features with that of a McMurray inverter.
2. Derive an expression for $i(t)$ for the parallel capacitor-quenched inverter.
3. List out the various assumptions made in the case of a parallel capacitor inverter and give arguments for their justification.
4. Give the advantages of VSIs over other types of inverters.
5. What are the advantages of the three-phase, six-step VSI?
6. With the help of waveforms, explain the part played by the diodes in a three-phase, six-step VSI.
7. Explain how the commutation of the main thyristor is achieved in a McMurray bridge inverter.
8. How does the auxiliary thyristor Th_{10} in a McMurray bridge inverter turn off and become ready for the next cycle?
9. Explain why the thyristor firing signals in a single-phase, full-bridge inverter have to be kept active throughout the half-cycle of the inverter.
10. Explain how the commutation of thyristors takes places in a single-phase bridge inverter (a) with a resistive load and (b) with an $R + L$ load.
11. Explain why energy flows from load to source for a fraction of the period in a single-phase bridge inverter with an $R + L$ load.
12. Explain with the help of waveforms the operation of (i) a single-phase, half-bridge inverter and (ii) a single-phase, full-bridge inverter.
13. How can the current and voltage ratings be decided for single-phase and three-phase VSIs?

14. What are the merits of the three-phase, six-step inverter?
15. Give the assumptions made before the analysis of a single-phase McMurray bridge inverter?
16. Explain how an alternating load current waveform is obtained for a single-phase McMurray inverter.
17. Explain the sequence of events that take place in a single-phase McMurray bridge inverter when turning on Th_{10} does not turn off Th_1. Suggest an alternative circuit that overcomes this drawback.
18. What conditions are necessary on the inductor pairs L_1, L_2 and L_3, L_4 in a single-phase McMurray–Bedford type of inverter for ensuring successful commutation of the thyristors?
19. Show the circuit turn-off time on the voltage waveform of a single-phase McMurray–Bedford type of inverter.
20. What is the current capability of the diode D_2 in a single-phase McMurray–Bedford type of inverter?
21. Draw the waveform of current through Th_1 in a McMurray–Bedford type of inverter and justify its shape.
22. Explain why the voltage through the capacitor C_1 in a McMurray–Bedford type of inverter exceeds the supply voltage E.
23. Explain how the operation of the McMurray bridge inverter gets modified if the diodes D_2 and D_3 were absent.
24. Explain the part played by the diodes in a 120°-mode VSI.
25. Explain the sequence of operations in an input circuit commutated inverter for commutating a set of thyristors, two of which are connected to the positive rail.
26. What are the advantages of the three-phase, input circuit commutated inverter over the three-phase VSI?
27. In an input circuit commutated inverter, explain how all the thyristors connected to the positive rail of the dc supply are turned off at the same time.
28. Justify the necessity of voltage control in a typical VSI.
29. Explain how the voltage control of a three-phase VSI is achieved (i) by controlling the dc link voltage, (ii) by using multiple inverters, and (iii) by using pulse width modulation.
30. Explain the principle of pulse width modulation with the help of waveforms.
31. Comment on the harmonic content of single and multiple pulse width modulated voltage waveforms.
32. For what operating conditions does the output current of a single PWM inverter become discontinuous?
33. What are the advantages of a PWM inverter as compared to a six-step VSI?
34. What are the merits and demerits of a six-step voltage source inverter?
35. Compare and contrast the features of square wave and sine wave PWMs.
36. Explain the technique of reduction of low-order harmonics using PWM.
37. How is constant torque operation achieved with a PWM inverter?

38. With the help of block diagrams, explain how regenerative braking can be achieved with the help of PWM inverters.
39. What are the advantages of CSIs as compared to VSIs?
40. Give the sequence of operations of a single-phase ASCI.
41. Give the sequence of operations of a three-phase bridge CSI.
42. How is regenerative braking of an ac motor achieved when it is fed by a three-phase CSI?
43. What are the disadvantages of the application of CSIs to drives?
44. Explain how commutation overlap occurs for upper and lower group thyristors in a three-phase CSI.
45. List out the merits of an LCI over a conventional six-step inverter.
46. What are the advantages and disadvantages of using a large dc link filter inductor in a three-phase CSI?
47. What are the functions of the dc link inductor in an LCI?
48. What are the applications of load commutated inverters?

Problems

1. The following data pertain to an input circuit commutated inverter: $E = 200$ V, $I_{ld} = 22$ A, $L_{c1} = 12$ mH, $C = 0.55$ μF. Determine the turn-off time of the thyristors.

   Ans. $t_{OFF} = 5$ μs

2. The following data are given for an input circuit commutated inverter: t_{OFF} of thyristors = 7.3 μs, $L_{c1} = 11$ mH, $I_{ld} = 19$ A, $E = 200$ V. Determine the value of C. [Hint: Use the trial and error method.]

   Ans. $C = 0.7$ μF

3. In the parallel capacitor-quenched inverter of Fig. 5.2(a), the data are $R_{ld} = 30\ \Omega$, turn ratio $(N_S/N_P) = 2$, t_{OFF} of the thyristor = 28 μs, $E = 150$ V, and $di/dt = 3$ A/μs. Determine the values of L and C.

   Ans. $L = 0.5$ mH, $C = 2.0$ μF

4. The following data pertain to the parallel capacitor-quenched inverter of Fig. 5.2(a): $R_{ld} = 20\ \Omega$, turn ratio $(N_S/N_P) = 2$, $C = 10$ μF, $E = 110$ V, and $(di/dt)_{rated} = 2.8$ A. Determine the values of L and t_{OFF}.

   Ans. $L = 0.8$ mH, $t_{OFF} = 100$ μs

5. The following data pertain to a single-phase McMurray bridge inverter. dc supply voltage $E = 160$ V, load resistance $R_{ld} = 10\ \Omega$, turn-off time $t_{OFF} = 40$ μs. Estimate the values of L and C.

   Ans. $C = 3.9$ μF, $L = 0.15$ mH

6. The following data pertain to an input circuit commutated inverter: $E = 210$ V, $C = 0.422$ μF, resistance R_{ld} of one phase winding is 10 Ω, $t_{OFF} = 6.3$ μs. Determine the value of L for successful turn-off of the thyristor. [Hint: At any instant two out of three phase windings are in parallel, and this parallel combination is in series with the third phase winding. The resulting circuit is connected across the battery.]

   Ans. $L = 10$ mH

7. The following data pertain to a single-phase McMurray bridge inverter: $E = 170$ V, turn-off time is 45 μs, $C = 4$ μF, load current is 0.7 times the peak

oscillatory current. Determine the approximate values of R_{ld} and L. [Hint: See Fig. 5.10.]

Ans. $R_{ld} = 10.2\ \Omega$, $L = 0.205$ mH

8. The following data pertain to a single-phase McMurray bridge inverter: $E = 120$ V, $L = 0.3$ mH, $C = 5\ \mu$F, $R_{ld} = 10\ \Omega$. Find the turn-off time of the thyristors and the ratio of the load current to the peak oscillatory current. [Hint: Use Fig. 5.10 to arrive at the values.]

Ans. $t_{OFF} = 53\ \mu$s, $I_{ld}/I_{cm} = 0.775$

9. The following data are given for a single-phase McMurray bridge inverter: $E = 130$ V, $C = 6\ \mu$F, $L = 0.35$ mH. The ratio I_{ld}/I_{cm} is 0.55. Determine the load resistance and the turn-off time of the thyristors. [Hint: Use Fig. 5.10 to arrive at the values.]

Ans. $R_{ld} = 13.9\ \Omega$, $t_{OFF} = 90.5\ \mu$s

10. A McMurray–Bedford inverter has the following data: $E = 140$ V, $t_{OFF} = 20\ \mu$s, load resistance is 10 Ω. Assume that $I_{cm} = 1.6\ I_{ld}$ and determine the values of the commutating elements.

Ans. $L = 1.29$ mH, $C = 1.26\ \mu$F

11. A McMurray–Bedford inverter has the following data: $E = 150$ V, $t_{OFF} = 25\ \mu$s, load current with an inductive load is 30 A. With $I_{cm} = 1.55 I_{ld}$, determine the values of the commutating elements.

Ans. $L = 0.442$ mH, $C = 6.2\ \mu$F

12. For a McMurray–Bedford inverter feeding a resistive load, the values of the commutating elements are $L = 0.6$ mH and $C = 11\ \mu$F. The dc source voltage $E = 180$ V. If the load current is 0.7 times the peak oscillatory current (I_{cm}), determine the turn-off time of the thyristor and the load resistance.

Ans. $t_{OFF} = 35\ \mu$s, $R_{ld} = 3.77\ \Omega$

13. For a McMurray–Bedford inverter feeding an inductive load, the values of the commutating elements are $L = 0.8$ mH and $C = 12\ \mu$F. The dc source voltage $E = 260$ V. If $I_{cm} = 1.3 I_{ld}$, determine t_{OFF} and the load current.

Ans. $t_{OFF} = 36\ \mu$s, $I_{ld} = 76.75$ A

14. For an input circuit commutated inverter, the following data are given: $E = 190$ V, $C = 0.8\ \mu$F, $L = 12$ mH, $t_{OFF} = 20.9\ \mu$s. Determine the value of I_{ld}.

Ans. $I_{ld} = 7$ A (approx)

15. The half-bridge inverter of Fig. 5.6(a) feeds a resistive load and has the following data: $E = 220$ V, $R_{ld} = 15\ \Omega$. Determine the (a) RMS output voltage at the fundamental frequency, (b) output power, (c) half-cycle average RMS, and peak currents of each thyristor, (d) distortion factor, and (e) input power factor.

Ans. (a) 99 V; (b) 654 W; (c) 5.94 A, 4.67 A, 9.33 A; (d) 0.9; (e) 0.81

16. Repeat the above calculations for the half-bridge inverter of Fig. 5.6(a) with $R_{ld} = 15\ \Omega$ and an inductive reactance of 12 Ω.

Ans. (a) 99 V; (b) 399 W; (c) 4.89 A, 3.69 A, 5.22 A; (d) 0.988; (e) 0.693

17. The three-phase, six-step inverter of Fig. 5.8(a) supplies a load of $R = 8\ \Omega$. The other data are $E = 200$ V and fundamental frequency = 50 Hz. Determine the

(a) RMS, average, and peak values of the fundamental current of the thyristors, (b) distortion factor, and (c) input power factor. [Hint: Each thyristor carries the line current.]

Ans. (a) 11.25 A, 5.85 A, 27.57 A; (b) 0.955; (c) 0.827

18. The three-phase, six-step inverter of Fig. 5.8(a) supplies a load of $R = 10\ \Omega$ and $L = 25$ mH. The inverter frequency is 50 Hz and the dc input voltage is 220 V. (a) Derive the expression for $i(t)$ as a Fourier series. Also determine the (b) RMS line voltage V_{RY}, (c) RMS phase voltage V_{RN}, (d) RMS line voltage at the fundamental frequency, $V_{RY(1)}$, (e) RMS phase voltage at the fundamental frequency $V_{RN(1)}$, (f) input power factor, (g) RMS line current, and (h) average and RMS thyristor currents.

Ans. (a) $i_R(t) = 11.02 \sin(\omega t - 38°) - 0.69 \sin(5\omega t - 75.7°) - 0.358 \times \sin(7\omega t - 79.7°) + 0.146 \sin(11\omega t - 83.4°) + \cdots$; (b) 179.63 V; (c) 103.71 V; (d) 171.53 V; (e) 99.03 V; (f) 0.749; (g) 7.81 A; (h) 2.77 A, 5.52 A

19. The single-phase, full-bridge inverter of Fig. 5.6(b) supplies a load of resistance 10 Ω and inductive reactance 12 Ω. If the supply voltage $E = 200$ V, determine the (a) RMS fundamental component of the output voltage, (b) output power, (c) half-cycle average, RMS, and peak currents of the thyristors, (d) input power factor, and (e) distortion factor. [Hint: See Worked Example 1. The difference between a half-bridge and a full-bridge inverter is that the output is a rectangular waveform of amplitude $E/2$ volts in the former and E volts in the latter.]

Ans. (a) 180 V; (b) 3240 W; (c) 20 A, 14.14 A, 20 A; (d) 0.81; (e) 0.9

20. The single-phase bridge inverter of Fig. 5.22 uses square wave PWM with two pulses per half-cycle as shown in Fig. 5.38. The modulation is so adjusted that the pulse width is $\pi/3$ and the distance between the pulses is $\pi/6$. The load is inductive with $R = 10\ \Omega$ and $L = 25$ mH. The dc input voltage is 220 V. (a) Determine the RMS output voltage and (b) express the instantaneous load current $i_{ld}(t)$ as a Fourier series. Take the frequency of the inverter to be 50 Hz.

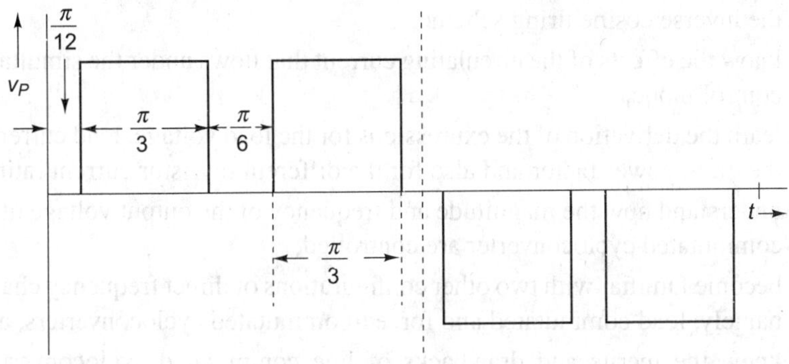

Fig. 5.38

Ans. (a) $V_{ld(RMS)} = 171.75$ V, (b) $i_{ld}(t) = 16.77 \sin(\omega t - 32.12°) + 6.19 \times \sin(3\omega t - 62.04°) - 1.2 \sin(5\omega t - 72.33°) - 0.63 \sin(7\omega t - 77.18°) - \cdots$

CHAPTER 6

Cycloconverters

6.1 Introduction

Though ac voltage controllers, described in Chapter 4, are useful as sources for ac drives, they suffer from the drawback that their output voltage waveforms have a high harmonic content. The voltage source and current source inverters dealt with in Chapter 5 are free from this disadvantage, but they consist of two or more power conversion stages and each of these stages contributes to the overall conversion losses. This prompted the development of the cycloconverter, which is an ac to ac converter or a direct frequency changer. This chapter deals with the various constructional and operational aspects of line commutated cycloconverters.

After going through this chapter the reader should

- become familiar with the principle of the line commutated cycloconverter,
- gain acquaintance with the simultaneous and non-simultaneous methods for the control of a cycloconverter and in the former case become familiar with the inverse cosine firing scheme,
- know the effects of the circulating current that flows under the simultaneous control mode,
- learn the derivation of the expressions for the load voltage, load current, and the input power factor and also for the different thyristor current ratings,
- understand how the magnitude and frequency of the output voltage of a line commutated cycloconverter are controlled,
- become familiar with two other configurations of direct frequency changers, namely, load commutated and forced-commutated cycloconverters, and
- know the merits and drawbacks of line commutated cycloconverters as compared to voltage source inverters (VSIs).

A single-phase cycloconverter consists of a thyristorized circuit whose output voltage waveform is fabricated by using sections of the input waveform. Figure 6.1(a) shows that a dual converter is employed in it for facilitating current flow in either direction; its rectifiers are denoted as *P* and *N* based on their current flow. The principle of the cycloconverter can be understood with the help of this dual converter with a resistive load. At t_1, when the terminal A is positive with respect to B, Th_1 and Th_3 are fired at a firing angle $\alpha_P > 0$. When the voltage v_{AB} crosses zero from the positive to the negative side (at t_2), the thyristors Th_2 and Th_4 are fired with $\alpha_P = 0$. v_{AB} goes down to zero at t_3 but is prevented by the devices to go negative because current can only flow when v_{AB} is positive. At t_4, Th_1 and Th_3 are fired again when line A is positive with $\alpha_P > 0$, and they stop conduction at t_5. At t_6, Th_5 and Th_7 are fired at the angle α_N $(= \alpha_P > 0)$. Similarly at t_7, Th_6 and Th_8 are fired with $\alpha_N = 0$, and finally at t_9, Th_5 and Th_7 are fired again at α_N $(= \alpha_P > 0)$. The cycle is then repeated by firing Th_1 and Th_3 at t_{11} (not shown) at an angle $\alpha_P > 0$.

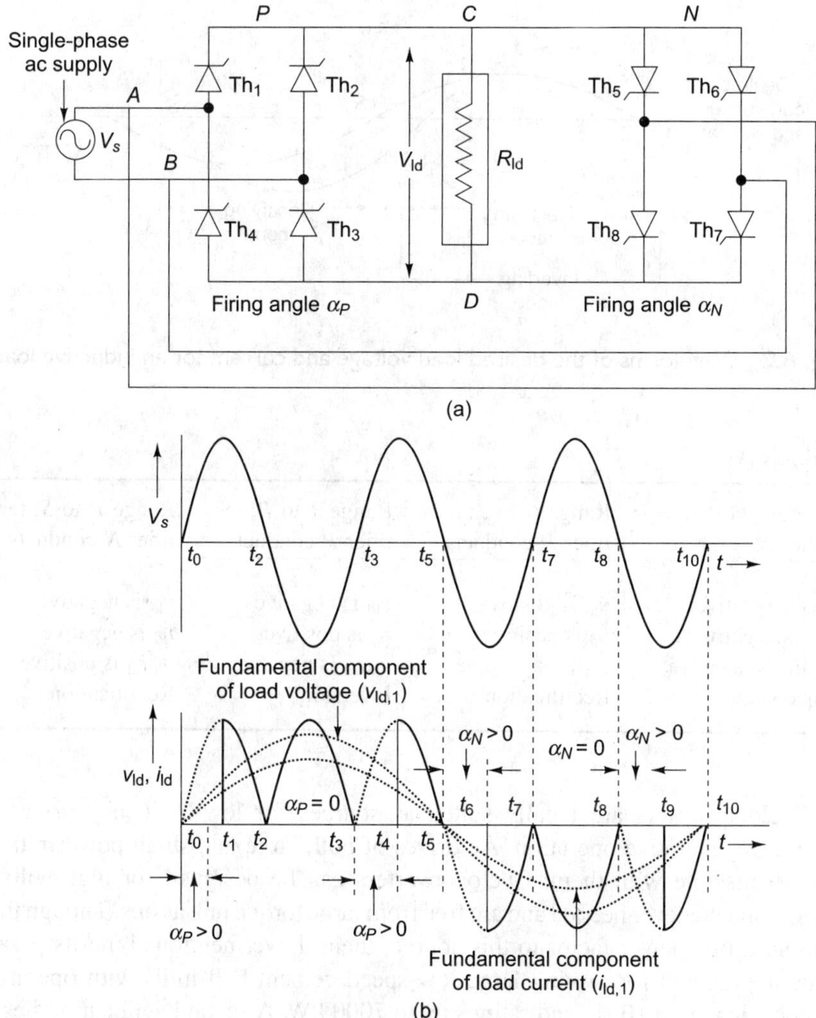

Fig. 6.1 Single-phase cycloconverter: (a) circuit diagram, (b) source and load waveforms with a resistive load

The load waveform [Fig. 6.1(b)] obtained is periodical and hence can be resolved into its harmonic components, with the fundamental component having the largest magnitude. As the load is resistive, the current waveform through the load is also of a similar shape, and its fundamental component will be in phase with the fundamental component of the voltage waveform. With an inductive load, however, the fundamental component of the load current lags that of the voltage as shown in Fig. 6.2. Recognizing that the rectifiers can conduct current through the load in only one direction, rectifier P conducts in the interval from Q to R and rectifier N conducts in the intervals R to S and O to Q. It is seen from Fig. 6.2 that the time period of the voltage waveform can be divided into rectifying and inverting modes of rectifiers P and N, as shown in Table 6.1.

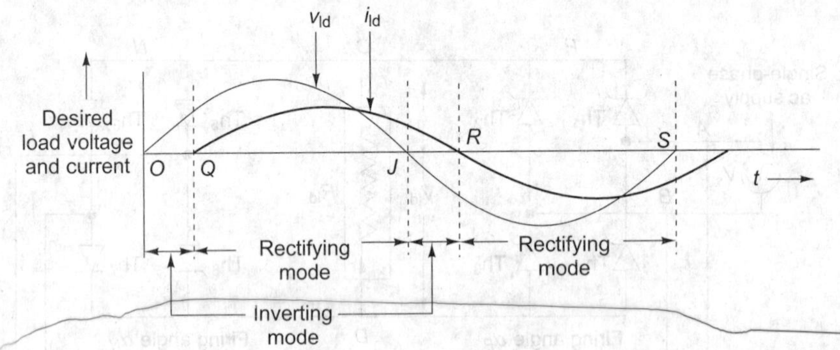

Fig. 6.2 Waveforms of the desired load voltage and current for an inductive load

Table 6.1

Range O to Q; rectifier N conducts	Range Q to J; rectifier P conducts	Range J to R; rectifier P conducts	Range R to S; rectifier N conducts
v_{ld} is positive i_{ld} is negative $v_{ld}i_{ld}$ is negative Inversion	v_{ld} is positive i_{ld} is positive $v_{ld}i_{ld}$ is positive Rectification	v_{ld} is negative i_{ld} is positive $v_{ld}i_{ld}$ is negative Inversion	v_{ld} is negative i_{ld} is negative $v_{ld}i_{ld}$ is positive Rectification

Cycloconverters are ideally suited as sources for low-frequency drives as well as regenerative operation. Ac drives of both large and small power ratings are permissible with them. Cycloconverters can be designed for high-voltage operation whenever needed and are free from large torque pulsations. Though they present a low power factor to this source, their above-mentioned merits permit wide application for driving large low-speed cement ball mills with operating frequencies up to 10 Hz and ratings up to 7000 kW. A second application lies in steel mill roller table drives where a cycloconverter makes it possible to directly drive rollers operating at low frequency, without any intermediary gear boxes. Finally, a cycloconverter can serve as a source for a number of ac motors operated in parallel.

6.2 Principle of the Cycloconverter

The following features are evident from the above discussion. First, a single-phase cycloconverter is just a dual converter which is operated in a particular sequence. Second, the load voltage waveform is composed of fragments of the supply voltage waveform obtained through the P and N rectifiers. If the firing angles α_P and α_N are increased, the RMS magnitude of v_{ld} decreases. Hence, from a fixed amplitude ac supply, variable magnitude ac can be obtained by means of a cycloconverter. It is also seen from Fig. 6.1(b) that the frequency of the load voltage is just one-third that of the ac supply. Now, if the load voltage is fabricated such that rectifiers P and N conduct for four half-cycles each, the output frequency will be one-fourth the ac supply frequency. It can therefore be concluded that a variable frequency output voltage can be obtained from a fixed frequency input voltage by suitably adjusting the firing angles of the rectifiers P and N. Therefore, a cycloconverter is a direct frequency converter which can give a variable magnitude, variable frequency output voltage from a fixed magnitude, fixed frequency ac input voltage. The appropriate values of firing angles at different instants can be arrived at with the help of the single-phase bridge of Fig. 6.1(a) redrawn as Fig. 6.3(a). The currents in the various parts of the circuit are also shown in this figure. In this circuit, called the *type S-S cycloconverter*, v_{AN} and v_{BN} are the voltages at the points A and B, respectively, with respect to the hypothetical centre point N of the transformer secondary. The difference $(v_{AN} - v_{BN})$ is rectified by rectifier P to constitute the positive load voltage v_{ld} for the first half of the duration shown in Fig. 6.3(b).

The load voltage waveform of Fig. 6.3(b) is constructed such that the rectifiers conduct for five half-cycles each, with $\alpha_P = 0$ and $\alpha_N = 0$, respectively, in all the positive and negative half-cycles. The fundamental component of the output will now have a frequency that is one-fifth the ac supply frequency. The shape of the output is nearly rectangular and has a large low-order harmonic content. The waveforms for the currents through the thyristor pairs and also the supply circuit are given in this figure, the current from the ac source is just a sinusoid.

The waveforms in Fig. 6.3(b) suggest that an output voltage which is nearly sinusoidal can be obtained by suitably adjusting the firing angles of the thyristors. This implies that the higher harmonics in such a revised output voltage waveform will be very small (less than 10%) and the fundamental component will be large (more than 90%). Figure 6.4 shows the output voltage of the cycloconverter of Fig. 6.3(a) for a typical sequence as given below. The firing angle applied to the first half-wave should be large, say 160°; the second one should be smaller, say 80°, and the middle one can be still smaller, say 0°. The fourth and fifth half-cycles should be 80° and 160°, respectively. Similarly, on the negative side, α should decrease from about 180° to 0° and then increase back to about 180°. With such phase control, the output consists of full half-cycles of the source voltage at its positive and negative peak regions, with the firing angle α increasing as the load voltage decreases to zero. Clearly, the harmonic content of this load voltage waveform will be much less than that of the load waveform given in Fig. 6.3(b). However, in this case, the supply current is very distorted and contains a harmonic component whose fundamental frequency is the same as the output frequency.

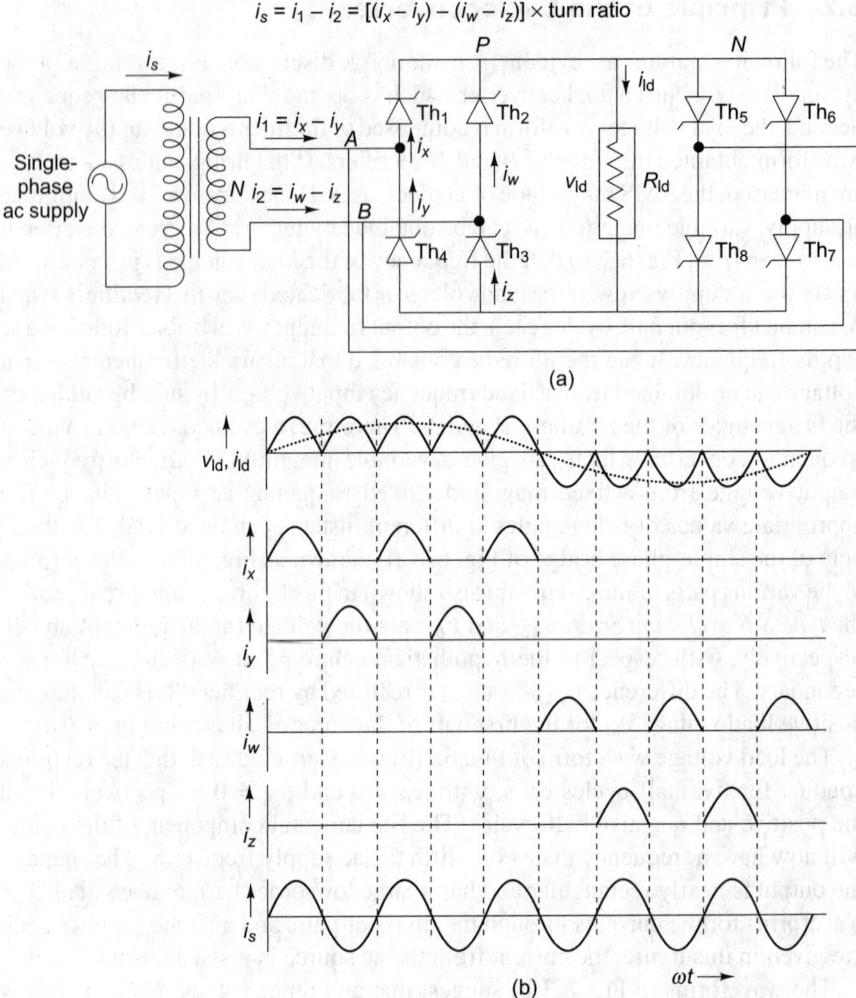

Fig. 6.3 The type S-S cycloconverter: (a) circuit diagram, (b) waveforms with zero firing angles for rectifier thyristors and with a resistive load

6.3 Non-simultaneous Control

Further examination of the circuit of Fig. 6.3(a) shows that if the thyristor pairs in rectifiers P and N conduct simultaneously at any instant, the ac source will be short-circuited through these devices. To obviate this, a reactor is interposed between the two rectifiers as shown in Fig. 6.5. It also helps in limiting the circulating current. At this point it is to be noted that both the rectifiers can be kept on, one (say P) carrying the load current and the other (N) carrying only the circulating current. The operating condition of the rectifier N can then be likened to that of a hot standby unit in a thermal power station. Such operation of the rectifiers is called *simultaneous control* and is dealt with in Section 6.4; a typical $s\phi$ cycloconverter is shown in Fig. 6.5. Though simultaneous control

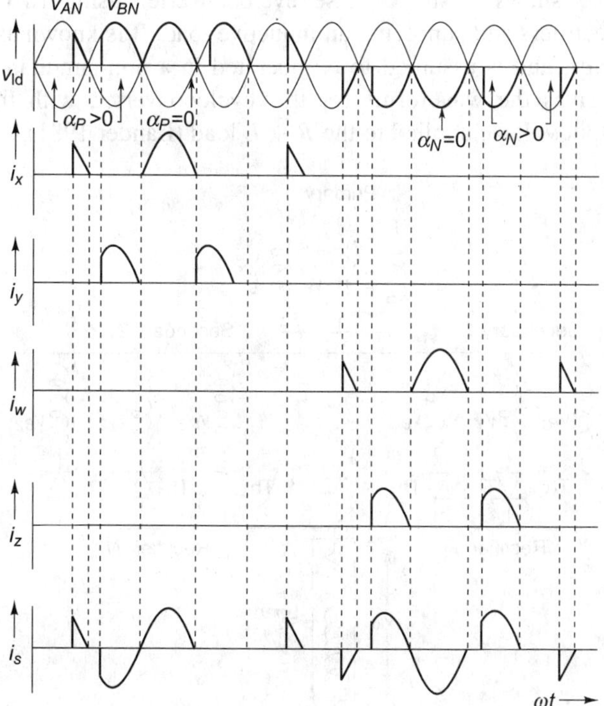

Fig. 6.4 Waveforms of the type S-S cycloconverter of Fig. 6.3(a) with finite firing angles for rectifier thyristors and with a resistive load

is advantageous from the viewpoint of quickness of response, it involves power losses due to the flow of circulating current through rectifiers. The other alternative for avoiding a short circuit through the two rectifiers is that of blocking the thyristors of the second rectifier (say N) when the first rectifier (P) is conducting the load current and vice versa. This type of control called *non-simultaneous control* is now discussed.

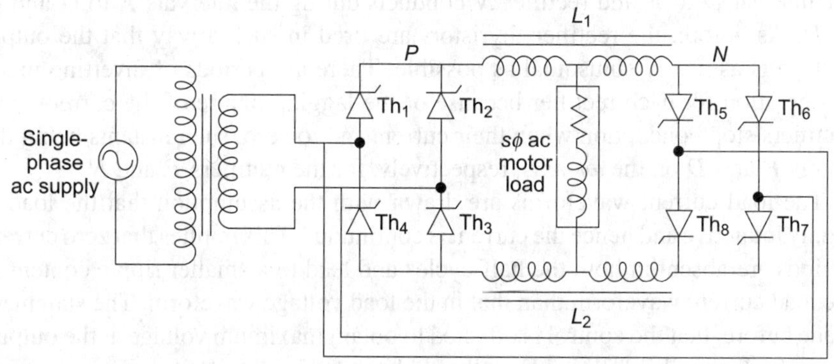

Fig. 6.5 A typical single-phase simultaneously controlled cycloconverter with an $s\phi$ ac motor load

Figure 6.6 shows a single-phase cycloconverter using two three-phase, half-wave rectifiers and supplying an inductive load. It is known as the *type T-S cycloconverter*, and is assumed to be operated in a non-simultaneous manner. Figure 6.7 gives the waveforms for this cycloconverter, with the maximum possible output voltage applied to the $R + L$ load (Lander 1981).

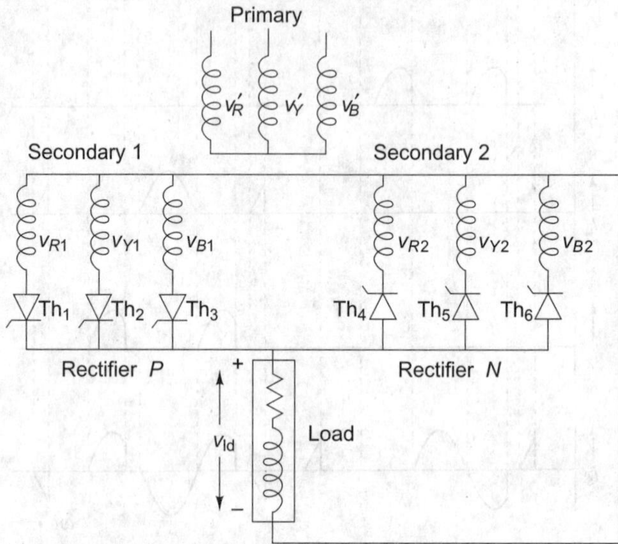

Fig. 6.6 Single-phase cycloconverter (with non-simultaneous control) supplied by two three-phase, half-wave controlled rectifiers and with an inductive load

The output frequency of the cycloconverter shown in Fig. 6.6 is such that the duration of one output cycle is slightly less than that of five cycles of the source voltage. The load current will lag the voltage and the rectifier conduction periods coincide with those of the current half-cycles. As shown by the desired and actual waveforms given, respectively, in Figs 6.7(a) and (b), rectifier P conducts during the interval D to F and rectifier N conducts during the intervals F to G and O to D. As before, the rectifier thyristors are fired in such a way that the output voltage is as nearly sinusoidal as possible. There are periods of inverting mode of operation for each rectifier because of the lagging nature of the current. The rectifiers stop conduction when their currents go to zero, such instants being the points F and D on the ωt-axis, respectively, for the rectifiers P and N.

The load current waveforms are drawn with the assumption that the load is highly inductive and hence the current is continuous. This implies that zero current periods are absent in both the half-cycles and lead to a smaller ripple content in the load current waveform than that in the load voltage waveform. The statement made before, that the control is effected to obtain maximum voltage at the output, can be understood as follows. For a three-phase, half-wave rectifier, the expression for the output voltage is given as

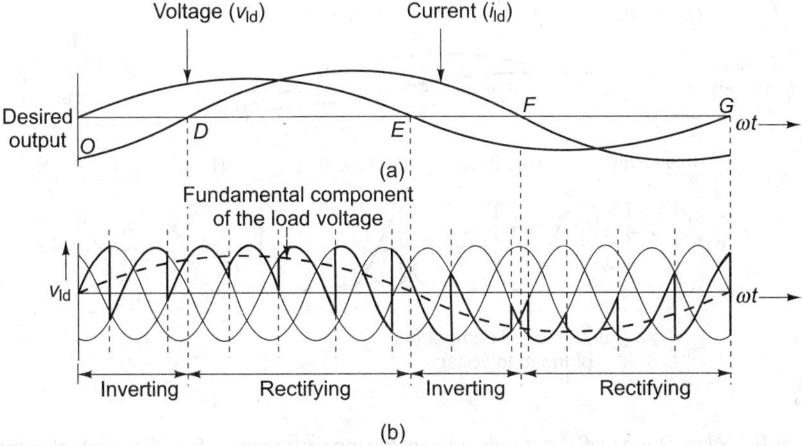

Fig. 6.7 Waveforms of the single-phase cycloconverter of Fig. 6.6 with maximum voltage to an inductive load: (a) desired waveforms of v_{ld} and i_{ld}, (b) actual waveform of v_{ld}

$$v_\alpha = \frac{3}{\pi} V_m \sin\frac{\pi}{n} \cos\alpha = V_0 \cos\alpha \tag{6.1}$$

where

$$V_0 = \frac{3}{\pi} V_m \sin\frac{\pi}{n}$$

Equation (6.1) shows that with $\alpha = 0$, the load gets the full half-cycle of the rectifier output voltage waveform, and hence the amplitude of the cycloconverter output will be V_0. This holds true for both the positive and negative half-cycles of the cycloconverter. If on the other hand it is desired to reduce the output amplitude of the load voltage, α has to be set at an appropriate non-zero value, say α_0. It is evident from our earlier discussion that the firing angles at other regions are to be progressively increased from α_0, so that at an angle of π on the ωt-axis the voltage magnitude will be zero. Figures 6.8(a) and (b) give the output voltage and current waveforms for the cycloconverter of Fig. 6.6 with the peak of the load voltage having a half-maximum amplitude.

An important operational aspect of a cycloconverter with non-simultaneous control is that of smooth transfer of load from one rectifier to the other at zero current instants. In practice a short time interval equal to the turn-off time of the thyristors is needed for the recovery of the outgoing rectifier thyristors before the incoming rectifier thyristors are turned on (Fig. 6.9). Employing a zero current detector, as shown in Fig. 6.10, is a foolproof method of ensuring the attainment of zero current by the outgoing thyristors.

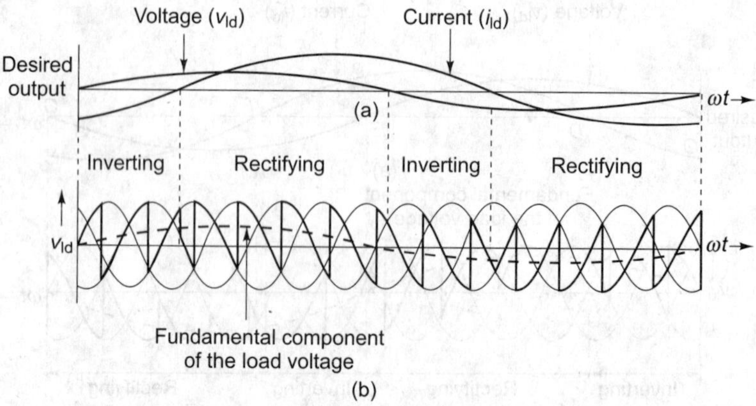

Fig. 6.8 Waveforms of the single-phase cycloconverter of Fig. 6.6 with the load voltage at half maximum to an inductive load: (a) desired waveforms of v_{ld} and i_{ld}, (b) actual waveform of v_{ld}

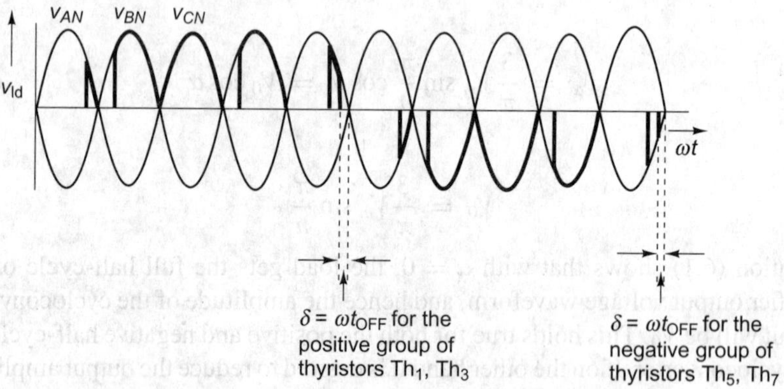

Fig. 6.9 Load voltage waveform of the single-phase cycloconverter of Fig. 6.5 showing the small time interval needed for forward recovery of thyristors

6.4 Simultaneous Control

As the name implies, in simultaneous control, both the rectifiers of the dual converter are maintained in the conducting state. Figure 6.10 shows the circuit of a single-phase bridge type of dual converter that can be used as a source with a reversible dc motor as its load. However, it can also be considered as that of a single-phase cycloconverter supplying a single-phase ac motor as shown in Fig. 6.5. All the single- and three-phase cycloconverters operated with non-simultaneous control can also be made to function with simultaneous control with the additional feature of an inductor interposed between the two rectifiers of the dual converter similar to that for the cycloconverter of Fig. 6.5. The dual converter introduced in Chapter 2 and the cycloconverter with simultaneous control have some common operational features. One such feature is the sequence of operations for changeover of conduction of the dual converter from one rectifier to another and is described below, *assuming that the load is a reversible dc motor.*

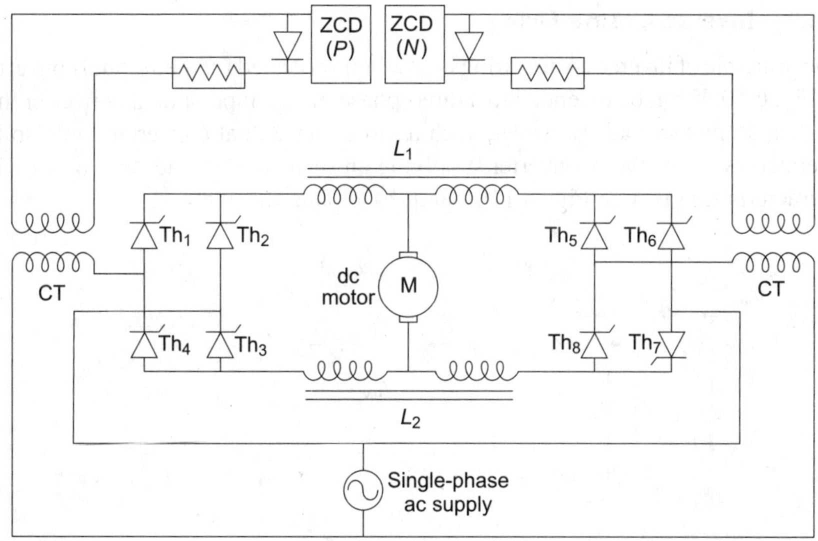

Fig. 6.10 Simultaneously controlled single-phase bridge type of dual converter with zero current detectors and a dc motor load

Rectifier P is assumed to initially be in the rectification mode, with α_P in the range $0° < \alpha_P < 90°$, and the rectifier N in the inversion mode, with α_N in the range $90° < \alpha_N < 180°$. Under these conditions, the motor of the dual converter is assumed to be in the positive motoring mode. Now, if the speed of the motor is to be reversed, the steps to be taken are as follows.

(a) α_P is increased and simultaneously α_N is decreased, ensuring that the following equation is satisfied.

$$\alpha_P + \alpha_N = 180° \quad (\text{or } \pi \text{ rad}) \tag{6.2}$$

As a consequence, the magnitude of the motor back emf exceeds both V_{α_P} and V_{α_N}.

(b) The armature current shifts to rectifier N, which works in the inverting mode, and the motor operates in the second quadrant. Thus, negative power is supplied to the motor, implying thereby that regenerative braking of the motor takes place. This condition continues as long as α_N is slowly reduced but remains in the range given by $90° < \alpha_P < 180°$.

(c) Just when the condition $\alpha_P = \alpha_N = 90°$ is attained, the motors stops.

(d) When α_N is in the range $0° < \alpha_N < 90°$, the rectifier N works in the rectifying mode and the motor reverses its direction of rotation. This completes the changeover of conduction from rectifier P to rectifier N.

Another common feature between the dual converter and the simultaneously controlled cycloconverter is the inverse cosine firing scheme. This scheme ensures that the firing angles α_P and α_N approximately obey the relationship given in Eqn (6.2), and is elaborated below.

6.4.1 Inverse Cosine Firing

The principle of inverse cosine firing is now demonstrated for the $s\phi$ dual converter of Fig. 6.10. It can be extended to a three-phase bridge type of dual converter that feeds a dc motor load. However, such a three-phase dual converter can also be operated as an $s\phi$ cycloconverter supplying an $s\phi$ ac motor load (Fig. 6.12). The characteristics given in Figs 6.11(a) and (b) explain this method.

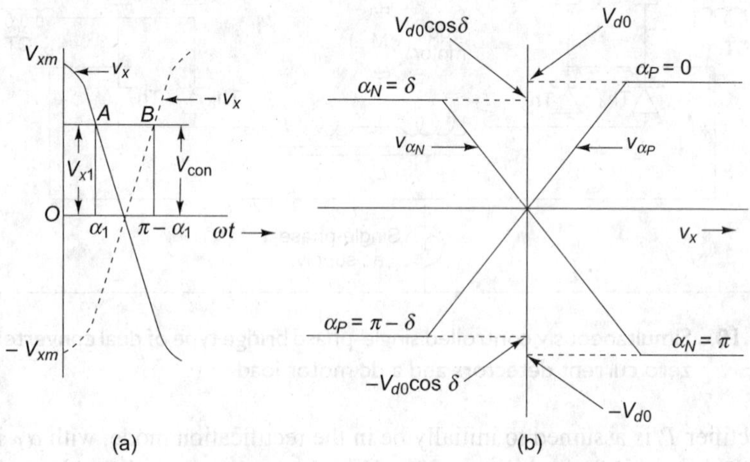

Fig. 6.11 (a) Waveform to illustrate the inverse cosine method of firing; (b) v_x vs $V_{d0}\cos\delta$ characteristic showing limits of α_P and α_N

The average voltage v_{α_1} of an n-phase, half-wave rectifier with a firing angle α_1 is

$$v_{\alpha_1} = \frac{n}{\pi} V_m \sin\frac{\pi}{n} \cos\alpha_1 \tag{6.3}$$

Defining

$$V_{d0} = \frac{n}{\pi} V_m \sin\frac{\pi}{n} \tag{6.4}$$

Eqn (6.3) can be written as

$$v_{\alpha_1} = V_{d0}\cos\alpha_1 \tag{6.5}$$

It is elaborated later that a cycloconverter can be constructed with a dual converter using either single-phase bridge rectifiers or three-phase, half-wave or full-wave rectifiers. The expressions for V_{d0} in these three cases are, respectively, $(2/\pi)V_m$, $(3/\pi)V_m\sin(\pi/3)$, and $(6/\pi)V_m\sin(\pi/3)$. Figure 6.11(a) consists of two sinusoidal waveforms denoted as v_x and $-v_x$. The equation for V_{x1} is taken as

$$V_{x1} = V_{xm}\cos\alpha_1 \tag{6.6}$$

Hence,

$$-V_{x1} = -V_{xm}\cos\alpha_1 \tag{6.7}$$

A dc control voltage V_{con} whose magnitude varies between 0 and V_{xm} is drawn so as to intersect the waveforms of v_x and $-v_x$ at A and B, respectively. V_{con} attains a value of $V_{xm}\cos\alpha_1$ at point A and $-V_{xm}\cos(\pi-\alpha_1)$ at point B. With the points A and B at the same level, V_{con} can be expressed as

$$V_{\text{con}} = V_{x1} = V_{xm}\cos\alpha_1 = -V_{xm}\cos(\pi-\alpha_1) \tag{6.8}$$

In the dual converter of Fig. 6.1(a), if the firing angle α_P of the devices of rectifier P which conduct current in the forward direction is made equal to α_1, and the firing angle α_N of the corresponding devices of rectifier N to $\pi-\alpha_1$, then the relation in Eqn (6.2) will be obeyed by α_P and α_N. It must be ensured that no direct circulating current flows through the rectifiers. Denoting the average voltages of the rectifiers P and N as V_{α_P} and V_{α_N}, respectively, their expressions with the above values for α_P and α_N become

$$V_{\alpha_P} = V_{d0}\cos\alpha_1 \tag{6.9}$$

and

$$V_{\alpha_N} = V_{d0}\cos(\pi-\alpha_1) = -V_{d0}\cos\alpha_1 \tag{6.10}$$

From Eqns (6.9) and (6.6), the ratio of V_{α_P} to V_{x1} can be written as

$$\frac{V_{\alpha_P}}{V_{x1}} = \frac{V_{d0}}{V_{xm}} = K_D \tag{6.11}$$

where K_D is a real number. With the ratio in Eqn (6.11) being independent of α_1, the characteristic v_{α_P} versus v_x is a straight line through the origin as shown in Fig. 6.11(b). v_{α_P} ranges between $-V_{d0}\cos\delta$ and V_{d0}. The reason for keeping the lower limit at $-V_{d0}\cos\delta$ and not at $-V_{d0}$ is as follows. In the actual operation of the dual converter, the upper limit of α_P cannot be increased to π radians but must be limited to $\pi-\delta$, where δ depends on the turn-off time of the thyristors. This is to allow enough margin for the commutation of the thyristors of rectifier P. Using Eqn (6.2), the lower limit of α_N becomes

$$\alpha_{N(\min)} = \pi - \alpha_{P(\max)} = \pi - (\pi-\delta) = \delta \tag{6.12}$$

Also, the upper limit of α_N can be written as

$$\alpha_{N(\max)} = \pi - \alpha_{P(\min)} = \pi - 0 = \pi \tag{6.13}$$

The plot v_{α_N} versus v_x is also a straight line [Fig. 6.11(b)], but with a negative slope. Its upper and lower limits are, respectively, $V_{d0}\cos\delta$ and $-V_{d0}$. Figure 6.11(a) suggests the following method for the firing of the thyristors in the two rectifiers of the dual converter. The straight line representing the dc control voltage V_{con} is drawn to intersect the two waveforms v_x and $-v_x$. The intersection with v_x gives for α_P the value α_1, which is the firing angle of the thyristors of rectifier P. Similarly the intersection of V_{con} with $-v_x$ gives α_N of rectifier N as $\pi-\alpha_1$. The control circuitry adjusts the firing angles of the two rectifiers accordingly. Since α_P equals $\cos^{-1}(V_{\text{con}}/V_{xm})$, this method is termed as the inverse cosine firing scheme and is adopted when the dual converter is operated with simultaneous control.

6.4.2 Firing Scheme for a Three-phase Dual Converter

Though a dual converter consisting of two single-phase bridge rectifiers has been considered in the previous section, a practical dual converter usually consists of two three-phase bridge rectifiers. Such a configuration has the merit of providing an output with minimum ripple content. Figure 6.12 shows a typical three-phase bridge type of dual converter that can be a source for a reversible dc motor as in Fig. 6.10. However, here it is operated as a cycloconverter (type DT-S) with an $s\phi$ ac motor as its load.

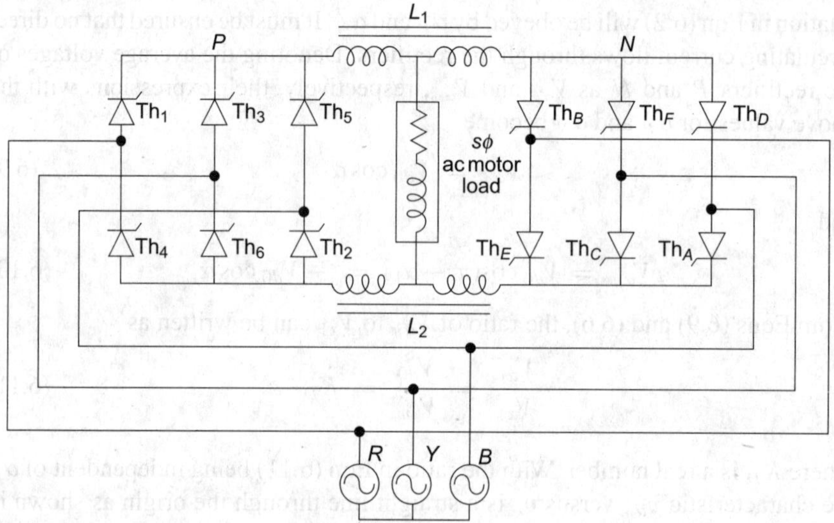

Fig. 6.12 Two three-phase, full-wave fully controlled rectifiers connected as a DT-S cycloconverter with an $s\phi$ ac motor load

In the firing scheme of Fig. 6.11(a), the peak of the waveform of v_{x1} is located at the origin of the single-phase ac supply so as to ensure a constant ratio between the waveforms of $v_{\alpha P}$ and v_{x1} as given in Eqn (6.11). On the other hand, in a three-phase dual converter, the peak of v_{x1} has to be located at the point of natural commutation of the supply waveform of v_R, which occurs at 30° on the ωt-axis, to ensure constancy between $v_{\alpha P}$ and v_{x1}. Similarly the peaks of the waveforms v_{x2} to v_{x6} have to be located, respectively, at the points of natural commutation of the supply waveforms $-v_B$, v_Y, $-v_R$, v_B, and $-v_Y$.

The following facts can be inferred from a scrutiny of the supply waveforms of Fig. 6.13. If the reference waveforms v_{x1} to v_{x6} are drawn as the scaled-down versions, respectively, of the waveforms of $-v_Y$, v_R, $-v_B$, v_Y, $-v_R$, and v_B, then the above condition will be fulfilled. For instance, in Fig. 6.13 the peak of the supply waveform $-v_Y$ occurs at an angle of 30° from the start of the waveform of v_R; hence v_{x1} can be obtained as a scaled-down version of v_Y. Accordingly, all the reference waveforms v_{x1} to v_{x6} shown in Fig. 6.14(a) are drawn in the order given above. The control voltage V_{con} and the reference waveform $-v_{x1}$

is also drawn in this figure. The thyristors Th_1 to Th_6 are fired at the angles of intersection of V_{con}, respectively, with the waveforms of v_{x1} to v_{x6}.

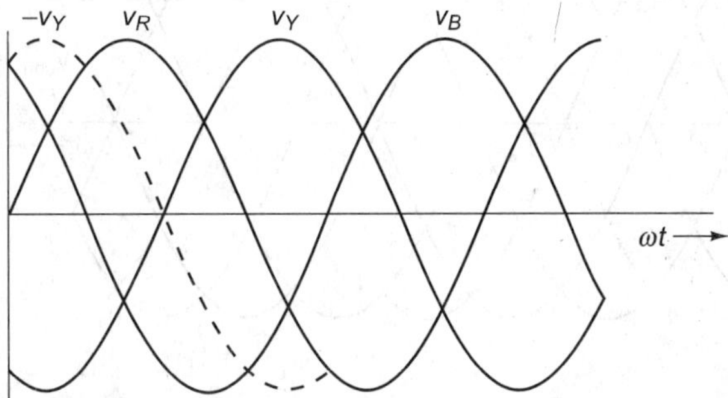

Fig. 6.13 Waveforms showing the method of obtaining a typical reference waveform

The thyristors Th_A to Th_F of rectifier N of Fig. 6.12 are to be fired at the angles given by the intersections of v_{con}, respectively, with the waveforms of $-v_{x1}$ to $-v_{x6}$, so as to ensure the relationship given in Eqn (6.2). For clarity, only the waveform of $-v_{x1}$ is shown in Fig. 6.14(a). Figure 6.14(b) gives the transfer characteristics of the rectifiers P and N of Fig. 6.12 and bears similarity with Fig. 6.11(b) (Dubey 1989) drawn for the single-phase bridge rectifier of Fig. 6.10.

6.4.3 Firing Scheme for a Single-phase, Line Commutated Cycloconverter

The inverse cosine firing scheme described in Section 6.4.1 is now modified so that the three-phase, full-wave dual converter of Fig. 6.12 can function as a *simultaneously controlled line commutated, single-phase cycloconverter*, called as the *type DT-S cycloconverter*. For this purpose, Fig. 6.14(a) has to be modified as follows. Instead of v_{con} having a flat control characteristic, it is now drawn as a slowly varying sine wave so as to ensure a sinusoidal output for the cycloconverter. Figure 6.15 is drawn accordingly with the reference voltage waveforms v_{x1} to v_{x6} drawn as before, a low-frequency sine wave v_{con} being superimposed thereon. The reference waveforms are assumed to be repeated over the entire duration of v_{con}. For the purpose of this discussion, the frequency of v_{con} is assumed to be nearly one-sixth of that of the reference voltage waveforms. It is evident that the intersection of v_{x1} with v_{con} occurs at an angle which is nearly 90° with respect to the angle of natural commutation of v_R. The output voltage of rectifier P of the dual converter now being proportional to $\cos \alpha_P$, its starting value with α_P = 90° will be nearly zero.

The intersection angles of the waveforms v_{x2} to v_{x6} progressively decrease because of the sinusoidal nature of v_{con}. Accordingly, the output voltage of rectifier P progressively increases. It follows that when v_{con} intersects any reference voltage v_{xn} (where n may have any value from 1 to 6) at its peak, then, as explained in the second paragraph of Section 6.4.2, the corresponding angle on the ωt-axis will be 0 and its cosine value becomes equal to 1. Also, the output voltage of

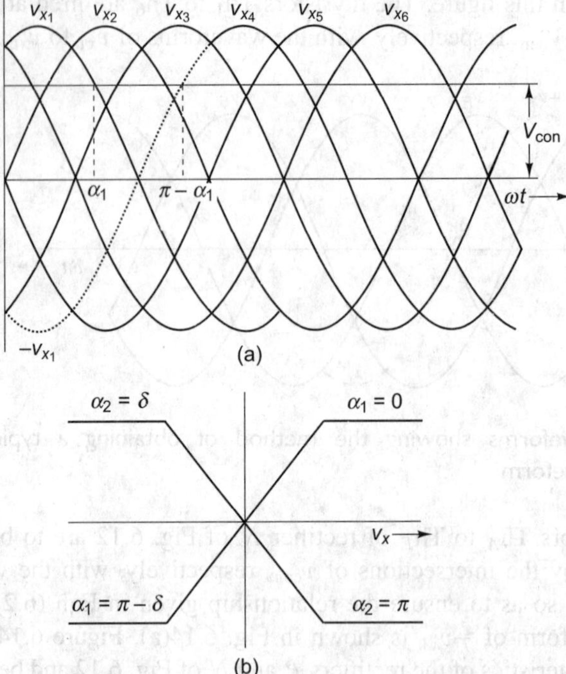

Fig. 6.14 Inverse cosine method: (a) waveforms to illustrate the firing angles of a three-phase dual converters, (b) α_1 and α_2 vs v_x

rectifier P will be maximum at this angle. The angle of intersection then gradually increases, becoming nearly 90° at the zero crossing of v_{con} (that is, the 180° point on the ωt-axis). Beyond this crossing, the inverse cosine firing scheme provides a negative output voltage of rectifier P with increasing magnitude, reaching a negative peak at the 270° point, and then slowly decreases, becoming zero at the 360° point on the ωt-axis. The modification of the inverse cosine firing scheme for the cycloconverter can be summed up by stating that the output voltage of rectifier P varies in the same manner as, and has the same frequency as the waveform of, v_{con}.

If instead of the reference waveforms v_{x1} to v_{x6}, the reference waveforms of $-v_{x1}$ to $-v_{x6}$ are drawn over the entire duration of v_{con}, their intersection with v_{con} will give the firing angles of rectifier N. The waveform so obtained will be nearly equal in magnitude but opposite in sign to that obtained for rectifier P.

The discussion so far has not dwelt upon the effect of the nature of the load on the output waveforms of the cycloconverter. With a resistive load, the load current waveform will be a scaled-down replica of the load voltage waveform. Accordingly, their zero crossings occur at $\omega t = 0°$ and $\omega t = 180°$ and also at angles which are multiples of 180°. With the rectifier thyristors conducting only in one direction, this implies that the rectifier P conducts from 0° to 180° on the ωt-axis and the rectifier N conducts from 180° to 360°. It is well known that for an inductive load, the current lags the voltage. Accordingly, Fig. 6.16(a) shows that the current is positive from A to C, and hence rectifier P conducts from A

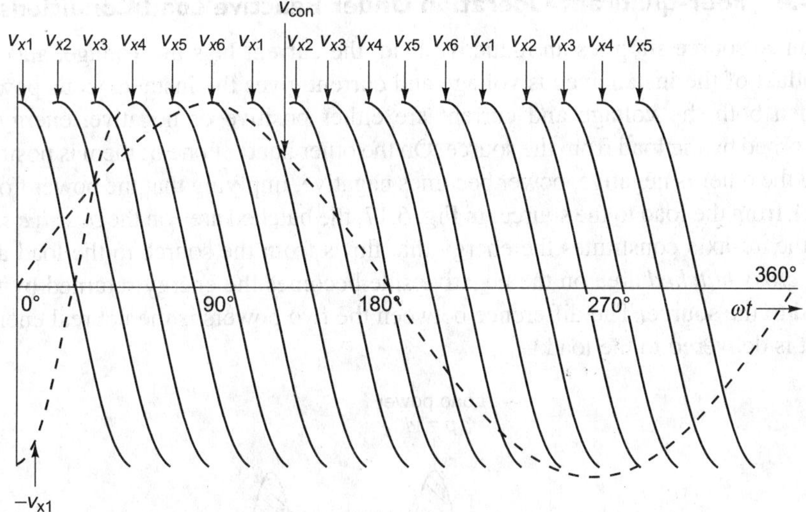

Fig. 6.15 Waveforms illustrating the method of obtaining a sinusoidal output for a single-phase cycloconverter

to C whereas rectifier N conducts from C to D and O to A. The intervals B to C and O to A are, respectively, the negative voltage region for rectifier P and the positive voltage region for rectifier N, and hence they are inversion regions. The dashed sinusoid in Fig. 6.16(b) shows one complete cycle of the cycloconverter output voltage waveform obtained as above (Lander 1981).

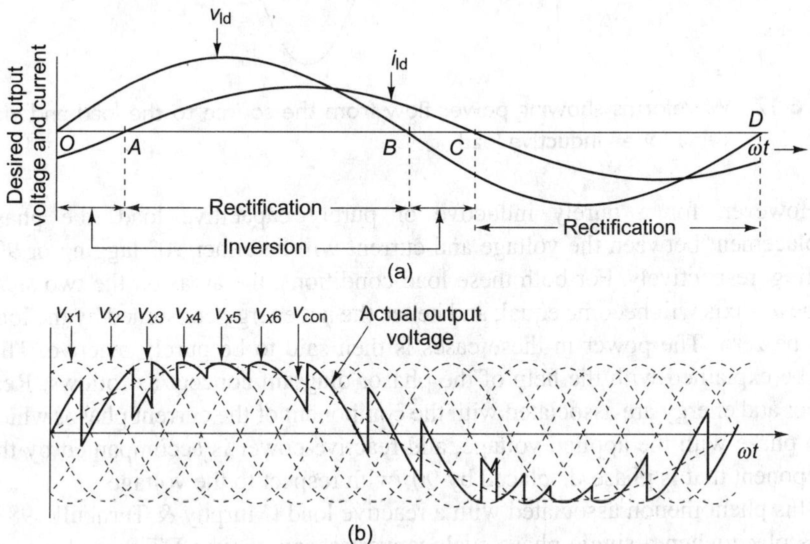

Fig. 6.16 Low-frequency output voltage waveform of a single-phase cycloconverter: (a) desired output voltage and current, (b) actual output voltage

6.4.4 Four-quadrant Operation Under Reactive Load Conditions

If an ac source supplies an inductive load, the current lags the voltage, and the product of the instantaneous voltage and current gives the instantaneous power. When both the voltage and current are either positive or negative, energy is absorbed by the load from the source. On the other hand, if one of them is positive and the other is negative, power becomes negative, implying that the power flows back from the load to the source. In Fig. 6.17, the hatched area on the *positive side* of the ωt-axis constitutes the energy that flows from the source to the load and the *cross-hatched* area on the negative side becomes the energy returned by the load to the source. The difference between the two powers is the net real energy that is delivered to the load.

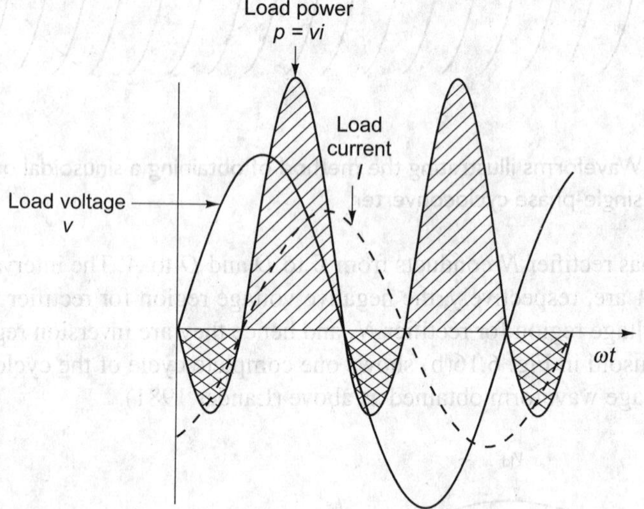

Fig. 6.17 Waveforms showing power flow from the source to the load and vice versa for an inductive load

However, for a purely inductive or purely capacitive load, the phase displacement between the voltage and current will be either 90° lagging or 90° leading, respectively. For both these load conditions, the areas on the two sides of the ωt-axis will become equal, and hence the net energy consumed by the load will be zero. The power in these cases is then said to be purely reactive. This can be explained with the help of the phasor diagram concept as follows. Real power and energy are associated with the component of the current phasor which is in phase with the applied voltage, and reactive power is accompanied by the component that is phase-displaced by 90° with respect to the voltage.

This phenomenon associated with a reactive load (Murphy & Turnbull 1988) takes place when a single-phase cycloconverter, say of type DT-S as shown in Fig. 6.12, feeds an inductive load. However, here the energy exchange from the source to load and vice versa will take place through the cycloconverter. This is demonstrated in Figs 6.16(a) and (b), which show the desired and actual waveforms of the cycloconverter. The source delivers power to the load through the cycloconverter in the intervals *AB* and *CD*. Also, power flows from the load back to the source during the intervals *OA* and *BC* and therefore the converters operate in the inverting mode during these intervals. These four modes of operation can also be represented in the four quadrants of the v-i plane (Fig. 6.18).

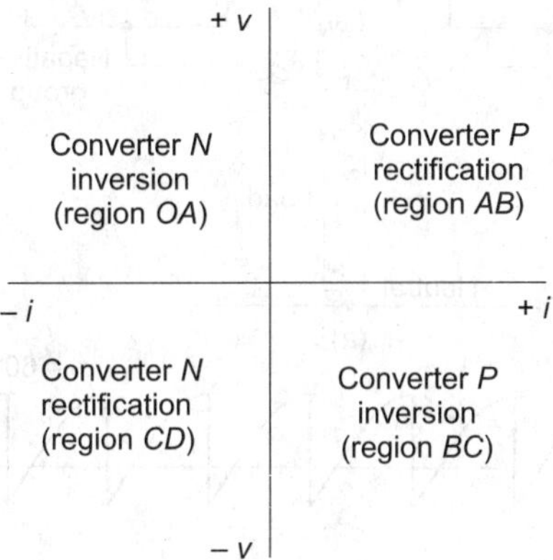

Fig. 6.18 Four-quadrant operation of a single-phase cycloconverter supplying an inductive load

6.4.5 Circulating Currents in a Simultaneously Controlled Single-phase Cycloconverter

In a simultaneously controlled single-phase cycloconverter, both the converters P and N of the dual converter will be in the conducting mode and hence a circulating current flows through them. The inductors L_1 and L_2, respectively, in the top and bottom lines of the dual converter (Fig. 6.10) serve to suppress this circulating current. This is true for the single-phase cycloconverter shown in Fig. 6.19(a) also, which consists of two three-phase, half-wave converters, and in which one centre-tapped inductor L_{PN} serves the purpose of suppressing the circulating current. The waveforms for the output voltages of the converters P and N, with α_P and α_N taken, respectively, as 60° and 120°, are drawn in Fig. 6.19(b) as also the waveforms of reactor voltage, circulating current, and load voltage. The ripple in the load voltage gets cancelled out when the mean of the voltages of the two converters is applied to the load. The thyristors of a group conduct current only in one direction; hence the circulating current also flows in one direction. Figure 6.19(b) also shows the following feature: the current builds up or falls down, respectively, when the reactor voltage is positive or negative. Hence, in the steady state, it has a fluctuating characteristic.

In the presence of load current, the operating conditions can be represented by the equivalent circuit of Fig. 6.19(c). It is seen that the load current is contributed by both the converters and hence can be written as

$$|i_{\mathrm{ld}}| = |i_P| + |i_N| \tag{6.14}$$

The currents i_P and i_N have a sinusoidal nature and Eqn (6.14) shows that the modulus of each has a mean value of $|i_{\mathrm{ld}}|/2$. A small circulating current

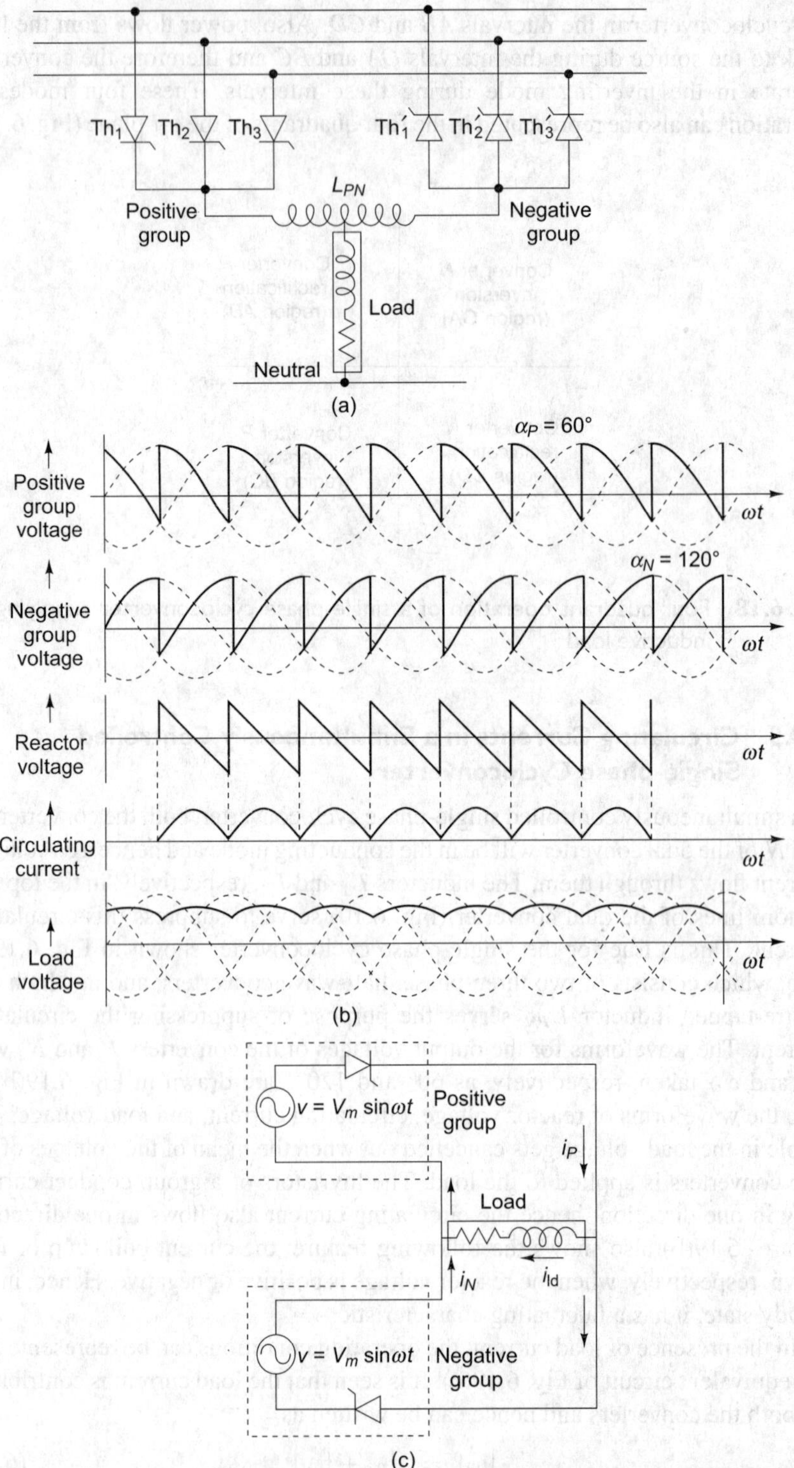

Fig. 6.19 Single-phase cycloconverter using two three-phase, half-wave rectifiers: (a) circuit diagram, (b) waveforms, (c) operating conditions when load current flows

of the ripple voltage and will be superimposed on the fundamental components of i_P and i_N.

One drawback of the simultaneously controlled cycloconverter is that the thyristors have to carry both the load as well as the circulating currents. The devices have therefore to be of a higher current rating than those used in a non-simultaneously controlled cycloconverter. A strategy for keeping down the currents flowing through the thyristors consists of operating the cycloconverters with non-simultaneous and simultaneous controls, respectively, under heavy- and light load conditions.

6.5 Circuit Analysis

The aim of circuit analysis is to (i) determine the total power factor presented to the source by the cycloconverter–motor (load) combination, (ii) arrive at the kilovolt-ampere (kVA) rating of the transformer, (iii) derive expressions for the RMS, half-cycle average and maximum current ratings of the thyristors, and (iv) determine the peak reverse- and forward blocking voltage ratings of the thyristors. This is done in six stages as detailed below.

6.5.1 Output Voltage

The output voltage of a single-phase cycloconverter is constructed with segments of the input ac voltage waveforms. Each of such segments is just the full-wave dual converter output with a firing angle α and can be expressed as

$$V_\alpha = 2\left(\frac{n}{\pi} V_m \sin\frac{\pi}{n} \cos\alpha\right) \tag{6.15}$$

where the expression within the brackets is that of the output voltage of a half-wave, n-phase rectifier. It is stated in Section 6.4.3 that a three-phase, full-wave dual converter is employed in a single-phase cycloconverter. It follows that three such dual converters constitute a three-phase cycloconverter.

The maximum voltage that can be obtained with a dual converter is given by Eqn (6.15) with α taken as zero. Accordingly, the peak of the output sinusoid of the single-phase cycloconverter, consisting of two three-phase rectifiers (Fig. 6.12), can be expressed as

$$\sqrt{2}V_{CC} = 2\left(\frac{3}{\pi} V_m \sin\frac{\pi}{3}\right) \tag{6.16}$$

where V_{CC} is the RMS output voltage of the cycloconverter. V_m can be expressed as $\sqrt{2}V_S$, where V_S is the line-to-neutral secondary voltage of a delta–star transformer. Also, recollecting that a minimum value of $\alpha = \delta$ has to be maintained to ensure successful commutation of the thyristor (Section 6.4.1), Eqn (6.16) has to be modified as

$$\sqrt{2}V_{CC} = 2 \times \frac{3}{\pi}\sqrt{2}V_S \sin\frac{\pi}{3} \cos\delta \tag{6.17}$$

Defining a voltage reduction factor $r = \cos\delta$, V_{CC} can be expressed as

$$V_{CC} = \frac{3\sqrt{3}}{\pi} r V_S \tag{6.18}$$

6.5.2 Input Displacement Factor

The ac supply currents in a cycloconverter have non-sinusoidal shapes, and hence the expressions for the input displacement factor, distortion factor, and input power factor need to be derived. The definitions of these quantities given in Section 2.9 have to be slightly modified here to suit the cycloconverter. The input displacement factor $\cos\phi_i$ (which is the same as the displacement factor $\cos\phi_1$ defined in Section 2.9) is the fundamental power factor, where ϕ_i is the phase displacement between the fundamental phase voltage and the fundamental phase current at the input of the cycloconverter. The distortion factor μ is the ratio of the fundamental RMS current to the total RMS current.

If the commutation overlap is negligible, the displacement factor $\cos\phi_i$, is equal to $\cos\alpha$, where α is the firing angle. In practice, some overlap exists and its effect is to increase ϕ_i, thus decreasing the displacement factor and total power factor of the cycloconverter–motor load combination.

In a cycloconverter system, the average phase displacement between the input current and voltage is large at low output voltages, thus making the input displacement factor low. With a purely resistive load, however, the displacement factor has its maximum value. A capacitive load has the same effect as an inductive load in reducing the displacement factor.

The input displacement factor $\cos\phi_i$ has been derived in classical analysis (Murphy & Turnbull 1988) with the assumption that the cycloconverter has an infinite number of input phases and negligible commutating reactance, and the output voltage is at its maximum. Thus, $\cos\phi_i$ is obtained as

$$\cos\phi_i = \frac{1}{\left\{1 + 4\left[\dfrac{\cos\phi_{CC} + \phi_{CC}\sin\phi_{CC}}{\pi\cos\phi_{CC}}\right]^2\right\}^{1/2}} \tag{6.19}$$

where $\cos\phi_{CC}$ is the power factor of the load. When the load voltage is varied by voltage control, the voltage reduction factor r (<1) also has to be taken into consideration and $\cos\phi_i$ gets reduced. Figure 6.20 shows plots of $\cos\phi_i$ versus ϕ_{CC} with r as a parameter, and $\cos\phi_i$ having a magnitude of 0.843 with $r = 1$ and $\cos\phi_{CC} = 1$. The curves are symmetrical about $\phi_{CC} = 0$ with the same value for $\cos\phi_i$ in the case of both lagging and leading power factors.

Again, classical analysis shows that the input power factor λ can be expressed as

$$\lambda = \frac{r\cos\phi_{CC}}{\sqrt{2}} \tag{6.20}$$

Thus, the distortion factor can be expressed as

$$\mu = \frac{\lambda}{\cos\phi_i} = \frac{r\cos\phi_{CC}}{\sqrt{2}\cos\phi_i} \tag{6.21}$$

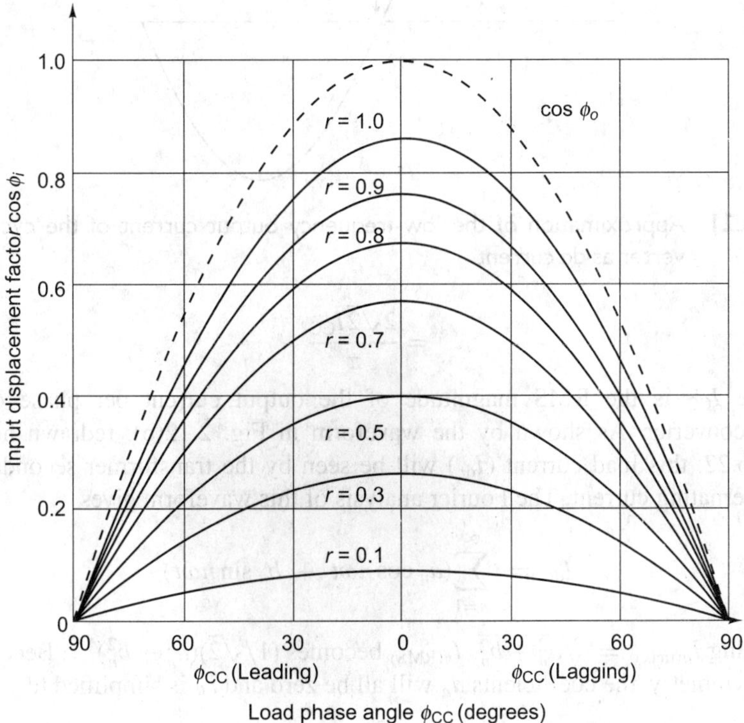

Fig. 6.20 Curves showing input displacement factor versus load phase angle for different values of *r*

6.5.3 Fundamental RMS Current

An expression for the fundamental RMS current I_1 at the input of the cycloconverter can be derived with the following assumptions.

(i) A single-phase cycloconverter is assumed to be constituted of a three-phase, full-wave fully controlled dual converter because of the fact that this ensures the absence of a dc component in the ac waveform at the secondary of the transformer. Accordingly, a three-phase cycloconverter will have three such dual converters.

(ii) The output frequency of the cycloconverter is assumed to be quite low so that it can be considered to be feeding a dc load with a current equal to the half-cycle average of the load current waveform, as illustrated in Fig. 6.21.

Denoting this half-cycle average of current as I_{dc}, it can be expressed as

$$I_{dc} = \frac{1}{\pi}\int_0^{\pi} \sqrt{2}I_{CC}\sin\omega t\, d(\omega t)$$

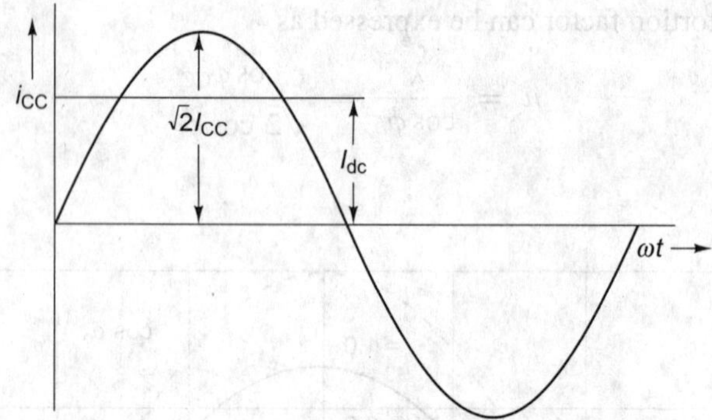

Fig. 6.21 Approximation of the low-frequency output current of the cycloconverter as dc current

$$= \frac{2\sqrt{2}I_{CC}}{\pi} \tag{6.22}$$

where I_{CC} is the RMS magnitude of the output current per phase of the cycloconverter. As shown by the waveform in Fig. 2.12(b), redrawn here as Fig. 6.22, this load current (I_{dc}) will be seen by the transformer secondary as an alternating current. The Fourier analysis of this waveform gives

$$I_{dc} = \sum_{n=1}^{\infty}(a_n \cos n\omega t + b_n \sin n\omega t) \tag{6.23}$$

Defining $I_{n(\max)} = \sqrt{a_n^2 + b_n^2}$, $I_{n(\text{RMS})}$ becomes $(1/\sqrt{2})(a_n^2 + b_n^2)^{1/2}$. Because of odd symmetry, the coefficients a_n will all be zero and i_R is simplified to

$$I_{dc} = \sum_{n=1}^{\infty} b_n \sin \omega t \tag{6.24}$$

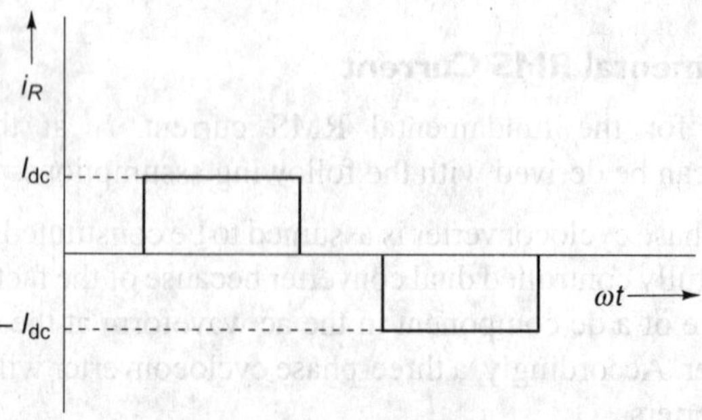

Fig. 6.22 Load current I_{dc} of the cycloconverter as seen from the transformer secondary

The fundamental component b_1, which is of interest, can be evaluated as

$$b_1 = \frac{2}{\pi}\int_{\theta}^{(\theta+120)} I_{dc} \sin \omega t \, d(\omega t) = \frac{2\sqrt{3}}{\pi} I_{dc} \sin(\theta + 60) \tag{6.25}$$

The maximum value of b_1 is

$$b_{1(\max)} = \frac{2\sqrt{3}}{\pi} I_{dc} \qquad (6.26)$$

and its RMS value is

$$I_1 = \frac{b_{1(\max)}}{\sqrt{2}} = \frac{\sqrt{6}}{\pi} I_{dc} \qquad (6.27)$$

Substituting from Eqn (6.22) for I_{dc}, the expression for I_1 becomes

$$I_1 = \frac{\sqrt{6}}{\pi} \times \frac{2\sqrt{2} I_{CC}}{\pi} = 0.7 I_{CC} \qquad (6.28)$$

6.5.4 Transformer Rating

With μ and I_1 determined, respectively, as in Eqns (6.21) and (6.28), the per phase RMS value of the transformer secondary current I_S is related to these quantities as

$$\mu = \frac{I_1}{I_S} \qquad (6.29)$$

This gives

$$I_S = \frac{I_1}{\mu} \qquad (6.30)$$

The kVA rating of the transformer feeding the cycloconverter is now obtained as

$$S = \frac{3 V_S I_S}{1000} \qquad (6.31)$$

Remarks:

(a) The expression for S in Eqn (6.31) holds good for any kind of load connected to the cycloconverter.

(b) When the cycloconverter–motor (load) combination presents a low power factor to the input supply, the expression on the right-hand side of Eqn (6.22) is taken as $\sqrt{2} I_{CC}$.

(c) The above derivation does not hold good for a cycloconverter fed by a single-phase bridge type of dual converter. In Worked Example 4, $I_{ld,1}$, which is the fundamental component, and I_{CC}, respectively, correspond to I_1 and I_S of Eqn (6.29).

6.5.5 Current Ratings

The current ratings of a cycloconverter depend on the load impedance. Denoting this impedance as Z_{ld}, the RMS rating of the output current can be written as

$$I_{CC} = \frac{V_{CC}}{Z_{ld}} \qquad (6.32)$$

where V_{CC} is given in Eqn (6.18).

The current ratings of the cycloconverter thyristors are obtained as follows. Each of the rectifiers conducts for a duration of one half-cycle of the output current, whose RMS value is I_{CC}. Denoting I_P and I_N, respectively, as the RMS magnitudes of the currents of the P and N rectifiers,

$$I_P = I_N = \frac{I_{CC}}{\sqrt{2}} \tag{6.33}$$

Also, the RMS current through each thyristor of a three-phase, full-wave rectifier is

$$I_{Th(RMS)} = \frac{I_P}{\sqrt{3}} = \frac{I_{CC}}{\sqrt{6}} \tag{6.34}$$

The maximum instantaneous thyristor current I_{ThM} is obtained as that of the cycloconverter. Thus,

$$I_{ThM} = \sqrt{2} I_{CC} \tag{6.35}$$

$$= \sqrt{2}\frac{V_{CC}}{Z_{ld}} \tag{6.36}$$

Denoting the half-cycle average current of the cycloconverter as I_{HA}, it is given as I_{dc} in Eqn (6.22). Thus,

$$I_{HA} = I_{dc} = \frac{2\sqrt{2} I_{CC}}{\pi} \tag{6.37}$$

With each thyristor conducting for one-third of a cycle, the average thyristor current rating can be written as

$$I_{HA(Th)} = \frac{I_{HA}}{3} = \frac{2\sqrt{2} I_{CC}}{3\pi} \tag{6.38}$$

6.5.6 Peak Reverse- and Forward Blocking Voltage Ratings

The peak reverse voltage rating of the thyristors can be obtained as that for the three-phase, full-wave fully controlled rectifiers discussed in Chapter 2. Its value is $\sqrt{3}V_m$, where V_m is the peak line-to-neutral magnitude of the input waveform. The peak voltage rating of the thyristor is arrived at as follows. Figure 6.23(b) gives the voltage waveforms at the two load terminals of the three-phase, full-wave rectifier shown in Fig. 6.23(a).

The voltage across thyristor Th_1 is shown in Fig. 6.23(c), from which it is seen that the peak forward voltage to be withstood by the thyristor occurs at an angle of $\pi/3$ from the origin. Alternatively, this can be obtained from the expressions for v_{Th_1} as follows:

$$v_{Th_1} = V_m \sin\omega t - V_m \sin\left(\omega t - \frac{2\pi}{3}\right)$$

$$= 2V_m \cos\left(\omega t - \frac{\pi}{3}\right)\sin\frac{\pi}{3} \tag{6.39}$$

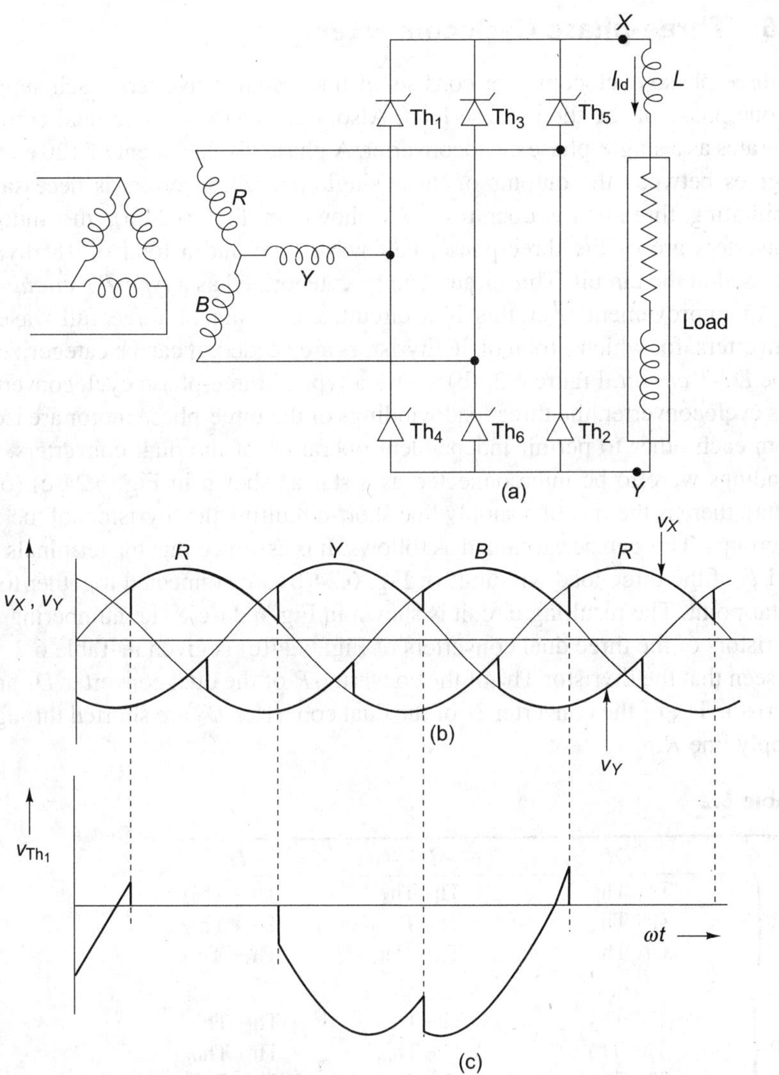

Fig. 6.23 Three-phase, full-wave fully controlled rectifier: (a) circuit diagram, (b) load voltage waveforms at *X* and *Y*, (c) voltage across Th_1

Differentiating v_{Th_1} with respect to ωt and equating it to zero gives the angle at which v_{Th_1} attains a maximum v_{Th_1M}:

$$\frac{dv_{Th}}{d(\omega t)} = -2V_m \sin\left(\omega t - \frac{\pi}{3}\right) \sin\frac{\pi}{3} = 0 \tag{6.40}$$

v_{Th_1M} occurs at $\omega t = 4\pi/3$ and the substitution of this value of ωt in Eqn (6.39) gives v_{Th_1M} as $-\sqrt{3}V_m$. Similarly, the maximum forward blocking voltage can be obtained as $\sqrt{3}V_m$, which is the same as the line-to-line voltage.

6.6 Three-phase Cycloconverters

A three-phase cycloconverter consists of three dual converters, each supplying to one phase of the three-phase load. Also, each of these three dual converters operates as a single-phase cycloconverter. A phase displacement of 120 electrical degrees between the outputs of these single-phase converters is necessary for facilitating three-phase operation. As shown in Fig. 6.24(a), the individual converters are of the three-phase, half-wave type and a total of 18 thyristors are used in the circuit. This circuit can be categorized as a *type T-T circuit*.

An improvement over this is a circuit, consisting of three full-wave dual converters, for which a total of 36 thyristors are needed. It can be categorized as a type *DT-T circuit*. Figure 6.24(b) shows a typical three-phase cycloconverter. In this cycloconverter, the three load windings of the three-phase motor are isolated from each other to permit independent operation of the dual converters. If the windings were to be interconnected as a star as shown in Fig. 6.24(c) (or as a delta), there is the risk of a supply line short-circuiting the thyristors of the P and N groups. This can be explained as follows. It is assumed that the terminals B, D, and F of the three load windings of Fig. 6.24(b) are connected together to form a star point. The resulting circuit is shown in Fig. 6.24(c). The numbering of the thyristors of the three dual converters of Fig. 6.24(c) is given in Table 6.2. It can be seen that the thyristor Th_2' of the converter P of the dual converter D_1 and the thyristor Th_8 of the converter N of the dual converter D_2 are shorted through the supply line R.

Table 6.2

	D_1	D_2	D_3
N	Th_1 Th_2	Th_7 Th_8	Th_{13} Th_{14}
	Th_3 Th_4	Th_9 Th_{10}	Th_{15} Th_{16}
	Th_5 Th_6	Th_{11} Th_{12}	Th_{17} Th_{18}
P	Th_1' Th_2'	Th_7' Th_8'	Th_{13}' Th_{14}'
	Th_3' Th_4'	Th_9' Th_{10}'	Th_{15}' Th_{16}'
	Th_5' Th_6'	Th_{11}' Th_{12}'	Th_{17}' Th_{18}'

Isolation of load windings is disadvantageous because all the six terminals of the motor have to be brought out, and such motors are rarely manufactured. If each of the dual converters is supplied through a separate transformer as shown in Fig. 6.25, a motor with conventional star- or delta-connected windings can be connected as a load for the cycloconverter. Such an isolation at the source end will eliminate the risk of shorting of thyristors.

The discussions made so far is with the assumption that the output frequency of the cycloconverter is less than the supply frequency. In the single-phase cycloconverter of Fig. 6.10, reasonable values for power output, efficiency, and harmonic content can be obtained, with an output frequency which bears a ratio of less than 1/2 with respect to the input frequency. On the other hand, with the single-phase cycloconverter of Fig. 6.19(a), acceptable performance is possible with an output frequency of 0 to 1/3 the supply frequency. If the output frequency in increased to more than 1/2 and 1/3, respectively, in the single-phase and three-phase cases, the output voltage waveform will have considerable harmonic distortion. This can be explained to be due to the fact that the output voltage is constituted of fewer segments of supply voltage. Such a distortion will get minimized only when the above-mentioned conditions on the frequency ratios are fulfilled.

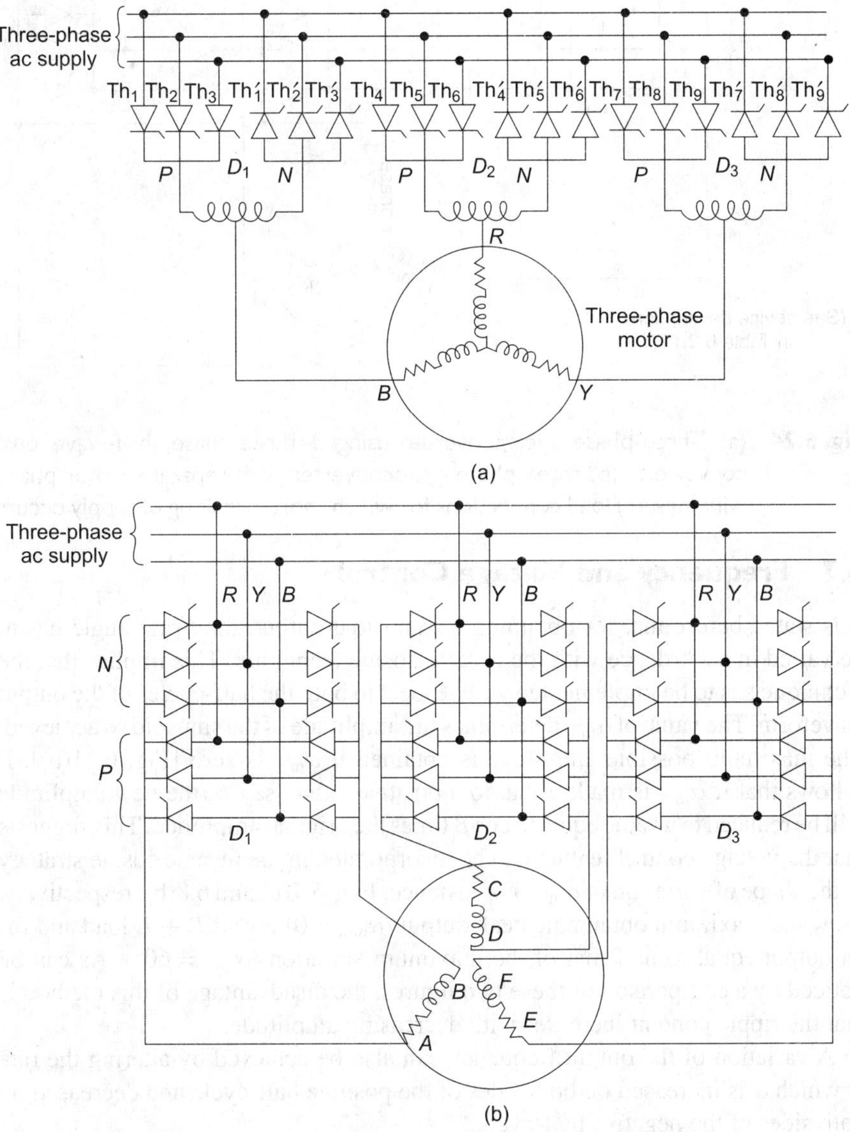

Figs 6.24(a) and (b)

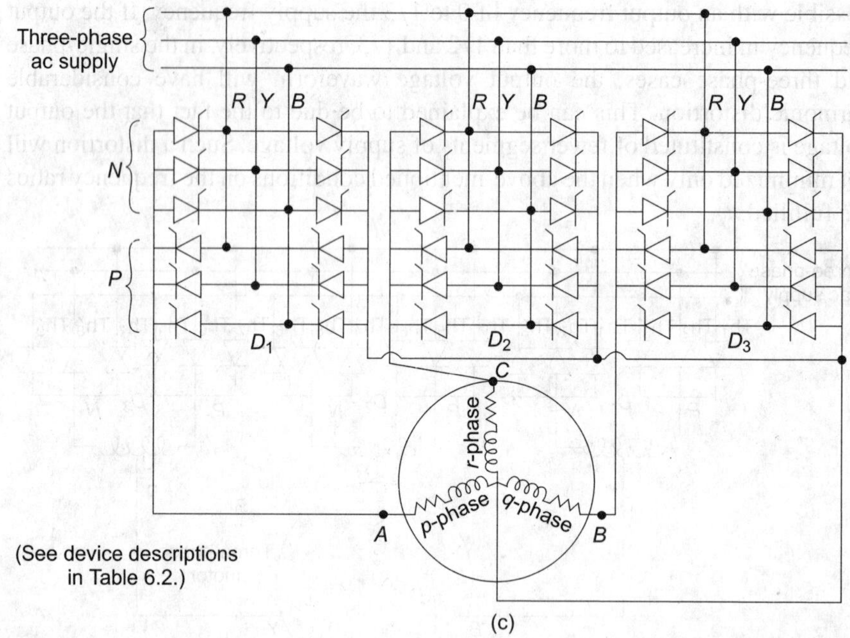

Fig. 6.24 (a) Three-phase cycloconverter using 3 three-phase, half-wave dual converters; (b) three-phase cycloconverter with separate motor phase windings; (c) load connections for which short-circuiting of supply occurs

6.7 Frequency and Voltage Control

It is stated before that for obtaining a sinusoidal output, the firing angle has to be varied in accordance with the inverse cosine technique. This implies that the technique has to be implemented with respect to both the half-cycles of the output waveform. The value of $\alpha_{\min}$ determines the amplitude of the sinusoid so achieved. The maximum possible amplitude is obtained if $\alpha_{\min}$ is zero [Fig. 6.11(b)]. It follows that if $\alpha_{\min}$ is made equal to a positive value, say β, the new amplitude will be reduced to a value equal to cos β times the original amplitude. This suggests that the voltage control feature can be incorporated in the inverse cosine strategy in the shape of variation of $\alpha_{\min}$. For instance, Figs 6.7(b) and 6.8(b), respectively, show the maximum obtainable peak output ($\alpha_{\min} = 0$) for an $R + L$ load and for an output equal to half that of the maximum variation ($\alpha_{\min} = 60°$). As can be noticed by a comparison of these two figures, the disadvantage of this method is that the ripple content increases with decreasing amplitude.

A variation of the output frequency can also be achieved by altering the rate at which α is increased on both sides of the positive half-cycle and decreased on both sides of the negative half-cycle.

A practical implementation of the two control features, namely, voltage and frequency, is shown with the help of the block diagram of Fig. 6.26. The output of a low-power master oscillator, which has the same voltage amplitude and frequency as the desired output of the cycloconverter, is connected to a block that incorporates a comparator and timing signal generator. The inverse

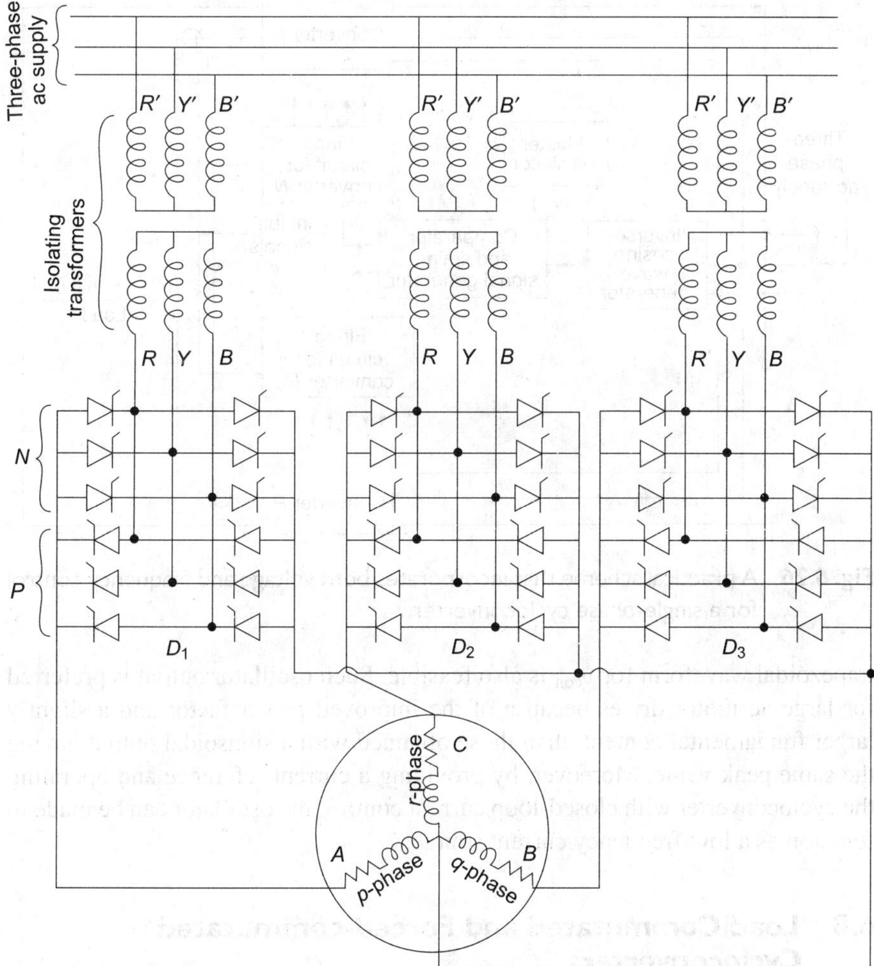

Fig. 6.25 Three-phase cycloconverter using isolating transformers for dual converters

cosine waveform generator gets its input from a three-phase supply; its output is connected to the comparator and compared with that of the master oscillator. The timing signal generator determines the firing angles α_P and α_N; accordingly, its two outputs are connected to the firing circuits of the P and N converters. As elaborated before, α_P and α_N are so adjusted that they approximately obey the relationship of Eqn (6.2). This approximation can be improved by reducing the ratio of the output frequency to that of the ac input frequency. Using the master oscillator has the merit that the direction of rotation of the drive motor can be reversed by just changing the phase sequence of the oscillator. Second, the voltage and frequency can be independently controlled. Third, control is also possible during the operation of the motor.

Though a sinusoidal output has been considered for the master oscillator in the above discussion, a reference oscillator and associated circuitry that give a

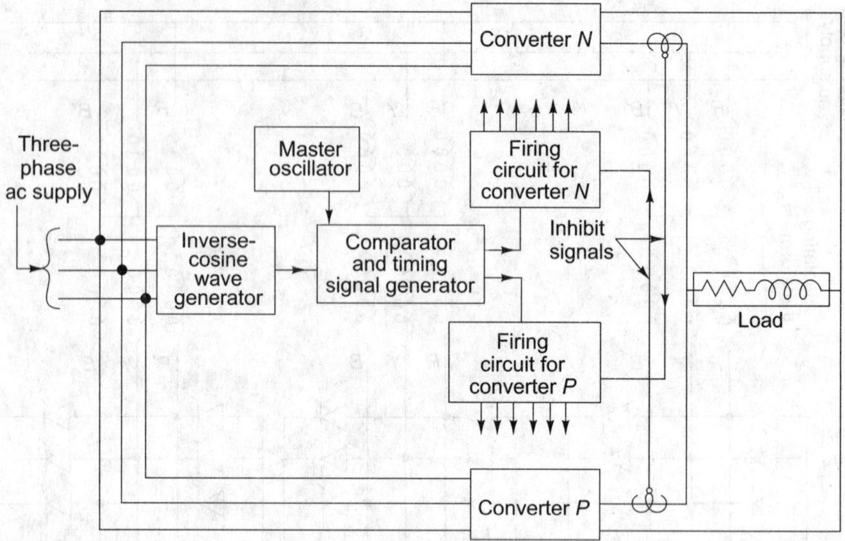

Fig. 6.26 A practical scheme that incorporates both voltage and frequency control for a single-phase cycloconverter

trapezoidal waveform for v_{con} is also feasible. Such oscillator output is preferred for large ac motor drives because of the improved power factor and a slightly larger fundamental content, than those obtained with a sinusoidal output having the same peak value. Moreover, by providing a current reference and operating the cycloconverter with closed-loop current control, the oscillator can be made to function as a low-frequency current source.

6.8 Load Commutated and Forced-commutated Cycloconverters

The converters dealt with so far are called *line commutated cycloconverters* because the commutation of the thyristors takes place naturally by the reversal of the ac line voltage. In a *load commutated cycloconverter*, commutation is achieved by the reversal of the load voltage. The requirement for this purpose is that the load must consist of a back emf which is independent of the source voltage, as in the case of the wound field and permanent magnet synchronous machines. Another constraint is that for deriving maximum torque per ampere in synchronous motors, the commutation should occur with a delay of 180 electrical degrees with respect to the machine voltage phasor. However, to provide a small margin for the turn-off of the thyristors and also commutation overlap, if any, commutation must be advanced with respect to this position. A shaft position sensor and associated control electronics are needed for this purpose. The main advantage of the load commutated cycloconverter is that the load frequency can be greater than that of the source. However, the provision of the shaft position sensor and associated electronic circuitry increases its cost.

It is stated above that the line commutated and load commutated cycloconverters are operated in such a way that the reversal of the line and load voltages, respectively, helps the natural commutation of the thyristors. A third method consists of providing forced commutation for the thyristors as in the case of an inverter. Its merit lies in the fact that the load frequency can be either higher of lower than the source frequency. Evidently the extra components needed for the commutation add to the cost and act as a deterrent for its use. However, with the advent of GTOs this drawback is no longer relevant and the cost has become competitive as compared to the line commutated type of cycloconverter.

6.9 Cycloconverter Versus Six-step VSI

A line commutated cycloconverter has the following advantages.

(a) Though the voltage source inverter (VSI) of Fig. 5.8(a) has been shown to have a battery as its dc source, it usually consists of an uncontrolled bridge rectifier and hence involves two stages of power conversion. The cycloconverter, on the other hand, has a single conversion stage, which implies lower conversion losses.

(b) Because of the adoption of natural commutation for the thyristors in a cycloconverter, similar to single- and three-phase rectifiers, simple conversion grade thyristors are permissible. As against this, forced commutation is employed in VSIs, and hence they need costly inverter grade thyristors.

(c) The cycloconverter can transfer power in either direction and hence handle reactive loads. It is also possible to obtain regenerative operation at full load torque over the complete speed range. On the other hand, regenerative braking of a VSI necessitates a second controlled rectifier.

(d) With the large number of thyristors (18 or 36) needed for three-phase cycloconverters, open-circuiting of any single thyristor does not lead to complete shutdown and the cycloconverter continues to operate with a slightly distorted output waveform. In addition, a balanced load is presented to the ac source even under unbalanced output conditions.

(e) The VSI gives a stepped output voltage waveform, which may cause non-uniform rotation of an ac motor, whereas the output voltage waveform of a cycloconverter is very nearly sinusoidal in nature.

The disadvantages of a line commutated cycloconverter are as follows.

(a) A commutation failure in any of the rectifiers of the dual converter may cause short-circuiting of the ac source, thus necessitating protection by means of intermediary fuses or circuit-breakers.

(b) A low-output-frequency waveform is obtained with the inverse cosine firing scheme. This frequency is usually limited to one-half or one-third of the input frequency. For example, with a source frequency of 50 Hz, the cycloconverter output frequency can at most be equal to 25 Hz. Such a limitation can be overcome by employing an engine-driven alternator as the ac source with a frequency of 100 Hz or more, so that the output frequency can be further raised. Alternatively, load commutated or forced-commutated cycloconverters can be employed with, of course, an increased cost. As a contrast, the frequency of the VSI can vary from any value below to any value above the base frequency. This is because its immediate input is dc, irrespective of the input supply being ac or dc.

(c) The control circuitry of a cycloconverter is quite complex in comparison to that of a VSI. It is therefore economical only for motor ratings exceeding 10 kV A. Using a two-phase ac motor as the load is one method of reducing this complexity.

(d) The cycloconverter presents a low input power factor, this being further decreased with a reduction of the output voltage. This contrasts with the high input power factor obtainable for a VSI consequent to use of an uncontrolled rectifier at its input side.

(e) The cycloconverter needs an ac source. Hence, with dc supply as the primary source, conversion to ac is a necessity. Such a conversion offsets the advantage of having a single conversion stage in a cycloconverter.

Summary

The cycloconverter is a static converter which directly converts a fixed voltage, fixed frequency input to variable voltage, variable frequency output. The output frequency of a line commutated cycloconverter is a submultiple of the input frequency. Its output can be varied by either simultaneous or non-simultaneous methods of control. Though the ohmic loss in simultaneous control adds to the total losses, it gives faster response and has the merit of smooth changeover from one mode of operation to another. The inverse cosine firing scheme ensures that there is negligible flow of dc circulating current through the dual converters. Voltage control of a cycloconverter is achieved by varying the α_{min} of the rectifiers whereas frequency control is implemented by varying the frequency of the master oscillator.

The limitation of the output frequency associated with a line commutated cycloconverter does not exist in the case of (i) forced-commutated and (ii) load commutated cycloconverters. However, the commutation circuitry for the first case and the additional features such as the shaft position sensor and electronic circuitry for the second case add to the overall cost of these cycloconverters.

The waveform of the line commutated cycloconverter is smoother than that of a VSI. It has good flexibility of operation by way of feasibility of regenerative braking, whereas the VSI needs additional equipment to incorporate this feature. The large number of thyristors used for a cycloconverter help in avoiding shutdown in the case of the open-circuiting of any single thyristor.

The drawbacks of a line commutated cycloconverter are (a) complexity of its control circuitry, which makes it economical only for ratings above 10 kV A, and (b) its low input power factor. However, with the merits elaborated above, it is ideally suited for low-frequency, high-capacity drives.

Worked Examples

1. A three-phase to single-phase cycloconverter is fed by a delta–star transformer and a three-phase, full-wave fully controlled dual converter. The RMS line-to-neutral voltage of the transformer secondary is 200 V. A load with $R = 1.2\ \Omega$ and $L = 10$ mH is connected at the output. The input and output frequencies of the cycloconverter are 50 Hz and 16.66 Hz, respectively. Assuming a voltage reduction factor (r) of 0.98, determine the (a) input displacement factor, (b) total power factor, (c) distortion factor, (d) kV A rating of the transformer, (e) half-cycle average current rating of the thyristors, and (f) RMS current rating of the thyristors.

Solution

(a) From Eqn (6.18),

$$V_{CC(RMS)} = \frac{3\sqrt{3}}{\pi} r V_S = \frac{3\sqrt{3}}{\pi} \times 0.98 \times 200 = 324.2 \text{ V}$$

$$I_{CC} = \frac{V_{CC}}{Z_{ld}} = \frac{V_{CC}}{\sqrt{R_{ld}^2 + (\omega_{CC} L_{ld})^2}}$$

$$= \frac{324.2}{\sqrt{(1.2)^2 + \left(\frac{314}{3} \times \frac{10}{1000}\right)^2}} = \frac{324.2}{1.59} = 203.9 \text{ A}$$

$$\phi_{CC} = \tan^{-1}\left(\frac{\omega_{CC} L_{ld}}{R_{ld}}\right) = \tan^{-1}\left(\frac{\frac{314}{3} \times \frac{10}{1000}}{1.2}\right) = 41^\circ$$

$$\cos \phi_{CC} = 0.754$$

For $r = 0.98$ and a load phase angle (ϕ_{CC}) of 41° (lag), the input displacement factor $\cos \phi_i$ is obtained from Eqn (6.19) as 0.71.

(b) From Eqn (6.20), the total power factor is

$$\lambda = \frac{r \cos \phi_{CC}}{\sqrt{2}} = \frac{0.98}{\sqrt{2}} \times 0.754 = 0.522$$

(c) From Eqn (6.21), the distortion factor is

$$\mu = \frac{\lambda}{\cos \phi_i} = \frac{0.522}{0.71} = 0.736$$

(d) From Eqn (6.28),

$$I_1 = 0.7 I_{CC} = 143 \text{ A}$$

The transformer secondary current

$$I_S = \frac{I_1}{\mu} = \frac{143}{0.736} = 194 \text{ A}$$

From the given data, the line-to-neutral voltage of the transformer is $V_S = 200$ V. Hence the transformer kVA rating is

$$S = \frac{3V_S I_S}{1000} = \frac{3 \times 200 \times 194}{1000} = 116.4 \text{ kV A}$$

From Eqn (6.35), the maximum current rating of the thyristor is

$$I_{\text{ThM}} = \sqrt{2} I_{\text{CC}} = \sqrt{2} \times 203.9 = 288.36 \text{ A}$$

(e) The half-cycle average current rating of the thyristor is

$$\frac{2\sqrt{2} I_{\text{CC}}}{3\pi} = \frac{2\sqrt{2} \times 203.9}{3\pi} = 61.2 \text{ A}$$

(f) The RMS current rating of the thyristor is obtained from Eqn (6.34) as

$$I_{\text{Th(RMS)}} = \frac{I_{\text{CC}}}{\sqrt{6}} = \frac{203.9}{\sqrt{6}} = 83.2 \text{ A}$$

2. A three-phase to single-phase cycloconverter which is fed by a three-phase dual converter has input and output frequencies of 50 Hz and 16.66 Hz, respectively. Its load consists of $R = 2.0\ \Omega$ and $L = 20$ mH. If the transformer secondary rating is 108 A, determine the (a) input displacement factor, (b) distortion factor, (c) fundamental component of the converter input current, (d) kVA rating of the transformer, (e) maximum and half-cycle average current ratings of the thyristors, and (f) RMS current rating of the thyristors.

Solution

(a) $$\cos\phi_{\text{CC}} = \frac{R_{\text{ld}}}{\sqrt{R_{\text{ld}}^2 + (\omega_{\text{CC}} L_{\text{ld}})^2}} = \frac{2}{\sqrt{2^2 + \left(\frac{314}{3} \times \frac{20}{1000}\right)^2}} = 0.69$$

This gives

$$\phi_{\text{CC}} = 46.37° \quad \text{or} \quad 0.81 \text{ rad}$$

Assuming that the voltage reduction factor r is unity and using Eqn (6.19), the input displacement factor is

$$\cos\phi_i = \left[1 + 4\left\{\frac{\cos\phi_{\text{CC}} + \phi_{\text{CC}} \sin\phi_{\text{CC}}}{\pi \cos\phi_{\text{CC}}}\right\}^2\right]^{-1/2} = 0.647$$

From Eqn (6.20),

$$\lambda = \frac{r}{\sqrt{2}} \cos\phi_{\text{CC}} = \frac{1}{\sqrt{2}} \times 0.69 = 0.488$$

(b) The distortion factor

$$\mu = \frac{\lambda}{\cos\phi_i} = \frac{0.488}{0.647} = 0.754$$

(c) The transformer secondary current rating $I_S = 108$ A. Hence the fundamental component of the converter input current

$$I_1 = \mu I_S = 0.754 \times 108 = 81.4 \text{ A}$$

(d) From Eqn (6.28),

$$I_{CC} = \frac{I_1}{0.7} = \frac{81.4}{0.7} = 116.3 \text{ A}$$

$$Z_{ld} = \sqrt{2^2 + \left(\frac{314}{3} \times \frac{20}{1000}\right)^2} = 2.9 \, \Omega$$

$$V_{CC} = I_{CC} Z_{ld} = 116.3 \times 2.9 = 337 \text{ V}$$

Using Eqn (6.18),

$$V_{CC(RMS)} = \frac{3\sqrt{3}}{\pi} r V_S$$

Substituting values for r and V_{CC} gives

$$\frac{3\sqrt{3}}{\pi} \times 1 \times V_S = 337$$

Hence,

$$V_S = \frac{337\pi}{3\sqrt{3}} = 204.0 \text{ V}$$

Finally, the transformer rating

$$S = \frac{3 V_S I_S}{1000} = \frac{3 \times 204 \times 108}{1000} = 66 \text{ kV A}$$

(e) The maximum thyristor current

$$I_{ThM} = \sqrt{2} I_{CC} = \sqrt{2} \times 116.3 = 164.5 \text{ A}$$

The half-cycle average of the thyristor current

$$I_{HA} = \frac{2\sqrt{2} I_{CC}}{3\pi} = \frac{2\sqrt{2} \times 116.3}{3\pi} = 34.9 \text{ A}$$

(f) The RMS current rating of the thyristors,

$$I_{Th(RMS)} = \frac{I_{CC}}{\sqrt{3}} = \frac{116.3}{\sqrt{3}} = 67.14 \text{ A}$$

3. A three-phase cycloconverter employing three full-wave bridge dual converters (type DT-T) supplies a low-speed synchronous motor that drives a cement mill. The data for the set-up are shaft power of the motor = 3000 kW, rated line-to-neutral voltage (V_{CC}) = 1000 V (RMS), rated frequency = 5 Hz, power factor of the motor, $\cos \phi_{CC} = 0.95$, efficiency of the motor = 90%, primary frequency of the three-phase transformer feeding the cycloconverter = 50 Hz, and voltage reduction factor (r) = 1.0. Compute the (a) kVA rating of the transformers and (b) maximum, average, and RMS current ratings of the thyristors.

Solution

The set-up is shown in Fig. 6.24(c).

(a) Input power to the motor = shaft power/η = 3000/0.9 = 3333 kW. Motor input per phase = 1111 kW. Motor current per phase

$$I_{CC} = \frac{\text{motor input per phase}}{V_{CC}\text{ per phase}} = \frac{1111 \times 1000}{1000 \times 0.95} = 1170\text{ A}$$

From Eqn (6.18),

$$V_{CC} = \frac{3\sqrt{3}}{\pi} r V_S$$

Substituting values gives

$$1000 = \frac{3\sqrt{3}}{\pi} \times 1 \times V_S$$

Thus,

$$V_S = \frac{1000\pi}{3\sqrt{3}} = 604.6\text{ V}$$

Also,

$$\cos\phi_{CC} = 0.95$$

From this,

$$\phi_{CC} = 0.318\text{ rad and } \sin\phi_{CC} = 0.312$$

From Eqn (6.19), the input displacement factor

$$\cos\phi_i = \left\{1 + 4\left[\frac{\cos\phi_{CC} + \phi_{CC}\sin\phi_{CC}}{\pi\cos\phi_{CC}}\right]^2\right\}^{-1/2}$$

$$= \left\{1 + \left[\frac{0.95 + 0.318 \times 0.312}{\pi \times 0.95}\right]^2\right\}^{-1/2}$$

$$= 0.818$$

Total power factor

$$\lambda = \frac{r}{\sqrt{2}}\cos\phi_{CC} = \frac{1}{\sqrt{2}} \times 0.95 = 0.672$$

Distortion factor

$$\mu = \frac{\lambda}{\cos\phi_i} = \frac{0.672}{0.818} = 0.822$$

From Eqn (6.28),

$$I_1 = 0.7 I_{CC} = 0.7 \times 1170 = 819\text{ A}$$

The transformer secondary current

$$I_S = \frac{I_1}{\mu} = \frac{819}{0.822} = 996.4 \text{ A}$$

The kVA rating of the transformer per phase of the cycloconverter is

$$\frac{3V_S I_S}{1000} = \frac{3 \times 604.6 \times 996.4}{1000} = 1807 \text{ kV A}$$

(b) The maximum instantaneous thyristor current

$$\sqrt{2} I_{CC} = 1170\sqrt{2} = 1654.6 \text{ A}$$

The half-cycle average of the thyristor current is

$$\frac{2\sqrt{2} I_{CC}}{3\pi} = \frac{2\sqrt{2} \times 1170}{3\pi} = 351 \text{ A}$$

(c) As each pair of thyristors conducts for one-third of the time, the RMS value of the current through a thyristor is

$$\frac{I_{CC}}{\sqrt{3}} = \frac{1170}{\sqrt{3}} = 675.5 \text{ A}$$

4. The single-phase to single-phase cycloconverter of Fig. 6.5 feeds an inductive load and operates with constant firing angle for the P and N rectifiers. It has the following data: input voltage = 180 V (RMS) at 50 Hz, output frequency = 16.66 Hz, the load is inductive with $R = 2\ \Omega$ and $L = 8$ mH, and the voltage reduction factor = 0.99. Both the positive- and negative-side firing angles are equal to 80°. Compute the (a) RMS value of the fundamental component of the output voltage, (b) input power factor, and (c) maximum, average, and RMS currents through the thyristors.

Solution

(a) The input and load voltage waveforms are given in Fig. 6.27. As the frequency of the cycloconverter output is 16.66 Hz, the load voltage waveform will have three input half-waves on the positive as well as negative sides, each half-wave starting from the appropriate firing angle.

The X-axis scale of the load voltage waveforms is set according to the frequency of the load, namely, 16.66 Hz. Hence, an α_P of 80° at a frequency of 50 Hz becomes 80°/3, which is just $\alpha_1/3$, with respect to this scale. Similarly, for the second and third positive half-cycles, the firing angles will, respectively, be

$$\frac{\alpha_2}{3} = \frac{80^\circ}{3} + 60^\circ = \frac{260^\circ}{3} \qquad \text{and} \qquad \frac{\alpha_3}{3} = \frac{80^\circ}{3} + 120^\circ = \frac{440^\circ}{3}$$

For the purpose of computing the RMS value of the fundamental component of the output voltage, one pair of waveforms (on the positive and negative sides) is considered at a time and the Fourier coefficients of the pair computed. The equation

$$v_{ld}(t) = V_m \sin \omega t = V_m \sin(314t)$$

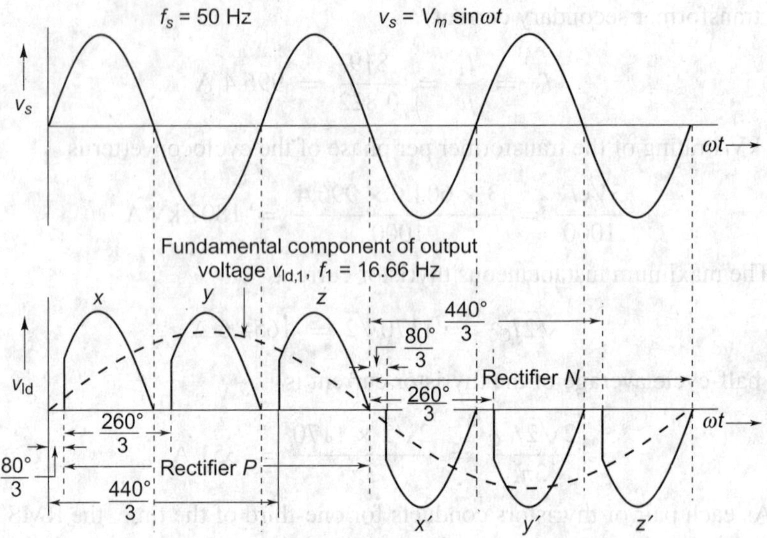

Fig. 6.27

with a frequency of 50 Hz now becomes

$$v_{ld}(t) = V_m \sin(3\omega_1 t - \theta) = V_m \sin\left\{3\frac{314}{3}t - \theta\right\}$$

where θ depends on the location of the starting point of the waveform.

The Fourier series for the load voltage is now written as

$$v_{ld}(t) = \sum_{n=1}^{\infty}(A_n \cos n\omega_1 t + B_n \sin n\omega_1 t)$$

The coefficients A_n will all be zero because of odd symmetry. Thus,

$$v_{ld}(t) = \sum_{n=1}^{\infty} B_n \sin n\omega_1 t$$

The fundamental component $B_1 \sin n\omega_1 t$ can now be written as

$$B_1 = b_1 + b_2 + b_3$$

where b_1, b_2, and b_3 are, respectively, the fundamental Fourier coefficients for the pairs of waveforms (x, x'), (y, y'), and (z, z'). Thus,

$$b_1 = \frac{2}{\pi}\int_{\alpha_1/3}^{\pi/3}(V_m \sin 3\omega_1 t)\sin\omega_1 t \, d(\omega_1 t)$$

where $\alpha_1 = 80\pi/180$ rad.

$$b_1 = \frac{2V_m}{\pi}\int_{\alpha_1/3}^{\pi/3}\frac{1}{2}(\cos 2\omega_1 t - \cos 4\omega_1 t)d(\omega_1 t)$$

$$= \frac{V_m}{\pi}\left[\frac{\sin 2\omega_1 t}{2} - \frac{\sin 4\omega_1 t}{4}\right]$$

with the lower limit $\alpha/3$ and upper limit $\pi/3$.

$$b_1 = \frac{V_m}{\pi}\left\{\left(\frac{\sin(2\pi/3)}{2} - \frac{\sin(4\pi/3)}{4}\right) - \left(\frac{\sin(2\alpha_1/3)}{2} - \frac{\sin(4\alpha_1/3)}{4}\right)\right\}$$

with $\alpha_1 = 80\pi/180$ rad. Substitution of values yields

$$b_1 = 0.489\frac{V_m}{\pi}$$

Now,

$$b_2 = \frac{2}{\pi}\int_{\alpha_2/3}^{2\pi/3} V_m \sin(3\omega_1 t - \pi)\sin\omega_1 t\, d(\omega_1 t)$$

where $\alpha_2 = 260\pi/180$ rad.

$$b_2 = \frac{2V_m}{\pi}\int_{\alpha_2/3}^{2\pi/3} \frac{1}{2}[\cos(2\omega_1 t - \pi) - \cos(4\omega_1 t - \pi)]d(\omega_1 t)$$

$$= \frac{V_m}{\pi}\left[\frac{\sin(2\omega_1 t - \pi)}{2} - \frac{\sin(4\omega_1 t - \pi)}{4}\right]$$

with the lower limit $\alpha_2/3$ and upper limit $2\pi/3$.

$$b_2 = \frac{V_m}{\pi}\left\{\left(\frac{\sin(4\pi/3 - \pi)}{2} - \frac{\sin(8\pi/3 - \pi)}{4}\right) - \left(\frac{\sin(2\alpha_2/3 - \pi)}{2} - \frac{\sin(4\alpha_2/3 - \pi)}{4}\right)\right\}$$

Substitution of values yields

$$b_2 = \frac{0.765V_m}{\pi}$$

Also,

$$b_3 = \frac{2}{\pi}\int_{\alpha_3/3}^{\pi} V_m \sin(3\omega_1 t - 2\pi)\sin\omega_1 t\, d(\omega_1 t)$$

where $\alpha_3 = 440\pi/180$ rad.

$$b_3 = \frac{2V_m}{\pi}\int_{\alpha_3/3}^{\pi} \frac{1}{2}[\cos(2\omega_1 t - 2\pi) - \cos(4\omega_1 t - 2\pi)]d(\omega_1 t)$$

$$= \frac{V_m}{\pi}\left[\frac{\sin(2\omega_1 t - 2\pi)}{2} - \frac{\sin(\omega_1 t - 2\pi)}{4}\right]$$

with the lower limit $\alpha_3/3$ and upper limit π.

$$b_3 = \frac{V_m}{\pi}\left[\left(\frac{\sin(2\pi - 2\pi)}{2} - \frac{\sin(4\pi - 2\pi)}{4}\right) - \left(\frac{\sin(2\alpha_3/3 - 2\pi)}{2} - \frac{\sin(4\alpha_3/3 - 2\pi)}{4}\right)\right]$$

Substitution of values gives

$$b_3 = \frac{0.277V_m}{\pi}$$

Finally,

$$B_1 = (0.489 + 0.765 + 0.277)\frac{V_m}{\pi} = \frac{1.531V_m}{\pi}$$

$$= \frac{1.531 \times 180\sqrt{2}}{\pi} = 124 \text{ V}$$

Thus the fundamental component of the load voltage waveform becomes

$$v_{\mathrm{ld},1}(t) = 124 \sin \omega_1 t \quad \text{with} \quad \omega_1 = 314/3$$

The RMS value of this fundamental component is

$$V_{\mathrm{ld},1} = \frac{124}{\sqrt{2}} = 87.7 \text{ V}$$

Likewise, the RMS value of the fundamental component of the load current is obtained as

$$I_{\mathrm{ld},1} = \frac{V_{\mathrm{ld},1}}{Z_{\mathrm{ld}}} = \frac{87.7}{\sqrt{2^2 + \left(\frac{314}{3} \times \frac{8}{1000}\right)^2}} = \frac{87.7}{2.168} = 40.45 \text{ A}$$

(b) The RMS value of the cycloconverter voltage (V_{CC}), which can be interpreted as the transformer secondary voltage, is obtained as

$$V_{\mathrm{CC}} = \frac{2}{2\pi}\left[\int_{\alpha_1/3}^{\pi/3} V_m^2 \sin^2(3\omega_1 t) d(\omega_1 t) + \int_{\alpha_2/3}^{2\pi/3} V_m^2 \sin^2(3\omega_1 t - \pi) d(\omega_1 t) + \int_{\alpha_3/3}^{\pi} V_m^2 \sin^2(3\omega_1 t - 2\pi) d(\omega_1 t)\right]$$

$$= \sqrt{\frac{2V_m^2}{2\pi}\left[\int_{\alpha_1/3}^{\pi/3}\left(\frac{1-\cos 6\omega_1 t}{2}\right) d(\omega_1 t) + \int_{\alpha_2/3}^{2\pi/3}\left(\frac{1-\cos(6\omega_1 t - \pi)}{2}\right) d(\omega_1 t) + \int_{\alpha_3/3}^{\pi}\left(\frac{1-\cos(6\omega_1 t - 2\pi)}{2}\right) d(\omega_1 t)\right]}$$

$$= \sqrt{\frac{V_m^2}{2\pi} \times 3 \times 0.639}$$

$$= \sqrt{\frac{\left(180\sqrt{2}\right)^2}{2\pi} \times 3 \times 0.639} = 140.6 \text{ V}$$

$$Z_{ld} = \sqrt{2^2 + \left(\frac{314}{3} \times \frac{8}{1000}\right)^2} = \sqrt{2^2 + (0.837)^2} = 2.168$$

The cycloconverter current I_{CC}, which can be interpreted as the transformer secondary current, becomes

$$I_{CC} = \frac{V_{CC}}{Z_{ld}} = \frac{140.6}{2.168} = 64.86 \text{ A}$$

The load power factor is

$$\cos\phi_{CC} = \frac{2}{\sqrt{2^2 + \left(\frac{314}{3} \times \frac{8}{1000}\right)^2}} = \frac{2}{2.168} = 0.923$$

So,

$$\phi_{CC} = 22.63° \text{ or } 0.395 \text{ rad}$$

and

$$\sin\phi_{CC} = 0.385$$

From Eqn (6.19), the input displacement factor $\cos\phi_i = 0.803$. The distortion factor becomes

$$\mu = \frac{\text{fundamental component of load current}}{\text{transformer secondary current}}$$

Thus,

$$\mu = \frac{I_{ld,1}}{I_{CC}} = \frac{40.45}{64.86} = 0.624$$

The total power factor is

$$\lambda = \mu \cos\phi_i = 0.624 \times 0.803 = 0.5$$

(c) The RMS and half-average of the thyristor pairs in the P and N rectifiers can be arrived at by an inspection of the single-phase cycloconverter of Fig. 6.5 and the waveforms given in Fig. 6.27.

The positive side waveforms x, y, and z are obtained as outputs of rectifier P. Here, waveforms x and z are obtained through the pair (Th_1, Th_3), whereas waveform y is obtained through the pair (Th_2, Th_4). Similarly, the negative side waveforms x' and z' are obtained through the pair (Th_5, Th_7) and waveform y' is obtained through the pair (Th_6, Th_8). Thus, each of the pairs, that is, (Th_1, Th_3) of rectifier P and (Th_5, Th_7) of rectifier N, conducts two input half-waves. Also, each of the thyristor pairs (Th_2, Th_4) and (Th_6, Th_8) conducts one input half-wave. Hence, $I_{Th(RMS)}$ of a pair conducting two half-waves [(Th_1, Th_3) or (Th_5, Th_7)]

$$= \sqrt{\frac{2}{\pi} \int_{\alpha_1/3}^{\pi/3} (\sqrt{2} I_{CC})^2 d(\omega t)}$$

where $\alpha_1 = 80°$. This works out to

$$I_{\text{Th(RMS)}}(\text{two half-waves}) = \sqrt{\frac{1}{\pi} \times 2I_{CC}^2\left[\frac{\pi}{3} - \frac{80\pi}{3 \times 180}\right]}$$

$$= 0.61I_{CC} = 0.61 \times 64.86 = 39.5 \text{ A}$$

Also, $I_{\text{Th(RMS)}}$ for a pair conducting one half-wave [(Th_2, Th_4) or (Th_6, Th_8)]

$$= \frac{39.5}{\sqrt{2}} = 27.9 \text{ A}$$

Considering the half-cycle average,

$$I_{\text{HA}}(\text{one of the first pairs}) = \frac{1}{2\pi}\int_{80°/3}^{60°} \sqrt{2}I_{CC}d(\omega t)$$

$$= \frac{\sqrt{2}I_{CC}}{2\pi}\left[\frac{\pi}{3} - \frac{80\pi}{3 \times 180}\right]$$

$$= \frac{\sqrt{2} \times 64.86}{2}\left[\frac{1}{3} - \frac{80}{540}\right]$$

$$= 8.5 \text{ A}$$

and

$$I_{\text{HA}}(\text{one of the second pairs}) = \frac{I_{\text{HA}}(\text{one of the first pairs})}{2}$$

$$= \frac{8.5}{2} = 4.25 \text{ A}$$

Exercises

Multiple Choice Questions

1. The purpose of reactors in a simultaneously controlled cycloconverter is ______.
   (a) to reduce the overcurrent during sudden load changes
   (b) to suppress current transients occurring during switching conditions
   (c) to keep down the circulating current that flows under normal operation
   (d) to suppress voltage spikes that occur during switching conditions
2. As compared to simultaneous control, non-simultaneous control has the drawback of ______.
   (a) high circuit cost
   (b) more I^2R losses
   (c) large current ratings for thyristors
   (d) large voltage ratings for thyristors
3. Not maintaining the relation $\alpha_P + \alpha_N = 180°$ in a simultaneously controlled cycloconverter leads to ______.
   (a) short-circuiting of the ac source (b) high voltage transients
   (c) high current transients (d) high dc circulating current
4. The peak inverse rating of a line commutated cycloconverter must be at least ______, where V_m is the peak line-to-neutral voltage of the ac supply.

(a) $\sqrt{2}V_m$
(b) $(\sqrt{3}/2)V_m$
(c) $V_m/2$
(d) $\sqrt{3}V_m$

5. The input power factor presented by a line commutated cycloconverter to the input supply is ______ that presented by a VSI.

(a) higher than
(b) lower than
(c) equal to
(d) much higher than

6. The RMS output voltage of a single-phase cycloconverter fed by a three-phase dual converter, supplied from an ac line-to-neutral voltage of 400 V (RMS), and having a voltage reduction factor of 0.96 will be nearly ______ V.

(a) 635.1
(b) 854.6
(c) 576.8
(d) 752.3

7. The torque pulsations in a line commutated, three-phase cycloconverter will be ______ those in a three-phase, six-step VSI.

(a) an order smaller than
(b) an order larger than
(c) much larger than
(d) nearly equal to

8. If the output voltage of a single-phase cycloconverter fed by a three-phase dual converter is 220 V (RMS) with a voltage reduction factor of 0.97, then the RMS magnitude of the input supply voltage will be nearly ______.

(a) 173.7 V
(b) 137.3 V
(c) 123.6 V
(d) 137.1 V

9. If the distortion factor is 0.9 and the total input power is 0.78, then the fundamental power factor angle ϕ_1 is nearly ______.

(a) 45°
(b) 60°
(c) 30°
(d) 22.5°

10. If a single-phase cycloconverter has an RMS output voltage of 250 V and feeds a load consisting of $R = 15\ \Omega$ and $X_L = 20\ \Omega$, the maximum instantaneous value of the current will be ______ A.

(a) 7.07
(b) 10.5
(c) 14.14
(d) 9.6

11. If a single-phase cycloconverter feeding a load composed of $R = 30\ \Omega$ and $X_L = 40\ \Omega$ has an output voltage of 250 V (RMS), the average current rating of the thyristors should be nearly ______ A.

(a) 1.5
(b) 5.1
(c) 3.5
(d) 4.2

12. If a single-phase, line commutated cycloconverter is fed by a three-phase, half-wave dual converter with a line-to-neutral voltage of 120 V (RMS), the forward blocking voltage of the thyristors should not be less than ______ V.

(a) 324
(b) 512
(c) 408
(d) 294

13. If the output current of a single-phase cycloconverter is 50 A, then the half-cycle average of the load current will be nearly ______ A.

(a) 36.4 (b) 45.0
(c) 54.1 (d) 40.8

14. If the output current of a single-phase cycloconverter is 40 A, then the average current through any of the thyristors will be nearly ______ A.

    (a) 12.0 (b) 14.2
    (c) 13.5 (d) 18.4

15. If the output current of a single-phase cycloconverter is 45 A, then the RMS value of the current through any of the thyristors of a three-phase, full-wave rectifier will be nearly ______ A.

    (a) 28.0 (b) 38.4
    (c) 36.7 (d) 18.4

16. The harmonic content in the output of a single-phase cycloconverter will be ______ that in a six-step inverter.

    (a) much more than (b) equal to
    (c) less than (d) slightly more than

17. The total power factor of a cycloconverter will always be ______ the distortion factor.

    (a) slightly more than (b) equal to
    (c) less than (d) much more than

18. During the simultaneous control of a cycloconverter, if rectifier P is initially conducting the load current and if it is desired that it should be transferred to rectifier N, the operations required are ______.

    (a) α_P is reduced to a minimum and then α_N is increased to a maximum
    (b) α_P is increased to a maximum and then α_N is increased to a maximum
    (c) α_P is increased to a maximum and then α_N is decreased to a minimum
    (d) α_P is decreased to a minimum and then α_N is decreased to a minimum

19. The output frequency of a three-phase cycloconverter should always be less than one-third the supply frequency, as otherwise ______.

    (a) there is likelihood of very high voltage peaks
    (b) it will not be possible to implement the inverse cosine scheme
    (c) firing angle control becomes more complicated
    (d) harmonic distortion in the output voltage increases

20. For a cycloconverter, if the voltage reduction factor $r = 0.95$ and $\phi_{CC} = 30°$, then the value of the input power factor λ will be nearly ______.

    (a) 1.23 (b) 0.58
    (c) 0.87 (d) 0.65

True or False

Indicate whether the following statements are *true* or *false*. Briefly justify your choice.

1. The purpose of reactors in a simultaneously controlled cycloconverter is to reduce the overcurrent during sudden load changes.
2. Non-simultaneous control of cycloconverters has the drawback of increasing the ohmic losses.

3. Not maintaining the relation $\alpha_N + \alpha_P = 180°$ in a simultaneously controlled cycloconverter leads to shorting of the ac source.
4. No separate ac or dc source is needed for implementing the inverse cosine firing scheme in a cycloconverter.
5. The peak inverse voltage rating of thyristors used in a cycloconverter should be at least $\sqrt{3}V_m$, where V_m is the peak line-to-neutral voltage of the ac supply.
6. Voltage control of a line commutated cycloconverter is achieved by changing the rate at which α is decreased as the peak is approached.
7. Regenerative braking is easier to implement with a line commutated cycloconverter than with a VSI.
8. Protection against commutation failure of the thyristors in a line commutated cycloconverter is necessary.
9. The relationship $\alpha_N + \alpha_P = 180°$ cannot be exactly satisfied in the dual converters used in single- and three-phase cycloconverters.
10. The power factor presented by a line commutated cycloconverter to the input supply is higher than that presented by a VSI.
11. Current sensing devices are necessary when non-simultaneous control is implemented for a cycloconverter.
12. The output voltage of a single-phase cycloconverter supplied from an ac input voltage of 350 V (RMS) and having a voltage reduction factor of 0.98 will be nearly 228 V (RMS).
13. For a single-phase cycloconverter, if the output voltage is nearly 353 V and the voltage reduction factor is 0.97, then the magnitude of the supply voltage will be 220 V.
14. If the distortion factor of a single-phase ac cycloconverter is 0.9 and the total power factor is 0.45, then the fundamental power factor angle ϕ_1 should be 60°.
15. If a single-phase cycloconverter that feeds a load consisting of $R = 15\ \Omega$ and $X_L = 20\ \Omega$ has an RMS output voltage of 250 V, the maximum and half-average current ratings of the thyristors will have to be kept, respectively, at 10 A and 6 A.
16. The shape of the fundamental component of the output voltage waveform in a cycloconverter is always an amplified version of the control voltage.
17. The maximum current rating of the thyristors used in a simultaneously controlled cycloconverter will be less than that used in a non-simultaneously controlled one.
18. The torque pulsations in a line commutated, three-phase cycloconverter will be larger than those in a three-phase, six-step VSI.

Short Answer Questions

1. Explain the principle of a single-phase cycloconverter.
2. How is a near sinusoidal output obtained for single-phase cycloconverters?
3. What are the merits and demerits of simultaneous controls when implemented in a cycloconverter?

4. What are the steps involved in the changeover of conduction from one rectifier to another in a simultaneously controlled cycloconverter?
5. How is the changeover of conduction from one rectifier to another effected in the case of a non-simultaneously controlled cycloconverter?
6. Explain the operation of a cycloconverter that feeds a reactive load.
7. Describe the firing scheme for a single-phase cycloconverter that is rigged up with a three-phase dual converter.
8. How does four-quadrant operation occur in a single-phase cycloconverter?
9. How does circulating current occur in a simultaneously controlled cycloconverter and what are its drawbacks?
10. Define distortion factor and displacement factor with regard to a cycloconverter and explain their significance.
11. How are frequency and voltage controls achieved for a cycloconverter?
12. Explain the principle of a load commutated cycloconverter and discuss its advantages and disadvantages over line commutated converters.
13. What are the applications of cycloconverters?
14. What are the advantages of a cycloconverter over the six-step VSI?
15. What are the drawbacks of a line commutated cycloconverter?
16. What are the events that take place when one of the thyristors gets open-circuited (a) in a cycloconverter using single-phase, half-wave rectifiers and (b) in a cycloconverter that consists of three-phase, full-wave rectifiers?
17. Explain why the dual converters used in a three-phase cycloconverter have to be isolated either at the input or at the load sides.

Problems

1. The type DT-S cycloconverter of Fig. 6.12 is fed by a delta–star transformer. The RMS line-to-neutral voltage of the secondary is 170 V. A load with $R = 1.6\ \Omega$ and $L = 15$ mH is connected at the output. The input and output frequencies of the cycloconverter are 50 Hz and 12.5 Hz, respectively. Assuming a voltage reduction factor of 0.96, determine the (a) transformer rating, (b) maximum, average, and RMS currents through the thyristors, and (c) input displacement factor, distortion factor, and total power factor.

   Ans. (a) 63.43 kV A; (b) 192.16 A, 40.78 A, 55.5 A; (c) 0.715, 0.765, 0.547

2. The DT-S cycloconverter of Fig. 6.12 has input and output frequencies of 50 Hz and 16.66 Hz, respectively. The following data are given: input displacement factor is 0.32, secondary current of transformer is 110 A, distortion factor, is 0.68, and phase voltage of transformer secondary is 550 V. Compute the following quantities: (a) load resistance R, (b) Load inductance L, (c) load current I_{CC}, (d) load voltage V_{CC} (e) maximum, average, and RMS currents through the thyristors.

   Ans. (a) 2.772 Ω; (b) 76.9 mH; (c) 106.86 A; (d) 909.69 V; (e) 151 A, 32 A, 43.6 A

3. A low-speed synchronous motor that is driving a steel mill has to supply a shaft power of 2800 kW to a load. The rated voltage and frequency of the motor

are 1200 V (RMS) and 10 Hz. The motor operates at a power factor of 0.92 and has an efficiency of 88%. If a three-phase cycloconverter is connected to drive this motor, specify the following quantities: (a) secondary current ratings, (b) secondary voltage rating, (c) and kVA rating of each of the three 50-Hz transformers feeding the cycloconverter. Assume 98% efficiency for the cycloconverter, which employs a three-phase, full-wave dual converter in each phase. Determine the following: (d) input displacement factor, (e) distortion factor, (f) total power factor, and (g) maximum, average, and RMS current ratings of the thyristors. [Hint: Take the efficiency of the cycloconverter as the voltage reduction factor. Assume that the dual converters constituting the cycloconverter are full-wave ones.]

Ans. (a) 841 A; (b) 740 V; (c) 1867 kV A; (d) 0.8; (e) 0.8; (f) 0.64; (g) 1359 A, 288 A, 392 A

4. The type S-S cycloconverter of Fig. 6.3(a) feeds an inductive load and operates with constant firing angle for the P and N rectifiers. It has the following data: input voltage V_s = 200 V, input frequency is 50 Hz, and output frequency is 16.66 Hz. The load consists of a 2.5 Ω resistance and 10 mH inductance. The voltage reduction factor is 0.95. Firing angles $\alpha_P = \alpha_N = 60°$. Compute the following: (a) RMS values of the output voltage and the fundamental component of the output voltage, (b) maximum, average, and RMS currents through the thyristors, and (c) input displacement factor, distortion factor, and input power factor. [Hint: See Worked Example 4.]

   Ans. (a) 194.2 V, 124.9 V; (b) First pair: 95.6 A, 22.5 A, 67.6 A; second pair: 67.6 A, 11.3 A, 47.8 A; (c) 0.923, 0.643, 0.594

5. The type T-S cycloconverter of Fig. 6.28 is fed by a delta–star transformer with a secondary line-to-neutral voltage of 120 V (RMS). The load is inductive and consists of a resistance of 0.5 Ω and an inductance of 5 mH. The input and output frequencies are 50 Hz and 12.5 Hz, respectively. Assuming a voltage reduction factor of 0.97, determine the (a) maximum, average, and RMS thyristor currents, (b) total power factor, and (c) transformer rating.

   Ans. (a) 214 A, 45.5 A, 62 A; (b) 0.54; (c) 50.76 kV A

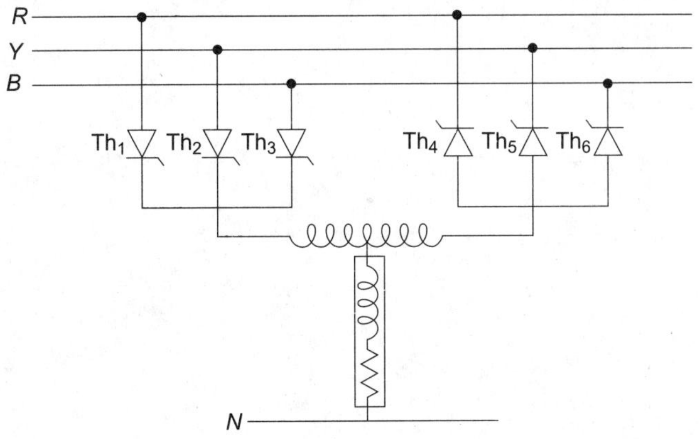

Fig. 6.28

6. The DT-S cycloconverter of Fig. 6.12 has input and output frequencies of 50 Hz and 16.66 Hz, respectively. The load consists of $R = 1.8\ \Omega$ and an inductance of 16 mH. If the transformer secondary current rating is 96 A, determine the (a) input displacement factor and distortion factor, (b) line-to-neutral voltage of the secondary winding, (c) transformer rating, and (d) maximum, average, and RMS thyristor currents.

   Ans. (a) 0.66, 0.78; (b) 160 V; (c) 46 kV A; (d) 152 A, 32 A, 44 A

7. A low-speed synchronous motor driving a cement mill has to supply a shaft power of 2000 kW to a load at 0.92 power factor (Fig. 6.29). The rated voltage and frequency of the motor are 800 V (RMS) and 10 Hz; the supply frequency is 50 Hz. The motor has an efficiency of 95%. If a three-phase cycloconverter running at 98% efficiency and employing three half-wave dual converters is to drive this motor, (a) specify the kVA rating of each of the three transformers, and compute the (b) input power factor and (c) maximum, average, and RMS current ratings of the thyristors. [Hint: Take the efficiency of the cycloconverter as the voltage reduction factor.]

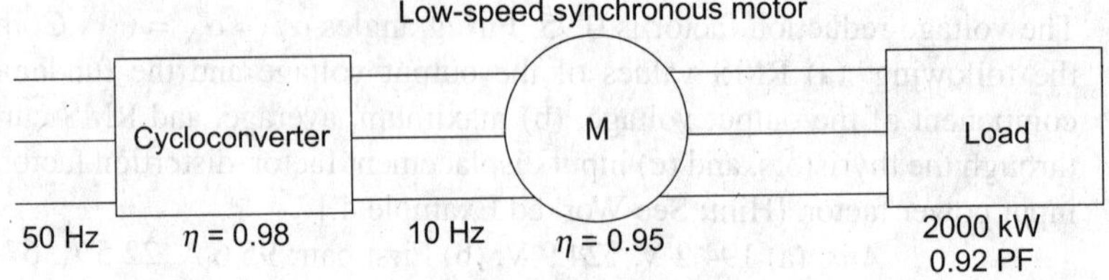

Fig. 6.29

   Ans. (a) 2472 kV A; (b) 0.64; (c) 1349 A, 286 A, 390 A

CHAPTER 7

Dc Drives

7.1 Introduction

Dc motors are preferred to their ac counterparts in drive applications that need speed control over a wide range. Dc drives have many advantages such as ease of frequent starting, braking, and reversing. Though the commutators used in these machines need considerable maintenance, the advantages of semiconductor-controlled operation outweigh this drawback. Dc drives therefore have wide-ranging applications such as electric traction, cranes, lathe machines, rolling mills, machine tools, etc.

The relationships that govern a separately excited dc motor are derived here and various methods of speed control based on these expressions are presented. Finally their implementation by means of closed-loop schemes is dealt with.

Hitherto, dc series motors had been preferred to separately excited motors in drives that require high starting torque, are subjected to frequent torque overloads, and need frequent starting. It will be shown later that separately excited drives can provide a performance comparable to that of the series motor with much simpler drive circuitry.

Shunt motors have a common source for the field and armature circuits, independent control of these two elements is not possible, except by means of conventional methods such as inserting a rheostat in the field circuit, etc. As they do not find use in closed-loop drives, dc shunt as well as compound motors are not considered here for discussion.

The variable dc voltage required for dc drives can be provided by single- or three-phase rectifiers as well as dc choppers. Figures 7.1(a) and (b), respectively, show these two configurations (Rashid 1994).

After going through this chapter the reader should

- become familiar with the different methods of speed control used in separately excited dc drives,
- get acquainted with the theory behind the speed–torque characteristic of separately excited dc drives,

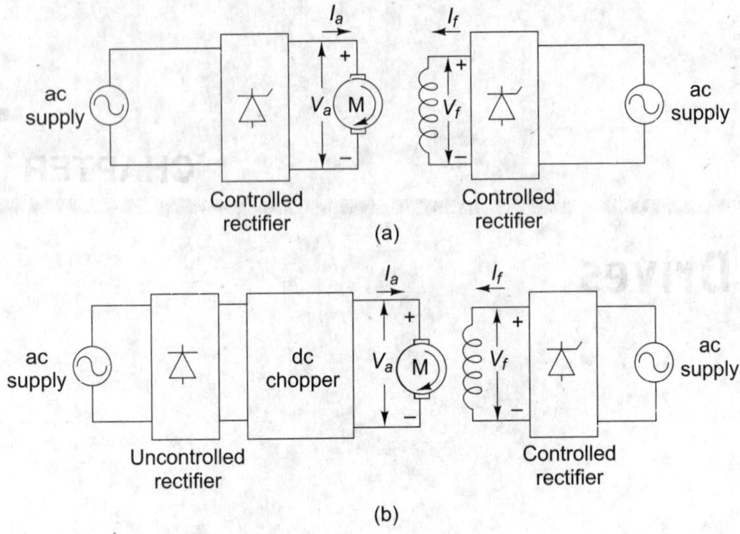

Fig. 7.1 Dc motor with armature fed by (a) a controlled rectifier, (b) an uncontrolled rectifier and dc chopper

- become familiar with the theory which forms the basis for the implementation of the armature and field control methods of speed control,
- understand the necessity of an inner current loop in addition to an outer speed loop,
- get acquainted with comprehensive control schemes,
- become familiar with the four-quadrant operation of a dc drive,
- understand how the approximate analysis of a dc series motor helps in plotting the speed–torque characteristic,
- understand the theory behind the speed control and regenerative braking of chopper-based drives, and
- become familiar with the problems associated with regenerative braking of chopper-based dc drives.

7.2 Steady-state Relationships of a Separately Excited Dc Motor

Figure 7.2 shows a separately excited dc motor with independent sources of supply for the armature and field circuits. The relationships governing these circuits (Sen 1981) are

$$V_a = E_b + R_a I_a \tag{7.1}$$

$$E_b = K_b \Phi_f \Omega_m \tag{7.2}$$

and

$$T_d = K_t \Phi_f I_a \tag{7.3}$$

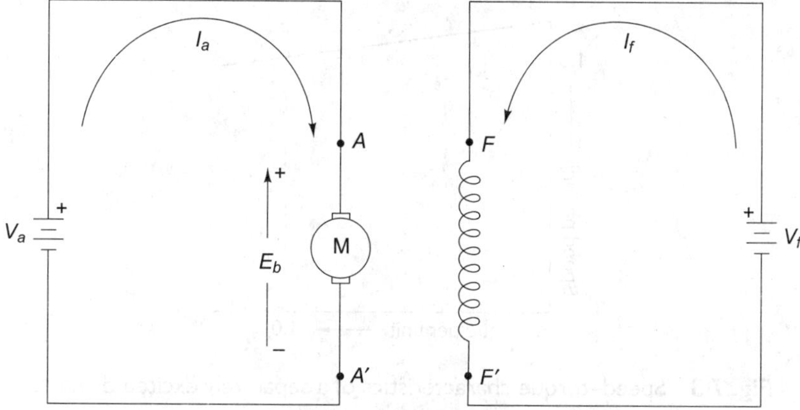

Fig. 7.2 Circuit diagram of a separately excited dc motor

where V_a is the constant applied armature voltage (in V), V_f is the constant applied field voltage (in V), I_a is the constant current in the armature circuit (in A), Φ_f is the constant field flux per pole (in Wb), ϕ_f is the field flux (variable) per pole (in Wb), R_a is the armature resistance (in Ω), K_b is the back emf constant [units as per Eqn (7.2)], $\mathcal{T}_d$ is the average torque (constant) developed by the motor (in N m), $\overline{\mathcal{T}}_d(t)$ is the average torque (variable) developed by the motor (in N m), $\omega_m(t)$ is the angular speed of the rotor (variable, in rad/s), Ω_m is the average value of $\omega_m(t)$ (constant), $\Omega_m(s)$ is the Laplace transform of $\omega_m(t)$, K_t is the torque constant [units as per Eqn (7.3) and numerically equal to K_b], $\overline{\mathcal{T}}_{\mathrm{ld}}(t)$ is the load torque (variable, in N m), i_f is the variable field current (in A), and I_f is the constant field current (in A). Also see the paragraph above Eqn (7.6) for the variable versions of the currents and voltages.

Substituting for E_b and I_a, respectively, from Eqns (7.2) and (7.3) in Eqn (7.1) and solving for Ω_m in terms of the torque $\mathcal{T}_d$ gives

$$\Omega_m = \frac{V_a}{K_b \Phi_f} - \frac{R_a}{(K_b \Phi_f)^2}\mathcal{T}_d \tag{7.4}$$

If the field voltage V_f is kept constant, the field current I_f and field flux Φ_f will also be constant. Also *if the speed and torque are considered as variables*, the former can be denoted as $\omega_m(t)$ and will become a function of $\overline{\mathcal{T}}_d(t)$, which will also be a function of time. With these considerations, Eqn (7.4) will get modified as follows:

$$\omega_m(t) = \frac{V_a}{K_1} - \frac{R_a}{K_1^2}\overline{\mathcal{T}}_d(t) \tag{7.5}$$

where $K_1 = K_b \Phi_f$, which in turn is numerically equal to $K_t \phi_f$.

Figure 7.3 gives the plot of speed per unit drawn versus average torque per unit as given by Eqn (7.5). With the assumption that any reduction of flux due to armature reaction is negligible, it will be a drooping straight line with a slope equal to $-R_a/K_1^2$. In a practical machine this assumption is not feasible; hence the actual slope of the speed–torque characteristic will be less than that shown in Fig. 7.3. The normal drop in speed from no load to full load is about 5%.

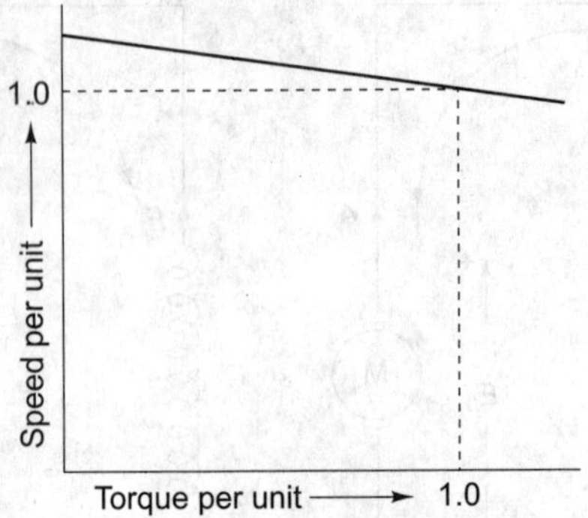

Fig. 7.3 Speed–torque characteristics of a separately excited dc motor

7.3 Speed Control of a Separately Excited Dc Motor

Dc motors are used for driving various kinds of loads, each with a unique speed–torque characteristic. The aim of speed control is to obtain a speed–torque relationship as demanded by the load. For instance, fans and centrifugal pumps have a nearly linear speed–torque relationship as shown in Fig. 7.4(a). On the other hand, coiler drives have the inverse characteristic given in Fig. 7.4(b).

Three methods of steady-state speed control are described in detail here:

(a) armature control,
(b) field control, and
(c) combined armature and field control.

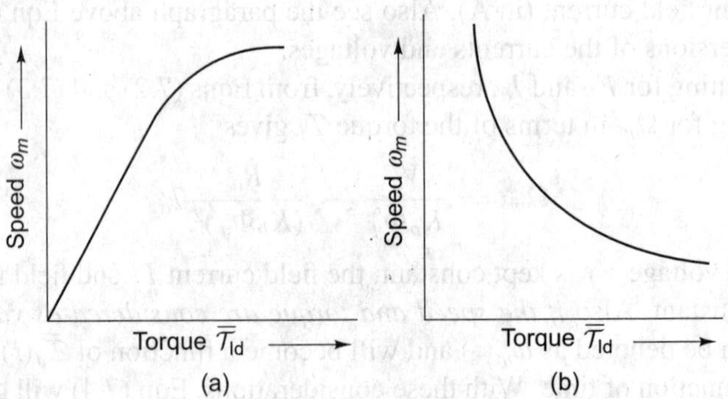

Fig. 7.4 Speed–torque characteristics of (a) fans and centrifugal pumps, (b) coiler drives

7.3.1 Armature Control Method

Figure 7.5 shows a set of speed–torque characteristics of a separately excited motor drawn for various values of kV_a, where k is a fraction. The plot with $k = 1$ corresponds to the rated armature voltage; the other characteristics with k having values less than 1 are parallel to this.

In a drive, all the currents and voltages become variables and are hence denoted by lower case letters. Accordingly, Eqn (7.1) can be written for a drive as

$$v_a = e_b + R_a i_a \tag{7.6}$$

This can be solved for i_a as

$$i_a = \frac{v_a - e_b}{R_a} = \frac{v_a - K_1\omega_m}{R_a} \tag{7.7}$$

If v_a is slightly increased, i_a will also increase, provided the field flux is kept constant. Consequently the developed torque $\overline{T}_d$ increases, thereby causing acceleration of the motor. Eventually the motor settles at a higher speed at which the developed torque $\overline{T}_d$ becomes equal the load torque $\overline{T}_{\text{ld}}$. The above steps are reversed for a decreasing v_a and the motor ultimately settles down to a lower steady-state speed. Thus, for meeting the load torque at any instant, the applied voltage has to be correspondingly varied. This method called *armature control* is convenient for constant torque control wherein the armature current is maintained at its rated value for all speeds. The constant torque characteristics drawn in Fig. 7.5 are typical armature control characteristics in which the torque is varied by changing v_a, the armature current being maintained constant throughout. Decreasing the voltage v_a by a large amount, however, leads to the following undesirable sequence of events:

(a) the back emf e_b exceeds the applied voltage v_a,

(b) the motor current gets reversed and also increased leading to heavy ohmic losses, and

(c) the motor will be constrained to work in the generation mode.

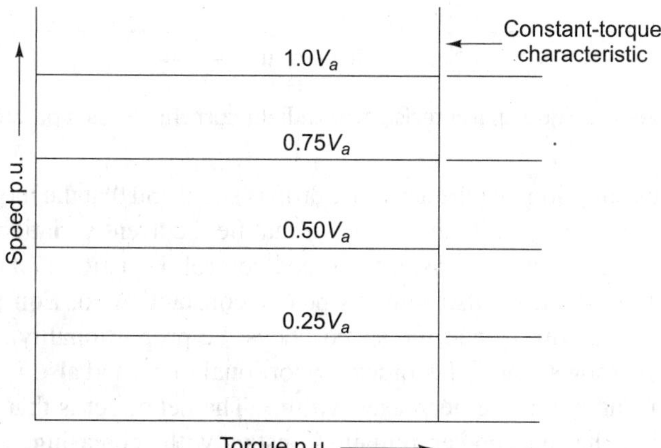

Fig. 7.5 Speed–torque characteristics with armature voltage (V_a) as a parameter

Hence it is necessary to decrease v_a only by a small amount. *It should also be ensured that the armature voltage does not exceed the rated value, as otherwise heavy current flows through the armature, which has a low resistance.*

7.3.2 Field Control Method

If the field current i_f of the separately excited motor shown in Fig. 7.1 is reduced, the field flux ϕ_f as well as back emf e_b $(= K_b i_f \omega_m)$ will decrease. As a consequence, the armature current i_a will increase. The armature resistance R_a being very small, this increase in armature current will be much higher than the reduction in field current. The developed torque is correspondingly increased to a value above the load torque and this accelerates the motor. The motor will finally settle down to a higher speed at which the developed torque will become equal to the load torque. The reverse operation of increasing the field current will make the motor settle down to a lower speed. Thus, any kind of load torque can be catered to by such field current variations. Figure 7.6 shows the speed–torque characteristics with the field current i_f as a parameter. They are seen to be drooping straight lines, the slope decreasing with increasing field current.

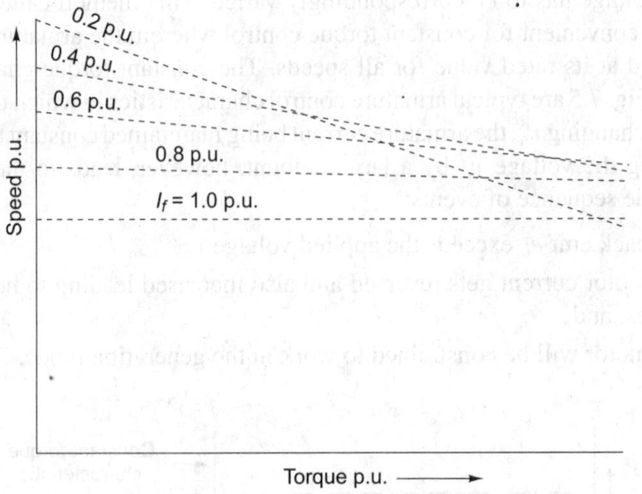

Fig. 7.6 Speed–torque characteristics with field current (I_f) as a parameter

With the assumption that the armature drop is small and that the motor operates at its rated current $I_{a(\max)}$, it can be shown that field current variation provides a constant horsepower characteristic for speed control. Equation (7.1) shows that with v_a kept constant, e_b also remains nearly constant. Also, Eqn (7.2) shows that ϕ_f decreases with i_f and the speed obeys the proportionality $\omega_m \propto 1/\phi_f$. Equation (7.3) shows that $\overline{T}_d$ becomes proportional to ϕ_f and also i_f at the rated armature current and hence decreases with i_f. The net effect is that the product $\overline{T}_d\omega_m$, which is the shaft power, remains constant with decreasing i_f. Figure 7.7 shows such a constant horsepower characteristic.

The assumption that the motor current remains constant at $I_{a(\max)}$ with reduction in field current is only approximate. In practice, the armature reaction has a pronounced effect when the field current (and hence the field flux) is reduced. Consequently, the maximum current that the motor can carry without sparking at the commutator will also decrease slightly.

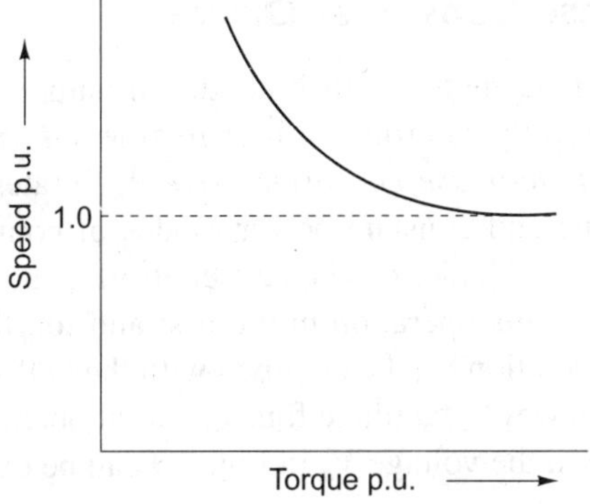

Fig. 7.7 Constant horsepower characteristic

7.3.3 Combined Armature and Field Control Method

Some drives require a wide variation of speed and the two methods described above have to be combined to cover this range. The merit of armature control (Fig. 7.5) is that the maximum torque capability of the motor can be maintained at all speeds, ranging from the standstill condition up to the rated (or base) speed, the field flux being kept constant throughout this range. For speeds above the rated speed, armature control is not permitted because the armature voltage cannot be increased beyond its rated value. Instead, the armature voltage is kept constant at the base value and field current control, usually called *field weakening*, is adopted, which is the same as constant horsepower control (Fig. 7.7). The speed increase so obtained is accompanied by a reduced torque and the resulting characteristic has the shape of a hyperbola.

The coiler drive [Fig. 7.4(b)] and electric traction (Fig. 7.8) are examples of loads that require wide speed variation. The latter shows how the torque and power vary for the whole range of speeds up to the maximum permissible speed, the armature current being maintained constant at its rated value throughout this range.

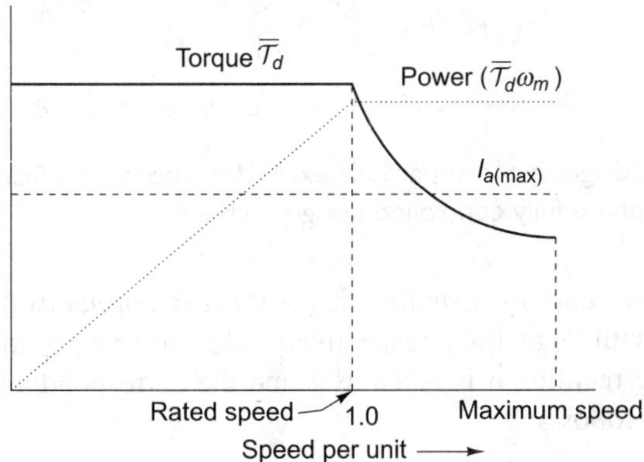

Fig. 7.8 Torque, power, and armature current vs speed

7.4 Single-phase Converter Drives

A separately excited dc motor with both the armature and the field fed by separate single-phase fully controlled bridge rectifiers is shown in Fig. 7.9. This circuit, called the *single-phase converter drive*, facilitates the implementation of the constant torque and constant-power modes of control discussed earlier. The rectifiers used on both sides can be half-controlled rectifiers as well. A fully controlled rectifier permits operation in the first and fourth quadrants, whereas only first-quadrant operation can be obtained with the half-controlled rectifier. A dual converter will, however, facilitate four-quadrant operation of the motor. For continuous conduction, the voltage V_a in Fig. 7.9 can be expressed as

$$V_a = \frac{2V_m}{\pi}\cos\alpha_1,\quad 0 < \alpha_1 < \frac{\pi}{2} \tag{7.8}$$

where V_m is the peak line-to-neutral (LN) value of the ac supply voltage and α_1 is the firing angle of the thyristors of rectifier A. Likewise, V_f can be expressed as

$$V_f = \frac{2V_m}{\pi}\cos\alpha_2,\quad 0 < \alpha_2 < \frac{\pi}{2} \tag{7.9}$$

where α_2 is the firing angle of rectifier B. Considering the field voltage V_f constant and substituting for V_a from Eqn (7.8) in Eqn (7.5), $\omega_m(t)$ becomes

$$\omega_m(t) = \frac{2V_m}{\pi K_1}\cos\alpha_1 - \frac{R_a}{K_1^2}T_d \tag{7.10}$$

where the expression given in Eqn (7.8) is substituted for V_a.

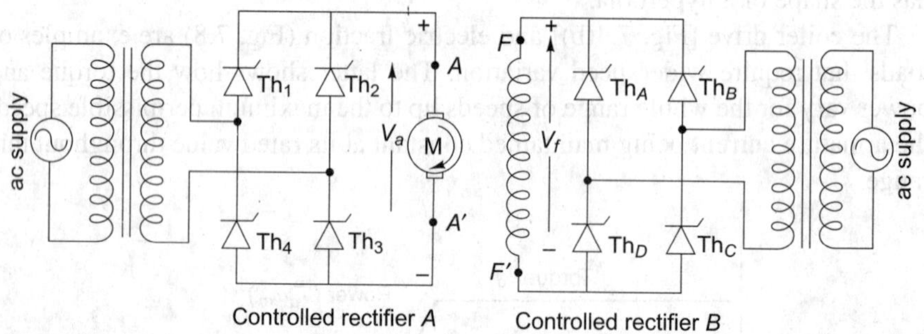

Fig. 7.9 Circuit diagram of a separately excited dc motor and field, each fed by a single-phase fully controlled bridge rectifier

If, on the other hand, the armature current is discontinuous, Eqns (2.14) and (2.15) are used with V_α and R_{ld}, respectively, taken to be equal to V_a and R_a. For this single-phase rectifier, n is taken as 2 and the corresponding expression for ω_m is derived as follows:

$$V_a = \frac{V_m}{\pi}\left[\sin\left(\alpha_e - \frac{\pi}{2}\right) - \sin\left(\alpha - \frac{\pi}{2}\right)\right]$$

which can be written as

$$V_a = \frac{V_m}{\pi}(\cos\alpha - \cos\alpha_e) \tag{7.11}$$

Also, E_b is expressed as $E_b^* = E_b(\alpha_e - \alpha)/\pi$, where $E_b = K_1\Omega_m$. Hence Eqn (7.1), for this discontinuous case, is written as

$$V_a = E_b^* + R_a I_a$$

with V_a and E_b^* as given above. Thus,

$$I_a = \frac{V_a - E_b^*}{R_a} = \frac{1}{R_a}\left[\frac{V_m}{\pi}(\cos\alpha - \cos\alpha_e) - \frac{K_1\Omega_m(\alpha_e - \alpha)}{\pi}\right] \tag{7.12}$$

If the speed, torque, and the armature current are treated as variables, they can be, respectively, denoted as ω_m, $\overline{T}_d$, and i_a. Equation (7.12) now becomes

$$i_a = \frac{1}{R_a}\left[\frac{V_m}{\pi}(\cos\alpha - \cos\alpha_e) - \frac{K_1\omega_m(\alpha_e - \alpha)}{\pi}\right]$$

Using Eqn (7.3) to express i_a as $\overline{T}_d/K_1$, where $K_1 = K_t\Phi_f$, the above equation changes to

$$\frac{\overline{T}_d}{K_1} = \frac{1}{R_a}\left[\frac{V_m}{\pi}(\cos\alpha - \cos\alpha_e) - \frac{K_1\omega_m(\alpha_e - \alpha)}{\pi}\right]$$

Solving this equation for ω_m now yields

$$\omega_m(t) = \frac{V_m(\cos\alpha - \cos\alpha_e)}{K_1(\alpha_e - \alpha)} - \frac{\pi R_a}{K_1^2(\alpha_e - \alpha)}\overline{T}_d \tag{7.13}$$

The speed–torque curves shown in Fig. 7.10 (Dubey 1989) are drawn using Eqn (7.13) with α as a parameter. Conduction may be continuous or discontinuous, and in the latter case α_e has to be read from the chart given in Fig. 2.21, with E_b taken to be equal to $(K_1\omega_m)$. The boundary between continuous and discontinuous conduction is shown by a dotted line in Fig. 7.10. Discontinuous conduction occurs for torque values below the rated value. Also, the speed–torque curves for continuous conduction are seen to be parallel straight lines with negative slopes; the slope of each line is determined by the armature resistance. This aspect can also be seen in Eqn (7.10), which is derived for continuous conduction. Figure 7.10 also shows that the speed regulation is poor (that is, the speed ω_m falls steeply) when the motor operates in the discontinuous conduction mode. The reasons for such an operation are as follows.

(a) In the continuous conduction mode, if the load torque increases suddenly, the speed goes down because the instantaneous developed torque $\overline{T}_d$ becomes less than the instantaneous load torque $\overline{T}_{\text{ld}}$ and deceleration occurs as per the equation

$$\frac{d\omega_m(t)}{dt} = \overline{T}_d - \overline{T}_{\text{ld}} \tag{7.14}$$

where the frictional torque is neglected. As a feedback action, a decrease in ω_m causes a decrease in e_b and an increase in i_a, as can be seen

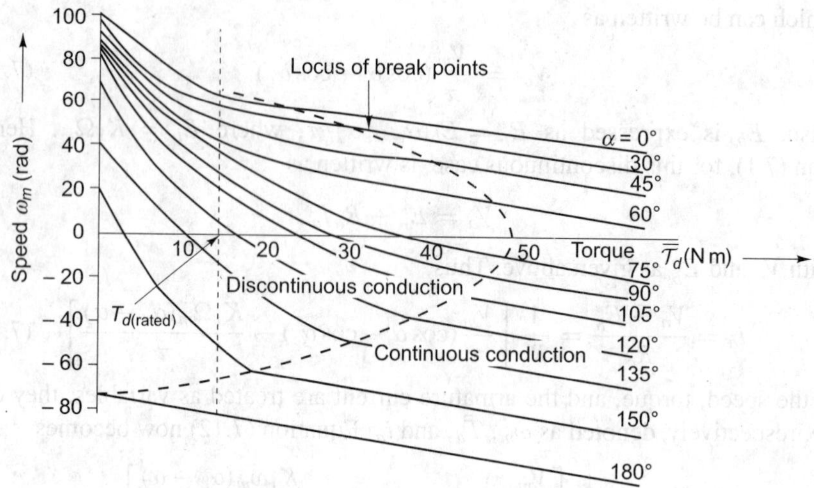

Fig. 7.10 Speed–torque characteristics of a typical single-phase rectifier drive with α (elec. degrees) as a parameter

from Eqn (7.7). The increase in i_a, in turn, leads to an increase in the developed torque $\overline{T}_d$, which now approaches $\overline{T}_{\text{ld}}$, and $\omega_m(t)$ decreases as per Eqn (7.14). A steady state is reached at a speed slightly less than the original speed, with $\overline{T}_d$ eventually becoming equal to $\overline{T}_{\text{ld}}$. Equation (7.10) reflects this net effect and provides a small slope for the $\omega_m(t)$-$\overline{T}_d$ characteristic.

(b) In the discontinuous conduction mode also, an increase in load torque causes a decrease in $\omega_m(t)$, which in turn leads to a decrease in E_b and an increase in α_e. The product $E_b\alpha_e$ remains nearly constant and the increase in current is quite less, as depicted by Eqn (7.12). The developed torque also increases marginally and there will be a very small increase in $\omega_m(t)$ *due to this feedback*. The net effect is that there is a large drop in $\omega_m(t)$ and hence the $\omega_m(t)$-$\overline{T}_d$ characteristic has a large negative slope. Equation (7.13) represents this behaviour, giving a steep slope for the $\omega_m(t)$-$\overline{T}_d$ curve.

The region of discontinuous conduction can be reduced by inserting an extra inductance in series with the motor armature.

7.5 Three-phase Converter Drives

The circuit of a three-phase, full-wave fully controlled bridge rectifier is given in Fig. 7.11, with a separately excited dc motor as the load. The field is assumed to be supplied through a similar controlled rectifier. Assuming that the field current is constant, the waveforms of load voltage and current for both continuous and discontinuous conduction are given in Figs 7.12(a) and (b), respectively. The speed–torque relationship for this case is arrived at as follows.

7.5.1 Discontinuous Conduction

Similar to the single-phase case, the angles α_1 and α_2 are taken, respectively, as the firing angles of the converters feeding the armature and field. The line-to-line

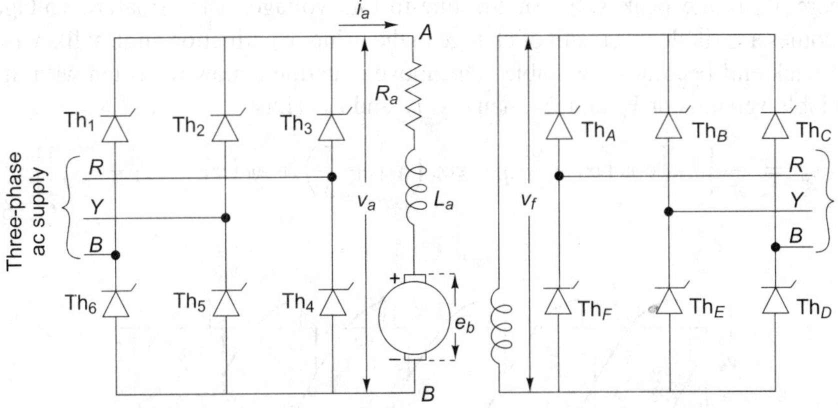

Fig. 7.11 Circuit diagram of a separately excited dc motor fed by a three-phase, full-wave fully controlled bridge rectifier

and the load voltage waveforms are given in Figs 7.12(a) and (b), respectively, for continuous and discontinuous conduction. As shown in these figures, the firing angle is measured from the point of natural commutation P, which is located at an angle of $\pi/3$ from the origin. The average voltage applied to the armature now becomes

$$V_a = \frac{3}{\pi}\int_{\alpha_1+\pi/3}^{\alpha_{e1}+\pi/3} V_p \sin(\omega t)d(\omega t) + \frac{3}{\pi}E_b\left(\alpha_1 + \frac{\pi}{3} - \alpha_{e1}\right) \tag{7.15}$$

The expression on the right-hand side of Eqn (7.15) can be evaluated as

$$V_a = \frac{3}{\pi}\left\{V_p\left[\cos\left(\alpha_1 + \frac{\pi}{3}\right) - \cos\left(\alpha_{e1} + \frac{\pi}{3}\right)\right] + E_b\left(\alpha_1 - \alpha_{e1} + \frac{\pi}{3}\right)\right\}$$

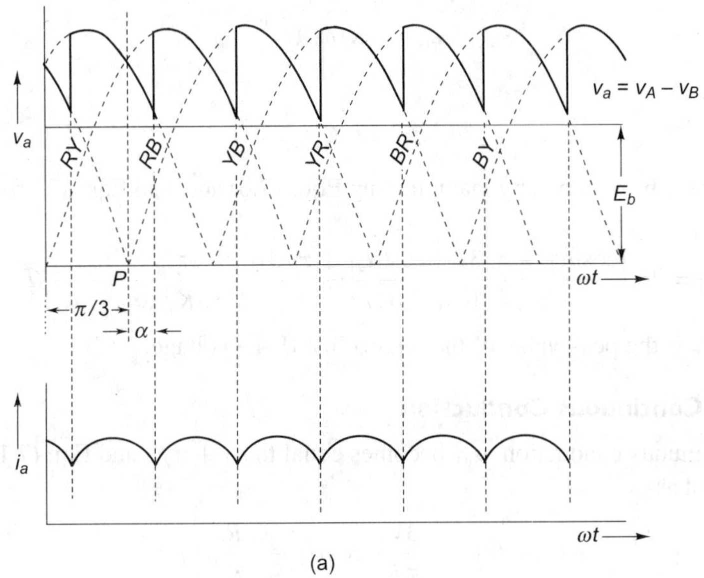

(a)

Fig. 7.12(a)

where V_p is the peak value of the line-to-line voltage. The armature voltage becomes a variable as a result of change in the firing or extinction angles; likewise the back emf becomes a variable. The above equation is now rewritten with the variable versions of V_a and E_b, namely, v_a and e_b. Thus,

$$v_a = \frac{3}{\pi}\left\{V_p\left[\cos\left(\alpha_1 + \frac{\pi}{3}\right) - \cos\left(\alpha_{e1} + \frac{\pi}{3}\right) + e_b\left(\alpha_1 - \alpha_{e1} + \frac{\pi}{2}\right)\right]\right\} \tag{7.16}$$

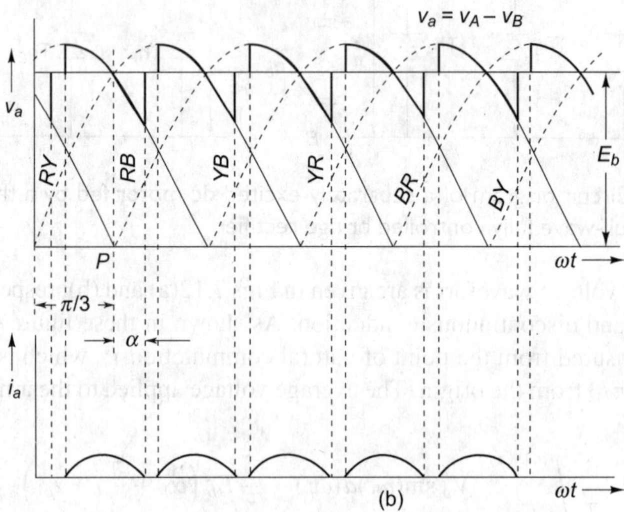

Fig. 7.12 Waveforms for the drive of Fig. 7.11 with (a) continuous conduction and (b) discontinuous conduction

Equations (7.2) and (7.3) can now be rewritten as

$$e_b = K_1\omega_m(t) \tag{7.17}$$

and

$$\overline{T}_d = K_1 i_a \tag{7.18}$$

ω_m can now be obtained by manipulating Eqns (7.6) and also Eqns (7.16)–(7.18) as

$$\omega_m(t) = V_p\frac{[\cos(\alpha_1 + \pi/3) - \cos(\alpha_{e1} + \pi/3)]}{K_1(\alpha_{e1} - \alpha_1)} - \frac{\pi}{3K_1^2}\frac{R_a}{(\alpha_{e1} - \alpha_1)}\overline{T}_d \tag{7.19}$$

where V_p is the peak value of the line-to-line (L-L) voltage.

7.5.2 Continuous Conduction

For continuous conduction, α_{e1} becomes equal to $\alpha_1 + \pi/3$ and Eqn (7.19) gets simplified as

$$\omega_m(t) = \frac{3V_p}{\pi K_1}\cos\alpha_1 - \frac{R_a}{K_1^2}\overline{T}_d \tag{7.20}$$

The $\omega_m(t)$-$\overline{T}_d$ curves shown in Fig. 7.13 are drawn with α as a parameter, based on the relationships given in Eqns (7.19) and (7.20) (Dubey 1989). As in the case of Fig. 7.10, the dashed line in Fig. 7.13 demarcates the regions of continuous and discontinuous conduction.

As stated earlier the region of discontinuous conduction can be reduced by the addition of an inductor in series with the motor armature.

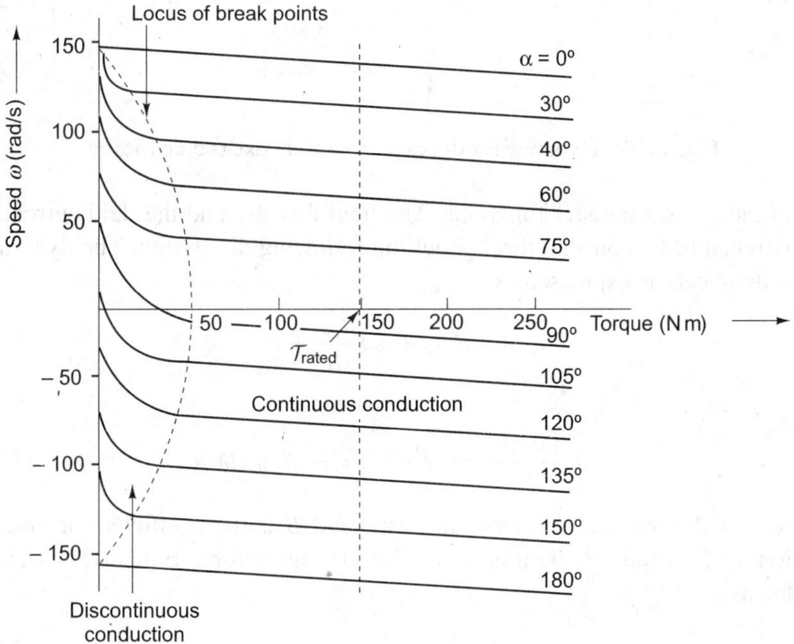

Fig. 7.13 Speed–torque characteristics of a typical three-phase rectifier drive with α (electrical degrees) as a parameter

7.6 Dynamic Behaviour of a Separately Excited Dc Motor Fed by Rectifiers

The response of a dc drive to suddenly applied reference inputs and sudden disturbances can be obtained with the help of equations that describe the dynamic behaviour of the machine. Such a model will be particularly useful for (a) determining the stability and (b) designing the components of closed-loop drives.

Figure 7.14 shows the schematic-cum-circuit diagram that serves as the dynamic model of a separately excited motor. The relevant mathematical relations for dynamic analysis are derived below.

7.6.1 Speed Control Using Armature Voltage Variation

The dynamic analysis of speed control using armature voltage variation is now taken up. The resulting dynamic equations will then be Laplace-transformed to

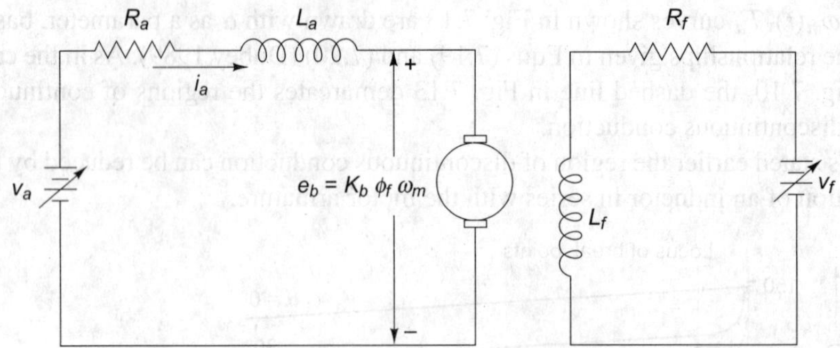

Fig. 7.14 Dynamic model of a separately excited dc motor

obtain the control transfer functions. The field flux Φ_f and the field current I_f are assumed to be constant throughout the following discussion. The dynamics of the drive can be expressed as

$$v_a = R_a i_a + L_a \frac{di_a}{dt} + e_b \tag{7.21}$$

and

$$J\frac{d\omega_m(t)}{dt} = \overline{T}_d - \overline{T}_{\mathrm{ld}} - B\omega_m(t) \tag{7.22}$$

where J is the moment of inertia in kg m^2 and B is the coefficient of viscous friction in N m/rad s^{-1}. Taking $e_b = K_1\omega_m(t)$ as before, Eqn (7.21) can be written as

$$v_a = R_a i_a + L_a \frac{di_a}{dt} + K_1\omega_m(t) \tag{7.23}$$

The instantaneous developed torque is expressed as

$$\overline{T}_d = K_1 i_a \tag{7.24}$$

Equation (7.22) can now be written as

$$J\frac{d\omega_m(t)}{dt} = K_1 i_a - \overline{T}_{\mathrm{ld}} - B\omega_m(t) \tag{7.25}$$

Laplace-transforming Eqn (7.23) and solving for $I_a(s)$ gives

$$I_a(s) = \frac{K_R V_a(s)}{1 + s\tau_a} - \frac{K_2}{1 + s\tau_a}\Omega_m(s) \tag{7.26}$$

where $\tau_a = L_a/R_a$, $K_R = 1/R_a$, and $K_2 = K_1/R_a = K_1 K_R$. Laplace-transforming Eqn (7.25) and solving for $\Omega_m(s)$ yields

$$\Omega_m(s) = \frac{K_3}{1 + s\tau_m} I_a(s) - \frac{K_4}{1 + s\tau_m}\overline{T}_{\mathrm{ld}}(s) \tag{7.27}$$

where $\tau_m = J/B$, $K_3 = K_1/B$, and $K_4 = 1/B$. Figure 7.15 shows the block diagram representing Eqns (7.26) and (7.27). If the Laplace transform of the load torque variable is taken as zero, the following equations can be obtained from Fig. 7.15:

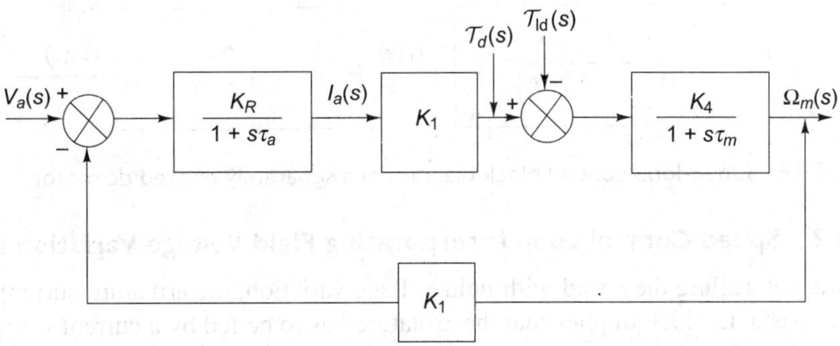

Fig. 7.15 Control block diagram for the dynamic model given in Fig. 7.14

$$\frac{[V_a(s) - K_1\Omega_m(s)]K_R}{1 + s\tau_a} = I_a(s) \tag{7.28}$$

Also,

$$\frac{K_1K_4I_a(s)}{1 + s\tau_m} = \frac{K_3I_a(s)}{1 + s\tau_m} = \Omega_m(s) \tag{7.29}$$

where K_3 is taken to be equal to K_1K_4. Substitution of the expression for Ω_m(s) from Eqn (7.29) in Eqn (7.28) gives the transfer function $I_a(s)/V_a(s)$ as

$$\frac{I_a(s)}{V_a(s)} = \frac{K_R(1 + s\tau_m)}{(1 + s\tau_m)(1 + s\tau_a) + K_1K_3K_R} \tag{7.30}$$

Equation (7.30) can be written in a compact form as

$$\frac{I_a(s)}{V_a(s)} = \frac{K_5(1 + s\tau_m)}{(1 + s\tau_1)(1 + s\tau_2)} \tag{7.31}$$

where

$$\tau_1 + \tau_2 = \frac{\tau_m + \tau_a}{1 + K_1K_3K_R}$$

and

$$\tau_1\tau_2 = \frac{\tau_m\tau_a}{1 + K_1K_3K_R}$$

and

$$K_5 = \frac{K_R}{1 + K_1K_3K_R}$$

Using Eqns (7.29) and (7.31), the closed-loop diagram of Fig. 7.15 can be redrawn as the open-loop diagram of Fig. 7.16. The resulting system can be considered as a cascade of two transfer function blocks.

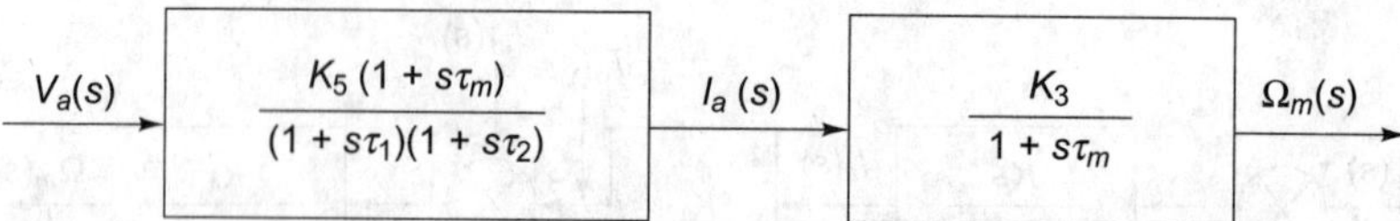

Fig. 7.16 Open-loop control block diagram of a separately excited dc motor

7.6.2 Speed Control Loop Incorporating Field Voltage Variation

While controlling the speed with field voltage variation, the armature current is kept constant, which implies that the armature has to be fed by a current source. It is assumed that any change in armature current due to variation of field current is negligible. This is justified by the fact that the time constant and hence the response time of the armature circuit are much smaller than the corresponding field quantities.

The field circuit dynamics can be expressed as

$$v_f = R_f i_f + L_f \frac{di_f}{dt} \tag{7.32}$$

With constant armature current, the instantaneous developed torque will be proportional to the instantaneous field current and is expressed as

$$\mathcal{T}_d = K_n i_f \tag{7.33}$$

where K_n is the constant of proportionality. The dynamics of the motor load system remains the same as given in Eqn (7.22). Using Eqn (7.33), this equation becomes

$$J\frac{d\omega_m(t)}{dt} = K_n i_f - \mathcal{T}_{\text{ld}} - B\omega_m(t) \tag{7.34}$$

Laplace-transforming Eqn (7.32) and solving for $I_f(s)$ gives

$$I_f(s) = \frac{K_6 V_f(s)}{1 + s\tau_f} \tag{7.35}$$

where $K_6 = 1/R_f$ and $\tau_f = L_f/R_f$. Also, taking the Laplace transform of Eqn (7.34) and solving for $\Omega_m(s)$ gives

$$\Omega_m(s) = \frac{K_7 I_f(s)}{1 + s\tau_m} - \frac{K_4 \mathcal{T}_{\text{ld}}(s)}{1 + s\tau_m} \tag{7.36}$$

where $K_4 = 1/B$, as defined before, $K_7 = K_n/B$, and $\tau_m = J/B$. Equations (7.35) and (7.36) are represented by the block diagram of Fig. 7.17.

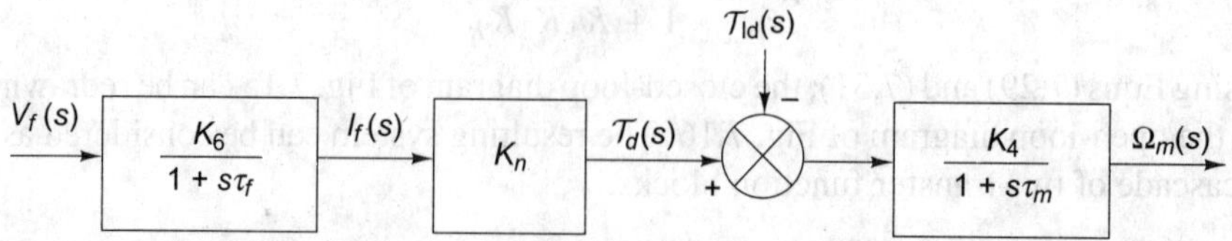

Fig. 7.17 Open-loop block diagram showing both field and motor load system transfer functions

7.6.3 Closed-loop Current Control

The block diagram of Fig. 7.16 shows that under dynamic conditions, the variation of the armature voltage causes changes in armature current as well as speed. However, under the transient conditions that occur during the starting or braking of the motor and also under the sudden speed variation caused due to sudden overload, etc., the current flowing through the rectifier thyristors may exceed the rated value. Such an overcurrent condition can be prevented by providing a feedback loop from the armature current variable $i_a(s)$. At the same time, it is desirable that the current be maintained at its maximum permitted value in order to exploit the full drive torque capability and facilitate a fast response. The schematic-cum-block diagram of Fig. 7.18 shows the usual method of implementing such a closed-loop current control. A voltage signal e_1, which sets the desired value of current, is passed through a current limiter, which outputs the reference signal I_{a1} for the current control loop.

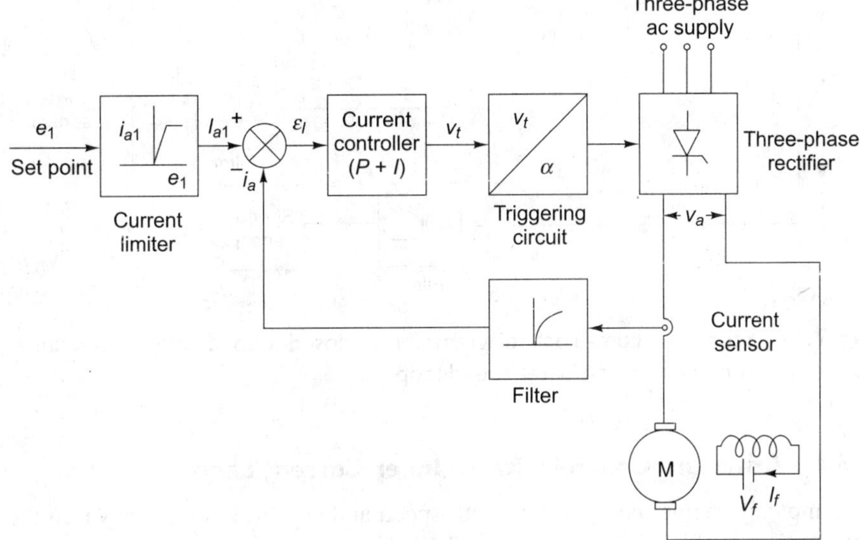

Fig. 7.18 Schematic-cum-block diagram for closed-loop current control of a separately excited dc drive

In a complete speed control scheme, however, the current and speed loops form, respectively, the inner and outer loops. Such a configuration is called a *cascade control scheme*. The real utility of the signal e_1 and the current limiter can be seen in this configuration. e_1 becomes the output of the speed (outer) controller. Under small load disturbances, the inner loop comes into action; the armature current i_a is made to track the desired value I_{a1}. On the other hand, under large load disturbances, which are the cause of transient conditions, both the loops come into play; e_1 may attain large values as a consequence of large variation in ω (speed). The current limiter then fixes the value of I_{a1} at the maximum permissible value I_{am}. This has a dual benefit: first, the current error ε_I acts on the current controller, whose output (v_t) adjusts the firing angle α of the rectifier so as to make the shaft speed approach the set speed; second, the motor also gets limited by the firing angle adjustment. If, however, it is desired to dispense with the outer speed loop, then the signal e_1 and the current limiter will not be used; the current reference signal I_{a1} can be set at its maximum value I_{am}. It is evident from the above discussion that such a configuration is sometimes used for open-loop speed control at the same time having a closed-loop configuration for current control. Figure 7.19 shows such a cascade control system whose detailed description follows.

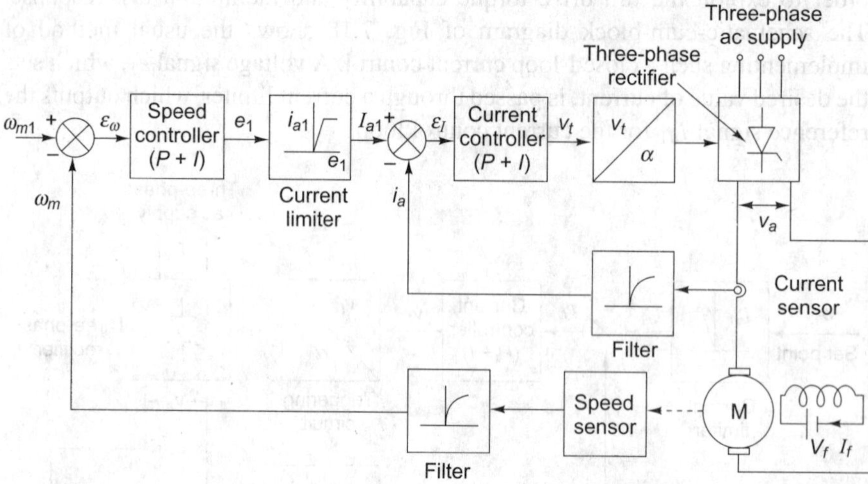

Fig. 7.19 Schematic-cum-block diagram for a closed-loop dc drive with inner current loop and outer speed loop

7.6.4 Armature Control with an Inner Current Loop

A complete scheme incorporating both speed and current loops is shown in the schematic-cum-block diagram of Fig. 7.19.

A signal proportional to the actual speed of the motor is obtained from a speed sensor and passed through the filter to remove any ripple in the output of the speed sensor. It is then compared with the signal ω_{m1}, which represents the desired speed. The error ε_ω so obtained is passed through a speed controller whose output signal e_1 adjusts the rectifier firing angle α as explained above. The speed controller, which is of the $P + I$ type, stabilizes the drive and helps in nullifying the steady-state error between the actual and desired speeds. As before, the output v_t of the current controller adjusts the rectifier firing angle to bring the shaft speed closer to the desired speed ω_{m0}.

The cascaded loops are designed in such a way that major system disturbances occur within the current (inner) loop and are immediately attended to by the corresponding inner loop controller. A typical transient disturbance occurring within the current loop can be in the form of a sudden change in the input ac voltage and is corrected dynamically as per the following sequence of operations.

A reduction in the ac supply voltage causes a reduction in the voltage applied to the motor. The armature current decreases as per the relation $i_a = (v_a - e_b)/R_a$, thereby reducing the developed motor torque and making the load torque exceed the developed torque. The retarding torque so developed slows down the motor, thus activating both the speed and current loops. The reduction in current is sensed by the inner current loop and the necessary reduction in the firing angle is made to initiate an increase in the current. A disturbance can also be in the form of a sudden load torque demand which drastically reduces the shaft speed. The large error in speed (ε_ω) so caused leads to the saturation of the current limiter. The resulting low speed provides a small e_b, which in turn leads to a large armature current. This current may be greater than i_{a1}; the negative current error acts on the system to restore the speed, eventually equalizing the developed and load torques. In both types of disturbances, it is the inner current loop which quickly corrects the situation, though the outer loop senses the reduction in speed. This is because its electrical time constant is much smaller than the mechanical time constant of the former. This feature also facilitates the quick restoration of steady-state conditions.

Yet another merit of the inner current loop is as follows. The drive response becomes sluggish in the discontinuous conduction mode because of the reduction in the combined gain of the firing circuit and rectifier blocks. Thus a drive designed for operation with continuous conduction may give an oscillatory response or even become unstable under discontinuous conduction. However, the fast action of the inner current loop, which encompasses both the firing circuit and the rectifier blocks, enables it to quickly set the situation right.

The current sensor can be placed either on the dc or on the ac side of the rectifier, the latter approach being preferred because of the facility of using current transformers.

7.6.5 A Comprehensive Control Scheme for Wide-range Speed Control

A scheme that provides speed control both below and above the base speed is shown in the block diagram of Fig. 7.20 (Rashid 1994). It incorporates field control in addition to the cascaded combination of armature and current loops. In the field loop a signal proportional to the actual back emf ($E_b = V_a - I_a R_a$) is compared with the desired value of the back emf (E_{b1}); the error signal passing through a field controller block is usually of the $P + I$ type. The output signal v_{ef} of the latter block controls the firing angle α_f of the field rectifier, thus causing a variation of its output voltage.

As emphasized earlier, the scheme given in Fig. 7.20 must be capable of maintaining constant field flux up to the base speed and constant rated voltage beyond this speed; this requirement is nearly achieved by the following procedure. For speeds below the base speed, the signal e_b ($= v_a - i_a R_a$) in the field control loop will be small and the error signal ε_f will be large, causing the saturation of the field controller. The firing angle α_f is set at its minimum value so that the field voltage and current are adjusted to be at their maximum values. The armature and inner current loops come into play as detailed above. On the other

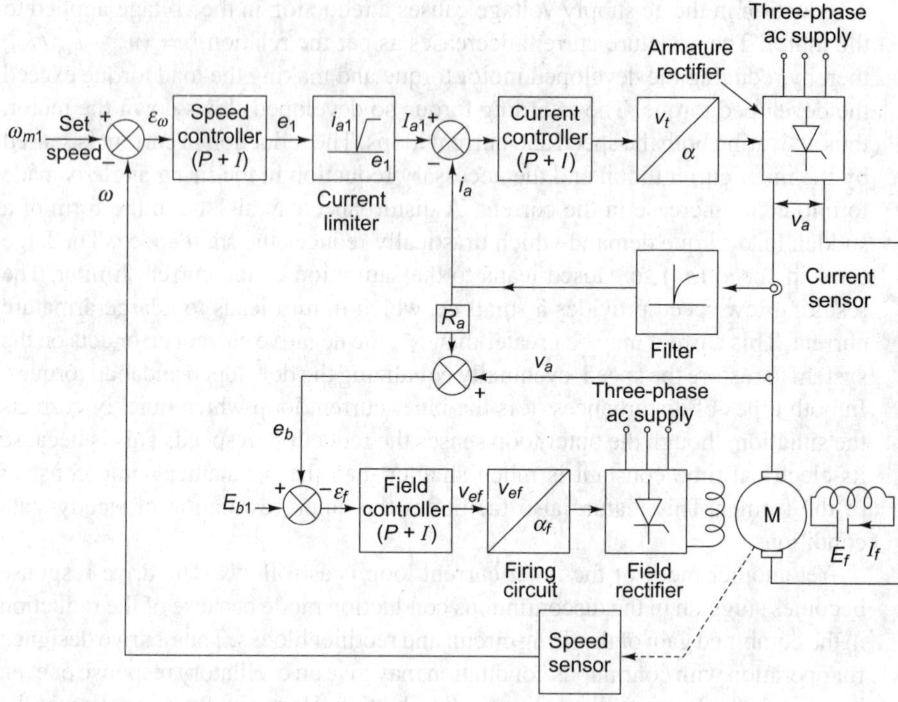

Fig. 7.20 Schematic-cum-block diagram of a comprehensive control scheme for a separately excited dc drive

hand, when the motor speed nearly equals the base speed, the error signal ε_f will be small and the field controller comes out of saturation, enabling v_{ef} to vary with ε_f. Now if operation above the base speed is desired, ω_{m1} is set to the desired value, making ε_ω positive as per the sequence explained in Section 7.6.4. The firing angle α_a of the rectifier thyristors is reduced, thus increasing the armature voltage v_a, initially close to its rated value. The motor accelerates and the signal proportional to the back emf increases, causing a decrease in ε_f and increase in α_f. The field current and voltage continue to decrease till the motor speed attains the value corresponding to ω_{m1}. This makes the speed error ε_ω small, causing restoration of v_a back to its original value, which is close to its rated value.

The inclusion of the field control loop slows down the response of the drive because of the large field time constant delaying the field reversal operation. This can be remedied by using a fully controlled rectifier in the speed loop and facilitating quick field reversal by means of a technique called *field forcing*.

7.6.6 Closed-loop Drive with Four-quadrant Operation

It is shown in Chapter 2 that hoist operation needs a variable speed drive with four-quadrant operation. Though such an operation can be obtained by other methods, a dual converter is preferred because of the smooth reversals facilitated by it. Figure 7.21 shows a schematic-cum-block diagram of a closed-loop drive using a dual converter with non-simultaneous control. The salient features of this drive (Dubey 1989) are

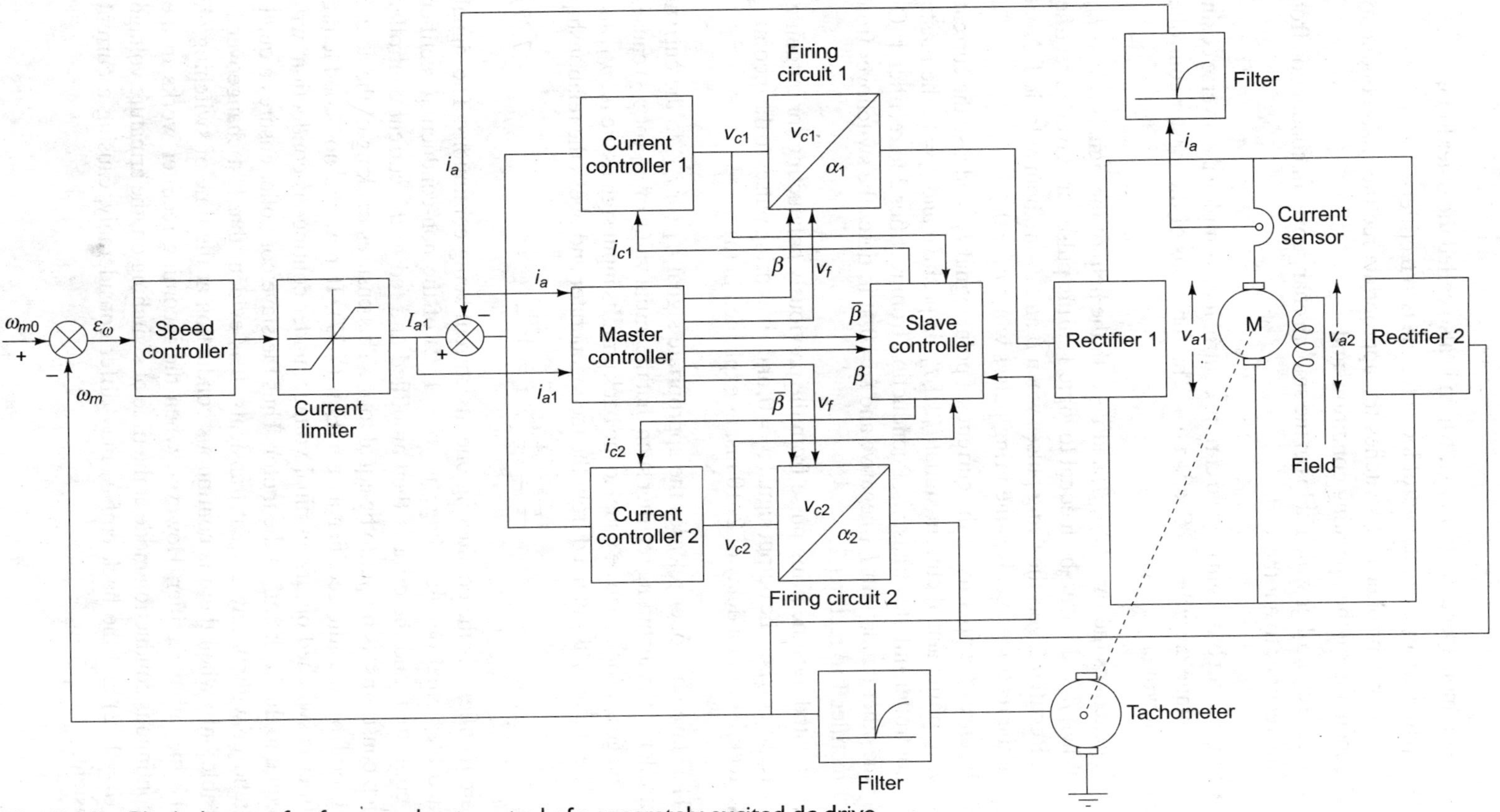

Fig. 7.21 Block diagram for four-quadrant control of a separately excited dc drive

(a) a common speed control loop, which is also called the outer loop,
(b) separate inner current control loops for the two rectifiers,
(c) a master controller that switches the appropriate rectifier into conduction depending on the operating condition, and
(d) a slave controller which helps in the smooth changeover of conduction from one rectifier to another.

The firing circuit blocks in the two current control loops employ the inverse cosine firing technique described in Section 6.4. The functions of the master controller are the following.

(a) It generates the signals β, $\overline{\beta}$, and v_f with the help of the signals i_a and I_{a1}. When β and v_f are both equal to logic 1, firing pulses are given to rectifier 1. Rectifier 2 gets the firing pulses when $\overline{\beta}$ and v_f are both at logic 1. None of the rectifiers receives the firing pulses when $v_f = 0$.
(b) It senses the polarity of the current set point signal I_{a1} and also the current-zero of the actual current signal i_a. If I_{a1} is positive and $i_a = 0$, the master controller initiates transfer of conduction from rectifier 2 to rectifier 1. On the other hand, with I_{a1} negative and $i_a = 0$, it initiates the switchover from rectifier 1 to rectifier 2.
(c) It withdraws the firing pulse from the previously conducting (or outgoing) rectifier when the output signal $v_f = 0$, and gives firing pulses to the incoming rectifier after a delay of 2–10 ms, as explained in (a).

The slave controller releases the appropriate signal i_{c1} (or i_{c2}) to the current controller of the incoming rectifier such that the initial value of v_{c1} (or v_{c2}) causes the voltage v_{a1} (or v_{a2}) to be equal to the back emf. Its input signals are v_r, v_{c1}, and v_{c2}. The signal v_r is obtained from the tachogenerator and obeys the relationship

$$\frac{v_r}{e_b} = \frac{v_{ci}}{v_a}, \quad i = 1, 2 \tag{7.37}$$

where rectifier i is the conducting one, its current being continuous. The signal v_r and the actual output voltage v_{ci} (i = 1, 2) of the non-conducting rectifier are compared, and the error is then amplified and fed to the current controller which constrains its output to be equal to v_r. This ensures tracking of the actual back emf by the idle rectifier as per Eqn (7.37). If now the non-conducting rectifier is switched on, its terminal voltage, under continuous conduction, will be equal to the back emf of the motor. Thus the slave controller ensures a quick switchover without any current transients. It is evident that fast changeover is possible only when there is continuous conduction at the time of switching on of the incoming rectifier. However, when the incoming rectifier works in the discontinuous conduction mode at the time of switching on, the armature voltage v_{ai} (i = 1, 2) and the back emf e_b may differ considerably, causing a current transient.

7.6.7 Speed Control of a Rectifier Controlled Dc Series Motor

A dc series motor whose speed is controlled by a single-phase half-controlled rectifier and its waveforms are shown, respectively, in Figs 7.22(a) and (b). The relation between the angular speed and torque of the motor is derived in Eqn (3.57) of Section 3.4 as

$$\omega_m(t) = \frac{v_a}{\sqrt{\overline{T}_d K_1 K_2}} - \frac{R}{K_1 K_2} \tag{7.38}$$

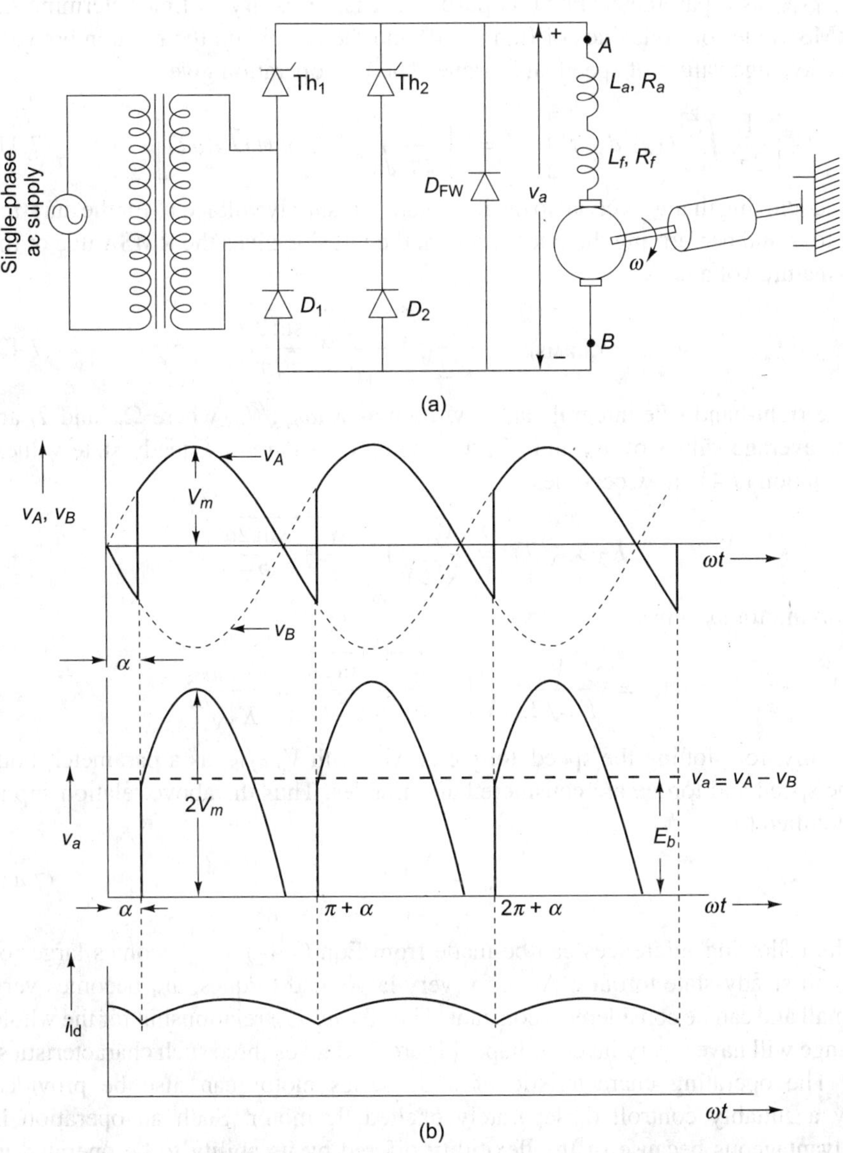

Fig. 7.22 Dc series motor drive fed by a half-controlled rectifier: (a) schematic diagram, (b) waveforms

For very small load torques, ω_m can be approximated as

$$\omega_m(t) \approx \frac{v_a}{\sqrt{\overline{T}_d K_1 K_2}} \tag{7.39}$$

Defining $K_3 = \sqrt{K_1 K_2}$, the armature voltage can be expressed as

$$v_a = K_3 \sqrt{\overline{T}_d}\,\omega_m(t) \tag{7.40}$$

It is now desired to obtain the speed–torque characteristics of the motor with $V_{a(\mathrm{RMS})}$ as a parameter. For this purpose, it is necessary to first determine the RMS values of both sides of Eqn (7.40) and then to obtain the relation between the average values of speed and torque. The first operation gives

$$\left[\frac{1}{2\pi}\int_0^{2\pi}(v_a)^2 d(\omega t)\right]^{1/2} = \left[\frac{1}{2\pi}\int_0^{2\pi} K_3^2 \omega_m^2(t)\overline{T}_d(\omega t)\right]^{1/2} \tag{7.41}$$

Recognizing that v_a is equal to the instantaneous supply voltage v_s for the duration α to π and performing the integration on the left side gives the RMS value of the armature voltage as

$$V_{a(\mathrm{RMS})} = \frac{V_m}{\sqrt{2}}\sqrt{1 - \frac{\alpha}{\pi} + \frac{\sin 2\alpha}{2\pi}} \tag{7.42}$$

The right-hand side integral can be written as $K_3\overline{\omega}_m\sqrt{T_d}$, where Ω_m and T_d are the average values of ω_m and $\overline{T}_d$, thus being considered as steady-state values. Equation (7.41) now becomes

$$K_3\overline{\omega}_m\sqrt{T_d} = \frac{V_m}{\sqrt{2}}\sqrt{1 - \frac{\alpha}{\pi} + \frac{\sin 2\alpha}{2\pi}} \tag{7.43}$$

Solving for $\overline{\omega}_m$ gives

$$\overline{\omega}_m = \frac{V_m}{K_3\sqrt{2T_d}}\sqrt{1 - \frac{\alpha}{\pi} + \frac{\sin 2\alpha}{2\pi}} = \frac{V_{a(\mathrm{RMS})}}{K_3\sqrt{T_d}}$$

Finally, for plotting the speed–torque curves with $V_{a(\mathrm{RMS})}$ as a parameter, both the speed and torque are considered as variables. Thus the above relationship is modified to

$$\omega_m = \frac{V_{a(\mathrm{RMS})}}{K_3\sqrt{T_d}} \tag{7.44}$$

The following inferences can be made from Eqn (7.44). ω_m becomes large for small steady-state torques. Also, for very large load torques, ω_m becomes very small and can be considered a constant. Thus the ω_m-$\overline{T}_d$ relationship for the whole range will have a very inverse shape. Figure 7.23 gives three such characteristics.

The operating characteristic of a dc series motor can also be provided by a suitably controlled, separately excited dc motor. Such an operation is advantageous because of the flexibility offered by its ability to be operated in the regenerative braking mode. However, this involves field reversal operations which are complicated.

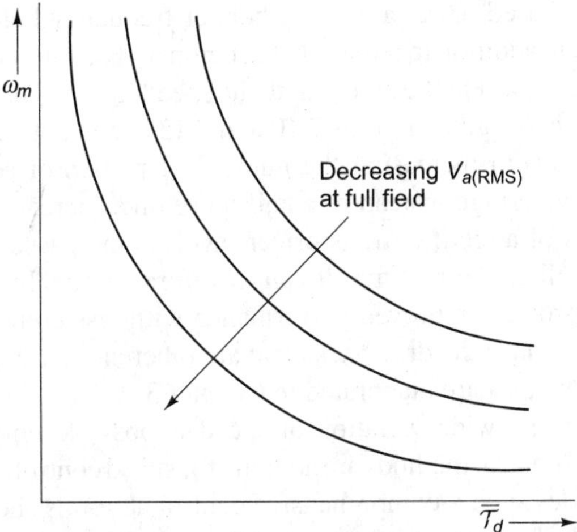

Fig. 7.23 Speed–torque characteristics of a dc series motor with armature voltage V_a as a parameter

7.7 Chopper-based Dc Drives

The analysis of a separately excited dc motor fed by a first-quadrant chopper is dealt with in Chapter 3. The speed–torque characteristics given in Figs 3.5 and 3.6 are reproduced here, respectively, as Figs 7.24(a) and (b).

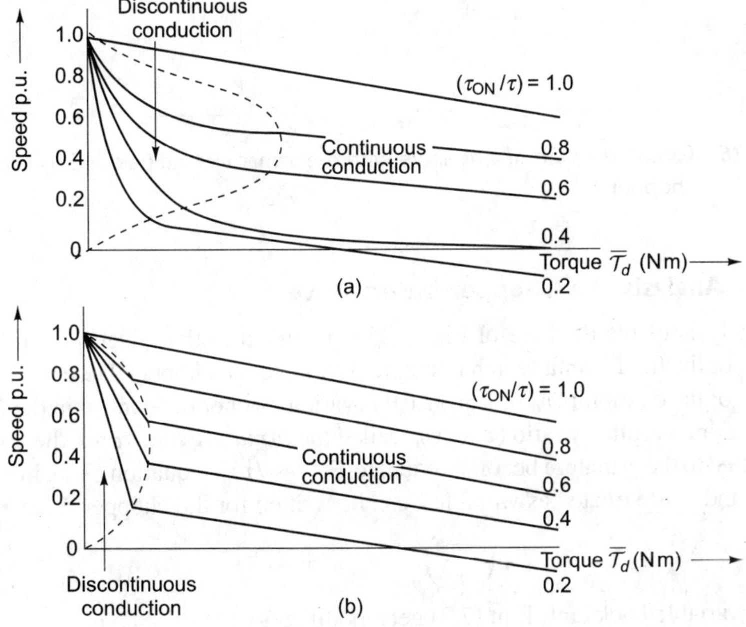

Fig. 7.24 Speed–torque characteristics of a chopper-based dc drive with (a) low and (b) high chopper frequency

For a chopper-based drive, a higher chopper frequency helps in obtaining a smoother output, in addition to reducing the region of discontinuous conduction in the speed–torque plane. On the other hand, the speed–torque characteristics of the converter-fed dc drives given in Figs 7.10 and 7.13 show that better performance can be obtained by (a) increasing the number of phases of ac supply and (b) using the full-wave bridge instead of a half-wave one. Increasing the number of secondary phases of a rectifier transformer involves increased cost both for the transformer as well as the rectifier. It can therefore be concluded that chopper-based dc drives provide improved performance with less complicated circuitry. However, the latter have the drawbacks that are inherent in a step-down chopper, and these disadvantages are elaborated in Chapter 3.

As stated before, a wide variation of speed is possible only when both the armature and field control methods are adopted for speed control. A chopper-based circuit that provides such a comprehensive control action is shown in Fig. 7.25, in which Ch_1 and Ch_2, respectively, control the armature and field voltages taken together. They provide the constant torque-cum-constant-power characteristic shown in Fig. 7.8.

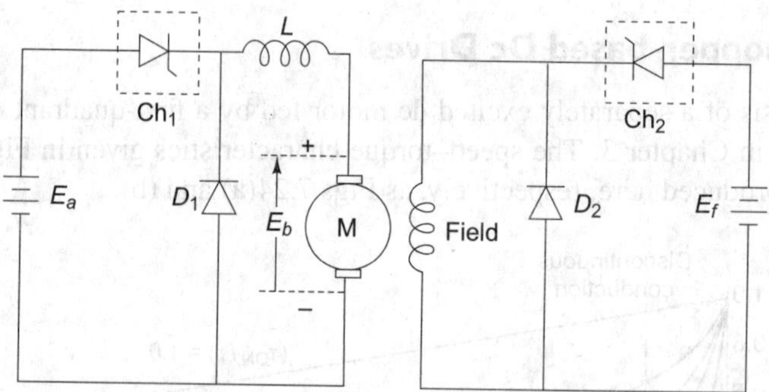

Fig. 7.25 Circuit diagram of a dc drive with the armature and field fed by separate choppers

7.7.1 Analysis of a Chopper-based Drive

The analysis of the dc drive of Fig. 7.25 is identical to that given in Chapter 3 except for the field circuit which is supplied by a second chopper. Here the supply voltage of the dc motor [v_a of Eqn (7.6)] is varied by chopper control in the range 0 to E. For an arbitrary ratio $(\tau_{ON}/\tau)_a$ called the *armature time ratio*, the voltage v_a applied to the armature becomes equal to $E_a(\tau_{ON}/\tau)_a$. Equation (7.1), in which v_a, e_b, and i_a are treated as variables, can be written for the chopper drive as

$$E_a\left(\frac{\tau_{ON}}{\tau}\right)_a = e_b + i_a R_a \tag{7.45}$$

With a variable back emf, Eqn (7.2) gets modified to

$$e_b = K_b \phi_f \omega_m(t) \tag{7.46}$$

ϕ_f in turn can be written as $K_f i_f$. Thus, e_b becomes

$$e_b = K_b K_f i_f \omega_m(t) \tag{7.47}$$

For the chopper-controlled field, $v_f = E_f(\tau_{ON}/\tau)_f$, where $(\tau_{ON}/\tau)_f$ is the *field time ratio*. Thus,

$$E_f\left(\frac{\tau_{ON}}{\tau}\right)_f = i_f R_f \tag{7.48}$$

where v_f is written as $i_f R_f$. Solving Eqn (7.48) for i_f gives

$$i_f = \frac{E_f}{R_f}\left(\frac{\tau_{ON}}{\tau}\right)_f \tag{7.49}$$

With ϕ_f expressed in terms of i_f as $K_f i_f$, as stated above, the average torque can be written as

$$\overline{T}_d = K_t K_f i_f i_a \tag{7.50}$$

Noting that the torque constant K_t is numerically equal to the back emf constant K_b and using the expression in Eqn (7.49) for i_f, $\overline{T}_d$ can be expressed as

$$\overline{T}_d = K_b K_f\left(\frac{E_f}{R_f}\right)\left(\frac{\tau_{ON}}{\tau}\right)_f i_a \tag{7.51}$$

The following facts can be inferred from the above relations. The supply voltages E_a, E_f, resistances R_a, R_f, and constants K_b, K_f are usually given as data. If the speed ω_m is specified, the armature current i_a as well as the developed torque $\overline{T}_d$ can be determined. Alternatively, if i_a is given, the torque and the speed can be computed. Up to the base speed, the field time ratio is kept constant and the armature time ratio is varied to obtain constant torque control. Above the base speed, the reverse operation is adopted so as to achieve constant-power control. The expression for the speed as a function of torque, given in Eqn (3.25), is reproduced here as

$$\omega = \frac{E}{K_1'}\left(\frac{\tau_{ON}}{\tau}\right) - \frac{R_a}{(K_1')^2}\overline{T}_d \tag{7.52}$$

where $\overline{T}_d$ is the torque (variable), τ_{ON}/τ is the armature time ratio, and $K_1' = K_t\phi_f$. Knowing E, K_1', and R_a, the critical ω versus $\overline{T}_d$ characteristics can be obtained as shown in Fig. 7.24(a).

7.7.2 Regenerative Braking of a Separately Excited Dc Motor

It is stated in Chapter 3 that regenerative braking of a separately excited dc motor is possible with a second-quadrant chopper. Such a chopper, shown in Fig. 3.8(b), is redrawn here as Fig. 7.26(a) with an input filter also incorporated in it and the dc motor taken as the load. The purpose of the filter, consisting of an inductor L_f and a capacitor C_f, is to reduce the voltage and current ripples in the source.

During regenerative braking the current i_a flows from the load (dc motor) to the battery source (E). As before, for facilitating chopper action, an additional inductor L_1 is added to the armature circuit of the motor of Fig. 7.26(a), making the total inductance L equal to $L_1 + L_a$. Assuming continuous conduction, the operation of this circuit can be explained as follows. The waveforms of the load quantities v_{ld}, i_a, and the source current i_s are shown in Fig. 7.26(b). The chopper Ch_2 is operated periodically with a period $\tau = 1/f_{Ch}$, where f_{Ch} is the chopper

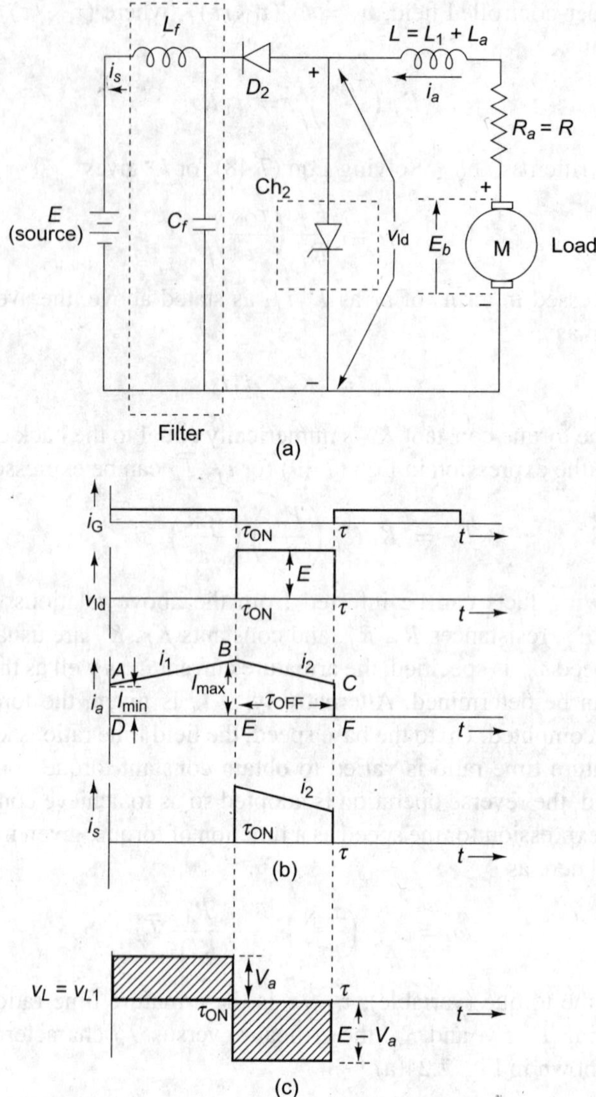

Fig. 7.26 Regenerative braking of a chopper-based dc drive: (a) circuit diagram, (b) waveforms, (c) voltage across inductor L for one period (τ)

frequency and $\tau = \tau_{ON} + \tau_{OFF}$. During the interval $0 \le t \le \tau_{ON}$, the chopper switch Ch_2 remains closed, making $v_{ld} = 0$. The current due to the back emf E_b flows through the armature resistance R_a, L, and Ch_2. The following discussion is made with the assumption that the system has attained steady state.

During the interval $0 \le t \le \tau_{ON}$, the current i_a increases from I_{min} to I_{max}. The combined kinetic energy of the rotor and shaft load is converted into electrical energy and stored in the inductance L. During the next interval $\tau_{ON} \le t \le \tau$, this $(1/2)Li^2$ energy is transferred from the inductance L to the battery (source) in the shape of current that flows towards the source. Consequently, the current i_a

decreases from I_{max} to I_{min}; a negligible fraction of this energy is dissipated in R_a.

The intervals $(0, \tau_{ON})$ and (τ_{ON}, τ) are therefore, respectively, called the *energy storage* and *energy transfer* periods. Over one period $(0, \tau)$, energy transfer takes place in two steps:

(a) from the load at low voltage to the inductor, and then
(b) from the inductor to the battery source at a higher voltage.

As the motor speed falls, the ON time τ_{ON} of the chopper is increased to maintain a nearly constant braking torque. When Ch_2 is just opened at the end of the interval $0 \le t \le \tau_{ON}$, the armature current I_a tries to rush through the source but the filter inductor L_f initially prevents this; however, the filter capacitor C_f provides an alternative path to this current.

The following is an analysis of the regenerative mode of operation. Since the dc machine is working as a generator, the terminal voltage becomes

$$V_a = E_b - I_a R_a \tag{7.53}$$

Assuming L_f and L_a to be very small as compared to the external inductor L_1, the voltage across L can be taken to be nearly equal to the drop across L_1. Thus the voltage profile across L_1 in the intervals $0 \le t \le \tau_{ON}$ and $\tau_{ON} \le t \le \tau$ will be as shown in the striped figure [Fig. 7.26(c)]. The relation

$$V_a \tau_{ON} = (E - V_a)\tau_{OFF} \tag{7.54}$$

holds good here. Writing τ_{OFF} as $(\tau - \tau_{ON})$ and rearranging the terms gives

$$V_a = E(1 - \delta) \tag{7.55}$$

where $\delta = \tau_{ON}/\tau$. Equating the right-hand sides of Eqns (7.53) and (7.55) gives

$$E(1 - \delta) = E_b - I_a R_a$$

or

$$I_a = \frac{E_b - E(1 - \delta)}{R_a} \tag{7.56}$$

With constant field current, the developed torque can be written as

$$T_d = K_m I_a \tag{7.57}$$

I_a being the average armature current.

Approximating the lines *AB* and *BC* of this waveform as straight lines, considering the areas of the trapeziums *ABED* and *BCFE*, and averaging over the period τ, the average armature current is

$$I_a = \frac{I_{max} + I_{min}}{2}$$

During the period $0 \le t \le \tau_{ON}$, the battery (source) is cut off from the load when the chopper is on. The circuit equation governing this period with $R = R_a$ can be written as

$$Ri_a + L\frac{di_a}{dt} = E_b \tag{7.58}$$

with the initial condition $i_a(0) = I_{min}$. As stated before, energy is stored in the inductor during this interval. The solution of Eqn (7.58) with the given initial condition is obtained as

$$i_a(t) = \frac{E_b}{R}(1 - e^{-t/T_a}) + I_{min}e^{-t/T_a}, \quad 0 \leq t \leq \tau_{ON} \tag{7.59}$$

where $T_a = L/R$. At $t = \tau_{ON}$, $i_a(t)$ attains a maximum I_{max}. Thus,

$$i_a(\tau_{ON}) = I_{max} = \frac{E_b}{R}(1 - e^{-\tau_{ON}/T_a}) + I_{min}e^{-\tau_{ON}/T_a} \tag{7.60}$$

In the second interval $\tau_{ON} \leq t \leq \tau$, the chopper is off and the energy stored in the inductor L flows back into the source. The circuit condition for this period is described by the differential equation

$$Ri'_a + L\frac{di'_a}{dt'} = E_b - E \tag{7.61}$$

with the initial condition

$$i'_a(t')_{t'=0} = I_{max}$$

and the time t' is related to t as

$$t' = t - \tau_{ON}$$

The solution of Eqn (7.61) is obtained as

$$i'_a(t') = \frac{E_b - E}{R}(1 - e^{-t'/T_a}) + I_{max}e^{-t'/T_a}, \quad 0 \leq t' \leq \tau_{OFF} \tag{7.62}$$

At $t' = \tau_{OFF}$, $i'_a(t')$ falls down to I_{min}. Thus,

$$i'_a(\tau_{OFF}) = I_{min} = E_b - \frac{(E_b - E)}{R}(1 - e^{-\tau_{OFF}/T_a}) + I_{max}e^{-\tau_{OFF}/T_a} \tag{7.63}$$

Equations (7.60) and (7.63) can be solved, respectively, for I_{max} and I_{min} as

$$I_{max} = \frac{E_b}{R} - \frac{E}{R}\frac{(e^{\tau_{OFF}/T_a} - 1)}{(e^{\tau/T_a} - 1)} \tag{7.64}$$

and

$$I_{min} = \frac{E_b}{R} - \frac{E}{R}\frac{(1 - e^{-\tau_{OFF}/T_a})}{(1 - e^{-\tau/T_a})} \tag{7.65}$$

As stated before, the average load current I_a can be approximated as the mean of I_{max} and I_{min}. The exact expression for I_a can, however, be determined as

$$I_a = \frac{1}{\tau}\left[\int_0^{\tau_{ON}} i_a(t)dt + \int_0^{\tau_{OFF}} i'_a(t')dt'\right] \tag{7.66}$$

Upon evaluating the integrals as indicated in Eqn (7.66), the expression for I_a is obtained as given in Eqn (7.56). The current ripple denoted as I_Δ can be expressed as

$$I_\Delta = \frac{I_{max} - I_{min}}{2} \tag{7.67}$$

Finally, the amount of power regenerated by the dc motor can be determined as

$$P_{\text{reg}} = \frac{1}{\tau}\int_0^{\tau_{\text{OFF}}} E i'_a(t')dt' \tag{7.68}$$

Evaluation of the integral in Eqn (7.68) gives

$$P_{\text{reg}} = \frac{E}{\tau}\left[\frac{(E_b - E)}{R}\{\tau_{\text{OFF}} + T_a(e^{-\tau_{\text{OFF}}/T_a} - 1)\} - T_a I_{\max}(e^{-\tau_{\text{OFF}}/T_a} - 1)\right] \tag{7.69}$$

with $I_{\max}$ as given in Eqn (7.64).

7.7.3 Regenerative Braking of a Dc Series Motor

The circuit of Fig. 7.26(a) can also be used for regenerative braking of a dc series motor with the following considerations. First, the inductance L of this figure must be replaced by $L_1 + L_a$, where L_a is now the inductance of the series field. Second, regeneration involves reversal of armature current, the direction of the field current remaining the same as in the motoring mode. However, when the reversal of the armature current of a series motor is attempted, the current through the series field also gets reversed. Therefore the field connections have to be reversed to maintain the original direction of current in the field circuit. A third feature which concerns the analysis of the series motor is as follows. When the voltage applied to a series motor is changed, both the back emf e_b as well as the current i_a in the series field vary as shown in Fig. 7.27(a). Clearly there is no fixed proportionality between these two variables, the ratio e_b/i_a typically changing as shown in Fig. 7.27(b). If this varying characteristic is approximated by a constant characteristic of magnitude K_s, then the original expression for E_b can be approximated by considering it as well as the speed in rad/s as variables [Eqn (7.2)]:

$$e_b = K_b\phi_f\omega_m = K_b\phi(i_a)\omega_m = K_s\omega_m \tag{7.70}$$

where $K_s \approx K_b\phi(i_a)$. Finally, any slight change in the field inductance due to the saturation of the magnetic circuit is neglected, because the effect of such a change on the steady-state performance is usually very small, thus justifying the assumption of constancy of the field inductance L_f.

The analysis of the series motor under regenerative operation can be done exactly as that of the separately excited motor after taking the above considerations into account. The expression for the regenerated power also remains the same as given by Eqn (7.69). Equations (7.56) and (7.70) can be combined to give ω_m as

$$\omega_m = \frac{(1-\delta)E + R_a I_a}{K_s} \tag{7.71}$$

Substitution for I_a from Eqn (7.57) in Eqn (7.71) yields

$$\omega_m = \frac{(1-\delta)E}{K_s} + \frac{R_a \mathcal{T}_{d(\text{av})}}{K_s K_m} \tag{7.72}$$

Equation (7.72) shows that the ω-$\overline{\mathcal{T}}_d$ curve will have a positive slope. This

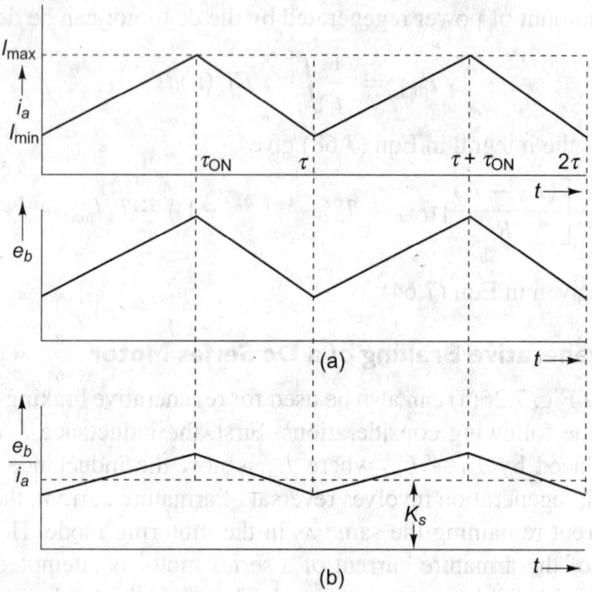

Fig. 7.27 Regenerative braking of dc series motor. Waveforms of (a) armature current i_a and back emf e_b; (b) plot of (e_b/i_a) vs time

feature is confirmed by Fig. 7.28, which is drawn for a typical value of δ, namely, δ_1. The positive slope of the ω-$\overline{T}_d$ curve indicates that a finite amount of load torque is always needed, as otherwise the drive may attain excessively high speeds leading to unstable operation. A problem that may arise at the time of reversal of the field is that the initial build-up of back emf in the generating mode may take considerable time.

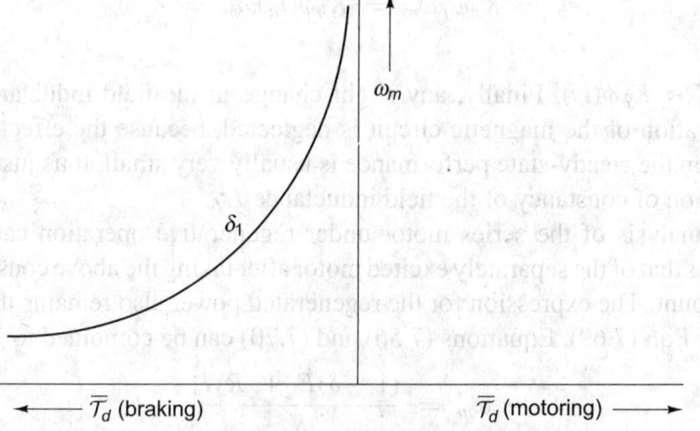

Fig. 7.28 Speed–torque characteristic of a dc series motor in the braking mode

Comments on regenerative braking of dc motors using chopper control If the input dc supply of a separately excited motor consists of a battery of fixed voltage E, regenerative braking is possible if in the equation

$$I_a = \frac{E_b - E}{R_a} \tag{7.73}$$

E_b exceeds E, because only then the motor operates as a generator and current flows from the dc machine to the source. This implies that the minimum speed ω_m at which regeneration is possible cannot go below E_b/K_1, where K_1 is the back emf constant [Eqn (7.17)]. On the other hand, Eqn (7.71) shows that the minimum speed can go down to zero, because under low-load conditions, ω_m can be approximated as $E(1 - \delta)/K_s$, in which δ can be increased to 1. Thus, in addition to providing improved performance, the dc chopper also makes regeneration possible at all speeds. In battery-operated vehicles, regenerated power can be stored in the battery and hence more mileage can be derived per charge on the battery. Secondly, regenerative braking of a series motor is possible only with a dc chopper though it is not as simple as in a separately excited dc motor.

Summary

Speed control over a wide range requires a combination of armature and field control strategies. This is possible with rectifiers and dc choppers, the latter being preferred for the following reasons. Higher chopper frequency provides a smoother output and also helps in reducing the region of discontinuous conduction in the speed–torque plane. The step-up and second-quadrant choppers facilitate regenerative braking of dc motors.

For dc drives, cascaded systems that incorporate an inner current loop together with an outer speed loop are very effective in quick restoration of steady-state conditions after a system disturbance.

The operation of a separately excited dc motor, in such a way as to provide the performance of a dc series motor, has the merit that the motor can be made to work in the regenerative braking mode also.

A dual converter with control features such as the master controller, slave controller, current- and speed control loops helps in smooth four-quadrant operation of the separately excited dc motor.

Worked Examples

1. A 200-V, 1500-rpm rated, separately excited dc motor is supplied by a single-phase controlled bridge rectifier and is connected to a load having a torque of 30 N m at a speed of 1000 rpm. The other data for the drive are $R_a = 0.20\ \Omega$, back emf constant $K_b = 68$ V/(Wb rad/s), $\phi_f = 12 \times 10^{-3}$ Wb, and ac supply voltage V_s = 150 V (RMS) at 50 Hz. Assuming that friction and no-load losses are negligible and that the load is highly inductive, compute (a) the torque developed by the motor and (b) the firing angle of the thyristors.

Solution

(a) As 1 h.p. is equal to 746 W, the rated current of the motor is computed as

$$I_{\text{rated}} = \frac{10 \times 746}{200} = 37.3 \text{ A}$$

The back emf constant K_b is numerically equal to the torque constant K_t. Torque developed by the motor = $K_b \phi_f I_a = 68 \times 12 \times 10^{-3} \times 37.3 = 30.44$ N m

(b) The back emf is obtained from Eqn (7.2) as

$$E_b = K_b \phi_f \bar{\omega}_m = 68 \times 12 \times 10^{-3} \times \frac{2\pi \times 1000}{60} = 85.5 \text{ V}$$

Equation (7.3) gives torque developed as

$$\mathcal{T}_d = K_b \phi_f I_a$$

Hence,

$$I_a = \frac{\mathcal{T}_d}{K_b \phi_f} = \frac{30}{68 \times 12 \times 10^{-3}} = 36.8 \text{ A}$$

The applied voltage is

$$V_a = E_b + I_a R_a = 85.5 + 36.8 \times 0.2 = 92.86 \text{ V}$$

As the load is highly inductive, the conduction will be continuous. Hence, V_a, the output voltage of a single-phase bridge rectifier, is given as

$$V_\alpha = V_m \left(\frac{n}{\pi}\right) \sin\left(\frac{\pi}{n}\right) \cos\alpha$$

where $V_m = 150\sqrt{2}$ and $n = 2$. Hence,

$$150\sqrt{2}\left(\frac{2}{\pi}\right) \sin\frac{\pi}{2} \cos\alpha = 92.86$$

This gives $\alpha = 46.6°$.

2. A 220-V, 1500-rpm rated, separately excited dc motor is supplied by a three-phase, half-wave controlled rectifier and runs at 700 rpm. The other data are $R_a = 0.22\ \Omega$, $L_a = 0.525$ mH, ac supply voltage $V_s = 100$ V (RMS) at 50 Hz, $\alpha = 60°$, $\phi_f = 10 \times 10^{-3}$ Wb, and $K_b = 68.75$ V/(Wb rad/s). Determine the (a) armature current and (b) load torque.

Solution

The circuit is shown in Fig. 7.29.

(a) It is necessary to first determine whether the conduction is continuous or discontinuous:

$$E_b = K_b \phi_f \omega_m = 68.75 \times 10 \times 10^{-3} \times \frac{2\pi \times 700}{60} = 50.4$$

$$I_a = \frac{V_a - E_b}{R_a} = \frac{V_a - 50.4}{R_a}$$

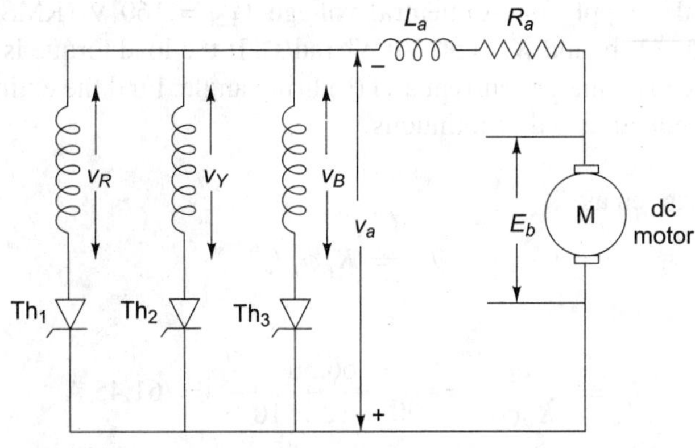

Fig. 7.29

The average voltage V_a is found from the expression

$$\frac{3V_m}{\pi}\frac{\sqrt{3}}{2}\cos\alpha$$

if the conduction is continuous, and from the expression

$$\frac{3V_m}{2\pi}(\cos\alpha \;-\; \cos\alpha_e)$$

if it is discontinuous.

$$\text{Load power factor} = \cos\phi \;=\; \cos\left[\tan^{-1}\left(\frac{\omega L_a}{R_a}\right)\right]$$

$$= \cos\left[\tan^{-1}\left(\frac{314 \times 0.525/100}{0.22}\right)\right]$$

$$= 0.8$$

The factor a is

$$a \;=\; \frac{E_b}{V_m} \;=\; \frac{79.2}{100\sqrt{2}} \;=\; 0.56$$

From the chart given in Fig. 2.21, with $\alpha = 60°$, $a = 0.56$, and $\cos\phi = 0.8$, $\alpha_e = 184°$. As $\alpha_e - \alpha = 124°$, which is more than 120°, the conduction is continuous. Hence,

$$V_a \;=\; V_m\frac{3\sqrt{3}}{2\pi}\cos\alpha \;=\; 100\sqrt{2} \times \frac{3\sqrt{3}}{2\pi}\cos 60° \;=\; 58.5\ \text{V}$$

$$I_a \;=\; \frac{58.5 \;-\; 50.4}{0.22} \;=\; 36.8\ \text{A}$$

(b) From Eqn (7.3)

$$\mathcal{T}_d \;=\; K_b\phi_f I_a \;=\; 68.75 \times 10 \times 10^{-3} \times 36.8 \;=\; 25.3\ \text{N m}$$

3. An 180-V, 1500-rpm rated, separately excited dc motor is supplied by a three-phase half-wave controlled rectifier and runs at 1200 rpm. Other data: $R_a = 0.24\ \Omega$,

L_a = 1.02 mH, supply line-to-neutral voltage V_{LN} = 160 V (RMS) at 50 Hz, $\phi_f = 12 \times 10^{-3}$ Wb, and K_b = 90 V/(Wb rad/s). If the load torque is 66.36 N m, determine the (a) armature current and (b) firing angle. Find the extinction angle also if the conduction is discontinuous.

Solution

(a) Torque is given as

$$T_d = K_b \phi_f I_a$$

Hence,

$$I_a = \frac{T_d}{K_b \phi_f} = \frac{66.36}{90 \times 12 \times 10^{-3}} = 61.45 \text{ A}$$

(b) $E_b = K_b \phi_f \omega_m = 90 \times 12 \times 10^{-3} \times \frac{2\pi \times 1200}{60} = 135.7 \text{ V}$

Hence,

$$V_a = E_b + I_a R_a = 135.7 + 61.45 \times 0.24 = 150.45 \text{ V}$$

$$a = \frac{E_b}{V_m} = \frac{135.7}{160\sqrt{2}} = 0.6$$

$$\tan\phi = \frac{\omega L}{R} = \frac{314 \times 1.02/1000}{0.24} = 1.33$$

$$\cos\phi = 0.6$$

V_a can also be written as

$$V_a = V_m \frac{3\sqrt{3}}{2\pi} \cos\alpha$$

if the conduction is continuous, and as

$$V_a = \frac{3V_m}{2\pi}(\cos\alpha - \cos\alpha_e)$$

if it is discontinuous. It is seen from the chart of Fig. 2.21 that with $a = 0.6$ and $\cos\phi = 0.6$, the conduction is discontinuous for almost all values of α. Hence the average voltage is

$$V_a = \frac{3V_m}{2\pi}(\cos\alpha - \cos\alpha_e)$$

Substituting the values of V_a and V_m gives

$$150.45 = \frac{3 \times 160\sqrt{2}}{2\pi}(\cos\alpha - \cos\alpha_e)$$

or

$$\cos\alpha - \cos\alpha_e = 1.393$$

By trial and error it is found from the chart that $\alpha = 55^\circ$ and the corresponding $\alpha_e = 145^\circ$ satisfy the above equality. Hence the firing angle $\alpha = 55^\circ$ and the extinction angle $\alpha_e = 145^\circ$.

4. A three-phase, full-wave bridge type of controlled rectifier supplies the armature of a 15-h.p., 315-V, separately excited dc motor. The field circuit gets its supply from a three-phase, full-wave diode rectifier bridge connected to the same supply. The data are supply $V_{LN} = 180$ V (RMS), $R_a = 0.26\ \Omega$, $L_a = \infty$, $R_f = 180\ \Omega$, $K_b = 80$ V/(Wb rad/s), ϕ_f is given as $K_1 I_f$, where $K_1 = 0.013$ Wb/A, $\alpha = 55^\circ$, and $I_a = 60$ A.

Compute the (a) torque developed by the motor, (b) speed, and (c) input power factor of the drive. If the motor is now made to work as a generator by reversing the field polarity, and carries the same armature current as before, find the new firing angle and the regenerated power.

Solution

(a) The torque developed is given by the expression

$$\mathcal{T}_d = K_b K_1 I_f I_a$$

All quantities in the right-hand side are known except I_f. As the field is connected to the output of an uncontrolled rectifier, with $n = 3$,

$$V_f = 180\sqrt{2} \times 2 \times \frac{3}{\pi} \times \frac{\sqrt{3}}{2} \cos 0^\circ = 421.0 \text{ V}$$

$$I_f = \frac{V_f}{R_f} = \frac{421}{180} = 2.34 \text{ A}$$

$$\mathcal{T}_d = 80 \times 0.013 \times 2.34 \times 60 = 146 \text{ N m}$$

(b) The back emf is given as

$$E_b = V_a - I_a R_a$$

As the inductance in the armature circuit is very large, conduction can be taken to be continuous. Hence

$$V_a = 2\left(\frac{n}{\pi}\right) V_m \sin\left(\frac{\pi}{n}\right) \cos\alpha$$

where V_m is the peak of the phase voltage $= V_s\sqrt{2} = 180\sqrt{2} = 255$ V. Thus,

$$V_a = 2 \times \frac{3\sqrt{3}}{2\pi} \times 180\sqrt{2} \cos 55^\circ = 241.5 \text{ V}$$

$$E_b = 241.5 - 60 \times 0.26 = 225.9 \text{ V}$$

$$\omega = \frac{E_b}{K_b K_1 I_f} = \frac{225.9}{80 \times 0.013 \times 2.34}$$

$$\frac{2\pi N}{60} = \omega$$

$$\text{Speed } N = \frac{\omega \times 60}{2\pi} = \frac{60}{2\pi} \times \frac{225.9}{80 \times 0.013 \times 2.34} = 886 \text{ rpm}$$

(c) The total power input to the unit is computed as follows:

$$I_{a1} = \frac{\sqrt{6}}{\pi} \times 60 = 46.78$$

$$I_{f1} = \frac{\sqrt{6}}{\pi} \times 2.34 = 1.82$$

$$\cos 55° = 0.5736$$

$$\sin 55° = 0.8192$$

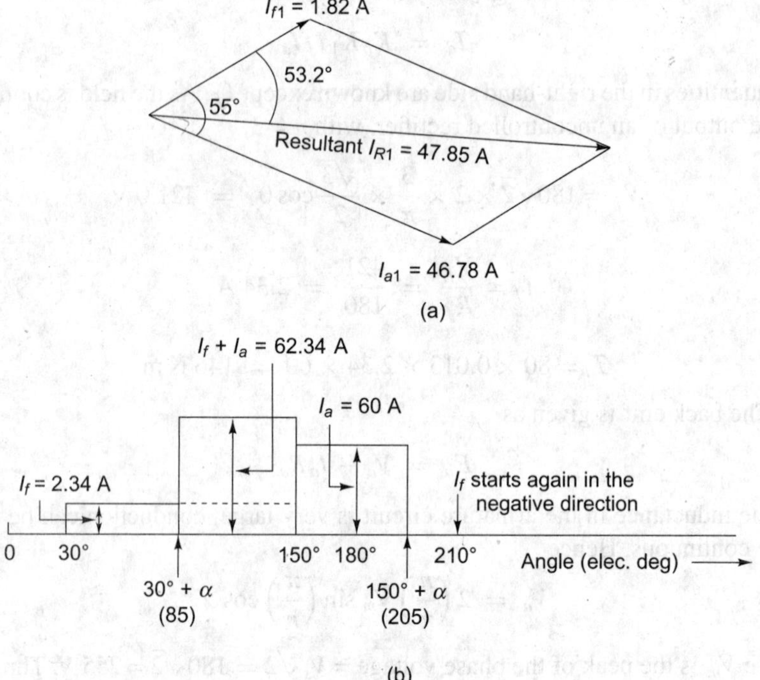

Fig. 7.30

From Fig. 7.30(a),

$$\begin{aligned} I_{R1}^2 &= I_{f1}^2 + I_{a1}^2 + 2I_{f1}I_{a1}\cos\alpha \\ &= 1.82^2 + 46.78^2 + 2 \times 1.82 \times 46.78 \times \cos 55° \\ &= 2289.365 \end{aligned}$$

$$I_{R1} = \sqrt{2289.365} = 47.85 \text{ A}$$

$$\tan\phi = \frac{I_{a1}\sin\alpha}{I_{f1}+I_{a1}\cos\alpha} = \frac{46.78\times 0.8192}{1.82+46.78\times 0.5736} = 1.3375$$

$$\beta = 53.2^\circ$$

From Fig. 7.30(b),

$$I_{\text{RMS}} = \left(\frac{2}{2\pi}\frac{\pi}{180}\left[\int_{30}^{85}(I_f)^2 d(\omega t) + \int_{85}^{150}(I_a+I_f)^2 d(\omega t) + \int_{150}^{205}(I_a)^2 d(\omega t) + \int_{205}^{210} 0^2 d(\omega t)\right]\right)^{1/2}$$

$$= \sqrt{\frac{1}{180}[I_f^2\times 55 + (I_a+I_f)^2\times 65 + I_a^2\times 55]}$$

$$= \sqrt{\frac{1}{180}(2.34^2\times 55 + 62.34^2\times 65 + 60^2\times 55)}$$

$$= 50$$

$$\text{Input power factor} = \frac{47.85\ \cos 53.2^\circ}{I_{\text{RMS}}} = \frac{28.653}{50} = 0.57$$

(d) Back emf just before polarity reversal = 225.9 V, back emf after reversal = −225.9 V, armature voltage after reversal = – 225.9 + 60 × 0.26 = −210.3 V. So,

$$2\times\frac{3}{\pi}\times\frac{\sqrt{3}}{2}\times 180\sqrt{2}\ \cos\alpha = -\ 210.3$$

$$\frac{540\sqrt{6}}{\pi}\cos\alpha = -\ 210.3$$

$$\cos\alpha = \frac{-210.3\pi}{540\sqrt{6}} = -\ 0.499,\quad \alpha = 120^\circ$$

Regenerated power = 210.3 × 60 = 12.62 kW.

5. A single-phase fully controlled bridge rectifier controls the input to a 10-h.p., 150-V, 1200-rpm, separately excited dc motor. The field of the dc motor is also supplied by a single-phase fully controlled bridge converter connected to the same ac source and is set to give three-fourths of its field current. The ac input is a single-phase, 220-V, 50-Hz supply. The armature resistance $R_a = 0.18\ \Omega$ and the field resistance $R_f = 170\ \Omega$. There are sufficient inductances in the armature and the field circuits to make the currents continuous and nearly constant. (a) Compute the firing angle of the rectifier when the motor supplies rated power at the rated speed. (b) When the firing angles of the armature and field circuit are kept the same as in (a) and the machine runs with no load, it takes an armature current equal to 8% of the full load value; with these conditions, determine the no-load speed. (c) Determine the percentage speed regulation assuming that the field flux is related to I_f as $\phi_f = K_1 I_f$, where $K_1 = 0.015$. Also $K_b = 70$ V/(Wb rad/s).

Solution

The circuit diagram of the drive is shown in Fig. 7.31.

(a) The maximum field current is obtained with a firing angle α_f of 0°.

$$V_f = \frac{2V_m}{\pi}\cos\alpha_f$$

$$I_f = \frac{2V_m}{R_f\pi}\cos\alpha_f$$

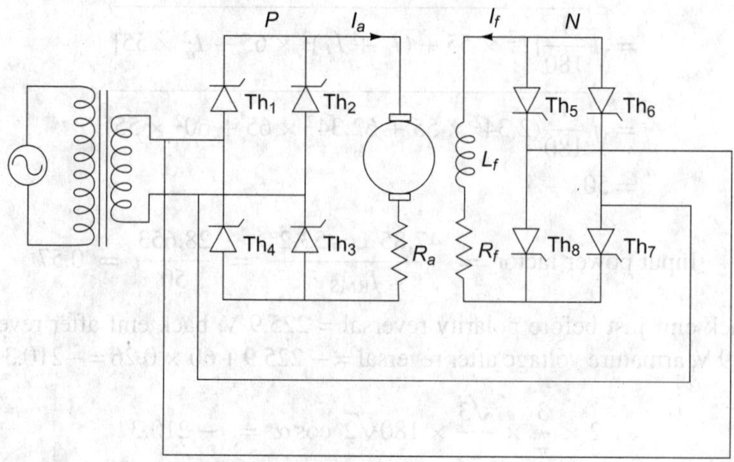

Fig. 7.31

$$I_{f(\max)} = \frac{2V_m}{R_f\pi} = \frac{2 \times 220\sqrt{2}}{\pi \times 170} = 1.17 \text{ A}$$

$$I_f = 0.75 I_{f(\max)} = 0.75 \times 1.17 = 0.88 \text{ A}$$

$\cos\alpha_f$ can be determined from the equation

$$\frac{2V_m}{R_f\pi}\cos\alpha_f = 0.75 I_{f(\max)} = 0.75\frac{2V_m}{R_f\pi}$$

Hence $\alpha_f = 41.4°$. At the rated speed,

$$E_b = K_b K_1 I_f \omega = \frac{70 \times 0.015 \times 1.17 \times 2\pi \times 1200}{60} = 115.8 \text{ V}$$

$$\text{Rated current} = \frac{10 \times 746}{150} \approx 50 \text{ A}$$

$$V_a = E_b + I_a R_a = 115.8 + 50 \times 0.18 = 124.8 \text{ V}$$

V_a is also given by

$$\frac{2V_m}{\pi} \cos \alpha_a = \frac{2 \times 220\sqrt{2} \cos \alpha}{\pi} = 124.8$$

or

$$\cos \alpha = \frac{\pi \times 124.8}{2 \times 220 \times \sqrt{2}} = 0.63$$

$$\alpha = 51°$$

(b) At no load, with α_a and α_f as before,

$$E_b = V_a - I_a R_a = 124.8 - \left(\frac{8}{100} \times 50\right) \times 0.18 = 124.8 - 4 \times 0.18 = 124.1 \text{ V}$$

$$\omega = \frac{E_b}{K_b K_1 I_f} = \frac{124.1}{70 \times 0.015 \times 0.88} = 134.3 \text{ rad/s}$$

Hence, no-load speed

$$N = \frac{134.3 \times 60}{2\pi} = 1283 \text{ rpm}$$

(c) Finally,

$$\text{Speed regulation} = \frac{\text{no-load speed} - \text{full-load speed}}{\text{full-load speed}}$$

$$= \frac{1283 - 1200}{1200}$$

$$= 0.0692$$

or the percentage speed regulation = 6.92%.

6. The armature of an 8-h.p., 140-V, 1500-rpm, separately excited dc drive is controlled by a single-phase fully controlled bridge rectifier. The field is also supplied by a single-phase fully controlled bridge from the same ac source. The circuit diagram is given in Fig. 7.31. The other data for the drive are as follows: the field current is set at its maximum value, ac source voltage = 200 V at 50 Hz, $R_a = 0.2\ \Omega$, $R_f = 170\ \Omega$, firing angle of the armature converter, $\alpha_a = 20°$, $d_f = 0°$, load torque = 55 N m, and $K_b = 75$ V/(Wb rad/s).

(a) Determine the speed of the motor under the above load conditions. (b) For the same load torque as in (a), determine the firing angle of the field rectifier if the speed has to be increased to 1600 rpm. The field flux is related to the field current as $\phi_f = K_1 I_f$, where $K_1 = 0.013$ A/Wb.

Solution

(a) As the field current is maximum, $\alpha_f = 0$.

$$V_f = \frac{2V_m}{\pi} \cos 0° = \frac{2\sqrt{2} \times 200}{\pi} = 180 \text{ V}$$

$$I_f = \frac{V_f}{R_f} = \frac{180}{170} = 1.06 \text{ A}$$

$$V_a = \frac{2V_m}{\pi} \cos \alpha_a = \frac{2\sqrt{2} \times 200}{\pi} \cos 20° = 169 \text{ V}$$

$$I_a = \frac{T_d}{K_b K_1 I_f} = \frac{55}{75 \times 0.013 \times 1.06} = 53.2 \text{ A}$$

$$E_b = V_a - I_a R_a = 169 - 53.2 \times 0.2 = 158.4 \text{ V}$$

$$\omega = \frac{E_b}{K_b K_1 I_f} = \frac{158.4}{75 \times 0.013 \times 1.06} = 153.3 \text{ rad/s}$$

Hence the speed with the given data = $(60/2\pi) \times 153.3 = 1464$ rpm.
(b) The new speed is 1600 rpm:

$$\omega = \frac{2 \times \pi \times 1600}{60} = 167.6 \text{ rad/s}$$

$$E_b = K_b K_1 I_f = 75 \times 0.013 \times 167.6 I_f$$

$$V_a = 169 = E_b + I_a R_a$$

$$169 = 75 \times 0.013 \times 167.6 I_f + \frac{55 \times 0.2}{75 \times 0.013 I_f}$$

This is of the form

$$ax^2 - 169x + b = 0$$

where $x = I_f$ and

$$a = (75 \times 0.013) \times 167.6 = 163.5$$

Also,

$$b = \frac{55 \times 0.2}{75 \times 0.013} = 11.28$$

and $x = I_f$. Therefore,

$$x = \frac{169 \pm \sqrt{169^2 - 4ab}}{2a} = \frac{169 \pm 145.54}{2 \times 163.5}$$

Taking the plus sign,

$$I_f = x = 0.962$$

$$V_f = I_f R_f = 0.962 \times 170 = 163.54$$

α_f is obtained from the equation $2V_m \cos \alpha_f / \pi = 163.54$. Hence, $\cos \alpha_f = 0.908$ and $\alpha_f = 24.7°$.

7. A 15-h.p., 480-V, 1000-rpm dc drive consists of a three-phase ac source connected to a three-phase, full-wave fully controlled bridge rectifier with the armature of a separately excited dc motor at its output. The field is also controlled by the same type of controlled rectifier and by the same ac supply. The data for this drive are ac supply $V_{LL} = 346$ V (RMS) at 50 Hz, $R_a = 0.25\ \Omega$, $R_f = 224\ \Omega$, $K_b = 80$ V/(Wb rad/s), developed torque = 80 N m at 1000 rpm. The firing angle α_f of the field rectifier is set to give maximum field current. (a) Find the firing angle α_a of the armature rectifier. (b) For the same load torque as in part (a), if the speed is to be increased to 1300 rpm, determine the new α_f. The field flux ϕ_f is expressed as $K_1 I_f$, where $K_1 = 0.014$ Wb/A. (c) Find the input power factor for the last operating condition.

Solution

(a) As the field current is set at its maximum, α_f of the field rectifier is 0°.

$$V_{f(\max)} = 2\left(\frac{n}{\pi}\right) V_m \sin\left(\frac{\pi}{n}\right) \cos 0^\circ$$

where $n = 3$ and

$$V_m = \sqrt{2} V_{\text{LN(RMS)}} = \sqrt{2}\frac{V_{\text{LL(RMS)}}}{\sqrt{3}} = \sqrt{2} \times \frac{346}{\sqrt{3}} = 282.5 \text{ V}$$

Thus,

$$V_{f(\max)} = \frac{2 \times 3\sqrt{3} \times 282.5}{2\pi} = 467.3$$

$$I_{f(\max)} = \frac{V_{f(\max)}}{R_f} = \frac{467.3}{224} = 2.09 \text{ A}$$

$$E_b = K_b K_1 I_f \omega = \frac{80 \times 0.014 \times 2.09 \times 2\pi \times 1000}{60} = 245 \text{ V}$$

$$I_a = \frac{T_d}{K_b K_1 I_f} = \frac{80}{80 \times 0.014 \times 2.09} = 34.2 \text{ A}$$

Now,

$$V_a = E_b + I_a R_a = 245 + 34.2 \times 0.25 = 253.6 \text{ V}$$

V_a is also evaluated as

$$V_a = 2 \times \frac{3\sqrt{3}}{2\pi} V_m \cos\alpha_a = \frac{2 \times 3\sqrt{3} \times 282.5}{2\pi} \cos\alpha_a$$

Hence,

$$\cos\alpha_a = \frac{253.6\pi}{3\sqrt{3} \times 282.5} = 0.543$$

$$\alpha_a = 57^\circ$$

(b) For the same load torque as before, armature current I_a is obtained as

$$I_a = \frac{T_d}{K_b K_1 I_f} = \frac{80}{80 \times 0.014 I_f}$$

$$\text{New speed} = 1300 \text{ rpm} = 1300 \times \frac{2\pi}{60} = 136.1 \text{ rad/s}$$

Also $E_b = 80 \times 0.014 \times 136.1 I_f$. The applied armature voltage remains the same as before.

$$V_a = E_b + I_a R_a$$

Substitution of values gives

$$253.6 = 80 \times 0.014 \times 136.1 I_f + \frac{80 \times 0.25}{80 \times 0.014 I_f}$$

$$= 152.5 I_f + \frac{17.86}{I_f}$$

This gives the quadratic equation

$$152.5 I_f^2 - 253.6 I_f + 17.86 = 0$$

$$I_f = \frac{253.6 \pm 231.1}{2 \times 152.5}$$

Taking the plus sign, $I_f = 1.59$ A. Field voltage = $I_f R_f = 1.59 \times 224 = 356$ V. $\cos\alpha_f$ is found from the equation

$$2 \times \frac{3\sqrt{3}}{2\pi} \times 282.5 \cos\alpha_f = 356$$

This gives $\cos\alpha_f = 0.76$. or $\alpha_f = 40.4°$.

(c) Input power factor calculation:

$$I_{a1} = \frac{\sqrt{6}}{\pi} I_a = \frac{\sqrt{6}}{\pi} \times 34.2 = 26.67 \text{ A}, \quad \theta_{a1} = -57° = -\alpha_a$$

$$I_{f1} = \frac{\sqrt{6}}{\pi} I_f = \frac{\sqrt{6}}{\pi} \times 1.59 = 1.24 \text{ A}, \quad \theta_{f1} = -40.4° = -\alpha_f$$

From Fig. 7.32(a), substitution of values gives

$$I_{R1}^2 = I_{f1}^2 + I_{a1}^2 + 2 I_{f1} I_{a1} \cos 16.6°$$

$$= 1.24^2 + 26.67^2 + 2 \times 1.24 \times 26.67 \times 0.9583$$

$$= 776.21$$

and

$$I_{R1} = 27.86$$

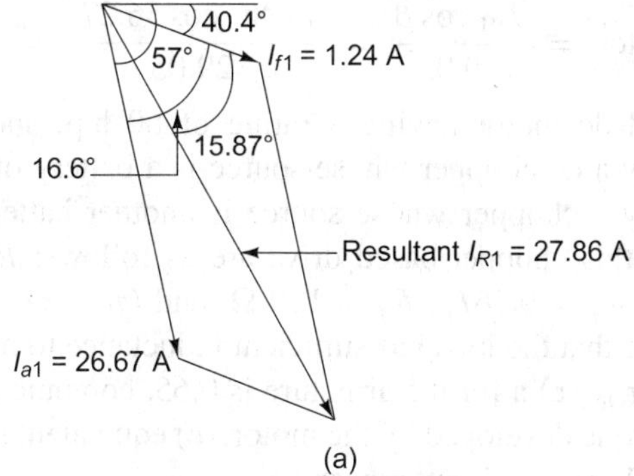

(a)

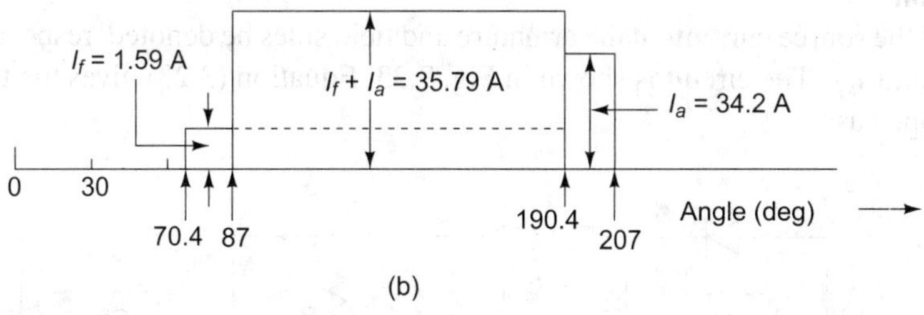

(b)

Fig. 7.32

$$\tan\beta = \frac{26.67\ \sin 16.6^\circ}{1.24 + 26.67\ \cos 16.6^\circ} = \frac{7.619}{26.798} = 0.2843$$

$$\beta = 15.87^\circ$$

$$\beta + 40.4 = 56.27^\circ$$

From Fig. 7.32(b),

$$I_{\text{RMS}} = \sqrt{\frac{\pi}{180} \times \frac{2}{2\pi}\left[\int_{70.4}^{87} I_f^2 d(\omega t) + \int_{87}^{190.4} (I_f + I_a)^2 d(\omega t) + \int_{190.4}^{207} I_a^2 d(\omega t)\right]}$$

$$= \left(\frac{1}{180}[(1.59)^2(87 - 70.4) + (34.2 + 1.59)^2(190.4 - 87) + (34.2)^2(207 - 190.4)]\right)^{1/2}$$

$$= \sqrt{\frac{1}{180}[41.966 + 132{,}447.55 + 19{,}416.024]}$$

$$= \sqrt{\frac{151{,}905.54}{180}}$$

$$= 29.05 \text{ A}$$

$$\text{Input power factor} = \frac{I_{R1}\cos\beta}{29.05} = \frac{27.86\ \cos 56.27^\circ}{29.05} = 0.53$$

8. A separately excited dc motor having a rating of 60 h.p. and running at 1200 rpm is supplied by a dc chopper whose source is a battery of 500 V. The field is also supplied by a chopper whose source is another battery of 300 V. The data pertaining to this chopper-based drive are as follows: $R_a = 0.18\ \Omega$, $K_b = 70$ V/(Wb rad/s), $\phi_f = 0.16I_f$, $R_f = 120\ \Omega$, and $(\tau_{\text{ON}}/\tau)_f$ for the field chopper is 0.85. Assume that the load has sufficient inductance to make the load current continuous. If (τ_{ON}/τ) a for the armature is 0.65, compute the (a) mean armature current, (b) torque developed by the motor, (c) equivalent resistance for the armature circuit, and (d) total input power.

Solution

(a) Let the source currents at the armature and field sides be denoted, respectively, as I_{sa} and I_{sf}. The circuit is shown in Fig. 7.33. Equation (3.23) gives the torque developed as

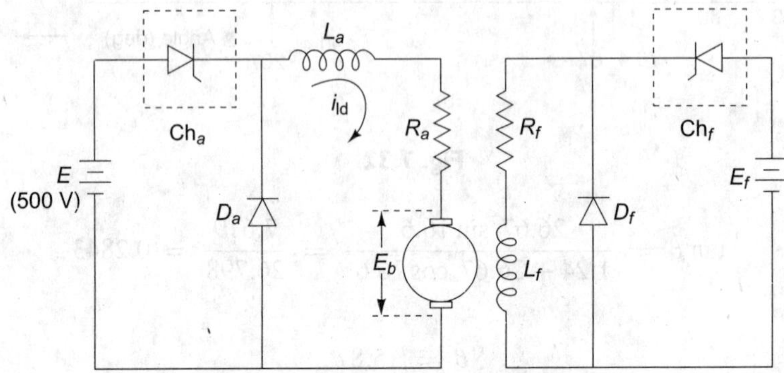

Fig. 7.33

$$\mathcal{T}_d = K_t\phi_f I_{\text{ld}}(\omega) = K_b K_1 I_f I_a$$

Here,

$$K_b K_1 = 70 \times 0.016 = 1.12$$

$$I_f = E_f\left(\frac{\tau_{\text{ON}}}{\tau}\right)_f \frac{1}{R_f} = \frac{300 \times 0.85}{120} = 2.125\ \text{A}$$

Hence the average torque is

$$\mathcal{T}_d = 1.12 \times 2.125 I_a = 2.38 I_a$$

$$\omega = \frac{2\pi \times 1200}{60} = 125.7$$

$$E_b = K_b K_1 I_f \omega = 2.38 \times 125.7 = 299\ \text{V}$$

$$V_a = E\left(\frac{\tau_{ON}}{\tau}\right)_a = 500 \times 0.65 = 325 \text{ V}$$

The mean armature current is

$$I_a = \frac{V_a - E_b}{R_a} = \frac{325 - 299}{0.18} = 144.4 \text{ A}$$

(b) Torque developed $T_d = 2.38 \times 144.4 = 343.8$ N m.
(c) Input (or source) current is

$$I_{sa} = I_a\left(\frac{\tau_{ON}}{\tau}\right)_a = 144.4 \times 0.65 = 93.9 \text{ A}$$

Also,

$$I_{sf} = I_f \times \left(\frac{\tau_{ON}}{\tau}\right)_f = 2.125 \times 0.85 = 1.81 \text{ A}$$

$$\text{Armature source equivalent resistance} = \frac{\text{armature source voltage}}{\text{armature source current}} = \frac{500}{93.9} = 5.32\ \Omega$$

(d) Total input power = power input to armature + power input to field $= E_a I_{sa} + E_f I_{sf}$. Hence, it is given as $P_i = 500 \times 93.9 + 300 \times 1.81 = 46{,}950 + 543 = 47{,}493 \text{ W} \approx 47.5$ kW.

9. A separately excited dc motor has a rating of 50 h.p. and when supplied by a battery of 480 V through a chopper, it has a mean armature current of 120 A. The field is also supplied by a chopper whose source is a battery of 250 V. Other data for this chopper-based drive are $R_a = 0.2\ \Omega$, $R_f = 125\ \Omega$, $K_b = 72$ V/(Wb rad/s), $\phi_f = 0.015 I_f$, $(\tau_{ON}/\tau)_a = 0.7$, and $(\tau_{ON}/\tau)_f = 0.9$. The armature circuit has sufficient inductance to make the current continuous. Compute the (a) speed of the motor, (b) torque developed by the motor, (c) equivalent resistance, and (d) total input power.

Solution

(a) The circuit is the same as that given in Fig. 7.33.

$$V_a = 480\left(\frac{\tau_{ON}}{\tau}\right)_a = 480 \times 0.7 = 336 \text{ V}$$

$$E_b = V_a - I_a R_a = 336 - 120 \times 0.2 = 312 \text{ V}$$

$$\omega = \frac{E_b}{K_b \phi_f} = \frac{E_b}{K_b \times 0.015 I_f}$$

Here,

$$I_{sf} = 250\left(\frac{\tau_{ON}}{\tau}\right)_f \frac{1}{R_f} = \frac{250 \times 0.9}{125} = 1.8 \text{ A}$$

Hence,

$$\omega = \frac{E_b}{K_b \times 0.015 I_f} = \frac{312}{72 \times 0.015 \times 1.8} = 160 \text{ rad/s}$$

Also,

$$\text{speed} = \frac{60\omega}{2\pi} = \frac{60 \times 160}{2\pi} = 1528 \text{ rpm}$$

(b)
$$\text{Torque developed} = K_b \times 0.015 I_f I_a$$
$$= 72 \times 0.015 \times 1.8 \times 120$$
$$= 233.3 \text{ N m}$$

The source current on the armature side is

$$I_{sa} = I_a \left(\frac{\tau_{ON}}{\tau}\right)_a = 120 \times 0.7 = 84 \text{ A}$$

(c)
$$\text{Armature source equivalent resistance} = \frac{\text{armature source voltage}}{\text{armature source current}}$$
$$= \frac{480}{84} = 5.7\ \Omega$$

(d) Total input power

$$P_i = \text{power input to armature} + \text{power input to field}$$
$$= E_a I_{sa} + E_f I_{sf}$$

where

$$I_{sf} = I_f \left(\frac{\tau_{ON}}{\tau}\right)_f = 1.8 \times 0.9 = 1.62 \text{ A}$$

Hence,

$$P_i = 480 \times 84 + 250 \times 1.62 = 40{,}320 + 405 = 40{,}725 \text{ W} \approx 40.7 \text{ kW}$$

10. A dc chopper is used for regenerative braking of a separately excited dc motor as shown in Fig. 7.24(b). The data are $E = 400$ V, $R_a = 0.2\ \Omega$, and $L_a = 0.2$ mH. The back emf constant K_b^1 ($= K_b\phi_f$), assuming ϕ_f to be constant, is equal to 1.96 V/rad s, and the average load current is 200 A. The frequency of the chopper is 1 kHz and $(\tau_{ON}/\tau)_a$ is 0.5. Compute the (a) average load voltage, (b) motor speed, (c) power regenerated and fed back to the battery, (d) equivalent resistance viewed from the motor side when it is working as a generator, and (e) minimum and maximum permissible speeds for regenerative braking.

Solution

(a) The circuit and waveforms are shown in Figs 7.26(a) and (b). The average load voltage is

$$V_a = E\left(1 - \frac{\tau_{ON}}{\tau}\right)_a = 400 \times 0.5 = 200 \text{ V}$$

(b)
$$I_a = \frac{I_{max} + I_{min}}{2} = 200 \text{ A}$$

or $I_{max} + I_{min} = 400$ A. From Eqns (7.64) and (7.65),

$$I_{max} + I_{min} = \frac{2E_b}{R_a} - \frac{E}{R_a}\left[\frac{1 - e^{-\tau_{OFF}/T_a}}{1 - e^{-\tau/T_a}} + \frac{e^{\tau_{OFF}/T_a} - 1}{e^{\tau/T_a} - 1}\right]$$

Here,

$$T_a = \frac{L_a}{R_a} = \frac{0.0002}{0.2} = 0.001$$

$$\tau = \frac{1}{1000} = 0.001$$

Hence,

$$\frac{\tau}{T_a} = 1 \text{ and } \frac{\tau_{ON}}{T_a} = \frac{0.0005}{0.001} = 0.5$$

Also

$$\tau_{OFF}/T_a = 0.5$$

Substituting the above values in the equation for $I_{max} + I_{min}$ gives

$$400 = \frac{2E_b}{0.2} - \frac{400}{0.2}\left[\frac{1 - 0.606}{1 - 0.368} + \frac{1.649 - 1}{2.718 - 1}\right]$$
$$= 10E_b - 2000(0.623 + 0.377)$$

This gives the value of the back emf as

$$E_b = 240$$

$$\text{Speed in rad/s} = \frac{240}{K_b^1} = \frac{240}{1.96} = 122.4$$

$$\text{Speed in rpm} = \frac{60}{2\pi} \times 122.4 = 1169 \text{ rpm}$$

(c) From Eqn (7.69), the power regenerated is

$$P_{reg} = \frac{E}{\tau}\left\{\frac{(E_b - E)}{R}[\tau_{OFF} + T_a(e^{-\tau_{OFF}/T_a} - 1)] - T_a I_{max}(e^{-\tau_{OFF}/T_a} - 1)\right\}$$

$$I_{max} = \frac{E_b}{R} - \frac{E}{R}\left[\frac{e^{\tau_{OFF}/T_a} - 1}{e^{\tau/T_a} - 1}\right]$$
$$= \frac{240}{0.2} - \frac{400}{0.2}\left[\frac{1.649 - 1}{2.718 - 1}\right]$$
$$= 1200 - 2000 \times \frac{0.649}{1.718} = 444 \text{ A}$$

Substitution of all values in the expression for P_{reg} gives

$$P_{reg} = 400\left\{\frac{240-400}{0.2}\left[\frac{\tau_{OFF}}{\tau} + \frac{T_a}{\tau}(e^{-0.5}-1) - \frac{T_a}{\tau}I_{max}(e^{-0.5}-1)\right]\right\}$$

$$= 400\left\{\frac{-160}{0.2}[0.5 + 1(0.606-1)] - 1 \times 444(0.606-1)\right\}$$

$$= 400\left\{\frac{(-160 \times 0.106)}{0.2} + 444 \times 0.394\right\}$$

$$= 400 \times 90 = 36{,}000 \text{ W or } 36 \text{ kW}$$

(d) Average generated voltage

$$E_b = E(1-\delta) + I_a R_a = 400(1-0.5) + 200 \times 0.2 = 240 \text{ V}$$

Average current through the motor, I_a = 200 A. Hence the equivalent resistance is

$$\frac{E_b}{I_a} = \frac{240}{200} = 1.2\ \Omega$$

(e) Minimum and maximum permissible speeds for regeneration are obtained from the inequality

$$0 \le \omega_m \le \frac{E}{K_b^1}$$

or

$$0 \le \omega_m \le \frac{400}{1.96} = 204$$

Thus the maximum permissible speed is 204 rad/s or 1949 rpm. Also the minimum permissible speed is zero rpm.

11. A dc drive consists of a dc series motor and a three-phase, full-wave fully controlled rectifier supplied by a three-phase ac supply V_{LL} of 270 V (RMS). The other data are R_a = 0.06 Ω, R_f = 0.04 Ω, and K_b = 0.7 V/(Wb rad/s). The field flux is expressed in terms of armature current as $K_1 I_a$, where K_1 = 0.015 Wb/A. The speed of the motor is 1200 rpm. If the firing angle α is 60°, determine the (a) armature current, (b) developed torque, and (c) input power factor.

Solution

(a)

$$V_{s(LN)(RMS)} = \frac{270}{\sqrt{3}} = 155.9 \text{ V}$$

$$V_{m(LN)(RMS)} = 155.9\sqrt{2} = 220.5 \text{ V}$$

$$E_b = K_b \phi_f \omega_m = K_b K_1 I_a \omega_m$$

Here

$$\omega_m = \frac{2\pi \times 1200}{60} = 125.67 \text{ rad/s}$$

$$E_b = 0.7 \times 0.015 \times 125.67 I_a = 1.32 I_a$$

The rectifier output V_a is given as

$$V_a = 2V_m \left(\frac{3}{\pi}\right) \sin\left(\frac{\pi}{3}\right) \cos\alpha$$

$$= 2 \times 220.5 \times \frac{3\sqrt{3}}{2\pi} \cos 60°$$

$$= 182.3 \text{ V}$$

From the motor point of view, V_a is given as

$$V_a = E_b + (R_a + R_f)I_a$$

$$= 1.32I_a + (0.06 + 0.04)I_a$$

$$= 1.42I_a$$

Equating both the expressions for V_a gives

$$182.3 = 1.42I_a$$

or

$$I_a = \frac{182.3}{1.42} = 128.4 \text{ A}$$

(b) The developed torque is

$$T_d = K_b K_1 (I_a)^2$$

$$= 0.7 \times 0.015 \times (128.4)^2$$

$$= 173 \text{ N m}$$

(c) From Fig. 2.12, the current in the secondary of the transformer is

$$I_{s(\text{RMS})} = \sqrt{\frac{2}{2\pi}\int_0^{2\pi/3} (I_a)^2 d(\omega t)} = I_a\sqrt{\frac{2}{3}} = 128.4\sqrt{\frac{2}{3}} = 104.8 \text{ A}$$

The fundamental component of the armature current is

$$I_{a1} = \frac{\sqrt{6}}{\pi} I_a = \frac{\sqrt{6}}{\pi} \times 128.4 = 100.1 \text{ A}$$

$$\phi_1 = -\alpha_a = -60°$$

$$\text{Input power factor} = \frac{I_{a1}\cos\phi_1}{I_{s(\text{RMS})}} = \frac{100.1 \cos 60°}{104.8} = 0.48$$

12. A dc series motor is supplied by a battery of 420 V with a dc chopper interposed between the battery and the motor. It has a mean armature current of 120 A. Other data for this chopper-based drive are $R_a = 0.05\ \Omega$, $R_f = 0.06\ \Omega$, $K_b = 0.72$ V/(Wb rad/s), and $\phi_f = 0.016\ I_a$. The duty ratio τ_{ON}/τ is 0.65. Compute the (a) speed of the motor, (b) torque developed by the motor, (c) equivalent input resistance, and (d) total input power.

Solution

The set-up is as shown in Fig. 7.25:

$$V_a = 420\frac{\tau_{ON}}{\tau} = 420 \times 0.65 = 273 \text{ V}$$

$$E_b = V_a - I_a(R_a + R_f)$$
$$= 273 - 120(0.05 + 0.06)$$
$$= 259.8 \text{ V}$$

$$\omega = \frac{E_b}{K_b \times 0.016 I_a}$$
$$= \frac{259.8}{0.72 \times 0.016 \times 120}$$
$$= 188 \text{ rad/s}$$

$$\text{Speed } N = \frac{188 \times 60}{2\pi}$$
$$= 1795 \text{ rpm}$$

$$\text{Torque developed by the motor} = K_b \times 0.016(I_a)^2$$
$$= 0.72 \times 0.016 \times (120)^2$$
$$= 165.9 \text{ N m}$$

The source current is

$$I_s = I_a\frac{\tau_{ON}}{\tau} = 120 \times 0.65 = 78 \text{ A}$$

$$\text{Input power} = EI_s = 420 \times 78 = 32{,}760 \text{ W} = 32.76 \text{ kW}$$

13. A dc chopper is used for regenerative braking of a dc series motor as shown in Fig. 7.26(a). The data are $E = 400$ V, $R_a = 0.05\ \Omega$, $R_f = 0.04\ \Omega$, $L = L_a + L_f = 0.09$ mH. The product $K_b\phi_f$ may be assumed to be constant at 1.5 V/rad s and the average load current is 200 A. The frequency of the chopper is 1 kHz and $\tau_{ON}/\tau = 0.6$. Compute the (a) average load voltage, (b) motor speed, (c) power regenerated and fed back to the battery, (d) equivalent resistance viewed from the motor side when it is working as a generator, and (e) minimum and maximum permissible speeds.

Solution

(a) $$V_a = E\left(1 - \frac{\tau_{ON}}{\tau}\right) = 400(1 - 0.6) = 160 \text{ V}$$

(b) The correct value of E_b is arrived at as follows:

$$I_a = \frac{I_{max} + I_{min}}{2} = 200 \text{ A}$$

or

$$I_{max} + I_{min} = 400 \text{ A}$$

$$T_a = \frac{L_a + L_f}{R_a + R_f} = \frac{0.09 \times 10^{-3}}{0.09} = 0.001 \text{ s}$$

$$\tau = \frac{1}{f_{\text{Ch}}} = \frac{1}{1000} = 0.001 \text{ s}$$

From Eqns (7.64) and (7.65),

$$2I_a = 400 = I_{\max} + I_{\min} = \frac{2E_b}{R_a} - \frac{E}{R_a}\left[\frac{1 - e^{-\tau_{\text{OFF}}/T_a}}{1 - e^{-\tau/T_a}} + \frac{e^{\tau_{\text{OFF}}/T_a} - 1}{e^{\tau/T_a} - 1}\right]$$

Thus,

$$400 = \frac{2E_b}{0.09} - \frac{400}{0.09}\left[\frac{0.33}{0.632} + \frac{0.492}{1.718}\right]$$

Rearranging terms gives

$$\frac{2E_b}{0.09} = 400 + \frac{400}{0.09} \times 0.808$$

and we obtain

$$E_b = 179.6 \text{ V}$$

$$\omega = \frac{179.6}{1.5} = 119.5 \text{ rad/s}$$

This gives

$$\text{Speed } N = 119.5 \times \frac{60}{2\pi} = 1141 \text{ rpm}$$

$$I_{\max} = \frac{E_b}{R} - \frac{E}{R}\left[\frac{e^{\tau_{\text{OFF}}/T_a} - 1}{e^{\tau/T_a} - 1}\right]$$
$$= \frac{179.6}{0.09} - \frac{400}{0.09} \times \frac{0.492}{1.728}$$
$$= 724.5 \text{ A}$$

$$I_{\min} = \frac{E_b}{R} - \frac{E}{R}\frac{(1 - e^{-\tau_{\text{OFF}}/T_a})}{(1 - e^{-\tau/T_a})}$$
$$= \frac{179.6}{0.09} - \frac{400}{0.09} \times \frac{0.33}{0.632}$$
$$= -325 \text{ A}$$

(c) $$P_{\text{reg}} = \frac{E}{\tau}\left\{\frac{(E_b - E)}{R}\left[\frac{\tau_{\text{OFF}}}{\tau} + \frac{T_a}{\tau}(e^{-\tau_{\text{OFF}}/T_a} - 1)\right] - \frac{T_a I_{\max}}{\tau}(e^{-\tau_{\text{OFF}}/T_a} - 1)\right\}$$

$$= 400\left\{\frac{(179.6 - 400)}{0.09}[0.4 + 1(0.67 - 1)] - 1 \times 724.5(0.67 - 1)\right\}$$

$$= 400\left\{\frac{-220.4}{0.09}[0.4 - 0.33] - 724.5(-0.33)\right\}$$

$$= 400\left\{\frac{-220.4}{0.09} \times 0.07 + 239\right\}$$

$$= 400\{-171.4 + 239\}$$

$$= 400 \times 67.6 = 27{,}031 \text{ W} \approx 27 \text{ kW}$$

(d) Equivalent resistance $=$ back emf/average current $=$ 179.6/200 $\approx$ 0.9 Ω
(e) The minimum and maximum speeds are given as

$$0 \leq N \leq \frac{E}{K_b \phi_f}\frac{60}{2\pi}$$

or

$$0 \leq N \leq \frac{400}{1.5}\frac{60}{2\pi}$$

This gives

$$0 \leq N \leq 2546 \text{ rpm}$$

Hence the minimum and maximum speeds are 0 and 2546 rpm, respectively.

Exercises

Multiple Choice Questions

1. In a separately excited dc motor, constant horsepower control of speed is achieved by ______ variation and constant torque control by ______ variation.
   (a) armature current, armature voltage
   (b) field current, armature voltage
   (c) field voltage, field current
   (d) armature voltage, field current
2. In the armature control method, if the applied armature voltage V_a is decreased by a large amount, then the motor ______.
   (a) speed gets reduced
   (b) gets overloaded
   (c) is constrained to work as a generator
   (d) quickly comes to a standstill
3. During the discontinuous conduction mode of a drive with armature control using a single-phase converter, the reduction in ω_m for an increase in I_a will be ______ that for the continuous mode.
   (a) more than (b) equal to
   (c) less than (d) independent of

4. An inner current control loop along with an outer speed loop gives ______ response for a change in the ac supply voltage as compared to a simple speed control loop.
   (a) slower (b) the same
   (c) quicker (d) much slower

5. In a closed-loop drive with four-quadrant operation, the master controller performs the function of ______ and the slave controller helps in ______.
   (a) smooth changeover of conduction from one rectifier to the other, switching the appropriate rectifier into conduction
   (b) smooth changeover of conduction from one rectifier to the other, sudden changeover of conduction from one rectifier to other
   (c) sudden changeover of conduction from one rectifier to the other, smooth changeover of conduction from one rectifier to the other
   (d) switching the appropriate rectifier into conduction, smooth changeover of conduction from one rectifier to the other

6. For a dc motor, the region of discontinuous conduction in the speed–torque plane is ______ with chopper control than that with rectifier control.
   (a) greater (b) much lesser
   (c) lesser (d) much greater

7. Regenerative braking of a separately excited dc motor using dc chopper control gives ______ that using rectifier control.
   (a) better performance than
   (b) inferior performance than
   (c) the same performance as
   (d) much inferior performance as compared to

8. In a speed control system incorporating an outer speed loop and an inner current loop, the drive response becomes sluggish in the discontinuous conduction mode because ______.
   (a) the field response is slower than that with continuous conduction
   (b) the output of the speed controller is smaller than that in continuous conduction
   (c) the combined gain of the firing circuitry and the rectifier is smaller than that with continuous conduction
   (d) the combined gain of the firing circuitry and the rectifier is larger than that with continuous conduction

9. During the regenerative braking operation of a chopper-based dc drive, energy transfer takes place ______.
   (a) from the load at high voltage to the source at low voltage
   (b) from the source at high voltage to the load at low voltage
   (c) from the source at low voltage to the load at high voltage
   (d) from the load at low voltage to the source at high voltage

10. Dc drives that use series motors are more advantageous than drives that employ separately excited motors because they facilitate ______.
    (a) high starting torque
    (b) frequent starting

(c) frequent torque overloads
(d) all of the above
(e) some of the above

11. In a dc drive supplied by a single-phase ac source and a rectifier, discontinuous conduction can be avoided by ________.
(a) decreasing the frequency of the ac supply
(b) adding a capacitor in series with the armature of the dc motor
(c) increasing the frequency of the ac supply
(d) adding an inductor in series with the armature of the dc motor.

12. A three-phase, full-wave controlled rectifier feeds a separately excited dc motor. The data are T_d = 140 N m, I_a = 60 A, and ω_m = 65 rad/s. With these data, the back emf of the motor will be nearly ________ V.
(a) 102 (b) 178
(c) 152 (d) 194

13. A three-phase, full-wave controlled rectifier feeds a separately excited dc motor and the data are E_b = 150 V, ω_m = 75 rad/s, and I_a = 56 A. With these data, the torque of the motor will be ________ N m.
(a) 168 (b) 108
(c) 184 (d) 112

14. A dc drive consisting of a three-phase, full-wave controlled rectifier together with a separately excited dc motor has the following data: R_a = 0.2 Ω, $K_b\phi_f$ = 0.8. With these data the slope of the speed–torque curve will be nearly ________ rad/N m s.
(a) −0.31 (b) −0.48
(c) −0.36 (d) −0.27

15. A dc drive consisting of a three-phase, full-wave controlled rectifier together with a separately excited dc motor has the following data: V_a = 160 V, $K_b\phi_f$ = 0.75. With these data the intercept of the speed–torque curve on the ω-axis will be nearly ________ rad/s.
(a) 254 (b) 213
(c) 198 (d) 176

True or False

Indicate whether the following statements are *true* or *false*. Briefly justify your choice.

1. In a separately excited dc motor, constant horsepower control of speed is achieved by armature voltage variation and constant torque control by field current variation.
2. During the speed control of a dc motor by armature control, if there is a large decrease in the applied armature voltage V_a, then the motor will be constrained to work as a generator.
3. During the discontinuous conduction mode of armature control using a single-phase converter drive, the reduction in ω_m for an increase in I_a will be less than that for the continuous mode.

4. A current control loop within a speed control loop gives quicker response for a change in the ac supply voltage as compared to a simple speed control loop.
5. In a comprehensive control scheme, the master controller performs the function of sudden changeover of conduction from one rectifier to the other and the slave controller helps in the smooth changeover of conduction from one rectifier to the other.
6. For a dc motor, the region of discontinuous conduction of the speed–torque curves with chopper control will be greater than that with rectifier control.
7. The regenerative braking of a separately excited dc motor using dc chopper control is advantageous compared to regenerative braking using rectifier control.
8. In a complete speed control system incorporating an outer speed loop and an inner current loop, the drive response becomes sluggish in the discontinuous conduction mode because the field response is slower than that with continuous conduction.
9. During the regenerative operation of a chopper-based dc drive, energy transfer takes place from the load at low voltage to the source at high voltage.
10. Dc series motors give better performance than separately excited motors for driving loads for which frequent starting is not needed and frequent torque overloads are absent.
11. If a dc drive consisting of a single-phase, full-wave controlled rectifier and a separately excited dc motor has a torque of 140 N m, armature current of 65 A, and ω_m of 60 rad/s, then the back emf of the motor will be nearly 151 V.
12. If a dc drive consisting of a single-phase, full-wave controlled rectifier and a separately excited dc motor has a back emf of 160 V, ω_m of 68 rad/s, and armature current of 62 A, then the torque will be nearly 146 N m.
13. If a dc drive consisting of a single-phase, full-wave controlled rectifier and a separately excited dc motor has an armature resistance of 0.25 Ω and $K_b\phi_f$ equal to 0.78, then the slope of the speed–torque curve will be −0.41 rad/N m s.
14. If a dc drive consisting of a single-phase, full-wave controlled rectifier and a separately excited dc motor has an armature voltage of 172 V and a value of 0.8 for the product $K_b\phi_f$, then the intercept of the speed–torque curve on the ω_m-axis will be nearly 256 rad/s.

Short Answer Questions

1. Derive an expression for ω_m as a function of torque for a separately excited dc motor.
2. How is speed control by armature voltage variation done, at the same time conforming to the required speed–torque characteristic?
3. What is the effect of a large decrease in V_a during the armature voltage control of a separately excited dc motor?
4. How is speed control achieved by field current variation, at the same time following the required speed–torque characteristic?
5. Explain how speed control with constant horsepower operation is obtained by field current variation.

6. Describe a scheme that provides wide variation of speed.
7. Explain the difference between the operating features of speed–torque curves obtained with continuous conduction and those obtained with discontinuous conduction.
8. For the case of armature control, derive the expressions which lead to a closed-loop block diagram.
9. In the case of armature control, give the steps that help in converting a closed control loop into an open one.
10. In the case of field control, derive the steps that lead to an open control loop.
11. Explain the operation of a separately excited dc motor with closed-loop current control.
12. Explain the necessity of cascading of the speed and current loops in the case of a separately excited dc motor.
13. Give the conditions under which drive response becomes sluggish in the case of speed control with an inner current loop.
14. Explain how cascaded speed and current loops provide quick response to inner loop disturbances.
15. Give the block diagram of a comprehensive control scheme that provides wide variation in speed for a separately excited dc motor.
16. What are the salient features of the closed-loop dc drive that provides four-quadrant operation with controlled rectifiers?
17. What are the functions of the master controller employed in a closed-loop dc drive with four-quadrant operation?
18. Explain why chopper-based dc drives give better performance than rectifier-controlled drives.
19. Describe a scheme that provides four-quadrant operation of a separately excited dc motor using dc choppers.
20. Derive expressions for I_{max}, I_{min}, and P_{reg} for the regenerative mode of operation of a separately excited dc motor supplied by a chopper circuit.
21. Why is the regenerative braking of a chopper-based dc series motor more complicated than that of a separately excited motor?
22. For a separately excited dc motor, explain why regenerative braking using a dc chopper is preferred to braking with a controlled rectifier.

Problems

1. A 220-V, 1500-rpm rated, separately excited dc motor is supplied by a single-phase controlled bridge rectifier and is connected to a load with a torque of 35 N m at a speed of 1100 rpm. Other data are $R_a = 0.22\ \Omega$, back emf constant $K_b = 72$ V/(Wb rad/s), $\phi_f = 14 \times 10^{-3}$ Wb. It is supplied with a single-phase ac supply voltage V_s of 160 V (RMS) at 50 Hz. Assuming that the load is highly inductive, compute the (a) armature current, (b) firing angle of the thyristors, and (c) input power factor of the rectifier.

   Ans. (a) 34.72 A, (b) 30.8°, (c) 0.77

2. A 200-V, 1500-rpm rated, separately excited dc motor is supplied by a three-phase, half-wave controlled rectifier and runs at 1000 rpm. The other data are $R_a = 0.18\ \Omega$, $L_a = 0.43$ mH, single-phase ac supply voltage $V_{s(\text{RMS})} = 130$ V at 50 Hz, $\alpha = 65°$, $\phi_f = 10 \times 10^{-3}$ Wb, and $K_b = 70$ V/(Wb rad/s). Compute the (a) armature current and (b) load torque. (c) State whether the conduction is continuous or otherwise.

   Ans. (a) 36.6 A, (b) 25.6 N m, (c) discontinuous

3. A 150-V, 1200-rpm rated, separately excited dc motor is supplied by a three-phase, half-wave controlled rectifier and runs at 1000-rpm. The other data are $R_a = 0.15\ \Omega$, $L_a = 0.64$ mH, supply voltage $V_{\text{LN}} = 138$ V (RMS) at 50 Hz, $\phi_f = 11.8 \times 10^{-3}$ Wb, and $K_b = 86$ V/(Wb rad/s). If the firing angle is 40°, determine (a) the armature current and (b) the extinction angle in case the conduction is discontinuous.

   Ans. (a) $I_a = 125.0$ A, (b) $\alpha_e = 148°$

4. A 180-V, 1300-rpm rated, separately excited dc motor is supplied by a three-phase, half-wave controlled rectifier, the load torque being 65 N m. Other data are $R_a = 0.20\ \Omega$, $\phi_f = 14 \times 10^{-3}$ Wb, $K_b = 81$ V/(Wb rad/s), and $\alpha = 50°$. If the ac supply voltage is 170 V (RMS) at 50 Hz and there is sufficient inductance in the circuit to give continuous conduction, determine the speed of the motor.

   Ans. $N = 980$ rpm

5. A three-phase, full-wave controlled rectifier bridge supplies the armature of a 15-h.p., 300-V, separately excited dc motor. The field circuit gets its supply from a three-phase, full-wave diode rectifier bridge connected to the same ac supply. The other data are supply voltage $V_{\text{LN}} = 180$ V (RMS) at 50 Hz, $R_a = 0.25\ \Omega$, $L_a = \infty$, $R_f = 182\ \Omega$, and $K_b = 82$ V/(Wb rad/s). ϕ_f is given as $K_1 I_f$, where $K_1 = 0.014$ Wb/A, $\alpha = 62°$, and $I_a = 58$ A. Compute the (a) torque developed by the motor and (b) the speed, and (c) input power factor of the drive.

   Ans. (a) 154 N m, (b) 659 rpm, (c) 0.48

6. A three-phase, full-wave controlled rectifier bridge supplies the armature of a 15-h.p., 280-V, separately excited dc motor. The field of the motor is supplied by a three-phase, full-wave diode rectifier bridge connected to the same ac supply. The other data are supply voltage $V_{\text{LN}} = 170$ V (RMS) at 50 Hz, $R_a = 0.21\ \Omega$, $L_a = \infty$, $R_f = 178\ \Omega$, $K_b = 80$ V/(Wb rad/s). ϕ_f is given as $K_1 I_f$, where $K_1 = 0.012$ Wb/A, speed is 670 rpm, and $I_a = 55$ A. Compute the (a) torque developed by the motor, (b) firing angle α, and (c) input power factor of the drive.

   Ans. (a) 118 N m, (b) 68°, (c) 0.42

7. A three-phase, full-wave controlled rectifier bridge supplies the armature of a 13-h.p., 280-V, separately excited dc motor. The field is also supplied by a three-phase, full-wave controlled rectifier with its firing angle adjusted to give maximum field current. Other data are supply voltage $V_{\text{LN}} = 175$ V (RMS) at 50 Hz, $R_a = 0.19\ \Omega$, $L_a = \infty$, $R_f = 175\ \Omega$, $K_b = 79$ V/(Wb rad/s). ϕ_f is given as $K_1 I_f$, where $K_1 = 0.013$ Wb/A, $\alpha = 60°$, and speed is 700 rpm. Determine the (a) armature current, (b) torque developed by the motor, and (c) input power factor. If the motor is now made to run as a generator by reversing the field

polarity and carries the same armature current as before, determine the (d) new firing angle and (e) regenerated power.

Ans. (a) 150 A, (b) 360 N m, (c) 0.49, (d) 68.9°, (c) 22.14 kW

8. A single-phase fully controlled bridge rectifier controls the input to a 12-h.p., 160-V, 1300-rpm, separately excited dc motor. The field of the dc motor is supplied by a single-phase fully controlled bridge rectifier connected to the same ac source and is set to give (7/8)th of its full voltage. Other data are the following: the ac input is a single-phase, 230-V, 50-Hz supply, the armature resistance $R_a = 0.19\ \Omega$, and the field resistance $R_f = 172\ \Omega$. There are sufficient inductances in the armature and field circuits to make the currents continuous and nearly constant. (a) Compute the firing angles of the armature and field rectifiers when the motor supplies rated power at the rated speed. (b) When the armature and field circuit firing angles are kept the same as in (a) and the machine runs with no load, it takes an armature current equal to 8.5% of the full load value; with these conditions determine the no-load speed. (c) Also determine the percentage speed regulation. The field flux is related to I_f as $\phi_f = K_1 I_f$, where $K_1 = 0.016$. Also $K_b = 68$ V/(Wb rad/s).

Ans. (a) $\alpha_a = 36°$, $\alpha_f = 29°$; (b) 1388 rpm; (c) 6.77%

9. The armature of a 150-V, 1500-rpm rated, separately excited dc drive is fed by a single-phase fully controlled bridge rectifier. The field is also supplied by a similar bridge from the same ac source. The other data are as follows: I_f is set to its maximum value, the ac supply voltage is 200 V at 50 Hz, $R_a = 0.21\ \Omega$, and $R_f = 175\ \Omega$. The firing angle of the armature converter $\alpha_a = 30°$, load torque is 53 N m, $K_b = 75$ V/(Wb rad/s). (a) Determine the speed of the motor under the above load conditions. (b) For the same load torque as in (a), determine the firing angle of the field rectifier if the speed has to be increased to 1650 rpm. The relation between ϕ_f and I_f is $\phi_f = K_1 I_f$, where $K_1 = 0.014$ A/Wb.

Ans. (a) 1287 rpm, (b) $\alpha_f = 39°$

10. A 460-V, 1000-rpm rated dc drive consists of a three-phase, full-wave fully controlled bridge rectifier with the armature of a separately excited dc motor at its output. The field is also controlled by a similar rectifier fed by the same ac supply. The data for this drive are as follows: ac supply voltage $V_{LL} = 350$ V at 50 Hz, $R_a = 0.24\ \Omega$, $R_f = 220\ \Omega$, $K_b = 80$ V/(Wb rad/s), $T_d = 78$ N m at 1050 rpm. The firing angle α_f of the field rectifier is set to give 90% of the field current. (a) Find the firing angles α_a and α_f of the armature and field rectifier thyristors. (b) For the same load torque and α_a as in part (a), if the speed is to be increased to 1350 rpm, determine the new α_f. The field flux ϕ_f is expressed as $K_1 I_f$, where $K_1 = 0.0145$ Wb/A. (c) Find the input power factor of the unit for the last operating condition.

Ans. (a) $\alpha_a = 57.4°$, $\alpha_f = 25.9°$; (b) 45.6°; (c) 0.51

11. A separately excited dc motor has a rating of 55 h.p. and runs at 1100 rpm. It is supplied by a dc chopper whose source is a battery of 480 V. The field is also supplied by a chopper whose source is another battery of 280 V. The other data for the drive are $R_a = 0.19\ \Omega$, $K_b = 72$ V/(Wb rad/s), $\phi_f = 0.017 I_f$, $R_f = 118\ \Omega$, and τ_{ON}/τ for the field chopper = 0.82. The load has sufficient inductance to make the load current continuous. If τ_{ON}/τ for the armature chopper is 0.60, compute for rated speed and power the (a) mean armature current, (b) torque

developed by the motor, (c) equivalent resistance for the armature current, and (d) total input power.

Ans. (a) 71.9 A, (b) 171.3 N m, (c) 11.13 Ω, (d) 21.2 kW

12. A separately excited dc motor has a rating of 48 h.p., and when supplied by a battery of 460 V through a chopper, it has a mean armature current of 115 A. The field is also supplied by a chopper whose source is a battery of 240 V. Other data for this chopper-based drive are $R_a = 0.2\ \Omega$, $R_f = 122\ \Omega$, $K_b = 70$ V/(Wb rad/s), $\phi_f = 0.014 I_f$, $(\tau_{ON}/\tau)_a = 0.72$, and $(\tau_{ON}/\tau)_f = 0.91$. The armature circuit has sufficient inductance to make the current continuous. Compute the (a) speed of the motor, (b) torque developed by the motor, (c) equivalent resistance, and (d) total input power.

Ans. (a) 1678 rpm, (b) 201.7 N m, (c) 5.56 Ω, (d) 38.5 kW

13. A dc chopper is used for regenerative braking of a separately excited dc motor as shown in Fig. 7.26(b). The data are as follows: E = 380 V, $R_a = 0.19\ \Omega$, L_a = 0.21 mH. The back emf constant K'_b (= $K_b\phi_f$) assuming ϕ_f to be constant is 1.98 V/rad s and the average load current is 204 A. The frequency of the chopper is 1.2 kHz and τ_{ON}/τ is 0.6. Compute the (a) average load voltage, (b) motor speed, (c) power regenerated and fed back to the battery, (d) equivalent resistance viewed from the motor side when it is working as a generator, and (e) minimum and maximum permissible speeds for regenerative braking.

Ans. (a) 152 V, (b) 925 rpm, (c) 30.0 kW, (d) 0.94 Ω, (e) $0 < N < 1833$ rpm

14. A dc drive consisting of a dc series motor is supplied by a three-phase, full-wave fully controlled rectifier. The data are as follows: the line-to-line voltage of ac supply is 255 V, K_b = 0.68 V/(Wb rad/s), speed N = 1250 rpm, armature current I_a = 85 A, and $R_a + R_f = 0.14\ \Omega$. The field flux is expressed in terms of armature current as $0.016 I_a$. Determine the (a) firing angle α_a and (b) input power factor. Assume that there is sufficient inductance in the motor circuit to cause continuous conduction.

Ans. (a) 67.3°, (b) 0.37

15. A dc series motor together with a three-phase, full-wave fully controlled rectifier forms a dc drive. The data are as follows: ac line voltage V_{LL} = 278 V, $R_a = 0.07\ \Omega$, $R_f = 0.05\ \Omega$, $K_b = 0.72$ V/(Wb rad/s). The field flux is expressed in terms of the armature current as $0.014 I_a$. If the firing angle α is 65° and the armature current is 86 A, determine the (a) speed of the motor and (b) input power factor.

Ans. (a) 1634 rpm, (b) 0.4

16. A dc series motor is supplied by a battery of 400 V through a dc chopper. The data are as follows: speed of the motor is 1650 rpm, the duty cycle τ_{ON}/τ is 0.62, $R_a = 0.04\ \Omega$, $R_f = 0.08\ \Omega$, $K_b = 0.75$ V/(Wb rad/s), and $\phi_f = 0.014 I_a$. If sufficient inductance has been included in the armature to make the motor current continuous, compute the (a) armature current, (b) torque developed, and (c) input power.

Ans. (a) 128.2 A, (b) 172.6 N m, (c) 31.8 kW

17. A dc series motor together with a chopper forms a dc drive. The data are as follows: input is a dc battery of 410 V, armature current is 110 A, $R_a = 0.05\ \Omega$,

$R_f = 0.07\ \Omega$, $L_a + L_f = 0.15$ mH, $K_b = 0.74$ V/(Wb rad/s), and field flux $\phi_f = 0.015 I_a$. If the speed of the motor is 1680 rpm, determine the (a) duty ratio, (b) torque developed, and (c) input power.

Ans. (a) 0.556, (b) 134.3 N m, (c) 25.0 kW

18. A dc chopper is used for the regenerative braking of a dc series motor as shown in Fig. 7.26(b). The data are $E = 420$ V, $R_a = 0.11\ \Omega$, $R_f = 0.07\ \Omega$, $L_a + L_f = 0.18$ mH, the product $K_b\phi_f$ may be assumed to be constant at 1.7 V/rad s, and the average load current is 210 A. If the frequency of the chopper is 0.8 kHz and $\tau_{ON}/\tau = 0.65$, compute the (a) average load voltage, (b) motor speed, (c) power regenerated and fed back to the battery, (d) equivalent resistance viewed from the motor side when it is working as a generator, and (e) minimum and maximum permissible speeds.

Ans. (a) 147 V, (b) 999 rpm, (c) 18.9 kW, (d) 0.85 Ω, (e) $0 \le N \le 2359$ rpm

CHAPTER 8

Ac Drives

8.1 Introduction

Dc drives have hitherto been used in various industrial applications because the variable voltage required by them can be obtained from dc choppers or controlled rectifiers, which are simple and inexpensive. Moreover, separately excited dc motors have a decoupled structure. However, the following facts emerge from a look at their disadvantages. A dc machine has a commutator which is expensive and needs constant maintenance. The brushes used with the commutator need periodical replacement because of constant wear and tear. In the recent past, ac drives have replaced dc drives in applications such as machine tool drives, paper mills, waste-water treatment, etc. Again, in high-horsepower applications, cycloconverter-fed induction motor drives and load commutated synchronous motor drives are preferred. This trend is continuing even though the power electronic portion of these ac drives constitutes 70% of their total cost. Ac machines exhibit highly coupled, non-linear, and multivariable structure in contrast to the simple structure of dc motors; consequently their control requires complex control algorithms. To implement closed-loop adjustable speed drives, it is important to know the static as well as dynamic behaviour of an ac machine. However, ac drives have the following overweighing advantages.

(a) Ac machines are 20–40 % lighter than dc motors. They are inexpensive and require less maintenance compared to dc motors.

(b) The advent of silicon controlled rectifiers with high di/dt and dv/dt ratings as well as IGBTs has given impetus to the use of ac drives in applications with power ratings of 100 h.p. and above.

(c) Easy implementation of complex control algorithms by means of microprocessors/microcomputers, and the consequent improved reliability and increased flexibility have popularized microprocessor-based ac drives. Flexibility can be attributed to the replaceability of hardware circuitry by microprocessor software. The development of software packages for optimum pulse width modulation has led to the increased use of PWM-based VSIs.

Hence, we can sum up by saying that recent device technology as well as high-speed microprocessors have tilted the balance, thus making ac drives quite viable.

After going through this chapter the reader should

- understand how the expression for developed torque of an induction motor is derived with the help of its equivalent circuit,
- know how the speed–torque curves can be split into the forward motoring, plugging, and regenerating modes,
- get acquainted with the different methods for speed control of the induction motor, namely, stator voltage control, rotor voltage control, stator frequency control, V/f control, E_s/f control, and stator current control,
- become familiar with the implementation of some closed-loop control schemes, namely, V/f control, slip regulation with the constant V/f feature, stator voltage control, and stator current control,
- understand the principles behind the rotor resistance control and slip energy recovery schemes, and know the derivation of the expressions for developed torque in these two cases,
- know the derivation of expressions for developed torque in the cylindrical and salient pole rotor versions of synchronous motor drives, and
- know the various applications of synchronous motor drives.

8.2 Induction Motor Drives

There are two types of ac drives: asynchronous and synchronous. This section is devoted to induction motor drives which come under the first category. These drives are finding increased usage in adjustable speed ac drive systems. Due to the transformer action between the stator winding (considered to be the primary) and the rotor winding (secondary), ac voltages are induced in the latter when the stator winding is supplied with balanced three-phase ac voltages. The stator windings are spatially distributed in such a manner that several cycles of magnetomotive force (mmf) are produced in the air gap consequent to the effect of multiple poles, and the mmf sets up a spatially distributed sinusoidal flux density in the air gap. The speed of the rotating field is known as the *synchronous speed* and is given as

$$N_s = \frac{120f}{P} \tag{8.1}$$

where N_s is the speed of the field in rpm, f is the stator frequency in hertz, and P is the number of poles. If n_s is the speed of the field in rps and p is the number of pole pairs, Eqn (8.1) can be written as

$$n_s = \frac{f}{p} \tag{8.2}$$

Multiplying Eqn (8.2) by 2π yields

$$\omega_s = \frac{\omega}{p} \tag{8.3}$$

where $\omega_s = 2\pi n_s$ is the synchronous speed in mechanical rad/s and $\omega = 2\pi f$ is the supply frequency in electrical rad/s. If ω_m is the angular speed of the rotor in mechanical rad/s, the slip is defined as

$$s = \frac{\omega_s - \omega_m}{\omega_s} \tag{8.4}$$

8.2.1 Equivalent Circuit and Analytical Relationships

A physical interpretation of the induction motor action leads to the per phase equivalent circuit shown in Fig. 8.1(a). It highlights the fact that the stator and rotor windings play the same role as the primary and secondary windings of a transformer. The applied voltage V_s is equal to the sum of the drop across the stator impedance $R_s + jX_s$ and a counter-emf E_s generated by the synchronously rotating air gap flux wave. Due to transformer action, E_s is converted to $E_r' = (N_r/N_s)(sE_s)$ in the rotor phase, where s is the slip per unit and N_r/N_s is the rotor-to-stator turn ratio. The no-load excitation current I_0 consists of two components, namely, the core loss component $I_c = E_s/R_m$ and the magnetizing component $I_m = E_s/jX_m$, where R_m is the resistive equivalent of the excitation loss and jX_m is the reactance due to the magnetizing inductance. The induced voltage E_r' of the rotor is equal to the product of the rotor current I_r' and the rotor impedance $R_r' + jX_r'$. Figure 8.1(b) shows the equivalent circuit in which

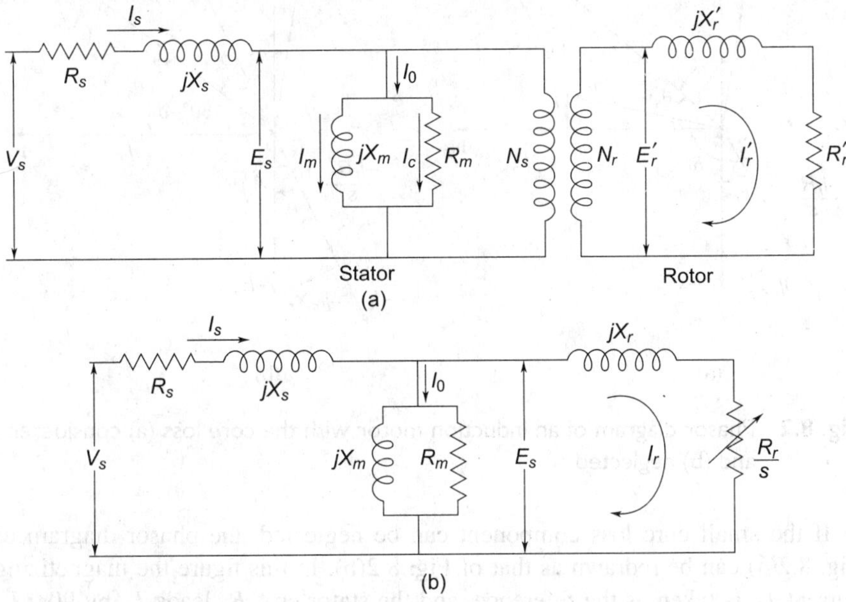

Fig. 8.1 Equivalent circuit of induction motor: (a) as originally derived, (b) with all quantities referred to the stator

all quantities are referred to the stator side. The stator current I_s is equal to the sum of the excitation component I_0 and the current I_r; the latter is the rotor current I_r' referred to the stator winding. The current I_r, which can be interpreted as the load component of the stator current, can be written as

$$I_r = \frac{N_r}{N_s} I'_r = \frac{(N_r/N_s)^2 s E_s}{R'_r + jX'_r} = \frac{E_s}{R_r/s + jX_r} \tag{8.5}$$

where the parameters R_r and X_r are, respectively, the rotor resistance and reactance referred to the stator. At standstill, the slip s is equal to 1; this condition corresponds to a short-circuited transformer. On the other hand, s becomes equal to zero when the motor runs at synchronous speed. With $s = 0$ the rotor current I_r also becomes zero; this motor condition is analogous to an open-circuited transformer. When the slip has a value between 0 and 1, the rotor current is governed by the term R_r/s because it is much greater than X_r. The phasor diagram of the equivalent circuit is shown in Fig. 8.2(a).

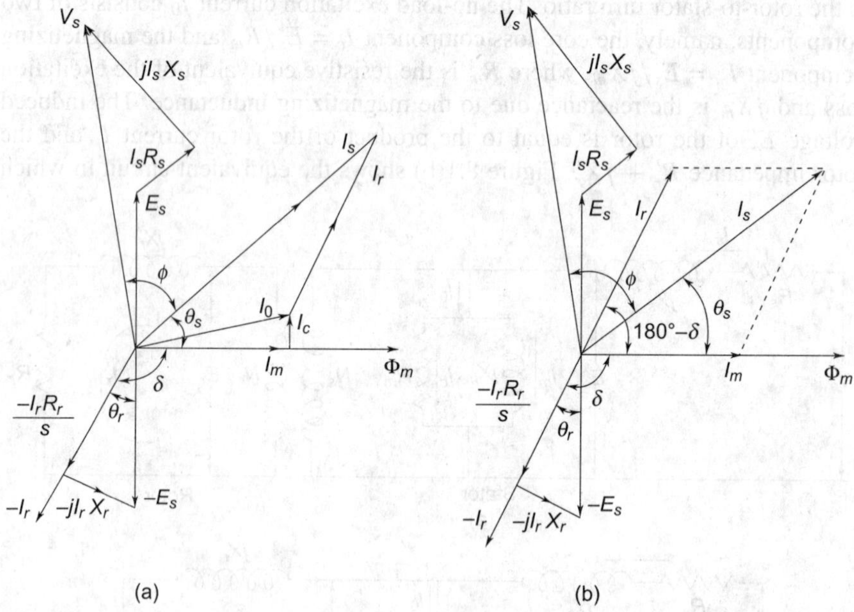

Fig. 8.2 Phasor diagram of an induction motor with the core loss (a) considered and (b) neglected

If the small core loss component can be neglected, the phasor diagram of Fig. 8.2(a) can be redrawn as that of Fig. 8.2(b). In this figure the magnetizing current I_m is taken as the reference, and the stator emf E_s leads I_m by 90°. I_r, which is considered to be the load component of the stator current, neutralizes the rotor mmf and lags E_s by the rotor power factor angle θ_r. The net stator current I_s is the phasor sum of I_m and I_r. The fundamental air gap flux ϕ_m is in phase with the magnetizing current I_m. If saturation is neglected, ϕ_m will be proportional to I_m.

The currents I_s, I_r, and I_m, respectively, represent the stator, rotor, and mutual mmf waves. They signify the continued effect of the three phases and rotate anticlockwise at a rate equal to ω_s mechanical rad/s. Likewise, ϕ_m represents the mutual air gap flux wave. The development of torque can be interpreted as the tendency of the mutual air gap flux and the winding mmf to align. The motor torque is proportional to the product of the amplitudes of ϕ_m and i_r and the sine of the angular displacement between them. Thus,

$$\mathcal{T}_d = k_1 \Phi_m I_r \sin(180^\circ - \delta) = k_1 \Phi_m I_r \sin\delta = k_1 \Phi_m I_s \sin\theta_s \tag{8.6}$$

where k_1 is a constant. Since δ and θ_s are constant, an anticlockwise steady torque is developed.

8.2.2 Circuit Analysis

The expressions for the input and output powers, power transferred from the stator to the rotor through the air gap, various losses, as well as developed torque can be obtained as follows. The input power is

$$P_{\text{in}} = 3V_s I_s \cos\phi \tag{8.7}$$

The stator copper loss and core loss are, respectively,

$$P_s = 3I_s^2 R_s \tag{8.8}$$

and

$$P_c = \frac{3E_s^2}{R_m} \tag{8.9}$$

The power transferred from the stator to the rotor through the air gap is

$$p_g = 3I_r^2 \frac{R_r}{s} \tag{8.10}$$

The rotor copper loss is

$$P_r = 3I_r^2 R_r \tag{8.11}$$

The output (or developed) power is

$$p_d = p_g - P_r = 3I_r^2 R_r \left(\frac{1-s}{s}\right) \tag{8.12}$$

If the power loss due to friction and windage is neglected, the torque (variable) developed at the shaft, $\overline{\mathcal{T}}_d$, can be written as

$$\overline{\mathcal{T}}_d = \frac{p_d}{\omega_m} = \frac{3I_r^2 R_r (1-s)}{\omega_m s} \tag{8.13}$$

Using Eqn (8.4), ω_m can be expressed as

$$\omega_m = \omega_s (1-s) \tag{8.14}$$

Substituting this expression for ω_m in Eqn (8.13) gives the developed torque as

$$\overline{\mathcal{T}}_d = \frac{3I_r^2 R_r}{s\omega_s} \tag{8.15}$$

Using Eqn (8.10), $\overline{T}_d$ can be expressed as

$$\overline{T}_d = \frac{P_g}{\omega_s} \tag{8.16}$$

where ω_s is the synchronous speed in mechanical rad/s as stated under Eqn (8.3). Equation (8.16) shows that the torque can be expressed as a ratio of the air gap power and the mechanical angular velocity of the stator field, with the unit N m. Since the torque is directly proportional to the air gap power irrespective of the speed, P_g is defined as the torque in synchronous watts.

If the core loss resistor, which usually has a high value, is omitted and the magnetizing reactance jX_m transferred to the input, the equivalent circuit of Fig. 8.1(b) gets simplified to that shown in Fig. 8.3.

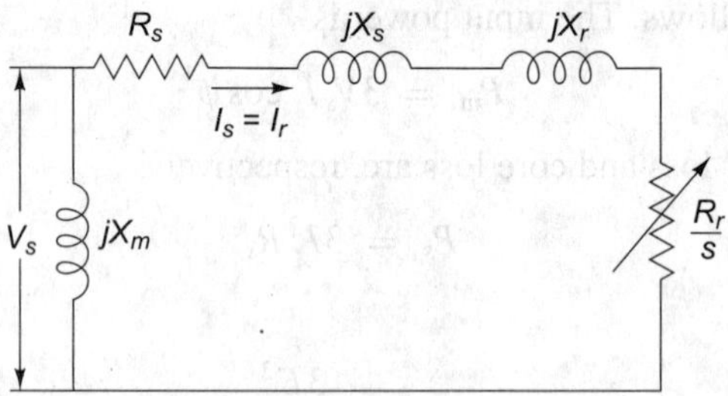

Fig. 8.3 Equivalent circuit of an induction motor with the magnetizing reactance shifted to the input side

This approximation is valid for an integral horsepower machine for which $|(R_s + jX_s)| << jX_m$. As shown in this figure, the stator current I_s is equal to I_r, which can be written as

$$I_r = \frac{V_s}{[(R_s + R_r/s)^2 + (X_s + X_r)^2]^{1/2}} \tag{8.17}$$

The expression for $\overline{T}_d$ in Eqn (8.15) can now be written as

$$\overline{T}_d = 3\frac{R_r}{s\omega_s}\frac{V_s^2}{[(R_s + R_r/s)^2 + (X_s + X_r)^2]} \tag{8.18}$$

8.2.3 Speed–torque Curves

If in Eqn (8.18) the supply voltage V_s and the supply frequency ω $(= p\omega_s)$ in electrical rad/s are considered constant and the torque is considered to be the variable $\overline{T}_d$, the latter becomes a function of the slip s and can be plotted as in Fig. 8.4 with the range of the slip as $-1 \le s \le 2$. The curve can also be interpreted as the speed–torque characteristic, with the speed taken as the rotor angular velocity ω_r in electrical rad/s. The relationship between ω_r and ω_m is

$$\omega_r = p\omega_m \tag{8.19}$$

where ω_m and p have the same meaning as before. The slip s can also be expressed in terms of electrical angular velocities as

$$s = \frac{\omega - \omega_r}{\omega} \tag{8.20}$$

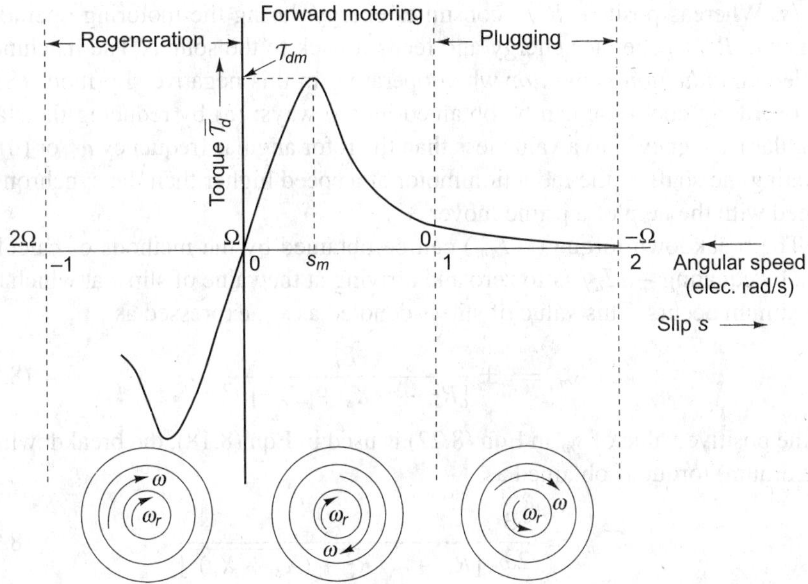

Fig. 8.4 Torque–slip curve of an induction motor

The total range of the slip axis can be divided into the following three regions:

(a) forward motoring ($0 \le s < 1$),

(b) plugging ($s > 1$), and

(c) regeneration ($s < 0$).

Considering the first slip condition, the motor rotates in the same direction as the air gap flux in this region; $\overline{\mathcal{T}}_d$ is zero at $s = 0$ and increases as s increases, with the air gap flux remaining constant. After reaching a maximum, called the *breakdown torque* ($\mathcal{T}_{dm}$), the torque decreases due to the reduction of the air gap flux because of the increased effect of the stator resistance. At $s = 1$ (or $\Omega_r = 0$), the machine is at a standstill and the corresponding torque is called the *starting torque*. Thus Eqn (8.18) becomes

$$\mathcal{T}_{\text{starting}} = 3\frac{R_r}{\omega_s}\frac{V_s^2}{(R_s + R_r)^2 + (X_s + X_r)^2} \tag{8.21}$$

In the plugging region, the rotor rotates in a direction opposite to that of the air gap flux. Such a condition occurs if the phase sequence of the stator supply is reversed when the rotor is rotating. The torque, which is always in the same direction as the rotating air gap flux, now acts as a braking torque. Though this braking torque is low, the current I_s ($= I_r$) is high. This can lead to heating due to the ohmic losses caused by dissipation within the motor. However, this type of braking is not normally resorted to in high power levels.

In the regeneration region, the rotor rotates at a speed higher than that of the air gap flux and the slip becomes negative, thus causing negative or braking torque. Second, the negative slip leads to negative equivalent rotor resistance

R_r/s. Whereas positive R_r/s consumes energy during the motoring operation, negative R_r/s generates energy and feeds it back to the source. The machine is called an *induction generator* when operating in this negative slip mode. Such an operating condition can be obtained in two ways: (a) by reducing the stator angular frequency ω to a value less than the rotor angular frequency ω_r or (b) by rotating the shaft of the induction motor at a speed higher than the synchronous speed with the help of a prime mover.

The breakdown torque $(-\mathcal{T}_{dm})$ can be obtained by the methods of calculus, i.e., by equating $-d\mathcal{T}_d/ds$ to zero and arriving at the value of slip s at which this maximum occurs. This value of slip is denoted as s_m, expressed as

$$s_m = \pm \frac{R_r}{[R_s^2 + (X_s + X_r)^2]^{1/2}} \tag{8.22}$$

If the positive value of s_m in Eqn (8.22) is used in Eqn (8.18), the breakdown (or maximum) torque is obtained as

$$\mathcal{T}_{dm} = \frac{3}{2\omega_s} \frac{V_s^2}{[R_s + \sqrt{R_s^2 + (X_s + X_r)^2}]} \tag{8.23}$$

The negative value of s_m in Eqn (8.22) gives the maximum regenerative torque as

$$\mathcal{T}_{rm} = \frac{3V_s^2}{2\omega_s[R_s - \sqrt{R_s^2 + (X_s + X_r)^2}]} \tag{8.24}$$

Usually, for motors other than fractional horsepower ones, the stator resistance is neglected because it is small compared to the other parameters. The expressions for $\overline{\mathcal{T}}_d$, s_m, $\mathcal{T}_{dm}$, and $\mathcal{T}_{rm}$ with this approximation are

$$\overline{\mathcal{T}}_d = \frac{3R_r}{s\omega_s} \frac{V_s^2}{[(R_r/s)^2 + (X_s + X_r)^2]} \tag{8.25}$$

$$s_m = \pm \frac{R_r}{X_s + X_r} \tag{8.26}$$

$$\mathcal{T}_{dm} = \frac{3V_s^2}{2\omega_s(X_s + X_r)} \tag{8.27}$$

and

$$\mathcal{T}_{rm} = \frac{-3V_s^2}{2\omega_s(X_s + X_r)} \tag{8.28}$$

The different portions of the torque–slip characteristic of Fig. 8.4 can be interpreted with the help of the following derivation. First the normalized torque is written as

$$\frac{\overline{\mathcal{T}}_d}{\mathcal{T}_{dm}} = \left(\frac{2R_r}{s}\right) \frac{X_s + X_r}{[(R_r/s)^2 + (X_s + X_r)^2]} \tag{8.29}$$

By substituting for s_m from Eqn (8.22), in which R_s is taken as zero, Eqn (8.29) becomes

$$\frac{\overline{T}_d}{T_{dm}} = \frac{2}{s/s_m + s_m/s} = \frac{2ss_m}{s_m^2 + s^2} \tag{8.30a}$$

Also, the normalized starting torque is evaluated as

$$\frac{T_s}{T_{dm}} = \frac{2R_r[R_s\sqrt{R_s^2 + (X_s + X_r)^2}]}{[(R_s + R_r)^2 + (X_s + X_r)^2]}$$

R_s is usually small compared to other parameters, this approximation being valid for motors of more than 1 kW rating. Thus, neglecting R_s, the above equation for normalized torque becomes

$$\frac{T_s}{T_{dm}} = \frac{2R_r(X_s + X_r)}{[(R_r^2 + (X_s + X_r)^2]} = \frac{2s_m}{s_m^2 + 1} \tag{8.30b}$$

Equation (8.30a) confirms that at $s = s_m, \overline{T}_d$ attains the value T_{dm} (Fig. 8.4). When $s << s_m$, the term s/s_m can be neglected and Eqn (8.30a) becomes

$$\frac{\overline{T}_d}{T_{dm}} = \frac{2s}{s_m} \tag{8.31}$$

Equation (8.31) shows that for small values of s, the normalized torque versus s curve is a straight line with slope $2/s_m$ and the speed decreases with s. As the rated torque occurs only for a small value of s, it can be concluded that the change of speed, corresponding to a change of torque from zero to the rated torque, is a small percentage of the rated speed. Finally, with $s >> s_m$, s_m/s becomes negligible and Eqn (8.30a) becomes

$$\frac{\overline{T}_d}{T_{dm}} = \frac{2s_m}{s} \tag{8.32}$$

Equation (8.32) reveals that for high values of s, the normalized torque has an inverse relationship with s, and the torque–slip curve is seen to be approaching that of a rectangular hyperbola. The motor operation in this region is unstable for most motors. The relationships in Eqns (8.31) and (8.32) are confirmed by Fig. 8.4.

8.2.4 Methods for Speed Control

Induction motors are required to drive various kinds of loads, each of them having a different torque versus speed characteristic. Among a number of methods that exist for this purpose, some commonly used ones are

(a) stator voltage control,
(b) stator frequency control,
(c) stator current control,
(d) controlling the stator voltage V_s and the frequency f with V_s/f kept constant,
(e) controlling the induced voltage E_s and the frequency f with E_s/f kept constant,

(f) rotor resistance control, and
(g) slip energy recovery scheme.

While the stator-related methods [(a)–(e)] are discussed below, the two rotor-related control strategies [(f) and (g)] are dealt with in Section 8.3.

8.2.4.1 Stator voltage control Figure 8.5 gives a set of torque–slip curves based on Eqn (8.18) with the stator voltage V_s as a parameter (Bose 1986), with V_{s5} as the 100% stator voltage. The constant torque characteristic of a typical load, namely, a machine tool, is also drawn in this figure. The points of intersection A, B, C, D, and E are the stable points of variable speed operation. The curves are seen to become smoother with a reduction in the stator voltage. The range of speed control depends upon the slip for maximum torque (s_m) given by Eqn (8.22), which is seen to be independent of V_s.

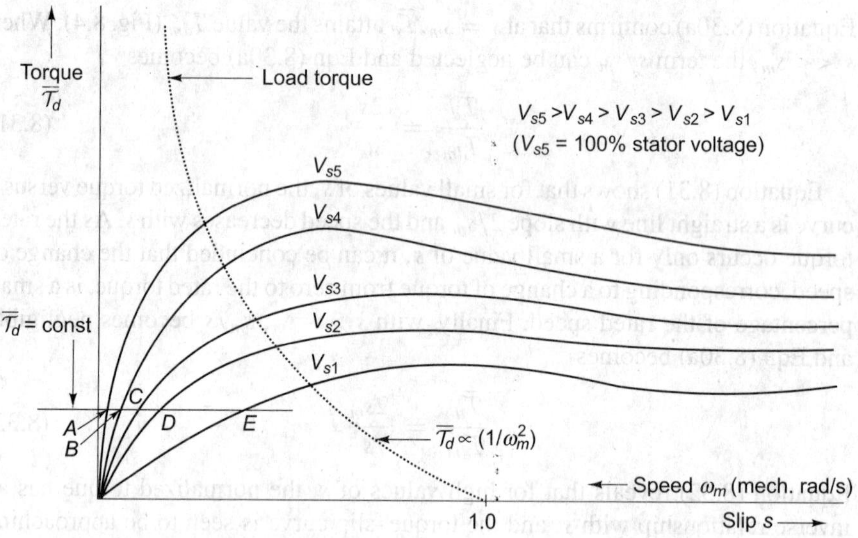

Fig. 8.5 Torque–slip curves of a typical induction motor with stator voltage as a parameter

For a machine with a relatively small s_m and a constant load torque characteristic, the range of speed control is small. It is seen from Fig. 8.5 that as the slip increases, the operating point shifts to a lower voltage curve but with an increase in the stator current. This can be verified with the help of Eqn (8.13) as follows. If originally $s = s_1$, the corresponding torque, denoted as $\mathcal{T}_{d1}$, becomes

$$\mathcal{T}_{d1} = \frac{3I_{r1}^2 R_r(1 - s_1)}{\omega_s \quad s_1} \tag{8.33}$$

The slip s_1 is usually much less than 1 and hence $1 - s_1$ can be approximated as 1. $\mathcal{T}_{d1}$ can now be written as

$$\mathcal{T}_{d1} = \frac{3I_{r1}^2}{\omega_s}\frac{R_r}{s_1} \tag{8.34}$$

If the slip increases to s_2 with the torque remaining at $\mathcal{T}_{d1}$, the torque equation can be expressed as

$$\mathcal{T}_{d1} = \frac{3I_{r2}^2}{\omega_s}\frac{R_r}{s_2} \tag{8.35}$$

If the two expressions on the RHSs of Eqns (8.34) and (8.35) are equated and the terms with currents taken to the left side, the equation

$$\frac{I_{r2}^2}{I_{r1}^2} = \frac{s_2}{s_1} > 1 \tag{8.36}$$

is obtained. That is,

$$I_{r2} > I_{r1} \tag{8.37}$$

The stator and rotor currents being equal as shown in Fig. 8.3, Eqn (8.37) confirms the statement made above regarding the increase of the stator current (I_s), and hence also the rotor current (I_r), with slip under the constant torque condition.

The range of speed is further reduced for a characteristic with the load torque inversely proportional to the square of the speed, as shown by the dotted curve drawn in Fig. 8.5. Hence this type of control is usually applied for (a) low-power applications requiring a low starting torque and a narrow range of speed at relatively low slip, for example in fractional horsepower drives such as fans, blowers, and centrifugal pumps. (b) It is also used to start high-power induction motors with a limit on the inrush of current. Stator voltage control is also used for cranes and hoists, where high slip is demanded only intermittently.

Stator voltage may be varied using any one of the following methods:

(a) connecting an autotransformer between the stator terminals and the ac supply terminals,
(b) using ac voltage controllers,
(c) using voltage-fed variable dc link inverters, or
(d) employing pulse width modulation (PWM) inverters

Out of these methods, ac voltage controllers are preferred because of their simplicity. They, however, suffer from the following drawbacks: (a) low input power factor, (b) high harmonic content, and (c) low operating efficiency. Motor derating is necessary at low speeds to avoid overheating due to excessive current and reduced ventilation. Hence their application is confined to drives with low power ratings and loads with a low speed range, which occur in steel mills, cement mills, etc.

8.2.4.2 Stator frequency control The speed of a three-phase induction motor can also be controlled by varying the input supply frequency (Bose 1986). Using the relation (8.14) for ω_m and noting that $\omega = p\omega_s$, the motor speed in mechanical rad/s can be expressed as

$$\omega_m = \omega_s(1-s) = \frac{\omega}{p}(1-s) = \frac{2\pi f}{p}(1-s) \tag{8.38}$$

Equation (8.38) shows that the mechanical angular velocity is approximately proportional to the frequency f, provided the slip s is much less than unity. The developed torque $\mathcal{T}_d$ given in Eqn (8.18) can be written as

$$\mathcal{T}_d = \frac{3pR_r}{s\omega} \frac{V_s^2}{(R_s + R_r/s)^2 + (X_s + X_r)^2} \tag{8.39}$$

If ω_0 is the nominal angular frequency of the input supply, any other angular frequency can be expressed as

$$\omega = k\omega_0 \tag{8.40}$$

where k can be less or greater than unity. Using this relation, $\mathcal{T}_d$ can be written as

$$\mathcal{T}_d = \frac{3pR_r}{sk\omega_0} \frac{V_s^2}{(R_s + R_r/s)^2 + k^2(X_s + X_r)^2} \tag{8.41}$$

The stator and rotor reactances in the denominator of Eqn (8.41), whose nominal values are computed with ω_0, also vary as k varies. Hence the torque becomes a function of k, provided the stator voltage V_s is kept constant. If the voltage drops across the stator resistance and reactance can be neglected, then V_s can be approximated as

$$V_s \approx E_s = 4.44 f N_s K_N \Phi_m \tag{8.42}$$

where f is the supply frequency, N_s is the number of stator turns, K_N is the winding factor (which is the product of the distribution factor and the pitch factor), and ϕ_m (which is the air gap flux) is the fundamental flux per pole of the rotating field. Equation (8.42) can now be written as

$$V_s \approx K_v f \Phi_m \tag{8.43}$$

where K_v incorporates all the constant terms in the RHS of Eqn (8.42). Equation (8.43) shows that, with V_s kept constant and f decreased from its nominal value f_0 $(= \omega_0/2\pi)$, the air gap flux ϕ_m increases; hence the degree of saturation of the magnetic circuit is increased, causing magnetic saturation. Consequently the original motor parameters will not hold good, and will lead to erroneous speed–torque characteristics. On the other hand, if the frequency is increased from f_0, the air gap flux ϕ_m and the developed torque $\mathcal{T}_d$ will decrease as illustrated in Fig. 8.6, in which the speed–torque curves are drawn as torque versus k curves with frequency as a parameter. The curve joining the peaks of the $\mathcal{T}_d$-k curves obeys the equation

$$\mathcal{T}_{dm} k^2 = \text{constant} \tag{8.44}$$

where $\mathcal{T}_{dm}$ is the maximum torque obtainable with the frequency $k\omega_0$. The relationship in Eqn (8.44) can be obtained from the expression for $\mathcal{T}_{dm}$ given in Eqn (8.23), which is rewritten here as

$$\mathcal{T}_{dm} = \frac{3}{2\omega_s} \frac{V_s^2}{[R_s + \sqrt{R_s^2 + (X_s + X_r)^2}]} \tag{8.45}$$

For frequencies above the rated frequency, R_s can be neglected as compared to the stator and rotor reactances. Using this approximation and also the relations

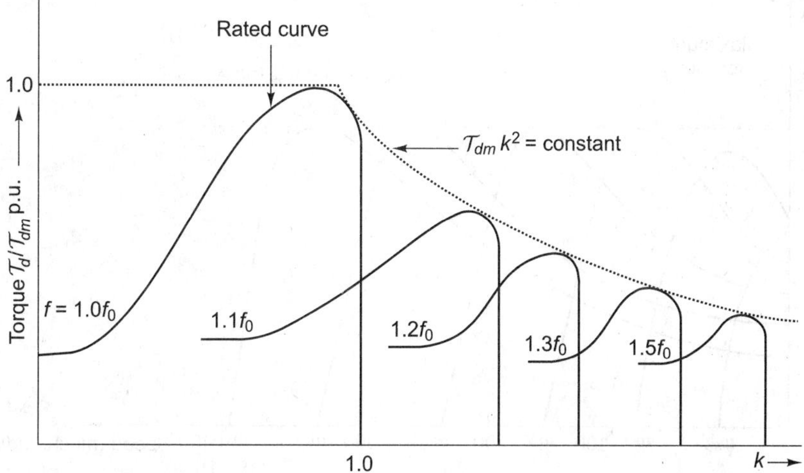

Fig. 8.6 Speed–torque curves of a typical induction motor with the frequency multiplier k as a parameter

$\omega_s = \omega/p$ and $\omega = k\omega_0$, T_{dm} can be expressed as

$$T_{dm} = \frac{3p}{2k\omega_0}\frac{V_s^2}{k(X_s + X_r)} = \frac{3pV_s^2}{2k^2\omega_0^2(L_s + L_r)} \tag{8.46}$$

where L_s and L_r are the leakage inductances of the stator and rotor, respectively. Equation (8.46) confirms the relationship in Eqn (8.44) because ω_0, p, V_s, L_s, and L_r are all constant. Another way of interpreting this equation is that the developed torque is inversely proportional to the square of the input frequency. For the range of f above f_0, the motor is said to be operating in the field weakening mode, and from Fig. 8.6 it is seen that the torque capability of the motor is reduced in this mode.

As stated above, the reduction of frequency below the nominal value f_0 leads to saturation of the magnetic circuit; hence frequency control is not permissible in this range of input frequency. Instead, if the ratio V_s/f (or equivalently V_s/k) is kept constant, the maximum torque as given by Eqn (8.46) remains approximately valid except in the low-frequency region, where the air gap flux is reduced by the stator impedance drop. Therefore, in this region the stator drop has to be compensated by an additional voltage boost so as to produce maximum torque. This type of control is known as *constant voltage/hertz control*. The speed–torque curves with ω as a parameter and with V/ω or V/k constant are given in Fig. 8.7.

It is evident from Eqn (8.43) that keeping V_s/ω constant throughout the range of speed is the same as keeping the air gap flux approximately constant at the value corresponding to the no-load operation, with the voltage and frequency at their rated values. The curves in Fig. 8.7 confirm that, in spite of the maximum (or breakdown) torque remaining constant for various values of ω, each of the characteristics falls down rapidly for very low values of ω_m. This decrease is caused by the reduction in the air gap flux consequent to the increased effect of the stator resistance. The stator drop $I_s R_s$ has the same value at all frequencies,

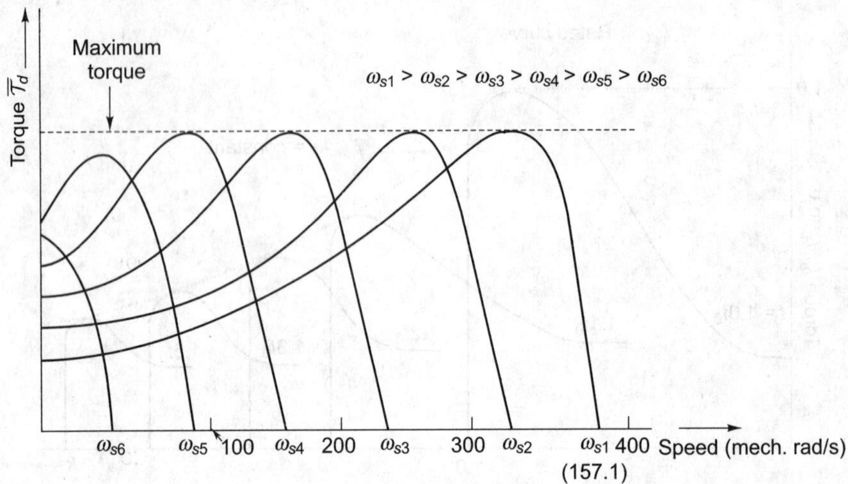

Fig. 8.7 Speed–torque curves of a typical induction motor with ω as a parameter and constant V/ω

whereas at very low frequencies, $I_s X_s$ is much less than that at the rated frequency and can be neglected. On the other hand, for frequency ranges above the rated frequency, $I_s R_s$ can be neglected as compared to the reactive drop.

One method of overcoming the effect of stator resistance drop in the low-frequency region is to give an additional boost to the input voltage so as to produce maximum torque as given at higher frequencies. This method is called the *voltage boost* method. A better method, known as *constant air gap flux operation*, is to keep the ratio E_s/ω constant instead of V_s/ω. Actually, V_s is the sum of the induced voltage E_s and the stator drop and is expressed as

$$V_s = E_s + I_s(R_s + jX_s) \tag{8.47}$$

The approximation $V_s \approx E_s$ does not hold good for low input frequencies. However, if E_s/Ω is kept constant, it can be ensured that the air gap flux is strictly constant. Since K_v is constant, the following proportionality is obtained from Eqn (8.42):

$$\Phi_m \propto \frac{|E_s|}{\Omega} \approx \frac{|V_s|}{\Omega} \tag{8.48}$$

Neglecting the core loss component R_m,

$$E_s = jX_m I_0 = j\Omega L_m I_0 \tag{8.49}$$

It follows from Eqns (8.48) and (8.49) that

$$\Phi_m \propto \frac{|E_s|}{\Omega} \propto L_m I_0 \tag{8.50}$$

Assuming that L_m remains constant and unaffected by saturation, Eqn (8.50) reveals that for maintaining the air gap flux constant, the magnetizing current I_0 has to be held constant at all speeds and loads. In practice, I_0 is held constant at the rated value corresponding to the normal full load operation at rated voltage and frequency.

From the equivalent circuit of Fig. 8.1(b), the rotor current can be written as

$$I_r = \frac{E_s}{\sqrt{(R_r/s)^2 + X_r^2}} \tag{8.51}$$

where $X_r = \omega L_r$. If $\omega_{R\omega}$ is taken as the actual angular frequency of the rotor voltage in electrical radians per second and ω_r is written as in Eqn (8.20), the slip can be expressed as

$$s = \frac{\omega - \omega_r}{\omega} = \frac{\omega_{R\omega}}{\omega} \tag{8.52}$$

Now I_r can be written as

$$I_r = \frac{E_s s}{\sqrt{R_r^2 + s^2\omega^2 L_r^2}} = \frac{E_s}{\omega}\frac{\omega_{R\omega}}{\sqrt{R_r^2 + \omega_{R\omega}^2 L_r^2}} \tag{8.53}$$

Substituting this expression for I_r in Eqn (8.13) gives

$$\mathcal{T}_d = \frac{3}{s\omega_s}\left(\frac{E_s}{\omega}\right)^2 \frac{\omega_{R\omega}^2 R_r}{R_r^2 + \omega_{R\omega}^2 L_r^2} \tag{8.54a}$$

Equations (8.3), (8.14), and (8.19) give the relation

$$s\omega_s = \omega_s - \omega_m = \frac{\omega}{p} - \frac{\omega_r}{p} = \frac{\omega_{R\omega}}{p} \tag{8.54b}$$

Thus, Eqn (8.54a) becomes

$$\mathcal{T}_d = 3p\left(\frac{E_s}{\omega}\right)^2 \frac{\omega_{R\omega} R_r}{R_r^2 + \omega_{R\omega}^2 L_r^2} \tag{8.55}$$

Equations (8.48) and (8.55) together imply that the torque is proportional to the square of the air gap flux at a given rotor frequency $\omega_{R\omega}$. Therefore, if ϕ_m is maintained constant under all operating conditions, *the induction motor torque will depend solely on* $\omega_{R\omega}$, which is the slip speed in electrical rad/s. The speed–torque curves in Fig. 8.8 are drawn with constant E/ω and, therefore, ensure that the air gap flux is maintained strictly constant at the value corresponding to normal full load operation at rated voltage and frequency. Here the maximum torque is seen to be the same for all supply frequencies including low frequencies. The deterioration in performance observed for the case of constant V/f for low frequency is seen to be absent here. It can be shown that the available torque will be much greater with constant air gap flux than that obtained for normal fixed frequency operation at rated voltage and frequency.

8.2.4.3 Stator current control The speed–torque characteristics of an induction motor can also be drawn with I_s as a parameter in order to obtain a relation between torque and stator current. This type of control is advantageous over stator voltage control, because it allows fast and effective control of the amplitude and the spatial phase angle of the stator mmf wave. Consequently, it provides a good dynamic response and high-quality torque control. To bring

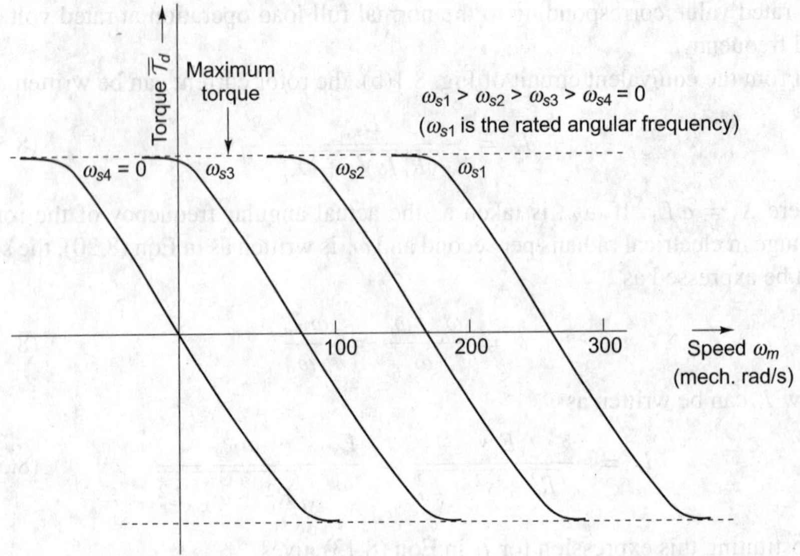

Fig. 8.8 Speed–torque curves of a typical induction motor with ω as a parameter and constant E/ω

out these features analytically, the torque is expressed in terms of stator current instead of stator voltage as follows.

The equivalent circuit of the induction motor given in Fig. 8.1(b) can be redrawn as in Fig. 8.9 with the assumption that the core loss is negligible. Equation (8.15) can be written using Eqn (8.54b) as

$$\overline{T}_d = \frac{3I_r^2 R_r}{s\omega_s} = \frac{3pI_r^2 R_r}{\omega_{R\omega}} \tag{8.56}$$

where ω is the synchronous speed in electrical rad/s.

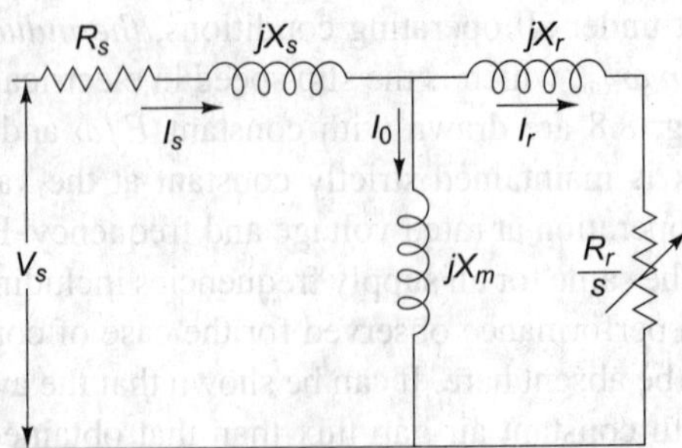

Fig. 8.9 Equivalent circuit of an induction motor with no core losses

I_r can be obtained from Fig. 8.9 as

$$|I_r| = \frac{s|I_s|\omega L_m}{\sqrt{R_r^2 + [s\omega(L_m + L_r)]^2}} \tag{8.57}$$

Using the relation $\omega_{R\omega} = s\omega$, Eqn (8.57) becomes

$$|I_r| = \frac{|I_s|\omega_{R\omega}L_m}{\sqrt{R_r^2 + [\omega_{R\omega}(L_r + L_m)]^2}} \tag{8.58}$$

Substituting this expression for I_r in Eqn (8.56), $\overline{T}_d$ can be expressed as

$$\overline{T}_d = 3pI_s^2 \frac{(\omega_{R\omega}L_m)^2 (R_r/\omega_{R\omega})}{R_r^2 + [\omega_{R\omega}(L_r + L_m)]^2} \tag{8.59}$$

The maximum torque and corresponding slip frequency can now be obtained as

$$T_{dm} = \pm 3pI_s^2 \frac{L_m^2}{2(L_r + L_m)} \tag{8.60}$$

and

$$\omega_{R\omega m} = \pm \frac{R_r}{L_r + L_m} \tag{8.61}$$

The magnetizing current I_0 can be written as

$$|I_0| = |I_s| \left[\frac{(R_r/s)^2 + X_r^2}{(R_r/s)^2 + (X_r + X_m)^2} \right]^{1/2} \tag{8.62}$$

$$= |I_s| \left[\frac{R_r^2 + (\omega_{R\omega}L_r)^2}{R_r^2 + [\omega_{R\omega}(L_r + L_m)]^2} \right]^{1/2} \tag{8.63}$$

If I_s and $\omega_{R\omega}$ are defined, the three quantities, I_r, T_d, and I_0 can be uniquely determined, respectively, from Eqns (8.58), (8.59), and (8.63). Particularly, if the air gap flux or equivalently I_0 is to be maintained constant, Eqn (8.63) shows that I_s should be related to $\omega_{R\omega}$ as

$$|I_s| = |I_0| \left[\frac{R_r^2 + [\omega_{R\omega}(L_r + L_m)]^2}{R_r^2 + (\omega_{R\omega}L_r)^2} \right]^{1/2} \tag{8.64}$$

where I_0 is kept constant at some desired value. Figure 8.10 shows the plot of I_s versus $\omega_{R\omega}$ as given by this equation (Murphy & Turnbull 1988). This plot shows that the magnitude of I_s attains the same values for both the motoring and regenerating modes of operation and becomes equal to I_0 when $\omega_{R\omega}$ is zero.

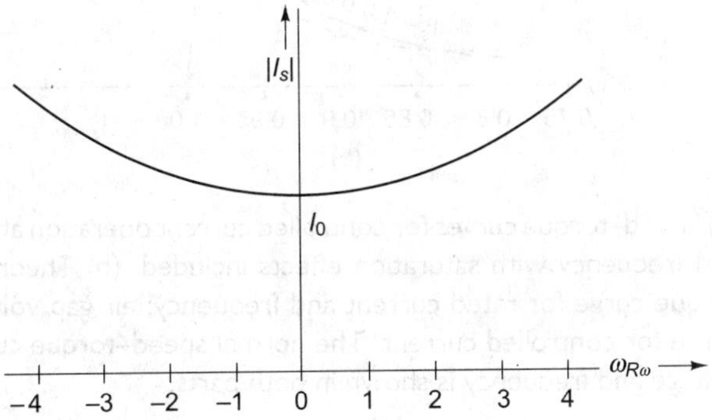

Fig. 8.10 Plot of I_s vs $\omega_{R\omega}$ for constant air gap flux

If the relationship between I_s and $\omega_{R\omega}$ is not maintained as in Fig. 8.10, the machine, which is operated with stator current control, is likely to be subjected to high magnetic saturation, and the machine inductance parameters (especially L_m) may vary.

The speed–torque characteristics, with stator current I_s as a parameter, are given in Fig. 8.11(a) with respect to a typical machine for which saturation is taken into consideration. The dotted curve in this figure shows the relationship with $V_s = 1.0$ p.u.

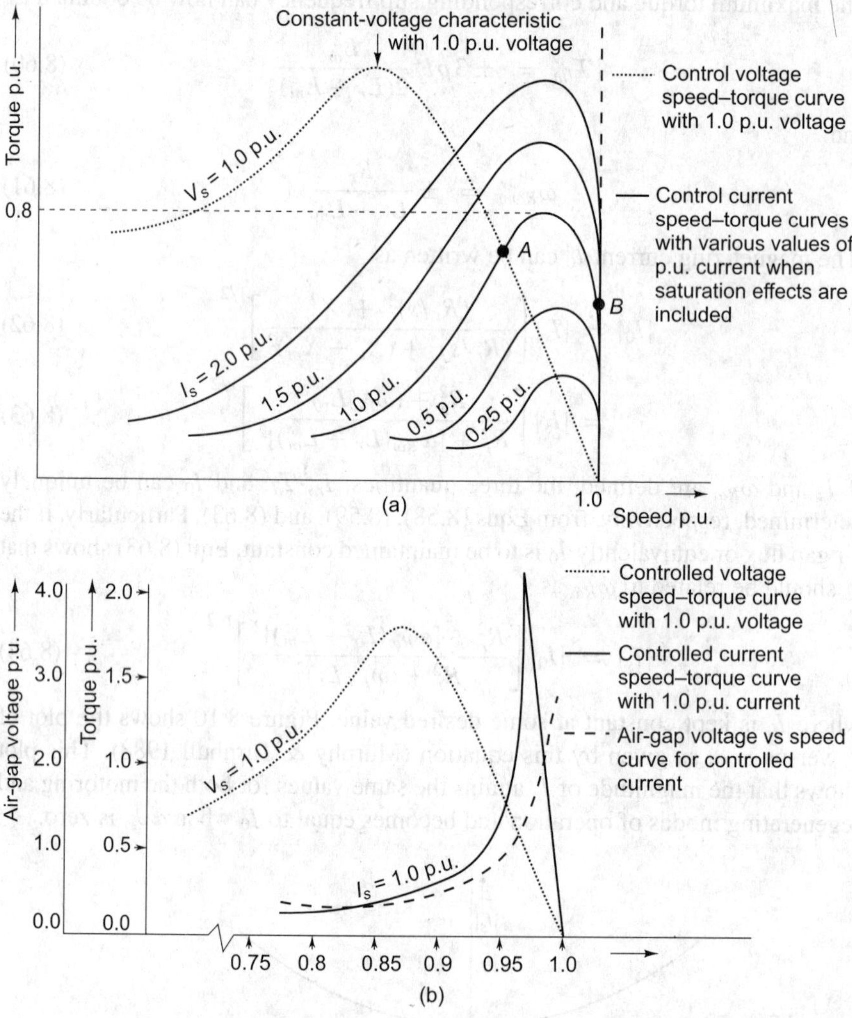

Fig. 8.11 (a) Speed–torque curves for controlled current operation at rated current and frequency with saturation effects included. (b) Theoretical speed–torque curve for rated current and frequency; air gap voltage vs speed curve for controlled current. The normal speed–torque curve for rated voltage and frequency is shown in both parts.

The speed–torque characteristic for $I_s = 1.0$ is given in Fig. 8.11(b) for the unsaturated condition of the machine as depicted by Eqn (8.59). A comparison of Figs 8.11(a) and (b) shows that

(a) the speed for maximum torque is slightly reduced under the saturated condition,

(b) the breakdown torque occurs at a higher value of slip (or rotor frequency), and

(c) the maximum per-unit (p.u.) torque is considerably reduced from about 2.0 to about 0.8, due to saturation.

In Fig. 8.11(a), the intersection point A is a possible operating point with the rated conditions $V_s = 1.0$ p.u., $I_s = 1.0$ p.u., and air gap flux $\phi = 1.0$ p.u. The same torque can be developed at the point B on the stable side of the speed–torque characteristic with $I_s = 1.0$ p.u., but at this point the air gap flux and the corresponding core losses will be high. The terminal voltage as given by Eqn (8.47) will also be increase. Controlled current operation at point A is preferred because of these reasons. However, this is on the unstable side of the speed–torque curve and it will not be possible to implement open-loop operation. A feedback loop is constructed to ensure that the operating point falls on the curve given in Fig. 8.10, thereby maintaining the rated air gap flux. This also ensures that the operating point falls on the speed–torque curve corresponding to the rated voltage.

8.2.4.4 Stator voltage, rotor slip, and stator frequency controls Induction motor drives may have to be operated at different speed ranges depending on the load–torque characteristics. Each of these ranges is governed by one or more of the control strategies discussed above. A comprehensive policy will, however, be helpful in covering the entire range of operation of the motor. For operation below base speed, it is necessary to avoid saturation of the core in this speed range. Stator frequency control has to be implemented by maintaining V/f or preferably E/f constant, which is known as constant flux (or constant torque) control. Stator frequency control for the region above the base frequency involves reduced air gap flux and torque capability (Fig. 8.6). In this mode, the envelope of the maximum (or breakdown) torque obeys the relation $\overline{T}_{dm}k^2 =$ constant. This is because $\overline{T}_{dm}$ becomes a variable above the base frequency [Eqn (8.46)]. The plots of Figs 8.6 and 8.7 are combined in Fig. 8.12 to give the two modes $\omega_s < \omega_0$ and $\omega_s > \omega_0$, where ω_0 is the base angular frequency. In the latter figure, the line AB corresponds to E/f (or V/f) control, whereas the portions BC and CD together correspond to the relation $\overline{T}_{dm}k^2 =$ constant.

8.2.4.5 Why and how of the constant horsepower mode It should, however, be noted that certain loads like traction drives have a constant horsepower characteristic and this feature has to be incorporated in the speed–torque characteristics of adjustable speed drives. The following derivation gives an insight into how such a characteristic is obtained. For this purpose, it is assumed that $\omega_{R\omega}$ is restricted to small values. Equation (8.6) gives the developed torque as

$$T_d = k_1 \Phi_m I_r \sin \delta \tag{8.65}$$

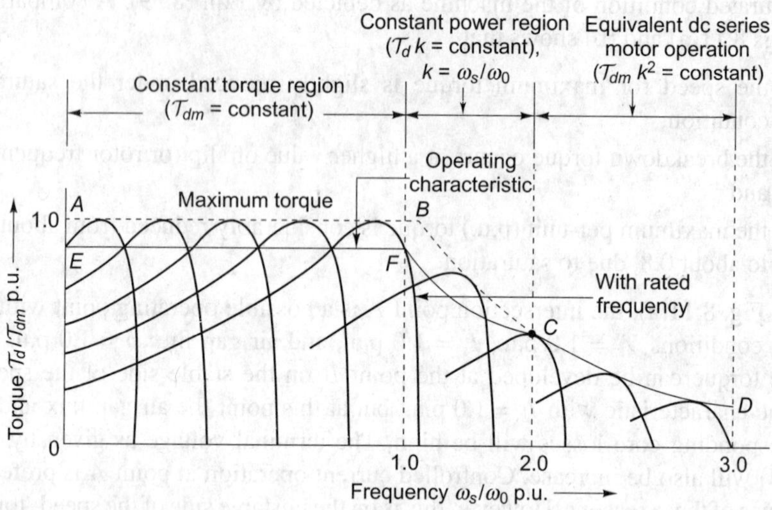

Fig. 8.12 Regions of speed–torque curves with variable voltage, variable frequency power supply

where I_r is taken to be equal to I_s as in Fig. 8.3. At low rotor frequencies, $\sin\delta$ is nearly unity and Eqn (8.65) can be approximated as

$$\mathcal{T}_d = k_1 \Phi_m I_r \tag{8.66}$$

I_r can be written from Eqn (8.5) as

$$|I_r| = \frac{|E_s|}{[(R_r/s)^2 + X_r^2]^{1/2}} \tag{8.67}$$

When s is small, $X_r << R_r/s$ and the rotor current I_r can be approximated as

$$|I_r| = \frac{s|E_s|}{R_r} = \frac{\omega_{R\omega}}{\omega}\frac{|E_s|}{R_r} \tag{8.68}$$

Since the air gap flux ϕ_m is proportional to $|E_m|/\omega$, Eqn (8.68) shows that the rotor current is proportional to the product of the air gap flux and the rotor slip frequency. Taking $1/R_r$ to be equal to a constant k_2, Eqn (8.68) can be written as

$$|I_r| = k_2\omega_{R\omega}\frac{|E_s|}{\omega} \tag{8.69}$$

Now taking $\phi_m = (k_3|E_s|)/\omega$ as per Eqn (8.50) and also using Eqn (8.69), Eqn (8.66) can be expressed, with the torque considered as a variable, as

$$\overline{\mathcal{T}}_d = k_4\left(\frac{|E_s|}{\omega}\right)^2 \omega_{R\omega} \tag{8.70}$$

where $k_4 = k_1 k_2 k_3$.

It can be seen from Eqn (8.70) that for constant stator voltage and slip angular frequency, the torque (under low-slip operation) is inversely proportional to ω^2. *Now, if $\omega_{R\omega}$ is increased linearly with ω, $\mathcal{T}_d$ varies as the inverse of ω*

or approximately as the inverse of speed, thus giving a constant horsepower characteristic. The upper speed limit for such an operating condition is governed by the maximum practical value of the rotor frequency. This constant horsepower mode of operation corresponds to the field weakening operation of a dc motor; it can be achieved for a limited range above base speed by supplying the induction motor with constant stator voltage (E_s) at higher frequencies.

8.2.4.6 High-speed motoring operation In general, it may be necessary to operate the motor at speeds higher than those obtained with the constant horsepower mode. In this case, the motor voltage E_s and the slip frequency $\omega_{R\omega}$ are maintained constant at their maximum values as the stator frequency is further increased; hence, the maximum output torque ($\overline{T}_{dm}$) varies as the inverse of the square of the speed (ω^2).

8.2.4.7 A composite picture of induction motor operation The above-discussed regions of the speed–torque curves of a *practical drive system* with a variable voltage, variable frequency supply are demarcated in Fig. 8.12 as *EF*, *FC*, and *CD*. *The corresponding voltage frequency relationship is shown in Fig. 8.13.* This figure also shows the developed torque, stator current, and slip frequency as functions of speed. In the constant torque region, the maximum available torque is shown to be somewhat lower than the breakdown torque due to the limited inverter current capability (portion *EF* in Fig. 8.12).

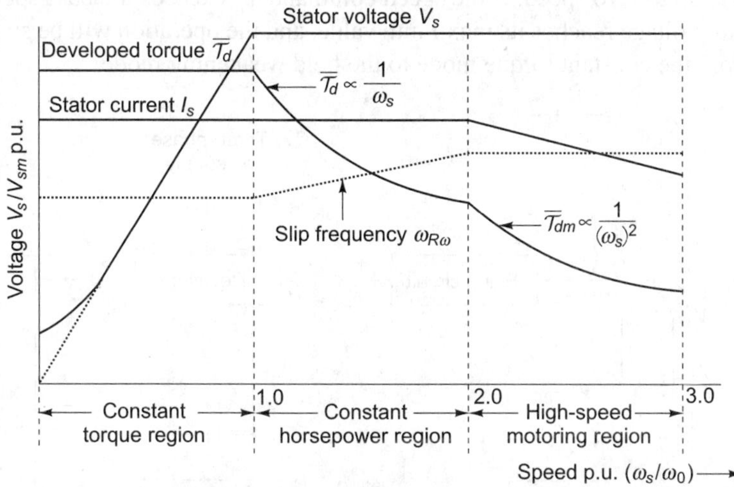

Fig. 8.13 Voltage, current, slip frequency, and torque of an induction motor as functions of speed for the operating characteristic of Fig. 8.12

On the right edge of the constant torque region, the stator voltage reaches the rated value and the machine enters the constant horsepower mode as elaborated above (portion *FC* in Fig. 8.12). In this mode, the stator current stays nearly constant but the rotor frequency increases linearly and the operating point moves closer to the breakdown point. Again, on the right of the constant horsepower region, the breakdown torque $\overline{T}_{dm}$ is reached (point *C* in Fig. 8.12) and the machine speed can be further increased by increasing the frequency, with reduction in the stator current (portion *CD* in Fig. 8.12). In this region, the slip frequency is held constant.

8.2.5 Control Schemes for Speed Control of Induction Motors

The implementation of some of the methods discussed above for the speed control of induction motors is now dealt with. The scheme may be either open loop or closed loop in nature; the latter is preferred from the point of view of better dynamic response and stability. Two broad categories of closed-loop control are (i) scalar control and (ii) vector control. Whereas in the former case, only the magnitude of the desired variable is changed, both the magnitude and phase are varied in the latter method. The following discussion will be confined to scalar control.

The block diagram of a typical open-loop speed control scheme is given in Fig. 8.14. The semiconductor-controlled converter can be a three-phase rectifier followed either by a dc link converter and a six-step or PWM type of VSI as shown in Fig. 8.14(a), or by a cycloconverter as shown in Fig. 8.14(b). The speed command ω_r^* is very close to the motor speed and the transfer function G_ϕ is a V/f gain constant which generates V_{s1}^* from the signal ω_r^*. V_{s1}^* is further boosted by the signal V_{s1} for the following reason. As the stator voltage V_s approaches zero, it will be absorbed by the stator resistance. The signal V_{s1} is introduced to overcome this resistance drop, so that the rated air gap flux and full torque become available up to zero speed. If the speed command ω_r^* exceeds the base speed ω_0, the stator voltage reaches its maximum value, and the operation will be switched over from the constant torque mode to the field weakening mode.

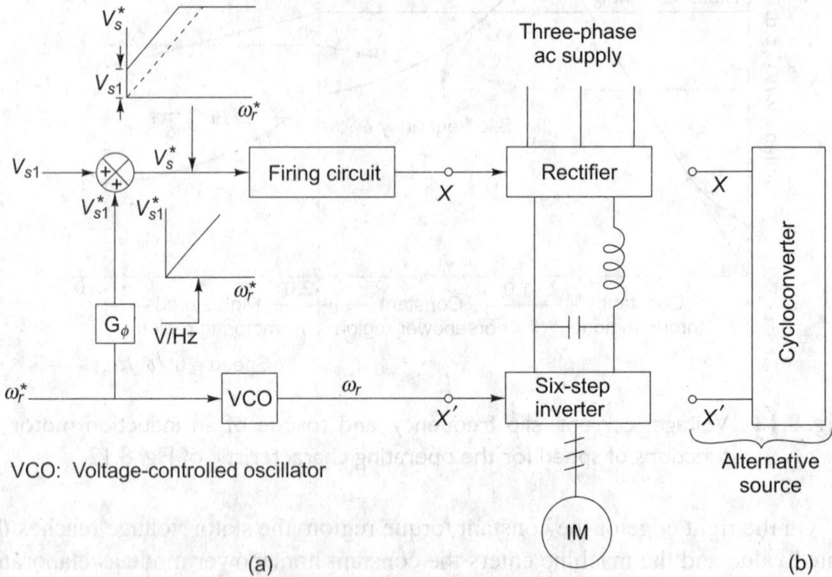

Fig. 8.14 Block-cum-schematic diagram of a typical open-loop speed control scheme

The open-loop scheme works quite satisfactorily in applications for which high dynamic performance is not required. Hence it is used for all general-purpose applications. Its cost is low because it does not involve either speed sensing or flux sensing apparatus and can be easily implemented on all commercially available induction motors. However, it suffers from the following disadvantages. (a) The ac line voltage fluctuations and the stator impedance drop cause corresponding fluctuations in the magnitude of the air gap flux. (b) The rotor slip increases and the motor slows down slightly when load torque is applied, thus giving a poor dynamic response; the system is also inherently less stable. (c) Quick changes in the frequency command signal may cause excessive current, thus warranting the introduction of current limit control.

8.2.5.1 Closed-loop volts/hertz control Figure 8.15 shows an induction motor drive with terminal volts/hertz control and speed feedback. The speed error is fed to a speed controller whose output determines the inverter frequency and voltage. A signal from the stator circuit measures the current and passes through a current limit signal, which comes into play only when the motor current rises to a value greater than a preset maximum value. The output of the current limit block controls the rate at which the inverter frequency and voltage are increased. For instance, if the speed command is given as a step input, the motor current rises to the previously adjusted limit. The desired output frequency and voltage are then gradually increased so that the motor speed follows the inverter frequency, at the same time ensuring that the rotor breakdown frequency is not exceeded.

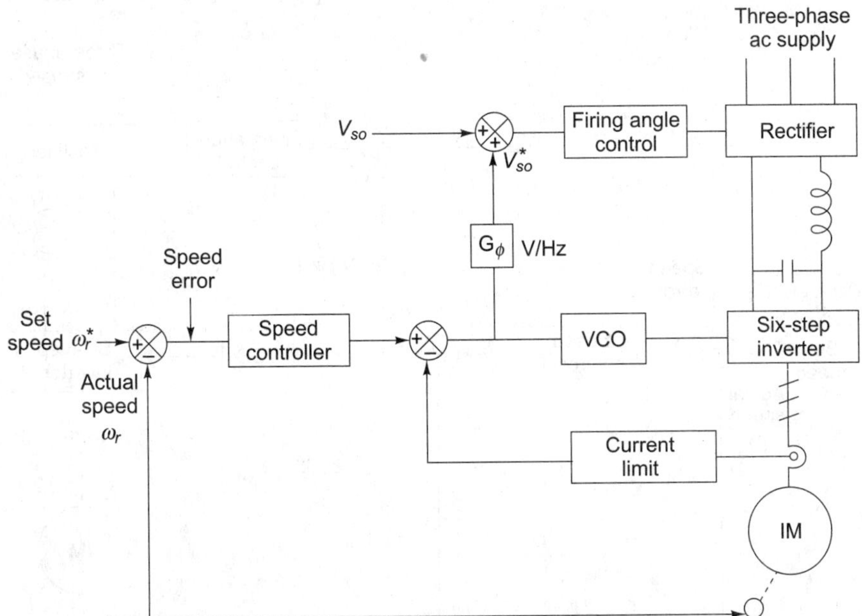

Fig. 8.15 Closed-loop speed control scheme with terminal V/ω control and inner current limit loop

As a result of the volts/hertz control the machine accelerates at constant torque and the current limit control ensures that the speed rises slowly till the set speed is reached. After this the stator current falls below the limit and steady-state operation is achieved. If the set speed is above the base speed value, the operation takes place at constant voltage. For any further increase of frequency, volts/hertz control cannot be implemented and the motor is operated in the field weakening region.

The advantage of this type of control is that the motor speed is adjusted to the desired value, giving better speed regulation, thus resulting in reduced speed sensitivity to fluctuations of shaft load. Second, speed feedback helps in uniform drive performance over the entire frequency range. As a consequence, any reduced stability feature that is present in open-loop control is eliminated.

8.2.5.2 Closed-loop slip regulation with volts/hertz control Figure 8.16 shows an alternative scheme in which current limit control is dispensed with, and in its place, the slip frequency $\omega_{R\omega}$ influences the rate of rise of inverter frequency and voltage. The speed error passes through the speed controller as before but the latter generates the slip frequency command $\omega^*_{R\omega}$. This signal when combined with the tachometer signal becomes the inverter frequency command signal ω^*_s. The signal ω^*_s passes through a function generator G_ϕ for the generation of the stator voltage command V^*_{so}. The function generator incorporates the low-frequency stator drop compensation so that the air gap flux remains approximately constant. As indicated by Eqn (8.55), the developed torque becomes a function of the slip frequency $\omega_{R\omega}$. For this reason, the slip control loop can be considered as an inner

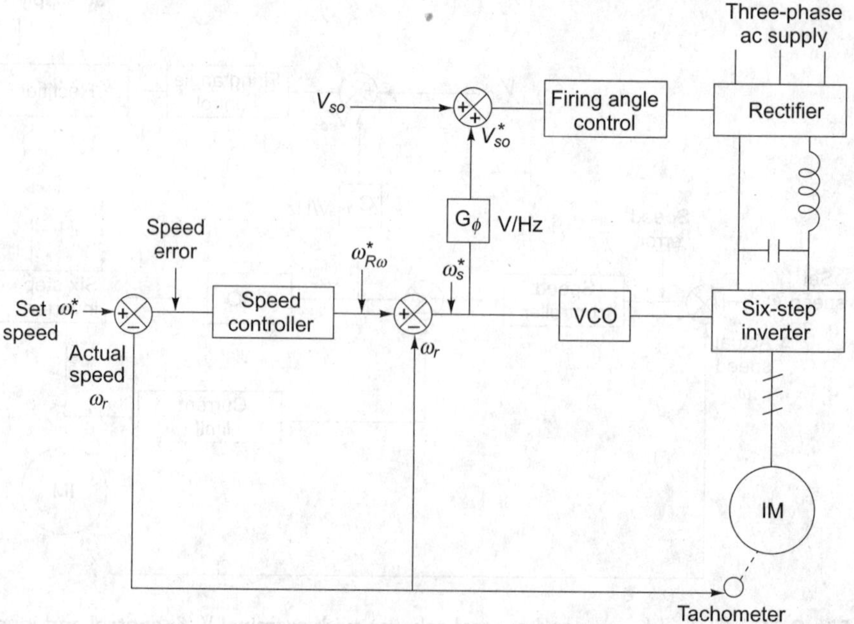

Fig. 8.16 Block-cum-schematic diagram of closed-loop slip regulation with V/ω control

torque loop. When a sudden speed increase is required, the speed controller keeps down the slip frequency demand so that the motor operates just below the rotor breakdown frequency and a large torque is developed by the motor. This helps in quick acceleration of the rotor to the desired speed. The slip frequency $\omega^*_{R\omega}$ then settles down to a steady-state value as dictated by the load torque. On the other hand, when the speed command is a negative step, the speed controller generates a negative slip, causing the induction motor to work in the induction generator mode. The energy so generated has to be dissipated in a dynamic braking resistor or returned back to the ac supply line.

Slip frequency control serves the same purpose as torque limit control and the expensive current sensor is avoided. The same tachometer signal is used as a feedback signal both for the inner stator frequency control loop as well as for the outer torque control loop.

8.2.5.3 Stator voltage control with constant slip frequency The control loop of Fig. 8.16 can also be used to provide constant slip control and has been employed in the past for this purpose due to ease of implementation. Stator frequency is defined by adding a constant predetermined value that constitutes the fixed slip frequency to the tachometer signal. The speed error signal now controls the stator voltage only. The ratio V/f is no longer constant and the air gap flux varies over a wide range. With $\phi_m = (k_3|E_s|)/\omega$ as defined below Eqn (8.69), Eqn (8.70) can be written as

$$\overline{T}_d = k_4 \left(\frac{\phi_m}{k_3}\right)^2 \omega_{R\omega} \tag{8.71}$$

and for constant $\omega_{R\omega}, \overline{T}_d$ becomes

$$\overline{T}_d = k_5 \phi_m^2 \tag{8.72}$$

where $k_5 = k_4/(k_3)^2$. Now Eqn (8.72) shows that the torque is proportional to the square of the air gap flux. The speed loop now controls the stator voltage, air gap flux, and developed torque. At light loads, the stator voltage is automatically reduced and the flux ϕ_m also falls to a low value. As a result, the core losses in the motor are reduced and the overall efficiency of the motor is improved. The disadvantage of this type of control is that a change in the load condition involves a change in the air gap flux level and the dynamic performance is very poor. The motor speed is highly sensitive to load fluctuations and the response to a sudden change in the set speed is very sluggish. Thus, this control is unsuitable for high-performance drives.

8.2.5.4 Closed-loop stator current control Because of the merits of closed-loop stator current control (Murphy & Turnbull 1988), enumerated in Section 8.2.4.3, this type of control is usually implemented for an induction motor drive fed by a CSI. The induction motor torque can be determined from a knowledge of the stator current I_s and the rotor frequency (or slip frequency) $\omega_{R\omega}$. In this scheme, which is illustrated in Fig. 8.17, the stator current I_s is determined from a signal proportional to the error between the actual dc link current I_d and its desired value I_d^*; the dc link current is regulated by the phase-controlled rectifier. The set point I_d^* for the dc link current is obtained as the output of the speed controller whose input is the speed error. The actual dc link current I_d is fed to a function generator which generates the slip frequency command $\omega_{R\omega}^*$. The relationship of $\omega_{R\omega}$ with the stator current I_s is maintained as demanded by the characteristic in Fig. 8.10 to ensure the operation of the motor with constant air gap flux. The signal $\omega_{R\omega}^*$ is added to the actual rotor frequency ω_r to give the stator frequency signal ω_s^*. The functional relationship between I_s and $\omega_{R\omega}$ involves the rotor resistance, rotor inductance, and magnetizing inductance, which must be accurately determined. When the motor operates under no load, $\omega_{R\omega}^*$ is zero but the dc link current I_d has a minimum value corresponding to the magnetizing current I_m. This minimum link current is also necessary for providing satisfactory commutation to the inverter. Apart from this method, which is termed as stator current versus slip frequency control, two other strategies available for flux control are (a) direct flux sensing and (b) voltage sensing (Sen et al. 1978).

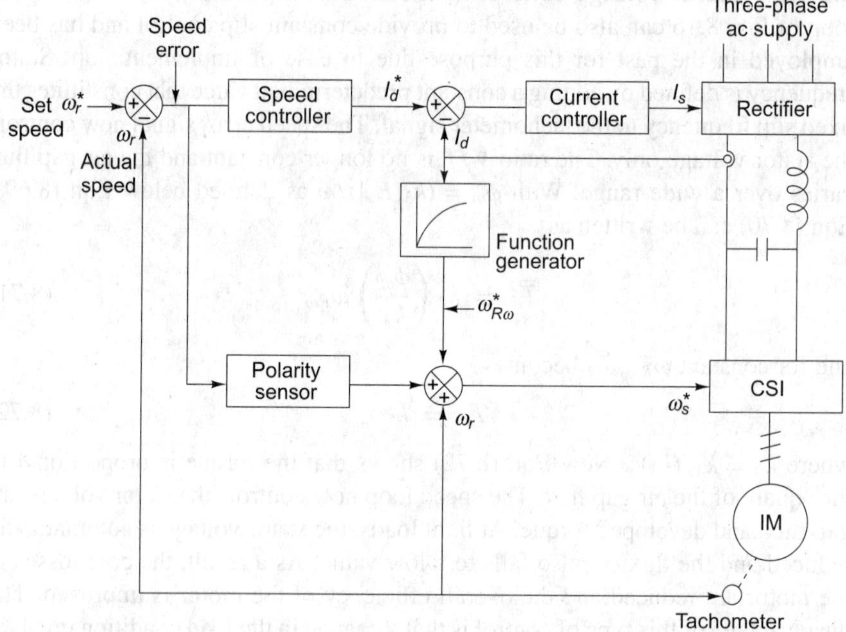

Fig. 8.17 Block-cum-schematic diagram of closed-loop stator current control

8.3 Rotor-related Control Systems for a Wound Rotor Induction Motor

All the control principles discussed in the previous section involve only stator-circuit quantities, namely, stator voltage, stator frequency, and stator current. These methods can be implemented for both the squirrel cage and wound rotor types of induction motors. In this section two schemes involving rotor-circuit quantities are presented. As the rotor bars of the squirrel cage induction motor are short-circuited, the manipulation of rotor-circuit quantities, namely, rotor voltage, current, or resistance, is not possible. On the other hand, the rotor winding of a wound rotor motor is accessible through the slip rings and hence the schemes discussed below can be applied to it. The two methods dealt with here for the latter type of motor are (a) inserting a variable external resistance in the rotor circuit and (b) extracting power from the rotor circuit and returning it to the supply. The latter method is called the *slip energy recovery scheme*.

8.3.1 Rotor Resistance Control

The principle of speed control by variation of rotor resistance (Rashid 1994) can be understood from Eqn (8.18), which gives the torque developed by the motor as

$$\overline{T}_d = 3\frac{R_r}{s\omega_s}\frac{V_s^2}{[(R_s + R_r/s)^2 + (X_s + X_r)^2]} \tag{8.73}$$

Equation (8.73) shows that, with the other quantities remaining constant, torque varies with rotor resistance. In a wound rotor induction machine, it is possible to connect a three-phase rheostat through the slip rings as shown in Fig. 8.18(a). Again Fig. 8.18(b) shows that different values of the resistance R_{st} of the three-phase rheostat connected to the rotor give rise to different normalized

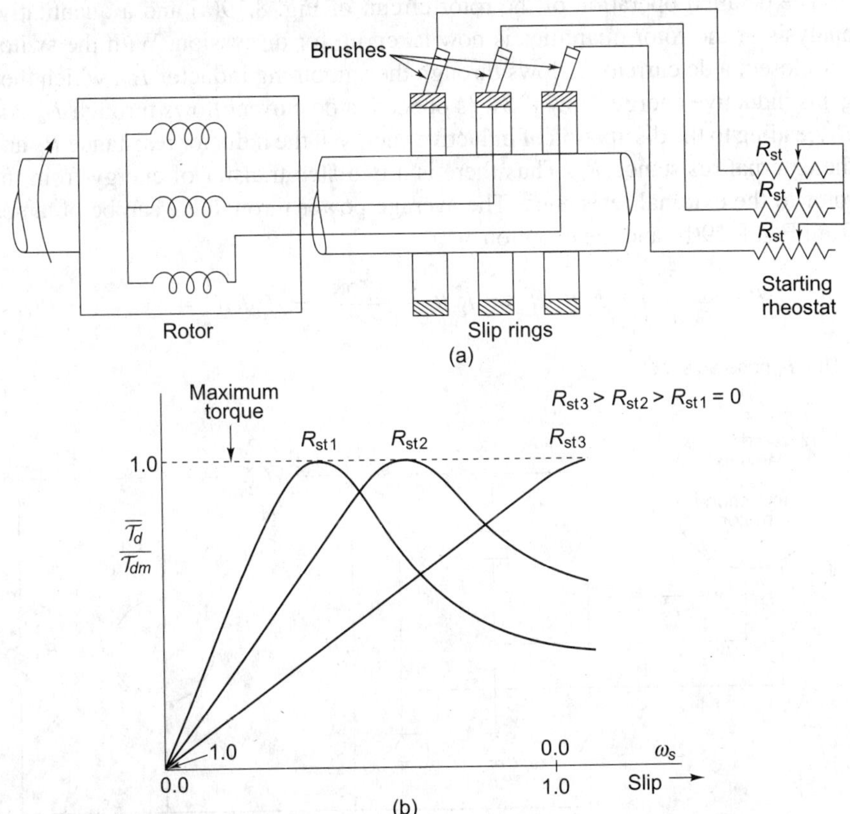

Fig. 8.18 Wound rotor induction motor with a variable three-phase rheostat: (a) schematic diagram, (b) normalized torque vs slip characteristics

torque versus slip characteristics. The increase in the rotor resistance does not affect the value of the maximum torque but increases the slip at maximum torque (s_m) and also the starting torque T_s. These facts are confirmed by Eqn (8.30b) and also Figs 8.18(b) and 8.19. Therefore there is considerable flexibility of operation in this method. This technique, however, suffers from the drawback that there is an increase in the rotor copper loss if only a higher starting torque is needed. A second disadvantage is that unbalanced currents flow through the rotor windings in the case of a mismatch amongst the resistors used for the three-phase rheostat. To overcome this mismatch, a diode rectifier, a smoothing inductor, a semiconductor switch (Sw), and an external resistor are connected in the rotor circuit as shown in Fig. 8.19(a). Figure 8.19(b) gives the same figure with the stator and rotor windings brought out and with the switch Sw replaced by a thyristor. The combination of the four additional elements gives the effect of a strictly balanced three-phase rheostat. Continuous variation of the rotor resistance can now be obtained by operating the thyristor as a dc chopper, to which the slip power is fed after rectification. The variation of the duty ratio of the chopper results in the variation of the effective resistance in the rotor circuit, the range so obtained being $0 \leq R_{\text{ext}} \leq R_e$. Figure 8.19(c) depicts the resulting torque–slip characteristics.

The detailed operation of the rotor circuit of Fig. 8.19(a) and a quantitative analysis of the rotor quantities is now taken up for discussion. With the switch Sw closed, a dc current I_d flows through the smoothing inductor L_d, which then gains inductive energy. Now if Sw is opened, a dc current flows through L_d and R_e, leading to the dissipation of inductive energy at the inductor resistance R_d and the external resistance R_e. Thus there is a two-step transfer of energy from the rotor to the external resistance. The average power through R_e can be obtained from Fig. 8.19(b) and the equation

$$P_{\text{av}} = \frac{1}{\tau}\int_{\tau_{\text{ON}}}^{\tau} I_d^2 R_e dt = I_d^2 R_e \frac{\tau - \tau_{\text{ON}}}{\tau} = I_d^2 R_e (1 - \delta) \tag{8.74}$$

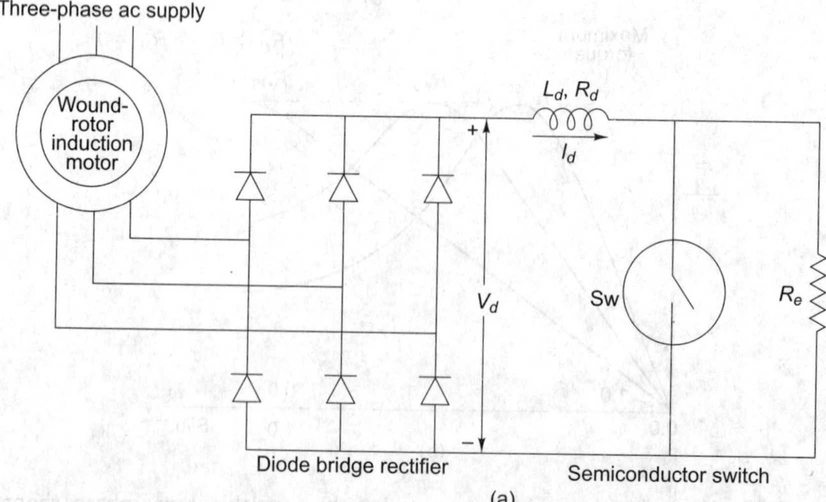

Fig. 8.19(a)

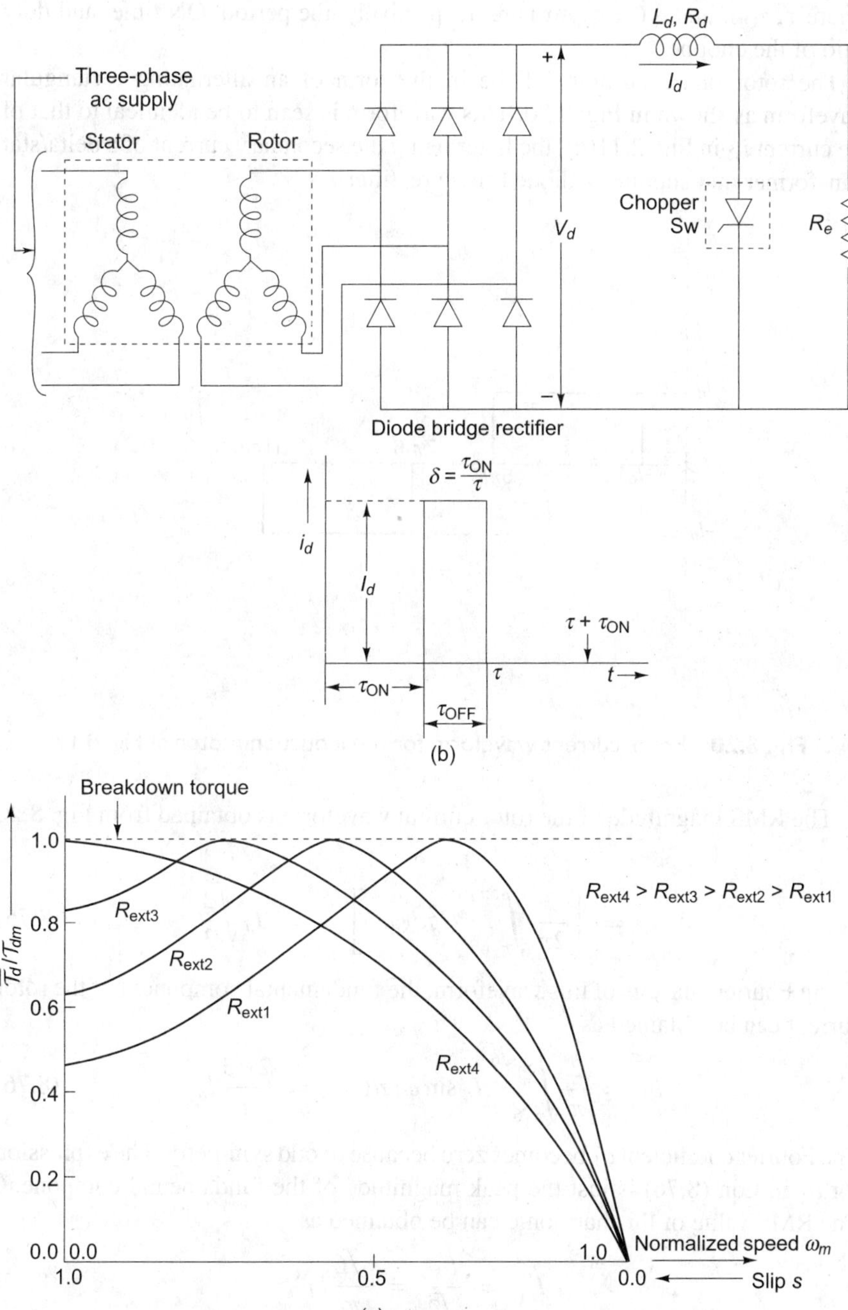

Fig. 8.19 Circuit diagram of a wound rotor induction motor with rotor resistance control with (a) a semiconductor switch and (b) a chopper (the duration of current I_d through L_d and R_e is also shown). (c) Speed–torque curves with R_{ext} as a parameter.

where τ, τ_{ON}, and δ (= τ_{ON}/τ) are, respectively, the period, ON time, and duty ratio of the chopper.

The rotor phase current will be in the form of an alternating rectangular waveform as shown in Fig. 8.20. This waveform is seen to be identical to that of the current i_R in Fig. 2.11(d), the latter being the secondary current of a delta/star transformer that supplies a diode bridge rectifier.

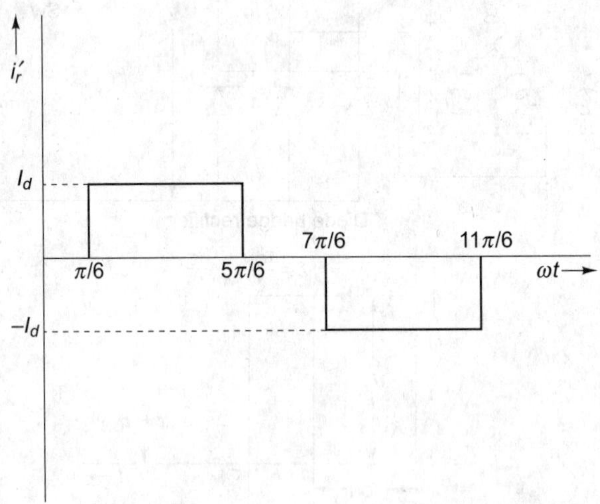

Fig. 8.20 Rotor current waveform for the induction motor of Fig. 8.19

The RMS magnitude of the rotor current waveform is obtained from Fig. 8.20 as

$$I_r' = \left[\frac{2}{2\pi}\int_{\pi/6}^{5\pi/6} I_d^2 d(\omega t)\right]^{1/2} = I_d\sqrt{\frac{2}{3}} \tag{8.75}$$

Using Fourier analysis of this waveform, the fundamental component of the rotor current can be obtained as

$$b_1 = \frac{2}{\pi}\int_{\pi/6}^{5\pi/6} I_d \sin\omega t\, d(\omega t) = \frac{2\sqrt{3}}{\pi} I_d \tag{8.76}$$

The Fourier coefficient a_1 becomes zero because of odd symmetry. The expression for b_1 in Eqn (8.76) is just the peak magnitude of the fundamental component. The RMS value of this harmonic can be obtained as

$$I_{r1}' = \frac{b_1}{\sqrt{2}} = \frac{\sqrt{6}}{\pi} I_d \tag{8.77}$$

Using Eqn (8.75) I_{r1}' can be expressed as

$$I_{r1}' = \frac{3}{\pi} I_r' \tag{8.78}$$

The average total resistance at the output of the diode bridge is

$$R_{out} = R_d + R_e(1 - \delta) \tag{8.79}$$

The power consumed per phase by this equivalent resistance (R_{out}) is

$$P_{out} = \frac{1}{3} I_d^2 R_{out} \tag{8.80}$$

Using Eqn (8.75), Eqn (8.80) can be written as

$$P_{out} = 0.5(I_r')^2 R_{out} = (I_r')^2(0.5R_{out}) \tag{8.81}$$

Equation (8.81) can be interpreted as power dissipation in a resistance of $0.5R_{out}$ ohms. The total copper loss in the rotor can be written, using Eqn (8.78), as

$$P_{loss} = 3(I_r')^2(R_r' + 0.5R_{out}) = \frac{\pi^2}{3}(I_{r1}')^2(R_r' + 0.5R_{out})$$

The fundamental slip power is

$$sP_{g1} = 3(I_{r1}')^2(R_r' + 0.5R_{out}) \tag{8.82}$$

or

$$P_{g1} = 3\frac{(I_{r1}')^2}{s}(R_r' + 0.5R_{out}) \tag{8.83}$$

The total mechanical power developed in the motor becomes

$$P_m = (1-s)P_{g1} = 3(I_{r1}')^2(R_r' + 0.5R_{out})\frac{(1-s)}{s} \tag{8.84}$$

The total power output per phase ($E_r' I_{r1}' \cos\phi$) now becomes the sum of the rotor copper loss per phase (which is one-third of the total copper loss) and one-third of the total mechanical power P_m. Thus,

$$\begin{aligned} E_r' I_{r1}' \cos\phi &= \frac{\pi^2}{9}(I_{r1}')^2(R_r' + 0.5R_{out}) + (I_{r1}')^2(R_r' + 0.5R_{out})\frac{(1-s)}{s} \\ &= \left[\left(\frac{\pi^2}{9} - 1\right)(R_r' + 0.5R_{out}) + \frac{(R_r' + 0.5R_{out})}{s}\right](I_{r1}')^2 \end{aligned} \tag{8.85}$$

If N_{sr} is the stator-to-rotor turn ratio, the right-hand quantities can be expressed as $I_R^2(R_1 + R_2/s)$, where

$$R_1 = N_{sr}^2\left(\frac{\pi^2}{9} - 1\right)(R_r' + 0.5R_{out})$$

and

$$R_2 = N_{sr}^2(R_r' + 0.5R_{out})$$

Also,

$$I_R^2 = \frac{1}{N_{sr}^2}(I_{r1}')^2$$

X_R can similarly be defined as

$$X_R = (N_{sr})^2 X_r'$$

where X_r' is the rotor resistance. Thus,

$$E_r I_R \cos\phi = I_R^2\left(R_1 + \frac{R_2}{s}\right) \tag{8.86}$$

where E_r and I_R are the rotor voltage and current, respectively, referred to the stator side. The equivalent circuit can now be constructed as in Fig. 8.21(a). Figure 8.21(b) shows its modified version, with X_m shifted to the left of the stator impedance.

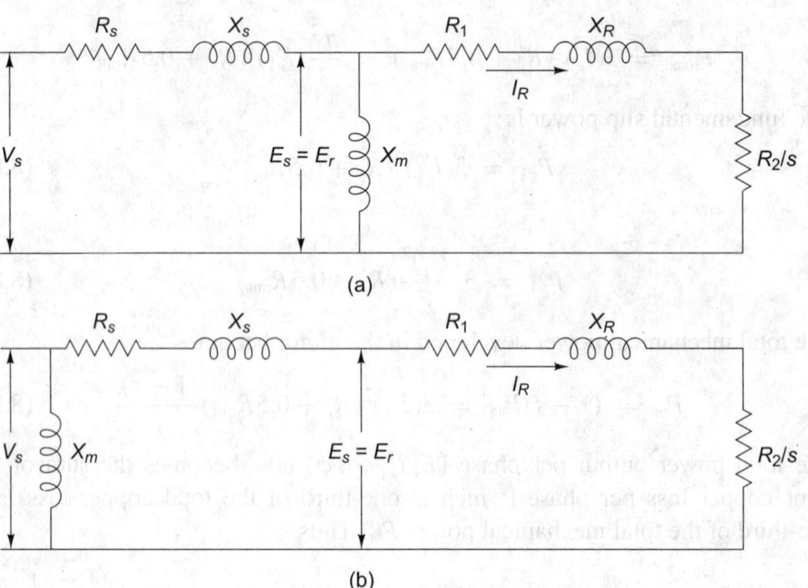

Fig. 8.21 Wound rotor induction motor: (a) equivalent circuit, (b) circuit with magnetizing reactance shifted to the input side

Now the phasor of the rotor current can be written as

$$\overline{I}_R = \frac{\overline{V}_s}{(R_s + R_1 + R_2/s) + j(X_s + X_R)} \tag{8.87}$$

The developed torque becomes

$$\overline{T}_d = \frac{P_m}{\omega_m} = \frac{P_{g1}(1-s)}{\omega_m} = \frac{P_{g1}(1-s)}{\omega_s(1-s)} = \frac{P_{g1}}{\omega_s} \tag{8.88}$$

$$= \frac{3}{\omega_s}|\overline{I}_R|^2\left(\frac{R_2}{s}\right) N_m \tag{8.89}$$

where $|\overline{I}_R|$ and R_2 are as defined above. Substituting for $\overline{I}_R$ from Eqn (8.87) gives

$$\overline{T}_d = \frac{3}{\omega_s}\frac{|\overline{V}_s|^2(R_2/s)}{(R_s + R_1 + R_2/s)^2 + (X_s + X_R)^2} \tag{8.90}$$

Equation (8.90) shows that the developed torque can be computed if the values of the rotor quantities, namely, the slip s, external resistance R_e, resistance R_d of the inductor, and duty ratio δ $(= \tau_{ON}/\tau)$, are available.

8.3.2 Slip Energy Recovery Scheme

The method of rotor resistance control suffers from the drawback that a large power loss occurs in the additional resistor. This method is therefore applicable to the cases where the extra loss in the resistor is acceptable. It can also be used for starting induction motors which operate as fixed-speed machines. The slip energy recovery scheme presented here is free from additional losses because it involves the return of electrical energy to the source. Its principle is to insert a variable back emf into the rotor circuit in such a way that the resultant energy can be fed back into the ac mains network that feeds the stator of the induction motor. The overall effect of such a system is efficient control of the motor speed.

The above-mentioned principle was initially implemented with a motor generator set, so that energy could be recovered and fed back to the dc motor either (a) by having its rotor on the same shaft as the induction motor (the Kramer system) or (b) by operating it as a prime mover for another ac generator from which ac power is returned to the supply network (the Scherbius scheme). These methods are illustrated in Figs 8.22(a) and (b), respectively. Due to the high cost and relatively high power losses in such installations, the principle of slip energy recovery is now implemented by means of semiconductor-controlled converters.

The modern Scherbius system is applied to a wound rotor induction motor with slip rings which provide the connection to the rotor circuit. The motor and static converters are usually designed to suit specific applications. The power capacity of drives which are controlled by Scherbius systems may range from hundreds of kilowatts to a few megawatts. The cost of these systems is therefore justified only for the control of high-capacity motors.

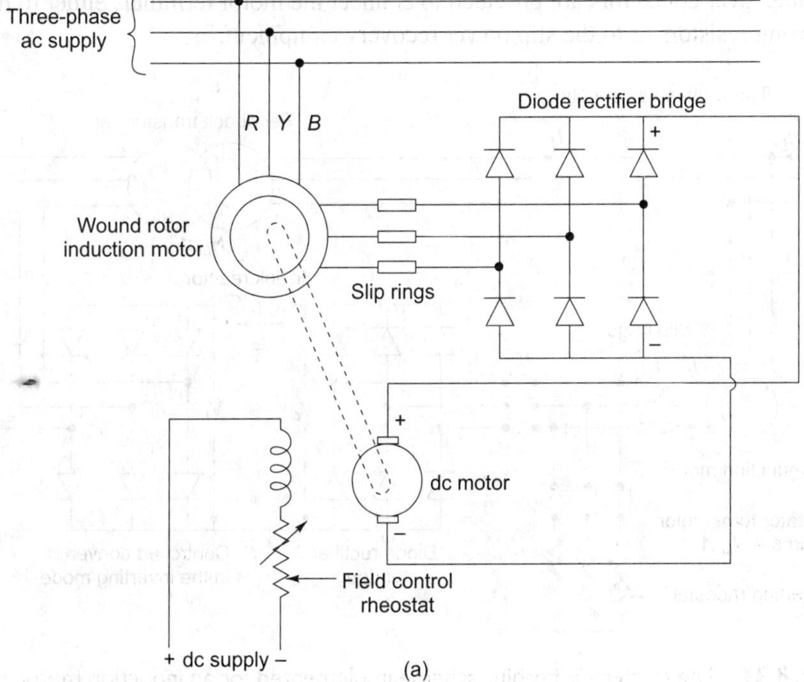

Fig. 8.22(a)

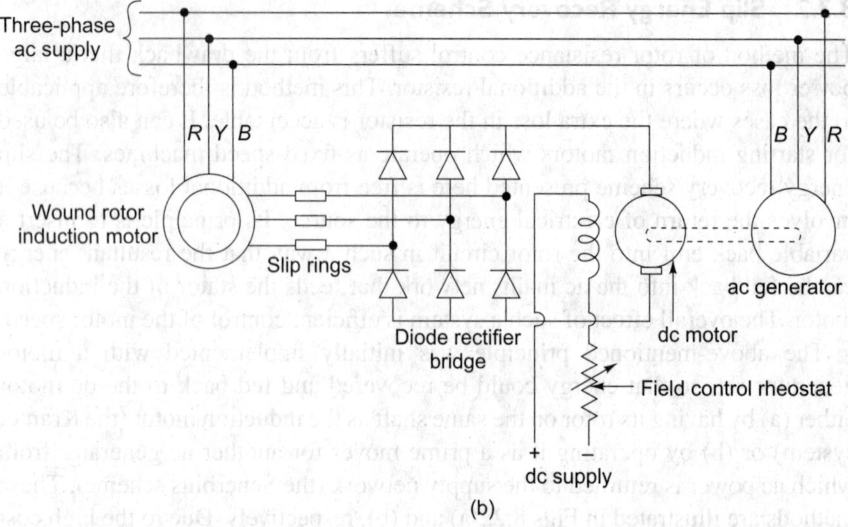

Fig. 8.22 Speed control of a wound rotor induction motor: (a) Kramer scheme, (b) Scherbius scheme

8.3.2.1 Modern Scherbius systems Figure 8.23 shows the basic scheme for a Scherbius system (Finney 1988), with which speed control is possible only in the subsynchronous range. The stator of the induction motor, which is directly connected to the supply network, produces a rotating field in the air gap. The starting resistors help in bringing the motor up to the specified speed range. Changeover contactors are provided to connect the motor terminals either to the starting resistors or to the slip power recovery equipment.

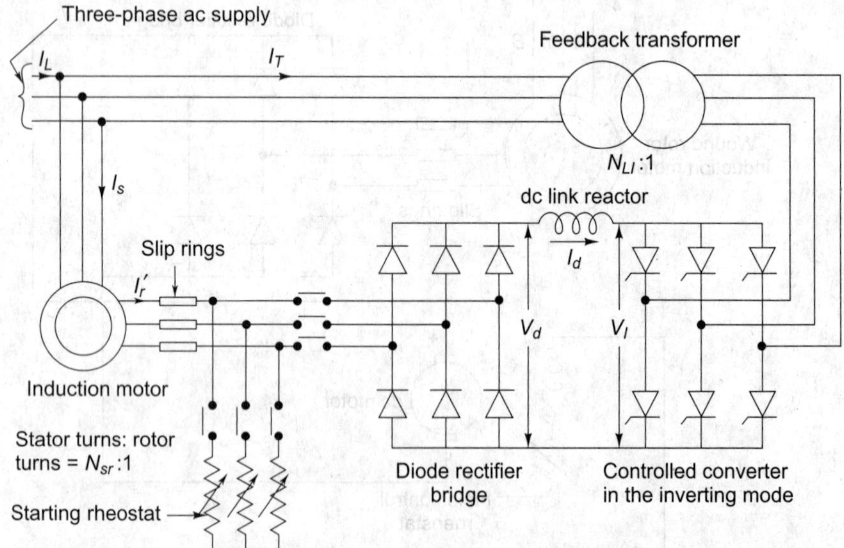

Fig. 8.23 The modern Scherbius scheme implemented for an induction motor

When slip recovery control is effected, the rotor voltage is rectified by a diode bridge whose output is applied against a dc link voltage. The magnitude of the link voltage is adjusted by changing the firing angle of a line commutated thyristor inverter which feeds the recovered power back to the mains network through a step-up transformer.

The principle of operation of the Scherbius system is as follows. The rotating field of the stator induces a voltage in the rotor, resulting in the flow of rotor current. This rotor current is rectified by the diode bridge to produce direct current which has to pass through the thyristor inverter to complete the dc circuit. While flowing through the inverter, it has to overcome the reverse voltage V_I set up in the inverter; this feature enables the inverter to extract energy from the rotor. In the steady state the rectified rotor voltage equals the reverse voltage of the inverter, with the assumption that the resistance of the link inductor is negligible. The circulating rotor current produces the necessary load torque as per the usual induction motor action. When a voltage is generated in the rotor, it has to slow down with respect to the stator rotating field. The motor speed can be controlled by varying the firing angle of the inverter thyristors as explained below. The input voltage of the diode bridge is varied in order to maintain a balance between V_d and V_I, and in this process the motor speed gets changed. In most systems, a thyristor rectifier which operates in the inverting mode (i.e., with $\alpha > 90°$) serves as the inverter.

The following features are important for the successful operation of the system, and their approximate fulfilment also facilitates the analysis of the circuit. A requirement for the correct operation of the inverter is that the smoothing reactor in the dc link should be sufficiently large so that the link current is fairly smooth and continuous. This implies that the current in the rotor will have a quasi-square wave shape.

The rotor frequency, which depends on the difference of the speeds of the stator rotating field and the rotor, is relatively low and may even become zero. The rotor voltage is also proportional to this speed difference and hence has a small magnitude. The speed of the induction motor is controlled by varying the dc link voltage. This is done by controlling the firing angle of the inverter. The rotor voltage has to match the dc link voltage and in this process the rotor slows down or speeds up, as is appropriate. The feedback transformer is of the step-down type and helps to match the supply voltage with the constant dc link voltage, which in turn depends upon the speed range over which the slip recovery system operates. As will be evident from the analysis, the torque developed by the motor is approximately proportional to the rotor current and, hence, the dc link current. Operation just below the synchronous speed corresponds to high converter currents at very low voltages, and operation at minimum speed corresponds to high voltages and low currents. This is with the understanding that torque reduces with speed.

The commutation overlaps, both in the diode bridge as well as in the inverter, are neglected; the losses in both the bridges are also ignored. As shown in Fig. 8.24, the rotor phase current waveform is symmetrical with respect to the slip frequency rotor phase voltage waveform. Hence the fundamental rotor current will be in phase with the phase voltage. Harmonics in the rotor current cause only a small harmonic current to flow in the stator. Therefore the machine-induced emf and, hence, flux can be assumed to be sinusoidal. It follows that the torque is produced by the fundamental component of the rotor current, whereas the harmonic currents produce only pulsating torques. For a given inverter firing angle α_i, the voltage V_I on the dc side of the inverter, the dc output voltage V_d of the diode bridge, and the input voltage of the diode bridge have fixed values. This shows that the variation of α_i changes all these voltages, which in turn causes a change in the motor speed. The power-flow diagram for the system is given in Fig. 8.25.

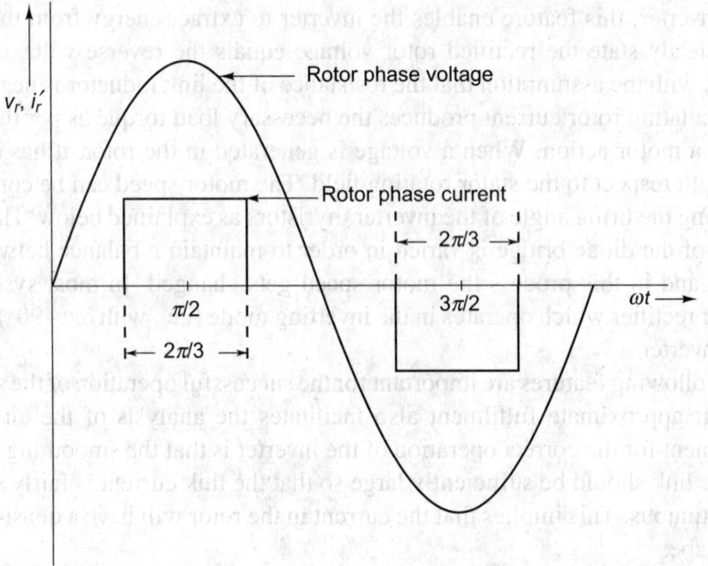

Fig. 8.24 Slip frequency rotor phase voltage and current waveforms

Analysis The output of the diode bridge rectifier is

$$V_d = \frac{3\sqrt{6}s\,V_s}{\pi N_{sr}} \tag{8.91}$$

where N_{sr} is the stator-to-rotor turn ratio and V_s is the input supply voltage. The dc side voltage of the inverter is

$$V_I = \frac{3\sqrt{6}V_s \cos\alpha_i}{\pi N_{LI}} \tag{8.92}$$

when N_{LI} is the ratio between the transformer line side voltage and the inverter ac-side voltage. If the resistance of the inductor L_d is negligibly small, the following relationship holds good:

$$V_d + V_I = 0 \tag{8.93}$$

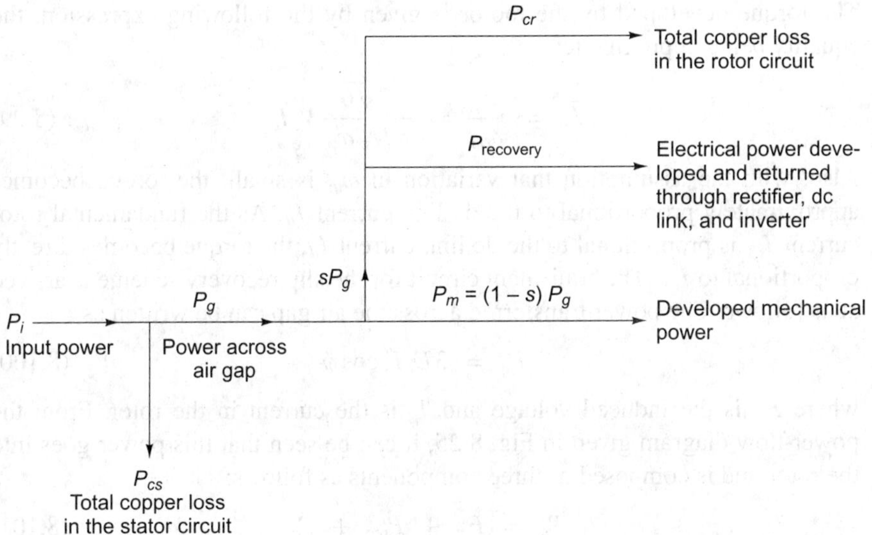

Fig. 8.25 Power-flow diagram of an induction motor with the modern Scherbius scheme

That is,

$$\frac{3\sqrt{6}sV_s}{\pi N_{sr}} + \frac{3\sqrt{6}}{\pi}\frac{V_s}{N_{LI}}\cos\alpha_i = 0 \tag{8.94}$$

This gives the slip as

$$s = -\frac{N_{sr}}{N_{LI}}\cos\alpha_i \tag{8.95}$$

Since the firing angle α_i of the inverter is greater than or equal to 90°, s is always positive and can be expressed as

$$s = \frac{N_{sr}}{N_{LI}}|\cos\alpha_i| \tag{8.96}$$

If N_{LI} is chosen to be equal to N_{sr}, then the slip will vary from 0 to 1, that is, the speed will vary from synchronous speed to zero speed (i.e., the standstill condition). Thus the motor speed can be controlled in the subsynchronous region simply by controlling the inverter firing angle. An expression for the torque is obtained as follows. Since the slip power flows through the dc link, it can be written as

$$sp_g = V_d I_d \tag{8.97}$$

This gives

$$p_g = \frac{V_d I_d}{s} \tag{8.98}$$

The torque developed by the motor is given by the following expression, the equality being approximate:

$$\overline{T}_d = \frac{p_g}{\omega_m} = \frac{3\sqrt{6}}{\pi N_{sr}\omega_m} V_s I_d \tag{8.99}$$

Also, with the assumption that variation in ω_m is small, the torque becomes approximately proportional to the dc link current I_d. As the fundamental rotor current I_{r1} is proportional to the dc link current I_d, the torque becomes directly proportional to I_{r1}. The equivalent circuit for the slip recovery scheme is arrived at as follows. The power transferred across the air gap can be written as

$$P_g = 3E_r I'_r \cos\phi \tag{8.100}$$

where E_r is the induced voltage and I'_r is the current in the rotor. From the power-flow diagram given in Fig. 8.25, it can be seen that this power goes into the rotor and is composed of three components as follows:

$$P_g = P_m + P_{\text{rec}} + P_{cr} \tag{8.101}$$

where P_m is the mechanical power at the shaft, P_{rec} is the power recovered and sent back to the ac supply, and P_{cr} is the total copper loss. The expressions for the three components are now derived. From Fig. 8.23, the power recovered can be written as

$$P_{\text{rec}} = -V_I I_d \tag{8.102}$$

The negative sign in Eqn (8.95) arises from Eqn (8.93), which shows that the polarity of the inverter output (V_I) is opposite to that of V_d. Similar to Eqn (8.77) derived under rotor resistance control, the RMS value of the fundamental component of the rotor current can be expressed as

$$I'_{r1} = \frac{\sqrt{6}I_d}{\pi} \tag{8.103}$$

This gives

$$I_d = \frac{\pi I'_{r1}}{\sqrt{6}} \tag{8.104}$$

Finally, the substitution of the expressions for I_d and V_I in Eqn (8.102) gives

$$P_{\text{rec}} = \frac{-3V_s I'_{r1}}{N_{LI}} \cos\alpha_i \tag{8.105}$$

P_{cr} denotes the total copper loss in the rotor circuit and consists of the loss in the rotor resistance R'_r and that in the link inductor resistance R_d. Thus,

$$P_{cr} = 3(I'_r)^2 R'_r + I_d^2 R_d \tag{8.106}$$

Similar to Eqn (8.75) derived under rotor resistance control, I'_r can be obtained as

$$I'_r = \sqrt{\frac{2}{3}} I_d$$

Thus, I_d is expressed as

$$I_d = \sqrt{\frac{3}{2}} I'_r \tag{8.107}$$

Using this expression for I_d, Eqn (8.106) can be written as

$$P_{cr} = 3(I'_r)^2 R'_r + \frac{3}{2}(I'_r)^2 R_d \tag{8.108}$$

$$= 3(I'_r)^2(R'_r + 0.5R_d) \tag{8.109}$$

Using Eqns (8.104) and (8.107), I'_r can be expressed as

$$I'_r = \frac{\pi}{3} I'_{r1} \tag{8.110}$$

Substitution for I'_r in Eqn (8.109) gives

$$P_{cr} = \frac{\pi^2}{3}(I'_{r1})^2(R'_r + 0.5R_d) \tag{8.111}$$

The total slip power with the fundamental component of the rotor current (denoted as sp_{g1}) is composed of the ohmic loss with the fundamental component I'_{r1} and the recovered power. Thus,

$$sp_{g1} = 3(I'_{r1})^2(R'_r + 0.5R_d) + P_{\text{rec}} \tag{8.112}$$

Substitution of the expression for P_{rec} given in Eqn (8.105) gives

$$sp_{g1} = 3(I'_{r1})^2(R'_r + 0.5R_d) - \frac{3V_s I'_{r1}}{N_{LI}} \cos\alpha_i \tag{8.113}$$

That is,

$$p_{g1} = \frac{3}{s}\left[(I'_{r1})^2(R'_r + 0.5R_d) - \frac{V_s I'_{r1}}{N_{LI}} \cos\alpha_i\right] \tag{8.114}$$

Noting that the mechanical power P_m is expressed as $P_m = P_{g1}(1 - s)$ and using Eqn (8.114) for P_{g1} gives P_m as

$$p_m = \left[3(I'_{r1})^2(R'_r + 0.5R_d) - \frac{3V_s I'_{r1}}{N_{LI}} \cos\alpha_i\right] \frac{(1-s)}{s} \tag{8.115}$$

As before [Eqn (8.85)], the power output per phase ($E'_r I'_{r1} \cos\phi$) is expressed as the sum of the rotor copper losses per phase and the mechanical power developed per phase, where E'_r and I'_{r1} are the rotor voltage and fundamental component of the rotor current referred to the stator, respectively. That is,

$$E'_r I'_{r1} \cos\phi = \frac{P_{cr}}{3} + \frac{p_m}{3} \tag{8.116}$$

Using the expressions for P_{cr} and p_m, respectively, from Eqns (8.111) and (8.115), Eqn (8.116) becomes

$$E'_r I'_{r1} \cos\phi = \frac{\pi^2}{9}(I'_{r1})^2(R'_r + 0.5R_d) + \left[(I'_{r1})^2(R'_r + 0.5R_d) - \frac{V_s I'_{r1}}{N_{LI}} \cos\alpha_i\right] \frac{(1-s)}{s} \tag{8.117}$$

Sorting out the terms containing $(R'_r + 0.5R_d)$ gives

$$E'_r I'_{r1} \cos\phi = \left(\frac{\pi^2}{9} - 1\right)(I'_{r1})^2(R'_r + 0.5R_d) + \frac{(I'_{r1})^2}{s}(R'_r + 0.5R_d)$$

$$-\frac{(1-s)}{s}\frac{V_s I'_{r1}}{N_{LI}}\cos\alpha_i \tag{8.118}$$

The first and second terms on the right-hand side of Eqn (8.118) can be considered, respectively, as the fixed and variable equivalent resistances on the rotor side. The inverter firing angle always being greater than 90°, the third term on the right-hand side is always positive and can be written as $+[(1-s)V_s I'_{r1}|\cos\alpha_i|]/sN_{LI}$. Defining V_{eq} as the voltage on the stator side corresponding to the expression $(1-s)V_s|\cos\alpha_i|/N_{LI}$, it is expressed as

$$V_{eq} = \frac{(1-s)V_s|\cos\alpha_i|}{N_{LI}}N_{sr}$$

Also, I_{r1} is related to I'_{r1} as

$$I_{r1} = \frac{I'_{r1}}{N_{sr}}$$

Using these relations, Eqn (8.118) can be expressed as a stator side quantity. Thus,

$$E_r I_{r1}\cos\phi = \left(\frac{\pi^2}{9} - 1\right)I_{r1}^2(R'_r + 0.5R_d)N_{sr}^2 + \frac{I_{r1}^2}{s}(R'_r + 0.5R_d)N_{sr}^2 + \frac{V_{eq}I_{r1}}{s} \tag{8.119}$$

where E_r is the rotor voltage referred to the stator side. Equation (8.119) is obtained with the assumption that the losses in the rectifier and inverter bridges are negligible. If R_1 and R_2 are denoted as the fixed and variable resistances on the stator side corresponding to the rotor side quantities $[(\pi^2/9) - 1](R'_r + 0.5R_d)$ and $(R'_r + 0.5R_d)/s$, respectively, they can be expressed as

$$R_1 = \left(\frac{\pi^2}{9} - 1\right)N_{sr}^2(R'_r + 0.5R_d)$$

and

$$R_2 = N_{sr}^2(R'_r + 0.5R_d)$$

Equation (8.119) can now be written in terms of R_1 and R_2 as

$$E_r I_{r1}\cos\phi = I_{r1}^2\left(R_1 + \frac{R_2}{s}\right) + \frac{V_{eq}I_{r1}}{s} \tag{8.120}$$

All the quantities in Eqn (8.120) now are referred to the stator side. The approximate equivalent circuit that represents this equation is given in Fig. 8.26, with $X_r = X'_r\,(N_{sr}^2)$, where X'_r is the rotor reactance. All the quantities in this figure are referred to the stator side.

The expression for the developed torque can be derived from the equivalent circuit as follows. The magnitude of the voltage source, given by V_{eq}/s, is a real

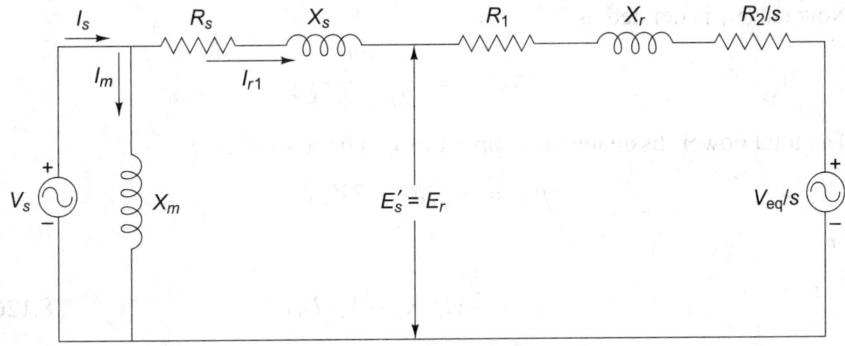

Fig. 8.26 Equivalent circuit of an induction motor with the modern Scherbius scheme

number because it is a dc voltage. It can therefore be treated like a voltage drop across the resistance. Also V_s^2 can be expressed as

$$V_s^2 = \left[\left(R_s + R_1 + \frac{R_2}{s}\right) I_{r1} + \frac{V_{eq}}{s}\right]^2 + I_{r1}^2 (X_s + X_r)^2 \qquad (8.121)$$

By defining

$$R = R_s + R_1 + \frac{R_2}{s}$$

$$X = X_s + X_r$$

and

$$V^* = \frac{V_{eq}}{s}$$

Eqn (8.121) can be written as

$$V_s^2 = (RI_{r1} + V^*)^2 + I_{r1}^2 X^2 \qquad (8.122)$$

Opening the bracket on the right-hand side and rearranging terms gives

$$V_s^2 = (X^2 + R^2) I_{r1}^2 + 2RV^* I_{r1} + (V^*)^2 \qquad (8.123)$$

Equation (8.123) can be expressed in the form of a quadratic equation in I_{r1} as

$$I_{r1}^2 + \frac{2RV^*}{R^2 + X^2} I_{r1} + \frac{(V^*)^2 - V_s^2}{R^2 + X^2} = 0 \qquad (8.124)$$

The two possible solutions of Eqn (8.124) are

$$\frac{1}{2}\left[-\frac{2RV^*}{R^2 + X^2} \pm \sqrt{\frac{4R^2(V^*)^2}{(R^2 + X^2)^2} - \frac{4[(V^*)^2 - V_s^2]}{R^2 + X^2}}\right]$$

I_{r1}, which the fundamental rotor current referred to the stator side, has to be positive. Hence

$$I_{r1} = -\frac{RV^*}{R^2 + X^2} + \sqrt{\frac{R^2(V^*)^2}{(R^2 + X^2)^2} - \frac{(V^*)^2 - V_s^2}{R^2 + X^2}} \qquad (8.125)$$

Now $\tan\phi_{r1}$ is defined as

$$\tan\phi_{r1} = \frac{XI_{r1}}{RI_{r1} + V^*}$$

The total power dissipated as copper loss can be written as

$$sp_g = 3I_{r1}^2R_2 + 3V_{eq}I_{r1}$$

or

$$p_g = \frac{3}{s}[I_{r1}^2R_2 + V_{eq}I_{r1}] \tag{8.126}$$

Now the mechanical power developed is given as

$$p_m = (1-s)p_g = 3[I_{r1}^2R_2 + V_{eq}I_{r1}]\frac{(1-s)}{s} \tag{8.127}$$

The developed torque is

$$\overline{T}_d = \frac{p_m}{\omega_m} = \frac{(1-s)p_g}{\omega_s(1-s)} = \frac{p_g}{\omega_s}$$

Using Eqn (8.126),

$$\overline{T}_d = \frac{3}{s\omega_s}[I_{r1}^2R_2 + V_{eq}I_{r1}] \tag{8.128}$$

where ω_s is the synchronous angular velocity in mechanical rad/s. The power factor of the drive is now arrived at from Figs 8.24 and 8.26.

The stator current vector is equal to the vectorial sum of the rotor current vector and the magnetizing current vector. Thus,

$$I_s\angle\phi_s = I_r\angle\phi_r + I_m\angle\phi_m \tag{8.129}$$

The line side current I_T of the feedback transformer can be expressed as

$$\frac{I_T}{I_r'}\frac{\angle\phi_T}{\angle\phi_r'} \approx \frac{1}{N_{LI}}$$

Also,

$$\frac{I_r'}{I_r}\frac{\angle\phi_r'}{\angle\phi_r} \approx N_{sr}$$

Hence,

$$\frac{I_T}{I_r}\frac{\angle\phi_T}{\angle\phi_r} \approx \frac{N_{sr}}{N_{LI}}$$

or

$$I_T\angle\phi_T \approx \frac{N_{sr}}{N_{LI}}I_r\angle\phi_r \tag{8.130}$$

Finally, the total line current can be written as

$$I_L\angle\phi_L = I_s\angle\phi_s + I_T\angle\phi_T \tag{8.131}$$

The drive power factor can now be expressed as

$$\text{Power factor} = \frac{V_s I_L \cos\phi_L}{V_s I_L} = \cos\phi_L \tag{8.132}$$

8.4 Synchronous Motors

This section is devoted to the theory behind different types of synchronous motors that are controlled by power electronic equipment. There are three types of synchronous machines, namely, the wound field, permanent magnet, and reluctance type. Here, only the wound field type of machines are dealt with because of their application in high and very high power drives.

The wound field synchronous motor can operate at unit power factor at full load. Hence, it has the benefit of operating with minimum stator current rating and can be supplied by an inverter of low volt-ampere rating. There are two subcategories of the wound field machine: (a) the cylindrical rotor machine, which has a round rotor, and (b) the salient pole machine, which has projecting rotor poles.

8.4.1 Cylindrical Rotor Machine

The cylindrical rotor type of synchronous machine has a uniform air gap between the slotted stator and rotor. The stator is constituted of iron laminations which are stacked together and the rotor is solid, with slots beneath its surface. The armature (or stator) has a distributed three-phase winding in its slots, whereas the rotor slots have a single winding, called the *field winding*. For the machine to operate as a synchronous motor, a balanced three-phase supply is given to its stator windings, and the rotor winding is provided with dc excitation. A balanced set of voltages, applied to the stator, gives rise to balanced three-phase armature currents and they, in turn, set up a flux wave in the air gap. This flux has an approximately sinusoidal distribution in space with constant amplitude. This wave rotates at a synchronous speed, and if the rotor were also to rotate at the synchronous speed, the magnetic fields of the stator and rotor would be stationary relative to one another. A steady electromagnetic torque is developed because the axis of the rotor magnetic field has a tendency to align with that of stator field. The speed N in rpm is given as $N = 60f/p$, where $f = 50$ Hz and p is the number of pole pairs.

The synchronous motor has the drawback that starting torque is not developed in it as in the case of the induction motor. An auxiliary machine like the induction motor or some other motor has to be connected to its shaft so as to bring up its speed to the synchronous speed.

The rotor field winding is excited with direct current supplied through slip rings and brushes by a controlled rectifier or a dc generator. The latter, known as an *exciter*, is usually mounted on the shaft of the synchronous machine. The field current of the synchronous machine can be varied by adjusting the firing angle of the semiconductor-controlled rectifier or the field rheostat of the exciter of a dc generator, as the case may be. If an ac generator together with a controlled rectifier can be placed on the same shaft as that of the synchronous machine, the rectifier plays the same part as that of an exciter. This type of set-up allows the rectified output to be connected directly to the field winding of the main motor and has the benefit of eliminating the slip rings and brushes.

8.4.1.1 The mechanism of the cylindrical rotor motor The currents in the three phases of the armature of a cylindrical rotor machine set up a synchronously rotating magnetic field in the air gap. With the rotor also rotating at the same speed, the armature mmf wave is stationary with respect to the rotor. The net effect is that a flux, which is the result of the field flux and an armature flux (caused by the armature current), is set up in the air gap. The effect of the armature flux on the resultant air gap flux is called the *armature reaction*. Figure 8.27 shows the phasor diagram of a synchronous generator in which the vectorial addition of the field flux ϕ_f and the armature flux ϕ_{arm} gives rise to ϕ_r, which causes the generation of the voltage E_r.

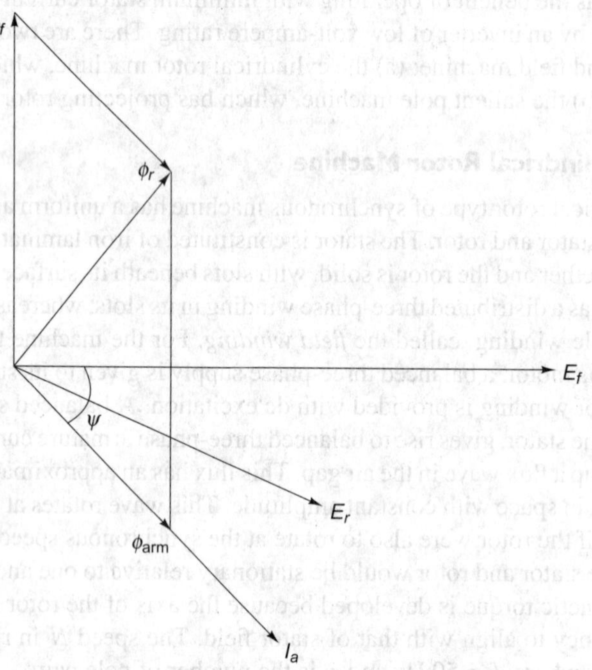

Fig. 8.27 Phasor diagram of fluxes in a cylindrical rotor synchronous motor

Figure 8.28(a) shows the equivalent circuit of one phase of the stator, where V is the applied voltage and E_r is the voltage generated as above. Also, E_f is the voltage due to the field current and corresponds to the back emf of the dc generator. Figure 8.28(b) shows the phasor diagram for the circuit of Fig. 8.28(a). ψ is the internal power factor angle between the armature current and the voltage E_f. The resultant voltage E_r and the applied voltage V as well as the flux phasors are also shown in this diagram. The armature current phasor I_a is drawn in phase opposition to the armature flux ϕ_{arm}. This is in variance with the standard notation shown in Fig. 8.27 for a synchronous generator and can be explained as follows.

When the synchronous machine operates as a motor, the ac supply voltage causes the armature current I_a to flow in opposition to the field voltage E_f. With generator convention, the phase displacement of I_a with respect to E_f exceeds 90° and hence the electrical power output becomes negative. This shows that the machine is running as a motor. However, it is more convenient to follow the motor convention by reversing the direction of the current phasor so that the motor power becomes positive. Figure 8.28(b) is drawn in conformity with this motor convention. The armature reaction is represented by the magnetizing reactance X_m, whereas the stator leakage flux is represented by X_l. The total reactance is denoted by X_s and hence the relation $X_s = X_m + X_l$. This figure is drawn with the assumption that the armature resistance is negligible.

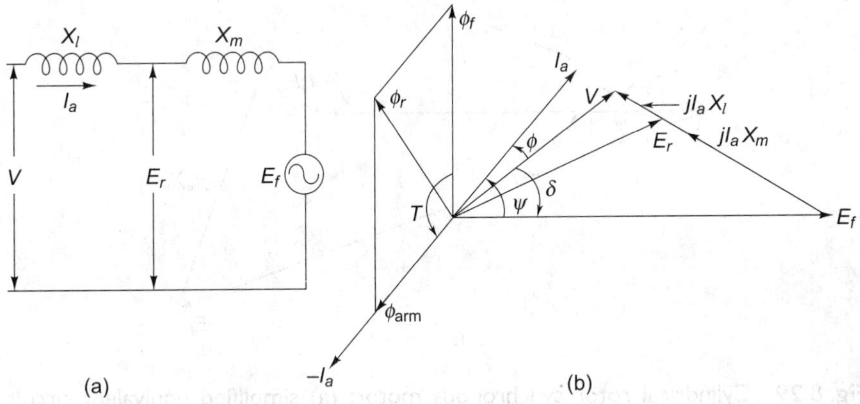

Fig. 8.28 Cylindrical rotor synchronous motor: (a) equivalent circuit, (b) complete phasor diagram

Figure 8.29(a) shows the circuit diagram in which the reactance X_s is as given above and R_a is the small resistance of the stator winding. The intermediate voltage E_r shown in Fig. 8.28(b) is eliminated, as it is not required in the following derivations. The corresponding phasor diagram is shown in Fig. 8.29(b), in which the flux vectors together with the vector E_r are eliminated. As before, ψ is the internal power factor angle; also, δ is termed the *load angle* and ϕ the *power factor angle*.

An expression for the shaft torque is now derived in terms of the load angle, applied voltage V, and field voltage E_f. The following relationships can be written with the help of the phasor diagram:

$$E_f \cos\delta = V - I_a R_a \cos\phi - I_a X_s \sin\phi \tag{8.133}$$

and

$$E_f \sin\delta = -I_a X_s \cos\phi + I_a R_a \sin\phi \tag{8.134}$$

These are now rewritten to bring all terms containing ϕ to the left-hand side. Thus,

$$I_a R_a \cos\phi + I_a X_s \sin\phi = V - E_f \cos\delta \tag{8.135}$$

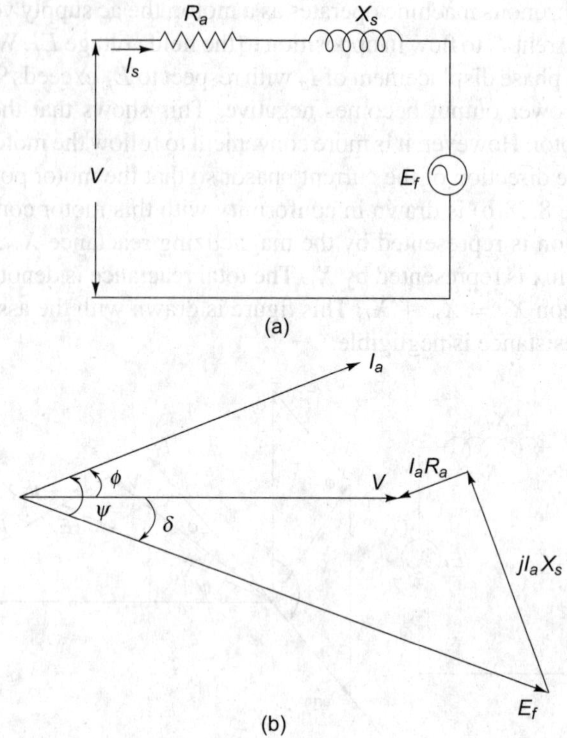

Fig. 8.29 Cylindrical rotor synchronous motor: (a) simplified equivalent circuit, (b) phasor diagram

and

$$I_a X_s \cos\phi - I_a R_a \sin\phi = -E_f \sin\delta \tag{8.136}$$

Equations (8.135) and (8.136) are solved to give $I_a \cos\phi$ as

$$I_a \cos\phi = \frac{VR_a - E_f R_a \cos\delta - X_s E_f \sin\delta}{R_a^2 + X_s^2} \tag{8.137}$$

The total electrical power input to the synchronous motor can be written as

$$P_i = 3VI_a \cos\phi \tag{8.138}$$

Substitution of the expression for $I_a \cos\phi$ from Eqn (8.137) in Eqn (8.138) now gives

$$P_i = \frac{3(V^2 R_a - VE_f R_a \cos\delta - VE_f X_s \sin\delta)}{R_a^2 + X_s^2} \tag{8.139}$$

If the term R_a, which is very small compared to X_s, is neglected, Eqn (8.139) now gets simplified as

$$P_i = \frac{-3VE_f \sin\delta}{X_s} \tag{8.140}$$

Since the field voltage E_f lags the applied voltage V_a, the load angle is negative. Therefore the right-hand side of Eqn (8.140) becomes negative. The electrical power provided at the motor input terminals is converted into mechanical power at the shaft terminals. If the copper losses in the stator are neglected (that is, with $R_a = 0$), the torque in N m developed by the motor can be obtained by dividing the electrical power by the synchronous angular velocity ω_s in mechanical rad/s, which in turn is equal to ω/p, where ω is the synchronous speed in electrical rad/s of the input supply and p is the number of pole pairs. The developed torque $\overline{T}_d$ can thus be written as

$$\overline{T}_d = \frac{P_i}{\omega_s} = \frac{P_i}{\omega/p} \qquad (8.141)$$

Using the expression in Eqn (8.140) for the power P_i, the expression for torque becomes

$$\overline{T}_d = \frac{-3p}{\omega}\frac{VE_f}{X_s}\sin\delta \qquad (8.142)$$

This equation implies that for a fixed field voltage, the developed torque is a function of $\sin\delta$. The maximum torque, called the *pull out torque*, occurs at $\delta = 90°$ and is given as

$$\overline{T}_{d(\max)} = \left[\frac{-3p}{\omega}\frac{VE_f}{X_s}\right] \qquad (8.143)$$

Figure 8.30 gives the plot of $\overline{T}_d$ versus δ with the regions of stability for the motoring and generating modes of operation marked therein. It is evident that the motor becomes unstable beyond the angles of pull out torque; these occur for

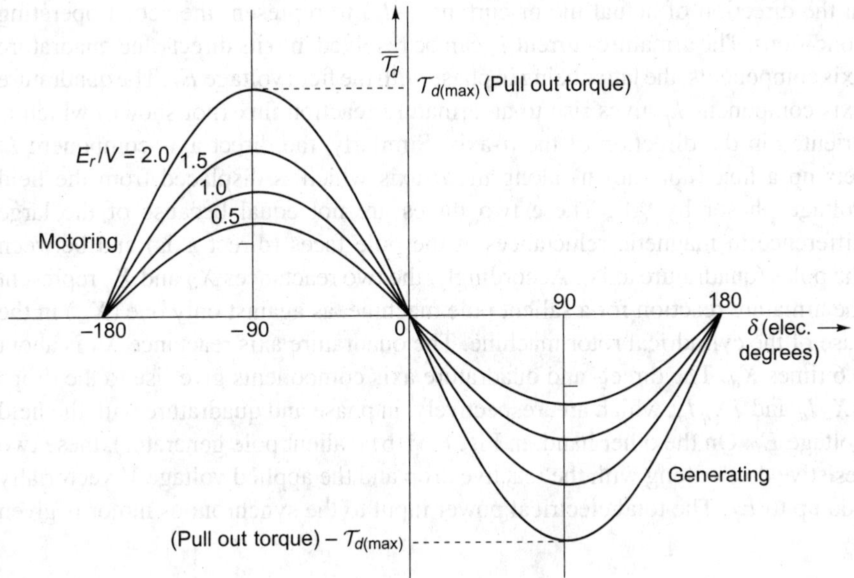

Fig. 8.30 Steady-state load angle curves of cylindrical rotor synchronous motor

$\delta = -90°$ in the motoring mode and $\delta = 90°$ in the generating mode. The region of stability can be extended by employing a semiconductor-controlled converter to generate the firing signals of the inverter, which, as stated in the beginning of this section, is the source for the stator. This converter in turn gets signals from a shaft position sensor. Chapter 9 gives a detailed discussion of such a converter.

8.4.2 Salient Pole Machine

The salient pole synchronous machine has a rotor with protruding poles, causing the air gap flux to be non-uniform. In a cylindrical rotor machine, the armature flux due to the armature ampere-turns depends on the electrical angle between the phase current I_a and the field voltage E_f. On the other hand, in a salient pole machine, this flux is a function of the displacement of the mmf wave from the salient poles. It follows that the magnetizing reactance X_m and the synchronous reactance given by the equation $X_s = X_m + X_l$ depend on the mmf orientation relative to the field poles. The following derivation is based on the two-axis theory, which postulates that the effects on the direct (or polar) axis are separate from those on the quadrature (or interpolar) axis.

The phasor diagram of a salient pole synchronous motor having negligible resistance and lagging current is shown in Fig. 8.31(a). The phasor diagram of a salient pole synchronous generator with a lagging current is also drawn in Fig. 8.31(b) for comparison. It is seen from the former figure that the actual armature current of the synchronous motor is nearly in phase opposition to the excitation or field voltage E_f. Using generator convention, the angle between E_f and I_a is seen to be greater than 90° and the electrical output is negative. Similar to the cylindrical rotor case, the motor current I_a in Fig. 8.31(a) has been reversed for convenience. However, the armature reaction flux ϕ_{arm} is drawn in the direction of actual motor current $(-I_a)$ to represent the actual operating conditions. The armature current I_a can be resolved into its direct- and quadrature axis components, the latter being in phase with the field voltage E_f. The quadrature axis component I_q gives rise to an armature reaction flux (not shown) which is oriented in the direction of the q-axis. Similarly, the direct-axis component I_d sets up a flux (not shown) along the d-axis which is displaced from the field voltage phasor by 90°. These two fluxes are not equal because of the large difference in magnetic reluctances at the pole faces (direct axis) and between the poles (quadrature axis). Accordingly, the two reactances X_d and X_q represent the armature reaction for a salient pole machine, as against only one (X_m) in the case of the cylindrical rotor machine. The quadrature axis reactance X_q is about 0.6 times X_d. The direct- and quadrature axis components give rise to the drops jX_dI_d and jX_qI_q, which are, respectively, in phase and quadrature with the field voltage E_f. On the other hand, in Fig. 8.31(b) (salient pole generator), these two resistive drops along with the reactive drop and the applied voltage V vectorially add up to E_f. The total electrical power input to the synchronous motor is given by

$$P_i = 3VI_a \cos\phi \tag{8.144}$$

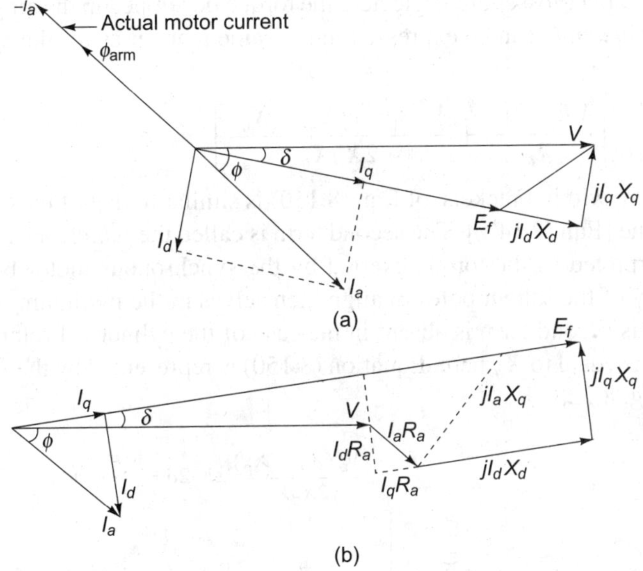

Fig. 8.31 Phasor diagrams of (a) a salient pole synchronous motor and (b) a salient pole synchronous generator

where V is the phase-to-neutral voltage and I_a is the phase current for, say, a star-connected three-phase armature for the motor. I_a is related to I_d and I_q as

$$I_a \cos(180° - \phi) = I_q \cos\delta - I_d \sin\delta$$

or

$$I_a \cos\phi = -I_q \cos\delta + I_d \sin\delta \tag{8.145}$$

Substituting this expression for $I_a \cos\phi$ in Eqn (8.144) gives

$$p_i = 3V(-I_q \cos\delta + I_d \sin\delta) \tag{8.146}$$

The following relations are also obtained from Fig. 8.31(a):

$$V\cos\delta - E_f = I_d X_d \tag{8.147}$$

and

$$V\sin\delta = I_q X_q \tag{8.148}$$

Equations (8.147) and (8.148) can be solved for I_d and I_q to give

$$I_d = \frac{V\cos\delta - E_f}{X_d} \quad \text{and} \quad I_q = \frac{V\sin\delta}{X_q}$$

By substituting these expressions for I_d and I_q in Eqn (8.146), the expression for the power input is obtained as

$$p_i = -3\left[\frac{VE_f\sin\delta}{X_d} + \frac{V^2\sin 2\delta(X_d - X_q)}{2X_dX_q}\right] \tag{8.149}$$

If the stator copper losses are neglected, the torque developed in the air gap of the *salient pole machine* can be expressed in the same manner as for the cylindrical rotor. Thus,

$$\overline{T}_d = -\frac{3p_i}{\omega}\left[\frac{VE_f \sin\delta}{X_d} + \frac{V^2 \sin 2\delta(X_d - X_q)}{2X_d X_q}\right] \tag{8.150}$$

The first term, within brackets, of Eqn (8.150) is similar to that of the cylindrical rotor machine [Eqn (8.142)]. The second term is called the *reluctance torque* and can be interpreted as the torque exerted by the synchronous motor because of the tendency of the salient poles to align themselves in the minimum reluctance position. This second term is absent in the case of the cylindrical rotor machine because X_d is equal to X_q here. Equation (8.150) is represented by the $\overline{T}_d$-δ curve shown in Fig. 8.32.

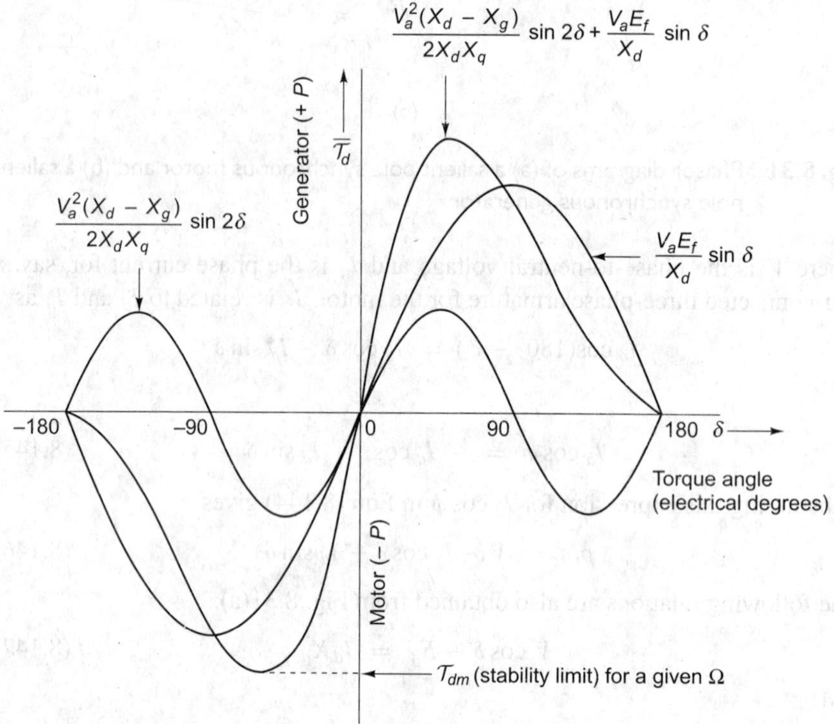

Fig. 8.32 T_d vs δ curve of a salient pole synchronous motor and its components

8.4.3 Speed and Torque Control

The speed of a synchronous motor is related to the input frequency as follows:

$$N = \frac{60f}{p} \tag{8.151}$$

where p is the number of pole pairs and N is the speed in rpm. This suggests that one method for the precise control of speed could be to use a crystal-controlled oscillator as an adjustable frequency reference. Such an oscillator has the advantages of stability and freedom from drift. Though using such a reference has the drawback of giving rise to considerable harmonic content, the resulting heating effects can be neglected. However, if the source is a six-step VSI, irregular motion or cogging can occur at low speeds. For motors that operate under such low-speed conditions it is preferable to employ sine wave PWM inverters.

The torque expressions given in Eqns (8.142) and (8.150), respectively, for the cylindrical and salient pole motors can be used for adjustable frequency operation of these motors. Considering the equation

$$|\mathcal{T}_{dm}| = \left|\frac{3p}{\omega}\frac{VE_f}{X_s}\right| \tag{8.152}$$

constant torque operation can be obtained as follows. With constant field excitation, both E_f and X_s are proportional to frequency. Hence the ratio E_f/X_s is independent of frequency. In addition, for constant volt/hertz or V/ω operation, the motor can be operated with constant maximum torque up to the base speed.

It must be kept in mind that Eqns (8.142) and (8.150) are derived under the assumption that the armature resistance can be neglected. Under high-speed conditions this assumption does not lead to serious errors, but at low speeds the armature resistance R_a becomes significant compared to X_s. Under constant V/ω control, this causes the available torque to reduce.

8.4.4 Open- and Closed-loop Control Schemes

The preceding discussion provides the basis for open- and closed-loop speed control of synchronous motors. Figure 8.33(a) shows an open-loop scheme in which a six-step inverter is employed for multiple-motor control. Figure 8.33(b) shows another open-loop scheme that employs a cycloconverter. Constant V/ω operation can also be implemented with the scheme shown in Fig. 8.33(a), but

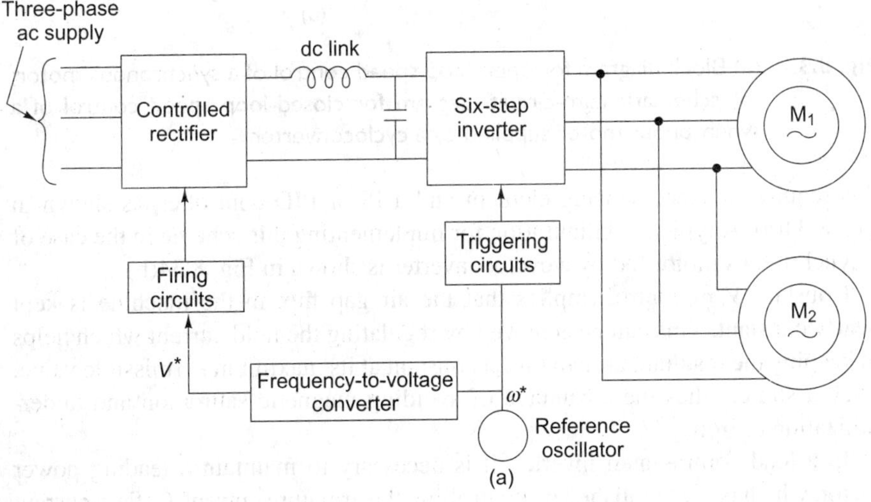

Fig. 8.33(a)

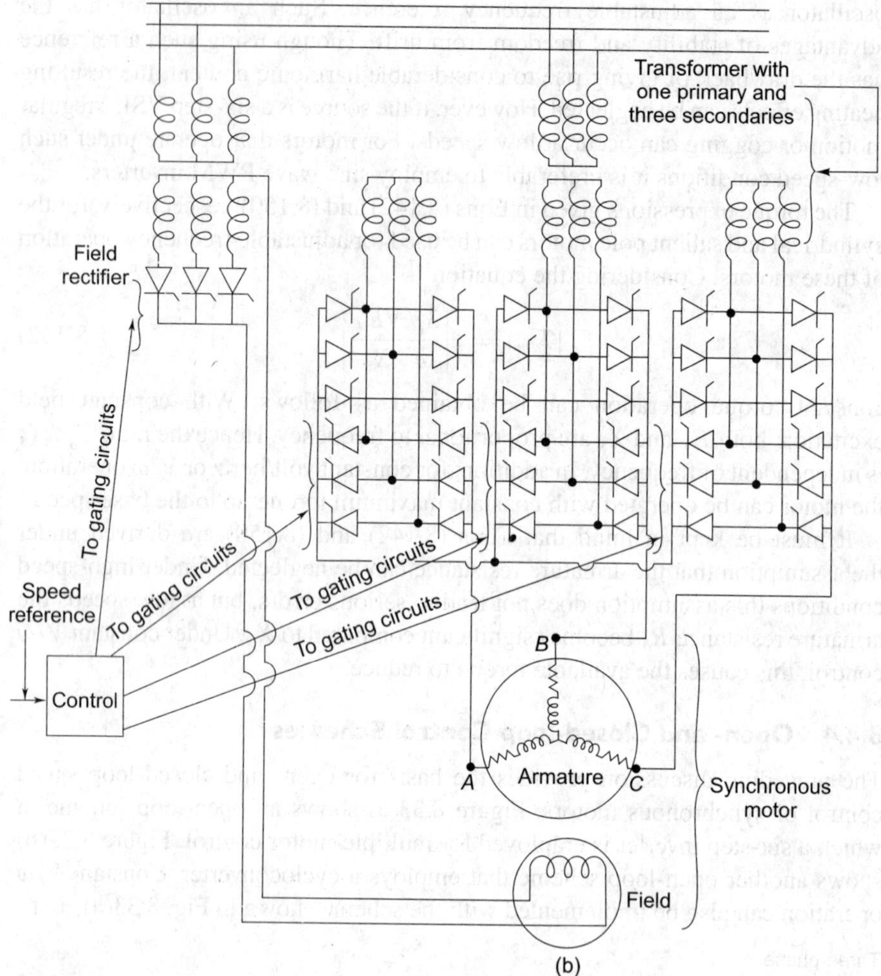

Fig. 8.33 (a) Block diagram for open-loop speed control of a synchronous motor; (b) schematic-cum-circuit diagram for closed-loop speed control of a synchronous motor supplied by a cycloconverter

this requires a speed sensing element and a PI or PID controller, as shown in Fig. 8.34(a). A typical configuration for implementing this scheme in the case of a synchronous motor fed by a cycloconverter is shown in Fig. 8.34(b).

Constant V/ω control implies that the air gap flux in the machine is kept nearly constant. This can be achieved by regulating the field current which helps in keeping the resultant air gap flux ϕ constant at its maximum permissible value. Such a strategy has the advantage of avoiding magnetic saturation and under-utilization of iron.

In a load commutated inverter, it is necessary to maintain a leading power factor, which is accomplished by controlling the armature current I_a, field current I_f, and torque angle ψ. The significance of the torque angle can be explained as follows. Torque is developed in a synchronous motor because of the tendency of the stator flux ϕ_{arm} and the air gap flux ϕ_f to align with each other; the torque so

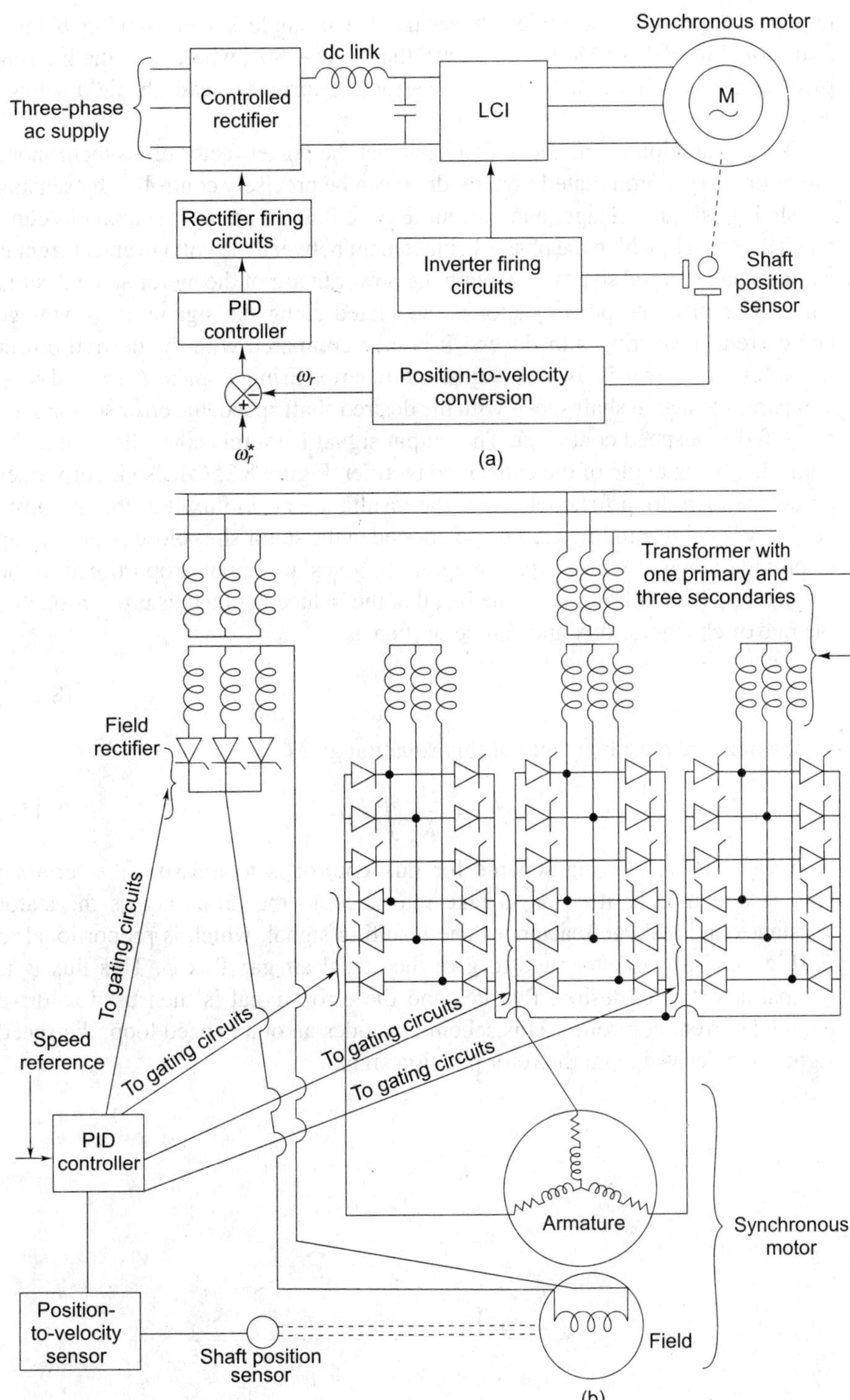

Fig. 8.34 (a) Block diagram for closed-loop control of a synchronous motor fed by a load commutated inverter (LCI); (b) schematic-cum-circuit diagram for closed-loop control of a synchronous motor fed by a cycloconverter

produced depends on the angle between them. This angle is seen from Fig. 8.35(a) to be equal to 360° – (180° + 90° – ψ), that is, ψ + 90°, where ψ is the internal phase angle, i.e., the angle by which the armature current I_a leads the field voltage E_f.

The explanation given above confirms that the power factor of a synchronous motor or a load commutated inverter drive can be precisely controlled by sensing the shaft position, voltage, and current. Figure 8.35(b) shows a schematic-cum-block diagram in which the phase displacement between the voltage and current is automatically varied so as to maintain the power factor of the motor at the desired value. The machine power factor is calculated using the signals from voltage and current-measuring transducers. It is then compared with the desired power factor, and the error is used to adjust the inverter firing signals. A second loop compares the actual shaft speed with the desired shaft speed, the error so obtained being fed to a speed controller. The output signal from this controller is used to adjust the firing angle of the controlled rectifier. Figure 8.35(b) also incorporates an independent loop for controlling the resultant air gap flux. For this purpose, auxiliary sensing windings can be positioned in the stator slots close to the air gap to provide a signal which, after integration, gives an output proportional to the air gap flux. This follows from the fact that the induced voltage is proportional to the rate of change of flux and can be written as

$$e = K_1 \frac{d\phi}{dt} \tag{8.153}$$

Integration and rearrangement of this equation gives

$$\phi = \frac{1}{K_1} \int e\, dt \tag{8.154}$$

An alternative sensing scheme for flux control is to measure the terminal voltage and modify it so as to take into account the drops across the stator resistance and leakage reactance. The resulting signal, which is proportional to E [Fig. 8.35(a)], is integrated to give the actual air gap flux ϕ. This flux ϕ is compared with the desired flux ϕ^*, and the error signal is then used to drive the field current controller. This scheme includes an outer speed loop; the speed signals are derived from the rotor position signal.

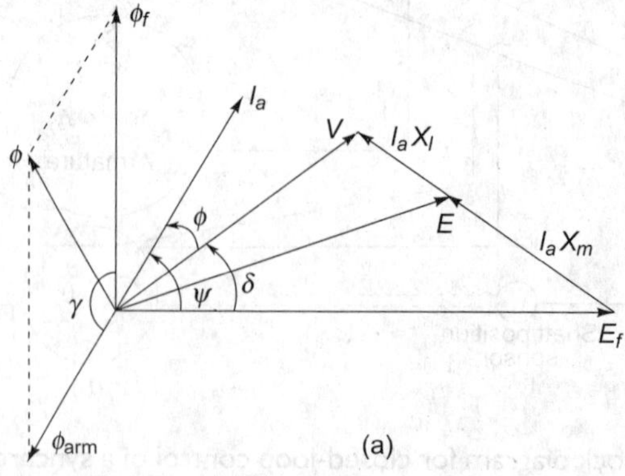

Fig. 8.35(a)

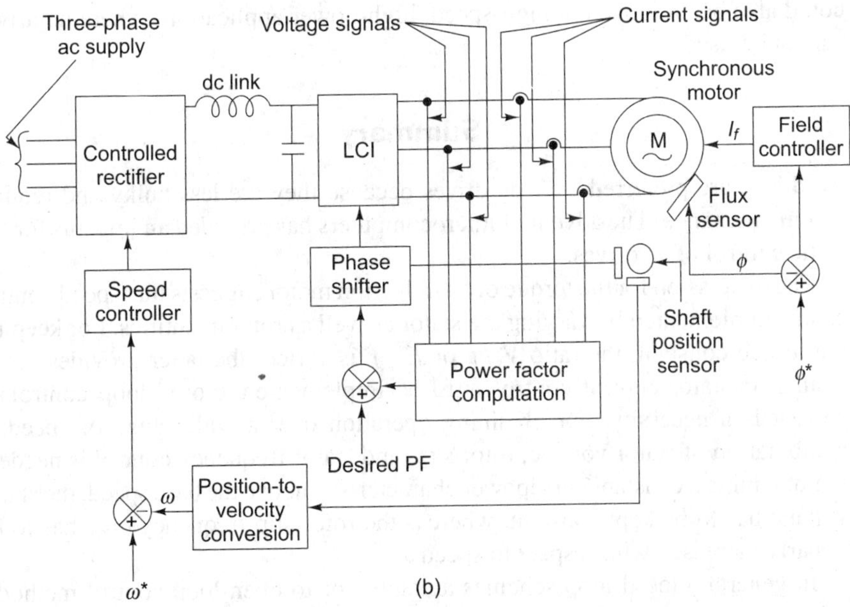

Fig. 8.35 (a) Complete phasor diagram of a synchronous motor; (b) block diagram for constant power factor and constant flux control of a synchronous motor

8.4.5 Applications of Synchronous Motor Drives

Variable speed synchronous drives have wide application in the following areas.

(a) Ac synchronous motors supplied by cycloconverters can be used for large low-speed reverse rolling mills, mine hoists, rotary cement kilns, low-power elevators, and direct-wheel drives in large earth-moving equipment. The rotor field winding of these motors is fed with direct current either through slip rings or from an ac exciter with rotating rectifiers. The drive permits four-quadrant operation with smooth transition from one quadrant to another and therefore provides rapid response. The torque harmonics due to the cycloconverter are small and of relatively high frequency.

(b) High-speed drives with capacities ranging from 10 MW to 100 MW in which dc link converters with natural commutation are used.

(c) Low-power servomotors with capacities up to 10 kW but with high dynamic response. The synchronous motors in this case are of the permanent magnet rotor type and are supplied by transistor inverters. As compared to induction motor drives, these drives have the advantage of eliminating rotor losses. However, a drawback associated with them is that field weakening is more difficult than in the case of induction motors.

With the recently developed signal processing techniques, wound rotor synchronous motor drives have the potential of becoming alternatives to dc drives, which are presently dominating the above-mentioned areas. In addition, they could also be employed in high-speed, high-power applications where dc drives cannot be used.

Summary

Ac drives are preferred over dc drives because they are less bulky and require less maintenance. The advent of microcomputers has provided an impetus for the static control of ac drives.

The expression for the torque of an induction motor suggests that speed control can be implemented by varying the stator as well as rotor quantities. For keeping the torque constant, the ratio V_s/f or E_s/f is varied; the latter provides better results. If stator current control is to be implemented, closed-loop control of current is a necessity. For obtaining operation over a wide range of speed, a combination of stator voltage, rotor slip, and stator frequency control is needed. For obtaining a constant horsepower characteristic above the base speed, the stator voltage has to be kept constant, whereas the rotor slip frequency $\omega_{R\omega}$ has to be linearly increased with respect to speed.

In general, closed-loop schemes are superior to open-loop control methods. Slip frequency control has the merit that the expensive current sensor, which is needed with closed-loop V/f control, is eliminated. The method of stator voltage control with constant slip frequency has the drawback that a change in the load condition changes the air gap flux level and hence deteriorates the dynamic performance. Closed-loop stator current control involves less hardware, because in this scheme a signal proportional to the magnitude of the stator current is obtained from a measurement of the dc link current. In this case, the slip frequency $\omega_{R\omega}$ is generated by a function generator. Stator current control provides good dynamic response and precise torque control.

Rotor resistance control can be implemented for a wound rotor induction motor; the variation of resistance is effected with the help of a diode rectifier and a semiconductor switch. A modern Scherbius system incorporates a thyristor inverter through which slip energy is fed back to the supply network through a transformer. This feature is helpful in increasing the efficiency of the system. However, for successful operation, it needs a large smoothing inductor in the dc link and a feedback transformer to match the supply voltage and the dc link voltage.

The torque developed by a cylindrical rotor type of synchronous motor depends on the angle δ between the applied voltage V_a and the field voltage E_f as well as on the synchronous reactance X_s. As compared to this, the torque developed by a salient pole motor is a function of both the direct- and quadrature axis reactances, in addition to the three quantities V_a, E_f, and δ. Constant torque operation can be obtained in both the cylindrical and salient pole cases by maintaining the ratio V_a/ω constant, where V_a is the applied voltage and ω is the rotor speed. A closed-loop scheme that incorporates V/f control involves a speed sensing element and a PID controller. Direct measurement of air gap flux is done either by using flux sensing apparatus or by measuring the terminal voltage and suitably manipulating it. Synchronous motor drives are used in large low-speed applications such as rotary cement kilns as well as for small servomotor purposes; the rotors in the latter case are of the permanent magnet type.

Worked Examples

1. A three-phase, 400-V, 500-Hz, four-pole, star-connected induction motor is supplied by a three-phase ac voltage controller with an input supply voltage of 440 V line-to-line. The data for the induction motor are $R_s = 0.35\ \Omega$, $R_r = 0.18\ \Omega$, $X_s = 0.9\ \Omega$, $X_r = 0.7\ \Omega$, and $X_m = 25\ \Omega$, all quantities being referred to the stator. The rotor speed is 1475 rpm. If the no-load losses are negligible, compute the (a) firing angle of the thyristors of the controller, (b) slip, (c) air gap power, (d) slip for maximum torque s_m, (e) maximum torque $\mathcal{T}_{dm}$ and, (f) efficiency.

Solution

(a) The line-to-neutral voltage of the input supply is $440/\sqrt{3} = 254$ V (RMS). The line-to-neutral voltage applied to the induction motor is $400/\sqrt{3}$ = 231 V (RMS). From Eqn (4.2),

$$V_{\text{RMS}} = \frac{V_m}{\sqrt{2}}\sqrt{1 - \frac{\alpha}{\pi} + \frac{\sin 2\alpha}{2\pi}}$$

$$231 = 254\sqrt{1 - \frac{\alpha}{\pi} + \frac{\sin 2\alpha}{2\pi}}$$

By trial and error, α is obtained as

$$\alpha \approx 1.0 \text{ rad or } 57.5^\circ$$

$$\text{(b) Synchronous speed} = \frac{120 f}{p} = \frac{120 \times 50}{4} = 1500 \text{ rpm}$$

$$\text{Slip} = \frac{1500 - 1475}{1500} = 0.0167$$

(c) The power transferred through the air gap is

$$P_g = 3I_r^2\left(\frac{R_r}{s}\right)$$

where

$$I_r = \frac{V_s}{[(R_s + R_r/s)^2 + (X_s + X_r)^2]^{1/2}}$$

$$= \frac{231}{[(0.35 + 0.18/0.0167)^2 + (0.9 + 0.7)^2]^{1/2}}$$

$$= \frac{231}{11.243} = 20.55 \text{ A}$$

Hence,

$$P_g = 3 \times (20.55)^2 \times \frac{0.18}{0.0167} = 13{,}650 \text{ W}$$

(d)
$$s_m = \frac{R_r}{[R_s^2 + (X_s + X_r)^2]^{1/2}}$$

$$= \pm \frac{0.18}{[0.35^2 + (0.9 + 0.7)^2]^{1/2}}$$

$$= \pm 0.11$$

(e)
$$T_{dm} = \frac{3}{2\omega_s} \frac{V_s^2}{\{R_s + \sqrt{R_s^2 + (X_s + X_r)^2}\}}$$

Here,

$$\omega_s = 2\pi \times \frac{1500}{60} = 50\pi = 157.1$$

$$T_{dm} = \frac{3}{2 \times 157.1} \times \frac{(231)^2}{1.988} = 256.3 \text{ N m}$$

(f)
$$\begin{aligned}\text{Power output} &= \text{air-gap power} - \text{rotor copper loss}\\ &= 13{,}650 - (20.55)^2 \times 0.18\\ &= 13{,}650 - 76\\ &= 13{,}574 \text{ W}\end{aligned}$$

$$\begin{aligned}\text{Power input} &= 3 \times \text{input voltage} \times \text{input current} \times \cos\phi_i\\ &= 3 \times \text{input voltage} \times \frac{\text{input voltage}}{\text{input impedance}} \cos\phi_i\\ &= \frac{3 \times 254^2}{Z_i} \times \cos\phi_i\end{aligned}$$

Z_i is the parallel combination of X_m and the combination of stator and rotor impedance. Here,

$$\begin{aligned}Z_i &= \frac{j25\,[(0.35 + 0.18/0.0167) + j(0.9 + 0.7)]}{(0.35 + 0.18/0.0167) + j(0.9 + 0.7 + 25)}\\ &= 25 \times \frac{11.243\angle 30.9^\circ}{28.83}\\ &= 9.75\angle 30.9^\circ\end{aligned}$$

Hence, the angle of the input current is a lagging one at 30.9°. Thus,

$$\begin{aligned}\text{Power input} &= 3 \times \text{input voltage} \times \text{input current} \times \cos\phi_i\\ &= \frac{3 \times 254^2}{9.75} \cos 30.9^\circ = 17{,}034\end{aligned}$$

$$\text{Efficiency} = \frac{\text{power output}}{\text{power input}} \times 100 = \frac{13{,}574}{17{,}034} \times 100 = 79.7\%$$

2. Stator voltage control is applied to a three-phase, 400-V, 50-Hz, four-pole, star-connected induction motor. The data for the motor are $R_s = 0.35\ \Omega$, $R_r = 0.18\ \Omega$, $X_s = 0.9\ \Omega$, $X_r = 0.7\ \Omega$, and $X_m = 25\ \Omega$, all quantities being referred to the stator. The no-load losses are negligible. A load with a constant torque characteristic

$T_{ld} = 60$ N m is supplied by the motor, with the base voltage equal to the motor terminal voltage. If the machine is first operated at the base voltage, and then at 0.8 times the base voltage, determine the following: (a) rotor current and (b) efficiency with both the voltages.

Solution

(i) *With base voltage*

(a) Here the synchronous speed is

$$\omega_s = \frac{2\pi \times 1500}{60} = 157.1 \text{ mechanical rad/s}$$

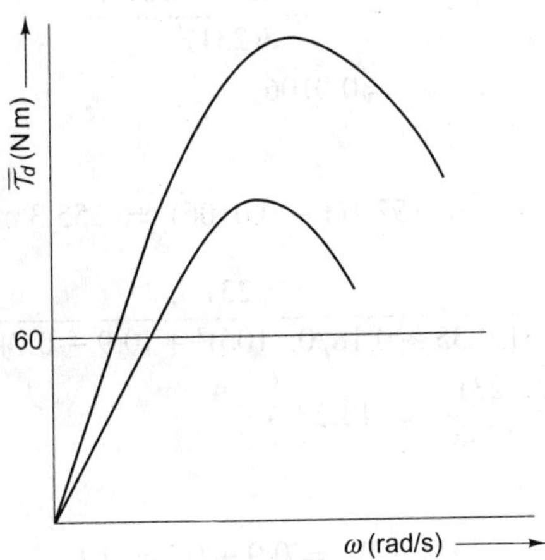

Fig. 8.36

Figure 8.36 gives the plots of the two curves with base voltage and 0.8 times the base voltage. The phase-to-neutral voltage is

$$V_s = \frac{400}{\sqrt{3}} = 231 \text{ V}$$

We now need the determine the approximate slip with base voltage. With a small slip s_1,

$$I_r \approx V_s \frac{s_1}{R_r}$$

Now

$$\begin{aligned}\mathcal{T}_d &= \frac{3I_r^2 R_r(1 - s_1)}{\omega_m s_1} \\ &= \frac{3I_r^2 R_r(1 - s_1)}{\omega_s(1 - s_1)s_1} \\ &= 3\left(V_s^2 \frac{s_1^2}{R_r^2}\right) \frac{R_r}{\omega_s s_1}\end{aligned}$$

$$= \frac{3V_s^2 s_1}{R_r \omega_s}$$

The slip s_1 with load torque T_{ld} = 60 N m is calculated as follows:

$$\frac{3V_s^2 s_1}{R_r \omega_s} = 60$$

The approximate slip

$$s_1 = \frac{60 R_r \omega_s}{3V_s^2}$$

$$= \frac{60 \times 0.13 \times 157.1}{3(231)^2}$$

$$= 0.0106$$

Now

$$\omega_m = \omega_s(1 - s_1) = 157.1(1 - 0.0106) = 155.3 \text{ mechanical rad/s}$$

$$I_r = \frac{231}{[(0.35 + 0.18/0.0106)^2 + (0.9 + 0.7)^2]^{1/2}}$$

$$= \frac{231}{17.40} = 13.28 \text{ A}$$

Also,

$$X_s + X_r = 0.9 + 0.7 = 1.6$$

and

$$[R_s + (R_r/s)] = 0.35 + \frac{0.18}{0.0106} = 0.35 + 16.98 = 17.33\ \Omega$$

(b) Output power

$$P_o = T_{ld}\omega_m = 60 \times 155.3 = 9320 \text{ W}$$

The stator current is given as

$$I_s \angle \phi_s = I_r \angle \phi_r + I_m \angle \phi_m$$

$$= 13.28 \angle -5.7^\circ - \frac{231}{25} \angle -90^\circ$$

$$= 16.9 \angle -33.6^\circ$$

Input power

$$P_i = 3I_s \times \text{supply voltage} \times \text{power factor}$$

Here,

$$\cos\phi_r = \cos\left[\tan^{-1}\frac{\text{total load reactance}}{\text{total equivalent load resistance}}\right]$$

Substitution of numerical values gives

$$\cos\phi_r = \cos\left[\tan^{-1}\left(\frac{1.6}{17.33}\right)\right] = 0.995$$

Here,

$$p_i = 3 \times V_s \times I_s \times \cos\phi$$

$$= 3 \times 231 \times 16.9 \times 0.995 = 11{,}653 \text{ W}$$

The approximate efficiency is

$$\frac{P_o \times 100}{P_i} = \frac{9320 \times 100}{11{,}653} = 80\%$$

(ii) *With 0.8 times the base voltage*
(a′) For the characteristic with voltage = $0.8V_s$, the torque is given as

$$\mathcal{T}_d = T_{\text{ld}}$$

$$= \frac{3(0.8V_s)^2 s_2}{R_r \omega_s}$$

$$= 60 \text{ N m}$$

The approximate slip $s_2 = 0.0166$.

$$I_r = \frac{0.8 \times 231}{[(0.35 + 0.18/0.0166)^2 + (0.9 + 0.7)^2]^{1/2}}$$

$$= \frac{0.8 \times 231}{11.31}$$

$$= 16.34 \text{ A}$$

(b′) $\omega_m = \omega_s(1 - s_2) = 157.1(1 - 0.0166) = 154.4$ mechanical rad/s
Output power

$$P_o = T_{\text{ld}}\omega_m = 60 \times 154.4 = 9264 \text{ W}$$

Approximate input power

$$P_i = 3I_s V_s \cos\phi_r$$

As before, the power factor is obtained from the ratio

$$\frac{\text{Total load reactance}}{\text{Total equivalent load resistance}}$$

Therefore,

$$\cos\phi_r = \cos\left[\tan^{-1}\left(\frac{1.6}{11.19}\right)\right] = 0.99$$

$$I_s\angle\phi_s = I_r\angle\phi_r + I_m\angle\phi_m$$

$$= 16.34\angle{-8.1^\circ} + \frac{231}{25}\angle{-90^\circ}$$

$$= 19.87\angle{-35.5^\circ}$$

Thus,

$$P_i = 3 \times 19.87 \times 231 \times 0.99 = 13{,}289 \text{ W}$$

The approximate efficiency is

$$\frac{P_o}{P_i} \times 100 = \frac{9264 \times 100}{13{,}289} = 69.7\%$$

3. A three-phase, 50-Hz, four-pole, star-connected induction motor is supplied by a three-phase ac voltage controller with an input voltage of 440 V (line-to-line) and $\alpha = 60°$; the corresponding output voltage is considered to be the base voltage V_s. The other data are $R_s = 0.4\ \Omega$, $R_r = 0.2\ \Omega$, $X_s = 0.85\ \Omega$, $X_r = 0.75\ \Omega$, and $X_m = 24\ \Omega$, all quantities being referred to the stator. Stator voltage control is applied to the drive, which has a load torque that obeys the equation $T_{ld}s^2 = 0.12$. Compute the following quantities for V_s and $0.6V_s$: (a) slip, (b) air gap power, (c) torque developed by the motor, (d) slip for maximum torque s_m, (e) maximum torque T_{dm}, and (f) efficiency.

Solution

With base voltage V_s

(a)
$$N = \frac{120 \times 50}{4} = 1500 \text{ rpm}$$

$$\omega_s = 2\pi\frac{N}{60} = 2\pi \times \frac{1500}{60} = 157.1 \text{ rad/s}$$

The line-to-neutral input voltage for the drive is

$$\frac{440}{\sqrt{3}} \text{ V (RMS)} = 254 \text{ V (RMS)} = V_i$$

The phase-to-neutral input voltage for the motor is

$$V_s = V_i\sqrt{1 - \frac{\alpha}{\pi} + \frac{\sin 2\alpha}{2\pi}}$$

$$= 254\sqrt{1 - \frac{\pi/3}{\pi} + \frac{\sqrt{3}/2}{2\pi}}$$

$$= 254\sqrt{1 - \frac{1}{3} + 0.1378} = 227.8 \text{ V}$$

For small slip, $\omega_m \approx \omega_s$.

$$T_d = \frac{3I_r^2 R_r}{\omega_s s}$$

where $I_r \approx V_s s/R_r$. Thus,

$$T_d \approx \frac{3R_r}{\omega_s s}\frac{V_s^2}{R_r^2}s^2$$

$$= \frac{3V_s^2 s}{\omega_s R_r}$$

As $\mathcal{T}_d = T_{ld}$,

$$\frac{3V_s^2 s}{\omega_s R_r} = \frac{0.12}{s^2}$$

Hence,

$$s^3 = \frac{0.12 \times \omega_s R_r}{3V_s^2} = \frac{0.12 \times 157.1 \times 0.2}{3 \times 227.8}$$

Thus,

$$s = 0.0289$$

(b) Power transferred through the air gap,

$$P_g = \frac{3I_r^2 R_r}{s}$$

where

$$I_r = \frac{V_s}{\sqrt{(R_s + R_r/s)^2 + (X_s + X_r)^2}}$$

where $R_s + R_r/s = 0.4 + 0.2/0.089$ and $X_s + X_r = 0.85 + 0.75 = 1.6\ \Omega$. Therefore,

$$I_r = \frac{227.8}{7.49} = 30.41$$

Hence,

$$P_g = 3 \times (30.41)^2 \times \frac{0.2}{0.0289} = 19{,}204\ \text{W}$$

(c) The torque developed by the motor is

$$\mathcal{T}_d = \frac{P_g}{\omega_s} = \frac{19{,}204}{157.1} = 122.2\ \text{N m}$$

(d) The slip for maximum torque is

$$s_m = \pm\frac{R_r}{X_s + X_r} = \pm\frac{0.2}{0.85 + 0.75} = \pm 0.125$$

(e) The maximum torque is

$$\mathcal{T}_{dm} = \frac{3V_s^2}{2\omega_s(X_s + X_r)} = \frac{3(227.8)^2}{2 \times 157.1 \times (0.85 + 0.75)} = 310\ \text{N m}$$

(f) The power output is

$$P_o = P_g - I_r^2 R_r = 19{,}204 - [3 \times (30.41)^2 \times 0.2] = 19{,}204 - 555 = 18{,}649\,\text{W}$$

The approximate power input = $3V_s I_r \cos\phi$, where

$$\phi = \tan^{-1}\left[\frac{X_s + X_r}{R_s + R_r/S}\right] = \tan^{-1}\left[\frac{1.6}{0.4 + 0.2/0.029}\right] = 12.3^\circ$$

From this,

$$\cos\phi = 0.977$$

Power input

$$P_i = 3 \times 227.8 \times 3.41 \times 0.977 = 20{,}304\ \text{W}$$

(g) Efficiency $= \dfrac{\text{power output } P_o}{\text{power input } P_i} \times 100$

$$= \frac{18{,}649}{20{,}304} \times 100$$

$$= 91.8\%$$

With $0.6V_s$

$$0.6V_s = 0.6 \times 227.8 = 136.68$$

(a′) $T_d = T_{\text{ld}}$ or

$$\frac{3I_r^2 R_r}{s\omega_s} = \frac{0.12}{s^2}$$

That is,

$$\frac{3R_r}{s\omega_s}\left[\frac{0.6V_s s}{R_r}\right]^2 = \frac{0.12}{s^2}$$

or

$$\frac{3(0.6V_s)^2 s^3}{\omega_s R_r} = 0.12$$

That is,

$$s^3 = \frac{0.12 \times 157.1 \times 0.2}{3(136.68)^2}$$

$$s = \sqrt[3]{\frac{0.12 \times 157.1 \times 0.2}{3(136.68)^2}}$$

$$= 0.0406$$

(b′) Power transferred through the air gap,

$$P_g = \frac{3I_r^2 R_r}{s}$$

$$I_r = \frac{0.6V_s}{\sqrt{(R_s + R_r/s)^2 + (X_S + X_r)^2}}$$

where s = 0.0406 and with R_s, R_r, X_s, and X_r as given data. Therefore,

$$I_r = \frac{136.68}{5.56}$$
$$= 24.58 \text{ A}$$

Hence,

$$P_g = \frac{3 \times (24.58)^2 \times 0.2}{0.0406} = 8929 \text{ W}$$

(c′) Torque developed

$$\mathcal{T}_d = \frac{P_g}{\omega_s} = \frac{8929}{157.1} = 56.84 \text{ N m}$$

(d′) Slip for maximum torque

$$s_m = \pm \frac{R_r}{X_s + X_r} = \pm 0.125 \text{ (same as before)}$$

(e′) Maximum torque

$$\mathcal{T}_{dm} = \frac{3(0.6V_s)^2}{2\omega_s(X_s + X_r)} = 111.5 \text{ N m}$$

Check: $\mathcal{T}_{dm}$ should be $(0.6)^2$ times the maximum torque with base voltage V_s:

$$(0.6)^2 \times 310 = 111.5 \text{ N m}$$

(f′) Power output

$$P_o = P_g - I_r^2 R_r = 8929 - 3 \times (24.58)^2 \times 0.2 = 8929 - 363 = 8566 \text{ W}$$

Approximate power input

$$P_i = 3 \times 0.6 V_s I_r \cos\phi = 3 \times 136.68 \times 24.58 \cos\phi$$

$$\phi = \tan^{-1}\left[\frac{X_s + X_r}{R_s + R_r/S}\right]$$
$$= \tan^{-1}\left[\frac{1.6}{0.4 + 0.2/0.0406}\right]$$
$$= 16.72°$$

This gives

$$\cos\phi = 0.958$$

Thus,

$$P_i = 3 \times 136.68 \times 24.58 \times 0.958 = 9655$$

$$\text{Efficiency} = \frac{8566}{9655} = 88.7\%$$

4. A three-phase, 50-kW, 1470-rpm, 400-V, 50-Hz, four-pole, star-connected induction motor has the following data: $R_s = 0.42\ \Omega$, $R_r = 0.23\ \Omega$, $X_s = 0.95\ \Omega$, $X_r = 0.85\ \Omega$, and $X_m = 28\ \Omega$, all quantities being referred to the stator side. The motor is operated with frequency control. If the slip for maximum torque at the given supply frequency is 0.12, determine the (a) supply frequency, (b) breakdown torque, and (c) speed at maximum torque.

Solution

(a) The given supply frequency is represented as

$$\omega = k\omega_0 = 2\pi f_0$$

where $f_0 = 50$ Hz. The slip for maximum torque is

$$s_m = \frac{R_r}{k\omega_0(L_s + L_r)} = \frac{R_r}{k(X_s + X_r)}$$

Thus,

$$\frac{R_r}{k(X_s + X_r)} = 0.12$$

$$k = \frac{R_r}{0.12(X_s + X_r)} = \frac{0.23}{0.12(0.95 + 0.85)} = 1.065$$

Therefore the supply frequency

$$k\omega_0 = 1.065 \times 314 = 334.4 \text{ electrical rad/s}$$

$$= 53.24 \text{ Hz}$$

(b) The phase-to-neutral voltage is

$$V_s = \frac{400}{\sqrt{3}} = 231 \text{ V}$$

The breakdown torque is given in Eqn (8.46) as

$$\left(\frac{3p}{2k^2\omega_0^2}\right)\left(\frac{V_s^2}{L_s + L_r}\right) = \left(\frac{3 \times 2 \times 231^2}{2 \times (1.065)^2 \times (314)^2 0.95 + 0.85/314}\right)\left(\frac{(231)^2}{0.95 + 0.85}\right)$$

$$= 249.7 \text{ N m}$$

(c) The synchronous speed (mechanical) is given in Eqn (8.3). Thus,

$$\omega_s = \frac{k\omega_0 \times 2}{P} = \frac{334.4 \times 2}{4} = 167.2 \text{ mechanical rad/s}$$

Angular speed at maximum torque = $167.2(1 - s) = 267.2(1 - 0.12) = 147.1$ mechanical rad/s. Finally, the speed at maximum torque is

$$N = 147.1 \times \frac{60}{2\pi} = 1405 \text{ rpm}$$

5. A three-phase, 50-kW, 1475-rpm, 400-V, 50-Hz, four-pole, star-connected induction motor has the following data: $R_s = 0.42\ \Omega$, $R_r = 0.23\ \Omega$, $X_s = 0.95\ \Omega$, $X_r = 0.85\ \Omega$, and $X_m = 30\ \Omega$, all quantities being referred to the stator side. The motor is operated with frequency control. If the breakdown torque is 225 N m at the supply frequency, determine the (a) supply frequency, (b) slip at maximum torque, and (c) speed at maximum torque.

Solution

(a) The supply frequency can be written as

$$\omega = k\omega_0 = k \times 314 \text{ electrical rad/s}$$

$$\text{Applied voltage (phase-to-neutral)} = \frac{400}{\sqrt{3}} = 231 \text{ V}$$

With the given frequency, $\mathcal{T}_{dm}$ from Eqn (8.46) is obtained as

$$\mathcal{T}_{dm} = \frac{3pV_s^2}{2k^2\omega_0^2(L_s + L_r)}$$
$$= \frac{3pV_s^2}{2k^2\omega_0^2(X_s + X_r)}$$
$$= \frac{3 \times 2 \times (231)^2}{2k^2 \times 314(0.95 + 0.85)}$$

$\mathcal{T}_{dm}$ is given as 225 N m. Hence,

$$\frac{3 \times 2 \times (231)^2}{2k^2 \times 314(0.95 + 0.85)} = 225$$

or

$$k^2 = \frac{3 \times 2 \times (231)^2}{2 \times 225 \times 314 \times 1.8}$$

Thus,

$$k = \sqrt{1.2588} = 1.122$$

The supply frequency

$$k\omega_0 = 1.122 \times 314 = 352.3 \text{ rad/s or } 56.07 \text{ Hz}$$

(b) Slip for maximum torque [Eqn (8.26)] is

$$\frac{R_r}{\omega(L_s + L_r)} = \frac{0.23}{k\omega_0(L_s + L_r)}$$
$$= \frac{0.23}{k(X_s + X_r)}$$
$$= \frac{0.23}{1.122(0.95 + 0.85)}$$
$$= 0.114$$

(c) The synchronous speed (mechanical) [Eqn (8.3)] is

$$k\omega_0 \times \frac{2}{P} = 352.3 \times \frac{2}{4}$$

$$= 176.15 \text{ mechanical rad/s}$$

$$= 176.15 \times \frac{30}{\pi} \text{ rpm} = 1682 \text{ rpm}$$

$$\text{Rotor speed} = 176.15(1 - s)$$

$$= 176.15(1 - 0.114)$$

$$= 156.07 \text{ mechanical rad/s} = 1490 \text{ rpm}$$

6. A three-phase, 50-kW, 1475-rpm, 420-V, 50-Hz, four-pole, star-connected induction motor has the following data: $R_s = 0.4\ \Omega$, $R_r = 0.21\ \Omega$, $X_s = 0.95\ \Omega$, $X_r = 0.85\ \Omega$, and $X_m = 32\ \Omega$, all quantities being referred to the stator side. If the frequency is increased to 58 Hz by frequency control, determine (a) the slip at maximum torque, (b) the speed at maximum torque, and (c) the breakdown torque.

Solution

(a) The phase-to-neutral supply voltage

$$V_s = \frac{420}{\sqrt{3}} = 242.5 \text{ V}$$

The new synchronous angular frequency is

$$k\omega_0 = 2\pi \times 58 = 364.4 \text{ electrical rad/s}$$

Here,

$$k = \frac{58}{50} = 1.16$$

The slip at maximum torque [Eqn (8.26)] is

$$s_m = \frac{R_r}{k\omega_0(L_s + L_r)}$$

$$= \frac{R_r}{k(X_s + X_r)}$$

$$= \frac{0.21}{1.16(0.95 + 0.85)}$$

$$= 0.1006$$

(b) The speed at maximum torque [Eqn (8.3)] is

$$\frac{k\omega_0 \times 2}{P} = \frac{364.4 \times 2}{4} = 182.2 \text{ mechanical rad/s}$$

or

$$\frac{182.2 \times 30}{\pi} = 1740 \text{ rpm}$$

(c) Breakdown torque $= \dfrac{3pV_s^2}{2(k\omega_0)^2(L_s + L_r)}$

$$= \frac{3pV_s^2}{2k^2\omega_0(X_s + X_r)}$$

$$= \frac{3pV_s^2}{2k(k\omega_0)(X_s + X_r)}$$

$$= \frac{3 \times 2 \times (242.5)^2}{2 \times 1.16 \times 364.4(0.95 + 0.85)}$$

$$= 231.9 \text{ N m}$$

7. A 30-kW, 400-V, 50-Hz, four-pole, star-connected induction motor is operated with constant volt/Hz control corresponding to the rated voltage and frequency. Other data are $R_s = 0.33\ \Omega$, $R_r = 0.22\ \Omega$, $X_s = 0.9\ \Omega$, and $X_r = 0.9\ \Omega$, all the quantities being referred to the stator side. If it has to supply a load torque of 100 N m, calculate the following quantities with rated frequency and with a frequency of 40 Hz: (a) slip for the given load torque, (b) slip for maximum torque, (c) breakdown torque, (d) breakdown rotor current, and (e) efficiency.

Solution

With rated frequency

(a) Supply voltage

$$V_s \text{ (line-to-neutral)} = \frac{400}{\sqrt{3}} = 231 \text{ V}$$

Supply angular frequency

$$\omega_0 \text{ (base frequency)} = 2\pi \times 50 = 314 \text{ electrical rad/s}$$

$$\frac{V_s}{\omega_0} = \frac{231}{314} = 0.736$$

The developed torque should be equal to the load torque. Hence,

$$T_d = \frac{3R_r}{s\omega_s}\left[\frac{V_s^2}{(R_s + R_r/s)^2 + (X_s + X_r)^2}\right] = T_{\text{ld}} = 100$$

The approximate value of s is first obtained by taking

$$\left(R_s + \frac{R_r}{s}\right)^2 + (X_s + X_r)^2 \approx \left(\frac{R_r}{s}\right)^2$$

Thus,

$$\left(\frac{3R_r}{s\omega_s}\right)\left(\frac{V_s^2}{(R_r/s)^2}\right) = 100$$

or

$$\frac{3V_s^2 s}{\omega_s R_r} = 100$$

The mechanical synchronous angular velocity (ω_s) is given as

$$\omega_s = \frac{\omega_0}{\text{number of pole pairs}} = \frac{314}{2} = 157 \text{ mech. rad/s}$$

Therefore,

$$\frac{3 \times (231)^2 \times s}{157 \times 0.22} = 100$$

$$s = \frac{100 \times 157 \times 0.22}{3 \times (231)^2} = 0.0215$$

The exact value of s is found by trial and error, by equating the exact expression for $\mathcal{T}_d$ with $T_{\text{ld}} = 100$ N m:

$$\mathcal{T}_d = \frac{3 \times 0.2}{s \times 157} \times \frac{(231)^2}{(0.33 + 0.22/s)^2} + (0.9 + 0.9)^2 = 100 \text{ N m}$$

By trial and error, the slip that gives the load torque with rated frequency is $s = 0.024$.

(b) The approximate slip for maximum torque [Eqn (8.26)] is

$$\frac{R_r}{X_s + X_r} = \frac{0.22}{1.80} = 0.122$$

The exact slip for maximum torque is

$$\frac{R_r}{\sqrt{R_s^2 + (X_s + X_r)^2}} = \frac{0.22}{\sqrt{(0.33)^2 + (1.8)^2}} = 0.120$$

(c) The breakdown torque is

$$\mathcal{T}_{dm} = \frac{3}{2\omega_s}\left\{\frac{(231)^2}{0.33 + \sqrt{(0.33)^2 + (1.8)^2}}\right\}$$

$$= \frac{3}{2 \times 157}\left\{\frac{(231)^2}{0.33 + \sqrt{(0.33)^2 + (1.8)^2}}\right\} = 236 \text{ N m}$$

(d) The breakdown rotor current is

$$I_r = \frac{V_s}{[(R_s + R_r/s)^2 + (X_s + X_r)^2]^{1/2}}$$

where $s = 0.12$. Thus, using the data given in the example,

$$I_r = \frac{231}{[(0.33 + 0.22/0.12)^2 + (1.8)^2]^{1/2}} = \frac{231}{2.81} = 82.1 \text{ A}$$

(e) Power input

$$P_i = 3V_s I_r \cos\phi$$

$$\cos\phi = \cos\left[\tan^{-1}\left(\frac{1.8}{0.33 + 0.22/0.024}\right)\right] = 0.714$$

$$P_i = 3 \times 231 \times 82.1 \times 0.714$$
$$= 40{,}598 \text{ W}$$

$$P_o = P_g \; - \; 3I_r^2 R_r$$
$$= 3I_r^2 \frac{R_r}{s} \; - \; 3I_r^2 R_r$$
$$= 3I_r^2 R_r \left[\frac{1}{s} - 1\right]$$
$$= 3 \times (23.9)^2 \times 0.22 \left[\frac{1}{0.12} - 1\right]$$
$$= 32{,}624 \text{ W}$$

$$\text{Efficiency} \; = \; \frac{P_o}{P_i} \times 100 \; = \; \frac{32{,}624}{40{,}598} \times 100 \; = \; 8.35\%$$

With a supply frequency of 40 Hz

(a) Supply voltage $= 231 \times \dfrac{40}{50} = 231 \times 0.8 = 184.8$ V

Angular frequency in electrical rad/s,

$$k\omega_0 \; = \; 2\pi \times 40 \; = \; 251.33$$

$$\omega_s \; = \; \frac{k\omega_0}{\text{number of pole pairs}} \; = \; \frac{251.33}{2} \; = \; 125.66 \text{ mechanical rad/s}$$

The approximate value of slip s_2^1 to meet a load demand of 100 N m is determined as before:

$$s_2^1 \; = \; \frac{100 \times 125.66 \times 0.22}{3 \times (184.8)^2} \; = \; 0.027$$

To determine the exact slip, T_{ld} is equated to the exact expression for $\mathcal{T}_d$. Here

$$X_{s2} \; = \; X_s \times 0.8 \; = \; 0.9 \times 0.8 \; = \; 0.72 \; \Omega$$

Similarly,

$$X_{r2} \; = \; 0.72 \; \Omega$$

$$\mathcal{T}_d \; = \; 100 \; = \; \frac{3 \times 0.22}{S_2 \times 125.66} \times \frac{(184.8)^2}{(0.33 + 0.22/s_2)^2 + (0.72 + 0.72)^2}$$

By trial and error, the slip is obtained as

$$s_2 \approx 0.0313$$

(b) The approximate slip for maximum torque is

$$\frac{R_r}{X_{s2}+X_{r2}} = \frac{0.22}{0.72+0.72} = 0.153$$

The exact slip for maximum torque is

$$\frac{R_r}{\sqrt{R_s^2+(X_{s2}+X_{r2})^2}} = \frac{0.22}{\sqrt{(0.33)^2+(1.44)^2}} = 0.149$$

(c) The breakdown torque is

$$\left(\frac{3}{2\omega_s}\right)\left[\frac{V_s^2}{R_s+\sqrt{R_s^2+(X_{s2}+X_{r2})^2}}\right]$$

$$= \frac{3}{2\times 125.66}\times\left[\frac{(184.8)^2}{0.33+\sqrt{(0.33)^2+(1.44)^2}}\right] = 225.6\text{ N m}$$

(d) The breakdown rotor current with $s = 0.149$ and $X_s + X_r = 1.44$ is

$$I_r = \frac{V_s}{[(R_s+R_r/s)^2+(X_{s2}+X_{r2})^2]^{1/2}} = \frac{184.8}{2.31} = 80\text{ A}$$

(e) Power input

$$P_i = 3V_sI_r\cos\phi$$

$$\cos\phi = \cos\left[\tan^{-1}\left(\frac{1.44}{0.33+0.22/0.149}\right)\right] = 0.782$$

$$P_i = 3\times 184.8\times 80.0\times 0.782$$

$$= 34{,}683\text{ W}$$

$$P_o = P_g - 3I_r^2R_r$$

$$= 3I_r^2R_r\left(\frac{1}{s}-1\right)$$

$$= 3\times(80.0)^2\times 0.22\left(\frac{1}{0.149}-1\right)$$

$$= 24{,}119\text{ W}$$

$$\text{Efficiency} = \frac{P_o}{P_i}\times 100 = \frac{24{,}119\times 100}{34{,}683} = 69.54\%$$

8. A 30-kW, 400-V, 50-Hz, four-pole, star-connected induction motor is operated with constant volts/Hz control corresponding to the rated voltage and frequency. Other data are $R_s = 0.34\ \Omega$, $R_r = 0.22\ \Omega$, $X_s = 1.0\ \Omega$, and $X_r = 0.8\ \Omega$, all parameters being referred to the stator side. If it has to operate at a reduced supply frequency such that the slip for maximum torque is 0.160, determine the (a) new supply frequency, (b) breakdown torque with the new frequency, (c) slip to provide a torque of 80 N m at this new frequency, (d) rotor current to provide this torque, and (e) efficiency.

Solution

(a) For small values of stator resistance, the expression $R_r/(X_{s2} + X_{r2})$ gives a nearly correct value of slip at maximum torque. Thus,

$$0.16 = \frac{R_r}{X_{s2} + X_{r2}} = \frac{R_r}{k\omega_0(L_{s2} + L_{r2})}$$

$$= \frac{R_r}{k(X_s + X_r)} = \frac{0.22}{k(1.0 + 0.8)}$$

Thus,

$$k = \frac{0.22}{0.16 \times 1.8} = 0.764$$

The new angular supply frequency is

$$k\omega_0 = 0.764 \times 314 = 240 \text{ electrical rad/s}$$

(b) Breakdown torque

$$\mathcal{T}_{dm} = \frac{3}{2\omega_s}\left\{\frac{(0.764 V_s)^2}{R_s + \sqrt{R_s^2 + k^2(X_s + X_r)^2}}\right\}$$

With

$$\omega_s = \frac{k\omega_0}{\text{number of pole pairs}} = \frac{240}{2} = 120 \text{ mechanical rad/s}$$

and

$$V_s \text{ (line-to-neutral)} = \frac{400}{\sqrt{3}} = 231 \text{ V}$$

The breakdown torque is

$$\mathcal{T}_{dm} = \frac{3}{2 \times 120}\left\{\frac{(0.764 \times 231)^2}{0.34 + \sqrt{(0.34)^2 + (0.764)^2(X_s + X_r)}}\right\} = 221.64 \text{ N m}$$

(c) The approximate slip required to provide a torque of 80 N m at the new frequency is determined as follows:

$$\mathcal{T}_d = \frac{3I_r^2 R_r}{sk\omega_s} = \frac{3R_r}{sk\omega_s}\left[\frac{(0.764 V_s)^2}{(R_r/s)^2}\right] = 80$$

or

$$s = \frac{80 \times 120 \times 0.22}{3 \times (0.764 \times 231)^2} = 0.0173$$

The exact slip required to give a load torque of 80 N m is found using Eqn (8.18) as

$$80 = \frac{3 \times 0.22}{s_2 \times 120} \times \frac{(231 \times 0.764)^2}{(0.34 + 0.22/s_2)^2 + [(1.0 + 0.8) \times 0.764]^2}$$

s_2 is found from trial and error as 0.025.
(d) The rotor current is

$$I_{r2} = \frac{0.764V_s}{[(R_s + R_r/s_2)^2 + 0.764(X_s + X_r)^2]^{1/2}}$$
$$= \frac{0.764 \times 231}{[(0.34 + 0.22/0.025)^2 + (0.764)^2(1.8)^2]^{1/2}}$$
$$= 19.09 \text{ A}$$

(e) Power input $= 3(0.764V_s)I_r \cos\phi$, where $\tan\phi = 0.764(X_s + X_r)/[R_s + (R_r/s_2)]$. Thus,

$$\cos\phi = \cos\left[\tan^{-1}\left(\frac{0.764 \times 1.8}{0.34 + 0.22/0.025}\right)\right] = 0.989$$

$$P_i = 3(0.764 \times 231) \times 19.09 \times 0.989$$
$$= 9996 \text{ W}$$

$$P_o = 3I_r^2 R_r\left(\frac{1}{s} - 1\right)$$
$$= 3 \times (19.09)^2 \times 0.22\left(\frac{1}{0.025} - 1\right)$$
$$= 9380 \text{ W}$$

$$\text{Efficiency} = \frac{P_o}{P_i} \times 100 = \frac{9380}{9996} \times 100 = 93.84\%$$

9. A 30-kW, 400-V, 50-Hz, four-pole, star-connected induction motor is operated with constant volts/Hz control corresponding to the rated voltage and frequency. Other data are $R_s = 0.33\ \Omega$, $R_r = 0.22\ \Omega$, and $X_s + X_r = 1.8\ \Omega$, all quantities being referred to the stator side. If it has to operate at a reduced supply frequency such that the breakdown torque is 210 N m, calculate the (a) new supply frequency, (b) slip to give maximum torque, (c) slip to provide a load torque of 90 N m at this frequency, (d) rotor current to provide this torque, and (e) efficiency.

Solution

This problem comes under stator current control.
(a) The breakdown torque is

$$T_{dm} = \frac{3}{2k\omega_s}\left\{\frac{(kV_s)^2}{R_s + \sqrt{R_s^2 + k^2(X_s + X_r)^2}}\right\}$$

$$210 = \frac{3}{2k\omega_s}\left\{\frac{(kV_s)^2}{R_s + \sqrt{R_s^2 + k^2(X_s + X_r)^2}}\right\}$$

Here, with p as the number of pole pairs, the synchronous speed in mech. rad/s is

$$\omega_s = \frac{314}{p} = \frac{314}{2} = 157$$

Also,

$$V_s = 231 \text{ V}$$

The breakdown torque is

$$210 = \frac{3}{2 \times 157} \left\{ \frac{k(231)^2}{0.33 + \sqrt{(0.33)^2 + k^2(1.8)^2}} \right\}$$

With trial and error, $k = 0.6$. The new supply frequency = 0.6 × 50 = 30 Hz. The new synchronous speed is given as

$$\frac{\text{New angular frequency}}{\text{Number of pole pairs}} = \frac{2 \times \pi \times 30}{2} = 94.25 \text{ rad/s}$$

(b) $$\text{Slip to give maximum torque} = \frac{R_r}{\sqrt{R_s^2 + (k\omega_0)^2(L_r + L_s)^2}}$$

$$= \frac{R_r}{\sqrt{R_s^2 + k^2(X_s + X_r)^2}}$$

$$= \frac{0.22}{\sqrt{(0.33)^2 + (0.6)^2(1.8)^2}}$$

$$= 0.195$$

(c) The approximate slip required to give a load torque of 90 N m is found from the equation

$$90 = \frac{3R_r}{s \times 0.6\omega_s} \frac{(0.6V_s)^2}{(R_r/s)^2}$$

that is,

$$90 = \frac{3(0.6V_s)^2 s}{0.6\omega_s R_r} = \frac{3 \times (0.6 \times 231)^2 \times s}{0.6 \times 94.25 \times 0.22}$$

or

$$s = \frac{90 \times 0.6 \times 94.25 \times 0.22}{(3 \times 231)^2} = 0.0194$$

The exact slip required to give a load torque of 90 N m is found from the equation

$$90 = \frac{3R_r}{sk\omega_s} \left[\frac{(0.6V_s)^2}{(R_s + R_r/s)^2 + k(X_s + X_r)^2} \right]$$

$$= \frac{3 \times 0.22}{s \times 0.6 \times 94.25} \times \left\{ \frac{(0.6 \times 231)^2}{(0.33 + 0.22/s)^2 + (0.6)(1.8)^2} \right\}$$

By trial and error, to give a torque of 90 N m at 30 Hz, s is found to be 0.0209.

(d) The rotor current required to provide this torque is

$$I_r = \frac{0.6V_s}{(R_s + R_r/s)^2 + [0.6(X_s + X_r)]^2}$$

Substituting the given values, the rotor current is obtained as

$$I_r = 12.72 \text{ A}$$

(e) The approximate power input

$$P_i = 3(0.6V_s)I_r \cos\phi$$

where

$$\cos\phi = \cos\left[\tan^{-1}\left(\frac{0.6 \times 1.8}{0.33 + 0.22/0.0209}\right)\right] = 0.995$$

$$P_i = 3 \times 0.6 \times 231 \times 12.72 \times 0.995$$
$$= 5263 \text{ W}$$

As per Eqn (8.12), the output power is

$$P_o = 3 \times (12.72)^2 \times 0.22\left(\frac{1}{0.0209} - 1\right)$$
$$= 5003 \text{ W}$$

$$\text{Efficiency} = \frac{P_o}{P_i} \times 100 = \frac{5003}{5263} \times 100 = 95.05\%$$

10. A 420-V, 50-Hz, four-pole, star-connected induction motor is operated with a current source of 30 A at its input. Other data are $R_s = 0.35\ \Omega$, $R_r = 0.18\ \Omega$, $L_s = 6.4$ mH, $L_r = 7.6$ mH, and $L_m = 0.08$ H, all parameters being referred to the stator side. It has to supply a load torque of 85 N m. Find (a) the slip, (b) the slip to give maximum torque, (c) maximum torque, and (d) magnetizing current to maintain the air gap flux constant at its no-load value with rated voltage and frequency; the slip remains the same as for a load torque of 85 N m.

Solution

(a) The slip required to give a torque of 85 N m is found from the equation

$$T_d = \frac{3pI_s^2(\omega_{R\omega}L_m)^2(R_r/\omega_{R\omega})}{R_r^2 + [\omega_{R\omega}(L_r + L_m)]^2}$$
$$= T_{\text{ld}} = 85 \text{ N m}$$

This equation can be written as

$$\frac{3pI_s^2L_m^2R_r/\omega_{R\omega}}{(R_r^2/\omega_{R\omega})^2 + (L_r + L_m)^2} = 85$$

Defining

$$x = \frac{R_r}{\omega_{R\omega}}$$

and substituting the values for the various quantities yields the quadratic equation

$$85x^2 - 34.56x + 0.652 = 0$$

giving

$$x = \frac{R_r}{\omega_{R\omega}} = 0.0198$$

Thus,

$$\omega_{R\omega} = \frac{0.18}{0.0198} = 9.08$$

and

$$s = \frac{9.08}{\omega} = 0.029$$

(b) The angular slip required to provide maximum torque is given by Eqn (8.61). Thus,

$$\omega_{R\omega m} = \frac{R_r}{L_r + L_m} = \frac{0.18}{0.0076 + 0.08} = 2.055$$

That is,

$$s_m = \frac{2.055}{314} = 0.0065$$

(c) The maximum torque is

$$\mathcal{T}_{dm} = (3pI_s^2)\left[\frac{L_m^2}{2(L_r + L_m)}\right]$$

Substitution of the values gives the maximum torque as

$$\mathcal{T}_{dm} = [3 \times 2 \times (30)^2]\left[\frac{(0.08)^2}{2(0.0076 + 0.08)}\right] = 197.26 \text{ N m}$$

(d) The magnetizing current at no-load condition with rated voltage and frequency to maintain the air gap flux constant is found as follows. At no load, $s = 0$ and the rotor impedance tends to infinity. Hence, for the no-load condition

$$I_s = \frac{V_s}{\sqrt{R_s^2 + [\omega(L_s + L_m)]^2}}$$

$$V_{s(\text{LN})} = \frac{420}{\sqrt{3}} = 242.5 \text{ V}$$

Here,

$$I_s = \frac{242.5}{\{(0.35)^2 + 314(0.08 + 0.0064)^2\}^{1/2}}$$
$$= 8.94 \text{ A}$$

As per Eqn (8.62), the magnetizing current is computed as

$$I_0 = \frac{I_s[R_r^2 + (\omega_{R\omega}L_r)^2]^{1/2}}{[R_r^2 + \{\omega_{R\omega}(L_r + L_m)\}^2]^{1/2}}$$

$$= \frac{8.94[(0.18)^2 + (9.08 \times 0.0076)^2]^{1/2}}{[(0.18)^2 + \{9.08(0.0076 + 0.08)\}^2]^{1/2}}$$

$$= \frac{8.94 \times 0.19277}{0.81552}$$

$$= 8.94 \times 0.2364 = 2.11 \text{ A}$$

11. A 415-V, 50-Hz, four-pole, star-connected induction motor operated with a constant current source has a breakdown torque, $\mathcal{T}_{dm}$ of 210 N m. Other data are $R_s = 0.04\ \Omega$, $R_r = 0.20\ \Omega$, $L_s = 8$ mH, $L_r = 8$ mH, and $L_m = 0.09$ H, all parameters being referred to the stator side. Determine the (a) the current supplied at the source, (b) slip to give a load torque of 78 N m, and (c) magnetizing current to maintain the air gap flux constant at its no-load value with rated voltage and frequency. Assume that the slip remains the same as that for a load torque of 78 N m.

Solution

(a) The maximum torque [Eqn (8.60)] is determined as

$$\mathcal{T}_{dm} = 3pI_s^2 \frac{L_m^2}{2(L_r + L_m)} = 210 \text{ N m}$$

That is,

$$210 = (3 \times 2 \times I_s^2) \times \frac{(0.09)^2}{2(0.098)} = \frac{3 \times 2 \times I_s^2 \times 0.0081}{2 \times 0.098}$$

giving

$$I_s^2 = 846.9$$

or

$$I_s = 29.1 \text{ A}$$

(b) The slip required to give a torque of 78 N m is determined from the equation

$$\frac{3pI_s^2 L_m^2 R_r/\omega_{R\omega}}{(R_r^2/\omega_{R\omega})^2 + (L_r + L_m)^2} = 78$$

Defining

$$x = \frac{R_r}{\omega_{R\omega}}$$

and substituting the values gives

$$78 = \frac{3 \times 2 \times (29.1)^2 \times (0.09)^2 \times x}{x^2 + (0.098)^2}$$

This results in the quadratic equation

$$78x^2 - 41.16x + 0.749 = 0$$

which gives x as

$$x = \frac{R_r}{\omega_{R\omega}} = 0.0189$$

Thus,

$$\omega_{R\omega} = \frac{R_r}{0.0189} = \frac{0.2}{0.0189} = 10.58$$

and

$$s = \frac{\omega_{R\omega}}{\omega} = \frac{10.58}{314} = 0.0337$$

(c) The stator current at no load with rated voltage and frequency is determined as follows. At no load, $s = 0$ and the rotor impedance tends to infinity. Hence I_s for the no-load condition is

$$\frac{V_s}{\sqrt{R_s^2 + \{\omega(L_s + L_m)\}^2}}$$

Here,

$$V_s = \frac{415}{\sqrt{3}} = 239.6 \text{ V}$$

Hence,

$$I_s = \frac{239.6}{\sqrt{(0.4)^2 + 314(0.098)^2}}$$
$$= 7.79 \text{ A}$$

Using Eqn (8.62), the magnitizing current is determined as

$$I_0 = 7.79 \frac{[R_r^2 + (\omega_{R\omega} L_r)^2]^{1/2}}{[R_r^2 + \{\omega_{R\omega}(L_r + L_m)\}^2]^{1/2}}$$
$$= 7.79 \frac{[(0.2)^2 + (10.58 \times 0.008)^2]^{1/2}}{[(0.2)^2 + (10.58 \times 0.098)]^{1/2}}$$
$$= \frac{7.79 \times 0.217}{1.056} = 1.6 \text{ A}$$

12. A three-phase, 400-V, 50-Hz, 1460-rpm, four-pole induction motor is subjected to rotor resistance control. If the maximum torque occurs for a chopper duty ratio (δ) of 0.2 and a slip of 0.13, determine (a) the value of the external resistance and (b) the torque for a chopper duty ratio (δ) of 0.6 and a slip of 0.04. The stator-to-rotor turn ratio is 2 and the resistance of the link inductor is 0.011 Ω. The motor parameters are $R_s = 0.08$ Ω, $X_s = 0.95$ Ω, $R_r' = 0.055$ Ω, and $X_r' = 0.5$ Ω.

Solution

(a) The line-to-neutral voltage is

$$V_{s(\text{LN})} = \frac{400}{\sqrt{3}} = 231 \text{ V}$$

The synchronous speed in rpm,

$$N = \frac{50 \times 120}{4} = 1500$$

The angular synchronous speed (mech.) of the rotor is

$$\omega_m = \frac{2\pi \times 1500}{60} = 157.1 \text{ rad/s}$$

The expression for the slip required to get maximum torque is obtained by differentiating the expression for torque with respect to s:

$$\mathcal{T}_d = \frac{(3/\omega_s)|\bar{V}_s|^2(R_2/s)}{[R_s + R_1 + R_2/s]^2 + (X_s + X_r)^2}$$

Setting

$$\frac{d\mathcal{T}_d}{ds} = 0$$

gives

$$s_m = \frac{R_2}{\sqrt{(R_s + R_1)^2 + (X_s + X_R)^2}}$$

The slip required to give maximum torque is given here as

$$s_m = 0.13$$

From the theory given in Section 8.3.1,

$$R_{\text{out}} = R_d + R_e(1 - \delta)$$
$$= 0.011 + R_e(1 - 0.2) = 0.011 + 0.8R_e$$

$$R_2 = N_{sr}^2[R_r' + 0.5R_{\text{out}}]$$
$$= 4[0.055 + 0.5(0.011 + 0.8R_e)]$$
$$= 4(0.0605 + 0.4R_e)$$

$$R_1 = N_{sr}^2\left[\frac{\pi^2}{9} - 1\right](R_r' + 0.5R_{\text{out}})$$
$$= 4\left[\frac{\pi^2}{9} - 1\right](0.0605 + 0.4R_e)$$

$$X_R = N_{sr}^2 X_r' = 4 \times 0.28 = 1.12\ \Omega$$

Substituting values in the expression for s_m gives

$$0.13 = \frac{4(0.0605 + 0.4R_e)}{\sqrt{0.08 + 4(\pi^2/9 - 1)(0.0605 + 0.4R_e)^2 + (0.95 + 1.12)^2}}$$

A fairly good approximate for R_e is obtained by neglecting the first squared term under the radical sign, which is small compared to the second squared term, in the denominator. Thus,

$$\frac{4(0.0605 + 0.4R_e)}{0.95 + 1.12} = 0.13$$

which gives $R_e = 0.0169$.

(b) The torque for $\delta = 0.6$ and $s = 0.04$ is computed as follows:

$$R_{\text{out}} = R_d + R_e(1 - 0.6) = 0.011 + 0.0169 \times 0.4 = 0.0178\ \Omega$$

Also,

$$\begin{aligned} R_2 &= N_{sr}^2(R_r' + 0.5R_{\text{out}}) \\ &= 4(0.055 + 0.5 \times 0.0178) = 0.2556\ \Omega \\ R_1 &= N_{sr}^2\left[\frac{\pi^2}{9} - 1\right](R_r' + 0.5R_{\text{out}}) \\ &= 4\left[\frac{\pi^2}{9} - 1\right](0.055 + 0.5 \times 0.0178) \\ &= 0.0247\ \Omega \end{aligned}$$

Finally,

$$\mathcal{T}_d = \frac{3}{157.1} \times \frac{(231)^2(0.2556/0.04)}{(0.08 + 0.0247 + 0.2556/0.04)^2 + (0.95 + 1.12)^2} = 140.13\,\text{N m}$$

13. Consider the induction motor of Example 12. For this motor, determine (a) the duty ratio for a speed of 750 rpm and a torque equal to the rated torque and also (b) the speed for a duty ratio of 0.7 and a torque equal to the rated torque.

Solution

(a) $$\text{Slip with a speed of 750 rpm} = \frac{1500 - 750}{1500} = 0.5$$

$$\text{Rated slip} = \frac{1500 - 1460}{1500} = 0.0267$$

The rated torque can be expressed as

$$\mathcal{T}_{\text{rated}} = \frac{3pR_r}{s\omega}\left[\frac{V_s^2}{(R_s + R_r/s)^2 + (X_s + X_r)^2}\right]$$

where

$$R_r = N_{sr}^2 R_r' = 4 \times 0.055 = 0.22\ \Omega$$

Also,

$$X_r = 4 \times 0.28 = 1.12\ \Omega$$

Substituting the values gives

$$\begin{aligned} \mathcal{T}_{\text{rated}} &= \left(\frac{3 \times 2 \times 0.22}{0.0267 \times 314}\right)\left[\frac{(231)^2}{(0.08 + 0.22/0.0267)^2 + (0.95 + 1.12)^2}\right] \\ &= 114.3\ \text{N m} \end{aligned}$$

The duty ratio can be obtained from the equation

$$T_{\text{rated}} = \frac{3}{\omega_m}\left\{\frac{|V_s|^2(R_2/s_1)}{[R_s + R_1 + R_2/s]^2 + (X_s + X_r)^2}\right\}$$

with ω_m = 157.1 rad/s and s_1 = 0.2.

$$R_{\text{out}} = R_d + R_e(1-\delta) = 0.011 + 0.0169(1-\delta)$$

$$R_1 = N_{sr}^2\left(\frac{\pi^2}{9} - 1\right)(R_r' + 0.5R_{\text{out}})$$

$$= 4\left(\frac{\pi^2}{9} - 1\right)\{0.055 + 0.5\,[0.011 + 0.0169(1-\delta)]\}$$

$$= 0.386\,[0.0605 + 0.00845(1-\delta)]$$

$$R_2 = N_{sr}^2[R_r' + 0.5R_{\text{out}}]$$

$$= 4\,[0.055 + 0.5\,\{0.011 + 0.0169(1-\delta)\}]$$

$$= 4\,[0.0605 + 0.00845(1-\delta)]$$

By substituting the values and defining $x = 0.0605 + 0.00845(1 - \delta)$, the torque equation becomes

$$114.3 = \frac{3}{157.1}\frac{(231)^2 \times (4/0.5)x}{\{0.08 + 0.386x + (4/0.5)x\}^2 + (0.95 + 1.12)^2}$$

Defining $x = 0.0605 + 0.00845\,(1 - \delta)$, the above torque equation can be written as a quadratic equation:

$$16.388x^2 - 16.57x + 1 = 0$$

The solutions of this quadratic equation are

$$x_1 = 0.947 \text{ and } x_2 = 0.0644$$

The higher value is inadmissible. Thus δ can be obtained from the lower value as

$$0.0605 + 0.00845(1-\delta) = 0.0644$$

giving δ = 0.54. Thus the duty ratio required to give the rated torque is 0.54.

(b) Let the speed for a duty ratio of 0.7 be N_2. Here

$$R_2 = 4\,[0.0605 + 0.00845(1 - 0.7)] = 0.252$$

$$R_1 = 0.386\,[0.0605 + 0.00845(1 - 0.7)] = 0.0243$$

Defining

$$s_2 = \frac{1500 - N_2}{1500}$$

the equation for the torque can be written as

$$114.3 = \left(\frac{3}{157.1}\right)\frac{(231)^2 \times (4/s_2) \times 0.252}{(0.08 + 0.0243 + 0.252/s_2)^2 + (0.95 + 1.12)^2}$$

This gives the quadratic equation

$$394.9s_2^2 - 821.1s_2 + 5.84 = 0$$

The solutions are

$$s_2 = 2.072 \text{ or } 0.0071$$

As the larger value is inadmissible, N_2 can be obtained from the smaller value of s_2 as

$$0.0071 = \frac{1500 - N_2}{1500}$$

This gives N_2 = 1489 rpm. Thus the speed required to give the rated torque with a duty ratio of 0.7 is 1489 rpm.

14. A 420-V, 50-Hz, four-pole, 1470-rpm induction motor is subjected to the Scherbius scheme of control. It has the following parameters: R_s = 0.12 Ω, X_s = 0.35 Ω, R_r and X_r (referred to the stator side) are, respectively, 0.09 Ω and 0.68 Ω, and X_m = 20 Ω. The drive has to provide a speed range that is 40% below the synchronous speed. The maximum value of α is 170°. Assume that the dc link inductor resistance R_d is zero. If N_{sr} = 1.8, determine (a) the ratio N_{sr}/N_{LI} and (b) the torque and drive power factor for α = 100° and a speed of 780 rpm. (c) If the transformer is now removed and the inverter connected directly to the ac mains with α altered to give 1.48 times the torque obtained in (b) but with the same slip as in (b), determine this new α and drive power factor.

Solution

(a) The maximum slip can be written as

$$s_m = -\left(\frac{N_{sr}}{N_{LI}}\right)\cos\alpha_{im}$$

where α_{im} is the maximum value of α_i. The synchronous speed N_s can be written as

$$N_s = \frac{f(120)}{P} = \frac{50 \times 120}{4} = 1500 \text{ rpm}$$

With s_m given as 0.4, actual speed is

$$N = 1500(1 - s_m) = 1500(1 - 0.4) = 900 \text{ rpm}$$

Thus,

$$0.4 = -\left(\frac{N_{sr}}{N_{LI}}\right)\cos 170° = 0.984\frac{N_{sr}}{N_{LI}}$$

This gives

$$\frac{N_{sr}}{N_{LI}} = \frac{0.4}{0.984} = 0.406$$

(b) $$V_s \text{ per phase} = \frac{420}{\sqrt{3}} = 242.5 \text{ V}$$

For $\alpha = 100°$,

$$V_{eq} = (1-s)\left(\frac{N_{sr}}{N_{LI}}\right) V_s |\cos\alpha_i|$$
$$= (1-0.48) \times 0.406 \times 242.5 \times |\cos 100°|$$
$$= 0.52 \times 17.1 = 8.89 \text{ V}$$

$$\text{Slip for a speed of 780 rpm} = \frac{1500-78}{1500} = 0.48$$

$$R_1 = (R'_r + 0.5R_d)N_{sr}^2\left[\frac{\pi^2}{9}-1\right]$$

Here $R_d = 0$ and $R_r = R'_r N_{sr}^2$. Thus,

$$R_1 = R'_r N_{sr}^2\left[\frac{\pi^2}{9}-1\right] = R_r\left[\frac{\pi^2}{9}-1\right]$$
$$= 0.09\left(\frac{\pi^2}{9}-1\right) = 0.0087$$
$$R_2 = (R'_r + 0.5R_d)N_{sr}^2$$
$$= R'_r N_{sr}^2 = R_r = 0.09$$

$$R = R_s + R_1 + \frac{R_2}{s}$$
$$= 0.12 + 0.0087 + \frac{0.09}{0.48} = 0.3162$$

$$X = X_s + X_r = 0.35 + 0.68 = 1.03$$

$$V^* = \frac{V_{eq}}{s} = \frac{-17.1 \times 0.52}{0.48}$$
$$= 35.625 \times 0.52 = 18.525$$

$$I_{r1} = -\frac{RV^*}{R^2+X^2} + \sqrt{\frac{R^2(V^*)^2}{(R^2+X^2)^2} - \frac{(V^*)^2 - V_s^2}{R^2+X^2}}$$

Here $R^2 + X^2 = (0.3162)^2 + (1.03)^2 = 1.1609$. Thus,

$$I_{r1} = -\frac{0.3162 \times 18.525}{1.1609} + \sqrt{\left(\frac{0.3162 \times 18.525}{1.1609}\right)^2 - \frac{(18.525)^2 - (242.5)^2}{1.1609}}$$
$$= -5.046 + \sqrt{25.16 + 50630.13}$$
$$= -5.046 + 224.47 = 219.42 \text{ A}$$

From Eqn (8.128),

$$\mathcal{T}_d = \frac{3}{s\omega_s}[I_{r1}^2 R_2 + V_{\text{eq}} I_{r1}]$$

Substituting the values,

$$\mathcal{T}_d = \frac{3}{0.48 \times 157.1}[(219.42)^2 \times 0.09 + 8.89 \times 219.42] = 250 \text{ N m}$$

From Eqn (8.122) and the expression for ϕ_r below Eqn (8.125),

$$\cos\phi_r = \frac{RI_{r1} + V^*}{V_s} = \frac{0.3162 \times 219.42 + 18.525}{242.5} = 0.3625$$

This gives $\phi_r = -68.75°$.

Check for I_{r1}

$$\begin{aligned}(I_{r1}X)^2 + (RI_{r1} + V^*)^2 &= (219.42 \times 1.03)^2 + (0.3162 \times 219.42 + 18.525)^2 \\ &= (226.0)^2 + (87.9)^2 \\ &= (242.5)^2\end{aligned}$$

Check for ϕ_r

$$\tan\phi_r = \frac{I_{r1}X}{R_{r1} + V^*} = \frac{226}{87.9} = 2.571$$

$$\phi_r = -68.75°$$

Now,

$$\begin{aligned}I_s\angle\phi_s &= I_{r1}\angle\phi_{r1} + I_m\angle{-90°} \\ &= 219.42\angle{-68.75°} + \frac{242.5}{20}\angle{-90°} \\ &= 79.53 - j204.5 - j12.13 = 79.53 - j216.63\end{aligned}$$

$$\begin{aligned}I_T\angle\phi_T &= I_{r1}\angle\phi_{r1}\frac{N_{sr}}{N_{LI}} \\ &= 0.406(79.53 - j204.5) = 32.29 - j83.03\end{aligned}$$

$$\begin{aligned}I_L\angle\phi_L &= I_S\angle\phi_s + I_T\angle\phi_T \\ &= 79.53 - j216.63 + 32.29 - j83.03 \\ &= 111.82 - j299.66\end{aligned}$$

$$\tan\phi_L = \frac{-299.66}{111.82} = -2.68$$

Thus $\phi_L = -69.54°$ and $\cos\phi_L = 0.35$.

(c) If the inverter is directly connected to the mains, $N_{LI} = 1$ and

$$\frac{N_{sr}}{N_{LI}} = \frac{1.8}{1.0} = 1.8$$

The values of the various quantities are calculated below. R_1 and R_2 remain the same as before because N_{sr} remains the same. R and X also remain unaltered.

$$V^* = \frac{V_{eq}}{s} = \left(\frac{N_{sr}}{N_{LI}}\right)\frac{V_s \cos\alpha_i(1-s)}{s}$$

$$= \frac{1.8}{1.0} \times \frac{242.5\cos\alpha_i}{0.48}(1-0.48)$$

$$= 0.52 \times 909.38\cos\alpha_i$$

$$= 472.88\cos\alpha_i$$

$$V_{eq} = (1-s)\frac{N_{sr}}{N_{LI}}V_s\cos\alpha_i = 0.52 \times \frac{1.8}{1.0} \times 242.5\cos\alpha_i = 226.98\cos\alpha_i$$

$$I_{r1} = -\frac{RV^*}{R^2+X^2} + \sqrt{\left(\frac{RV^*}{R^2+X^2}\right)^2 - \frac{(V^*)^2 - V_s^2}{R^2+X^2}}$$

Here,

$$R^2 + X^2 = (0.3162)^2 + (1.03)^2 = 1.1609$$

Thus,

$$(R^2+X^2)^2 = (1.1609)^2 = 1.3476$$

$$I_{r1} = \frac{-0.3162 \times 472.88\cos\alpha_i}{1.1609} + \sqrt{\left(\frac{0.3162 \times 909.33\cos\alpha_i}{1.1609}\right)^2 - \frac{(472.88\cos\alpha_i)^2 - (242.5)^2}{1.1609}}$$

$$= -247.68\cos\alpha_i + \sqrt{(247.68)^2\cos^2\alpha_i + \frac{(242.5)^2 - (472.88)^2\cos\alpha_i}{1.1609}}$$

Choosing $\cos\alpha_i$ as 0.12 to give 1.48 times the original torque,

$$I_{r1} = -247.68 \times 0.12 + \sqrt{(247.68)^2(0.12)^2 + \frac{(242.5)^2 - (472.88 \times 0.12)^2}{1.1609}}$$

$$= -15.46 + 219.36 = 203.9$$

Check for I_{r1}

$$(I_{r1}X)^2 + (I_{r1}R + V^*)^2 = (210.02)^2 + (121.22)^2 = (242.49)^2 = V_s^2$$

Hence I_{r1} is correct.

Now

$$\mathcal{T}_d = \frac{3}{0.48 \times 157.1} \times 203.9 \times (203.9 \times 0.09 + 226.98 \times 0.12) = 369.8$$

$$1.48 \text{ times the original torque} = 1.48 \times 250 = 370.0$$

Hence the value of $\cos \alpha_i$ is correct.

$$\tan \phi_{r1} = \frac{203.9 \times 1.03}{203.9 \times 0.3162 + 56.75} = \frac{210.02}{121.22} = 1.7325$$

This gives

$$\phi_{r1} = -60^\circ$$

$$I_{r1} \angle \phi_{r1} = 203.9 \angle -60^\circ = 101.93 - j176.59$$

$$I_m \angle \phi_m = \frac{242.5}{20} \angle -90^\circ = -j12.13$$

$$\begin{aligned} I_s \angle \phi_s &= I_{r1} \angle \phi_{r1} + I_m \angle \phi_m \\ &= 101.93 - j176.59 - j12.13 \\ &= 101.93 - j188.72 \end{aligned}$$

$$\begin{aligned} I_T \angle \phi_T &= \frac{N_{sr}}{N_{LI}} I_{r1} \angle \phi_{r1} \\ &= \frac{1.8}{1} \times 203.9 \angle -60^\circ \\ &= 1.8(101.93 - j176.59) \\ &= 183.47 - j317.86 \end{aligned}$$

$$\begin{aligned} I_L \angle \phi_L &= I_s \angle \phi_s + I_T \angle \phi_T \\ &= 101.93 - j188.72 + 183.47 - j317.86 \\ &= 285.4 - j506.58 \end{aligned}$$

$$\tan \phi_L = 1.775$$

$$\phi_L = -60.6^\circ$$

$$\cos \phi_L = 0.491$$

15. A 440-V, 50-Hz, 1465-rpm, four-pole, star-connected, wound rotor induction motor has the following parameters referred to the stator: $R_s = 0.04\ \Omega$, $X_s = 0.28\ \Omega$, $X_m = 12\ \Omega$, $R'_r = 0.048\ \Omega$, $X'_r = 0.105\ \Omega$, $R_d = 0.1\ \Omega$, and $N_{sr} = 2$. The motor is controlled by a static Scherbius scheme. The set-up is designed for a speed range that is 35% below synchronous speed with a maximum firing angle of 168°. The derived parameter R_1 is to be neglected. Compute the (a) ratio N_{sr}/N_{LI}, (b) motor speed for 1.738 times the rated torque and the inverter firing angle α of 93°, (c) α_i for 2.7 times the rated torque and a speed 1200 rpm, and (d) drive power factor with the last condition.

Solution

(a) The maximum slip occurs at maximum firing angle and minimum speed. Thus,

$$s_m = 0.35 = \frac{N_{sr}}{N_{LI}} \times \cos 168°$$

Hence,

$$\frac{N_{sr}}{N_{LI}} = \frac{0.35}{\cos 168°} = 0.358$$

$$\text{(b) Rated slip} = \frac{1500 - 1465}{1500} = \frac{35}{1500} = 0.0233$$

$$V_{s(\text{LN})} = \frac{440}{\sqrt{3}} = 254 \text{ V}$$

$$\text{Rated torque } \mathcal{T}_d = \frac{3R_r}{s\omega_s} \frac{V_s^2}{(R_s + R_r/s)^2 + (X_s + X_r)^2}$$

where $R_r = N_{sr}^2 R_r' = 4 \times 0.048 = 0.192\ \Omega$ and $X_r = N_{sr}^2 X_r' = 4 \times 0.105 = 0.42\ \Omega$. Substituting the values gives

$$\mathcal{T}_d = \frac{3 \times 0.192}{0.0233 \times 157.11} \times \frac{(254)^2}{(0.04 + 0.192/0.0233)^2 + (0.28 + 0.42)^2} = 147.03$$

Let the motor speed be N_1. Then the corresponding slip is

$$s_1 = \frac{1500 - N_1}{1500}$$

$$R_2 = (R_r' + 0.5R_d)N_{sr}^2 = 4(0.048 + 0.5 \times 0.1) = 0.392$$

$$R_1 = (R_r' + 0.5R_d)N_{sr}^2 \left[\frac{\pi^2}{9} - 1\right]$$

R_1 can be neglected.

$$V_{\text{eq}} = (1 - s_1)\frac{N_{sr}}{N_{LI}} V_s \cos\alpha = (1 - s_1) \times 0.358 \times 254 \times |\cos 93°| = 4.76(1 - s_1)$$

$$R = R_s + R_1 + \frac{R_2}{s_1}$$

$$= R_s + 0 + \frac{R_2}{s_1} = 0.04 + \frac{0.394}{s_1}$$

$$X = X_s + X_r' N_{sr}^2 = 0.28 + 0.105 \times 4 = 0.70\ \Omega$$

$$V^* = \frac{-4.76(1 - s_1)}{s_1}$$

$$I_{r1} = \frac{-RV^*}{R^2 + X^2} + \sqrt{\left(\frac{RV^*}{R^2 + X^2}\right)^2 + \frac{V_s^2 - (V^*)^2}{R^2 + X^2}}$$

$$\mathcal{T}_d = 147.03 = \frac{3}{s_1 \omega_s}[I_{r1}^2 R_2 + |V_{eq}| I_{r1}]$$

$$= \frac{3}{s_1 \times 157.1}[I_{r1}^2 \times 0.392 - 4.76 I_{r1}]$$

Assuming $s_1 = 0.05$, I_{r1} works out to 35.95 A and $\mathcal{T}_d$ works out to 255.52, which is nearly 1.738 times the rated torque (that is, 147.03 N m). Hence the motor speed N_x is obtained from the equation

$$\frac{1500 - N_x}{1500} = 0.05$$

giving $N_x = 1425$. Hence the motor speed for 1.738 times the rated torque and an α_i of 93° is 1425 rpm.

(c)

$$\text{Slip} = \frac{1500 - 1200}{1500} = \frac{300}{1500} = 0.2$$

$$V_{eq} = (1 - 0.2) \times 0.358 \times 254 \cos \alpha_i = 72.74 \cos \alpha_i$$

$$R = R_s + R_1 + \frac{R_2}{s} = 0.04 + 0 + \frac{0.392}{0.2}$$

$$= 0.04 + 1.96 = 2.0$$

$$X = X_s + X_r' \times 4 = 0.28 + 0.105 \times 4 = 0.7$$

$$V^* = \frac{V_{eq}}{s} = \frac{72.74}{0.2} \cos \alpha_i = 363.7 \cos \alpha_i$$

Assuming $\alpha_i = 102°$,

$$\cos \alpha_i = 0.2079$$

$$I_{r1} = \frac{-2 \times 363.7 \cos \alpha_i}{2^2 + (0.7)^2} + \sqrt{\left(\frac{2 \times 363.7 \cos \alpha_i}{2^2 + (0.7)^2}\right)^2 + \frac{254^2 - (363.7 \cos \alpha_i)^2}{2^2 + (0.7)^2}}$$

Substituting the value of $\cos \alpha_i$ gives

$$I_{r1} = -33.68 + 119.29 = 85.61$$

$$V_{eq} = 72.74 \times 0.2079 = 15.12$$

$$\mathcal{T}_d = \frac{3 \times 85.61}{0.2 \times 157.1}(85.61 \times 0.392 + 15.12) = 397.9$$

Check for T_d

2.7 times the rated torque is

$$2.7 \times 147.03 = 397.9 \text{ N m}$$

Check for I_{r1}

$$V^* = 363.7 \cos \alpha_i = 363.7 \times 0.2079 = 75.61$$

$$(RI_{r1} + V^*)^2 + (XI_{r1})^2 = (85.61 \times 2 + 75.61)^2 + (85.61 \times 0.7)^2 = 254^2$$

$$\tan \phi_r = \frac{59.93}{246.83} = 0.2428$$

$$\phi_r = -13.65°$$

$$\begin{aligned} I_s \angle \phi_s &= I_r \angle \phi_r + I_m \angle \phi_m \\ &= 85.61 \angle -13.65° + \frac{254}{12} \angle -90° \\ &= 85.61(0.9718 - j0.2359) - j21.17 \\ &= 83.20 - j41.37 \end{aligned}$$

$$I_L \angle \phi_L = I_s \angle \phi_s + I_T \angle \phi_T$$

Here,

$$\begin{aligned} I_T \angle \phi_T &= \frac{N_{sr}}{N_{LI}} I_r \angle \phi_r \\ &= 0.358 \times 85.61(0.9718 - j0.2359) \\ &= 29.79 - j7.23 \end{aligned}$$

So

$$\begin{aligned} I_L \angle \phi_L &= 83.20 - j41.37 + 29.79 - j7.23 \\ &= 112.99 - j48.6 \end{aligned}$$

$$\tan \phi_L = \frac{-48.6}{112.99} = -0.43$$

$$\phi_L = -23.27°$$

$$\cos \phi_L = 0.919$$

Hence the drive power factor is 0.919.

Exercises

Multiple Choice Questions

1. Even though the control of ac drives is more complex than that of dc drives, they prove to be cheaper and more reliable, and their operation is more flexible than dc drives because ________.
   (a) their current-carrying capacity is small
   (b) their voltage rating is small
   (c) of the absence of commutators
   (d) using a microprocessor allows one to replace hardware components with software, thus reducing the cost of circuitry
2. If the slip of an induction motor is 0.04 and the synchronous speed is 1500, then the rotor speed is ________ rpm.
   (a) 1560 (b) 1440
   (c) 1320 (d) 1280
3. Neglecting the friction and windage losses, the torque developed in an induction motor is proportional to ________.
   (a) $s/(s-1)$ (b) $1/(s-1)$,
   (c) $(s-1)/s$ (d) $(1-s)/s$
4. The ranges of slip for forward motoring and plugging are ________ and ________, respectively.
   (a) $0 \le s < 1, s < 0$ (b) $0 \le s < 1, s > 0$
   (c) $0 \le s < 1, s > 1$ (d) $0 \le s < 1, s \ge 0$
5. The range of rotor frequency in electrical rad/s for the regenerative mode of operation of an induction motor is ________, where ω is the input supply frequency in electrical rad/s.
   (a) $-\omega$ to ω (b) 0 to ω
   (c) ω to 2ω (d) $-\omega$ to 0
6. For a value of slip much greater than s_m, the ratio $\mathcal{T}_d/\mathcal{T}_{dm}$ will approximate to ________.
   (a) $s/2s_m$ (b) $2s/s_m$
   (c) $2/s_m s$ (d) $2s_m/s$
7. For a value of slip much less than s_m, the plot of $\mathcal{T}_d/\mathcal{T}_{dm}$ versus the slip is a straight line and has a slope equal to ________.
   (a) $4/s_m$ (b) $2/s_m$
   (c) $1/2s_m$ (d) $s_m/2$
8. Considering stator voltage control of an induction motor, if the operating point moves from a curve of higher voltage to one of lower voltage at the same torque, it leads to ________.
   (a) voltage spikes in the stator winding
   (b) lower I^2R losses in the stator winding and higher losses in the rotor winding
   (c) voltage spikes in the rotor winding
   (d) higher I^2R losses in the stator and rotor

9. In the speed control scheme for an induction motor using frequency control with constant stator voltage, decreasing the frequency below base frequency leads to ______.

   (a) increased I^2R losses in the machine
   (b) an increased degree of saturation of the magnetic circuit
   (c) decreased I^2R losses in the machine
   (d) a decreased degree of saturation of the magnetic circuit

10. In the speed control scheme for an induction motor using frequency control, considering the torque versus k plots (where $k = \omega/\omega_0$, with ω_0 as the base speed), if the peaks of these plots are joined by a smooth curve, this curve obeys the relationship ______.

    (a) $\mathcal{T}_{dm}\sqrt{k}$ = constant (b) $\mathcal{T}_{dm}/k$ = constant
    (c) $k/\mathcal{T}_{dm}$ = constant (d) $\mathcal{T}_{dm}k^2$ = constant

11. When the ratio E_s/ω is kept constant during the frequency control of an induction motor under low-slip operation, then the torque $\mathcal{T}_d$ ______.

    (a) depends only on the slip frequency $\omega_{R\omega}$ and is independent of ω
    (b) depends only on ω and is independent of $\omega_{R\omega}$
    (c) depends on both ω and $\omega_{R\omega}$
    (d) is independent of both $\omega_{R\omega}$ and ω

12. Stator current control of an induction motor is superior to stator voltage control due to the fact that ______.

    (a) it leads to low magnetic saturation
    (b) it has less torque pulsations
    (c) it gives high values for the amplitude and phase of the stator mmf wave
    (d) it provides fast and effective control of the amplitude and phase of the stator mmf wave

13. With stator current control of an induction motor, the breakdown torque ______ and the maximum per-unit torque ______.

    (a) occurs at a lower value of slip, increases
    (b) occurs at a lower value of slip, decreases
    (c) occurs at a higher value of slip, increases
    (d) occurs at a higher value of slip, decreases

14. If the stator resistance of an induction motor is negligible, then the slip for maximum torque is ______ and the maximum torque is ______.

    (a) $\pm X_s/(R_r + X_r)$, $2V_s^2\omega_s/(3X_s - X_r)$
    (b) $\pm X_r/(R_r + X_s)$, $2V_s^2/[3\omega_s^2(X_r + X_s)]$
    (c) $\pm R_r/(X_s + X_r)$, $3V_s^2/[2\omega_s(X_s + X_r)]$
    (d) $\pm R_r/(X_s + X_r)$, $3V_s^2/[4\omega_s(X_s - X_r)]$

15. Considering stator voltage control of an induction motor, as the magnitude of the stator voltage is decreased, the speed–torque curve ______.

    (a) becomes more steep at the origin and the peak is more sharply defined
    (b) becomes less steep at the origin and the peak is more sharply defined
    (c) becomes less steep at the origin and the peak portion becomes flatter
    (d) becomes more steep at the origin and the peak portion becomes flatter

16. Considering the speed variation of an induction motor using frequency control, it is better to keep E_s/ω constant rather than keeping V_s/ω constant for obtaining constant flux operation, because in the latter case, ______.
    (a) at high frequency both drops $I_s R_s$ and $I_s X_s$ are negligible
    (b) at high frequency the drop $I_s X_s$ is negligible but the drop $I_s R_s$ is predominant
    (c) at low frequency both the drops $I_s R_s$ and $I_s X_s$ are negligible
    (d) at low frequency the drop $I_s X_s$ is negligible but the drop $I_s R_s$ is predominant

17. Considering comprehensive control of an induction motor with constant stator voltage E_s and constant slip frequency $\omega_{R\omega}$, the torque developed is proportional to ______, where ω is the input frequency.
    (a) $\omega/2$ (b) $1/\omega^2$
    (c) ω (d) $1/\omega$

18. With constant counter-emf E_s and constant slip frequency, the torque developed in an induction motor will be ______.
    (a) directly proportional to $\omega^{1/2}$ (b) directly proportional to $\omega^{3/2}$
    (c) inversely proportional to ω^2 (d) inversely proportional to ω

19. For a cylindrical rotor synchronous motor, if the armature resistance is neglected, then the expression for maximum developed torque is given as ______.
    (a) $(-3p/\omega)(V_a X_s/E_f)$ (b) $(-3p/\omega)(V_a E_f/X_s)$
    (c) $(-3p/\omega)(E_f X_s/V_a)$ (d) $(-3p/\omega)(X_s/V_a E_f)$

20. For a salient pole synchronous machine, the drop $I_d X_d$ is ______ and the drop $I_q X_q$ is ______.
    (a) in phase with the armature voltage V_a, in quadrature with the field voltage E_f
    (b) in phase with the field voltage E_f, in quadrature with the armature voltage V_a
    (c) in phase with the armature voltage V_a, in quadrature with the armature voltage V_a
    (d) in phase with the armature voltage V_a, in quadrature with the field voltage E_f

21. Keeping the air gap flux constant for a synchronous motor has twin effects, namely, ______.
    (a) avoiding high I^2R losses as well as high voltage spikes
    (b) avoiding magnetic saturation as well as high I^2R losses
    (c) avoiding high voltage spikes and making better use of iron
    (d) avoiding magnetic saturation and making better use of iron

22. An induction motor is normally not operated in the plugging region because ______.
    (a) it leads to unstable operation
    (b) it leads to high voltage in the motor
    (c) it leads to high I^2R heating in the motor

(d) there is risk of the negative resistance causing a short circuit in the rotor circuit

23. In the regeneration region, the value of the slip s of an induction motor will be between ________.

(a) 1 and 2 (b) 0 and 1
(c) 0 and -1 (d) -1 and -2

24. Considering the two methods for varying the stator voltage of an induction motor, namely, (i) using ac voltage controllers and (ii) employing square wave PWM inverters, ________.

(a) the first method leads to high I^2R heating and the second to high harmonic content in the output
(b) the first method leads to high content of all harmonics and the second to high I^2R
(c) the first method gives more high-order harmonics whereas the second method has both high- and low-order harmonics
(d) the first method gives a large amount of both high- and low-order harmonics whereas the second method gives a high low-order harmonic content

25. Considering the frequency control of an induction motor, keeping V/ω constant throughout the speed range has the disadvantage that ________.

(a) the magnetic circuit gets saturated
(b) the speed–torque characteristic falls down rapidly for low values of ω_m
(c) high I^2R losses occur both in the stator and rotor
(d) the permitted power limit of the motor may be exceeded

26. In the stator current control of an induction motor, if the prescribed relationship between the stator current I_s and the slip frequency $\omega_{R\omega}$ is not maintained, it leads to ________.

(a) high I^2R losses in the stator and rotor
(b) a drastic reduction in the developed torque
(c) high magnetic saturation of the machine
(d) the introduction of high harmonic content

27. Increasing the value of the synchronous reactance in a cylindrical rotor synchronous motor leads to ________.

(a) a decrease in the power output
(b) an increase in the power output
(c) an increase in the reactive power with the power output remaining unaffected
(d) no change in the real and reactive power outputs

28. Assuming that the direct-axis reactance of a salient pole synchronous motor is equal to the synchronous reactance of a cylindrical rotor motor, and the other quantities remain the same in both cases, the power generated in a salient pole motor is ________ that generated in a cylindrical rotor type of motor.

(a) less than (b) equal to
(c) more than (d) much less than

29. Considering the rotor resistance control of an induction motor, if $N_{sr} = 2$, $R'_r = 0.05\ \Omega$, and $R_{out} = 0.03\ \Omega$, then the values of R_1 and R_2 are ________ and ________, respectively.
    (a) 0.10, 0.38 (b) 0.05, 0.29
    (c) 0.025, 0.26 (d) 0.005, 0.052

30. In the case of Scherbius control of an induction motor, the speed of the induction motor varies with the ________.
    (a) firing angle α_i of the inverter
    (b) resistance R_1
    (c) resistance R_2/s
    (d) input voltage of the diode rectifier

31. In the case of Scherbius control of an induction motor, if the ratios N_{sr} and N_{LI} are 2 and 4, respectively, and α is 150° then the slip will be nearly equal to ________.
    (a) 0.54 (b) 0.43
    (c) 0.74 (d) 0.48

32. In a Scherbius system, if the drive has to provide a speed range that is 30% below synchronous speed and α is 120°, then the ratio N_{sr}/N_{LI} will be ________.
    (a) 0.6 (b) 0.8
    (c) 0.45 (d) 0.30

33. In the Scherbius scheme, if $R'_r = 0.08\ \Omega$, $R_d = 0$, and $N_{sr} = 3$, the values of R_1 and R_2 will be nearly ________ and ________, respectively.
    (a) 0.03, 1.45 (b) 0.10, 0.90
    (c) 0.07, 0.72 (d) 0.12, 0.83

34. For the Scherbius scheme, the expression for the power recovered and fed back to the mains network is ________.
    (a) $(-1/3)(V_s/N_{sr})I'_{r1} \sin \alpha_i$ (b) $-3(V_s/N_{LI})I'_{r1} \sin \alpha_i$
    (c) $-3(V_s/N_{LI})I'_{r1} \cos \alpha_i$ (d) $-3(V_s/N_{sr})I'_{r1} \sin \alpha_i$

35. In the Scherbius scheme, if the dc link current is doubled, then the total copper loss P_{cr} ________ that of the original value.
    (a) becomes three times (b) becomes four times
    (c) becomes half (d) becomes double

36. A requirement for the successful operation of the Scherbius scheme is that the inductance of the dc link inductor should be ________.
    (a) large (b) small
    (c) very small (d) zero

37. In the case of rotor resistance control of an induction motor, if the stator-to-rotor turn ratio is increased, then the torque ________.
    (a) decreases (b) increases
    (c) remains unaffected (d) increases to a large value

38. In the case of rotor resistance control of an induction motor, if the value of the dc link current is 10 A, then the RMS value of the rotor current will be nearly ________ A.

(a) 7.5 (b) 9.6
(c) 8.2 (d) 10.3

39. In the case of rotor resistance control of an induction motor, if the duty ratio increases, then the average power through the external resistor ________.
(a) increases (b) remains the same
(c) decreases (d) increases to a very large value

40. In the case of rotor resistance control of an induction motor, if the rotor resistance varies, the slip for maximum torque ________ and the maximum torque ________.
(a) remains constant, varies
(b) varies, varies
(c) remains constant, remains constant
(d) varies, remains constant

41. In the case of a modern Scherbius system, the purpose of the three resistors at the three terminals of the rotor of the motor is to ________.
(a) increase or decrease the current during normal operation
(b) limit the current in the dc link inductor
(c) dissipate the energy when high voltages occur in the system
(d) start the motor in a smooth manner

True or False

Indicate whether the following statements are *true* or *false*. Briefly justify your choice.

1. The absence of commutators makes ac drives cheap and facilitates more reliable and flexible performance as compared to dc drives.
2. If the slip of an induction motor is 0.06 and the synchronous speed is 1500, then the rotor speed is 1425 rpm.
3. If the friction and windage losses in an induction motor are neglected, then the torque developed by it will be proportional to $s/(1-s)$.
4. The ranges of slip for forward motoring and plugging for an induction motor are $0 \leq s < 1$ and $s > 1$, respectively.
5. The range of rotor frequency in electrical rad/s, for the regenerative mode of an induction motor is $-\omega$ to ω, where ω is the input supply frequency in electrical rad/s.
6. When the stator resistance of an induction motor is neglected, then the slip for maximum torque is given as $\pm X_r/(R_r + X_s)$, whereas the maximum torque is given as $3V_s^2/[4\omega_s/2(X_r + X_s)]$.
7. Considering the stator voltage control of an induction motor, for a decrease in the magnitude of the stator voltage, the speed–torque characteristic becomes less steep at the origin and flatter at the peak region.
8. For the constant flux operation of an induction motor using frequency control, it is better to keep E_s/ω constant rather than V_s/ω, because in the latter case, at low frequencies, the drop $I_s X_s$ is negligible but the drop $I_s R_s$ is predominant, causing the flux and torque to reduce.
9. With the stator voltage E_s as well as slip frequency kept constant, the torque developed by an induction motor is proportional to $1/\omega$, where ω is the input electrical frequency.

10. Considering an induction motor operating in the field weakening region, with constant slip frequency and constant induced voltage E_s, the torque developed will be inversely proportional to the input frequency ω.
11. For a cylindrical rotor synchronous motor with negligible armature resistance, the expression for maximum developed torque will be $-(3p/\omega)(VE_f)/X_s$.
12. For a salient pole synchronous machine, the phasor for the drop $I_d X_d$ is in phase with the armature voltage V_a and that for the drop $I_q X_q$ is in quadrature with the field voltage E_f.
13. Keeping the air gap flux constant for a salient pole synchronous motor has twin advantages of avoiding magnetic saturation as well as making better use of the iron lamination.
14. An induction motor is normally not operated in the plugging region because there is risk of the negative rotor resistance causing a short circuit in the rotor circuit.
15. In the regeneration region, the value of the slip s of an induction motor will be between 1 and 2.
16. Considering the two methods for varying the stator voltage of an induction motor, namely, (a) using ac voltage controllers and (b) employing square wave PWM inverters, the latter method has the drawback of having high harmonic content of all orders in the output whereas the former has the drawback of high I^2R heating.
17. Considering frequency control of an induction motor, keeping V/ω constant throughout the range has the disadvantage that it leads to the saturation of the magnetic circuit.
18. In the stator current control of an induction motor, if the prescribed relationship between the stator current I_s and the slip frequency $\omega_{R\omega}$ is not maintained, then it leads to the introduction of a high harmonic content in the machine.
19. In a cylindrical rotor synchronous motor, increasing the value of the synchronous reactance decreases the power output.
20. Assuming that the direct-axis reactance of a salient pole synchronous motor is equal to the synchronous reactance of a cylindrical rotor motor, with the other quantities remaining the same in both the cases, the total power generated in a salient pole motor is less than that generated in a cylindrical rotor type of motor.
21. For a value of slip much greater than s_m, the ratio $\mathcal{T}_d/\mathcal{T}_{dm}$ will approximate to $2s_m/s$.
22. For a value of slip much less than s_m, the plot of $\mathcal{T}_d/\mathcal{T}_{dm}$ versus the slip will be a straight line with a slope equal to $1/2s_m$.
23. Considering the stator voltage control of an induction motor, if the operating point moves from a curve of higher voltage to one of lower voltage at the same torque, it leads to higher I^2R losses in the stator and rotor.
24. Considering a speed control scheme for an induction motor using frequency control, if the peaks of all the torque versus k plots are joined together by a smooth curve, this curve obeys the expression $\mathcal{T}_{dm}k^2 = \text{constant}$.
25. When the ratio E_s/ω is kept constant during the frequency control of an induction motor, the torque $\mathcal{T}_d$ depends only on ω and is independent of $\omega_{R\omega}$.
26. Stator current control of an induction motor is superior to stator voltage control because it gives high values for the amplitude and phase of the stator mmf wave.

27. With stator current control of an induction motor, the breakdown torque occurs at a higher value of slip and the maximum per-unit torque decreases.
28. Considering comprehensive control of an induction motor with constant stator voltage E_s as well as constant slip frequency, the torque developed will be proportional to $\omega/2$, where ω is the input frequency.
29. With constant slip frequency and constant induced voltage E_s the torque developed in an induction motor will be inversely proportional to $\omega^{1/2}$.
30. Considering the rotor resistance control of an induction motor, if $N_{sr} = 2$, $R_r' = 0.05\ \Omega$, and $R_{\text{out}} = 0.03\ \Omega$, then the values of R_1 and R_2 will be 0.05 and 0.29, respectively.
31. Considering Scherbius control of an induction motor, the speed of the induction motor varies with the firing angle α_i of the inverter.
32. In the case of Scherbius control of an induction motor, if the ratios N_{sr} and N_{LI} are 2 and 4, respectively, and α is 150°, then the slip will be nearly equal to 0.74.
33. In a Scherbius system, if the drive has to provide a speed range that is 30% below synchronous speed and the firing angle α_i is 120°, then the ratio N_{sr}/N_{LI} will be 0.6.
34. In the Scherbius scheme if $R_r' = 0.08\ \Omega$, $R_d = 0$, and $N_{sr} = 3$, the values of R_1 and R_2 will be nearly 0.12 and 0.83, respectively.
35. For the Scherbius scheme, the expression for the power recovered and fed back to the mains networks is $-3(V_s/N_{sr})I_{r1}' \sin \alpha_i$.
36. In the Scherbius scheme, if the dc link current is doubled with respect to its original value, then the total copper loss P_{cr} becomes four times its original value.
37. A requirement for the successful operation of the Scherbius scheme is that the inductance of the dc link inductor should be large.
38. In the case of rotor resistance control of an induction motor, if the stator-to-rotor turn ratio is increased, then the torque decreases.
39. During the rotor resistance control of an induction motor, if the value of dc link current is 10 A, then the RMS value of the rotor current will be nearly 7.5 A.
40. During the rotor resistance control of an inductance motor, if the duty ratio increases, then the average power through the external resistor decreases.
41. In the case of rotor resistance control of an induction motor, if the rotor resistance varies, then the slip for maximum torque remains constant.
42. In a modern Scherbius system, the three resistors located at the three terminals of the rotor of the motor are meant to facilitate the smooth starting of the motor.

Short Answer Questions

1. What are the advantages of ac drives over dc drives?
2. Derive an expression for torque with the help of an approximate equivalent circuit of the induction motor.
3. With the help of a diagram show the regions of forward motoring, plugging, and regeneration in the characteristic of an induction motor.
4. With the help of the approximation that the stator resistance can be neglected as compared to the other parameters, derive an expression for the ratio T_d/T_{dm} as a function of s and s_m. Give a suitable interpretation of this expression.

5. In stator voltage control, explain how the I^2R losses increase when the operating point is shifted to a lower voltage curve.
6. Considering speed variation using frequency control for an induction motor, explain why voltage/hertz control is preferred over simple frequency control in the low-frequency range?
7. Discuss the merits of constant E_s/ω control over constant V_s/ω control.
8. Derive an expression for developed torque as a function of speed and stator current.
9. How is constant torque operation below base speed achieved with the help of stator current control?
10. Explain how the torque developed in an induction motor is reduced under saturated conditions.
11. Describe a comprehensive speed control method both below and above base speed.
12. Describe the closed-loop speed control scheme for an induction motor using (a) a six-step inverter and (b) a cycloconverter.
13. Give a closed-loop slip regulation scheme with constant volts/hertz control for an induction motor.
14. Discuss stator voltage control with constant slip frequency as applied to an induction motor. What are its merits and demerits?
15. What are the advantages of stator current control over stator voltage control?
16. What are the different methods of providing excitation to a cylindrical rotor type of synchronous motor?
17. Explain the effects of armature reaction on the operation of a synchronous motor.
18. Derive an expression for the shaft torque developed by the cylindrical rotor type of synchronous motor.
19. Derive an expression for the shaft torque developed by the salient pole type of synchronous motor.
20. List out the assumptions made while arriving at an expression for the torque for a salient pole synchronous motor. Under what conditions do these assumptions fail to hold good?
21. List out the advantages of V/f control as applied to a synchronous motor.
22. Give a scheme for closed-loop speed regulation of an induction motor with the constant volts/hertz feature incorporated.
23. List out the various applications of synchronous drives.
24. Discuss the effect of rotor resistance variation on the speed–torque curves of an induction motor.
25. How is rotor resistance control implemented using a static converter?
26. Explain how the slip s depends upon the N_{sr}/N_{LI} ratio and firing angle α in the case of rotor resistance control.

27. Describe the conventional Scherbius scheme of slip energy recovery and enumerate its drawbacks.
28. Give the circuit of a modern Scherbius system and explain the functions of each of its modules.
29. Explain how speed is varied in the modern Scherbius scheme.
30. Draw the equivalent circuit of an induction motor for which the Scherbius scheme is applied and derive expressions for R_1, R_2, and V_1'/s.
31. List out the important assumptions that are made in connection with a static Scherbius system.
32. Justify the following statement for stator voltage control: 'The range of speed is further reduced for a characteristic with the load torque proportional to the square of speed, as could be seen from the dashed line in Fig. 8.5.'

Problems

1. A three-phase, 400-V, 50-Hz, four-pole, star-connected induction motor is connected to an ac mains network through a three-phase ac voltage controller. The data are ac mains voltage = 460 V, $R_s = 0.32\ \Omega$, $R_r = 0.16\ \Omega$, $X_s = 0.85\ \Omega$, $X_r = 0.75\ \Omega$, and $X_m = 24\ \Omega$, all parameters being referred to the stator. The rotor speed is 1465 rpm. Assuming negligible no-load losses, compute the (a) firing angle of the ac controller, (b) slip, (c) air gap power, (d) slip for maximum torque s_m, (e) maximum torque T_{dm}, and (f) efficiency.

   Ans. (a) 65°, (b) 0.0233, (c) 20.52 kW, (d) 0.098, (e) 261 N m, (f) 92.3%

2. A three-phase, 420-V, 50-Hz, four-pole, star-connected induction motor operated with stator voltage control has the following data: $R_s = 0.3\ \Omega$, $R_r = 0.17\ \Omega$, $X_s = 0.93\ \Omega$, $X_r = 0.67\ \Omega$, and $X_m = 22\ \Omega$, all parameters being referred to the stator. A load with a constant torque characteristic is supplied by the motor with $T_{ld} = 63$ N m. Considering the base voltage to be the motor terminal voltage, compute the following quantities with base voltage and 0.75 times the base voltage: (a) rotor current and (b) efficiency. [Hint: First find the approximate slip and then obtain the correct value by trial and error.]

   Ans. With base voltage: (a) 13.96 A, (b) 96.9%; with 0.75 times the base voltage: (a) 19.0 A, (b) 93.7%

3. A three-phase, 50-Hz, four-pole, star-connected induction motor is supplied by a three-phase ac voltage controller with an input voltage of 460 V (line-to-line) and $\alpha = 60°$, the corresponding output being considered as the base voltage V_s. The motor supplies a load torque that obeys the equation $T_{ld}s^2 = 0.15$. If stator voltage control is implemented for this motor, determine the approximate values of the (a) slip, (b) torque developed by the motor, (c) slip for maximum torque, (d) maximum torque, and (e) efficiency both with base voltage V_s and 0.65 times the base voltage. Other data are $R_s = 0.28\ \Omega$, $R_r = 0.16\ \Omega$, $X_s = 0.91\ \Omega$, $X_r = 0.68\ \Omega$, and $X_m = 23\ \Omega$, all parameters being referred to the stator.

   Ans. With base voltage: (a) 0.03, (b) 169.8 N m, (c) 0.099, (d) 286 N m, (e) 88.68%; with 0.65 times the base voltage: (a) 0.0374, (b) 84.0 N m, (c) 0.099, (d) 120.8 N m, (e) 85.3%

4. A three-phase, 50-kW, 1475-rpm, 410-V, 50-Hz, four-pole, star-connected induction motor has the following parameters: $R_s = 0.41\ \Omega$, $R_r = 0.24\ \Omega$, $X_s = 1.05\ \Omega$, $X_r = 0.75\ \Omega$, and $X_m = 29\ \Omega$, all parameters being referred to the stator. The motor is operated with frequency control. If the slip for maximum torque at an unknown supply frequency is 0.127, determine the (a) supply frequency, (b) breakdown torque, and (c) speed at maximum torque.

   Ans. (a) 51.28 Hz, (b) 226.8 N m, (c) 1343 rpm

5. A three-phase, 52-kW, 1470-rpm, 420-V, 50-Hz, four-pole, star-connected induction motor has the following data: $R_s = 0.4\ \Omega$, $R_r = 0.25\ \Omega$, $X_s = 0.9\ \Omega$, $X_r = 0.9\ \Omega$, and $X_m = 27\ \Omega$, all parameters being referred to the stator. The motor is operated with frequency control. If the breakdown torque is 216 N m at an unknown supply frequency, determine (a) this frequency, (b) the slip for maximum torque, and (c) the speed for maximum torque.

   Ans. (a) 54.25 Hz, (b) 0.125, (c) 1424 rpm

6. A three-phase, 50-kW, 986-rpm, 400-V, 50-Hz, six-pole, star-connected induction motor has the following parameters: $R_s = 0.38\ \Omega$, $R_r = 0.22\ \Omega$, $X_s = 1.00\ \Omega$, $X_r = 0.8\ \Omega$, and $X_m = 25\ \Omega$, all parameters being referred to the stator. The motor is subjected to frequency control. If the breakdown torque at the supply frequency is 206 N m, determine (a) this frequency, (b) the speed for maximum torque, and (c) the efficiency with a slip of 0.025.

   Ans. (a) 53 Hz, (b) 940 rpm, (c) 93.5%

7. A three-phase, 400-V, 1480-rpm, 50-Hz, four-pole, star-connected induction motor has the following data: $R_s = 0.4\ \Omega$, $R_r = 0.20\ \Omega$, $X_s = 0.85\ \Omega$, $X_r = 0.95\ \Omega$, and $X_m = 29\ \Omega$ all parameters being referred to the stator. Using frequency control the frequency is increased to 55 Hz. Determine the (a) speed at breakdown torque, (b) value of breakdown torque, and (c) stator and rotor copper losses for a slip of 0.02.

   Ans. (a) 1487 rpm, (b) 191.3 N m, (c) stator Cu loss 571 W, (d) rotor Cu loss 285 V

8. A 30-kHz, 400-V, 50-Hz, four-pole, star-connected induction motor is operated with constant volts/Hz control corresponding to the rated voltage and frequency. The parameters of the machine are $R_s = 0.31\ \Omega$, $R_r = 0.20\ \Omega$, $X_s = 0.95\ \Omega$, $X_r = 0.85\ \Omega$, and $X_m = 26\ \Omega$, all parameters being referred to the stator. If it has to supply a constant load torque of 110 N m, compute for the rated frequency the (a) slip for given T_{ld}, (b) slip for $\mathcal{T}_{dm}$, (c) breakdown rotor current, and (d) efficiency. Repeat the calculations for a frequency of 45 Hz.

   Ans. With rated frequency: (a) 0.0242, (b) 0.109, (c) 82.5 A, (d) 94.4%; with a frequency of 45 Hz: (a) 0.027, (b) 0.121, (c) 81.7 A, (d) 93.3%

9. A 420-V, 50-Hz, six-pole, star-connected induction motor is subjected to constant volts/Hz control corresponding to rated voltage and frequency. The data are $R_s = 0.25\ \Omega$, $R_r = 0.18\ \Omega$, $X_s = 0.85\ \Omega$, $X_r = 0.95\ \Omega$, and $X_m = 32\ \Omega$, all parameters being referred to the stator. If it has to supply a constant load torque of 105 N m, determine the following quantities at 50 Hz and 42 Hz: (a) speed for the given T_{ld}, (b) slip for maximum torque, (c) air gap power, and (d) efficiency.

   Ans. With 50 Hz: (a) 982 rpm, (b) 0.99, (c) 16.53 kW, (d) 95.7%; with 42 Hz: (a) 978 rpm, (b) 0.117, (c) 13.88 kW, (d) 94.9%

10. A 400-V, 50-Hz, four-pole, star-connected induction motor is operated with constant volts/Hz control corresponding to the rated voltage and frequency. Other data are $R_s = 0.32\ \Omega$, $R_r = 0.20\ \Omega$, $X_s = 1.2\ \Omega$, and $X_r = 0.7\ \Omega$, all parameters being referred to the stator. If it has to operate at a reduced frequency such that the slip for maximum torque is 0.15, determine the (a) new supply frequency, (b) breakdown torque with the new frequency, (c) speed when a load torque of 75 N m is to be supplied, and (d) rotor current to provide this torque.

Ans. (a) 34 Hz, (b) 209.6 N m, (c) 976 rpm, (d) 18.0 A

11. A 25-kW, 400-V, 50-Hz, six-pole, star-connected induction motor is operated with constant volts/Hz corresponding to rated voltage and frequency. The motor parameters are $R_s = 0.28\ \Omega$, $R_r = 0.26\ \Omega$, $X_s = 1.2\ \Omega$, $X_r = 0.68\ \Omega$, and $X_m = 21\ \Omega$, all of them being referred to the stator. At a reduced supply frequency, the slip for maximum torque is 0.20. Determine the (a) new supply frequency, (b) breakdown torque with this frequency, (c) slip required to provide a torque of 100 N m at this new frequency, (d) stator current with the slip as in (c), and (e) efficiency with the slip as in (c).

Ans. (a) 22.8 Hz, (b) 196.6 N m, (c) 0.0675, (d) 29.26, (e) 86.9%

(12) A 30-kW, 420-V, 50-Hz, four-pole, star-connected induction motor is operated with constant volts/Hz control corresponding to rated voltage and frequency. The parameters are $R_s = 0.31\ \Omega$, $R_r = 0.25\ \Omega$, $X_s = 1.2\ \Omega$, $X_r = 0.8\ \Omega$, and $X_m = 30\ \Omega$, all of them being referred to the stator. If it has to operate at a reduced frequency such that the breakdown torque is 225 N m, calculate the (a) new supply frequency, (b) slip required to give maximum torque, (c) slip to provide a load torque of 105 N m at this frequency, (d) stator and rotor copper losses at this slip, and (e) efficiency at this slip.

Ans. (a) 34.75 Hz, (b) 0.176, (c) 0.0385, (d) rotor Cu loss 442 W, stator Cu loss 683 W, (e) 95%

12. A 32-kW, 440-V, 50-Hz, 980-rpm, six-pole, star-connected induction motor is operated with constant volts/Hz control corresponding to rated voltage and frequency. The parameters are $R_s = 0.25\ \Omega$, $R_r = 0.28\ \Omega$, $X_s = 1.1\ \Omega$, $X_r = 0.9\ \Omega$, and $X_m = 31\ \Omega$, all of them being referred to the stator. At a reduced frequency the breakdown torque is given as 216 N m. Compute the (a) new supply frequency, (b) slip required to provide load torque that obeys the relation $T_{ld}s^2 = 0.11$, and (c) efficiency with the slip as in (b).

Ans. (a) 17.25 Hz, (b) 0.043, (c) 92.2%

13. A 416-V, 50-Hz, four-pole, star-connected induction motor is operated with a current source of 32 A as its input. The parameters are $R_s = 0.32\ \Omega$, $R_r = 0.16\ \Omega$, $L_s = 5.5$ mH, $L_r = 7.5$ mH, and $L_m = 0.09$ H, all of them being referred to the stator. If it has to supply a load torque of 95 N m, find the (a) slip required to provide this torque, (b) slip required to give maximum torque, (c) breakdown torque, and (d) magnetizing current required to maintain the air gap flux at its no-load value with rated voltage and frequency.

Ans. (a) 0.027, (b) 0.00523, (c) 255.2 N m, (d) 1.63 A

14. A 420-V, 50-Hz, six-pole, star-connected induction motor is operated with a constant current source of 29 A as its input. The motor data are $R_s = 0.30\ \Omega$, $R_r = 0.18\ \Omega$, $L_s = 6.2$ mH, $L_r = 6.9$, and $L_m = 0.1$ H, all parameters being

referred to the stator. If the slip for a particular load torque is 0.010 H, determine the (a) torque, (b) slip required to provide maximum torque, and (c) breakdown torque.

Ans. (a) 63.35 N m, (b) 0.00536, (c) 236 N m

15. A 420-V, 50-Hz, four-pole, star-connected induction motor operated with a constant current source has a breakdown torque of 215 N m. Other data are $R_s = 0.35\ \Omega$, $R_r = 0.18\ \Omega$, $L_s = 8.5$ mH, $L_r = 0.6$ mH, and $L_m = 0.08$ H, all parameters being referred to the stator. Determine the (a) current source value, (b) slip required to give a load torque of 100 N m, and (c) magnetizing current required to maintain the air gap flux constant at its no-load value with rated voltage and frequency.

Ans. (a) 30 A, (b) 0.0287, (c) 7.23

16. A 440-V, 50-Hz, six-pole, star-connected induction motor supplied by a constant current source of 31 A operates with a slip value that gives a maximum torque of 0.0055. Determine the (a) rotor resistance and (b) slip required to give a load torque of 110 N m. Other data are $R_s = 0.34\ \Omega$, $L_s = 9.0$ mH, $L_r = 0.9$ mH, and $L_m = 0.1$ H.

Ans. (a) 0.174 Ω, (b) 0.0274

17. A three-phase, 400-V, 50-Hz, 1472-rpm, four-pole, wound rotor induction motor is to be provided with rotor resistance control. If maximum torque occurs for a chopper duty ratio of 0.25 and a slip of 0.31, determine (a) the external resistance and (b) torque for a chopper duty ratio of 0.55 and a slip of 0.05. The stator-to-rotor turn ratio is 1.8 and the resistance of the link inductor is 0.015. The motor parameters are $R_s = 0.10\ \Omega$, $X_s = 0.29\ \Omega$, $R_r = 0.06\ \Omega$, and $X_r = 0.45\ \Omega$, all of them being referred to the stator.

Ans. (a) 0.0025 Ω, (b) 335 N m

18. A three-phase, 420-V, 50-Hz, 1468-rpm, four-pole, wound rotor induction motor is to be equipped with the rotor resistance control feature. Maximum torque occurs for a chopper duty ratio of 0.32 and a slip of 0.57. Other data are $R_s = 0.12\ \Omega$, $X_s = 0.30$, $R_r = 0.08\ \Omega$, and $X_r = 0.48$, all the parameters being referred to the stator side, the stator-to-rotor turn ratio is 2.2, and the resistance of the link inductor is 0.018 Ω. Determine (a) the external resistance and (b) torque for a duty ratio of 0.42 and a slip of 0.08.

Ans. (a) 0.202, (b) 247.3 N m

19. Consider the induction motor of Problem 18. For this motor determine (a) the duty ratio for a speed of 1410 rpm and a torque equal to 1.24 times the rated torque and (b) the speed for a duty ratio of 0.65 and a torque equal to 0.895 times the rated torque.

Ans. (a) 0.4, (b)1440 rpm

20. Consider the induction motor of Problem 18. For this motor determine (a) the duty ratio for a speed of 1420 rpm and a torque equal to 0.584 times the rated torque and (b) the speed for a duty ratio of 0.68 and a torque equal to 0.5 times the rated torque.

Ans. (a) 0.6, (b) 1440 rpm

21. A 400-V, 50-Hz, four-pole, 1472-rpm, wound rotor induction motor is subjected to Scherbius control. The data are $R_s = 0.13\ \Omega$, $X_s = 0.33\ \Omega$, and

X_m = 23 Ω; R_r and X_r (both referred to the stator side) are, respectively, 0.10 Ω and 0.159 Ω. The drive has to provide a speed range that is 35% below synchronous speed. Assume that R_d = 0.009 Ω. If N_{sr} = 2.1, determine (a) the ratio N_{sr}/N_{LI} with α_m equal to 165° and (b) the torque and drive power factor for α_i = 105° and a speed of 800 rpm. The transformer is now removed and the inverter connected directly to the ac mains; also, α_i is altered to give the same torque and slip as in (b). (c) Determine this new α_i and the drive power factor.

Ans. (a) 0.362; (b) 290.7 N m, 0.446; (c) 92.9°, 0.457

22. A 440-V, 50-Hz, six-pole, 964-rpm, wound rotor induction motor is equipped with the Scherbius control feature. The data are R_s = 0.12 Ω, X_s = 0.31 Ω, and the rotor parameters R'_r = 0.12 Ω and X'_r = 0.129 Ω, and X_m = 28 Ω. The drive has to provide a speed range that is 25% below synchronous speed. Assume that R_d is negligible. If N_{sr} = 2.3, determine (a) the ratio N_{sr}/N_{LI} with α_m equal to 168° and (b) the torque and drive power factor for α_i = 98° and a speed of 550 rpm. The transformer is now removed and the inverter connected directly to the ac mains; also, α_i is adjusted to give 1.11 times the torque but the same slip as in (b). (c) Determine this new α_i and the drive power factor.

Ans. (a) 0.256; (b) 328.1 N m, 0.399; (c) 91.7°, 0.439

23. A 420-V, 50-Hz, 1470-rpm, four-pole, star-connected induction motor has Scherbius control applied to it. The stator parameters are R_s = 0.06 Ω, X_s = 0.28 Ω, X_m = 12 Ω and the rotor parameters are R'_r = 0.02 Ω, X'_r = 0.10 Ω. The drive is designed to give a speed range that is 32% below synchronous speed and the maximum firing angle α_{im} is 155°. R_d can be taken as zero and N_{sr} = 1.9. Determine (a) the ratio N_{sr}/N_{LI}, (b) α_i for obtaining 1.4 times the rated torque at a speed of 780 rpm, and (c) the drive power factor for the conditions in (b). The friction, windage, and core losses are negligible.

Ans. (a) 0.353, (b) 101°, (c) 0.337

24. A 440-V, 50-Hz, 980-rpm, six-pole, star-connected induction motor is provided with Scherbius speed control. The stator parameters are R_s = 0.08 Ω, X_s = 0.29 Ω, X_m = 15 Ω and the rotor parameter are R'_r = 0.02 Ω and X'_r = 0.09 Ω. The drive is designed to give a speed range that is 45% below synchronous speed and the maximum firing angle α_{im} is 163°. If N_{sr} = 1.6 and R_d = 0.01 Ω, compute (a) the ratio N_{sr}/N_{LI}, (b) α_i for obtaining 2.29 times the rated current at a speed of 760 rpm, and (c) the torque and drive power factor for the conditions in (b). The core, friction, and windage losses are negligible.

Ans. (a) 0.471; (b) 113.6° (approx); (c) 863.17 N m, 0.875

CHAPTER 9

Brushless Dc Motors

9.1 Introduction

The dc motor is widely used because of its merits, namely, (i) ease of variation of speed and (ii) the linear speed–torque characteristics in its range of operation except in the high-torque region. Its main drawback is the commutator, which makes it bulky and costly, and this imposes limitations on the design. Its field poles are situated on the stator, making the field stationary. Though the armature conductors carry alternating currents, the commutator and brushes are so arranged that the spatial distribution of the armature current direction as well as armature magnetomotive force (mmf) remain constant for all speeds. The stator field and armature mmf waves interact to provide a steady torque at any permissible speed. To obtain sparkless commutation, the brushes are placed in the neutral axis, thus giving a displacement of 90° between the field and armature mmf waves.

Ac motors are preferred in applications where the commutator is undesirable, but there are still some areas in which they cannot replace dc motors. This gave the impetus for the development of the *brushless dc motor*, which is a combination of (a) an ac machine, (b) a solid-state inverter, (c) electronic control circuitry, and (d) a rotor position sensor. These modules constitute a drive system that provides a linear speed–torque characteristic similar to that of the conventional dc motor. The construction of this ac motor is the same as that of a self-controlled synchronous machine with a permanent magnet rotor. Here, the inverter is the power converter, with the ac machine connected at its output. The frequency of the inverter is changed in proportion to the speed of the ac motor so that the armature and the rotor mmf waves revolve at the same speed. This results in the production of a steady torque at all speeds as in a dc motor. The three components, namely, the rotor position sensor, the electronic control circuitry, and the inverter, perform the same function as the brushes and the commutator in a dc motor. A slight difference between the brushless dc motor and the conventional dc motor is that the former need not operate at a displacement of 90° between the field and armature mmf waves, as it is free from the problem of sparking. This displacement is usually set at a suitable value so that the performance requirements are satisfied.

Brushless dc motors are suited for applications requiring high starting torque, good efficiency at low speeds, and continuous speed variation from standstill to full speed. With its low torque ripple, it is a strong contender for high-performance servo drives used in the machine tool industry as well as those employed in robotics.

After going through this chapter the reader should

- understand the principle of the brushless dc motor,
- know the constructional features of the electronic commutator, which is the heart of the brushless dc motor,
- get acquainted with the various types of sensors used with the electronic commutator,
- become familiar with the mechanism of torque production in the sinusoidal-type two-phase, brushless dc motor and also in the three-phase, half- and full-wave configurations,
- know the method of implementation of the closed-loop control of a brushless dc drive,
- get acquainted with the different types of current controllers that can be used in the closed-loop scheme,
- become familiar with the recent developments in this field, and
- know how the performance of the brushless dc motor differs from those of the three conventional motors, namely, the dc motor, the induction motor, and the brushless synchronous motor.

Brushless dc motors are suited for applications requiring high starting torque, good efficiency at low speeds, and continuous speed variation from standstill to full speed. With its low torque ripple, it is a strong contender for high-performance servo drives used in the machine tool industry as well as those employed in robotics.

9.2 Sinusoidal and Trapezoidal Brushless Dc Motors

A distinction is made in the industry between the following two types of brushless motors. The self-controlled synchronous machine with a standard permanent magnet rotor, whose air gap flux distribution and back emf waveform are both sinusoidal, is called a *sinusoidal brushless dc machine*. On the other hand, the term *brushless dc motor* is usually assigned to a self-synchronous machine, in which these two waveforms are trapezoidal in nature. These two types of brushless dc motors have the following common features: the drive characteristics and the control methods are the same, and in both cases the motor must be energized with controlled currents that are synchronized with the rotor position. They, however, differ in the following aspects of construction and operation. The standard synchronous motor requires sinusoidal current excitation, whereas the trapezoidal brushless dc machine is supplied with a square-wave or quasi-square wave current. The rotor position sensor for the trapezoidal brushless dc motor usually consists of simple position detectors such as magnetic position sensors,

Hall effect devices, or optical position sensors. By sensing the rotor magnetic field, these sensors determine the phase-switching points. On the other hand, more precise position information has to be given to the sinusoidal brushless dc motor, so that the resulting armature mmf waveforms are accurately positioned. The trapezoidal brushless dc motor will henceforth be called the *brushless dc motor*.

9.3 Electronic Commutator

The most important part of a brushless dc motor is the electronic commutator, in which the following events take place. The signals from the rotor position sensor act on the electronic control circuitry and the resulting output signals operate appropriate switches in the inverter circuit. The development of the brushless dc motor is explained below. In a dc motor, the stator contains salient poles, and sometimes the interpoles also, which are energized by a field coil. The rotor, usually called the armature, carries a current which is supplied by an external dc source. The rotor currents have to be switched periodically in order to keep the two fields as nearly perpendicular to each other as possible for maintaining unidirectional torque between the rotor and stator. In a conventional dc motor, this switching is done with the help of the mechanical commutator and the brushes. Figures 9.1(a) and (b), respectively, show a two-segment commutator for a two-pole machine and the connection of the coil sides to the commutator segments. The external dc supply is connected to these segments through fixed brushes. Figure 9.2 shows the waveform of the torque, as a function of the electrical position, for this simple two-commutator segment. It is seen that the torque is unidirectional but pulsating in nature.

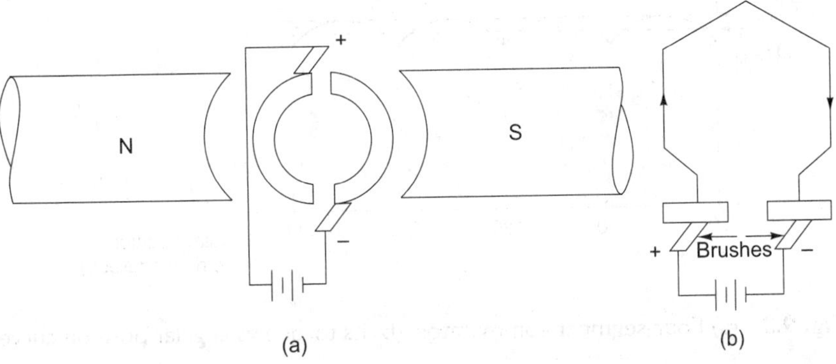

Fig. 9.1 Connection of (a) a two-segment commutator and (b) coil sides

The electrical angles are measured with respect to the neutral axis. Maximum torque occurs when the coil sides are situated at the centres of the north and south poles, respectively. Now, if the number of coil sides, and hence the number of commutator segments, is doubled, as shown in Fig. 9.3(a), the torque waveform will become fairly smoother [Fig. 9.3(b)]. The mean torque in this case can be calculated to be 0.9 times the maximum torque.

It is possible to design a dc motor in which the rotor has permanent magnet poles and the stator current is switched. However, in this case the brushes have

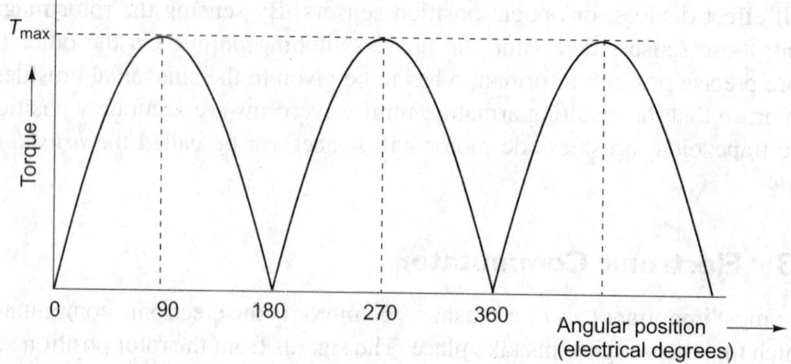

Fig. 9.2 Torque vs angular position curve for a two-segment commutator

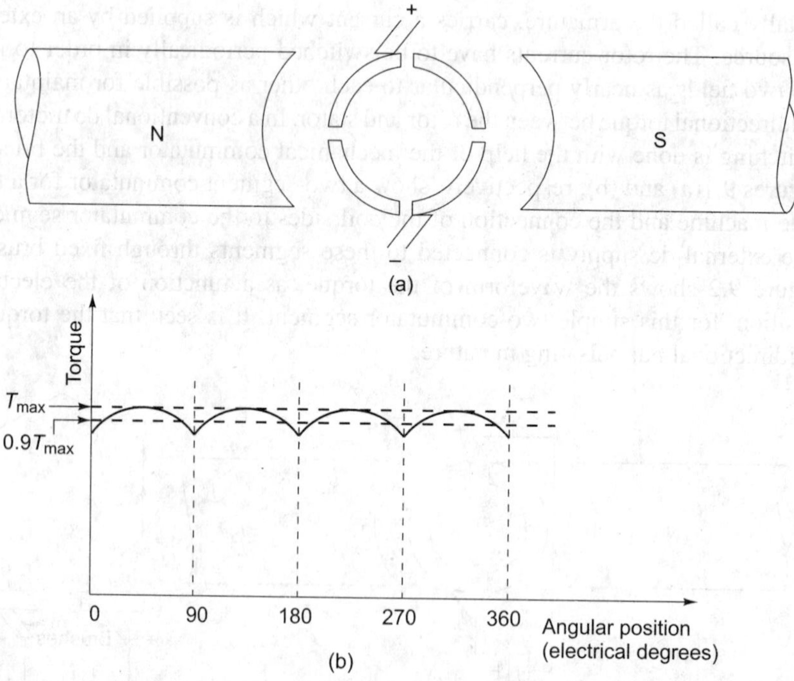

Fig. 9.3 (a) Four-segment commutator; (b) its torque vs angular position curve

to rotate at the same speed as the rotor to maintain unidirectional torque. Though such a rotor–stator combination may not give optimum performance, as in the conventional machine, it can be advantageously used in electronic commutator circuits. This is because with a stationary commutator it is easier to operate the electronic devices that are used as switches. The spread out of a typical inverted dc machine with coils and switches is shown in Fig. 9.4. The stator flux can be controlled as required, by means of the switches connected to the positive or negative buses, at the appropriate moments.

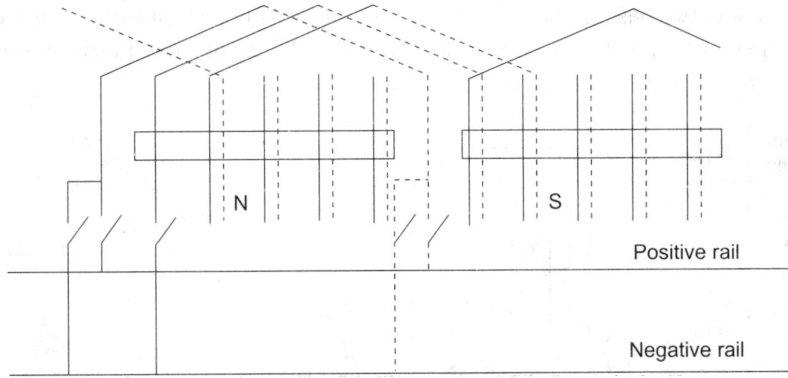

Fig. 9.4 Switching circuit for an electronic commutator

This inverted machine will have a rotor of the salient-pole type. Hence, the field flux constitutes the major portion of the air gap flux, and the effect of armature reaction can be neglected.

Though the circuit of Fig. 9.4 has been used in some electronic commutator motors, its drawback is that it employs a large number of switching devices and is hence expensive. It is stated above that a four-coil-side armature (stator) which is accompanied with a four-segment commutator gives a fairly high mean torque and will be optimum from the point of view of the number of commutator segments, because it gives a fairly smooth torque function with minimum number of switches. Figure 9.5(a) shows one such four-segment electronic commutator, which is meant to be used with a two-phase ac synchronous motor having windings A and B. The switch positions that give appropriate polarities for the windings are given in Fig. 9.5(b).

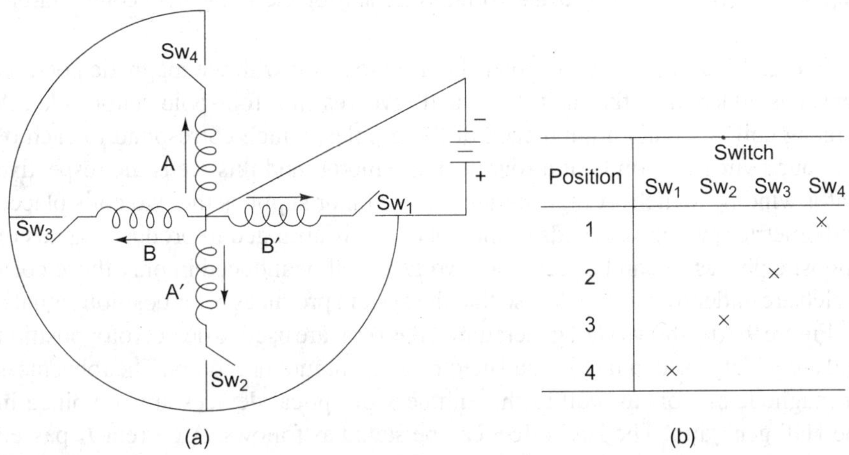

Position	Switch Sw_1	Sw_2	Sw_3	Sw_4
1				×
2			×	
3		×		
4	×			

Fig. 9.5 (a) Four-segment electronic commutator (b) table of electronic switching

Figure 9.6 shows the implementation of the electronic commutator for a two-phase ac motor. With the *p-n-p* and *n-p-n* transistors playing the roles of switches, the base signals for the transistors of Fig. 9.6 are obtained from a magnetic rotor position sensor as shown in Fig. 9.7(a), or from a Hall sensor as shown in Fig. 9.7(b) (Mazda 1990).

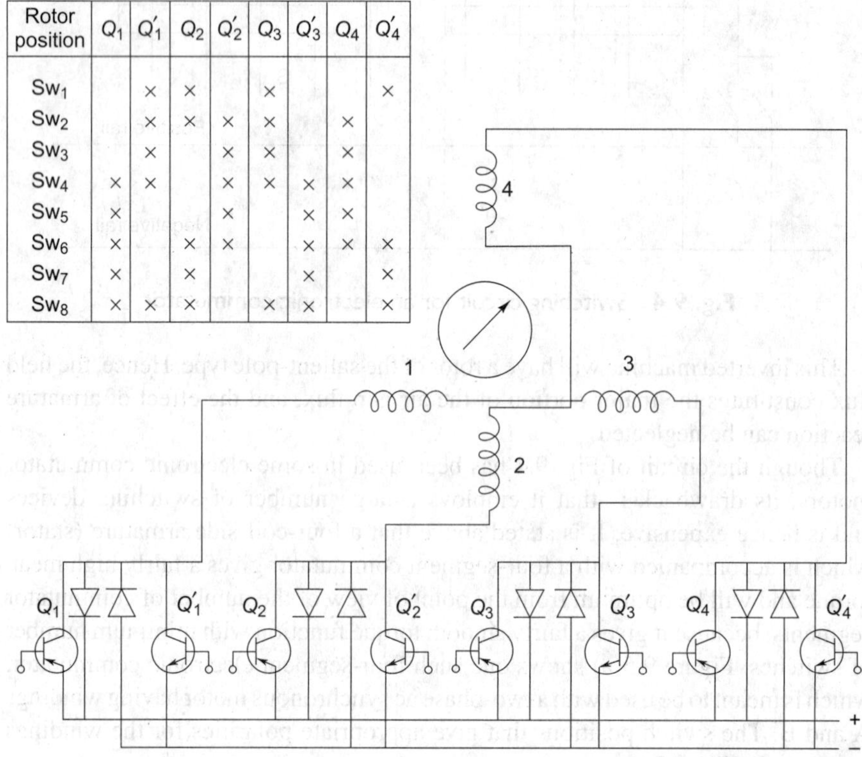

Rotor position	Q_1	Q_1'	Q_2	Q_2'	Q_3	Q_3'	Q_4	Q_4'
Sw_1		×	×		×			×
Sw_2		×	×	×	×		×	
Sw_3		×		×	×		×	
Sw_4	×	×		×	×	×	×	
Sw_5	×			×		×	×	
Sw_6	×		×	×		×	×	×
Sw_7	×		×			×		×
Sw_8	×	×	×		×	×		×

Fig. 9.6 Detailed circuitry of electronic switching for a four-segment commutator

In Fig. 9.7(a), the auxiliary rotor that is associated with the magnetic position sensor is attached to the main rotor and revolves in a four-pole stator yoke. A sensing coil is wound around each of these poles, which correspond to each of the four switching points of a four-segment motor, and this feeds the respective stator winding switch. A high-frequency oscillator supplies the two coils placed at diametrically opposite ends of the yoke, and is arranged to produce signals of opposite phases. It can be seen that a voltage will be induced in only those coils which are under the rotor poles, so that the system produces rotor position signals.

Figure 9.7(b) shows Hall generators (HGs) that are used to detect rotor position and are widely used in many small electronic commutator motors. The robustness of magnetic sensors as well as the lightness of optical devices are combined in the Hall generator. The Hall effect can be stated as follows. A current I_c passed between the two faces of a thin conductor or a semiconductor placed in a transverse magnetic field B would result in a redistribution of the charge carried within the device and the induction of a voltage across it in a direction perpendicular to both the current flow and the magnetic field. This voltage, termed as the *Hall voltage*, is proportional to both I_c and B.

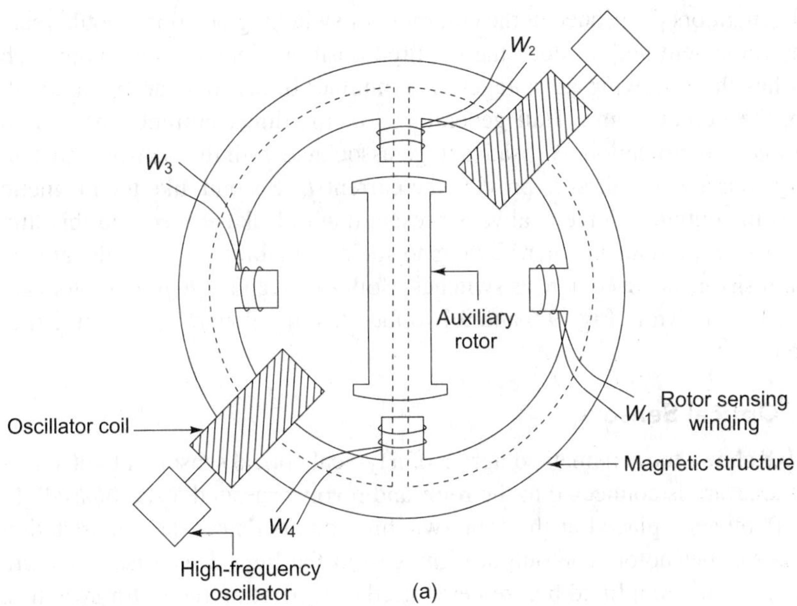

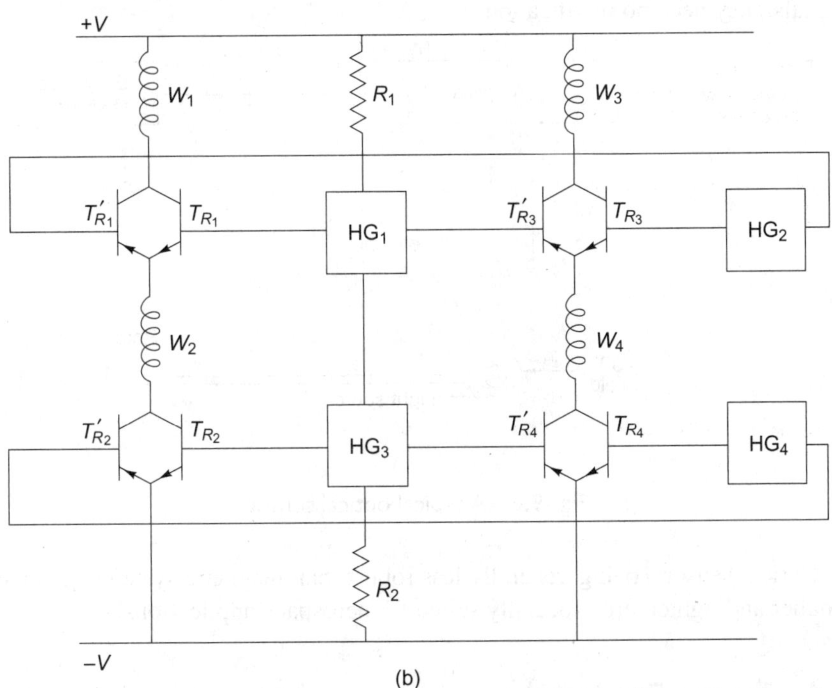

Fig. 9.7 (a) Magnetic rotor position sensor; (b) a typical Hall sensor consisting of Hall generators, transistors, resistors, and inductors

Hall generators positioned at the commutator switching positions would react with the rotor field and produce the required position signals. Employing such sensors has the following advantages. The number of sensors can be minimal. Second, the output from a Hall generator is of the direct current type and so does not need rectification. A disadvantage associated with this sensor is that an auxiliary source is required to provide the current I_c. Second, like the magnetic sensor, some output voltage is always present due to leakage flux, and this flux builds up in a gradual manner. Owing to their availability with high ratings, power transistors are used here as switches. Both n-p-n and p-n-p transistors are connected as shown in Figs 9.6 and 9.7 to facilitate the flow of current in either direction.

9.3.1 Optical Sensors

Figure 9.8 shows the schematic diagram of a typical optical sensor. A light shield with an aperture is connected to the rotor and revolves around a stationary light source. Photocells placed at the four switching points detect the commutation positions for the motor. The output signals from the light detectors, which are usually weak, are amplified before being used for operating the stator switches. The signals can, however, be made to rise sharply and, being unidirectional signals, they need no rectification.

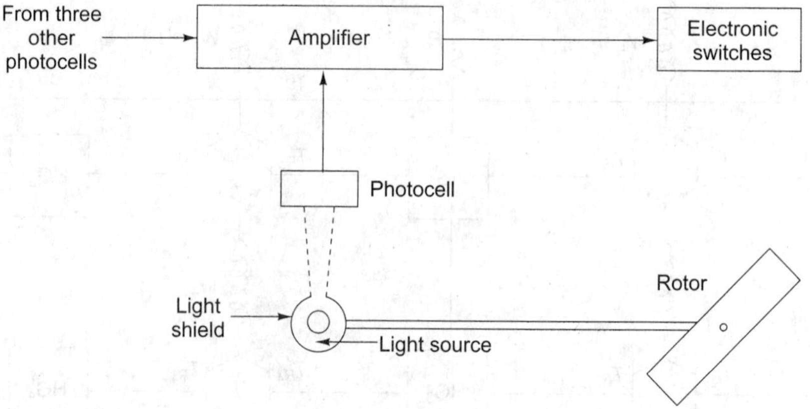

Fig. 9.8 A typical optical sensor

Optical sensors, being generally less robust than magnetic systems but much smaller and lighter, are especially suited for aerospace applications.

9.4 Torque Production

In this section, the unidirectional torque production feature in the case of a brushless dc motor is first examined for a sinusoidally operated two-phase, brushless motor, and then extended to the three-phase, brushless, half-wave and full-wave configurations.

9.4.1 The Sinusoidal-type Two-phase, Brushless Dc Motor

The sinusoidal-type two-phase, brushless dc motor (Murphy & Turnbull 1988) is just a self-controlled synchronous motor with a permanent magnet rotor, as shown in Fig. 9.9(a). It is designed in such a way that the torques generated in the two phases are functions, respectively, of the sine and cosine of the rotor angle [Fig. 9.9(b)].

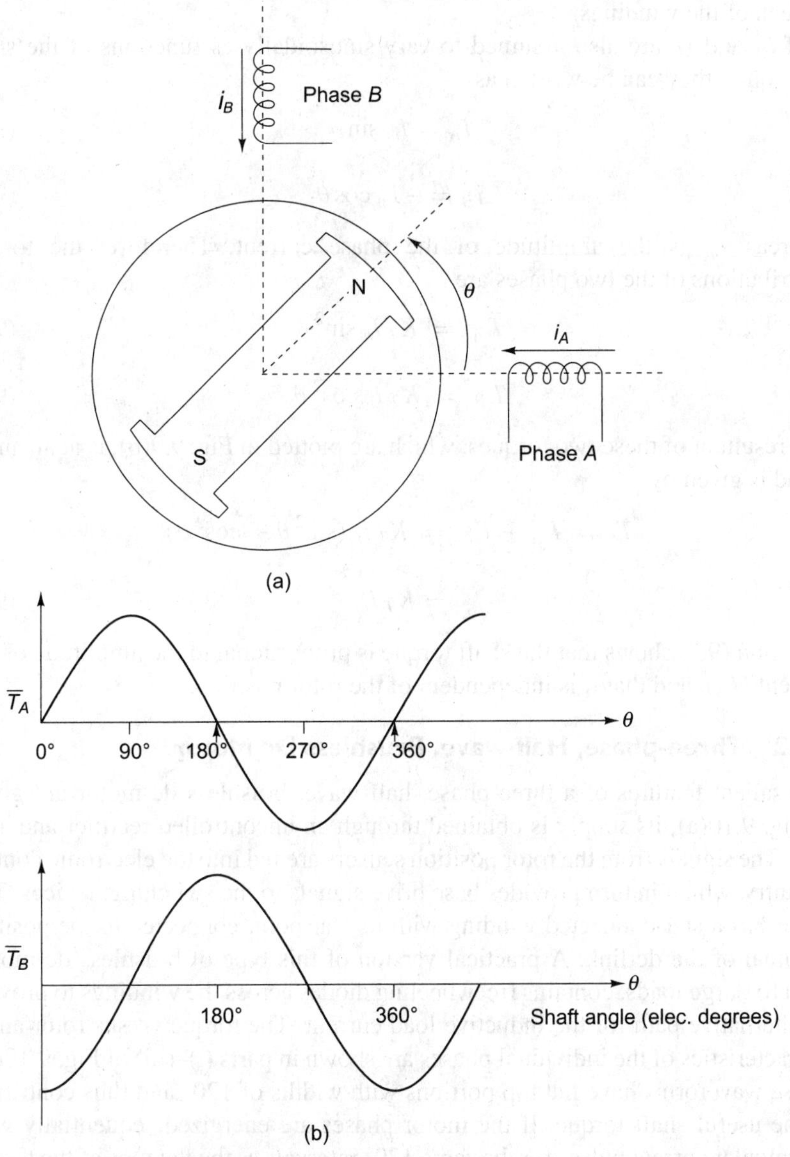

Fig. 9.9 (a) Schematic diagram of a two-phase, brushless dc motor; (b) torque–position waveforms

Thus, they are expressed as

$$\overline{T}_A = K_T i_A \sin\theta \tag{9.1}$$

$$\overline{T}_B = -K_T i_B \cos\theta \tag{9.2}$$

where i_A and i_B are the instantaneous phase currents and K_T is the torque constant of each of the windings.

If i_A and i_B are also assumed to vary sinusoidally as functions of the shaft position θ, they can be written as

$$i_A = I_m \sin\theta \tag{9.3}$$

$$i_B = -I_m \cos\theta \tag{9.4}$$

whereas I_m is the amplitude of the phase current. Therefore, the torque contributions of the two phases are

$$\overline{T}_A = K_T I_m \sin^2\theta \tag{9.5}$$

$$\overline{T}_B = K_T I_m \cos^2\theta \tag{9.6}$$

The resultant of these two torques, which are plotted in Fig. 9.9(b), is at an angle θ and is given by

$$T = \overline{T}_A + \overline{T}_B = K_T I_m(\sin^2\theta + \cos^2\theta)$$

$$= K_T I_m \tag{9.7}$$

Equation (9.7) shows that the shaft torque is proportional to the amplitude of the current (I_m), and that it is independent of the rotor position.

9.4.2 Three-phase, Half-wave, Brushless Dc Motor

The salient features of a three-phase, half-wave, brushless dc motor are given in Fig. 9.10(a); its supply is obtained through an uncontrolled rectifier and a dc link. The signals from the rotor position sensors are fed into the electronic control circuitry, which in turn provides base drive signals to the switching devices. The stator has a star-connected winding with its star point connected to the positive terminal of the dc link. A practical version of this type of brushless dc motor, used for large loads, contains freewheeling diodes across the windings to provide an alternative path for the inductive load current. The torque versus rotor angle characteristics of the individual phases are shown in parts (i)–(iii) of Fig. 9.10(b). These waveforms have flat top portions with widths of 120° and thus contribute to the useful shaft torque. If the motor phases are energized sequentially with rectangular current pulses for the same 120° intervals as the flat tops of the torque waveform, the motor develops a steady positive torque that is independent of the shaft position. The phase currents for positive torque production in the three phases are also shown in parts (iv)–(vi) of Fig. 9.10(b) and for negative torque production in parts (vii)–(ix). Each motor phase current has a conduction angle

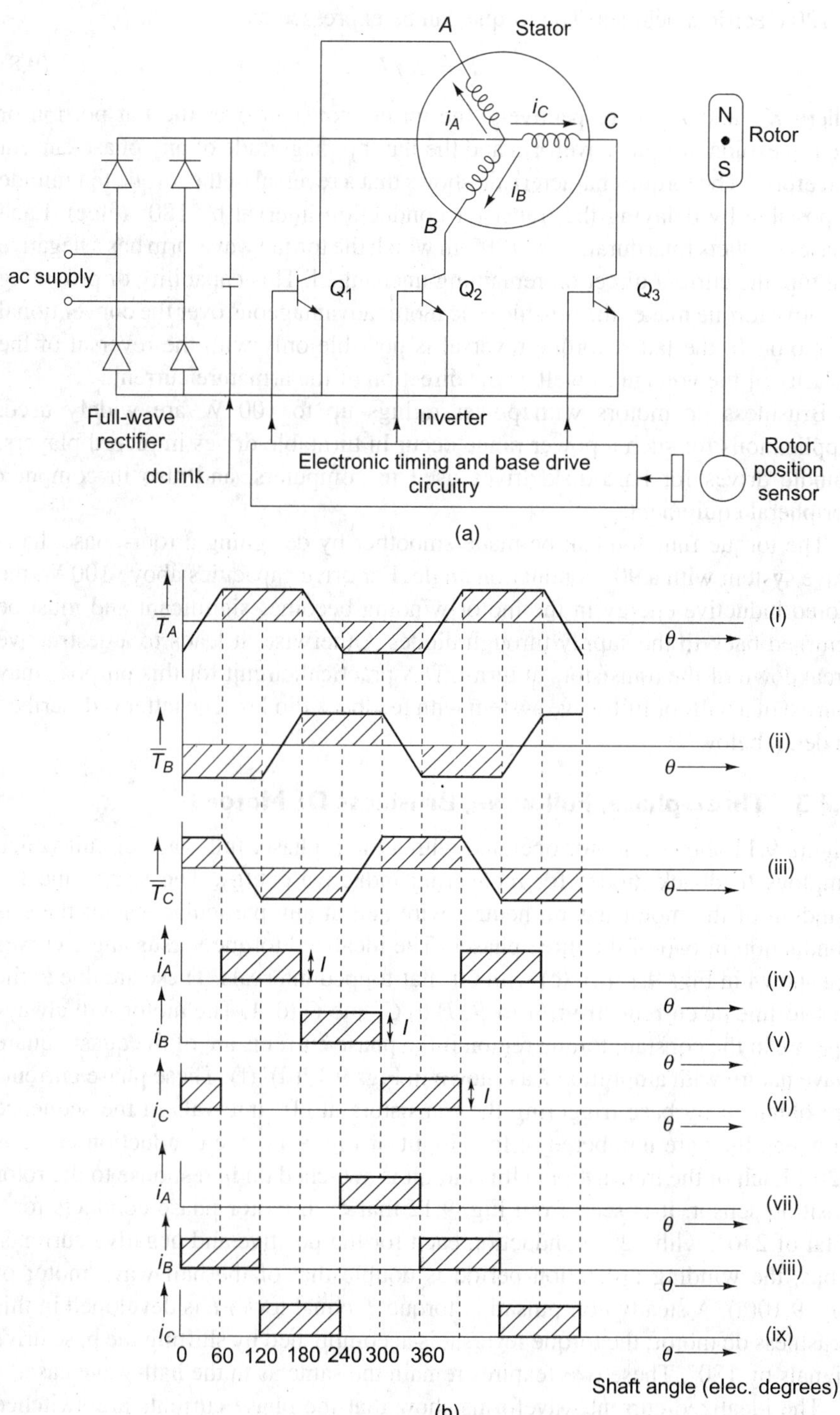

Fig. 9.10 (a) Circuit of a three-phase, half-wave, brushless dc motor; (b) idealized waveforms giving torque vs shaft angle characteristics for *A*, *B*, *C* phases; currents in *A*, *B*, *C* phases for positive and negative torque

of 120 electrical degrees. The torque can be expressed as

$$T = K_T I \tag{9.8}$$

where K_T and I are, respectively, the torque constant over the flat portion of the trapezoidal torque waveform and the flat top magnitude of any phase current waveform. The torque characteristic shows that a reversal of the torque magnitude is possible by delaying the transistor conduction interval by 180° (elec). Each phase conducts for a duration of 120°, in which the torque waveform has a negative flat top, the current direction remaining unchanged. This capability of providing negative torque makes the brushless dc motor advantageous over the conventional dc motor. In the latter, torque reversal is possible only with the reversal of the polarity of the voltage as well as the direction of the armature current.

Brushless dc motors with power ratings up to 100 W are widely used. Applications for such a power range occur in turntable drives in record players, spindle drives for hard-disk drives used in computers, and also in computer peripheral equipment.

The torque function can be made smoother by designing a four-phase, half-wave system with a 90° conduction angle. For drive capacities above 100 W, the stored inductive energy in the motor winding becomes significant and must be returned back to the supply through diodes. Otherwise, it leads to a destructive breakdown of the transistors at turn-off. A practical circuit for this purpose may consist of a half- or full-wave system with feedback diodes. The latter is described in detail below.

9.4.3 Three-phase, Full-wave, Brushless Dc Motor

Figure 9.11 shows a motor operated with a three-phase, full-wave circuit which employs feedback diodes for recovering inductive energy. The star-connected winding of the motor has no neutral wire and at any particular instant there is conduction in two of the three phases. The idealized torque versus angle curves are shown in Figs 9.12(a)–(c) with 60° flat-topped intervals. These are due to the line-to-line dc currents from A to B, B to C, and C to A. The motor will always operate in the constant torque region if the phase currents are of the quasi-square wave nature with amplitude I as shown in Figs 9.12(d)–(f). These phase currents are obtained by base triggering the transistors at 60° intervals in the sequence in which they are numbered in the circuit of Fig. 9.11; the conduction angle is 120°. Each of the transistors in this circuit is switched on in response to the rotor position sensor. It is seen from Fig. 9.12 that each motor phase conducts for a total of 240°, with 120° conduction each for the positive and negative currents. Thus, the winding utilization period is double that of the half-wave motor of Fig. 9.10(a). A steady non-pulsating torque T equal to $K_T I$ is developed in this brushless dc motor; the torque reversal is accomplished by shifting the base drive signals by 180°. These two features remain the same as in the half-wave case.

The idealized current waveforms show that the phase currents are switched instantly from one phase to the other. In practice, however, the inductive nature of the load will delay both the build-up as well as the collapse of the currents.

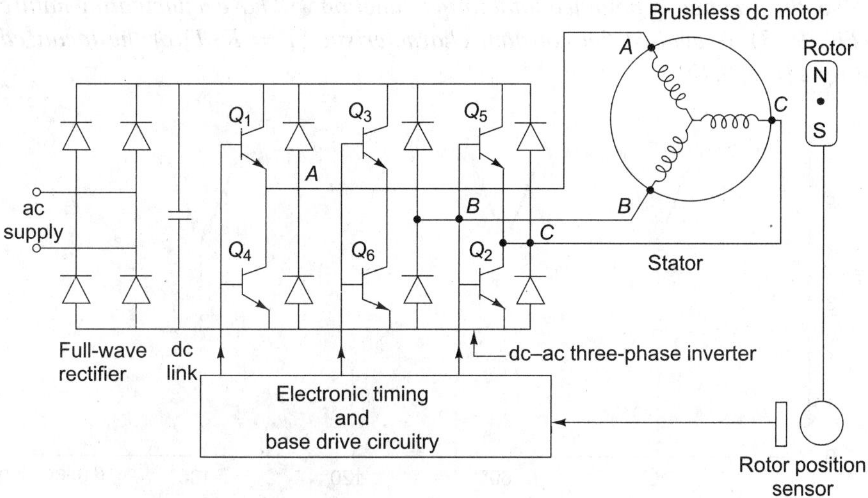

Fig. 9.11 Circuit of a three-phase, full-wave, brushless dc motor

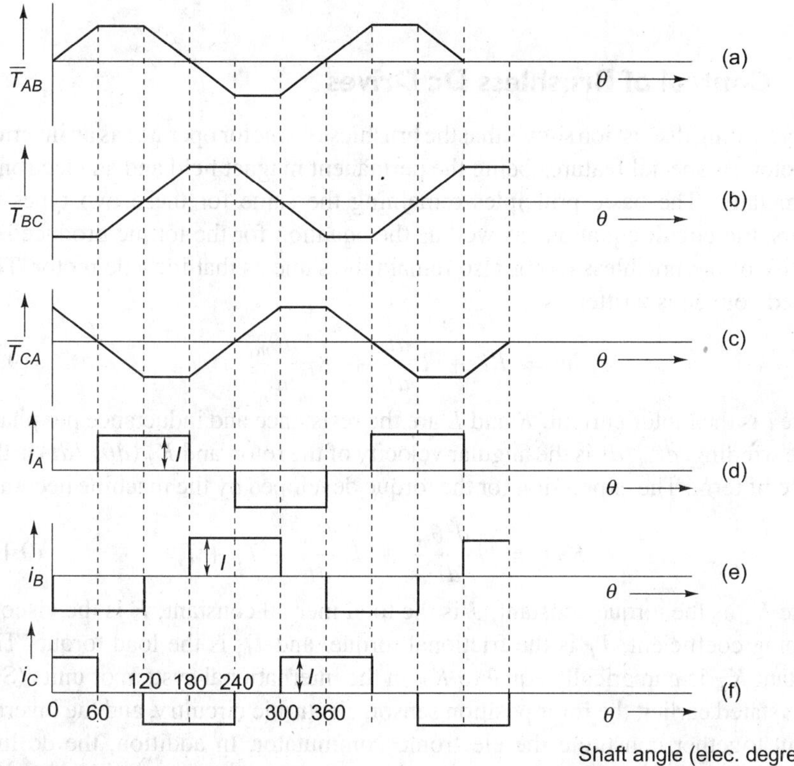

Fig. 9.12 Idealized waveforms for a three-phase, full-wave, brushless ac motor consisting of static torque vs shaft angle characteristics $\overline{T}_{AB}$, $\overline{T}_{BC}$, and $\overline{T}_{CA}$ and phase currents i_A, i_B, and i_C

Also, *the practically obtained total torque function will have a fluctuating nature (Fig. 9.13) as against the constant characteristic* $[T = K_T I]$ *of the idealized torque (Fig. 9.12).*

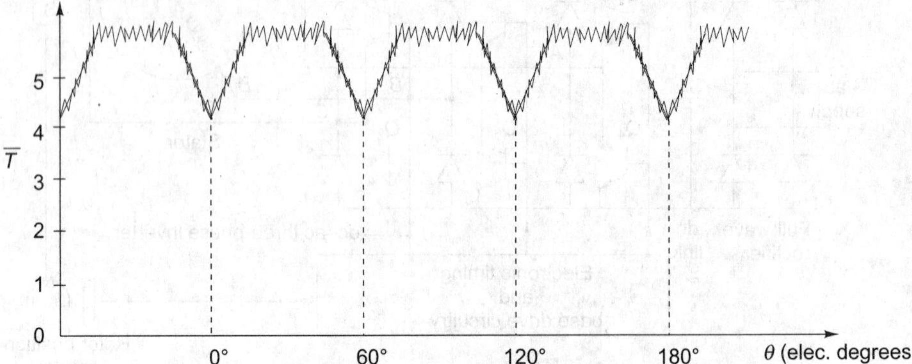

Fig. 9.13 A typical practical waveform of total torque vs shaft angle for the three-phase, full-wave brushless dc motor

9.5 Control of Brushless Dc Drives

The preceding discussion shows that the brushless dc motor operates as an inverted dc motor, its special features being the permanent magnet field and an electronic commutator. The basic principles remaining the same for these two types of motors, the circuit equations as well as the equation for the torque produced in the case of the brushless motor also remain the same as that for a dc motor. The applied voltage is written as

$$v = Ri + L\frac{di}{dt} + K_E\frac{d\theta_m}{dt} \tag{9.9}$$

where i is the motor current, R and L are the resistance and inductance per phase of the winding, $d\theta_m/dt$ is the angular velocity of the rotor, and $K_E(d\theta_m/dt)$ is the back emf term. The expression for the torque developed by the machine becomes

$$\overline{T} = K_T i = J\frac{d^2\theta_m}{dt^2} + F\frac{d\theta_m}{dt} + T_f + T_{\text{ld}} \tag{9.10}$$

where K_T is the torque constant, J is the total inertial constant, F is the viscous damping coefficient, T_f is the frictional torque, and T_{ld} is the load torque. The constant K_T is numerically equal to K_E in the international system of units (SI).

As stated earlier, the rotor position sensor, electronic circuitry, and the inverter all put together constitute the electronic commutator. In addition, the dc link voltage and link current correspond to the armature voltage and current in the conventional dc motor. Speed control of the brushless dc motor can be effected by varying the dc link voltage with the help of a dc chopper or a pulse width modulated controller. The input voltage to the electronic commutator is thereby varied and, in turn, this voltage causes variation of the motor speed.

The inner current control loop module that serves as a torque control feature in conventional dc drives can also be implemented in brushless dc drives by operating the dc link in the current controlled PWM mode. A better method is to utilize the devices used in the inverter to regulate the amplitude of the motor current by PWM control and effect the changeover from one phase to another as per the signals obtained from the rotor position sensor. For this purpose, the actual current signal is obtained either from the motor leads or from the dc link. Current feedback can also be incorporated in a conventional PWM current control loop. Current controller IC chips are now available, their circuit consisting of six output power transistors, a PWM current control circuit, and a Hall logic decoder. Thermal and under-voltage protection features are incorporated in such circuits.

9.5.1 A Typical Brushless Dc Drive

Figure 9.14 shows the schematic-cum-block diagram of a typical dc brushless drive consisting of the inner current loop and the outer speed loop. The drive consists of a three-phase uncontrolled bridge rectifier, a dc link together with an LC filter, and a MOSFET-based voltage source inverter.

The shaft speed of the motor is measured using a tachometer and compared with a reference speed to give the error in speed. The latter is fed into a PID (proportional integral and derivative) controller, which generates the torque signals. A limiter is provided to restrict this torque to the value of the reference torque T^*. In addition, it limits the winding currents to the maximum permissible value and also provides protection to the inverter switches. The reference current generator then takes the T^* signal as well as the rotor position signals and produces the signals I_R^*, I_Y^*, and I_B^* based on the output of the limiter. The hysteresis current controller uses these signals to give out control signals that operate the inverter switches. The net effect is that this controller forces the winding current to remain close to the reference current signals. Two salient features of this drive are described below.

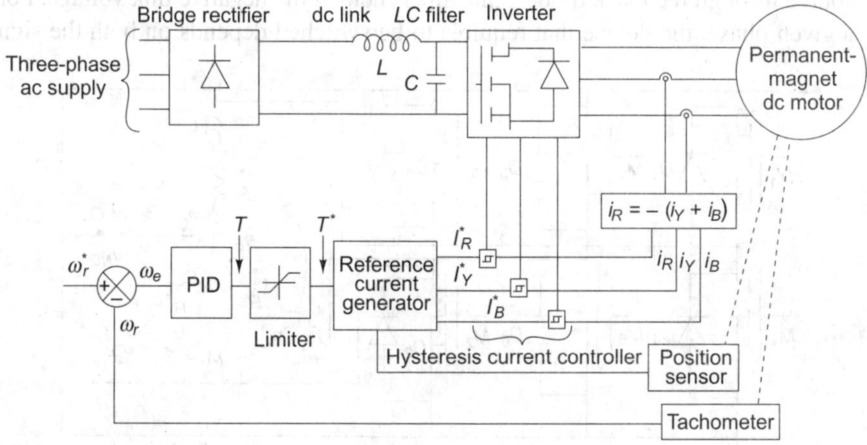

Fig. 9.14 A typical closed-loop brushless dc drive

9.5.1.1 Hysteresis Current Controller With the hysteresis controller comparing the actual current with the command signal as described earlier, if the magnitude of the error is greater than the present tolerance value, the inverter is switched appropriately. There is a dead band, called the hysteresis band, within which the inverter switches do not operate. Typical single-phase reference and actual hysteresis currents are shown in Fig. 9.15. The following discussion holds good for a trapezoidal current waveform also. The hysteresis current controller operates separately on each phase, Fig. 9.15 showing the current in one phase.

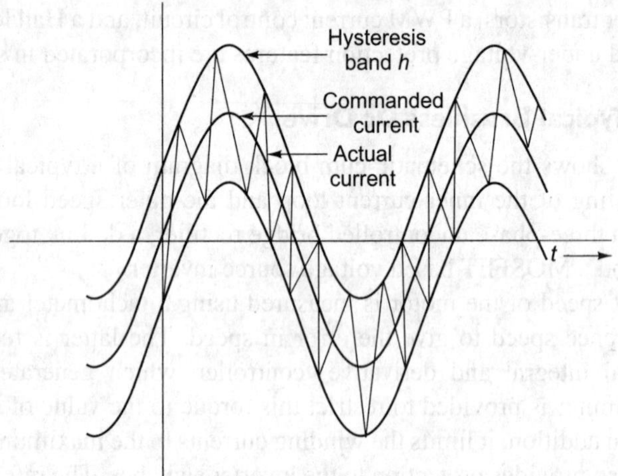

Fig. 9.15 Waveforms for hysteresis control

9.5.1.2 Inverter In the three-phase inverter circuit shown in Fig. 9.16, each output line is held at the positive dc link voltage when the MOSFETs are switched into conduction. On the other hand, when the inductive energy flows back into the source through feedback diodes, this line is held at the negative link voltage. For a given phase, the device that requires to be switched depends on both the sign

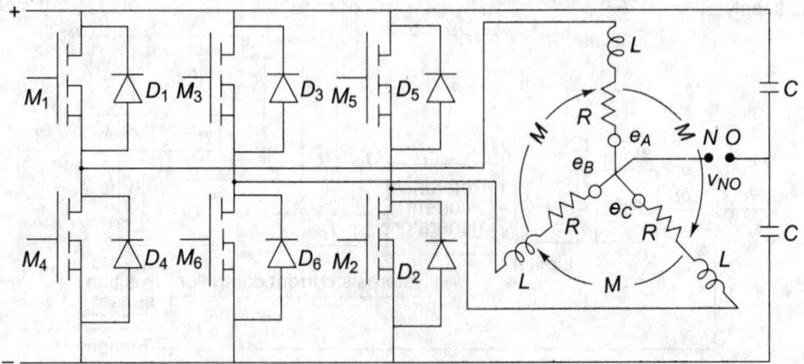

Fig. 9.16 MOSFET-based six-step inverter

of the error and that of the reference current I^*. For a hysteresis band of width h, the algorithm for the current control of the inverter leg having MOSFETs M_1 and M_4 is as follows.

- If $I^* > 0$ and $I > (I^* + h/2)$, turn off M_1.
- If $I^* > 0$ and $I < (I^* - h/2)$, turn on M_1.
- If $I^* < 0$ and $I > (I^* + h/2)$, turn on M_4.
- If $I^* < 0$ and $I < (I^* - h/2)$, turn off M_4.

These steps hold good for other inverter legs also. Thus, the hysteresis current regulator keeps down the machine winding currents so that they lie within the specified band. A drawback of this controller is that the switching frequency of the hysteresis is unknown and depends on the motor parameters, namely, speed, and also the dc bus voltage. The frequency also varies over an electrical cycle.

The ideal shapes of the induced emfs and currents in the three phases are given in Fig. 9.17.

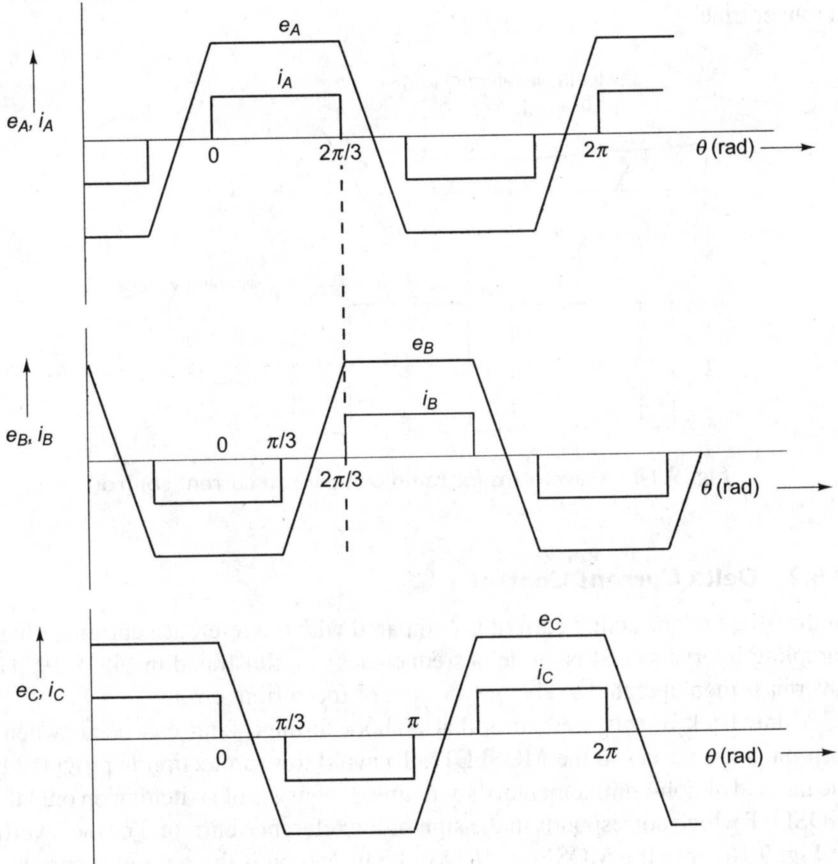

Fig. 9.17 Ideal wave shapes of induced emfs and currents

9.6 Other Current Controllers

Apart from the hysteresis current controller, some other types of current controllers, detailed below, can be employed in the brushless dc drive.

9.6.1 Ramp Comparison Current Control

In this method, the current error is compared with a triangular reference waveform. When the error is greater than the instantaneous magnitude of the triangular waveform, the control signal switches the inverter devices (MOSFETs) on. Again, when the error magnitude is lower, the devices are switched off. Figure 9.18 shows a typical switching voltage waveform. The switching algorithm is as follows, with the notation I meant for the reference current.

- If $I^* > 0$ and $i_{\text{error}} > i_{\text{triangle}}$, turn M_1 on or else turn M_1 off.
- If $I^* < 0$ and $i_{\text{error}} > i_{\text{triangle}}$, turn M_4 on or else turn M_4 off.

With an increase in the error, the appropriate MOSFET remains in conduction for a longer time.

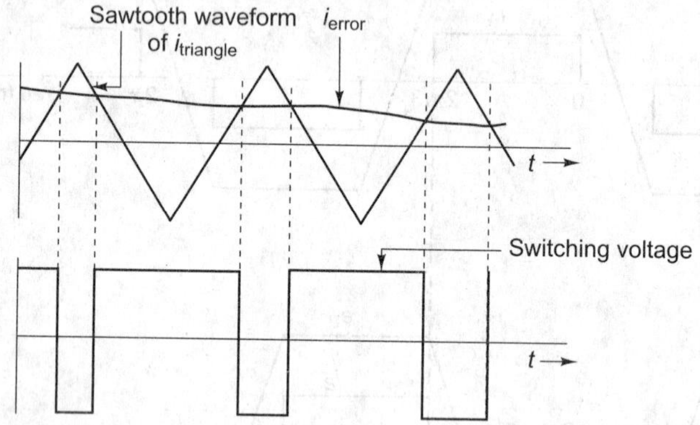

Fig. 9.18 Waveforms for ramp comparison current control

9.6.2 Delta Current Control

In this strategy, the actual current is compared with the reference current at fixed sampling intervals (that is, at delta frequencies), as illustrated in Fig. 9.19. The inverter is then operated to change the sign of the current error.

A drawback of this method is that a shoot through fault can occur when a turn-on delay occurs in the MOSFETs. To avoid this, protection is provided by the method of non-complementary switching. It consists of switching on only that MOSFET which corresponds to the sign of the reference current. For the inverter of Fig. 9.16, only the MOSFET M_1 will be turned on if the reference current is positive. The other device on the leg (M_4) will not be turned on as long as this condition remains.

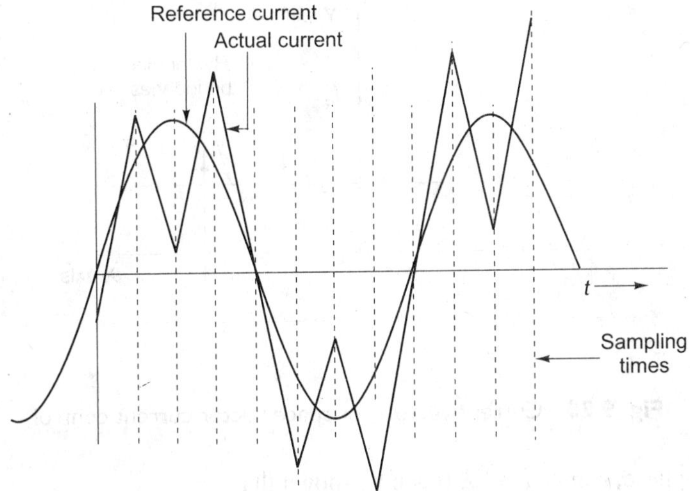

Fig. 9.19 Waveforms for delta current control

9.6.3 Space Vector Current Control

This method consists of using the space vector current error, which is defined *in the case of loads with the isolated neutral*, to determine the inverter gating signals. The space current vector is defined with reference to a two-dimensional space as the complex quantity $\overline{I}$, where

$$\overline{I} = \overline{I}_R + a\overline{I}_Y + a^2\overline{I}_B \tag{9.11}$$

and $a = e^{j(2\pi/3)}$.

The two axes used are the real axis, which is coincident with the axis of the R-phase, and an axis perpendicular to the R-axis. The effect is the same as vectorially adding the phasors. The resultant can be used to decide about the inverter gating signals, taking into account all the current errors simultaneously.

With an earthed neutral unfaulted system, $\overline{I}$ in Eqn (9.11) will be equal to zero, whereas it is non-zero for an isolated neutral system. Its error with respect to the reference current signal is denoted as $\Delta\overline{I}$, and is termed as the *space vector current error*. Thus,

$$\Delta\overline{I} = \overline{I} - \overline{I}^* \tag{9.12}$$

where $\overline{I}^*$ is the reference current vector.

If this vector $\Delta\overline{I}$ is drawn, then a line drawn parallel to the X-axis at its tip defines one of the hysteresis boundaries (Fig. 9.20). The mirror reflection of this line about the X-axis constitutes the second hysteresis boundary. It is shown below that due to the absence of the neutral, the dependency between the phases causes the space current error vector to exceed the hysteresis boundary *whenever the largest individual phase error exceeds two-thirds of the hysteresis boundary*.

Apart from the space current error vector defined above, the two other variables of interest for the following derivation are (i) the component of $\Delta\overline{I}$ in the direction of the R-phase, namely, $(\Delta\overline{I})_R$, and (ii) the individual phase error in the direction

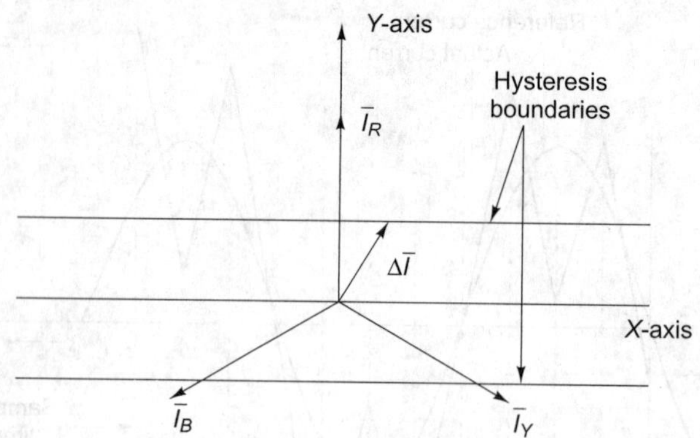

Fig. 9.20 Current vectors for space vector current control

of the R-phase, namely, $\Delta\bar{I}_R$. It can be shown that

$$\Delta\bar{I} = \Delta\bar{I}_R - \frac{1}{2}(\Delta\bar{I}_Y + \Delta\bar{I}_B) + j\frac{\sqrt{3}}{2}(\Delta\bar{I}_Y - \Delta\bar{I}_B) \tag{9.13}$$

The component of $\Delta\bar{I}$ in the R-axis direction is

$$(\Delta\bar{I})_R = \Delta\bar{I}_R - \frac{1}{2}(\Delta\bar{I}_Y + \Delta\bar{I}_B) \tag{9.14}$$

Noting that

$$\Delta\bar{I}_R + \Delta\bar{I}_Y + \Delta\bar{I}_B = 0 \tag{9.15}$$

we get,

$$\Delta\bar{I}_R = -(\Delta\bar{I}_Y + \Delta\bar{I}_B) \tag{9.16}$$

Therefore,

$$(\Delta\bar{I})_R = \Delta\bar{I}_R + \frac{1}{2}\Delta\bar{I}_R = \frac{3}{2}\Delta\bar{I}_R \tag{9.17}$$

or

$$\Delta\bar{I}_R = \frac{2}{3}[(\Delta\bar{I})_R] \tag{9.18}$$

Similarly, for $(\Delta\bar{I})_Y$ and $(\Delta\bar{I})_B$, Eqn (9.18) confirms the above proposition.

After checking the above condition simultaneously in all phases, the space current error is used to determine the inverter gating signals. That is, after comparing the space current error vector with the hysteresis band, the controller applies the voltage vector which is nearest to the current error vector to the corresponding machine winding.

9.7 Recent Trends

The design of brushless dc motors is an area towards which considerable research and development effort has been directed. Three aspects in which progress has been made are elaborated below.

9.7.1 Materials used for the Permanent Magnet

The brushless dc motor consists of a permanent magnet rotor and hence the material used for this magnet requires special attention. Research in magnet technology has led to the use of samarium–cobalt rare-earth material as a magnetic material because it has a very high energy product and high coercivity. A consequence of this is that smaller sizes of magnets can be placed on the rotor, making the inertia of the rotor smaller compared to that of a dc brush motor of the same capacity. Thus, dc brushless motors can have higher torque/inertia ratios as compared to their counterparts of the same capacity.

9.7.2 Alternative Methods for Rotor Position Sensing

The rotor position sensor is one of the costlier components of the brushless motor, and hence its elimination reduces the complexity of construction of the motor. Using the machine terminal voltages for estimating the rotor position is one such method that has been developed and found to give fairly accurate results.

9.7.3 Estimation of Winding Currents

Instead of direct measurement, winding currents are estimated by measuring the dc link current and constructing an observer. The observer also utilizes information on power device switching patterns to estimate the dc link current. At appropriate intervals in the switching cycle, information regarding the estimated as well as the actual dc link currents is used to correct any error in observer estimation.

9.8 Brushless Dc Motors Compared with Other Motors

The brushless dc motor has some features that are common with each of the following machines:

(a) the conventional brush dc motor,
(b) the inverter-driven induction motor, and
(c) the brushless sinusoidally operated synchronous motor.

To highlight its merits as well as demerits, a comparison of the brushless dc motor with these machines is carried out below.

9.8.1 Brushless Dc Motor Versus Brush Dc Motor

The disadvantages of the mechanical commutator used with a conventional dc motor are as follows.

(i) The commutator along with its brushes is subjected to sparking, causing interference with electrical signals. This also leads to a reduction of its lifespan.
(ii) The commutator is mounted on the shaft, alongside the rotor so as to permit the brushes to be stationary. This limits the speed of the motor and also its power output.

(iii) It needs a high ratio of diameter to length of the armature iron, thus limiting the machine design.

(iv) Copper, which is the material used for the commutator, softens at higher temperatures and this feature limits the current rating of the machine.

(v) It is subject to environmental features such as dirt and moisture.

(vi) A substantial amount of heat dissipation occurs in the rotor because of the presence of windings in it.

The brushless dc machine, on the other hand, does not have any heat-producing elements. The 'inside-out' construction of the machine permits increased heat dissipation, because the heat-generating windings are close to the stator surface, thus facilitating natural cooling. These positive features permit higher current ratings and greater steady-torque outputs than the traditional dc machines of the same size. Moreover, it retains the advantages of the conventional dc motor, namely, high starting torque and fairly high efficiency at all speeds. An advantageous feature present in small-sized, brushless dc motors is that they have smoother torque versus angle characteristics. This can be explained as follows. Small-sized mechanical commutator dc motors have a smaller number of commutator segments and consequently give a fluctuating torque versus angle characteristic. As against this, the brushless dc motor with a four-segment electronic commutator gives a much smoother torque function, using its switches in an amplification mode. Any residual torque ripple can be suppressed by closed-loop operation by means of speed feedback, which also enables the motor to have excellent low-speed performance.

9.8.2 Brushless Dc Motor Versus Induction Motor

The brushless dc motor gives a performance similar to that of an induction motor driven by an inverter. These two machines have similar constructional features also, the induction motor being simpler due to the absence of rotor position sensors. The most important difference is as follows: the frequency at which the inverter (that is, the stator) switches operate in the brushless dc motor is determined by the rotor speed, which is sensed by the rotor position sensors; this rotor speed in turn depends on the winding currents. In the inverter-controlled induction motor, the switching frequency totally relies on the frequency of the oscillator present in the inverter. This controls the rotating field in the air gap; the rotor merely follows this field. It can be inferred from this that the speed of an induction motor can therefore be controlled by varying the inverter frequency. This feature simplifies the operation of the induction motor. Other disadvantages of the brushless dc motor are as follows.

(i) In addition to the motor leads (as in the induction motor), the leads from the position sensor have also to be brought out, and they may be subjected to interference from external signals.

(ii) It takes a larger starting current and this current surge may demagnetize the permanent magnet. The current surge magnitudes are, however, less significant in smaller motors which have higher impedances for their windings.

(iii) Since the operation of the inverter that feeds an induction motor is independent of the motor, several motors can be connected at its output. The implementation of this feature is not feasible in a brushless dc motor.

9.8.3 Brushless Dc Motor Versus Brushless Synchronous Motor

Whereas the rotor sensing apparatus in the case of the trapezoidal brushless dc machine consists of simple position sensing devices like the Hall effect devices, the sinusoidal type of brushless synchronous motor needs more precise position transducers so as to accurately synthesize the sinusoidal current waveforms. The total torque in the latter is directly proportional to the current, but the torque function in the brushless dc machine is slightly fluctuating. It can, however, be improved by closed-loop control. Except for these differences, the operational aspect of both these machines can be considered to be similar.

Summary

The brushless dc motor is an ac motor with operation similar to that of a conventional dc motor, with the exception that the mechanical commutator in the latter machine is replaced by an electronic commutator. This commutator consists of a rotor position sensor, electronic switching circuitry, and an inverter. This combination helps in the production of a unidirectional torque similar to that in a dc motor. A three-phase, full-wave inverter gives better performance than a half-wave circuit. In a typical closed-loop brushless dc drive, speed control is achieved by varying the dc link voltage, whereas current can be controlled by operating the dc link in a current controlled PWM mode.

The hysteresis as well as some other types of current controllers help in maintaining the machine winding currents close to their specified values.

Since the rotor position sensor is a costly module, indirect methods of estimating this position have been found to give fairly accurate results. Likewise, indirect methods for measuring winding currents have proved to be reliable.

While the brushless dc motor gives a performance comparable to the conventional dc motor, the cumbersome commutator of the latter can be replaced by the more compact electronic commutator. Thus, for the same size of the machine, the brushless motor has higher current ratings and can hence provide greater steady-torque output.

The brushless dc drive system broadly conforms to the induction motor drive. The two drives differ in the following feature. Whereas the speed of the conventional induction motor drive can be controlled by varying the inverter frequency, the signals from the rotor position sensor decide the inverter switching instants and hence the speed of the brushless dc drive.

The brushless dc motor is superior to the brushless synchronous motor because of the simplicity of its rotor position sensor.

Exercises

Multiple Choice Questions

1. The construction of a brushless dc motor is identical to that of a ______.
   (a) brush dc motor
   (b) stepper motor
   (c) induction motor
   (d) self-controlled synchronous motor with a permanent magnet rotor
2. The self-controlled synchronous motor is supplied with ______ excitation, whereas the brushless dc motor is supplied with ______ excitation.
   (a) sinusoidal current, sinusoidal current
   (b) quasi-square wave current, sinusoidal current
   (c) sinusoidal current, quasi-square wave current
   (d) quasi-square wave current, quasi-square wave current
3. For optimum performance, a ______ electronic commutator has to be used in a brushless dc motor.
   (a) six-segment (b) two-segment
   (c) four-segment (d) eight-segment
4. The ______ sensor is preferred for rotor position sensing in a brushless dc motor.
   (a) optical sensor (b) Hall effect sensor
   (c) magnetic sensor (d) electromagnetic
5. In a three-phase, full-wave, brushless dc motor, the conduction period of the transistors is ______, and each motor phase winding conducts for ______.
   (a) 240°, 240° (b) 240°, 120°
   (c) 120°, 240° (d) 120°, 120°
6. In ramp comparison current control, the output goes ______ when the error exceeds the instantaneous value of the triangular waveform, whereas it goes ______ when the error is less than the instantaneous value.
   (a) high, low (b) low, low
   (c) low, high (d) high, high
7. The conventional brush dc machine has a limited current rating because ______.
   (a) the commutator along with its brushes is subjected to sparking
   (b) the material used for the commutator softens at higher temperature
   (c) the commutator is subjected to environmental features such as dirt and moisture
   (d) the commutator is mounted on the shaft by the side of the rotor
8. Any slight torque ripple at the shaft of the brushless dc motor can be suppressed by ______.
   (a) closed-loop operation by means of speed feedback
   (b) inserting an inductor in the stator circuit
   (c) using damper windings on the rotor
   (d) using special magnetic materials for the rotor

9. The inverter supplying a brushless dc motor cannot have a multimotor output because ______.
   (a) the leads from the position sensor may be subjected to interference from external signals
   (b) the motor load takes a larger starting current and this current may demagnetize the permanent magnet
   (c) the inverter cannot operate independently of the rotor
   (d) the heat-generating windings are close to the stator surface
10. In a brushless dc motor the total, torque is ______, and in a brushless synchronous motor the total torque is ______.
    (a) directly proportional to current, slightly fluctuating in nature
    (b) inversely proportional to current, slightly fluctuating in nature
    (c) slightly fluctuating in nature, highly fluctuating in nature
    (d) slightly fluctuating in nature, directly proportional to current
11. In an earthed neutral unfaulted three-phase system, the current given by the expression $I = (2/3)(I_R + aI_Y + a^2 I_B)$, where $a = e^{j(2\pi/3)}$, is ______, and in an isolated neutral unfaulted three-phase system it is ______.
    (a) non-zero, non-zero (b) zero, non-zero
    (c) zero, zero (d) non-zero, zero

True or False

Indicate whether the following statements are *true* or *false*. Briefly justify your choice.

1. The construction of a brushless dc motor is identical to that of a self-controlled synchronous motor.
2. The self-controlled synchronous motor is supplied with quasi-square wave current and the brushless dc motor is supplied with sinusoidal current.
3. For optimum performance, a four-segment electronic commutator has to be used in a brushless dc motor.
4. The electromagnetic sensor is preferred for rotor position sensing in a brushless dc motor.
5. In a three-phase, full-wave, brushless dc motor, the conduction period of the transistors is 240°, whereas each motor phase winding conducts for 120°.
6. In ramp-comparison current control, the output goes high when the error exceeds the instantaneous values of the triangular waveform, whereas it goes low when the error is less than the instantaneous values.
7. The conventional brush dc machine has a limited current rating because the commutator, along with its brushes, is subjected to sparking.
8. Any slight torque ripple occurring in the brushless dc motor can be suppressed by closed-loop operation using speed feedback.
9. The inverter supplying a brushless dc motor cannot have a multimotor output because the motor load takes a large starting current and this current may demagnetize the permanent magnet.
10. In a brushless dc motor the total torque is directly proportional to current, whereas in a brushless synchronous motor the total torque is slightly fluctuating in nature.

11. In an earthed neutral unfaulted three-phase system, the current given by the expression $I = (2/3)(I_R + aI_Y + a^2 I_B)$, where $a = e^{j(2/3)}$, is non-zero, whereas in an isolated neutral unfaulted three-phase system it is zero.

Short Answer Questions

1. Describe the constructional features of a brushless dc motor.
2. In what aspects does the sinusoidal brushless dc motor differ from the trapezoidal motor.
3. Explain how the supply to the stator winding of a brushless dc motor is switched with electronic components so as to develop a unidirectional torque.
4. Explain the principle of the Hall effect and describe a method for rotor position sensing using Hall generators.
5. With the help of a schematic diagram and waveforms, explain the operation of (i) a three-phase, half-wave, brushless dc motor and (ii) a three-phase, full-wave, brushless dc motor.
6. Describe the different modules of a typical closed-loop brushless dc drive system.
7. Explain the operation of a hysteresis current controller.
8. Explain the principle of (a) a ramp comparison current controller, (b) a delta current controller, and (c) a space vector current controller.
9. What are the three important aspects of the modern brushless dc drive which have been improved by R & D effort?
10. Compare and contrast a brush dc motor with a brushless dc motor.
11. Compare and contrast a brushless dc motor with an induction motor.
12. Compare and contrast a brushless synchronous motor with a brushless dc motor.

CHAPTER 10

Control Circuits for Electronic Equipment

10.1 Introduction

The power electronic devices introduced in Chapter 1 need control circuits to enable them to provide a variable output. Such control circuits form the brain of various power converters such as controlled rectifiers, inverters, ac controllers, and cycloconverters. These circuits involve devices such as the DIAC, UJT, and PUT. Other components accompanying power converters are opto-isolators, pulse transformers, digital IC chips, pulse train generators, zero crossing detectors, etc.

After going through this chapter the reader should

- get acquainted with pulse transformers that help in passing square waves and trigger pulses, which are used for firing thyristors and TRIACs, and also how they help in isolating the control circuit from the load circuit,
- know the shape of the pulse that reliably turns on a thyristor, at the same time facilitating reduction in turn-on losses,
- become familiar with a pulse generating circuit that gives a pulse train as its output,
- get acquainted with various types of opto-isolator circuits and their specifications,
- realize that thyristor circuits that have inductive loads need either a long-duration pulse or a pulse train for reliable firing,
- know some commonly used phase delay circuits and their phasor diagrams,
- understand the construction of zero voltage and zero current detection circuits and their applications,
- get acquainted with the ramp-comparator and digital firing schemes for firing of thyristors, and
- become familiar with drive circuits that are used for thyristors, power transistors, GTOs, and MOSFETs.

10.2 Pulse Transformers

Various types of transformers are used in power circuits; namely, current transformers for stepping down current and facilitating current measurements, autotransformers for getting variable voltage supply, Scott connected transformers for converting a three-phase supply into a two-phase one, and so on. Pulse transformers provide isolation between the low-voltage control circuits that handle voltages in the range of 5–20 V and the power semiconductor devices used in circuits that deal with voltages of the order of 100 V and above.

With the primary voltage kept constant, the maximum secondary voltage of a pulse transformer for full load current will be slightly less than that for a no-load condition. This reduction is due to potential drops in the primary and secondary windings. This voltage drop determines the regulation of the transformer, which is defined as the change in the output voltage between the no-load and full load conditions, expressed as a percentage of the full load voltage. On the other hand, the 'efficiency' of the transformer is defined as the ratio of the output power to the input power, the difference between them being dissipated in the form of core and ohmic losses. The insulation between the primary and secondary windings should be such as to withstand the large voltage difference between them. High-voltage transients due to external or internal causes result in phenomena like corona, dielectric failure, surface creepage, and flashover between points.

'Corona' is a partial discharge within the transformer. It leads to damage of the insulations and causes radio frequency interference which affects all adjacent equipment and circuits. 'Flashover' is said to have occurred when arcing takes place between the parts of a transformer, and 'creepage' is flashover across the surface of the insulation. Both these phenomena can cause high voltages in secondary circuits. The dielectric strength of the insulation between the primary and secondary windings is usually measured as the maximum withstanding voltage per unit thickness of insulation. Solids have a higher 'dielectric strength' than liquids and gases. In a transformer the presence of gas causes reduction in the dielectric strength and also results in corona. It is therefore important to avoid air gaps across the insulating system.

Electrostatic shielding provides protection against voltage transfer through the interwinding capacitance. It is necessary to prevent the transfer of transient voltages or high-frequency noise from the primary circuit to the secondary circuit. The shield is usually a grounded metallic plate between the primary and secondary windings. Electromagnetic shielding is used to attenuate the magnetic field which leaks from the magnetic circuit of the transformer and causes induction of voltages in the nearby circuitry.

Pulse transformers are designed to pass square waves and also waveforms with short rise and fall times without distortion. Figure 10.1 gives a typical voltage waveform obtained as the output of a pulse transformer with a square wave as its input (Mazda 1990). The size of the transformer should be small.

With 'span ratio' defined as the ratio of pulse width to pulse rise time, a pulse transformer should be able to pass pulses having small rise times and high span ratios. Another desirable feature is the resolution of high-frequency pulse trains.

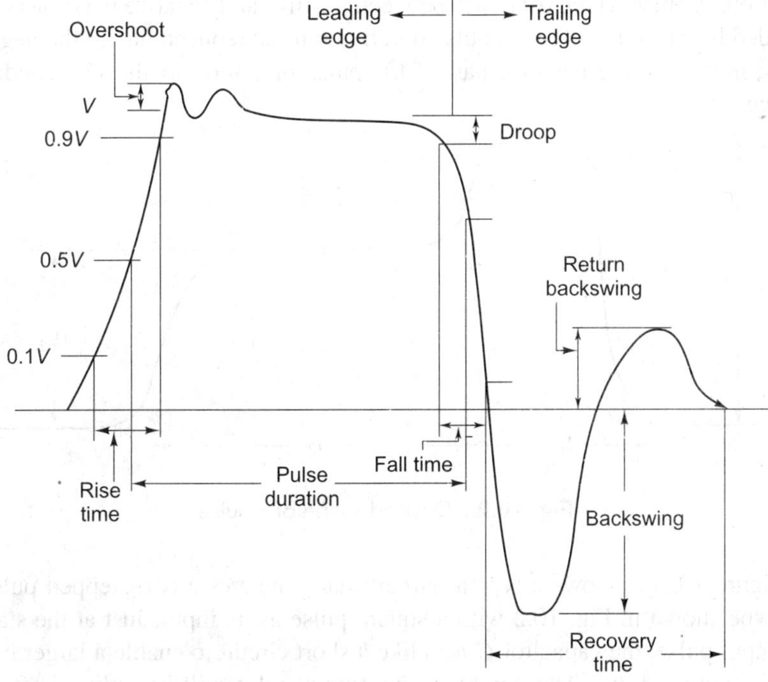

Fig. 10.1 Output voltage waveform of a pulse transformer

Power semiconductors, namely, thyristors, GTOs, and TRIACs, are turned on by trigger pulses applied at the control terminals. In contrast, the transistor requires a current control signal at its base during the whole conduction period. A common feature between the power transistor and the GTO is that both can be turned on and off by appropriate control signals. On the other hand, devices of the thyristor family depend on a momentary discontinuation of the load current for their turn-off.

The initial amplitude of the control pulse should be high (V_1 in Fig. 10.2) to provide the overdrive required just at turn-on. This initial amplitude should be confined to a short duration t_1 to reduce the turn-on time. This is because a longer duration leads to excessively high power dissipation in the control circuitry; the height of the pulse is therefore reduced to V_2 for a time t_2 after the time t_1. The thyristor, transistor, and other semiconductor devices being charge operated, the total height of the corresponding control current pulse given by

$$i_{\text{control}} = \frac{v_{\text{pulse}}}{\text{resistance of the control circuit}}$$

should be at least equal to the charge required to turn on the device. In the case of a transistor the time $t_1 + t_2$ is equal to the total conduction period, whereas for a thyristor it should be large enough to ensure that the current through the device attains the latching current value. However, the pulse duration should be longer in the case of a thyristor controlling an $R + L$ load. As explained later, a pulse train applied for the duration $t_1 + t_2$ is as good as a rectangular pulse having the same duration. A pulse which oscillates between positive and negative maxima is to be avoided because the effective pulse duration will be reduced due to the negative parts; moreover, the negative part of the pulse may turn off the semiconductor device.

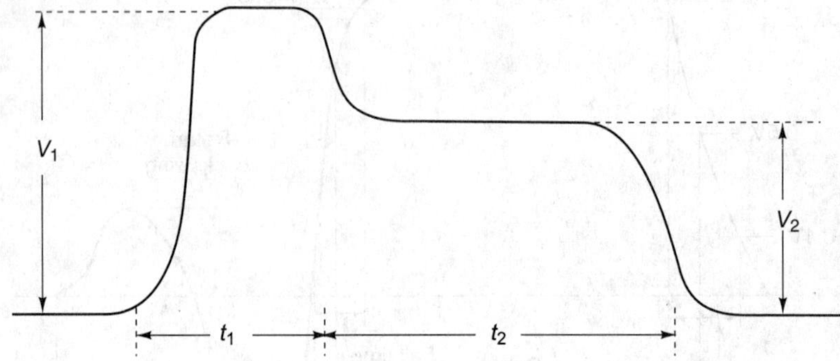

Fig. 10.2 Desired shape of a pulse

Figure 10.3(a) shows a typical circuit that generates a two-stepped pulse of the type shown in Fig. 10.2 with a square pulse as its input. Just at the start of the input pulse, the capacitor C acts like a short circuit to enable a larger height for the output pulse. The height of the output pulse will be reduced after the capacitor is charged, which now presents a high impedance. The resistance R_1, which provides an alternative path, also reduces the height of the output pulse.

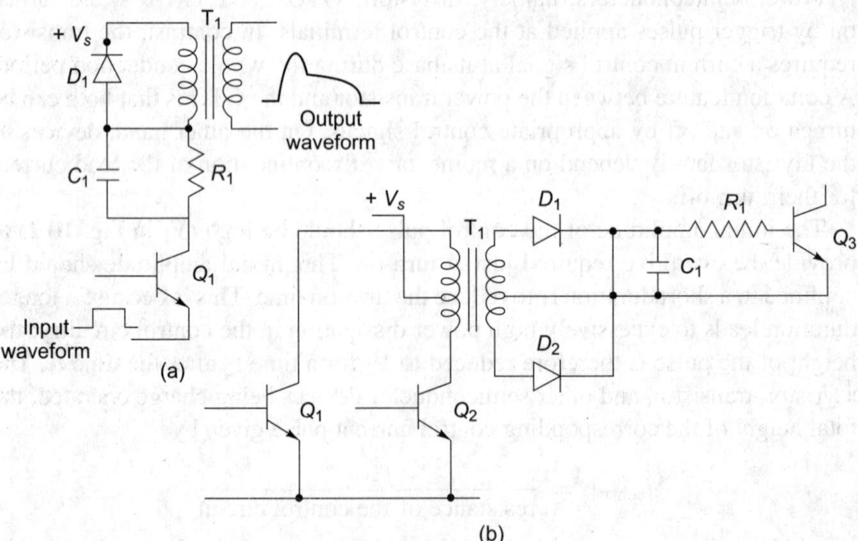

Fig. 10.3 (a) Typical pulse generating circuit using a pulse transformer; (b) pulse transformer used in an oscillatory circuit

Optical couplers (or opto-isolators), considered in the next section, are more suited than pulse transformers for applications requiring continuous drive (like those using transistors), as they do not exhibit the phenomenon of saturation.

However, as shown in Fig. 10.3(b), pulse transformers can be used by operating them in an oscillatory circuit, the output being rectified to cater to a dc load.

Figure 10.4 shows a circuit that generates a pulse train using a pulse transformer whose primary winding is in the collector circuit of transistor Q_1. The operation of this circuit starts with the turn-on of Q_1 and the consequent induction of voltage in the auxiliary winding N_3 (connected in the base circuit of Q_1); the polarity of the voltage will be such that diode D_1 gets reverse biased and Q_1 turns off. In the mean time, C_1 charges through R_1, and Q_1 gets turned on again. This process of turning on and off of Q_1 continues as along as the voltage V_1 continues to be at the input of the pulse generator circuit.

The purpose of the diode D_m across the winding N_1 is as follows. During a no-pulse interval at the base of Q_1, the transistor turns off and a voltage of opposite polarity is induced across the primary. As a consequence, the freewheeling diode D_m conducts and the magnetic energy of the transformer decays through it to zero.

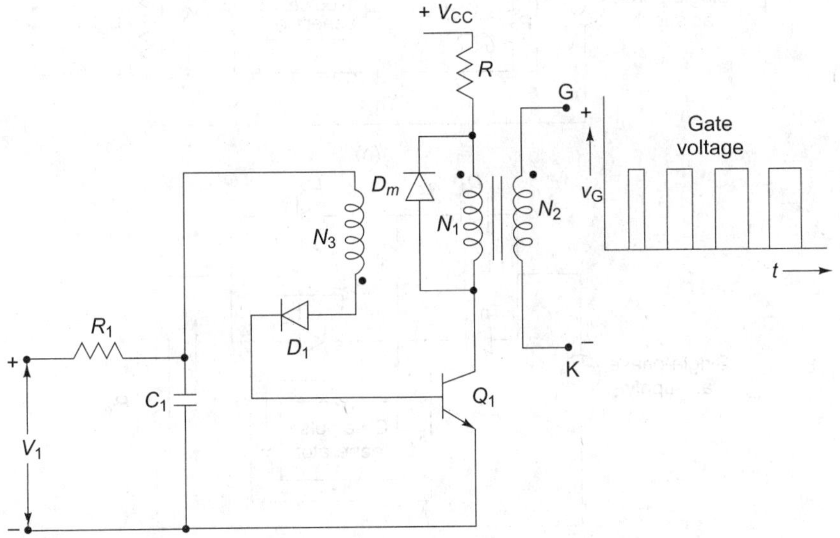

Fig. 10.4 A typical circuit that generates a pulse train

The ac controller shown in Fig. 10.5(a) needs isolation in its gate circuit, because with the common gate pulse generator, the points A, B, and C form a common-cathode connection and short-circuit the two thyristors. This can be prevented by employing two pulse transformers as shown in Fig. 10.5(b). The alternative connection shown in Fig. 10.5(c), however, does not require the pulse transformers because it employs two power diodes and makes use of the common-cathode connection for the thyristors.

As the cost of the two pulse transformers is less than that of the power diodes, the connection of Fig. 10.5(b) is usually preferred, its operation remaining the same as that of Fig. 10.5(a).

Another application where isolation is necessary is the three-phase, full-wave rectifier given in Fig. 10.6 in which the pulse transformers T_1, T_2, and T_3 isolate

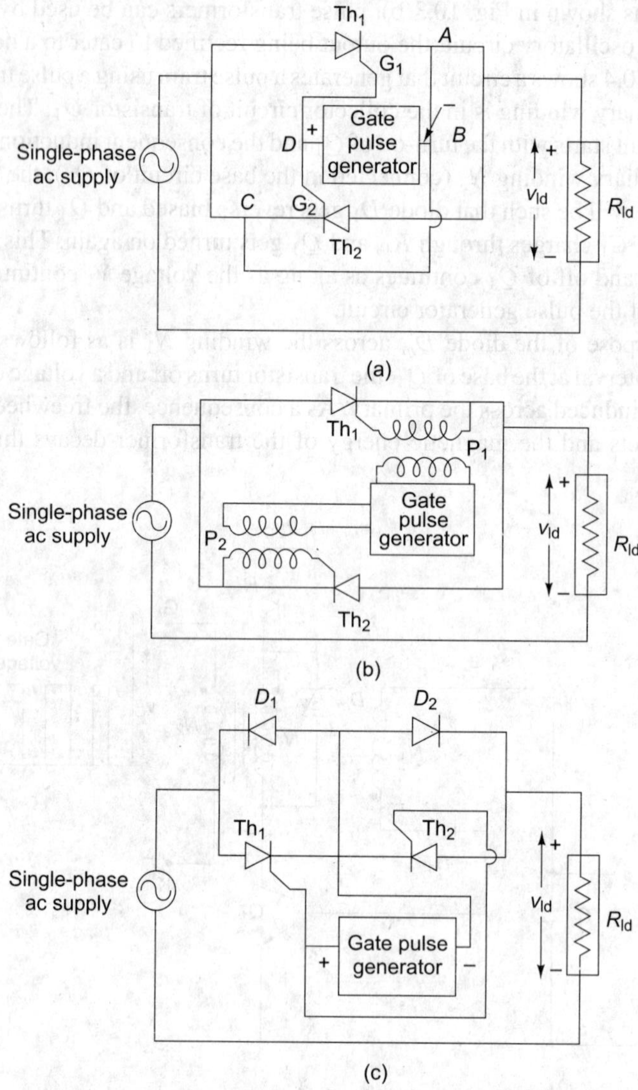

Fig. 10.5 Firing circuits for an ac controller (a) with cathodes connected together, (b) using pulse transformers P_1 and P_2 for isolating the cathodes, and (c) using diode and thyristor pairs connected as a bridge

the control circuitry from the main power circuit consisting of thyristors Th_1 to Th_6. In this case the pulse transformers will have ratings of a few watts and all switching is done on the low-voltage side. The circuits on the secondary side of the pulse transformers will consist of only passive components such as resistors, capacitors, diodes, etc.

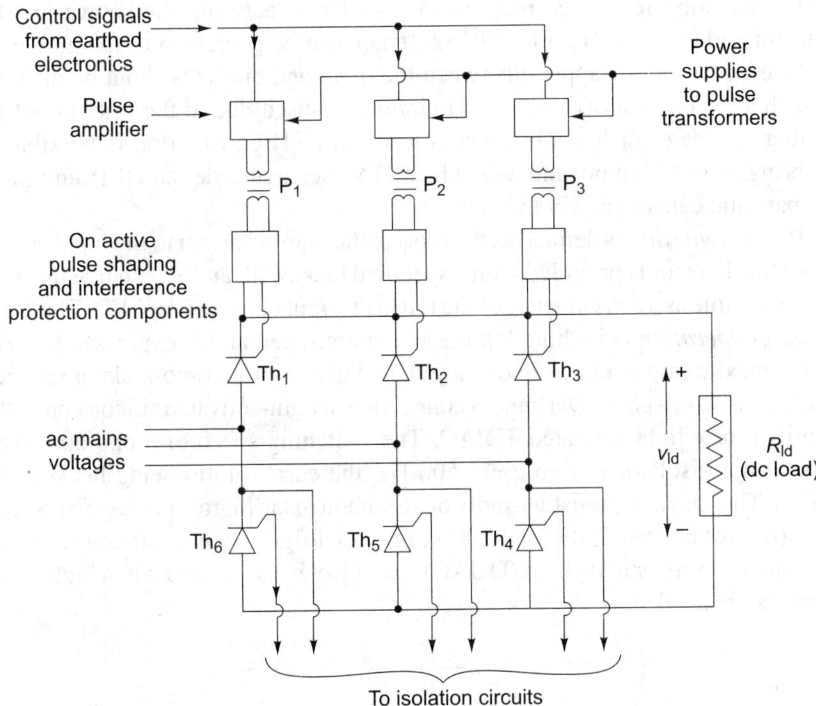

Fig. 10.6 Pulse transformers P_1, P_2, and P_3 used in a three-phase, full-wave controlled rectifier

10.3 Opto-isolators

An opto-isolator, like the pulse transformer, is another module that is employed to provide electrical isolation between logic level signals and drive level signals. Opto-isolators, also called optical couplers, are made from a combination of an optical source and an optical detector, both being housed in a single package. When used for power semiconductor drive applications, the gap between the source and the detector is completely filled with glass or plastic separators. Such an arrangement enables the components to be quite close, thus improving the coupling efficiency and at the same time providing good isolation between the two elements. Light-emitting diodes (LEDs) are commonly used as optical sources. At present, many types of LEDs are available, two of the more common ones being those made of GaAs with emission in the infrared (940 mm) range and those made of GaAlAs with emission in the vicinity of the infrared (750–850 mm) range. In a combination of a silicon photodetector and LED material, the latter should match well with the spectral response of the former. The LED should also have good efficiency, implying that it should have good light emission for the given current input. The performance of the GaAlAs type of LED is satisfactory with regard to both these aspects.

From the point of view of the overall performance of an opto-isolator, the following parameters are important: (a) isolation between the source and detector, (b) input–output current transformation ratio, and (c) speed of operation.

The isolation resistance measures the isolation between the source and the detector and is of the order of 10^{11} Ω. It can also be expressed as the maximum voltage which can be applied between the input and output without breakdown. If such a breakdown occurs, a short circuit is caused due to the bridging of the emitter and detector leads by molten lead wires. High isolation is possible by (a) bringing out the input and output pairs from separate sides and (b) minimizing the parasitic capacitance in the dielectric.

The *current ratio* is defined as the ratio of the output current to the input current when the detector is provided with a specified bias. With an LED source this ratio falls with time as a consequence of the fall in the light output of the LED. The factor *speed of operation* is nothing but the switching speed and is expressed in terms of the maximum operating frequency. The detector of an opto-isolator can be a phototransistor, a photo Darlington connection, a light-activated silicon controlled recitifier, or a light-activated TRIAC. The switching speed of a phototransistor type of opto-isolator can go up to 500 Hz, the current ratio being between 0.1 and 1. The current (transfer) ratio of the photo Darlington package shown in Fig. 10.7 is between 100% and 600%, the exact figure being difficult to assess. The photo-thyristor and photo-TRIAC type of isolators are used when high output currents are needed.

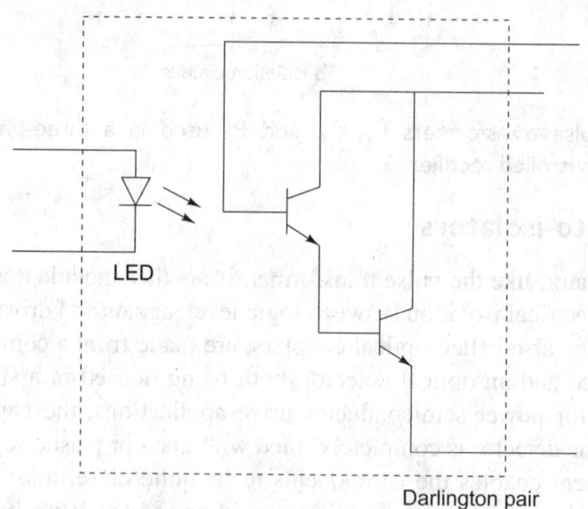

Fig. 10.7 Photo Darlington package

Figure 10.8 shows three types of applications for opto-isolators. The power semiconductor devices that are controlled in the three cases are a power transistor, a MOSFET, and a TRIAC. In Fig. 10.8(a) a separate low-voltage supply is used for the opto-isolator (Mazda 1990) and hence the output voltage rating is low. No separate V_{GG} is needed for the connections shown in Figs 10.8(b). In these cases the necessary supply is derived from the same source as for the main semiconductor device. The optocoupler output stages of Figs 10.8(b) (Rashid 1994) and (c) will see the full main supply voltage when the power MOSFET and power TRIAC are, respectively, off. Some types of optocouplers have a zero crossing detector built into the package so that the power device, which is usually

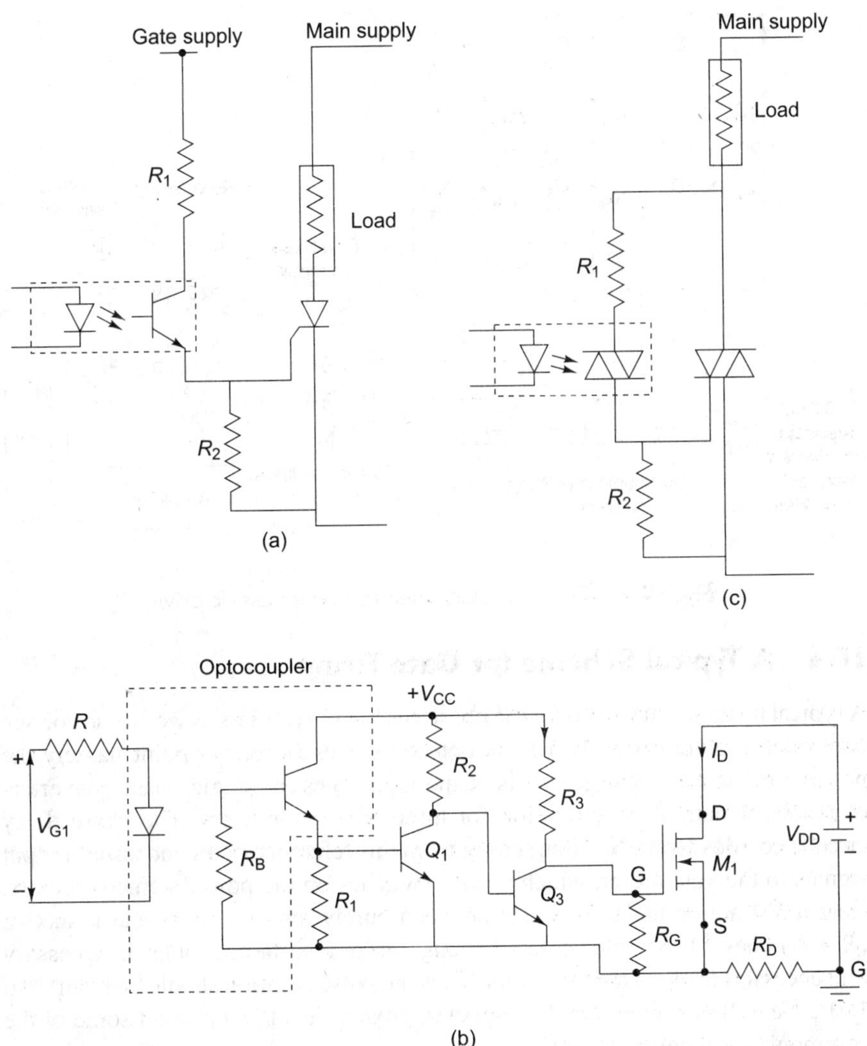

Fig. 10.8 Opto-isolators used for the triggering of (a) a power transistor, (b) a MOSFET, and (c) a TRIAC

a thyristor or a TRIAC, can be switched on at, or with a delay from, the zero crossing points of the ac mains. Radio frequency interference is minimized with such an integrated package.

Another application of opto-isolators is in the measurement of rotor position sensing for brushless dc motors, a typical configuration being that shown in Fig. 10.9. Light from the sources S_1 to S_4 passes through slotted wheels onto the photodetectors D_1 to D_4, which then give electrical signals that are amplified by the amplifier. The output of the amplifier provides the input to the electronic switching sequencer. Finally, the sequencer output signals operate the switches of the inverter, with the result that the stator winding currents are switched synchronously with the rotor position.

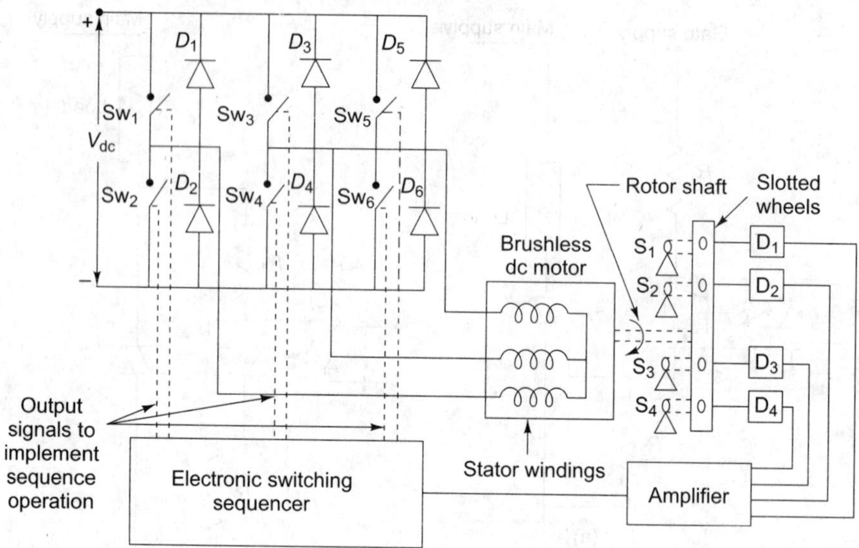

Fig. 10.9 Opto-isolators used in a brushless dc drive

10.4 A Typical Scheme for Gate Firing

A typical firing scheme used for the phase control of rectifiers as well as ac voltage controllers is given in Fig. 10.10. The detector senses a reference point, namely, the positive-going zero voltage points of the input lines for single-phase converters or points of natural commutation for three-phase converters. The phase delay circuit provides a variable time setting from this reference point and issues output signals to the gate driver, which in turn switches on the power semiconductors. Here a distinction has to be made between purely resistive loads and inductive $(R + L)$ ones. In the latter case a rectangular, or a sustained, pulse is necessary for successful firing of the thyristors. Thus the pulse duration should be from 0° to 180°. We will now examine this aspect in greater detail and present some of the commonly used delay circuits.

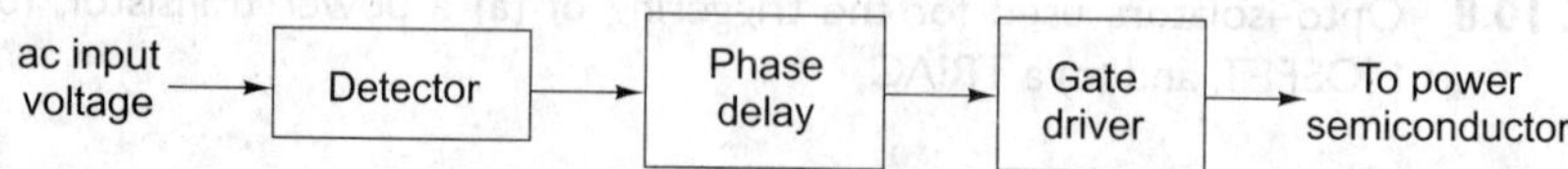

Fig. 10.10 Typical scheme for the gate firing of a controlled semiconductor device

10.4.1 Firing Pulse Requirements for Inductive Loads

If the load of an ac voltage controller is resistive and the semiconductor device is a thyristor, a single trigger pulse would be sufficient either at the start of each half-cycle (zero firing angle) or with a delay of α electrical degrees (finite firing angle). However, it should be ensured that the pulse width is larger than the turn-on time of the thyristor. The waveforms for the single-phase ac controller of Fig. 10.11(a) are given in Fig. 10.11(b) for zero firing angle, the load being

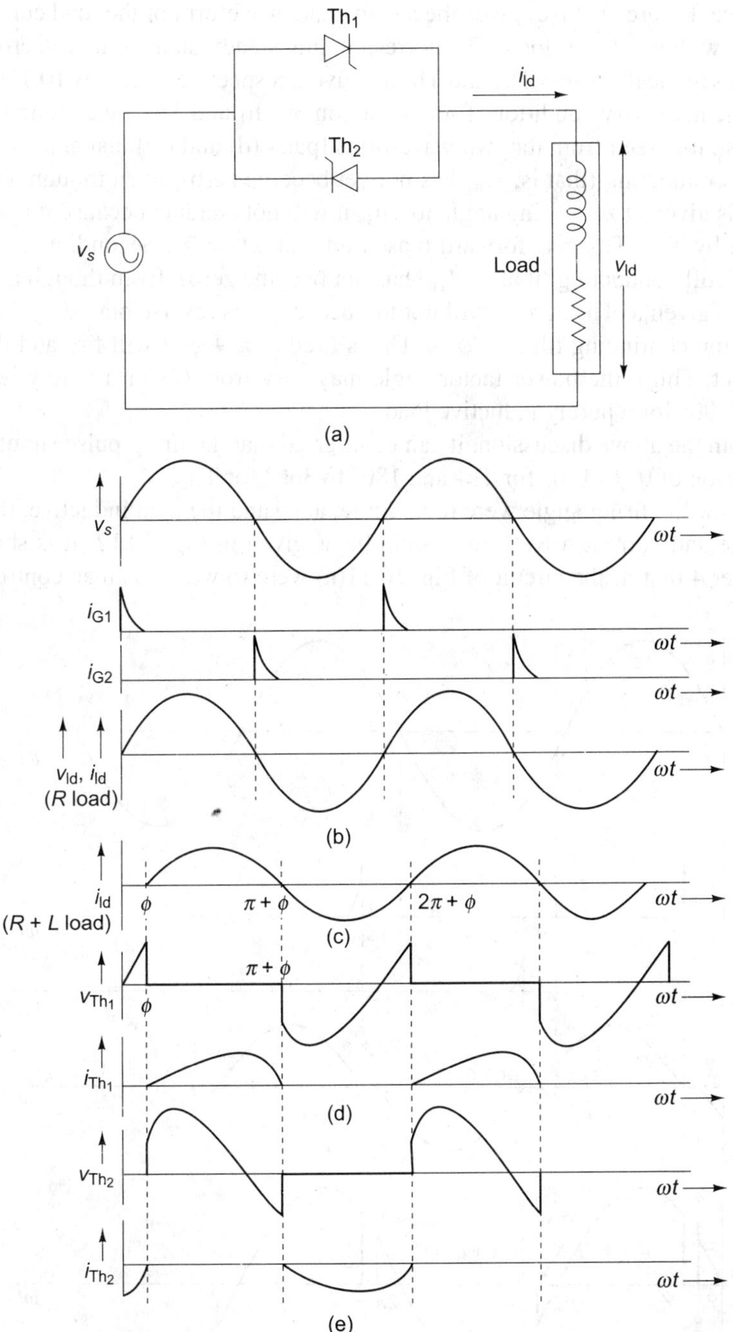

Fig. 10.11 (a) Single-phase ac controller circuit; waveforms of (b) the ac controller with resistive load, (c) load current with $R + L$ load, (d) v_{Th_1}, i_{Th_1}, (e) v_{Th_2}, i_{Th_2}

resistive. Figure 10.11(c) gives the steady-state waveform of the load current for $\alpha = 0°$ with an $R + L$ load. The corresponding steady-state voltages across and currents in the thyristors Th_1 and Th_2 are given, respectively, in Figs 10.11(d) and (e). The necessary conditions for conduction of Th_1 and Th_2 are as follows.

First, it is seen from the two waveforms [parts (d) and (e)] that at $\omega t = 0$, Th_2 is still conducting (that is, I_{Th_2} has not yet become zero). Even though a trigger pulse is given at zero firing angle to Th_1, it will not conduct because it is reverse biased by Th_2. Th_1 gets forward biased only at $\omega t = \phi$. Second, at $\omega t = \pi$, Th_1 is still conducting (that is, I_{Th_1} has not become zero). Even though a trigger pulse is given to Th_2 at π, it will not fire because it is reverse biased by Th_1, this condition continuing till $\pi + \phi$. If Th_2 is fired at $\pi + \phi$ it will fire and thereon conduct. Third, the power factor angle may vary from 0° for a purely resistive load to 90° for a purely inductive load.

From the above discussion, it can be inferred that the firing pulse should have a duration of 0° to 180° for Th_1 and 180° to 360° for Th_2.

If now the firing angle were to be finite, at α, and the load inductive, the load voltage and current waveforms would be as given in Fig. 10.12. It is shown in Chapter 4 that, if the circuit of Fig. 10.11(a) were to work as an ac controller, α

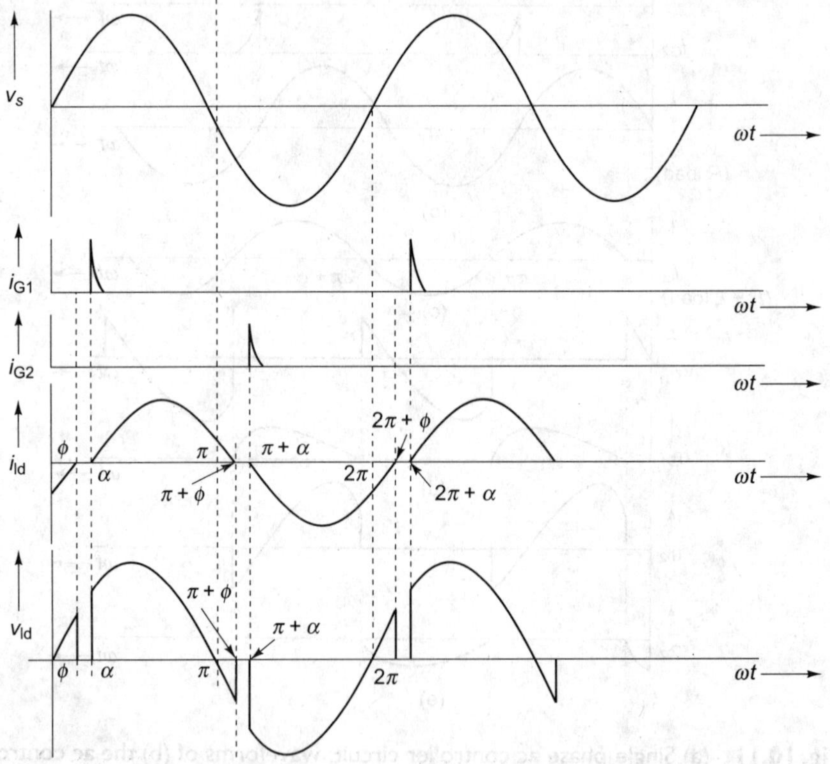

Fig. 10.12 Waveforms of load voltage and current for an ac controller supplying an inductive load

should satisfy the inequality

$$\alpha > \phi \tag{10.1}$$

The above statements imply that the firing angle may also range up to 90°. This feature together with the fact that the turn-on time is higher for an inductive load necessitate keeping the firing pulse duration at 180° for both the thyristors. This discussion holds good for the single-phase bridge rectifier also. It follows that for an n-phase rectifier, the duration of the firing pulse for the thyristor should be $360/n$ degrees or $2\pi/n$ rad. The pulse for this purpose can be rectangular as shown for G_2 and G_5 in Fig. 10.19 for a three-phase rectifier or a series of pulses in quick succession, called the pulse train, as shown in Fig. 10.4.

10.4.1.1 Phase delay circuits Because of the necessity of varying the firing angle α in all thyristorized rectifiers, phase delay circuits have to be employed with them. Figures 10.13(b), (c), and (d) give three phase delay circuits, the outputs of which can be applied to the gate–cathode circuit of the single-phase, half-wave controlled rectifier of Fig. 10.13(a). The phasor diagrams and other features of these phase delay circuits are considered for discussion here.

The circuit and phasor diagram given in Fig. 10.13(b) show that the output voltage of the RC phase shifting circuit has a smaller magnitude but the same frequency as that of Fig. 10.13(a). The voltage vector $V_{out} = V_c$ lags that of V_s by α. This angle is varied by changing the resistance R, its theoretical range of variation is $0° < \alpha < 90°$.

The centre-tapped circuit and its phasor diagram given in Fig. 10.13(c) show that the phase angle of the phasor V_{PN} can be varied in the range of $0° < \alpha < 180°$ with respect to V_s by varying the resistance R. This circuit is therefore preferred to that of 10.13(b).

The circuit of Fig. 10.13(d), called a ramp and pedestal circuit, also provides more delay than that of Fig. 10.13(b). When the input voltage V_s is in its negative half-cycle, the switch Sw is kept open. Just when V_s crosses zero to become positive, Sw is closed so that the capacitor C charges to the battery voltage V_R with a steep slope, say, from t_0 to t_1. Theoretically, C should charge instantaneously to V_R, but due to the presence of internal resistance of the battery and other resistive elements in the circuit, C takes a finite time from t_0 to t_1 up to the value V_R; this time can be further increased by increasing the battery voltage V_R. The capacitor charges to its final value V_{out} at t_2, the slope depending on the resistance R. Thus the total delay α is given by

$$\alpha = \omega(t_1 - t_0 + t_2 - t_1) = \omega(t_2 - t_0) \tag{10.2}$$

The nomenclature of this circuit is derived from the fact that the capacitor charges quickly to the dc voltage V_R and then slowly to V_{out}. It is stated in Chapter 1 that time delay can also be generated by transistorized circuits such as multivibrators. A typical circuit employed for a single-phase, half-wave rectifier is that shown in Fig. 1.28, which is reproduced here as Fig. 10.14. The detailed operation of this circuit is also described in Chapter 1. As stated therein, it can be used to control the firing angle of the thyristor by varying the time period of the astable multivibrator.

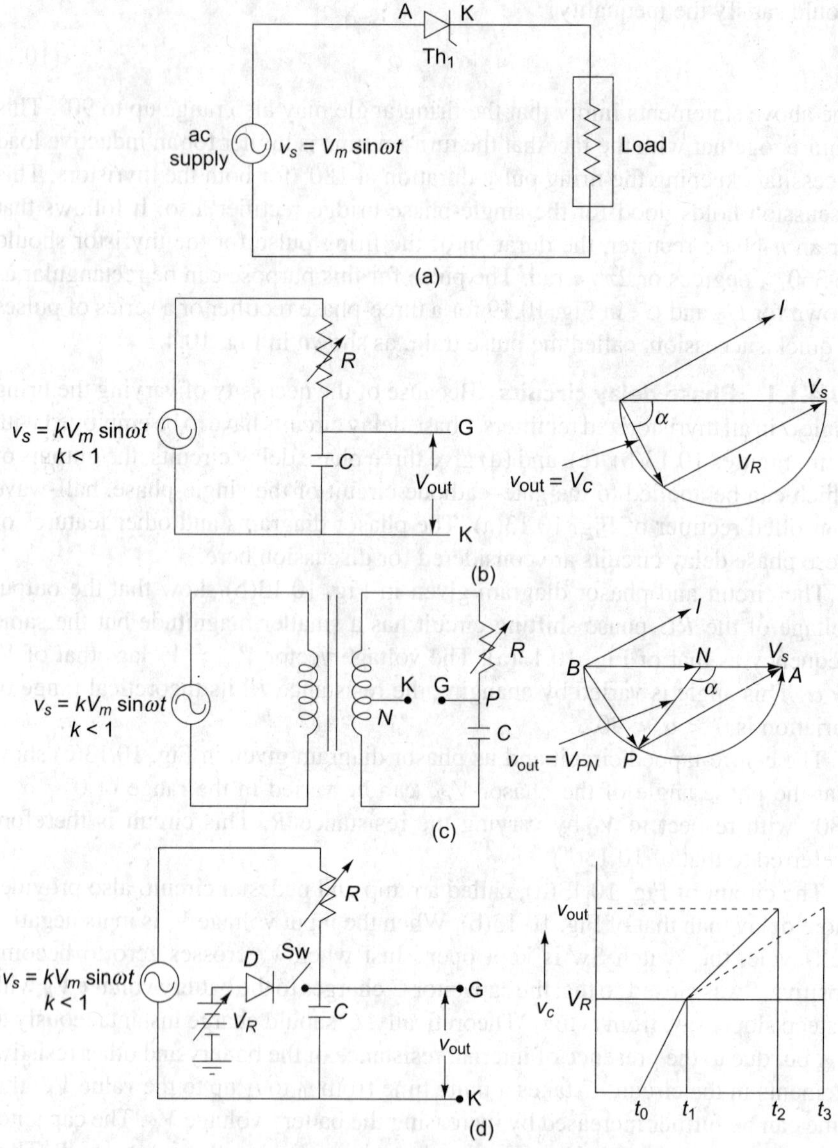

Fig. 10.13 (a) Circuit of a single-phase, half-wave controlled rectifier; (b) *RC* phase shifting circuit fed from an ac source and phasor diagram; (c) *RC* phase shifting circuit fed from a centre-tapped secondary and phasor diagram; (d) ramp and pedestal circuit and phasor diagram

10.5 Zero Crossing Detection

A detector to sense the zero crossing point or some other reference point on the voltage waveform is required in many power electronic circuits, such as rectifiers, ac controllers, and cycloconverters. In the case of rectifiers, the firing angle has to be measured from the positive-going zero of the voltage, whereas in the case of ac

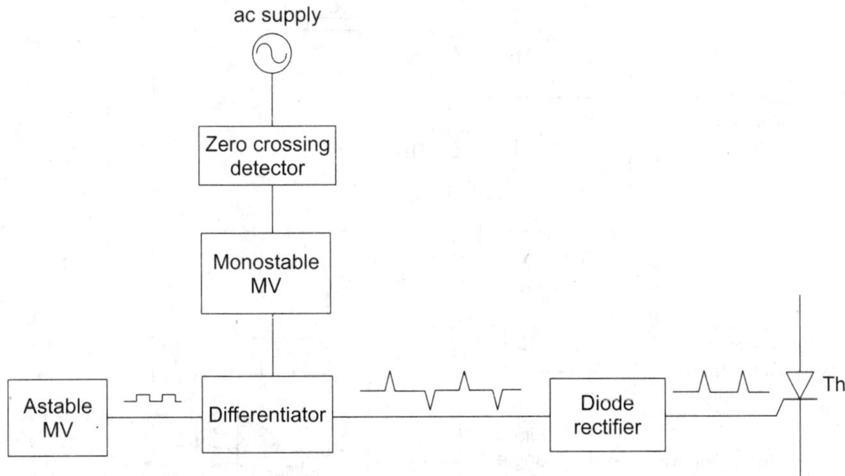

Fig. 10.14 Time delay circuit using a monostable MV

controllers detection is necessary at both the positive-going and the negative-going zero crossings. Typical circuits for zero voltage and zero current detection are presented here.

10.5.1 Zero Voltage Detection

A typical detection scheme for the zero voltage crossing in a single-phase bridge converter is shown in Fig. 10.15(a). The isolation transformer T_1 serves the following purposes.

(a) The main circuit of the thyristor is at the supply potential whereas the trigger circuit pulses need to be generated with reference to the logical ground of the control circuitry. It follows that the detection of the zero crossing of the supply voltage as well as the trigger pulses must be isolated from the line potential.

(b) The centre-tapped secondary of the isolation transformer together with diodes D_1 and D_2 serve as a dc power supply for the gate triggering circuit, with the logical ground potential as its reference.

The detection of zero crossing points of the input voltage waveform is implemented as per the following sequence. In the firing angle block, the ac supply voltage is converted into a periodical ramp voltage with a constant peak-to-peak amplitude; the latter gets synchronized to the zero crossing of the line voltage as shown by the waveforms of Fig 10.15(b). It is then compared with a control voltage V_{con}; the required trigger pulses for the thyristor pairs are then generated in alternate half-cycles as shown by these waveforms. The triggering angle, which can be varied from 0° to nearly 180°, will be proportional to the control voltage. IC chips, which perform all the above-mentioned operations, considerably simplify the circuitry.

The modification of this scheme for triggering the thyristors of an ac controller is straightforward and shown in Fig. 10.16(a). Here the thyristors Th_1 and Th_2

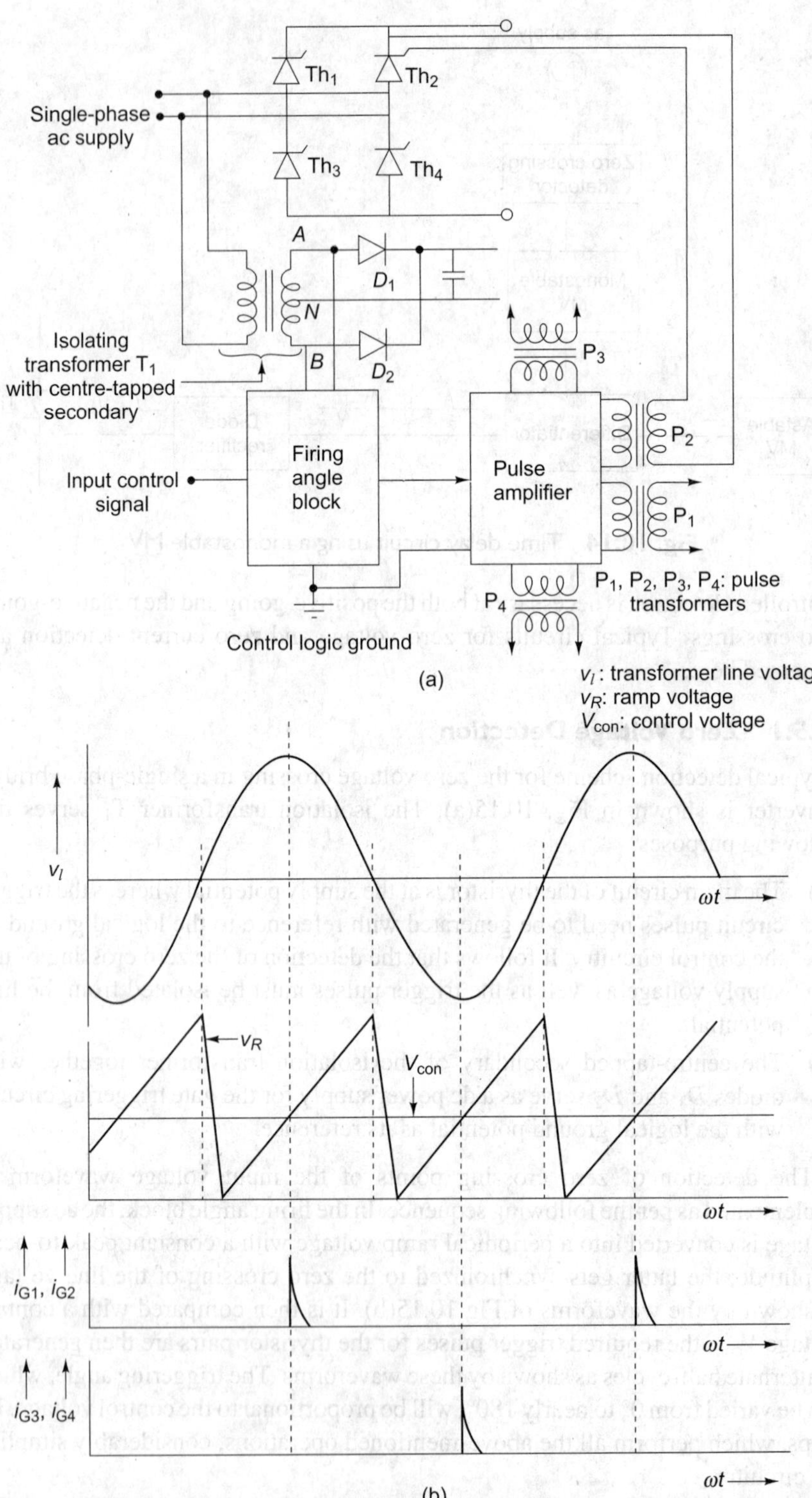

Fig. 10.15 (a) Zero voltage detector circuit; (b) waveforms

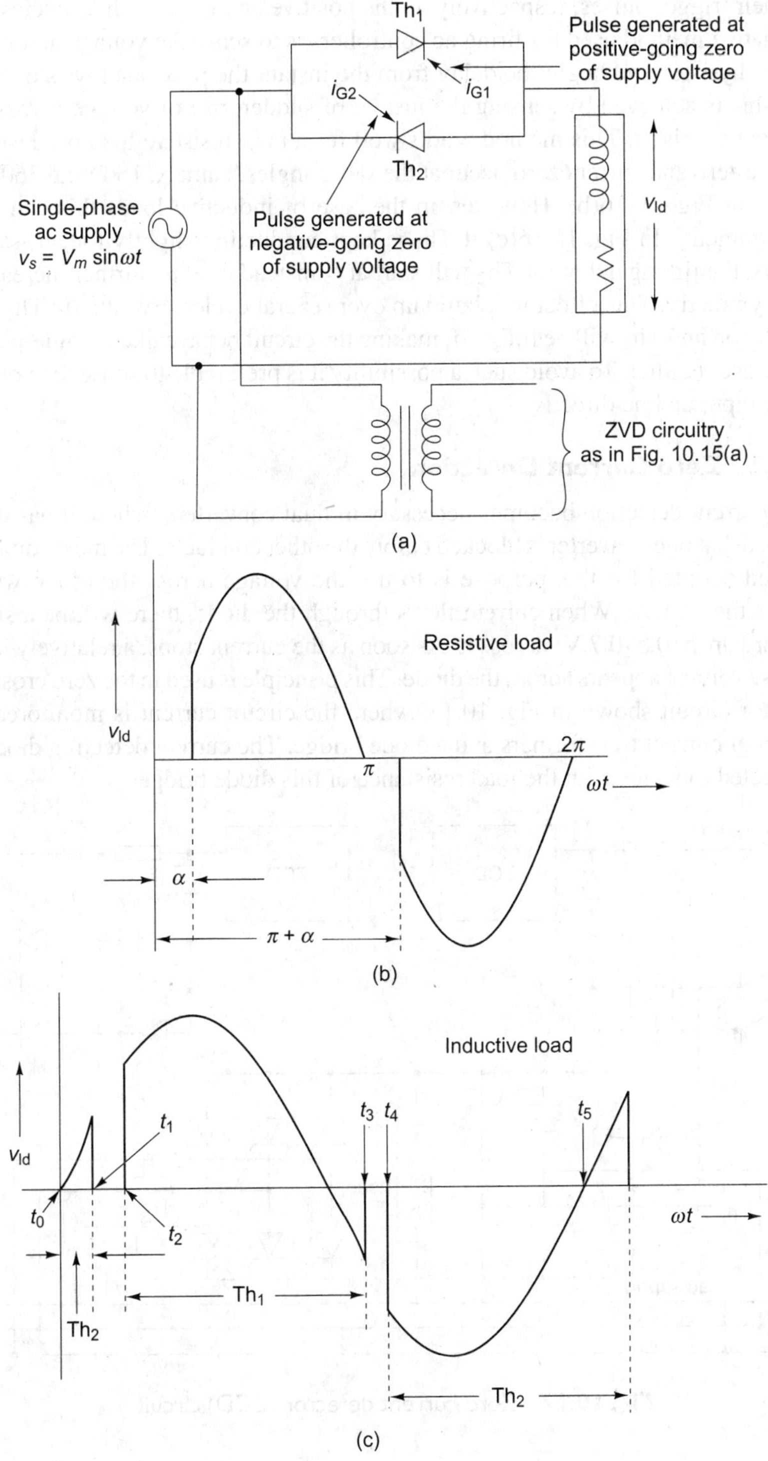

Fig. 10.16 Zero voltage detector (ZVD): (a) circuit for an ac controller, (b) load voltage waveform with resistive load, (c) load voltage waveform with inductive load

get their trigger pulses, respectively, in the positive and negative half-cycles. An alternative method used for firing ac controllers is to sense the voltage across the single thyristor and begin the delay from the instant the previous thyristor turns off. This is achieved by sensing the instant of sudden rise of voltage across the previous thyristor. This method works well for a pure resistive load because the voltage zero and current zero occur at the same angles, namely, 180° and 360°, as shown in Fig. 10.16(b). However, in the case of inductive loads this can lead to asymmetry. In Fig. 10.16(c) if Th_1 is kept conducting slightly longer, say, if $t_3' > t_3$, the timing delay for Th_2 will start at t_3' instead of at t_3, further increasing the asymmetry. This effect will build up over several cycles. Eventually Th_1 will be fully on and Th_2 will be fully off, making the circuit behave like a single-phase, half-wave rectifier. To avoid such a possibility it is preferable to sense the voltage of the input ac line directly.

10.5.2 Zero Current Detection

Zero current detection becomes necessary in dual converters, where it has to be ensured that one converter is blocked before the other conducts. The most common method adopted for this purpose is to use the voltage across the diode which carries the current. When current flows through the diode, there will be a small forward drop (0.5–0.7 V) across it. As soon as the current stops, a relatively large reverse voltage appears across the diode. This principle is used in the zero crossing detector circuit shown in Fig. 10.17, where the circuit current is monitored by means of current transformers and a diode bridge. The current detecting diode is connected in series with the load resistance of this diode bridge.

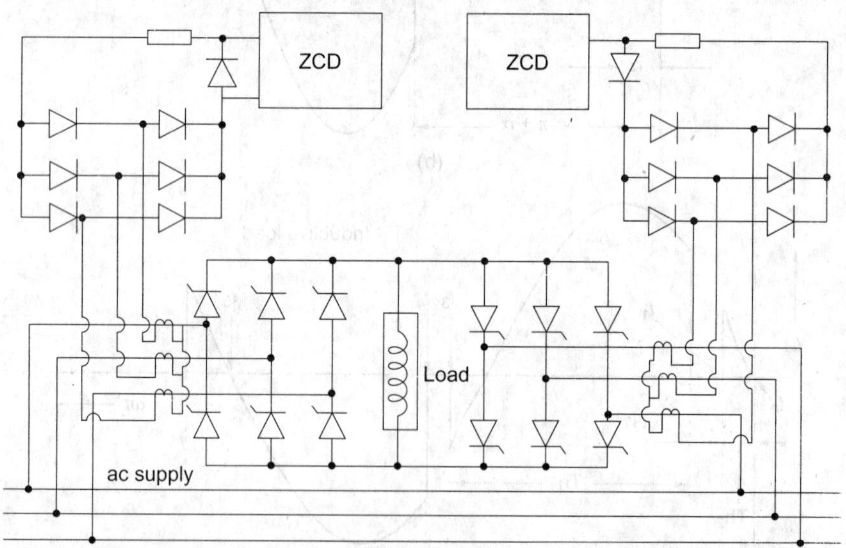

Fig. 10.17 Zero current detector (ZCD) circuit

10.6 Gate Firing Schemes for a Three-phase Bridge Converter

The general gate firing scheme presented in the block diagram of Fig. 10.15(a) for a single-phase, full-wave rectifier can be extended to the three-phase bridge rectifier shown in Fig. 10.18. Two methods of implementation are discussed below, one with a combination of analog and digital components and the other with purely digital chips.

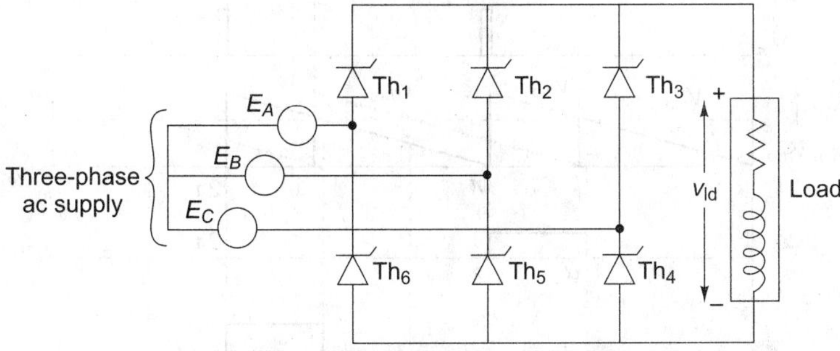

Fig. 10.18 Three-phase, full-wave controlled bridge rectifier

10.6.1 Ramp-comparator Method

Figures 10.19(a) and (b) show the block diagram and waveforms for the trigger circuitry of the three-phase, full-wave rectifier shown in Fig. 10.19(c). The line voltages are obtained through a synchronizing transformer. A scheme for the line

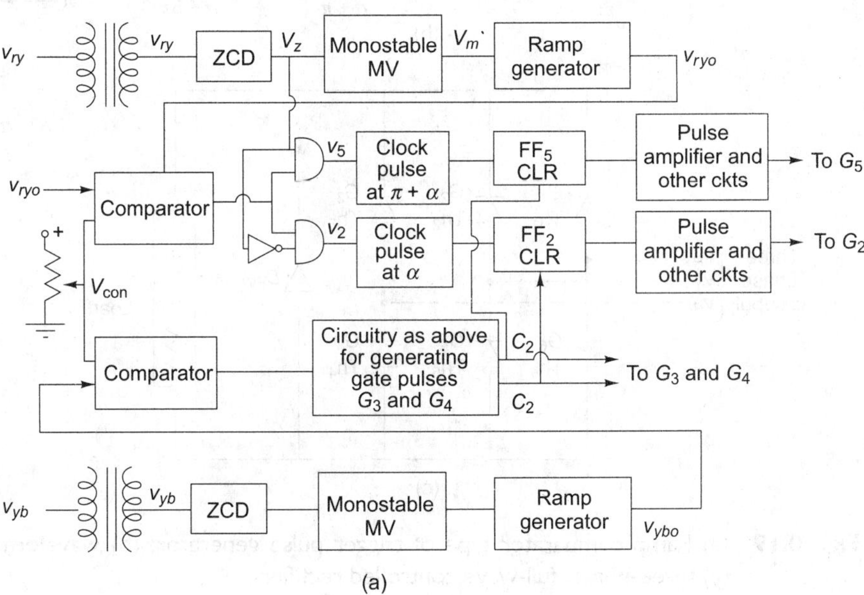

Fig. 10.19(a)

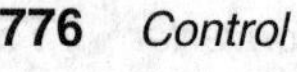

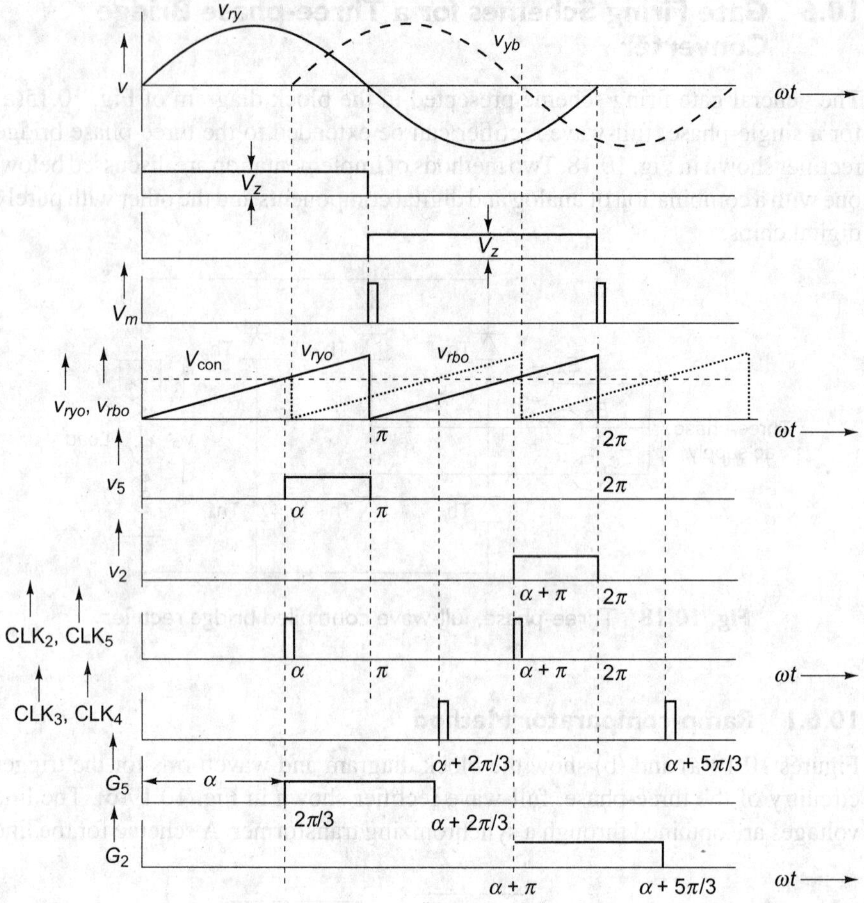

(b)

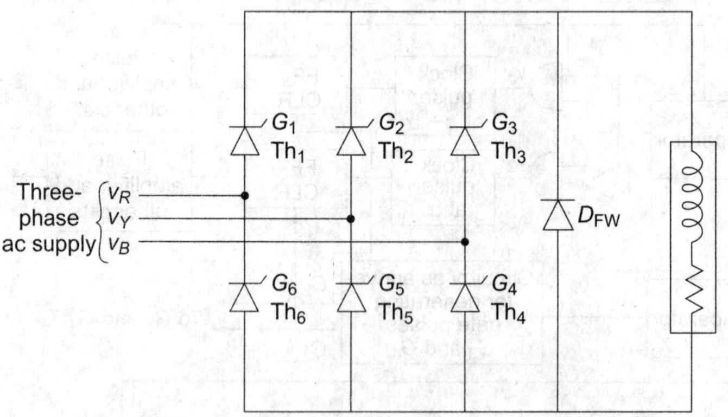

(c)

Fig. 10.19 (a) Ramp-comparator type of trigger pulse generator; (b) waveforms; (c) three-phase, full-wave controlled rectifier

voltage v_{RY} is demonstrated here. The stepped down voltage signal passes through a zero crossing detector circuit, which converts both halves of the sinusoidal wave into a rectified rectangular pulse of width V_z. At the zero crossing of the synchronizing signal, a pulse train of fixed amplitude V_m is generated by a monostable multivibrator. These pulses are used to generate a periodic ramp voltage which starts at 0 and ends at π. This ramp voltage generator is just an integrator, with V_m as its input. The ramp voltage v_{ryo} is then compared with a dc variable control voltage V_{con}, which determines the firing angle α. With the help of AND gates, two pulses V_2 and V_5 are generated. Again, in the blocks named 'Clock pulse at α' and 'Clock pulse at $\pi + \alpha$', small pulses are generated at α and $\pi + \alpha$ with the help of monostable multivibrators. The output of these two blocks then sets the flip-flops FF_2 and FF_5. The drive circuits (pulse amplifier, etc.) now generate the firing pulses for the thyristors Th_2 and Th_5.

The flip-flops FF_2 and FF_5 are cleared by signals C_2 obtained from another channel where the line voltage v_{YB} lags v_{RY} by 120° (elec.). The bottom portion of the figure shows a block containing circuitry, which is similar to the top portion, for generating gate pulses for thyristors Th_3 and Th_4. The pulse widths of the pulses G_2 and G_5 are maintained to have a duration of 120°, or 360°/3, for the reasons explained in Section 10.4. The complete scheme consists of the circuitry necessary for the voltages v_{RY}, v_{YB}, and v_{BR}.

The output voltage of a three-phase converter is given as

$$v_\alpha = \frac{3}{\pi} V_m \sin\frac{\pi}{3} \cos\alpha = V_0 \cos\alpha \tag{10.3}$$

where $V_0 = (3/\pi)V_m \sin(\pi/3)$, and α is proportional to the control voltage V_{con} and can be expressed as

$$\alpha = kV_{\text{con}} \tag{10.4}$$

Thus,

$$v_\alpha = V_0 \cos(kV_{\text{con}}) \tag{10.5}$$

The variation of the average voltage v_α of the bridge converter is thus a cosine function of the control voltage V_{con} and is not linear. Though the output obtained by this method is not affected by line voltage fluctuations, it is quite sensitive to variations in component values, particularly those (components) in the ramp generators.

10.6.2 Digital Firing Scheme

Figure 10.20(a) gives the block diagram of a digital firing scheme (Dubey et al. 1986) used for a three-phase, full-wave rectifier whose operation is explained below. Figure 10.20(b) shows the waveforms of the circuit.

The phase lock loop (PLL) takes in the ac supply with frequency f and gives an ac output equal to $6\times2^{N+1}f$, thus functioning as a frequency multiplier. The ring counter generates the required three-phase synchronizing waveforms spaced at 60°. The PLL output is passed through a $\div$ 6 ring counter R_B to generate a frequency of $2^{N+1}f$. This is used as a clock for the three sets of N-bit down-counters 1, 2, and 3. The preset enable (PE) points of these down-counters are

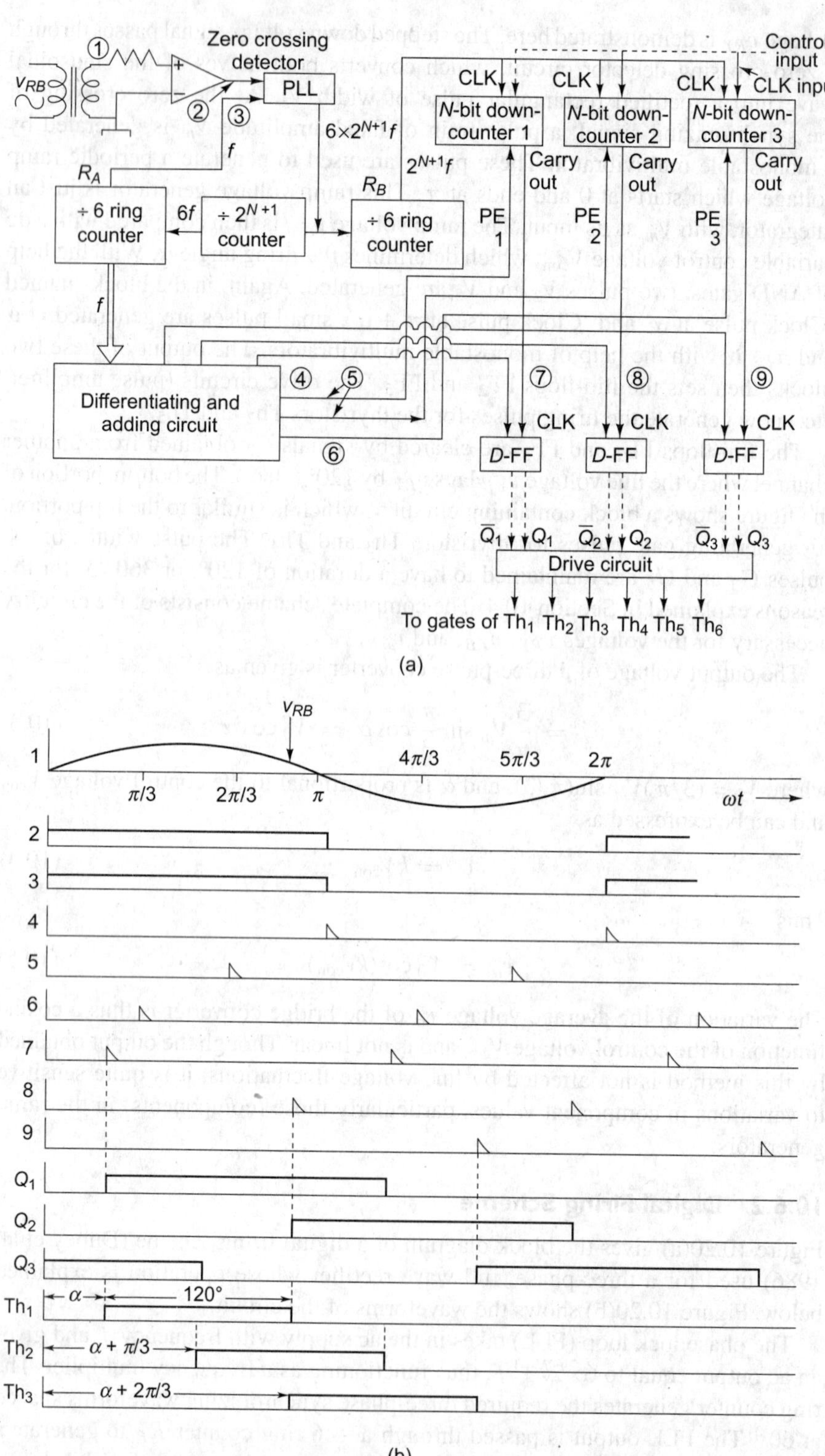

Fig. 10.20 (a) Digital type of trigger pulse generator; (b) waveforms

made high at the zero crossings of v_{RB}, v_{BY}, and v_{YR}. These zero crossings are obtained by differentiating and adding the appropriate synchronizing waveforms obtained from the output of the ring counter R_A. An external signal acts as the digital control input to adjust the firing angles of the three phases through the three down-counters. The 'carry out' signals of the down-counters consist of the output pulses obtained after the counting is completed. These signals, along with the synchronizing waveforms from the differentiating and adding circuit, constitute the input signals for the D flip-flops, whose outputs are fed to the drive circuit. This last circuit generates the required 120°-wide gating pulses for the thyristors.

The merits of the digital firing scheme are as follows: (a) the resolution of the firing angle is as low as $180°/2^N$ and thus precise firing angle control is possible and (b) the circuit is insensitive to input waveform distortion and transients.

10.7 Gate Drive Circuits

A drive circuit forms the final module of any control scheme. A typical circuit that switches on a semiconductor device and its output waveform are shown in Fig. 10.21. The characteristics of the various devices, namely, the thyristor, GTO, power transistor, and MOSFET, being different from each other, their respective gate- or base drive circuits do not resemble each other. Typical drive circuits for each of these devices are presented here and their special features highlighted.

10.7.1 Pulse Amplifier Circuit for Thyristors

The gate triggering pulses needed for low-power thyristors have quite small amplitudes and hence these can be provided by commonly available IC chips without any pulse amplifiers. On the other hand, the trigger circuits of high-power thyristors are set to give a high output at the manufacturing stage so that they may be immune to any noise signals. The initial and final heights of these gate triggering pulses that correspond to the durations t_1 and t_2 of Fig. 10.2 should be of the order of 3 A and 0.5 A, respectively. Hence the output current pulses of pulse generating circuits, which are of the order of a few milliamperes, need amplification. A typical pulse amplifier circuit, also called a *drive circuit*, is shown in Fig. 10.21(a). The input pulse, which is obtained from a typical firing angle module, turns on the MOSFET M, which in turn supplies an amplified gate current pulse to the thyristors through the pulse transformer. The diode D_1 is necessary to prevent the negative gate current pulse caused by the transformer magnetizing current when the MOSFET M turns off. The diode D_2 provides a path for the transformer magnetizing current, so as to dissipate the energy in the magnetic core by means of the resistance R_1. Six such circuits are needed for a three-phase, full-wave bridge converter, and the six thyristors in this converter are fired in sequence with an interval of 60° between any two thyristors. At the moment of starting and also when current is discontinuous, two thyristors, one from the top group and one from the bottom group, have to be fired as elaborated in Chapter 2. The six pairs thus selected are to be fired with intervals of 60°, as shown in Fig. 10.21(b).

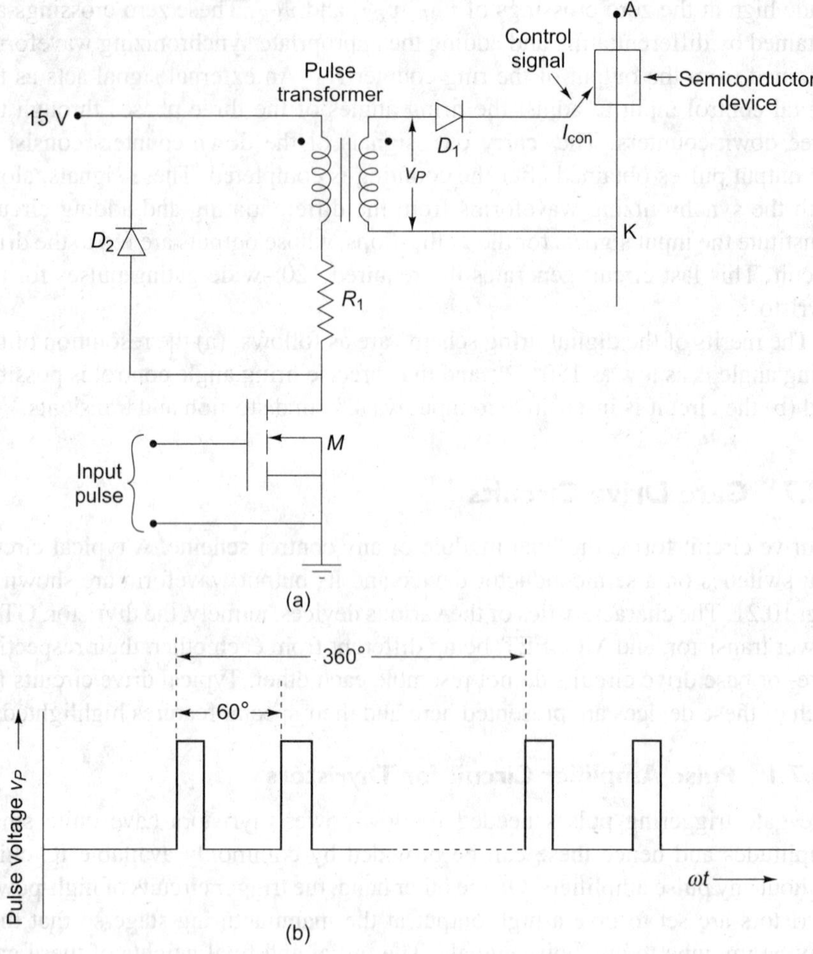

Fig. 10.21 (a) Pulse amplifier circuit for triggering a thyristor; (b) waveforms

10.7.2 Drive Circuit for Power Transistors

A typical drive circuit for switching a power transistor on and off is given in Fig. 10.22(a) (Mohan et al. 1989) and its waveforms are shown in Fig. 10.22(b).

During turn-on, the *p-n-p* driver transistor Q_1 is turned on by saturating one of the internal transistors in the op-amp comparator so as to facilitate the flow of base current for the power transistor Q_M. The base current can be calculated as follows:

$$V_{AA} = V_{CE(sat)}|_{Q_1} + R_A I_A + V_{BE(ON)}|_{Q_M} \tag{10.6}$$

$$I_{B(ON)}|_{Q_M} = I_A - \frac{V_{BE(ON)}|_{Q_M}}{R_B} \tag{10.7}$$

With respect to the collector current I_C required for a specific application, the corresponding $I_{B(ON)}$ and $V_{BE(ON)}$ are obtained from the power transistor data sheets. The $V_{CE(sat)}$ for Q_1 can also be obtained from these data sheets. For the selection of R_B, a trade-off is made as follows: a small R_B will allow a faster turn-off but will also lead to higher power dissipation and vice versa. A suitable value can be estimated from the equation

$$R_B = \frac{V_{BE}(\text{storage})}{I_B(\text{storage})} \tag{10.8}$$

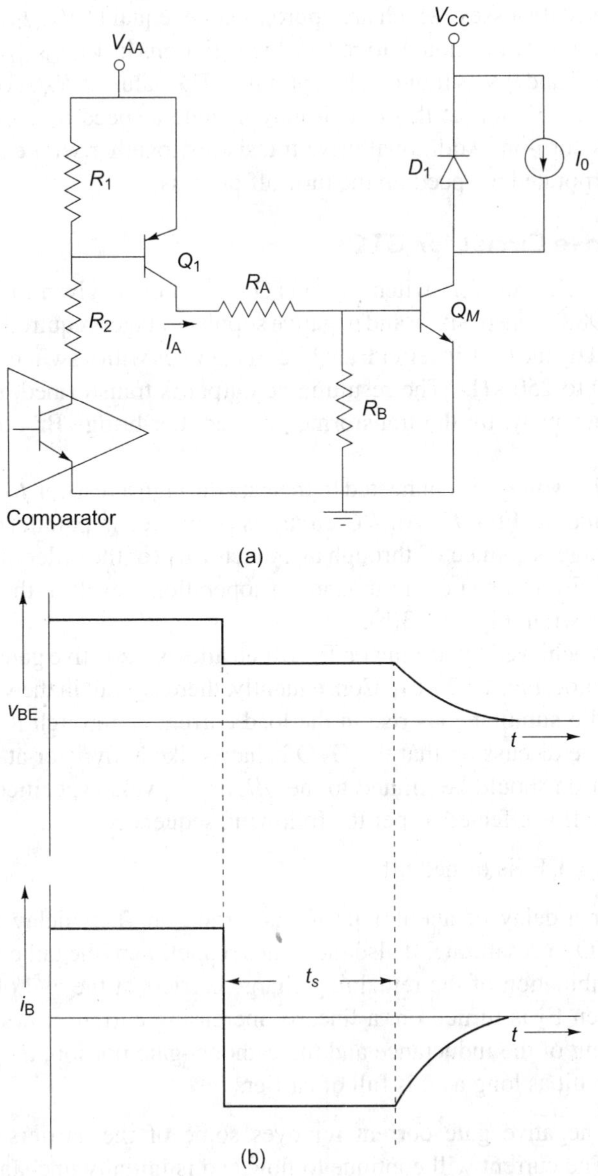

Fig. 10.22 (a) Base drive circuit for a BJT; (b) waveforms

where the numerator and denominator values are the estimated values for the storage time interval (t_s). The current I_A is obtained from Eqn (10.7) as

$$I_A = I_{B(ON)} + \frac{V_{BE(ON)}|_{Q_M}}{R_B} \tag{10.9}$$

where $I_{B(ON)}$ and $V_{BE(ON)}$ are known ON state values. With I_A being fixed thus, only the value of V_{AA} needs to be determined. For this purpose, a trade-off is necessary between the following two requirements: V_{AA} should be small to keep down the ON state losses, which are approximately equal to $V_{AA}I_A$. On the other hand, it should be large enough to reduce the variations in $V_{BE(ON)}|_{Q_M}$. A value of V_{AA} between 7 and 9 V is found to be optimum. The value of R_A is obtained from Eqn (10.6). A variation of this circuit may include a speed up capacitor across R_A for quick turn-on. Additional drive transistors together with capacitors may also be incorporated to speed up the turn-off process.

10.7.3 Drive Circuit for GTOs

A typical drive circuit for switching a GTO on and off is given in Fig. 10.23(a) (Thorborg 1988). The positive and negative supply voltages required in this circuit are generated by the FET inverter F_3, while F_4 operates with a switching frequency between 150 to 250 kHz. The resulting ac output is transformed, rectified, and filtered, respectively, by the transformer T_1, rectifier bridge Br_1, and capacitor pair C_1, C_2.

The positive voltage is connected to the gate through a resistor R_1 (of the order of 8–12 Ω) and the FET F_1. An RC circuit is connected in parallel with R_1. The negative voltage is connected through an inductor L_1 (of the order of 0.4–0.6 μH) and the FET F_2. The switching on and -off operations result in the gate current waveform shown in Fig. 10.23(b).

Turn-on is achieved by turning on F_1, which allows a positive gate current with a high overshoot [Fig. 10.23(b)]. Consequently, there is a fall in the voltage across the GTO and a simultaneous rise in the load current i_T through it. It is evident from the above discussion that the GTO behaves like a thyristor at turn-on. The di/dt at turn-on should be limited to the $(di/dt)_{max}$ value specified by the data sheets. Turn-off is effected as per the following sequence.

Step 1 The FET F_1 is turned off.

Step 2 After a delay of about 1 μs, F_2 is turned on. This delay is needed to allow the GTO to desaturate. It also facilitates reduction of the tail current caused by the recombination of the remaining charge carriers at the end of the turn-on interval. When F_2 is turned on, a linearly increasing current flows through the path consisting of the inductance and the cathode–gate portion, the latter acting as a short circuit as long as it is full of carriers.

Step 3 The negative gate current removes some of the carriers while some recombine. The current will continue to flow and is initially unchanged because of the inductance. The cathode–gate voltage will rise, causing a short avalanche breakdown which accelerates the removal of the carriers. Simultaneously, the current decays to zero, indicating that the GTO has attained its OFF state. As

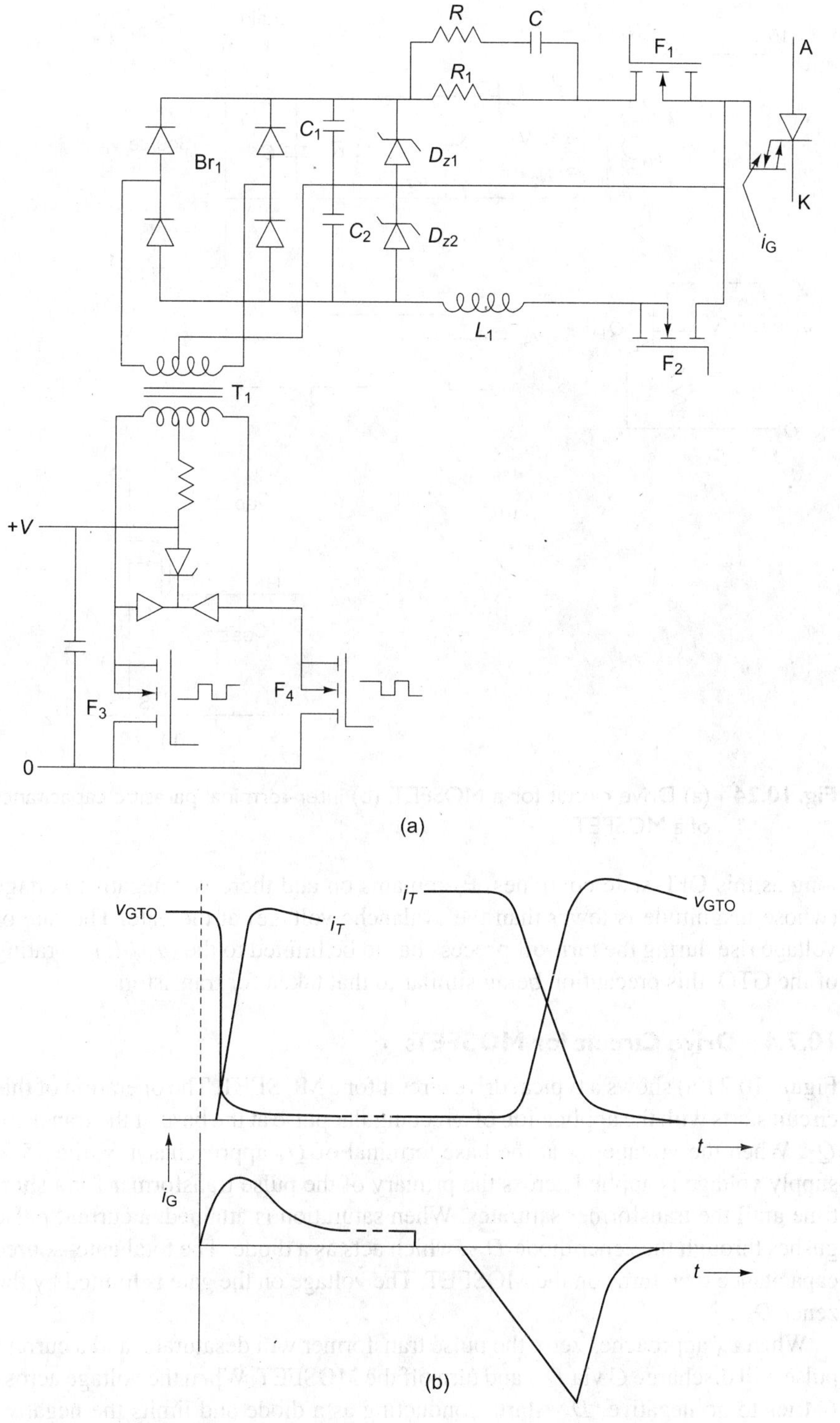

(a)

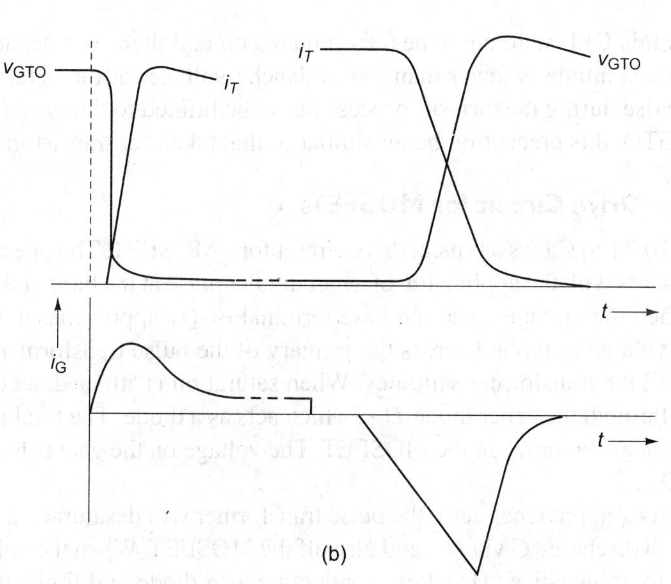

(b)

Fig. 10.23 (a) Drive circuit for a GTO; (b) waveforms

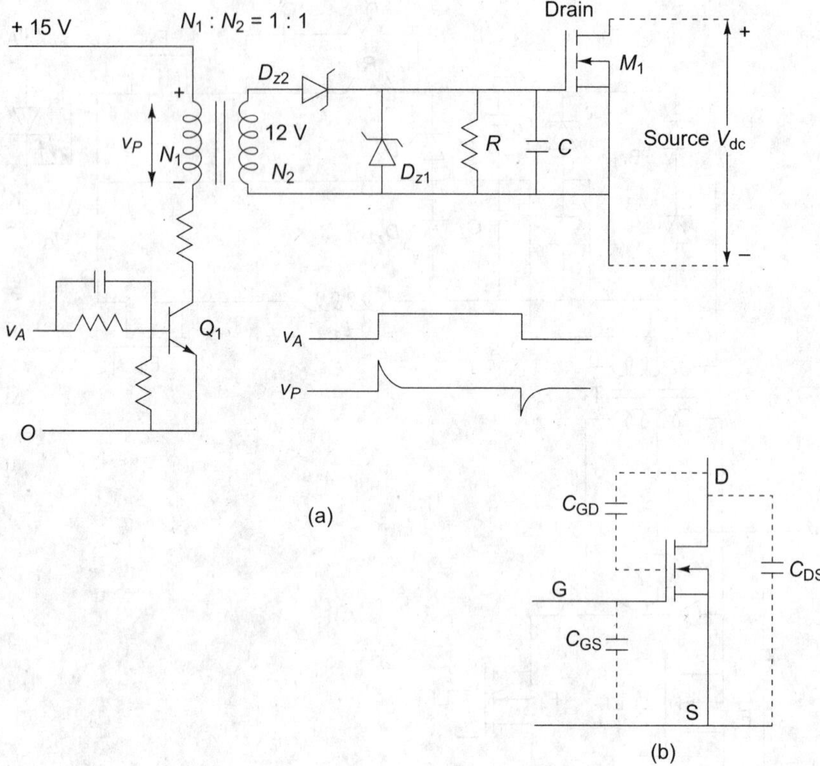

Fig. 10.24 (a) Drive circuit for a MOSFET; (b) inter-terminal parasitic capacitances of a MOSFET

long as this OFF state continues, F_2 remains on and there is a negative voltage (whose magnitude is lower than the avalanche voltage) at the gate. The rate of voltage rise during the turn-off process has to be limited to the $(dv/dt)_{max}$ rating of the GTO, this precaution being similar to that taken for transistors.

10.7.4 Drive Circuit for MOSFETs

Figure 10.24(a) shows a typical drive circuit for a MOSFET. The operation of this circuit starts with the application of a rectangular pulse at the base of the transistor Q_1. When the voltage v_A at the base terminal of Q_1 approaches 1 V, the 15 V supply voltage is applied across the primary of the pulse transformer for a short time until the transformer saturates. When saturation is attained, a current pulse gushes through the zener diode D_{z2}, which acts as a diode. The total gate–source capacitance then turns on the MOSFET. The voltage on the gate is limited by the zener D_{z1}.

When v_A approaches zero, the pulse transformer will desaturate, and a current pulse will discharge C via D_{z2} and turn off the MOSFET. When the voltage across C tries to go negative, D_{z1} starts conducting as a diode and limits the negative voltage at the gate to less than 1 V. The zener diode D_{z1} is necessary to protect the gate against voltages that exceed the maximum rated gate–source voltage, which is ±20 V. An overvoltage fault, which may come from the drive circuit, is otherwise likely to occur. It can also come from the main circuit as follows. When the dc (15 V) supply at the left is turned on, there will be a voltage division between the parasitic capacitances C_{GD} and C_{GS} [Fig. 10.24(b)], thus leading to a fault.

10.8 Transformer-isolated Circuits for Driving MOSFETs and IGBTs

In the recent past, rapid developments have taken place in the area of driver circuit technology to keep in step with the emerging device technology. Driver ICs are now routinely used for switching on MOSFETs and IGBTs. The following is a brief introduction to this aspect of control circuits used for MOSFETs and IGBTs.

10.8.1 Multipurpose Use of the Isolating Transformer

It is stated in Sections 10.2 and 10.3 that pulse transformers and opto-isolators are employed to provide electrical isolation between logic level as well as drive level signals. The former equipment are found to introduce a great deal of flexibility in the design of the drive circuit. Further benefits are accrued from the following modification. The same isolating transformer that is used to transfer the control signal from the control circuits to the isolated drive circuit can also be employed to provide isolated dc bias power as well, thus doing away with a separate transformer for the isolated dc supplies. If, in an application, the MOSFET to be controlled is to be kept on for a long time, the circuit shown in Fig. 10.25(a) can be used. The control voltage in this circuit is modulated by a high-frequency oscillator output before being applied to the buffer circuits. This results in a high-frequency ac signal appearing across the transformer primary when the control voltage is high

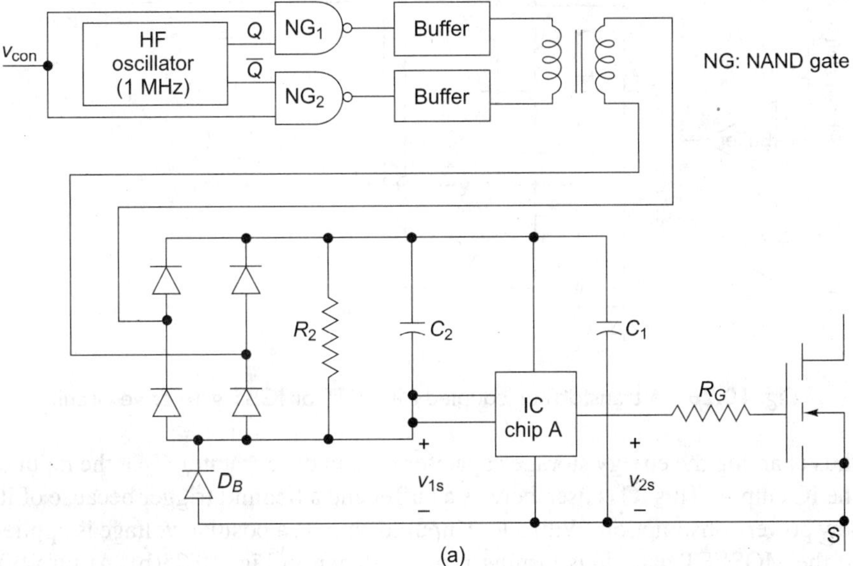

Fig. 10.25(a)

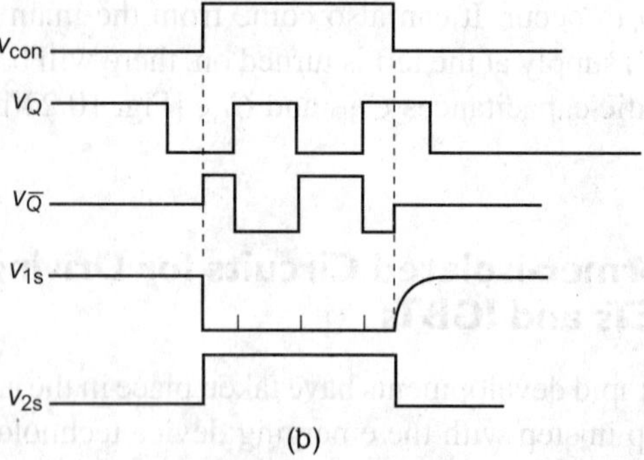

Fig. 10.25 (a) A transformer-isolated MOSFET gate drive using a high-frequency modulated carrier; (b) waveforms

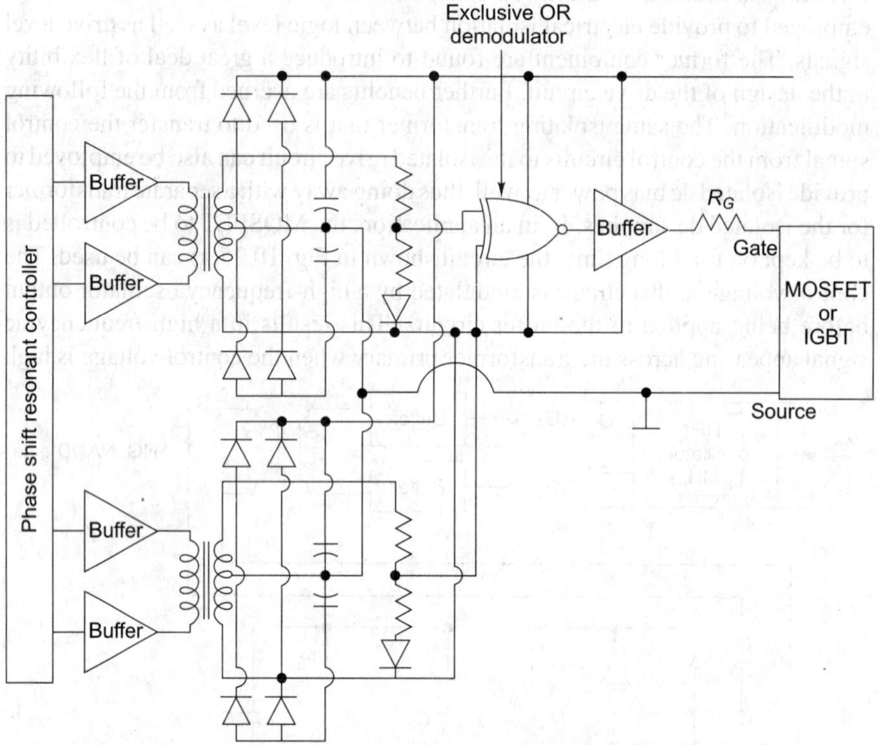

Fig. 10.26 A transformer-coupled MOSFET or IGBT gate drive circuit

thus charging the energy storage capacitor C_1 and the capacitor C_2 at the input of the IC chip A. This IC is used here as a buffer and a Schmitt trigger because of its low power consumption. With a low input to chip A, a positive voltage is applied to the MOSFET gate, thus turning it on as shown in Fig. 10.25(b). At turn-off, the control voltage goes low and the voltage across the transformer primary goes to zero. Now C_2 discharges through R_2 and the input voltage to the chip A goes high; this causes the output voltage to go low, thus turning the MOSFET off. The diode D_B is used to prevent the energy stored in the capacitor C_1 from discharging into the resistor R_2.

10.8.2 A Circuit with Smooth Duty Ratio Transition

In PWM inverters such as motor drives and UPSs, there is a need for a smooth transistion in the duty ratio from a finite value to either 0 or 1. The use of a phase shift resonant controller combined with two transformers, a demodulator, and some buffers will make an efficient gate drive circuit for a MOSFET or an IGBT, as shown in the drive circuit of Fig. 10.26. The transformers may have a relatively large leakage reactance, making it easy to produce noise-immune transformers with a high isolation test voltage.

Summary

Isolation of the power and the control circuits is necessary in most power electronic circuits to avoid exposing the low-power control components to high voltages and high currents. Pulse transformers and opto-isolators are two such isolating components. The latter are emerging as strong contenders in this field because of the high coupling efficiency provided by them. Including a zero crossing detector in opto-isolator ICs increases their utility.

Pulse-generator circuits should be designed to produce pulses of appropriate shape and size so as to reliably fire a thyristor. At the same time it should be ensured that there is minimum power loss in such pulses.

Phase delay circuits provide variable firing angles for thyristors. The ramp and pedestal circuit gives a wide range of delay as compared to other delay circuits. Zero voltage detectors are needed to provide a reference point for firing angle measurement in the case of single- and three-phase rectifiers as well as ac voltage controllers. Zero current detectors are essential for successful operation of non-simultaneously controlled dual converters.

Converters that employ high-current thyristors need pulse amplifier circuits because pulses of a few milliamperes cannot trigger such high amperage devices. The pulse frequency should conform to the time sequences needed in a particular rectifier or inverter. Drive circuits for power transistors should be properly designed to minimize the ON state losses, at the same time ensuring flexibility of operation. A typical control circuit for a GTO consists of a combination of both turn-on and turn-off circuits. Similarly, a typical control circuit of a MOSFET should be capable of generating both positive and negative voltages.

Pulse transformers are found to introduce a great deal of flexibility in the design of the drive circuitry. This feature has been advantageously applied for gate circuits used for MOSFETs and IGBTs.

Exercises

Multiple Choice Questions

1. The dielectric strength of insulation between the primary and secondary windings of a pulse transformer is usually measured as ______.
   (a) the maximum capacitance per unit thickness of insulation between the windings
   (b) the maximum distance between the windings
   (c) the maximum voltage that can be withstood per unit thickness of insulation between the windings
   (d) the maximum charge per unit thickness of insulation between the windings
2. Electrostatic shielding safeguards a pulse transformer against ______.
   (a) high current transients being transferred through the windings
   (b) magnetic field leakage through the interwinding insulation
   (c) high-frequency noise being transferred through the interwinding capacitances
   (d) voltage transfer through the interwinding capacitances
3. The span ratio of a pulse transformer is defined as the ratio of ______.
   (a) the pulse width to the pulse fall time
   (b) the pulse height to the pulse rise time
   (c) the pulse width to the pulse rise time
   (d) the pulse rise time to the pulse height
4. A firing pulse which oscillates between positive and negative maxima should be avoided because ______.
   (a) the effective pulse duration will be reduced due to the negative parts
   (b) the negative part of the pulse may turn off the semiconductor device
   (c) the positive and negative parts nullify each other
   (d) all of the above
   (e) some of the above
5. Pulse transformers are necessary to isolate the firing signals of a thyristorized single-phase ac controller because ______.
   (a) in their absence, high current transients may pass from the cathode of one device to that of another
   (b) in their absence, high-frequency noise may pass from the cathode of one device to that of another
   (c) the common-cathode connection may short the ac source
   (d) a fault in the circuit may damage both the devices
6. The current transfer ratio of an opto-isolator is defined as ______.
   (a) the ratio of output current to input current when the detector is subjected to a specific bias
   (b) the ratio of input current to output current when the detector is subjected to a specific bias
   (c) the ratio of output current to input current when the detector is not subjected to any bias

(d) the ratio of input current to output current when the detector is not subjected to any bias

7. When the single-phase, centre-tapped secondary type of rectifier, shown in Fig. 10.27, is used to supply an inductive load, a firing signal in the shape of a single trigger pulse fails to trigger the thyristors because ______.

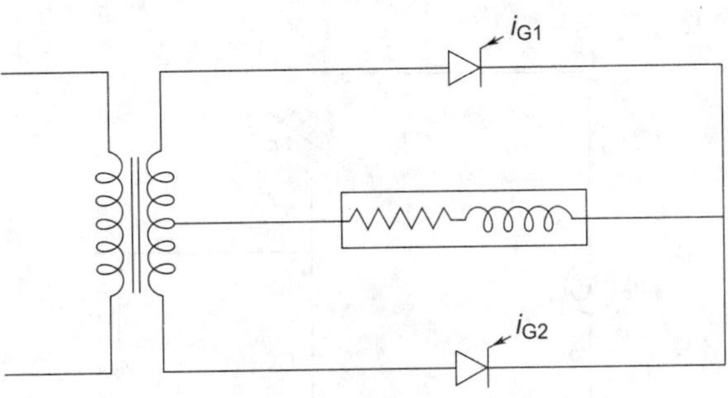

Fig. 10.27

(a) the inductive nature of the load attenuates the firing signals
(b) the charge associated with such a trigger pulse is insufficient to trigger the thyristor
(c) due to the inductive nature of the load, the voltage of the conducting thyristor goes negative for some time, preventing the other thyristor from firing
(d) the inductive nature of the load offers high impedance, preventing the thyristors from reaching the latching current value

8. In an ac controller ______ detection is needed, and in a cycloconverter ______ detection is needed.

(a) zero voltage, zero voltage
(b) zero current, zero voltage
(c) zero current, zero current
(d) zero voltage, zero current

9. The phase locked loop in a typical digital firing scheme functions as a ______.

(a) frequency modulator
(b) frequency multiplier
(c) frequency adder
(d) frequency divider

10. In a MOSFET-based pulse amplifier circuit for thyristor triggering, the purpose of D_1 is ______, and the function of D_2 is ______.

(a) to provide a path for the transformer magnetizing current, to prevent high negative current transients
(b) to prevent negative gate signals when the thyristor is off, to provide a path for the magnetizing current
(c) to prevent a negative gate current caused by the transformer magnetizing current when the MOSFET turns off, to dissipate the energy in the magnetic core by means of the resistor R_1
(d) to provide a path for the transformer magnetizing current, to prevent a negative gate signal when the thyristor is off

11. In a MOSFET drive circuit, the zener diode D_{z1} is necessary to ______.

(a) prevent negative gate currents flowing from the gate to the source

(b) discharge a current pulse across C
(c) protect the gate against voltages exceeding the maximum negative gate–source voltage
(d) allow transformer current to flow for a short time until the transformer saturates

12. The range of phase delay that can be obtained with the phase delay circuit of Fig. 10.28 is ________.

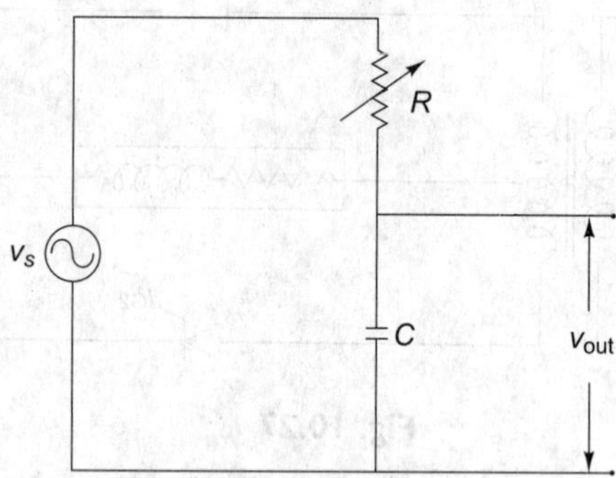

Fig. 10.28

(a) 30°–120° (b) 0°–180°
(c) 30°–180° (d) 0°–90°

13. The range of phase delay that can be obtained with the phase delay circuit of Fig. 10.29 is ________.

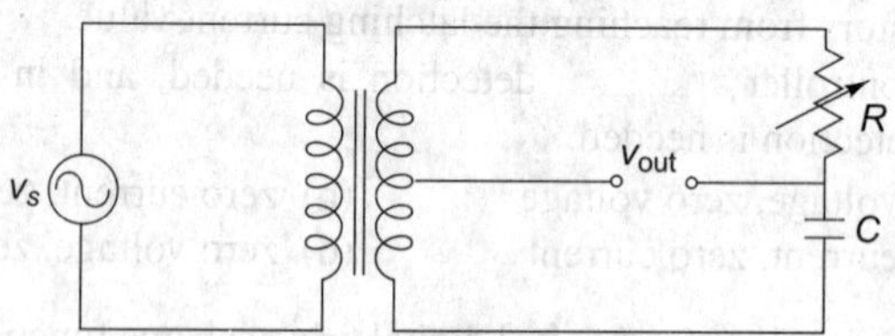

Fig. 10.29

(a) 60°–180° (b) 0°–180°
(c) 90°–180° (d) 30°–150°

14. The ramp and pedestal type of delay circuit of Fig. 10.13(d) gives ________ the RC phase shifting circuit of Fig. 10.13(b).

(a) less delay than (b) more delay than
(c) a delay equal to (d) a delay which is much less than

15. In single- and three-phase controlled rectifiers ________ detection is needed, and in a dual converter used as a cycloconverter ________ detection is needed.

(a) zero voltage, zero voltage (b) zero current, zero current
(c) zero current, zero voltage (d) zero voltage, zero current

16. Considering the ramp-comparator method of getting a delayed phase angle, if a dc voltage of 2 V is given at the input of the ramp voltage generator, a voltage ramp having a slope of nearly ________ is obtained at its output.

(a) 45° (b) 30°
(c) 63° (d) 75°

17. In the digital firing scheme employed for a three-phase, full-wave rectifier, ______ ring counters and ______ N-bit down-counters are used.
    (a) two, three (b) three, four
    (c) three, two (d) four, three

18. Considering the pulse amplifier type of drive circuit for thyristors, the trigger current of high-power thyristors is kept at a high value for ______.
    (a) facilitating reliable firing of the thyristors
    (b) providing immunity from temperature variations
    (c) providing immunity from noise signals
    (d) providing immunity from voltage transients

19. Considering the drive circuit for power transistors, the value of the resistance R_B should be large enough ______, and small enough ______.
    (a) to avoid slow turn-off, to avoid high power dissipation
    (b) to avoid slow turn-off, to avoid leakage of current
    (c) to avoid leakage of current, to avoid electromagnetic interference
    (d) to avoid high power dissipation, to avoid slow turn-off

20. Considering the drive circuit for power transistors, the power supply voltage V_{AA} should not be too small ______, and should not be too large ______.
    (a) so as to keep the ON state losses small, so as to reduce the variation of $V_{BE(ON)}$
    (b) so as to facilitate noise immunity, so as to keep down the ON state losses
    (c) so as to facilitate reduction in the variation of $V_{BE(ON)}$, so as to avoid leakage of current
    (d) so as to facilitate reduction in the variation of $V_{BE(ON)}$, so as to keep down the ON state losses

21. Considering the firing circuit for a GTO, ______ in order to turn it on, and ______ in order to turn it off.
    (a) F_2 is turned on, F_1 is turned on
    (b) F_1 is turned on, F_2 is turned off and after a small delay F_2 is turned on
    (c) F_1 is turned on, F_1 is turned off and after a small delay F_2 is turned on
    (d) F_1 is turned on, F_2 is turned off and then F_1 is turned on

22. Considering the firing circuit of a MOSFET, the zener diode D_{z1} is necessary ______, and the zener diode D_{z2} is necessary ______.
    (a) to provide a path for the discharging of capacitor C, to provide a path for the charging of capacitor C
    (b) to protect the gate against high voltages, to provide a path for the charging of capacitor C
    (c) to protect the gate against high voltages, to provide immunity to noise signals
    (d) to protect the gate against high voltages, to provide a path for the discharging of capacitor C

23. Considering the firing circuit of a MOSFET, the voltage V_A should be ______ for it to turn on and ______ for it to turn off the device.

(a) 1 V, 0 V (b) 0 V, 1 V
(c) 2 V, 1 V (d) 2 V, 0 V

True or False

Indicate whether the following statements are *true* or *false*. Briefly justify your choice.

1. The dielectric strength of the insulation between the primary and secondary windings of a pulse transformer is usually measured as the maximum voltage that can be withstood per unit thickness of insulation between the windings.
2. Electrostatic shielding safeguards a pulse transformer against magnetic field leakage through the interwinding insulation.
3. The span ratio of a pulse transformer is defined as the ratio of the pulse width to the pulse fall time.
4. A firing pulse which oscillates between positive and negative maxima should be avoided because the positive and negative parts nullify each other.
5. Pulse transformers are necessary to isolate the firing signals of a thyristorized single-phase ac controller because the common-cathode connection may short the ac source.
6. The current transfer ratio of an opto-isolator is defined as the ratio of input current to output current when the detector is not subjected to any bias.
7. When a single-phase, centre-tapped secondary type of rectifier, shown in Fig. 10.30, is used to supply an inductive load, a firing signal in the shape of a single trigger pulse fails to trigger the thyristors because, due to the inductive nature of the load, the voltage of the conducting thyristor goes negative for some time, preventing the other thyristor from firing.

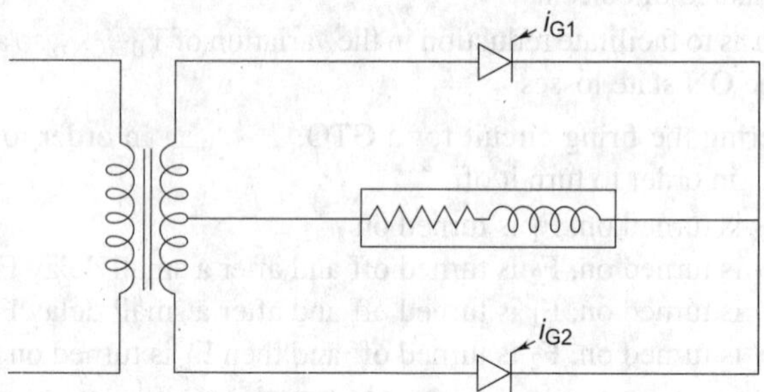

Fig. 10.30

8. Zero voltage detection is employed in both the ac controller and the cycloconverter.
9. The phase-locked loop used in a typical digital firing scheme functions as a frequency modulator.
10. In a MOSFET-based pulse amplifier circuit for thyristor triggering, the purpose of D_1 is to prevent a negative gate current caused by the transformer magnetizing current when the MOSFET turns off, and the function of D_2 is to provide a path for the transformer magnetizing current.

11. In a MOSFET drive circuit, the zener diode D_{z1} is necessary to protect the gate against voltages exceeding the maximum negative gate–source voltage.
12. The range of phase delay that can be obtained with the phase delay circuit of Fig. 10.31 is 0°–90°.

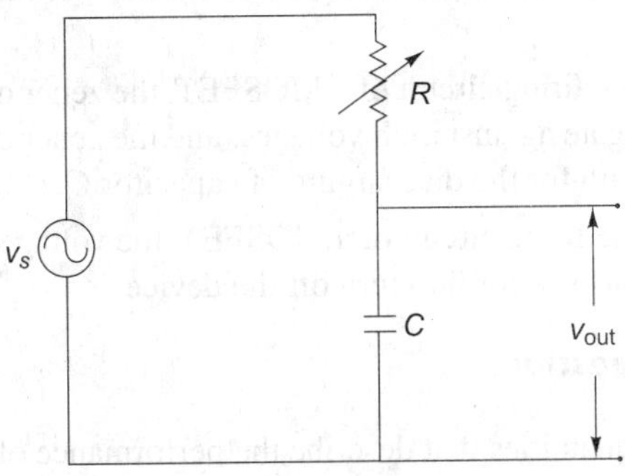

Fig. 10.31

13. The range of phase delay that can be obtained with the phase delay circuit of Fig. 10.32 is 30°–180°.

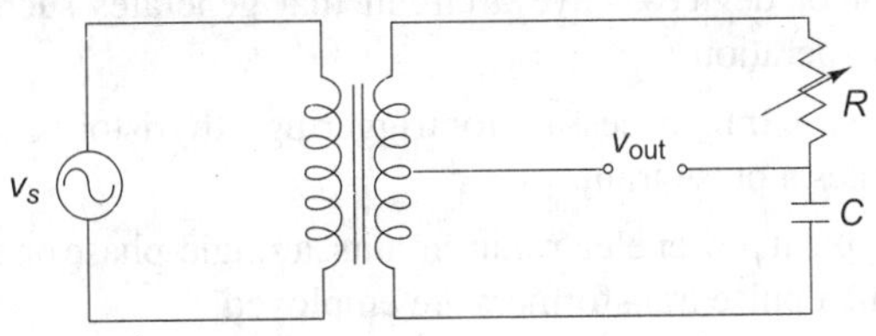

Fig. 10.32

14. The ramp and pedestal type of delay circuit of Fig. 10.13(d) gives more delay than the *RC* phase shifting circuit of Fig. 10.13(b).
15. Zero current detection is needed in single- and three-phase controlled rectifiers, and zero voltage detection is needed in a dual converter used as a cycloconverter.
16. Considering the ramp-comparator method of getting a delayed phase angle, if a dc voltage of 2 V is given at the input of the ramp voltage generator, a voltage ramp having a slope of nearly 63° is obtained at its output.
17. In the digital firing scheme employed for a three-phase, full-wave rectifier, three ring counters and two *N*-bit down-counters are used.
18. Considering the pulse amplifier type of drive circuit for thyristors, the trigger current of high-power thyristors is kept at a high value for facilitating reliable firing of the thyristors.
19. Considering the drive circuit for power transistors, the value of the resistance R_B should be large enough to avoid high power dissipation, and small enough to avoid slow turn-off.

20. Considering the drive circuit for power transistors, the power supply voltage V_{AA} should not be too small so as to facilitate reduction in the variation of $V_{BE(ON)}$, and should not be too large so as to keep down the ON state losses.
21. Considering the firing circuit for a GTO, the MOSFET F_1 is switched on at first in order to turn it on; in order to turn it off, F_2 is turned off and then F_1 is turned on.
22. Considering the firing circuit of a MOSFET, the zener diode D_{z1} is necessary to protect the gate against high voltages, and the zener diode D_{z2} is necessary to provide a path for the discharging of capacitor C.
23. Considering the firing circuit of a MOSFET, the voltage V_A should be 0 V for it to turn on and 1 V for it to turn off the device.

Short Answer Questions

1. What are the quantities that describe the performance of a pulse transformer?
2. Explain the terms (a) corona, (b) flashover, (c) creepage, and (d) dielectric strength as applied to a pulse transformer.
3. Why are electrostatic and electromagnetic shieldings necessary for a pulse transformer? What materials are used for these shieldings?
4. Draw a typical pulse voltage waveform required for the triggering of semiconductor devices. Give a circuit that generates such a waveform and explain its operation.
5. When is a pulse train necessary for triggering a thyristor? Give a typical circuit that generates a pulse train.
6. Give two typical power electronic circuits, a single-phase one and a three-phase one, in which pulse transformers are employed.
7. Describe the various constructional features of opto-isolators.
8. Explain the various performance features associated with opto-isolators.
9. Give three circuits, one with a thyristor, one with a transistor, and one with a TRIAC, in which opto-isolators are employed.
10. Explain why the firing signal required for inductive loads should be either rectangular in shape or in the form of a pulse train.
11. Give any two phase shifting circuits and explain their operation.
12. Give a typical circuit for zero voltage crossing detection and explain its operation.
13. Give a typical circuit for zero current crossing detection and explain its operation.
14. Give a typical scheme that employs both analog and digital components for the firing of a three-phase bridge converter.
15. Give a typical digital-electronic scheme for the firing of a three-phase bridge converter.
16. Describe typical drive circuits that are used for (a) a thyristor and (b) a GTO.
17. Describe typical drive circuits that are used for (a) a power transistor and (b) a MOSFET.

CHAPTER 11

Industrial Applications

11.1 Introduction

The inception of the thyristor during the 1960s as a power device led to the development of compact, reliable, and maintenance-free drive circuits. This application got a further boost with the advent of the gate turn-off thyristor, power transistor, and MOSFET, among others, as power electronic devices. The utility of these devices spread to other areas as well, such as uninterruptible power supplies (UPSs), induction heating, high-voltage dc (HVDC) transmission, electrical welding, static reactive volt-ampere (VAR) control of power systems, etc. This chapter is devoted to a detailed discussion of these industrial applications.

After going through this chapter the reader should

- know the role of power electronic converters in UPSs,
- get acquainted with (i) converters used in HVDC transmission, (ii) the derivation of expressions for the ac source current in terms of the dc current and for the reactive power drawn by these converters, and (iii) the various modes of converter control,
- understand the principles underlying induction heating and welding, and know the detailed circuitry for implementing the same,
- become familiar with thyristor switched capacitors (TSCs) and thyristor controlled inductors (TCIs) that are widely used for static VAR control of power systems, and
- know the constructional and operational details of switch mode power supplies (SMPSs).

11.2 Uninterruptible Power Supplies

In applications such as chemical processes, computers, and life support systems, alternative supplies at the mains frequency are needed when the mains supply fails. It is important that the changeover be effected instantaneously in respect of these loads. The *uninterruptible power supply* is therefore used to describe such a system. These power supplies usually include an inverter operated from a dc storage battery or some other dc source. Their name originated from the fact that they can be pressed into service in a very short time and also switched off as soon as (a) the mains supply gets restored or (b) a motor generator set is started and brought up to the speed required to supply the required power. Figure 11.1 shows the block diagram of a typical UPS system in which switches Sw_1 and Sw_2 are, respectively, kept closed and open to ensure that the load, called the *critical load*, gets its power directly from the ac mains. The dc battery is trickle-charged from this mains under normal conditions. The switch conditions will, however, be reversed, that is, Sw_1 is opened and Sw_2 is closed, when there is a mains supply failure.

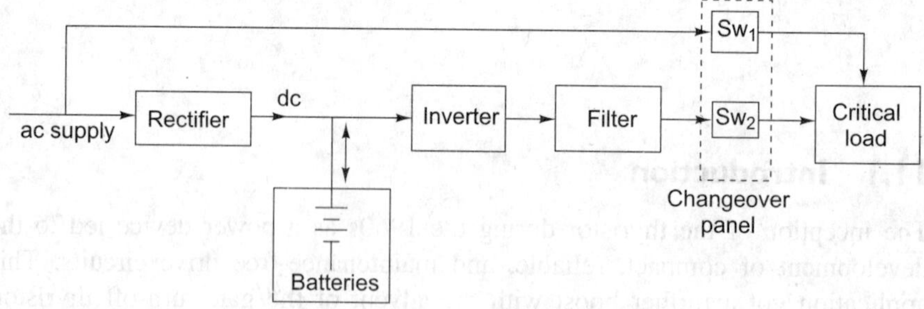

Fig. 11.1 Block diagram of a UPS

The inverter in the UPS is operated at a fixed frequency and its nearly sinusoidal output is obtained by means of optimized pulse width modulation (PWM) or waveform synthesis followed by a filter to further improve it. The inverter needs to be synchronized with the mains to ensure that no waveform distortion occurs upon mains failure. Important parts of the UPS, namely, the batteries, inverter, and rectifier are described below.

11.2.1 Batteries

Among the many types of battery systems that exist, two of the secondary cell type, namely, the lead acid type and the alkaline type, are commonly used. However, because of its higher ampere-hour efficiency and lower cost as compared to the latter, the lead acid cell is preferred for UPS applications. The battery draws a small amount of current during trickle-charging, which helps in keeping it in a fully charged state. The capacity of the battery is expressed in ampere-hours; this rating is equal to the product of the constant discharge current and the duration above which the battery voltage falls below a certain level called the final discharge voltage. When the battery supplies the load current, a check is kept under the mains shutdown condition and its magnitude constantly monitored so that it does not fall below the *final discharge level*. This is because any further use of the battery after it attains this level will only shorten the life of the battery.

When the battery is released from load, it is kept charged with constant current I_c till the voltage goes up to the trickle-charge voltage level V_{TC}. After this, it is kept under a constant voltage state as shown in Fig. 11.2. The current now goes down to its trickle-charge value and remains at that level.

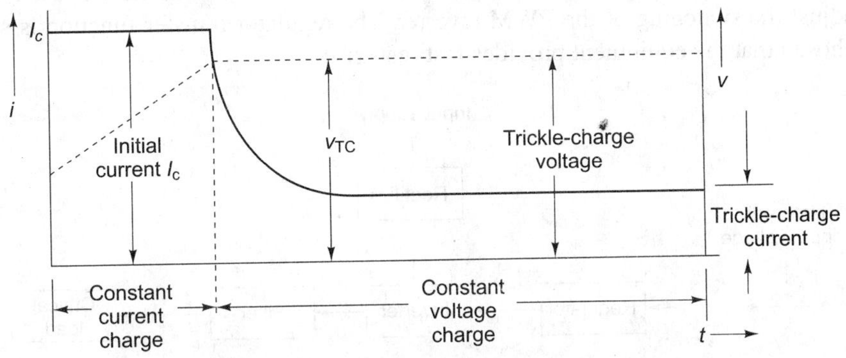

Fig. 11.2 Battery charging characteristic

The battery charging characteristic can be programmed so as to bring it to the fully charged state in minimum time.

11.2.2 Inverters

The main problem encountered with inverters is that of the introduction of harmonics into the UPS due to the non-linear nature of most of the loads. One method of minimizing these harmonics is by assessing a factor called the *per cent total harmonic distortion* (PTHD) factor and keeping it below 5%; also, the magnitude of each harmonic component as a ratio of V_1 is kept below 3% of the fundamental. The harmonic content of the output voltage can be assessed by means of the PTHD factor, which is defined as

$$\text{PTHD} = \frac{\left(\sum_{k=2}^{\infty} V_k^2\right)^{1/2}}{V_1} \times 100 \tag{11.1}$$

where V_1 and V_k are, respectively, the RMS values of the fundamental and kth harmonic components. For inverters used in UPSs, this factor is required to be less than 5%; also, each harmonic component has to have a magnitude less than 3% of the fundamental.

A single- or three-phase PWM inverter is used in a modern UPS, because it permits selective harmonic elimination as explained in Chapter 5. An isolation transformer is also used at the output of the inverter as shown in Fig. 11.3. For high-capacity UPSs two or more inverters may be used, which are paralleled through

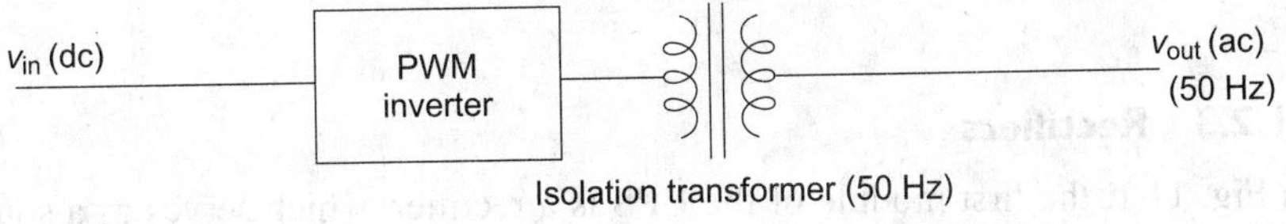

Fig. 11.3 Inverter with isolation transformer

transformers so as to feed a common loop output. Another way to minimize the harmonic content is to use closed-loop control as shown in Fig. 11.4. Here the actual output is compared with the sinusoidal reference and the error is used to adjust the switching of the PWM inverter. The regulator transfer function is so chosen that the equipment provides fast response.

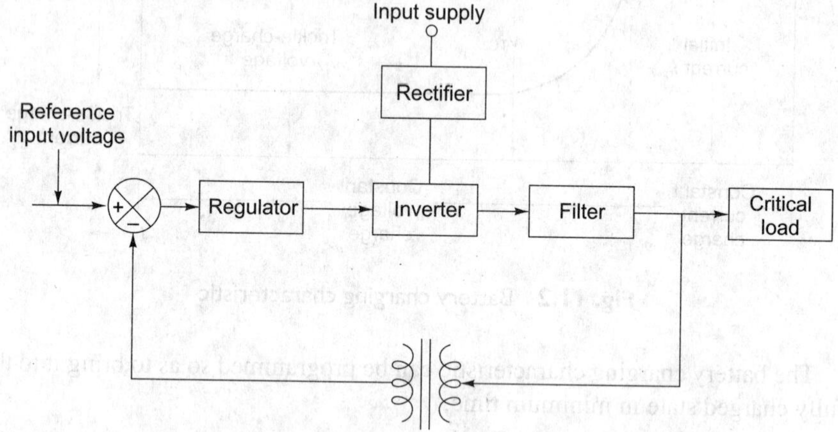

Fig. 11.4 Closed-loop-controlled UPS

For increased reliability the supply line is itself used as a backup to the UPS, and an ac controller is used to transfer the load from the UPS directly to the supply. A block diagram for such a scheme is shown in Fig. 11.5.

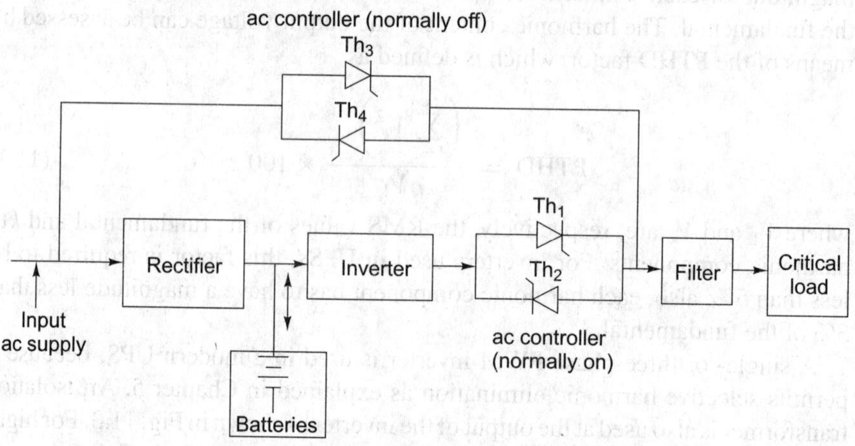

Fig. 11.5 A typical UPS with mains as the backup

11.2.3 Rectifiers

In Fig. 11.1, the first module of the UPS is a rectifier, which serves as a source of dc supply to the inverter and also keeps the battery unit under trickle charge.

Though this can be implemented in a number of ways, the arrangement shown in Fig. 11.6 has some merits as follows. A major portion of the power goes to the inverter through a three-phase diode bridge, the small quantity of power for battery charging being fed through a single-phase, centre-tapped type of controlled rectifier. This allows the charging voltage of the battery to be controlled with respect to both magnitude and polarity. The thyristor Th_3 is normally off and is turned on only when there is a supply failure, so that the battery serves its purpose.

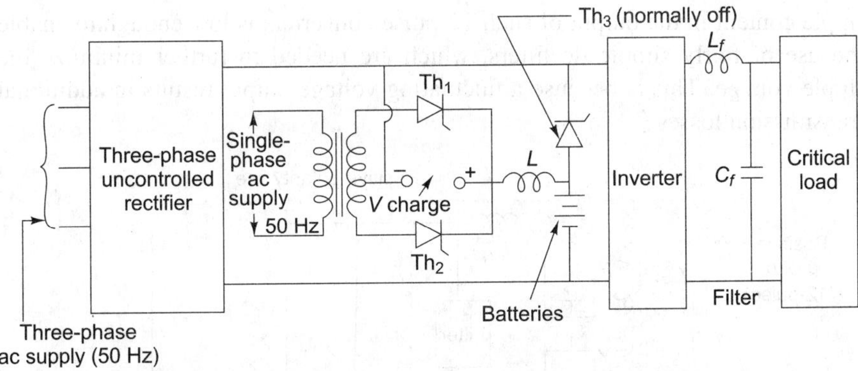

Fig. 11.6 A typical UPS using a three-phase rectifier

11.3 High-voltage Dc Transmission

Alternating current power is used as a power source as well as for transmission purposes because it can be conveniently generated and also converted from one voltage to another. Transmission of ac power over long distances can, however, result in relatively high transmission losses. Transient stability problems and operational requirements such as dynamic damping of electrical systems may also arise with long transmission lines. Dc transmission is an alternative which overcomes most of these problems. However, it is economically feasible only when the transmission distance exceeds 500–600 km; this distance is smaller in the case of underwater cables.

Figure 11.7 shows the following features: At the sending end of the HVDC (Mohan et al. 1989) system, the converter *S* transforms the low ac voltage to a higher ac voltage and then rectifies it to a dc voltage for transmission. At the receiving end the HVDC is converted back to HVAC and then stepped down to low-voltage ac. When the operating conditions demand that power be transmitted from system *B* to system *A*, the roles of the converters *R* and *S* are reversed.

The power source used for HVDC transmission is usually a thyristor converter which is made up of serially connected devices to obtain high-voltage capability. As elaborated later, two 12-pulse fully controlled bridges are usually employed for supplying power through the positive and negative poles (Fig. 11.8). The converters are designed to operate in both rectification and inversion modes. The

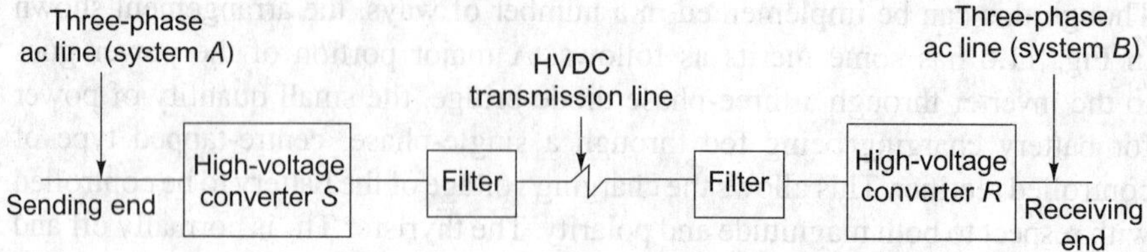

Fig. 11.7 Block diagram of a two-terminal HVDC system

ripple content in the output of such 12-pulse converters is low enough to enable the use of fairly simple dc filters, which are needed to further minimize the ripple voltage. This is because a fluctuating voltage output results in additional transmission losses.

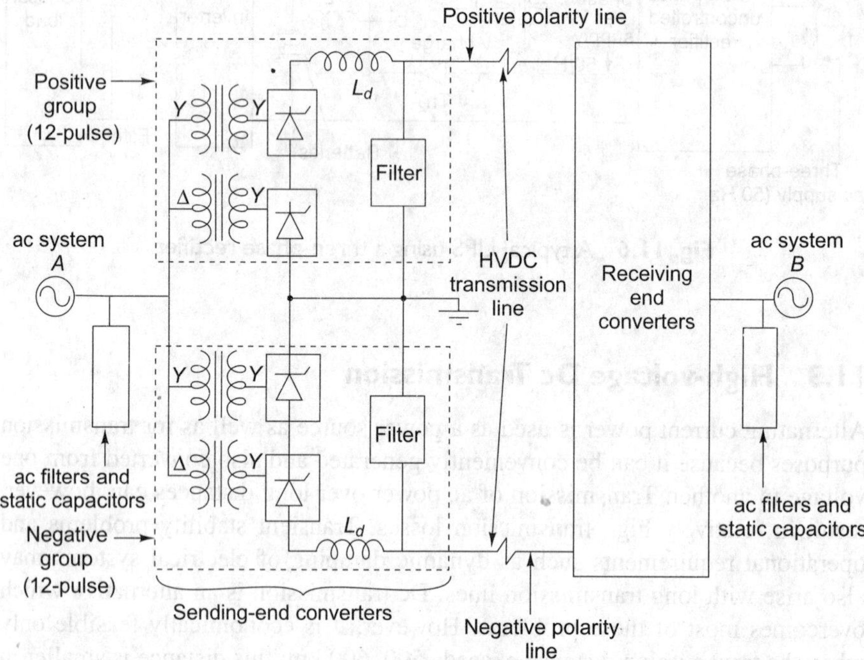

Fig. 11.8 A typical HVDC system with detailed circuit connections of the sending-end converters

Figure 11.8 shows that each converter terminal consists of positive and negative poles, each pole in turn consisting of two six-pulse input line frequency bridge converters connected through Y-Y and Δ-Y transformers to give a 12-pulse output. Ac filters are employed to reduce the current harmonics generated by the converters and also to prevent their entry into the ac system. Capacitors are installed at the sending-end side in line with ac filters for supplying lagging reactive power to the rectifier; similar capacitors are also required at the receiving end. The smoothing reactor L_d at the sending end ensures that the ripple in the dc voltage is not carried over as a current ripple into the dc line and the dc side filter banks.

11.3.1 Twelve-pulse Line frequency Converters

Two six-pulse controlled converters are combined into a 12-pulse converter for dc transmission. The two six-pulse converters are connected through star-star and Δ-star transformers (Fig. 11.9) so as to facilitate reduction of the current harmonics generated on the ac side, as well as the voltage ripple produced on the dc side of the converter. The series connection of the two six-pulse converters is necessary to ensure that the specifications associated with HVDC transmission are fulfilled.

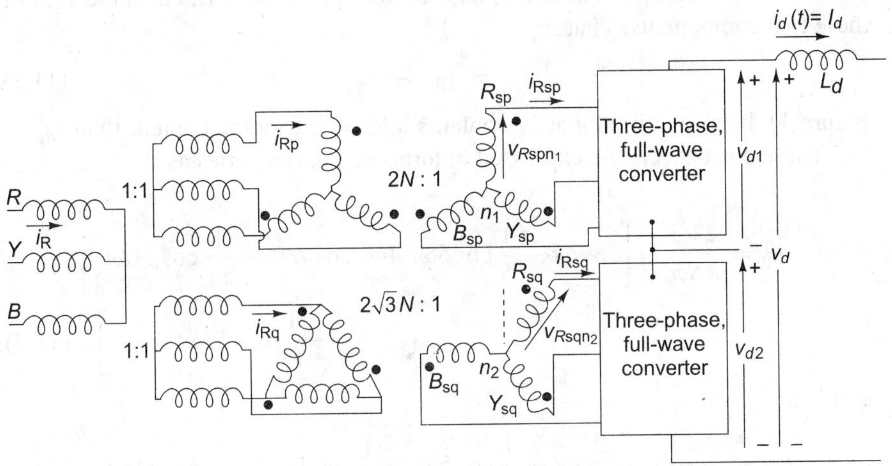

Fig. 11.9 A 12-pulse converter circuit using star-star and Δ-star transformers

The following notations pertain to the HVDC system under consideration.

i_{Rp}, i_{Rq}: Instantaneous R-phase primary currents of the Y-Y and Δ-Y transformers

i_R: Sum of i_{Rp} and i_{Rq}

i_{Rsp}, i_{Rsq}: R-phase secondary currents of the Y-Y and Δ-Y transformers

$i_{(Rsp)_1}$, $i_{(Rsq)_1}$: Fundamental components of i_{Rsp} and i_{Rsq}, respectively

i_d, I_d: Instantaneous and average output currents of the 12-pulse converter combination

v_{Rspn_1}, V_{Rspn_1}: Instantaneous and RMS values of R-phase line-to-neutral voltages on the secondary side of the Y-Y transformer

v_{Rsqn_2}, V_{Rsqn_2}: Instantaneous and RMS values of R-phase line-to-neutral voltages on the secondary side of the Δ-Y transformer

v_{d1}, V_{d1}: Instantaneous and average output voltages of the rectifier connected to the Y-Y transformer

v_{d2}, V_{d2}: Instantaneous and average output voltages of the rectifier connected to the Δ-Y transformer

v_d, V_d: Sums of instantaneous and average voltages of the rectifiers connected to the Y-Y and Δ-Y transformers

The different vector groups for the two transformers are such that, on the secondary side, the voltage V_{Rspn} leads V_{Rsqn} by 30° (Fig. 11.9). The smoothing reactor L_d is assumed to give a direct current characteristic to the current i_d. Assuming that the per phase leakage inductances of the transformers are negligible, and that the current is in the form of the rectangular pulses, the current waveforms of the individual currents, i_{Rp} and i_{Rq} will have stepped shapes as shown in Fig. 11.10(a). The total phase current i_{R} can be written as the sum of these two components. That is,

$$i_{\text{R}} = i_{\text{Rp}} + i_{\text{Rq}} \tag{11.2}$$

Figure 11.10(a) confirms that i_{R} contains a lower harmonic content than i_{Rp} or i_{Rq}. These currents can be expressed in terms of Fourier series as

$$i_{\text{Rp}} = \frac{2\sqrt{3}}{2N\pi} I_d \left[\cos\omega t - \frac{1}{5}\cos 5\omega t + \frac{1}{7}\cos 7\omega t - \frac{1}{11}\cos 11\omega t + \frac{1}{13}\cos 13\omega t + \cdots \right] \tag{11.3}$$

and

$$i_{\text{Rq}} = \frac{2\sqrt{3}}{2N\pi} I_d \left[\cos\omega t + \frac{1}{5}\cos 5\omega t - \frac{1}{7}\cos 7\omega t - \frac{1}{11}\cos 11\omega t + \frac{1}{13}\cos 13\omega t + \cdots \right] \tag{11.4}$$

The transformer ratios are taken as $2N$:1 and $2\sqrt{3}N$:1, and i_{Rp} and i_{Rq} are referred to the secondary side. i_{R} can be expressed as

$$i_{\text{R}} = i_{\text{Rp}} + i_{\text{Rq}} = \frac{2\sqrt{3}}{N\pi} I_d \left[\cos\omega t - \frac{1}{11}\cos 11\omega t + \frac{1}{13}\cos 13\omega t + \cdots \right] \tag{11.5}$$

Equation (11.5) shows that the total current i_{R} has harmonics of the order of $12k \pm 1$, where k is an integer, with the 11th harmonic being the lowest order harmonic. The expression for i_{R} also confirms that there is much less ripple content with the resulting 12-pulse operation. The dc side voltage waveforms v_{d1} and v_{d2} shown in Fig. 11.10(b) are shifted by 30° with respect to each other.

With the two six-pulse converters connected in series on the dc side, the total dc voltage v_d can be written as

$$v_d = v_{d1} + v_{d2} \tag{11.6}$$

As can be expected, v_d has 12 ripple pulsations per cycle of ac line frequency. Fourier analysis also confirms that the voltage harmonics are of the order of $12k$, where k is an integer, confirming that the 12th harmonic is the lowest order

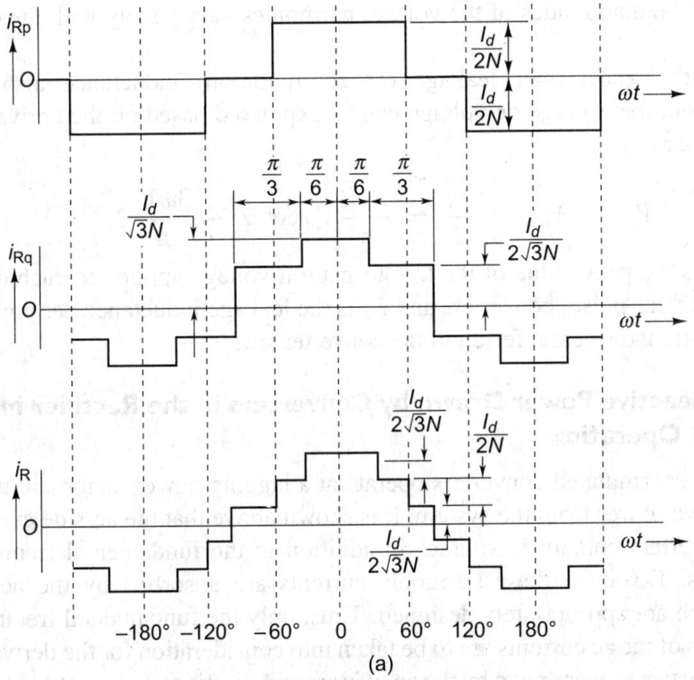

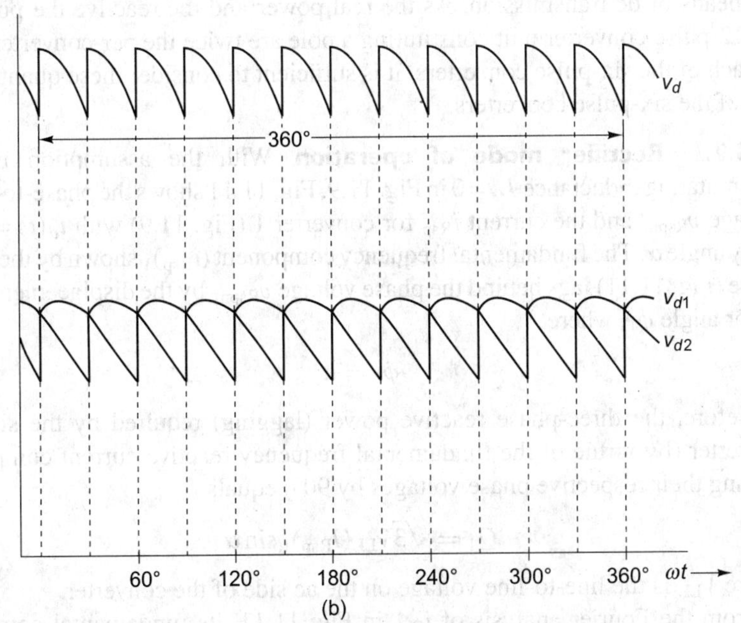

Fig. 11.10 Sending end waveforms assuming $L_s = 0$: (a) currents i_{Rp}, i_{Rq}, and i_R; (b) voltages v_{d1}, v_{d2}, and v_d

harmonic. The magnitudes of the voltage harmonics vary widely with the delay angle α.

Taking the transformer leakage (or commutation) inductance also into consideration, the average dc voltage can be expressed based on the derivations in Chapter 2 as

$$V_{d1} = V_{d2} = \frac{V_d}{2} = \frac{3\sqrt{3}}{\pi} V_m \cos\alpha - \frac{3\omega L_s}{\pi} I_d \tag{11.7}$$

where V_m is the peak value of the line-to-neutral voltage applied to each of the full-wave (or six-pulse) converters and L_s is the leakage inductance per phase of each of the transformers referred to the converter side.

11.3.2 Reactive Power Drawn by Converters in the Rectifier Mode of Operation

The ac line commutated converters operate at a lagging power factor and hence draw reactive power from the system. It is shown above that the ac-side currents of the converters contain harmonics in addition to the fundamental frequency components. However, these harmonic currents are absorbed by the ac-side filters, which are appropriately designed. Thus, only the fundamental frequency components of the ac currents are to be taken into consideration for the derivation of (i) the reactive power drawn by the converters and (ii) the real power transferred by means of dc transmission. As the real power and the reactive the power for the 12-pulse converter unit constituting a pole are twice the per-converter values of each of the six-pulse converters, it is sufficient to consider these quantities for one of the six-pulse converters.

11.3.2.1 Rectifier mode of operation With the assumption that the commutating inductance $L_s = 0$ in Fig. 11.9, Fig. 11.11 shows the phase-to-neutral voltage v_{Rspn_1} and the current i_{Rsp} for converter 1 (Fig. 11.9) with $i_d(t) = I_d$ at a delay angle α. The fundamental frequency component $(i_{\mathrm{Rsp}})_1$ shown by the dashed curve (Fig. 11.11) lags behind the phase voltage v_{Rspn_1} by the displacement power factor angle ϕ_1, where

$$\phi_1 = \alpha \tag{11.8}$$

Therefore, the three-phase reactive power (lagging) required by the six-pulse converter (by virtue of the fundamental frequency reactive current components lagging their respective phase voltages by 90°) equals

$$Q_1 = \sqrt{3} V_{\mathrm{LL}} (I_{\mathrm{Rsp}})_1 \sin\alpha \tag{11.9}$$

where V_{LL} is the line-to-line voltage on the ac side of the converter.

From the Fourier analysis of i_{Rsp} in Fig. 11.11, its fundamental component $(i_{\mathrm{Rsp}})_1$ can be written in terms of I_d as

$$(I_{\mathrm{Rsp}})_1 = \frac{\sqrt{6}}{\pi} I_d = 0.78 I_d \tag{11.10}$$

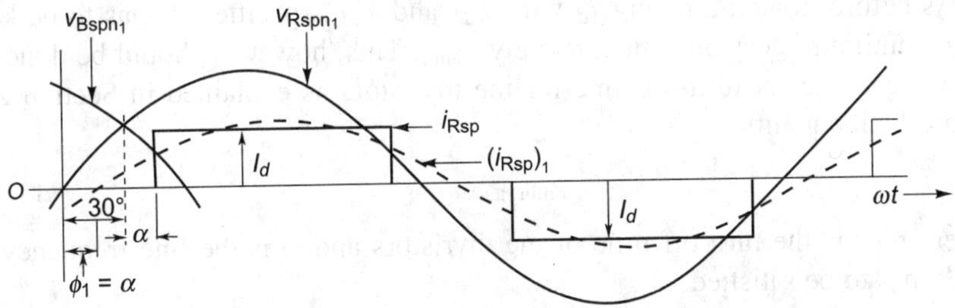

Fig. 11.11 Idealized waveforms of v_{Rspn_1}, i_{Rsp}, and $(i_{Rsp})_1$ assuming $L_s = 0$

Equations (11.9) and (11.10) can be combined to give

$$Q_1 = \sqrt{3}V_{LL}\left(\frac{\sqrt{6}}{\pi}\right) I_d \sin\alpha = 1.35 V_{LL} I_d \sin\alpha \tag{11.11}$$

The real power transmitted through each of the six-pulse converters can be obtained from Eqn (11.7), with the assumption that the leakage inductance L_s of the transformer is negligible, as

$$P_1 = V_{d1} I_d = 1.35 V_{LL} I_d \cos\alpha \tag{11.12}$$

The first consideration in the design of converters is the minimization of Q_1 for a desired power transfer P_1. Second, I_d should also be kept as small as possible to minimize the ohmic losses on the dc transmission line. Equation (11.12) shows that with P_1 and V_{LL} having specified values, the minimization of I_d is possible only by choosing a small value for the firing angle α in the rectifier mode of operation. Taking practical considerations also into account, α is maintained in the range $10° \le \alpha \le 20°$.

11.3.2.2 Inverter mode of operation For arriving at the real and reactive powers for inverter operation, the extinction angle γ has to be considered, where

$$\gamma = 180° - (\alpha + \mu)$$

With the assumption that $L_s \approx 0$, the commutation angle μ also becomes negligible, and thus

$$\gamma = 180 - \alpha$$

Proceeding along lines similar to those for the rectifier, the various quantities at the inverter end can be derived after taking the above approximations into account (Mohan et al. 1989) as

$$V_{d1} = V_{d2} = \frac{V_d}{2} = 1.35 V_{LL} \cos\gamma \tag{11.13}$$

$$P_1' = 1.35 V_{LL} I_d \cos\gamma \tag{11.14}$$

and

$$Q_1' = 1.35 V_{LL} I_d \sin\gamma \tag{11.15}$$

As before, for minimizing I_d with P_{d1} and V_{LL}, specified, γ has to be kept at its minimum possible value, namely, γ_{min}. This, however, should be done by allowing sufficient turn-off time for the thyristors as explained in Section 2.5. Thus, the inequality

$$\gamma_{min} \geq \omega t_{OFF} \tag{11.16}$$

where t_{OFF} is the turn-off time of the thyristors and ω is the line frequency in rad/s, has to be satisfied.

11.3.3 Control of HVDC Converters

Control of HVDC consists of placing the operating point (v_d, i_d) at an appropriate location in the v_d-i_d plane. Considering rectifier operation, this point falls in the first quadrant of this plane, whereas with inverter operation, it is located in the fourth quadrant. The following discussion is confined to per-pole operation because the positive and negative poles are operated under identical conditions. Figure 11.12(a) shows the positive pole consisting of a sending-end 12-pulse converter operating as a rectifier and a receiving-end 12-pulse converter providing inverter operation. The control of the converter A at the sending end can be understood with the help of the static characteristic given in Fig. 11.12(b).

The first line AB is that of a constant firing angle characteristic drawn between v_d and i_d. The equation of this line will be that given by Eqn (11.7), which is reproduced here, in its variable version:

$$v_{d1} = V_0 \cos\alpha - \frac{3\omega L_s}{\pi} i_d$$

where $V_0 = (3\sqrt{3}/\pi)V_m$. With constant α, the first term on the right-hand side becomes constant; thus Eqn (11.7) is of the form $y = -mx + c$, with $y = v_{d1}$, $x = i_d$, $c = V_0 \cos\alpha$, and the slope $m = -3\omega L_s/\pi$. The quantity $|3\omega L_s/\pi|$ can therefore be interpreted as an internal resistance of the converter A and denoted as R_s. With a load condition that demands high value for i_d, the converter operates at the minimum firing angle α_{min} (say, 5°). Denoting the load resistance as R_{ld} and neglecting the drop $R_s i_d$, the load current can be written as

$$i_{ld} = \frac{v_{d1}}{R_{ld}} = \frac{V_0 \cos\alpha_{min}}{R_{ld}} \tag{11.17}$$

Equation (11.17) shows that i_{ld} depends on the ac input voltage and the load resistance. The line BC represents the constant current control characteristic and this is the normal mode of operation of the rectifier. For maintaining constant current, the firing angle α is adjusted so as to make i_{ld} equal to $i_{ld(Ref)}$. If $i_{ld} > i_{ld(Ref)}$ due to any reason, the controller increases the firing angle α, thus keeping down v_{d1} and restoring i_d. The operating point now moves down from B to C. At point C, the firing angle reaches 90° and, neglecting the overlap angle (which is equivalent to neglecting $R_s i_d$), the voltage v_d changes polarity; the converter then becomes an inverter. The line CD thus falls in the region of inverter operation.

At point D, the inverter firing angle is at its maximum value and any further increase can cause commutation failure. For safe operation of the inverter, it should be ensured that the extinction angle γ [$= 180° - (\alpha + \mu)$] is greater than

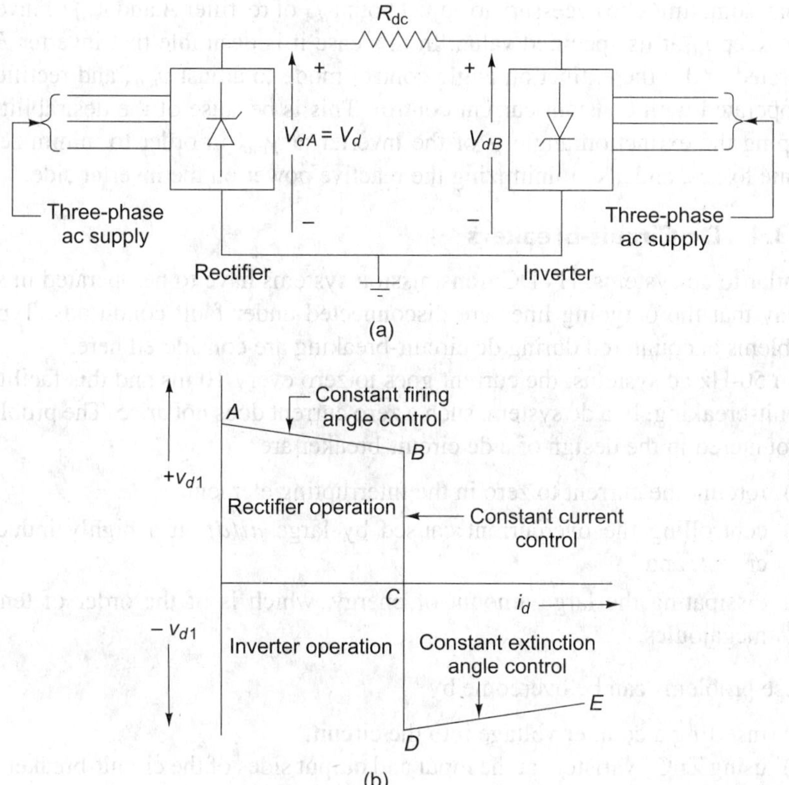

Fig. 11.12 (a) Schematic of a two-terminal HVDC connection; (b) control characteristic

$\gamma_{\min} = \omega\tau_{\text{OFF}}$. Typically, $\gamma_{\min}$ is of the order of 15°–30° for a thyristor. In the region *DE*, γ is kept constant at this minimum value and hence the name *extinction angle control*. In this range any increase in the load current increases the overlap angle μ and reduces $|v_{d1}|$, this being the negative resistance characteristic of an inverter. The aims of converter control are the following.

(i) Minimization of generation of non-characteristic harmonics.

(ii) Safe inverter operation with the fewest possible commutation failures, even with distorted ac voltages.

(iii) Lowest possible consumption of reactive power. From Eqns (11.11) and (11.15) it is seen that this requires operation with $\alpha = \alpha_{\min}$ and $\gamma = \gamma_{\min}$, respectively, in order to eliminate the risk of commutation failure.

(iv) Smooth transition from constant current control to constant extinction angle control.

(v) Sufficient stability margins and response time when the short-circuit ratio is low, where

$$\text{Short-circuit ratio} = \frac{\text{ac system short-circuit strength in MW}}{\text{dc power in MW}} \tag{11.18}$$

It may sometimes be necessary to adjust both v_{dA} of rectifier A and v_{dB} of inverter B to keep i_d at its specified value. In this case it is desirable that inverter B be operated under the extinction angle control mode to adjust v_{dB}, and rectifier A be operated with constant current control. This is because of the desirability of keeping the extinction angle γ of the inverter at γ_{min} in order to minimize the ohmic losses, and also minimizing the reactive power on the inverter side.

11.3.4 Dc Circuit-breakers

Similar to ac systems, HVDC transmission systems have to be operated in such a way that the outgoing lines are disconnected under fault conditions. Typical problems encountered during dc circuit-breaking are considered here.

In 50-Hz ac systems, the current goes to zero every 10 ms and this facilitates circuit-breaking. In a dc system, such a zero current does not arise. The problems encountered in the design of a dc circuit-breaker are

(a) forcing the current to zero in the interrupting element,

(b) controlling the overcurrent caused by large di/dt in a highly inductive circuit, and

(c) dissipating the large amount of energy, which is of the order of tens of megajoules.

These problems can be overcome by

(a′) inserting a counter voltage into the circuit,

(b′) using ZnO_2 varistors at the input and output sides of the circuit-breaker, and

(c′) inserting a resistor in the circuit to dissipate the energy.

All these features are incorporated in the dc circuit-breaker shown in Fig. 11.13. It is of the air blast type and the current is commutated by connecting a series combination of an inductor L and a capacitor C across it. The commutation circuit is thus oscillatory and leads to zero current in the circuit-breaker. When the opening of the circuit-breaker increases the voltage across the commutation circuit, it is limited by the three ZnO varistors, which enter into conduction due to the high-voltage conditions of the circuit.

A two-terminal dc system does not need a dc circuit-breaker because, with a fast converter control response, it is possible to bring the current quickly to zero. Dc circuit-breakers can, however, provide additional flexibility of operation in multiterminal systems.

11.3.5 Advantages of HVDC Transmission

The advantages of HVDC transmission can be summarized as follows:

(a) Trouble-free long-distance transmission is made possible.

(b) By providing an asynchronous interconnection of two ac systems, all interconnection problems such as stability, power swings, etc., which exist in an ac transmission system, are eliminated.

(c) It is possible to precisely control the power flow without inadvertent loop flow, which is unavoidable in interconnected ac systems.

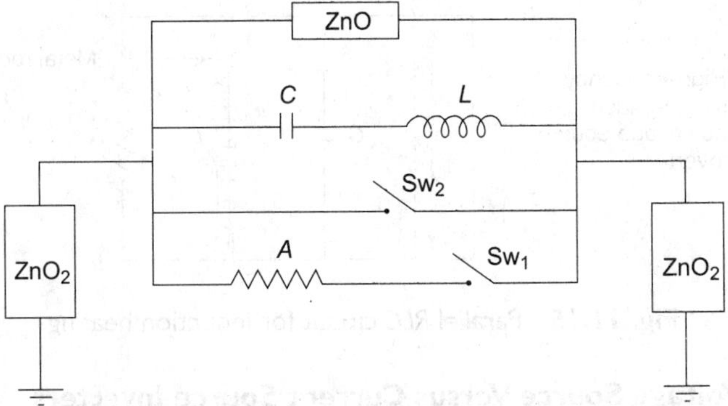

Fig. 11.13 Circuit diagram of a dc circuit-breaker

11.4 Induction Heating

Line commutated controlled rectifiers and inverters find industrial applications in induction heating.

11.4.1 Principle of Induction Heating

The principle of induction heating can be explained with the help of a hypothetical single-phase circuit (Fig. 11.14), where the load consists of a thick metal rod, which is actually like a short circuit for the secondary winding. As a result, heavy current flows through the rod, which gets heated up. With such a dead short circuit, however, the secondary winding of the transformer may get unduly heated and thus damaged. This circuit, though impractical, helps in understanding the principle of induction heating. A more practical circuit is described below.

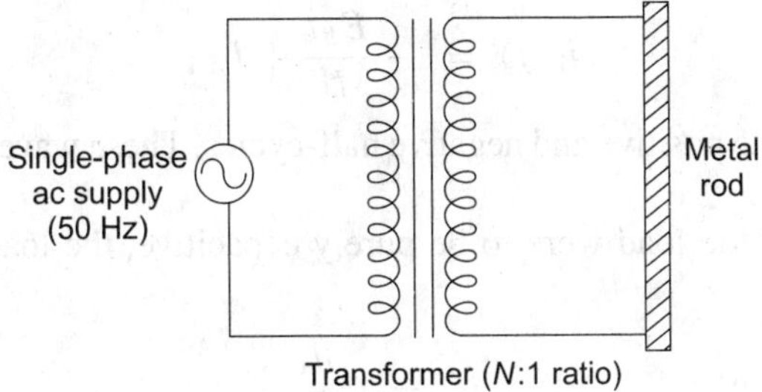

Fig. 11.14 Hypothetical circuit for induction heating

Figure 11.15 shows a parallel *RLC* circuit supplied by a high-frequency, single-phase ac voltage source inverter. The inductance coil wound round the metal rod acts like the primary of a transformer with the metal rod (workpiece) acting as the secondary; consequently, circulating currents flow through the metal rod, thereby heating it. The performance of this circuit vis-à-vis a circuit supplied by a current source inverter is now examined in detail.

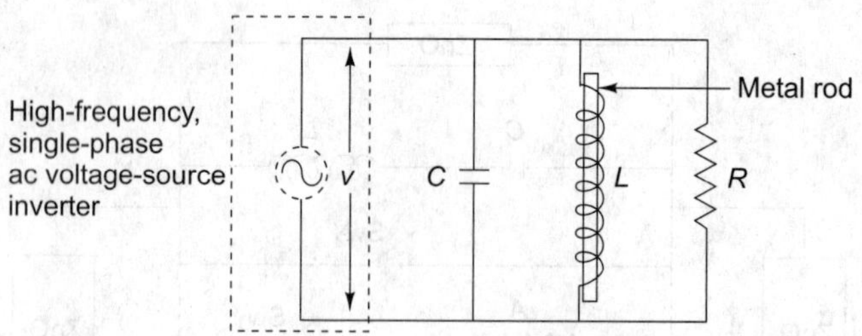

Fig. 11.15 Parallel *RLC* circuit for induction heating

11.4.2 Voltage Source Versus Current Source Inverters

A single-phase transistorized inverter feeding a load, which can be either inductive or capacitive, is given in Fig. 11.16(a). The primary voltage is assumed to have a square waveform with amplitude E_b. If the load is purely inductive, the load current will be in the shape of ramps with positive and negative slopes, respectively, in the positive and negative half-cycles of the primary voltage waveform, the overall load waveform having a triangular shape. This can be seen from the expression of the load current, written as

$$i_{\text{ld}} = \frac{1}{L}\int v\,dt \tag{11.19}$$

where v will be equal to the battery voltages $+E_B$ and $-E_B$, respectively, in the positive and negative half-cycles. Denoting the positive and negative peaks of the load current waveform as $I_{\text{ld},1}$ and $-I_{\text{ld},2}$, the two ramps can be expressed as

$$i_{\text{ld}}(t) = \frac{E_B t}{L} - I_{\text{ld},2} \tag{11.20}$$

$$i_{\text{ld}}(t) = -\frac{E_B t}{L} + I_{\text{ld},1} \tag{11.21}$$

respectively, in the positive and negative half-cycles. These waveforms are shown in Fig. 11.16(b).

In contrast, if the load were to be purely capacitive, the load current can be expressed as

$$i_{\text{ld}} = C\frac{dv}{dt} \tag{11.22}$$

With v having a square waveform, the load current will have the shape of alternately positive- and negative-going sharp current pulses.

If the source in Fig. 11.16(a) is now taken to be a current source having a square output waveform of amplitude I_B, the load currents with pure inductive and pure capacitive loads will be, respectively, in the form of positive- and negative-going triangular waveforms. Thus the roles of the pure inductive and capacitive loads will be reversed. The already drawn waveforms for the voltage source can now be properly interpreted. For example, the voltage waveform with a purely capacitive load will have the equations

$$v_{ld}(t) = \frac{I_B t}{C} - V_{ld,2} \tag{11.23}$$

and

$$v_{ld}(t) = \frac{-I_B t}{C} + V_{ld,1} \tag{11.24}$$

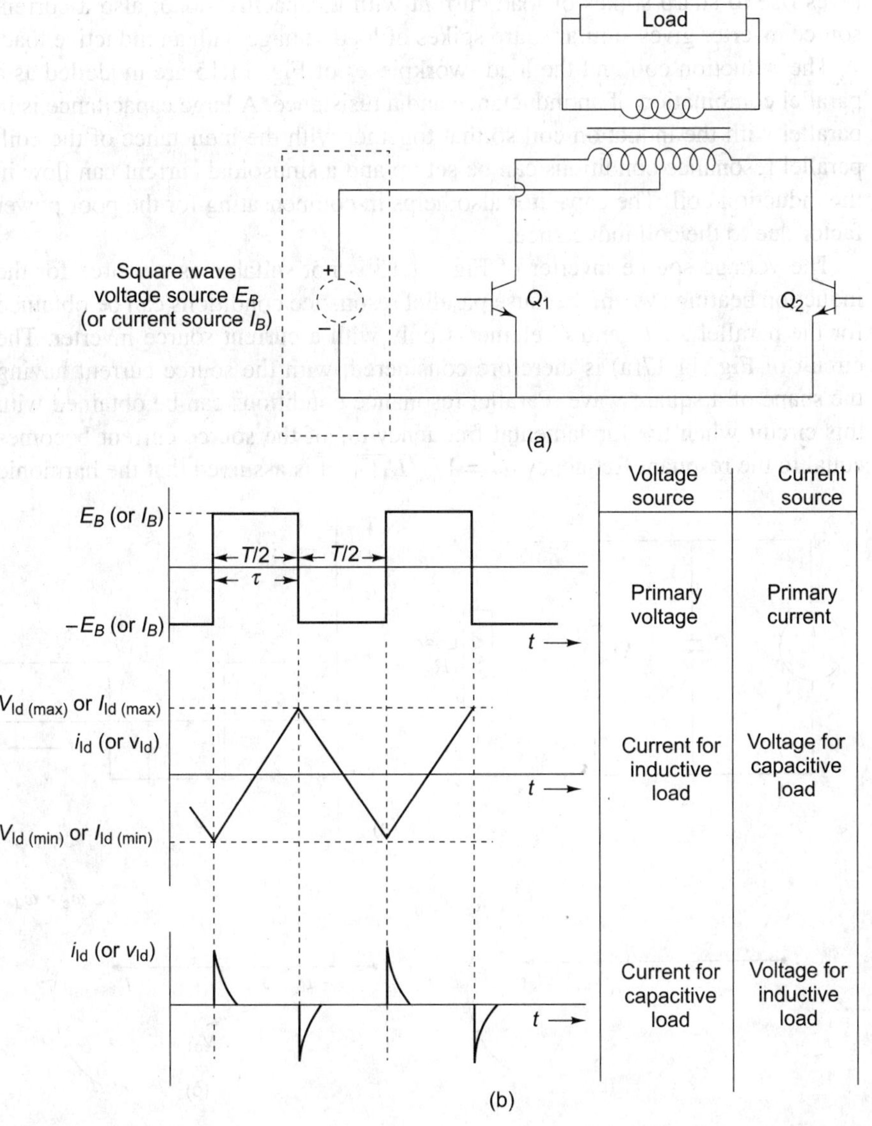

Fig. 11.16 (a) A single-phase inverter; (b) inverter waveforms with voltage and current sources

for the positive- and negative sloped ramps, where $V_{\text{ld},1}$ and $-V_{\text{ld},2}$ are the positive and negative peaks of the load voltage with an inductive load; the voltage waveform will have the shape of alternately positive- and negative-going sharp voltage pulses. This is because of a relation $v_{\text{ld}} = L(di/dt)$; its interpretation is given in Fig. 11.16(b).

11.4.3 Need for a Current Source Inverter Established

The above discussion can be summed up by stating that a voltage source inverter gives rise to sharp spikes of load current with a capacitive load; also a current source inverter gives similar sharp spikes of load voltage with an inductive load.

The induction coil and the load (workpiece) of Fig. 11.15 are modelled as a parallel combination of an inductance and a resistance. A large capacitance is in parallel with the induction coil so that together with the inductance of the coil, parallel resonance conditions can be set up and a sinusoidal current can flow in the induction coil. The capacitor also helps in compensating for the poor power factor due to the coil inductance.

The voltage source inverter of Fig. 11.15 is not suitable as a source for the induction heating system, because parallel resonance conditions can be obtained for the parallel R, L, and C elements only with a current source inverter. The circuit of Fig. 11.17(a) is therefore considered, with the source current having the shape of a square wave. Parallel resonance conditions can be obtained with this circuit when the fundamental frequency ω_1 of the source current becomes equal to the resonant frequency $\omega_r = 1/\sqrt{L_1 C_1}$. It is assumed that the harmonic

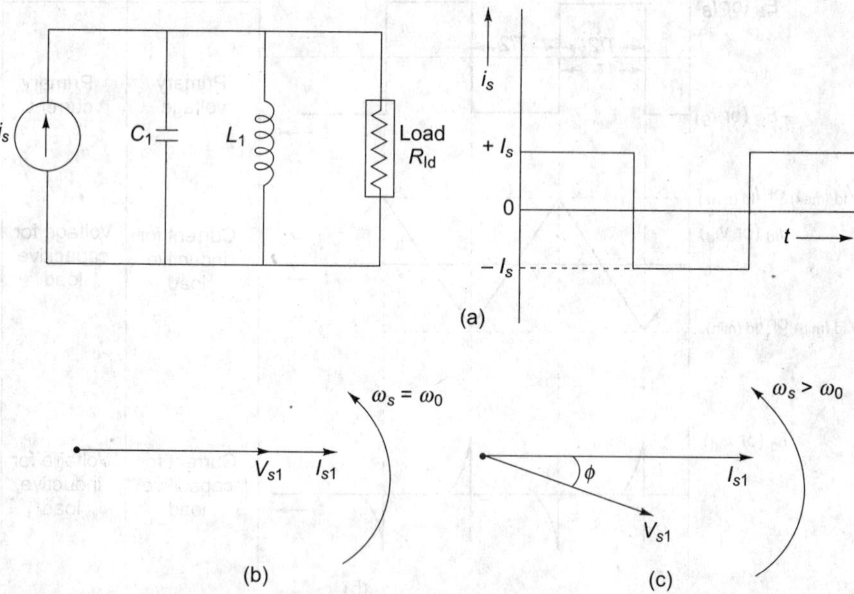

Fig. 11.17 (a) A parallel *RLC* circuit with a current source, and its i_s vs t waveform; (b) phasor diagram with V_{s1} in phase with I_{s1}; (c) phasor diagram with V_{s1} lagging I_{s1}

impedance of the parallel resonant load at the harmonic frequencies of the input current source is negligibly small, and thus the resulting voltage V_{s1} at the load can be considered to be sinusoidal. As illustrated in Fig. 11.17(b) the load voltage under resonance conditions will be in phase with the fundamental component of the source current.

The current source inverter is usually implemented by means of a three-phase, full-wave controlled rectifier and a single-phase bridge converter operating in the inverting mode with a large series inductor interposed between them; a typical configuration for this combination is given in Fig. 11.18(a). However, such a thyristorized circuit consumes reactive power; if the resonant load were to supply this reactive power, it will greatly reduce the complexity of the circuit. To facilitate this the load voltage is made to lag the input current [Fig. 11.17(c)]. This operation is possible with $\omega_1 > \omega_r$. But these conditions imply that the load as a whole will have a capacitive character. It is shown above that the load voltage for a circuit with a current source inverter feeding a capacitive load will have a triangular shape and, therefore, will be smooth. This feature confirms that the current source inverter is more appropriate as a source for induction heating purposes.

11.4.4 A Practical Circuit for Induction Heating

The circuit of a current source parallel resonant inverter, which is constituted of (i) a line commutated converter operating as a rectifier, (ii) a series inductor, and (iii) a single-phase converter operating in the inverting mode, is given in Fig. 11.18(a), and its waveforms are shown in Fig. 11.18(b). A small inductance L_1 is inserted in series with the resonant load so as to suppress high di/dt through the inverter thyristor during commutation. The output current of the current source inverter will therefore deviate from the ideal square wave shape; instead it will have a trapezoidal shape as shown in Fig. 11.18(b). The figure also shows that a reverse voltage is impressed across Th_A for a duration equal to γ/ω_s, where γ is the extinction angle of the line commutated inverter. Evidently, γ/ω_s should be greater than the turn-off time of the thyristor.

The power output of the inverter of Fig. 11.18(a) can be controlled by one of the following methods: (a) The operating frequency ω_s is increased above the natural frequency ω_r; consequently the power output decreases if I_d is held constant by

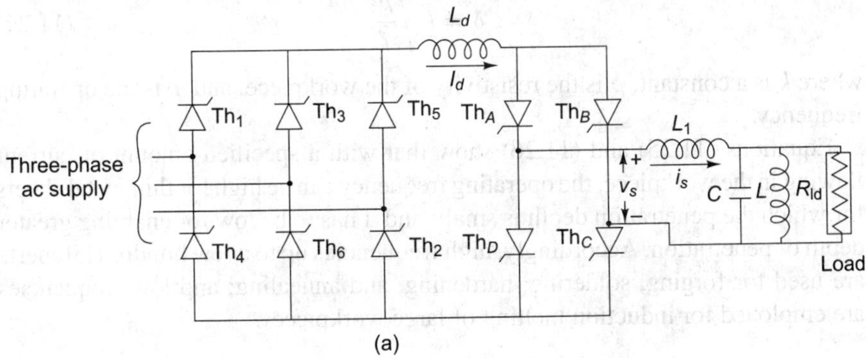

Fig. 11.18(a)

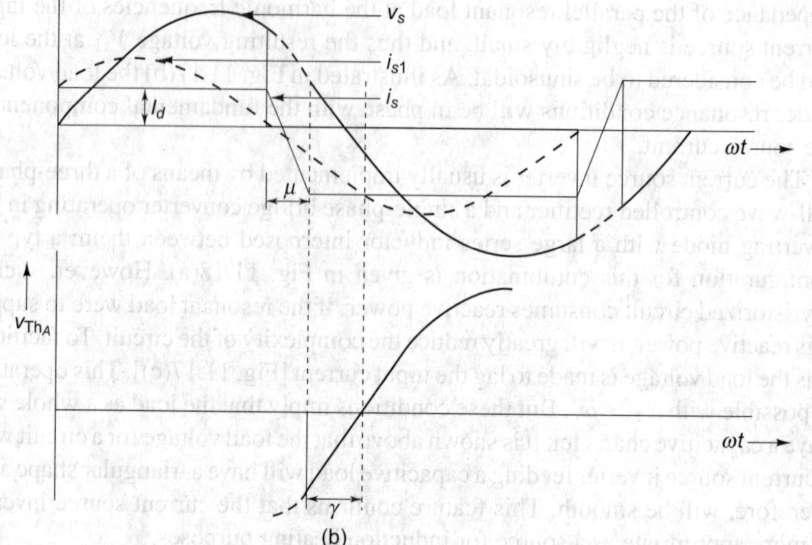

Fig. 11.18 (a) A practical circuit of a current source parallel resonant inverter; (b) waveforms for induction heating

means of a controlled dc source. (b) The current I_d is controlled, keeping the frequency of the inverter constant at ω_s.

11.4.5 Choice of Frequency

The workpiece in an induction heating set-up consists of a thick metal plate; the heating effect is proportional to the square of the current passing through it. This current depends upon the frequency of operation as follows:

$$I(x) = I_0 e^{-x/\delta} \tag{11.25}$$

where I_0 is the current at the surface and x is the depth of penetration from the surface. The factor δ is like a time constant, as seen by the relation $I(x) = I_0/e$ when $x = \delta$. This δ is given as

$$\delta = k\sqrt{\frac{\rho}{f}} \tag{11.26}$$

where k is a constant, ρ is the resistivity of the workpiece, and f is the operating frequency.

Equations (11.25) and (11.26) show that with a specified amount of current flowing in the workpiece, the operating frequency can be high for thin metal sheets for which the penetration depth is small; and it has to be low for enabling greater depth of penetration. Accordingly, high frequencies up to a few hundred kilohertz are used for forging, soldering, hardening, and annealing; and low frequencies are employed for induction melting of large workpieces.

11.5 Welding

The welding together of metal parts by heating them up to melting point is one area where conventional control equipment, such as mercury-vapour valves of the grid-control or the ignitron type, has given place to thyristorized equipment. Welding is of two kinds: (a) resistance welding and (b) arc welding. Both types require a heavy current at low voltage. Whereas alternating current is used in the first case, direct current is employed for arc welding.

11.5.1 Resistance Welding

Alternating current is used for resistance welding because of the ease with which the required voltage and current can be obtained with a transformer. Resistance welding can be implemented for iron, steel, brass, copper, aluminium, and other such metals. It may be required at a particular location, as in the case of (a) butt jointing, where the two ends of rods and wires are joined, and (b) spot welding, where two sheets of metal are joined with an overlapping joint. When this type of welding is done along a particular length of a sheet it is termed as *seam welding*.

Figure 11.19 shows that an ac controller interposed between the input ac supply and the primary winding of a step-down transformer serves as a source of controlled power to the welding unit.

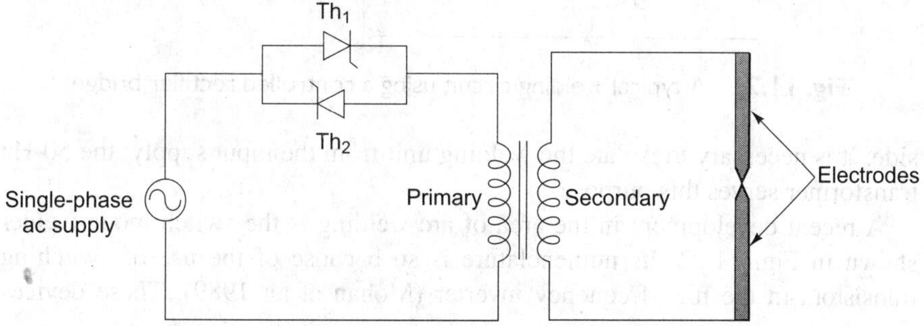

Fig. 11.19 Circuit diagram for resistance welding

Water cooling is provided for the thyristors in the primary circuit and also for the electrodes on the secondary side. Thyristors with high peak inverse voltage are used in ac controllers.

11.5.2 Arc Welding

In arc welding, the heat necessary for melting is obtained by an arc struck between the two electrodes, one of them being the workpiece, as illustrated in Fig. 11.20.

Direct voltage of the order of 40–60 V and direct current of the order of 500–800 A are required for this type of welding. As shown in Fig. 11.21 a single-phase step-down (50 Hz) transformer and a single-phase thyristor bridge are used to provide the required load voltage.

An alternative scheme may consist of a single-phase transformer followed by a diode bridge and a dc chopper. It is desirable that the load current have very low ripple content. This requirement is met by connecting an inductor on the load

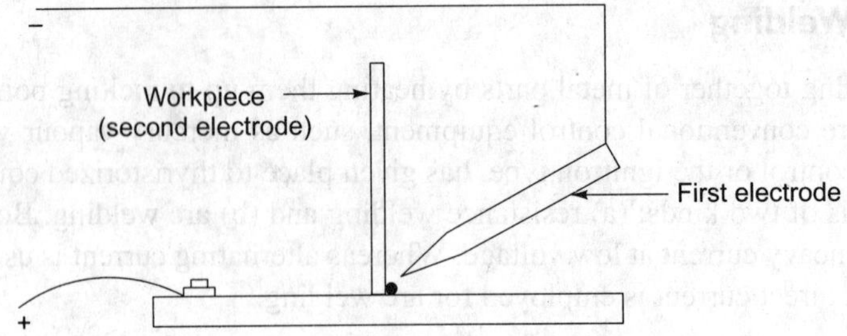

Fig. 11.20 A typical setup for arc welding

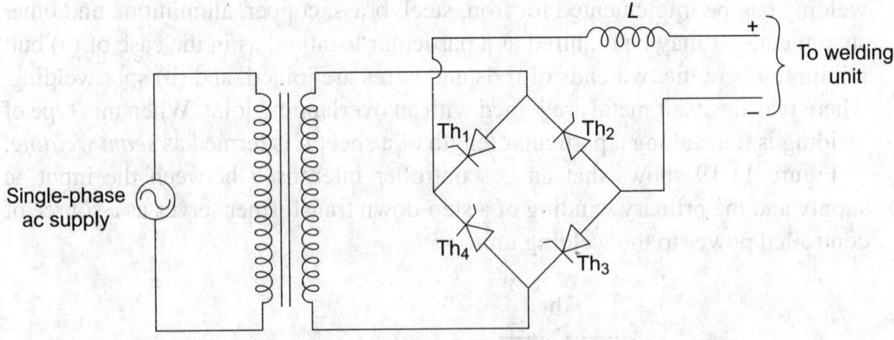

Fig. 11.21 A typical welding circuit using a controlled rectifier bridge

side. It is necessary to isolate the welding unit from the input supply; the 50-Hz transformer serves this purpose.

A recent development in the area of arc welding is the switch mode welder shown in Fig. 11.22. Its nomenclature is so because of the use of switching transistors in the high-frequency inverter (Mohan et al. 1989). These devices

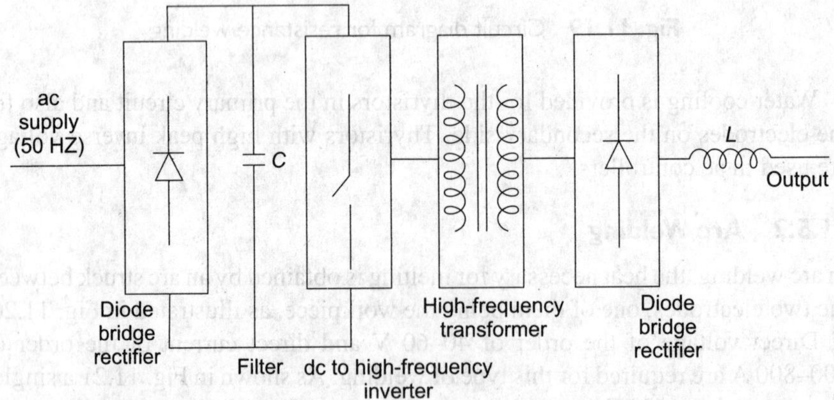

Fig. 11.22 Block diagram of a switch mode welder circuit

consume much less power than those operating in the active region of i_C-v_{CE} characteristics. As before, the small inductor L on the output side serves to minimize the ripple content.

There is a substantial reduction in the size and weight of this welding unit because of the use of the high-frequency transformer, its efficiency being above 85%. This transformer also serves the purpose of providing electrical isolation of the load from the input supply. As can be expected, the size of the unit shown in Fig. 11.21 will be much larger than the switch mode welder because of the 50-Hz transformer.

11.6 Static VAR Control

The loads that are supplied by electric utility systems consist of (a) induction motors and induction furnaces which operate at a lagging power factor and hence account for most of the reactive power consumed in the system and (b) domestic loads such as fans, lights, and other appliances which operate at a power factor near unity. The power and distribution transformers act as reactive power loads because of the magnetizing current drawn by them. The loads in a three-phase network are usually unbalanced and give rise to negative sequence currents which also contribute to the reactive power consumed by the system. The overall effect is that the system (as a whole) operates at a low power factor. Such an operation has the following disadvantages.

(a) With a highly lagging power factor, a large current flows through the system and causes considerable I^2R losses.

(b) The kilovolt-ampere (kVA) loading on the system, which is given by

$$\text{kVA} = \sqrt{(\text{kW})^2 + (\text{kVAR})^2} \tag{11.27}$$

increases because of high kVAR (reactive kilovolt-ampere) rating. Consequently, lower kW capacity of the generator will be available.

(c) The voltage regulation at the load end, which is defined as

$$\%\ \text{regulation} = 100 \times \frac{V_{\text{no load}} - V_{\text{full load}}}{V_{\text{full load}}} \tag{11.28}$$

increases, leading to high overvoltages in the system and violation of the normally specified regulation whose range is from − 10% to + 5%.

The above aspects can be seen with the help of (a) the circuit as well as (b) the phasor diagram given in Fig. 11.23. It is seen from the phasor diagram that with I_p remaining constant, if I_q increases by ΔI_q, the terminal voltage V_{ld} decreases by ΔV_{ld}. This in turn results in a reduction of the real power P, in spite of I_p remaining constant.

The disadvantages detailed above can be overcome by installing either capacitor banks or synchronous motors at the load end of the system. The capacitor banks can provide the required amount of capacitance to the system, thereby supplying the required amount of kVAR to the system. But these banks suffer from the demerit that they are bulky; the bulk increases with further augmentation

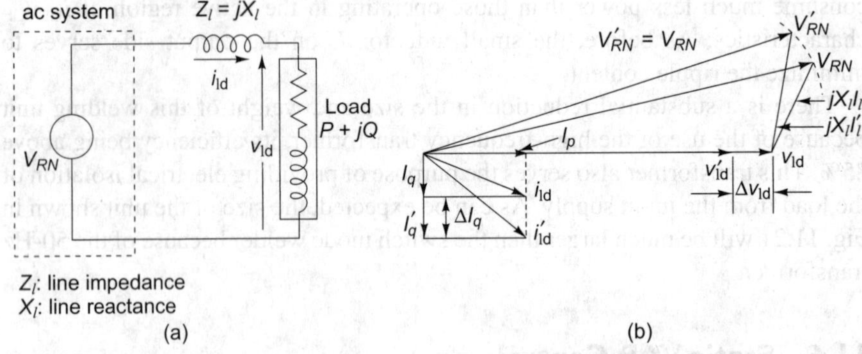

Fig. 11.23 (a) A simple ac system; (b) its phasor diagram

consequent to the increase in the substation load. Moreover, installation of additional banks proves to be cumbersome for the consumer.

Synchronous motors supply reactive power to the system when operated at a leading power factor. Such configurations are called *synchronous condensers*. However, the synchronous condenser is a very costly machine and installing one such machine at every load centre is not a feasible solution.

Thyristorized VAR controllers, called *static VAR controllers* (Dubey et al. 1986), have many merits as compared to the capacitor banks and synchronous condensers considered above. These are as follows.

(a) They form a power electronic interface between the system and the load, and provide quick control over the reactive power.

(b) They are much cheaper and more compact than the conventional equipment.

(c) They prevent voltage flickers as well as rapid changes in the reactive power caused by industrial loads such as arc furnaces.

(d) Such loads also present an unbalanced load condition for the three-phase system, and static VAR controllers can compensate for such imbalances.

(e) Static VAR controllers can also provide dynamic voltage regulation to improve the stability of a two-generator system. They are categorized as

(i) thyristor controlled inductors,
(ii) thyristor switched capacitors, and
(iii) switching converters with minimum energy storage elements.

The first two types of controllers are dealt with in detail here since these are widely used.

11.6.1 Thyristor Controlled Inductors

A thyristor controlled inductor (TCI) can be treated as a variable inductor whose reactive power consumption can be controlled by changing the firing angle of the thyristors. This can be explained with the help of the two-terminal power system shown in Fig. 11.24. The generating station *A* supplies real power P_{ld} through the transmission line to meet the load power at station *B* as well as I^2R losses. The reactive power Q_{ld} required by the load has also to be transmitted through the same line, thus making the total power equal to $P_{\text{ld}} + jQ_{\text{ld}}$. Under full load conditions, the load power factor will be low, because the reactive power consumed as well as the load current will be quite high. It was elaborated before that the load voltage falls down with an increase of current. The load voltage at the receiving end can also be restored by taking measures such as the provision of capacitor banks for power factor correction. On the other hand, under light load conditions, the transmission line capacitance comes into picture (called the *Ferranti effect*) and acts as a source of reactive power.

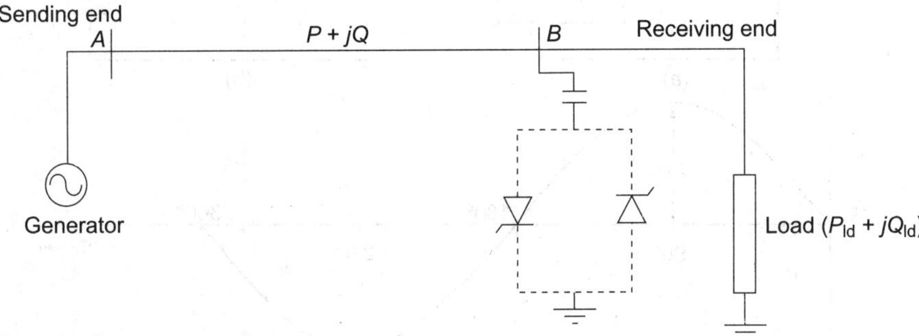

Fig. 11.24 A typical two-terminal power system with a TSC at the load end

Because of inadequate load inductance for the consumption of this reactive power, the load voltage magnitude may even exceed that of the sending end voltage. The load power factor will also be low, but this time it will be a leading one because of the flow of capacitive reactive power. Thyristor controlled inductors can absorb this reactive power through the transmission line and restore normal conditions for the voltage as well as the power factor. The lower the load power under light load conditions, the greater will be the capacitive reactive power flow and the larger will be the magnitude of inductance required from the thyristor controlled inductor. It is clear from the above discussion that TCIs are appropriate only under light load conditions. A practical circuit, which can take care of both light- and full load conditions, is considered later. Figure 11.25(a) shows the circuit of a thyristor controlled inductor and Fig. 11.25(b) shows its equivalent. In the latter the inductor is shown as a load fed from a single-phase ac controller and the analysis of such a circuit is dealt with in Chapter 4. Figure 11.25(c) gives the waveforms of the TCI currents.

Equation (4.16) can be modified as

$$L\frac{di_{\mathrm{ld}}}{dt} = V_m \sin \omega t \tag{11.29}$$

with the current i_{ld} starting with a zero value at a firing angle α and again becoming zero at α_e. Thus $i_{\mathrm{ld}} = 0$ at $\omega t = \alpha$ and also at $\omega t = \alpha_e$, the latter serving as the final condition.

The ranges for current conduction in the positive and negative half-cycles are, respectively, $\alpha < \omega t < \alpha_e$ and $\pi + \alpha < \omega t < \pi + \alpha_e$. The solution of Eqn (11.29), with the initial condition $i_{\mathrm{ld}} = 0$ at $\omega t = \alpha$ and the final value ωt, is

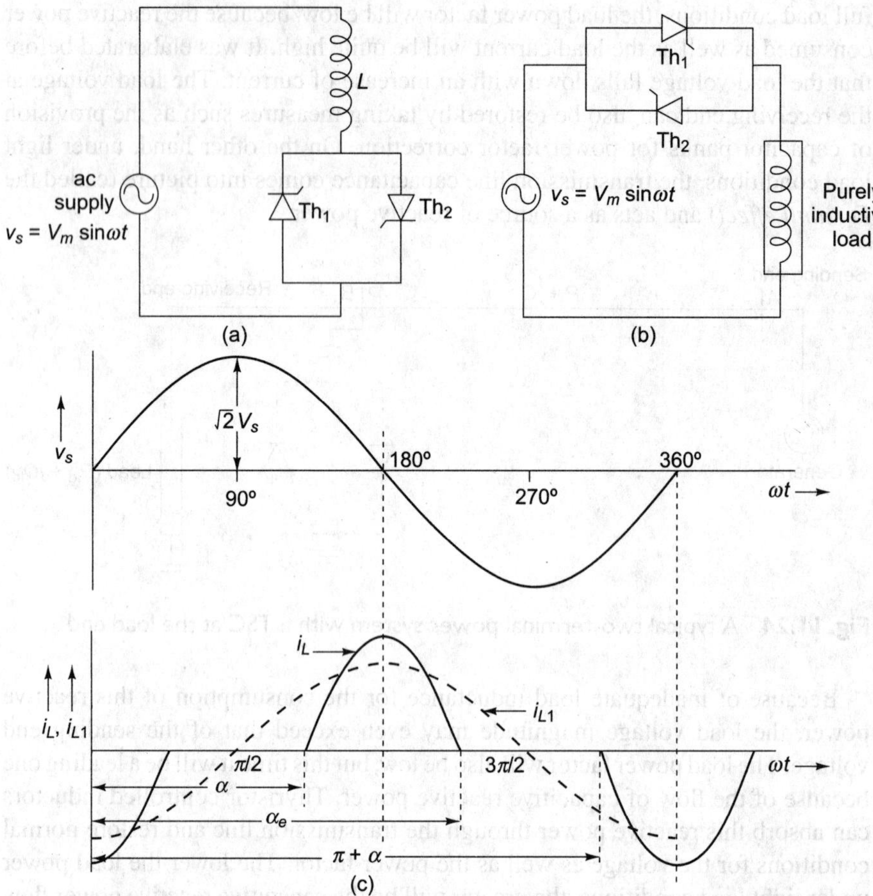

Fig. 11.25 (a) An ac system with a TCI; (b) equivalent circuit of a TCI; (c) waveforms of the TCI current and its fundamental component

obtained as

$$\int di_{\text{ld}} = \frac{V_m}{L}\int_{\alpha}^{\omega t} \sin \omega t \, dt \tag{11.30}$$

Evaluation of the integral and substitution of the initial and final values gives

$$i_{\text{ld}} = \frac{V_m}{\omega L}[\cos\alpha - \cos\omega t]$$

Also, the final condition on i_{ld} gives

$$\cos\alpha = \cos\alpha_e \tag{11.31}$$

This gives

$$\alpha_e = 2\pi - \alpha \tag{11.32}$$

because $\alpha = \alpha_e$ is inadmissible. From the arguments of Section 4.3 it is evident that for the circuit to work as an ac controller the inequality

$$\alpha > \frac{\pi}{2} \text{ rad} \tag{11.33}$$

has to be fulfilled. With $\alpha = \pi/2$, however, the circuit will behave as though the ac supply is directly connected to the inductor, with the current lagging the supply voltage by 90° (or $\pi/2$ rad) as shown in Fig. 11.26. This is also true for $\alpha < \pi/2$.

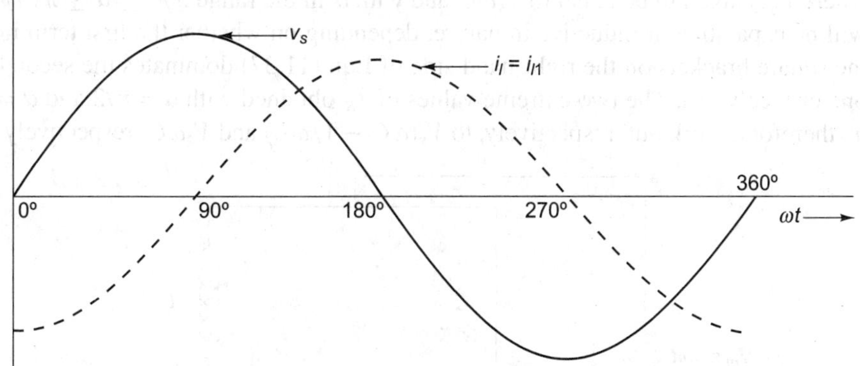

Fig. 11.26 Waveforms of the source voltage and lagging current

Hence, for the circuit to function as a variable inductor it is necessary that the firing angle α should be greater than 90° (electrical). The chart in Fig. 4.4(c) shows that the current is always discontinuous with this condition on α. If a Fourier analysis is done for this discontinuous but periodic waveform of i_L, the fundamental component i_{L1} appears as in Fig. 11.25(c); its RMS magnitude $I_{L1(\text{eq})}$ can be worked out as

$$I_{L1(\text{eq})} = \frac{V_s}{\pi\omega L}[2\pi - 2\alpha + \sin 2\alpha], \quad \frac{\pi}{2} < \alpha < \pi \tag{11.34}$$

where V_s is the RMS magnitude of the ac supply voltage, equal to $(V_m/\sqrt{2})$. The equivalent inductance with a firing angle $\alpha > 0$ can now be written as

$$L_{1(\text{eq})} = \frac{V_s}{\omega I_{L1(\text{eq})}} \tag{11.35}$$

All the other harmonics of i_{ld} are odd, that is, 3, 5, 7, 9, etc. By connecting a three-phase TCI in a delta, the triplen harmonics 3, 9, etc. circulate in the delta connectors and do not enter the system. Other harmonics, namely, the fifth, seventh, etc., can also be eliminated by connecting series *LCR* filters across the system.

11.6.2 Practical Realization of TCI

Figure 11.27 gives a practical circuit for implementing the TCI on a single-phase system (Dubey et al. 1986). The capacitor draws a fixed amount of current from the system and the net current can be varied by phase control of the thyristors. As the TCI and the capacitor *C* are connected in parallel, the total fundamental component of the reactive current drawn becomes

$$I_R = I_c - I_{L1(eq)} \tag{11.36}$$

Writing I_R as a function of α and $I_{L1(eq)}$ as given in Eqn (11.34) gives I_R as

$$I_R(\alpha) = V_s\left[\omega C - \frac{1}{\omega L}\left(2 - \frac{2\alpha}{\pi} + \frac{1}{\pi}\sin 2\alpha\right)\right] \tag{11.37}$$

where I_c is taken to be equal to $V_s\omega C$ and with α in the range $\pi/2 \le \alpha \le \pi$. I_R will be capacitive or inductive in nature, depending on whether the first term in the square brackets on the right-hand side of Eqn (11.37) dominates the second one or vice versa. The two extreme values of I_R obtained with $\alpha = \pi/2$ and $\alpha = \pi$, therefore, work out, respectively, to $V_s(\omega C - 1/\omega L)$ and $V_s\omega C$, respectively.

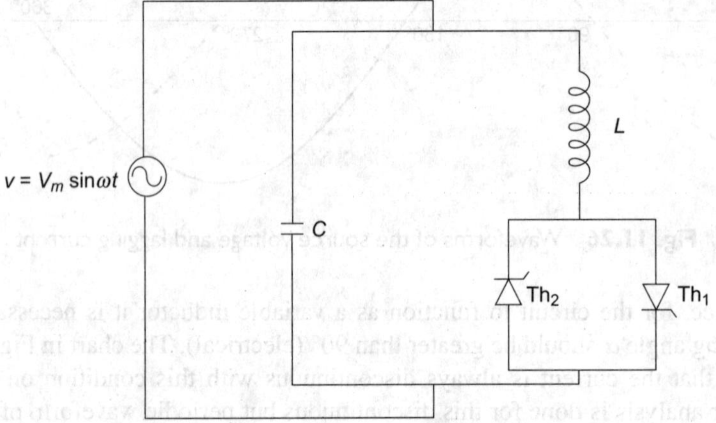

Fig. 11.27 Realization of a TCI with a fixed capacitor in parallel with a variable inductor

11.6.3 Control of Bus Voltage Using TCI

The voltage control for a low-voltage bus in a power station can be understood by considering the transmission line AB shown in Fig. 11.28(a) with its sending end and receiving end voltages denoted, respectively, as V_s and V_r. The phasor diagrams of this two-bus system with the load current lagging by an angle ϕ_1 and leading by an angle ϕ_2 are, respectively, given in Figs 11.28(b) and (c).

Figure 11.28 shows that with a fixed magnitude receiving end voltage V_r, the magnitude of the sending end voltage V_s will be greater or smaller than that of V_r for the transmission line current, respectively, lagging or leading the receiving end voltage. If now V_s were to have a fixed magnitude, the inequalities

$$|V_r| < |V_s| \qquad \text{with a lagging current}$$

and

$$|V_r| > |V_s| \qquad \text{with a leading current}$$

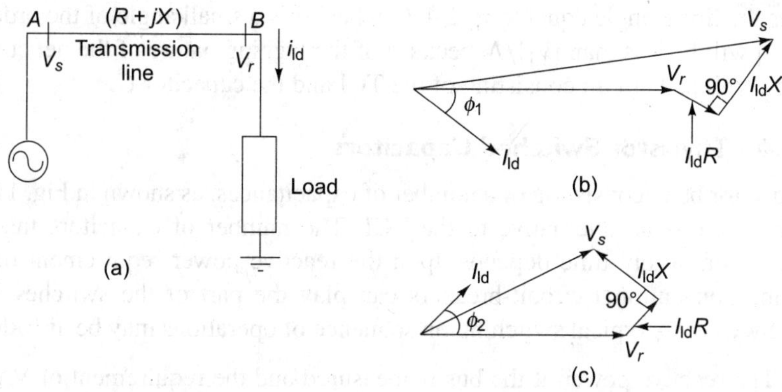

Fig. 11.28 (a) A two-terminal transmission line; (b) phasor diagram with lagging current; (c) phasor diagram with leading current

hold good. This suggests a method of controlling the magnitude of the LV (low-voltage) bus voltage using a TCI. Figure 11.29 shows such a scheme. If V_1 in this figure is considered as the fixed sending end voltage because of its interconnection with the large system, V_2 on the LV (low-voltage) bus becomes the LV end voltage. The equivalent resistance and leakage reactance of the transformer T_1 plays the same part as the R and X of the transmission line of Fig. 11.28(a). The magnitude $|V_2|$ of the low-voltage bus, which depends on the load conditions, is measured using a potential transformer and compared with the required reference voltage V_{ref}. If the regulator adjusts the firing angles α of Th_1 to be nearly equal to π, L_{eq} will be equal to zero and the current drawn will be of a purely capacitive nature. $|V_2|$ is thus regulated to be greater than $|V_1|/N$; on the other hand if the regulator

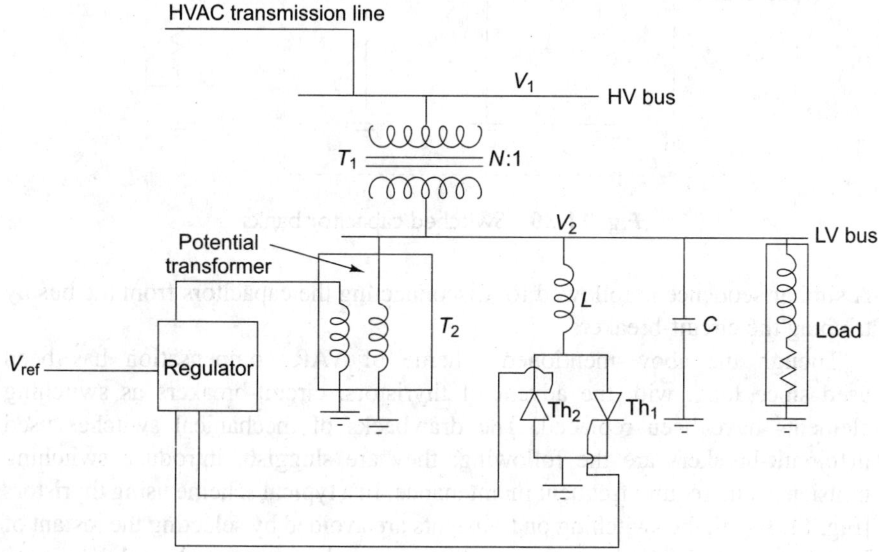

Fig. 11.29 Schematic diagram for bus voltage control

makes the firing angle equal to $\pi/2 + \theta$, where θ is a small angle of the order of $5°$, $|V_2|$ will be less than $|V_1|/N$ because of the lagging nature of the net current in the combined circuit consisting of the TCI and the capacitor C.

11.6.4 Thyristor Switched Capacitors

A capacitor bank consisting of a number of capacitances, as shown in Fig. 11.30, can be used as an alternative to the TCI. The number of capacitors that are switched on at any time depends upon the reactive power requirement of the system. Conventional circuit-breakers can play the part of the switches Sw_1, Sw_2, Sw_3, etc. A typical switching on sequence of operations may be as follows:

(a) The reactive power at the bus is measured and the requirement of VAR is estimated.

(b) The required number of relays are energized.

(c) The corresponding number of circuit-breakers are operated and their contacts closed; that is, the switches Sw_1, Sw_2, etc., which represent the circuit-breakers, are closed.

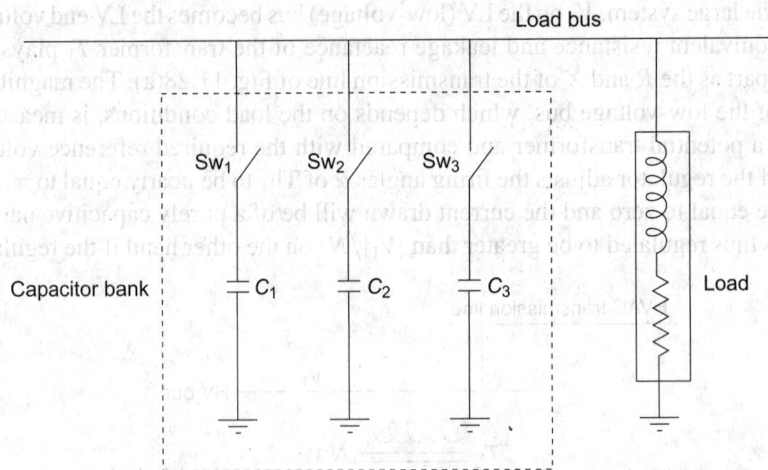

Fig. 11.30 Switched capacitor banks

A similar sequence is followed for disconnecting the capacitors from the bus by tripping the circuit-breakers.

Though the above-mentioned scheme of VAR compensation has been used since long, with the advent of thyristors, circuit-breakers as switching elements have been replaced. The drawbacks of mechanical switches used in circuit-breakers are the following: they are sluggish, introduce switching transients, and require frequent maintenance. In a typical scheme using thyristors [Fig. 11.31(a)], the switching on transients are avoided by selecting the instant of switching such that the system voltage corresponds in magnitude and polarity to the capacitor voltage. The problem of switching off transients also does not occur, because the thyristors are turned off at the next instant of zero current after the

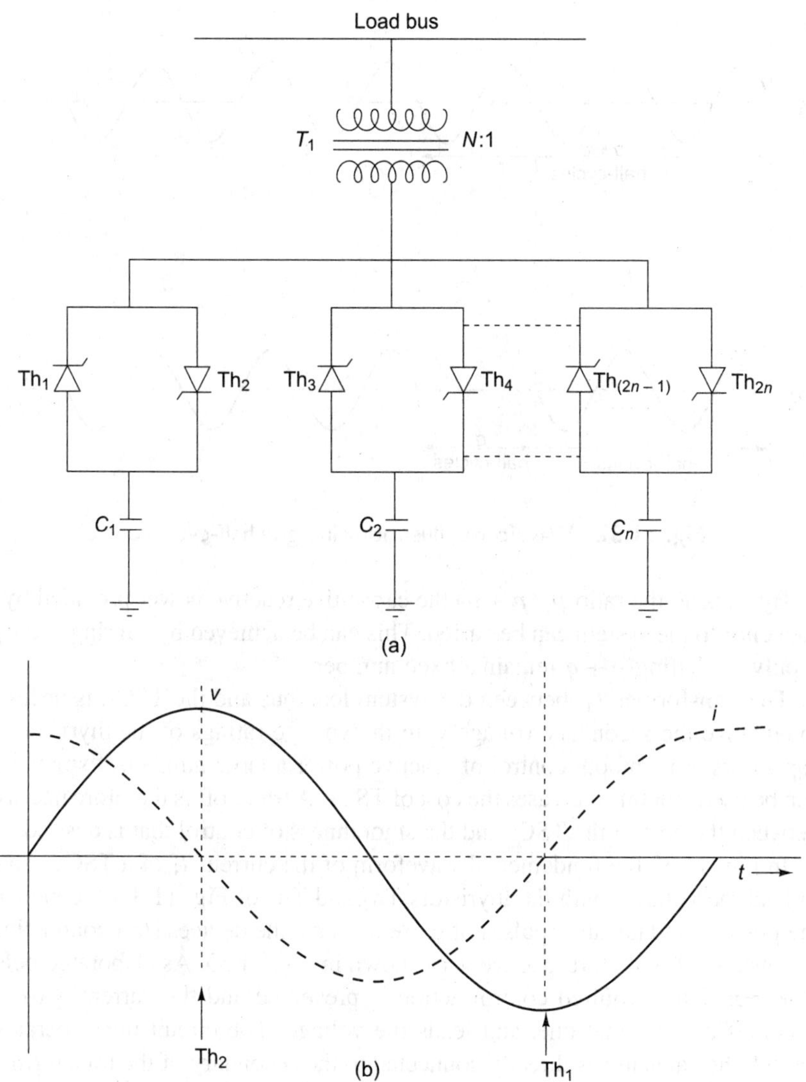

Fig. 11.31 Thyristor switched capacitors

removal of the gate pulse. The system voltage, at the instant of zero current, will be at its peak as shown in Fig. 11.31(b). The capacitors therefore get charged to this positive or negative peak voltage, thus becoming ready for the next switching.

The time required for the switch on or off of the thyristors can be split into two parts, namely, the time required to detect the required change in the reactive power and time required to wait for the favourable switching condition. Each of these takes a half-cycle, thus making a minimum of two half-cycles for switching on or off. Another feature of the TSC is that the integral half-cycle method of control is used as against phase control used in TCIs. It consists of switching the thyristor on for p half-cycles and switching it off for q half-cycles as shown in Fig. 11.32, where both p and q are integers. Both p and q should be greater than 2 for the reasons stated above.

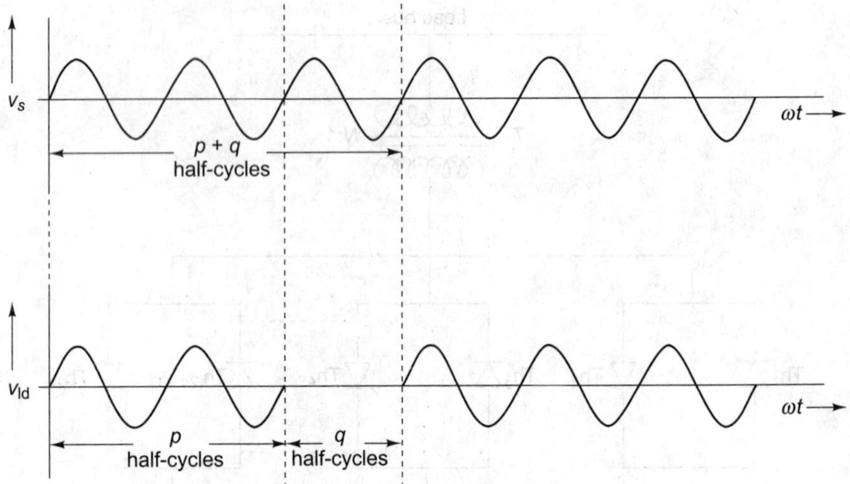

Fig. 11.32 Waveforms illustrating integral half-cycle control

By varying the ratio $p/(p+q)$ the capacitive reactive power provided by the capacitor to the system can be varied. This can be achieved by varying p only or q only but letting $p+q$ remain a fixed number.

The transformer T_1, between the system load bus and the TSCs, is necessary to equalize the secondary voltage with the voltage ratings of the thyristors and capacitors. For smooth control of reactive power a large number of small TSCs can be used, but this increases the cost of TSCs. A trade-off is therefore necessary between the cost of the TSCs and the smoothness of control that is desired.

In Fig. 11.33 the fundamental waveform of the current I_1 of a TSC is shown to lead the voltage, with the thyristors Th_2 and Th_1 of Fig. 11.31(a) conducting the positive and negative pulses of current. Thus, the devices start conduction at the instants A and B, respectively, as shown in Fig. 11.33. As elaborated before, if $\alpha < \pi/2$ the required control action is prevented and the current waveform becomes a sinusoidal one, and leads the voltage. The circuit then operates as though the capacitor is directly connected to the secondary of the transformer.

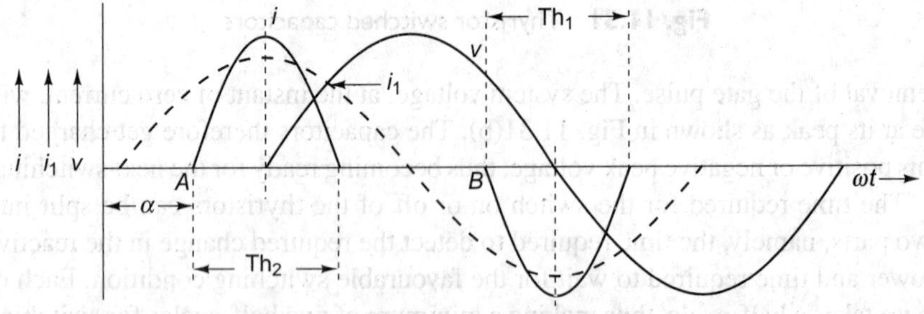

Fig. 11.33 Waveforms for TSC operation

11.7 Desirable Features of Dc Power Supplies

Almost all analog and digital electronic systems need regulation of the dc power supplied to them. For this purpose, the power supplies must be operated in such a manner that (a) the load voltage is maintained within specified tolerance in spite of load changes and supply voltage fluctuations, (b) the output (or load) is electrically isolated from the input, and (c) a provision is made for multiple outputs, each of which have different voltage and current ratings. Other requirements are (i) the power supply should have minimum size and weight and (ii) it should operate with high efficiency. It is shown below that switch mode power supplies fulfil most of the above-mentioned requirements.

11.7.1 Linear Power Supplies

Linear power supplies have long been used as dc power supplies. An understanding of this equipment will help one to better appreciate switch mode power supplies.

Figure 11.34(a) shows a schematic-cum-block diagram of a typical linear power supply system. A 50-Hz transformer is interposed in order to provide electrical isolation between the input and the output and also to give the desired voltage at the output. A transistor that operates in its active region is connected in series with the supply and helps in the regulation.

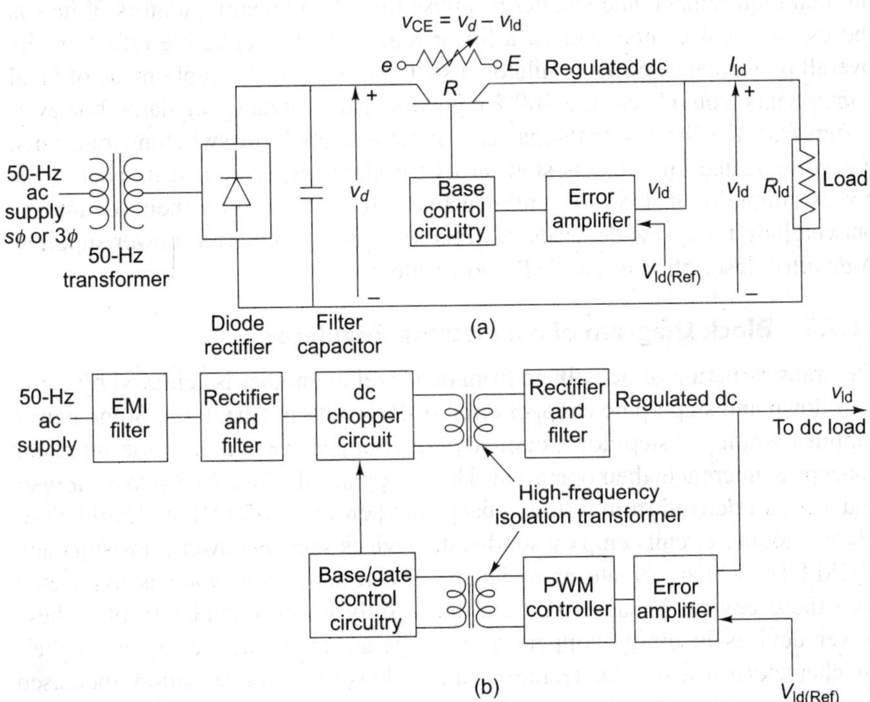

Fig. 11.34 Block diagram of (a) a linear power supply and (b) a switch mode power supply (SMPS)

By comparing the load voltage v_{ld} with a reference voltage $V_{ld(Ref)}$, the circuit adjusts the transistor base current such that v_{ld} (= $v_d - v_{CE}$) becomes equal to $V_{ld(Ref)}$. The transistor acts as an adjustable resistor, where the difference $v_d - v_{ld}$ between the input and the output voltages appears across the device, causing power losses in it. To minimize these losses, the turn ratio of the transformer is adjusted such that the minimum value of v_d is slightly greater than v_{ld} but does not exceed the load voltage by a large margin.

The linear supplies described above suffer from the following disadvantages: (i) they require a low-frequency (50-Hz) transformer which has a large size as compared to a typical high-frequency transformer, (ii) the operation of the transistor in its active region leads to power loss. On the positive side, the total cost of linear supplies will be less for power ratings below 25 W.

11.7.2 Comparison of Linear and Switch Mode Power Supplies

While linear regulated power supplies had been widely used in the past, the need for more efficient and compact regulated dc power supplies led to the development of switch mode power supplies (SMPS) [Fig. 11.34(b)]. The availability of fast switching high-voltage power transistors and low-loss ferrite and metallic glass materials made the SMPS more reliable and practicable. Some important differences between the linear and switch mode regulators are as follows. Switch mode regulators have higher efficiency, smaller size and weight, lower thermal requirement, and smaller response time than linear regulators. Whereas the use of ideal components in a linear regulator has negligible effect on the overall performance of the regulator, a switch mode power supply made of ideal components would become a 100% regulator. The switching regulator, however, suffers from the disadvantages that it is noisy, associated with switching transients, has complicated circuitry, is slow in its transient response, and its analysis is not as simple as that of conventional linear regulators. With their advantages outweighing these few demerits, SMPSs are now the preferred power supplies. A detailed description of an SMPS now follows.

11.7.3 Block Diagram of a Switching Regulator

The transformation of dc voltage from one level to another is achieved by using step-down and step-up dc chopper circuits dealt with in Chapter 3. Some power supplies employ a step-down-cum-step-up chopper also because the regulator concept is inherent in their operation. This chopper, called the *buck boost chopper* and a circuit derived from it (the flyback chopper) are dealt with in detail below. These chopper circuits employ solid-state devices such as power transistors and MOSFETs, which operate as switches. The term *switch mode* is associated with these devices because they are either fully on on completely off. These power devices in these choppers do not operate in the active region of their i-v characteristics, and this feature results in lower power dissipation. Increased switching speeds and higher voltage and current ratings are some of their positive features (Mohan et al. 1989).

A comparison of Figs 11.34(a) and (b) shows that the input rectifier and the filter following it are common for both the circuits. The transistor, acting as a variable resistor, does the necessary regulation in the power supply, shown in Fig. 11.34(a). On the other hand, the voltage adjustment in an SMPS is done by the combined action of the dc chopper circuit, isolating transformer, output rectifier, and filter of Fig. 11.34(b).

The electromagnetic interference (EMI) filter at the input of the switch mode supply is needed to prevent any conducted EMI. The dc chopper block of Fig. 11.34(b) converts the input dc voltage from one level to another using the technique of high-frequency switching, a consequence of which is the generation of ac across the isolation transformer. The secondary output of the transformer is rectified and filtered, the final output being the load voltage.

In the linear regulator of Fig. 11.34(a) regulation is achieved by obtaining the error by comparing it with a reference load voltage, amplifying it, and then using this signal as an input to the base drive circuitry. For the switch mode supply the output of the error amplifier operates a PWM controller. The output of this passes through an HF transformer (which provides isolation) to the base or gate control circuit (Fig. 11.35). An optocoupler may as well be used in place of the high-frequency (HF) transformer for electrical isolation. Some of the loads may be of the unregulated type and may need electrical isolation.

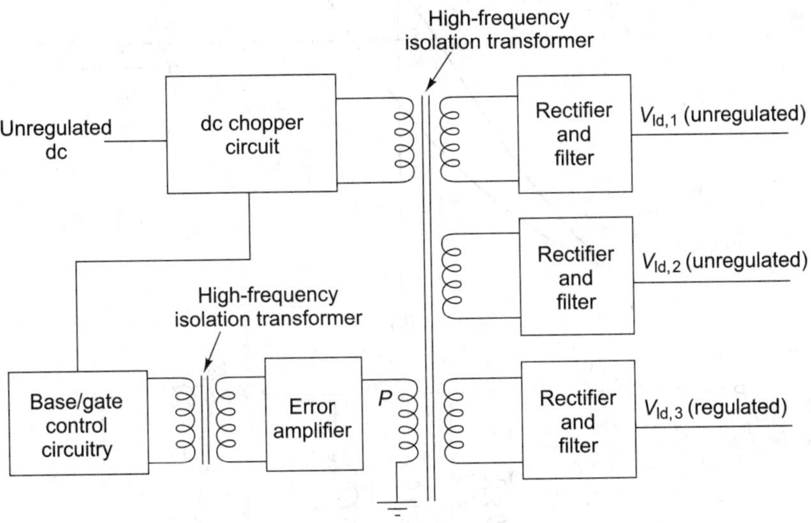

Fig. 11.35 Isolation transformer in an SMPS

11.7.4 Isolation Transformer

Figure 11.36(a) shows the schematic diagram of a typical high-frequency transformer which provides isolation of the dc chopper from the output stage consisting of the rectifier and filter. The transformer core characteristic in terms of its *B-H* loop is shown in Fig. 11.36(b), in which B_1 is the maximum flux density above which saturation occurs and B_2 is the remnant flux density. The dc choppers used in the SMPS utilize the isolation transformer either for unidirectional core excitation, where only the first quadrant part of the *B-H* loop is needed, or for

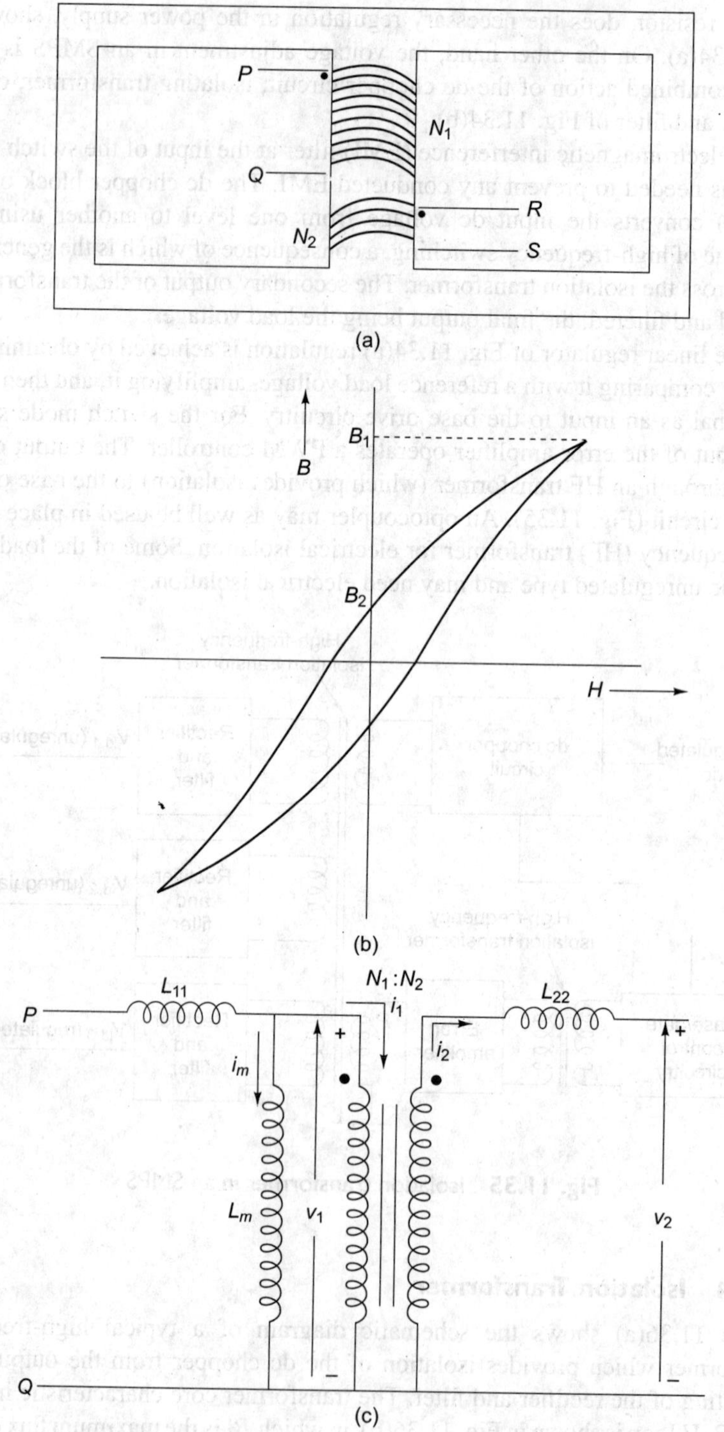

Fig. 11.36 (a) Core of a high-frequency transformer; (b) B-H diagram of the core material; (c) equivalent circuit for transformer windings

bidirectional core excitation, where the entire B-H loop is needed. The buck boost dc chopper can be modified to provide electrical isolation by means of unidirectional core excitation. This modification is called the *flyback chopper*.

In Fig. 11.36(c), N_1:N_2 is the turn ratio of the transformer and L_m is the magnetizing inductance referred to the primary side. L_{11} and L_{22} are, respectively, the leakage inductances of the primary and secondary windings. The inductive energy of these inductances is required to be absorbed by the semiconductor switches and their snubber circuits. It is therefore necessary to minimize their sizes. The magnetizing current i_m flows through the semiconductor switches and increases their current ratings; hence, the magnetizing inductance L_m is usually made as high as possible to keep down these ratings. However, for the flyback chopper considered below, this is not necessary because the transformer there has to perform two functions, namely, (i) storing energy as an inductor and (ii) providing electrical isolation as a transformer.

11.7.5 Buck Boost Chopper

The buck boost chopper used in an SMPS provides an output voltage of negative polarity with respect to the common terminal of the input voltage. The magnitude of this output can be higher or lower in magnitude than that of the input voltage. It can be considered to be a cascade connection of the step-down and step-up choppers considered in Chapter 3. Accordingly, its output to input voltage ratio is the product of the output-to-input ratios of these choppers and is expressed as

$$\frac{v_{\mathrm{ld}}}{v_d} = \frac{\delta}{1-\delta} \tag{11.38}$$

where the duty ratio is defined as (Chapter 3)

$$\delta = \frac{\tau_{\mathrm{ON}}}{\tau}$$

This equation implies that the output voltage can be greater or lower than the input voltage. For example, if $\delta = 0.4$, this ratio becomes $0.4/0.6 = 2/3$; on the other hand, with $\delta = 0.6$, it becomes $0.6/0.4 = 3/2$. The circuit of Fig. 11.37 gives the buck boost characteristic discussed above. When the switch Sw is closed, the diode D is reverse biased and provides energy to the inductor L; and upon the opening of this switch, the energy stored in the inductor is transferred to the output. In the steady-state analysis presented below the output capacitor is considered to be sufficiently large so as to result in a steady load voltage $v_{\mathrm{ld}}(t) = V_{\mathrm{ld}}$.

Figure 11.38(a) shows the waveforms of the voltage and current of the inductor L in the continuous conduction mode of the inductor current. Figures 11.38(b) and (c), respectively, show the circuit conditions during the ON and OFF conditions of the switch.

Equation (3.4) holds good in this case and is reproduced below:

$$\int_0^{\tau} v_L dt = 0 \tag{11.39}$$

Application of Eqn (11.39) to the inductor voltage in Fig. 11.38 gives

$$v_d \tau_{\mathrm{ON}} - v_{\mathrm{ld}}(\tau - \tau_{\mathrm{ON}}) = 0 \tag{11.40}$$

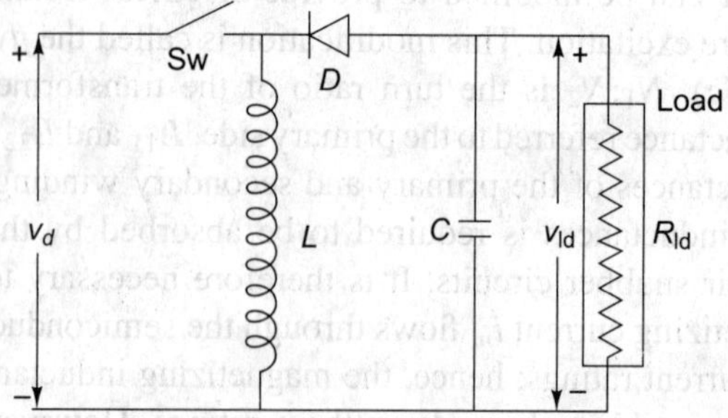

Fig. 11.37 Circuit of a buck boost chopper

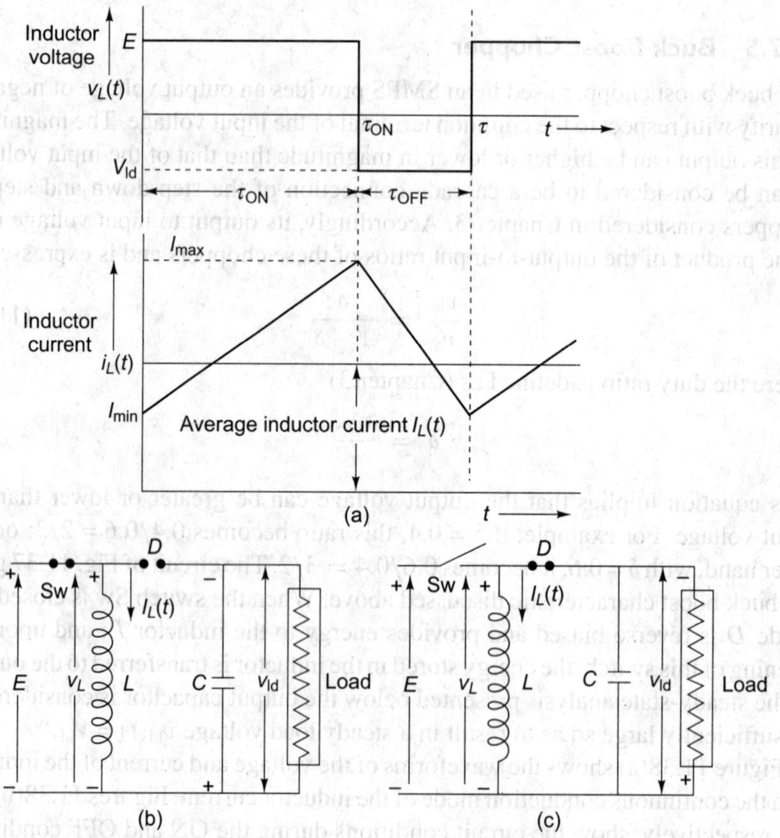

Fig. 11.38 (a) Waveforms, (b) switch-closed circuit, and (c) switch-open circuit of a buck boost chopper

or
$$\frac{v_{\mathrm{ld}}}{v_d} = \frac{\tau_{\mathrm{ON}}}{\tau - \tau_{\mathrm{ON}}} = \frac{\delta}{1-\delta} \tag{11.41}$$

Assuming that the input power P is equal to the output power P_{ld}, and denoting i_s and i_{ld}, respectively, as the mean source and load currents,

$$\frac{i_{\mathrm{ld}}}{i_s} = \frac{P_{\mathrm{ld}}}{v_{\mathrm{ld}}} \div \frac{P}{v_d} = \frac{v_d}{v_{\mathrm{ld}}} = \frac{1-\delta}{\delta} \tag{11.42}$$

11.7.6 Principle of the Flyback Chopper

Dc choppers having different topologies can be employed in the corresponding block of the SMPS. Some commonly used converters as well as the dc chopper topologies from which their circuits are derived are given below.

Derived choppers	Basic chopper circuits
Flyback chopper	Buck boost chopper
Forward chopper	Step-down chopper
Push-pull chopper	Step-down chopper
Half-bridge chopper	Step-down chopper
Full-bridge chopper	Step-down chopper

It is seen that with the exception of the flyback chopper all the above-derived choppers use the step-down chopper topology. Only the flyback chopper circuit and an SMPS based on it are considered here; the other derived choppers are described in Appendix C. Figure 11.37 gives the circuit of a buck boost chopper and can be redrawn as Fig. 11.39(a) after reversal of the input terminals. If a second winding is placed on the inductor core so as to electrically isolate the load from the source, with the polarity of the second inductor winding made opposite to that of a conventional transformer, Fig. 11.39(a) in turn can be redrawn as Fig. 11.39(b), which is just the figure of the flyback chopper.

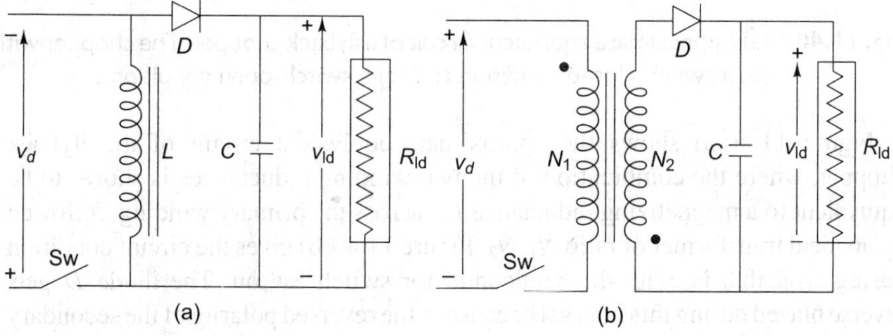

Fig. 11.39 Circuit of (a) a buck boost chopper with terminals reversed and (b) a flyback chopper

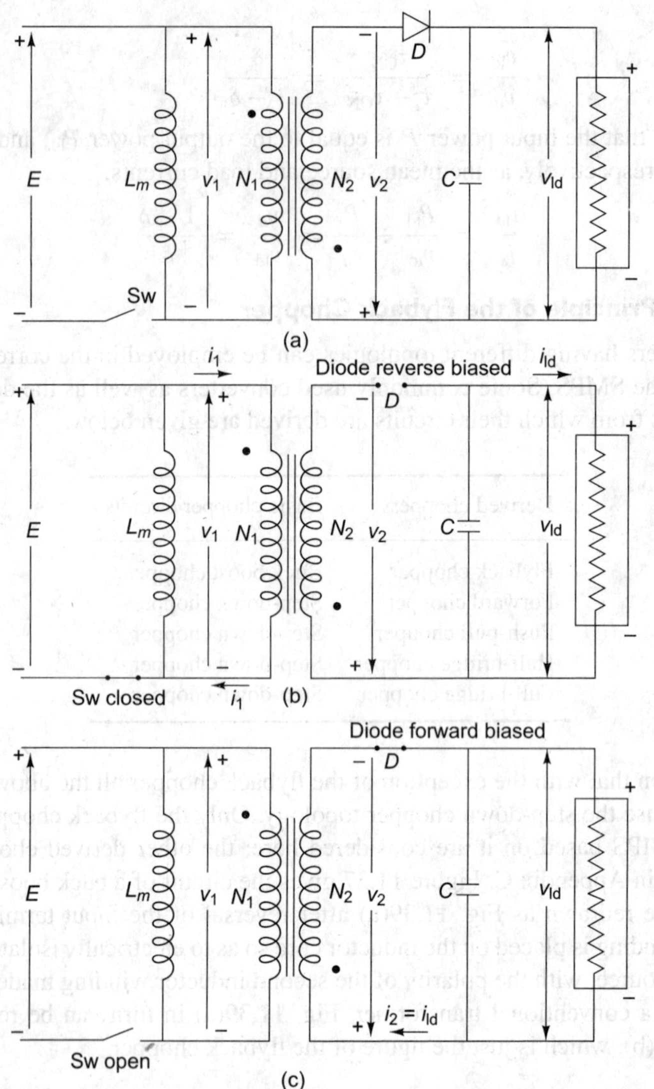

Fig. 11.40 (a) Approximate equivalent circuit of a flyback chopper. The chopper with (b) a switch-closed condition and (c) a switch-open condition

Figure 11.40(a) shows the approximate equivalent circuit of the flyback chopper, where the combination of the two winding inductances is shown to be equivalent to a magnetizing inductance L_m across the primary winding, followed by an ideal transformer of ratio N_1:N_2. Figure 11.40(b) gives the circuit condition during τ_{ON}, that is, with the semiconductor switch Sw on. The diode D gets reverse biased during this interval because of the reversed polarity of the secondary winding and the flux in the core builds up from ϕ_{min} to ϕ_{max}. Figure 11.40(c) gives the circuit for the interval τ_{OFF} (the semiconductor switch is off). In this interval the current i_2 in the secondary of the transformer maintains the magnetization of the core. Thus it flows in the loop consisting of the dot, secondary winding, diode

D, load, and back to the dot. In this period the core flux will get reduced from ϕ_{max} to ϕ_{min} and the voltage across the load will become equal to $-V_{ld}$. In the steady state, the flux attains the values ϕ_{min} at $t = 0$, ϕ_{max} at $t = \tau_{ON}$, again ϕ_{min} at $t = \tau$, and so on.

Figure 11.41 shows the voltage waveform across the magnetizing inductance L_m, flux ϕ in the core, the source and load currents, and finally the equivalent current across L_m referred to the primary.

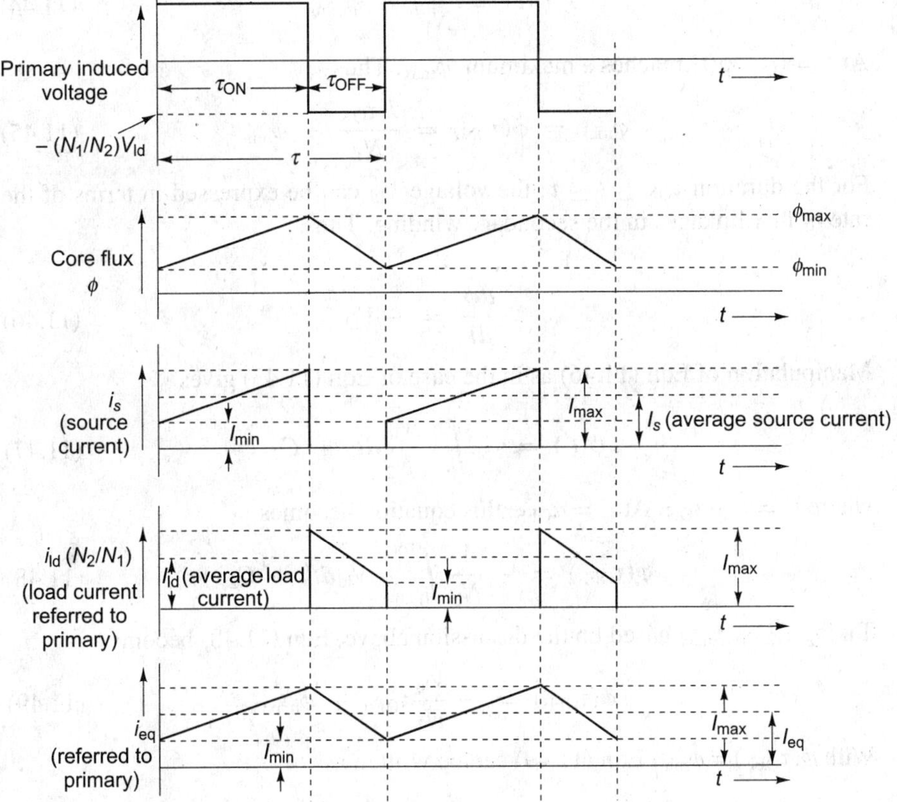

Fig. 11.41 Waveforms of a flyback chopper

The output-to-input voltage ratio (V_{ld}/E) is arrived at with the help of appropriate expressions for flux as detailed below. Neglecting losses, the applied voltage E will be equal to the induced emf, which can be expressed as the rate of flux linkages in the primary winding. This can be written as

$$N_1 \frac{d\phi}{dt} = E \tag{11.43}$$

Solving this equation for $d\phi$ gives

$$d\phi = \frac{E}{N_1} dt$$

Integration of both sides will result in the following expression for $\phi(t)$:

$$\phi(t) = \int_0^t \frac{E}{N_1} dt + C_1$$

From the sequence of events given earlier for the steady state, C_1 is taken as $\phi_{\min}$ as per the ϕ vs t plot of Fig. 11.41 and $\phi(t)$ can be written for the interval $0 \le t \le \tau_{\text{ON}}$ as

$$\phi(t) = \frac{E}{N_1} t + \phi_{\min} \tag{11.44}$$

At $t = \tau_{\text{ON}}$, $\phi(t)$ reaches a maximum, $\phi_{\max}$. Thus

$$\phi_{\max} = \phi(\tau_{\text{ON}}) = \frac{E\tau_{\text{ON}}}{N_1} + \phi_{\min} \tag{11.45}$$

For the duration $\tau_{\text{ON}} \le t \le \tau$, the voltage V_{ld} can be expressed in terms of the rate of flux linkages in the secondary winding. Thus,

$$N_2 \frac{d\phi}{dt} = -V_{\text{ld}} \tag{11.46}$$

Manipulation of Eqn (11.46) as in the case of Eqn (11.43) gives

$$\phi(t') = -\int_0^{t'} \frac{V_{\text{ld}}}{N_2} dt' + C_2 \tag{11.47}$$

where $t' = t - \tau_{\text{ON}}$. At $t' = \tau_{\text{OFF}}$, this equation becomes

$$\phi(\tau_{\text{OFF}}) = -\frac{1}{N_2}\int_0^{\tau_{\text{OFF}}} V_{\text{ld}} dt' + C_2 \tag{11.48}$$

Taking C_2 as $\phi_{\max}$ based on the discussion above, Eqn (11.48) becomes

$$\phi(\tau_{\text{OFF}}) = -\frac{V_{\text{ld}}}{N_2}\tau_{\text{OFF}} + \phi_{\max} \tag{11.49}$$

With $\phi(\tau_{\text{OFF}}) = \phi_{\min}$, Eqn (11.49) can be written as

$$\phi_{\min} = -\frac{V_{\text{ld}}}{N_2}\tau_{\text{OFF}} + \phi_{\max} \tag{11.50}$$

A combination of Eqns (11.45) and (11.50) gives

$$\phi_{\max} = \frac{E\tau_{\text{ON}}}{N_1} - \frac{V_{\text{ld}}}{N_2}\tau_{\text{OFF}} + \phi_{\max} \tag{11.51}$$

Equation (11.51) can be solved for V_{ld}/E as

$$\frac{V_{\text{ld}}}{E} = \frac{N_2}{N_1}\frac{\tau_{\text{ON}}}{\tau_{\text{OFF}}} = \frac{N_2}{N_1}\frac{\delta}{1-\delta} \tag{11.52}$$

with δ equal to τ_{ON}/τ. Equation (11.52) is similar to Eqn (11.41), which pertains to the buck boost chopper. The equivalent inductor current referred to the primary, which is the sum of the source and load currents (both referred to the primary), can be obtained by starting from the following two equations which are, respectively, similar to Eqns (11.43) and (11.46):

$$L_m \frac{di_{eq}}{dt} = E \text{ for the period } 0 \le t \le \tau_{ON} \tag{11.53}$$

and

$$L_m \frac{di_{eq}}{dt} = -V_{ld} \frac{N_1}{N_2} \text{ for the period } \tau_{ON} \le t \le \tau \tag{11.54}$$

Writing Eqn (11.53) as an integral gives

$$i_{eq}(t) = \int_0^t \frac{E}{L_m} dt + I_{min}, \quad 0 \le t \le \tau_{ON} \tag{11.55}$$

$$= \frac{E}{L_m} t + I_{min} \tag{11.56}$$

where the constant of integration is written as I_{min} as per the i_{eq} vs t plot of Fig. 11.41. $i_{eq}(t)$ reaches a maximum I_{max} at $t = \tau_{ON}$. Hence,

$$I_{max} = \frac{E}{L_m} \tau_{ON} + I_{min} \tag{11.57}$$

For the interval $\tau_{ON} \le t \le \tau$, the equivalent current i_{eq} can be obtained from Eqn (11.54) as

$$i_{eq}(t') = -\int_0^{t'} \frac{V_{ld}}{L_m} \frac{N_1}{N_2} dt' + I_{max} \tag{11.58}$$

where $t' = t - \tau_{ON}$. Evaluation of the integral gives

$$i_{eq}(t') = I_{max} - \frac{V_{ld}}{L_m} \frac{N_1}{N_2} (t') \tag{11.59}$$

The load current i_{ld} can be expressed as

$$i_{ld}(t') = \frac{N_1}{N_2} i_{eq}(t') \tag{11.60}$$

Substituting for $i_{eq}(t')$ from Eqn (11.59) gives

$$i_{ld}(t') = \frac{N_1}{N_2} I_{max} - \frac{V_{ld}}{L_m} \left(\frac{N_1}{N_2}\right)^2 (t') \tag{11.61}$$

The average of the load current i_{ld} can be written as

$$I_{ld} = \frac{1}{\tau} \int_{\tau_{ON}}^{\tau} i_{ld}(t') dt' \tag{11.62}$$

Substitution of the expression for $i_{ld}(t')$ from Eqn (11.61) in Eqn (11.62) gives

$$I_{ld} = \frac{1}{\tau} \int_{\tau_{ON}}^{\tau} \left[\frac{N_1}{N_2} I_{max} - \frac{V_{ld}}{L_m} \left(\frac{N_1}{N_2}\right)^2 (t') \right] dt' \tag{11.63}$$

Evaluating the integral and solving for I_{max} gives

$$I_{max} = \frac{I_{ld}}{1-\delta}\frac{N_2}{N_1} + \frac{V_{ld}}{2L_m}\frac{N_1}{N_2}(1-\delta)\tau \tag{11.64}$$

I_{min} is obtained from Eqn (11.57) as

$$I_{min} = I_{max} - \frac{E}{L_m}\tau_{ON} \tag{11.65}$$

Using Eqn (11.52), I_{min} can be expressed as

$$I_{min} = I_{max} - \frac{V_{ld}}{L_m}\frac{N_1}{N_2}\frac{(1-\delta)}{\delta}\tau_{ON} \tag{11.66}$$

Finally, substituting the expression for I_{max} from Eqn (11.64), and then simplifying using the relation $\tau_{ON} = \delta\tau$ gives

$$I_{min} = \frac{I_{ld}}{1-\delta}\frac{N_2}{N_1} - \frac{V_{ld}}{2L_m}\frac{N_1}{N_2}(1-\delta)\tau \tag{11.67}$$

11.7.7 A Practical Flyback Chopper Circuit

Figure 11.42(a) shows a practical flyback chopper circuit where the transistor Q plays the same part as the switch Sw of Fig. 11.40(a) (Kitsum 1984). Also, Fig. 11.42(b) shows the complete flyback chopper circuit including that of the pulse width modulator which performs the regulation process of the SMPS, with a control voltage V_{con} derived from the output of the flyback chopper.

The modulator is assumed to be the linear-ramp-and-comparator type where a linear ramp voltage is compared with the control voltage V_{con} [Fig. 11.42(c)], which is derived from the output voltage v_{ld}. A_1 and A_2 are, respectively, a comparator and an error amplifier. A decrease in v_{ld} will cause V_{con}, which is the output of A_2, to increase; this in turn will lead to the triggering of the comparator at a higher potential point on the ramp, thus resulting in a wider output pulse. This wide pulse in turn adjusts the currents in the output circuit such that a higher output voltage v_{ld} is obtained. If v_{ld} increases, each of the events in the sequence mentioned above will be reversed, with the result that the flyback chopper operates as a voltage-regulator.

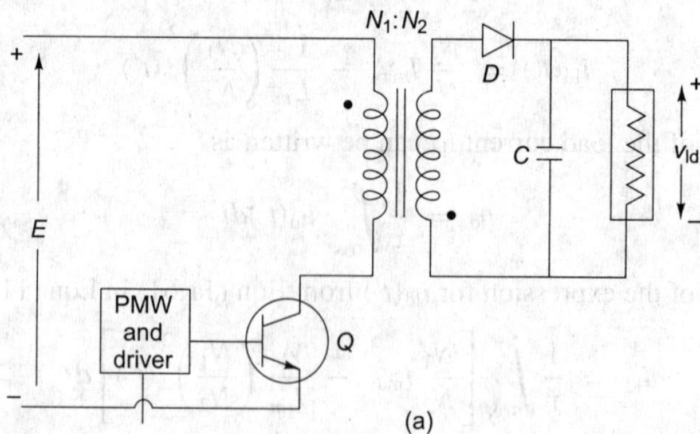

Fig. 11.42(a)

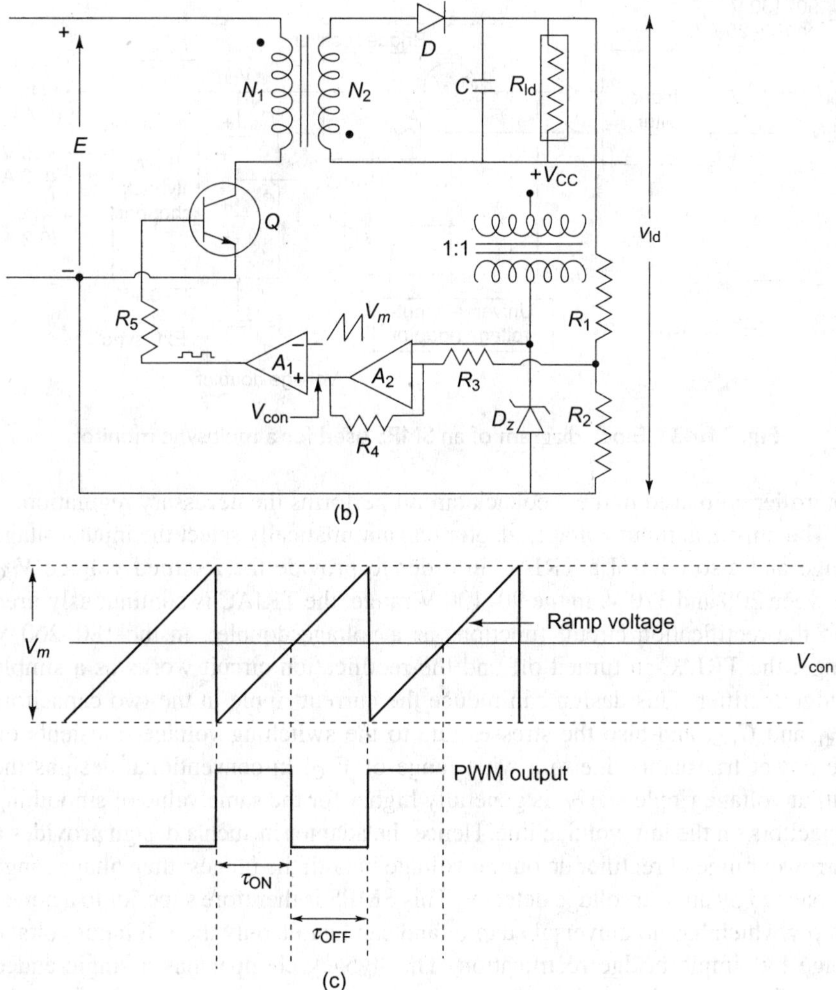

Fig. 11.42 (a) Outline of a practical flyback chopper circuit; (b) complete flyback chopper circuit; (c) comparison of linear ramp voltage with V_{con}

11.7.8 A Typical SMPS Using a Flyback Chopper

All the individual modules dealt with above are now integrated to constitute an SMPS (Tong et al. 1991). Figure 11.43 gives the block diagram of such a typical low-cost SMPS which is used in multisync colour monitors; its specifications are the following:

Inputs: 90–130 V ac or 180–260 V ac at 50 Hz

Outputs:

+110 V, 0.7 A for HV, drivers, and deflection

+12 V, 0.3 A for auxiliary use

+5 V, 0.2 A for logic ICs

Power: 90 W with overload protection

The salient features of this power supply are the rectification circuit, the universal input voltage adaptor, and the 90-W flyback chopper. A current mode

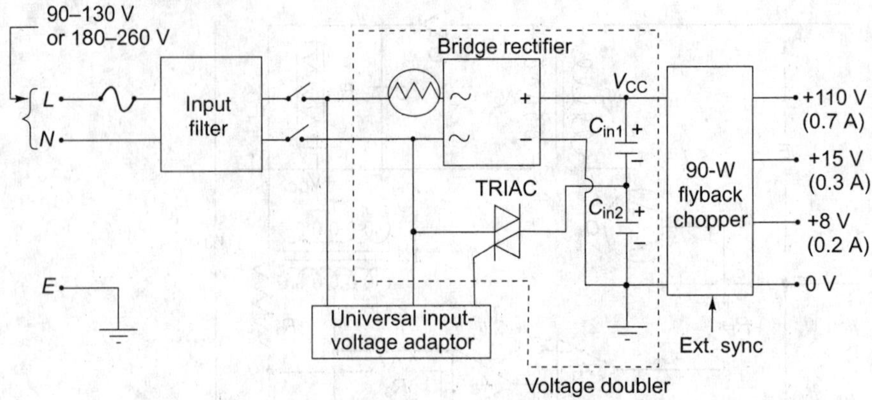

Fig. 11.43 Block diagram of an SMPS used for a multisync monitor

controller chip used in the feedback circuit performs the necessary regulation.

The universal input voltage adaptor can automatically select the input voltage range and also control a TRIAC in order to provide the rectified voltage V_{CC} between 200 and 370 V. In the 90–130 V range, the TRIAC is continuously fired and the rectification circuit functions as a voltage doubler. In the 180–260 V range, the TRIAC in turned off and the rectification circuit works as a simple bridge rectifier. This design can reduce the current ripple in the two capacitors C_{in1} and C_{in2}, and also the stresses due to the switching voltage transients on the power transistors due to a wide range of V_{CC}. In conventional designs the output voltage ripple at V_{CC} is generally higher for the same value of smoothing capacitors on the low-voltage line. Hence the adaptor in such a design provides a narrower range of rectifier dc output voltage in both the ranges; the voltage range is selected by an overvoltage detector. This SMPS is therefore superior to a power supply which has no universal adaptor and can handle only the full input voltage range by simple bridge rectification. The flyback chopper has a single-ended discontinuous-mode topology and performs the major power transfer from the dc (at V_{CC}) to the load. Figure 11.44 gives the complete circuit of the flyback chopper.

11.7.9 Control of SMPSs

An important aspect of the SMPS is the closed-loop control of its output voltages, which must be regulated to be within a tolerance band, say, $\pm 1\%$ about its nominal value in response to changes (i) in the load voltage and (ii) in the input line voltage. This is achieved by means of a closed-loop control system with negative feedback, as shown in Fig. 11.45.

The load voltage v_{ld} is compared with its reference value $V_{ld(Ref)}$ in the error amplifier, thereby resulting in the control voltage v_{con}. This voltage in turn adjusts the duty ratio of one or more switches in the converter.

11.7.9.1 Duty ratio PWM control A modification of the load voltage control scheme presented above is the *direct duty ratio PWM*. Here $v_{con}(t)$,

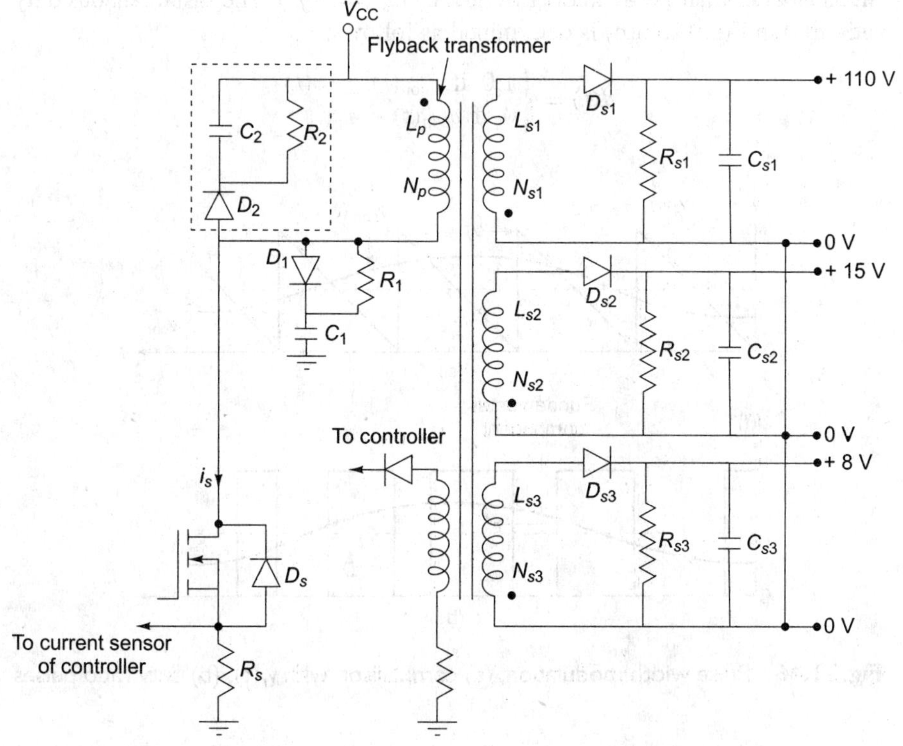

Fig. 11.44 Complete circuit of a flyback chopper

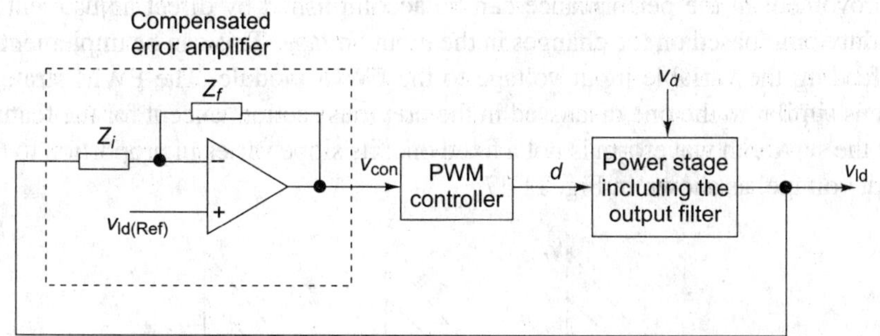

Fig. 11.45 Voltage regulation using a feedback control system

which is the output of the error amplifier, is compared with a sawtooth waveform $v_r(t)$ [Fig. 11.46(a)], thereby determining the switching fequency f_s. The control voltage $v_{\text{con}}(t)$ consists of a dc component and a small ac component $\tilde{v}_{\text{con}}(t)$. Thus,

$$v_{\text{con}}(t) = v_{\text{con}} + \tilde{v}_{\text{con}}(t)$$

where $v_{\text{con}}(t)$ varies from 0 to $\hat{V}_r$ as shown in Fig. 11.46(a). Here $\tilde{v}_{\text{con}}(t)$ is a sinusoidal ac perturbation in the control voltage at a frequency ω_p, where ω_p is much smaller than the switching frequency ω_s (= $2\pi f_s$). The instantaneous duty ratio $d(t)$ in Fig. 11.46(b) is determined as follows:

$$d(t) = \begin{cases} 1.0 \text{ if } v_{\text{con}}(t) \geq v_r(t) \\ 0 \text{ if } v_{\text{con}}(t) < v_r(t) \end{cases}$$

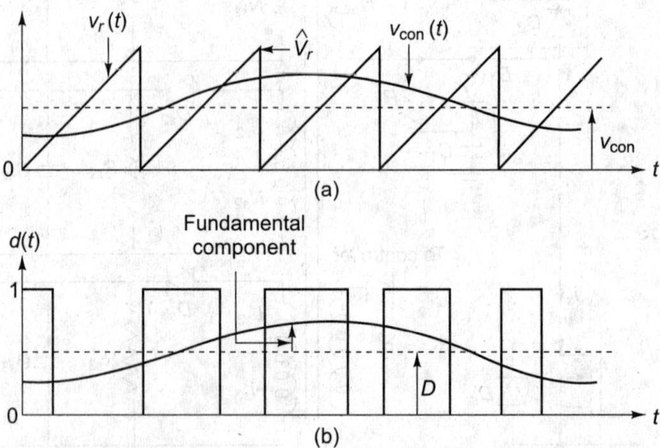

Fig. 11.46 Pulse width modulation: (a) comparison with $v_r(t)$, (b) duty ratio pulses

11.7.9.2 Voltage feed-forward PWM control Though closed-loop direct duty ratio control is satisfactory, its dynamic performance in regulating the load voltage in response to the changes in the input supply voltage is slow. A further improvement in the performance can be accomplished by direct adjustment of the duty ratio based on the changes in the input voltage. This can be implemented by feeding the variable input voltage to the PWM module. The PWM strategy here is similar to the one discussed in the previous section, except for the feature that the sawtooth waveform is not a fixed one; its slope varies in proportion to the input voltage, as shown in Fig. 11.47.

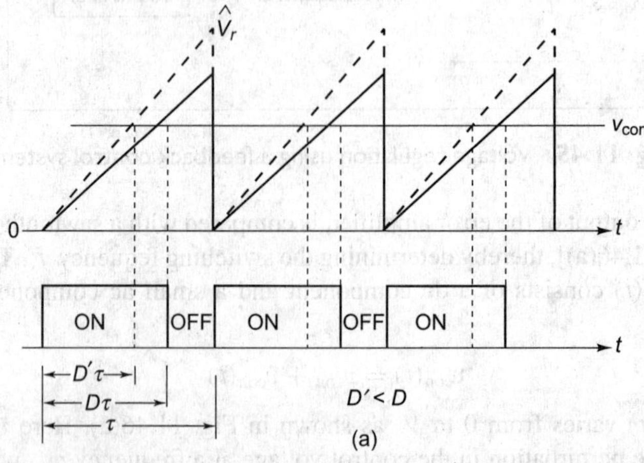

Fig. 11.47(a)

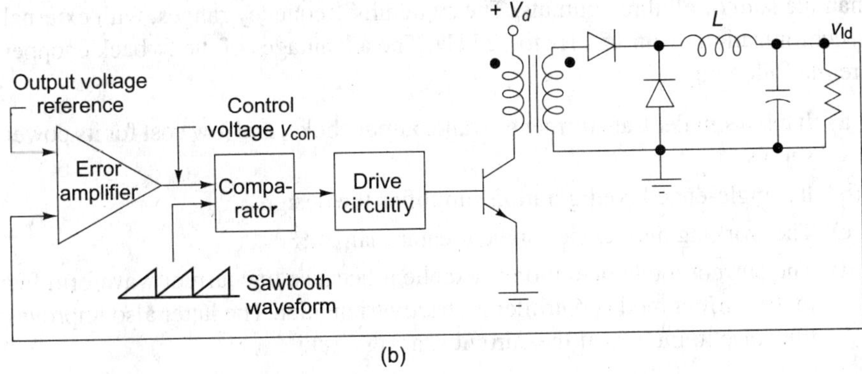

Fig. 11.47 (a) Effect on duty ratio with voltage feed-forward; (b) PWM duty ratio control

It is seen from Fig. 11.47(a) that an increase in the slope of the sawtooth waveform leads to a decrease in the pulse width. This in turn will effectively compensate the load voltage changes. The complete closed-loop block diagram for this strategy is shown in Fig. 11.47(b). Here the control voltage v_{con} adjusts the duty ratio of the transistor switch by comparing the control voltage with a fixed frequency, but a variable slope, sawtooth waveform. Controlling at the duty ratio adjusts the voltage across the inductor and also its current. Eventually the load voltage is adjusted back to its reference value.

11.7.9.3 Current mode control In this scheme, which is a modification of that shown in Fig. 11.47(b), an additional inner loop is constructed as shown in Fig. 11.48. This new loop directly controls the inductor current and accordingly adjusts the load voltage. This modification of the variable duty ratio has a significant effect on the dynamic performance of the outer feedback loop (Mohan et al. 1994).

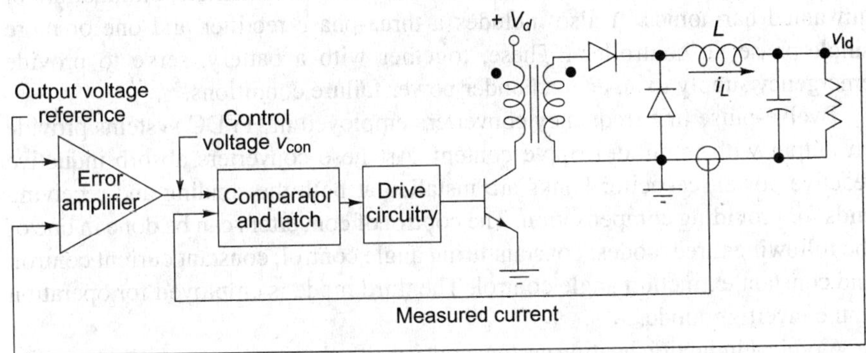

Fig. 11.48 Current mode control

11.7.10 Merits and Demerits of the Flyback Chopper

The flyback chopper is designed to operate in the voltage range 200–370 V dc. The total power that can be supplied by it is 90 W, which is slightly higher than the sum of all three outputs. The switching frequency ranges, with external synchronization, from 15 kHz to 32 kHz. The advantages of the flyback chopper are the following.

(a) It has a smaller transformer size and output choke, and low cost for its power supply.
(b) Its single-ended configuration simplifies the design.
(c) The working duty cycle can be greater than 50%.
(d) The current mode operation is excellent because the current waveform fed to the current mode controller is strictly triangular. The latter also improves the noise immunity of the current sensing circuit.

Its disadvantages are the following.

(a) Its high RMS and peak transformer currents result in larger losses in the power switch and windings.
(b) The higher ripple current appearing in the output capacitors produces greater output ripple voltage, which may cause screen interference. However, as the switching frequency of the power supply is designed in synchronization with the horizontal frequency, this adverse effect is considerably minimized.

Multisync colour monitors have the advantage of adapting to several modes of computer displays such as CGA, EGA, and VGA, which are used in IBM PCs. The three display modes have different horizontal resolutions and scanning frequencies ranging from 15.7 kHz to 31.5 kHz. The SMPS described above is useful for synchronizing the horizontal scanning frequencies of such multisync colour monitors.

Summary

A typical UPS system employs PWM inverters to facilitate elimination of unwanted harmonics. It also includes a three-phase rectifier and one or more single-phase ac controllers. These, together with a battery, serve to provide emergency supply to an ac load under power failure conditions.

Twelve-pulse line frequency converters employed in HVDC systems provide an output with minimum ripple content. As these converters absorb inductive reactive power, capacitor banks are installed at both the sending and receiving ends for providing compensation. The control of converters can be done in one of the following three modes: constant firing angle control, constant current control, and constant extinction angle control. The third mode is employed for operation in the inverting mode.

A typical induction heating system consists of a three-phase controlled rectifier and a single-phase line commutated inverter with a series inductor between them. Switch mode welders are becoming popular because of their reduced size and high efficiency.

Thyristor controlled inductors and thyristor switched capacitors incorporate single-phase ac controllers so as to provide variable reactive power as per the demands of the power system. TCIs are also useful in controlling the voltage of a low-voltage bus.

Switch mode power supplies (SMPSs) give superior performance as compared to linear regulated power supplies. A variation of the buck boost chopper called the flyback chopper forms the heart of the SMPS. This chopper gives an excellent current mode operation, but may give rise to high ripple content and increased switching losses. A universal input voltage adaptor serves to increase the input voltage range of the SMPS.

Current mode control is by and large superior to the various methods for control of SMPSs

Worked Examples

1. An inductive load is supplied from a single-phase, 230-V, 50-Hz ac supply. The load current varies between the two extreme limits $(4 - j0)$ A and $(6 - j12)$ A. A thyristor controlled inductor (TCI) is installed to compensate for the reactive power, thus making the power factor equal to unity. Find the values of L and C of the TCI.

Solution

From Eqn (11.37), the expression for reactive current flowing through the TCI is

$$I_R(\alpha) = V_s\left\{\omega C - \left(\frac{1}{\omega L}\right)\left[2 - \frac{2\alpha}{\pi} + \frac{1}{\pi}\sin 2\alpha\right]\right\}, \qquad \frac{\pi}{2} \le \alpha \le \pi$$

With $\alpha = \pi$, the reactive current I_R works out to $I_R(\pi) = V_s\omega C$. This capacitive reactive current has to be equal to the reactive part of the current in the second limit. Thus,

$$V_s\omega C = 12$$

and

$$C = \frac{12}{V_s\omega} = \frac{12}{230 \times 314}\text{ F} = \frac{12 \times 10^6}{230 \times 314}\ \mu\text{F} = 166\ \mu\text{F}$$

For the first limit, namely, $4 - j0$, no compensation is needed. Hence the expression for $I_R(\alpha)$ should be zero when $\alpha = \pi/2$.

$$I_R\left(\frac{\pi}{2}\right) = V_s\left\{\omega C - \left(\frac{1}{\omega L}\right)\left[2 - \frac{2}{\pi} \times \frac{\pi}{2} + \frac{1}{\pi}\sin\pi\right]\right\}$$

$$= V_s\left(\omega C - \frac{1}{\omega L}\right)$$

This current should be zero, giving

$$\omega C = \frac{1}{\omega L}$$

This gives

$$L = \frac{1}{\omega^2 C} = \frac{10^6}{(314)^2 \times 166}\text{ H} = 61\text{ mH}$$

Thus the required values of L and C are 61 mH and 166 μF, respectively.

2. If the TCI of Example 1 is used to improve the power factor of a load current of $(4 - j4)$ A to unity, determine the firing angle α of the TCI.

Solution

The concerned equation is

$$V_s \left\{\omega C - \left(\frac{1}{\omega L}\right)\left[2 - \frac{2\alpha}{\pi} + \frac{1}{\pi}\sin 2\alpha\right]\right\} = 4 \qquad \text{(i)}$$

As per Example 1,

$$\omega C = \frac{1}{\omega L}$$

Thus, taking $\omega c = 1/\omega L$, an approximate value of α can be found from the equation

$$V_s \left\{\omega C - \omega C\left[2 - \frac{2\alpha}{\pi}\right]\right\} = 4$$

or

$$V_s \omega C \left\{1 - \left[2 - \frac{2\alpha}{\pi}\right]\right\} = 4$$

Substituting the value of $V_s\omega C$,

$$12\left\{1 - 2 + \frac{2\alpha}{\pi}\right\} = 4$$

giving

$$\alpha = \frac{2\pi}{3} \text{ rad or } 120° \text{ (approx)}$$

By trial and error, the exact value of α that satisfies Eqn (i) is found to be 122°.

3. A TCI installed in a bus with a voltage of 240 V has values for C and L of 200 μF and 0.0507 H, respectively. The in-phase component of the current remains at 12 A throughout the range. Determine (a) the range of current that can be compensated and (b) the value of the firing angle required to compensate a current of $(12 - j11)$ A.

Solution

(a) When $\alpha = \pi$,

$$\begin{aligned} I_R(\pi) &= V_s\omega C \\ &= 240 \times 314 \times 200 \times 10^{-6} \\ &= 15.07 \text{ A} \end{aligned}$$

When $\alpha = \pi/2$, the current provided by the TCI is

$$I_R\left(\frac{\pi}{2}\right) = V_s\left(\omega C - \frac{1}{\omega L}\right) = 240\,(0.0628 - 0.0628) = 0$$

Hence the range of current that can be compensated is (12 + $j0$) A to (12 $-j15.07$) A.

(b)
$$V_s\left\{\omega C - \left(\frac{1}{\omega L}\right)\left[2 - \frac{2\alpha}{\pi} + \frac{1}{\pi}\sin 2\alpha\right]\right\} = 1$$

As a first approximation, that is, neglecting the sin 2α term,

$$V_s\left\{\omega C - \left(\frac{1}{\omega L}\right)\left[2 - \frac{2\alpha}{\pi}\right]\right\} = 11$$

Substitution of values gives

$$240\left\{0.0628 - 0.0628\left(2 - \frac{2\alpha}{\pi}\right)\right\} = 11$$

which gives

$$\alpha = 2.717 \text{ rad or } 155.7^\circ$$

By trial and error, the exact value of α is found to be 176°.

4. An inductive load supplied from a single-phase ac supply of 220 V (50 Hz) has current ranging from (5 + $j2$) A to (8 $-j10$) A. A thyristor controlled inductor is to be installed to make the power factor unity throughout the range. Determine the values of L and C of the TCI.

Solution

For $\alpha = \pi$, compensation is to be provided for the reactive component of $j10$. Hence,

$$I_R(\pi) = V_s\omega C = 10$$

This gives

$$\omega C = \frac{10}{V_s} = \frac{10}{220}$$

This gives

$$C = \frac{10}{220 \times 314} \text{ F or } \frac{10 \times 10^6}{220 \times 314} \text{ μF}$$

or

$$C = 145 \text{ μF}$$

For $\alpha = \pi/2$, TCI can provide a current equal to $V_s\,[\omega C - 1/\omega L]$. Thus,

$$V_s\left[\omega C - \frac{1}{\omega L}\right] = -2$$

or

$$220\left[\frac{10}{220} - \frac{1}{\omega L}\right] = -2$$

This gives $L = 58.4$ mH.

5. If the TCI of Example 4 of Chapter 6 is used to improve the power factor of a load current of $(6 - j6)$ A, determine the firing angle α of the TCI.

Solution

$$I_R(\alpha) = V_s\left\{\omega C - \left(\frac{1}{\omega L}\right)\left[2 - \frac{2\alpha}{\pi} + \frac{1}{\pi}\sin 2\alpha\right]\right\}$$

Substitution of the given values yields

$$6 = 220\left\{\frac{10}{220} - \left(\frac{10^3}{314 \times 58.4}\right)\left[2 - \frac{2\alpha}{\pi} + \frac{1}{\pi}\sin 2\alpha\right]\right\}$$

$$6 = 220\left\{0.0454 - 0.0545\left[2 - \frac{2\alpha}{\pi} + \frac{1}{\pi}\sin 2\alpha\right]\right\}$$

That is,

$$6 = 10 - 12\left(2 - \frac{2\alpha}{\pi} + \frac{1}{\pi}\sin 2\alpha\right) \tag{11.68}$$

or

$$4 = 12\left(2 - \frac{2\alpha}{\pi} + \frac{1}{\pi}\sin 2\alpha\right)$$

As a first approximation

$$2 - \frac{2\alpha}{\pi} = 0.333$$

Thus,

$$\alpha = 2.62 \text{ rad or } 150°$$

An exact value of α is found by trial and error as 2.16 rad or 123.5°.

Exercises

Multiple Choice Questions

1. The capacity of a battery is expressed in ______.
   (a) kilowatts (b) kilovolt-amperes
   (c) watt-hours (d) ampere-hours
2. When a battery is released from load, it is first kept charged with constant ______, and after a certain time, with constant ______.
   (a) voltage, current (b) current, voltage
   (c) high current, low current (d) high voltage, low voltage
3. For an inverter the per cent total harmonic distortion is given by the expression ______.
   (a) $\sum_{k=1}^{\infty} V_k^2/\sqrt{V_1}$ (b) $\left(\sum_{k=3}^{\infty} V_k^2\right)^{1/2}/(V_1^2 + V_2^2)^{1/2}$

(c) $\sum_{k=2}^{\infty} V_k^2/\sqrt{V_1}$ (d) $\sum_{k=2}^{\infty} V_k^2/V_1$

4. The Harmonic content at the output of an inverter is minimized by the ______.
   (a) series connection of two or more inverters through transformers
   (b) open-loop control of two or more inverters
   (c) parallel connection of two or more inverters through transformers
   (d) open-loop control of each inverter, individually
5. High-voltage dc transmission is economical only for distances ______.
   (a) below 200 km (b) above 300 km
   (c) below 500 km (d) above 500 km
6. For HVDC transmission, 12-pulse converters are rigged up by the series connection of two six-pulse converters through Y-Y and Δ-Y transformers. This type of connection helps in the reduction of ______ and ______.
   (a) current harmonics generated on the ac side, the voltage ripple produced on the dc side
   (b) voltage harmonics generated at the ac side, the current ripple produced on the dc side
   (c) voltage harmonics generated on the ac side, the voltage ripple produced on the dc side
   (d) current harmonics generated on the ac side, the current ripple produced on the dc side
7. The reactive power drawn by converters is given by the expression ______ and the real power transmitted by them is given by ______.
   (a) $3.15 V_{LL} I_d \cos\alpha$, $1.35 V_{LL} I_d \cos\alpha$
   (b) $1.35 V_{LL} I_d \cos\alpha$, $1.35 V_{LL} I_d \sin\alpha$
   (c) $1.35 V_{LL} I_d \cos\alpha$, $3.15 V_{LL} I_d \sin\alpha$
   (d) $1.35 V_{LL} I_d \sin\alpha$, $1.35 V_{LL} I_d \cos\alpha$
8. In the V_d-I_d diagram given in Fig. 11.49, the characteristic AB pertains to constant ______ control, BCD to constant ______ control, and DE to constant ______ control.
   (a) extinction angle, firing angle, current
   (b) extinction angle, current, firing angle
   (c) firing angle, current, extinction angle
   (d) firing angle, extinction angle, current
9. Dc circuit-breakers are usually of the ______ type.
   (a) bulk oil (b) air blast
   (c) minimum oil (d) air break
10. For the input current waveform given in Fig. 11.50, the voltage waveform for a capacitive load consists of ______ and that for inductive load consists of ______.
    (a) alternate positive- and negative-going trigger pulses, alternate positive- and negative-going trigger pulses
    (b) alternate positive- and negative sloped ramps, alternate positive- and negative sloped ramps

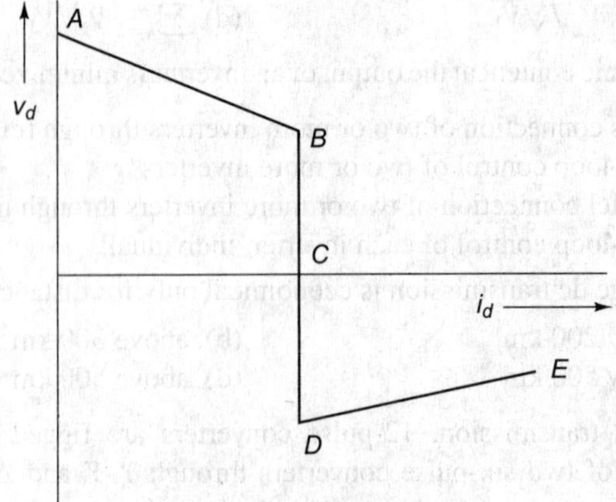

Fig. 11.49

(c) alternate positive- and negative sloped ramps, alternate positive- and negative-going trigger pulses
(d) alternate positive- and negative-going trigger pulses, alternate positive- and negative sloped ramps

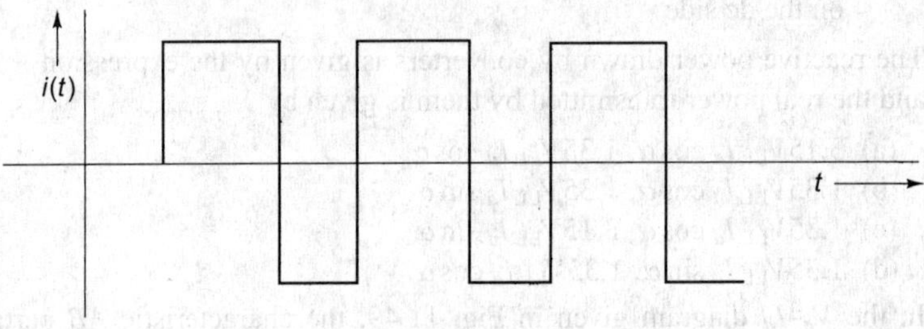

Fig. 11.50

11. Considering induction heating, the inductance coil and workpiece combination is modelled as ________.
    (a) a capacitance in series with a resistance
    (b) an inductance in series with a resistance
    (c) an inductance in parallel with a resistance
    (d) a capacitance in parallel with a resistance
12. For induction heating purposes the operating frequency is chosen such that it is ________ for thin metal sheets and ________ for thick metal sheets.
    (a) low, very low (b) high, low
    (c) very high, low (d) low, very high
13. For the purpose of arc welding the voltage should not be less than ________ and the current should not be less than ________.
    (a) 100 V, 400 A (b) 40 V, 500 A
    (c) 80 V, 300 A (d) 200V, 400 A

14. The set-up required for power electronic control of welding is either _______ or _______.
    (a) a transformer followed by a diode bridge rectifier, a transformer followed by a fully controlled rectifier and a dc chopper
    (b) a transformer followed by a half-controlled bridge rectifier, a transformer followed by a diode bridge rectifier and a dc chopper
    (c) a transformer followed by a fully controlled bridge rectifier, a transformer followed by a diode bridge rectifier and a dc chopper
    (d) a transformer followed by a fully controlled bridge rectifier, a transformer followed by a half-controlled bridge rectifier and a dc chopper
15. With the notation NL indicating no load and FL indicating full load, the percentage voltage regulation at the load end of a transmission line is given as _______.
    (a) $100(V_{FL} - V_{NL})/V_{FL}$ (b) $100(V_{NL} - V_{FL})/V_{NL}$
    (c) $100(V_{FL} - V_{NL})/V_{NL}$ (d) $100(V_{NL} - V_{FL})/V_{FL}$
16. Under full load and no-load conditions, a power system requires _______ reactive power and _______ reactive power, respectively.
    (a) capacitive, inductive (b) inductive, inductive
    (c) capacitive, capacitive (d) inductive, capacitive
17. The two extreme values of the total fundamental component of reactive current drawn by a thyristor controlled inductor are _______ and _______, respectively, where V_s is the supply voltage.
    (a) $V_s(\omega L - 1/\omega C)$, $V_s \omega L$ (b) $V_s(\omega C - 1/\omega L)$, $V_s \omega L$
    (c) $V_s(\omega C - 1/\omega L)$, $V_s \omega C$ (d) $V_s(\omega L - 1/\omega C)$, $V_s \omega C$
18. For a single-phase transmission line from station A (sending end) to station B (receiving end) and for a lagging load at the receiving end, $|V_A|$ will be _______ than $|V_B|$, and for a leading load, $|V_A|$ will be _______ than $|V_B|$.
    (a) less, less (b) more, less
    (c) more, more (d) less, more
19. Considering integral cycle control of thyristor switched capacitors, if the semiconductor is switched on for p half-cycles and off for q half-cycles, the capacitive reactive power provided to the system varies as the ratio _______.
    (a) $(p+q)/q$ (b) $(p+q)/p$
    (c) $p/(p+q)$ (d) $q/(p+q)$
20. A transformer used with the flyback chopper of an SMPS has a twofold purpose: _______ and _______.
    (a) as an inductor to smoothen the current, to provide electrical isolation
    (b) as a voltage balancer, as protective equipment for faults
    (c) to provide energy storage as an inductor, to provide electrical isolation
    (d) to safeguard against voltage transients, as a voltage balancer
21. If the ratio τ_{ON}/τ for a buck boost chopper is 0.7, then the ratio V_{ld}/V_{input} is nearly _______.
    (a) 2.33 (b) 2.66
    (c) 1.33 (d) 3.00

22. If the ratio τ_{ON}/τ for a buck boost chopper is 0.4, then the ratio I_{ld}/I_{input} is ________.
    (a) 2.5 (b) 1.5
    (c) 0.75 (d) 1.25

23. The leakage inductances L_{11} and L_{22} of an isolation transformer have to be minimized because ________.
    (a) the power losses due to them will considerably reduce the efficiency
    (b) the inductive energy of these inductances flows through the circuit and causes voltage transients
    (c) electromagnetic interference will increase because of increase in inductive energy
    (d) the inductive energy has to be absorbed by the semiconductor switches and their snubber circuits

24. For a flyback chopper with turn ratio $N_2{:}N_1 = 300{:}200$ and τ_{ON} and τ_{OFF} of 5 μs and 10 μs, respectively, the ratio V_{ld}/E is ________.
    (a) 0.84 (b) 0.58
    (c) 0.33 (d) 0.27

25. For the input voltage waveform given in Fig. 11.51, the current waveform for a capacitive load consists of ________ and that for an inductive load consists of ________.
    (a) alternate positive- and negative-going trigger pulses, alternate positive- and negative-going trigger pulses
    (b) alternate positive- and negative-going trigger pulses, alternate positive- and negative sloped ramps
    (c) alternate positive- and negative sloped ramps, alternate positive- and negative sloped ramps
    (d) alternate positive- and negative sloped ramps, alternate positive- and negative-going trigger pulses

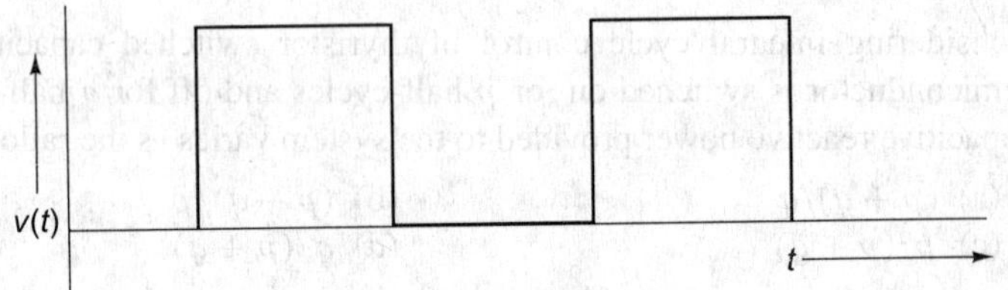

Fig. 11.51

26. When a synchronous motor is under-excited it ________.
    (a) supplies capacitive reactive power to the system
    (b) supplies inductive reactive power to the system
    (c) takes capacitive reactive power from the system
    (d) does not take any reactive power

27. A synchronous condenser is not preferred to the static VAR reactor for power factor improvement because ________.
    (a) its I^2R losses are very high
    (b) it leads to high-voltage peaks in the circuit
    (c) the circuit is subjected to more noise content
    (d) it is not feasible to install it at all stations
28. Dc circuit-breakers are not needed in an HVDC transmission system because ________.
    (a) there is no possibility for overcurrent in such a system
    (b) overcurrent is associated with under-voltage, their product being quite small
    (c) with fast converter control response, it is possible to reduce the current
    (d) the inductance in an HVDC system will protect it against overcurrents
29. In a circuit with current source input, if the load voltage phasor lags the source current, the circuit will behave like ________.
    (a) a source of inductive reactive power
    (b) a source of capacitive reactive power
    (c) a load that consumes capacitive reactive power
    (d) a purely resistive circuit
30. A thyristor controlled inductor will work as a variable inductor for α ________ and as a fixed inductor for α ________.
    (a) $\leq \pi/2, > \pi/2$ (b) $< \pi/2, \geq \pi/2$
    (c) $> \pi/2, \leq \pi/2$ (d) $< \pi/2, > \pi/2$
31. For implementation of constant extinction angle control, the extinction angle γ is kept constant at its minimum value, namely, ________.
    (a) 15°–30° (b) 60°–90°
    (c) 30°–45° (d) 25°–40°
32. For obtaining a capacitive reactive current of 10 A at a bus voltage of 230 V, a capacitor with a capacitance of ________ μF has to be connected to the bus.
    (a) 98 (b) 156
    (c) 214 (d) 138
33. For acting as a sink for a capacitive reactive current of 12 A at a voltage of 240 V, an inductor with an inductance of nearly ________ mH has to be connected to the bus.
    (a) 55.5 (b) 63.7
    (c) 44.3 (d) 28.7

True or False

Indicate whether the following statements are *true* or *false*. Briefly justify your choice.

1. The capacity of a battery is expressed in kilovolt-amperes.
2. When a battery is released from load, it is first kept charged with constant current, and after a certain time, with constant voltage.

3. For an inverter the per cent total harmonic distortion is given by the expression

$$\frac{\sum_{k=1}^{\infty}(V_k)^2}{(V_1)^{1/2}}$$

4. The harmonic content at the output of an inverter is minimized by the parallel connection of two or more inverters.
5. High-voltage dc transmission is economical only for distances below 500 km.
6. For HVDC transmission, 12-pulse converters are rigged up by the series connection of two six-pulse converters through Y-Y and Δ-Y transformers. This type of connection helps in the reduction of current harmonics generated on the ac side, and the voltage ripple produced on the dc side.
7. The reactive power drawn by the converters is given by the expression $1.35 V_{LL} I_d \cos\alpha$ and the real power transmitted by them is given by $1.35 V_{LL} I_d \sin\alpha$.
8. In the v_d-i_d diagram given in Fig. 11.52, the characteristic AB pertains to constant firing angle control, BCD to constant current control, and DE to constant extinction angle control.

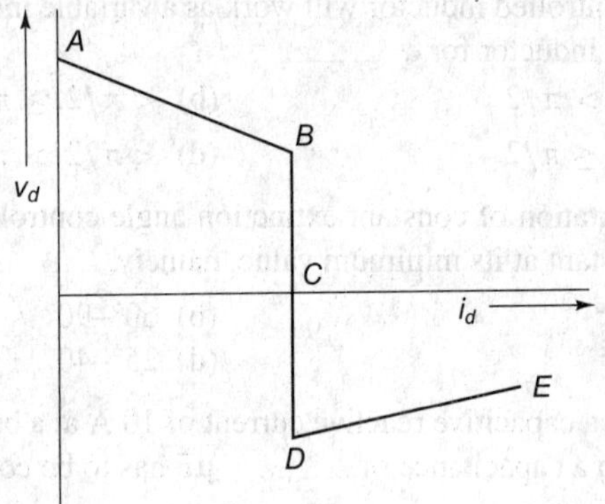

Fig. 11.52

9. Dc circuit-breakers are usually of the air blast type.
10. For the input current source waveform given in Fig. 11.53, the voltage waveform for a capacitive load consists of alternate positive- and negative-going trigger pulses, and that for inductive load consists of alternate positive- and negative sloped ramps.
11. Considering induction heating, the inductance coil and workpiece combination is modelled as an inductance in parallel with a resistance.
12. For induction heating purposes the operating frequency is chosen such that it is low for thin metal sheets and very low for thick metal sheets.

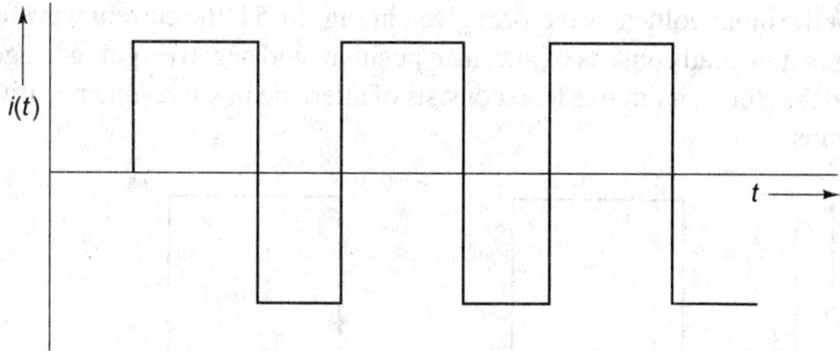

Fig. 11.53

13. For the purpose of arc welding the voltage should not be less than 40 V and the current should not be less than 500 A.
14. The set-up required for power electronic control of welding is either a transformer followed by a diode bridge rectifier or a transformer followed by a fully controlled rectifier and chopper.
15. With the notation NL indicating no load and FL indicating full load, the percentage voltage regulation at the load end of a transmission line is given as

$$100 \times \frac{V_{FL} - V_{NL}}{V_{FL}}$$

16. Under full load and no-load conditions, a power system requires capacitive and inductive reactive power, respectively.
17. The two extreme values of the total fundamental component of reactive current drawn by a thyristor controlled inductor are $V_s(\omega C - 1/\omega L)$ and $V_s\omega C$, respectively, where V_s is the supply voltage.
18. For a single-phase transmission line from station A (sending end) to station B (receiving end) and for a lagging load at the receiving end, $|V_A|$ will be smaller than $|V_B|$; on the other hand, for a leading load $|V_A|$ will be larger than $|V_B|$.
19. Considering integral cycle control of thyristor switched capacitors, if the semiconductor is switched on for p integral half-cycles and off for q integral half-cycles, the capacitive reactive power provided to the system varies as the ratio $(q + p)/q$.
20. The transformer used with the flyback converter of an SMPS has a twofold purpose: (i) to provide energy storage as an inductor and (ii) to provide electrical isolation.
21. If the ratio τ_{ON}/τ for a buck boost chopper is 0.7, then the ratio V_{ld}/V_{input} is nearly 2.66.
22. If the ratio τ_{ON}/τ for a buck boost chopper is 0.4, then the ratio I_{ld}/I_{input} is 1.5.
23. The leakage inductances L_{11} and L_{22} of an isolation transformer have to be minimized because the inductive energy has to be absorbed by the semiconductor switches and their snubber circuits.
24. For a flyback chopper with turn ratio N_2:N_1 = 300:200 and τ_{ON} and τ_{OFF} of 5 μs and 10 μs, respectively, the ratio V_{ld}/E is 0.27.

25. For the input voltage waveform given in Fig. 11.54, the current waveform for a capacitive load consists of alternate positive- and negative-going trigger pulses and that for an inductive load consists of alternate positive- and negative sloped ramps.

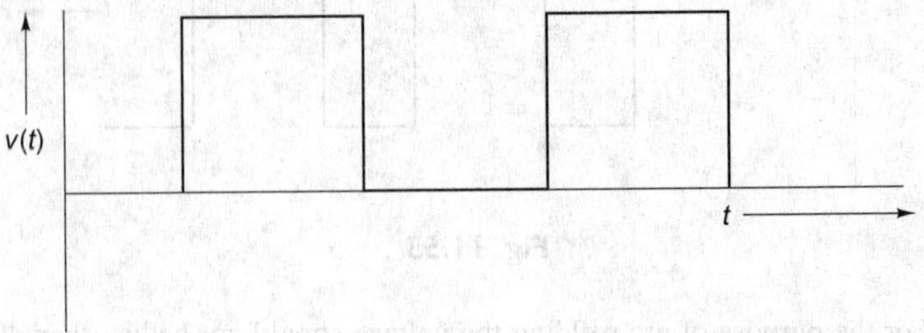

Fig. 11.54

26. When a synchronous motor is under-excited it supplies inductive reactive power to the system.
27. A synchronous condenser is not preferred to the static VAR reactor for power factor improvement because it is not feasible to install it at all stations.
28. Dc circuit-breakers are not needed in an HVDC transmission system because overcurrent is associated with under-voltage, their product being quite small.
29. In a circuit with current source input, if the load voltage phasor lags the source current, the circuit will behave like a source of capacitive reactive power.
30. A thyristor controlled inductor will work as a variable inductor for $\alpha > \pi/2$ and as a fixed inductor for $\alpha \leq \pi/2$.
31. For implementation of constant extinction angle control, the extinction angle is kept constant at its minimum value, namely, 15°–30°.
32. For obtaining a capacitive reactive current of 10 A at a bus voltage of 230 V, a capacitor with a capacitance of 13.8 μF has to be connected to the bus.
33. For acting as a sink for a capacitive reactive current of 12 A at a voltage of 240 V, an inductor with an inductance of 44.3 mH has to be connected to the bus.

Short Answer Questions

1. Discuss the various aspects of the following parts of a UPS: (a) battery, (b) inverter, and (c) rectifier.
2. Briefly describe the salient features of a typical HVDC system.
3. Considering HVDC transmission, derive the expression for the total current per phase of the transformer secondary.
4. Considering HVDC transmission, derive an expression for the average dc voltage obtained at the output of the converter.
5. Derive expressions for the real power transmitted and reactive power taken by converters used in HVDC transmission.

6. List out the various aims of converter control in HVDC transmission.
7. With the help of a (v_d, i_d) diagram discuss the various types of control of HVDC converters.
8. List out the advantages of HVDC transmission.
9. Write short notes on dc circuit-breakers.
10. What are the desirable features of a circuit used for induction heating? How is such a circuit rigged up?
11. Describe a practical circuit used for induction heating.
12. What are the considerations that go into the choice of suitable frequency for induction heating.
13. Write short notes on the use of controlled rectifiers for (a) resistance welding and (b) arc welding.
14. What are the advantages of thyristor VAR controllers over other types of VAR controllers?
15. Why are TCIs necessary on transmission lines?
16. Explain how a TCI circuit is rigged up.
17. Discuss the various aspects of bus control using a TCI.
18. How is a thyristor switched capacitor practically realized?
19. How does the switch mode regulator differ from a linear regulator? Use block diagrams to support your answer.
20. With the help of a block-cum-schematic diagram, discuss the different features of an SMPS.
21. With the help of suitable circuit diagrams, explain the principle of a flyback chopper.
22. Discuss the features of a practical flyback chopper.
23. Give the block diagram of an SMPS that uses a flyback chopper. Describe the various modules that go into its construction.
24. What are the advantages of a flyback chopper?
25. Derive an expression for the ratio of output and input voltages for a buck boost chopper.
26. Derive an expression for the ratio of output and input voltages for a flyback chopper.
27. Explain how the flyback chopper is derived from a buck boost chopper.

Problems

1. An inductive load is supplied from a single-phase ac source with a voltage of 235 V at 50 Hz. The load current varies from $(5-j0)$ A to $(7-j13)$ A. (a) Determine the values of L and C for a thyristor controlled inductor which has to improve the power factor to unity throughout the range of load current. (b) If the TCI of part (a) is used to improve the power factor of a load current of $(6-j7)$ A to unity, determine the firing angle α of the TCI.

   Ans. (a) $C = 176\ \mu\text{F}$, $L = 57.6$ mH; (b) $\alpha = 116°$

2. A TCI installed at a bus with a voltage of 225 V at 50 Hz has the values for C and L as 190.2 μF and 53.35 mH, respectively. The in-phase component of the load current remains at 10 A throughout the range. Determine (a) the range of load current that can be compensated so as have a power factor of unity and (b) the value of α required to compensate a current of $(10 - j8)$ A.

   Ans. (a) $(10 + j0)$ A to $(10 - j13.44)$ A, (b) $\alpha = 119°$

3. An inductive load supplied from a single-phase ac supply of 236 V, 50 Hz has a current ranging from $(6 + j3)$ A to $(7 - j12)$ A. A thyristor controlled inductor is to be installed to improve the power factor to unity throughout the range. Determine (a) the values of L and C of the TCI and (b) the firing angle α required to improve the power factor of a load current of $(6.5 - j7.1)$ A to unity.

   Ans. (a) C = 162 μF, L = 50 mH; (b) 124°

4. A thyristor controlled inductor installed at a bus with a voltage of 231 V at 50 Hz has the values for C and L as 180 μF and 51 mH, respectively. The in-phase component of the load current remains at 9 A throughout the range. Determine (a) the range of load current that can be compensated so as to have a power factor of unity and (b) the value of α to compensate for a load current of $(9 - j9)$ A.

   Ans. (a) $(9 + j1.42)$ A to $(9 - j13)$ A, (b) 127.5°

5. A single-phase system bus of voltage 240 V at 50 Hz has a load current that varies in the range $I_1 = (8 + j2)$ A to $I_2 = (8 - j14)$ A with the real part remaining at 8 throughout the range. A thyristor controlled reactor with C = 186 μF and L = 52 mH is installed to improve the power factor. (a) Examine the correctness of the values of L and C for I_1 and I_2; express their errors (if any) as percentages. (b) For the case of incorrect compensation, determine the net power factor of the current.

   Ans. (a) C is adequate, L is higher by 8.8%; (b) 0.996

CHAPTER 12

Microprocessor Fundamentals

12.1 Introduction

A microprocessor chip is a digital electronic circuit that becomes a versatile digital system called the *microcomputer* when it is accompanied by memories, input/output devices, and other accessory chips. Thus, a study of the microcomputer involves understanding the microprocessor, also called the central processing unit, and a variety of IC chips, together with the techniques required to interconnect (or interface) them with the processor. This chapter gives a brief description of both 8-bit and 16-bit microprocessors as well as their supporting chips so as to, respectively, constitute the 8-bit and 16-bit microcomputers. It provides a background for Chapter 13, which deals with applications of microcomputers to power electronic drives.

After going through this chapter the reader should

- become familiar with digital circuits such as latches, buffers, encoders, decoders, etc.
- know the structure and operation of flip-flops, which are the basic building blocks of digital circuits such as shift registers,
- gain familiarity with the architectures and instruction sets of 8-bit and 16-bit microprocessors,
- know how arithmetic computations can be carried out using assembly language programs, and
- become familiar with the hardware as well as software aspects of interfacing the 8-bit microprocessor (8085A) with peripherals.

12.2 Digital Systems

A typical digital system is constructed from a few basic electronic devices, which are used repeatedly in various topological combinations. A memory cell called a flip-flop stores binary digits (0 and 1). This flip-flop and another circuit called the NAND gate form the basic building blocks of any digital system. IC chips that perform the following functions are available in the market: binary addition, decoding (demultiplexing), data selection (multiplexing), counting, storage of binary data (memories and registers), digital-to-analog (D/A) and analog to-digital (A/D) conversion, etc. Large-scale integration of MOSFETs is also done for storage and conversion of data; these chips are called *memories*.

Standard combinations of basic electronic modules with less than about 12 gates on a chip are classified as small-scale integration (SSI). The flip-flops discussed here are also available as SSI packages. Most other functions using bipolar junction transistors are examples of medium-scale integration (MSI); each such chip has between 12 and 100 gates.

Developments in semiconductor technology have made it feasible to put several hundreds or even thousands of transistors and resistors on a single chip. Whereas high-density chips having more than 100 gates come under the category of large-scale integration (LSI), those having more than 1000 gates are classified as very large scale integration (VLSI) chips. It is possible to design either extremely complex and powerful chips that can perform specialized tasks or general-purpose chips that can be programmed to perform many such tasks. The microprocessor comes under the latter category and is programmable in the same sense as a general-purpose computer. The microprocessor is especially attractive for use in a large number of applications including power electronic circuits because of the flexibility available to change or expand the functions performed by it, without any rewiring or redesigning of the system.

12.2.1 Latches, Tristate Logic, and Buffers

One special feature of microprocessors is that only clocked circuits are used in them. Therefore, to analyse such circuits, timing diagrams (or waveforms) are needed along with truth tables. These timing diagrams are just drawings of the input and output waveforms of the voltages at the nodes of the circuit, as observed in a multichannel oscilloscope. Figures 12.1(a) and (b), respectively, show the truth table and its timing diagram equivalent for a NOR gate. Figure 12.1(c) shows the same timing diagram with a magnified time scale so as to clearly bring out the rise times and propagation delays.

Timing diagrams are necessary to observe any glitches or unwanted signals that may creep into the output waveform of a digital circuit. Such erroneous output waveforms [Fig. 12.1(d)] may occur when the signal has to pass through a number of devices. Figure 12.1(e) shows a typical three-device circuit wherein propagation delay in the digital devices may lead to glitches.

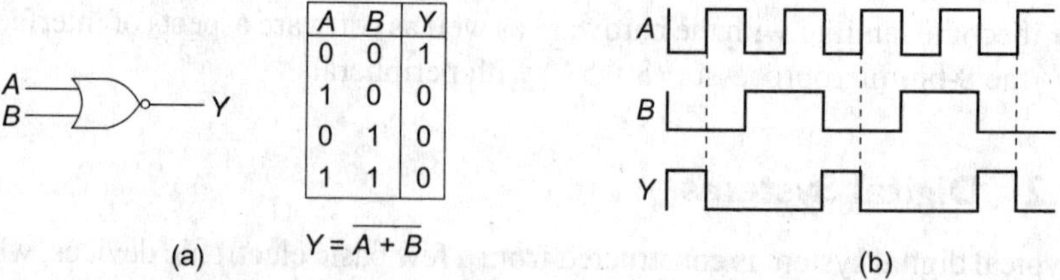

A	B	Y
0	0	1
1	0	0
0	1	0
1	1	0

Figs 12.1(a) and (b)

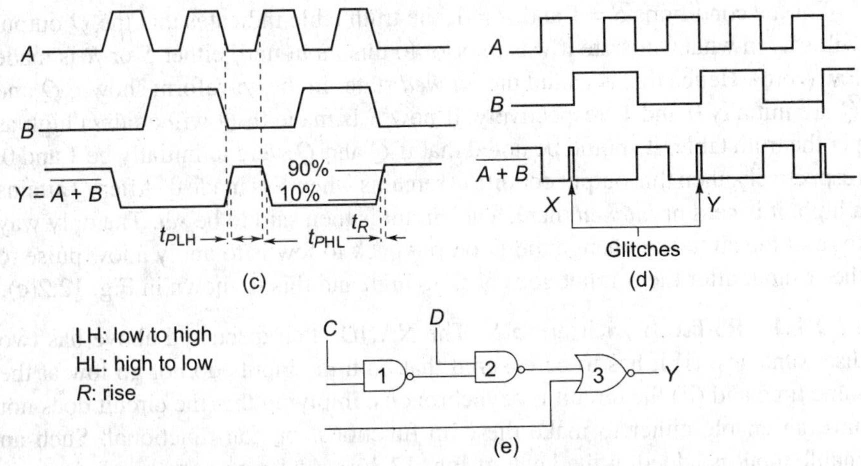

Fig. 12.1 NOR gate: (a) truth table, (b) ideal timing diagram, (c) timing diagram showing rise times (t_R) and propagation delays (t_P), (d) timing diagram showing glitches, (e) circuit that can produce glitches

Latches and flip-flops are memory circuits which store either a 1 or a 0 and whose outputs change only when a clock (or enable) input is pulsed. These chips are therefore classified as *clocked digital circuits*. Figures 12.2(a), (b), and (c), respectively, show the schematic diagram, truth table, and timing diagrams of a simple NAND latch, in which Q and $\overline{Q}$ are complementary outputs, S is the set or preset input, and R is reset or clear input. From the truth table it is seen that for $S = R = 0$ the output is indeterminate; this can be explained as follows. For a NAND gate, if one of the inputs is low, a high output is obtained irrespective of the other inputs. This means that with $S = R = 0$, both Q and $\overline{Q}$ will be high. But this violates the definition of Q and $\overline{Q}$ being complementary. Hence, such a state is called as *indeterminate* or *prohibited* state and is indicated by an asterisk in the truth table.

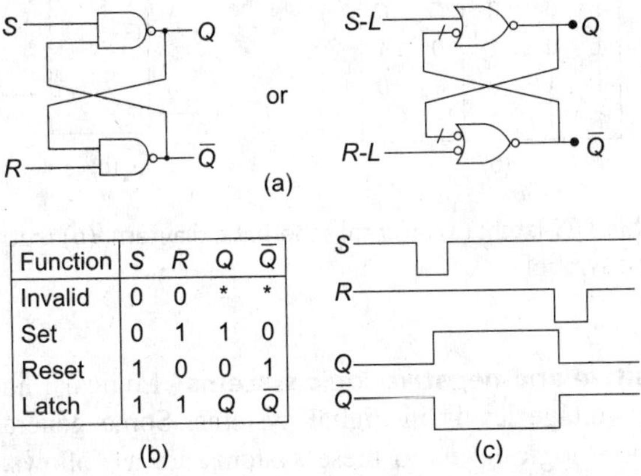

Function	S	R	Q	$\overline{Q}$
Invalid	0	0	*	*
Set	0	1	1	0
Reset	1	0	0	1
Latch	1	1	Q	$\overline{Q}$

Fig. 12.2 Cross-coupled NAND gate latch: (a) schematic diagram, (b) truth table, (c) waveform showing the effect of *S* and *R* pulses on the outputs

For the conditions $S = 1$ and $R = 1$, the truth table indicates that the Q output will stay in whatever state it was in prior to this input until either S or R is made low (zero). Hence this is called the *latched* state. In the waveform shown, Q and $\overline{Q}$ are initially 0 and 1, respectively. If now S is made 0, Q will attain a high as per the truth table. It should be noted that if Q and $\overline{Q}$ were to initially be 1 and 0, respectively, then this output condition remains when S is made 0. After Q attains a high, it is held or *latched* there. The circuit is then said to be *set*. The only way to reset the circuit or change the Q output back to low is to apply a low pulse to the R input after the S input goes back to high and this is shown in Fig. 12.2(c).

12.2.1.1 *RS* latch with enable The NAND latch discussed above has two disadvantages: (i) it has to be ensured that both the inputs do not go low at the same time and (ii) the circuit is asynchronous, implying that the circuit does not have an enable either to make the chip functional or non-functional. Such an enable input is added in the latch of Fig. 12.3(a). As long as the enable input is low in this circuit, the outputs of the input NAND gates are high and the Q and $\overline{Q}$ outputs are latched. Only when the enable input is high can the R and S inputs affect Q and $\overline{Q}$. The states of the R and S inputs just before the enable goes low will determine the states latched on the Q and $\overline{Q}$ outputs. The input NAND gates invert the polarity of the pulse required at S or R to set or reset the latch. Figures 12.3(b) and (c), respectively, show the truth table and schematic symbol of this latch.

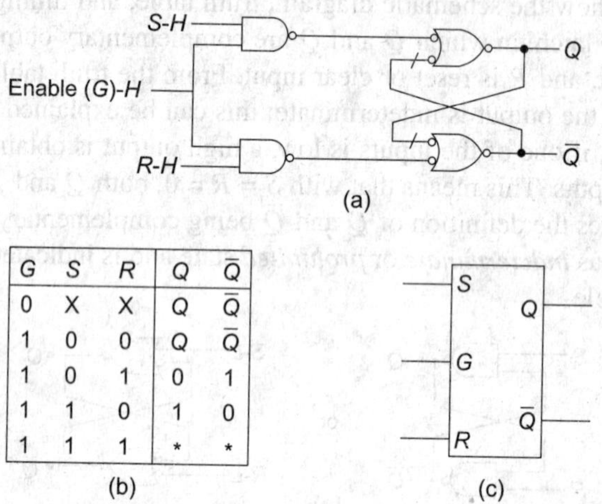

G	S	R	Q	$\overline{Q}$
0	X	X	Q	$\overline{Q}$
1	0	0	Q	$\overline{Q}$
1	0	1	0	1
1	1	0	1	0
1	1	1	*	*

Fig. 12.3 Enabled *RS*-latch: (a) internal schematic diagram, (b) truth table, (c) schematic symbol

12.2.1.2 Positive and negative logic systems Logical 1 and 0 are usually represented by voltage levels in digital systems. Some general rules for the definition of these logic levels in these systems are as follows. *Positive logic* (active high levels) means that the most positive logic voltage level (also called the high or *H* level) is defined to be the logic 1 state, and the most negative logic voltage level (also called the low level) is defined to be the logic 0 state. *Negative* *logic* (active low levels) is just the opposite. Here the most positive (or high) level is a logic 0 state and the most negative (or low) level is a logic 1 state.

The use of H and L in a schematic diagram eliminates the requirement of specifying whether the associated truth table is written in positive or negative logic. The alternate schematic diagram in Fig. 12.2(a) and the diagram in Fig. 12.3(a) can to be understood accordingly.

12.2.1.3 D latch In the circuit of the RS latch of Fig. 12.3, there is still a set of indeterminate output conditions which occur when $S = 1$ and $R = 1$. This problem can be overcome in many applications by connecting an inverter between the S and R inputs [Fig. 12.4(a)]. The resulting circuit is called a D latch. As seen from the truth table in Fig. 12.4(b) or from a scrutiny of the circuit, the Q output follows the D input as long as the enable input at terminal G is high. The logic state of the D input just before the enable state goes low will be latched on the output when G is made 0. From the point of view of a latch, the enable input is called the *strobe* or *gate*.

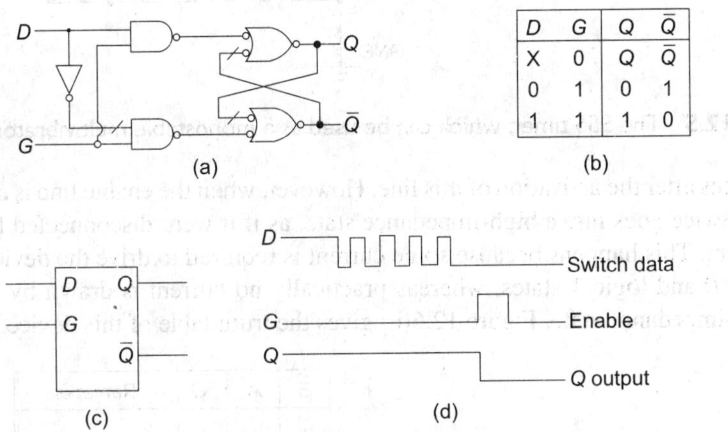

Fig. 12.4 D-latch: (a) internal schematic diagrams, (b) truth table, (c) schematic symbol, (d) timing for debouncing switch contacts

Figures 12.4(c) and (d) show the circuit symbol and the timing diagram of the D latch. For this purpose the bouncing output of a switch is the data input to the D latch. If the latch is enabled only after the bouncing stops, as shown in the figure, then none of the bounces will appear at the Q output; this is one of the methods to debounce switch signals. The enable pulse can be produced from the switch transition with two 555 timers used as monostables. Figure 12.5 shows one such timer. A negative switch transition triggers one 555 timer, which is set for an output time of 10–15 ms; the falling edge of this output then triggers a second 555 timer to produce an enable pulse of the desired width.

12.2.1.4 Tristate (or high-impedance) devices Tristate logic devices have three states, namely, logic 1, logic 0, and high impedance. Figure 12.6(a) shows such a device with three logic states; apart from the input and the output lines, the *enable* is also shown. The device functions in the same way as ordinary logic

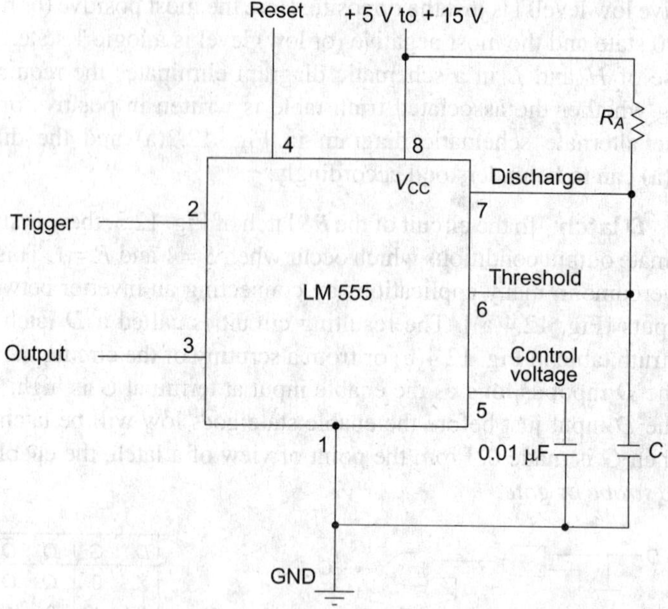

Fig. 12.5 The 555 timer, which can be used as a monostable multivibrator

devices after the activation of this line. However, when the enable line is disabled, the device goes into a high-impedance state, as if it were disconnected from the system. This happens because some current is required to drive the device in the logic 0 and logic 1 states, whereas practically no current is drawn by it in the high-impedance state. Figure 12.6(b) gives the truth table of this device.

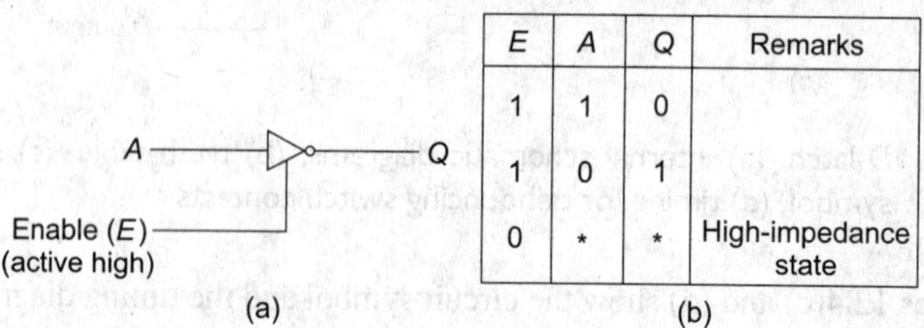

E	A	Q	Remarks
1	1	0	
1	0	1	
0	*	*	High-impedance state

(b)

Fig. 12.6 Tristate inverter and its truth table

12.2.1.5 Buffer A buffer [Fig. 12.7(a)] is a logic circuit which amplifies current or power, the logic level of its output being the same as that at its input. This chip is used primarily to increase the driving capability of a logic circuit and is hence also called a *driver*.

The tristate chip shown in Fig. 12.7(a) functions as a buffer when the enable line is low and stays in the high-impedance state when this line is high. This type of buffer is commonly used to increase the driving capability of the data and address buses of a microcomputer, which is described in Section 12.4. Figure 12.7(b) gives the truth table of this buffer.

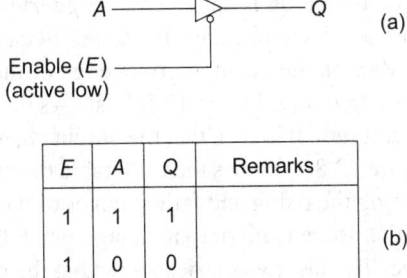

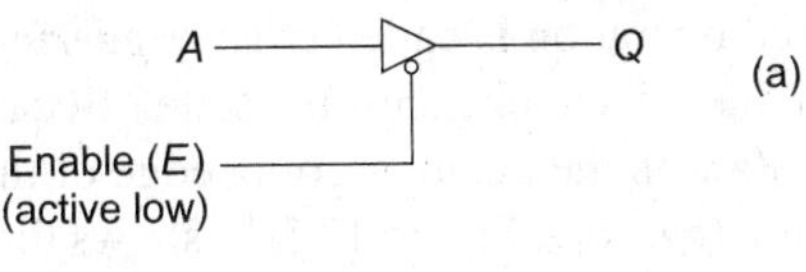

E	A	Q	Remarks
1	1	1	
1	0	0	
0	*	*	High-impedance state

(b)

Fig. 12.7 Tristate buffer and its truth table

12.2.2 Flip-flops

A flip-flop is a device which remains in one of the two logic states until triggered into another. This feature facilitates its use as a memory element. A large number of these flip-flop elements are used in a computer to constitute its memory.

12.2.2.1 Edge-triggered *D* flip-flop Figure 12.8(a) shows a *D* flip-flop with an *RC* circuit whose time constant is deliberately made much smaller than the width of the clock pulse. Consequently the capacitor charges fully when the CLK signal goes high and a narrow positive voltage spike is produced across the resistor. Slightly later, the trailing edge of the clock pulse results in a narrow negative spike. The narrow positive spike enables the input gate for an instant, thus activating it. This is equivalent to sampling the *D* output and forcing the *Q* output either to be set or reset. The narrow negative spike, however, does not serve any purpose.

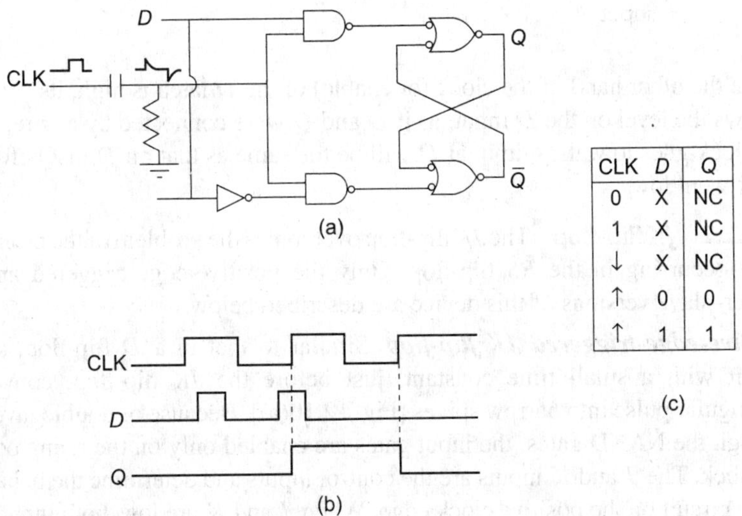

CLK	D	Q
0	X	NC
1	X	NC
↓	X	NC
↑	0	0
↑	1	1

Fig. 12.8 Edge-triggered *D* flip-flop: (a) circuit, (b) waveforms, (c) truth table

The above-described operation is called *edge-triggering* because the flip-flop responds only when the clock changes its state. Because of the triggering occurring, in Fig. 12.8(a), on the positive-going edge of the clock, the operation is called *positive-edge-triggering*. Figure 12.8(b) shows the action of the flip-flop by means of a timing diagram. It is seen that the output changes only on the rising edge of the clock. Figure 12.8(c) shows the truth table of this flip-flop, the up and down arrows representing the rising and falling edges of the clock. It is seen from the first three entries that there is no output change when the clock is low, high, or on its negative edge. The last two entries show that the output changes on the positive edge of the clock.

Figure 12.9 shows the output waveforms of the *D* latch and the positive-edge-triggered *D* flip-flop for the same *D* input and clock waveforms. The *D* flip-flop updates its *Q* output only at the moment of the occurrence of the positive edge. This implies that the *D* flip-flop takes a snapshot of the *D* signal at each positive edge of the clock and displays it at *Q*.

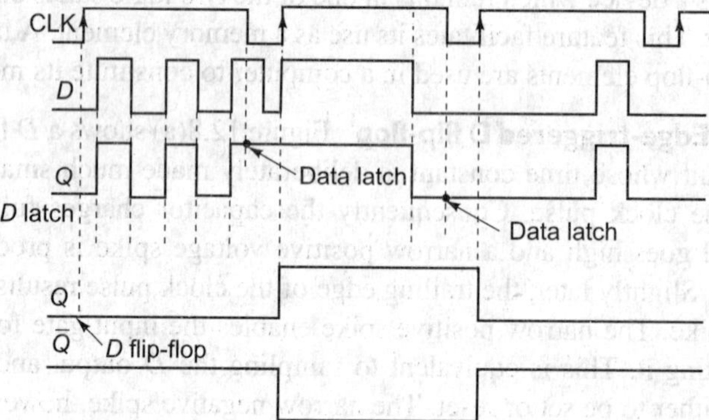

Fig. 12.9 Waveforms of the *Q* outputs for a *D* latch and a *D* flip-flop for the same *D* input

On the other hand, if the clock (or enable) of the *D* latch is high, its *Q* output follows the level on the *D* input, as if *D* and *Q* were connected by a wire. When the clock goes low, the output at *Q* will be the same as that on *D* just before the clock went low.

12.2.2.2 *JK* flip-flop The *JK* flip-flop overcomes the problem of the prohibited input occurring in the *RS* flip-flop. Only the positive-edge-triggered and the master–slave versions of this device are described below.

Positive-edge-triggered JK flip-flop Similar to that in a *D* flip-flop, an *RC* circuit with a small time constant, just before the *JK* flip-flop, converts a rectangular pulse into narrow spikes [Fig. 12.10(a)]. Because of double inversion through the NAND gates, the input gates are enabled only on the rising edge of the clock. The *J* and *K* inputs are the control inputs and determine the behaviour of the circuit on the positive clock edge. When *J* and *K* are low, both input gates are disabled and the circuit is inactive at all times including the rising edge of the clock.

When J is low (0) and K is high (1), the upper gate is disabled; so there is no means for the flip-flop to be set, the only possibility being to reset it. When Q is high, the lower gate passes a reset trigger as soon as the positive clock edge arrives. This forces Q to become low. Therefore, with $J = 0$ and $K = 1$, a rising clock edge resets the flip-flop. On the other hand with $J = 1$ and $K = 0$, the lower gate is disabled and it becomes impossible to reset the flip-flop. But the flip-flop can be set when Q is low and $\overline{Q}$ is high; therefore the upper gate passes a set trigger on the positive clock edge. This drives Q into the high state. The final result is that the next positive clock edge sets the flip-flop.

Depending on the current state of the output, it is possible to either set or reset the flip-flop when both J and K are high. If Q is high, the lower gate passes a reset trigger on the next positive clock edge. On the other hand, when Q is low, the upper gate passes a set trigger on the next positive edge. In either case, Q changes to the complement of the last state. It can therefore be concluded that the combination $J = 1$ and $K = 1$ makes the flip-flop toggle on the next positive clock edge. The word *toggle* means switching to the opposite state.

The timing diagram given in Fig. 12.10(b) shows the action described above. When J is high and K is low, the rising clock edge sets Q high. On the other hand, when J is low and K is high, the rising clock edge sets Q low. It is also seen that when J and K are simultaneously high, the output toggles on each rising clock edge.

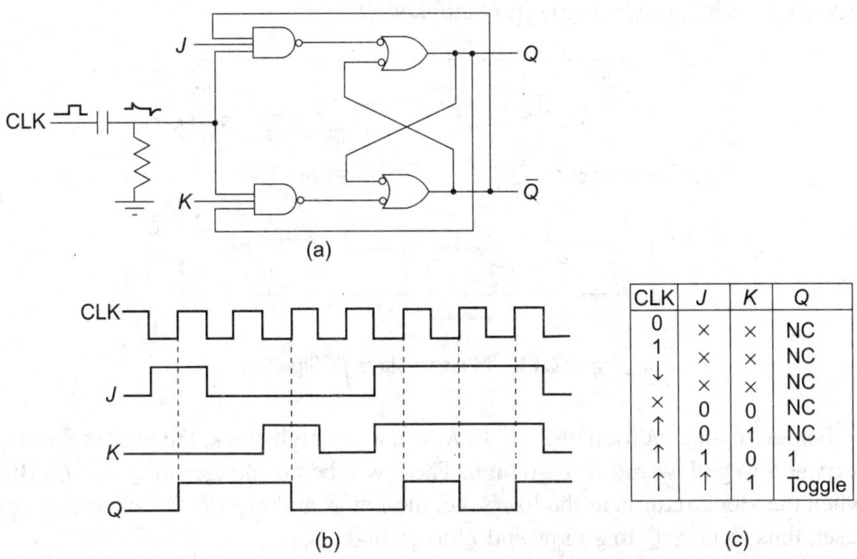

CLK	J	K	Q
0	×	×	NC
1	×	×	NC
↓	×	×	NC
×	0	0	NC
↑	0	1	NC
↑	1	0	1
↑	↑	1	Toggle

Fig. 12.10 Edge-triggered *JK* flip-flop: (a) circuit, (b) waveforms, (c) truth table

The truth table in Fig. 12.10(c) summarizes the operation. The circuit is inactive when the clock is low, high, or on its negative edge, and also when both J and K are low. The output Q is seen to change only on the rising edge of the clock.

The problem of racing The *JK* flip-flop shown in Fig. 12.10(a) has to be edge-triggered to avoid oscillations. If the *RC* circuit is removed and the clock connected straight to the gates, then, with $J = K = 1$ and a high clock, the output will toggle. New outputs are fed back to the input gates under this condition. After an interval equal to twice the propagation time (time taken by the output to change after the inputs have changed), the output toggles again, and once more new outputs return to the input states. The output can toggle repeatedly in this manner as long as the clock is high. This means that the output consists of oscillations during the positive half-cycle of the clock. Toggling more than once during a clock cycle is called *racing* and it occurs when the clock pulse duration is greater than the propagation delay of the chip.

JK master–slave flip-flop The *JK* master–slave flip-flop shown in Fig. 12.11 is a flip-flop circuit in which the problem of racing is eliminated. It is a combination of two clocked latches; the first is called the *master* and the second the *slave*. The master is seen to be positively clocked whereas the slave is negatively clocked. This implies that

(a) when the clock is high, the master is active and the slave is inactive, and

(b) when the clock is low, the master is inactive and the slave is active.

To start with, the Q output is taken to be low and $\overline{Q}$ high. With $J = 1$, $K = 0$, and a high clock, the master goes into the set state producing a high S and a low R. Nothing happens to the Q and $\overline{Q}$ outputs because the slave is inactive while the clock is high. If now the clock goes low, the high S and low R force the slave into the set state, producing high Q and low $\overline{Q}$.

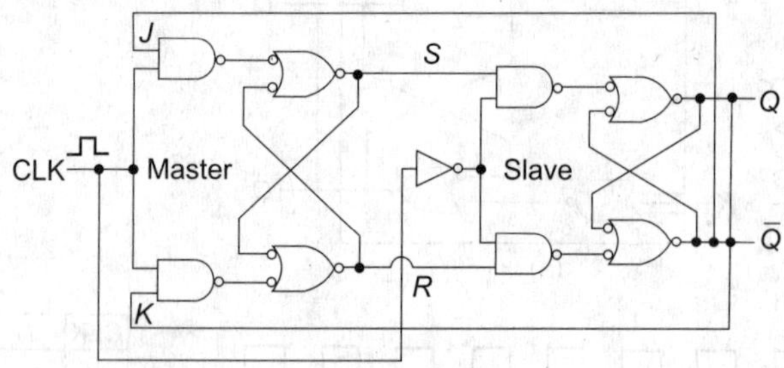

Fig. 12.11 Master–Slave *JK* flip-flop

For the input condition of $J = 0$, $K = 1$, and a high clock, the master is reset, forcing S to go low and R to go high. There will be no changes in Q and $\overline{Q}$. But when the clock returns to the low state, the low S and high R force the slave to reset, thus forcing Q to go low and $\overline{Q}$ to go high.

If the J and K inputs are both high, the master toggles once while the clock is high, the slave then toggles once when the clock goes low. It is seen from the above discussion that, irrespective of what the master does, the slave copies it.

The master–slave flip-flop is shown to be level-clocked in Fig. 12.11. Therefore, while the clock is high, any changes in J and K can affect the S and R outputs. Because of this feature, the J and K inputs are normally kept constant during the positive half-cycle of the clock. After the clock goes low, the master becomes inactive and J and K are allowed to change.

12.2.3 Registers and Shift Registers

A register is used for the temporary storage of data bits. The size of a register may be 4 bits or its multiples; the 8-bit and 16-bit combinations are used in the 8-bit microprocessor. In a shift register, the output of each flip-flop is connected to the input of the adjacent flip-flop, and so each successive clock pulse moves the data bits one flip-flop to the left, or right, depending on the order of the connection. Four basic types of shift registers, which are commonly used, are described below.

12.2.3.1 Serial-in serial-out shift register Figure 12.12(a) shows four *D* flip-flops connected to form a serial-in serial-out (SISO) shift register, and Fig. 12.12(b) shows the clock, data, and output waveforms for this circuit. The waveforms show that each clock pulse moves the data pulse one flip-flop to the right. Hence the output is delayed four clock pulses from the moment of giving the input to the first flip-flop.

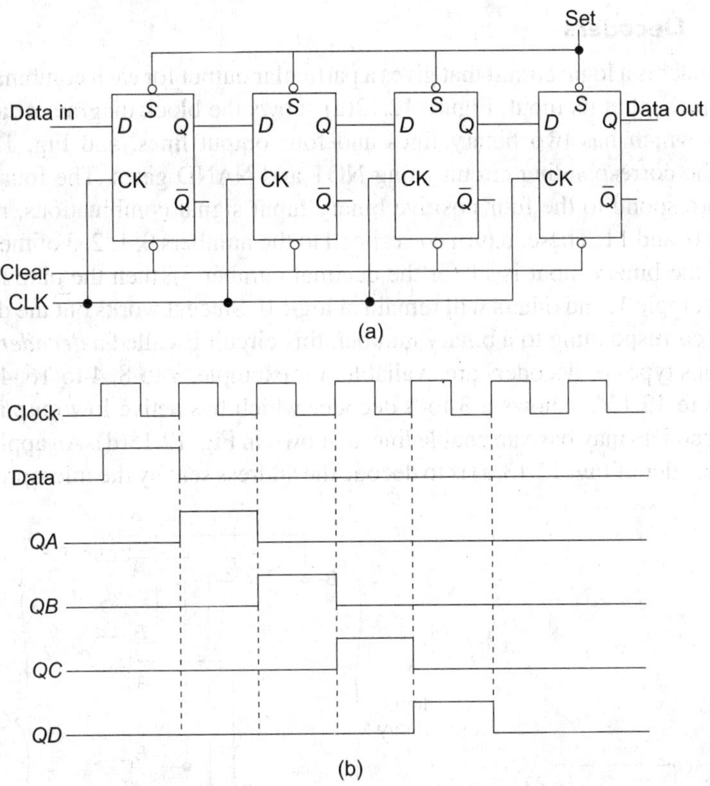

Fig. 12.12 Serial-in serial-out shift register: (a) logic diagram, (b) clock, data, and output waveforms

The SISO shift register is used for delaying digital signals, the amount of delay depending on the clock frequency and the number of flip-flops in the register.

12.2.3.2 Serial-in parallel-out shift register A serial-in parallel-out (SIPO) shift register is connected in the same way as the SISO shift register of Fig. 12.12(a), but the outputs of each flip-flop are available on the pins of the IC package.

12.2.3.3 Parallel-in serial-out shift register Parallel-in serial-out (PISO) shift registers allow for parallel inputting of a number into the register and then shifting it out to the end of the register. This is equivalent to causing a serial output. This is widely used to convert parallel format data to serial format. A start bit, seven data bits, one parity bit, and two stop bits can be loaded into an 11-bit PISO shift register and clocked to produce a serial data character.

12.2.3.4 Parallel-in parallel-out shift register As stated earlier, a parallel-in shift register allows a binary number, placed on the parallel input pins, to be loaded into the internal registers. With some parallel load shift registers, these parallel input data appear immediately at the parallel outputs, but with others, a clock pulse is required to move the parallel loaded data to the outputs.

12.2.4 Decoders

The decoder is a logic circuit that gives a particular output for each combination of signals present at its input. Figure 12.13(a) shows the block diagram of a 2-to-4 decoder which has two binary lines and four output lines, and Fig. 12.13(b) shows the corresponding circuit using NOT and NAND gates. The four output lines correspond to the four positive binary input signal combinations, namely, 00, 01, 10, and 11. These in turn correspond to the numbers 0, 1, 2, 3 of the output lines. If the binary input is 11 (or the decimal number 3), then the output line 3 will be at logic 1, and others will remain at logic 0. Since it works out the decimal number corresponding to a binary number, this circuit is called a *decoder*.

Various types of decoders are available; for example, 3-to-8, 4-to-16, 4-to-10, etc. Figure 12.13(c) shows a 3-to-8 decoder which has active low output lines. Some decoders may have an enable line as shown in Fig. 12.13(d). An application of the decoder of Fig. 12.13(a) is to decode the address sent by the microprocessor

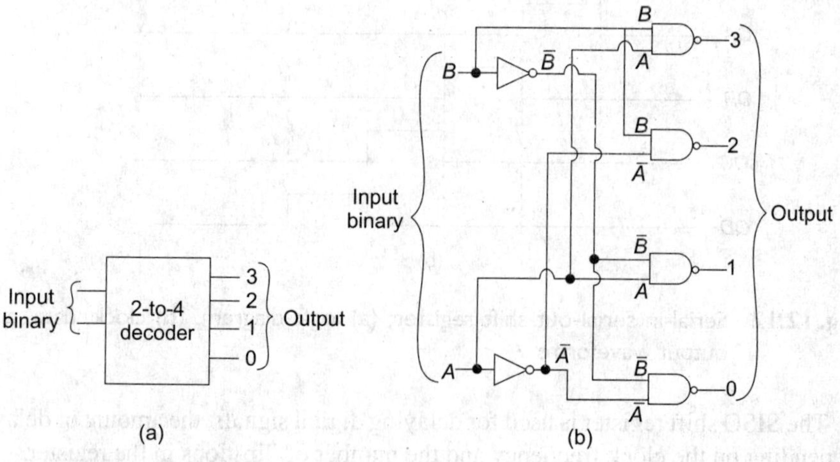

Figs 12.13(a), (b)

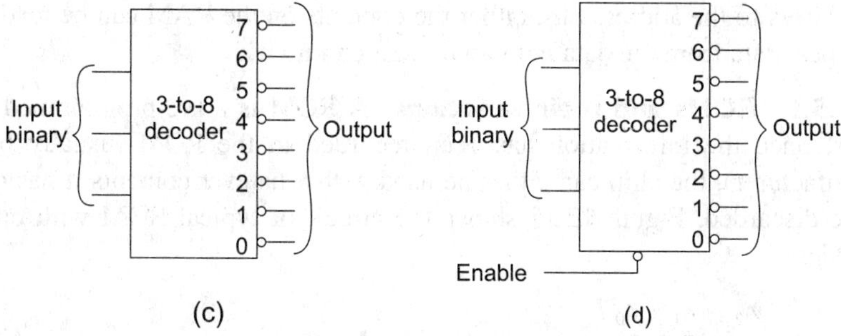

Fig. 12.13 (a) Block diagram of 2-to-4 decoder; (b) circuit of 2-to-4 decoder; (c) block diagram of 3-to-8 decoder; (d) 3-to-8 decoder with enable

on the address bus, and go to the corresponding memory location, as shown in Fig. 12.14.

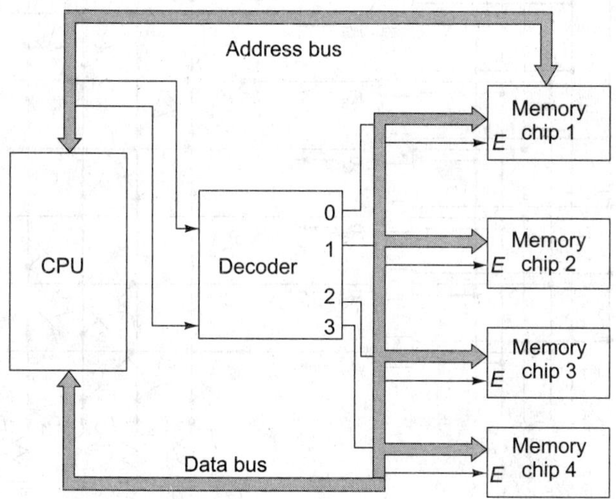

Fig. 12.14 A simple tristate memory section

12.2.5 Memories

The memory of a microcomputer is an IC chip where programs and data are stored, and from where the data are retrieved whenever the programs need them. This type of memory is called *read-only memory* (ROM) and is the simplest of its kind. It is equivalent to a group of registers, each of which stores a byte (a group of 8 bits) permanently. By applying control signals, a byte can be read from any memory location. The word 'read' means 'to make the contents of the memory location appear at the output terminals of the ROM'.

Another type of memory that is needed for temporarily storing partial answers during a computer run is called *random access memory* (RAM) or *read–write memory*, and is also equivalent to a group of addressable registers. After supplying an address to the address bus, either the contents on the RAM can be read from it or new data from the data bus can be written into it.

12.2.5.1 ROMs and their variations A ROM is a pre-programmed chip. Thus, once the information has been recorded in the ROM (usually by the manufacturer), the chip can either be used with whatever contents it has or has to be discarded. Figure 12.15 shows the circuit of typical ROM with on-chip decoders.

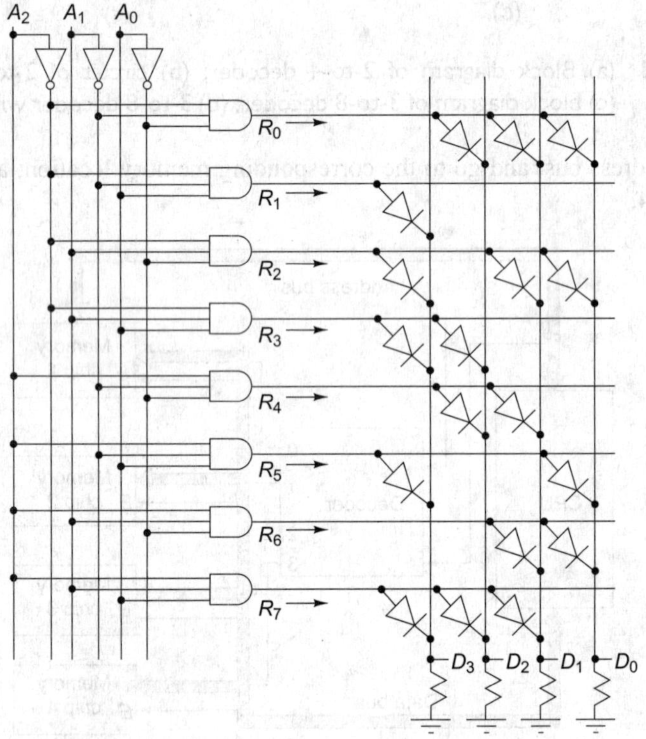

Fig. 12.15 A ROM with an on-chip decoder

A user-programmable ROM, also known as a PROM, is a one-time programmable ROM; it can be programmed by the user just once. After being programmed, the PROM behaves just like the ROM. Both in the ROM and the PROM there is a fuse in each cell. Burning or retaining a fuse decides whether the cell contains a '1' or a '0'. An instrument called the *PROM programmer* does the storing by burning.

A PROM that can be erased by ultraviolet light and then reprogrammed is known as an EPROM. In order to erase the EPROM, it has to be taken out of its normal circuit and placed in front of a special ultraviolet eraser for a period of 15–60 minutes depending on the intensity of the ultraviolet light and time of exposure. The inconvenience and other technical problems due to the removal of an EPROM from its normal circuit of operation are eliminated in electrically erasable and programmable ROMs or EEPROMs. These chips can be erased and programmed during normal operation, the erasure time varying from 1 to 10 s.

12.2.5.2 Static and dynamic RAMs Two types of RAMS, namely, static RAM (or SRAM) and dynamic RAM (or DRAM), are used in a microprocessor system. Once a bit of information is written into a cell of an SRAM, the cell retains it until it is overwritten or electrical power is taken off the chip. The cell itself is a flip-flop as shown in Fig. 12.16(a) and may consist of 4, 6, or 8 transistors or MOSFETS.

Figure 12.16(b) shows a DRAM cell chip which has a much smaller cell than an SRAM. Here a bit of information is stored in the form of charge in a capacitor. Because of its smaller cell structure, a DRAM can store about four times as much information as an SRAM in the same area. However, as the information is just a charge in the capacitor, it requires refreshing once every few milliseconds in order to retain the stored information. This refreshing needs extra circuitry and makes the interfacing of DRAMs to microprocessors more complex than the interfacing of SRAMs. Generally, in systems that require large memory capacity, DRAMs are used to lower the memory cost. On the other hand, SRAMs are used where speed is an important consideration, memory size is not large, and there is no limitation on the cost.

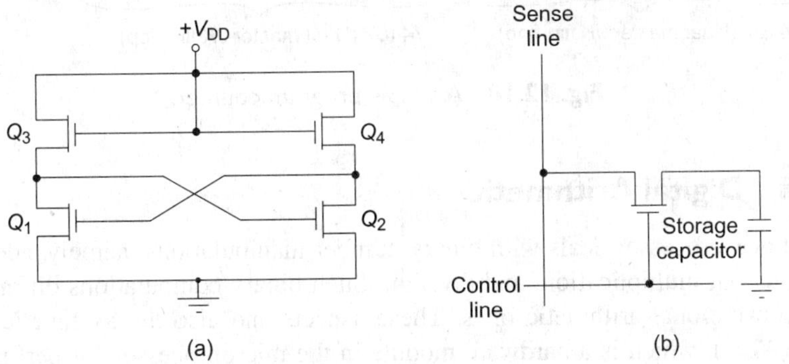

Fig. 12.16 (a) Static RAM cell; (b) dynamic RAM cell

12.2.6 Counters

A counter is a register that can count the number of clock pulses arriving at its input.

12.2.6.1 Controlled counter A controlled ripple counter counts clock pulses only when directed to do so. Figure 12.17 shows such a counter whose count signal can be low or high. Since it conditions the J and K inputs, the signal COUNT controls the action of the counter, forcing it either to do nothing or to count clock pulses.

When COUNT is low, the J and K inputs are low; therefore all flip-flops remain latched in spite of the clock pulses arriving at the counter. When COUNT and also the J and K inputs are high, the counter successively increases its

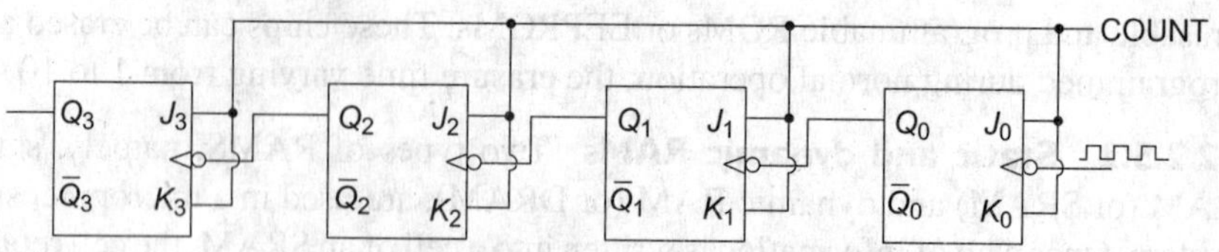

Fig. 12.17 Controlled ripple counter

count from 0000 to 1111, each negative clock pulse incrementing the stored count by 1. A microprocessor includes a program counter that keeps track of the instruction being executed. Figure 12.18 shows part of such a program counter, whose principle of operation is similar to that of the controlled counter.

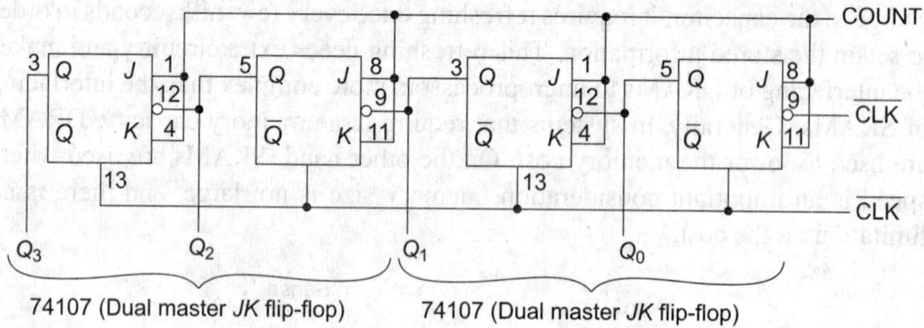

Fig. 12.18 A simple program counter

12.3 Digital Arithmetic

The microprocessor deals with binary number manipulations, namely, addition, subtraction, multiplication, and division. Such binary computations differ from the conventional arithmetic ones. These aspects and also the arithmetic logic unit (ALU), which is a hardware module in the microprocessor for performing computations, are detailed below.

12.3.1 Binary Addition

Figure 12.19(a) shows the truth table for the addition of two single-bit numbers with the half adder circuits shown in Fig. 12.19(b). Here A and B are the numbers to be added and SUM and CARRY are the outputs of the half adder. The logic relation

$$A \oplus B = S$$

holds good here with $\oplus$ denoting the exclusive-OR (XOR) operation. As shown in Fig. 12.19(b), this function is implemented with one XOR and one AND gate.

The circuit of Fig. 12.19(b) is only a half adder because it does not take into account a carry-over from a previous digit. Figure 12.20(a) shows the truth table for a full adder, which takes into account a carry-over (C_{in}) from a previous digit. It is seen from the last entry that when A, B, and C_{in} are all 1's, then both S and

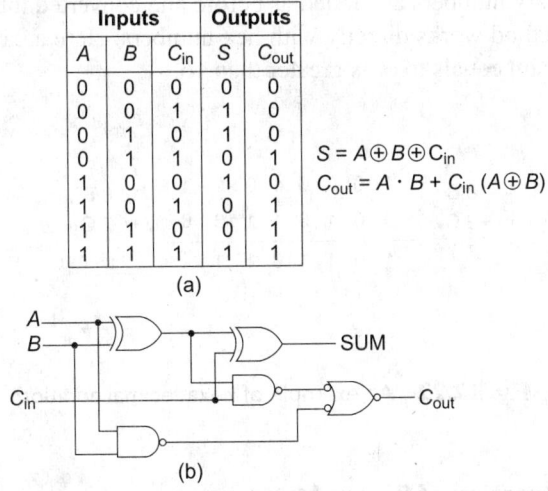

Inputs		Outputs	
A	B	S	C
0	0	0	0
0	1	1	0
1	0	1	0
1	1	0	1

(a)

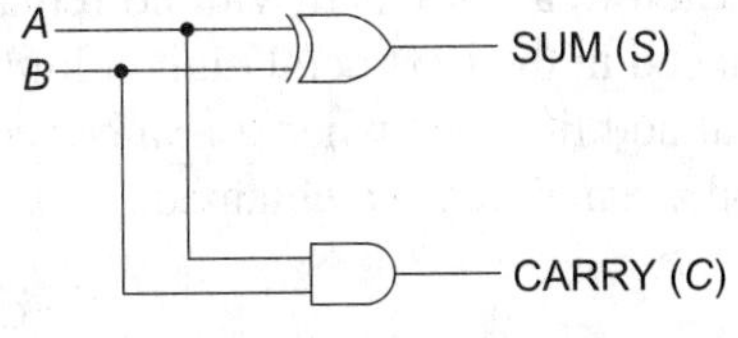

(b)

Fig. 12.19 Half adder: (a) truth table, (b) circuit

C_{out} are 1's. It can easily be verified that the logic expressions for S and C_{out} are

$$S = A \oplus B \oplus C_{in}$$

$$C_{out} = A \cdot B + C_{in}(A \oplus B)$$

Figure 12.20(b) shows the schematic diagram of a full adder circuit built with AND, OR, and XOR gates.

Inputs			Outputs	
A	B	C_{in}	S	C_{out}
0	0	0	0	0
0	0	1	1	0
0	1	0	1	0
0	1	1	0	1
1	0	0	1	0
1	0	1	0	1
1	1	0	0	1
1	1	1	1	1

$S = A \oplus B \oplus C_{in}$

$C_{out} = A \cdot B + C_{in}\ (A \oplus B)$

(a)

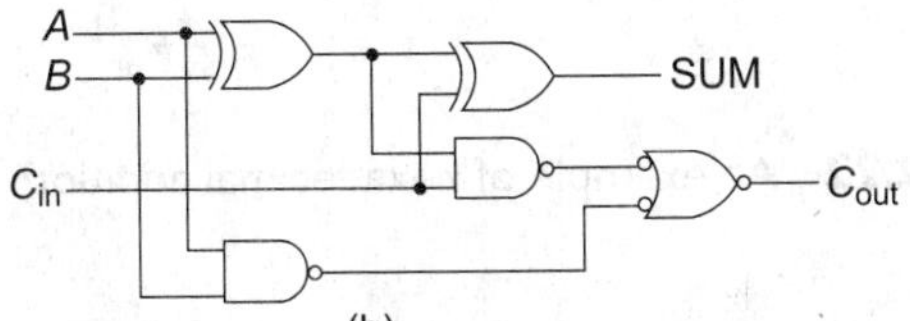

(b)

Fig. 12.20 Full adder: (a) truth table, (b) circuit

12.3.2 Octal and Hexadecimal Addition

Octal and hexadecimal addition are used as a shorthand to represent long binary numbers such as ROM addresses. Figure 12.21 shows two ways of adding the octal numbers 46 and 35. The first method is to convert both numbers to their binary equivalents. The binary numbers are then added using the rules for binary addition; the resulting sum is the binary sum of the two octal numbers. The second method works directly with the octal form: 6 added to 5 gives 11; this number is in decimal code. If the decimal number (11_{10}) is viewed from the octal code, there will be a carry of 1 to the next digit and a remainder of 3. This 3 is written down and the carry is added to the next digit column. Then 4 plus 3 plus carry gives 8, which is equal to a carry of 1 with no remainder. The 0 is written down and the carry is placed in the next octal digit column. The procedure is the same as used for decimal addition, but when a number equal to or greater than 8 is produced, a carry and a remainder are obtained.

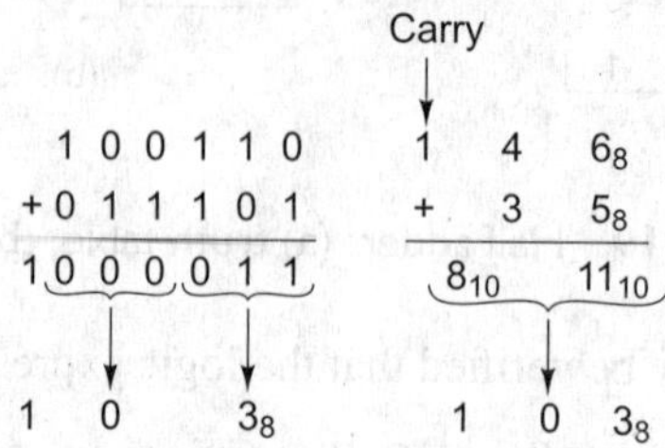

Fig. 12.21 An example of octal addition

Figure 12.22 shows the addition of two hexadecimal numbers 8B and 2C. It is seen that a hexadecimal digit, also called a hex digit, represents four binary digits. The binary numbers are added as before and converted into hexadecimal. The second method works directly with hex numbers. Here a carry is produced whenever the sum equals to or is greater than 16.

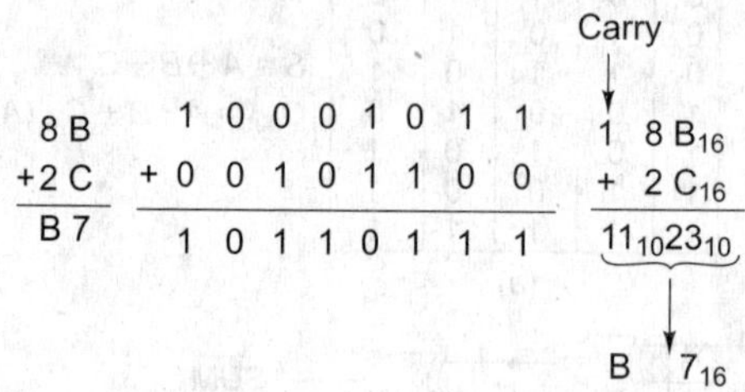

Fig. 12.22 An example of hexadecimal addition

12.3.3 Subtraction of Binary Numbers

Figure 12.23(a) shows the truth table for subtraction of B from A. When B is larger than A, a borrow is made from the previous digit as shown in Fig. 12.23(b). Figure 12.23(c) shows an example of binary subtraction using these rules. The logical expression for the difference output for subtraction and that for borrow out are given as

$$D = A \oplus B \oplus B_{\text{in}} \tag{12.1}$$

and

$$B_{\text{out}} = \bar{A} \cdot D + (A \oplus B) \cdot B_{\text{in}} \tag{12.2}$$

Since these expressions are very similar to those for the full adder which is considered for addition, it is quite easy to construct a circuit that can add as well as subtract two binary digits. Figure 12.24 shows such a circuit which can be shown to add the two inputs with carry-in if the control input is high and subtract *B* from *A* with borrow if the control input is low.

12.3.4 2's Complement Subtraction

The 2's complement of a binary number is formed by inverting each bit in the number and adding 1 to the least significant digit. Subtraction of one binary number from another can be effected by adding the minuend and the 2's complement of the subtrahend as shown in Fig. 12.23(d). A carry (that is, a 1 in the box) from the most significant bit indicates that the result is positive. If a 0 appears in the box, it indicates that the result is negative and is obtained in the 2's complement form. To get back the number which is the result of the subtraction, each bit of this 2's complement is inverted and 1 added to the result. This method is easy to implement on a microprocessor because it needs only simple operations, namely, inversion, increment, and addition.

Inputs		Outputs	
A	*B*	*D*	(*B*)
0	0	0	0
0	1	1	1
1	0	1	0
1	1	0	0

Difference = $A \oplus B$

Borrow = $\overline{A} \cdot B$

(a)

Inputs			Outputs	
A	*B*	B_{in}	*D*	B_{out}
0	0	0	0	0
0	0	1	1	1
0	1	0	1	1
0	1	1	0	1
1	0	0	1	0
1	0	1	0	0
1	1	0	0	0
1	1	1	1	1

Difference = $A \oplus B \oplus B_{in}$

Borrow = $\overline{A} \cdot B + \overline{(A \oplus B)} \cdot B_{in}$

(b)

$$\begin{array}{r} 12 \\ -2 \\ \hline 10 \end{array} \qquad \begin{array}{r} 1\,1\,0\,0 \\ -0\,0\,1\,0 \\ \hline 1\,0\,1\,0 \end{array}$$

(c)

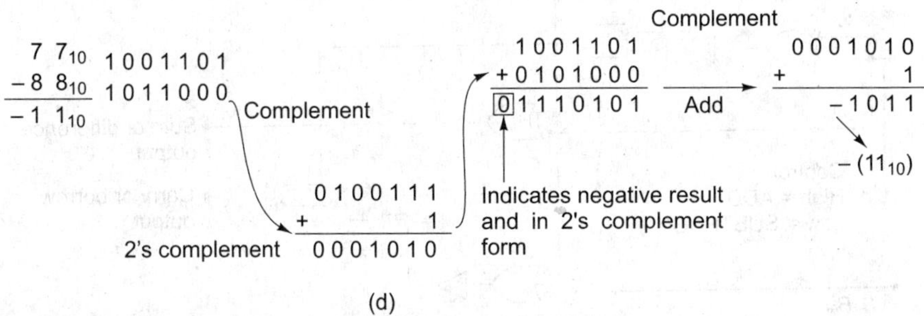

(d)

Fig. 12.23 Binary subtraction: (a) half subtractor truth table, (b) full subtractor truth table, (c) example of subtraction, (d) example of 2's complement subtraction

To detect whether a number stored in an 8-bit RAM is positive or negative, the usual method is to use the most significant bit as a sign bit. Positive numbers are represented in their true binary form with a 0 sign bit. Negative numbers are represented in their 2's complement form with the sign bit set, that is, equal to 1.

Examples of numbers expressed using the sign bit, and how they can be stored in an 8-bit RAM, are given below:

	sign bit ↓	
+6	0 0000110	
+45	0 0101101	
+104	0 1101000	
−13	1 1110011	Signs and two's complement of magnitude
−55	1 1001001	
−116	1 0001100	
−45	1 1010011	

The method of obtaining the 2's complement sign-and-magnitude form is identical to that for obtaining a 2's complement binary number given above.

12.3.5 Arithmetic and Logic Unit

The arithmetic and logic unit (ALU) is an important part of the microprocessor which performs addition, subtraction, as well as the logic functions AND, OR, and XOR.

Figure 12.24 shows an ALU circuit, constructed with simple gates, that can add or subtract two single-bit binary numbers with carry or borrow. The operation that is executed is determined by a high or a low on a select input. The TTL integrated circuit chip 74181 is an expanded version of this chip and can perform many logic and arithmetic operations on two 4-bit binary numbers. Evidently the ALU that is used in 8085A will, however, be an 8-bit chip that can do these operations.

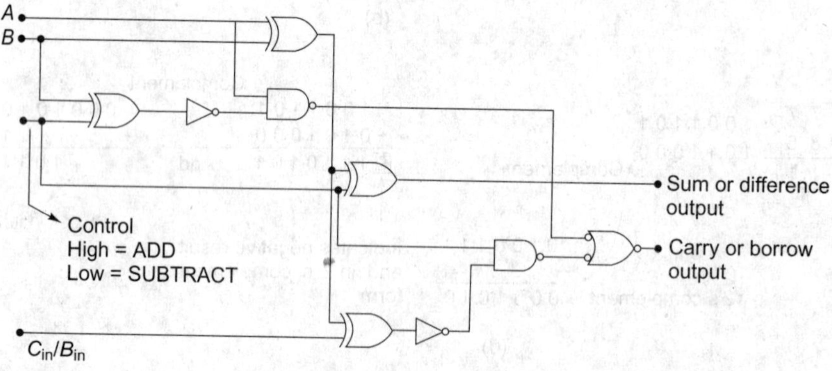

Fig. 12.24 Circuit to add or subtract two 1-bit binary numbers on command

12.4 Microprocessor Structure and Programming

The interconnections of the digital chips introduced in the earlier sections to make a complete microprocessor system are now dealt with. The ALU introduced in the previous section is a part of the microprocessor. A RAM is employed to store the intermediate results and data, whereas a ROM is used to store a set of binary coded instructions. If the control circuitry is also incorporated, then the ALU can be made to perform a sequence of operations as per the program. The sequence of binary coded instructions, which is just a sequence of specific operations to be performed, is called a machine language program. The following is a very brief explanation of the structure and operation of the microprocessor.

12.4.1 Salient Features of a Microcomputer/Microprocessor

Figure 12.25 shows the three important modules that constitute a digital microcomputer, namely, the central processing unit (microprocessor), memory, and the I/O ports. The memory section consists of ROM, RAM, and magnetic discs or tapes and is used to store not only programs or instruction sequences (ROM) but also intermediate results (RAM). Another use of the memory is to store the data that are processed by the computer and also the output data that result after processing. The central processing unit (CPU) contains the ALU, the control circuitry, some registers, and an address/program counter. To execute a program, the address of the location of the relevant instruction is sent out by the CPU together with the necessary memory control signal. Thus the instruction code goes from the memory to the CPU, gets decoded, and then executed. After each operation the program counter gets incremented to show the address of the location of the next instruction in memory that is to be executed.

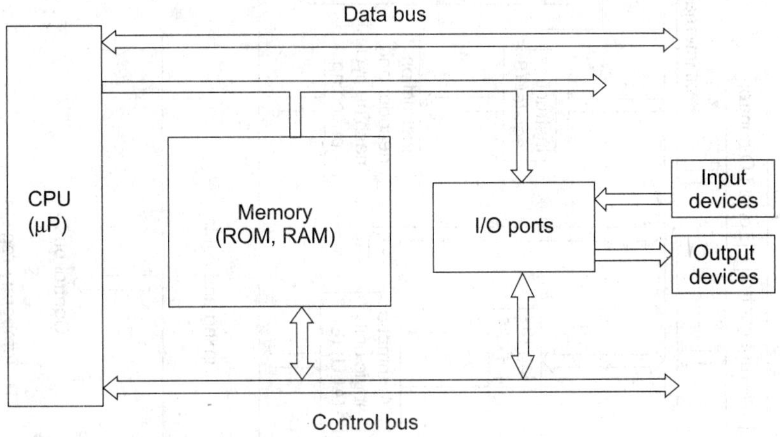

Fig. 12.25 A microcomputer system

The input–output section is constituted of the input–output (I/O) ports, which form the interface of the computer with the outside world. An input port, which is an IC chip, allows the data entered from a keyboard, digitized from a pressure sensor, or in some other form, to be read into the computer under CPU control. An output port is also a chip that is used to send data from the CPU to some output device such as a teletypewriter, video screen display, or motor-speed control unit. Both these ports are just parallel *D* flip-flops which pass data upon receiving a signal (called a strobe) from the CPU.

12.4.2 Architecture of the 8085A Microprocessor

A microprocessor is classified as an *n*-bit processor which implies that its ALU has a width of *n* bits. The 8085A CPU is an 8-bit microprocessor; Fig. 12.26 shows its functional block diagram. The salient features of the 8085A microprocessor are

- the arithmetic and logic group of circuits,
- registers, program counter, and address latch, and
- instruction register, instruction decoder, and control circuits.

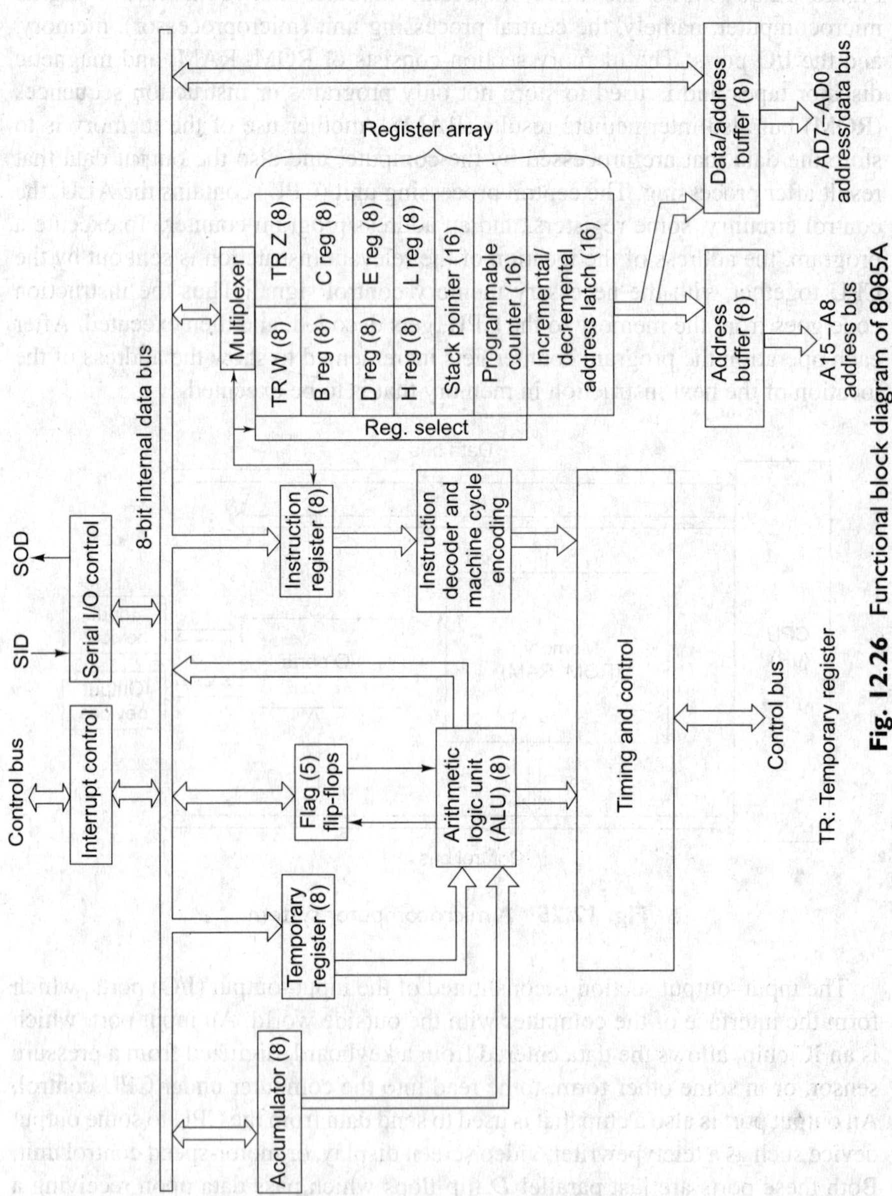

TR: Temporary register

Fig. 12.26 Functional block diagram of 8085A

12.4.2.1 Arithmetic–logic group of circuits This group includes the accumulator, ALU, temporary register, decimal adjust circuitry, and flag flip-flops.

The accumulator is a special 8-bit register that holds one of the quantities to be operated upon by the ALU and also receives the results of an operation performed by the ALU. For example, an instruction, when executed, might add the contents of register *B* to the contents of the accumulator and place the result in the accumulator itself.

The ALU can increment a binary quantity, decrement a binary quantity, and perform logic operations such as AND, OR, XOR, addition, and subtraction. The temporary register receives data from the internal data bus and holds it for manipulation by the ALU. The decimal adjust circuitry is used when binary coded decimal (BCD) addition or subtraction is performed, so as to maintain the correct BCD format in the result.

There are five flip-flops called *flags*, which by means of set/reset conditions, indicate conditions which result from computations. They are the carry flag bit (C), parity flag bit (P), auxiliary carry flag bit (AC), zero flag bit (Z), and sign flag bit (S). The carry flag is set to 1 when there is a carry. The parity flag is set to 1 if the binary number in the accumulator has an even number of bits. The zero flag is set if an instruction execution result in all 0's in the accumulator. If this bit is 1 then the number in the accumulator is negative. The auxiliary carry bit will be set to 1 if there is a carry from bit 3 of the accumulator. When BCD addition or subtraction is being performed this flag indicates that a decimal adjust must be done.

12.4.2.2 General-purpose registers, program counter, and address latch The 8085A contains six 8-bit general-purpose registers which are arranged in pairs. These are B and C, D and E, and H and L. Any of these registers can be used alone or two of them as a pair. For example, register B alone can be used to store a byte (8 bits) of data, or together with register C it can be used to store 16-bit data or a 16-bit address. The temporary register pair W and Z is not available to the user.

The stack pointer (SP), program counter (PC), and address latch (AL) are special-purpose 16-bit registers. The SP is used to store an address and is useful when the program includes one or more subroutines. The PC stores the memory address currently being read from or written into by the CPU. After each instruction it is incremented automatically so that it points to the location of the next instruction or data in memory. The AL selects an address to be sent out from the PC, SP, or one of the 16-bit general-purpose register pairs. Also, it latches this address on to the address lines for the required length of time. The 16-bit address bus from the 8085A allows addressing up to 2^{16} or 65,536 memory locations. An incrementer/decrementer allows the contents of any of the 16-bit registers to be incremented/decremented.

12.4.2.3 Instruction register, instruction decoder, and control circuits All programs gets executed sequentially. An instruction goes from the memory to the internal data bus; from this bus it goes to the instruction register, which holds the instruction for use by the instruction decoder. The instruction decoder interprets the instruction and issues appropriate control signals to carry it out. For example, if an 'ADD C' instruction comes from the memory into the decoder, the decoder will recognize that a second fetch operation is needed so as to fetch the number from register C to be added to the number in the accumulator. The first fetch operation, of course, is the one needed to fetch the instruction itself from the memory. The instruction decoder accordingly directs the control circuitry to send out another memory read signal, and then a signal to shift the data in register C into the temporary register, so that it can be added to the number in the accumulator. The actual addition takes place in the ALU, and when it is completed, the control circuits steer the result from the ALU into the accumulator. The control signals then increment the program counter to the next memory address and send out another memory read signal to get (that is, fetch) the next instruction from memory.

The 8085A requires a + 5 V supply and also an external crystal to be connected to the X_1 and X_2 inputs, the latter being used for generating clock signals.

12.4.3 Assembly Language for 8085A

A microcomputer program is stored in a binary coded form in successive memory locations. As it is in a form that can be directly used by the machine, it is called *machine language*. However, it is unintelligible to the human programmer because of the maze of 1's and 0's contained in it. Using a hexadecimal version of these binary codes is convenient, and about 256 individual instruction codes are needed to cover the entire range.

An easily understandable language called the *assembly language* has been devised for the human programmer. Microprocessor manufacturers devise symbolic codes called *mnemonics* for the instructions. A mnemonic is a combination of letters suggesting the operation to be performed by the instruction it represents. Each one is a two- to four-letter word which is similar to a truncated English word and has to be translated into machine language for being run on the microcomputer. For instance, the instruction 'move the contents of the accumulator to a memory location' is written in mnemonics as 'MOV M,A', the memory location M is defined in a previous instruction.

12.4.3.1 The instruction set of 8085A Appendix E shows the mnemonics of all assembly language instructions and the hexadecimal codes represented by them. An instruction usually consists of the operation code or *opcode* and the operand; these are separated by a single space. The opcode is the mnemonic for a particular operation and the operand is the register or data to be acted upon. The various instructions given in Appendix E can be broadly grouped into the following types:

(a) Data transfer group: move, move immediate, and load immediate

(b) Arithmetic and logic group: add, subtract, increment, decrement, logical, arithmetic, and logical immediate

(c) Branch control group: jump and conditional jump

(d) I/O machine control group: input–output and HLT

12.4.3.2 Restart instructions The 8085A has eight restart instructions, namely, RST0, RST1, RST2, RST3, RST4, RST5, RST6, and RST7. Each of them is a 1-byte CALL instruction and provides an efficient method for calling subroutines which are used repeatedly in the main program. Table 12.1 gives these RST instructions, their binary as well as hex codes, and the locations to which the microprocessor jumps due to their insertion in the program.

Table 12.1 Restart instructions

Mnemonic	Binary code	Hex code	Call location in hex code
RST0	11000111	C7	0000
RST1	11001111	CF	0008
RST2	11010111	D7	0010
RST3	11011111	DF	0018
RST4	11100111	E7	0020
RST5	11101111	EF	0028
RST6	11110111	F7	0030
RST7	11111111	FF	0038

Incorporating a typical RST instruction, say, RST0, in the main program has the following effects. Because it is equivalent to a CALL instruction, the microprocessor first pushes the contents of the program counter onto the stack. Then the program branches to the address 0000H as shown in Fig. 12.27. The subroutine written at locations 0000H to 0007H will then be executed, with RET causing a return to the instruction in the main program which is next to the RST0 instruction.

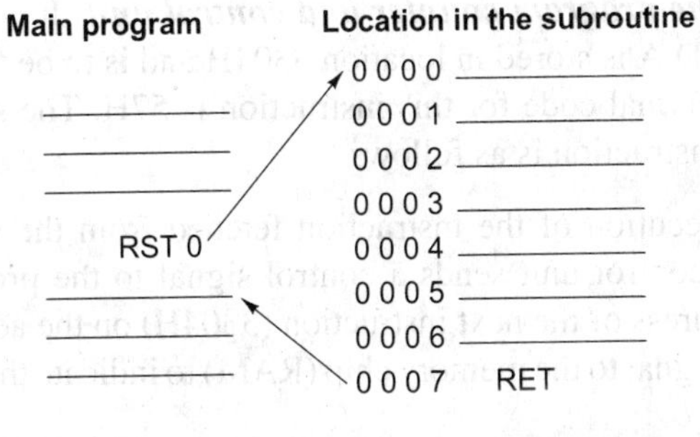

Fig. 12.27 Example showing the use of the RST0 instruction

12.4.3.3 Various steps of instruction execution The following examples illustrate how data transfer actually takes place in a microprocessor.

(a) Decoding and executing an instruction Let us assume that the accumulator contains the data byte 84H, and the instruction MOV D,A is fetched from the memory. The hex code for this instruction is 57H. The sequence of steps for decoding and executing this instruction are as follows.

(i) The contents of the data bus (57H) are placed in the instruction register and decoded.

(ii) The contents of the accumulator (84H) are transferred to the temporary register in the ALU.

(iii) The contents of the temporary register are transferred to register D.

Figure 12.28 illustrates these operations (Gaonkar 1986).

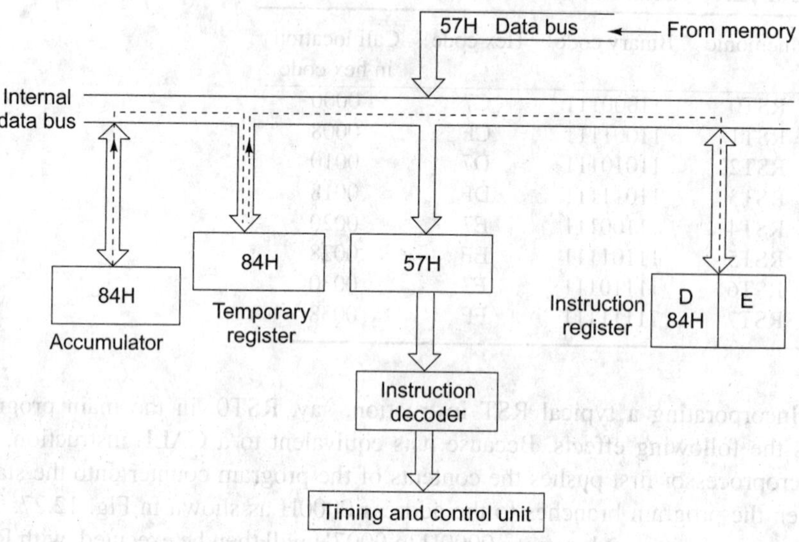

Fig. 12.28 Decoding and executing of an instruction

(b) The role of the program counter and control unit It is assumed that the instruction MOV D,A is stored in location 3304H and is to be fetched. As stated above, the hexadecimal code for this instruction is 57H. The sequence of steps for fetching this instruction is as follows.

(i) After the execution of the instruction fetched from the memory location 3303H, the control unit sends a control signal to the program counter to place the address of the next instruction (3304H) on the address bus. It also sends a $\overline{RD}$ signal to the memory chip (RAM) to indicate that the data should be read.

(ii) The program counter places the address (3304H) on the address bus, which is decoded by the memory decoder.

(iii) The memory chip places the data 57H present at location 3304H on the data bus.

(iv) This data then goes from the external data bus to the internal data bus and thereon to the instruction decoder.

(v) The instruction is decoded and executed as in (a) above.

Steps (i)–(v) are illustrated in Fig. 12.29.

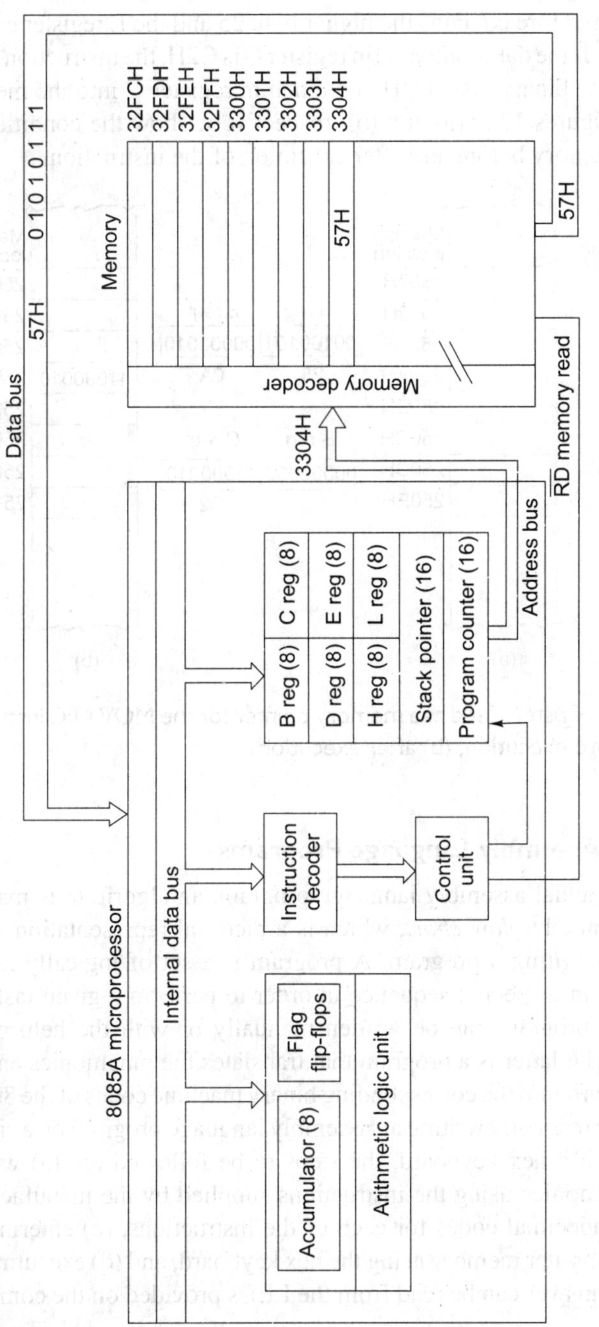

Fig. 12.29 Illustration of the functions of a program counter and an instruction decoder

12.4.3.4 Illustration of the use of the memory pointer The instruction that involves a memory move is written as MOV M,C and is assumed to be in the memory location 250AH. In the 8085A microprocessor, the HL register is used as a memory pointer (M).

The H register therefore contains the higher byte 25 and the L register contain the lower byte 0AH. If the data contained in register C is C2H, the instruction MOV M,C will transfer the binary code C2H contained in register C into the memory location 250AH. Figures 12.30(a) and (b), respectively, show the conditions of the registers and memory before and after execution of the instruction.

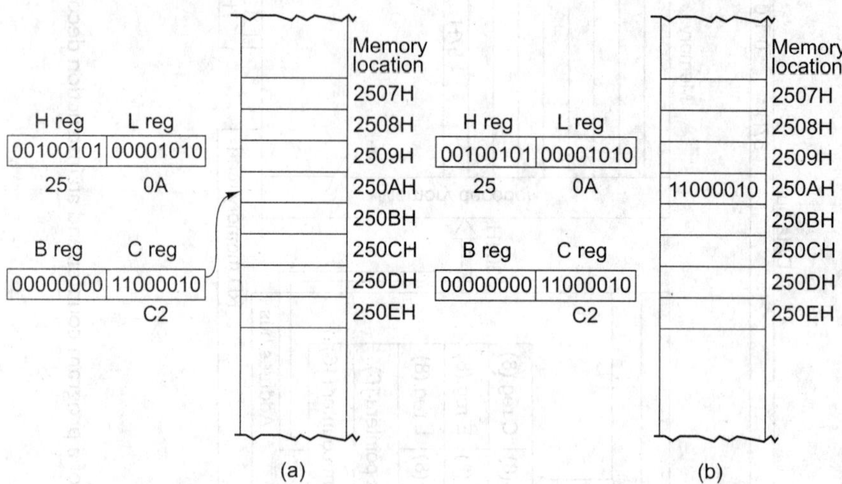

Fig. 12.30 H and L registers used as a memory pointer for the MOV M,C instruction: (a) before execution, (b) after execution

12.4.4 Writing Assembly Language Programs

Before writing the actual assembly language program, an algorithm is made in a schematic form called a *flow chart*, which is a pictorial representation of the steps necessary for writing a program. A program is a set of logically related instructions written in a specific sequence in order to perform a given task. An assembly language program can be written manually or with the help of the *8085A assembler*. The latter is a program that translates the mnemonics entered by the ASCII keyboard into the corresponding binary machine codes of the 8085A microprocessor. For manually writing an assembly language program on a single-board computer with a hex keyboard, the steps to be followed are (a) writing instructions in mnemonics using the instructions supplied by the manufacturer, (b) finding the hexadecimal codes for each of the instructions, (c) entering the program into the computer memory using the hex keyboard, and (d) executing the program. The final answer can be read from the LEDs provided on the computer panel.

Manual writing and execution is done only for small programs. This is because the process of picking up the corresponding machine codes and then entering them is tedious and subject to errors. Figure 12.31 shows a flow chart made to take in 25 samples of data from an input port and store them in successive locations of memory.

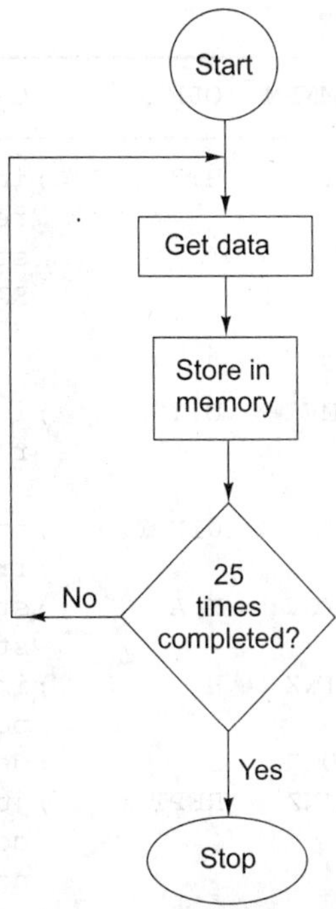

Fig. 12.31 Flow chart for storing 25 samples of data from the input port

The format for writing an assembly language program is given below:

PROGRAM TITLE					
ADDRESS	INSTRUCTION	LABEL	MNEMONIC	OPERAND	COMMENTS
1	2	3	4	5	6

The first column contains the memory address in which the instruction is stored. The hex code of the instruction is written in the second column. Whenever branching or looping has to be done, some instructions need to be labelled. This label represents the address of the instruction that is labelled. Whenever it is needed to jump to this address a branching instruction is written with the name of the label. The next two fields contain the mnemonic and the operand, respectively. The comments field, which is usually started with a semicolon, is an important part of the assembly language program. Comments explain the operation in the instruction and do not form a part of the instruction. These are written for the benefit of the programmer and/or the user. Three typical assembly language programs are given below to illustrate the above-mentioned aspects. Program 12.1 shows an assembly language program for the flow chart given in Fig. 12.31.

Programming Example 12.1 To take in 25 data samples from port 01 and store them in consecutive locations.

Assembly language program

ADDR	INSTR	LABEL	MNEM	OPERAND	COMMENTS
2800	21		LXI	H,3260H	;initialize the H and L registers to the first storage address in the RAM
2801	60				
2802	32				
2803	16		MVI	D,19H	;initialize counter register to 25
2804	19				
2805	DB	REPT	IN	01H	;take in data
2806	01				;from port 1
2807	77		MOV	M,A	;store the data in RAM storage
2808	23		INX	H	;increment H and L to point to next address
2809	15		DCR	D	;decrement counter by 1
280A	C2		JNZ	REPT	;jump to REPT if counter register is not zero
280B	05				
280C	28				
280D	76		HLT		;stop

Programming Example 12.2 To add the elements of an array of five numbers stored in locations 2400_H to 2404_H in the memory. Data: 2400 38H, 2401 39H, 2402 3AH, 2403 3BH, 2404 3CH.

Assembly language program

ADDR	INSTR	LABEL	MNEM	OPERAND	COMMENTS
2000	21		LXI	H, 2400	;initialize the H and L registers to first storage address in the RAM
2001	00				
2002	24				
2003	0E		MVI	C,05H	;initialize counter
2004	05				;register to 5
2005	3E		MVI	A,00H	;clear the
2006	00				;accumulator
2007	86	RAMA	ADD	M	;add the contents of memory to the accumulator

(contd)

(contd)

ADDR	INSTR	LABEL	MNEM	OPERAND	COMMENTS
2008	23		INX	H	;increment H and L to point to the next location
2009	0D		DCR	C	;decrement counter by 1
200A	CA		JZ	SITA	;jump to location 2010_H if counter
200B	10				;register is
200C	20				;zero
200D	C3		JMP	RAMA	;otherwise jump
200E	07				;to location 2007_H
200F	20				
2010	77	SITA	MOV	M,A	;store final result in the next memory location
2011	76		HLT		;stop

Programming Example 12.3 To arrange a given array of 10 numbers in descending order and store the resulting array in the same memory locations. Memory location 2600H contains the array length (10) and the array starts at location 2601H. Data: 2600 0AH, 2601 15H, 2602 20H, 2603 25H, 2604 2AH, 2605 31H, 2606 38H, 2607 3CH, 2608 41H, 2609 45H, 260A 4FH.

Assembly language program

ADDR	INSTR	LABEL	MNEM	OPERAND	COMMENTS
2000	21		LXI	H,2600	;initialize the HL register pair as a pointer to the first storage address in the RAM (array length)
2001	00				
2002	26				
2003	4E		MOV	C,M	;transfer the array length to register C, which is the counter
2004	23		INX	H	;increment the memory pointer
2005	11		LXI	D,2601	;put the memory location number 2601 in the DE pair
2006	01				
2007	26				

(contd)

(contd)

ADDR	INSTR	LABEL	MNEM	OPERAND	COMMENTS
2008	1A		LDAX	D	; load the accumulator with the contents of the memory location in the DE pair
2009	OD	BELL	DCR	C	; decrement the counter
200A	CA		JZ	TREE	; check for zero, if zero go to location TREE
200B	1D				
200C	20				
200D	23	JAR	INX	H	; increment the HL register pair by 1
200E	1A		LDAX	D	; load the accumulator with the contents of memory
200F	46		MOV	B,M	; transfer memory contents to register B
2010	B8		CMP	B	; compare contents of B with those of accumulator
2011	D2		JNC	BELL	; if there is no carry go to location BELL
2012	09				
2013	20				
2014	EB		XCHG		; exchange the contents of H with D and those of L with E
2015	7E		MOV	A,M	; transfer memory contents to accumulator
2016	70		MOV	M,B	; transfer contents of register B to memory
2017	12		STAX	D	; store contents of register A in the memory location contained in the DE pair.
2018	EB		XCHG		; exchange contents of H with D and those of L with E
2019	OD		DCR	C	; decrement counter
201A	C2		JNZ	JAR	; check for zero, if zero go to location JAR.
201B	OD				
201C	20				

(contd)

(contd)

ADDR	INSTR	LABEL	MNEM	OPERAND	COMMENTS
201D	13	TREE	INX	D	; increment DE pair by 1
201E	21		LXI	H,2600	; load HL pair with the number 2600
201F	00				
2020	26				
2021	35		DCR	M	; decrement the HL register
2022	7E		MOV	A,M	; transfer memory contents to accumulator
2023	FE		CPI	01	; compare accumulator contents with 1
2024	01				
2025	CA		JZ	FINAL	; if result is zero go to location 202C
2026	2C				
2027	20				
2028	4E		MOV	C,M	; transfer memory contents to register C
2029	62		MOV	H,D	; transfer register D contents to register H
202A	6B		MOV	L,E	; transfer register E contents to register L
202B	C3		JMP	BELL	; go to location 2009
202C	76		FINAL	HLT;	; stop

12.4.5 Stack and Subroutines

The stack is a set of memory locations in the RAM of the microcomputer system which is defined by the programmer at the beginning of the main program. Its purpose is to temporarily store either the contents of some registers or a particular part of the program, called the subroutine, which is used repeatedly in the main program. The four instructions which are connected with the stack are PUSH, POP, CALL, and RET. For the purpose of stack operations, a 16-bit word, called the *program status word* (PSW) is defined as

$$\text{PSW} = |\underline{\text{A}}|\underline{\text{F}}|$$

where A is the contents of the accumulator and F is the contents of the flag register.

The starting location of the stack is defined by means of the instruction 'LXI SP, $m_1m_2m_3m_4$', where $m_1m_2m_3m_4$ is the 16-bit address of a memory location and is called the *top of the stack*. If the contents of a register pair, say, the BC pair, are to be stored on the stack, a 'PUSH B' instruction has to be given. Then the contents of register B will be stored at a location which is one less than the top of the stack and those of C at the next lower location, which is two less than the top of the stack. For example, if the top of the stack is defined by the instruction LXI

SP, 3550H, and the contents of the BC pair are 1245H, a PUSH B instruction will result in the following operations.

(a) The stack pointer, which originally has the number 3550H, is reduced by 1 and the contents of register B (12H) are stored in location 354FH.

(b) The stack pointer is again reduced by 1 and the contents of register C (45H) are stored in location 354EH. Figure 12.32 illustrates these operations.

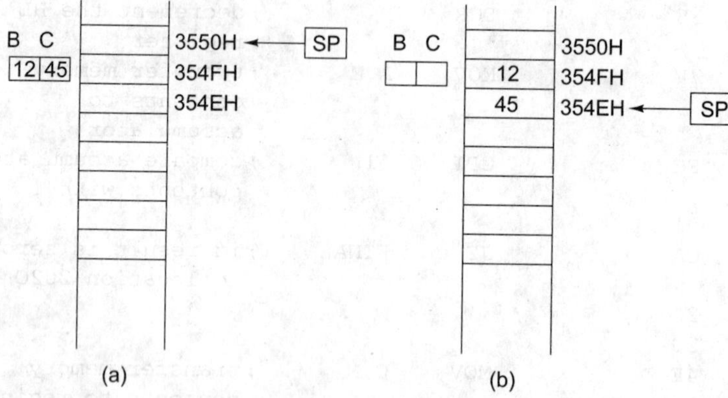

Fig. 12.32 Stack contents (a) before PUSH instruction, (b) after PUSH instruction

The rule to be followed is that the higher byte is stored at the higher location on the stack. If now a POP B instruction is given, the contents of the location 354EH (on the stack) will be restored back to register C and the stack pointer will be incremented by 1. Again the contents of location 354FH will be restored back to register B and the stack pointer will again be incremented by 1, thus pointing to the top of the stack (3550H). These operations are illustrated in Fig. 12.33.

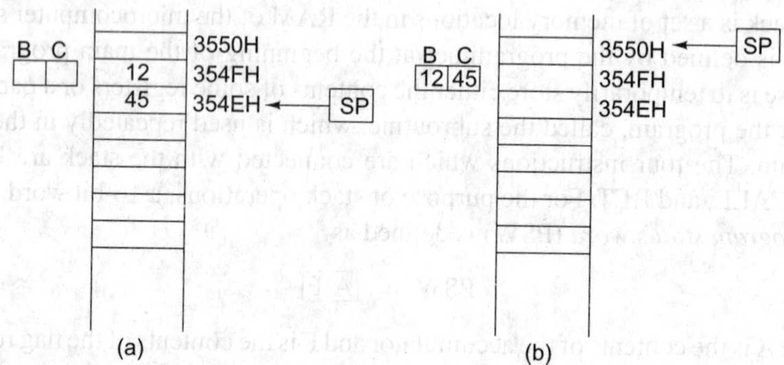

Fig. 12.33 Stack contents (a) before POP instruction, (b) after POP instruction

The CALL and RET instructions are two other instructions that are used with the stack. The 'CALL $n_1n_2n_3n_4$' instruction is a 3-byte instruction which has to be included in the main program whenever a subroutine is to be called. The subroutine is a separate program module which can be repeatedly called by the main program to do any computation, say, multiplication of two numbers. On encountering the relevant CALL instruction in the main program, the microprocessor saves the contents of the program counter, (the address of the next instruction in the main program), on the stack and decrements the stack pointer by 2 as detailed above. Then it jumps unconditionally to the memory location specified by the second and third bytes of the CALL instruction. A RET instruction, which is a 1-byte instruction, causes an unconditional return from the subroutine to the main program, and is the last instruction in the subroutine. It inserts the two bytes from the top of the stack into the program counter and increments the stack pointer by two.

12.5 Timing Waveforms

Both truth tables and timing diagrams are needed for the analysis of digital circuits. Timing diagrams are just plots of the input and output waveforms. For example, the timing diagram in Fig. 12.1(c) for a NOR gate shows the rise times as well as propagation delays of the gate.

In the 8085A microprocessor the execution of an instruction is timed by a sequence of machine cycles, each of which is divided into clock periods. One to five machine cycles are needed for this purpose; these are denoted as MC1, MC2, etc. The first machine cycle of an instruction is the opcode fetch (or instruction fetch) machine cycle, which consists of four clock cycles. The other machine cycles usually have a duration of three clock cycles. In some special cases, however, a machine cycle may have five or even six clock cycles, with subsequent ones having three clock cycles. A typical opcode fetch machine cycle followed by a memory read machine cycle is illustrated in Fig. 12.34.

Though the 8085A has no SYNC signal to signify the start of a new machine cycle, it uses the ALE (address latch enable) signal for this purpose. This signal is output true during the first clock period of every machine cycle, at which time the AD0–AD7 lines are outputting the lower order address. The symbol AD stands for address/data, which implies that the lower order address bus is used for data after the first clock period of any machine cycle. Releasing the lower order address bus for data purposes is called *demultiplexing* of the address bus. Accordingly it is saved from clock period 2 onwards by using a latch. The status signals S_0 and S_1 identify various operations depending on their binary condition (high/low). For example, the opcode fetch machine cycle is identified by $S_0 = 1$ and $S_1 = 1$, as shown in Fig. 12.34.

Timing diagrams can be drawn for the memory read, memory write, I/O read, I/O write, interrupt acknowledge, halt, hold, and reset machine cycles. They clearly indicate the precise timings of the address, data, and control signals. It should be emphasized that it is the control signals $IO/\overline{M}$, $\overline{RD}$, and $\overline{WR}$ which control the various memory and I/O operations.

If the actual timing diagram is faulty as could be seen with the help of a logic analyser, the exact location of the fault can be detected from this diagram. Thus any hardware fault can be located with the help of these diagrams and the logic analyser.

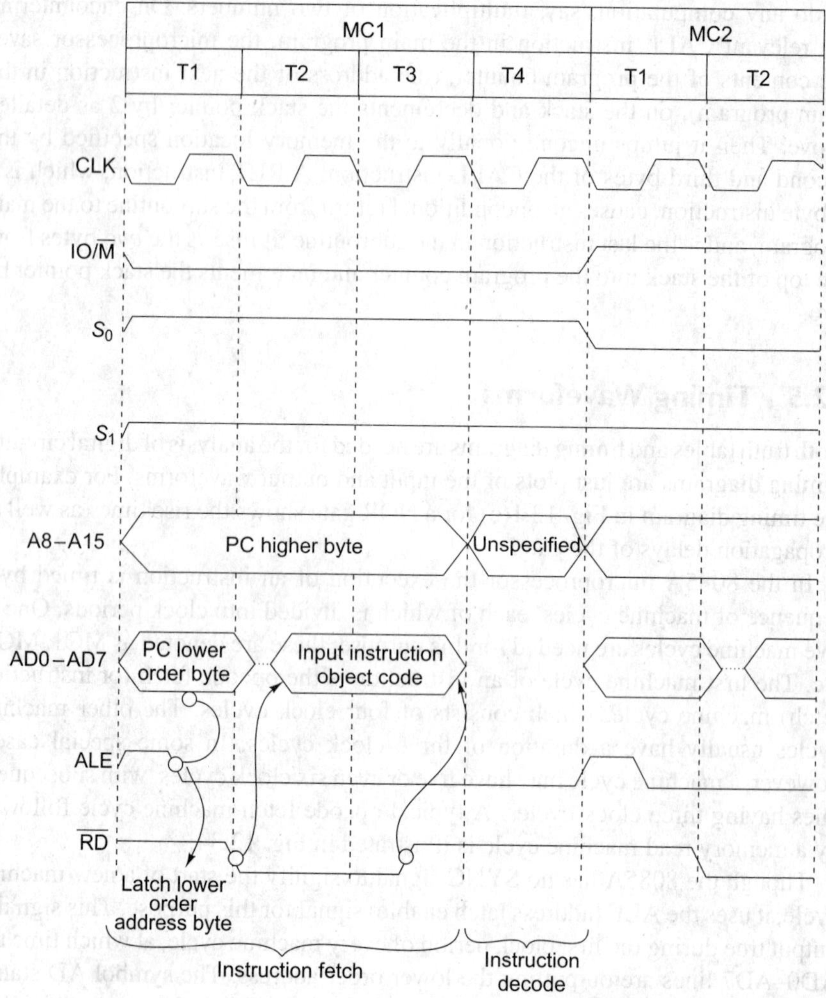

Fig. 12.34 A four-clock-period instruction fetch machine cycle

12.6 Interfacing the Microprocessor with Peripherals

The microprocessor has to (i) take data from input devices such as keyboards and A/D converters, (ii) read instructions from memory, (iii) process data according to instructions, and (iv) send the results to output devices such as D/A converters, LEDs, video monitors, and printers. These equipment, which are external to the microprocessor, are called peripherals or I/O's. The memory can also be considered as a special type of I/O. The technology of interfacing consists of designing hardware circuits and writing software programs so as to facilitate communication with these peripherals. The associated logic circuits are called I/O ports or interfacing devices.

12.6.1 Interfacing I/O devices

I/O devices can be connected as either peripherals or memory locations; these techniques are, respectively, called peripheral I/O or memory-mapped I/O. In the peripheral type, the IN or OUT instructions are used, which are 2-byte ones. In the memory-mapped type, the I/O is identified by a 16-bit address as associated instructions.

Data transfer is effected in this case by means of (a) STA, (b) LDA, (c) MOV M,R, and (d) MOV R,M instructions. Parallel data transfer to I/O devices can be done in a number of ways. As shown by the tree diagram of Fig. 12.35, data transfer can be broadly categorized into microprocessor-controlled and device-controlled types. In the former case the microprocessor sets up the conditions for data transfer to ensure that no data is lost in transfer from/to the peripherals which respond with much lesser speed. Memory-mapped I/O is resorted to when the peripheral is much faster than the microprocessor and the volume of data is quite large. In the direct memory access (DMA) type of data transfer, which comes under this category, a device called the DMA controller sends a HOLD signal to the microprocessor, and the microprocessor releases the data and address buses to this controller. Such a procedure facilitates transfer of data with high speed.

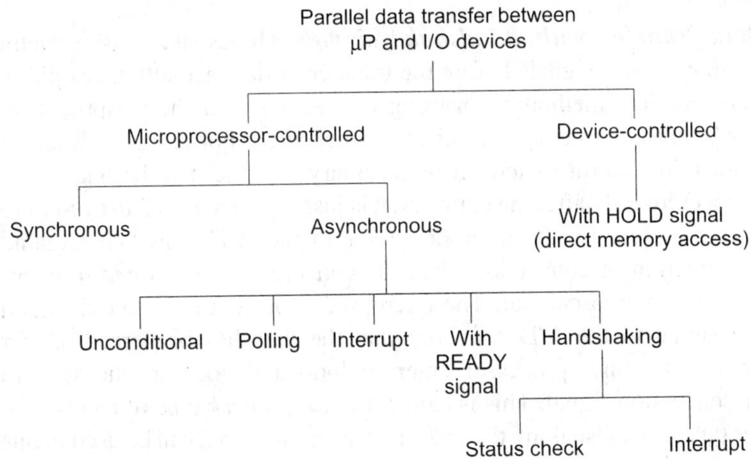

Fig. 12.35 Methods for parallel data transfer

The following are different microprocessor-controlled data transfers.

(a) Synchronous data transfer When the speed of the I/O devices matches that of the microprocessor, this type of data transfer can be adopted. The IN and OUT instructions are used for this purpose with the understanding that the I/O device is ready for transferring the data.

(b) Unconditional data transfer For this type of data transfer, it is assumed that the peripheral is always available and no matching of speeds is needed. An example is that of relays connected to the output port which can be switched on or off by outputting a 1 or a 0. It is evident that synchronous data transfer can also be considered to be of this category.

(c) Data transfer with polling or status check In this case it is assumed that data is not readily available and therefore the microprocessor goes on checking (or polling for) the availability of data by means of a loop. An example is that of inputting data from a keyboard, in which case the microprocessor goes on polling till the key is pressed.

(d) Data transfer with interrupt The 8085A microprocessor has five interrupt pins, namely, TRAP, RST 7.5, RST 6.5, RST 5.5, and INT, in the order of their priority. An interrupt signal can be given at any of these pins when data is available at the peripheral. The microprocessor then completes the execution of the current instruction under consideration, takes the data from the peripheral, and returns back to the next instruction of the program. Data transfer with polling [method (a)] is disadvantageous because the microprocessor wastes its time in executing a loop; on the other hand, in the interrupt case it can execute other programs till the interrupt occurs.

(e) Data transfer with ready signal This method is useful whenever the peripheral is much slower than the microprocessor. Here a READY signal goes low and causes the addition of clock cycles (T states) allowing more time for the peripheral to receive or send data. This technique is usually adopted with slow memory chips.

(f) Data transfer with handshake signals Handshaking is a method of exchanging control signals before the transfer of data actually takes place. This is a very efficient method of ensuring the readiness of the peripheral for data transfer because perfect synchronization is ensured. For instance, when an A/D converter is used in a system, it is necessary that the data be taken in by the microprocessor only after the conversion is just completed. A *start-of-conversion* signal is first sent by the microprocessor to the A/D converter. Again, soon after completion of conversion, the A/D converter sends an *end-of-conversion* signal to the microprocessor. The microprocessor then reads the data and then sends a signal to the A/D converter that the data has been read. During the conversion, the microprocessor enters a loop and goes an checking for the end-of-conversion signal. This is called the *status check* type of handshaking. In another type of handshaking the end-of-conversion signal can be used to interrupt the microprocessor; this is called the interrupt type of handshaking.

12.6.2 I/O Ports

An *input port* is a device in which data from the peripheral is unloaded and the microprocessor takes it from this port through the accumulator. Similarly, data to be sent out from the microprocessor is transferred through the accumulator to the *output port*; the peripheral then takes in the data from this port. Figure 12.36 shows a schematic diagram of an I/O port which acts as an interface between the input/output peripherals and the microprocessor. An I/O port may be programmable or non-programmable, the latter behaving as an input port if it is connected in the input mode.

Figure 12.37 shows an 8-key input port in which the address lines are decoded using an 8-input NAND gate. When the address lines A7–A0 are high (FFH), the output of the NAND gate G_A goes low and this is combined with the control

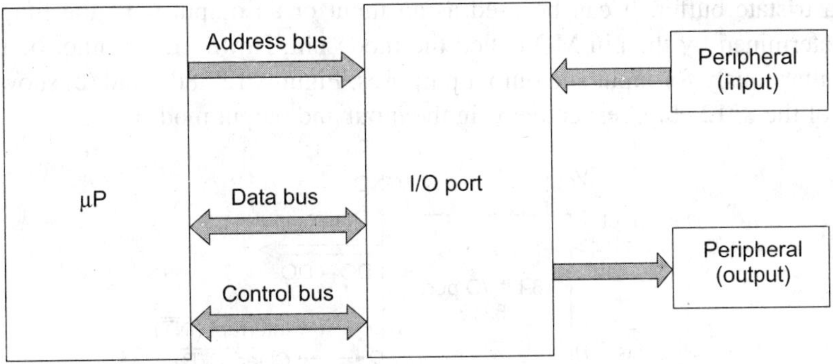

Fig. 12.36 I/O port for the 8085 chip

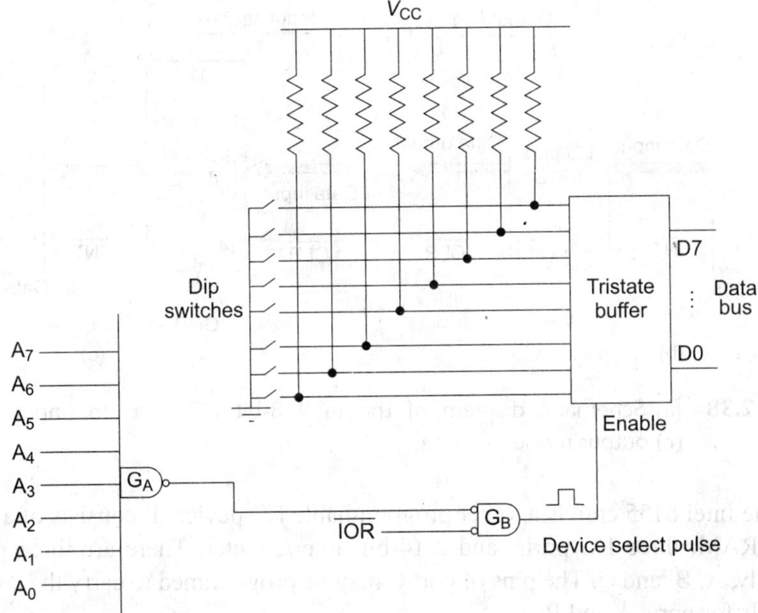

Fig. 12.37 Decode logic for a dip switch input port

signal $\overline{\text{IOR}}$ at the input of the gate G_B. When the microprocessor executes the instruction IN FFH, G_B generates a device select pulse which serves as the enable pulse of the tristate buffer. The data from the keys is then put on the data bus D7–D0 and then taken into the accumulator. Similar to this non-progammable input port, a port connected in the output mode works as a non-programmable output port. But a programmable I/O port has the flexibility of being programmed as an input or output port; the external electrical connections remain the same in both cases.

The Intel 8212 chip, which is an 8-bit non-programmable I/O port, is shown in Fig. 12.38(a). It is a 24-pin I/O device with eight *D* latches, each followed by a tristate buffer. It can be used as an input or an output port; the function is determined by the pin MD called the mode pin. However, it cannot be used simultaneously for input and output purposes. Figures 12.38(b) and (c) show the use of the 8212 chip, respectively, in the input and output modes.

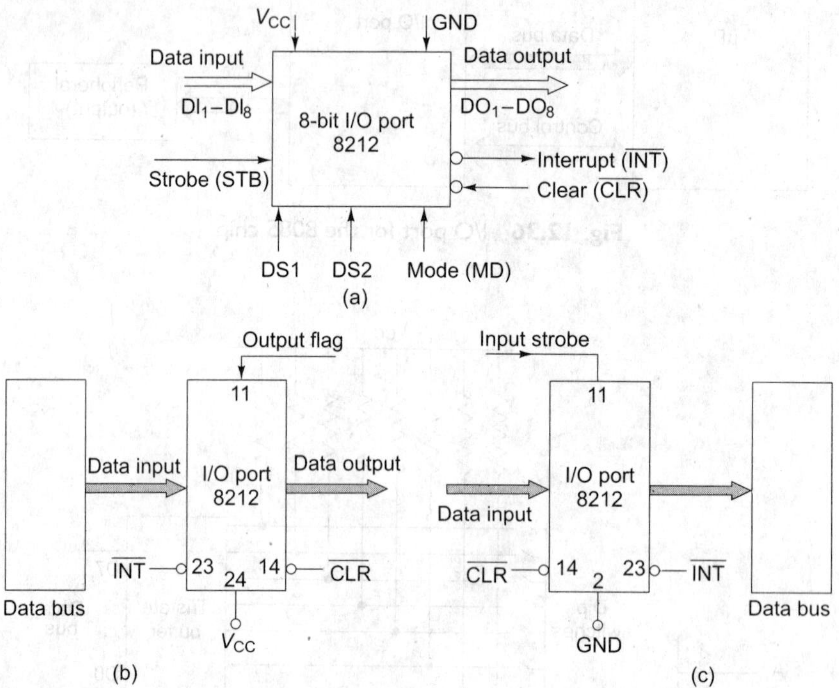

Fig. 12.38 (a) Schematic diagram of the Intel 8-bit I/O port; (b) input mode; (c) output mode

The Intel 8155 chip is another programmable I/O device. It consists of a 256-byte RAM, three I/O ports, and a 14-bit timer/counter. There are three ports, namely, A, B, and C. The pins of port C may be programmed to carry the control signals for ports A and B.

A third chip, namely, the Intel 8255A, functions as a programmable peripheral interface (PPI). Its main function is to act as an interface between the microprocessor and the peripherals. It has three 8-bit ports, namely, port A, port B, and port C. The last port is capable of being split up into port C (upper) and port C (lower). Each bit of port C can also be programmed individually. For programming purposes, these ports are numbered 00H, 01H, and 02H, respectively. This interface chip contains a register called the *control word register* whose 8 bits can be programmed so that the three ports can function either as input or output ports. The pin diagram and functional block diagram of the 8255A are, respectively, given in Figs 12.39(a) and (b). A program to interface a process with a microprocessor is given below.

Programming Example 12.4 It is desired to write an assembly language program for the 8085A to acquire data, namely, two analog process variables

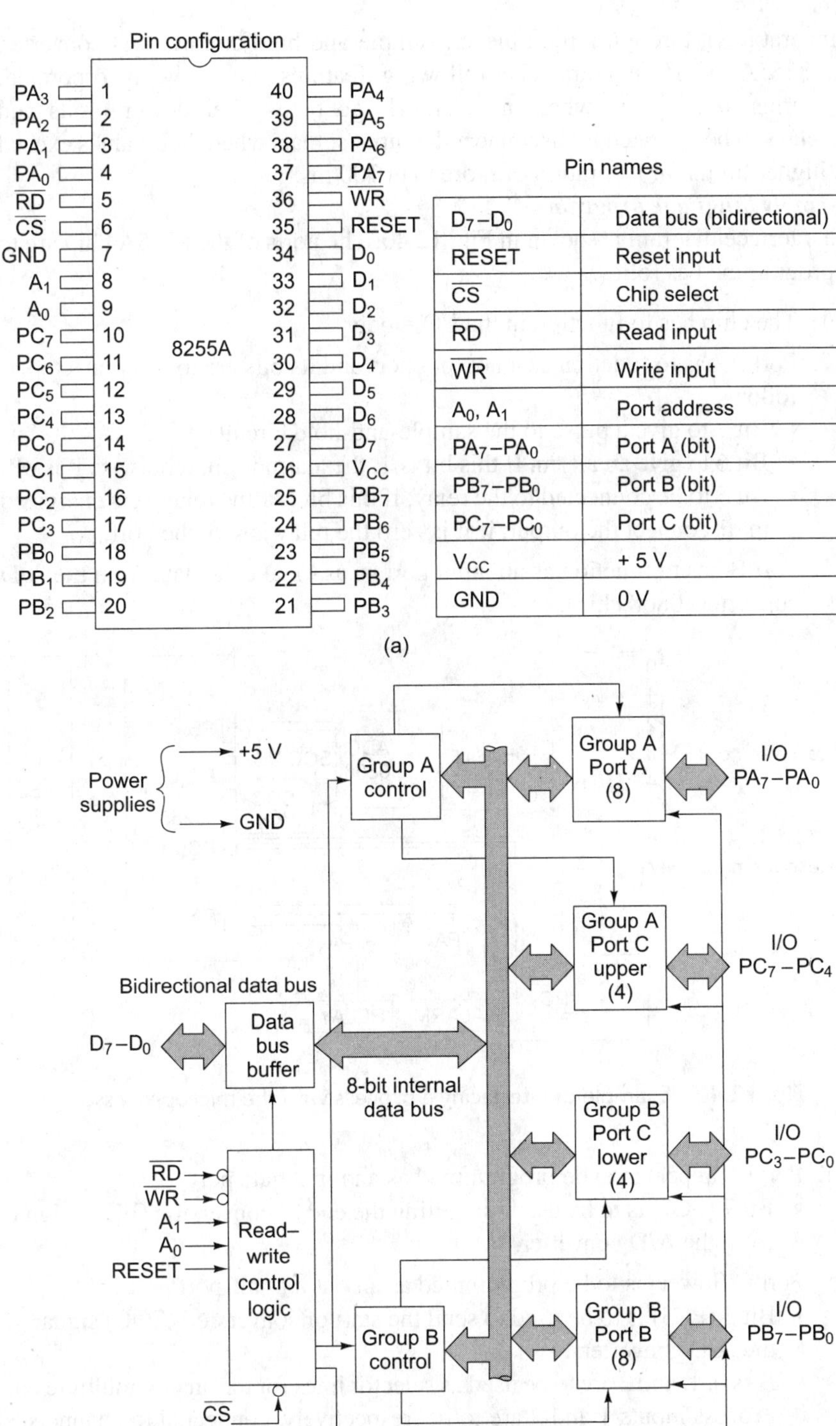

Pin names

D_7–D_0	Data bus (bidirectional)
RESET	Reset input
$\overline{CS}$	Chip select
$\overline{RD}$	Read input
$\overline{WR}$	Write input
A_0, A_1	Port address
PA_7–PA_0	Port A (bit)
PB_7–PB_0	Port B (bit)
PC_7–PC_0	Port C (bit)
V_{CC}	+ 5 V
GND	0 V

Fig. 12.39 8255A programmable peripheral interface chip: (a) pin diagram, (b) functional diagram

(temperatures), through a multiplexer, sample-and-hold circuit, A/D converter, and 8255A interface chip. The following features are to be incorporated: (a) alarms to be given when input signals go below their lower limits and (b) relays to be operated to disconnect the input signals when their values exceed the higher limits for a duration of more than 100 ms.

Assembly language program

The interface diagram is shown in Fig. 12.40. The ports of the 8255A chip are to be programmed as follows.

1. The chip has to function in the I/O mode.
2. Port A has to function as an output port and its bits are to be connected as follows.
   - Bit 7 to give a pulse to the sample-and-hold circuit.
   - Bit 6 to give an alarm. If this bit is 1, the alarm is on, otherwise it is off.
   - Bit 5 to be connected to the relay; if this bit is 1 the relay gets energized to disconnect the output; if it is zero the relay is switched off.
3. Port B has to function as an input port so as to take the data from the A/D converter (0808 chip).

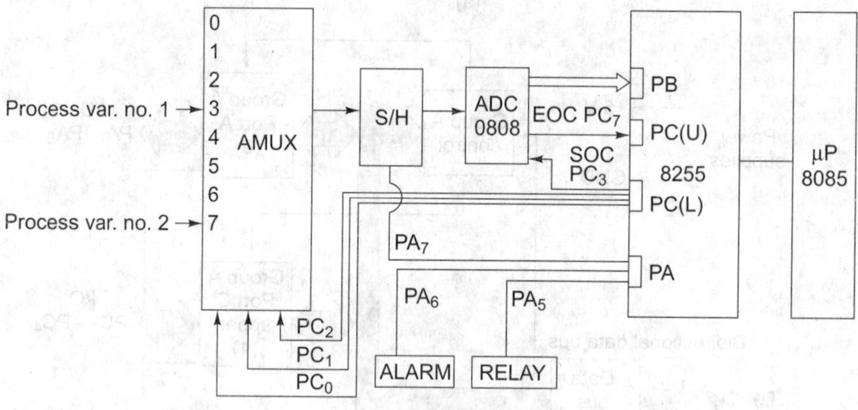

Fig. 12.40 Example of interfacing a process with the microprocessor

4. Port C (upper) is to be programmed as an input half-port.
   - Bit 7 (PC7) is to be used for getting the end of conversion (EOC) signal from the A/D converter.
5. Port C (lower) is to be programmed as an output half-port.
   - Bit 3 (PC3) is to be used to send the start of conversion (SOC) signal to the A/D converter.
   - Bits 0, 1, and 2 are to be used as selector lines for the analog multiplexer. Process inputs 1 and 2 are to be, respectively, connected to channels 3 and 7 of the analog multiplexer AMUX.

The low (LLM) and high (HLM) limits, respectively, are 36H and 50H. The assembly program for this example is given below.

ADDR	INSTR	LABEL	MNEM	OPERAND	COMMENTS
3000	31		LXI	SP,27FF	;Initialize stack pointer
3001	FF				
3002	27				
3003	3E		MVI	A,8A	;Load accumulator with control word and send to control word register
3004					
3005	D3		OUT	03H	
3006	03				
3007	0E	P4	MVI	C,01	;set counter to 1
3008	01				
3009	CD	P3	CALL	P1	;call the subroutine at P1
300A					
300B					
300C	CD		CALL	P2	;call the subroutine for performing limit check
300D					
300E					
300F	0C		INR	C	;increment counter
3010	79		MOV	A,C	;transfer counter number to accumulator
3011	FE		CPI	02H	;compare with the number of inputs
3012	02				
3013	CA		JZ	P3	;jump to P3 if not zero
3014					
3015					
3016	C3		JMP	P4	;jump to the beginning for initializing the counter
3017					
3018					
3019	79	P1	MOV	A,C	;transfer counter contents to accumulator
301A	FE		CPI	01H	;compare with 01
301B	01				
301C	C2		JNZ	LINE	;if not zero go to location LINE
301D	26				
301E	30				

(contd)

(contd)

ADDR	INSTR	LABEL	MNEM	OPERAND	COMMENTS
301F	3E		MVI	A,02H	;transfer first input channel
3020	02				;to accumulator and send through
3021	D3		OUT	02H	;port C(L)
3022	02				
3023	CD		CALL	PULSE	;call subroutine that sends the
3024	73				;pulse to S/H
3025	30				
3026	3E		MVI	A,0AH	;start of conversion pulse
3027	0A				;without affecting the
3028	D3		OUT	02H	;channel address
3029	02				
302A	3E		MVI	A,02H	;end SOC pulse sent
302B	02				;through
302C	D3		OUT	02H	;port 2
302D	02				
302E	C3		JMP	P6	;jump to P6 for checking
302F	40				;EOC pulse
3030	30				
3031	3E	LINE	MVI	A,06H	;transfer second input
3032	06				;channel to accumulator and
3033	D3		OUT	02H	;send through port C
3034	02				
3035	CD		CALL	PULSE	;call the subroutine
3036	73				;that sends the pulse to S/H
3037	30				
3038	3E		MVI	A,0EH	;start of conversion
3039	0E				;pulse without affecting channel
303A	D3		OUT	02H	;address
303B	02				
303C	3E		MVI	A,06H	;end of conversion pulse
303D	06				
303E	D3		OUT	02H	
303F	02				
3040	DB	P6	IN	02H	;check for
3041	02				;EOC pulse

(contd)

(contd)

ADDR	INSTR	LABEL	MNEM	OPERAND	COMMENTS
3042	E6		ANI	80H	;by noting the
3043	80				;setting of
3044	C2		JNZ	P6	;zero flag
3045	40				
3046	30				
3047	DB		IN	01H	;input the process
3048	01				;variable
3049	C9		RET		
304A	FE	P2	CPI	HLM	;check for high limit
304B	50				;and operate relay
304C	D2		JNC	RELAY	;if carry flag shows zero
304D	55				
304E	30				
304F	FE		CPI	LLM	;check for low limit
3050	3C				;and operate alarm
3051	DA		JC	ALARM	;if carry flag shows zero
3052	68				
3053	30				
3054	C9		RET		
3055	11	RELAY	LXI	D,4000H	;set time
3056	00				;delay to
3057	40				;200 ms
3058	06		MVI	B,20H	;make bit 5 high
3059	20				;to send through Port A
305A	78	P7	MOV	A,B	;to relay
305B	D3		OUT	00H	
305C	00				
305D	1B		DCX	D	
305E	7B		MOV	A,E	;check for completion
305F	B2		ORA	D	;of time delay
3060	C2		JNZ	P7	
3061	5A				
3062	30				
3063	3E		MVI	A,00H	;transfer 00H into
3064	00				;accumulator and output through
3065	D3		OUT	00H	;port A to indicate end
3066	00				;of pulse to relay
3067	C9		RET		
3068	11	ALARM	LXI	D,FFFFH	;set DE pair for an
3069	FF				;interval of
306A	FF				;0.5 s

(contd)

(contd)

ADDR	INSTR	LABEL	MNEM	OPERAND	COMMENTS
306B	06		MVI	B,40H	;make bit 6 high to
306C	40				;send signal to alarm
306D	CD		CALL	P7	
306E	5A				
306F	30				
3070	C3		JMP	FINAL	;jump to end of
3071	82				;program
3072	30				
3073	3E	PULSE	MVI	A,80H	;to sample and
3074	80				;hold
3075	D3		OUT	00H	
3076	00				
3077	16		MVI	D,03H	;delay for
3078	03				;acquisition
3079	15	P5	DCR	D	;time
307A	C2		JNZ	P5	
307B	79				
307C	30				
307D	3E		MVI	A,00H	;transfer 00H into
307E	00				;accumulator and output through
307F	D3		OUT	00H	;port A to indicate end
3080	00				;of pulse to alarm
3081	C9		RET		
3082	76	FINAL	HLT		;stop

12.7 The 8086 Microprocessor

The 8086 chip is an Intel 16-bit microprocessor which is a successor of the 8085A chip. Its data bus has a width of 16 bits whereas its address bus is 20-bit wide. The basic structure of the Intel 8086 microprocessor is shown in Fig. 12.41. The CPU is divided into two independent functional parts, namely, the bus interface unit (BIU) and the execution unit (EU). These two units are interconnected by a 16-bit data bus, called the A bus, and an 8-bit Q bus. The functions of the BIU are

(a) sending out addresses and fetching instructions from memory,
(b) reading data from memory and ports,
(c) writing data to memory and ports,
(d) fetching six instruction bytes ahead of time from memory for the instruction fetch queue (IQ), and
(e) keeping the EU busy with prefetched instructions under normal conditions.

The functions of the EU are as follows:

(a) telling the BIU where to fetch instruction or data from,

(b) obtaining instructions from the IQ maintained by the BIU,
(c) decoding instructions, and
(d) executing instructions.

These two units operate independently of each other, enabling the 8086 microprocessor to simultaneously fetch instructions (BIU) and decode and execute them (EU).

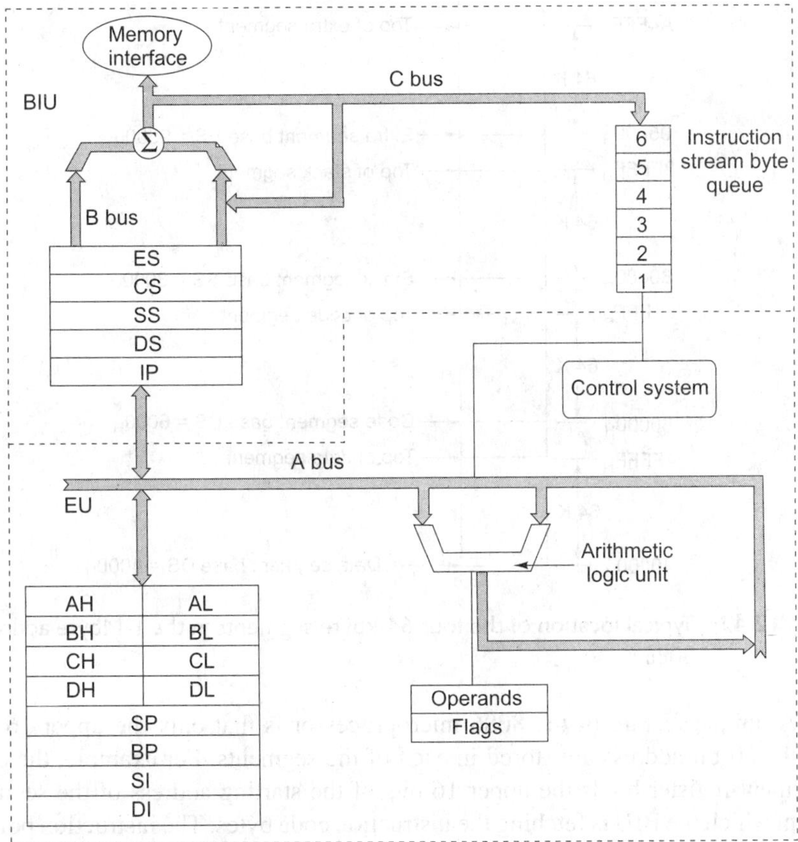

Fig. 12.41 Basic internal structure of 8086

12.7.1 Bus Interface Unit

The salient features of the bus interface unit are the queue, segment registers, and instruction pointer. The prefetched instruction bytes are held in the queue, which is a group of registers arranged in a first-in-first-out manner. Because of the overlapping of fetching and execution, a program runs much faster on the 8086 than on the 8085. Fetching the next instruction while the current instruction is being executed, is called *pipelining*.

There are four segment registers in the 8086, namely, the code segment (CS), stack segment (SS), extra segment (ES), and data segment (DS) registers. Each segment register is of length 64 kbytes (65,536 bytes), whereas the total memory capacity of the microprocessor is 2^{20} or 1,048,576 bytes, which is 16 times the 64 kbyte length. This total length is also 1 Mbyte because it is equal to $2^{10} \times 2^{10}$ bytes, 2^{10} bytes being 1 kbyte. These four segments can be located anywhere in the memory as shown in Fig. 12.42.

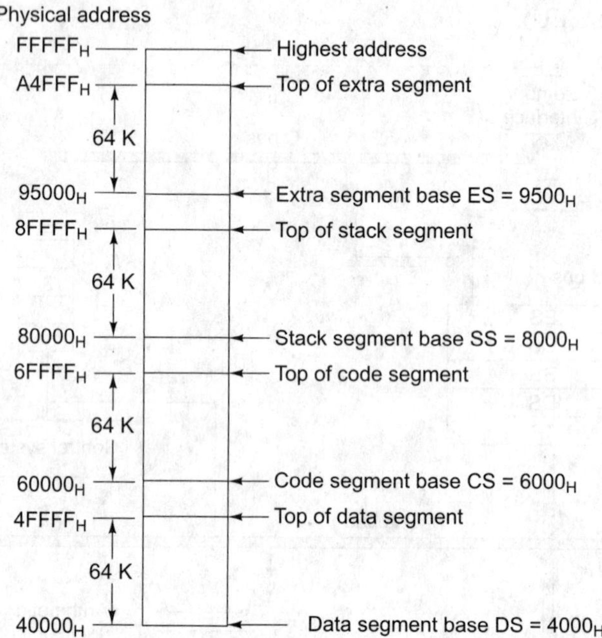

Fig. 12.42 Typical location of the four 64-kbyte segments in the 1-Mbyte address of 8086

A unique feature of the 8086 microprocessor is that only the upper 16 bits of the 20-bit address are stored in each of the segments. For example, the code segment register holds the upper 16 bits of the starting address of the segment from which the BIU is fetching the instruction code bytes. The instruction pointer (IP) register holds the 16-bit address of the next code byte within this segment. This value in the IP is called the *offset* because it must be added to the segment base address in the CS to obtain the 20-bit physical address.

Figures 12.43(a) and (b) show the locations of the physical addresses in the portion of the memory that contains the code segment and the method of arriving at the physical address. In a similar manner, the physical address of a stack (memory) location is split up into two parts, the upper 16 bits being stored in the stack segment, the 16-bit offset being kept in the stack pointer (SP) register. The method of arriving at the physical address is illustrated in Figs 12.43(c) and (d). Third, to access data in the memory, the 8086 microprocessor stores the upper 16 bits of the address in the data segment. The 16-bit offset, also called the *effective address*, is added to this as in the case of the code and stack segments to obtain the 20-bit physical address. In all these cases the upper 16 bits in the segments are shifted to the left by four bits before adding the offset. For example, in Fig. 12.43(d) the contents of the stack segment are shifted to the left by four bits, these bits being taken as zero. The 16-bit number in the SP register is then added to this shifted number to obtain the physical address of the stack segment.

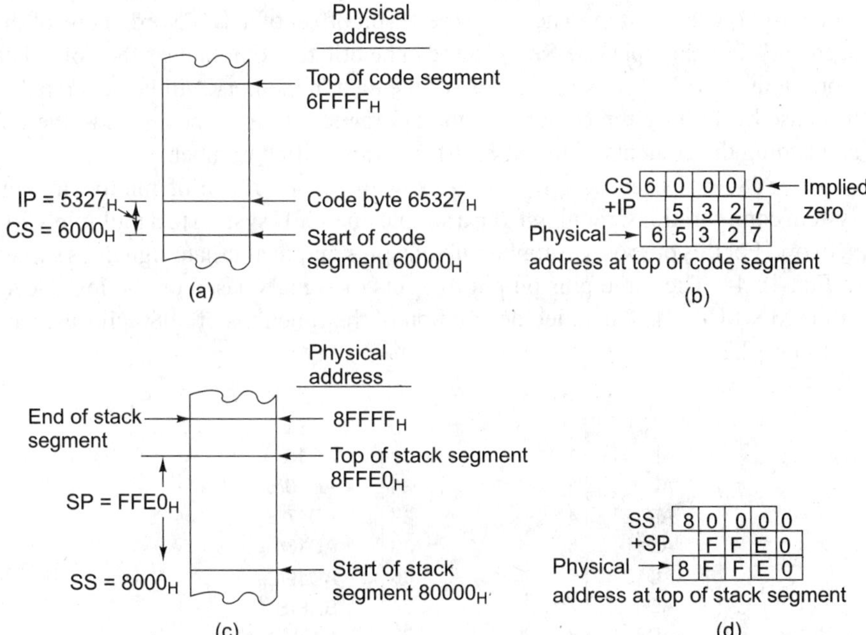

Fig. 12.43 Addition of IP to CS to produce the physical address of the code byte: (a) memory map of addition of IP to CS, (b) calculation of physical address at the top of the code segment, (c) memory map of addition of SP to SS, (d) calculation of physical address at the top of the stack segment

12.7.2 Execution Unit

The execution unit contains the following subunits:

(a) general-purpose registers,
(b) stack pointer, base pointer, source index and destination index registers,
(c) flag register, and
(d) control circuitry, instruction decoder, and ALU.

There are eight general-purpose registers, namely, AH, AL, BH, BL, CH, CL, DH, and DL. These can be used separately to temporarily store 8-bit data or together to store 16-bit data. For example, the combination of BH and BL is called the BX register and can be used to store a 16-bit number. The 16-bit register AX is called the accumulator for 16-bit operations whereas the register AL acts as the accumulator for 8-bit operations.

As in the 8085A microprocessor, the stack is a section of the memory kept apart to store addresses and data during the execution of a subprogram (here called a procedure). The stack pointer register contains the 16-bit offset from the start of the segment (which is one location higher than the memory location where a word has been stored most recently on the stack); this location is called the top of stack. This has been explained before and illustrated in Fig. 12.43(c). The BP, SI, and DI registers can be used for temporary storage of data in the same way as the general-purpose registers. They hold the 16-bit offset of a data word in one of the segments. For example, the SI register can be utilized to hold the offset of a data word in the DS. The physical address of the data in memory will be generated in this case by shifting the contents of the DS register four bit positions to the left and adding the contents of the SI register to this shifted number.

The 8086 has been designed to work with a wide range of microcomputer system configurations, ranging from a simple one-CPU system to a multiple-CPU network. For this purpose, a number of 8086 pins output alternate signals as shown in Fig. 12.44. The same pins output two sets of signals based on the logic level of the MN/$\overline{MX}$ pin. A detailed description of the functions of 8086 pins is given in Table 12.2.

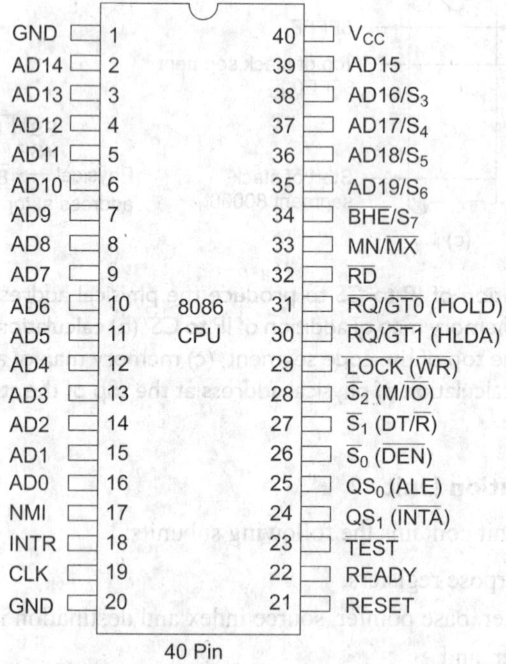

Fig. 12.44 8086 pin diagram

12.7.3 Programming on the 8086

The 8086 has its own assembly language like the 8085A microprocessor. In addition, it has an additional facility, namely, programmability in a high-level language such as BASIC, FORTRAN, and PASCAL. The statements in these languages are more English-like and this makes them easier than assembly language. The drawback with these languages is that a compiler program has also to be run to convert them into machine code, and this takes its own time

Table 12.2 Detailed functional description of the pins of the 8086 chip

Pin	Description	Type
AD0–AD15	Data/address bus	Bidirectional, tristate
A16/S3, A17/S4	Address/segment identifier	Output, tristate
A18/S5	Address/interrupt enable status	Output, tristate
A19/S6	Address/status	Output, tristate
$\overline{BHE}/\overline{S7}$	High-order byte/status	Output, tristate
$\overline{RD}$	Read control	Output, tristate
READY	Wait state request	Input
$\overline{TEST}$	Wait for test control	Input
INTR	Interrupt request	Input
NMI	Non-maskable interrupt request	Input
RESET	System reset	Input
CLK	System clock	Input
MN/$\overline{MX}$	= GND for maximum system	
$\overline{S0}$, $\overline{S1}$, $\overline{S2}$	Machine cycle status	Output, tristate
$\overline{R0}/\overline{GT0}/\overline{RQ}/\overline{GT1}$	Local bus priority control	Bidirectional
QS0, QS1	Instruction queue status	Output, tristate
$\overline{LOCK}$	Bus hold control	Output, tristate
MN/$\overline{MX}$	= V_{CC} for minimum system	
M/$\overline{IO}$	Memory or I/O access	Output, tristate
$\overline{WR}$	Write control	Output, tristate
ALE	Address latch enable	Output, tristate
DT/$\overline{R}$	Data transmit/receive	Output, tristate
$\overline{DEN}$	Data enable	Output, tristate
$\overline{INTA}$	Interrupt acknowledge	Output
HOLD	Hold request	Input
HLDA	Hold acknowledge	Output
V_{CC}, GND	Power, ground	

for conversion. Execution of these programs therefore takes more time than those written in assembly language. Programs which involve process control or mechanical controls (for example, the operation of a robot) must run fast and hence are written in assembly language. On the other hand, programs that involve large amounts of data, such as insurance company records and payrolls, are written in high-level languages. However, recently introduced high-level languages have assembly language features also. Due to space limitations, only assembly language programs are dealt with here.

12.7.3.1 Addressing modes of 8086 The different ways in which the processor accesses data are called addressing modes. The MOV instruction can be used to explain this. It is written in the form

$$\text{MOV } d, s$$

where s is the source and d is the destination. When this instruction is executed, a byte from the source location is copied into the destination location. The source can be an 8-bit or a 16-bit number written directly in the instruction, an 8-bit or a 16-bit register as given in column 5 of Table 12.3, or a memory location specified in 24 different ways as shown in columns 2–4 of Table 12.3. The destination can be any of the above except an 8-bit or a 16-bit number. The different combinations can be classified under the following three modes:

(a) immediate addressing mode,
(b) register addressing mode, and
(c) direct addressing mode.

Table 12.3 MOD R/M bit patterns for 8086 instructions

MOD	Memory mode			Register mode	
R/M	00	01	10	11	
				W = 0	W = 1
000	(BX) + (SI)	(BX) + (SI) + d8	(BX) + (SI) + d16	AL	AX
001	(BX) + (DI)	(BX) + (DI) + d8	(BX) + (DI) + d16	CL	CX
010	(BP) + (SI)	(BP) + (SI) + d8	(BP) + (SI) + d16	DL	DX
011	(BP) + (DI)	(BP) + (DI) + d8	(BP) + (DI) + d16	BL	BX
100	(SI)	(SI) + d8	(SI) + d16	AH	SP
101	(DI)	(DI) + d8	(DI) + d16	CH	BP
110	d16(direct address)	(BP) + d8	(BP) + d16	DH	SI
111	(BX)	(BX) + d8	(BX) + d16	BH	DI

An example of mode (a) is MOV DX,2458H. When executed, the immediately available 16-bit number, 2458H, will be put into the 16-bit register DX. Likewise, the instruction MOV DL,58H will transfer the 8-bit number into the 8-bit register DL. An example of mode (b) is MOV DX,BX. When executed, it copies the contents of the 16-bit BX register into the 16-bit DX register; the contents of the BX register remain intact. As the source is a register, this instruction comes under the register addressing mode. The instruction MOV DL,BL similarly copies 8-bit data contained in BL to the DL register. The instruction MOV DL,[48AEH] is an example of mode (c). The square brackets around the number 48AEH mean 'the contents of the memory location/locations at a displacement from the segment base of'. The actual 20-bit address is obtained by shifting the data segment base contained in the DS segment register four bits to the left and adding the effective address 48AEH as explained earlier. It should be noted that the instruction MOV DX, 48AEH puts the 16-bit number into the 16-bit register DX. Hence the presence or absence of square brackets around a 16-bit number makes a difference in the operation.

It is always helpful to make a flow chart before actually writing an assembly level program. A typical flow chart is given in Fig. 12.45 and programming example 12.5 corresponds to this chart. This program reads the temperature of a tank which is sensed by a resistance temperature detector (RTD) or a thermocouple 100 times, and checks the value for its high limit. If the limit is exceeded, an alarm is sent out and then the temperature is stored in memory; otherwise after

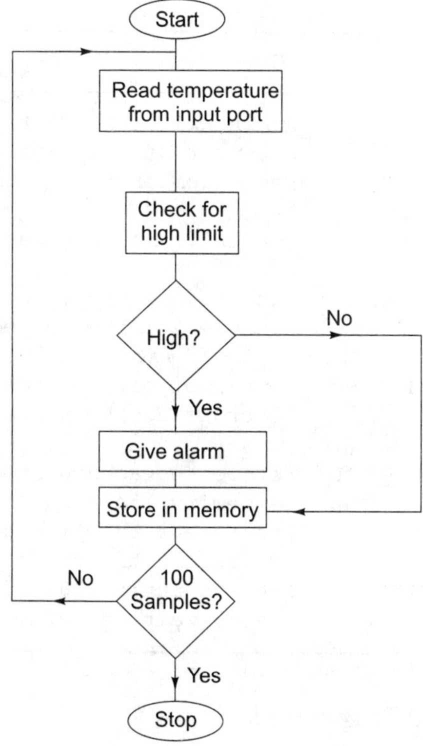

Fig. 12.45 Flow chart for programming example 12.5

comparison it is directly stored in memory. The start and stop blocks are usually circles. Operations like reading, checking, storing, adding, and subtracting are shown as rectangular blocks. The diamond-shaped block indicates a decision making stage. Depending on whether the decision is yes or no, the program branches to the appropriate location.

Programming Example 12.5 To read 100 samples of the temperature of a tank, check for its high limit (150 °C), send an alarm if it exceeds this limit, and then store it in memory.

Assembly language program

LABEL	MNEM	OPERAND	COMMENTS
	MOV	DX,FFFEH	;initialize the 8255 interface chip with
	MOV	AL,99H	;ports A and B, respectively, as input and
	MOV	DX,AL	;output ports
	MOV	SI,55H	;initialize the SI register

(contd)

(contd)

LABEL	MNEM	OPERAND	COMMENTS
	MOV	BL,64H	;initialize the BL register as a counter of 100 samples
START	MOV	DX,FFF8H	;read temperature from port A
	IN	AL,DX	
	CMP	AL,96H	;check if it is greater than 150 °C
	JB	STORE	;if less, store it
ALARM	MOV	AL,01	;if greater than or equal to 150 °C, make
	MOV	DX,FFFAH	;bit 1 of port B high, thus energizing the alarm
	OUT	DX,AL	
STORE	MOV	45H SI,AL	;store the accumulator contents in a memory location
	INC	SI	;increment the SI register by 1
	DEC	BL	;decrement the counter by 1
	JNZ	START	;repeat until 100 values are read
	INT	3	;stop, wait for command from user

12.7.4 Structured Programming

The development of a program is facilitated by programming tools called *standard programming structures*. There are three basic operations, namely, sequence, decision (or selection), and repetition (or iteration). A sequence consists of a series of actions, a decision involves choosing between two alternative actions, and repetition means repeating a series of actions until some condition is satisfied. Standard programming structures can be constructed based on these three operations. Three typical structures, namely, the simple sequence chart, the if-then-else flow chart, and the while-do flow chart, are shown in Fig. 12.46.

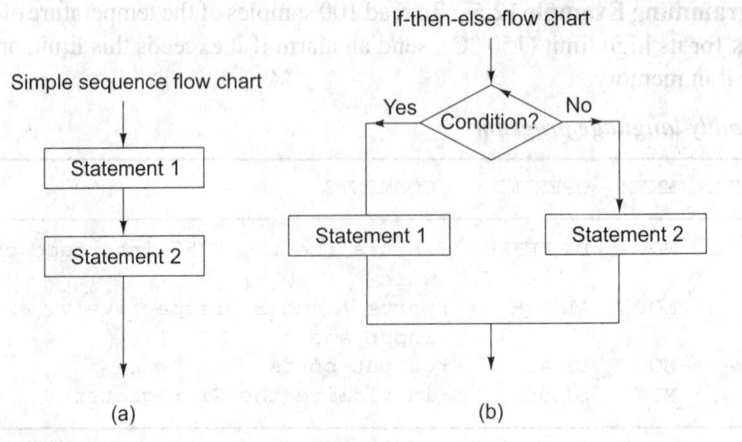

Figs 12.46(a), (b)

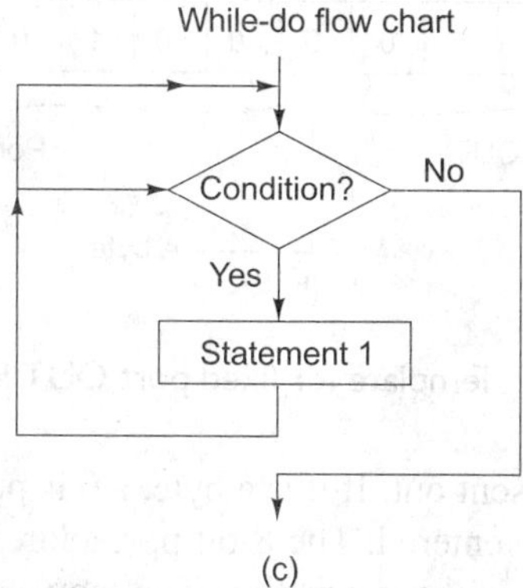

Fig. 12.46 Standard programming structures: (a) sequence, (b) if-then-else, (c) while-do

The structure shown in Fig. 12.46(a) is that of the simple sequence in which the desired actions to be taken are written in the desired order. The if-then-else structure shown in Fig. 12.46(b) is a decision making operation and directs the operation one way or another based on some condition. The while-do structure shown in Fig. 12.46(c) is just one form of repetition; it indicates that some action or sequence of actions has to be taken as long as some condition is fulfilled.

An entire program can be divided into the above and other basic programming structures; each of them is considered to be a module. Hence this approach is termed the *modular approach*, and simplifies program writing and facilitates checking at a later stage.

12.7.5 Writing Machine Code for Instructions

For simple programs, hand coding can be done for the 8086 assembly language instructions. The intel SDK-86 prototyping board can be used to enter and run such programs. Such an approach would be helpful in understanding the manner of coding of programs with instruction templates. Figure 12.47(a) shows the template for the fixed port OUT instruction and Fig. 12.47(b) shows an example (OUT 2CH) which asks the microprocessor to copy the contents of the 16-bit AL register to the port 2CH.

It is evident that two bytes are required for the instruction. Seven out of the eight upper-byte bits tell the 8086 that this is an 'output to a fixed port' instruction. The eighth upper-byte bit W in the template tells the 8086 whether a word (16 bits)

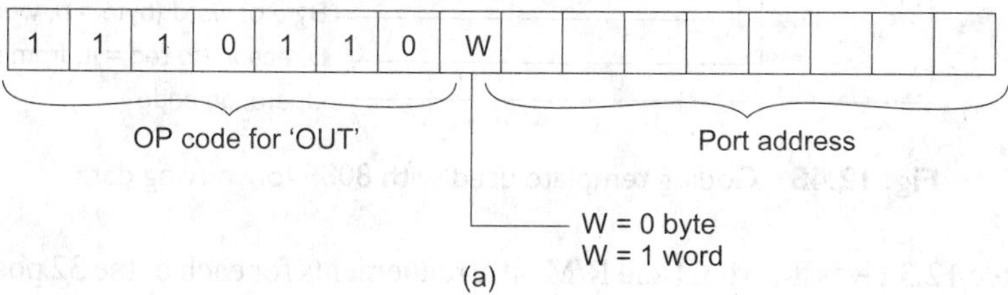

Fig. 12.47(a)

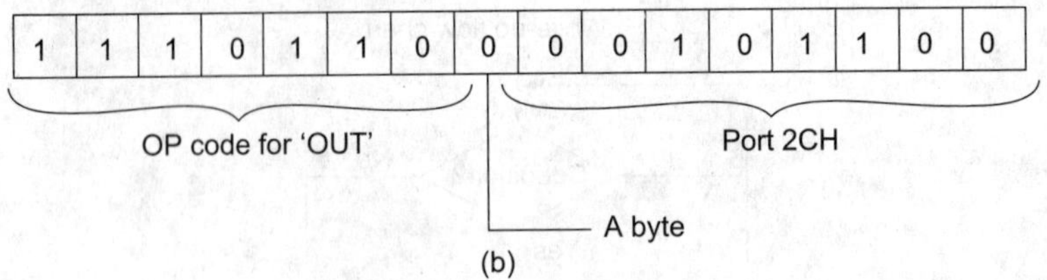

Fig. 12.47 Template for fixed port OUT instruction

or a byte (8 bits) is to be sent out. If it is a byte, a 0 is put at this location, and if it is a word, a 1 has to be entered. The 8-bit port address 2CH or 00101100_2 is written in the second byte of the instruction. The sequence for entering the above instruction is that the first instruction byte is loaded in one memory location and the second instruction is put in the next.

Template filling for the MOV instruction is shown in the coding template of Fig. 12.48 for the 8086 processor. As stated before, the MOV instruction copies from a register to a register, from a register to a memory location, or from a memory location to a register. This figure shows that at least two code bytes are required for the MOV instruction. The upper six bits have the opcode which signifies the general type of instruction, say, adding, subtracting, copying data, etc. The 6-bit opcode for the MOV instruction is 100010. The W bit in the first word (16 bit) is used to indicate whether a word or a byte is being moved. For the former a 1 and for the latter a 0 is put in this location. In this instruction, one of the operands has to be a register; so 3 bits in the second byte are earmarked for the register code. These register codes are given in Fig. 12.49. The D bit in the first byte of the MOV instruction shows whether the data is being moved 'to' the register whose code is written in the REG field of the second byte or 'from' that register. In the former case, D is taken as 1 and for the latter D is taken as 0. The remaining fields in the second byte of the instruction code, namely, MOD and R/M, are used to specify the desired addressing mode for the other operand.

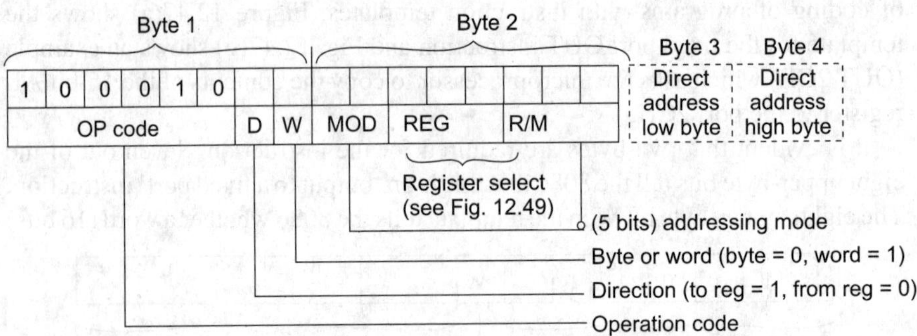

Fig. 12.48 Coding template used with 8086 for moving data

Table 12.3 gives the MOD and R/M bit arrangements for each of the 32 possible addressing modes. If the second operand in the instruction is also one of the eight registers, the binary number 11 is put in the MOD field and the 3-bit code for that

Register field bit assignments
(If W = 1, 16-bit register; if W = 0, 8-bit register)

Register code		Register code		Segment code register	
AX	0 0 0	AL	0 0 0	ES	0 0
CX	0 0 1	CL	0 0 1	CS	0 1
DX	0 1 0	DL	0 1 0	SS	1 0
BX	0 1 1	BL	0 1 1	DS	1 1
SP	1 0 0	AH	1 0 0		
BP	1 0 1	CH	1 0 1		
SI	1 1 0	DH	1 1 0		
DI	1 1 1	BH	1 1 1		

Fig. 12.49 Instruction codes for 8086 registers

register is entered in R/M field. If the second operand is a memory location, then there are 24 ways of obtaining the effective address of the operand in memory. As stated earlier, the effective address can be given directly in the instruction or it can be contained in a register. A third alternative is that it can be the sum of the contents of two registers and a displacement. All the possible alternatives are given in Table 12.3. Coding of some commonly occurring MOV instructions are given below:

Coding MOV SI, CX This instruction will copy a word from the CX register to the SI register. As per the 8086 instruction codes, the 6-bit opcode for the MOV instruction is 100010. The W bit is 1 because a word is being moved. The D bit is filled up with the consideration that the move is 'from' the register CX or, alternatively, it is 'to' the register SI. If it is considered that the move is to SI, the D bit is 1. Also, 110 is put in the REG field to represent the register code for SI. The MOD field will be 11 to represent the move as from one register to another (register addressing mode). The R/M field will be filled with the code of the CX register, namely, 001. The resulting code for the instruction MOV SI, CX is shown below:

1000101111110001

If the move is considered to be from CX, then D will be zero and the contents of the REG and R/M fields will have to be swapped.

Coding MOV DL, [BX] This instruction will copy a byte from the memory location whose effective address is contained in BX to the DL register. The effective address will be added to the data segment base in DS to produce the physical address.

As before, the 6-bit opcode for the MOV instruction is 100010. The D bit is 1 because data is moved to the register DL. The W bit is 0 because the instruction is moving a byte into DL. The 3-bit register code for DL (010) is put in the REG field. The remaining fields, namely, MOD and R/M, are filled as follows. First the box containing the desired addressing mode is to be located. The box containing [BX] is in the bottom row of the second column. The MOD bit pattern for this is read from the top of the column as 00. Its R/M bit pattern is in the bottom row of the first column and is given as 111. Assembling all the above bits together gives the resulting code as shown below:

1000101000010111

Coding MOV 56H [DI], CH This instruction will copy a byte from the CH register to a memory location, the effective address of which will be computed by adding the indicated displacement 56H to the contents of the DI register. The actual physical address will be produced by shifting the contents of the data segment base in DS four bits to the left and adding this effective address to the result.

The 6-bit opcode is 100010. The D bit is 0 because the move is 'from' a register and the W bit is 0 as a byte is moved. As the move is from the register CH, the binary digits 101 are to be entered in the REG field. From column 3 and row 6 of Table 12.3 the MOD bits for [DI] + D8 are 01 and the R/M bits are 101. The binary code for the displacement 56H is 01010110 and this forms the third byte of the template of Fig. 12.48. Thus the 3-byte code for this instruction is as given below:

100010000110110101010110

Coding MOV BX, [8A72H] This instruction copies the contents of two memory locations into the register CX. The template for this instruction is arrived as follows. The opcode is 100010. As the move is to a register, D is 1. A word is copied into the BX register, hence a 1 is to be entered in the W column. The REG field should be filled as 011, which is the code for the BX register. The displacement, which is just a 16-bit number, is located in column 2 and row 7 of Table 12.3. Hence the MOD bits are 00 and the R/M bits are 110. Thus the first two code bytes for the instruction are shown in the first half of the binary numbers given below:

1000101100011110 0111001010001010

The two code bytes are followed by the lower byte of the direct address 72H (01110010 binary). The higher byte of the direct address 8AH (10001010 binary) is placed after that; the two bytes put together will be as shown in the second half of the number given above.

Coding MOV CS: [BX], CH This instruction copies a byte from the CH register to a memory location whose effective address is contained in the BX register. In the normal course an effective address in BX will be added to the data segment base in DS to generate the physical memory address. CS, in this instruction, indicates that the BIU should add the effective address to the code segment base in CS to produce the physical address. The combination CS: is called *segment override prefix*. This is a unique case in which the 8-bit code for the segment override prefix has to be located in the memory prior to the code for the rest of the instruction. In the present case the code byte for the segment override prefix is 001XX110, where in the field XX the code for CS (01) is to be inserted. Thus the prefixed byte is 00101110. This is then followed by the code for the other part of the instruction as shown below. The codes for the other segment registers are 00 for ES, 10 for SS, and 11 for DS.

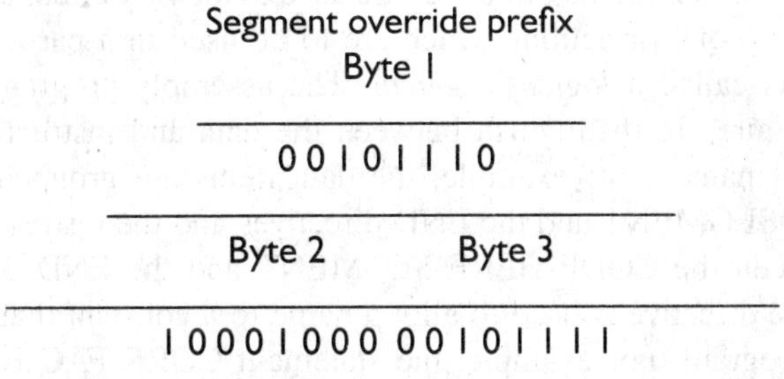

The computation of memory address for the various combinations given in Table 12.3 can also be represented by the block diagram of Fig. 12.50.

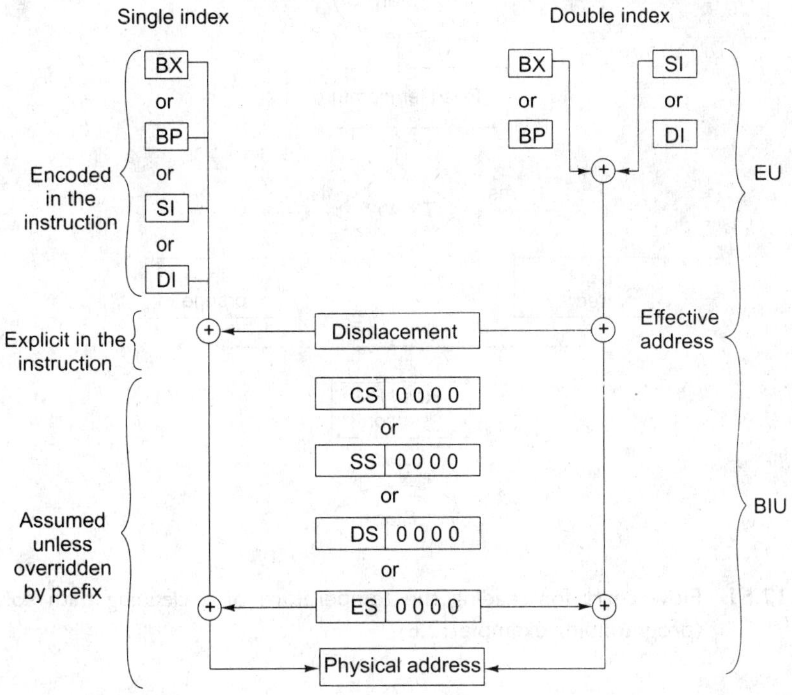

Fig. 12.50 Memory address computation

12.7.6 Necessity of an Assembler

The examples in Section 12.7.5 explain how the 8086 microprocessor arrives at the effective address and also how instructions can be coded by constructing the template step by step. Whereas hand coding is possible for short programs, large programs take considerable time, and become tedious and prone to error with such coding. The 8086 assembler, like the 8085A assembler, removes the tediousness of programming and makes programs much more readable.

Using an assembler which accompanies assembly language program involves assembler directives. Some of the most important directives are now dealt with.

The SEGMENT and END directives are used to identify either a group of data items or a group of instructions which are to be used in a particular segment; this segment is called a *logical segment*. The assembly program that follows shows this feature. To distinguish between the data and instructions, they are given different names. For example, the data items are grouped between the DATA-HERE-SEGMENT and the END directives and the instruction codes are grouped between the CODE-HERE-SEGMENT and the END directives. The EQU or equate directive is used to allot a name to a constant that is repeatedly used in the program. For example, the statement CORR-FAC EQU 15H in a program tells the assembler to insert the value 15H every time it finds the name CORR-FAC.

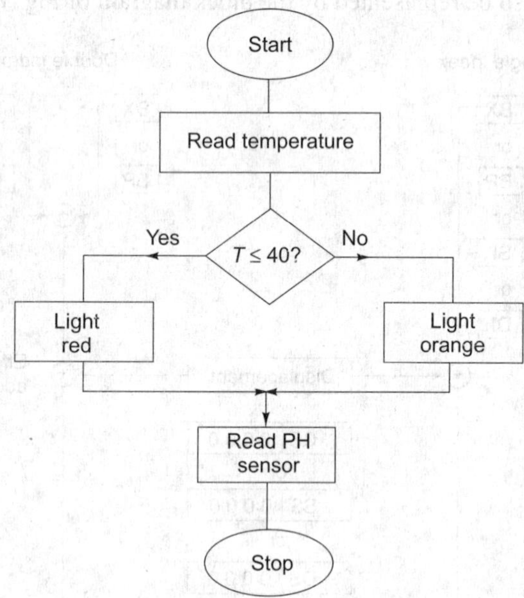

Fig. 12.51 Flow chart for reading the temperature of a cleaning bath solution (programming example 12.6)

The DB, DW, and DD directives give names to the variables in the program. For example, the statement TANK-TEMP DB 27H declares a variable of type byte (8 bits) and gives it the name TANK-TEMP. Similarly, the statement MUL DW 3527H declares a variable of type word (16 bits) with the name MUL. Finally, the directive DD is used to declare a variable of type double word (32 bits). If a number is written after DB, DW, or DD, the data item would be initialized with that value when the program is loaded from the disc into the RAM.

Programming Example 12.6 The flow chart in Fig. 12.51 is made for an assembler. The sequence of operations are reading the temperature of a cleaning bath solution and lighting one of the two lamps according to the temperature read, i.e., if it is below or equal to 40 °C, a red lamp is to be lighted, and if it is above 40 °C, an orange lamp is to be lighted.

Assembly language program

Registers used: AL DX; Ports used: FFFE4; port control register: FFF8H (port A) as temperature input, FFFAH (port B) as lamp control output (red = bit 0, orange = bit 1), FFFCH (port C) as *p*H sensor input; Procedures used: none.

```
CODE-HERE SEGMENT
          ASSUME CS: CODE-HERE

; initialize port FFFAH as an output port
          MOV DX,FFFEH  ; point DX to port control register
          MOV AL,99H    ; load control word to set up output port
          OUT DX,AL     ; send control word to control register
; initialization complete
          MOV DX,FFF8H  ; point DX at input port
          IN AL,DX      ; read temp. from sensor on input port
          CMP AL,40     ; compare temp. with 40 deg. C
          JBE RED       ; if temp <= 40 go to light red lamp
          JMP ORANGE    ; else go to light orange lamp
RED:      MOV AL,01     ; load code to light red lamp
          MOV DX,FFFAH  ; point DX at output port
          OUT DX,AL     ; send code to light red lamp
          JMP EXIT      ; go to next main line
ORANGE:   MOV AL,02     ; load code to light orange lamp
          MOV DX,FFFAH  ; point DX at output port
          OUT DX,AL     ; send code to light orange lamp
EXIT:     MOV DX,FFFCH  ; next main line instruction
CODE-HERE ENDS
          END
```

12.7.7 The 8086 Instruction Set

The instructions of the 8086 are categorized into seven types as follows:

(a) data transfer instructions,
(b) arithmetic instructions,
(c) logical instructions,
(d) execution transfer instructions,
(e) string instructions,
(f) interrupt instructions, and
(g) processor control instructions.

The 8086 instruction set is longer and involves more operations than that of the 8085A microprocessor. For example, the feature of string instructions is not available in the latter. Other additional features of the 8086 microprocessor are (a) multiplication and division operations between 8-bit as well as 16-bit numbers, (b) availability of a number of conditional jump instructions, and (c) programmability of a port number as a fixed or variable one.

12.7.7.1 Data transfer instructions The XLAT instruction is meant for the translation of a byte in the AL register using a lookup table in memory. Also the instruction LAHF is meant for copying the lower byte of the flag register into

AH. The XCHG instruction is more flexible to the extent that it causes either an 8- or a 16-bit value to be exchanged either between two registers or between a register and a memory location. The AL and AX registers act as accumulators, respectively, for 8-bit and 16-bit operations. Other data transfer instructions are similar to those in the 8085A microprocessor.

12.7.7.2 Arithmetic instructions If the 8086 is operating on numbers other than binary numbers such as ASCII (American standard code for information interchange), BCD (binary coded decimal), etc., the following instructions have to be executed in order to adjust the result back to the proper numeric format. Some special instructions which perform this kind of adjustment are detailed below.

AAA (ASCII adjust for addition)—When two ASCII codes are added without masking the 'three' in the upper nibble of each, this instruction makes sure that the result is in the correct unpacked BCD.

AAD (BCD-to-binary conversion before division)—The 'three' in the upper nibble of the ASCII codes is masked out before AAD. AAD converts two unpacked BCD digits in AH and AL to the equivalent binary number in AL. This adjustment must be made before dividing the two unpacked BCD digits in AX by an unpacked BCD byte. After division, AL will contain the BCD quotient and AH will contain the unpacked BCD remainder.

AAM (BCD adjust after multiply)—The first four bits of the ASCII digits are masked before the multiplication of two numbers. After the two resulting unpacked BCD digits are multiplied, the AAM instruction is used to adjust their product to two unpacked BCD digits in AX.

AAS (ASCII adjust for subtraction)—The 8086 allows subtraction of the ASCII codes for two decimal digits without masking the 'three' in the upper nibble of each. The AAS instruction is then used to make sure that the result is the correct unpacked BCD.

12.7.7.3 Logical instructions By means of the shift (SAL, SAR) and rotate (ROL, ROR) instructions, respectively, a byte or word specified by the instruction can be shifted and rotated to the left or the right by the number of bits specified by the CL register. The TEST instruction performs the logical AND of bytes or words and the result is used to set the flags. Neither operand is altered by this instruction.

12.7.7.4 Program execution transfer instructions The 8086 has 18 conditional jump instructions. A jump is executed only if the flag(s) specified in the instructions are in the proper state. The CALL and RET instructions can be of the intrasegment (within the same segment) type or the intersegment (between two different segments) type.

The following three loop instructions are specially used for the 8086 microprocessor: LOOP, LOOPZ/LOOPE, and LOOPNZ/LOOPNE. All these instructions cause the repetition of a group of instructions a number of times as indicated by the count register CX. This register is decremented by 1 after the group is executed once. After completion, the execution will jump to a destination within the range of −128 bytes and +127 bytes depending on the conditions of the CX register and the zero flag.

12.7.7.5 String instructions The 8086 can move, compare a string byte to check a password, and scan a data string using its string instructions. The strings that are operated upon can be numeric, alphanumeric (ASCII, EBCDIC, etc.), or in any format that can be stored in the memory in the form of binary data. These instructions may use just the SI and DS registers for an address or SI and DS registers with DI and ES registers. The LODS instruction can be used to load the AL (AX) register with a byte (word) from a string. The STOS instruction is used to store the byte (word) contained in the AL (AX) register in memory.

12.7.7.6 Interrupt instructions The 8086 has three different types of interrupt instructions: INT, INTO, and IRET. INT can be used to call a far procedure (i.e., a subroutine in another segment). The second byte of the instruction specifies which of the 256 possible interrupt service subroutines are to be executed. For example, with the INT35 instruction the 8086 will get a new value for the IP from the address 008CH (which is four times 35) and a new value for the CS from the address 008EH (which is four times 35 + 2).

The INTO instruction will cause the 8086 to make an indirect far call to a procedure written by the programmer to handle the overflow condition, provided the overflow flag is set.

The significance of the IRET instruction is as follows. When it is executed, the 8086 pushes the flags and the current IP and CS registers into the stack. It then loads a new value for the CS and loads IP with the starting address of the procedure which the programmer writes for the response to that interrupt. Once the interrupt service subroutine has been executed, the flags and the values of CS and IP have to be popped out of the stack, so that the 8086 can return to the section of the code that was being executed when the interrupt occurred. The IRET (or return from the interrupt) instruction must be placed at the end of interrupt service subroutines.

12.7.7.7 Processor control instructions These instructions consist mostly of flag set and clear instructions. Some of these instructions are used in multiprocessor environments (ESC and LOCK) and some others cause an instruction to use a segment register other than the one normally used. For instance, if an instruction normally uses the data segment register as part of a memory address, a segment override prefix instruction can cause the extra segment register to be used instead.

The detailed set of instructions of the 8086 microprocessor can be found in the book by D.V. Hall (1986).

Summary

A microprocessor is a digital system that consists of basic digital circuits such as latches, shift registers, decoders, counters, tristate buffers, and the ALU. In the Intel 8085A microprocessor, arithmetic and other data manipulations are done by 8-bit binary numbers, and it is therefore categorized as an 8-bit microprocessor. Familiarity with binary, octal, and hexadecimal numbers and their addition/subtraction helps in the understanding of the internal processor operations.

Only the processing part is handled by the microprocessor chip. To convert it into a full-fledged microcomputer, additional circuits such as I/O and memory chips are necessary.

There are two types of memory: ROM for permanent storage of programs and RAM for temporary storage of programs and data. More sophisticated ROMs have the capability to get the old programs erased so that new programs can be written thereon.

The stack is a part of the memory that is set aside to facilitate calling and execution of subroutines.

Familiarity with interfacing techniques is necessary for writing programs when the microprocessor is used for instrumentation and process control applications.

Compared to the 8085A microprocessor, the Intel 8086 chip (which is a 16-bit microprocessor) is more powerful and has a relatively large memory capacity of 1 Mbyte. Also its instruction set is more versatile, which facilitates quicker execution. Accordingly, mastering assembly level programming for the 8086 microprocessor needs more skill and practice. From the hardware point of view there is considerable flexibility in this 16-bit microprocessor: (i) both fetching and execution of instructions can be done simultaneously and (ii) the microprocessor can work either in the maximum or in the minimum mode. Other features of the 8086 microprocessor are, however, similar to those of the 8-bit microprocessor.

Exercises

Multiple Choice Questions

1. IC chips having more than ________ gates but less than ________ gates come under the category of LSI.
   (a) 500, 1500 (b) 200, 1200
   (c) 100, 1000 (d) 100, 2000
2. Glitches in the output waveform of IC chips are caused due to ________.
   (a) defective gates
   (b) fluctuation in the input signals
   (c) defective connection in the circuit
   (d) propagation delay in the circuit.
3. In the *RS* latch with a clock signal, the input condition S = ________, R = ________ gives rise to indeterminate output.
   (a) 1, 0 (b) 1, 1
   (c) 0, 1 (d) 0, 0
4. The *D* latch acts as ________ for the *D* input when the clock is high.
   (a) an inverter (b) a buffer
   (c) a window (d) a high-impedance circuit
5. The problem of racing occurs in a *JK* flip-flop when the ________ the clock pulse duration.
   (a) input pulse duration is less than

(b) input pulse duration is more than
(c) propagation delay is more than
(d) propagation delay is less than

6. A decoder chip helps in converting the input ________ into the equivalent ________ at the output.

(a) trigger pulse, level pulse (b) binary code, decimal code
(c) decimal code, binary code (d) binary code, octal code

7. In the controlled counter of Fig. 12.17, when COUNT is ________ all flip-flops remain ________ in spite of the clock pulses arriving at the counter.

(a) low, high (b) high, latched
(c) high, low (d) low, latched

8. The result of the addition of the two octal numbers 46_8 and 32_8 is ________.

(a) 100_8 (b) 87_8
(c) 110_8 (d) 64_8

9. The 2's complement, sign-and-magnitude form of −24 is ________.

(a) 11100111 (b) 11101000
(c) 11000101 (d) 00011000

10. If the number 9_{10} is subtracted from the number 16_{10} by the binary 2's complement method, then the result will be ________.

(a) 10000111 (b) 11000110
(c) 01000111 (d) 01110100

11. If the number 13_{10} is subtracted from the number 9_{10} by the binary 2's complement method, then the result will be ________.

(a) 00111100 (b) 11000100
(c) 11111100 (d) 00000011

12. The auxiliary carry bit of the 8085A microprocessor will be set when there is a carry from bit ________ of the accumulator.

(a) 5 (b) 7
(c) 4 (d) 3

13. In the 8085A microprocessor, the ________ holds the instruction to enable the ________ to decode the instruction and cause the issuing of the appropriate control signal.

(a) accumulator, instruction decoder
(b) instruction decoder, instruction register
(c) control unit, instruction decoder
(d) instruction register, instruction decoder

14. In the 8085A microprocessor, the ________ flag will be set to ________ if the binary number in the accumulator has even number of bits.

(a) auxiliary carry, 1 (b) parity, 1
(c) parity, 0 (d) zero, 1

15. The restart instruction RST3 is equivalent to a CALL instruction that takes the microprocessor to the location ________.

(a) 0030 (b) 0010
(c) 0018 (d) 0008

16. The number of machine cycles for an OPCODE FETCH instruction will be ______.

    (a) less than 3 but more than 1 (b) between 2 and 5
    (c) between 4 and 7 (d) between 4 and 6

17. In the following program, if the initial data in the accumulator is 00110101 and the flag register contents are 0011 1100, then the contents of the 16-bit registers after the execution of the program will be ______.

```
2100    LXI SP,21EFH
2103    LXI H,2280H
2106    LXI B,2144H
2109    LXI D,1823H
210C    PUSH H
210D    PUSH D
210E    PUSH PSW
210F    PUSH B
:
2120    POPD
2121    POPB
2122    POP PSW
2123    POP H
2124    RET
```

    (a) BC 2280H; DE 1823H; HL 353CH; PSW 2144H
    (b) BC 2144H; DE 2280H; HL 1823H; PSW 353CH
    (c) BC 353CH; DE 2144H; HL 2280H; PSW 1823H
    (d) BC 1823H; DE 353CH; HL 2144H; PSW 2280H

18. The control word 10011001 gives the following port assignments to the 8255 ports: port A ______, port B ______, port C(L) ______, port C(U) ______.

    (a) output, input, input, output (b) input, output, input, input
    (c) output, output, input, input (d) input, input, output, output

19. The exchange of the signals SOC and EOC between the microprocessor and the A/D converter can be categorized as data transfer ______.

    (a) with ready signal (b) with interrupt signal
    (c) with handshaking (d) which is unconditional

20. The 8185 chip has ______ number of ports ______ the 8255 chip.

    (a) as many, as (b) more, than
    (c) less, than (d) much more, than

21. The number of flags in the 8086 microprocessor are ______ whereas those in the 8085 microprocessor are ______.

(a) 5, 9 (b) 6, 6
(c) 9, 5 (d) 5, 5

22. If in the 8086 microprocessor, the stack segment base is defined as 5120H and the stack pointer contents are 89FDH, then the physical address at the top of the stack is ______.
    (a) 85FCDH (b) 58CDFH
    (c) 59FBDH (d) 59BFDH

23. The EQU directive is used to ______.
    (a) allot a fixed number to a variable that is repeatedly used in the program
    (b) allot a name for the starting address of the program
    (c) allot a name for a subroutine used in the program
    (d) allot a name to a constant that is repeatedly used in the program

24. The SEGMENT and END directives are used to identify ______.
    (a) an array of numbers that are to be shifted from a particular location in a segment to a new location
    (b) a group of instructions that are to be kept together in a particular segment
    (c) an array of numbers whose members are to be added
    (d) a sequence of instructions that form a subroutine

True or False

Indicate whether the following statements are *true* or *false*. Briefly justify your choice.

1. IC chips having more than 500 gates but less than 1500 gates come under the category of LSI.
2. Glitches in the output waveforms of IC chips are caused due to propagation delay.
3. In the *RS* latch with an enable signal, the input condition $S = 0$, $R = 0$ gives rise to an indeterminate output condition.
4. The *D* latch acts a window for the *D* input when the clock is high.
5. The problem of racing occurs in a *JK* flip-flop when the propagation delay is greater than the clock pulse duration.
6. The decoder chip helps in converting an input decimal number into the equivalent binary code in the output.
7. In the controlled ripple counter of Fig. 12.17, when COUNT is high, all flip-flops remain latched in spite of the clock pulses arriving at the counter.
8. The result of the addition of the two octal numbers 46_8 and 32_8 is 78_8.
9. The 2's complement, sign-and-magnitude form of -24_{10} is 11101000.
10. If the number 9_{10} is subtracted from the number 16_{10} by the 2's complement method, then the result will be 10000111.
11. The auxiliary carry bit of the 8085A microprocessor will be set when there is a carry from bit 7 of the accumulator.
12. The parity flag will be set to 1 if a binary number has even number of bits.
13. In the 8085A, the instruction decoder holds the instruction to enable the instruction register to issue the necessary control signals.

14. The restart instruction RST3 is equivalent to a CALL instruction that takes the microprocessor to the location 0028H.
15. An OPCODE FETCH machine cycle can have more than four clock cycles.
16. In the following program, if the initial data in the accumulator is 00110101 and the flag register contents are 0011 1100, then the contents of the 16-bit registers after the execution of the program will be BC 353CH; HL 2280H; DE 2144H; PSW 1823H.

```
2100  LXI SP,21EFH
2103  LXI H,2280H
2106  LXI B,2144H
2109  LXI D,1823H
210C  PUSH H
210D  PUSH D
210E  PUSH PSW
210F  PUSH B
 :
2120    POP D
2121    POP B
2122    POP PSW
2123    POP H
2124    RET
```

17. The control word 10011001 gives the following port assignment to the 8255 ports: port A output, port B input, port C(U) input, and port C(L) output.
18. The exchange of the SOC and EOC signals between the microprocessor and the A/D converter can be categorized under data transfer instructions with interrupt.
19. The 8155 chip has as many number of ports as the 8255 chip; in addition it has some extra features also.
20. The number of flags in the 8086 microprocessor is 9, whereas that in the 8085 microprocessor is only 5.
21. If in the 8086 microprocessor the stack segment base is defined as 5120H and the stack pointer contents are 89FDH, then the physical address at the top of the stack is 58CFDH.
22. The coding for the instruction MOV DX, [238 FH] is

```
10001011 00010110 10001111 00100011
 Byte 1   Byte 2   Byte 3   Byte 4
```

23. The EQU directive is used to allot a name to a variable that is repeatedly used in a program of 8086.
24. The SEGMENT and END directives are used to identify a group of data items as well as a group of instruction codes.

Short Answer Questions

1. Give the circuit and truth table of an enabled *RS* latch and explain its operation. Why is the input $R = S = 1$ prohibited?
2. Distinguish between a *D* latch and a positive-edge-triggered *D* flip-flop. Taking a typical input waveform, give the corresponding outputs in both cases.
3. Explain the functions of (a) a buffer and (b) a tristate device.
4. What is racing? Give a flip-flop circuit in which racing occurs.
5. Explain the working of (a) a SIPO shift register and (b) a negative-edge-triggered *D* flip-flop.
6. Explain the part played by a decoder circuit when it is used with a memory chip.
7. Write short notes on (a) EPROMS and (b) dynamic versus static ROMS.
8. Draw the schematic of a BCD counter and explain its operation.
9. Explain the difference between half and full adders.
10. With the help of an example, explain how subtraction is done in the 8085A microprocessor.
11. Explain the part played by the ALU in the 8085A microprocessor. With what circuits are the arithmetic type of instructions implemented?
12. List out the various 16-bit general-purpose registers in the 8085A microprocessor and briefly explain how they are used.
13. Give the sequence of steps in fetching an instruction from the memory of the 8085A microprocessor.
14. Write an assembly language program for the 8085A microprocessor to arrange the following numbers in ascending order: 12H, 25H, 08H, 3FH, 02H.
15. Write a program for the 8085A microprocessor to divide all the following numbers by 5_{10}, add the resulting numbers, and then store the sum in memory: 14H, 28H, 0FH, 2DH, 32H.
16. List out all the stack operations and give examples for the same.
17. Give the timing diagram for the OPCODE FETCH machine cycle.
18. List out the various kinds of data transfer used with the 8085A microprocessor.
19. List out the important parts of the 8255 interface chip.
20. List out the various parts of the (a) bus interface unit and (b) execution unit of the 8086 microprocessor.
21. Work out the physical address in the following cases.
    (a) The starting address of the CS and the offset in the IP are 5410H and 042AH, respectively.
    (b) The starting address of the SS and the offset in the SP are given as 7320H and 1154H, respectively.
22. Give typical instructions for the three addressing modes of the 8086 microprocessor.
23. List out the basic programming structures of the 8086 microprocessor and give their flow charts.

24. Give the template for the MOV instruction and explain how the various bit locations in it are filled.
25. Using Table 12.3, give the coding for the following instructions.
    (a) MOV DX, [7342H]
    (b) MOV 56H [SI], DH
    (c) MOV SS: [CX], BH
26. Write notes on the following instructions of the 8086 microprocessor.
    (a) Execution transfer instructions
    (b) String instructions
    (c) Interrupt instructions

CHAPTER 13

Microcomputer Control of Industrial Equipment

13.1 Introduction

Microprocessors and microcomputers have made their impact on power electronics based industrial equipment since the 1970s. In the early stages the microcomputer was used to acquire data from various plants and perform statistical computations. A Boolean function synthesizer for programmable logic applications constituted the second application.

A typical data acquisition system has transducers as primary sensors. After appropriate signal conditioning, the transducer outputs are converted into digital signals by an A/D converter. The microcomputer processes these signals and sends the resulting outputs to actuators through D/A converters either for giving high and low alarms or for tripping circuit-breakers. The general block diagram of a data acquisition system which can represent either a process plant or a power electronic drive is shown in Fig. 13.1.

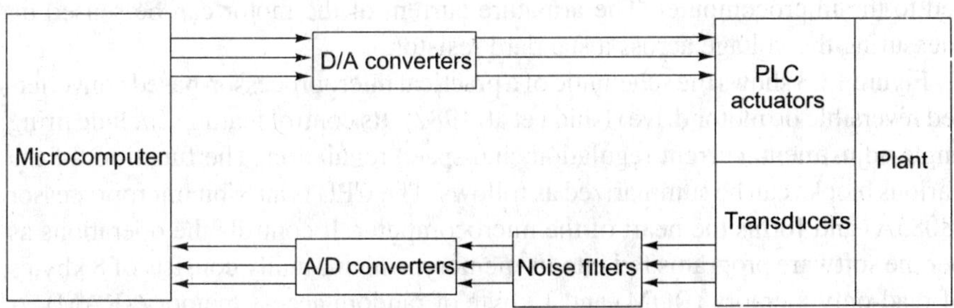

Fig. 13.1 A data-acquisition-cum-programmable logic synthesizer

With the evolution of more powerful microcomputers, their applications have gradually expanded to the closed-loop control of drives and other areas such as heating controls, uninterrupted power supply systems, and control of photovoltaic power conversion. Other functions performed by microcomputers include (a) feedback controllers, (b) providing gate firing signals for phase-controlled conversion, (c) PWM wave generation for inverters, (d) optimal and adaptive control, (e) diagnostics, and (f) miscellaneous computations.

The previous chapter introduced two early Intel microprocessors, namely, 8085A and 8086. Some typical programming examples including an interfacing program have also been presented therein. This chapter is devoted to a detailed discussion of the application of the microcomputer to two of the most commonly used electrical drives, namely, a reversible dc drive and an induction motor drive. A few other related topics are also dealt with.

After going through this chapter the reader should

- get acquainted with the role of microcomputers in various types of dc drives,
- understand how microcomputers can provide precise and quick control of induction motor drives,
- become familiar with the uniform sampling method used for pulse width modulation waveform generation and recognize the benefits arising out of the microcomputer-based implementation of it, and
- know the various benefits accruing out of microcomputer-based control.

13.2 Closed-loop Control of Dc Drives

The closed-loop control of dc drives, discussed in Chapter 7, usually involves one or more feedback loops. Process signals are sampled through transducers at discrete instants for computational purposes, and switching of power semiconductor devices is accomplished by gate stroke base control. Therefore, a microcomputer-based closed-loop system is essentially a discrete control system. Figure 13.2 shows a block-cum-schematic diagram of a typical separately excited dc drive which has an outer speed loop and an inner current control loop. The motor speed is sensed by a tachometer, the output of which is fed through an A/D converter to the microcomputer. However, a shaft encoder can also be used for this purpose, and in that case the digital output of the shaft encoder can be directly fed to the microcomputer. The armature current of the motor can be sensed by measuring the voltage across a standard resistor.

Figure 13.3 shows the schematic of a practical microprocessor-based converter-fed reversible dc motor drive (Ishida et al. 1982). Its control features include firing angle adjustment, current regulation, and speed regulation. The functions of the various blocks can be summarized as follows. The CPU is an 8-bit microprocessor (8085A) and forms the heart of the microcomputer. It controls the operations as per the software programs fed into its memory, which usually consists of 8 kbytes of read-only memory (ROM) and 1 kbyte of random access memory (RAM). A synchronous signal detector/counter is used for checking the phase of the firing pulse and for controlling the firing period. The current is sensed on the ac side by means of a current transformer (CT). The output of the CT secondary is passed through a diode bridge and a current detector (which includes an A/D converter) and is finally fed into the microcomputer. Finally the signal is fed back into the microcomputer. The actual speed is obtained by counting the pulses from a pulse generator and passing its output through the speed detector to the microcomputer.

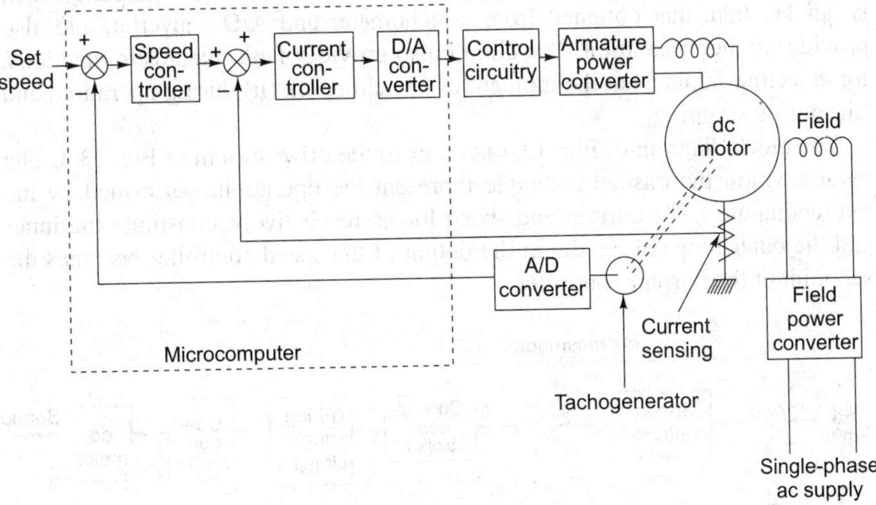

Fig. 13.2 Block diagram of a separately excited dc drive

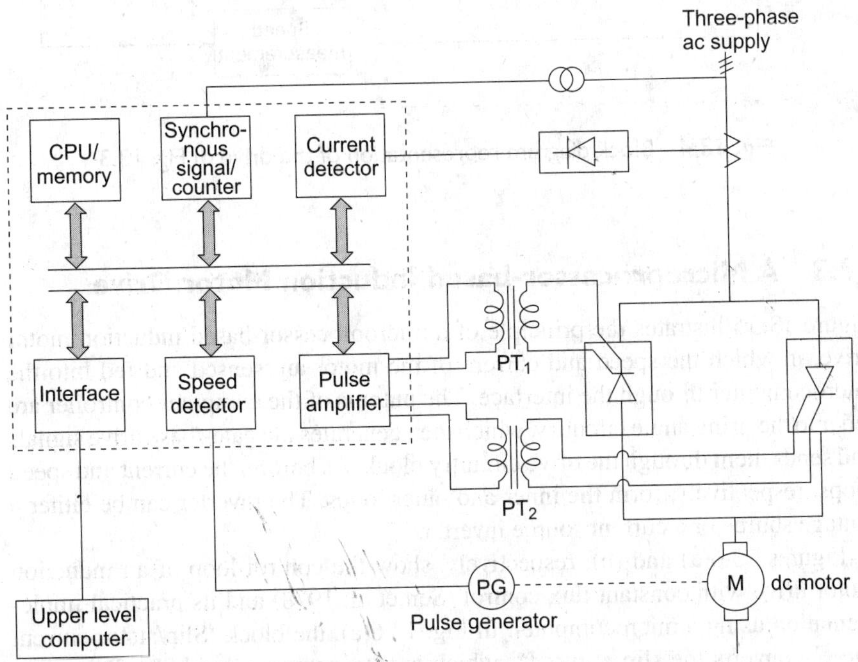

Fig. 13.3 Microprocessor-based converter-fed reversible dc drive

The power electronic converter is a three-phase, dual converter in which each of the converters is a three-phase, full-wave rectifier. The firing pulses from the firing angle module are amplified in the pulse amplifier before being fed to the gates of the thyristors. The pulse transformers PT_1 and PT_2 provide isolation for the control circuits from the power circuits.

The pulse generator, used in the speed feedback loop, provides sampling which is quicker than that obtained from a tachometer and A/D converter, and also provides higher resolution. The circuit also provides fault diagnostic functions, for detecting faults (i) at the moment of switching on, (ii) during operation, and (iii) during stopping.

The block diagram of Fig. 13.4 pertains to the drive system of Fig. 13.3. The blocks within the dashed rectangle represent the operations performed by the microcomputer. The current and speed loops, respectively, constitute the inner and the outer loops of the drive; the output of the speed controller becomes the set point of the current controller.

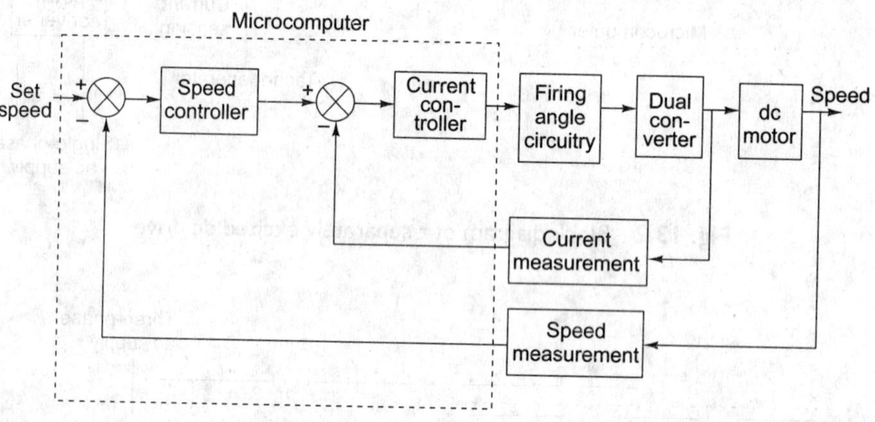

Fig. 13.4 Block diagram representation of the drive of Fig. 13.3

13.3 A Microprocessor-based Induction Motor Drive

Figure 13.5 illustrates the principle of a microprocessor-based induction motor drive, in which the speed and current of the motor are sensed and fed into the microcomputer through the interface. The outputs of the computer controller are fed into the firing angle circuitry, which then generates the gate-/base drive signals and sends them through the drive circuitry block. As before, the current and speed loops, respectively, form the inner and outer loops. The inverter can be either a voltage source or a current source inverter.

Figures 13.6(a) and (b), respectively, show the control loop of an induction motor drive with constant flux control (Sen et al. 1978) and its practical implementation using a microcomputer. In Fig. 13.6(a), the block 'Slip/stator current curve' converts the slip s into I_s^*, which is then compared with I_L. This latter current is a measure of the actual induction motor current. The error operates on the rectifier to adjust the firing angle of its thyristors. In Chapter 8, three methods are suggested for flux control (Sen et al. 1978) namely, (a) direct flux sensing,

(b) voltage sensing, and (c) stator current versus slip frequency control. The last method is employed here because the motor current can be easily controlled by the current source inverter. The stator current versus slip frequency characteristic [Fig. 13.7(b)] for the flux regulation method can be obtained from the equivalent circuit of the induction motor [Fig. 13.7(a)], with all parameters referred to the stator.

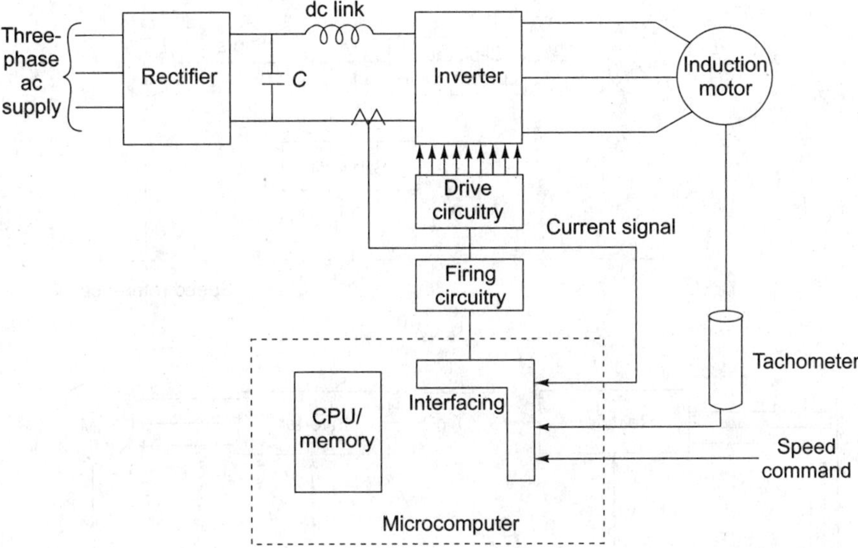

Fig. 13.5 Schematic-cum-block diagram of a microprocessor-based induction motor drive

Referring to Fig. 13.7(a), the stator current phasor $\overline{I}_s$ is the phasor sum of the magnetizing current phasor $\overline{I}_m$ and the rotor-reflected current phasor $\overline{I}_r$. Thus,

$$\overline{I}_s = \overline{I}_m + \overline{I}_r \tag{13.1}$$

In the flux control scheme the magnetizing inductance L_m is kept constant. For a particular supply frequency f, the induced emf E_s can be computed from the equation

$$\overline{E}_s = 2\pi f L_m \overline{I}_m \tag{13.2}$$

This gives

$$\frac{\overline{E}_s}{f} = 2\pi L_m \overline{I}_m \tag{13.3}$$

This gives $\overline{I}_m$ as

$$\overline{I}_m = \frac{\overline{E}_s}{2\pi f L_m} \tag{13.4}$$

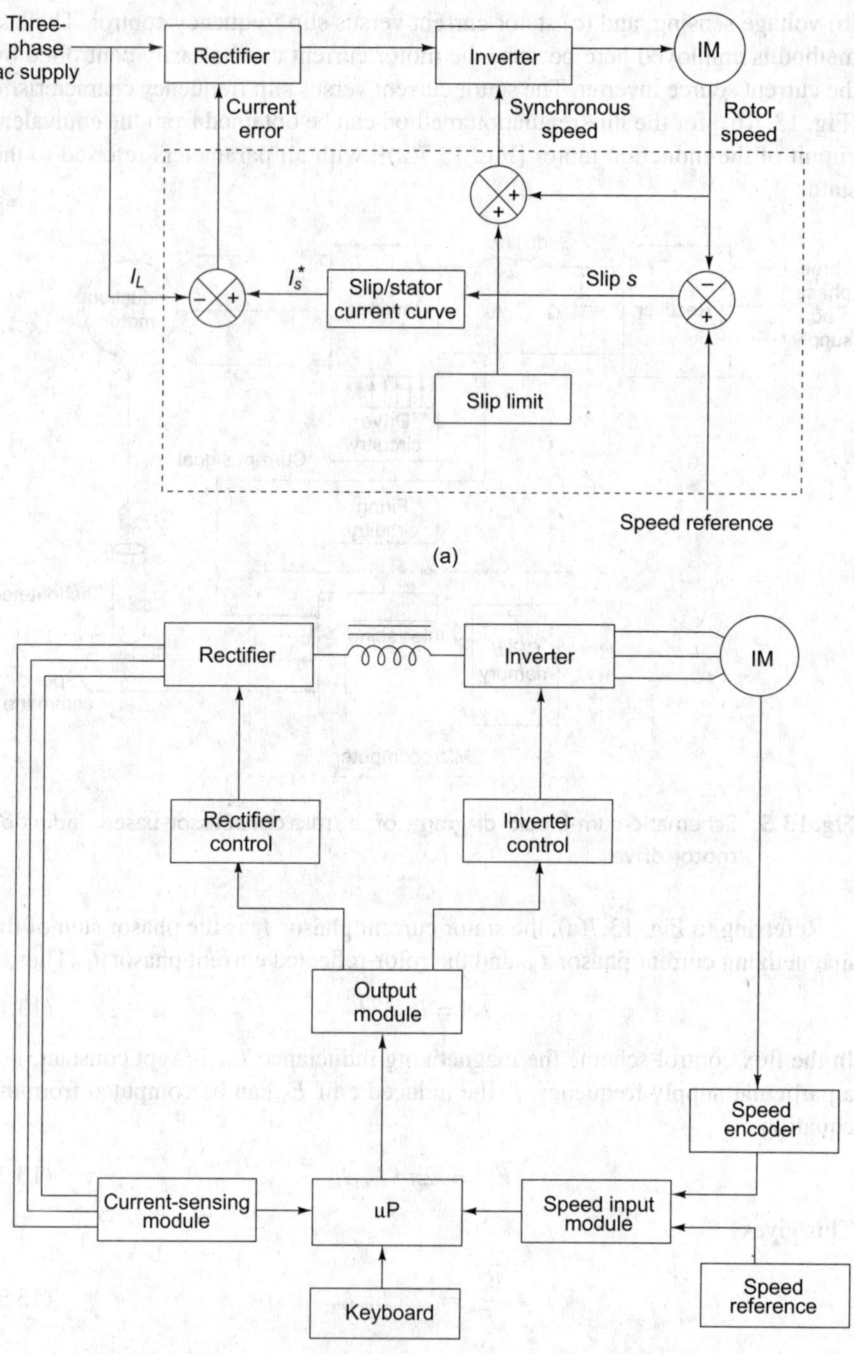

Fig. 13.6 (a) Block diagram of stator current vs slip frequency control of an induction motor (IM); (b) microprocessor (μP) implementation

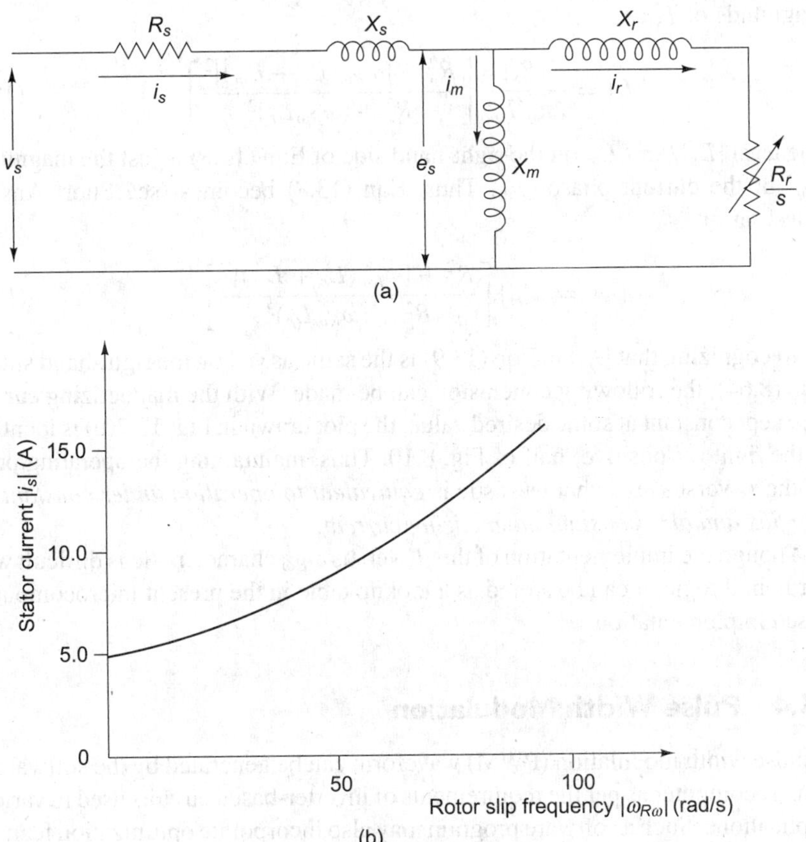

Fig. 13.7 (a) Equivalent circuit of an induction motor; (b) stator current vs slip frequency characteristic for constant air gap flux operation

The phasor $\overline{I}_r$ can be computed from the expression

$$\overline{I}_r = \frac{\overline{E}_s}{R_r/s + j2\pi f L_r} \tag{13.5}$$

Taking s as the ratio of the slip frequency $f_{R\omega}$ and the supply frequency f, Eqn (13.5) becomes

$$\overline{I}_r = \frac{\overline{E}_s}{\dfrac{R_r}{f_{R\omega}/f} + j2\pi f L_r} \tag{13.6}$$

$\overline{I}_r$ can also be written as

$$\overline{I}_r = \frac{\overline{E}_s/f}{R_r/f_{R\omega} + j2\pi L_r} \tag{13.7}$$

where R_r and L_r are, respectively, the rotor resistance and inductance, both referred to the stator side. Adding the expressions for the phasors $\overline{I}_m$ and $\overline{I}_r$ given, respectively, in Eqns (13.4) and (13.7) and then taking the magnitude yields the magnitude of $\overline{I}_s$ as

$$|I_s| = \frac{|E_s|}{2\pi f L_m}\left[\frac{R_r^2 + [\omega_{R\omega}(L_r + L_m)]^2}{R_r^2 + (\omega_{R\omega} L_r)^2}\right]^{1/2} \tag{13.8}$$

The term $|E_s|/2\pi f L_m$ on the right-hand side of Eqn (13.8) is just the magnitude $|I_m|$ of the current phasor $\overline{I}_m$. Thus, Eqn (13.8) becomes (see Short Answer Question 9)

$$|I_s| = |I_m|\left[\frac{R_r^2 + [\omega_{R\omega}(L_r + L_m)]^2}{R_r^2 + (\omega_{R\omega} L_r)^2}\right]^{1/2} \tag{13.9}$$

Recognizing that $|I_m|$ in Eqn (13.9) is the same as $|I_0|$ on the right-hand side of Eqn (8.64), the following conclusion can be made. With the magnetizing current $|I_m|$ kept constant at some desired value, the plot drawn in Fig. 13.7(b) is identical to the right (or positive) half of Fig. 8.10. Thus, maintaining the operating point on the I_s versus $\omega_{R\omega}$ characteristic is *equivalent to operation under constant air gap flux and also constant magnetizing current.*

Though the implementation of this I_s versus $\omega_{R\omega}$ characteristic is difficult with hardwired logic, it can be stored as a lookup table in the present microcomputer-based implementation.

13.4 Pulse Width Modulation

A pulse width modulation (PWM) waveform can be generated by the software of a microcomputer as per the requirements of inverter-based circuits used in various applications. Such a software program may also incorporate optimization features such as elimination of unwanted harmonics. A technique for the generation of PWM waveforms is to store the lookup table of digital words corresponding to pulse and notch widths at different magnitudes of the fundamental voltage in memory for later conversion into PWM waveforms. An example is that of a commonly used PWM scheme called the *regular sampled PWM*.

In the natural sampling method shown in Fig. 13.8(a), the carrier waveform is compared directly with the modulating (or reference) wave to determine the switching instants. As against this, the regular (or uniform) sampling method given in Fig. 13.8(b) is based on the sample-and-hold principle and has certain advantages as elaborated below (Bose et al. 1983).

In the uniform sampling method, the sinusoidal modulating wave is sampled at regular intervals corresponding to the positive peaks of the synchronized carrier wave. The sample-and-hold circuit maintains a constant level for the waveform between two sampling instants. This process results in a stepped (or amplitude modulated) version of the reference waveform, which is always symmetric with respect to the carrier waveform, whereas the naturally sampled pulse has the disadvantage of being asymmetrical about the trough of the carrier. The points of intersection of this waveform with the triangular carrier waveform are taken as the inverter switching instants. The resulting waveform consists of pulses whose widths are proportional to the step height of the reference waveform, with the centres of these pulses occurring at uniformly spaced sampling times.

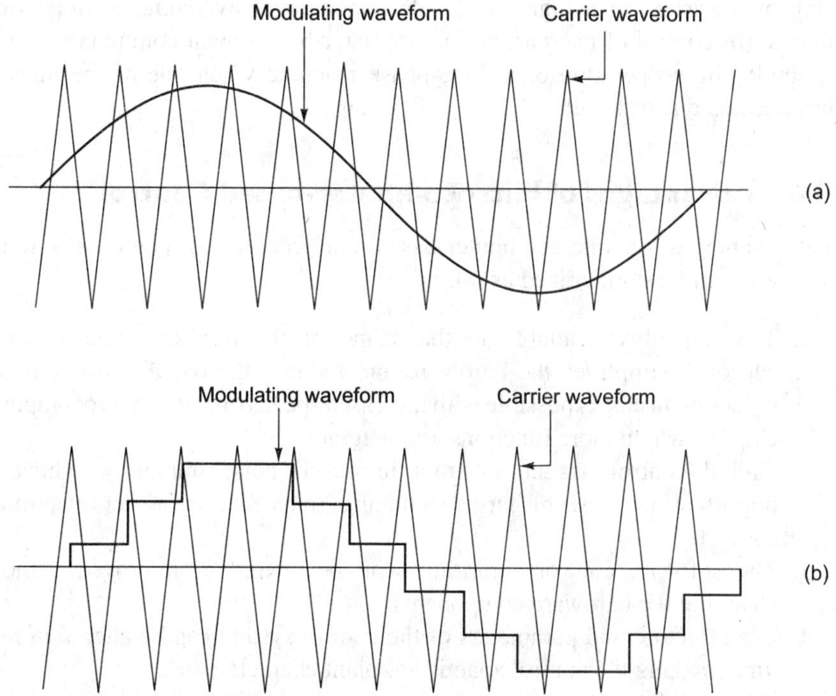

Fig. 13.8 (a) Natural sampling method and (b) uniform sampling method for a sinusoidal modulating waveform

The uniform sampling method has been found to show considerable improvement in performance by way of eliminating the subharmonics when non-integer frequency ratios are used. Further, it is easily adaptable to microprocessor implementation. Yet another merit is that there is a considerable reduction of memory requirement as compared to natural sampling. This is because the number of sine values needed to fully define one cycle of the stepped version of the modulating waveform is equal to the carrier ratio used. In contrast, the number of samples required for a normal sine wave to describe the complete cycle at intervals of 0.5° is 720. A digital pulse width modulator based on regular sampling is commercially available as a large-scale integration (LSI) chip.

13.5 Protection Including Fault-monitoring and Control

System protection under faulty conditions as well as fault preventing control are now assigned to the microcomputer. When, under emergency conditions, the circuits have to be switched off by way of disconnecting the base-/gate drive signals, dedicated hardware may also be employed. For example, the operating conditions of high pressure and low water level in a boiler are translated into Boolean expressions based on the corresponding transducer signals. Examples of such logic involving situations in power electronic applications are (a) selective firing of thyristors in a high-voltage dc converter to overcome commutation failures, (b) control of overcurrent in a thyristor by resonant commutation, and (c) single-phase operation of a three-phase machine when one of the lines is disconnected due to a fault.

13.6 Advantages of Microcomputer-based Control

The advantages of microcomputer-based applications to power electronic converters can be summarized as follows.

1. The flexibility available in the shape of the hardware-cum-software approach simplifies the hardware and reduces the overall cost. Further reduction in cost is possible with increased speeds of newer microcomputer chips in which more functions are integrated.
2. Digital control has an inherent improved noise immunity, which is important in view of large switching transients in power electronic converters.
3. The software control algorithms can be altered or improved without changing the hardware components.
4. The structure and parameters of the control system can be altered in real time, making the control adaptive to plant characteristics.
5. Powerful diagnostics can be incorporated in the software, thus facilitating the reduction of system down time. The related software can be designed for use by semi-skilled technicians.
6. The complex computational and decision making capabilities of the microcomputer facilitate the application of optimal and adaptive control strategies, which help in the optimization of the drive system.
7. The reliability of an LSI or a very large scale integration (VLSI) chip is much higher than that of analog electronic circuits that involve large number of components. Thus, system reliability, which is estimated as the mean time between failure (MTBF) can be increased. Reliability is further improved by temperature derating and by employing high-quality components.
8. A high level of integration of IC chips avoids large voltage and current transients in a microcomputer-based system and helps in the reduction of electromagnetic interference (EMI) problems. Only nominal shielding on controller hardware is sufficient. Noise coupling through the input supply and other power supplies can be minimized by providing good analog or digital filtering.
9. With digital signal processing, the problems of signal drifts and parameter variations present in analog controllers can be eliminated.
10. The current trend is that of adopting hierarchical control, which involves a minicomputer at a higher level, with the microcomputer located at the drive level. A mainframe is normally installed at the highest level to prepare management-related tabulations such as production schedules, etc. In the microcomputer-based control of drives, compatibility with the same level or higher level of computers does not pose any problems.

13.7 Limitations of Microcomputer-based Control

Because of the execution of the control functions in a serial manner, a microcomputer-based system proves to be slower than a dedicated analog or digital hardware control system. Dedicated analog hardware signals are processed in parallel paths. A microcomputer used for multitasking gets further slowed down due to time-sharing. These problems can, however, be solved by multiprocessor configurations. The other problems are the following.

1. Quantization error is introduced as a consequence of the use of A/D and D/A converters. This can be minimized by increasing the bit sizes of the microcomputer and signal converters.
2. Monitoring of intermediate signals is not possible due to signal processing by software. As against this, monitoring of a hardware-controlled system proves to be simpler and does not present any problems.
3. Software development, especially when assembly language is used, may be time-consuming and expensive, but this additional cost is justified by the outweighing advantages.

Summary

The use of microcomputers in the field of dc and ac drives as also in other industrial applications involving power electronic converters has come to stay. For dc drive control, the speed and current signals are fed into the microcomputer, which then compares the output signals with set values and implements the desired control strategy by means of software. In the current versus slip frequency technique employed for speed control, and other similar controls, the characteristics can be stored as lookup tables in the memory of the microcomputer. The PWM technique can be implemented by way of microcomputer software, thus providing considerable flexibility of operation. Microcomputers have also been successfully used for fault-monitoring and prevention.

The software-cum-hardware approach increases the flexibility and decreases the overall cost in most applications. The microcomputer also facilitates real-time control. Immunity from noise signals and reliability of LSI or VLSI chips are the other merits of microcomputer-based control. Though software development is time-consuming and proves to be slower than hardwired analog and digital schemes, microcomputer-based control has become popular because of its other outweighing merits.

Exercises

Multiple Choice Questions

1. A dc drive has the speed feedback in the ______ loop and current feedback in the ______ loop.

   (a) open, outer
   (b) inner, outer
   (c) inner, open
   (d) outer, inner

2. In a closed-loop dc drive, the functions performed by the microcomputer are ________.
   (a) speed measurement, current control, and error detection
   (b) current measurement, speed control, and error detection
   (c) speed control, current control, and error detection
   (d) speed control, current control, and current measurement
3. In the scheme for slip current control of an induction motor drive, the current error signal is fed into the ________ and the synchronous speed signal is fed into the ________.
   (a) rectifier, inverter
   (b) rectifier, rectifier
   (c) inverter, rectifier
   (d) inverter, inverter
4. For implementing constant flux control using the current versus slip frequency method, the rotor current is expressed as $I_r =$ ________, where f is the supply frequency and $f_{R\omega}$ is the slip frequency.
   (a) $(E_s/f)/(R_r/f + j2\pi L_r)$
   (b) $(E_s f)/(R_r f_{R\omega} + j2\pi L_r)$
   (c) $(E_s/f)/(R_r f_{R\omega} + j2\pi L_r)$
   (d) $(E_s/f)/(R_r/f_{R\omega} + j2\pi L_r)$
5. In the scheme for microcomputer-based control of an induction motor, the input signals to the microcomputer are ________.
   (a) the terminal voltage of the motor, actual speed signal, and actual current signal
   (b) the reference speed signal, actual speed signal, and actual current signal
   (c) the reference speed signal, reference current signal, and actual speed signal
   (d) the reference speed signal, terminal voltage of the motor, and reference current signal
6. In the scheme for microcomputer-based control of an induction motor, the current signal is sensed ________.
   (a) at the input supply lines of the rectifier
   (b) at the dc link
   (c) at the input lines of the induction motor
   (d) at the windings of the induction motor
7. In the implementation of the PWM technique, the ________ signal is asymmetrical about the trough of the carrier; also the memory requirement with the ________ PWM is smaller.
   (a) uniformly sampled, naturally sampled
   (b) naturally sampled, naturally sampled
   (c) naturally sampled, uniformly sampled
   (d) uniformly sampled, uniformly sampled
8. As compared to dedicated digital hardware implementation of a drive system, the speed of a microcomputer-based system is ________ and the monitoring of intermediate signals ________.
   (a) high, does not pose problems
   (b) slow, poses problems
   (c) slow, does not pose problems
   (d) high, poses problems

True or False

Indicate whether the following statements are *true* or *false*. Briefly justify your choice.

1. A dc drive has the speed feedback in the inner loop and the current feedback in the outer loop.
2. In a closed-loop dc drive, the functions performed by the microcomputer are speed control, current control, and error detection.
3. In the scheme for slip-current control of an induction motor drive, the current error signal is fed into the inverter and the synchronous speed signal is fed into the rectifier.
4. For implementing constant flux control using the current versus slip frequency method, the rotor current is expressed as $I_r = (E_s/f)/(R_r f_{R\omega} + j2\pi L_r)$.
5. In the scheme for microcomputer-based control of an induction motor, the input signals to the microcomputer are the reference speed signal, actual speed signal, and current signal.
6. In the scheme for microcomputer-based control of an induction motor, the current signal is sensed at the dc link.
7. A uniformly sampled pulse is asymmetrical about the trough of the carrier and the memory requirement with a naturally sampled PWM is smaller.
8. As compared to dedicated digital hardware implementation of a drive system, the speed of a microcomputer-based system is high and the monitoring of intermediate signals poses problems.

Short Answer Questions

1. What are the various industrial applications of microcomputers?
2. Describe a microcomputer-based closed-loop reversible dc drive. What are the functions performed by the microcomputer in it?
3. Describe the uniform sampling method of PWM wave generation.
4. Describe a typical induction motor drive that implements constant flux control.
5. What are the advantages of the uniform sampling method of PWM wave generation over the natural sampling method?
6. Give examples of microcomputer-based protection of equipment and explain the part played by the microcomputer in them.
7. List out the advantages of microcomputer-based control used for drive systems.
8. Give the drawbacks of microcomputer-based control of drive systems. What are the techniques that minimize/eliminate them?
9. Show that adding Eqns (13.4) and (13.7) and then taking the magnitude results in the equation

$$|I_s| = |I_m| \left[\frac{R_r^2 + \omega_{R\omega}^2 (L_r + L_m)^2}{R_r^2 + \omega_{R\omega}^2 L_r^2} \right]^{1/2}$$

CHAPTER 14

Field Oriented Control of Ac Motors

14.1 Introduction

The dc motor has been widely used because the simplicity of its structure makes the implementation of speed control quite straightforward and easy. The complex structure of an ac motor, on the other hand, makes its control complicated.

Field oriented control is a technique that consists of transforming the dynamic structures of both synchronous and asynchronous types of ac motors to that of a separately excited, compensated dc motor. This helps in considerable improvement of their performance throughout the speed range, including fast torque response under four-quadrant operation.

After going through this chapter the reader should

- understand the mechanical analogies of (a) a dc motor and (b) a three-phase induction motor,
- get acquainted with the broad steps that are involved in field oriented control using a mathematical model of the induction motor,
- understand the general block diagram that explains the principle of field oriented control using the mathematical model of the induction motor,
- become familiar with the implementation of field oriented control for the position control of a wound rotor induction motor fed by a voltage source pulse width modulator (PWM),
- know how the microprocessor implementation of speed control of an induction motor fed by a current source inverter is achieved using the principle of field oriented control,
- understand the application of field oriented control for a synchronous motor fed by a three-phase, line commutated cycloconverter, and
- get acquainted with some other aspects of field oriented control, namely, (a) direct and indirect field oriented control, (b) implementation using stator currents, (c) problems associated with flux acquisition, etc.

A comparison of the torque production in dc and ac motors is made below using mechanical analogies. A strategy is then evolved for the development of the field oriented control of ac motors.

14.1.1 Torque Production in Electrical Motors

14.1.1.1 Dc motor A tangential force is produced in a dc motor due to the interaction of the radial magnetic flux density B_f, which is set up by the field current i_f flowing through the windings of the stationary poles, and the axial currents i_a flowing in the conductors situated on the periphery of the rotating armature. In the two-pole dc set-up shown in Fig. 14.1, the forces F_i and F_i' experienced by the conductors in slots A and B, respectively, form a couple. The torque T_d developed by the motor is the sum of the torques produced by the conductors in all such pairs of slots. Current is supplied to the armature through the commutator brushes so that the armature magnetomotive force (mmf) can get fixed up in space and displaced by 90° from the stationary main field axis. This perpendicular relationship is independent of the speed of rotation, and hence the electromagnetic torque T_d of the dc motor is proportional to the product of the field flux and the armature current. Assuming that the magnetic saturation is negligible, the field flux is proportional to the field current and is unaffected by the armature current because of the orthogonal (or 90°) displacement. Considering a

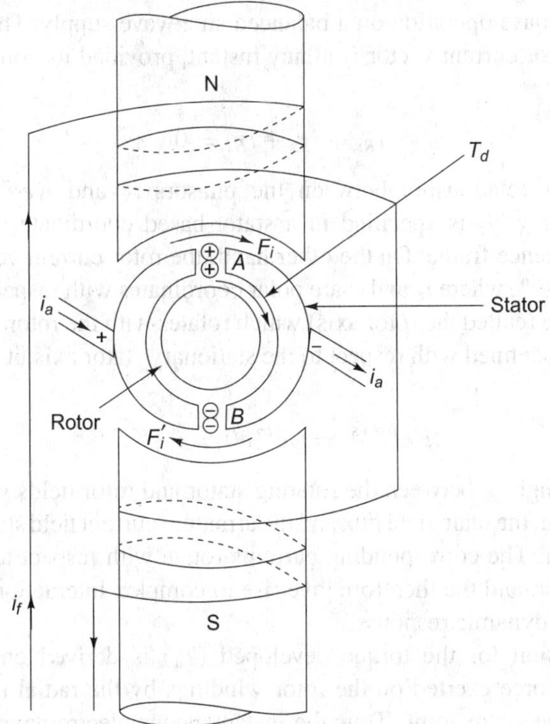

Fig. 14.1 Torque production in a two-pole dc motor

separately excited dc motor, if the field current is kept constant, the torque will be proportional to the armature current. The field flux and the developed torque can therefore be independently controlled.

14.1.1.2 Induction motor A three-phase, wound rotor induction motor consists of three stator windings and three rotor windings [Fig. 14.2(a)] (Murphy & Turnbull 1988). The following discussion is applicable to the squirrel cage type of induction motor also, with the understanding that the cage rotor can be represented by an equivalent three-phase wound rotor. The stator currents in the three-phases i_{Rs}, ai_{Ys}, and $a^2 i_{Bs}$ ($a = e^{j2\pi/3}$) can be combined to give the complex stator current vector shown in Fig. 14.2(b) as follows:

$$\bar{i}_s = i_{Rs} + i_{Ys}e^{j(2\pi/3)} + i_{Bs}e^{j(4\pi/3)} \tag{14.1}$$

This complex stator current vector can be interpreted as a space vector related to the resultant stator current distribution or fundamental mmf wave due to the three phases, as shown in Fig. 14.2(b). The three vectors i_{Rs}, ai_{Ys}, and $a^2 i_{Bs}$ are stationary vectors representing the sinusoidal spatial mmf contributions of the individual stator phases, with the spatial maximum of each phase mmf on the axis of the phase winding. The vector $\bar{i}_s$ therefore indicates the instantaneous magnitude and angular position of the resultant stator mmf wave. For balanced sinusoidal three-phase currents, this vector has a constant amplitude and rotates with constant angular velocity, and represents the uniformly rotating field for normal three-phase operation on a balanced sine wave supply. Thus Eqn (14.1) defines the stator current vector $\bar{i}_s$ at any instant, provided its components obey the relation

$$i_{Rs} + i_{Ys} + i_{Bs} = 0$$

The angular relationship between the phasors $\bar{i}_s$ and $\bar{i}_r e^{j\varepsilon}$ is shown in Fig. 14.2(d) as χ. i_s is specified in a stator-based coordinate system with a stationary reference frame. On the other hand, the rotor current vector $\bar{i}_r$ can be expressed as $i_r e^{j\alpha}$, where i_r and α are polar coordinates with respect to a moving reference frame (called the rotor axis) which rotates with the rotor [Fig. 14.2(c)]. If this vector is defined with respect to the stationary stator axis, it is represented as

$$i_r e^{j(\alpha+\varepsilon)} = i_r e^{j\alpha} e^{j\varepsilon} = \bar{i}_r e^{j\varepsilon}$$

The space angle χ between the rotating stator and rotor fields varies with the load. In addition, the main field flux and the armature current field strongly interact with each other. The corresponding currents rotate with respect to the stator as well as the rotor, and the therefore give rise to complex interactions that lead to an osciallatory dynamic response.

The expression for the torque developed (T_d) is derived on the basis of the tangential force exerted on the rotor windings by the radial magnetic field produced by the stator mmf. Thus the instantaneous electromagnetic torque is obtained as (Leonhard 1985)

$$\bar{T}_d = \frac{2}{3} M \Im[\bar{i}_s(\bar{i}_r e^{j\varepsilon})^*] \tag{14.2}$$

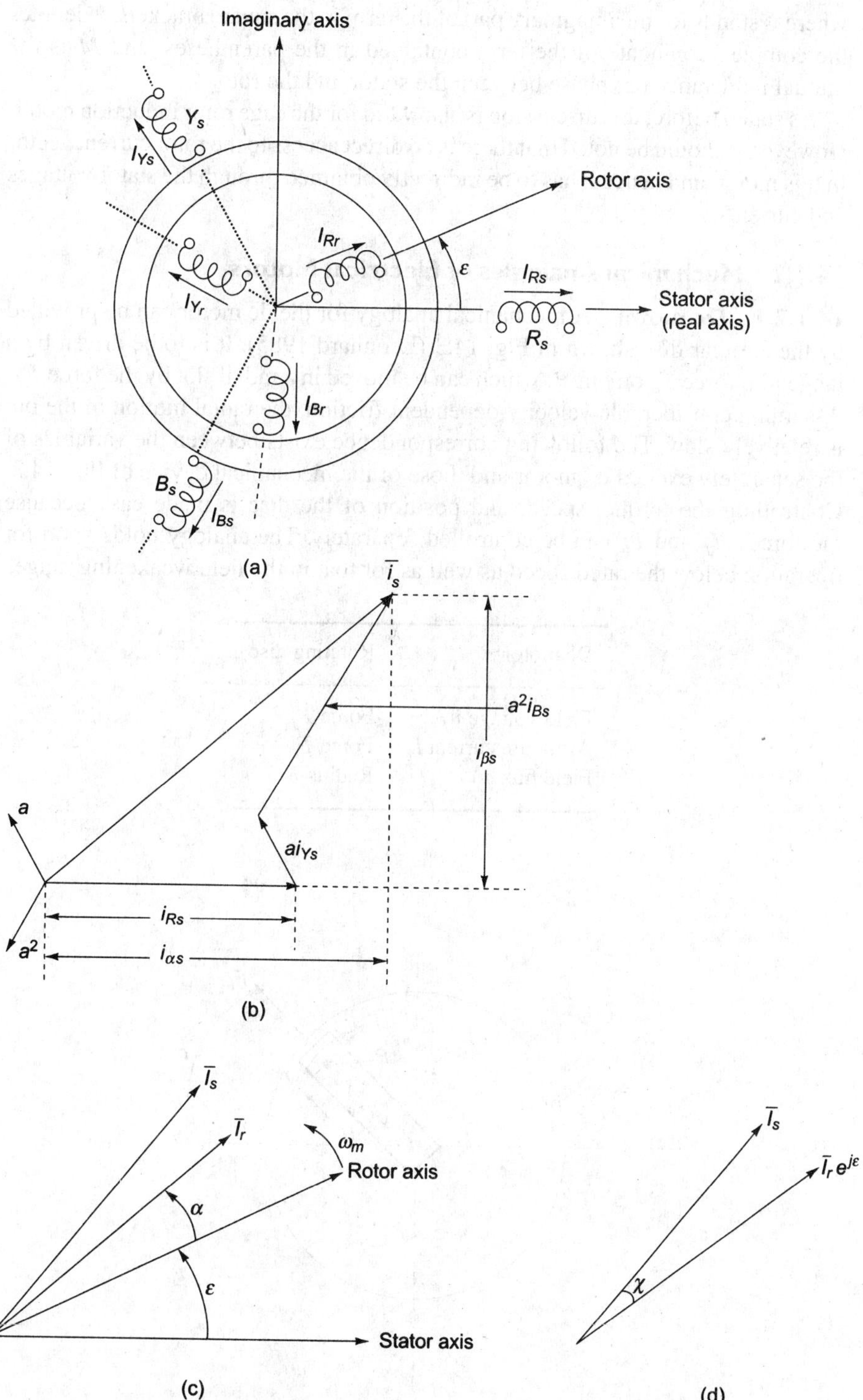

Fig. 14.2 (a) Schematic diagram of a three-phase, wound rotor induction motor; (b) components of the stator current phasor; (c) relative positions of stator- and rotor current phasors (d) rotor current vector with respect to rotor axis is $\bar{I}_r = I_r e^{j\alpha}$ and with respect to stator axis it is $\bar{I}_r e^{j\varepsilon}$

where $\Im$ stands for the imaginary part of the term in the square brackets, * denotes the complex conjugate of the term contained in the parentheses, and M is the mutual inductance per phase between the stator and the rotor.

As stated before, this discussion is also valid for the cage rotor induction motor. However, it should be noted that there is no direct access to the rotor current vector in this motor, and hence it has to be indirectly obtained through the stator voltages and currents.

14.1.2 Mechanical Analogies of Electrical Motors

14.1.2.1 Dc motor A mechanical analogy for the dc motor can be provided by the circular disc shown in Fig. 14.3 (Leonhard 1985). It is to be driven by a tangential force F_a on pin P, which can be moved in a radial slot by the force F_f. Assuming considerable velocity-dependent friction, the radial motion of the pin is relatively slow. The following correspondence exists between the variables of the separately excited dc motor and those of the mechanical device of Fig. 14.3. Controlling the torque, speed, and position of the disc is made easy because the forces F_f and F_a can be controlled separately. The analogy holds good for operation below the rated speed as well as for that in the field weakening range.

Dc motor	Rotating disc
Field voltage v_f	Force F_f
Armature current i_a	Force F_a
Field flux ϕ_f	Radius R

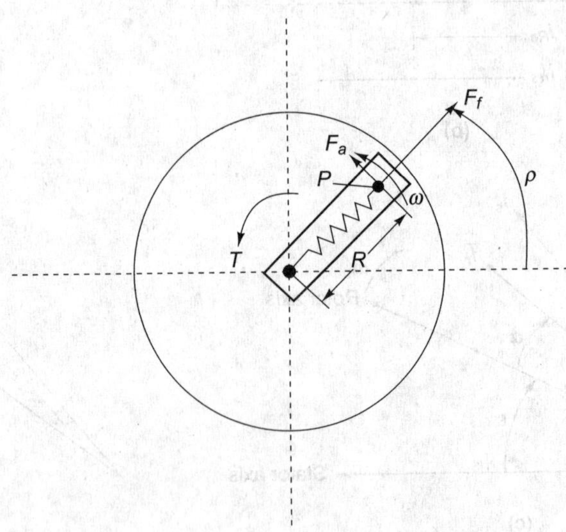

Fig. 14.3 Mechanical analogy of a dc motor

14.1.2.2 Induction motor The circular disc of Fig. 14.4 provides a mechanical analogy for the cage rotor induction motor. In this case the pin is driven by the three connecting rods which apply forces f_R, f_Y, and f_B representing the three stator currents. In order to produce smooth circular motion keeping the pin at a given radius, it is necessary to generate constant radial and tangential forces. This in turn needs a well-coordinated set of alternating forces f_R, f_Y, and f_B to be generated. The analogy will be complete if the polar coordinates R and ρ of the pin are assumed to correspond to the polar position of the fundamental flux wave.

A comparison of Figs 14.3 and 14.4 suggests the following steps for simplification of disc control in the latter.

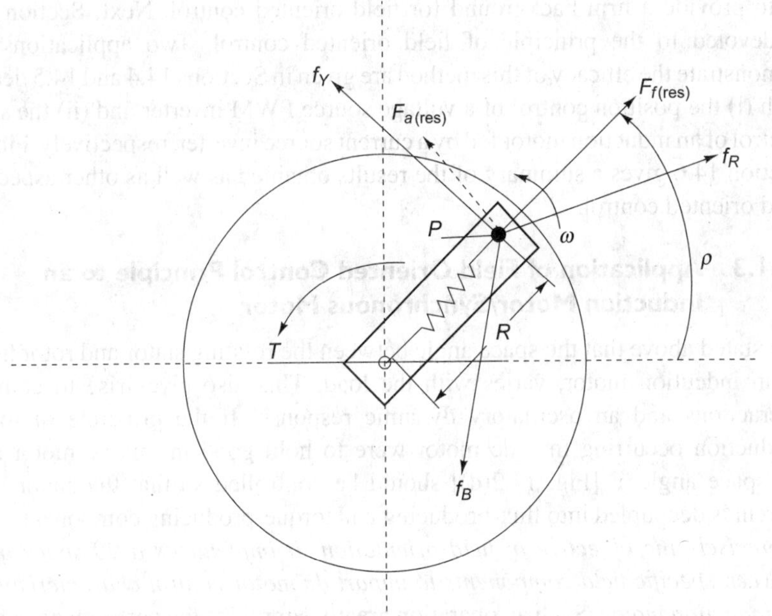

Fig. 14.4 Mechanical analogy of a three-phase induction motor

(a) The speed controller determines the requisite driving force to maintain a specified speed for the disc of Fig. 14.4.

(b) The two orthogonal forces $F_{a(\text{res})}$ and $F_{f(\text{res})}$ are computed.

(c) By a two- to three-coordinate transformation, the forces $F_{a(\text{res})}$ and $F_{r(\text{res})}$ are converted to the forces f'_R, f_Y, and f'_B based on the instantaneous angular position ρ of the pin. Such conversion implies the necessity of flux measurement in a field oriented control scheme.

Based on the steps described above for the mechanical device of Fig. 14.4, *the concept of field oriented control can be evolved as follows:*

(a) Similar to the two-axis theory of synchronous machines, a variable frame of reference can be defined based on the rotor flux which rotates around the rotor with slip speed.

(b) The three stator currents can be considered to be composed of two orthogonal components, namely, a current i_{ds} in the direction of the flux axis, representing the field current of a dc motor and another current i_{qs} perpendicular to the flux axis, representing the armature current of the dc motor, which produces torque. These two components can be computed with the help of flux measurement as well as the information about the torque required to maintain a specific speed.

(c) By a two- to three-coordinate transformation, the currents i_{ds} and i_{qs} are transformed into the stator currents i_{Rs}, i_{Ys}, and i_{Bs} which are actually supplied to the motor.

A mathematical model of an induction motor is developed in Section 14.2 so as to provide a firm background for field oriented control. Next, Section 14.3 is devoted to the principle of field oriented control. Two applications that demonstrate the efficacy of this method are given in Sections 14.4 and 14.5 dealing with (i) the position control of a voltage source PWM inverter and (ii) the speed control of an induction motor fed by a current source inverter, respectively. Finally, Section 14.6 gives a summary of the results obtained as well as other aspects of field oriented control.

14.1.3 Application of Field Oriented Control Principle to an Induction Motor/Synchronous Motor

It is stated above that the space angle between the rotating stator and rotor fields, in an indcution motor, varies with the load. This also gives rise to complex interactions and an oscillatory dynamic response. If the principle of torque production occurring in a dc motor were to hold good in this ac motor also, the space angle χ [Fig. 14.2(d)] should be controlled so that the stator input current is decoupled into flux-producing and torque-producing components. *This is precisely the objective of field orientation: it implements a 90° space angle between specific field components to impart dc motor control characteristics to the induction motor*. Such an operation greatly simplifies the performance of this ac motor, which is inherently complex as emphasized above.

The practical implementation of the principle of field orientation requires information regarding the magnitude and position of the rotor flux vector. The control action takes place in a field coordinate system using the rotor flux vector as a reference frame for the stator currents (and voltages). The transformed stator currents are then resolved into direct and quadrative axis components, which are, respectively, similar to the field and armature currents of a compensated dc motor.

14.2 Mathematical Model of an Induction Motor

The analysis presented below is based on the assumption that the machine is a three-phase, two-pole induction motor with a short-circuited wound or cage rotor. It is further assumed that (a) the spatial harmonics, slot effects, eddy currents, and iron losses can be neglected, (b) the iron core is of infinite permeability, and (c) the stator and rotor have equal number of turns. The following equations can be written for describing the dynamics of the motor in terms of the stator- and rotor current phasors:

$$R_s\bar{i}_s + L_s\frac{d\bar{i}_s}{dt} + M\frac{d}{dt}(\bar{i}_r e^{j\varepsilon}) = \bar{v}_s \tag{14.3}$$

$$R_r\bar{i}_r + L_r\frac{d\bar{i}_r}{dt} + M\frac{d}{dt}(\bar{i}_s e^{-j\varepsilon}) = 0 \tag{14.4}$$

$$J\frac{d\omega_m}{dt} = \overline{T}_d - \overline{T}_{\mathrm{ld}} = \frac{2}{3}M\Im[\bar{i}_s(\bar{i}_r e^{j\varepsilon})^*] - \overline{T}_{\mathrm{ld}} \tag{14.5}$$

$$\frac{d\varepsilon}{dt} = \omega_m \tag{14.6}$$

The expression given for the torque (T_d) developed in the motor [Eqn (14.5)] is the instantaneous electromagnetic torque of the motor. J, ω_m, and ε are, respectively, the inertia constant, angular speed, and angular position of the motor. T_{ld} is the effective load torque. Similar to the complex stator current vector $\bar{i}_s$ [Eqn (14.1)], the complex rotor current vector $\bar{i}_r$ is constituted of the rotor current phasors in the three phases. Thus,

$$\bar{i}_r(t) = i_{Rr}(t) + i_{Yr}(t)e^{j(2\pi/3)} + i_{Br}(t)e^{j(4\pi/3)} \tag{14.7}$$

The stator current phasor can be written in rectangular coordinates as [Fig. 14.2(b)].

$$\bar{i}_s(t) = i_{\alpha s} + j i_{\beta s} \tag{14.8}$$

where $i_{\alpha s}$ and $i_{\beta s}$ can be considered as the instantaneous currents in an equivalent two-phase stator winding that establishes the same resultant mmf as the three-phase winding. This pair can also be interpreted as the currents obtained after a three- to two-coordinate transformation on the currents i_{Rs}, i_{Ys}, and i_{Bs}. The stator voltage phasor is also defined as a combination of phase-to-neutral voltages and is written as

$$\bar{v}_s(t) = v_{Rs}(t) + v_{Ys}(t)e^{j(2\pi/3)} + v_{Bs}(t)e^{j(4\pi/3)} \tag{14.9}$$

Because of the isolated neutral and the symmetry of the three-phase windings, the sum of the phase voltages and phase currents will be equal to zero. R_s and L_s, are respectively, the resistance and leakage inductance per phase of the stator winding. Likewise R_r and L_r are the parameters of the rotor winding. M is the mutual inductance per phase. σ_s, σ_r, and σ are, respectively, the stator, rotor, and total leakage factors. With the assumption of equal stator and rotor windings, these three leakage factors are defined by the following equations:

$$L_s = (1 + \sigma_s)M \tag{14.10}$$

$$L_r = (1 + \sigma_r)M \tag{14.11}$$

and

$$\sigma = 1 - \frac{1}{(1+\sigma_s)(1+\sigma_r)} \tag{14.12}$$

The mutual air gap flux is a measure of the magnetizing current and can be detected by the stator search coils or Hall effect sensors. This measurement defines the magnetizing current vector $\bar{i}_m$, which is the sum of the stator- and rotor current vectors with respect to the stator axis. Thus,

$$\bar{i}_m = \bar{i}_s + \bar{i}_r e^{j\varepsilon} \tag{14.13}$$

Equation (14.13) can be used to eliminate the rotor current term $\bar{i}_r e^{j\varepsilon}$ in Eqn (14.3) and also the term $\bar{i}_r$ in Eqn (14.4). However, a better choice is to define a modified magnetizing current phasor representing the rotor flux, including the leakage flux. The reason for this can be understood from the discussion that follows the development of the model of the induction motor in field coordinates. This phasor is denoted by $\bar{i}_{mr}$ with reference to the stator axis and can be written as

$$\bar{i}_{mr} = \bar{i}_s + (1+\sigma_r)\bar{i}_r e^{j\varepsilon} \tag{14.14}$$

Referring to the phasor diagram of Fig. 14.5, $\bar{i}_{mr}$ can also be written as

$$\bar{i}_{mr} = i_{mr} e^{j\rho} \tag{14.15}$$

where i_{mr} is the magnitude of the phasor $\bar{i}_{mr}$ and ρ is the phase angle (called the flux angle) with respect to the stator axis. The instantaneous angular velocity of this phasor is given as

$$\omega_{mr} = \frac{d\rho}{dt} \tag{14.16}$$

which in the steady state corresponds to the angular velocity ω_1 of the stator current phasor. The angular displacements between the various phasors as well as the angular velocities are given in Fig. 14.5. If the stator voltages and currents are referred to this moving frame of coordinates, the corresponding direct- and

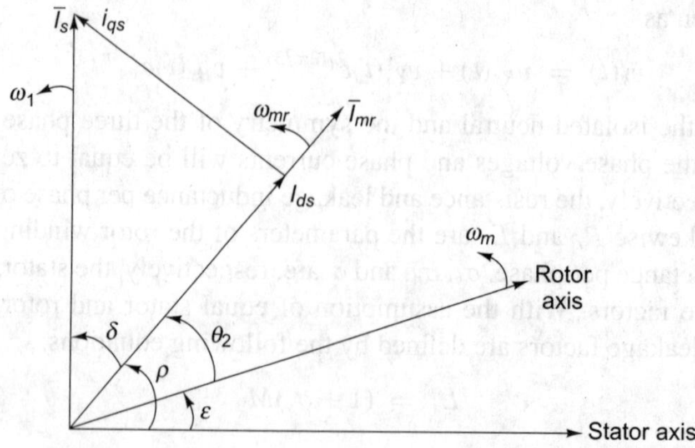

Fig. 14.5 Relative positions of the stator current and rotor flux phasors

quadrature axis components are obtained. If the harmonics generated in the inverter are neglected, these orthogonal components are constant in the steady state. These are given as

$$\overline{v}_s e^{-j\rho} = v_{ds} + j v_{qs} \tag{14.17}$$

where

$$\overline{v}_s = v_{\alpha s} + j v_{\beta s} \tag{14.18}$$

The phasor $e^{-j\rho}$ causes a coordinate transformation from stator-based coordinates to the moving frame of reference defined by the phasor $\overline{i}_{mr}$, and

$$\overline{i}_s e^{-j\rho} = i_{ds} + j i_{qs} = \overline{i}_s e^{j\delta} \tag{14.19}$$

If $\overline{i}_{mr}$ is introduced into Eqn (14.5), the revised expression for $\overline{T}_d$ can be written as

$$\begin{aligned}\overline{T}_d &= \frac{2}{3}\frac{M}{(1+\sigma_r)}\Im[\overline{i}_s(\overline{i}_{mr} - \overline{i}_s)^*] \\ &= \frac{2}{3}\frac{M}{(1+\sigma_r)}\Im[\overline{i}_s\,\overline{i}_{mr}^{\,*}]\end{aligned} \tag{14.20}$$

Using Eqn (14.15) for $\overline{i}_{mr}$, Eqn (14.20) can be written as

$$\overline{T}_d = \cdot\frac{2}{3}\frac{M}{(1+\sigma_r)} i_{mr}\Im[\overline{i}_s e^{-j\rho}] \tag{14.21}$$

From Eqn (14.19), as well as from Fig. 14.5, the term $\Im[\overline{i}_s e^{-j\rho}]$ can be seen to be equal to i_{qs}. Thus Eqn (14.21) can be written as

$$\overline{T}_d = \frac{2}{3}\frac{M}{(1+\sigma_r)} i_{mr} i_{qs} = k i_{mr} i_{qs} \tag{14.22}$$

where $k = (2/3)[M/(1+\sigma_r)]$.

Equation (14.22) shows that the developed torque of the induction motor is proportional to the product of the magnitudes of the magnetizing current phasor $\overline{i}_{mr}$ and the quadrature component of the stator current phasor $\overline{i}_{qs}$. This bears analogy to the torque of an unsaturated, separately excited dc motor, which is proportional to the product of field and armature currents. The quadrature axis component i_{qs} is analogous to the armature current of the dc machine and can be quickly varied to provide a quick response to a sudden torque demand.

Equations (14.10)–(14.12), (14.14), and (14.15) may be inserted into Eqns (14.1), (14.3)–(14.7), and (14.9) to obtain the transformed model of the induction motor. First, the stator-related differential equation [Eqn (14.3)] is taken up for transformation into the new frame of reference and is reproduced here:

$$R_s\overline{i}_s + L_s\frac{d\overline{i}_s}{dt} + M\frac{d}{dt}(\overline{i}_r e^{j\varepsilon}) = \overline{v}_s \tag{14.23}$$

Writing the phasor $\bar{i}_s$ in terms of the new coordinates as in Eqn (14.19) and eliminating $\bar{i}_r e^{j\varepsilon}$ from Eqn (14.14) gives

$$R_s[i_{ds}e^{j\rho} + ji_{qs}e^{j\rho}] + \left(L_s - \frac{M}{1+\sigma_r}\right)\frac{d}{dt}[i_{ds}e^{j\rho} + ji_{qs}e^{j\rho}]$$
$$+ \frac{M}{(1+\sigma_r)}\frac{d}{dt}(i_{mr}e^{j\rho}) = v_{ds}e^{j\rho} + jv_{qs}e^{j\rho} \tag{14.24}$$

Using Eqns (14.10)–(14.12), $L_s - M/(1+\sigma_r)$ can be written as σL_s. Also performing the differentiations indicated, Eqn (14.24) becomes

$$e^{j\rho}R_s[i_{ds} + ji_{qs}] + \sigma L_s\left[\frac{di_{ds}}{dt}e^{j\rho} - \frac{d\rho}{dt}i_{qs}e^{j\rho}\right] + j\sigma L_s\left[\frac{d\rho}{dt}i_{ds} + \frac{di_{qs}}{dt}\right]e^{j\rho}$$
$$+ \frac{Me^{j\rho}}{1+\sigma_r}\left(\frac{di_{mr}}{dt} + j\frac{d\rho}{dt}i_{mr}\right) = v_{ds}e^{j\rho} + jv_{qs}e^{j\rho} \tag{14.25}$$

Multiplying throughout by $e^{-j\rho}$ gives

$$R_s[i_{ds} + ji_{qs}] + \sigma L_s\left[\frac{di_{ds}}{dt} - \frac{d\rho}{dt}i_{qs}\right] + j\sigma L_s\left[\frac{d\rho}{dt}i_{ds} + \frac{di_{qs}}{dt}\right]$$
$$+ \frac{M}{1+\sigma_r}\left(\frac{di_{mr}}{dt} + j\frac{d\rho}{dt}i_{mr}\right) = v_{ds} + jv_{qs} \tag{14.26}$$

Separating the real and imaginary parts in Eqn (14.26) and taking L_s/R_s as the time constant τ_s gives

$$\sigma\tau_s\frac{di_{ds}}{dt} + i_{ds} = \frac{v_{ds}}{R_s} + \sigma\tau_s\omega_{mr}i_{qs} - (1-\sigma)\tau_s\frac{di_{mr}}{dt} \tag{14.27}$$

$$\sigma\tau_s\frac{di_{qs}}{dt} + i_{qs} = \frac{v_{qs}}{R_s} - \sigma\tau_s\omega_{mr}i_{ds} - (1-\sigma)\tau_s\omega_{mr}i_{mr} \tag{14.28}$$

Equations (14.27) and (14.28) describe the interactions due to stator voltages. If fast control loops are available for the variables i_{ds} and i_{qs}, then the undesirable terms $\sigma\tau_s\omega_{mr}i_{qs}$, $-(1-\sigma)\tau_s(di_{mr}/dt)$, $-\sigma\tau_s\omega_{mr}i_{ds}$, and $-\tau_s(1-\sigma)\omega_{mr}i_{mr}$ can be cancelled by the application of compensating signals to the reference inputs. If an impressed current scheme involving i_{ds} and i_{qs} is used, these two equations can be neglected because the two components i_{ds} and i_{qs} are then the control inputs to the transformed plant.

In the next step, the dynamic equation for rotor current is taken up for transformation. Eliminating the terms in $\bar{i}_r$ from Eqn (14.4) with the help of Eqn (14.14) gives

$$\frac{R_r}{1+\sigma_r}[\bar{i}_{mr}e^{-j\varepsilon} - \bar{i}_s e^{-j\varepsilon}] + L_r\frac{d}{dt}\left[\frac{\bar{i}_{mr}e^{-j\varepsilon} - \bar{i}_s e^{-j\varepsilon}}{1+\sigma_r}\right] + M\frac{d}{dt}(\bar{i}_s e^{-j\varepsilon}) = 0 \tag{14.29}$$

The terms $[-L_r/(1+\sigma_r)](d/dt)(\bar{i}_s e^{-j\varepsilon})$ and $M(d/dt)(\bar{i}_s e^{-j\varepsilon})$ cancel each other by virtue of Eqn (14.11). Equation (14.29) can now be written as

$$\frac{R_r\bar{i}_{mr}e^{-j\varepsilon}}{1+\sigma_r} + \frac{L_r}{1+\sigma_r}\frac{d}{dt}[\bar{i}_{mr}e^{-j\varepsilon}] = \frac{R_r}{1+\sigma_r}\bar{i}_s e^{-j\varepsilon} \tag{14.30}$$

Multiplying throughout by $(1 + \sigma_r)e^{j\varepsilon}$ and performing the differentiation gives

$$R_r\bar{i}_{mr} + L_r\left[\frac{d\bar{i}_{mr}}{dt} - j\frac{d\varepsilon}{dt}\bar{i}_{mr}\right] = R_r\bar{i}_s \tag{14.31}$$

Writing $\bar{i}_{mr}$ as $i_{mr}e^{j\rho}$, $\bar{i}_s$ as $i_s e^{j(\delta+\rho)}$ and dividing throughout by R_r gives

$$i_{mr}e^{j\rho} + \tau_r\frac{d(i_{mr}e^{j\rho})}{dt} - j\omega_m\tau_r i_{mr}e^{j\rho} = i_s e^{j(\delta+\rho)} \tag{14.32}$$

Multiplying throughout by $e^{-j\rho}$ and performing the differentiation on the left-hand side gives

$$\tau_r\frac{di_{mr}}{dt} + i_{mr} + j\tau_r\omega_{mr}i_{mr} - j\tau_r\omega_m i_{mr} = i_{ds} + ji_{qs} \tag{14.33}$$

where L_r/R_r and $d\rho/dt$ are, respectively, taken as τ_r and ω_{mr} [Eqn (14.16)]. Also, $d\varepsilon/dt$ is taken as ω_m [Eqn (14.6)] after performing the differentiation of the second term in Eqn (14.32). Separating out the real and imaginary parts gives the final equation as

$$\tau_r\frac{di_{mr}}{dt} + i_{mr} = i_{ds} \tag{14.34}$$

and

$$\omega_{mr} - \omega_m = \frac{i_{qs}}{\tau_r i_{mr}} = \omega_2 \tag{14.35}$$

Equations (14.5), (14.22), (14.27), (14.28), (14.34), and (14.35) define the model of the induction motor in field coordinates.

In the induction motor, i_{mr} is analogous to the main field flux of the dc machine and is controlled by i_{ds}, the direct component of the stator current vector [Eqn (14.34)]. However, the rotor time constant τ_r introduces a fairly large time lag in the response of i_{mr} to a change in i_{ds}. This time lag may be around 1 s for a large machine and it plays the same part as the time lag in the response of the field flux of a dc machine to a variation in field voltage.

All these equations except Eqns (14.27) and (14.28) can be represented by the block diagram of Fig. 14.6 (Murphy & Turnbull 1988). According to the earlier statement, the three- to two- phase transformation gives the orthogonal two-phase currents ($i_{\alpha s}$ and $i_{\beta s}$). The second conversion from ($i_{\alpha s}$, $i_{\beta s}$) to (i_{ds}, i_{qs}) is based on the rotor flux angle ρ. The vector $\bar{i}_s$ is rotated by an angle $-\rho$ to obtain these currents, which are in field oriented coordinates. Thus,

$$\begin{aligned}\bar{i}_s e^{-j\rho} &= (i_{\alpha s} + ji_{\beta s})(\cos\rho - j\sin\rho)\\ &= (i_{\alpha s}\cos\rho + i_{\beta s}\sin\rho) + (i_{\beta s}\cos\rho - i_{\alpha s}\sin\rho)\\ &= i_{ds} + ji_{qs}\end{aligned} \tag{14.36}$$

The use of the dynamic flux model given in Fig. 14.6 has the following advantages.

(a) The flux signals are not based on local measurements and are not affected by slot harmonics and stator resistance.
(b) Using a standard motor without an additional sensor is possible.
(c) Flux sensing is operative down to zero frequency because no open-ended integration, which would be subjected to drift, is needed.

However, there are stringent accuracy requirements, for example with respect to modulo 2π integration of ρ; this can be best solved by digital computation in a microprocessor as will be shown later. The dependence of the rotor model on the time constant τ_r still constitutes a source of error because a false position ρ of the flux wave would lead to undesirable coupling between the d- and q-axis and could eventually invalidate the idea of control in field coordinates, even resulting in instability.

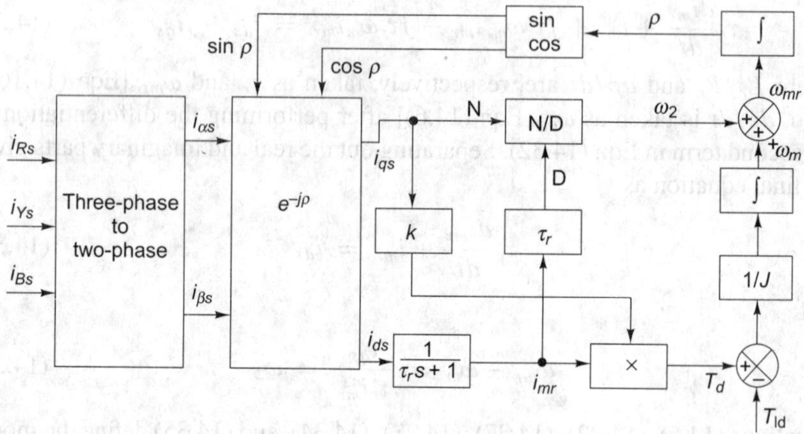

Fig. 14.6 Block diagram representation of an induction motor in field coordinates

14.3 Implementation of Field Oriented Control for a VSI Controlled Ac Motor

Figure 14.7 is a general block diagram explaining the implementation of field oriented control with the example of the speed control of an ac motor fed by a voltage source inverter (Gabriel & Leonhard 1982). Beginning with the components i_{ds} and i_{qs}, Eqns (14.5), (14.22), (14.34), and (14.35) of the motor are represented in detail, indicating that the magnetizing current i_{mr} (rotor flux) can only be changed through a large lag τ_r as in the case of a dc machine. The quadrature component, being delayed by the leakage time constant τ_s [Eqn (14.28)], can be changed quickly, thus providing a means of rapidly controlling the torque. This offers a clue as to how the motor ought to be operated: (a) the magnetizing current i_{mr} should be maintained at maximum value, below rated speed, limited

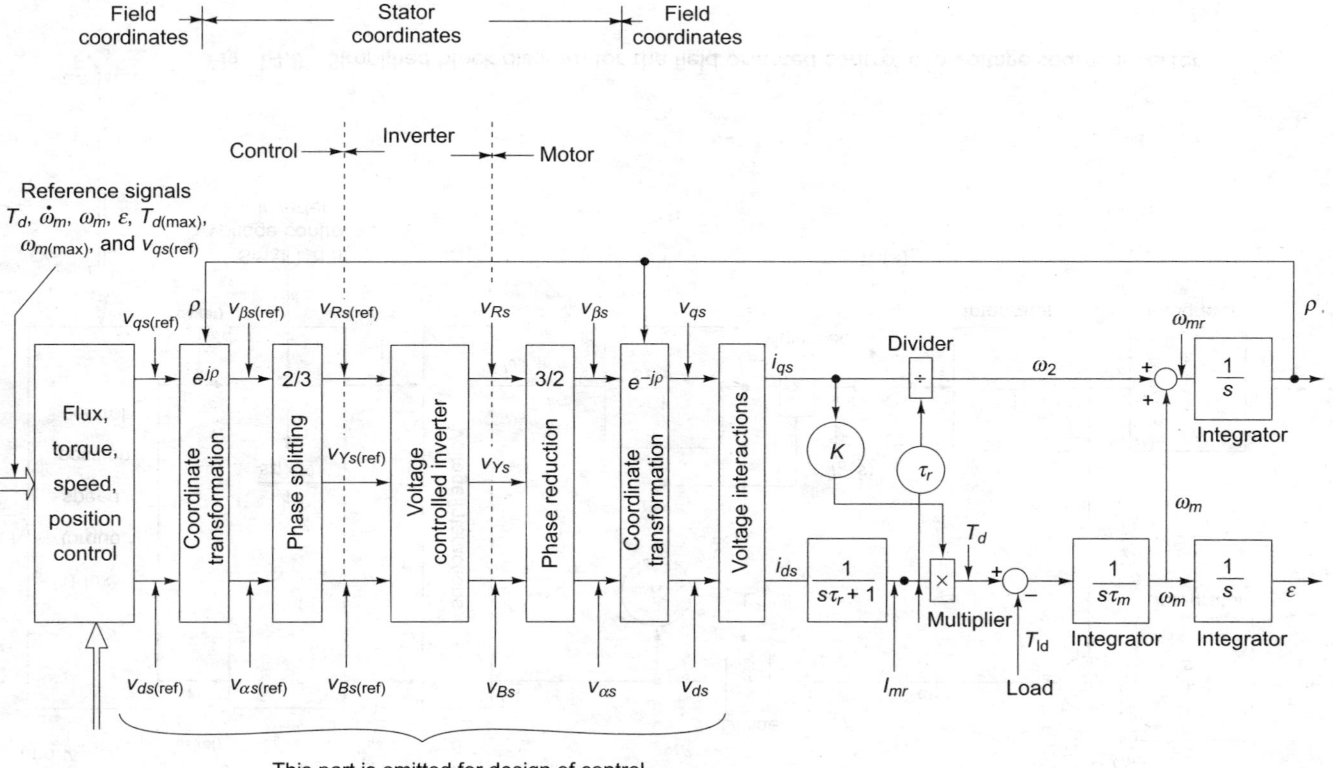

Fig. 14.7 General block diagram for the field oriented control of an a.c. motor fed by a voltage source inverter

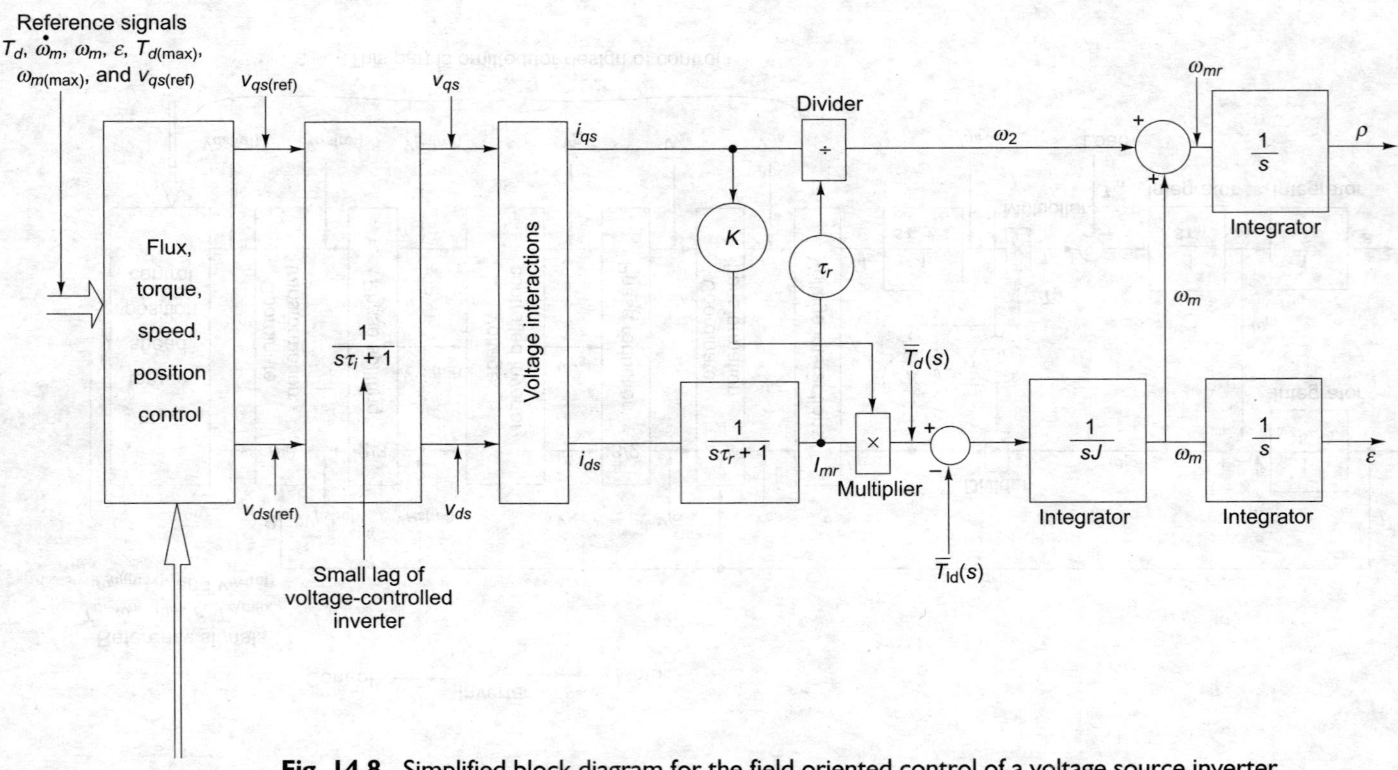

Fig. 14.8 Simplified block diagram for the field oriented control of a voltage source inverter

by the saturation of the iron; (b) in the field weakening range it should be kept at the voltage ceiling of the inverter; and (c) the torque should be controlled by the quadrature current i_{qs}. Proceeding to the left of Fig. 14.7, the voltage interactions of Eqns (14.27) and (14.28) are encountered as also the coordinate transformation using $e^{-j\rho}$ of the two-phase orthogonal ac phase components $v_{\alpha s}$ and $v_{\beta s}$, which constitute the vector $\overline{v}_s$ and correspond to demodulation. These are given by Eqns (14.17) and (14.18). The two-phase voltages follow immediately from the three-phase terminal voltages v_{Rs}, v_{Ys}, and v_{Bs} of the motor [Eqn (14.9)]. The voltage source inverter is assumed to have three voltage reference inputs $v_{Rs(\text{ref})}$, $v_{Ys(\text{ref})}$, and $v_{Bs(\text{ref})}$, which are produced from the voltages $v_{ds(\text{ref})}$ and $v_{qs(\text{ref})}$ by the inverse operations described above, including a modulation with the flux angle ρ.

By assuming that (a) all the operations on the reference variables are executed correctly and (b) the inverter has only a small control delay that is negligible in comparison with the motor dynamics, the middle portion of Fig. 14.7 can be eliminated, thus simplifying the scheme to that of Fig. 14.8. Therefore the design of field oriented control will be greatly simplified by adjoining the dc quantities $v_{ds(\text{ref})}$ and $v_{qs(\text{ref})}$ through a small equivalent lag to the quantities v_{ds} and v_{qs} within the motor. The control scheme now becomes very simple, containing the field lag, the torque generation, and the mechanical part of the drive.

The modulation of the control signals, using $e^{j\rho}$, is based on the actual flux. This implies that the motor is self-controlled and cannot go out of step. If overloaded, the speed controller will attain saturation and confine the operation to correspond to constant maximum torque, down to standstill, and also to regeneration depending on the operating conditions.

14.4 Position Control with a Voltage Source PWM Inverter

The general concepts developed in the preceding sections are implemented on a voltage source inverter (VSI) drive used for the position control of a wound rotor induction motor as shown in Fig. 14.9 (Gabriel & Leonhard 1982). The VSI is assumed to be of constant voltage and incorporates a standard PWM scheme. The inverter is controlled by two voltage references according to the principle laid down in Fig. 14.8. Two of the stator currents are measured with Hall transducers and fed back for closed-loop speed control purposes. An optical speed sensor is used for speed measurement. The output of this sensor is integrated and combined with positional reference so as to enable the scheme to serve as a position control scheme also. In contrast with Fig. 14.7, here the field is weakened by a limiting controller for the magnitude of the stator voltages.

The following groups of functions are performed by the microprocessor shown in Fig. 14.9.

(i) Flux computation, coordinate transformation, torque control, analog output
(ii) Speed control, τ_r identification
(iii) Voltage limit, field and position control

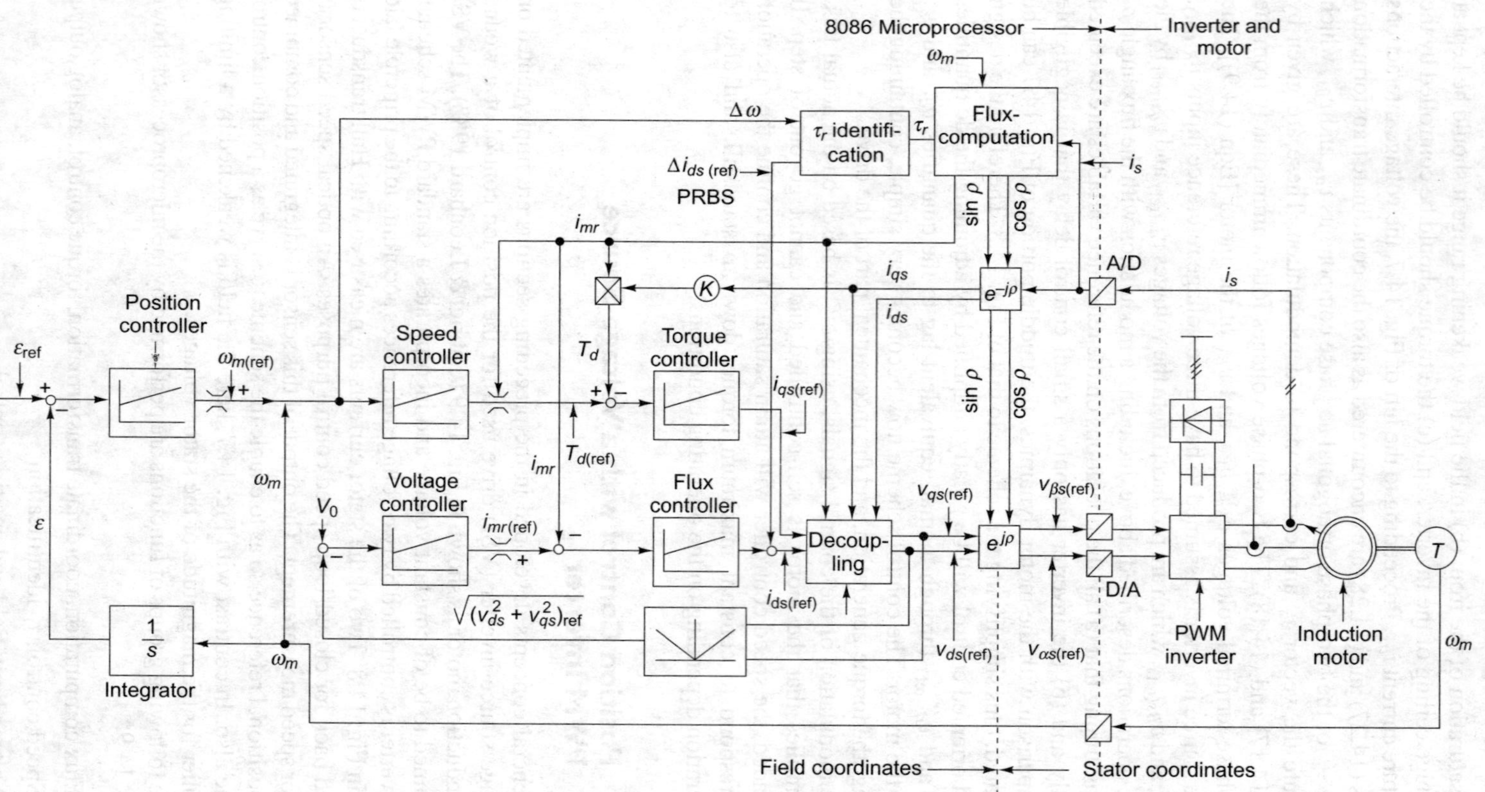

Fig. 14.9 Position control with a voltage source PWM inverter

14.4.1 τ_r Identification

An error in the estimation of the time constant τ_r may be caused due to the temperature dependence of the rotor resistance R_r, this in turn leading to imperfect field oriented control. To detect such an error, an adaptive feature τ_r *identification* has been incorporated into the control structure. It adjusts the flux computation to the changing conditions as follows. A pseudo-random binary sequence (PRBS) generator, which is programmed to function as a shift register with feedback, produces a low-level binary signal which is injected into the d-channel. The response, at the output of the q-channel measured as noise at the input of the speed controller, is correlated with the injected noise signal. If there is a clearly detectable correlation, it is concluded that the d- and q-axis are not orthogonal and that this is a consequence of the error in the value of the rotor time constant τ_r. The sign of the correlation function is utilized for the correction of the parameter τ_r in the flux model so that the two axes are correctly displaced from each other.

14.5 Speed Control of an Induction Motor Fed by a Current Source Inverter

For medium-rated drives, say up to a few hundred kilowatts, current source inverters are chosen because they employ only 12 line grade thyristors; they also give a four-quadrant, six-pulse operation. The cost of such a control-circuit is comparable with a dc drive that uses a dual converter. The main circuit of a current source inverter, shown in Fig. 14.10, consists of a two-quadrant, line commutated converter which supplies a forced-commutated machine side inverter through a large smoothing reactor. A characteristic feature of this circuit is that two alternate phases of the motor carry the dc link current during periods other than commutation intervals. Hence, neglecting commutation, the motor currents are of a six-step form, where 120° block currents are supplied through the motor terminals.

The control task is split into two parts: (a) that of controlling the link current by means of the line side converter and (b) that of commutating the dc current to the appropriate motor phase. By writing the stator current vector in polar form as

$$\bar{i}_s(t) = i_s(t)e^{j\zeta(t)} \tag{14.37}$$

its magnitude (outside the commutation intervals) is determined only by the link current $i_d(t)$ as given by the equation

$$i_s(t) = \sqrt{3}i_d(t) \tag{14.38}$$

The angle ζ is given by the switching state of the inverter, which is commanded by a control circuit containing a six-step ring counter. The state of the ring counter may be represented by a complex number $e^{j\xi(t)}$, where $\xi(t)$ advances discontinuously in either the forward or reverse direction in steps of 60°. Due to the link reactor and reactances of the motor, $i_s(t)$ and $\zeta(t)$ are both continuous functions. Though in the steady state, outside the commutation intervals, $\zeta(t)$ follows the switching state $\xi(t)$ of the ring counter, it is associated with continuous

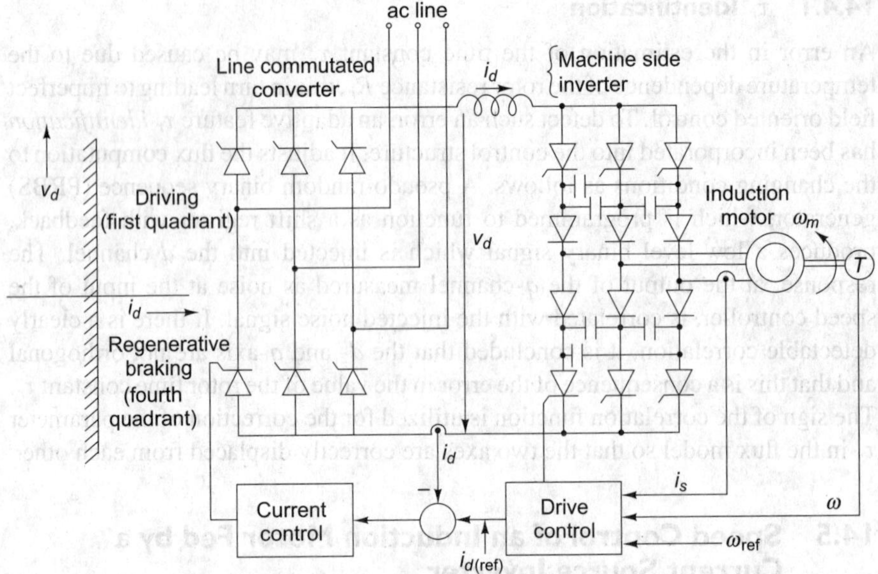

Fig. 14.10 A typical induction motor drive fed by a current source inverter

commutation transients that occur just after switching. This problem is solved by an angle control loop which is described later.

The above-mentioned argument suggests that a current source inverter (CSI) can be controlled in polar, rather than rectangular, coordinates. This is achieved as described below. First, the flux and torque controllers determine the desired stator current in field coordinates. These are combined into the polar form as

$$i_{ds(\text{ref})} + j i_{qs(\text{ref})} = i_{s(\text{ref})} e^{j\delta_{\text{ref}}} \tag{14.39}$$

By adding the flux angle $\rho(\text{mod } 2\pi)$ to the angle δ_{ref}, the stator current vector is obtained as

$$\bar{i}_s(\text{ref}) = i_{s(\text{ref})} e^{j(\delta_{\text{ref}} + \rho)} = i_{s(\text{ref})} e^{j\zeta_{\text{ref}}} \tag{14.40}$$

$i_s(\text{ref})$ and ζ_{ref} of Eqn (14.40) provide the command signals, respectively, for the line side converter and machine side CSI.

14.5.1 Application of Field Oriented Control

The principle of field orientation (Leonhard 1985) is now applied to the basic control structure inherent in a power circuit (Fig. 14.11) consisting of the model of the induction motor in field coordinates in the upper right-hand corner, assuming impressed stator current in polar coordinates i_s and ζ. The direct and quadrature current components i_{ds} and i_{qs} are obtained from i_s and $\delta = \zeta - \rho$ by a polar to rectangular transformation (P/R). The inverse operation is performed at the output of the flux and torque controllers. The transformation into stator coordinates is achieved by adding the angle $\rho(\text{mod } 2\pi)$ of the flux wave determined by some method of flux acquisition. The output signals of the drive control are the reference

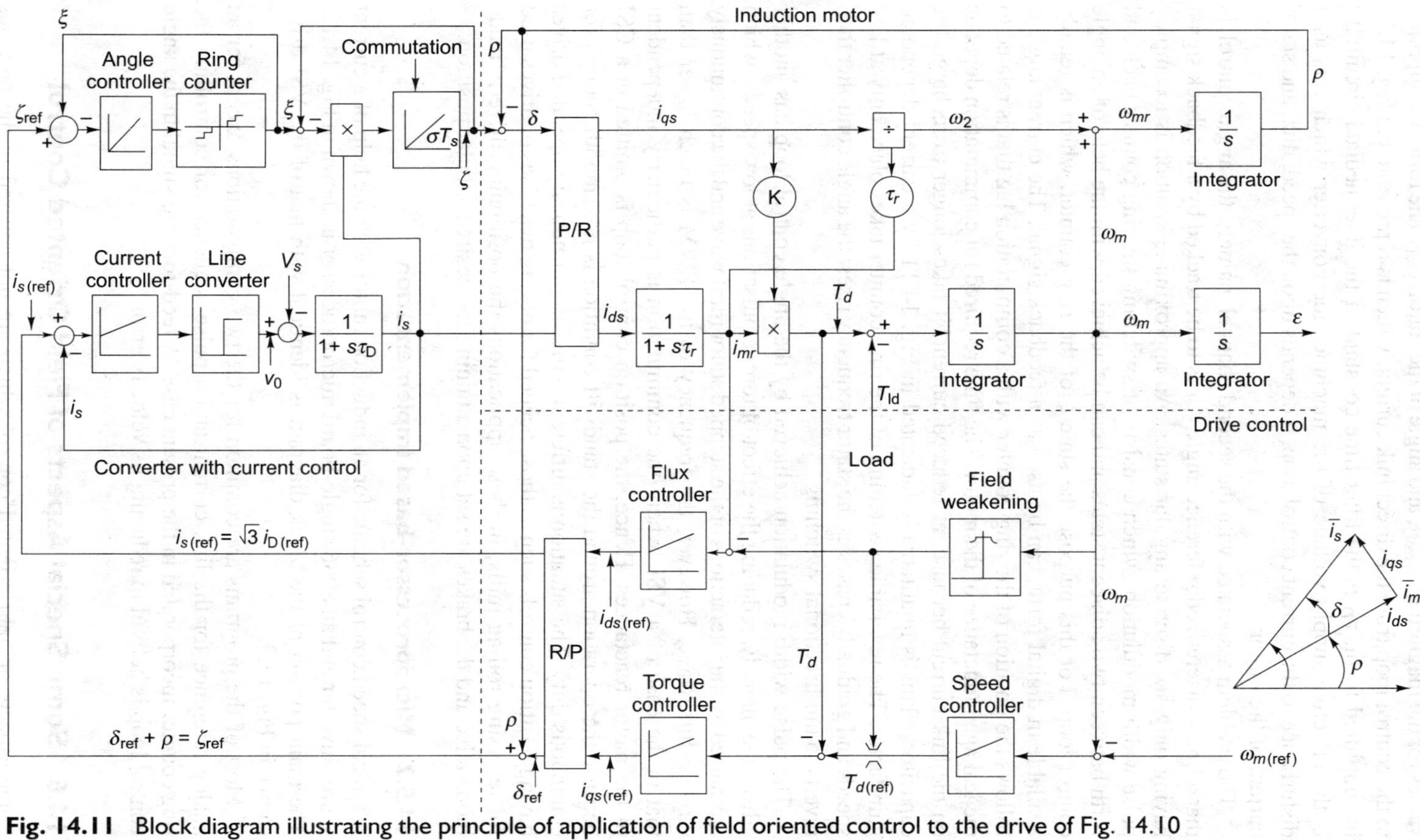

Fig. 14.11 Block diagram illustrating the principle of application of field oriented control to the drive of Fig. 14.10

values for the magnitude, $i_{s(\text{ref})}$, and angle of the stator, ζ_{ref}. The former is supplied to the control loop from the direct link current, which is represented in Fig. 14.11 in simplified form. τ_D is the filter time constant of the intermediate circuit; V_s is the effective supply voltage of the machine side converter which acts as a disturbance to the current control loop, it depends upon the speed, flux, and stator current of the motor.

The problem associated with the second channel, namely, the angle control, is that a finely discretized reference angle ζ_{ref} is to be tracked by a feedback signal having only six discrete angular states. An approximate solution is to employ pulse width modulation, which consists of switching the ring counter back and forth between two adjacent states and can be achieved with the help of an angle control loop. For this purpose the state ξ of the ring counter, which is readily available in digital form, can be used as a feedback signal. The current angle ζ follows the position of the ring counter with a commutation lag that is related to the leakage inductance of the motor. The time required for commutation depends on the link current because, at reduced current, it takes longer to recharge the capacitors. This is qualitatively indicated in Fig. 14.11 by a current dependent gain factor. The reversible switching of the ring counter takes place only at low speeds and gradually ceases as the stator frequency rises; the angle controller then reverts to unidirectional switching.

The pulse width modulation achieved by the angle control loop has another advantage, namely, reducing the effect of torque pulsations at low speeds, which are caused by the interactions of the stepped stator mmf wave and the continuously moving flux wave. However, the frequency of the PWM is much lower than that in the case of a VSI, where the commutation is particularly independent of the motor reactances. Hence, the position control loop is omitted in a CSI because the operation around the standstill condition is not smooth enough for continuous duty. The situation is different with discontinuously operated drives having position controls such as those needed for boring machines or drives used for reversing rolling mills; in these applications the positioning drive is made inoperative and the brakes are set upon attaining the desired angular position.

14.5.2 Microprocessor-based Implementation

A typical speed control scheme for an induction motor with the help of a current source inverter and an 8086 single-board microprocessor is shown in Fig. 14.12. The main portion of the block diagram is identical with that of the VSI drive shown in Fig. 14.9.

Most of the programs are common for the two configurations except for an analog reference for the link current and a pulse sequence for controlling the ring counter incorporated in the present case. A feed-forward signal in the angle control loop is helpful in reducing its velocity error.

14.6 Some Special Aspects of Field Oriented Control

Some aspects of field oriented control which are not covered in the previous sections are discussed here.

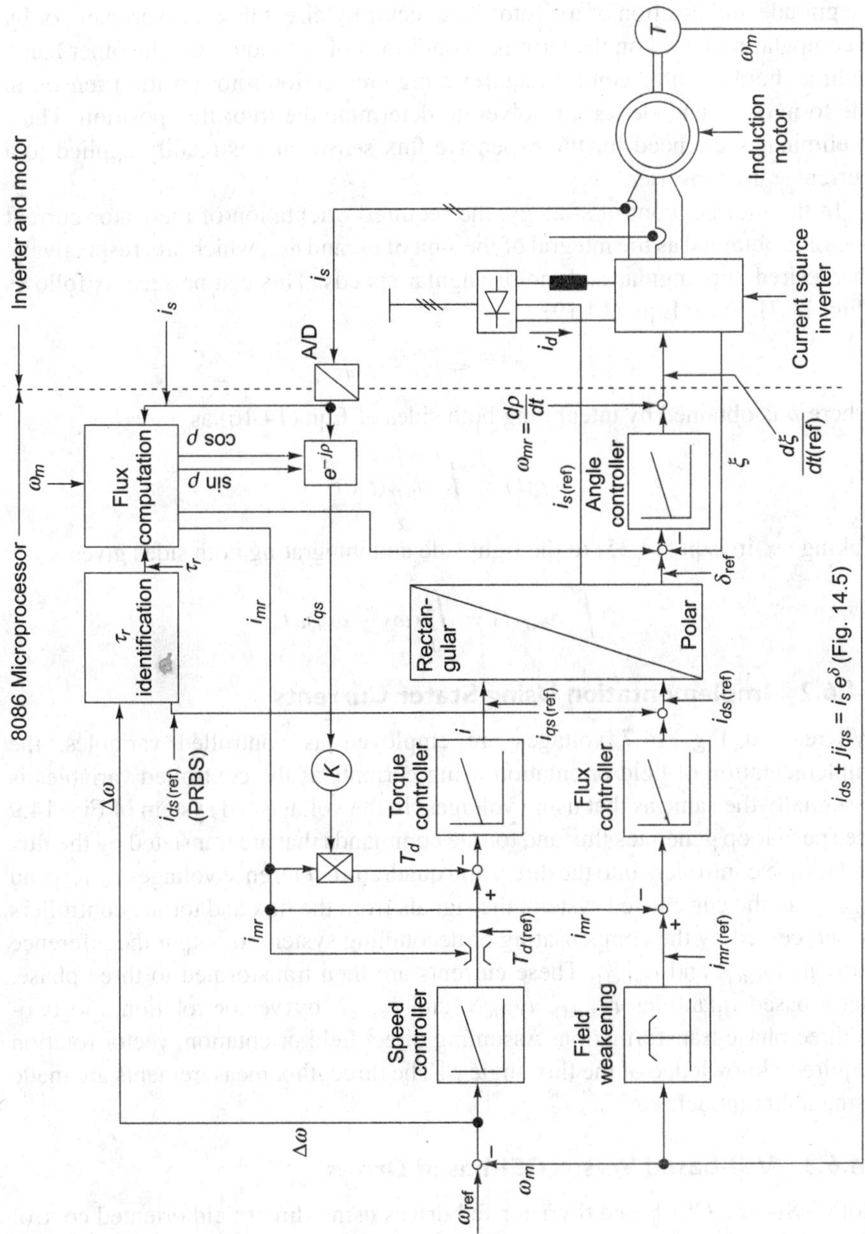

Fig. 14.12 Microprocessor implementation of the scheme of Fig. 14.11 for the drive of Fig. 14.10

14.6.1 Direct and Indirect Field Oriented Control

The implementation of field oriented control can be classified into two groups: (a) direct control and (b) indirect control. The classification is based on the method that is employed to determine the rotor flux vector. All the methods described above are direct control methods which involve determination of the magnitude and position of the rotor flux vector by direct flux measurement or by a computation based on the terminal conditions of the motor. On the other hand, indirect field oriented control requires a high resolution rotor position sensor, in the form of an encoder or a resolver, to determine the rotor flux position. Thus, it eliminates the need for the expensive flux sensor and is readily applied to a current source inverter.

In the indirect control strategy, the required orientation of the stator current vector is obtained as the integral of the sum of ω_2 and ω_m, which are, respectively, the desired slip angular and motor angular speeds. This can be seen as follows [Fig. 14.7]. From Eqn. (14.19),

$$\bar{i}_s e^{-j\rho} = i_{ds} + j i_{qs}$$

where ρ is obtained by integrating both sides of Eqn (14.16) as

$$\rho(t) = \int_0^t \omega_{mr}(t)dt$$

Taking ω_m in Eqn (14.35) to the right side and integrating both sides gives

$$\int_0^t \omega_{mr}(t) = \int_0^t (\omega_2 + \omega_m)dt$$

14.6.2 Implementation Using Stator Currents

Whereas in Fig. 14.7 voltages are employed as controlled variables, the implementation of field orientation using currents as the controlled variables is essentially the same as that using voltages. In the voltage-fed system of Fig. 14.9 the speed loop generates flux and torque commands that are translated by the flux and torque controllers into the direct and quadrature reference voltages $v_{ds(\text{ref})}$ and $v_{qs(\text{ref})}$. In the current-fed system, the signals from the flux and torque controllers are processed by the compensating or decoupling system to output the reference currents $i_{ds(\text{ref})}$ and $i_{qs(\text{ref})}$. These currents are then transformed to three-phase, stator-based quantities $i_{Rs(\text{ref})}$, $i_{Ys(\text{ref})}$, and $i_{Bs(\text{ref})}$ by vector rotation and two- to three-phase transformation. Assuming direct field orientation, vector rotation requires a knowledge of the flux angle ρ. The direct flux measurements are made using a flux model.

14.6.3 VSI-based Versus CSI-based Drives

Both VSI- and CSI-based thyristor-fed drives using direct field oriented control show similar behaviour in steady-state operations as well as for large-amplitude transients. A difference in operation exists, however, at low speeds, with which the CSI causes a slightly cogging motion while the voltage source PWM inverter drive provides a perfectly smooth operation up to and including the standstill condition. Hence, the VSI-type PWM inverter drive is suited for sensitive loads such as elevators, winders, paper mills, or auxiliary steel-mill drives.

The CSI drive has the merit of performing with very little audible noise. With the VSI-based PWM inverter on the other hand, noise is audible but can be minimized using special switching sequences.

14.6.4 Problems Associated with Flux Acquisition

The vector rotation in Fig. 14.7 is based on the rotor flux angle, and hence the precise value of the rotor flux position is of great importance for the success of the scheme. One method is to measure the flux density in the air gap of the machine directly by positioning appropriately spaced magnetoresistive devices such as Hall sensors on the face of the stator teeth. Signals from local samples can be interpolated to get an estimate of the exact magnitude and angular position of the air gap flux wave. The signal that represents the magnetizing current $\bar{i}_{mr}$ can then be obtained by adding a voltage component proportional to the stator current vector. This method suffers from the drawback that Hall sensors are mechanically fragile and cannot withstand severe vibrations and thermal stresses. Harmonics caused by rotor slots further contribute to the inaccuracy in the estimate. A motor on which these sensors are mounted has to be specially manufactured and cannot be a readily available standard motor.

A better method, which also helps in suppressing undesirable slot harmonics, is to place sensing coils, each with a width equal to the full pole pitch, on the stator; one arrangement could be to enclose them in the wedges covering the stator slots. The signals obtained in this manner are proportional to the change of flux, which after integration gives an estimate of the main flux. Then, by adding voltages proportional to the stator currents, signals proportional to $\bar{i}_{mr}$ can be obtained. This method, though satisfactory from normal frequency to a low frequency, say 0.5 Hz, has got the defect that the drift of the integrators becomes large at very low frequencies. Hence it is unsuitable for drives used for position control. The motor will again be of the non-standard type.

Measuring the terminal voltage implies using the stator windings as flux sensing coils, but in this case the resistive drop is predominant, especially at low frequencies. It has to be compensated for before integrating the voltage signals. An added complication is that of change of stator resistance with temperature.

Yet another approach is to use the model equations of the motor. Equation (14.30), which is a vectorial first-order differential equation, may be employed for estimation of the phasor $\bar{i}_{mr}(t)$ in stator coordinates. The other data required for computing $\bar{i}_{mr}(t)$ are the rotor time constant τ_r, the direct-axis and quadrature axis components i_{ds} and i_{qs} of the stator current, and finally the angular velocity ω_2 [Eqns (14.34) and (14.35)]. A rotation using $e^{-j\rho}$ converts the stator current phasor $\bar{i}_s = i_{\alpha s} + ji_{\beta s}$ to the oriented phasor $i_{ds} + ji_{qs}$.

The last alternative is to compute the magnitude and position of the fundamental flux wave based on the flux model in field coordinates, which is represented by Eqns (14.34) and (14.35). As these equations are derived from Eqn (14.30), the stator current vector as well as the angular velocity ω_{mr} are required as inputs, as in the previous case. The flux model in field coordinates shown in Fig. 14.6 represents these equations. This method is preferred to others because of its simplicity and accuracy of computation. Its other merits have been elaborated in Section 14.2. The rotor time constant τ_r, which is a parameter in Eqn (14.34), has to be periodically updated as it is a source of inaccuracy. This is because of the variation of the rotor resistance with temperature.

14.6.5 Field Oriented Control of a Synchronous Motor

Figure 14.13 shows the power circuit of an ac synchronous motor drive in which the stator windings are fed by a three-phase, line commutated cycloconverter (Leonhard 1985). The field winding is fed by another fully controlled thyristor rectifier which gives two-quadrant operation. The output frequency of the

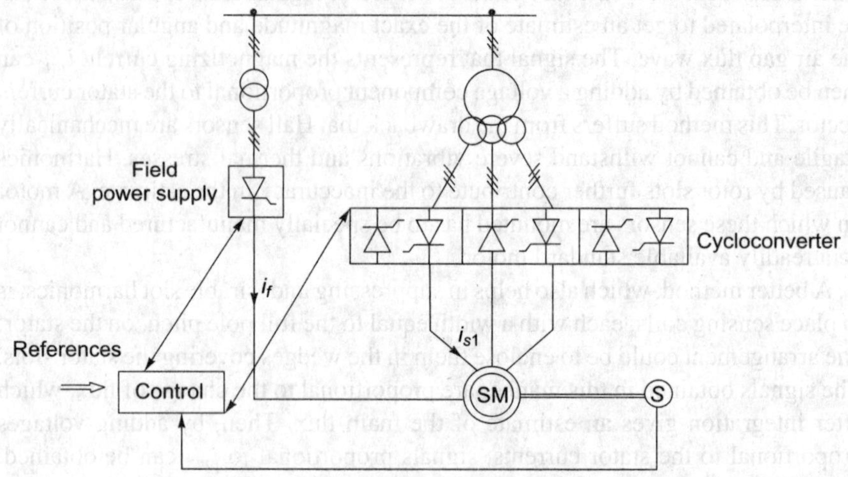

Fig. 14.13 Block-cum-schematic diagram of a synchronous motor drive fed by a cycloconverter

cycloconverter being low, it can only be used for a low-speed drive, say a mine hoist drive. The cycloconverter forms a part of a current control loop, thus tracking the sinusoidal current reference signals. If the current control is considered to be fast, the motor can be viewed as being fed by a current source. This drive differs from that of an induction motor in the aspect of dc excitation given to the rotor from a separate source.

A microprocessor implementation for this drive consists of torque-, speed-, and position control loops, which could be specified for the mine hoist drive under consideration. The field can be weakened by a separate control loop which limits the stator voltage.

Summary

The principle of field oriented control can be applied to the speed and position control of a large number of ac drives. Its adaptation to different constraints becomes easier by considering a simplified model of the symmetrical ac machine.

Though the implementation of field oriented control is seen to pose some complicated problems, it has been demonstrated that these problems can be handled conveniently and effectively with the help of a micro computer and other hardware. The cost of a power converter, however, seems to be a deterrent, and the benefits that accrue out of it need to be carefully assessed prior to its adoption.

It is hoped that the development of low-cost power semiconductor devices with better performance features will lead to a wider use of this control strategy.

Exercises

Multiple Choice Questions

1. The τ_r identification block in field oriented control with a voltage source inverter helps in checking the orthogonality of the phasors ______.

   (a) $i_{\alpha s}$ and $i_{\beta s}$ (b) v_{qs} and v_{ds}
   (c) i_{qs} and i_{ds} (d) $v_{\alpha s}$ and $v_{\beta s}$

2. In the angle control loop of the field control scheme using a current source inverter, reversible switching occurs only in the ______ range.

   (a) low-speed (b) medium-speed
   (c) high-speed (d) very high speed

3. In the mechanical analogy of a separately excited dc motor, the force F_f and radius R correspond to the ______ and ______, respectively.

   (a) field flux ϕ_f, armature current i_a
   (b) field voltage v_f, field flux ϕ_f
   (c) armature current i_a, field voltage v_f
   (d) field flux ϕ_f, field voltage v_f

4. In the block diagram that illustrates the principle of field oriented control using a voltage source inverter, the ______ block follows the *phase reduction* block and the ______ block precedes the *phase splitting* block.

   (a) voltage interaction, coordinate transformation
   (b) voltage controlled inverter, coordinate transformation
   (c) coordinate transformation with $e^{j\rho}$, coordinate transformation with $e^{-j\rho}$
   (d) coordinate transformation with $e^{-j\rho}$, coordinate transformation with $e^{j\rho}$

5. The rotor time constant τ_r makes the response of ______ slow, whereas the stator time constant $\sigma\tau_s$ makes the response of ______ fast.

   (a) i_{mr}, i_{qs} (b) i_{mr}, i_{ds}
   (c) i_{ds}, i_{qs} (d) i_{ds}, i_{mr}

6. The dynamic flux model used for the computation of the flux vector has the drawback that ______.

   (a) it is affected by slot harmonics and stator resistance
   (b) it depends on the stator time constant τ_s, which may vary
   (c) it depends on the rotor time constant τ_r, which may vary
   (d) it requires a costly sensor that measures the magnitude and angle of the flux vector

7. The expression for the torque T_d developed in an induction motor is given as ______.

   (a) $(2/3)M\Im[\bar{i}_r(\bar{i}_s e^{-j\varepsilon})^*]$ (b) $(2/3)M\Im[\bar{i}_s(\bar{i}_r e^{-j\varepsilon})^*]$
   (c) $(2/3)M\Im[\bar{i}_s(\bar{i}_r e^{j\varepsilon})^*]$ (d) $(2/3)M\Im[\bar{i}_r(\bar{i}_s e^{j\varepsilon})^*]$

8. If the phasor $\bar{i}_s$ is to be expressed with respect to the phasor i_{mr}, then it has to be multiplied by ______.

   (a) $e^{-j\varepsilon}$ (b) $e^{j\delta}$
   (c) $e^{j\varepsilon}$ (d) $e^{-j\delta}$

9. In the block diagram that explains the principle of field control of a VSI, a ______ transformation is required before the inverter and a ______ transformation is required after it.
   (a) three- to two-phase, two- to three-phase
   (b) two- to three-phase, two- to three-phase
   (c) three- to two-phase, three- to two-phase
   (d) two- to three-phase, three- to two-phase

10. With field oriented control, an induction motor cannot go out of step and can be termed as a self-controlled motor because ______.
    (a) its actual speed is fed back and keeps it in step
    (b) its terminal voltage is fed back and does the necessary regulation
    (c) the modulation of the control signals using $e^{j\rho}$ is based on the actual flux position
    (d) the flux computation is based on its actual terminal voltage

11. In the field control scheme using a voltage source PWM inverter, the output of the flux controller is ______ and that of the torque controller is ______.

    (a) $v_{qs(\text{ref})}$, $v_{ds(\text{ref})}$ (b) $i_{ds(\text{ref})}$, $i_{qs(\text{ref})}$
    (c) $i_{\alpha s(\text{ref})}$, $i_{\beta s(\text{ref})}$ (d) $i_{qs(\text{ref})}$, $i_{ds(\text{ref})}$

12. Field weakening with respect to field oriented control using a VSI and a CSI is achieved, respectively, by a ______ and a ______.
    (a) function generator, voltage controller
    (b) function generator, function generator
    (c) voltage controller, voltage controller
    (d) voltage controller, function generator

13. In the case of direct field control, the magnitude and position of the rotor flux vector is obtained by the ______, whereas in the case of indirect field control the flux vector details are obtained from the ______.
    (a) rotor position sensor, computations based on terminal conditions
    (b) direct flux measurement or computation based on terminal conditions, rotor position sensor
    (c) direct flux measurement or by means of a rotor position sensor, computations based on terminal conditions
    (d) computation based on terminal conditions or a rotor position sensor, computations based on a rotor position sensor

14. Field oriented control using a current source inverter has the drawback that it has a slightly cogging motion in the ______.
    (a) medium- and high-speed regions
    (b) standstill and high-speed regions
    (c) standstill and medium-speed regions
    (d) standstill and low-speed regions

15. The disadvantage of field control using a voltage source PWM inverter over that with a CSI is that its performance is accompanied with ______.
    (a) audible noise (b) torque pulsations
    (c) cogging (d) crawling
16. The stator currents $i_{\alpha s}$ and $i_{\beta s}$ are obtained from the currents i_{ds} and i_{qs} by multiplying the latter by ______.
    (a) $e^{j\varepsilon}$ (b) $e^{-j\varepsilon}$
    (c) $e^{j\rho}$ (d) $e^{-j\rho}$
17. The basic difference between the structure of a field-controlled synchronous motor and that of a field-controlled induction motor is that in the former ______.
    (a) the rotor needs a separate ac supply
    (b) the rotor needs both ac and dc supplies
    (c) the rotor has a small battery mounted on it
    (d) the rotor needs a separate dc source
18. Direct flux measurement by placing sensing coils, each with a width equal to the full pole pitch, on the stator is advantageous over that done by placing Hall sensors on the face of the stator teeth, but it suffers from the drawback that ______.
    (a) stray flux lines induce unwanted voltages at very low frequencies
    (b) the drift of the integrators is too large at very low frequencies
    (c) stray flux lines induce unwanted voltages at very high frequencies
    (d) the drift of the integrators is too large at very high frequencies
19. Obtaining an estimate of the flux vector by measuring the terminal voltages is superior to that by placing sensing coils of width equal to a pole pitch on the stator, but it has the disadvantage that ______.
    (a) the *IR* drop in the stator at low frequencies may cause errors due to the temperature dependence of the resistance
    (b) the set-up needs a specially manufactured machine to facilitate voltage sensing
    (c) the drift of the integrators is large in the low-speed range
    (d) stray flux lines induce unwanted voltages at very high speeds
20. Though obtaining an estimate of the magnitude and position of the flux density by means of a flux model is the best method to obtain the flux vector, it has the drawback that ______.
    (a) the flux model depends on the stator resistance, which varies with temperature
    (b) the flux model depends on the stator inductance, which may vary due to linkage of stray flux lines
    (c) the rotor time constant τ_r has to be periodically updated because the rotor resistance varies with temperature
    (d) the value of i_{mr} obtained from the equation

$$\tau_r \frac{di_{mr}}{dt} + i_{mr} = i_{qs}$$

    may not be correct because of the incorrectness of i_{qs}.

True or False

Indicate whether the following statements are *true* or *false*. Briefly justify your choice.

1. Considering the mechanical analogy of a separately excited dc motor, the force F_f represents the field flux ϕ_f and the radius R represents the field voltage v_f.
2. The expression for the instantaneous electromagnetic torque developed in an induction motor is given as

$$T_d = 2/3M\Im[\bar{i}_r(\bar{i}_s e^{-j\varepsilon})^*]$$

3. If the stator current phasor $\bar{i}_s$ is to be expressed with respect to the magnetizing current phasor $\bar{i}_{mr}$, then it has to be multiplied by the term $e^{-j\rho}$.
4. In the block diagram that explains the principle of field control of a VSI controlled induction motor, the coordinate transformation block follows the VSI block and the phase reduction block follows the coordinate transformation block.
5. The rotor time constant $\tau_r = L_r/R_r$ is high as compared to the stator time constant $\tau_s = L_s/R_s$.
6. The flux signal obtained as an output of the dynamic flux model has the drawback that it is affected by slot harmonics and stator resistance.
7. In the block diagram that illustrates the principle of field control, the reference signals $v_{qs(\mathrm{ref})}$ and $v_{ds(\mathrm{ref})}$ are subject to coordinate transformation by $e^{j\rho}$ before the VSI and the signals $v_{\alpha s}$ and $v_{\beta s}$ are subject to coordinate transformation by $e^{-j\rho}$ after the VSI.
8. A field oriented control induction motor cannot be pulled out of step because the speed is fed back to the input side, and this makes it a self-controlled motor.
9. In a microprocessor-based voltage source PWM inverter, the signal $i_{qs(\mathrm{ref})}$ is the output of the flux controller whereas the signal $i_{ds(\mathrm{ref})}$ is the output of the torque controller.
10. In field control using a voltage source PWM inverter, field weakening is affected by a voltage controller, whereas in field control using a current source inverter it is done by a function generator.
11. The τ_r identification block in the field-control of a voltage source PWM inverter helps in checking the orthogonality of the stator current vectors $i_{\alpha s}$ and $i_{\beta s}$.
12. In the angle controller used in field control with a current source inverter, reversible switching of the ring counter takes place only at low speeds.
13. In direct field oriented control, the magnitude and position of the rotor flux vector is determined either by direct flux measurement or by rotor position measurement with an encoder or a resolver.
14. Field oriented control with a current source inverter has the disadvantage that the motor rotation is not smooth at low speeds.
15. The advantage of field oriented control with a VSI is that it is accompanied by very little noise.
16. The basic difference between the implementation of field control using stator voltages and that using stator currents is that in the former case the flux and

torque controllers generate $v_{\alpha s}$ and $v_{\beta s}$, whereas in the latter case they generate $i_{\alpha s}$ and $i_{\beta s}$.

17. Direct flux measurement by placing sensing coils, with a width equal to the full pole pitch, on the stator is advantageous over that by placing Hall sensors on the face of the stator teeth, but it suffers from the drawback that stray flux lines induce unwanted voltages at very low speeds.
18. The method of obtaining an estimate of the flux vector directly by measuring the terminal voltages is superior to that of placing sensing coils on the stator, but has the disadvantage that the *iR* drop in the stator at low frequencies causes errors due to the temperature dependence of the resistance.
19. Though obtaining the magnitude and position of flux directly by means of the flux model is better than all other methods, it has the major drawback that the computation of the current i_{mr} from the equation

$$\tau_r \frac{di_{mr}}{dt} + i_{mr} = i_{qs}$$

may pose some problems.

Short Answer Questions

1. Give the mechanical analogy of an induction motor and the steps required to convert this into the mechanical analogy of a dc motor.
2. Give the sequence of steps that is to be followed in connection with field oriented control.
3. Starting from the expression for the developed torque given by

$$T_d = \frac{2}{3} M \Im[\bar{i}_s(\bar{i}_r e^{j\varepsilon})^*]$$

derive the expression for torque in field coordinates.

4. Draw the block diagram of the dynamic flux model and list out its advantages as well as possible sources of error.
5. Which part of the general block diagram illustrating the concept of field oriented control can be simplified into a single block? Give reasons.
6. Give the block diagram of the field oriented control of an induction motor supplied by a PWM type of voltage source inverter. Discuss the results obtained with this scheme.
7. Explain the usefulness of the τ_r identification block in the microcomputer implementation of an ac motor using a PWM type of voltage source inverter.
8. Based on the results obtained with simulation, compare the field oriented control of an ac motor using a current source inverter with that using a voltage source inverter.
9. Bring out the similarities that exist in the voltage and current models of field oriented control implementation.
10. Highlight the special features of field oriented control of an ac motor using a current source inverter.

11. How does an indirect field oriented control scheme differ from a direct control scheme?
12. Discuss three methods for flux computation along with their merits and demerits.
13. Give a typical block diagram that illustrates the field oriented control of a synchronous motor.
14. Give the set of equations that describes the model of an induction motor in field coordinates and explain the significance of each of them.

Appendix A

Fourier Series Expansion of Periodic Quantities

Power electronic converters and inverters give output voltage and -current waveforms which are periodic in nature. Hence, if a voltage $v(t)$ is periodic with time T, then the equality

$$v_1(t) = v_1(t+T) \tag{A1}$$

holds good. If the frequency of this voltage is f, then T can be written as

$$T = \frac{2\pi}{\omega} = \frac{1}{f}$$

Further, if v_1 is expressed as a function of angle $\theta(=\omega t)$, then Eqn (A1) can be rewritten as

$$v_1\left(\frac{\theta}{\omega}\right) = v_1\left(\frac{\theta}{\omega} + \frac{2\pi}{\omega}\right)$$

or

$$v(\omega t) = v(\omega t + 2\pi) \tag{A2}$$

where $v(\omega t) = v_1(t)$. $v(\omega t)$ can be expressed in terms of a Fourier series as

$$v(\omega t) = a_0 + \sum_{n=1}^{\infty}(a_n \cos n\omega t + b_n \sin n\omega t) \tag{A3}$$

a_0 can be determined as follows:

$$a_0 = \frac{1}{2\pi}\int_{-\pi}^{\pi} v(\omega t)\, d(\omega t) \tag{A4}$$

The coefficients a_n and b_n can be determined as

$$a_n = \frac{1}{\pi}\int_{-\pi}^{\pi} v(\omega t)\cos n\omega t \; d(\omega t) \tag{A5a}$$

and

$$b_n = \frac{1}{\pi}\int_{-\pi}^{\pi} v(\omega t)\sin n\omega t \; d(\omega t) \tag{A5b}$$

where $n = 1, 2, 3, \ldots$.

Because of the periodicity of the integrands, the interval of integration in Eqns (A5a) and (A5b) may be replaced by any other interval of length 2π. Thus,

$$a_n = \frac{1}{\pi}\int_{0}^{2\pi} v(\omega t)\cos n\omega t \; d(\omega t) \tag{A6a}$$

and

$$b_n = \frac{1}{\pi}\int_{0}^{2\pi} v(\omega t)\sin n\omega t \; d(\omega t) \tag{A6b}$$

If there is half-wave symmetry, as in the case of the output voltage [Fig. A.1(b)] of the ac voltage controller of Fig. A.1(a), the dc component a_0 and also the even harmonics will be absent. The expression in Eqn (A3) now becomes

$$v(\omega t) = \sum_{n=1,3,5,\ldots}^{\infty} (a_n \cos n\omega t + b_n \sin n\omega t) \tag{A7}$$

where a_n and b_n are given as in Eqns (A6a) and (A6b), respectively. The nth component of voltage can be written as

$$\begin{aligned} v_n &= a_n \cos n\omega t + b_n \sin n\omega t \\ &= (a_n^2 + b_n^2)^{1/2}\left[\frac{a_n}{\sqrt{a_n^2+b_n^2}}\cos n\omega t + \frac{b_n}{\sqrt{a_n^2+b_n^2}}\sin n\omega t\right] \\ &= V_{nm}\sin(n\omega t + \beta_n) \end{aligned} \tag{A8}$$

where $\beta_n = \tan^{-1}(a_n/b_n)$ and $V_{nm} = \sqrt{a_n^2 + b_n^2}$.

The Fourier coefficients of a periodic waveform of current $i(\omega t)$ with half-wave symmetry can be determined by proceeding along similar lines and can be written in terms of its Fourier coefficients c_n and d_n as

$$\begin{aligned} i(\omega t) &= \sum_{n=1,3,5,\ldots}^{\infty} (c_n \cos n\omega t + d_n \sin n\omega t) \\ &= \sum_{n=1,3,5,\ldots}^{\infty} i_n \quad \text{(say)} \end{aligned} \tag{A9}$$

Thus,

$$i_n = c_n \cos n\omega t + d_n \sin n\omega t = I_{nm}\sin(n\omega t + \phi_n) \tag{A10}$$

with $\phi_n = \tan^{-1}(c_n/d_n)$ and $I_{nm} = \sqrt{c_n^2 + d_n^2}$.

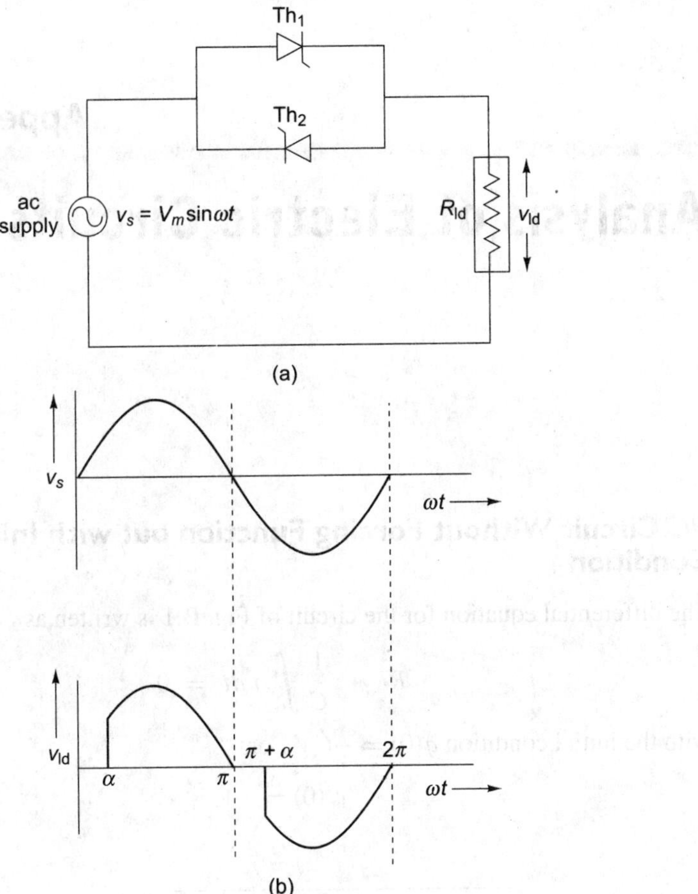

Fig. A.1

Appendix B

Analysis of Electric Circuits

RC Circuit Without Forcing Function but with Initial Condition

The differential equation for the circuit of Fig. B.1 is written as

$$Ri + \frac{1}{C}\int_0^t i\,dt = 0 \tag{B1}$$

with the initial condition $q(0) = -CV_o$, or

$$v_c(0) = -V_o$$

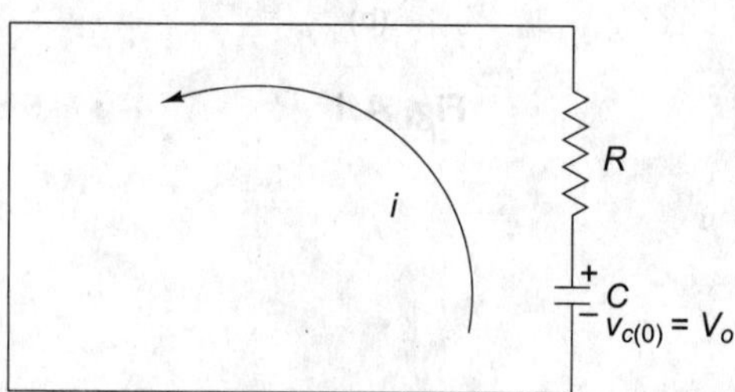

Fig. B.1

Equation (B1) can be written as

$$R\frac{dq}{dt} + \frac{q}{C} = 0$$

or

$$\frac{dq}{dt} + \frac{q}{RC} = 0 \tag{B2}$$

Laplace transformation gives

$$sQ(s) - q(0) + \frac{Q(s)}{RC} = 0 \tag{B3}$$

Rearranging Eqn (B3) and solving for $Q(s)$ gives

$$Q(s) = \frac{q(0)}{s + 1/RC} \tag{B4}$$

Inverse Laplace transformation of Eqn (B4) gives

$$\begin{aligned} q(t) &= q(0)e^{-t/RC} + K_1 \\ &= -CV_o e^{-t/RC} + K_1 \end{aligned}$$

where K_1 is the constant of integration. Applying the initial condition $q(0) = -CV_o$ again gives

$$-CV_o = -CV_o + K_1$$

or

$$K_1 = 0$$

Hence the solution can be written as

$$q(t) = -CV_o e^{-t/RC} \tag{B5a}$$

Also,

$$v_c(t) = -V_o e^{-t/RC} \tag{B5b}$$

Differentiation of $q(t)$ gives

$$i(t) = (V_o/R)e^{-t/RC} \tag{B5c}$$

Equation (B5c) confirms the fact that the initial current is due to the initial charge on the capacitor and it flows in the same direction as $i(t)$.

RC Circuit Connected to a Battery of Voltage E with Initial Condition on the Capacitor

The differential equation governing the circuit of Fig. B.2 when the switch Sw is turned on at $t = 0$ is given as

$$E = Ri + \frac{1}{C}\int_0^t i\,dt \tag{B6}$$

With the initial condition $v_c(0) = -V_o$ or

$$q(0) = -CV_o$$

Equation (B6) can be rewritten as

$$E = R\frac{dq}{dt} + \frac{q}{C}$$

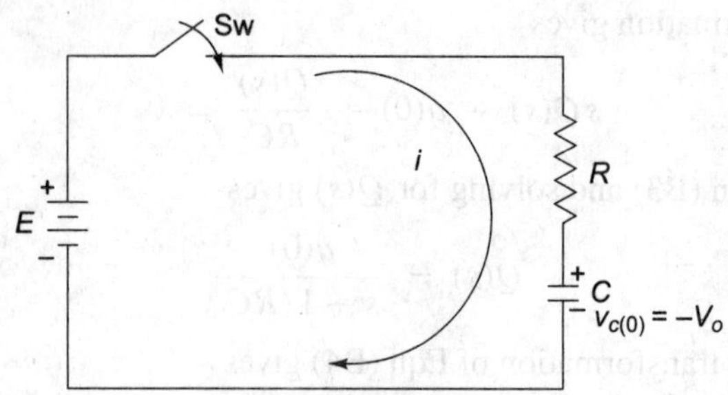

Fig. B.2

or

$$\frac{E}{R} = \frac{dq}{dt} + \frac{q}{RC} \tag{B7}$$

Laplace transformation gives

$$\left(\frac{E}{sR}\right) = Q(s)\left(s + \frac{1}{RC}\right) - q(0) \tag{B8}$$

This can be solved for $Q(s)$ as

$$Q(s) = \frac{E/R}{s(s+1/RC)} + \frac{q(0)}{s+1/RC} \tag{B9}$$

The right-hand side of Eqn (B9) can be split up into partial fractions as

$$Q(s) = \frac{CE}{s} - \frac{CE}{s+1/RC} + \frac{q(0)}{s+1/RC} \tag{B10}$$

Inverse Laplace transformation now gives

$$\begin{aligned} q(t) &= CE(1-e^{-t/RC}) + q(0)e^{-t/RC} \\ &= CE(1-e^{-t/RC}) - CV_o e^{-t/RC} \end{aligned} \tag{B11a}$$

$v_c(t)$ is obtained as

$$v_c(t) = \frac{1}{C}q(t) = E(1-e^{-t/RC}) - V_o e^{-t/RC} \tag{B11b}$$

Differentiation of Eqn (B11a) gives $i(t)$ as

$$i(t) = \frac{1}{R}(E+V_o)e^{-t/RC} \tag{B11c}$$

LC Circuit Without Forcing Function but with Non-zero Initial Condition

The equation that governs the circuit of Fig. B.3 is

$$L\frac{di}{dt} + \frac{1}{C}\int_0^t i\,dt = 0 \tag{B12}$$

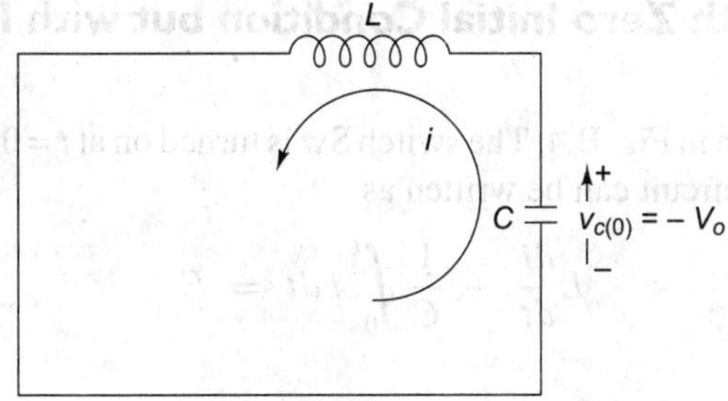

Fig. B.3

with $v_c(0) = -V_o$ or $q(0) = -CV_o$ and $i(0) = 0$. Equation (B12) can be written as

$$L\frac{d^2q}{dt^2} + \frac{q}{C} = 0$$

or

$$\frac{d^2q}{dt^2} + \frac{q}{LC} = 0 \tag{B13}$$

This is the equation of an unforced harmonic oscillator and its solution is given as

$$q(t) = A\cos\omega t + B\sin\omega t \tag{B14}$$

with $\omega = 1/\sqrt{LC}$.

The constants A and B now have to be evaluated. Using the initial condition on charge, $q(0) = -CV_o$, gives

$$q(0) = A = -CV_o$$

Thus,

$$q(t) = -CV_o\cos\omega t + B\sin\omega t \tag{B15}$$

Differentiation of $q(t)$ gives

$$i(t) = \omega CV_o\sin\omega t + \omega B\cos\omega t \tag{B16}$$

Using the initial condition on current, $i(0) = 0$, gives

$$i(0) = \omega B = 0 \text{ or } B = 0$$

Thus,

$$q(t) = -CV_o\cos\omega t \tag{B17a}$$

Also,

$$i(t) = V_o\sqrt{\frac{C}{L}}\sin\omega t \tag{B17b}$$

and

$$v_c(t) = -V_o\cos\omega t \tag{B17c}$$

LC Circuit with Zero Initial Condition but with Forcing Function E

The circuit is given in Fig. B.4. The switch Sw is turned on at $t = 0$. The differential equation for this circuit can be written as

$$L\frac{di}{dt} + \frac{1}{C}\int_0^t i\,dt = E \tag{B18}$$

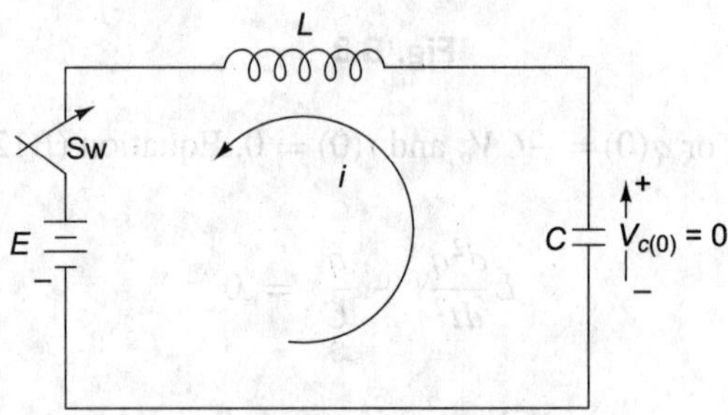

Fig. B.4

with $q(0) = 0$ and $i(0) = 0$. Equation (B18) can be written in terms of the charge q as

$$\frac{d^2q}{dt^2} + \frac{q}{LC} = \frac{E}{L} \tag{B19}$$

with $q(0) = 0$ and $i(0) = \dot{q}(0) = 0$. Equation (B19) is that of a forced harmonic oscillator. Laplace transformation and the application of initial conditions gives

$$\left(s^2 + \frac{1}{LC}\right) Q(s) = \frac{E}{Ls}$$

Thus,

$$Q(s) = \frac{E/L}{s(s^2 + \omega^2)} \tag{B20}$$

Splitting Eqn (B20) (with $\omega = 1/\sqrt{LC}$) into partial fractions and then performing inverse Laplace transformation gives $q(t)$, $v_c(t)$, and $i(t)$ as

$$q(t) = CE(1 - \cos\omega t) \tag{B21a}$$

$$v_c(t) = E(1 - \cos\omega t) \tag{B21b}$$

$$i(t) = E\sqrt{\frac{C}{L}}\sin\omega t \tag{B21c}$$

Unforced *RLC* Circuit with Initial Condition

The equation for the circuit of Fig. B.5 is

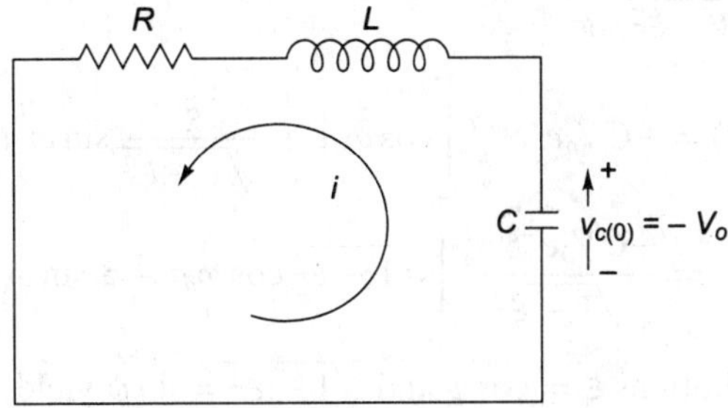

Fig. B.5

$$Ri + L\frac{di}{dt} + \frac{1}{C}\int_0^t i\,dt = 0 \tag{B22}$$

with $v_c(0) = -V_o$ or $q(0) = -CV_o$ and $i(0) = 0$. Equation (B22) can be rewritten as

$$\frac{d^2q}{dt^2} + \frac{R}{L}\frac{dq}{dt} + \frac{q}{LC} = 0 \tag{B23}$$

with $q(0) = -CV_o$ and $i(0) = \dot{q}(0) = 0$. Laplace transformation of Eqn (B23) gives

$$[s^2Q(s) - sq(0) - \dot{q}(0)] + \frac{R}{L}[sQ(s) - q(0)] + \frac{Q(s)}{LC} = 0 \tag{B24}$$

Rearrangement of terms gives

$$\left[s^2 + \left(\frac{R}{L}\right)s + \frac{1}{LC}\right]Q(s) = q(0)\left[s + \frac{R}{L}\right] = -CV_o\left(s + \frac{R}{L}\right)$$

where the initial condition $\dot{q}(0) = 0$ is taken into consideration. Now $Q(s)$ is obtained as

$$Q(s) = \frac{-CV_o(s + R/L)}{s^2 + (R/L)s + 1/LC} \tag{B25}$$

Equation (B25) can be written as

$$Q(s) = \frac{-CV_o(s + 2\xi\omega_n)}{s^2 + 2\xi\omega_n s + {\omega_n}^2} \tag{B26}$$

where $2\xi\omega_n = R/L$ and $\omega_n^2 = 1/LC$. $Q(s)$ of Eqn (B26) can now be split up as

$$Q(s) = \frac{(-CV_o)(s + \xi\omega_n)}{(s + \xi\omega_n^2)^2 + \omega_n^2(1 - \xi^2)} + \frac{(-CV_o)\xi\omega_n}{(s + \xi\omega_n)^2 + \omega_n^2(1 - \xi^2)} \tag{B27}$$

Inverse Laplace transformation gives

$$q(t) = -CV_oe^{-\xi\omega_n t}\cos\omega_d t - CV_o\frac{\xi}{\sqrt{1 - \xi^2}}e^{-\xi\omega_n t}\sin\omega_d t$$

where $\omega_d = \omega_n\sqrt{1-\xi^2}$, or

$$q(t) = -CV_o e^{-\xi\omega_n t}\left[\cos\omega_d t + \frac{\xi}{\sqrt{1-\xi^2}}\sin\omega_d t\right]$$

$$= \frac{-CV_o e^{-\xi\omega_n t}}{\sqrt{1-\xi^2}}\left[\sqrt{1-\xi^2}\cos\omega_d t + \xi\sin\omega_d t\right]$$

Finally, the substitutions $\xi = \cos\phi$ and $\sqrt{1-\xi^2} = \sin\phi$ yield

$$q(t) = \frac{-CV_o e^{-\xi\omega_n t}\sin(\omega_d t + \phi)}{\sqrt{1-\xi^2}} \tag{B28a}$$

Also,

$$v_c(t) = \frac{-V_o e^{-\xi\omega_n t}\sin(\omega_d t + \phi)}{\sqrt{1-\xi^2}} \tag{B28b}$$

Finally, differentiation of $q(t)$ with respect to t gives

$$i(t) = \frac{CV_o\omega_n e^{-\xi\omega_n t}\sin\omega_d t}{\sqrt{1-\xi^2}} \tag{B28c}$$

Forced *RLC* Circuit with Zero Initial Condition

The circuit starts functioning when the switch Sw of Fig. B.6 is turned on at $t = 0$. It can be described by the equation

$$Ri + L\frac{di}{dt} + \frac{1}{C}\int_0^t i\,dt = E \tag{B29}$$

with $i(0) = 0$ and $v_c(0) = 0$, which can be rewritten as

$$\frac{d^2q}{dt^2} + \frac{R}{L}\frac{dq}{dt} + \frac{q}{LC} = \frac{E}{L} \tag{B30}$$

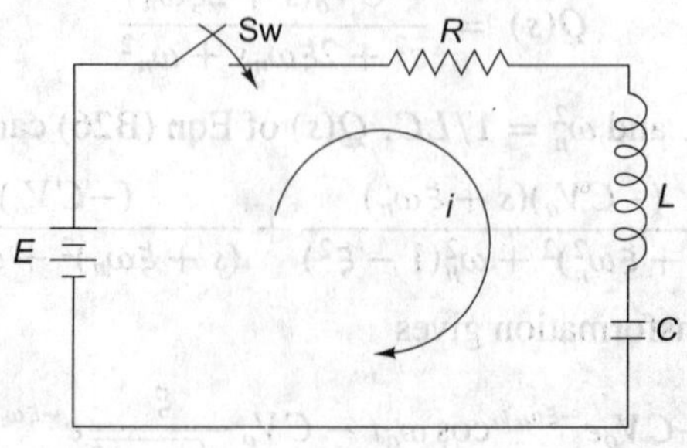

Fig. B.6

Equation (B30) is that of a standard second-order equation with zero initial condition and its solution can be written as

$$q(t) = CE\left[1 - \frac{e^{-\xi\omega_n t}}{\sqrt{1-\xi^2}}\sin(\omega_d t + \phi)\right] \tag{B31a}$$

$$v_c(t) = E\left[1 - \frac{e^{-\xi\omega_n t}}{\sqrt{1-\xi^2}}\sin(\omega_d t + \phi)\right] \tag{B31b}$$

and

$$i(t) = CE\frac{\omega_n e^{-\xi\omega_n t}}{\sqrt{1-\xi^2}}\sin\omega_d t \tag{B31c}$$

Forced *RLC* Circuit with Non-zero Initial Condition

The switch Sw is turned on at t = 0. The equation that describes the circuit of Fig. B.7 is

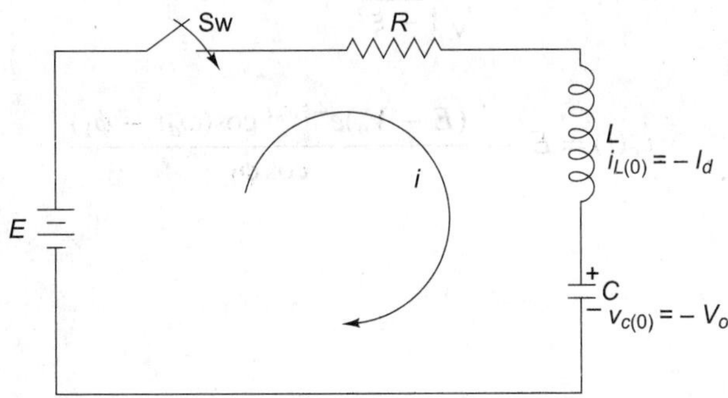

Fig. B.7

$$Ri + L\frac{di}{dt} + \frac{1}{C}\int_0^t i\,dt = E \tag{B32}$$

with initial conditions $q(0) = CV_o$ and $\dot{q}(0) = i_L(0) = -I_d$. Equation (B32) can be written as

$$\frac{d^2q}{dt^2} + \frac{R}{L}\frac{dq}{dt} + \frac{q}{LC} = \frac{E}{L} \tag{B33}$$

with $q(0) = CV_o$ and $i(0) = \dot{q}(0) = -I_d$. Equation (B33) is just a standard second-order one. Thus it can be written as

$$\frac{d^2q}{dt^2} + 2\xi\omega_n\frac{dq}{dt} + \omega_n^2 q = \frac{E}{L} \tag{B34}$$

where $2\xi\omega_n = R/L$, and $\omega_n^2 = 1/LC$. Laplace transformation of Eqn (B31) gives

$$[s^2Q(s) - sq(0) - \dot{q}(0)] + 2\xi\omega_n[sQ(s) - q(0)] + \omega_n^2 Q(s) = \frac{E}{sL} \tag{B35}$$

Equation (B35) can be solved for $Q(s)$ and then split into partial fractions as

$$Q(s) = \frac{E/L}{s(s^2 + 2\xi\omega_n s + \omega_n^2)} + \frac{(s + 2\xi\omega_n)CV_o}{s^2 + 2\xi\omega_n s + \omega_n^2} - \frac{I_d}{s^2 + 2\xi\omega_n s + \omega_n^2} \tag{B36}$$

Inverse Laplace transformation of Eqn (B36) gives

$$q(t) = CE - e^{-\xi\omega_n t}\sin\omega_d t\left[\frac{\xi}{\sqrt{1-\xi^2}}(CE - CV_o) + \frac{I_d}{\omega_n\sqrt{1-\xi^2}}\right]$$

$$-e^{-\xi\omega_n t}\cos\omega_d t(CE - CV_o)$$

Thus,

$$q(t) = CE - \frac{(CE - CV_o)e^{-\xi\omega_n t}\cos(\omega_d t - \phi_1)}{\cos\phi_1} \tag{B37a}$$

where

$$\tan\phi_1 = \frac{\xi + \dfrac{I_d}{\omega_n(CE - CV_o)}}{\sqrt{1-\xi^2}}$$

and

$$v_c(t) = E - \frac{(E - V_o)e^{-\xi\omega_n t}\cos(\omega_d t - \phi_1)}{\cos\phi_1} \tag{B37b}$$

Appendix C

Choppers for Switch Mode Power Supplies

In Chapter 11, the advantages of switch mode power supplies (SMPSs) over linear regulators have been enumerated, and a typical SMPS based on the flyback chopper has been dealt with in detail. Here the buck chopper is taken up at first for discussion, followed by the choppers derived from it, namely, the push–pull, half-bridge, full-bridge, and forward choppers. Switch mode power supplies using these derived choppers are widely used.

Buck Chopper

In a *buck chopper*, also called a *buck regulator*, the average load voltage V_{ld} is less than the input voltage E. It is indeed a practical version of the step-down chopper in which the closed-loop regulation feature is also incorporated. Figure C.1 gives the complete circuit including the pulse width modulator as in Fig. 11.42(b). It is assumed that the inductance L is sufficiently large so as to make the load current fairly constant.

The waveforms for the continuous mode of this regulator are shown in Fig. C.1(d). The circuit operation consists of two stages. During 0 to T_{ON}, the transistor Q is on and the input current flows through the inductor L, filter capacitor C, and the load resistor. The current i_L increases to I_{max} at T_{ON} and consequently the inductive energy in L increases. When Q is switched off at time T_{ON}, the inductive current continues to flow through L, C, the load, and also the freewheeling diode D_1, dissipating the inductive energy through R_{ld}; consequently i_L falls to I_{min} at τ. When Q is switched on to start the next cycle, the sequence described above is repeated. The equivalent circuits for the periods 0 to τ_{ON} and τ_{ON} to τ are given in Figs C.1(b) and (c), respectively.

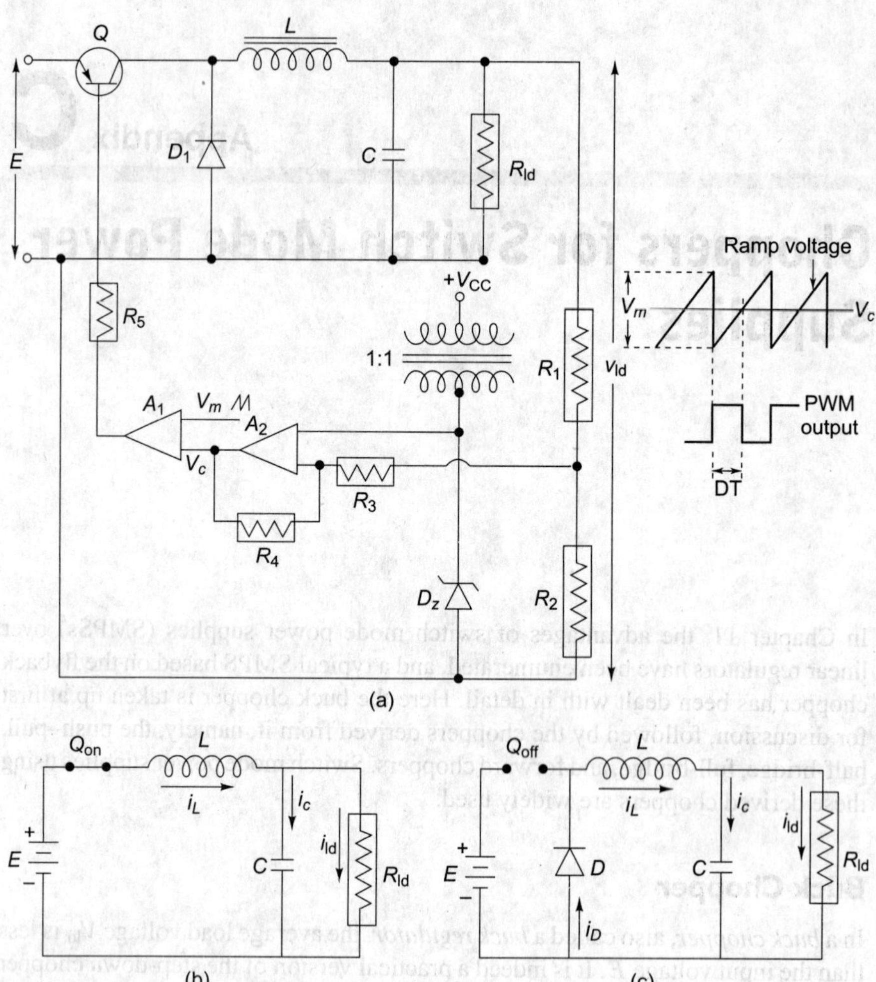

Figs C.1(a)–(c)

As per the voltage–time diagram given in Fig. C.1(d),

$$(E - V_{ld})\tau_{ON} = V_{ld}(\tau - \tau_{ON})$$

This gives

$$V_{ld} = E\frac{\tau_{ON}}{\tau} \tag{C1}$$

Setting $\delta = \tau_{ON}/\tau$, this can be written as

$$V_{ld} = E\delta \tag{C2}$$

An expression for ΔI ($= I_{max} - I_{min}$) can be derived as follows. The voltage across the inductor L can be written as

$$v_L = L\frac{di}{dt} \tag{C3}$$

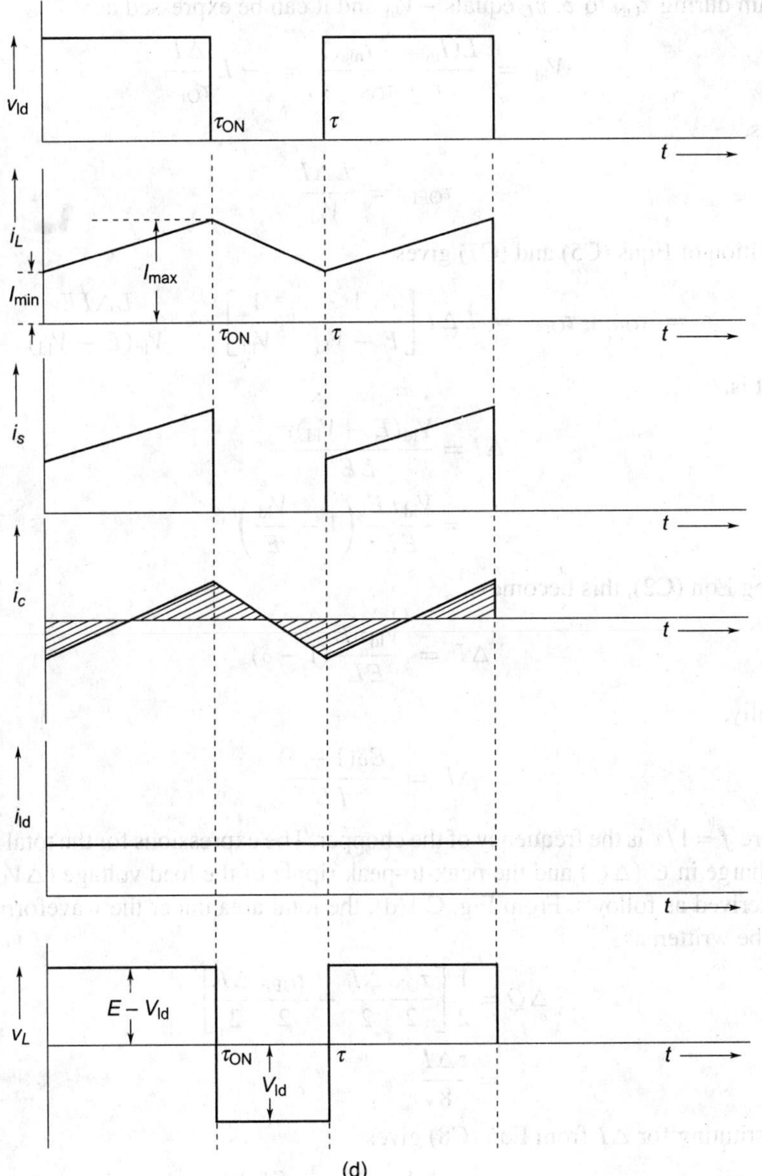

Fig. C.1

During the interval 0 to τ_{ON}, this can be written as

$$E - V_{ld} = \frac{L(I_{max} - I_{min})}{\tau_{ON}} = \frac{L\Delta I}{\tau_{ON}} \tag{C4}$$

where $\Delta I = I_{max} - I_{min}$. That is,

$$\tau_{ON} = \frac{L\Delta I}{E - V_{ld}} \tag{C5}$$

Again during τ_{ON} to τ, v_L equals $-V_{ld}$ and it can be expressed as

$$-V_{ld} = \frac{L(I_{min} - I_{max})}{\tau - \tau_{ON}} = -L\frac{\Delta I}{\tau_{OFF}} \tag{C6}$$

Thus,

$$\tau_{OFF} = \frac{L\Delta I}{V_{ld}} \tag{C7}$$

Addition of Eqns (C5) and (C7) gives

$$\tau = \tau_{ON} + \tau_{OFF} = L\Delta I\left[\frac{1}{E - V_{ld}} + \frac{1}{V_{ld}}\right] = \frac{L\Delta I E}{V_{ld}(E - V_{ld})}$$

That is,

$$\Delta I = \frac{V_{ld}(E - V_{ld})\tau}{LE}$$
$$= \frac{V_{ld}\tau E}{EL}\left(1 - \frac{V_{ld}}{E}\right)$$

Using Eqn (C2), this becomes

$$\Delta I = \frac{V_{ld}\tau E}{EL}(1 - \delta)$$

Finally,

$$\Delta I = \frac{E\delta(1 - \delta)}{fL} \tag{C8}$$

where $f = 1/\tau$ is the frequency of the chopper. The expressions for the total ripple of charge in C (ΔQ) and the peak-to-peak ripple of the load voltage (ΔV_{ld}) can be derived as follows. From Fig. C.1(d), the total area under the waveform of i_c can be written as

$$\Delta Q = \frac{1}{2}\left[\frac{\tau_{ON}}{2}\frac{\Delta I}{2} + \frac{\tau_{OFF}}{2}\frac{\Delta I}{2}\right]$$
$$= \frac{\tau\Delta I}{8} \tag{C9}$$

Substituting for ΔI from Eqn (C8) gives

$$\Delta Q = \frac{\tau}{8}E\frac{\delta(1 - \delta)}{fL} = \frac{E\delta(1 - \delta)}{8f^2L} \tag{C10}$$

Also, ΔV_{ld} becomes

$$\Delta V_{ld} = \frac{\Delta Q}{C} = E\frac{\delta(1 - \delta)}{sf^2LC} \tag{C11}$$

Certain variations in the circuit of the buck chopper lead to the derived choppers, namely, the push–pull chopper, half-bridge chopper, full-bridge chopper, and the forward chopper. These are described below.

Push–Pull Chopper

The basic circuit of a push–pull chopper is given in Fig. C.2. It is just a buck converter with a transformer isolation. This kind of isolation provides step-up, step-down, or reversed polarity for the input voltage. The transistors Q_1 and Q_2 are controlled to alternately switch on and off. The ratio τ_{ON}/τ for each transistor is regulated and attains a theoretical maximum of 0.5. When Q_1 is turned on, the diode D_1 is forward biased and part of the load current passes through it. In the next half-cycle Q_2 is turned on and the diode D_2 conducts the other part of the load current. When both Q_1 and Q_2 are off, both D_1 and D_2 are forward biased to maintain a continuous inductor current. The ratio V_{ld}/E for this case can be obtained as

$$\frac{V_{ld}}{E} = \frac{2N_2}{N_1}\tau_{ON} \tag{C12}$$

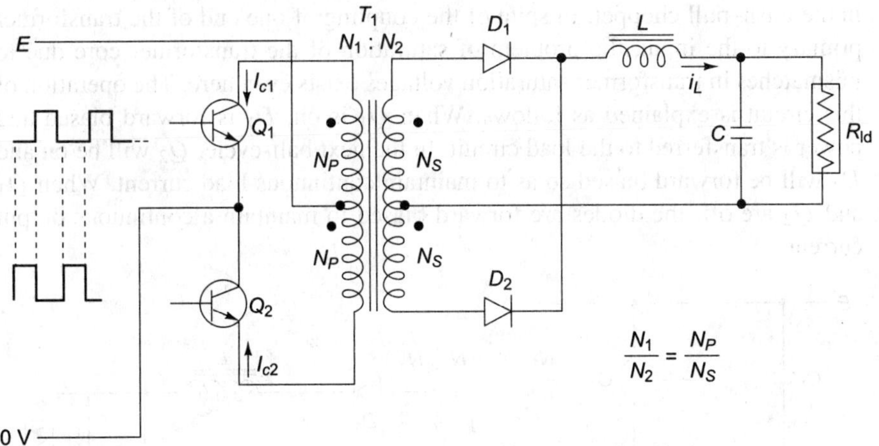

Fig. C.2

The collector voltage of Q_1 is zero during the ON time of Q_1 and increases to E during the OFF time of this transistor and to $2E$ during the ON time of Q_2. This shows that Q_1 and Q_2 must be rated to a voltage of at least $2E$. The collector current is given by

$$I_{max} = \frac{N_2}{N_1}\left(I_{ld} - \frac{I_L}{2}\right) + I_m \tag{C13}$$

where I_{ld}, I_L, and I_m are, respectively, the load current, inductor current, and the magnetizing current of the transformer. The reverse voltage capacity of the diode must be $2(N_2/N_1)E$, because when one diode is conducting, the whole of the secondary voltage appears against the other (non-conducting) diode.

The push–pull chopper is suitable for supplying large load currents, because the transistors Q_1 and Q_2 share this current effectively. During the changeover from one transistor to the other, however, the sudden power demand is likely to upset the balanced load sharing of the transistors, causing one side of the primary winding to carry a higher input current. If this effect continues over some cycles of operation, the transformer core will attain saturation. A non-magnetic gap in the transformer core is sometimes inserted in this type of chopper to remedy this situation.

Half-bridge Chopper

Like the push–pull chopper, the half-bridge chopper has two transistor switches that operate alternately. Figure C.3 gives the circuit for this chopper. The capacitance C is made equal to C_2 so that the input voltage is divided between them, thus enabling the common connecting point of these capacitors to have an average voltage of $E/2$. This arrangement facilitates the reduction of the peak collector voltage of the transistors to E, rather than $2E$, as in the case of the push–pull chopper. However, for a given output power, this chopper has an average primary current which is twice that of the average primary current in the push–pull chopper. In spite of the coupling of one end of the transformer primary to the input, the problem of saturation of the transformer core due to mismatches in transformer saturation voltages exists even here. The operation of this circuit is explained as follows. When Q_1 is on, D_1 is forward biased and power is transferred to the load circuit. In the next half-cycle, Q_2 will be on and D_2 will be forward biased so as to maintain continuous load current. When Q_1 and Q_2 are off, the diodes are forward biased to maintain a continuous output current.

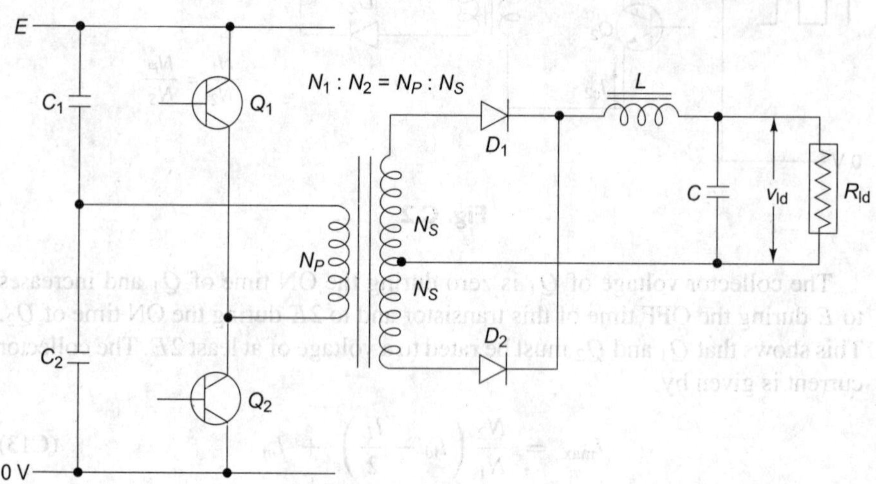

Fig. C.3

The steady-state ratio of the output to the input is given by

$$\frac{V_{\text{ld}}}{E} = \frac{N_2}{N_1}\delta \tag{C14}$$

The maximum collector current is given by

$$I_{c(\max)} = \frac{N_2}{N_1}\left(I_{\rm ld} + \frac{I_L}{2}\right) + I_m \tag{C15}$$

where I_m is the magnetizing current of the transformer. Since the capacitors C_1 and C_2 have to conduct high RMS current, they are usually bulky.

Full-bridge Chopper

The full-bridge chopper is shown in Fig. C.4. In this chopper, the transistor pairs Q_1, Q_4 and Q_2, Q_3 conduct in alternate intervals. The ON time of the transformers is pulse width controlled. It has the same collector voltage rating (E) as that of the half-bridge chopper. Also, it has imbalance conditions similar to the push–pull chopper and hence has the same transformer saturation problem. The steady-state ratio of the load voltage to the input voltage is

$$\frac{V_{\rm ld}}{E} = \frac{2N_2}{N_1}\delta \tag{C16}$$

The maximum collector current is given by

$$I_{c(\max)} = \frac{N_2}{N_1}\left(I_{\rm ld} + \frac{I_L}{2}\right) + I_m \tag{C17}$$

In all the choppers described above, it is necessary that the alternating nature of the transistor switching operation be maintained and any overlap of conduction be avoided. This is because overlapping of the duty ratio lends to an effect that is equivalent to shorting the primary winding of the transformer; this will cause high input currents to flow with consequent destruction of the switching transistors.

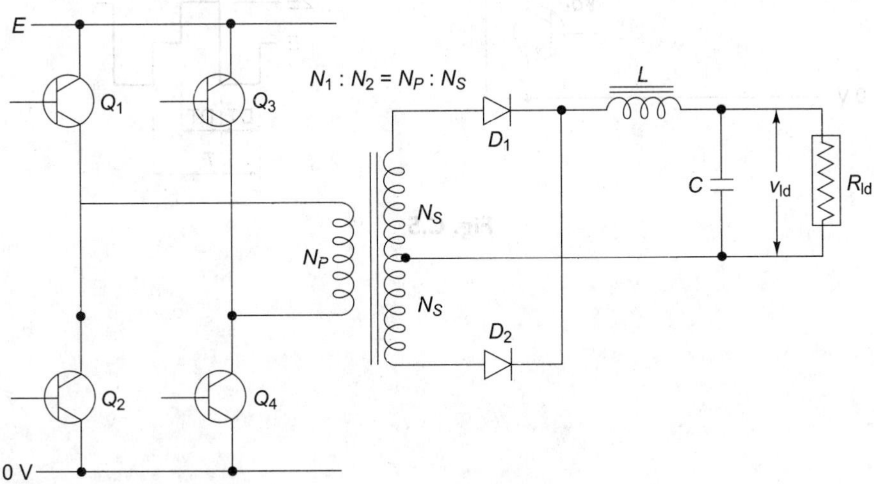

Fig. C.4

Forward Chopper

The circuit for the forward chopper, which is also a buck-derived chopper, is given in Fig. C.5. This circuit operates as follows. When Q_1 is on, diode D_1 is forward biased and energy is transferred to the load. When Q_1 turns off, D_1 is reverse biased but D_2 is forward biased to maintain a continuous current in the load circuit. Diode D_3 in series with winding W_2 is also forward biased to allow magnetic resetting of the core. This demagnetizing winding usually has the same number of turns as in the primary winding (W_1). Thus it is necessary to see that the collector of the transistor is rated to a value which is at least $2E$. The transistor duty cycle should not exceed 0.5 so as to maintain the transformer volt-second balance and also to prevent the saturation of the transformer core. If the ratio of W_2 to W_1 is varied, so that the winding W_2 has fewer turns, the transistor duty cycle may exceed 0.5, but the collector flyback voltage rating becomes higher. The ratio of V_{ld} to E is given for this chopper as

$$\frac{V_{ld}}{E} = \frac{N_2}{N_1}\delta \tag{C18}$$

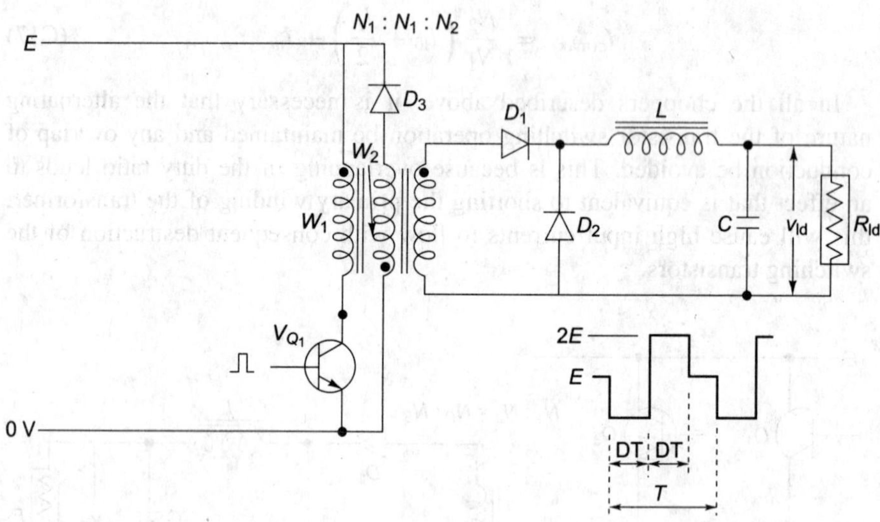

Fig. C.5

Appendix D

Resonant Pulse Inverters

Inverters employing the PWM technique have been successfully used to obtain any desired voltage or current waveform. However, they suffer from the drawback that the devices in them are switched on and off while conducting the load current and therefore subjected to high di/dt values. They are also subjected to high voltage stresses and consequent high switching power losses. The high di/dt and dv/dt also result in electromagnetic interference. These disadvantages inherent in the implementation of the PWM method will be considerably reduced if the switching devices are switched under zero current conditions. In an LC resonant circuit, current is forced to pass through zero. This suggests that connecting the devices in resonant circuits will help overcome the above-described problems. Inverters rigged up in this manner are known as *resonant pulse inverters*. The following discussion centres on some of the most commonly used resonant inverter circuits and their operation.

Series Resonant Inverters

As the name suggests, these inverters are based on the occurrence of series resonance in an RLC circuit. As the current goes to zero, the thyristor turns off and is said to be self-commutated. The output waveform is approximately sinusoidal with its frequency ranging from 200 Hz to 1000 kHz. Some of the applications of such inverters are induction heating, fluorescent lighting, and ultrasonic generators. One advantage of high-frequency switching is that small-sized inductors can be used.

Figure D.1 shows the circuit and waveform of a series inverter using two thyristors. The operation of the circuit can be explained as follows. To start with, Th_2 is assumed to be on. Now turning on Th_1 causes a damped oscillatory current to flow through the load, and also affects the commutation of Th_2. The current becomes zero at t_1 and Th_1 is self-commutated; also the capacitor C attains a value V_{C_1} at t_1, with the X-plate having positive polarity. For a short time, up to t_2, both the thyristors are off and the capacitor voltage remains constant at V_{C_1}. When Th_2 is fired at t_2, a reverse damped oscillatory current flows through the load. At t_3, this reverse current becomes zero and the capacitor voltage attains the value $-V_{C_2}$. The cycle repeats when Th_1 is fired at t_4.

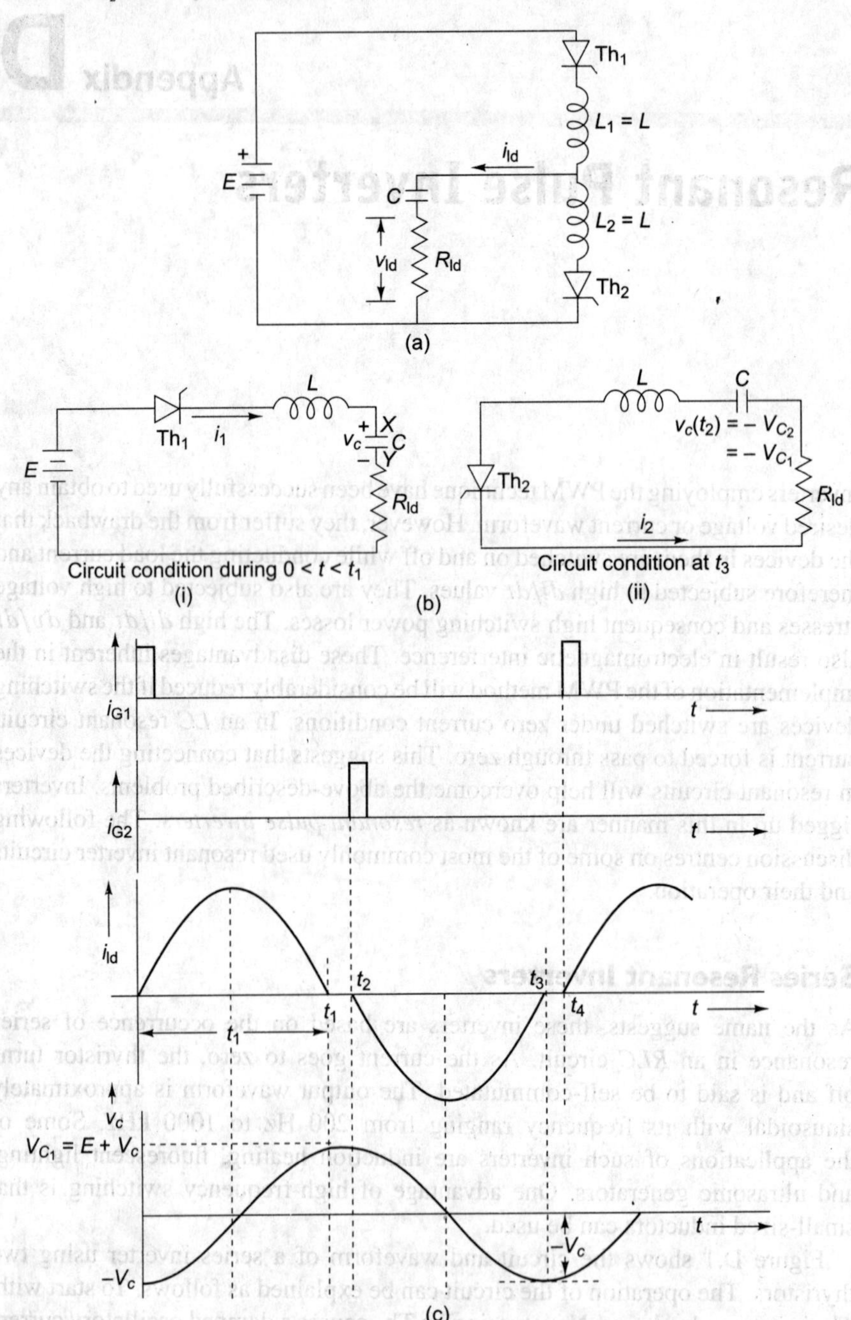

Fig. D.1

Analysis of the Circuit in Steady State

When Th_1 is fired at time $t = 0$, the equation describing the circuit is

$$L\frac{di_1}{dt} + Ri_1 + \frac{1}{C}\int_0^t i_1 dt = E \tag{D1}$$

The initial condition on v_c is written as

$$v_c(0) = \frac{1}{C}\int_{-\infty}^0 i_1 dt = -V_c$$

or

$$q_1(0) = -CV_c$$

The second initial conduction is $i_1(0) = dq_1/dt|_{t=0} = 0$. Equation (D1) can be written as

$$L\frac{d^2q_1}{dt^2} + R\frac{dq_1}{dt} + \frac{q_1}{C} = E \tag{D2}$$

with initial conditions $q_1(0) = -CV_c$ and $dq_1/dt|_{t=0} = 0$. Equation (D2) is a forced *RLC* circuit with non-zero initial conditions. From Appendix B, the solution of this equation can be written as

$$q_1(t) = CE - (CE + CV_c)e^{-\xi\omega_n t}\frac{\sin(\omega_d t + \phi)}{\sqrt{1-\xi^2}} \tag{D3a}$$

where $\sin\phi = \sqrt{1-\xi^2}$ and $\cos\phi = \xi$

$$v_c(t) = E - (E + V_c)e^{-\xi\omega_n t}\frac{\sin(\omega_d t + \phi)}{\sqrt{1-\xi^2}} \tag{D3b}$$

Also,

$$i_1(t) = \frac{dq_1}{dt} = \frac{(CE + CV_c)}{\sqrt{1-\xi^2}}e^{-\xi\omega_n t}\omega_n \sin\omega_d t \tag{D3c}$$

The voltage V_{C_1} attained at t_1 by the capacitor can be computed by substituting $t = \pi/\omega_d$ in Eqn (D3b). Thus,

$$V_{C_1} = \frac{E - (E + V_c)e^{-\xi\omega_n(\pi/\omega_d)}\sin(\pi + \phi)}{\sqrt{1-\xi^2}}$$

$$= E + (E + V_c)e^{-\xi\pi/\sqrt{1-\xi^2}} \tag{D4}$$

This value of capacitor voltage remains till t_2, when Th_2 is fired. The circuit equation for this condition (that is, just after firing of Th_2) becomes

$$L\frac{di_2^2}{dt'} + Ri_2 + \frac{1}{C}\int_0^t i_2 dt' = 0 \tag{D5}$$

with the initial condition $i_2(t')|_{t'=0} = 0$ and with $t' = t - t_2$. As i_2 flows in a direction opposite to that of i_1, the initial condition on the capacitor becomes

$$v_c(t')|_{t'=0} = -V_{C_1}$$

Equation (D5) is that of the unforced *RLC* circuit with an initial condition of Appendix B. From Eqn (B28b), the solution can be written as

$$v_c(t') = -V_{C_1} e^{-\xi\omega_n t'} \frac{\sin(\omega_d t' + \phi)}{\sqrt{1-\xi^2}} \tag{D6a}$$

and

$$i_2(t') = C V_{C_1} e^{-\xi\omega_n t'} \frac{\sin \omega_d t'}{\sqrt{1-\xi^2}} \omega_n \tag{D6b}$$

where

$$\phi = \tan^{-1} \frac{\sqrt{1-\xi^2}}{\xi} \tag{D6c}$$

At $t' = t_3 = \pi/\omega_d$, the current $i_2(t')$ becomes zero. At t_3, $v_c(t')$ becomes

$$v_c(t_3) = -V_{C_1} \frac{e^{-[\xi\omega_n(\pi/\omega_d)]} \sin(\pi + \phi)}{\sqrt{1-\xi^2}}$$

Simplification gives

$$v_c(t_3) = V_{C_1} e^{-\xi\pi/\sqrt{1-\xi^2}} \tag{D7a}$$

The expression on the right-hand side of Eqn (D7a) is just that of V_c, because in the steady state, the capacitor voltage is periodically repeated. Hence,

$$V_c = V_{C_1} e^{-\xi\pi/\sqrt{1-\xi^2}} \tag{D7b}$$

Setting $y = \xi\pi/\sqrt{1-\xi^2}$, Eqns (D4) and (D7b) can be solved for V_c as follows:

$$V_c = \frac{E(1+e^{-y})}{e^y - e^{-y}} = \frac{E(e^y+1)}{e^{2y}-1} = \frac{E}{e^y - 1} \tag{D8a}$$

Also,

$$V_{C_1} = V_c e^y = \frac{E e^y}{e^y - 1} \tag{D8b}$$

Adding E to both sides of Eqn (D8a) gives

$$E + V_c = \frac{E e^y}{e^y - 1} = V_{C_1} \tag{D9}$$

It can be seen from Eqns (D3c), (D6b), and (D9) that the peak values of the positive current pulse $i_1(t)$ and the negative current pulse $i_2(t')$ are equal. The thyristor Th_1 must be turned off before t_2. This implies that $t_2 - t_1$ must be greater than the turn-off time of this thyristor. t_2 is the half-cycle time of the inverter frequency, whereas t_1 is the half-cycle time of the damped oscillation of

i_1. Hence the inequality

$$\frac{\pi}{\omega} - \frac{\pi}{\omega_d} = t_{\mathrm{OFF}} \geq t_q \tag{D10}$$

should hold good, where ω is the angular frequency of the inverter, t_{OFF} is the available OFF time, and t_q is the turn-off time of the thyristor. A second condition is that Th_1 must be turned off before Th_2 is refired at t_2, otherwise the battery may be short-circuited through Th_1, L_1, L_2, and Th_2. The inverter frequency f can obtained from Eqn (D10) as

$$f \leq \frac{1}{2(t_q + \pi/\omega_d)} \tag{D11}$$

From Eqn (D11), the maximum inverter frequency is

$$f_{\max} = \frac{1}{2(t_q + \pi/\omega_d)} \tag{D12}$$

In spite of its simplicity, the series resonant inverter has the drawback of the source current not being continuous. The inverter has to be turned on during every half-cycle of the output voltage. This imposes a limit on the inverter frequency and the amount of energy transfer from the source to load. The thyristors are also subject to a high peak reverse voltage. The interval between t_1 and t_2 should exceed the turn-off time of the device.

The bidirectional series resonant inverter of Fig. D.2 overcomes most of these disadvantages. The operation of this circuit starts with the firing of the thyristor Th_1. A resonant current flows through Th_1, L, C, and the battery. Th_1 is commutated when the current reverses at t_1. The oscillatory current continues to flow, but now through the diode D, until the current again becomes zero at the end of the negative half-cycle. Now, if Th_1 is fired again, the cycle will repeat and a pure sinusoidal voltage will be applied to the load.

The merits of the bidirectional circuit are as follows. The reverse voltage of the thyristor is limited to the forward drop of the diode, which is slightly less than 1 V. Moreover, if the conduction time of the diode is greater than the turn-off time of the thyristor, no zero current interval is needed. As a strictly sinusoidal voltage appears across the output, the harmonic content is minimized. Here the output frequency is equal to the resonant frequency f_r. Thus,

$$f = f_r = \frac{\omega_r}{2\pi} \tag{D13}$$

where f_r is the resonant frequency of the series circuit in Hz and ω_r is the resonant angular frequency in rad/s. The maximum inverter frequency can be assessed from the following inequality:

$$2f \leq \frac{1}{t_q}$$

where t_q is the turn-off time of the thyristor. That is,

$$f_{\max} = \frac{1}{2t_q} \tag{D14}$$

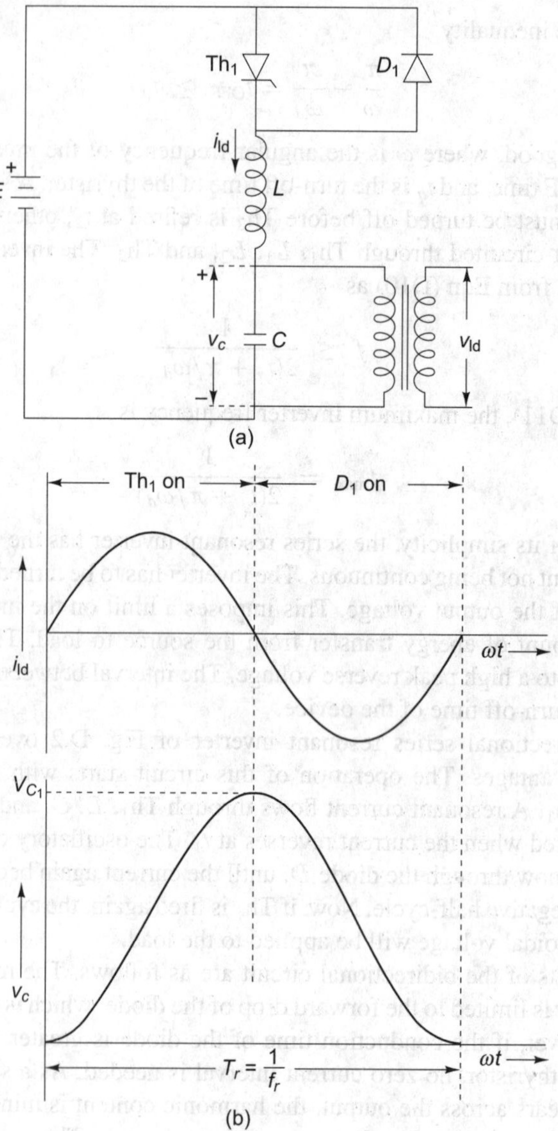

Fig. D.2

Evidently, f should be less than f_{max}.

Parallel Resonant Inverter

The parallel resonant inverter is the dual of the series inverter. Figure D.3(a) shows a parallel resonant circuit with a constant current source. The current is continuously controlled so as to provide short-circuit protection under fault conditions. The equation governing this current is written as

$$C\frac{dv_{\mathrm{ld}}}{dt} + \frac{v_{\mathrm{ld}}}{R} + \frac{1}{L}\int_0^t v_{\mathrm{ld}}dt = I_s \tag{D15}$$

with the initial load voltage $v_{\mathrm{ld}}(0) = 0$ and the initial inductor current $i_L(0) = 0$. The solution of the equation is obtained as

$$v_{\mathrm{ld}}(t) = \frac{I_s}{\omega_n C}e^{-\xi\omega_n t}\sin\omega_d t \tag{D16}$$

where $\omega_n^2 = 1/LC$ and $2\xi\omega_n = 1/RC$.

The factors which are to be considered for the parallel circuit are the input impedance $Z(j\omega)$ and the quality factor Q:

$$Z(j\omega) = \frac{V_{\mathrm{ld}}}{I_s}(j\omega) = \frac{R}{1 + j(R/\omega L) + j\omega CR} \tag{D17}$$

and

$$Q = \omega CR = \frac{R}{\omega L} = R\sqrt{\frac{C}{L}}$$

The plot of the current source waveform of the parallel circuit of Fig. D.3(a) is shown in Fig. D.3(b). This parallel circuit can be considered as an equivalent circuit of Fig. D.3(c) as follows: R_{ld} and L_m in the latter, respectively, correspond to R_1 and L_1 in the former. Also, the inductor L together with the battery in the

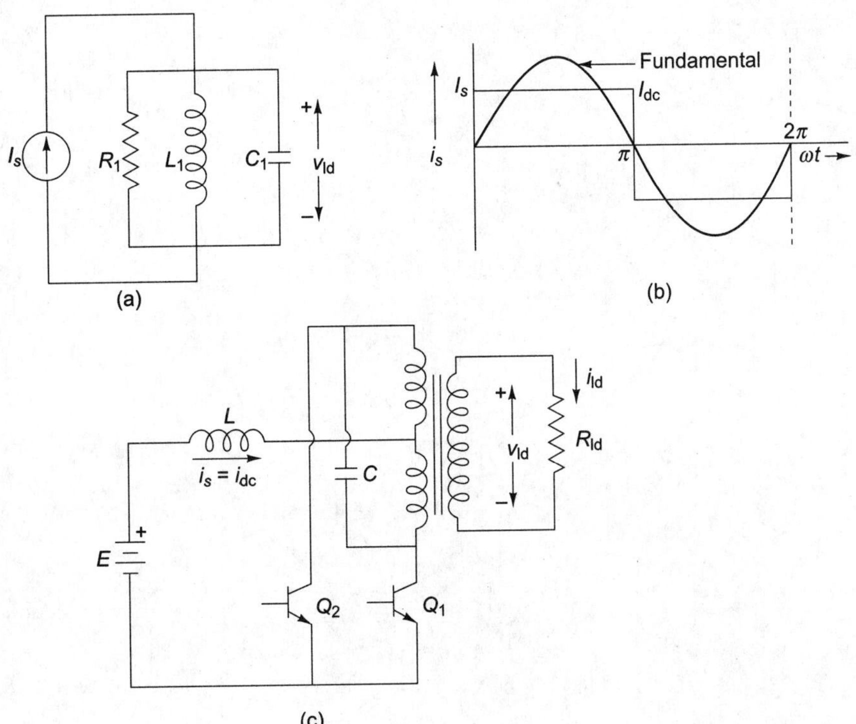

Fig. D.3

latter play the same part as the current source I_s in the former. Thus, L_m and C in the latter act as resonating elements. It can also be seen that the current source in (c) is switched alternately into the resonating circuit by transistors Q_1 and Q_2. The leakage inductance of the transformer in (c) is usually neglected.

Appendix **E**

8085/8080A Instruction Summary

Hex	Mnemonic	Hex	Mnemonic	Hex	Mnemonic	Hex	Mnemonic
CE	ACI 8-bit	3F	CMC	2B	DCX H	01	LXI B,16-bit
8F	ADC A	BF	CMP A	3B	DCX SP	11	LXI D,16-bit
88	ADC B	B8	CMP B	F3	DI	21	LXI H,16-bit
89	ADC C	B9	CMP C	FB	EI	31	LXI SP,16-bit
8A	ADC D	BA	CMP D	76	HLT	7F	MOV A,A
8B	ADC E	BB	CMP E	DB	IN 8-bit	78	MOV A,B
8C	ADC H	BC	CMP H	3C	INR A	79	MOV A,C
8D	ADC L	BD	CMP L	04	INR B	7A	MOV A,D
8E	ADC M	BE	CMP M	0C	INR C	7B	MOV A,E
87	ADD A	D4	CNC 16-bit	14	INR D	7C	MOV A,H
80	ADD B	C4	CNZ 16-bit	1C	INR E	7D	MOV A,L
81	ADD C	F4	CP 16-bit	24	INR H	7E	MOV A,M
82	ADD D	EC	CPE 16-bit	2C	INR L	47	MOV B,A
83	ADD E	FE	CPI 8-bit	34	INR M	40	MOV B,B
84	ADD H	E4	CPO 16-bit	03	INX B	41	MOV B,C
85	ADD L	CC	CZ 16-bit	13	INX D	42	MOV B,D
86	ADD M	27	DAA	23	INX H	43	MOV B,E
C6	ADI 8-bit	09	DAD B	33	INX SP	44	MOV B,H
A7	ANA A	19	DAD D	DA	JC 16-bit	45	MOV B,L
A0	ANA B	29	DAD H	FA	JM 16-bit	46	MOV B,M
A1	ANA C	39	DAD SP	C3	JMP 16-bit	4F	MOV C,A
A2	ANA D	3D	DCR A	D2	JNC 16-bit	48	MOV C,B
A3	ANA E	05	DCR B	C2	JNZ 16-bit	49	MOV C,C
A4	ANA H	0D	DCR C	F2	JP 16-bit	4A	MOV C,D
A5	ANA L	15	DCR D	EA	JPE 16-bit	4B	MOV C,E
A6	ANA M	1D	DCR E	E2	JPO 16-bit	4C	MOV C,H
E6	ANI 8-bit	25	DCR H	CA	JZ 16-bit	4D	MOV C,L
CD	CALL 16-bit	2D	DCR L	3A	LDA 16-bit	4E	MOV C,M
DC	CC 16-bit	35	DCR M	0A	LDAX B	57	MOV D,A
FC	CM 16-bit	0B	DCX B	1A	LDAX D	50	MOV D,B
2F	CMA	1B	DCX D	2A	LHLD 16-bit	51	MOV D,C

(contd)

(contd)

Hex	Mnemonic	Hex	Mnemonic	Hex	Mnemonic	Hex	Mnemonic
52	MOV D,D	71	MOV M,C	E5	PUSH H	9E	SBB M
53	MOV D,E	72	MOV M,D	F5	PUSH PSW	DE	SBI 8-bit
54	MOV D,H	73	MOV M,E	17	RAL	22	SHLD 16-bit
55	MOV D,L	74	MOV M,H	1F	RAR	30	SIM
56	MOV D,M	75	MOV M,L	D8	RC	F9	SPHL
5F	MOV E,A	3E	MVI A,8-bit	C9	RET	32	STA 16-bit
58	MOV E,B	06	MVI B,8-bit	20	RIM	02	STAX B
58	MOV E,C	0E	MVI C,8-bit	07	RLC	12	STAX D
5A	MOV E,D	16	MVI D,8-bit	F8	RM	37	STC
5B	MOV E,E	1E	MVI E,8-bit	D0	RNC	97	SUB A
5C	MOV E,H	26	MVI H,8-bit	C0	RNZ	90	SUB B
5D	MOV E,L	2E	MVI L,8-bit	F0	RP	91	SUB C
5E	MOV E,M	36	MVI M,8-bit	E8	RPE	92	SUB D
67	MOV H,A	00	NOP	E0	RPO	93	SUB E
60	MOV H,B	B7	ORA A	0F	RRC	94	SUB H
61	MOV H,C	B0	ORA B	C7	RST 0	95	SUB L
62	MOV H,D	B1	ORA C	CF	RST 1	96	SUB M
63	MOV H,E	B2	ORA D	D7	RST 2	D6	SUI 8-bit
64	MOV H,H	B3	ORA E	DF	RST 3	EB	XCHG
65	MOV H,L	B4	ORA H	E7	RST 4	AF	XRA A
66	MOV H,M	B5	ORA L	EF	RST 5	A8	XRA B
6F	MOV L,A	B6	ORA M	F7	RST 6	A9	XRA C
68	MOV L,B	F6	ORI 8-bit	FF	RST 7	AA	XRA D
69	MOV L,C	D3	OUT 8-bit	C8	RZ	AB	XRA E
6A	MOV L,D	E9	PCHL	9F	SBB A	AC	XRA H
6B	MOV L,E	C1	POP B	98	SBB B	AD	XRA L
6C	MOV L,H	D1	POP D	99	SBB C	AE	XRA M
6D	MOV L,L	E1	POP H	9A	SBB D	EE	XRI 8-bit
6E	MOV L,M	F1	POP PSW	9B	SBB E	E3	XTHL
77	MOV M,A	C5	PUSH B	9C	SBB H		
70	MOV M,B	D5	PUSH D	9D	SBB L		

References

Berendson, C.S. et al. 1993, 'Commutation strategies for brushless DC motors. Influence on instant torque', *IEEE Trans. Power Electron.*, vol. 8, no. 3.

Bimbhra, P.S. 1991, *Power Electronics*, Khanna Publishers, New Delhi.

Bose, B.K. 1982, 'Adjustable speed ac drives—A technology status review', *Proc. IEEE*, vol. 70, no. 2, pp. 116–35.

Bose, B.K. and H. Sutherland 1983, 'A high performance pulsewidth modulator for an inverter fed drive system using a microcomputer', *IEEE IAS Trans.*, vol. 19, no. 2, pp. 235–43.

Bose, B.K. 1986, *Power Electronics and AC Drives*, Prentice Hall, Englewood Cliffs, New Jersey.

Csaki, F., K. Ganzsky, and S. Marti 1975, *Power Electronics*, Akademiai Kiado, Budapest.

Dewan, S.B. and Straughen A. 1975, *Power Semiconductor Circuits*, Wiley, New York.

Dubey, G.K., S.R. Doradla, A. Joshi, and R.M.K. Sinha 1986, *Thyristorized Power Controllers*, Wiley Eastern Ltd, New Delhi.

Dubey, G.K. 1989, *Power Semiconductor Controlled Drives*, Prentice Hall, Englewood Cliffs, New Jersey.

Finney, D. 1988, *Variable Frequency AC Motor Drive Systems*, Peter Peregrinus Ltd, London.

Fransua, A. and R. Magureanu 1984, *Electrical Machines and Drive Systems*, Technical Press, Oxford, England.

Gabriel R. and W. Leonhard 1982, 'Microprocessor control of induction motor', *Proceedings of IEEE/IAS 7th International Semiconductor Power Converter Conference, Orlando, FL*, pp. 385–95.

Gaonkar, R.S. 1986, *Microprocessor Architecture, Programming and Applications with the 8085/8080A*, Wiley Eastern Ltd.

Gentry, F.E., F.W. Gutzwiller, F.W. Holonyak, and E.E. Von Zastrov 1964, *Semiconductor Controlled Rectifier*, Prentice Hall, Englewood Cliffs, London.

Hall, D.V. 1986, *Microprocessors and Interfacing: Programming and Hardware*, McGraw-Hill, New Delhi.

Ishida, K. et al. 1982, 'Microprocessor Control of Converter-fed DC Motor Drives', *Conf. Rec. - IAS Annu. Meet.*, pp. 619–23.

Kitsum, K. 1984, *Switch Mode Power Conversion—Basic Theory and Design*, Marcel Dekker, New York and Basel.

Lander, C.W. 1981, *Power Electronics*, 1st edn, McGraw-Hill, New York.

Leonhard, W. 1985, *Control of Electric Drives*, Narosa Publishing House, New Delhi.

Mazda, F.F. 1990, *Power Electronics Handbook*, Butterworths, London.

Murphy, J.M.D. and F.G. Turnbull 1988, *Power Electronic Control of AC Motors*, Pergamon Press, Oxford, England.

Mohan, Ned, T.M. Undeland, and W.P. Robbins 1989, *Power Electronics: Converters, Applications, and Design*, John Wiley, New York.

Mohan, Ned, T.M. Undeland, and W.P. Robbins 1994, *Power Electronics: Converters, Applications, and Design*, 2nd edn, John Wiley, New York.

Puchstein, A.F., T.C. Lloyd, and A.G. Conrad 1954, *Alternating Current Machines*, Wiley, New York.

Rajput, R.N. 1993, 'An analytical study of the McMurray–Bedford inverter', ME (Control and Instrumentation) dissertation, University of Delhi.

Ramshaw, R.S. 1975, *Power Electronics*, Chapman and Hall, London.

Rashid, M.H. 1994, *Power Electronics: Circuits, Devices, and Applications*, Prentice Hall of India, New Delhi, 1994.

Sen, P.C. et al. 1978, 'Microprocessor control of an induction motor with flux regulation', *Conference on Industrial Applications of Microprocessors (IECI '78), Philadelphia, PA*.

Sen, P.C. 1981, *Thyristorized DC Drives*, John Wiley, New York.

Thorborg, K. 1988, *Power Electronics*, Prentince Hall International, UK.

Tong, S.K. et al. 1991, 'External synchronous power supply with universal input voltage range for monitoring', *Collaborative Workshop on Microprocessors and Other Related Semiconductor Devices (Motorola Ltd and IIT Delhi)*, New Delhi, November 7–9.

Index

Q

R

S